Advanced Modern Algebra (0-13-087868-5)
Joseph Rotman

Algebra: Pure and Applied (0-13-088254-2)
Aigli Papantonopoulou

Applied Algebra: Codes, Ciphers, and Discrete Algorithms (0-13-067464-8)
Darel W. Hardy, Carol W. Walker

Introduction to Cryptography with Coding Theory (0-13-061814-4)
Wade Trappe, Lawrence C. Washington

Algebra: Abstract and Concrete (Stressing Symmetry), 2/E (0-13-067342-0)
Frederick M. Goodman

Invitation to Cryptology (0-13-088976-8)
Thomas H. Barr

Making, Breaking Codes: Introduction to Cryptology (0-13-030369-0)
Paul Garrett

Fundamentals of Complex Analysis with Applications to Engineering, Science, and Mathematics, 3/E (0-13-907874-6)
Edward Saff, Arthur Snider

Applied Complex Analysis with Partial Differential Equations (0-13-089239-4)
Nakhle H. Asmar

An Introduction to Mathematical Thinking: Algebra and Number Theory (0-13-184868-2)
Will Gilbert, Scott Vanstone

Introduction to Advanced Mathematics, 2/E (0-13-016750-9)
William J. Barnier, Norman Feldman

Analysis with an Introduction to Proof, 4/E (0-13-148101-0)
Steven R. Lay

The Structure of Proof: With Logic and Set Theory (0-13-019077-2)
Michael L. O'Leary

A Journey into Mathematics: An Introduction to Proofs (0-13-842360-1)
Joseph J. Rotman

Mathematical Reasoning: Writing and Proof (0-13-061815-2)
Ted A. Sundstrom

Differential Geometry of Curves and Surfaces (0-13-212589-7)
Manfredo DoCarmo

Differential Geometry and Its Applications, 2/E (0-13-065246-6)
John Oprea

Vector Calculus

Vector Calculus

Third Edition

Susan Jane Colley

Oberlin College

Upper Saddle River, NJ 07458

Library of Congress Cataloging-in-Publication Data

Colley, Susan Jane.
Vector calculus / Susan Jane Colley. -- 3rd ed.
p. cm.
Includes bibliographical references and index.
ISBN 0-13-185874-2
1. Vector analysis. I. Title.

QA433.C635 2006
515′.63--dc22

2005004445

Executive Acquisitions Editor: *George Lobell*
Editor in Chief: *Sally Yagan*
Production Editor: *Lori Hazzard*
Senior Managing Editor: *Linda Mihatov Behrens*
Executive Managing Editor: *Kathleen Schiaparelli*
Manufacturing Buyers: *Michael Bell and Ilene Kahn*
Director of Marketing: *Patrice Jones*
Marketing Assistant: *Rebecca Alimena*
Editorial Assistant/Print Supplements Editor: *Jennifer Urban*
Art Director: *Heather Scott*
Interior Designer: *Tamara Newnam*
Cover Designer: *Tamara Newnam*
Art Editor: *Thomas Benfatti*
Director of Creative Services: *Paul Belfanti*
Cover Image: *Digital Vision/Getty Images, Inc.*
Manager, Cover Visual Research & Permissions: *Karen Sanatar*
Art Studio: *Laserwords*

Pearson Prentice Hall
Pearson Education, Inc.
Upper Saddle River, New Jersey 07458

Printed in the United States of America

10 9 8 7 6 5 4

ISBN: 0-13-185874-2

Pearson Education LTD., *London*
Pearson Education Australia PTY, Limited, *Sydney*
Pearson Education Singapore, Pte. Ltd.
Pearson Education North Asia Ltd, *Hong Kong*
Pearson Education Canada, Ltd, *Toronto*
Pearson Educación de Mexico, S. A. de C.V.
Pearson Education—Japan, *Tokyo*
Pearson Education Malaysia, Pte. Ltd

To Will and Diane,

with love

About the Author

John Seyfried

Susan Colley is the Andrew and Pauline Delaney Professor of Mathematics at Oberlin College, having previously served as Chair of the Department.

She received S.B. and Ph.D. degrees in mathematics from the Massachusetts Institute of Technology prior to joining the faculty at Oberlin in 1983.

Her research focuses on enumerative problems in algebraic geometry, particularly concerning multiple-point singularities and higher-order contact of plane curves.

Professor Colley has published papers on algebraic geometry and commutative algebra, as well as articles on other mathematical subjects. She has lectured internationally on her research and has taught a wide range of subjects in undergraduate mathematics.

Professor Colley is a member of several professional and honorary societies, including the American Mathematical Society, the Mathematical Association of America, Phi Beta Kappa, and Sigma Xi.

Contents

Preface

Physical and natural phenomena depend on a complex array of factors. The sociologist or psychologist who studies group behavior, the economist who endeavors to understand the vagaries of a nation's employment cycles, the physicist who observes the trajectory of a particle or planet, or indeed anyone who seeks to understand geometry in two, three, or more dimensions recognizes the need to analyze changing quantities that depend on more than a single variable. Vector calculus is the essential mathematical tool for such analysis. Moreover, it is an exciting and beautiful subject in its own right, a true adventure in many dimensions.

The only technical prerequisite for this text, which is intended for a sophomore-level course in multivariable calculus, is a standard course in the calculus of functions of one variable. In particular, the necessary matrix arithmetic and algebra (not linear algebra) are developed as needed. Although the mathematical background assumed is not exceptional, the reader will still need to "think hard" in places.

My own objectives in writing the book are simple ones: to develop in students a sound conceptual grasp of vector calculus and to help them begin the transition from first-year calculus to more advanced technical mathematics. I maintain that the first goal can be met, at least in part, through the use of vector and matrix notation, so that many results, especially those of differential calculus, can be stated with reasonable levels of clarity and generality. Properly described, results in the calculus of several variables can look quite similar to those of the calculus of one variable. Reasoning by analogy will thus be an important pedagogical tool. I also believe that a conceptual understanding of mathematics can be obtained through the development of a good geometric intuition. Although I state many results in the case of n variables (where n is arbitrary), I recognize that the most important and motivational examples usually arise for functions of two and three variables, so these concrete and visual situations are emphasized to explicate the general theory. Vector calculus is in many ways an ideal subject for students to begin exploration of the interrelations among analysis, geometry, and matrix algebra.

Multivariable calculus, for many students, represents the beginning of significant mathematical maturation. Consequently, I have written a rather expansive text so that they can see that there is a "story" behind the results, techniques, and examples—that the subject coheres and that this coherence is important for problem solving. To indicate some of the power of the methods introduced, a number of topics, not always discussed very fully in a first multivariable calculus course, are treated here in some detail:

- an early introduction of cylindrical and spherical coordinates (§1.7);
- the use of vector techniques to derive Kepler's laws of planetary motion (§3.1);
- the elementary differential geometry of curves in $\mathbf{R}^3$, including discussion of curvature, torsion, and the Frenet–Serret formulas for the moving frame (§3.2);
- Taylor's formula for functions of several variables (§4.1);

- the use of the Hessian matrix to determine the nature (as local extrema) of critical points of functions of n variables (§§4.2 and 4.3);
- an extended discussion of the change of variables formula in double and triple integrals (§5.5);
- applications of vector analysis to physics (§7.4);
- an introduction to differential forms and the generalized Stokes's theorem (Chapter 8).

Included are a number of proofs of important results. The more technical proofs are collected as addenda at the end of the appropriate sections so as not to disrupt the main conceptual flow and to allow for greater flexibility of use by the instructor and student. Nonetheless, some proofs (or sketches of proofs) embody such central ideas that they are included in the main body of the text.

New in the Third Edition

I have retained the overall structure and tone of the first two editions. New features in this edition include the following:

- 216 additional exercises;
- 230 true/false exercises collected at the end of each chapter;
- a presentation of Newton's method for approximating solutions to systems of n equations in n unknowns (in §2.4);
- a discussion of numerical approximations of line integrals (in §6.1);
- various refinements throughout the text, including new examples and explanations.

How to Use this Book

There is more material in this book than can be covered comfortably during a single semester. Hence, the instructor will wish to eliminate some topics or subtopics—or to abbreviate the rather leisurely presentations of limits and differentiability. Since I frequently find myself without the time to treat surface integrals in detail, I have separated all material concerning parametrized surfaces, surface integrals, and Stokes's and Gauss's theorems (Chapter 7), from that concerning line integrals and Green's theorem (Chapter 6). In particular, in a one-semester course for students with little or no experience with vectors or matrices, instructors can probably expect to cover most of the material in Chapters 1–6, although no doubt it will be necessary to omit some of the optional subsections and to downplay many of the proofs of results. A rough outline for such a course, allowing for some instructor discretion, could be the following:

Chapter 1	8–9	lectures
Chapter 2	9	lectures
Chapter 3	4–5	lectures
Chapter 4	5–6	lectures
Chapter 5	8	lectures
Chapter 6	4	lectures
	38–41	lectures

If students have a richer background (so that much of the material in Chapter 1 can be left largely to them to read on their own), then it should be possible to treat a good portion of Chapter 7 as well. For a two-quarter or two-semester course, it should be possible to work through the entire book with reasonable care and rigor, athough coverage of Chapter 8 should depend on students' exposure to introductory linear algebra, as somewhat more sophistication is assumed there.

The exercises vary from relatively routine computations to more challenging and provocative problems, generally (but not invariably) increasing in difficulty within each section. In a number of instances, groups of problems serve to introduce supplementary topics or new applications. Each chapter concludes with a set of miscellaneous exercises that review and extend the ideas introduced in the chapter.

A word about the use of technology. The text was written without reference to any particular computer software or graphing calculator. Most of the exercises can be solved by hand, although there is no reason not to turn over some of the more tedious calculations to a computer. Those exercises that *require* a computer for computational or graphical purposes are marked with the symbol , and should be amenable to software such as *Mathematica*®, Maple®, or MATLAB.

Ancillary Materials

In addition to this text a **Student Solutions Manual** is available. An **Instructor's Solutions Manual** containing complete solutions to all of the exercises is available to course instructors from Prentice Hall. The reader can find errata for the text and accompanying solutions manuals at the following address:
www.oberlin.edu/math/faculty/colley/VCErrata.html.

Acknowledgments

I am very grateful to many individuals for sharing with me their thoughts and ideas about multivariable calculus. I would like to express particular appreciation to my Oberlin colleagues (past and present) Bob Geitz, Kevin Hartshorn, Michael Henle (who, among other things, carefully read the draft of Chapter 8), Gary Kennedy, Dan King, Greg Quenell, Daniel Steinberg, Daniel Styer, Jim Walsh, and Elizabeth Wilmer for their conversations with me. I am also grateful to Henry C. King, University of Maryland; Karen Saxe, Macalester College; David Singer, Case Western Reserve University; and Mark R. Treuden, University of Wisconsin at Stevens Point, for their helpful comments. Several colleagues reviewed various versions of the manuscript, and I am happy to acknowledge their efforts and many fine suggestions. In particular, for the first edition, I thank the following reviewers:

Marcel A. F. Déruaz, *University of Ottawa*;
Christopher C. Leary, *State University of New York, College at Geneseo*;
David C. Minda, *University of Cincinnati*;
Jeffrey Morgan, *University of Houston*;
Jeffrey L. Nunemacher, *Ohio Wesleyan University*;
Florin Pop, *Wagner College*;
John T. Scheick, *University of North Carolina at Chapel Hill*;
Theodore B. Stanford, *New Mexico State University*;
James Stasheff, *University of North Carolina at Chapel Hill (now emeritus)*.

For the second edition, I thank the following individuals:

Stanley Chang, *Wellesley College*;
Krzysztof Galicki, *University of New Mexico*;
Isom H. Herron, *Rensselaer Polytechnic Institute*;
Ashwani K. Kapila, *Rensselaer Polytechnic Institute*;
Monika Nitsche, *University of New Mexico*;
Leonard M. Smiley, *University of Alaska, Anchorage*;
Saleem Watson, *California State University, Long Beach*.

Finally, for the third edition, I thank:

Raymond J. Cannon, *Baylor University*;
Richard D. Carmichael, *Wake Forest University*;
Gabriel Prajitura, *State University of New York, College at Brockport*;
Dmitry Gokhman, *University of Texas at San Antonio*;
Mark Schwartz, *Ohio Wesleyan University*;
Floyd L. Williams, *University of Massachusetts, Amherst*.

Many people have been of invaluable assistance throughout the production process. For the first edition, I would like to thank Ben Miller for his hard work establishing the format for the initial drafts. I am very grateful to Linda Miller and Michael Bastedo for their numerous typographical contributions and to Joshua Davis for his assistance with proofreading. Special thanks go to Stephen Kasperick-Postellon for his manifold contributions to the typesetting, indexing, proofreading, and friendly critiquing of the manuscript. For the second edition, I express my appreciation to Joaquin Espinoza Goodman for his assistance with proofreading. And, for both the second and third editions, I am most grateful to Catherine Murillo for her help with any number of tasks. Without the efforts of these individuals, this project might never have taken physical form.

The staff at Prentice Hall have been extraordinarily kind and helpful to me. In particular, I am very grateful to my editor, George Lobell; to Gale Epps, Melanie Van Benthuysen, and Jennifer Urban, his editorial assistants; to Nicholas Romanelli, my production editor for the first edition (whose gentle suggestions improved the original manuscript immeasurably); and to Barbara Mack, my efficient and cheerful production editor for the second edition. For the first two editions, my sincere appreciation goes to Ron Weickart and the staff at Network Graphics for their fine rendering of the figures, and to Dennis Kletzing for his careful and enthusiastic composition work. For the third edition, I thank Lori Hazzard and the staff at Interactive Composition Corporation who did a fine job with the production of this new edition and I express my appreciation to Tom Benfatti of Prentice Hall for his effort with the figures.

Finally, I thank the many Oberlin students who had the patience to listen to me lecture and who inspired me to write and improve this volume.

SJC
sjcolley@math.oberlin.edu

To the Student: Some Preliminary Notation

Here are the ideas that you need to keep in mind as you read this book and learn vector calculus.

Given two sets A and B, I assume that you are familiar with the notation $A \cup B$ for the **union** of A and B—those elements that are in either A or B (or both):

$$A \cup B = \{x \mid x \in A \text{ or } x \in B\}.$$

Similarly, $A \cap B$ is used to denote the **intersection** of A and B—those elements that are in both A and B:

$$A \cap B = \{x \mid x \in A \text{ and } x \in B\}.$$

The notation $A \subseteq B$, or $A \subset B$, indicates that A is a **subset** of B (possibly empty or equal to B).

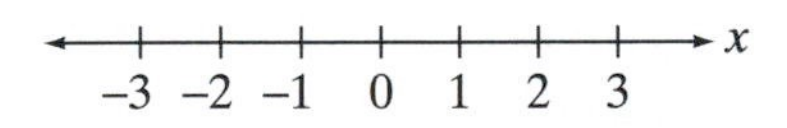

Figure 1 The coordinate line **R**.

One-dimensional space (also called the **real line** or **R**) is just a straight line. We put real number coordinates on this line by placing negative numbers on the left and positive numbers on the right. (See Figure 1.)

Two-dimensional space, denoted $\mathbf{R}^2$, is the familiar Cartesian plane. If we construct two perpendicular lines (the x- and y-**coordinate axes**), set the **origin** as the point of intersection of the axes, and establish numerical scales on these lines, then we may locate a point in $\mathbf{R}^2$ by giving a pair of numbers (x, y), the **coordinates** of the point. Note that the coordinate axes divide the plane into four **quadrants.** (See Figure 2.)

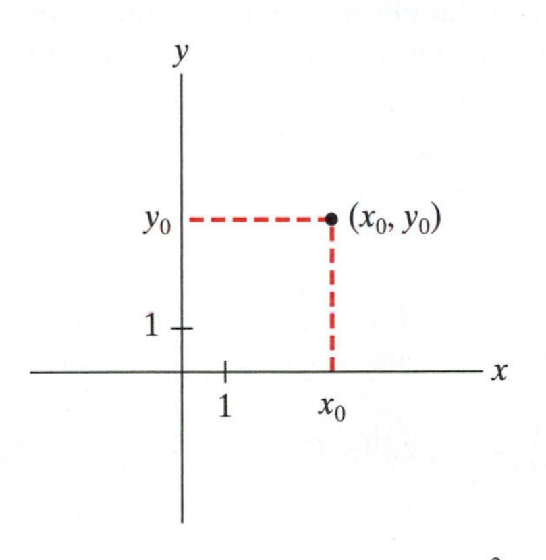

Figure 2 The coordinate plane $\mathbf{R}^2$.

Three-dimensional space, denoted $\mathbf{R}^3$, requires three mutually perpendicular coordinate axes (called the x-, y- and z-**axes**) that meet in a single point (called the **origin**) in order to locate an arbitrary point. Analogous to the case of $\mathbf{R}^2$, if we establish scales on the axes, then we can locate a point in $\mathbf{R}^3$ by giving a triple of numbers (x, y, z). The coordinate axes divide three-dimensional space into eight **octants.** It takes some practice to get your sense of perspective correct when sketching points in $\mathbf{R}^3$. (See Figure 3.) Sometimes we draw the coordinate axes in $\mathbf{R}^3$ in different orientations in order to get a better view of things. However, we always maintain the axes in a **right-handed configuration.** This means that if you curl the fingers of your right hand from the positive x-axis to the positive y-axis, then your thumb will point along the positive z-axis. (See Figure 4.)

Although you need to recall particular techniques and methods from the calculus you have already learned, here are some of the more important concepts to keep in mind: Given a function $f(x)$, the **derivative** $f'(x)$ is the limit (if it exists)

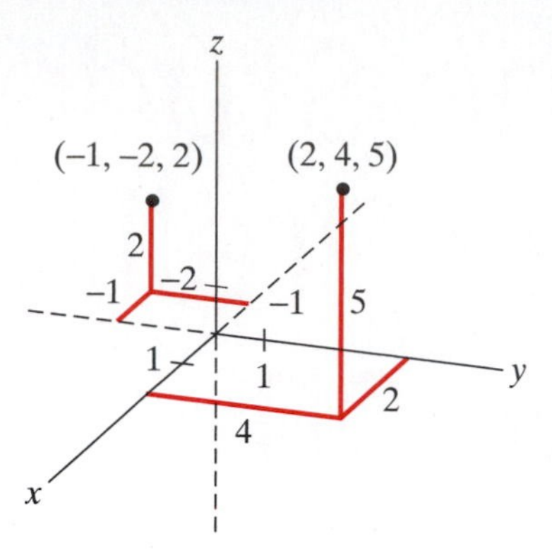

Figure 3 Three-dimensional space $\mathbf{R}^3$. Selected points are graphed.

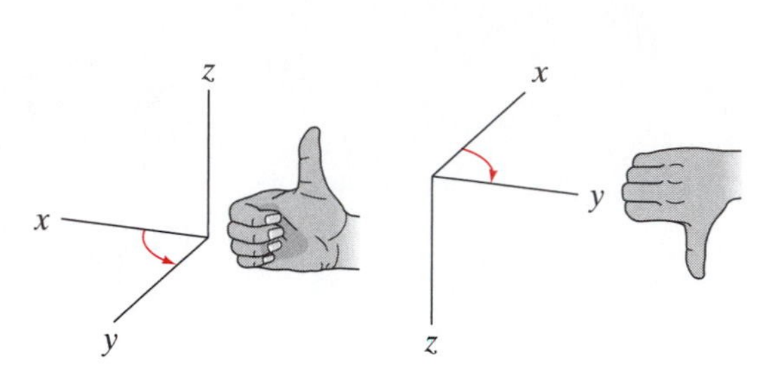

Figure 4 The x-, y-, and z-axes in $\mathbf{R}^3$ are always drawn in a right-handed configuration.

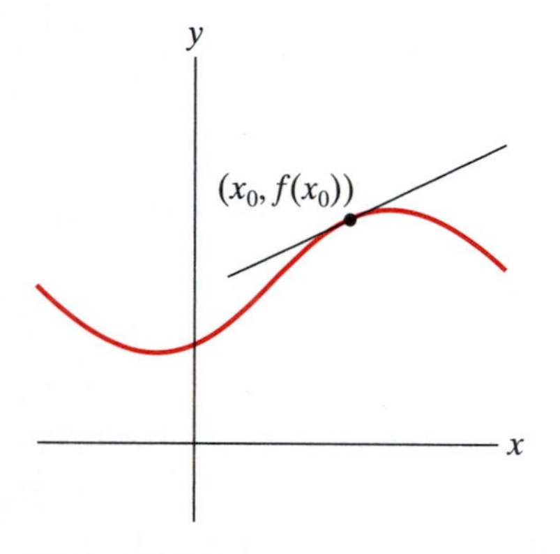

Figure 5 The derivative $f'(x_0)$ is the slope of the tangent line to $y = f(x)$ at $(x_0, f(x_0))$.

of the difference quotient of the function:

$$f'(x) = \lim_{h \to 0} \frac{f(x+h) - f(x)}{h}.$$

The significance of the derivative $f'(x_0)$ is that it measures the slope of the line tangent to the graph of f at the point $(x_0, f(x_0))$. (See Figure 5.) The derivative may also be considered to give the instantaneous rate of change of f at $x = x_0$. We also denote the derivative $f'(x)$ by df/dx.

The **definite integral** $\int_a^b f(x)\,dx$ of f on the closed interval $[a, b]$ is the limit (provided it exists) of the so-called **Riemann sums** of f:

$$\int_a^b f(x)\,dx = \lim_{\text{all } \Delta x_i \to 0} \sum_{i=1}^{n} f(x_i^*)\Delta x_i.$$

Here $a = x_0 < x_1 < x_2 < \cdots < x_n = b$ denotes a **partition** of $[a, b]$ into subintervals $[x_{i-1}, x_i]$, the symbol $\Delta x_i = x_i - x_{i-1}$ (the length of the subinterval), and x_i^* denotes any point in $[x_{i-1}, x_i]$. If $f(x) \geq 0$ on $[a, b]$, then each term $f(x_i^*)\Delta x_i$ in the Riemann sum is the area of a rectangle related to the graph of f. The Riemann sum $\sum_{i=1}^{n} f(x_i^*)\Delta x_i$ thus approximates the total area under the graph of f between $x = a$ and $x = b$. (See Figure 6.)

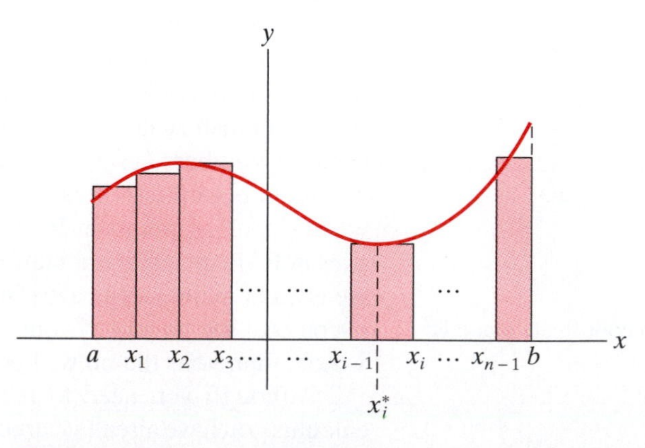

Figure 6 If $f(x) \geq 0$ on $[a, b]$, then the Riemann sum approximates the area under $y = f(x)$ by giving the sum of areas of rectangles.

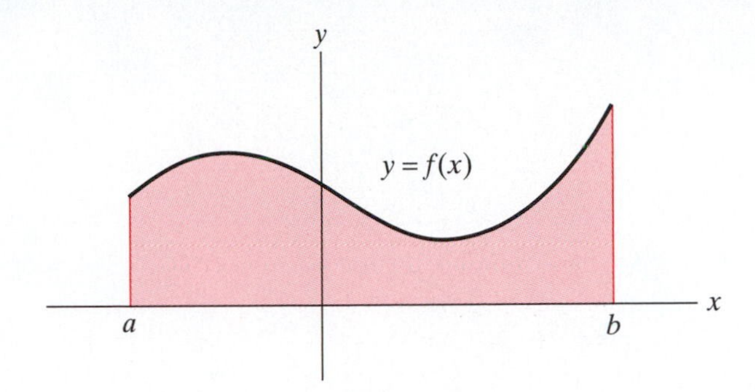

Figure 7 The area under the graph of $y = f(x)$ is $\int_a^b f(x)\,dx$.

The definite integral $\int_a^b f(x)\,dx$, if it exists, is taken to represent the area under $y = f(x)$ between $x = a$ and $x = b$. (See Figure 7.)

The derivative and the definite integral are connected by an elegant result known as **the fundamental theorem of calculus.** Let $f(x)$ be a continuous function of one variable, and let $F(x)$ be such that $F'(x) = f(x)$. (The function F is called an **antiderivative** of f.) Then

1. $\displaystyle\int_a^b f(x)\,dx = F(b) - F(a)$;

2. $\displaystyle\frac{d}{dx}\int_a^x f(t)\,dt = f(x)$.

Finally, the end of an example is denoted by the symbol ◆ and the end of a proof by the symbol ■.

Vector Calculus

CHAPTER 1

Vectors

1.1 Vectors in Two and Three Dimensions

For your study of the calculus of several variables, the notion of a vector is fundamental. As is the case for many of the concepts we shall explore, there are both *algebraic* and *geometric* points of view. You should become comfortable with both perspectives in order to solve problems effectively and to build on your basic understanding of the subject.

Vectors in $\mathbf{R}^2$ and $\mathbf{R}^3$: The Algebraic Notion

■ **Definition 1.1** A **vector** in $\mathbf{R}^2$ is simply an ordered pair of real numbers. That is, a vector in $\mathbf{R}^2$ may be written as

$$(a_1, a_2) \quad \text{(e.g., } (1, 2) \text{ or } (\pi, 17)\text{)}.$$

Similarly, a **vector** in $\mathbf{R}^3$ is simply an ordered triple of real numbers. That is, a vector in $\mathbf{R}^3$ may be written as

$$(a_1, a_2, a_3) \quad \text{(e.g., } (\pi, e, \sqrt{2})\text{)}.$$

To emphasize that we want to consider the pair or triple of numbers as a single unit, we will use **boldface** letters; hence $\mathbf{a} = (a_1, a_2)$ or $\mathbf{a} = (a_1, a_2, a_3)$ will be our standard notation for vectors in $\mathbf{R}^2$ or $\mathbf{R}^3$. Whether we mean that $\mathbf{a}$ is a vector in $\mathbf{R}^2$ or in $\mathbf{R}^3$ will be clear from context (or else won't be important to the discussion). When doing handwritten work, it is difficult to "boldface" anything, so you'll want to put an arrow over the letter. Thus, $\vec{a}$ will mean the same thing as $\mathbf{a}$. Whatever notation you decide to use, it's important that you distinguish the *vector* $\mathbf{a}$ (or $\vec{a}$) from the *single real number* a. To contrast them with vectors, we will also refer to single real numbers as **scalars.**

In order to do anything interesting with vectors, it's necessary to develop some arithmetic operations for working with them. Before doing this, however, we need to know when two vectors are equal.

Definition 1.2 Two vectors $\mathbf{a} = (a_1, a_2)$ and $\mathbf{b} = (b_1, b_2)$ in $\mathbf{R}^2$ are **equal** if their corresponding components are equal, that is, if $a_1 = b_1$ and $a_2 = b_2$. The same definition holds for vectors in $\mathbf{R}^3$: $\mathbf{a} = (a_1, a_2, a_3)$ and $\mathbf{b} = (b_1, b_2, b_3)$ are **equal** if their corresponding components are equal, that is, if $a_1 = b_1$, $a_2 = b_2$, and $a_3 = b_3$.

EXAMPLE 1 The vectors $\mathbf{a} = (1, 2)$ and $\mathbf{b} = \left(\frac{3}{3}, \frac{6}{3}\right)$ are equal in $\mathbf{R}^2$, but $\mathbf{c} = (1, 2, 3)$ and $\mathbf{d} = (2, 3, 1)$ are *not* equal in $\mathbf{R}^3$. ◆

Next, we discuss the operations of vector addition and scalar multiplication. We'll do this by considering vectors in $\mathbf{R}^3$ only; exactly the same remarks will hold for vectors in $\mathbf{R}^2$ if we simply ignore the last component.

Definition 1.3 (VECTOR ADDITION) Let $\mathbf{a} = (a_1, a_2, a_3)$ and $\mathbf{b} = (b_1, b_2, b_3)$ be two vectors in $\mathbf{R}^3$. Then the **vector sum $\mathbf{a} + \mathbf{b}$** is the vector in $\mathbf{R}^3$ obtained via componentwise addition: $\mathbf{a} + \mathbf{b} = (a_1 + b_1, a_2 + b_2, a_3 + b_3)$.

EXAMPLE 2 We have $(0, 1, 3) + (7, -2, 10) = (7, -1, 13)$ and (in $\mathbf{R}^2$):

$$(1, 1) + (\pi, \sqrt{2}) = (1 + \pi, 1 + \sqrt{2}).$$ ◆

Properties of vector addition. We have

1. $\mathbf{a} + \mathbf{b} = \mathbf{b} + \mathbf{a}$ for all $\mathbf{a}$, $\mathbf{b}$ in $\mathbf{R}^3$ (commutativity);
2. $\mathbf{a} + (\mathbf{b} + \mathbf{c}) = (\mathbf{a} + \mathbf{b}) + \mathbf{c}$ for all $\mathbf{a}$, $\mathbf{b}$, $\mathbf{c}$ in $\mathbf{R}^3$ (associativity);
3. a special vector, denoted $\mathbf{0}$ (and called the **zero vector**), with the property that $\mathbf{a} + \mathbf{0} = \mathbf{a}$ for all $\mathbf{a}$ in $\mathbf{R}^3$.

These three properties require proofs, which, like most facts involving the algebra of vectors, can be obtained by explicitly writing out the vector components. For example, for property 1, we have that if

$$\mathbf{a} = (a_1, a_2, a_3) \quad \text{and} \quad \mathbf{b} = (b_1, b_2, b_3),$$

then

$$\begin{aligned} \mathbf{a} + \mathbf{b} &= (a_1 + b_1, a_2 + b_2, a_3 + b_3) \\ &= (b_1 + a_1, b_2 + a_2, b_3 + a_3) \\ &= \mathbf{b} + \mathbf{a}, \end{aligned}$$

since real number addition is commutative. For property 3, the "special vector" is just the vector whose components are all zero: $\mathbf{0} = (0, 0, 0)$. It's then easy to check that property 3 holds by writing out components. Similarly for property 2, so we leave the details as exercises.

Definition 1.4 (SCALAR MULTIPLICATION) Let $\mathbf{a} = (a_1, a_2, a_3)$ be a vector in $\mathbf{R}^3$ and let $k \in \mathbf{R}$ be a scalar (real number). Then the **scalar product** $k\mathbf{a}$ is the vector in $\mathbf{R}^3$ given by multiplying each component of $\mathbf{a}$ by k: $k\mathbf{a} = (ka_1, ka_2, ka_3)$.

EXAMPLE 3 If $\mathbf{a} = (2, 0, \sqrt{2})$ and $k = 7$, then $k\mathbf{a} = (14, 0, 7\sqrt{2})$. ◆

The results that follow are not difficult to check—just write out the vector components.

> **Properties of scalar multiplication.** For all vectors **a** and **b** in $\mathbf{R}^3$ (or $\mathbf{R}^2$) and scalars k and l in **R**, we have
>
> **1.** $(k + l)\mathbf{a} = k\mathbf{a} + l\mathbf{a}$ (distributivity);
>
> **2.** $k(\mathbf{a} + \mathbf{b}) = k\mathbf{a} + k\mathbf{b}$ (distributivity);
>
> **3.** $k(l\mathbf{a}) = (kl)\mathbf{a} = l(k\mathbf{a})$.

It is worth remarking that none of these definitions or properties really depends on dimension, that is, on the number of components. Therefore we could have introduced the algebraic concept of a vector in $\mathbf{R}^n$ as an **ordered n-tuple** $(a_1, a_2, \ldots, a_n)$ of real numbers and defined addition and scalar multiplication in a way analogous to what we did for $\mathbf{R}^2$ and $\mathbf{R}^3$. Think about what such a generalization means. We will discuss some of the technicalities involved in §1.6.

Vectors in $\mathbf{R}^2$ and $\mathbf{R}^3$: The Geometric Notion

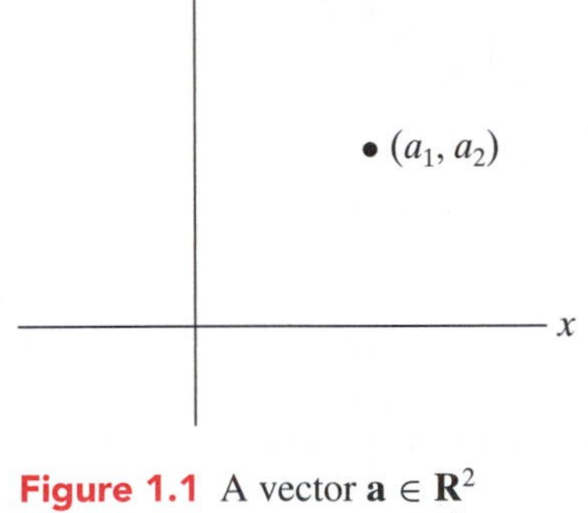

Figure 1.1 A vector $\mathbf{a} \in \mathbf{R}^2$ corresponds to a point in $\mathbf{R}^2$.

Although the algebra of vectors is certainly important and you should become adept at working algebraically, the formal definitions and properties tend to present a rather sterile picture of vectors. A better motivation for the definitions just given comes from geometry. We explore this geometry now. First of all, the fact that a vector **a** in $\mathbf{R}^2$ is a pair of real numbers (a_1, a_2) should make you think of the coordinates of a point in $\mathbf{R}^2$. (See Figure 1.1.) Similarly, if $\mathbf{a} \in \mathbf{R}^3$, then **a** may be written as (a_1, a_2, a_3), and this triple of numbers may be thought of as the coordinates of a point in $\mathbf{R}^3$. (See Figure 1.2.)

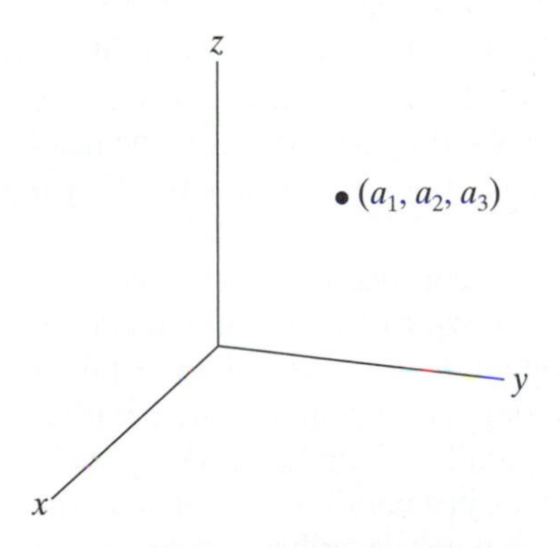

Figure 1.2 A vector $\mathbf{a} \in \mathbf{R}^3$ corresponds to a point in $\mathbf{R}^3$.

All of this is fine, but the results of performing vector addition or scalar multiplication don't have very interesting or meaningful geometric interpretations in terms of points. As we shall see, it is better to visualize a vector in $\mathbf{R}^2$ or $\mathbf{R}^3$ as an arrow that begins at the origin and ends at the point. (See Figure 1.3.) Such a depiction is often referred to as the **position vector** of the point (a_1, a_2) or (a_1, a_2, a_3).

If you've studied vectors in physics, you have heard them described as objects having "magnitude and direction." Figure 1.3 demonstrates this concept, provided that we take "magnitude" to mean "length of the arrow" and "direction" to be the

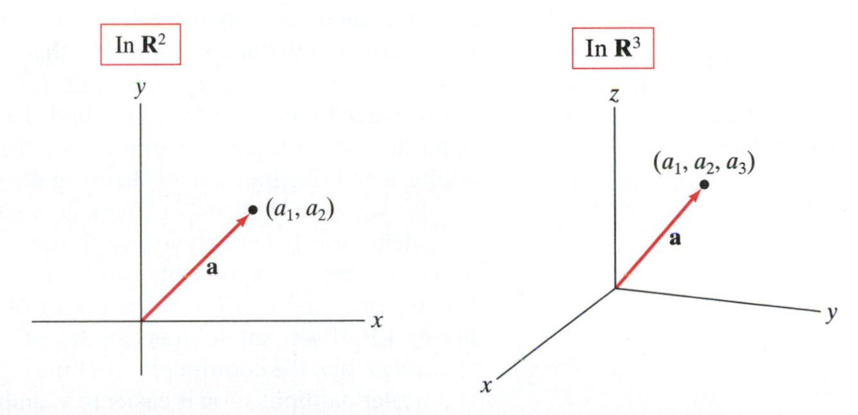

Figure 1.3 A vector **a** in $\mathbf{R}^2$ or $\mathbf{R}^3$ is represented by an arrow from the origin to **a**.

orientation or sense of the arrow. (Note: There is an exception to this approach, namely, the zero vector. The zero vector just sits at the origin, like a point, and has no magnitude and, therefore, an indeterminate direction. This exception will not pose much difficulty.) However, in physics, one doesn't demand that all vectors be represented by arrows having their tails bound to the origin. One is free to "parallel translate" vectors throughout $\mathbf{R}^2$ and $\mathbf{R}^3$. That is, one may represent the vector $\mathbf{a} = (a_1, a_2, a_3)$ by an arrow with its tail at the origin (and its head at (a_1, a_2, a_3)) or with its tail at any other point, so long as the length and sense of the arrow are not disturbed. (See Figure 1.4.) For example, if we wish to represent $\mathbf{a}$ by an arrow with its tail at the point (x_1, x_2, x_3), then the head of the arrow would be at the point $(x_1 + a_1, x_2 + a_2, x_3 + a_3)$. (See Figure 1.5.)

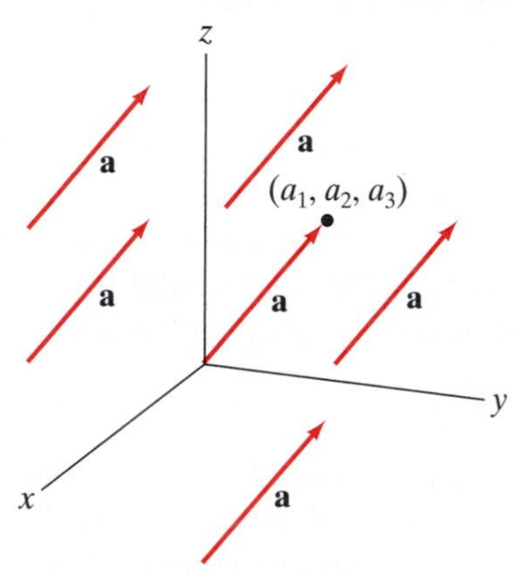

Figure 1.4 Each arrow is a parallel translate of the position vector of the point (a_1, a_2, a_3) and represents the same vector.

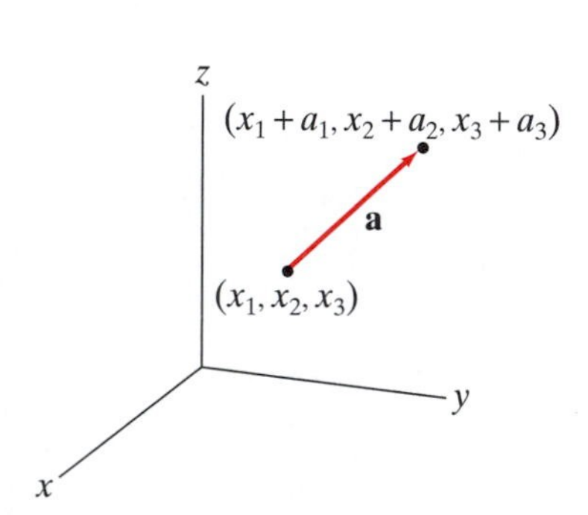

Figure 1.5 The vector $\mathbf{a} = (a_1, a_2, a_3)$ represented by an arrow with tail at the point (x_1, x_2, x_3).

With this geometric description of vectors, vector addition can be visualized in two ways. The first is often referred to as the "head-to-tail" method for adding vectors. Draw the two vectors $\mathbf{a}$ and $\mathbf{b}$ to be added so that the tail of one of the vectors, say $\mathbf{b}$, is at the head of the other. Then the vector sum $\mathbf{a} + \mathbf{b}$ may be represented by an arrow whose tail is at the tail of $\mathbf{a}$ and whose head is at the head of $\mathbf{b}$. (See Figure 1.6.) Note that it is *not* immediately obvious that $\mathbf{a} + \mathbf{b} = \mathbf{b} + \mathbf{a}$ from this construction!

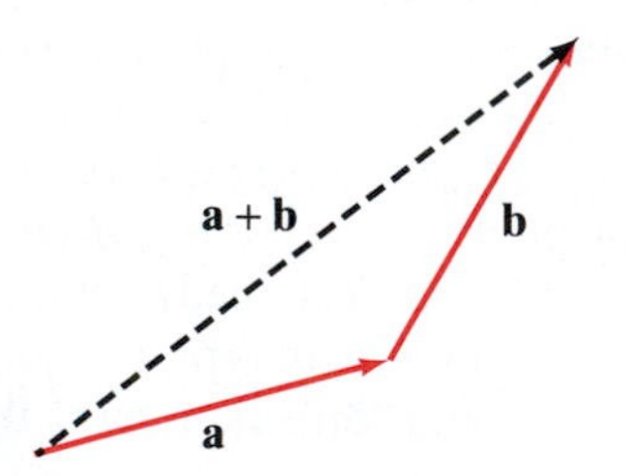

Figure 1.6 The vector $\mathbf{a} + \mathbf{b}$ may be represented by an arrow whose tail is at the tail of $\mathbf{a}$ and whose head is at the head of $\mathbf{b}$.

The second way to visualize vector addition is according to the so-called **parallelogram law**: If $\mathbf{a}$ and $\mathbf{b}$ are nonparallel vectors drawn with their tails emanating from the same point, then $\mathbf{a} + \mathbf{b}$ may be represented by the arrow (with its tail at the common initial point of $\mathbf{a}$ and $\mathbf{b}$) that runs along the diagonal of the parallelogram determined by $\mathbf{a}$ and $\mathbf{b}$ (Figure 1.7). The parallelogram law is completely consistent with the head-to-tail method. To see why, just parallel translate $\mathbf{b}$ to the opposite side of the parallelogram. Then the diagonal just described is the result of adding $\mathbf{a}$ and (the translate of) $\mathbf{b}$, using the head-to-tail method. (See Figure 1.8.)

We still should check that these geometric constructions agree with our algebraic definition. For simplicity, we'll work in $\mathbf{R}^2$. Let $\mathbf{a} = (a_1, a_2)$ and $\mathbf{b} = (b_1, b_2)$ as usual. Then the arrow obtained from the parallelogram law addition of $\mathbf{a}$ and $\mathbf{b}$ is the one whose tail is at the origin O and whose head is at the point P in Figure 1.9. If we parallel translate $\mathbf{b}$ so that its tail is at the head of $\mathbf{a}$, then it is immediate that the coordinates of P must be $(a_1 + b_1, a_2 + b_2)$, as desired.

Scalar multiplication is easier to visualize: The vector $k\mathbf{a}$ may be represented by an arrow whose length is $|k|$ times the length of $\mathbf{a}$ and whose direction is the same as that of $\mathbf{a}$ when $k > 0$ and the opposite when $k < 0$. (See Figure 1.10.)

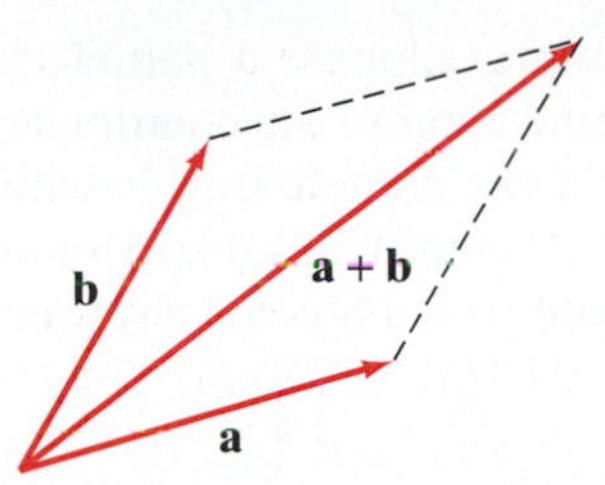

Figure 1.7 The vector $\mathbf{a}+\mathbf{b}$ may be represented by the arrow that runs along the diagonal of the parallelogram determined by **a** and **b**.

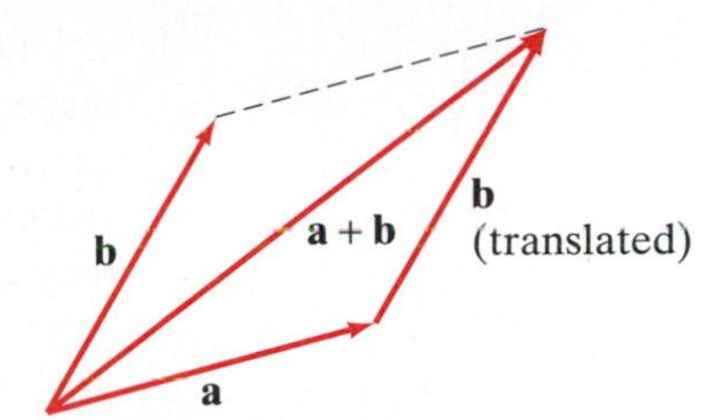

Figure 1.8 The equivalence of the parallelogram law and the head-to-tail methods of vector addition.

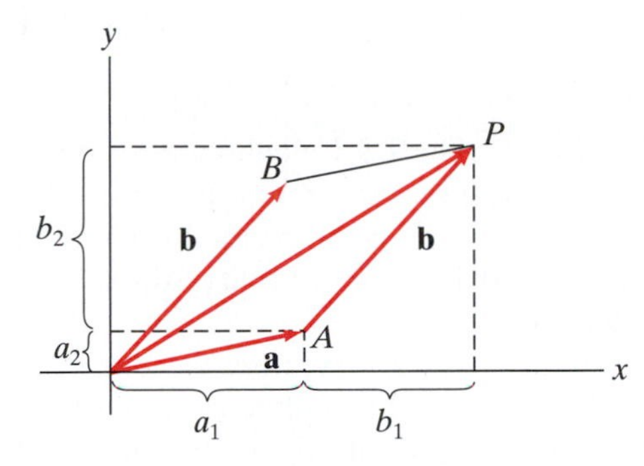

Figure 1.9 The point P has coordinates (a_1+b_1, a_2+b_2).

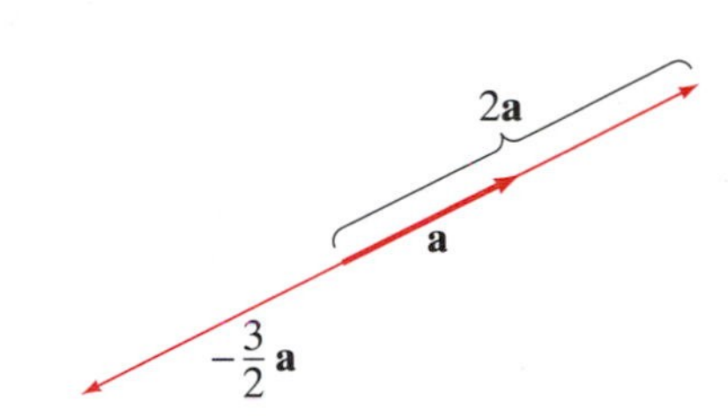

Figure 1.10 Visualization of scalar multiplication.

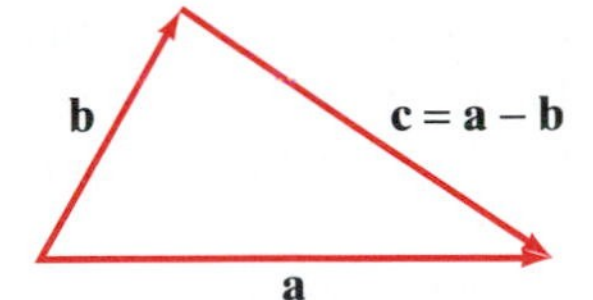

Figure 1.11 The geometry of vector subtraction. The vector **c** is such that $\mathbf{b}+\mathbf{c}=\mathbf{a}$. Hence, $\mathbf{c}=\mathbf{a}-\mathbf{b}$.

It is now a simple matter to obtain a geometric depiction of the **difference** between two vectors. (See Figure 1.11.) The difference $\mathbf{a}-\mathbf{b}$ is nothing more than $\mathbf{a}+(-\mathbf{b})$ (where $-\mathbf{b}$ means the scalar -1 times the vector **b**). The vector $\mathbf{a}-\mathbf{b}$ may be represented by an arrow pointing from the head of **b** toward the head of **a**; such an arrow is also a diagonal of the parallelogram determined by **a** and **b**. (As we have seen, the other diagonal can be used to represent $\mathbf{a}+\mathbf{b}$.)

Here is a construction that will be useful to us from time to time.

> ■ **Definition 1.5** Given two points $P_1(x_1, y_1, z_1)$ and $P_2(x_2, y_2, z_2)$ in $\mathbf{R}^3$, the **displacement vector from** P_1 **to** P_2 is
>
> $$\overrightarrow{P_1P_2} = (x_2 - x_1, y_2 - y_1, z_2 - z_1).$$

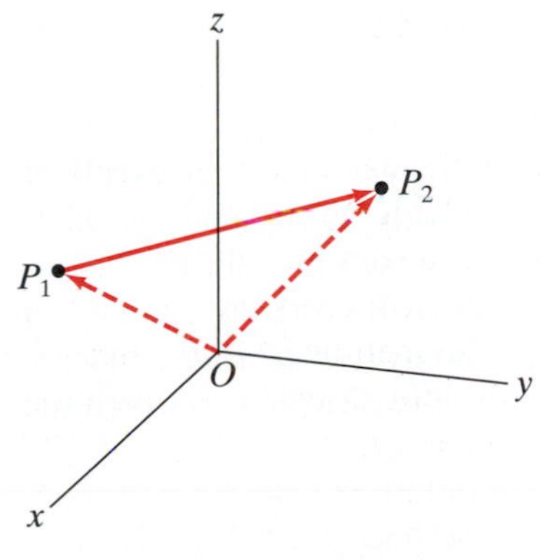

Figure 1.12 The displacement vector $\overrightarrow{P_1P_2}$, represented by the arrow from P_1 to P_2, is the difference between the position vectors of these two points.

This construction is not hard to understand if we consider Figure 1.12. Given the points P_1 and P_2, draw the corresponding position vectors $\overrightarrow{OP_1}$ and $\overrightarrow{OP_2}$. Then we see that $\overrightarrow{P_1P_2}$ is precisely $\overrightarrow{OP_2}-\overrightarrow{OP_1}$. An analogous definition may be made for $\mathbf{R}^2$.

In your study of the calculus of one variable, you no doubt used the notions of derivatives and integrals to look at such physical concepts as velocity, acceleration, force, etc. The main drawback of the work you did was that the techniques involved allowed you to study only *rectilinear,* or straight-line, activity. Intuitively, we all understand that motion in the plane or in space is more complicated than straight-line motion. Because vectors possess direction as well as magnitude, they are ideally suited for two- and three-dimensional dynamical problems.

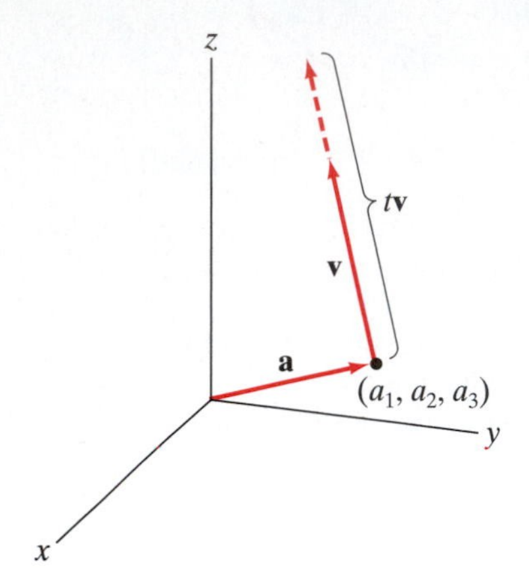

Figure 1.13 After t seconds, the point starting at $\mathbf{a}$, with velocity $\mathbf{v}$, moves to $\mathbf{a} + t\mathbf{v}$.

For example, suppose a particle in space is at the point (a_1, a_2, a_3) (with respect to some appropriate coordinate system). Then it has position vector $\mathbf{a} = (a_1, a_2, a_3)$. If the particle travels with constant velocity $\mathbf{v} = (v_1, v_2, v_3)$ for t seconds, then the particle's displacement from its original position is $t\mathbf{v}$, and its new coordinate position is $\mathbf{a} + t\mathbf{v}$. (See Figure 1.13.)

EXAMPLE 4 If a spaceship is at position (100, 3, 700) and is traveling with velocity (7, −10, 25) (meaning that the ship travels 7 mi/sec in the positive x-direction, 10 mi/sec in the negative y-direction, and 25 mi/sec in the positive z-direction), then after 20 seconds, the ship will be at position

$$(100, 3, 700) + 20(7, -10, 25) = (240, -197, 1200),$$

and the displacement from the initial position is (140, −200, 500). ◆

EXAMPLE 5 The S.S. Calculus is cruising due south at a rate of 15 knots (nautical miles per hour) with respect to still water. However, there is also a current of $5\sqrt{2}$ knots southeast. What is the total velocity of the ship? If the ship is initially at the origin and a lobster pot is at position (20, −79), will the ship collide with the lobster pot?

Since velocities are vectors, the total velocity of the ship is $\mathbf{v}_1 + \mathbf{v}_2$, where $\mathbf{v}_1$ is the velocity of the ship with respect to still water and $\mathbf{v}_2$ is the southeast-pointing velocity of the current. Figure 1.14 makes it fairly straightforward to compute these velocities. We have that $\mathbf{v}_1 = (0, -15)$. Since $\mathbf{v}_2$ points south-eastward, its direction must be along the line $y = -x$. Therefore, $\mathbf{v}_2$ can be written as $\mathbf{v}_2 = (v, -v)$, where v is a positive real number. By the Pythagorean theorem, if the length of $\mathbf{v}_2$ is $5\sqrt{2}$, then we must have $v^2 + (-v)^2 = (5\sqrt{2})^2$ or $2v^2 = 50$, so that $v = 5$. Thus, $\mathbf{v}_2 = (5, -5)$, and, hence, the net velocity is

$$(0, -15) + (5, -5) = (5, -20).$$

After 4 hours, therefore, the ship will be at position

$$(0, 0) + 4(5, -20) = (20, -80)$$

and thus will miss the lobster pot. ◆

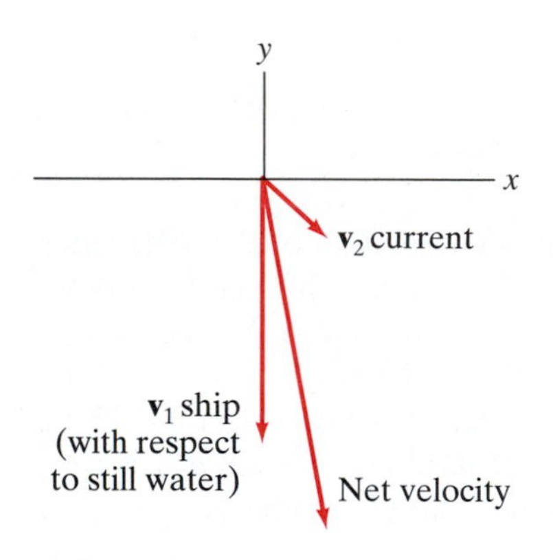

Figure 1.14 The length of $\mathbf{v}_1$ is 15, and the length of $\mathbf{v}_2$ is $5\sqrt{2}$.

EXAMPLE 6 The theory behind the venerable martial art of judo is an excellent example of vector addition. If two people, one relatively strong and the other relatively weak, have a shoving match, it is clear who will prevail. For example, someone pushing one way with 200 lb of force will certainly succeed in overpowering another pushing the opposite way with 100 lb of force. Indeed, as Figure 1.15 shows, the net force will be 100 lb in the direction in which the stronger person is pushing.

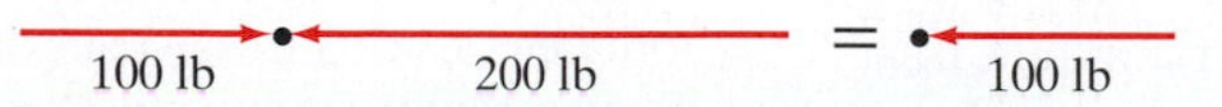

Figure 1.15 A relatively strong person pushing with a force of 200 lb can quickly subdue a relatively weak one pushing with only 100 lb of force.

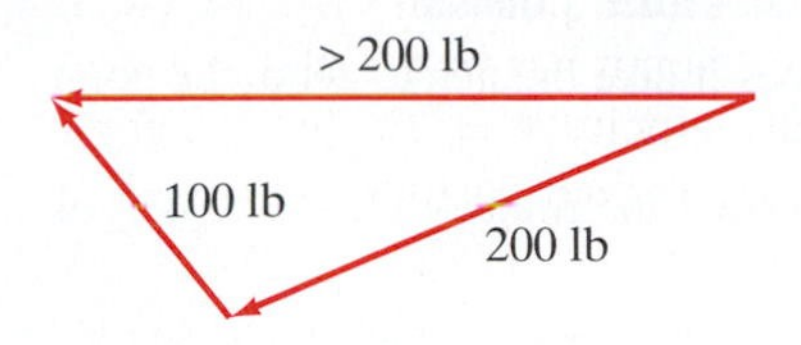

Figure 1.16 Vector addition in judo.

Dr. Jigoro Kano, the founder of judo, realized (though he never expressed his idea in these terms) that this sort of vector addition favors the strong over the weak. However, if the weaker participant applies his or her 100 lb of force in a direction only slightly different from that of the stronger, he or she will effect a vector sum of length large enough to surprise the opponent. (See Figure 1.16.) This is the basis for essentially all of the throws of judo, and what is meant when judo is described as the art of "using a person's strength against himself or herself." In fact, the word "judo" means "the giving way." One "gives in" to the strength of another by attempting only to redirect his or her force, rather than to oppose it. ◆

Exercises

1. Sketch the following vectors in $\mathbf{R}^2$:
(a) $(2, 1)$ (b) $(3, 3)$ (c) $(-1, 2)$

2. Sketch the following vectors in $\mathbf{R}^3$:
(a) $(1, 2, 3)$ (b) $(-2, 0, 2)$ (c) $(2, -3, 1)$

3. Perform the indicated algebraic operations. Express your answers in the form of a single vector $\mathbf{a} = (a_1, a_2)$ in $\mathbf{R}^2$.

(a) $(3, 1) + (-1, 7)$

(b) $-2(8, 12)$

(c) $(8, 9) + 3(-1, 2)$

(d) $(1, 1) + 5(2, 6) - 3(10, 2)$

(e) $(8, 10) + 3\,((8, -2) - 2(4, 5))$

4. Perform the indicated algebraic operations. Express your answers in the form of a single vector $\mathbf{a} = (a_1, a_2, a_3)$ in $\mathbf{R}^3$.

(a) $(2, 1, 2) + (-3, 9, 7)$

(b) $\frac{1}{2}(8, 4, 1) + 2(5, -7, \frac{1}{4})$

(c) $-2\left((2, 0, 1) - 6(\frac{1}{2}, -4, 1)\right)$

5. Graph the vectors $\mathbf{a} = (1, 2)$, $\mathbf{b} = (-2, 5)$, and $\mathbf{a} + \mathbf{b} = (1, 2) + (-2, 5)$, using both the parallelogram law and the head-to-tail method.

6. Graph the vectors $\mathbf{a} = (3, 2)$ and $\mathbf{b} = (-1, 1)$. Also calculate and graph $\mathbf{a} - \mathbf{b}$, $\frac{1}{2}\mathbf{a}$, and $\mathbf{a} + 2\mathbf{b}$.

7. Let A be the point with coordinates $(1, 0, 2)$, let B be the point with coordinates $(-3, 3, 1)$, and let C be the point with coordinates $(2, 1, 5)$.

(a) Describe the vectors $\overrightarrow{AB}$ and $\overrightarrow{BA}$.

(b) Describe the vectors $\overrightarrow{AC}$, $\overrightarrow{BC}$, and $\overrightarrow{AC} + \overrightarrow{CB}$.

(c) Explain, with pictures, why $\overrightarrow{AC} + \overrightarrow{CB} = \overrightarrow{AB}$.

8. Graph $(1, 2, 1)$ and $(0, -2, 3)$, and calculate and graph $(1, 2, 1) + (0, -2, 3)$, $-1(1, 2, 1)$, and $4(1, 2, 1)$.

9. If $(-12, 9, z) + (x, 7, -3) = (2, y, 5)$, what are x, y, and z?

10. What is the length (magnitude) of the vector $(3, 1)$? (Hint: A diagram will help.)

11. Sketch the vectors $\mathbf{a} = (1, 2)$ and $\mathbf{b} = (5, 10)$. Explain why $\mathbf{a}$ and $\mathbf{b}$ point in the same direction.

12. Sketch the vectors $\mathbf{a} = (2, -7, 8)$ and $\mathbf{b} = \left(-1, \frac{7}{2}, -4\right)$. Explain why $\mathbf{a}$ and $\mathbf{b}$ point in opposite directions.

13. How would you add the vectors $(1, 2, 3, 4)$ and $(5, -1, 2, 0)$ in $\mathbf{R}^4$? What should $2(7, 6, -3, 1)$ be? In general, suppose that

$$\mathbf{a} = (a_1, a_2, \ldots, a_n) \quad \text{and} \quad \mathbf{b} = (b_1, b_2, \ldots, b_n)$$

are two vectors in $\mathbf{R}^n$ and $k \in \mathbf{R}$ is a scalar. Then how would you define $\mathbf{a} + \mathbf{b}$ and $k\mathbf{a}$?

14. Find the displacement vectors from P_1 to P_2, where P_1 and P_2 are the points given. Sketch P_1, P_2, and $\overrightarrow{P_1P_2}$.

(a) $P_1(1, 0, 2)$, $P_2(2, 1, 7)$

(b) $P_1(1, 6, -1)$, $P_2(0, 4, 2)$

(c) $P_1(0, 4, 2)$, $P_2(1, 6, -1)$

(d) $P_1(3, 1)$, $P_2(2, -1)$

15. Let $P_1(2, 5, -1, 6)$ and $P_2(3, 1, -2, 7)$ be two points in $\mathbf{R}^4$. How would you define and calculate the displacement vector from P_1 to P_2? (See Exercise 13.)

16. If A is the point in $\mathbf{R}^3$ with coordinates $(2, 5, -6)$ and the displacement vector from A to a second point B is $(12, -3, 7)$, what are the coordinates of B?

17. Suppose that you and your friend are in New York talking on cellular phones. You inform each other of your own displacement vectors from the Empire State Building to your current position. Explain how you can use this information to determine the displacement vector from you to your friend.

18. Give the details of the proofs of properties 2 and 3 of vector addition given in this section.

19. Prove the properties of scalar multiplication given in this section.

20. (a) If $\mathbf{a}$ is a vector in $\mathbf{R}^2$ or $\mathbf{R}^3$, what is $0\mathbf{a}$? Prove your answer.

(b) If $\mathbf{a}$ is a vector in $\mathbf{R}^2$ or $\mathbf{R}^3$, what is $1\mathbf{a}$? Prove your answer.

21. (a) Let $\mathbf{a} = (2, 0)$ and $\mathbf{b} = (1, 1)$. For $0 \leq s \leq 1$ and $0 \leq t \leq 1$, consider the vector $\mathbf{x} = s\mathbf{a} + t\mathbf{b}$. Explain why the vector $\mathbf{x}$ lies in the parallelogram determined by $\mathbf{a}$ and $\mathbf{b}$. (Hint: It may help to draw a picture.)

(b) Now suppose that $\mathbf{a} = (2, 2, 1)$ and $\mathbf{b} = (0, 3, 2)$. Describe the set of vectors $\{\mathbf{x} = s\mathbf{a} + t\mathbf{b} \mid 0 \leq s \leq 1, 0 \leq t \leq 1\}$.

22. Let $\mathbf{a} = (a_1, a_2, a_3)$ and $\mathbf{b} = (b_1, b_2, b_3)$ be two nonzero vectors such that $\mathbf{b} \neq k\mathbf{a}$. Use vectors to describe the set of points inside the parallelogram with vertex $P_0(x_0, y_0, z_0)$ and whose adjacent sides are parallel to $\mathbf{a}$ and $\mathbf{b}$ and have the same lengths as $\mathbf{a}$ and $\mathbf{b}$. (See Figure 1.17.) (Hint: If $P(x, y, z)$ is a point in the parallelogram, describe $\overrightarrow{OP}$, the position vector of P.)

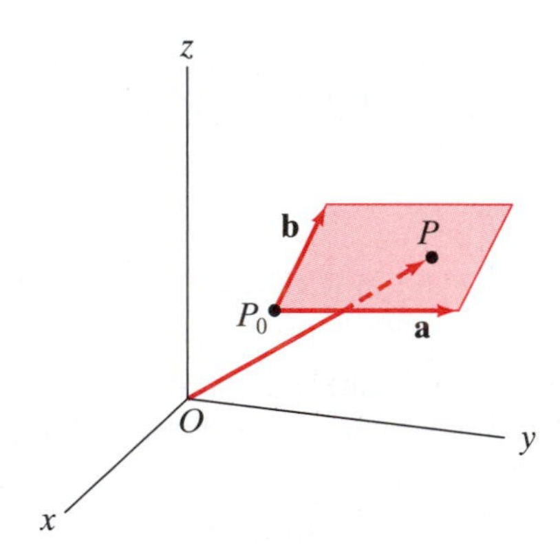

Figure 1.17 Figure for Exercise 22.

23. A flea falls onto marked graph paper at the point $(3, 2)$. She begins moving from that point with velocity vector $\mathbf{v} = (-1, -2)$ (i.e., she moves 1 graph paper unit per minute in the negative x-direction and 2 graph paper units per minute in the negative y-direction).

(a) What is the speed of the flea?

(b) Where is the flea after 3 minutes?

(c) How long does it take the flea to get to the point $(-4, -12)$?

(d) Does the flea reach the point $(-13, -27)$? Why or why not?

24. A plane takes off from an airport with velocity vector $(50, 100, 4)$. Assume that the units are miles per hour, that the positive x-axis points east, and that the positive y-axis points north.

(a) How fast is the plane climbing vertically at take-off?

(b) Suppose the airport is located at the origin and a skyscraper is located 5 miles east and 10 miles north of the airport. The skyscraper is 1,250 feet tall. When will the plane be directly over the building?

(c) When the plane is over the building, how much vertical clearance is there?

25. As mentioned in the text, physical forces (e.g., gravity) are quantities possessing both magnitude and direction and therefore can be represented by vectors. If an object has more than one force acting on it, then the **resultant** (or **net**) **force** can be represented by the sum of the individual force vectors. Suppose that two forces, $\mathbf{F}_1 = (2, 7, -1)$ and $\mathbf{F}_2 = (3, -2, 5)$, act on an object.

(a) What is the resultant force of $\mathbf{F}_1$ and $\mathbf{F}_2$?

(b) What force $\mathbf{F}_3$ is needed to counteract these forces (that is, so that *no* net force results and the object remains at rest)?

26. A 50 lb sandbag is suspended by two ropes. Suppose that a three-dimensional coordinate system is introduced so that the sandbag is at the origin and the ropes are anchored at the points $(0, -2, 1)$ and $(0, 2, 1)$.

(a) Assuming that the force due to gravity points parallel to the vector $(0, 0, -1)$, give a vector $\mathbf{F}$ which describes this gravitational force.

(b) Now, use vectors to describe the forces along each of the two ropes. Use symmetry considerations and draw a figure of the situation.

1.2 More About Vectors

The Standard Basis Vectors

In $\mathbf{R}^2$, the vectors $\mathbf{i} = (1, 0)$ and $\mathbf{j} = (0, 1)$ play a special notational role. Any vector $\mathbf{a} = (a_1, a_2)$ may be written in terms of $\mathbf{i}$ and $\mathbf{j}$ via vector addition and scalar multiplication:

$$(a_1, a_2) = (a_1, 0) + (0, a_2) = a_1(1, 0) + a_2(0, 1) = a_1\,\mathbf{i} + a_2\,\mathbf{j}.$$

(It may be easier to follow this argument by reading it in reverse.) Insofar as notation goes, the preceding work simply establishes that one can write either (a_1, a_2) or $a_1\mathbf{i} + a_2\mathbf{j}$ to denote the vector $\mathbf{a}$. It's your choice which notation

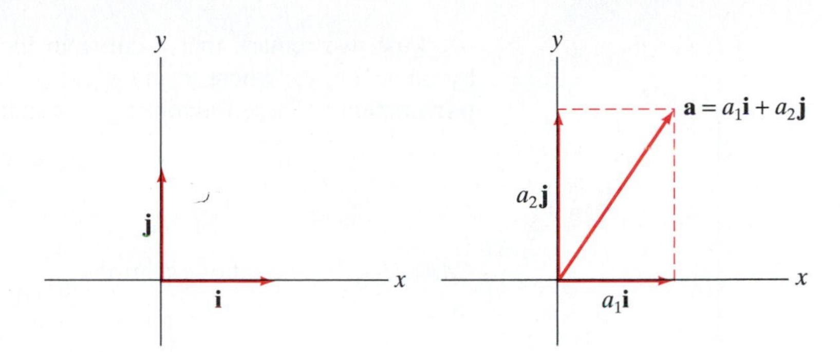

Figure 1.18 Any vector in $\mathbf{R}^2$ can be written in terms of $\mathbf{i}$ and $\mathbf{j}$.

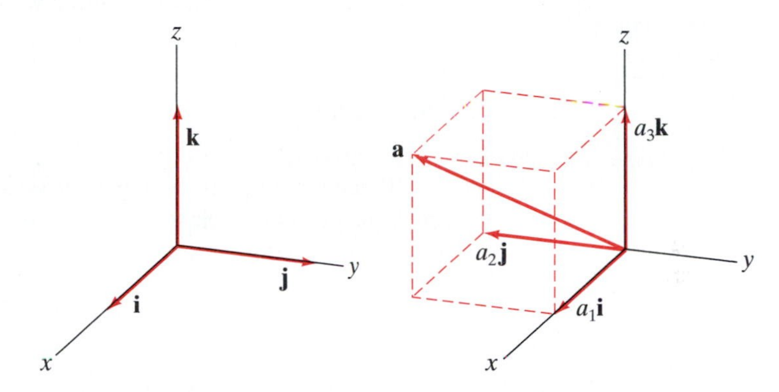

Figure 1.19 Any vector in $\mathbf{R}^3$ can be written in terms of $\mathbf{i}$, $\mathbf{j}$, and $\mathbf{k}$.

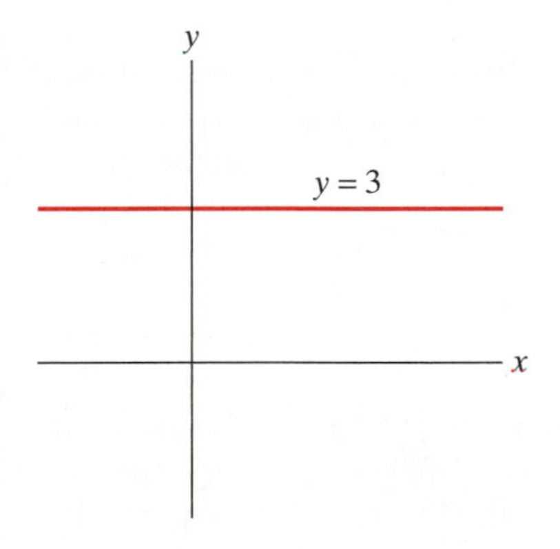

Figure 1.20 In $\mathbf{R}^2$, the equation $y = 3$ describes a line.

to use (as long as you're consistent), but the **ij**-notation is generally useful for emphasizing the "vector" nature of **a**, while the coordinate notation is more useful for emphasizing the "point" nature of **a** (in the sense of **a**'s role as a possible position vector of a point). Geometrically, the significance of the **standard basis vectors i and j** is that an arbitrary vector $\mathbf{a} \in \mathbf{R}^2$ can be decomposed pictorially into appropriate **vector components** along the x- and y-axes, as shown in Figure 1.18.

Exactly the same situation occurs in $\mathbf{R}^3$, except that we need three vectors, $\mathbf{i} = (1, 0, 0)$, $\mathbf{j} = (0, 1, 0)$, and $\mathbf{k} = (0, 0, 1)$, to form the standard basis. (See Figure 1.19.) The same argument as the one just given can be used to show that any vector $\mathbf{a} = (a_1, a_2, a_3)$ may also be written as $a_1\,\mathbf{i} + a_2\,\mathbf{j} + a_3\,\mathbf{k}$. We shall use both coordinate and standard basis notation throughout this text.

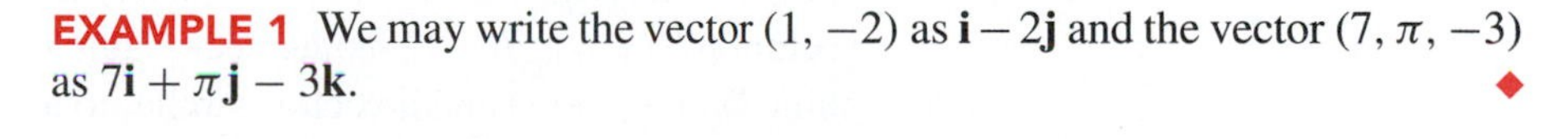

EXAMPLE 1 We may write the vector $(1, -2)$ as $\mathbf{i} - 2\mathbf{j}$ and the vector $(7, \pi, -3)$ as $7\mathbf{i} + \pi\mathbf{j} - 3\mathbf{k}$. ◆

Parametric Equations of Lines

In $\mathbf{R}^2$, we know that equations of the form $y = mx + b$ or $Ax + By = C$ describe straight lines. (See Figure 1.20.) Consequently, one might expect the same sort of equation to define a line in $\mathbf{R}^3$ as well. Consideration of a simple example or two (such as in Figure 1.21) should convince you that a single such linear equation describes a plane, not a line. A pair of simultaneous equations in x, y, and z is required to define a line.

We postpone discussing the derivation of equations for planes until §1.5 and concentrate here on using vectors to give sets of parametric equations for lines in $\mathbf{R}^2$ or $\mathbf{R}^3$ (or even $\mathbf{R}^n$).

Figure 1.21 In $\mathbf{R}^3$, the equation $y = 3$ describes a plane.

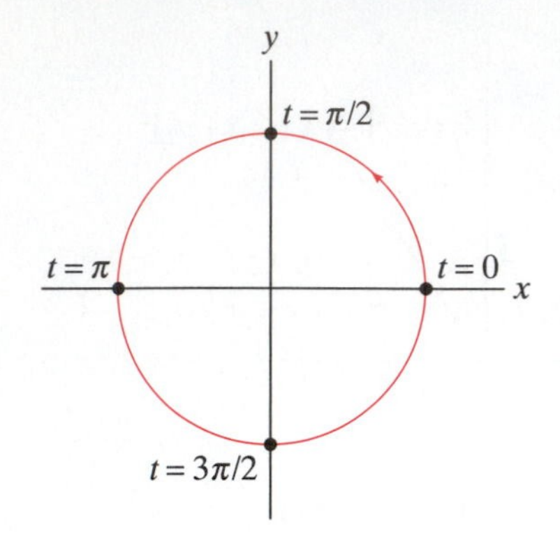

Figure 1.22 The graph of the parametric equations $x = 2\cos t$, $y = 2\sin t$, $0 \leq t < 2\pi$.

First, we remark that a curve in the plane may be described analytically by points (x, y), where x and y are given as functions of a third variable (the **parameter**) t. These functions give rise to **parametric equations** for the curve:

$$\begin{cases} x = f(t) \\ y = g(t) \end{cases}.$$

EXAMPLE 2 The set of equations

$$\begin{cases} x = 2\cos t \\ y = 2\sin t \end{cases} \quad 0 \leq t < 2\pi$$

describes a circle of radius 2, since we may check that

$$x^2 + y^2 = (2\cos t)^2 + (2\sin t)^2 = 4.$$

(See Figure 1.22.) ◆

Parametric equations may be used as readily to describe curves in $\mathbf{R}^3$; a curve in $\mathbf{R}^3$ is the set of points (x, y, z) whose coordinates x, y, and z are each given by a function of t:

$$\begin{cases} x = f(t) \\ y = g(t) \\ z = h(t) \end{cases}.$$

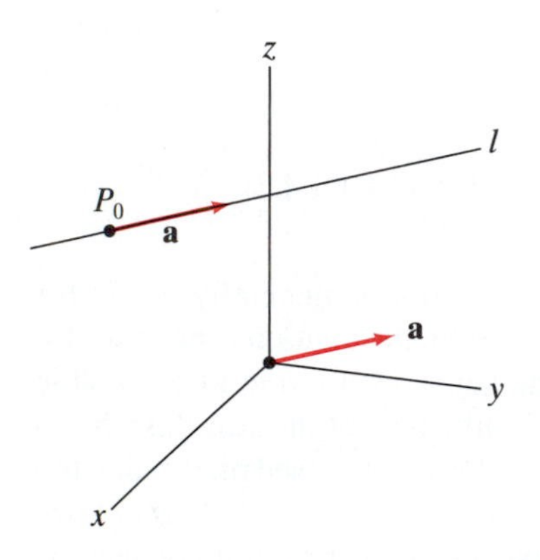

Figure 1.23 The line l is the unique line passing through P_0 and parallel to the vector **a**.

The advantages of using parametric equations are twofold. First, they offer a uniform way of describing curves in any number of dimensions. (How would you define parametric equations for a curve in $\mathbf{R}^4$? In $\mathbf{R}^{128}$?) Second, they allow you to get a dynamic sense of a curve if you consider the parameter variable t to represent time and imagine that a particle is traveling along the curve with time according to the given parametric equations. You can represent this geometrically by assigning a "direction" to the curve to signify increasing t. Notice the arrow in Figure 1.22.

Now, we see how to provide equations for lines. First, convince yourself that a line in $\mathbf{R}^2$ or $\mathbf{R}^3$ is uniquely determined by two pieces of geometric information: (1) a vector whose direction is parallel to that of the line and (2) any particular point lying on the line—see Figure 1.23. In Figure 1.24, we seek the vector

$$\mathbf{r} = \overrightarrow{OP}$$

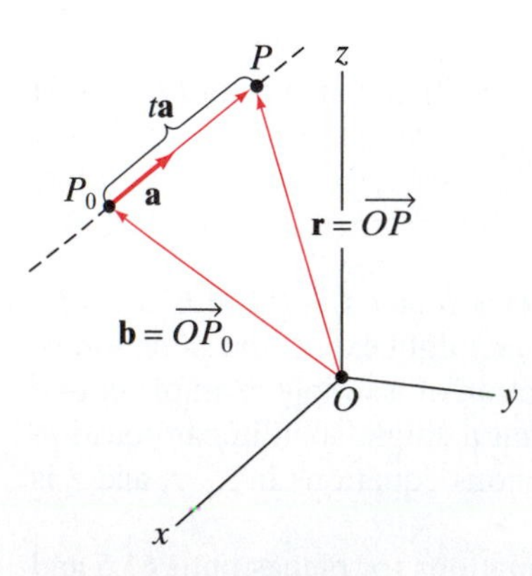

Figure 1.24 The graph of a line in $\mathbf{R}^3$.

between the origin O and an arbitrary point P on the line l (i.e., the position vector of $P(x, y, z)$). $\overrightarrow{OP}$ is the vector sum of the position vector **b** of the given point P_0 (i.e., $\overrightarrow{OP_0}$) and a vector parallel to **a**. Any vector parallel to **a** must be a scalar multiple of **a**. Letting this scalar be the parameter variable t, we have

$$\mathbf{r} = \overrightarrow{OP} = \overrightarrow{OP_0} + t\mathbf{a},$$

and we have established the following proposition:

■ **Proposition 2.1** The vector parametric equation for the line through the point $P_0(b_1, b_2, b_3)$, whose position vector is $\overrightarrow{OP_0} = \mathbf{b} = b_1\mathbf{i} + b_2\mathbf{j} + b_3\mathbf{k}$, and parallel to $\mathbf{a} = a_1\mathbf{i} + a_2\mathbf{j} + a_3\mathbf{k}$ is

$$\mathbf{r}(t) = \mathbf{b} + t\mathbf{a}. \tag{1}$$

Expanding formula (1),

$$\begin{aligned}\mathbf{r}(t) = \overrightarrow{OP} &= b_1\mathbf{i} + b_2\mathbf{j} + b_3\mathbf{k} + t(a_1\mathbf{i} + a_2\mathbf{j} + a_3\mathbf{k}) \\ &= (a_1t + b_1)\mathbf{i} + (a_2t + b_2)\mathbf{j} + (a_3t + b_3)\mathbf{k}.\end{aligned}$$

Next, write $\overrightarrow{OP}$ as $x\mathbf{i}+y\mathbf{j}+z\mathbf{k}$ so that P has coordinates (x, y, z). Then, extracting components, we see that the coordinates of P are $(a_1t + b_1, a_2t + b_2, a_3t + b_3)$ and our parametric equations are

$$\begin{cases} x = a_1t + b_1 \\ y = a_2t + b_2 \\ z = a_3t + b_3 \end{cases}, \tag{2}$$

where t is any real number.

These parametric equations work just as well in $\mathbf{R}^2$ (if we ignore the z-component) or in $\mathbf{R}^n$ where n is arbitrary. In $\mathbf{R}^n$, formula (1) remains valid, where we take $\mathbf{a} = (a_1, a_2, \dots, a_n)$ and $\mathbf{b} = (b_1, b_2, \dots, b_n)$. The resulting parametric equations are

$$\begin{cases} x_1 = a_1t + b_1 \\ x_2 = a_2t + b_2 \\ \quad\vdots \\ x_n = a_nt + b_n \end{cases}.$$

EXAMPLE 3 To find the parametric equations of the line through $(1, -2, 3)$ and parallel to the vector $\pi\mathbf{i} - 3\mathbf{j} + \mathbf{k}$, we have $\mathbf{a} = \pi\mathbf{i} - 3\mathbf{j} + \mathbf{k}$ and $\mathbf{b} = \mathbf{i} - 2\mathbf{j} + 3\mathbf{k}$ so that formula (1) yields

$$\begin{aligned}\mathbf{r}(t) &= \mathbf{i} - 2\mathbf{j} + 3\mathbf{k} + t(\pi\mathbf{i} - 3\mathbf{j} + \mathbf{k}) \\ &= (1 + \pi t)\mathbf{i} + (-2 - 3t)\mathbf{j} + (3 + t)\mathbf{k}.\end{aligned}$$

The parametric equations may be read as

$$\begin{cases} x = \pi t + 1 \\ y = -3t - 2 \\ z = t + 3 \end{cases}.$$

◆

EXAMPLE 4 From Euclidean geometry, two distinct points determine a unique line in $\mathbf{R}^2$ or $\mathbf{R}^3$. Let's find the parametric equations of the line through the points $P_0(1, -2, 3)$ and $P_1(0, 5, -1)$. The situation is suggested by Figure 1.25. To use formula (1), we need to find a vector $\mathbf{a}$ parallel to the desired line. The vector with tail at P_0 and head at P_1 is such a vector. That is, we may use for $\mathbf{a}$ the vector

$$\overrightarrow{P_0P_1} = (0 - 1, 5 - (-2), -1 - 3) = -\mathbf{i} + 7\mathbf{j} - 4\mathbf{k}.$$

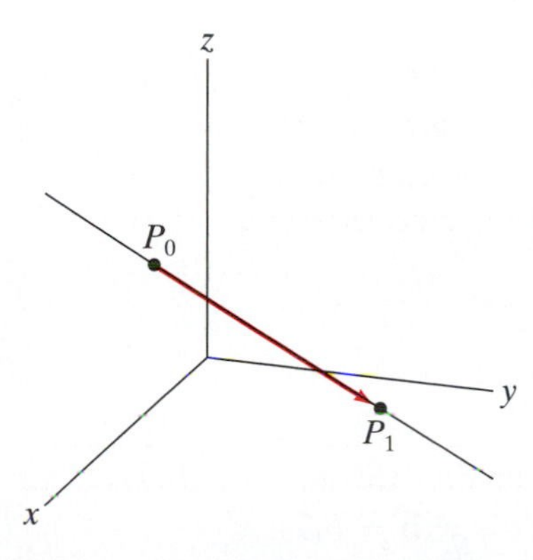

Figure 1.25 Finding equations for a line through two points in Example 4.

For $\mathbf{b}$, the position vector of a particular point on the line, we have the choice of taking either $\mathbf{b} = \mathbf{i} - 2\mathbf{j} + 3\mathbf{k}$ or $\mathbf{b} = 5\mathbf{j} - \mathbf{k}$. Hence, the equations in (2) yield parametric equations

$$\begin{cases} x = 1 - t \\ y = -2 + 7t \\ z = 3 - 4t \end{cases} \quad \text{or} \quad \begin{cases} x = -t \\ y = 5 + 7t \\ z = -1 - 4t \end{cases}.$$

◆

In general, given two *arbitrary* points

$$P_0(a_1, a_2, a_3) \quad \text{and} \quad P_1(b_1, b_2, b_3),$$

the line joining them has vector parametric equation

$$\mathbf{r}(t) = \overrightarrow{OP_0} + t\overrightarrow{P_0P_1}. \tag{3}$$

Equation (3) gives parametric equations

$$\begin{cases} x = a_1 + (b_1 - a_1)t \\ y = a_2 + (b_2 - a_2)t \\ z = a_3 + (b_3 - a_3)t \end{cases}. \tag{4}$$

Alternatively, in place of equation (3), we could use the vector equation

$$\mathbf{r}(t) = \overrightarrow{OP_1} + t\overrightarrow{P_0P_1}, \tag{5}$$

or perhaps

$$\mathbf{r}(t) = \overrightarrow{OP_1} + t\overrightarrow{P_1P_0}, \tag{6}$$

each of which gives rise to somewhat different sets of parametric equations. Again, we refer you to Figure 1.25 for an understanding of the vector geometry involved.

Example 4 brings up an important point, namely, that parametric equations for a line (or, more generally, for any curve) are *never* unique. In fact, the two sets of equations calculated in Example 4 are by no means the only ones; we could have taken $\mathbf{a} = \overrightarrow{P_1P_0} = \mathbf{i} - 7\mathbf{j} + 4\mathbf{k}$ or any nonzero scalar multiple of $\overrightarrow{P_0P_1}$ for $\mathbf{a}$.

If parametric equations are not determined uniquely, then how can you check your work? In general, this is not so easy to do, but in the case of lines, there are two approaches to take. One is to produce two points that lie on the line specified by the first set of parametric equations and see that these points lie on the line given by the second set of parametric equations. The other approach is to use the parametric equations to find what is called the **symmetric form** of a line in $\mathbf{R}^3$. From the equations in (2), assuming that each a_i is nonzero, one can eliminate the parameter variable t in each equation to obtain:

$$\begin{cases} t = \dfrac{x - b_1}{a_1} \\ t = \dfrac{y - b_2}{a_2} \\ t = \dfrac{z - b_3}{a_3} \end{cases}.$$

The symmetric form is

$$\frac{x - b_1}{a_1} = \frac{y - b_2}{a_2} = \frac{z - b_3}{a_3}. \tag{7}$$

In Example 4, the two sets of parametric equations give rise to corresponding symmetric forms

$$\frac{x-1}{-1} = \frac{y+2}{7} = \frac{z-3}{-4} \quad \text{and} \quad \frac{x}{-1} = \frac{y-5}{7} = \frac{z+1}{-4}.$$

It's not difficult to see that adding 1 to each "side" of the second symmetric form yields the first one. In general, symmetric forms for lines can differ only by a constant term or constant scalar multiples (or both).

The symmetric form is really a set of two simultaneous equations in $\mathbf{R}^3$. For example, the information in (7) can also be written as

$$\begin{cases} \dfrac{x-b_1}{a_1} = \dfrac{y-b_2}{a_2} \\[2ex] \dfrac{x-b_1}{a_1} = \dfrac{z-b_3}{a_3} \end{cases}.$$

This illustrates that we require two "scalar" equations in x, y, and z to describe a line in $\mathbf{R}^3$, although a single vector parametric equation, formula (1), is sufficient.

The next two examples illustrate how to use parametric equations for lines to identify the intersection of a line and a plane, or of two lines.

EXAMPLE 5 We find where the line with parametric equations

$$\begin{cases} x = t + 5 \\ y = -2t - 4 \\ z = 3t + 7 \end{cases}$$

intersects the plane $3x + 2y - 7z = 2$.

To locate the point of intersection, we must find what value of the parameter t gives a point on the line that also lies in the plane. This is readily accomplished by substituting the parametric values for x, y, and z from the line into the equation for the plane

$$3(t+5) + 2(-2t-4) - 7(3t+7) = 2. \tag{8}$$

Solving equation (8) for t, we find that $t = -2$. Setting t equal to -2 in the parametric equations for the line yields the point $(3, 0, 1)$, which, indeed, lies in the plane as well. ◆

EXAMPLE 6 We determine whether and where the two lines

$$\begin{cases} x = t + 1 \\ y = 5t + 6 \\ z = -2t \end{cases} \quad \text{and} \quad \begin{cases} x = 3t - 3 \\ y = t \\ z = t + 1 \end{cases}$$

intersect.

The lines intersect provided that there is a specific value t_1 for the parameter of the first line and a value t_2 for the parameter of the second line that generate the same point. In other words, we must be able to find t_1 and t_2 so that, by equating the respective parametric expressions for x, y, and z, we have

$$\begin{cases} t_1 + 1 = 3t_2 - 3 \\ 5t_1 + 6 = t_2 \\ -2t_1 = t_2 + 1 \end{cases}. \tag{9}$$

The last two equations of (9) yield

$$t_2 = 5t_1 + 6 = -2t_1 - 1 \quad \Rightarrow \quad t_1 = -1.$$

Using $t_1 = -1$ in the second equation of (9), we find that $t_2 = 1$. Note that the values $t_1 = -1$ and $t_2 = 1$ also satisfy the first equation of (9); therefore, we have solved the system. Setting $t = -1$ in the set of parametric equations for the first line gives the desired intersection point, namely, (0, 1, 2). ◆

Parametric Equations in General

Vector geometry makes it relatively easy to find parametric equations for a variety of curves. We provide two examples.

EXAMPLE 7 If a wheel rolls along a flat surface without slipping, a point on the rim of the wheel traces a curve called a **cycloid**, as shown in Figure 1.26.

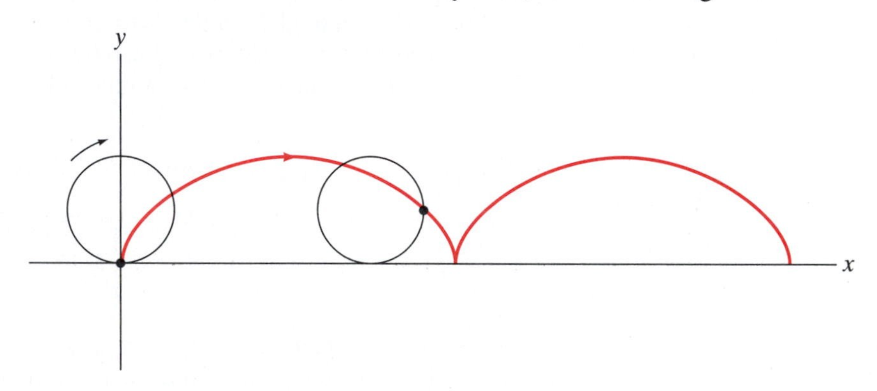

Figure 1.26 The graph of a cycloid.

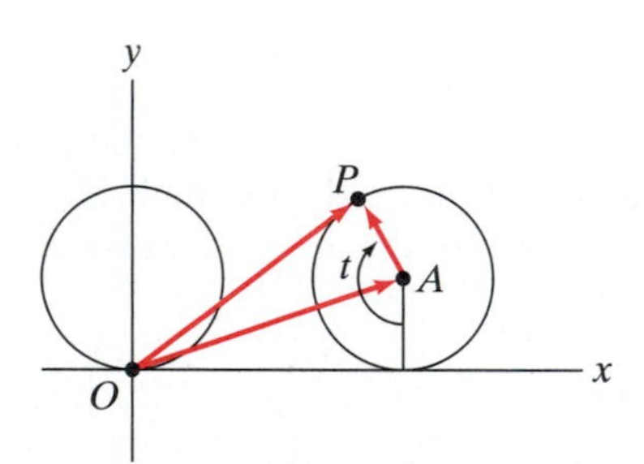

Figure 1.27 The result of the wheel in Figure 1.26 rolling through a central angle of t.

Suppose that the wheel has radius a and that coordinates in $\mathbf{R}^2$ are chosen so that the point of interest on the wheel is initially at the origin. After the wheel has rolled through a central angle of t radians, the situation is as shown in Figure 1.27. We seek the vector $\overrightarrow{OP}$, the position vector of P, in terms of the parameter t. Evidently, $\overrightarrow{OP} = \overrightarrow{OA} + \overrightarrow{AP}$, where the point A is the center of the wheel. The vector $\overrightarrow{OA}$ is not difficult to determine. Its **j**-component must be a, since the center of the wheel does not vary vertically. Its **i**-component must equal the distance the wheel has rolled; if t is measured in radians, then this distance is at, the length of the arc of the circle having central angle t. Hence, $\overrightarrow{OA} = at\mathbf{i} + a\mathbf{j}$.

The value of vector methods becomes apparent when we determine $\overrightarrow{AP}$. Parallel translate the picture so that $\overrightarrow{AP}$ has its tail at the origin, as in Figure 1.28. From the parametric equations of a circle of radius a,

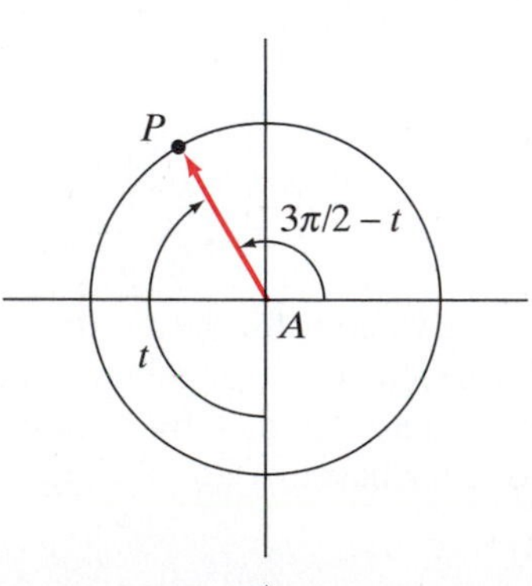

Figure 1.28 $\overrightarrow{AP}$ with its tail at the origin.

$$\overrightarrow{AP} = a\cos\left(\frac{3\pi}{2} - t\right)\mathbf{i} + a\sin\left(\frac{3\pi}{2} - t\right)\mathbf{j} = -a\sin t\,\mathbf{i} - a\cos t\,\mathbf{j},$$

from the addition formulas for sine and cosine. We conclude that

$$\begin{aligned}\overrightarrow{OP} = \overrightarrow{OA} + \overrightarrow{AP} &= (at\mathbf{i} + a\mathbf{j}) + (-a\sin t\mathbf{i} - a\cos t\mathbf{j}) \\ &= a(t - \sin t)\mathbf{i} + a(1 - \cos t)\mathbf{j},\end{aligned}$$

so the parametric equations are

$$\begin{cases} x = a(t - \sin t) \\ y = a(1 - \cos t) \end{cases}.$$

◆

EXAMPLE 8 If you unwind adhesive tape from a nonrotating circular tape dispenser so that the unwound tape is held taut and tangent to the dispenser roll, then the end of the tape traces a curve called the **involute** of the circle. Let's find the parametric equations for this curve, assuming that the dispensing roll has constant radius a and is centered at the origin. (As more and more tape is unwound, the radius of the roll will, of course, decrease. We'll assume that little enough tape is unwound so that the radius of the roll remains constant.)

Considering Figure 1.29, we see that the position vector $\overrightarrow{OP}$ of the desired point P is the vector sum $\overrightarrow{OB} + \overrightarrow{BP}$. To determine $\overrightarrow{OB}$ and $\overrightarrow{BP}$, we use the angle θ between the positive x-axis and $\overrightarrow{OB}$ as our parameter. Since B is a point on the circle,

$$\overrightarrow{OB} = a\cos\theta\,\mathbf{i} + a\sin\theta\,\mathbf{j}.$$

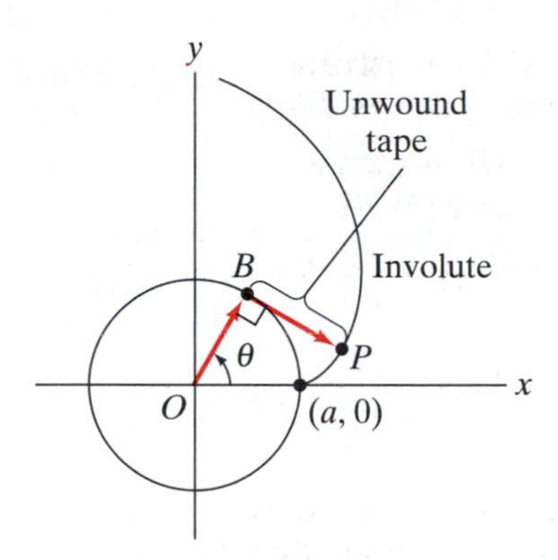

Figure 1.29 Unwinding tape, as in Example 8. The point P describes a curve known as the **involute** of the circle.

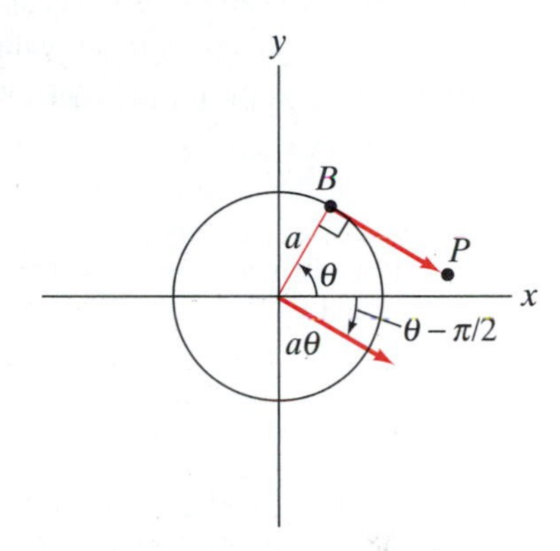

Figure 1.30 The vector $\overrightarrow{BP}$ must make an angle of $\theta - \pi/2$ with the positive x-axis.

To find the vector $\overrightarrow{BP}$, parallel translate it so that its tail is at the origin. Figure 1.30 shows that $\overrightarrow{BP}$'s length must be $a\theta$, the amount of unwound tape, and its direction must be such that it makes an angle of $\theta - \pi/2$ with the positive x-axis. From our experience with circular geometry and, perhaps, polar coordinates, we see that $\overrightarrow{BP}$ is described by

$$\overrightarrow{BP} = a\theta\cos\left(\theta - \frac{\pi}{2}\right)\mathbf{i} + a\theta\sin\left(\theta - \frac{\pi}{2}\right)\mathbf{j} = a\theta\sin\theta\,\mathbf{i} - a\theta\cos\theta\,\mathbf{j}.$$

Hence,

$$\overrightarrow{OP} = \overrightarrow{OB} + \overrightarrow{BP} = a(\cos\theta + \theta\sin\theta)\,\mathbf{i} + a(\sin\theta - \theta\cos\theta)\,\mathbf{j}.$$

So

$$\begin{cases} x = a(\cos\theta + \theta\sin\theta) \\ y = a(\sin\theta - \theta\cos\theta) \end{cases}$$

are the parametric equations of the involute, whose graph is pictured in Figure 1.31.

◆

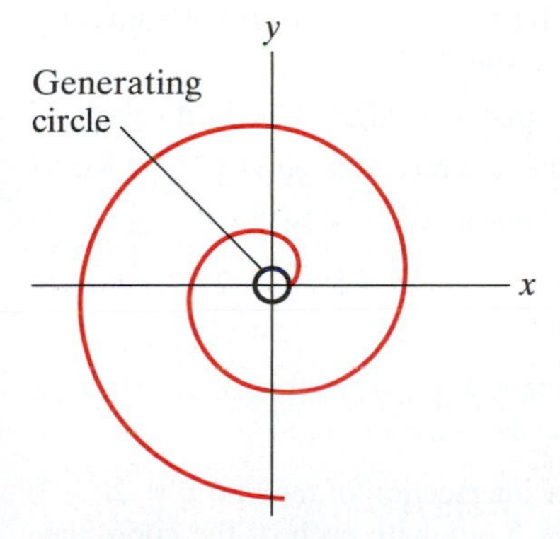

Figure 1.31 The involute.

Exercises

In Exercises 1–5, write the given vector by using the standard basis vectors for $\mathbf{R}^2$ *and* $\mathbf{R}^3$.

1. $(2, 4)$ **2.** $(9, -6)$ **3.** $(3, \pi, -7)$

4. $(-1, 2, 5)$ **5.** $(2, 4, 0)$

In Exercises 6–10, write the given vector without using the standard basis notation.

6. $\mathbf{i} + \mathbf{j} - 3\mathbf{k}$

7. $9\mathbf{i} - 2\mathbf{j} + \sqrt{2}\,\mathbf{k}$

8. $-3(2\mathbf{i} - 7\mathbf{k})$

9. $\pi\mathbf{i} - \mathbf{j}$ (Consider this to be a vector in $\mathbf{R}^2$.)

10. $\pi\mathbf{i} - \mathbf{j}$ (Consider this to be a vector in $\mathbf{R}^3$.)

11. Let $\mathbf{a}_1 = (1, 1)$ and $\mathbf{a}_2 = (1, -1)$.

(a) Write the vector $\mathbf{b} = (3, 1)$ as $c_1\mathbf{a}_1 + c_2\mathbf{a}_2$, where c_1 and c_2 are appropriate scalars.

(b) Repeat part (a) for the vector $\mathbf{b} = (3, -5)$.

(c) Show that *any* vector $\mathbf{b} = (b_1, b_2)$ in $\mathbf{R}^2$ may be written in the form $c_1\mathbf{a}_1 + c_2\mathbf{a}_2$ for appropriate choices of the scalars c_1, c_2. (This shows that $\mathbf{a}_1$ and $\mathbf{a}_2$ form a basis for $\mathbf{R}^2$ that can be used instead of $\mathbf{i}$ and $\mathbf{j}$.)

12. Let $\mathbf{a}_1 = (1, 0, -1)$, $\mathbf{a}_2 = (0, 1, 0)$, and $\mathbf{a}_3 = (1, 1, -1)$.

(a) Find scalars c_1, c_2, c_3, so as to write the vector $\mathbf{b} = (5, 6, -5)$ as $c_1\mathbf{a}_1 + c_2\mathbf{a}_2 + c_3\mathbf{a}_3$.

(b) Try to repeat part (a) for the vector $\mathbf{b} = (2, 3, 4)$. What happens?

(c) Can the vectors $\mathbf{a}_1$, $\mathbf{a}_2$, $\mathbf{a}_3$ be used as a basis for $\mathbf{R}^3$, instead of $\mathbf{i}, \mathbf{j}, \mathbf{k}$? Why or why not?

In Exercises 13–18, give a set of parametric equations for the lines so described.

13. The line in $\mathbf{R}^3$ through the point $(2, -1, 5)$ that is parallel to the vector $\mathbf{i} + 3\mathbf{j} - 6\mathbf{k}$.

14. The line in $\mathbf{R}^3$ through the point $(12, -2, 0)$ that is parallel to the vector $5\mathbf{i} - 12\mathbf{j} + \mathbf{k}$.

15. The line in $\mathbf{R}^2$ through the point $(2, -1)$ that is parallel to the vector $\mathbf{i} - 7\mathbf{j}$.

16. The line in $\mathbf{R}^3$ through the points $(2, 1, 2)$ and $(3, -1, 5)$.

17. The line in $\mathbf{R}^3$ through the points $(1, 4, 5)$ and $(2, 4, -1)$.

18. The line in $\mathbf{R}^2$ through the points $(8, 5)$ and $(1, 7)$.

19. Write a set of parametric equations for the line in $\mathbf{R}^4$ through the point $(1, 2, 0, 4)$ and parallel to the vector $(-2, 5, 3, 7)$.

20. Write a set of parametric equations for the line in $\mathbf{R}^5$ through the points $(9, \pi, -1, 5, 2)$ and $(-1, 1, \sqrt{2}, 7, 1)$.

21. (a) Write a set of parametric equations for the line in $\mathbf{R}^3$ through the point $(-1, 7, 3)$ and parallel to the vector $2\mathbf{i} - \mathbf{j} + 5\mathbf{k}$.

(b) Write a set of parametric equations for the line through the points $(5, -3, 4)$ and $(0, 1, 9)$.

(c) Write different (but equally correct) sets of equations for parts (a) and (b).

(d) Find the symmetric forms of your answers in (a)–(c).

22. A certain line in $\mathbf{R}^3$ has symmetric form

$$\frac{x-2}{5} = \frac{y-3}{-2} = \frac{z+1}{4}.$$

Write a set of parametric equations for this line.

23. Show that the two sets of equations

$$\frac{x-2}{3} = \frac{y-1}{7} = \frac{z}{5} \quad \text{and} \quad \frac{x+1}{-6} = \frac{y+6}{-14} = \frac{z+5}{-10}$$

actually represent the same line in $\mathbf{R}^3$.

24. Determine whether the two lines l_1 and l_2 defined by the sets of parametric equations l_1: $x = 2t - 5$, $y = 3t + 2$, $z = 1 - 6t$, and l_2: $x = 1 - 2t$, $y = 11 - 3t$, $z = 6t - 17$ are the same. (Hint: First find two points on l_1 and then see if those points lie on l_2.)

25. Do the parametric equations l_1: $x = 3t + 2$, $y = t - 7$, $z = 5t + 1$, and l_2: $x = 6t - 1$, $y = 2t - 8$, $z = 10t - 3$ describe the same line? Why or why not?

26. Do the parametric equations $x = 3t^3 + 7$, $y = 2 - t^3$, $z = 5t^3 + 1$ determine a line? Why or why not?

27. Do the parametric equations $x = 5t^2 - 1$, $y = 2t^2 + 3$, $z = 1 - t^2$ determine a line? Explain.

28. A bird is flying along the straight-line path $x = 2t + 7$, $y = t - 2$, $z = 1 - 3t$, where t is measured in minutes.

(a) Where is the bird initially (at $t = 0$)? Where is the bird 3 minutes later?

(b) Give a vector that is parallel to the bird's path.

(c) When does the bird reach the point $\left(\frac{34}{3}, \frac{1}{6}, -\frac{11}{2}\right)$?

(d) Does the bird reach $(17, 4, -14)$?

29. Find where the line $x = 3t - 5$, $y = 2 - t$, $z = 6t$ intersects the plane $x + 3y - z = 19$.

30. Where does the line $x = 1 - 4t$, $y = t - 3/2$, $z = 2t + 1$ intersect the plane $5x - 2y + z = 1$?

31. Find the points of intersection of the line $x = 2t - 3$, $y = 3t + 2$, $z = 5 - t$ with each of the coordinate planes $x = 0$, $y = 0$, and $z = 0$.

32. Show that the line $x = 5 - t$, $y = 2t - 7$, $z = t - 3$ is contained in the plane having equation $2x - y + 4z = 5$.

33. Does the line $x = 5 - t$, $y = 2t - 3$, $z = 7t + 1$ intersect the plane $x - 3y + z = 1$? Why?

34. Find the point of intersection of the two lines $l_1: x = 2t + 3$, $y = 3t + 3$, $z = 2t + 1$ and $l_2: x = 15 - 7t$, $y = t - 2$, $z = 3t - 7$.

35. Do the lines $l_1: x = 2t + 1$, $y = -3t$, $z = t - 1$ and $l_2: x = 3t + 1$, $y = t + 5$, $z = 7 - t$ intersect? Explain your answer.

36. (a) Find the distance from the point $(-2, 1, 5)$ to any point on the line $x = 3t - 5$, $y = 1 - t$, $z = 4t + 7$ (Your answer should be in terms of the parameter t.)

(b) Now find the distance between the point $(-2, 1, 5)$ and the line $x = 3t - 5$, $y = 1 - t$, $z = 4t + 7$. (The distance between a point and a line is the distance between the given point and the *closest* point on the line.)

37. (a) Describe the curve given parametrically by

$$\begin{cases} x = 2\cos 3t \\ y = 2\sin 3t \end{cases} \quad 0 \le t < \frac{2\pi}{3}.$$

What happens if we allow t to vary between 0 and 2π?

(b) Describe the curve given parametrically by

$$\begin{cases} x = 5\cos 3t \\ y = 5\sin 3t \end{cases} \quad 0 \le t < \frac{2\pi}{3}.$$

(c) Describe the curve given parametrically by

$$\begin{cases} x = 5\sin 3t \\ y = 5\cos 3t \end{cases} \quad 0 \le t < \frac{2\pi}{3}.$$

(d) Describe the curve given parametrically by

$$\begin{cases} x = 5\cos 3t \\ y = 3\sin 3t \end{cases} \quad 0 \le t < \frac{2\pi}{3}.$$

38. Suppose that a bicycle wheel of radius a rolls along a flat surface without slipping. If a reflector is attached to a spoke of the wheel at a distance b from the center, the resulting curve traced by the reflector is called a **curtate cycloid**. One such cycloid appears in Figure 1.32, where $a = 3$ and $b = 2$.

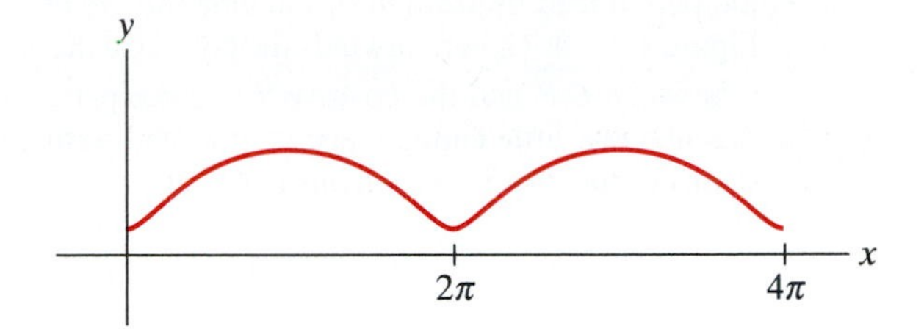

Figure 1.32 A curtate cycloid.

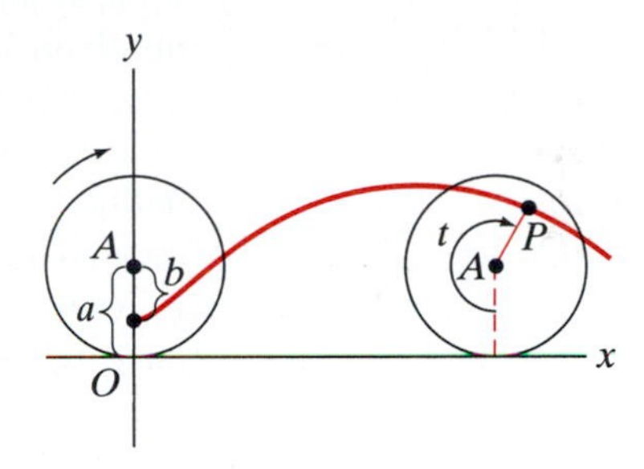

Figure 1.33 The point P traces a curtate cycloid.

Using vector methods or otherwise, find a set of parametric equations for the curtate cycloid. Figure 1.33 should help. (Take a low point of the cycloid to lie on the y-axis.) There is no theoretical reason that the cycloid just described cannot have $a < b$, although in such case the bicycle-wheel–reflector application is no longer relevant. (When $a < b$, the parametrized curve that results is called a **prolate cycloid**.) Your parametric equations should be such that the constants a and b can be chosen independently of one another. An example of a prolate cycloid, with $a = 2$ and $b = 4$, is shown in Figure 1.34. Try to think

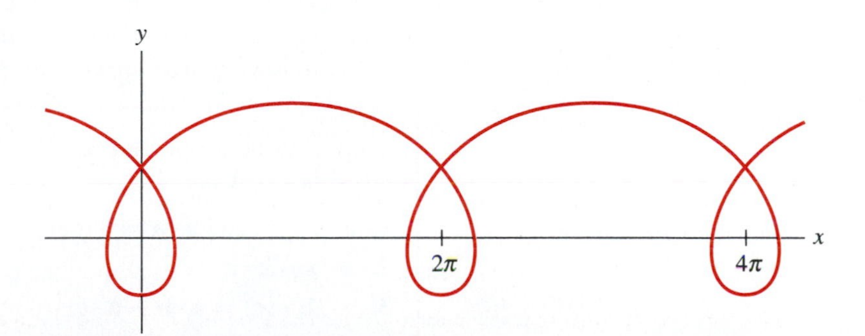

Figure 1.34 A prolate cycloid.

of a physical situation in which such a curve would arise.

39. Egbert is unwinding tape from a circular dispenser of radius a by holding the tape taut and perpendicular to the dispenser. Find a set of parametric equations for the path traced by the end of the tape (the point P in Figure 1.35) as Egbert unwinds the tape. Use the angle θ between $\overrightarrow{OP}$ and the positive x-axis for parameter. Assume that little enough tape is unwound so that the radius of the dispenser remains constant.

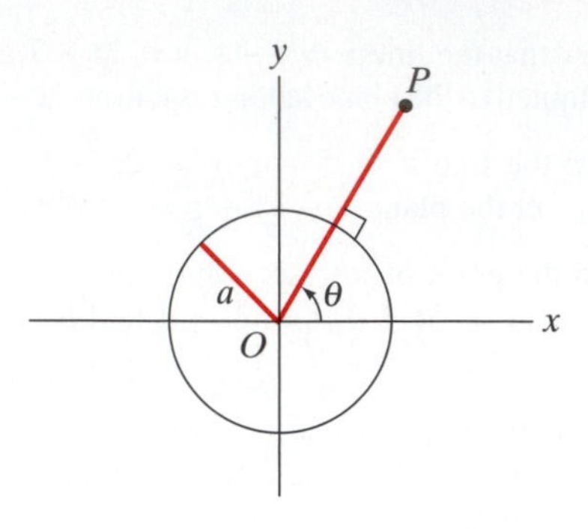

Figure 1.35 Figure for Exercise 39.

1.3 The Dot Product

When we introduced the arithmetic notions of vector addition and scalar multiplication, you may well have wondered why the product of two vectors was not defined. You might think that "vector multiplication" should be defined in a manner analogous to the way we defined vector addition (i.e., by componentwise multiplication). However, such a definition is not very useful. Instead, we shall define, and use, two different concepts of a product of two vectors: (1) the Euclidean inner product, or "dot" product, which may be defined for two vectors in $\mathbf{R}^n$ (where n is arbitrary) and (2) the "cross" or vector product, which is defined *only* for vectors in $\mathbf{R}^3$.

The Dot Product of Two Vectors

■ **Definition 3.1** Let $\mathbf{a} = (a_1, a_2, a_3)$ and $\mathbf{b} = (b_1, b_2, b_3)$ be two vectors in $\mathbf{R}^3$. The **dot** (or **inner** or **scalar**) **product of a and b**, denoted $\mathbf{a} \cdot \mathbf{b}$, is

$$\mathbf{a} \cdot \mathbf{b} = a_1 b_1 + a_2 b_2 + a_3 b_3.$$

In $\mathbf{R}^2$, the analogous definition is

$$\mathbf{a} \cdot \mathbf{b} = a_1 b_1 + a_2 b_2,$$

where $\mathbf{a} = (a_1, a_2)$ and $\mathbf{b} = (b_1, b_2)$.

EXAMPLE 1 In $\mathbf{R}^3$, we have

$$(1, -2, 5) \cdot (2, 1, 3) = (1)(2) + (-2)(1) + (5)(3) = 15.$$
$$(3\mathbf{i} + 2\mathbf{j} - \mathbf{k}) \cdot (\mathbf{i} - 2\mathbf{k}) = (3)(1) + (2)(0) + (-1)(-2) = 5.$$ ◆

In accordance with its name, the dot—or scalar—product takes two vectors and produces a *single real number* (*not* a vector).

The following facts are consequences of Definition 3.1:

Properties of dot products. If **a**, **b**, and **c** are any vectors in $\mathbf{R}^3$ (or $\mathbf{R}^2$) and $k \in \mathbf{R}$ is any scalar, then

1. $\mathbf{a} \cdot \mathbf{a} \geq 0$, and $\mathbf{a} \cdot \mathbf{a} = 0$ if and only if $\mathbf{a} = \mathbf{0}$;
2. $\mathbf{a} \cdot \mathbf{b} = \mathbf{b} \cdot \mathbf{a}$;
3. $\mathbf{a} \cdot (\mathbf{b} + \mathbf{c}) = \mathbf{a} \cdot \mathbf{b} + \mathbf{a} \cdot \mathbf{c}$;
4. $(k\mathbf{a}) \cdot \mathbf{b} = k(\mathbf{a} \cdot \mathbf{b}) = \mathbf{a} \cdot (k\mathbf{b})$.

Proof of Property 1 If $\mathbf{a} = (a_1, a_2, a_3)$, then we have

$$\mathbf{a} \cdot \mathbf{a} = a_1 a_1 + a_2 a_2 + a_3 a_3 = a_1^2 + a_2^2 + a_3^2.$$

This last expression is evidently nonnegative, since it is a sum of squares of real numbers. Moreover, such an expression is zero exactly when each of the terms is zero, that is, if and only if $a_1 = a_2 = a_3 = 0$. ■

We leave the proofs of properties 2, 3, and 4 as exercises.

Thus far, we have introduced the dot product of two vectors as a purely algebraic construction. It is the geometric interpretation of the definition that is really interesting. To establish this interpretation, we begin with the following:

■ **Definition 3.2** If $\mathbf{a} = (a_1, a_2, a_3)$, then the **length** of $\mathbf{a}$ (also called the **norm** or **magnitude**), denoted $\|\mathbf{a}\|$, is $\sqrt{a_1^2 + a_2^2 + a_3^2}$.

The motivation for this definition is evident if we draw $\mathbf{a}$ as the position vector of the point (a_1, a_2, a_3). Then the length of the arrow from the origin to (a_1, a_2, a_3) is

$$\sqrt{(a_1 - 0)^2 + (a_2 - 0)^2 + (a_3 - 0)^2},$$

as given by the distance formula, which is nothing more than an extension of the Pythagorean theorem in the plane. As we just saw, $\mathbf{a} \cdot \mathbf{a} = a_1^2 + a_2^2 + a_3^2$, and we have

$$\|\mathbf{a}\| = \sqrt{\mathbf{a} \cdot \mathbf{a}}$$

or, equivalently,

$$\mathbf{a} \cdot \mathbf{a} = \|\mathbf{a}\|^2. \tag{1}$$

Now we're ready to state the main result concerning the geometry of the dot product. If $\mathbf{a}$ and $\mathbf{b}$ are two nonzero vectors in $\mathbf{R}^3$ (or $\mathbf{R}^2$) drawn with their tails at the same point, let θ, where $0 \leq \theta \leq \pi$, be the angle between $\mathbf{a}$ and $\mathbf{b}$. If either $\mathbf{a}$ or $\mathbf{b}$ is the zero vector, then θ is indeterminate (i.e., can be any angle).

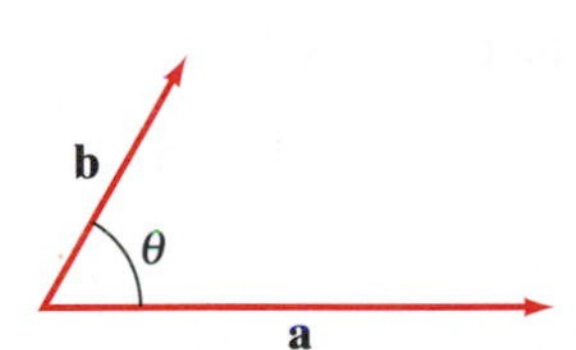

Figure 1.36 The dot product of $\mathbf{a}$ and $\mathbf{b}$ is $\|\mathbf{a}\| \, \|\mathbf{b}\| \cos\theta$.

■ **Theorem 3.3** If $\mathbf{a}$ and $\mathbf{b}$ are any two vectors in either $\mathbf{R}^2$ or $\mathbf{R}^3$, then

$$\mathbf{a} \cdot \mathbf{b} = \|\mathbf{a}\| \, \|\mathbf{b}\| \cos\theta.$$

(See Figure 1.36.)

Proof If either $\mathbf{a}$ or $\mathbf{b}$ is the zero vector, say $\mathbf{a}$, then $\mathbf{a} = (0, 0, 0)$ and so

$$\mathbf{a} \cdot \mathbf{b} = (0)(b_1) + (0)(b_2) + (0)(b_3) = 0.$$

Also, $\|\mathbf{a}\| = 0$ in this case, so the formula in Theorem 3.3 holds. In this case, the angle θ is indeterminate.

Now suppose that neither $\mathbf{a}$ nor $\mathbf{b}$ is the zero vector. Let $\mathbf{c} = \mathbf{b} - \mathbf{a}$. Then we may apply the law of cosines to the triangle whose sides are $\mathbf{a}$, $\mathbf{b}$, and $\mathbf{c}$ (Figure 1.37) to obtain

$$\|\mathbf{c}\|^2 = \|\mathbf{a}\|^2 + \|\mathbf{b}\|^2 - 2\|\mathbf{a}\| \, \|\mathbf{b}\| \cos\theta.$$

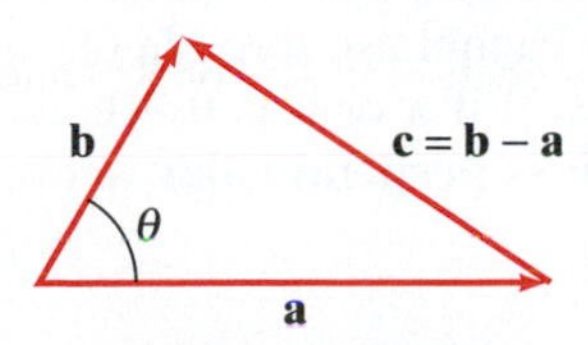

Figure 1.37 The vector triangle used in the proof of Theorem 3.3.

Thus,

$$2\|\mathbf{a}\|\,\|\mathbf{b}\|\cos\theta = \|\mathbf{a}\|^2 + \|\mathbf{b}\|^2 - \|\mathbf{c}\|^2 = \mathbf{a}\cdot\mathbf{a} + \mathbf{b}\cdot\mathbf{b} - \mathbf{c}\cdot\mathbf{c}, \tag{2}$$

from equation (1). Now, use the properties of the dot product. Since $\mathbf{c} = \mathbf{b} - \mathbf{a}$,

$$\begin{aligned}\mathbf{c}\cdot\mathbf{c} &= (\mathbf{b}-\mathbf{a})\cdot(\mathbf{b}-\mathbf{a})\\ &= (\mathbf{b}-\mathbf{a})\cdot\mathbf{b} - (\mathbf{b}-\mathbf{a})\cdot\mathbf{a}\\ &= \mathbf{b}\cdot\mathbf{b} - \mathbf{a}\cdot\mathbf{b} - \mathbf{b}\cdot\mathbf{a} + \mathbf{a}\cdot\mathbf{a},\end{aligned} \tag{3}$$

by properties 3 and 4 of the dot product. If we use equation (3) to substitute for $\mathbf{c}\cdot\mathbf{c}$ in equation (2), then

$$\begin{aligned}2\|\mathbf{a}\|\,\|\mathbf{b}\|\cos\theta &= \mathbf{a}\cdot\mathbf{a} + \mathbf{b}\cdot\mathbf{b} - (\mathbf{b}\cdot\mathbf{b} - \mathbf{a}\cdot\mathbf{b} - \mathbf{b}\cdot\mathbf{a} + \mathbf{a}\cdot\mathbf{a})\\ &= \mathbf{a}\cdot\mathbf{b} + \mathbf{b}\cdot\mathbf{a}\\ &= 2\mathbf{a}\cdot\mathbf{b},\end{aligned}$$

by property 2 of the dot product. By canceling the factor of 2 on both sides, the desired result is obtained. ■

Angles Between Vectors

Theorem 3.3 may be used to find the angle between two nonzero vectors **a** and **b**—just solve for θ in the formula in Theorem 3.3 to obtain

$$\theta = \cos^{-1}\frac{\mathbf{a}\cdot\mathbf{b}}{\|\mathbf{a}\|\,\|\mathbf{b}\|}. \tag{4}$$

The use of the inverse cosine is unambiguous, since we take $0 \le \theta \le \pi$ when defining angles between vectors.

EXAMPLE 2 If $\mathbf{a} = \mathbf{i} + \mathbf{j}$ and $\mathbf{b} = \mathbf{j} - \mathbf{k}$, then formula (4) gives

$$\theta = \cos^{-1}\frac{(\mathbf{i}+\mathbf{j})\cdot(\mathbf{j}-\mathbf{k})}{\|\mathbf{i}+\mathbf{j}\|\,\|\mathbf{j}-\mathbf{k}\|} = \cos^{-1}\frac{1}{(\sqrt{2}\cdot\sqrt{2})} = \cos^{-1}\frac{1}{2} = \frac{\pi}{3}.$$ ◆

If **a** and **b** are nonzero, then Theorem 3.3 implies

$$\cos\theta = 0 \quad \text{if and only if} \quad \mathbf{a}\cdot\mathbf{b} = 0.$$

We have $\cos\theta = 0$ just in case $\theta = \pi/2$. (Remember our restriction on θ.) Hence, it makes sense for us to call **a** and **b** **perpendicular** (or **orthogonal**) when $\mathbf{a}\cdot\mathbf{b} = 0$. If either **a** or **b** is the zero vector, then we cannot use formula (4), and the angle θ is undefined. Nonetheless, since $\mathbf{a}\cdot\mathbf{b} = 0$ if **a** or **b** is **0**, we adopt the standard convention and say that **the zero vector is perpendicular to every vector**.

EXAMPLE 3 The vector $\mathbf{i} + \mathbf{j}$ is orthogonal to the vector $\mathbf{i} - \mathbf{j} + \mathbf{k}$, since

$$(\mathbf{i}+\mathbf{j})\cdot(\mathbf{i}-\mathbf{j}+\mathbf{k}) = (1)(1) + (1)(-1) + (0)(1) = 0.$$ ◆

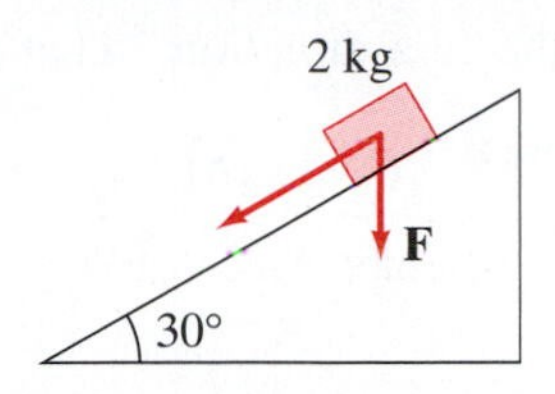

Figure 1.38 An object sliding down a ramp. The force due to gravity is downward, but the direction of travel of the object is inclined 30° to the horizontal.

Vector Projections

Suppose that a 2 kg object is sliding down a ramp having a 30° incline with the horizontal as in Figure 1.38. If we neglect friction, the only force acting on the object is gravity. What is the component of the gravitational force in the direction of motion of the object?

To answer questions of this nature, we need to find the projection of one vector on another. The general idea is as follows: Given two nonzero vectors **a** and **b**, imagine dropping a perpendicular line from the head of **b** to the line through **a**. Then the **projection of b onto a**, denoted $\text{proj}_{\mathbf{a}}\mathbf{b}$, is the vector represented by the arrow in Figure 1.39.

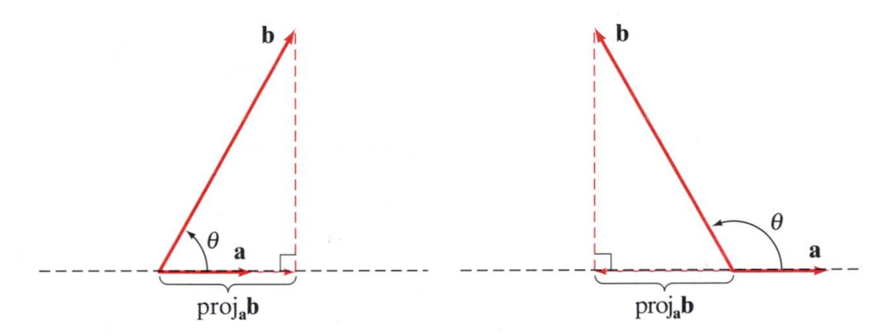

Figure 1.39 Projection of the vector **b** onto the vector **a**.

Given this intuitive understanding of the projection, we find a precise formula for it. Recall that a vector is determined by magnitude (length) and direction. It follows by definition that the direction of $\text{proj}_{\mathbf{a}}\mathbf{b}$ is either the same as that of **a**, or opposite to **a** if the angle θ between **a** and **b** is more than $\pi/2$. Trigonometry then tells us

$$|\cos\theta| = \frac{\|\text{proj}_{\mathbf{a}}\mathbf{b}\|}{\|\mathbf{b}\|}.$$

(The absolute value sign around $\cos\theta$ is needed in case $\pi/2 \leq \theta \leq \pi$.) Hence, with a bit of algebra, we have

$$\|\text{proj}_{\mathbf{a}}\mathbf{b}\| = \|\mathbf{b}\|\,|\cos\theta| = \frac{\|\mathbf{a}\|\,\|\mathbf{b}\|\,|\cos\theta|}{\|\mathbf{a}\|} = \frac{|\mathbf{a}\cdot\mathbf{b}|}{\|\mathbf{a}\|}$$

by Theorem 3.3. Thus, we know the magnitude and direction of $\text{proj}_{\mathbf{a}}\mathbf{b}$. To obtain a compact formula for $\text{proj}_{\mathbf{a}}\mathbf{b}$, note the following:

■ **Proposition 3.4** Let k be any scalar and **a** any vector. Then

1. $\|k\mathbf{a}\| = |k|\,\|\mathbf{a}\|$.
2. A **unit vector** (i.e., a vector of length 1) in the direction of a nonzero vector **a** is given by $\mathbf{a}/\|\mathbf{a}\|$.

Proof Part 1 is left as an exercise. (Write out $k\mathbf{a}$ and $\|k\mathbf{a}\|$ in terms of components.) For part 2, we must check that the length of $\mathbf{a}/\|\mathbf{a}\|$ is 1:

$$\left\|\frac{\mathbf{a}}{\|\mathbf{a}\|}\right\| = \left\|\frac{1}{\|\mathbf{a}\|}\mathbf{a}\right\| = \frac{1}{\|\mathbf{a}\|}\|\mathbf{a}\| = 1,$$

by part 1 (since $1/\|\mathbf{a}\|$ is a positive scalar). ■

Now $\text{proj}_{\mathbf{a}}\mathbf{b}$ is a vector of length $|\mathbf{a}\cdot\mathbf{b}|/\|\mathbf{a}\|$ in the "$\pm\mathbf{a}$-direction." That is,

$$\text{proj}_{\mathbf{a}}\mathbf{b} = \pm \underbrace{\left(\frac{|\mathbf{a}\cdot\mathbf{b}|}{\|\mathbf{a}\|}\right)}_{\substack{\text{length of}\\ \text{proj}_{\mathbf{a}}\mathbf{b}}} \times \underbrace{\frac{\mathbf{a}}{\|\mathbf{a}\|}}_{\substack{\text{unit vector}\\ \text{in direction of } \mathbf{a}}} = \pm \frac{\|\mathbf{a}\|\,\|\mathbf{b}\|\,|\cos\theta|}{\|\mathbf{a}\|}\frac{\mathbf{a}}{\|\mathbf{a}\|}.$$

Note that the angle θ keeps track of the appropriate sign of $\text{proj}_{\mathbf{a}}\mathbf{b}$; that is, when $0 \le \theta < \pi/2$, $\cos\theta$ is positive and $\text{proj}_{\mathbf{a}}\mathbf{b}$ points in the direction of $\mathbf{a}$, and when $\pi/2 < \theta \le \pi$, $\cos\theta$ is negative and $\text{proj}_{\mathbf{a}}\mathbf{b}$ points in the direction opposite to that of $\mathbf{a}$. Thus, we can eliminate *both* the $\pm$ sign and the absolute value, and we find that

$$\text{proj}_{\mathbf{a}}\mathbf{b} = \frac{\|\mathbf{a}\|\,\|\mathbf{b}\|\,\cos\theta}{\|\mathbf{a}\|}\frac{\mathbf{a}}{\|\mathbf{a}\|} = \frac{\mathbf{a}\cdot\mathbf{b}}{\|\mathbf{a}\|^2}\mathbf{a}$$

by Theorem 3.3, so that

$$\text{proj}_{\mathbf{a}}\mathbf{b} = \left(\frac{\mathbf{a}\cdot\mathbf{b}}{\mathbf{a}\cdot\mathbf{a}}\right)\mathbf{a} \qquad (5)$$

by equation (1). Formula (5) is concise and not difficult to remember.

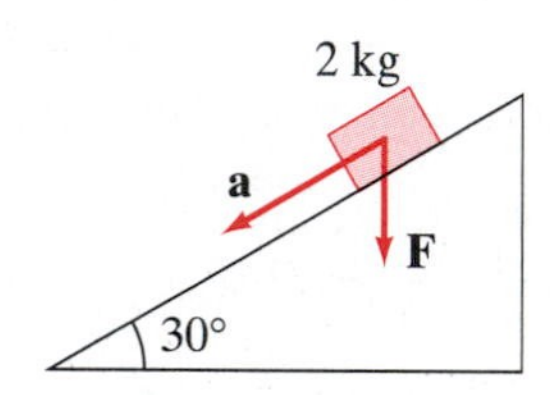

Figure 1.40 The 2 kg object sliding down a ramp in Example 4.

EXAMPLE 4 To answer the question posed at the beginning of this subsection, we need to calculate $\text{proj}_{\mathbf{a}}\mathbf{F}$, where $\mathbf{F}$ is the gravitational force vector and $\mathbf{a}$ points along the ramp as shown in Figure 1.40. We have a coordinate situation as shown in Figure 1.41. From trigonometric considerations, we must have $\mathbf{a} = a_1\mathbf{i} + a_2\mathbf{j}$ such that $a_1 = -\|\mathbf{a}\|\cos 30^\circ$ and $a_2 = -\|\mathbf{a}\|\sin 30^\circ$. Since we are really only interested in the *direction* of $\mathbf{a}$, there is no loss in assuming that $\mathbf{a}$ is a unit vector. Thus,

$$\mathbf{a} = -\cos 30^\circ\,\mathbf{i} - \sin 30^\circ\,\mathbf{j} = -\tfrac{\sqrt{3}}{2}\mathbf{i} - \tfrac{1}{2}\mathbf{j}.$$

Taking $g = 9.8$ m/sec^2, we have $\mathbf{F} = -2g\mathbf{j} = -19.6\mathbf{j}$. Therefore, formula (5)

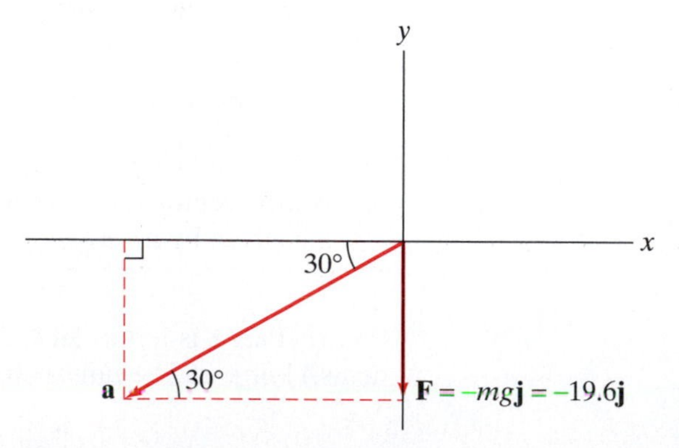

Figure 1.41 The vectors **a** and **F** in Example 4, realized in a coordinate system.

implies

$$\text{proj}_{\mathbf{a}}\mathbf{F} = \left(\frac{\mathbf{a}\cdot\mathbf{F}}{\mathbf{a}\cdot\mathbf{a}}\right)\mathbf{a} = \frac{(-\frac{\sqrt{3}}{2}\mathbf{i} - \frac{1}{2}\mathbf{j})\cdot(-19.6\,\mathbf{j})}{1}\left(-\frac{\sqrt{3}}{2}\mathbf{i} - \frac{1}{2}\mathbf{j}\right)$$

$$= 9.8\left(-\frac{\sqrt{3}}{2}\mathbf{i} - \frac{1}{2}\mathbf{j}\right)$$

$$\approx -8.49\,\mathbf{i} - 4.9\,\mathbf{j},$$

and the component of **F** in this direction is

$$\|\text{proj}_{\mathbf{a}}\mathbf{F}\| = \|-8.49\,\mathbf{i} - 4.9\,\mathbf{j}\| = 9.8\text{ N}.$$ ◆

Unit vectors—that is, vectors of length 1—are important in that they capture the idea of direction (since they all have the same length). Part 2 of Proposition 3.4 shows that every nonzero vector **a** can have its length adjusted to give a unit vector $\mathbf{u} = \mathbf{a}/\|\mathbf{a}\|$ that points in the same direction as **a**. This operation is referred to as **normalization** of the vector **a**.

EXAMPLE 5 A fluid is flowing across a plane surface with uniform velocity vector **v**. If **n** is a unit vector perpendicular to the plane surface, let's find (in terms of **v** and **n**) the volume of the fluid that passes through a unit area of the plane in unit time. (See Figure 1.42.)

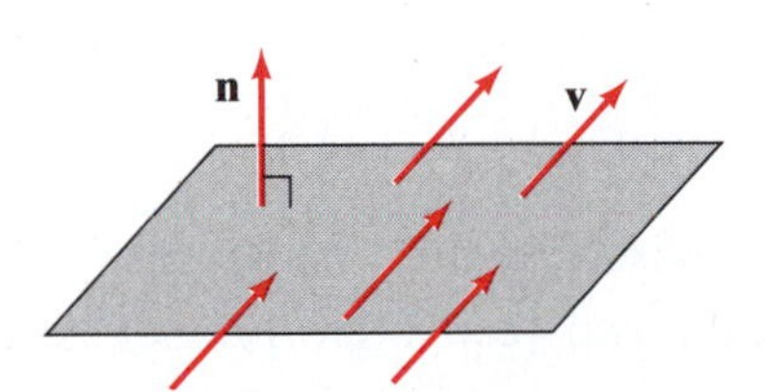

Figure 1.42 Fluid flowing across a plane surface.

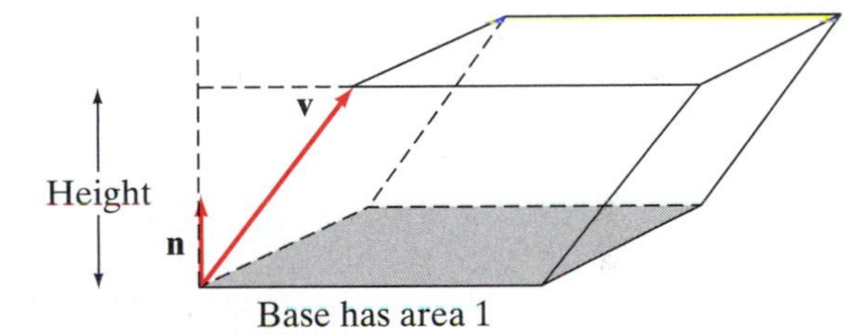

Figure 1.43 After one unit of time, the fluid passing across a square will have filled the box.

First, imagine one unit of time has elapsed. Then over a unit area of the plane (say over a unit square), the fluid will have filled a "box" as in Figure 1.43. The box may be represented by a parallelepiped (a three-dimensional analogue of a parallelogram). The volume we seek is the volume of this parallelepiped and is

$$\text{Volume} = (\text{area of base})(\text{height}).$$

The area of the base is 1 unit by construction. The height is given by $\|\text{proj}_{\mathbf{n}}\mathbf{v}\|$. From formula (5),

$$\text{proj}_{\mathbf{n}}\mathbf{v} = \left(\frac{\mathbf{n}\cdot\mathbf{v}}{\mathbf{n}\cdot\mathbf{n}}\right)\mathbf{n} = (\mathbf{n}\cdot\mathbf{v})\mathbf{n},$$

since $\mathbf{n}\cdot\mathbf{n} = \|\mathbf{n}\|^2 = 1$. Hence,

$$\|\text{proj}_{\mathbf{n}}\mathbf{v}\| = \|(\mathbf{n}\cdot\mathbf{v})\mathbf{n}\| = |\mathbf{n}\cdot\mathbf{v}|\,\|\mathbf{n}\| = |\mathbf{n}\cdot\mathbf{v}|,$$

by part 1 of Proposition 3.4. ◆

Vector Proofs

We conclude this section with two illustrations of how wonderfully well vectors are suited to providing elegant proofs of geometric results.

EXAMPLE 6 In an arbitrary triangle, show that the line segment joining the midpoints of two sides is parallel to and has half the length of the third side. (See Figure 1.44.) In other words, if M_1 is the midpoint of side $\overline{AB}$ and M_2 is the midpoint of side $\overline{AC}$, we wish to show that $\overline{M_1M_2}$ is parallel to $\overline{BC}$ and has half its length.

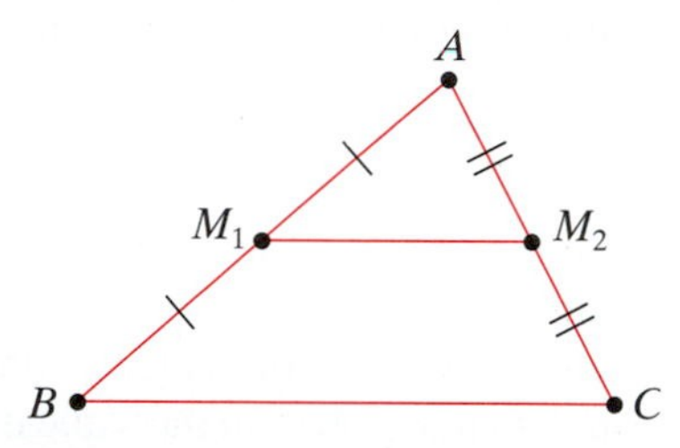

Figure 1.44 In triangle ABC, $\overline{M_1M_2}$ is parallel to $\overline{BC}$ and has half its length.

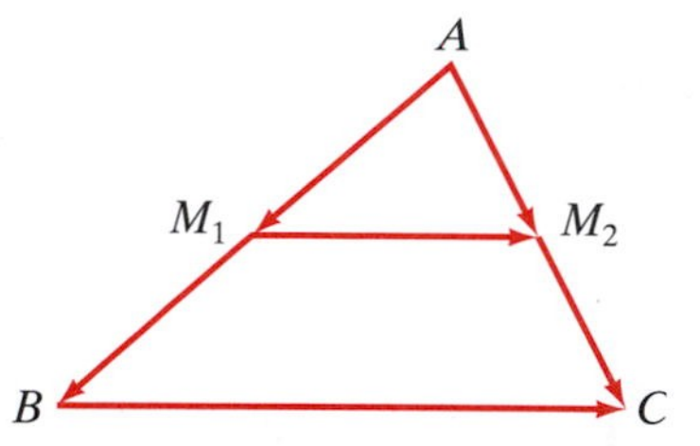

Figure 1.45 The vector version of triangle ABC in Example 6.

For a vector proof, we use the diagram in Figure 1.45, a slightly modified version of Figure 1.44. The midpoint conditions translate to the following statements about vectors:

$$\overrightarrow{AM_1} = \tfrac{1}{2}\overrightarrow{AB}, \qquad \overrightarrow{AM_2} = \tfrac{1}{2}\overrightarrow{AC}.$$

Now,

$$\overrightarrow{M_1M_2} = \overrightarrow{AM_2} - \overrightarrow{AM_1} = \tfrac{1}{2}\overrightarrow{AC} - \tfrac{1}{2}\overrightarrow{AB} = \tfrac{1}{2}(\overrightarrow{AC} - \overrightarrow{AB}) = \tfrac{1}{2}\overrightarrow{BC}.$$

But $\overrightarrow{M_1M_2} = \frac{1}{2}\overrightarrow{BC}$ is precisely what we wish to prove: To say $\overrightarrow{M_1M_2}$ is a scalar times $\overrightarrow{BC}$ means that the two vectors are parallel. Moreover, from part 1 of Proposition 3.4,

$$\|\overrightarrow{M_1M_2}\| = \|\tfrac{1}{2}\overrightarrow{BC}\| = \tfrac{1}{2}\|\overrightarrow{BC}\|,$$

so that the length condition also holds. ◆

EXAMPLE 7 Show that every angle inscribed in a semicircle is a right angle, as suggested by Figure 1.46.

To prove this remark, we'll make use of Figure 1.47, where **a** and **b** are "radius vectors" with tails at the center of the circle. We need only show that $\mathbf{a} - \mathbf{b}$ (a vector along one ray of the angle in question) is perpendicular to $-\mathbf{a} - \mathbf{b}$ (a vector along the other ray). In other words, we wish to show that

$$(\mathbf{a} - \mathbf{b}) \cdot (-\mathbf{a} - \mathbf{b}) = 0.$$

We have

$$(\mathbf{a} - \mathbf{b}) \cdot (-\mathbf{a} - \mathbf{b}) = (-1)(\mathbf{a} - \mathbf{b}) \cdot (\mathbf{a} + \mathbf{b}),$$

by property 4 of dot products,

$$\begin{aligned} &= (-1)\left((\mathbf{a} - \mathbf{b}) \cdot \mathbf{a} + (\mathbf{a} - \mathbf{b}) \cdot \mathbf{b}\right) \\ &= (-1)\left(\mathbf{a} \cdot \mathbf{a} - \mathbf{b} \cdot \mathbf{a} + \mathbf{a} \cdot \mathbf{b} - \mathbf{b} \cdot \mathbf{b}\right) \\ &= (-1)(\|\mathbf{a}\|^2 - \|\mathbf{b}\|^2), \end{aligned}$$

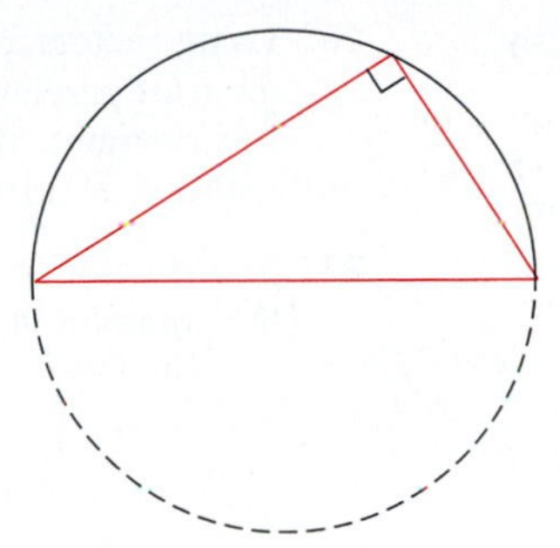

Figure 1.46 Every angle inscribed in a semicircle is a right angle.

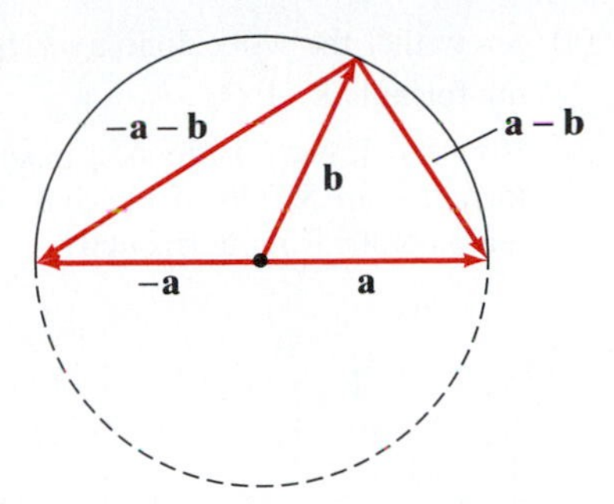

Figure 1.47 **a** and **b** are "radius vectors."

by properties 2 and 4,

$$= 0,$$

since both **a** and **b** are radius vectors (and therefore have the same length, namely, the radius of the circle). ◆

Vector proofs as in Examples 6 and 7 are elegant and sometimes allow you to write shorter and more direct proofs than those from your high school geometry days.

Exercises

Compute $\mathbf{a}\cdot\mathbf{b}$, $\|\mathbf{a}\|$, $\|\mathbf{b}\|$ *for the vectors listed in Exercises 1–6.*

1. $\mathbf{a} = (1, 5)$, $\mathbf{b} = (-2, 3)$

2. $\mathbf{a} = (4, -1)$, $\mathbf{b} = \left(\frac{1}{2}, 2\right)$

3. $\mathbf{a} = (-1, 0, 7)$, $\mathbf{b} = (2, 4, -6)$

4. $\mathbf{a} = (2, 1, 0)$, $\mathbf{b} = (1, -2, 3)$

5. $\mathbf{a} = 4\mathbf{i} - 3\mathbf{j} + \mathbf{k}$, $\mathbf{b} = \mathbf{i} + \mathbf{j} + \mathbf{k}$

6. $\mathbf{a} = \mathbf{i} + 2\mathbf{j} - \mathbf{k}$, $\mathbf{b} = -3\mathbf{j} + 2\mathbf{k}$

In Exercises 7–9, find the angle between each of the pairs of vectors.

7. $\mathbf{a} = \sqrt{3}\,\mathbf{i} + \mathbf{j}$, $\mathbf{b} = -\sqrt{3}\,\mathbf{i} + \mathbf{j}$

8. $\mathbf{a} = \mathbf{i} + \mathbf{j} - \mathbf{k}$, $\mathbf{b} = -\mathbf{i} + 2\mathbf{j} + 2\mathbf{k}$

9. $\mathbf{a} = (1, -2, 3)$, $\mathbf{b} = (3, -6, -5)$

In Exercises 10–12, calculate $\text{proj}_{\mathbf{a}}\mathbf{b}$.

10. $\mathbf{a} = \mathbf{i} + \mathbf{j}$, $\mathbf{b} = 2\mathbf{i} + 3\mathbf{j} - \mathbf{k}$

11. $\mathbf{a} = (\mathbf{i} + \mathbf{j})/\sqrt{2}$, $\mathbf{b} = 2\mathbf{i} + 3\mathbf{j} - \mathbf{k}$

12. $\mathbf{a} = \mathbf{i} + \mathbf{j} + 2\mathbf{k}$, $\mathbf{b} = 2\mathbf{i} - 4\mathbf{j} + \mathbf{k}$

13. Give a unit vector that points in the same direction as the vector $2\mathbf{i} - \mathbf{j} + \mathbf{k}$.

14. Give a unit vector that points in the direction opposite to the vector $-\mathbf{i} + 2\mathbf{k}$.

15. Give a vector of length 3 that points in the same direction as the vector $\mathbf{i} + \mathbf{j} - \mathbf{k}$.

16. Find three nonparallel vectors that are perpendicular to $\mathbf{i} - \mathbf{j} + \mathbf{k}$.

17. Is it ever the case that $\text{proj}_{\mathbf{a}}\mathbf{b} = \text{proj}_{\mathbf{b}}\mathbf{a}$? If so, under what conditions?

18. Prove properties 2, 3, and 4 of dot products.

19. Prove part 1 of Proposition 3.4.

20. Suppose that a force $\mathbf{F} = \mathbf{i} - 2\mathbf{j}$ is acting on an object moving parallel to the vector $\mathbf{a} = 4\mathbf{i} + \mathbf{j}$. Decompose $\mathbf{F}$ into a sum of vectors $\mathbf{F}_1$ and $\mathbf{F}_2$, where $\mathbf{F}_1$ points along the direction of motion and $\mathbf{F}_2$ is perpendicular to the direction of motion. (Hint: A diagram may help.)

21. In physics, when a constant force acts on an object as the object is displaced, the **work** done by the force is the product of the length of the displacement and the component of the force in the direction of the displacement. Figure 1.48 depicts an object acted upon by a constant force $\mathbf{F}$, which displaces it from the point P to the point Q. Let θ denote the angle between $\mathbf{F}$ and the direction of displacement.

(a) Show that the work done by $\mathbf{F}$ is determined by the formula $\mathbf{F} \cdot \overrightarrow{PQ}$.

(b) How much work is done in pushing a handtruck loaded with 500 lb of bananas 40 ft up a ramp inclined 30° from horizontal?

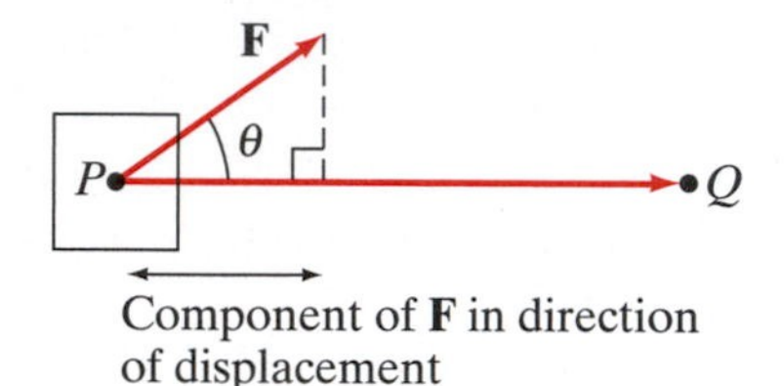

Figure 1.48 A constant force $\mathbf{F}$ displaces the object from P to Q. (See Exercise 21.)

22. Let $\mathbf{a}$ be a vector in $\mathbf{R}^3$. The **direction cosines** of $\mathbf{a}$ are the three numbers $\cos\alpha$, $\cos\beta$, $\cos\gamma$ determined by the angles α, β, γ between $\mathbf{a}$ and, respectively, the positive x-, y-, and z-axes. If $\mathbf{a} = a_1\mathbf{i} + a_2\mathbf{j} + a_3\mathbf{k}$, give expressions for the direction cosines of $\mathbf{a}$ in terms of the components of $\mathbf{a}$.

23. Let A, B, and C denote the vertices of a triangle. Let $0 < r < 1$. If P_1 is the point on $\overline{AB}$ located r times the distance from A to B and P_2 is the point on $\overline{AC}$ located r times the distance from A to C, use vectors to show that $\overline{P_1P_2}$ is parallel to $\overline{BC}$ and has r times the length of $\overline{BC}$. (This result generalizes that of Example 6 of this section.)

24. Let A, B, C, and D be four points in $\mathbf{R}^3$ such that no three of them lie on a line. Then $ABCD$ is a quadrilateral, though not necessarily one that lies in a plane. Denote the midpoints of the four sides of $ABCD$ by M_1, M_2, M_3, and M_4. Use vectors to show that, amazingly, $M_1M_2M_3M_4$ is always a parallelogram.

25. Use vectors to show that the diagonals of a parallelogram have the same length if and only if the parallelogram is a rectangle. (Hint: Let $\mathbf{a}$ and $\mathbf{b}$ be vectors along two sides of the parallelogram. Express vectors running along the diagonals in terms of $\mathbf{a}$ and $\mathbf{b}$. See Figure 1.49.)

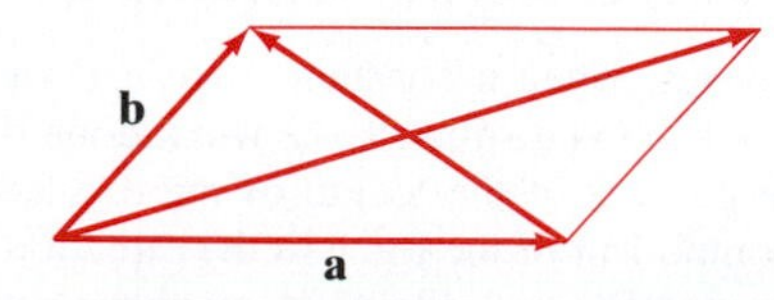

Figure 1.49 Diagram for Exercise 25.

26. Using vectors, prove that the diagonals of a parallelogram are perpendicular if and only if the parallelogram is a rhombus. (Note: A **rhombus** is a parallelogram whose four sides all have the same length.)

27. This problem concerns three circles of equal radius r that intersect in a single point O. (See Figure 1.50.)

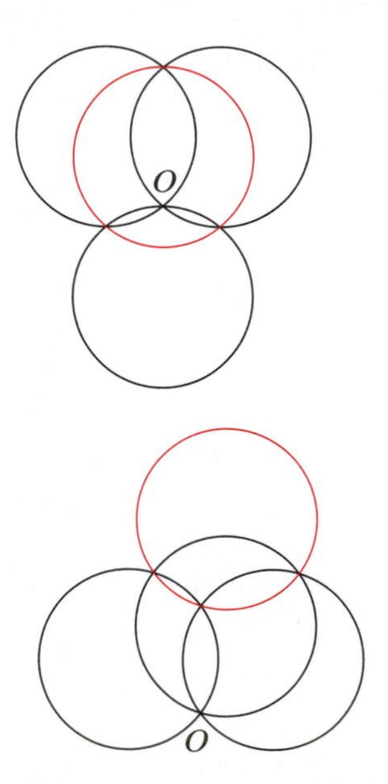

Figure 1.50 Two examples of three circles of equal radius intersecting in a single point O. (See Exercise 27.)

(a) Let W_1, W_2, and W_3 denote the centers of the three circles and let $\mathbf{w}_i = \overrightarrow{OW_i}$ for $i = 1, 2, 3$. Similarly, let A, B, and C denote the remaining intersection points of the circles and set $\mathbf{a} = \overrightarrow{OA}$, $\mathbf{b} = \overrightarrow{OB}$, and $\mathbf{c} = \overrightarrow{OC}$. By numbering the centers of the circles appropriately, write $\mathbf{a}$, $\mathbf{b}$, and $\mathbf{c}$ in terms of $\mathbf{w}_1$, $\mathbf{w}_2$, and $\mathbf{w}_3$.

(b) Show that A, B, and C lie on a circle of the same radius r as the three given circles. (Hint: The center of the circle is at the point P, where $\overrightarrow{OP} = \mathbf{w}_1 + \mathbf{w}_2 + \mathbf{w}_3$.)

(c) Show that O is the orthocenter of triangle ABC. (The **orthocenter** of a triangle is the common intersection point of the altitudes perpendicular to the edges.)

28. (a) Show that the vectors $\|\mathbf{b}\|\mathbf{a} + \|\mathbf{a}\|\mathbf{b}$ and $\|\mathbf{b}\|\mathbf{a} - \|\mathbf{a}\|\mathbf{b}$ are orthogonal.

(b) Show that $\|\mathbf{b}\|\mathbf{a} + \|\mathbf{a}\|\mathbf{b}$ bisects the angle between $\mathbf{a}$ and $\mathbf{b}$.

1.4 The Cross Product

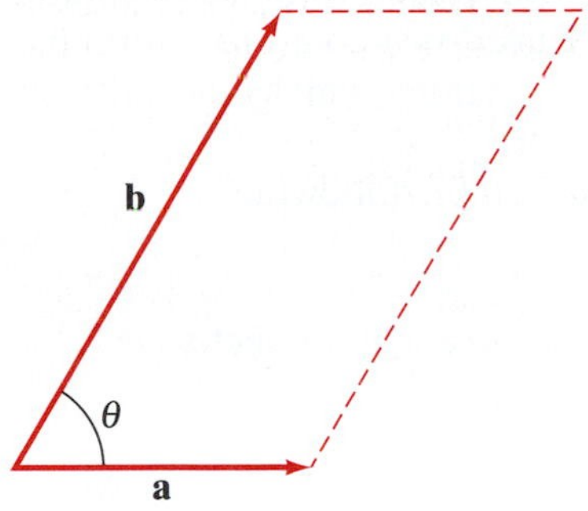

Figure 1.51 The area of this parallelogram is $\|\mathbf{a}\|\,\|\mathbf{b}\|\sin\theta$.

The cross product of two vectors in $\mathbf{R}^3$ is an "honest" product, in the sense that it takes two vectors and produces a third one. However, the cross product possesses some curious properties (not the least of which is that it *cannot* be defined for vectors in $\mathbf{R}^2$ without first embedding them in $\mathbf{R}^3$ in some way) making it less "natural" than may at first seem to be the case.

When we defined the concepts of vector addition, scalar multiplication, and the dot product, we did so algebraically (i.e., by a formula in the vector components) and then saw what these definitions meant geometrically. In contrast, we will define the cross product first *geometrically,* and then deduce an algebraic formula for it. This technique is more convenient, since the coordinate formulation is fairly complicated (although we will find a way to organize it so as to make it easier to remember).

The Cross Product of Two Vectors in $\mathbf{R}^3$

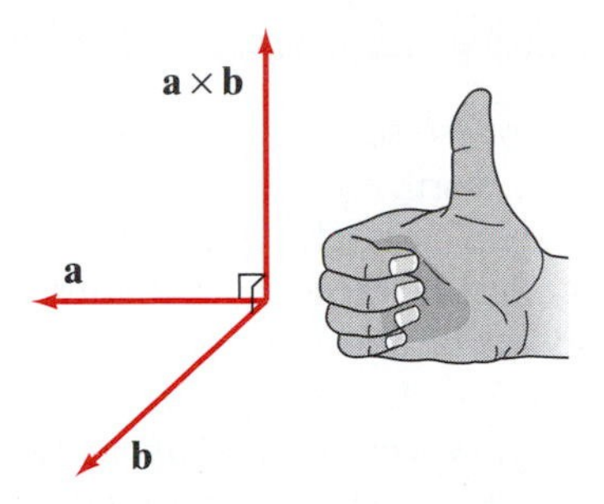

Figure 1.52 The right-hand rule for finding $\mathbf{a} \times \mathbf{b}$.

■ **Definition 4.1** Let $\mathbf{a}$ and $\mathbf{b}$ be two vectors in $\mathbf{R}^3$ (*not* $\mathbf{R}^2$). The **cross product** (or **vector product**) of $\mathbf{a}$ and $\mathbf{b}$, denoted $\mathbf{a} \times \mathbf{b}$, is the vector whose length and direction are given as follows:

- The length of $\mathbf{a} \times \mathbf{b}$ is the area of the parallelogram spanned by $\mathbf{a}$ and $\mathbf{b}$ or is $\mathbf{0}$ if either $\mathbf{a}$ is parallel to $\mathbf{b}$ or if $\mathbf{a}$ or $\mathbf{b}$ is $\mathbf{0}$. Alternatively, the following formula holds:

$$\|\mathbf{a} \times \mathbf{b}\| = \|\mathbf{a}\|\,\|\mathbf{b}\|\sin\theta,$$

  where θ is the angle between $\mathbf{a}$ and $\mathbf{b}$. (See Figure 1.51.)
- The direction of $\mathbf{a} \times \mathbf{b}$ is such that $\mathbf{a} \times \mathbf{b}$ is perpendicular to both $\mathbf{a}$ and $\mathbf{b}$ (when both $\mathbf{a}$ and $\mathbf{b}$ are nonzero) and is taken so that the ordered triple $(\mathbf{a}, \mathbf{b}, \mathbf{a} \times \mathbf{b})$ is a right-handed set of vectors, as shown in Figure 1.52. (If either $\mathbf{a}$ or $\mathbf{b}$ is $\mathbf{0}$, or if $\mathbf{a}$ is parallel to $\mathbf{b}$, then $\mathbf{a} \times \mathbf{b} = \mathbf{0}$ from the aforementioned length condition.)

By saying that $(\mathbf{a}, \mathbf{b}, \mathbf{a} \times \mathbf{b})$ is right-handed, we mean that if you let the fingers of your right hand curl *from* $\mathbf{a}$ *toward* $\mathbf{b}$, then your thumb will point in the direction of $\mathbf{a} \times \mathbf{b}$.

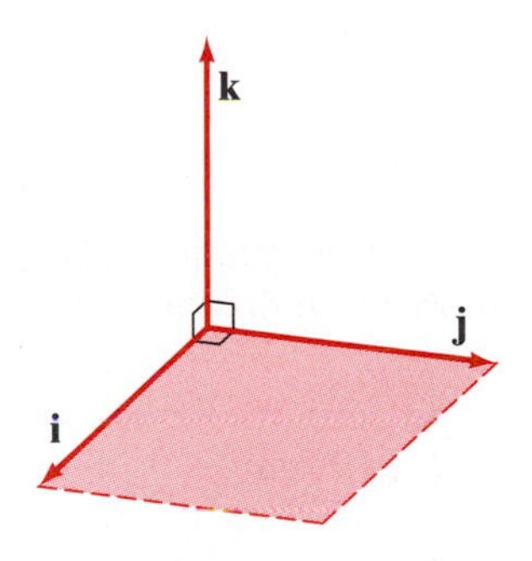

Figure 1.53 $\mathbf{i} \times \mathbf{j} = \mathbf{k}$.

EXAMPLE 1 Let's compute the cross product of the standard basis vectors for $\mathbf{R}^3$. First consider $\mathbf{i} \times \mathbf{j}$ as shown in Figure 1.53. The vectors $\mathbf{i}$ and $\mathbf{j}$ determine a square of unit area. Thus, $\|\mathbf{i} \times \mathbf{j}\| = 1$. Any vector perpendicular to *both* $\mathbf{i}$ and $\mathbf{j}$ must be perpendicular to the plane in which $\mathbf{i}$ and $\mathbf{j}$ lie. Hence, $\mathbf{i} \times \mathbf{j}$ must point in the direction of $\pm\mathbf{k}$. The "right-hand rule" implies that $\mathbf{i} \times \mathbf{j}$ must point in the positive $\mathbf{k}$ direction. Since $\|\mathbf{k}\| = 1$, we conclude that $\mathbf{i} \times \mathbf{j} = \mathbf{k}$. The same argument establishes that $\mathbf{j} \times \mathbf{k} = \mathbf{i}$ and $\mathbf{k} \times \mathbf{i} = \mathbf{j}$. To remember these basic equations, you can draw $\mathbf{i}$, $\mathbf{j}$, and $\mathbf{k}$ in a circle, as in Figure 1.54. Then the relations

$$\mathbf{i} \times \mathbf{j} = \mathbf{k}, \qquad \mathbf{j} \times \mathbf{k} = \mathbf{i}, \qquad \mathbf{k} \times \mathbf{i} = \mathbf{j} \tag{1}$$

may be read from the circle by beginning at any vector and then proceeding clockwise. ◆

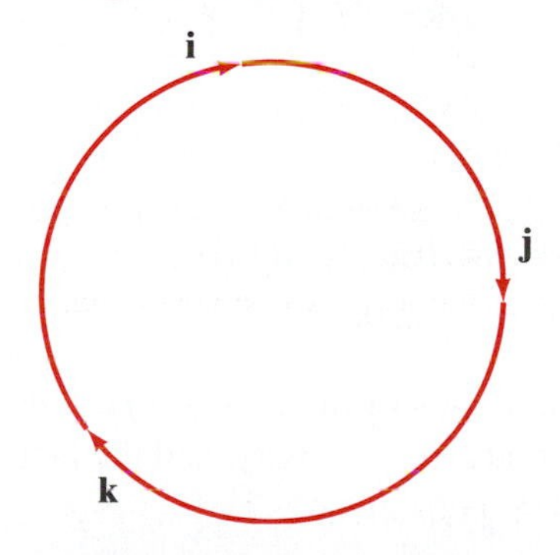

Figure 1.54 A mnemonic for finding the cross product of the unit basis vectors.

Properties of the Cross Product; Coordinate Formula

Example 1 demonstrates that the calculation of cross products from the geometric definition is not entirely routine. What we really need is a coordinate formula, analogous to that for the dot product or for vector projections, which is not difficult to obtain.

From our Definition 4.1, it is possible to establish the following:

Properties of the Cross Product. Let $\mathbf{a}$, $\mathbf{b}$, and $\mathbf{c}$ be any three vectors in $\mathbf{R}^3$ and let $k \in \mathbf{R}$ be any scalar. Then

1. $\mathbf{a} \times \mathbf{b} = -\mathbf{b} \times \mathbf{a}$ (anticommutativity);
2. $\mathbf{a} \times (\mathbf{b} + \mathbf{c}) = \mathbf{a} \times \mathbf{b} + \mathbf{a} \times \mathbf{c}$ (distributivity);
3. $(\mathbf{a} + \mathbf{b}) \times \mathbf{c} = \mathbf{a} \times \mathbf{c} + \mathbf{b} \times \mathbf{c}$ (distributivity);
4. $k(\mathbf{a} \times \mathbf{b}) = (k\mathbf{a}) \times \mathbf{b} = \mathbf{a} \times (k\mathbf{b})$.

We provide proofs of these properties at the end of the section, although you might give some thought now as to why they hold. It's worth remarking that these properties are entirely reasonable, ones which we would certainly want a product to have. However, you should be clear about the fact that the cross product *fails* to satisfy other properties which you might also consider to be eminently reasonable. In particular, since property 1 holds, we see that $\mathbf{a} \times \mathbf{b} \neq \mathbf{b} \times \mathbf{a}$ in general (i.e., the cross product is *not* commutative). Consequently, *be very careful about the order in which you write cross products.* Another property that the cross product does *not* possess is associativity. That is,

$$\mathbf{a} \times (\mathbf{b} \times \mathbf{c}) \neq (\mathbf{a} \times \mathbf{b}) \times \mathbf{c},$$

in general. For example, let $\mathbf{a} = \mathbf{b} = \mathbf{i}$ and $\mathbf{c} = \mathbf{j}$. Then

$$\mathbf{i} \times (\mathbf{i} \times \mathbf{j}) = \mathbf{i} \times \mathbf{k} = -\mathbf{k} \times \mathbf{i} = -\mathbf{j},$$

from properties 1 and 4, but $(\mathbf{i} \times \mathbf{i}) \times \mathbf{j} = \mathbf{0} \times \mathbf{j} = \mathbf{0} \neq -\mathbf{j}$. (The equation $\mathbf{i} \times \mathbf{i} = \mathbf{0}$ holds because $\mathbf{i}$ is, of course, parallel to $\mathbf{i}$.) Make sure that you do *not* try to use an associative law when working problems.

We now have the tools for producing a coordinate formula for the cross product. Let $\mathbf{a} = a_1\mathbf{i} + a_2\mathbf{j} + a_3\mathbf{k}$ and $\mathbf{b} = b_1\mathbf{i} + b_2\mathbf{j} + b_3\mathbf{k}$. Then

$$\begin{aligned}\mathbf{a} \times \mathbf{b} &= (a_1\mathbf{i} + a_2\mathbf{j} + a_3\mathbf{k}) \times (b_1\mathbf{i} + b_2\mathbf{j} + b_3\mathbf{k}) \\ &= (a_1\mathbf{i} + a_2\mathbf{j} + a_3\mathbf{k}) \times b_1\mathbf{i} + (a_1\mathbf{i} + a_2\mathbf{j} + a_3\mathbf{k}) \times b_2\mathbf{j} \\ &\quad + (a_1\mathbf{i} + a_2\mathbf{j} + a_3\mathbf{k}) \times b_3\mathbf{k},\end{aligned}$$

by property 2,

$$\begin{aligned}&= a_1b_1\mathbf{i} \times \mathbf{i} + a_2b_1\mathbf{j} \times \mathbf{i} + a_3b_1\mathbf{k} \times \mathbf{i} + a_1b_2\mathbf{i} \times \mathbf{j} + a_2b_2\mathbf{j} \times \mathbf{j} \\ &\quad + a_3b_2\mathbf{k} \times \mathbf{j} + a_1b_3\mathbf{i} \times \mathbf{k} + a_2b_3\mathbf{j} \times \mathbf{k} + a_3b_3\mathbf{k} \times \mathbf{k},\end{aligned}$$

by properties 3 and 4. These nine terms may look rather formidable at first, but we can simplify by means of the formulas in (1), anticommutativity, and the fact that $\mathbf{c} \times \mathbf{c} = \mathbf{0}$ for any vector $\mathbf{c} \in \mathbf{R}^3$. (Why?) Thus,

$$\begin{aligned}\mathbf{a} \times \mathbf{b} &= -a_2b_1\mathbf{k} + a_3b_1\mathbf{j} + a_1b_2\mathbf{k} - a_3b_2\mathbf{i} - a_1b_3\mathbf{j} + a_2b_3\mathbf{i} \\ &= (a_2b_3 - a_3b_2)\mathbf{i} + (a_3b_1 - a_1b_3)\mathbf{j} + (a_1b_2 - a_2b_1)\mathbf{k}. \qquad (2)\end{aligned}$$

EXAMPLE 2 Formula (2) gives

$$\begin{aligned}(\mathbf{i}+3\mathbf{j}-2\mathbf{k})\times(2\mathbf{i}+2\mathbf{k}) &= (3\cdot 2-(-2)\cdot 0)\mathbf{i}+(-2\cdot 2-1\cdot 2)\mathbf{j} \\ &\quad +(1\cdot 0-3\cdot 2)\mathbf{k} \\ &= 6\mathbf{i}-6\mathbf{j}-6\mathbf{k}.\end{aligned}$$

◆

Formula (2) is more complicated than the corresponding formulas for all the other arithmetic operations of vectors that we've seen. Moreover, it is a rather difficult formula to remember. Fortunately, there is a more elegant way to understand formula (2). We explore this reformulation next.

Matrices and Determinants: A First Introduction

A **matrix** is a rectangular array of numbers. Examples of matrices are

$$\begin{bmatrix}1 & 2 & 3\\ 4 & 5 & 6\end{bmatrix}, \quad \begin{bmatrix}1 & 3\\ 2 & 7\\ 0 & 0\end{bmatrix}, \quad \text{and} \quad \begin{bmatrix}1 & 0 & 0 & 0\\ 0 & 1 & 0 & 0\\ 0 & 0 & 1 & 0\\ 0 & 0 & 0 & 1\end{bmatrix}.$$

If a matrix has m rows and n columns, we call it "$m \times n$" (read "m by n"). Thus, the three matrices just mentioned are, respectively, 2×3, 3×2, and 4×4. To some extent, matrices behave algebraically like vectors. We discuss some elementary matrix algebra in §1.6. For now, we are mainly interested in the notion of a **determinant**, which is a real number associated to an $n \times n$ (square) matrix. (There is no such thing as the determinant of a nonsquare matrix.) In fact, for the purposes of understanding the cross product, we need only study 2×2 and 3×3 determinants.

■ **Definition 4.2** Let A be a 2×2 or 3×3 matrix. Then the **determinant** of A, denoted $\det A$ or $|A|$, is the real number computed from the individual entries of A as follows:

- 2×2 case

  If $A = \begin{bmatrix}a & b\\ c & d\end{bmatrix}$, then $|A| = \begin{vmatrix}a & b\\ c & d\end{vmatrix} = ad - bc.$

- 3×3 case

  If $A = \begin{bmatrix}a & b & c\\ d & e & f\\ g & h & i\end{bmatrix}$, then

$$\begin{aligned}|A| = \begin{vmatrix}a & b & c\\ d & e & f\\ g & h & i\end{vmatrix} &= aei + bfg + cdh - ceg - afh - bdi \\ &= a\begin{vmatrix}e & f\\ h & i\end{vmatrix} - b\begin{vmatrix}d & f\\ g & i\end{vmatrix} + c\begin{vmatrix}d & e\\ g & h\end{vmatrix}\end{aligned}$$

in terms of 2×2 determinants.

Perhaps the easiest way to remember and compute 2×2 and 3×3 determinants (but **not** higher-order determinants) is by means of a "diagonal approach." We write (or imagine) diagonal lines running through the matrix entries. The determinant is the sum of the products of the entries that lie on the same diagonal, where negative signs are inserted in front of the products arising from diagonals going from lower left to upper right:

- 2×2 case

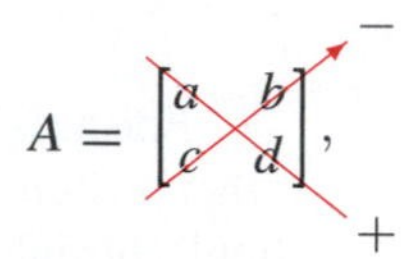

and

$$|A| = ad - bc.$$

- 3×3 case (we need to repeat the first two columns for the method to work)

$$A = \begin{bmatrix} a & b & c \\ d & e & f \\ g & h & i \end{bmatrix}.$$

Write

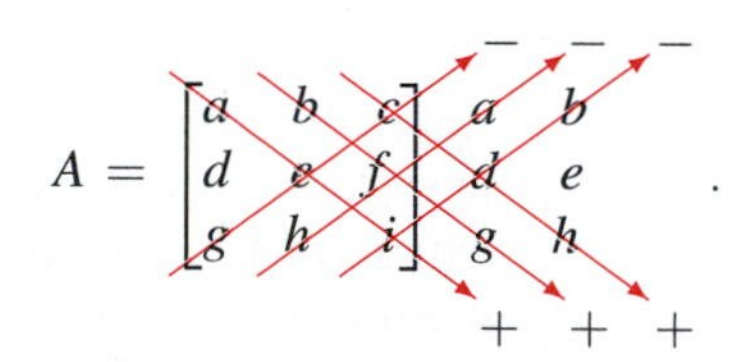

Then

$$|A| = aei + bfg + cdh - gec - hfa - idb.$$

IMPORTANT WARNING This mnemonic device does *not* generalize beyond 3×3 determinants.

We now state the connection between determinants and cross products.

Key Fact. If $\mathbf{a} = a_1\mathbf{i} + a_2\mathbf{j} + a_3\mathbf{k}$ and $\mathbf{b} = b_1\mathbf{i} + b_2\mathbf{j} + b_3\mathbf{k}$, then

$$\mathbf{a} \times \mathbf{b} = \begin{vmatrix} a_2 & a_3 \\ b_2 & b_3 \end{vmatrix}\mathbf{i} - \begin{vmatrix} a_1 & a_3 \\ b_1 & b_3 \end{vmatrix}\mathbf{j} + \begin{vmatrix} a_1 & a_2 \\ b_1 & b_2 \end{vmatrix}\mathbf{k} = \begin{vmatrix} \mathbf{i} & \mathbf{j} & \mathbf{k} \\ a_1 & a_2 & a_3 \\ b_1 & b_2 & b_3 \end{vmatrix}. \quad (3)$$

The determinants arise from nothing more than rewriting formula (2). Note that the 3×3 determinant in formula (3) needs to be interpreted by using the 2×2 determinants that appear in formula (3). (The 3×3 determinant is sometimes referred to as a "symbolic determinant.")

EXAMPLE 3

$$(3\mathbf{i} + 2\mathbf{j} - \mathbf{k}) \times (\mathbf{i} - \mathbf{j} + \mathbf{k}) = \begin{vmatrix} \mathbf{i} & \mathbf{j} & \mathbf{k} \\ 3 & 2 & -1 \\ 1 & -1 & 1 \end{vmatrix}$$

$$= \begin{vmatrix} 2 & -1 \\ -1 & 1 \end{vmatrix}\mathbf{i} - \begin{vmatrix} 3 & -1 \\ 1 & 1 \end{vmatrix}\mathbf{j} + \begin{vmatrix} 3 & 2 \\ 1 & -1 \end{vmatrix}\mathbf{k}$$

$$= \mathbf{i} - 4\mathbf{j} - 5\mathbf{k}.$$

We may also calculate the 3×3 determinant as

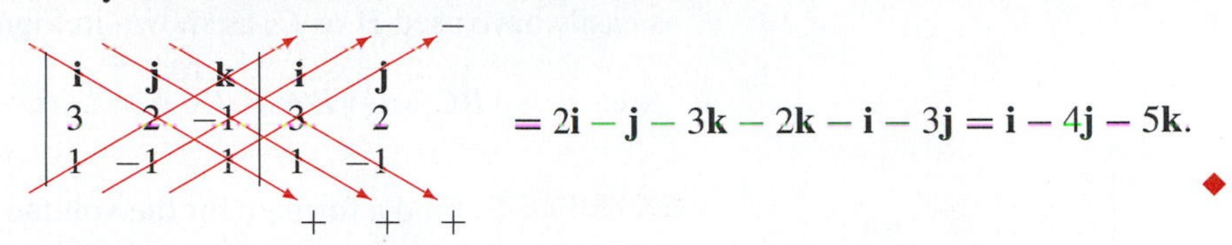

$$= 2\mathbf{i} - \mathbf{j} - 3\mathbf{k} - 2\mathbf{k} - \mathbf{i} - 3\mathbf{j} = \mathbf{i} - 4\mathbf{j} - 5\mathbf{k}.$$

◆

Areas and Volumes

Cross products are used readily to calculate areas and volumes of certain objects. We illustrate the ideas involved with the next two examples.

EXAMPLE 4 Let's use vectors to calculate the area of the triangle whose vertices are $A(3, 1)$, $B(2, -1)$, and $C(0, 2)$ as shown in Figure 1.55.

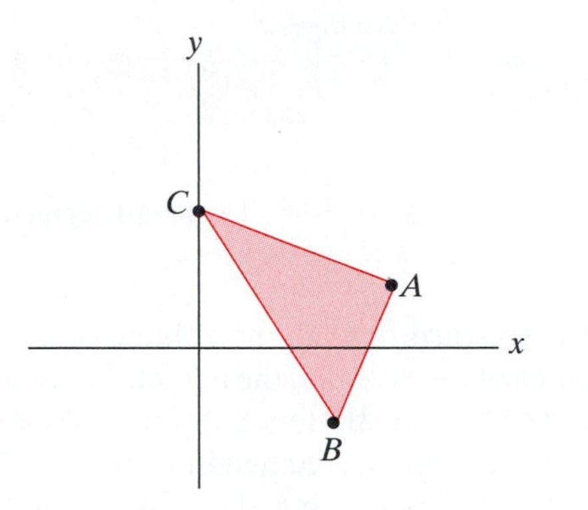

Figure 1.55 Triangle ABC in Example 4.

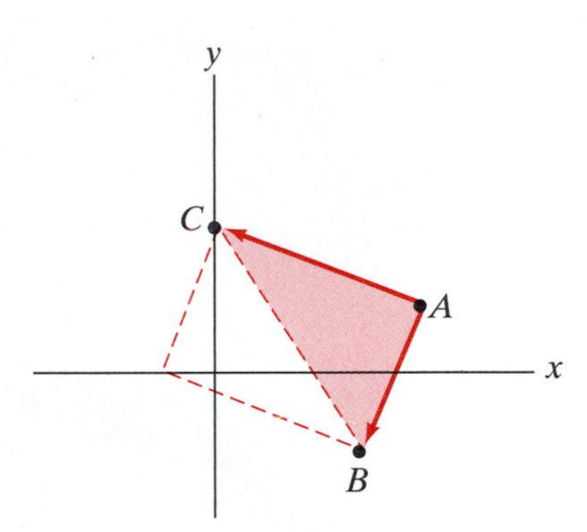

Figure 1.56 Any triangle may be considered to be half of a parallelogram.

The trick is to recognize that any triangle can be thought of as half of a parallelogram (see Figure 1.56), and that the area of a parallelogram is obtained from a cross product. In other words, $\overrightarrow{AB} \times \overrightarrow{AC}$ is a vector whose length measures the area of the parallelogram determined by $\overrightarrow{AB}$ and $\overrightarrow{AC}$, and so

$$\text{Area of } \Delta \text{ ABC} = \tfrac{1}{2}\|\overrightarrow{AB} \times \overrightarrow{AC}\|.$$

To use the cross product, we must consider $\overrightarrow{AB}$ and $\overrightarrow{AC}$ to be vectors in $\mathbf{R}^3$. This is straightforward: We simply take the **k**-components to be zero. Thus,

$$\overrightarrow{AB} = -\mathbf{i} - 2\mathbf{j} = -\mathbf{i} - 2\mathbf{j} - 0\mathbf{k},$$

and

$$\overrightarrow{AC} = -3\mathbf{i} + \mathbf{j} = -3\mathbf{i} + \mathbf{j} + 0\mathbf{k}.$$

Therefore,

$$\overrightarrow{AB} \times \overrightarrow{AC} = \begin{vmatrix} \mathbf{i} & \mathbf{j} & \mathbf{k} \\ -1 & -2 & 0 \\ -3 & 1 & 0 \end{vmatrix} = -7\mathbf{k}.$$

Hence,

$$\text{Area of } \Delta ABC = \tfrac{1}{2}\| -7\mathbf{k}\| = \tfrac{7}{2}.$$

◆

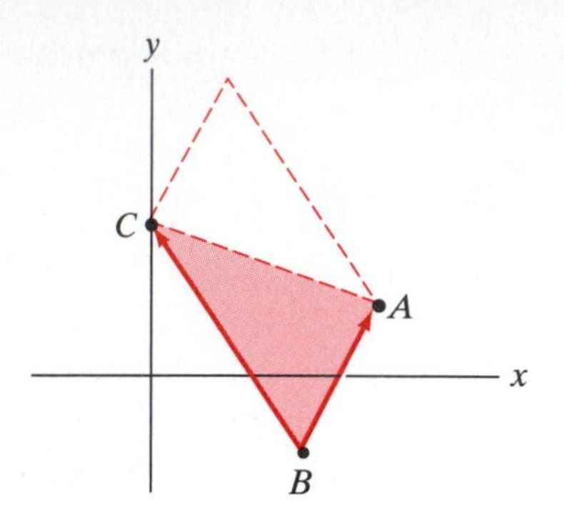

Figure 1.57 The area of ΔABC is $7/2$.

There is nothing sacred about using A as the common vertex. We could just as easily have used B or C, as shown in Figure 1.57. Then

$$\text{Area of } \Delta ABC = \tfrac{1}{2}\|\overrightarrow{BA} \times \overrightarrow{BC}\| = \tfrac{1}{2}\|(\mathbf{i}+2\mathbf{j}) \times (-2\mathbf{i}+3\mathbf{j})\| = \tfrac{1}{2}\|7\mathbf{k}\| = \tfrac{7}{2}.$$

EXAMPLE 5 Find a formula for the volume of the parallelepiped determined by the vectors **a**, **b**, and **c**. (See Figure 1.58.)

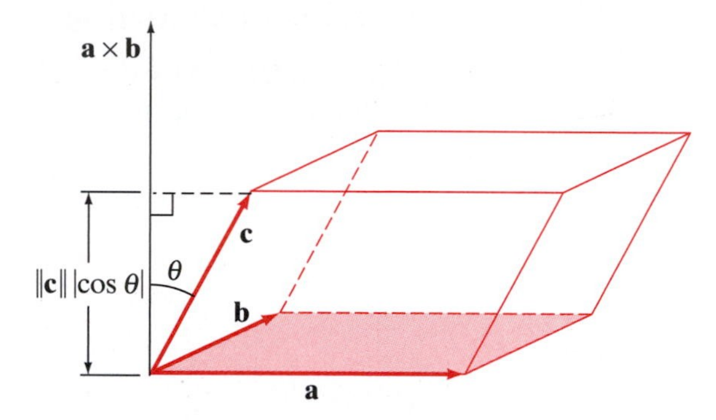

Figure 1.58 The parallelepiped determined by **a**, **b**, and **c**.

As explained in §1.3, the volume of a parallelepiped is equal to the product of the area of the base and the height. In Figure 1.58, the base is the parallelogram determined by **a** and **b**. Hence, its area is $\|\mathbf{a} \times \mathbf{b}\|$. The vector $\mathbf{a} \times \mathbf{b}$ is perpendicular to this parallelogram; the height of the parallelepiped is $\|\mathbf{c}\|\ |\cos\theta|$, where θ is the angle between $\mathbf{a} \times \mathbf{b}$ and $\mathbf{c}$. (The absolute value is needed in case $\theta > \pi/2$.) Therefore,

$$\begin{aligned}\text{Volume of parallelepiped} &= (\text{area of base})(\text{height}) \\ &= \|\mathbf{a} \times \mathbf{b}\|\ \|\mathbf{c}\|\ |\cos\theta| \\ &= |(\mathbf{a} \times \mathbf{b}) \cdot \mathbf{c}|.\end{aligned}$$

(The appearance of the $\cos\theta$ term should alert you to the fact that dot products are lurking somewhere.)

For example, the parallelepiped determined by the vectors

$$\mathbf{a} = \mathbf{i} + 5\mathbf{j}, \quad \mathbf{b} = -4\mathbf{i} + 2\mathbf{j}, \quad \text{and} \quad \mathbf{c} = \mathbf{i} + \mathbf{j} + 6\mathbf{k}$$

has volume equal to

$$\begin{aligned}|((\mathbf{i}+5\mathbf{j}) \times (-4\mathbf{i}+2\mathbf{j})) \cdot (\mathbf{i}+\mathbf{j}+6\mathbf{k})| &= |22\mathbf{k} \cdot (\mathbf{i}+\mathbf{j}+6\mathbf{k})| \\ &= |22(6)| \\ &= 132.\end{aligned}$$

◆

The real number $(\mathbf{a} \times \mathbf{b}) \cdot \mathbf{c}$ appearing in Example 5 is known as the **triple scalar product** of the vectors **a**, **b**, and **c**. Since $|(\mathbf{a} \times \mathbf{b}) \cdot \mathbf{c}|$ represents the volume of the parallelepiped determined by **a**, **b**, and **c**, it follows immediately that

$$|(\mathbf{a} \times \mathbf{b}) \cdot \mathbf{c}| = |(\mathbf{b} \times \mathbf{c}) \cdot \mathbf{a}| = |(\mathbf{c} \times \mathbf{a}) \cdot \mathbf{b}|.$$

In fact, if you are careful with the right-hand rule, you can convince yourself that the absolute value signs are not needed, that is,

$$(\mathbf{a} \times \mathbf{b}) \cdot \mathbf{c} = (\mathbf{b} \times \mathbf{c}) \cdot \mathbf{a} = (\mathbf{c} \times \mathbf{a}) \cdot \mathbf{b}. \tag{4}$$

This is a nice example of how the geometric significance of a quantity can provide an extremely brief proof of an algebraic property the quantity must satisfy. (Try proving it by writing out the expressions in terms of components to appreciate the value of geometric insight.)

We leave it to you to check the following beautiful (and convenient) formula for calculating triple scalar products:

$$(\mathbf{a} \times \mathbf{b}) \cdot \mathbf{c} = \begin{vmatrix} a_1 & a_2 & a_3 \\ b_1 & b_2 & b_3 \\ c_1 & c_2 & c_3 \end{vmatrix},$$

where $\mathbf{a} = a_1\mathbf{i} + a_2\mathbf{j} + a_3\mathbf{k}$, $\mathbf{b} = b_1\mathbf{i} + b_2\mathbf{j} + b_3\mathbf{k}$, and $\mathbf{c} = c_1\mathbf{i} + c_2\mathbf{j} + c_3\mathbf{k}$.

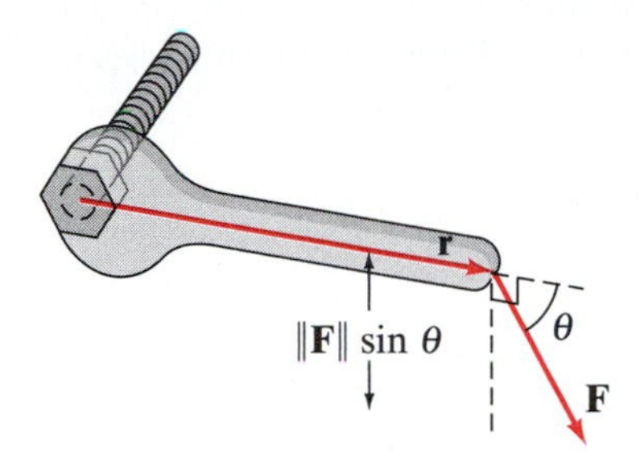

Figure 1.59 Turning a bolt with a wrench. The torque on the bolt is the vector $\mathbf{r} \times \mathbf{F}$.

Torque

Suppose you use a wrench to turn a bolt. What happens is the following: You apply some force to the end of the wrench handle farthest from the bolt and that causes the bolt to move in a direction perpendicular to the plane determined by the handle and the direction of your force (assuming such a plane exists). To measure exactly how much the bolt moves, we need the notion of **torque** (or twisting force).

In particular, letting $\mathbf{F}$ denote the force you apply to the wrench, we have

$$\text{Amount of torque} = (\text{length of wrench})(\text{component of } \mathbf{F} \perp \text{wrench}).$$

Let $\mathbf{r}$ be the vector from the center of the bolt head to the end of the wrench handle. Then

$$\text{Amount of torque} = \|\mathbf{r}\|\,\|\mathbf{F}\| \sin\theta,$$

where θ is the angle between $\mathbf{r}$ and $\mathbf{F}$. (See Figure 1.59.) That, is, the amount of torque is $\|\mathbf{r} \times \mathbf{F}\|$, and it is easy to check that the direction of $\mathbf{r} \times \mathbf{F}$ is the same as the direction in which the bolt moves (assuming a right-handed thread on the bolt). Hence it is quite natural to *define* the **torque vector T** to be $\mathbf{r} \times \mathbf{F}$. The torque vector $\mathbf{T}$ is a concise way to capture the physics of this situation.

Note that if $\mathbf{F}$ is parallel to $\mathbf{r}$, then $\mathbf{T} = \mathbf{0}$. This corresponds correctly to the fact that if you try to push or pull the wrench, the bolt does not turn.

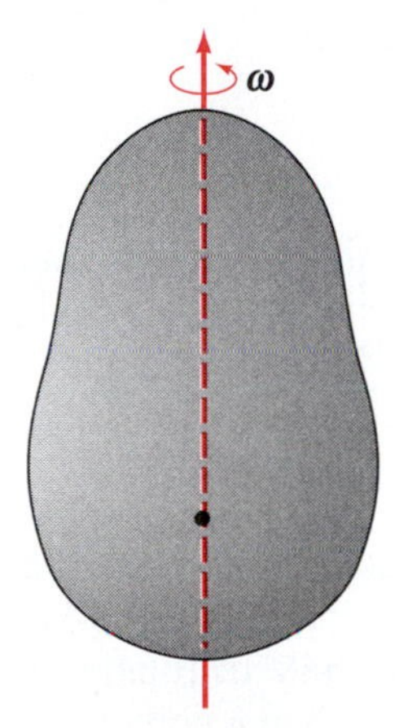

Figure 1.60 A potato spinning about an axis.

Rotation of a Rigid Body

Spin an object (a rigid body) about an axis as shown in Figure 1.60. What is the relation between the (linear) velocity of a point of the object and the rotational velocity? Vectors provide a good answer.

First we need to define a vector $\boldsymbol{\omega}$, the angular velocity vector of the rotation. This vector points along the axis of rotation, and its direction is determined by the right-hand rule. The magnitude of $\boldsymbol{\omega}$ is the angular speed (measured in radians per unit time) at which the object spins. Assume that the angular speed is constant in this discussion. Next, fix a point O (the origin) on the axis of rotation, and let $\mathbf{r}(t) = \overrightarrow{OP}$ be the position vector of a point P of the body, measured as a function of time, as in Figure 1.61. The velocity $\mathbf{v}$ of P is defined by

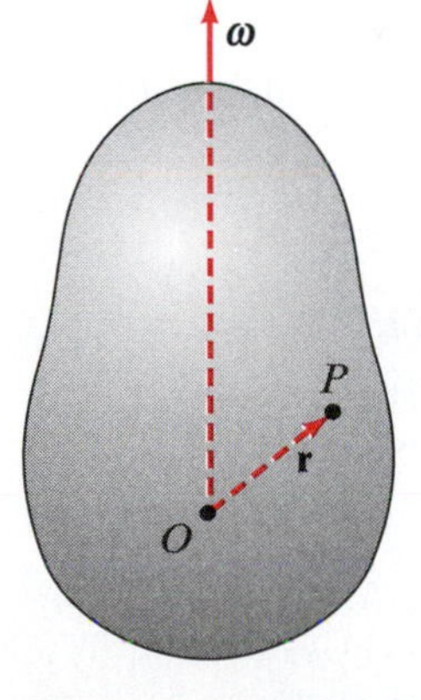

Figure 1.61 The angular velocity vector $\boldsymbol{\omega}$.

$$\mathbf{v} = \lim_{\Delta t \to 0} \frac{\Delta \mathbf{r}}{\Delta t},$$

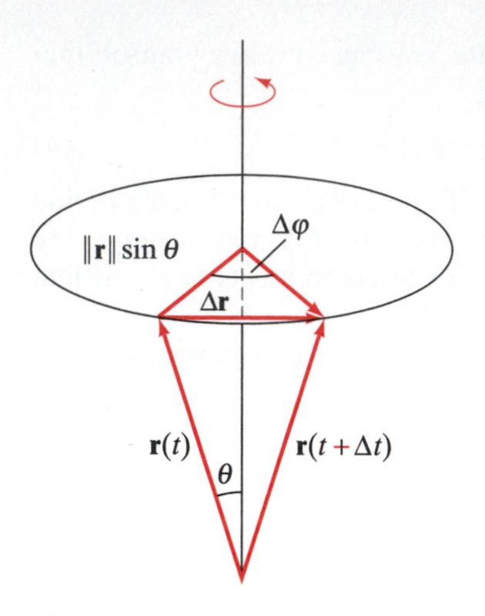

Figure 1.62 A spinning rigid body.

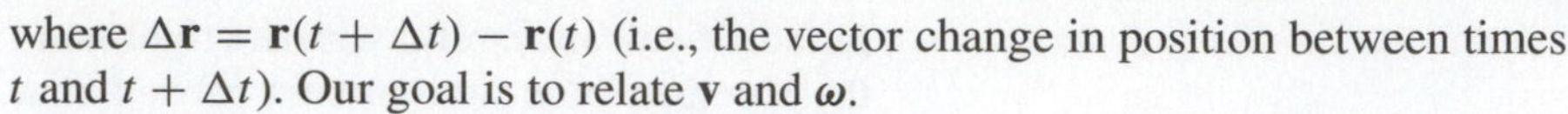
where $\Delta\mathbf{r} = \mathbf{r}(t + \Delta t) - \mathbf{r}(t)$ (i.e., the vector change in position between times t and $t + \Delta t$). Our goal is to relate $\mathbf{v}$ and $\boldsymbol{\omega}$.

As the body rotates, the point P (at the tip of the vector $\mathbf{r}$) moves in a circle whose plane is perpendicular to $\boldsymbol{\omega}$. (See Figure 1.62, which depicts the motion of such a point of the body.) The radius of this circle is $\|\mathbf{r}(t)\| \sin\theta$, where θ is the angle between $\boldsymbol{\omega}$ and $\mathbf{r}$. Both $\|\mathbf{r}(t)\|$ and θ must be constant for this rotation. (The direction of $\mathbf{r}(t)$ may change with t, however.) If $\Delta t \approx 0$, then $\|\Delta\mathbf{r}\|$ is approximately the length of the circular arc swept by P between t and $t + \Delta t$. That is,

$$\begin{aligned}\|\Delta\mathbf{r}\| &\approx (\text{radius of circle})(\text{angle swept through by } P)\\ &= (\|\mathbf{r}\| \sin\theta)(\Delta\varphi)\end{aligned}$$

from the preceding remarks. Thus,

$$\left\|\frac{\Delta\mathbf{r}}{\Delta t}\right\| \approx \|\mathbf{r}\| \sin\theta\, \frac{\Delta\varphi}{\Delta t}.$$

Now, let $\Delta t \to 0$. Then $\Delta\mathbf{r}/\Delta t \to \mathbf{v}$ and $\Delta\varphi/\Delta t \to \|\boldsymbol{\omega}\|$ by definition of the angular velocity vector $\boldsymbol{\omega}$, and we have

$$\|\mathbf{v}\| = \|\boldsymbol{\omega}\|\, \|\mathbf{r}\| \sin\theta = \|\boldsymbol{\omega} \times \mathbf{r}\|. \tag{5}$$

It's not difficult to see intuitively that $\mathbf{v}$ must be perpendicular to both $\boldsymbol{\omega}$ and $\mathbf{r}$. A moment's thought about the right-hand rule should enable you to establish the vector equation

$$\mathbf{v} = \boldsymbol{\omega} \times \mathbf{r}. \tag{6}$$

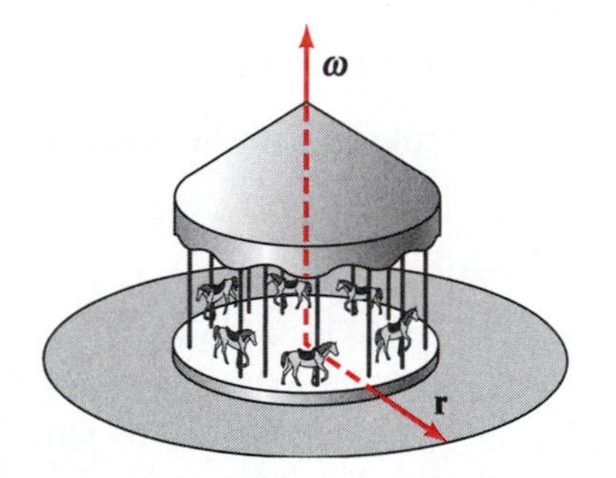

Figure 1.63 A carousel wheel.

If we apply formula (5) to a bicycle wheel, it tells us that the speed of a point on the edge of the wheel is equal to the product of the radius of the wheel and the angular speed (θ is $\pi/2$ in this case). Hence, if the rate of rotation is kept constant, a point on the rim of a large wheel goes faster than a point on the rim of a small one. In the case of a carousel wheel, this result tells you to sit on an outside horse if you want a more exciting ride. (See Figure 1.63.)

Summary of Products Involving Vectors

Following is a collection of some basic information concerning scalar multiplication of vectors, the dot product, and the cross product:

Scalar Multiplication: $k\mathbf{a}$

Result is a vector in the direction of $\mathbf{a}$.
Magnitude is $\|k\mathbf{a}\| = |k|\, \|\mathbf{a}\|$.
Zero if $k = 0$ or $\mathbf{a} = \mathbf{0}$.
Commutative: $k\mathbf{a} = \mathbf{a}k$.
Associative: $k(l\mathbf{a}) = (kl)\mathbf{a}$.
Distributive: $k(\mathbf{a} + \mathbf{b}) = k\mathbf{a} + k\mathbf{b}$; $(k + l)\mathbf{a} = k\mathbf{a} + l\mathbf{a}$.

Dot Product: $\mathbf{a} \cdot \mathbf{b}$

Result is a scalar.
Magnitude is $\mathbf{a} \cdot \mathbf{b} = \|\mathbf{a}\|\,\|\mathbf{b}\| \cos\theta$; θ is the angle between $\mathbf{a}$ and $\mathbf{b}$.
Magnitude is maximized if $\mathbf{a} \parallel \mathbf{b}$.
Zero if $\mathbf{a} \perp \mathbf{b}$, $\mathbf{a} = \mathbf{0}$, or $\mathbf{b} = \mathbf{0}$.
Commutative: $\mathbf{a} \cdot \mathbf{b} = \mathbf{b} \cdot \mathbf{a}$.
Associativity is irrelevant, since $(\mathbf{a} \cdot \mathbf{b}) \cdot \mathbf{c}$ doesn't make sense.
Distributive: $\mathbf{a} \cdot (\mathbf{b} + \mathbf{c}) = \mathbf{a} \cdot \mathbf{b} + \mathbf{a} \cdot \mathbf{c}$.
If $\mathbf{a} = \mathbf{b}$, then $\mathbf{a} \cdot \mathbf{a} = \|\mathbf{a}\|^2$.

Cross Product: $\mathbf{a} \times \mathbf{b}$

Result is a vector perpendicular to both $\mathbf{a}$ and $\mathbf{b}$.
Magnitude is $\|\mathbf{a} \times \mathbf{b}\| = \|\mathbf{a}\|\,\|\mathbf{b}\| \sin\theta$; θ is the angle between $\mathbf{a}$ and $\mathbf{b}$.
Magnitude is maximized if $\mathbf{a} \perp \mathbf{b}$.
Zero if $\mathbf{a} \parallel \mathbf{b}$, $\mathbf{a} = \mathbf{0}$, or $\mathbf{b} = \mathbf{0}$.
Anticommutative: $\mathbf{a} \times \mathbf{b} = -\mathbf{b} \times \mathbf{a}$.
Not associative: In general, $\mathbf{a} \times (\mathbf{b} \times \mathbf{c}) \neq (\mathbf{a} \times \mathbf{b}) \times \mathbf{c}$.
Distributive: $\mathbf{a} \times (\mathbf{b} + \mathbf{c}) = \mathbf{a} \times \mathbf{b} + \mathbf{a} \times \mathbf{c}$ and
$(\mathbf{a} + \mathbf{b}) \times \mathbf{c} = \mathbf{a} \times \mathbf{c} + \mathbf{b} \times \mathbf{c}$.
If $\mathbf{a} \perp \mathbf{b}$, then $\|\mathbf{a} \times \mathbf{b}\| = \|\mathbf{a}\|\,\|\mathbf{b}\|$.

Addendum: Proofs of Cross Product Properties

Proof of Property 1 To prove the anticommutativity property, we use the right-hand rule. Since

$$\|\mathbf{a} \times \mathbf{b}\| = \|\mathbf{a}\|\,\|\mathbf{b}\| \sin\theta,$$

we obviously have that $\|\mathbf{a} \times \mathbf{b}\| = \|\mathbf{b} \times \mathbf{a}\|$. Therefore, we need only understand the relation between the direction of $\mathbf{a} \times \mathbf{b}$ and that of $\mathbf{b} \times \mathbf{a}$. To determine the direction of $\mathbf{a} \times \mathbf{b}$, imagine curling the fingers of your right hand from $\mathbf{a}$ toward $\mathbf{b}$. Then your thumb points in the direction of $\mathbf{a} \times \mathbf{b}$. If instead you curl your fingers from $\mathbf{b}$ toward $\mathbf{a}$, then your thumb will point in the opposite direction. This is the direction of $\mathbf{b} \times \mathbf{a}$, so we conclude that $\mathbf{a} \times \mathbf{b} = -\mathbf{b} \times \mathbf{a}$. (See Figure 1.64.) ■

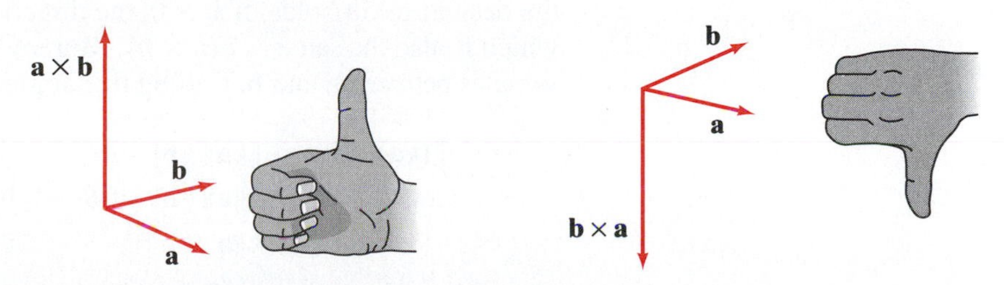

Figure 1.64 The right-hand rule shows why $\mathbf{a} \times \mathbf{b} = -\mathbf{b} \times \mathbf{a}$.

Proof of Property 2 First, note the following general fact:

■ **Proposition 4.3** Let $\mathbf{a}$ and $\mathbf{b}$ be vectors in $\mathbf{R}^3$. If $\mathbf{a}\cdot\mathbf{x}=\mathbf{b}\cdot\mathbf{x}$ for all vectors $\mathbf{x}$ in $\mathbf{R}^3$, then $\mathbf{a}=\mathbf{b}$.

To establish Proposition 4.3, write $\mathbf{a}$ as $a_1\mathbf{i}+a_2\mathbf{j}+a_3\mathbf{k}$ and $\mathbf{b}$ as $b_1\mathbf{i}+b_2\mathbf{j}+b_3\mathbf{k}$ and set $\mathbf{x}$ in turn equal to $\mathbf{i}$, $\mathbf{j}$, and $\mathbf{k}$. Proposition 4.3 is valid for vectors in $\mathbf{R}^2$ as well as $\mathbf{R}^3$.

To prove the distributive law for cross products (property 2), we show that, for any $\mathbf{x}\in\mathbf{R}^3$,

$$(\mathbf{a}\times(\mathbf{b}+\mathbf{c}))\cdot\mathbf{x}=(\mathbf{a}\times\mathbf{b}+\mathbf{a}\times\mathbf{c})\cdot\mathbf{x}.$$

By Proposition 4.3, property 2 follows.

From the equations in (4),

$$\begin{aligned}(\mathbf{a}\times(\mathbf{b}+\mathbf{c}))\cdot\mathbf{x}&=(\mathbf{x}\times\mathbf{a})\cdot(\mathbf{b}+\mathbf{c})\\&=(\mathbf{x}\times\mathbf{a})\cdot\mathbf{b}+(\mathbf{x}\times\mathbf{a})\cdot\mathbf{c},\end{aligned}$$

from the distributive law for dot products,

$$\begin{aligned}&=(\mathbf{a}\times\mathbf{b})\cdot\mathbf{x}+(\mathbf{a}\times\mathbf{c})\cdot\mathbf{x}\\&=(\mathbf{a}\times\mathbf{b}+\mathbf{a}\times\mathbf{c})\cdot\mathbf{x},\end{aligned}$$

again using (4) and the distributive law for dot products. ■

Proof of Property 3 Property 3 follows from properties 1 and 2. We leave the details as an exercise. ■

Proof of Property 4 The second equality in property 4 follows from the first equality and property 1:

$$\begin{aligned}k(\mathbf{a}\times\mathbf{b})&=-k(\mathbf{b}\times\mathbf{a}) &&\text{by property 1}\\&=-(k\mathbf{b})\times\mathbf{a} &&\text{by the first equality of property 4}\\&=\mathbf{a}\times(k\mathbf{b}) &&\text{by property 1.}\end{aligned}$$

Hence we need only prove the first equality.

If either $\mathbf{a}$ or $\mathbf{b}$ is the zero vector or if $\mathbf{a}$ is parallel to $\mathbf{b}$, then the first equality clearly holds. Otherwise, we divide into three cases: (1) $k=0$, (2) $k>0$, and (3) $k<0$. If $k=0$, then both $k\mathbf{a}$ and $k(\mathbf{a}\times\mathbf{b})$ are equal to the zero vector and the desired result holds. If $k>0$, the direction of $(k\mathbf{a})\times\mathbf{b}$ is the same as $\mathbf{a}\times\mathbf{b}$, which is also the same as $k(\mathbf{a}\times\mathbf{b})$. Moreover, the angle between $k\mathbf{a}$ and $\mathbf{b}$ is the same as between $\mathbf{a}$ and $\mathbf{b}$. Calling this angle θ, we check that

$$\begin{aligned}\|(k\mathbf{a})\times\mathbf{b}\|&=\|k\mathbf{a}\|\,\|\mathbf{b}\|\sin\theta\\&=k\|\mathbf{a}\|\,\|\mathbf{b}\|\sin\theta &&\text{by part 1 of Proposition 3.4}\\&=k\|\mathbf{a}\times\mathbf{b}\| &&\text{by Definition 4.1}\\&=\|k(\mathbf{a}\times\mathbf{b})\| &&\text{by part 1 of Proposition 3.4.}\end{aligned}$$

We conclude $(k\mathbf{a})\times\mathbf{b}=k(\mathbf{a}\times\mathbf{b})$ in this case.

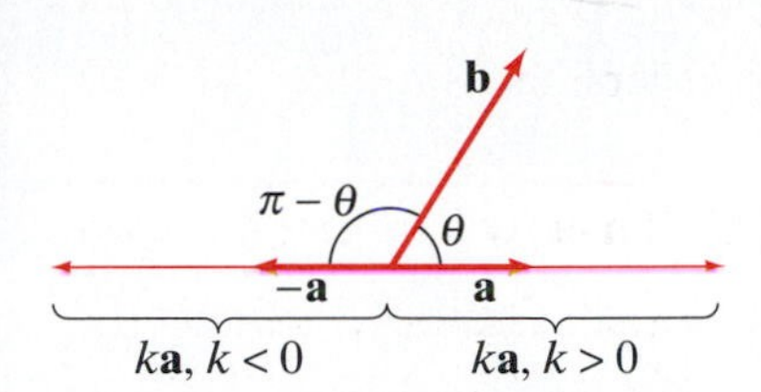

Figure 1.65 If the angle between **a** and **b** is θ, then the angle between $k\mathbf{a}$ and **b** is either θ (if $k > 0$) or $\pi - \theta$ (if $k < 0$).

If $k < 0$, then the direction of $(k\mathbf{a}) \times \mathbf{b}$ is the same as that of $(-\mathbf{a}) \times \mathbf{b}$, which is seen to be the same as that of $-(\mathbf{a} \times \mathbf{b})$ and thus the same as that of $k(\mathbf{a} \times \mathbf{b})$. The angle between $k\mathbf{a}$ and $\mathbf{b}$ is therefore $\pi - \theta$, where θ is the angle between **a** and **b**. (See Figure 1.65.) Thus,

$$\|(k\mathbf{a}) \times \mathbf{b}\| = \|k\mathbf{a}\|\,\|\mathbf{b}\| \sin(\pi - \theta) = |k|\,\|\mathbf{a}\|\,\|\mathbf{b}\| \sin\theta = \|k(\mathbf{a} \times \mathbf{b})\|.$$

So, again, it follows that $(k\mathbf{a}) \times \mathbf{b} = k(\mathbf{a} \times \mathbf{b})$. ■

Exercises

Evaluate the determinants in Exercises 1–4.

1. $\begin{vmatrix} 2 & 4 \\ 1 & 3 \end{vmatrix}$

2. $\begin{vmatrix} 0 & 5 \\ -1 & 6 \end{vmatrix}$

3. $\begin{vmatrix} 1 & 3 & 5 \\ 0 & 2 & 7 \\ -1 & 0 & 3 \end{vmatrix}$

4. $\begin{vmatrix} -2 & 0 & \frac{1}{2} \\ 3 & 6 & -1 \\ 4 & -8 & 2 \end{vmatrix}$

In Exercises 5–7, calculate the indicated cross products, using both formulas (2) and (3).

5. $(1, 3, -2) \times (-1, 5, 7)$

6. $(3\mathbf{i} - 2\mathbf{j} + \mathbf{k}) \times (\mathbf{i} + \mathbf{j} + \mathbf{k})$

7. $(\mathbf{i} + \mathbf{j}) \times (-3\mathbf{i} + 2\mathbf{j})$

8. Prove property 3 of cross products, using properties 1 and 2.

9. If $\mathbf{a} \times \mathbf{b} = 3\mathbf{i} - 7\mathbf{j} - 2\mathbf{k}$, what is $(\mathbf{a} + \mathbf{b}) \times (\mathbf{a} - \mathbf{b})$?

10. Calculate the area of the parallelogram having vertices $(1, 1)$, $(3, 2)$, $(1, 3)$, and $(-1, 2)$.

11. Calculate the area of the parallelogram having vertices $(1, 2, 3)$, $(4, -2, 1)$, $(-3, 1, 0)$, and $(0, -3, -2)$.

12. Find a unit vector that is perpendicular to both $2\mathbf{i} + \mathbf{j} - 3\mathbf{k}$ and $\mathbf{i} + \mathbf{k}$.

13. If $(\mathbf{a} \times \mathbf{b}) \cdot \mathbf{c} = 0$, what can you say about the geometric relation between **a**, **b**, and **c**?

Compute the area of the triangles described in Exercises 14–17.

14. The triangle determined by the vectors $\mathbf{a} = \mathbf{i} + \mathbf{j}$ and $\mathbf{b} = 2\mathbf{i} - \mathbf{j}$

15. The triangle determined by the vectors $\mathbf{a} = \mathbf{i} - 2\mathbf{j} + 6\mathbf{k}$ and $\mathbf{b} = 4\mathbf{i} + 3\mathbf{j} - \mathbf{k}$

16. The triangle having vertices $(1, 1)$, $(-1, 2)$, and $(-2, -1)$

17. The triangle having vertices $(1, 0, 1)$, $(0, 2, 3)$, and $(-1, 5, -2)$

18. Find the volume of the parallelepiped determined by $\mathbf{a} = 3\mathbf{i} - \mathbf{j}$, $\mathbf{b} = -2\mathbf{i} + \mathbf{k}$, and $\mathbf{c} = \mathbf{i} - 2\mathbf{j} + 4\mathbf{k}$.

19. What is the volume of the parallelepiped with vertices $(3, 0, -1)$, $(4, 2, -1)$, $(-1, 1, 0)$, $(3, 1, 5)$, $(0, 3, 0)$, $(4, 3, 5)$, $(-1, 2, 6)$, and $(0, 4, 6)$?

20. Verify that $(\mathbf{a} \times \mathbf{b}) \cdot \mathbf{c} = \begin{vmatrix} a_1 & a_2 & a_3 \\ b_1 & b_2 & b_3 \\ c_1 & c_2 & c_3 \end{vmatrix}$.

21. Show that $(\mathbf{a} \times \mathbf{b}) \cdot \mathbf{c} = \mathbf{a} \cdot (\mathbf{b} \times \mathbf{c})$ using Exercise 20.

22. Use geometry to show that $|(\mathbf{a} \times \mathbf{b}) \cdot \mathbf{c}| = |\mathbf{b} \cdot (\mathbf{a} \times \mathbf{c})|$.

23. (a) Show that the area of the triangle with vertices $P_1(x_1, y_1)$, $P_2(x_2, y_2)$, and $P_3(x_3, y_3)$ is given by the absolute value of the expression

$$\frac{1}{2}\begin{vmatrix} 1 & 1 & 1 \\ x_1 & x_2 & x_3 \\ y_1 & y_2 & y_3 \end{vmatrix}.$$

(b) Use part (a) to find the area of the triangle with vertices $(1, 2)$, $(2, 3)$, and $(-4, -4)$.

24. Suppose that **a**, **b**, and **c** are noncoplanar vectors in $\mathbf{R}^3$, so that they determine a tetrahedron as in Figure 1.66.

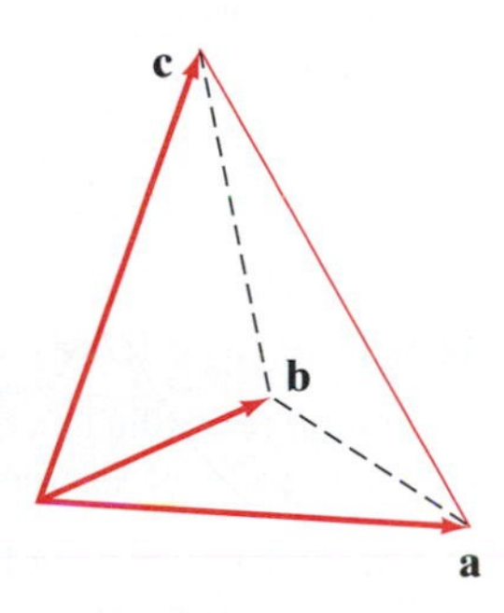

Figure 1.66 The tetrahedron of Exercise 24.

Give a formula for the surface area of the tetrahedron in terms of **a**, **b**, and **c**. (Note: More than one formula is possible.)

25. Suppose that you are given nonzero vectors **a**, **b**, and **c** in $\mathbf{R}^3$. Use dot and cross products to give expressions for vectors satisfying the following geometric descriptions:

(a) A vector orthogonal to **a** and **b**

(b) A vector of length 2 orthogonal to **a** and **b**

(c) The vector projection of **b** onto **a**

(d) A vector with the length of **b** and the direction of **a**

(e) A vector orthogonal to **a** and $\mathbf{b} \times \mathbf{c}$

(f) A vector in the plane determined by **a** and **b** and perpendicular to **c**.

26. Suppose **a**, **b**, **c**, and **d** are vectors in $\mathbf{R}^3$. Indicate which of the following expressions are vectors, which are scalars, and which are nonsense (i.e., neither a vector nor a scalar).

(a) $(\mathbf{a} \times \mathbf{b}) \times \mathbf{c}$ (b) $(\mathbf{a} \cdot \mathbf{b}) \cdot \mathbf{c}$

(c) $(\mathbf{a} \cdot \mathbf{b}) \times (\mathbf{c} \cdot \mathbf{d})$ (d) $(\mathbf{a} \times \mathbf{b}) \cdot \mathbf{c}$

(e) $(\mathbf{a} \cdot \mathbf{b}) \times (\mathbf{c} \times \mathbf{d})$ (f) $\mathbf{a} \times [(\mathbf{b} \cdot \mathbf{c})\mathbf{d}]$

(g) $(\mathbf{a} \times \mathbf{b}) \cdot (\mathbf{c} \times \mathbf{d})$ (h) $(\mathbf{a} \cdot \mathbf{b})\mathbf{c} - (\mathbf{a} \times \mathbf{b})$

Exercises 27–32 concern several identities for vectors **a**, **b**, **c**, *and* **d** *in* $\mathbf{R}^3$. *Each of them can be verified by hand by writing the vectors in terms of their components and by using formula (2) for the cross product and Definition 3.1 for the dot product. However, this is quite tedious to do. Instead, use a computer algebra system to define the vectors* **a**, **b**, **c**, *and* **d** *in general and to verify the identities.*

27. $(\mathbf{a} \times \mathbf{b}) \times \mathbf{c} = (\mathbf{a} \cdot \mathbf{c})\mathbf{b} - (\mathbf{b} \cdot \mathbf{c})\mathbf{a}$

28. $$\begin{aligned}\mathbf{a} \cdot (\mathbf{b} \times \mathbf{c}) &= \mathbf{b} \cdot (\mathbf{c} \times \mathbf{a}) = \mathbf{c} \cdot (\mathbf{a} \times \mathbf{b}) \\ &= -\mathbf{a} \cdot (\mathbf{c} \times \mathbf{b}) = -\mathbf{c} \cdot (\mathbf{b} \times \mathbf{a}) \\ &= -\mathbf{b} \cdot (\mathbf{a} \times \mathbf{c})\end{aligned}$$

29. $$(\mathbf{a} \times \mathbf{b}) \cdot (\mathbf{c} \times \mathbf{d}) = (\mathbf{a} \cdot \mathbf{c})(\mathbf{b} \cdot \mathbf{d}) - (\mathbf{a} \cdot \mathbf{d})(\mathbf{b} \cdot \mathbf{c}) = \begin{vmatrix} \mathbf{a} \cdot \mathbf{c} & \mathbf{a} \cdot \mathbf{d} \\ \mathbf{b} \cdot \mathbf{c} & \mathbf{b} \cdot \mathbf{d} \end{vmatrix}$$

30. $(\mathbf{a} \times \mathbf{b}) \times \mathbf{c} + (\mathbf{b} \times \mathbf{c}) \times \mathbf{a} + (\mathbf{c} \times \mathbf{a}) \times \mathbf{b} = \mathbf{0}$ (this is known as the **Jacobi identity**).

31. $(\mathbf{a} \times \mathbf{b}) \times (\mathbf{c} \times \mathbf{d}) = [\mathbf{a} \cdot (\mathbf{c} \times \mathbf{d})]\mathbf{b} - [\mathbf{b} \cdot (\mathbf{c} \times \mathbf{d})]\mathbf{a}$

32. $(\mathbf{a} \times \mathbf{b}) \cdot (\mathbf{b} \times \mathbf{c}) \times (\mathbf{c} \times \mathbf{a}) = [\mathbf{a} \cdot (\mathbf{b} \times \mathbf{c})]^2$

33. Establish the identity

$$(\mathbf{a} \times \mathbf{b}) \cdot (\mathbf{c} \times \mathbf{d}) = (\mathbf{a} \cdot \mathbf{c})(\mathbf{b} \cdot \mathbf{d}) - (\mathbf{a} \cdot \mathbf{d})(\mathbf{b} \cdot \mathbf{c})$$

of Exercise 29 *without* resorting to a computer algebra system by using the results of Exercises 27 and 28.

34. Egbert applies a 20 lb force at the edge of a 4 ft wide door that is half-open in order to close it. (See Figure 1.67.) Assume that the direction of force is perpendicular to the plane of the doorway. What is the torque about the hinge on the door?

35. Gertrude is changing a flat tire with a tire iron. The tire iron is positioned on one of the bolts of the wheel so that it makes an angle of 30° with the horizontal. (See Figure 1.68.) Gertrude exerts 40 lb of force straight down to turn the bolt.

(a) If the length of the arm of the wrench is 1 ft, how much torque does Gertrude impart to the bolt?

(b) What if she has a second tire iron whose length is 18 in?

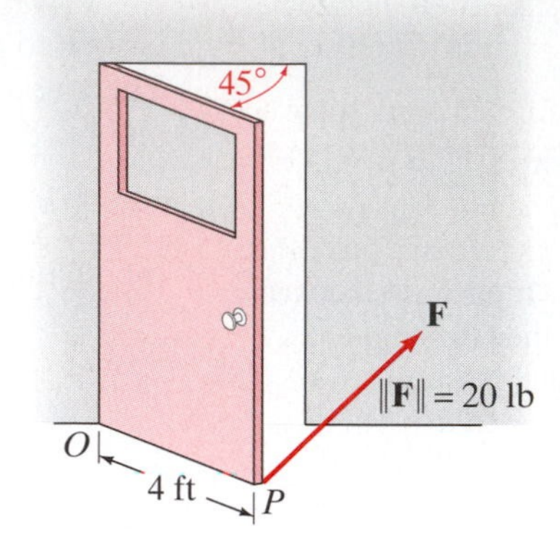

Figure 1.67 Figure for Exercise 34.

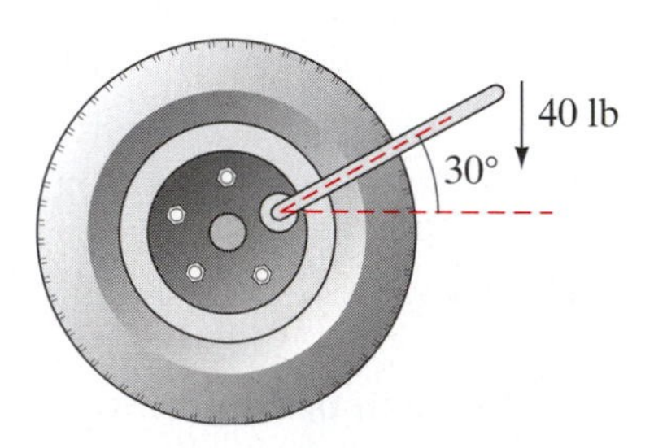

Figure 1.68 The configuration for Exercise 35.

36. Egbert is trying to open a jar of grape jelly. The radius of the lid of the jar is 2 in. If Egbert imparts 15 lb of force tangent to the edge of the lid to open the jar, how many ft-lb, and in what direction, is the resulting torque?

37. A 50 lb child is sitting on one end of a seesaw, 3 ft from the center fulcrum. (See Figure 1.69.) When she is

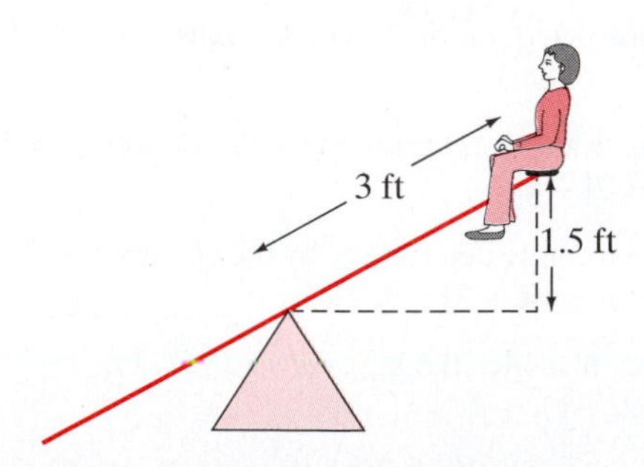

Figure 1.69 The seesaw of Exercise 37.

1.5 ft above the horizontal position, what is the amount of torque she exerts on the seesaw?

38. For this problem, note that the radius of the earth is approximately 3960 miles.

(a) Suppose that you are standing at 45° north latitude. Given that the earth spins about its axis, how fast are you moving?

(b) How fast would you be traveling if, instead, you were standing at a point on the equator?

39. Archie, the cockroach, and Annie, the ant, are on an LP record. Archie is at the edge of the record (approximately 6 in from the center) and Annie is 2 in closer to the center of the record. How much faster is Archie traveling than Annie? (Note: A record playing on a turntable spins at a rate of $33\frac{1}{3}$ revolutions per minute.)

40. A top is spinning with a constant angular speed of 12 radians/sec. Suppose that the top spins about its axis of symmetry and we orient things so that this axis is the z-axis and the top spins counterclockwise about it.

(a) If, at a certain instant, a point P in the top has coordinates $(2, -1, 3)$, what is the velocity of the point at that instant?

(b) What are the (approximate) coordinates of P one second later?

41. There is a difficulty involved with our definition of the angular velocity vector $\boldsymbol{\omega}$, namely, that we cannot properly consider this vector to be "free" in the sense of being able to parallel translate it at will. Consider the rotations of a rigid body about each of two parallel axes. Then the corresponding angular velocity vectors $\boldsymbol{\omega}_1$ and $\boldsymbol{\omega}_2$ are parallel. Explain, perhaps with a figure, that even if $\boldsymbol{\omega}_1$ and $\boldsymbol{\omega}_2$ are equal as "free vectors," the corresponding rotational motions that result must be different. (Therefore, when considering more than one angular velocity, we should always assume that the axes of rotation pass through a common point.)

1.5 Equations for Planes; Distance Problems

In this section, we use vectors to derive analytic descriptions of planes in $\mathbf{R}^3$. We also show how to solve a variety of distance problems involving "flat objects" (i.e., points, lines, and planes).

Coordinate Equations of Planes

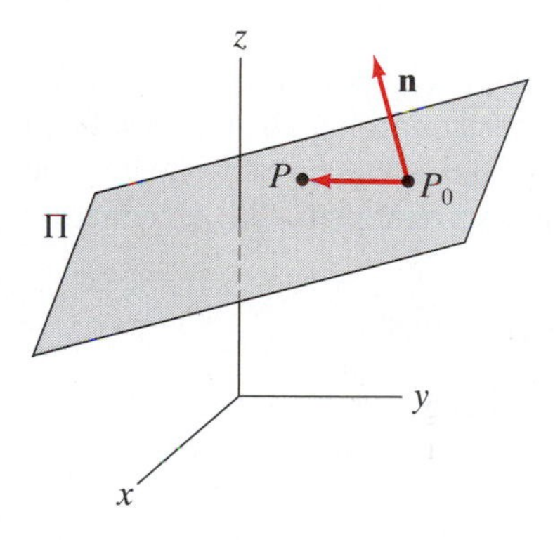

Figure 1.70 The plane in $\mathbf{R}^3$ through the point P_0 and perpendicular to the vector $\mathbf{n}$.

A plane Π in $\mathbf{R}^3$ is determined uniquely by the following geometric information: a particular point $P_0(x_0, y_0, z_0)$ in the plane and a particular vector $\mathbf{n} = A\mathbf{i}+B\mathbf{j}+C\mathbf{k}$ that is normal (perpendicular) to the plane. In other words, Π is the set of all points $P(x, y, z)$ in space such that $\overrightarrow{P_0P}$ is perpendicular to $\mathbf{n}$. (See Figure 1.70.) This means that Π is defined by the vector equation

$$\mathbf{n} \cdot \overrightarrow{P_0P} = 0. \tag{1}$$

Since $\overrightarrow{P_0P} = (x - x_0)\mathbf{i} + (y - y_0)\mathbf{j} + (z - z_0)\mathbf{k}$, equation (1) may be rewritten as

$$(A\mathbf{i} + B\mathbf{j} + C\mathbf{k}) \cdot ((x - x_0)\mathbf{i} + (y - y_0)\mathbf{j} + (z - z_0)\mathbf{k}) = 0$$

or

$$A(x - x_0) + B(y - y_0) + C(z - z_0) = 0. \tag{2}$$

This is equivalent to

$$Ax + By + Cz = D,$$

where $D = Ax_0 + By_0 + Cz_0$.

EXAMPLE 1 The plane through the point (3,2,1) with normal vector $2\mathbf{i}-\mathbf{j}+4\mathbf{k}$ has equation

$$\begin{aligned}(2\mathbf{i}-\mathbf{j}+4\mathbf{k})\cdot((x-3)\mathbf{i}+(y-2)\mathbf{j}+(z-1)\mathbf{k})&=0\\ \Longleftrightarrow 2(x-3)-(y-2)+4(z-1)&=0\\ \Longleftrightarrow 2x-y+4z&=8.\end{aligned}$$

◆

Not only does a plane in $\mathbf{R}^3$ have an equation of the form given by equation (2), but, conversely, any equation of this form must describe a plane. Moreover, it is easy to read off the components of a vector normal to the plane from such an equation: They are just the coefficients of x, y, and z.

EXAMPLE 2 Given the plane with equation $7x+2y-3z=1$, find a normal vector to the plane and identify three points that lie on that plane.

A possible normal vector is $\mathbf{n}=7\mathbf{i}+2\mathbf{j}-3\mathbf{k}$. However, any nonzero scalar multiple of $\mathbf{n}$ will do just as well. Algebraically, the effect of using a scalar multiple of $\mathbf{n}$ as normal is to multiply equation (2) by such a scalar.

Finding three points in the plane is not difficult. First, let $y=z=0$ in the defining equation and solve for x:

$$7x+2\cdot 0-3\cdot 0=1 \quad\Longleftrightarrow\quad 7x=1 \quad\Longleftrightarrow\quad x=\tfrac{1}{7}.$$

Thus $\left(\frac{1}{7},0,0\right)$ is a point on the plane. Next, let $x=z=0$ and solve for y:

$$7\cdot 0+2y-3\cdot 0=1 \quad\Longleftrightarrow\quad y=\tfrac{1}{2}.$$

So $\left(0,\frac{1}{2},0\right)$ is another point on the plane. Finally, let $x=y=0$ and solve for z. You should find that $\left(0,0,-\frac{1}{3}\right)$ lies on the plane. ◆

EXAMPLE 3 Put coordinate axes on $\mathbf{R}^3$ so that the z-axis points vertically. Then a plane in $\mathbf{R}^3$ is vertical if its normal vector $\mathbf{n}$ is horizontal (i.e., if $\mathbf{n}$ is parallel to the xy-plane). This means that $\mathbf{n}$ has no $\mathbf{k}$-component, so $\mathbf{n}$ can be written in the form $A\mathbf{i}+B\mathbf{j}$. It follows from equation (2) that a vertical plane has an equation of the form

$$A(x-x_0)+B(y-y_0)=0.$$

Hence, a nonvertical plane has an equation of the form

$$A(x-x_0)+B(y-y_0)+C(z-z_0)=0,$$

where $C\neq 0$. ◆

EXAMPLE 4 From high school geometry, you may recall that a plane is determined by three (noncollinear) points. Let's find an equation of the plane that contains the points $P_0(1,2,0)$, $P_1(3,1,2)$, and $P_2(0,1,1)$.

There are two ways to solve this problem. The first approach is algebraic and rather uninspired. From the aforementioned remarks, any plane must have an equation of the form $Ax+By+Cz=D$ for suitable constants A, B, C, and D. Thus, we need only to substitute the coordinates of P_0, P_1, and P_2 into this equation and solve for A, B, C, and D. We have that

- substitution of P_0 gives $A+2B=D$;
- substitution of P_1 gives $3A+B+2C=D$; and
- substitution of P_2 gives $B+C=D$.

Hence, we must solve a system of three equations in four unknowns:

$$\begin{cases} A + 2B \qquad\quad = D \\ 3A + \;\; B + 2C = D \\ \qquad\;\; B + \;\; C = D \end{cases}. \tag{3}$$

In general, such a system has either no solution or else infinitely many solutions. We must be in the latter case, since we know that the three points P_0, P_1, and P_2 lie on some plane (i.e., that some set of constants A, B, C, and D must exist). Furthermore, the existence of infinitely many solutions corresponds to the fact that any particular equation for a plane may be multiplied by a nonzero constant without altering the plane defined. In other words, we can choose a value for one of A, B, C, or D, and then the other values will be determined. So let's multiply the first equation given in (3) by 3, and subtract it from the second equation. We obtain

$$\begin{cases} A + \;\; 2B \qquad\quad = \;\; D \\ \quad -5B + 2C = -2D \\ \qquad\;\; B + \;\; C = \;\; D \end{cases}. \tag{4}$$

Now, multiply the third equation in (4) by 5 and add it to the second:

$$\begin{cases} A + 2B \qquad\quad = D \\ \qquad\qquad\; 7C = 3D \\ \qquad\;\; B + \;\; C = D \end{cases}. \tag{5}$$

Multiply the third equation appearing in (5) by 2 and subtract it from the first:

$$\begin{cases} A \qquad\quad -2C = -D \\ \qquad\qquad\;\; 7C = \;\; 3D \\ \qquad B + \quad\; C = \;\; D \end{cases}. \tag{6}$$

By adding appropriate multiples of the second equation to both the first and third equations of (6), we find that

$$\begin{cases} A \qquad\qquad = -\frac{1}{7}D \\ \qquad\quad 7C = \;\; 3D \\ \qquad B \qquad = \;\; \frac{4}{7}D \end{cases}. \tag{7}$$

Thus, if in (7) we take $D = -7$ (for example), then $A = 1$, $B = -4$, $C = -3$, and the equation of the desired plane is

$$x - 4y - 3z = -7.$$

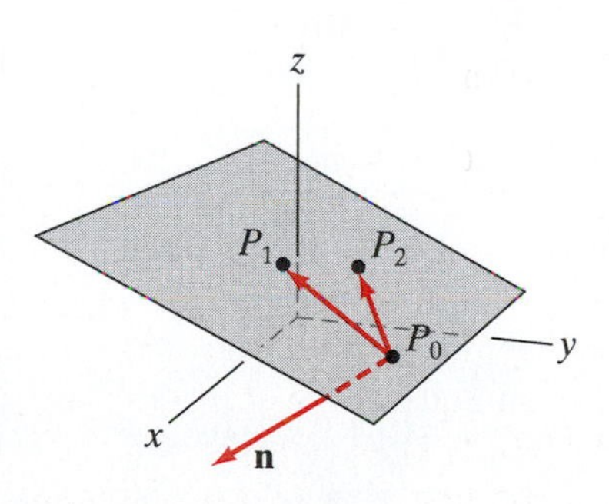

Figure 1.71 The plane determined by the points P_0, P_1, and P_2 in Example 4.

The second method of solution is cleaner and more geometric. The idea is to make use of equation (1). Therefore, we need to know the coordinates of a particular point on the plane (no problem—we are given three such points) and a vector $\mathbf{n}$ normal to the plane. The vectors $\overrightarrow{P_0P_1}$ and $\overrightarrow{P_0P_2}$ both lie in the plane. (See Figure 1.71.) In particular, the normal vector $\mathbf{n}$ must be perpendicular to them both. Consequently, the cross product provides just what we need. That is, we may take

$$\mathbf{n} = \overrightarrow{P_0P_1} \times \overrightarrow{P_0P_2} = (2\mathbf{i} - \mathbf{j} + 2\mathbf{k}) \times (-\mathbf{i} - \mathbf{j} + \mathbf{k})$$

$$= \begin{vmatrix} \mathbf{i} & \mathbf{j} & \mathbf{k} \\ 2 & -1 & 2 \\ -1 & -1 & 1 \end{vmatrix} = \mathbf{i} - 4\mathbf{j} - 3\mathbf{k}.$$

If we take $P_0(1, 2, 0)$ to be the particular point in equation (1), we find that the equation we desire is

$$(\mathbf{i} - 4\mathbf{j} - 3\mathbf{k}) \cdot ((x-1)\mathbf{i} + (y-2)\mathbf{j} + z\mathbf{k}) = 0$$

or

$$(x-1) - 4(y-2) - 3z = 0.$$

This is the same equation as the one given by the first method. ◆

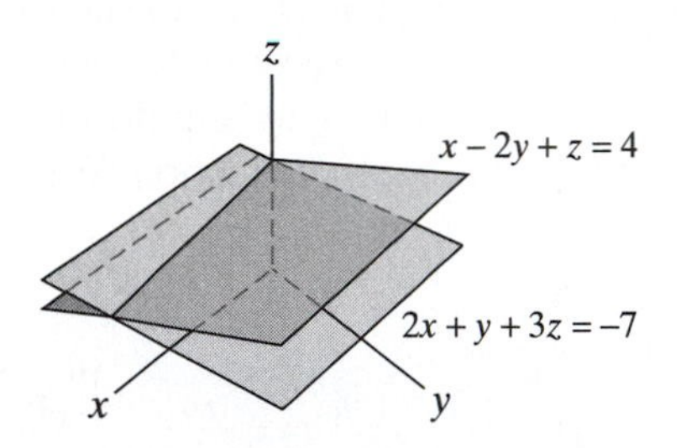

Figure 1.72 The line of intersection of the planes $x - 2y + z = 4$ and $2x + y + 3z = -7$ in Example 5.

EXAMPLE 5 Consider the two planes having equations $x - 2y + z = 4$ and $2x + y + 3z = -7$. We determine a set of parametric equations for their line of intersection. (See Figure 1.72.) We use Proposition 2.1. Thus, we need to find a point on the line and a vector parallel to the line. To find the point on the line, we note that the coordinates (x, y, z) of any such point must satisfy the system of simultaneous equations given by the two planes

$$\begin{cases} x - 2y + z = 4 \\ 2x + y + 3z = -7 \end{cases}. \tag{8}$$

From the equations given in (8), it is not too difficult to produce a single solution (x, y, z). For example, if we let $z = 0$ in (8), we obtain the simpler system

$$\begin{cases} x - 2y = 4 \\ 2x + y = -7 \end{cases}. \tag{9}$$

The solution to the system of equations (9) is readily calculated to be $x = -2$, $y = -3$. Thus, $(-2, -3, 0)$ are the coordinates of a point on the line.

To find a vector parallel to the line of intersection, note that such a vector must be perpendicular to the two normal vectors to the planes. The normal vectors to the planes are $\mathbf{i} - 2\mathbf{j} + \mathbf{k}$ and $2\mathbf{i} + \mathbf{j} + 3\mathbf{k}$. Therefore, a vector parallel to the line of intersection is given by

$$(\mathbf{i} - 2\mathbf{j} + \mathbf{k}) \times (2\mathbf{i} + \mathbf{j} + 3\mathbf{k}) = -7\mathbf{i} - \mathbf{j} + 5\mathbf{k}.$$

Hence, Proposition 2.1 implies that a vector parametric equation for the line is

$$\mathbf{r}(t) = (-2\mathbf{i} - 3\mathbf{j}) + t(-7\mathbf{i} - \mathbf{j} + 5\mathbf{k}),$$

and a standard set of parametric equations is

$$\begin{cases} x = -7t - 2 \\ y = -t - 3 \\ z = 5t \end{cases}.$$

◆

Parametric Equations of Planes

Another way to describe a plane in $\mathbf{R}^3$ is by a set of parametric equations. First, suppose that $\mathbf{a} = (a_1, a_2, a_3)$ and $\mathbf{b} = (b_1, b_2, b_3)$ are two nonzero, nonparallel vectors in $\mathbf{R}^3$. Then $\mathbf{a}$ and $\mathbf{b}$ determine a plane in $\mathbf{R}^3$ that passes through the origin. (See Figure 1.73.) To find the coordinates of a point $P(x, y, z)$ in this plane, draw a parallelogram whose sides are parallel to $\mathbf{a}$ and $\mathbf{b}$ and that has two opposite vertices at the origin and at P, as shown in Figure 1.74. Then there must exist scalars s and t so that the position vector of P is $\overrightarrow{OP} = s\mathbf{a} + t\mathbf{b}$. The plane may

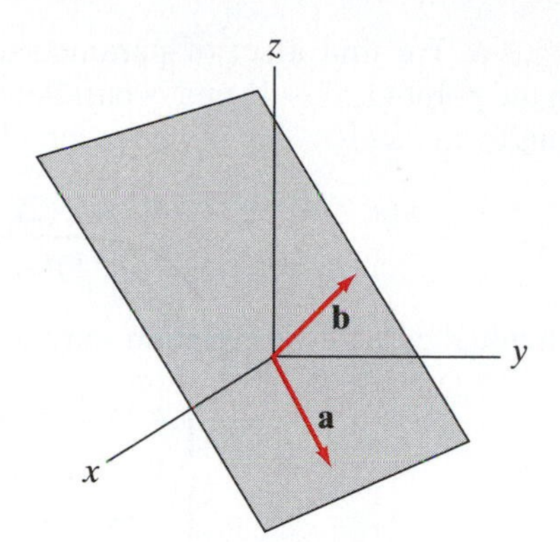

Figure 1.73 The plane through the origin determined by the vectors **a** and **b**.

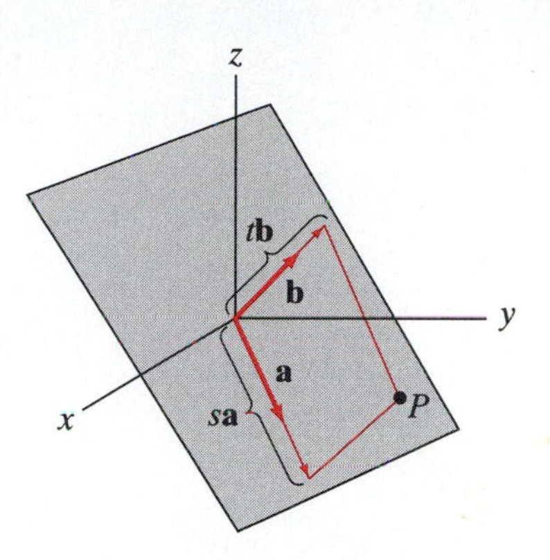

Figure 1.74 For the point P in the plane shown, $\overrightarrow{OP} = s\mathbf{a} + t\mathbf{b}$ for appropriate scalars s and t.

be described as

$$\{\mathbf{x} \in \mathbf{R}^3 \mid \mathbf{x} = s\mathbf{a} + t\mathbf{b};\ s, t \in \mathbf{R}\}.$$

Now, suppose that we seek to describe a general plane Π (i.e., one that does not necessarily pass through the origin). Let

$$\mathbf{c} = (c_1, c_2, c_3) = \overrightarrow{OP_0}$$

denote the position vector of a particular point P_0 in Π and let **a** and **b** be two (nonzero, nonparallel) vectors that determine the plane through the origin parallel to Π. By parallel translating **a** and **b** so that their tails are at the head of **c** (as in Figure 1.75), we adapt the preceding discussion to see that the position vector of any point $P(x, y, z)$ in Π may be described as

$$\overrightarrow{OP} = s\mathbf{a} + t\mathbf{b} + \mathbf{c}.$$

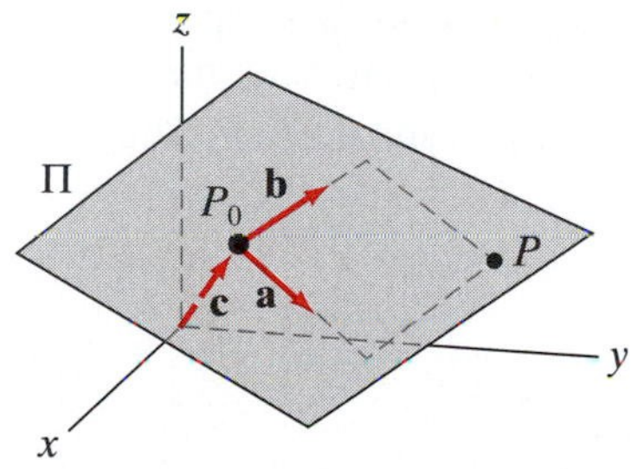

Figure 1.75 The plane passing through $P_0(c_1, c_2, c_3)$ and parallel to **a** and **b**.

To summarize, we have shown the following:

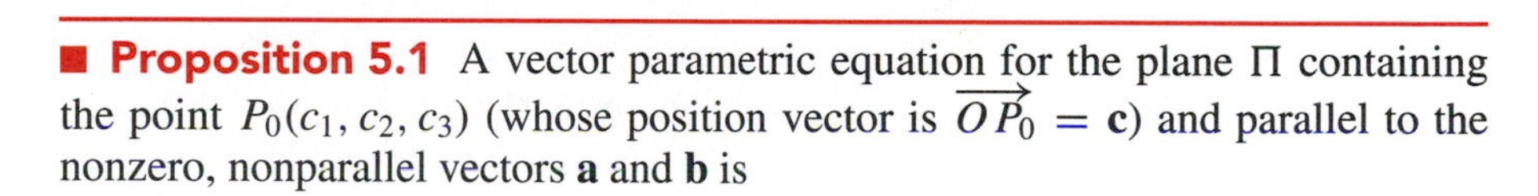

Proposition 5.1 A vector parametric equation for the plane Π containing the point $P_0(c_1, c_2, c_3)$ (whose position vector is $\overrightarrow{OP_0} = \mathbf{c}$) and parallel to the nonzero, nonparallel vectors **a** and **b** is

$$\mathbf{x}(s, t) = s\mathbf{a} + t\mathbf{b} + \mathbf{c}. \tag{10}$$

By taking components in formula (10), we readily obtain a set of parametric equations for Π:

$$\begin{cases} x = sa_1 + tb_1 + c_1 \\ y = sa_2 + tb_2 + c_2 \\ z = sa_3 + tb_3 + c_3 \end{cases}. \tag{11}$$

Compare formula (10) with that of equation (1) in Proposition 2.1. We need to use two parameters s and t to describe a plane (instead of a single parameter t that appears in the vector parametric equation for a line) because a plane is a two-dimensional object.

EXAMPLE 6 We find a set of parametric equations for the plane that passes through the point $(1, 0, -1)$ and is parallel to the vectors $3\mathbf{i} - \mathbf{k}$ and $2\mathbf{i} + 5\mathbf{j} + 2\mathbf{k}$.

From formula (10), any point on the plane is specified by

$$\begin{aligned}\mathbf{x}(s, t) &= s(3\mathbf{i} - \mathbf{k}) + t(2\mathbf{i} + 5\mathbf{j} + 2\mathbf{k}) + (\mathbf{i} - \mathbf{k}) \\ &= (3s + 2t + 1)\mathbf{i} + 5t\mathbf{j} + (2t - s - 1)\mathbf{k}.\end{aligned}$$

The individual parametric equation may be read off as

$$\begin{cases} x = 3s + 2t + 1 \\ y = 5t \\ z = 2t - s - 1 \end{cases}.$$

◆

Distance Problems

Cross products and vector projections provide convenient ways to understand a range of distance problems involving lines and planes: Several examples follow. What is important about these examples are the vector techniques for solving geometric problems that they exhibit, not the general formulas that may be derived from them.

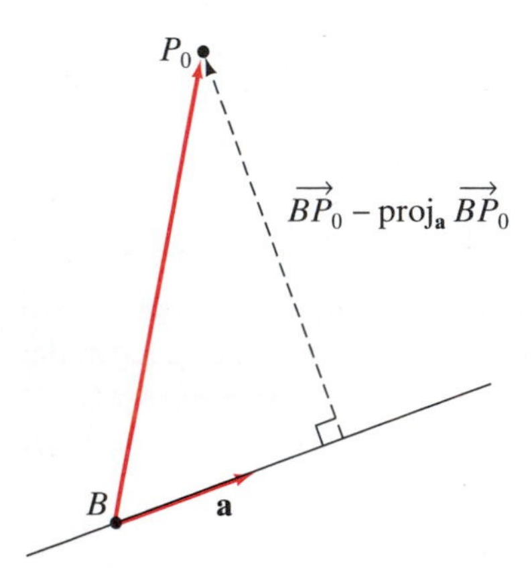

Figure 1.76 A general configuration for finding the distance between a point and a line, using vector projections.

EXAMPLE 7 (Distance between a point and a line) We find the distance between the point $P_0(2, 1, 3)$ and the line $\mathbf{l}(t) = t(-1, 1, -2) + (2, 3, -2)$ in two ways.

METHOD 1. From the vector parametric equations for the given line, we read off a point B on the line—namely, $(2, 3, -2)$—and a vector $\mathbf{a}$ parallel to the line—namely, $\mathbf{a} = (-1, 1, -2)$. Using Figure 1.76, the length of the vector $\overrightarrow{BP_0} - \text{proj}_{\mathbf{a}} \overrightarrow{BP_0}$ provides the desired distance between P_0 and the line. Thus, we calculate that

$$\begin{aligned}\overrightarrow{BP_0} &= (2, 1, 3) - (2, 3, -2) \\ &= (0, -2, 5); \\ \text{proj}_{\mathbf{a}} \overrightarrow{BP_0} &= \left(\frac{\mathbf{a} \cdot \overrightarrow{BP_0}}{\mathbf{a} \cdot \mathbf{a}} \right) \mathbf{a} \\ &= \left(\frac{(-1, 1, -2) \cdot (0, -2, 5)}{(-1, 1, -2) \cdot (-1, 1, -2)} \right) (-1, 1, -2) \\ &= (2, -2, 4).\end{aligned}$$

The desired distance is

$$\|\overrightarrow{BP_0} - \text{proj}_{\mathbf{a}} \overrightarrow{BP_0}\| = \|(0, -2, 5) - (2, -2, 4)\| = \|(-2, 0, 1)\| = \sqrt{5}.$$

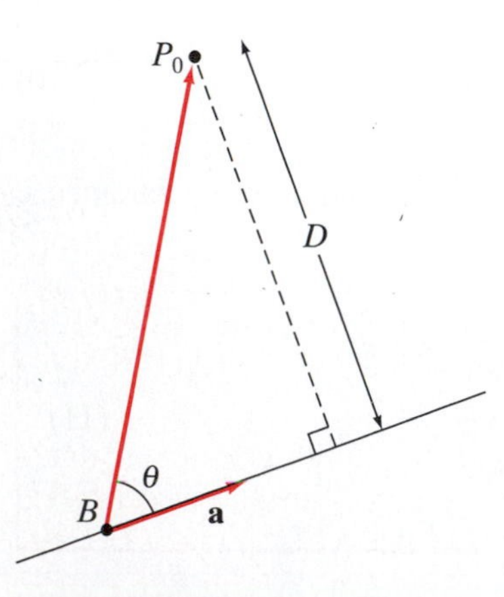

Figure 1.77 Another general configuration for finding the distance between a point and a line.

METHOD 2. In this case, we use a little trigonometry. If θ denotes the angle between the vectors $\mathbf{a}$ and $\overrightarrow{BP_0}$ as in Figure 1.77, then

$$\sin\theta = \frac{D}{\|\overrightarrow{BP_0}\|},$$

where D denotes the distance between P_0 and the line. Hence,

$$D = \|\overrightarrow{BP_0}\| \sin\theta = \frac{\|\mathbf{a}\| \, \|\overrightarrow{BP_0}\| \sin\theta}{\|\mathbf{a}\|} = \frac{\|\mathbf{a} \times \overrightarrow{BP_0}\|}{\|\mathbf{a}\|}.$$

Therefore, we calculate

$$\mathbf{a} \times \overrightarrow{BP_0} = \begin{vmatrix} \mathbf{i} & \mathbf{j} & \mathbf{k} \\ -1 & 1 & -2 \\ 0 & -2 & 5 \end{vmatrix} = \mathbf{i} + 5\mathbf{j} + 2\mathbf{k},$$

so that the distance sought is

$$D = \frac{\|\mathbf{i} + 5\mathbf{j} + 2\mathbf{k}\|}{\|-\mathbf{i} + \mathbf{j} - 2\mathbf{k}\|} = \frac{\sqrt{30}}{\sqrt{6}} = \sqrt{5},$$

which agrees with the answer obtained by Method 1. ◆

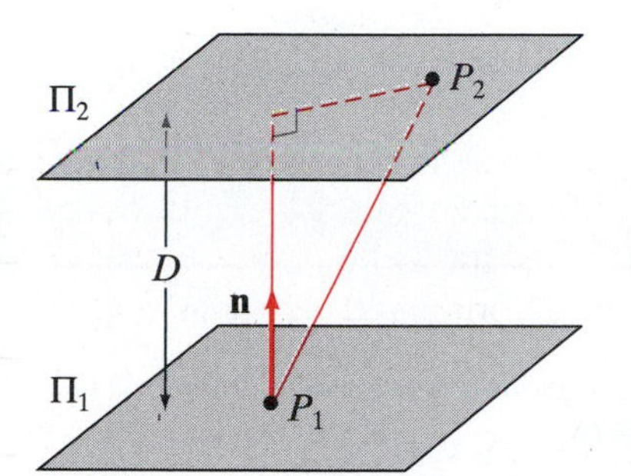

Figure 1.78 The general configuration for finding the distance D between two parallel planes.

EXAMPLE 8 (Distance between parallel planes) The planes

$$\Pi_1\colon\ 2x - 2y + z = 5 \quad \text{and} \quad \Pi_2\colon\ 2x - 2y + z = 20$$

are parallel. (Why?) We see how to compute the distance between them.

Using Figure 1.78 as a guide, we see that the desired distance D is given by $\|\text{proj}_{\mathbf{n}}\overrightarrow{P_1P_2}\|$, where P_1 is a point on Π_1, P_2 is a point on Π_2, and $\mathbf{n}$ is a vector normal to both planes.

First, the vector $\mathbf{n}$ that is normal to both planes may be read directly from the equation for either Π_1 or Π_2 as $\mathbf{n} = 2\mathbf{i} - 2\mathbf{j} + \mathbf{k}$. It is not hard to find a point P_1 on Π_1: the point $P_1(0, 0, 5)$ will do. Similarly, take $P_2(0, 0, 20)$ for a point on Π_2. Then

$$\overrightarrow{P_1P_2} = (0, 0, 15),$$

and calculate

$$\begin{aligned} \text{proj}_{\mathbf{n}}\overrightarrow{P_1P_2} = \left(\frac{\mathbf{n} \cdot \overrightarrow{P_1P_2}}{\mathbf{n} \cdot \mathbf{n}}\right)\mathbf{n} &= \left(\frac{(2, -2, 1) \cdot (0, 0, 15)}{(2, -2, 1) \cdot (2, -2, 1)}\right)(2, -2, 1) \\ &= -\tfrac{15}{9}(2, -2, 1) \\ &= -\tfrac{5}{3}(2, -2, 1). \end{aligned}$$

Hence, the distance D that we seek is

$$D = \|\text{proj}_{\mathbf{n}}\overrightarrow{P_1P_2}\| = \tfrac{5}{3}\sqrt{9} = 5. \quad ◆$$

EXAMPLE 9 (Distance between two skew lines) Find the distance between the two skew lines

$$\mathbf{l}_1(t) = t(2, 1, 3) + (0, 5, -1) \quad \text{and} \quad \mathbf{l}_2(t) = t(1, -1, 0) + (-1, 2, 0).$$

(Two lines in $\mathbf{R}^3$ are said to be **skew** if they are neither intersecting nor parallel. It follows that the lines must lie in parallel planes and that the distance between the lines is equal to the distance between the planes.)

To solve this problem, we need to find $\|\text{proj}_{\mathbf{n}}\overrightarrow{B_1B_2}\|$, the length of the projection of the vector between a point on each line onto a vector $\mathbf{n}$ that is perpendicular to both lines, hence, also perpendicular to the parallel planes that contain the lines. (See Figure 1.79.)

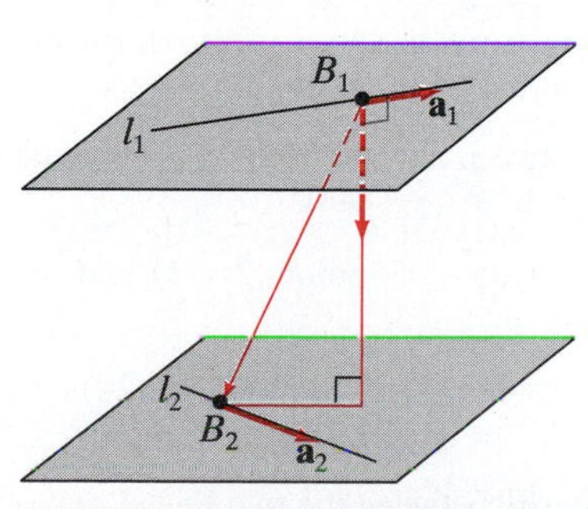

Figure 1.79 Configuration for determining the distance between two skew lines in Example 9.

From the vector parametric equations for the lines, we read that the point $B_1(0, 5, -1)$ is on the first line and $B_2(-1, 2, 0)$ is on the second. Hence,

$$\overrightarrow{B_1B_2} = (-1, 2, 0) - (0, 5, -1) = (-1, -3, 1).$$

For a vector $\mathbf{n}$ that is perpendicular to both lines, we may use $\mathbf{n} = \mathbf{a}_1 \times \mathbf{a}_2$, where $\mathbf{a}_1 = (2, 1, 3)$ is a vector parallel to the first line and $\mathbf{a}_2 = (1, -1, 0)$ is parallel to the second. (We may read these vectors from the parametric equations.) Thus,

$$\mathbf{n} = \mathbf{a}_1 \times \mathbf{a}_2 = \begin{vmatrix} \mathbf{i} & \mathbf{j} & \mathbf{k} \\ 2 & 1 & 3 \\ 1 & -1 & 0 \end{vmatrix} = 3\mathbf{i} + 3\mathbf{j} - 3\mathbf{k},$$

and so,

$$\begin{aligned} \text{proj}_{\mathbf{n}} \overrightarrow{B_1 B_2} &= \left(\frac{\mathbf{n} \cdot \overrightarrow{B_1 B_2}}{\mathbf{n} \cdot \mathbf{n}} \right) \mathbf{n} = \left(\frac{(-1, -3, 1) \cdot (3, 3, -3)}{(3, 3, -3) \cdot (3, 3, -3)} \right) (3, 3, -3) \\ &= -\tfrac{15}{27}(3, 3, -3) \\ &= -\tfrac{5}{3}(1, 1, -1). \end{aligned}$$

The desired distance is $\|\text{proj}_{\mathbf{n}} \overrightarrow{B_1 B_2}\| = \frac{5}{3}\sqrt{3}$. ◆

Exercises

1. Calculate an equation for the plane containing the point $(3, -1, 2)$ and perpendicular to $\mathbf{i} - \mathbf{j} + 2\mathbf{k}$.

2. Find an equation for the plane containing the point $(9, 5, -1)$ and perpendicular to $\mathbf{i} - 2\mathbf{k}$.

3. Find an equation for the plane containing the points $(3, -1, 2)$, $(2, 0, 5)$, and $(1, -2, 4)$.

4. Find an equation for the plane containing the points $(A, 0, 0)$, $(0, B, 0)$, and $(0, 0, C)$. Assume that at least two of A, B, and C are nonzero.

5. Give an equation for the plane that is parallel to the plane $5x - 4y + z = 1$ and that passes through the point $(2, -1, -2)$.

6. Find an equation for the plane that contains the line $x = 2t - 1$, $y = 3t + 4$, $z = 7 - t$, and the point $(2, 5, 0)$.

7. Find an equation for the plane that is perpendicular to the line $x = 3t - 5$, $y = 7 - 2t$, $z = 8 - t$ and that passes through the point $(1, -1, 2)$.

8. Find an equation for the plane that contains the two lines $l_1: x = t + 2$, $y = 3t - 5$, $z = 5t + 1$ and $l_2: x = 5 - t$, $y = 3t - 10$, $z = 9 - 2t$.

9. Give a set of parametric equations for the line of intersection of the planes $x + 2y - 3z = 5$ and $5x + 5y - z = 1$.

10. Give a set of parametric equations for the line through $(5, 0, 6)$ that is perpendicular to the plane $2x - 3y + 5z = -1$.

11. Find a value for A so that the planes $8x - 6y + 9Az = 6$ and $Ax + y + 2z = 3$ are parallel.

12. Find values for A so that the planes $Ax - y + z = 1$ and $3Ax + Ay - 2z = 5$ are perpendicular.

Give a set of parametric equations for each of the planes described in Exercises 13–18.

13. The plane that passes through the point $(-1, 2, 7)$ and is parallel to the vectors $2\mathbf{i} - 3\mathbf{j} + \mathbf{k}$ and $\mathbf{i} - 5\mathbf{k}$

14. The plane that passes through the point $(2, 9, -4)$ and is parallel to the vectors $-8\mathbf{i} + 2\mathbf{j} + 5\mathbf{k}$ and $3\mathbf{i} - 4\mathbf{j} - 2\mathbf{k}$

15. The plane that contains the lines l_1: $x = 2t + 5$, $y = -3t - 6$, $z = 4t + 10$ and l_2: $x = 5t - 1$, $y = 10t + 3$, $z = 7t - 2$

16. The plane that passes through the three points $(0, 2, 1)$, $(7, -1, 5)$, and $(-1, 3, 0)$

17. The plane that contains the line $l: x = 3t - 5$, $y = 10 - 3t$, $z = 2t + 9$ and the point $(-2, 4, 7)$

18. The plane determined by the equation $2x - 3y + 5z = 30$

19. Find a single equation of the form $Ax + By + Cz = D$ that describes the plane given parametrically as $x = 3s - t + 2$, $y = 4s + t$, $z = s + 5t + 3$. (Hint: Begin by writing the parametric equations in vector form and then find a vector normal to the plane.)

20. Find the distance between the point $(1, -2, 3)$ and the line $l: x = 2t - 5$, $y = 3 - t$, $z = 4$.

21. Find the distance between the point $(2, -1)$ and the line $l: x = 3t + 7$, $y = 5t - 3$.

22. Find the distance between the point $(-11, 10, 20)$ and the line $l: x = 5 - t$, $y = 3$, $z = 7t + 8$.

23. Determine the distance between the two lines $\mathbf{l}_1(t) = t(8, -1, 0) + (-1, 3, 5)$ and $\mathbf{l}_2(t) = t(0, 3, 1) + (0, 3, 4)$.

24. Compute the distance between the two lines $\mathbf{l}_1(t) = (t-7)\mathbf{i} + (5t+1)\mathbf{j} + (3-2t)\mathbf{k}$ and $\mathbf{l}_2(t) = 4t\mathbf{i} + (2-t)\mathbf{j} + (8t+1)\mathbf{k}$.

25. (a) Find the distance between the two lines $\mathbf{l}_1(t) = t(3, 1, 2) + (4, 0, 2)$ and $\mathbf{l}_2(t) = t(1, 2, 3) + (2, 1, 3)$.

(b) What does your answer in part (a) tell you about the relative positions of the lines?

26. (a) The lines $\mathbf{l}_1(t) = t(1, -1, 5) + (2, 0, -4)$ and $\mathbf{l}_2(t) = t(1, -1, 5) + (1, 3, -5)$ are parallel. Explain why the method of Example 9 cannot be used to calculate the distance between the lines.

(b) Find another way to calculate the distance. (Hint: Try using some calculus.)

27. Find the distance between the two planes given by the equations $x - 3y + 2z = 1$ and $x - 3y + 2z = 8$.

28. Calculate the distance between the two planes

$$5x - 2y + 2z = 12 \quad \text{and} \quad -10x + 4y - 4z = 8.$$

29. Show that the distance d between the two parallel planes determined by the equations $Ax + By + Cz = D_1$ and $Ax + By + Cz = D_2$ is

$$d = \frac{|D_1 - D_2|}{\sqrt{A^2 + B^2 + C^2}}.$$

30. Two planes are given parametrically by the vector equations

$$\mathbf{x}_1(s, t) = (-3, 4, -9) + s(9, -5, 9) + t(3, -2, 3)$$
$$\mathbf{x}_2(s, t) = (5, 0, 3) + s(-9, 2, -9) + t(-4, 7, -4).$$

(a) Give a convincing explanation for why these planes are parallel.

(b) Find the distance between the planes.

31. Write equations for the planes that are parallel to $x + 3y - 5z = 2$ and lie three units from it.

32. Suppose that $\mathbf{l}_1(t) = t\mathbf{a} + \mathbf{b}_1$ and $\mathbf{l}_2(t) = t\mathbf{a} + \mathbf{b}_2$ are parallel lines in either $\mathbf{R}^2$ or $\mathbf{R}^3$. Show that the distance D between them is given by

$$D = \frac{\|\mathbf{a} \times (\mathbf{b}_2 - \mathbf{b}_1)\|}{\|\mathbf{a}\|}.$$

(Hint: Consider Example 7.)

33. Let Π be the plane in $\mathbf{R}^3$ with normal vector $\mathbf{n}$ that passes through the point A with position vector $\mathbf{a}$. If $\mathbf{b}$ is the position vector of a point B in $\mathbf{R}^3$, show that the distance D between B and Π is given by

$$D = \frac{|\mathbf{n} \cdot (\mathbf{b} - \mathbf{a})|}{\|\mathbf{n}\|}.$$

34. Show that the distance D between parallel planes with normal vector $\mathbf{n}$ is given by

$$\frac{|\mathbf{n} \cdot (\mathbf{x}_2 - \mathbf{x}_1)|}{\|\mathbf{n}\|},$$

where $\mathbf{x}_1$ is the position vector of a point on one of the planes, and $\mathbf{x}_2$ is the position vector of a point on the other plane.

35. Suppose that $\mathbf{l}_1(t) = t\mathbf{a}_1 + \mathbf{b}_1$ and $\mathbf{l}_2(t) = t\mathbf{a}_2 + \mathbf{b}_2$ are skew lines in $\mathbf{R}^3$. Use the geometric reasoning of Example 9 to show that the distance D between these lines is given by

$$D = \frac{|(\mathbf{a}_1 \times \mathbf{a}_2) \cdot (\mathbf{b}_2 - \mathbf{b}_1)|}{\|\mathbf{a}_1 \times \mathbf{a}_2\|}.$$

1.6 Some n-dimensional Geometry

Vectors in $\mathbf{R}^n$

The algebraic idea of a vector in $\mathbf{R}^2$ or $\mathbf{R}^3$ is defined in §1.1, in which we asked you to consider what would be involved in generalizing the operations of vector addition, scalar multiplication, etc., to n-dimensional vectors, where n can be arbitrary. We explore some of the details of such a generalization next.

> **Definition 6.1** A **vector** in $\mathbf{R}^n$ is an ordered n-tuple of real numbers. We use $\mathbf{a} = (a_1, a_2, \ldots, a_n)$ as our standard notation for a vector in $\mathbf{R}^n$.

EXAMPLE 1 The 5-tuple $(2, 4, 6, 8, 10)$ is a vector in $\mathbf{R}^5$. The $(n+1)$-tuple $(2n, 2n-2, 2n-4, \ldots, 2, 0)$ is a vector in $\mathbf{R}^{n+1}$, where n is arbitrary. ◆

Exactly as is the case in $\mathbf{R}^2$ or $\mathbf{R}^3$, we call two vectors $\mathbf{a} = (a_1, a_2, \ldots, a_n)$ and $\mathbf{b} = (b_1, b_2, \ldots, b_n)$ **equal** just in case $a_i = b_i$ for $i = 1, 2, \ldots, n$. Vector addition and scalar multiplication are defined in complete analogy with Definitions 1.3

and 1.4: If $\mathbf{a} = (a_1, a_2, \ldots, a_n)$ and $\mathbf{b} = (b_1, b_2, \ldots, b_n)$ are two vectors in $\mathbf{R}^n$ and $k \in \mathbf{R}$ is any scalar, then

$$\mathbf{a} + \mathbf{b} = (a_1 + b_1, a_2 + b_2, \ldots, a_n + b_n)$$

and

$$k\mathbf{a} = (ka_1, ka_2, \ldots, ka_n).$$

The properties of vector addition and scalar multiplication given in §1.1 hold (with proofs that are no different from those in the two- and three-dimensional cases). Similarly, the dot product of two vectors in $\mathbf{R}^n$ is readily defined:

$$\mathbf{a} \cdot \mathbf{b} = a_1 b_1 + a_2 b_2 + \cdots + a_n b_n.$$

The inner product properties given in §1.3 continue to hold in n dimensions; we leave it to you to check that this is so.

What we *cannot* do in dimensions larger than three is to develop a pictorial representation for vectors as arrows. Nonetheless, the power of our algebra and analogy does allow us to define a number of geometric ideas. We define the **length** of a vector in $\mathbf{a} \in \mathbf{R}^n$ by using the dot product:

$$\|\mathbf{a}\| = \sqrt{\mathbf{a} \cdot \mathbf{a}}.$$

The **distance** between two vectors **a** and **b** in $\mathbf{R}^n$ is

$$\text{Distance between } \mathbf{a} \text{ and } \mathbf{b} = \|\mathbf{a} - \mathbf{b}\|.$$

We can even define the **angle** between two nonzero vectors by using a generalized version of equation (4) of §1.3:

$$\theta = \cos^{-1} \frac{\mathbf{a} \cdot \mathbf{b}}{\|\mathbf{a}\| \, \|\mathbf{b}\|}.$$

Here $\mathbf{a}, \mathbf{b} \in \mathbf{R}^n$ and θ is taken so that $0 \leq \theta \leq \pi$. (Note: At this point in our discussion, it is not clear that we have

$$-1 \leq \frac{\mathbf{a} \cdot \mathbf{b}}{\|\mathbf{a}\| \, \|\mathbf{b}\|} \leq 1,$$

which is a necessary condition if our definition of the angle θ is to make sense. Fortunately, the Cauchy–Schwarz inequality—formula (1) that follows—takes care of this issue.) Thus, even though we are not able to draw pictures of vectors in $\mathbf{R}^n$, we can nonetheless talk about what it means to say that two vectors are perpendicular or parallel, or how far apart two vectors may be. (Be careful about this business. We are *defining* notions of length, distance, and angle entirely in terms of the dot product. Results like Theorem 3.3 have no meaning in $\mathbf{R}^n$, since the ideas of angles between vectors and dot products are not independent.)

There is no simple generalization of the cross product. However, see Exercises 39–42 at the end of this section for the best we can do by way of analogy.

We can create a **standard basis** of vectors in $\mathbf{R}^n$ that generalize the **i**, **j**, **k**-basis in $\mathbf{R}^3$. Let

$$\begin{aligned} \mathbf{e}_1 &= (1, 0, 0, \ldots, 0), \\ \mathbf{e}_2 &= (0, 1, 0, \ldots, 0), \\ &\vdots \\ \mathbf{e}_n &= (0, 0, \ldots, 0, 1). \end{aligned}$$

Then it is not difficult to see (check for yourself) that

$$\mathbf{a} = (a_1, a_2, \ldots, a_n) = a_1\mathbf{e}_1 + a_2\mathbf{e}_2 + \cdots + a_n\mathbf{e}_n.$$

Here are two famous (and often handy) inequalities:

Cauchy–Schwarz inequality. For all vectors $\mathbf{a}$ and $\mathbf{b}$ in $\mathbf{R}^n$, we have

$$|\mathbf{a} \cdot \mathbf{b}| \leq \|\mathbf{a}\| \, \|\mathbf{b}\|. \tag{1}$$

Proof If $n = 2$ or 3, this result is virtually immediate in view of Theorem 3.3. However, in dimensions larger than three, we do not have independent notions of inner products and angles, so a different proof is required.

First note that the inequality holds if either $\mathbf{a}$ or $\mathbf{b}$ is $\mathbf{0}$. So assume that $\mathbf{a}$ and $\mathbf{b}$ are nonzero. Then we may define the projection of $\mathbf{b}$ onto $\mathbf{a}$ just as in §1.3:

$$\operatorname{proj}_{\mathbf{a}}\mathbf{b} = \left(\frac{\mathbf{a} \cdot \mathbf{b}}{\mathbf{a} \cdot \mathbf{a}}\right) \mathbf{a} = k\mathbf{a}.$$

Here k is, of course, the scalar $\mathbf{a} \cdot \mathbf{b}/\mathbf{a} \cdot \mathbf{a}$. Let $\mathbf{c} = \mathbf{b} - k\mathbf{a}$ (so that $\mathbf{b} = k\mathbf{a} + \mathbf{c}$). Then we have $\mathbf{a} \cdot \mathbf{c} = 0$, since

$$\begin{aligned}
\mathbf{a} \cdot \mathbf{c} &= \mathbf{a} \cdot (\mathbf{b} - k\mathbf{a}) \\
&= \mathbf{a} \cdot \mathbf{b} - k\mathbf{a} \cdot \mathbf{a} \\
&= \mathbf{a} \cdot \mathbf{b} - \left(\frac{\mathbf{a} \cdot \mathbf{b}}{\mathbf{a} \cdot \mathbf{a}}\right) \mathbf{a} \cdot \mathbf{a} \\
&= \mathbf{a} \cdot \mathbf{b} - \mathbf{a} \cdot \mathbf{b} \\
&= 0.
\end{aligned}$$

We leave it to you to check that the "Pythagorean theorem" holds, namely, that the following equation is true:

$$\|\mathbf{b}\|^2 = k^2 \|\mathbf{a}\|^2 + \|\mathbf{c}\|^2.$$

Multiply this equation by $\|\mathbf{a}\|^2 = \mathbf{a} \cdot \mathbf{a}$. We obtain

$$\begin{aligned}
\|\mathbf{a}\|^2 \, \|\mathbf{b}\|^2 &= \|\mathbf{a}\|^2 k^2 \|\mathbf{a}\|^2 + \|\mathbf{a}\|^2 \, \|\mathbf{c}\|^2 \\
&= \|\mathbf{a}\|^2 \left(\frac{\mathbf{a} \cdot \mathbf{b}}{\mathbf{a} \cdot \mathbf{a}}\right)^2 \|\mathbf{a}\|^2 + \|\mathbf{a}\|^2 \, \|\mathbf{c}\|^2 \\
&= (\mathbf{a} \cdot \mathbf{a}) \left(\frac{\mathbf{a} \cdot \mathbf{b}}{\mathbf{a} \cdot \mathbf{a}}\right)^2 (\mathbf{a} \cdot \mathbf{a}) + \|\mathbf{a}\|^2 \, \|\mathbf{c}\|^2 \\
&= (\mathbf{a} \cdot \mathbf{b})^2 + \|\mathbf{a}\|^2 \, \|\mathbf{c}\|^2.
\end{aligned}$$

Now, the quantity $\|\mathbf{a}\|^2 \|\mathbf{c}\|^2$ is nonnegative. Hence,

$$\|\mathbf{a}\|^2 \, \|\mathbf{b}\|^2 \geq (\mathbf{a} \cdot \mathbf{b})^2.$$

Taking square roots in this last inequality yields the result desired. ■

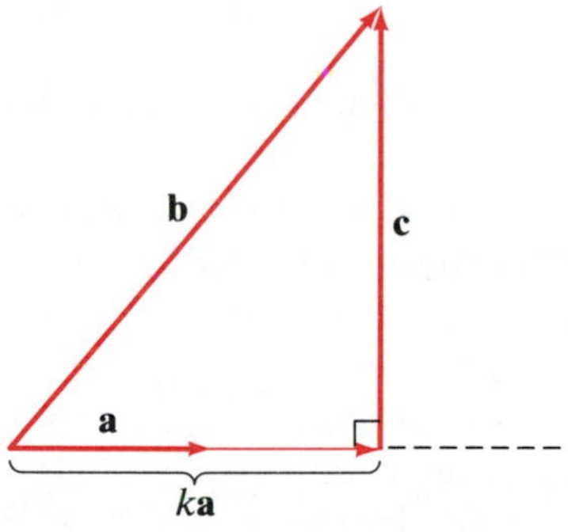

The geometric motivation for this proof of the Cauchy–Schwarz inequality comes from Figure 1.80.[1]

Figure 1.80 The geometry behind the proof of the Cauchy–Schwarz inequality.

The triangle inequality. For all vectors $\mathbf{a}, \mathbf{b} \in \mathbf{R}^n$ we have

$$\|\mathbf{a} + \mathbf{b}\| \leq \|\mathbf{a}\| + \|\mathbf{b}\|. \tag{2}$$

[1] See J. W. Cannon, *Amer. Math. Monthly* **96** (1989), no. 7, 630–631.

Proof Strategic use of the Cauchy–Schwarz inequality yields

$$
\begin{aligned}
\|\mathbf{a}+\mathbf{b}\|^2 &= (\mathbf{a}+\mathbf{b})\cdot(\mathbf{a}+\mathbf{b}) \\
&= \mathbf{a}\cdot\mathbf{a} + 2\mathbf{a}\cdot\mathbf{b} + \mathbf{b}\cdot\mathbf{b} \\
&\le \mathbf{a}\cdot\mathbf{a} + 2\|\mathbf{a}\|\,\|\mathbf{b}\| + \mathbf{b}\cdot\mathbf{b} && \text{by (1)} \\
&= \|\mathbf{a}\|^2 + 2\|\mathbf{a}\|\,\|\mathbf{b}\| + \|\mathbf{b}\|^2 \\
&= (\|\mathbf{a}\| + \|\mathbf{b}\|)^2 .
\end{aligned}
$$

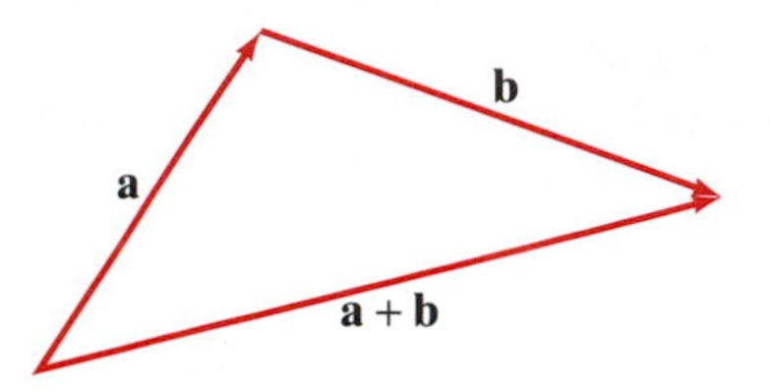

Figure 1.81 The triangle inequality visualized.

Thus, the result desired holds by taking square roots, since the quantities on both sides of the inequality are nonnegative. ■

In two or three dimensions the triangle inequality has the following obvious proof from which the inequality gets its name: Since $\|\mathbf{a}\|$, $\|\mathbf{b}\|$, and $\|\mathbf{a}+\mathbf{b}\|$ can be viewed as the lengths of the sides of a triangle, inequality (2) says nothing more than that the sum of the lengths of two sides of a triangle must be at least as large as the length of the third side, as demonstrated by Figure 1.81.

Matrices

We had a brief glance at matrices and determinants in §1.4 in connection with the computation of cross products. Now it's time for another look.

A matrix is defined in §1.4 as a rectangular array of numbers. To extend our discussion, we need a good notation for matrices and their individual entries. We used the upper case Latin alphabet to denote entire matrices and will continue to do so. We shall also adopt the standard convention and use the lower case Latin alphabet and two sets of indices (one set for rows, the other for columns) to identify matrix entries. Thus, the general $m \times n$ matrix can be written as

$$A = \begin{bmatrix} a_{11} & a_{12} & \cdots & a_{1n} \\ a_{21} & a_{22} & \cdots & a_{2n} \\ \vdots & \vdots & \ddots & \vdots \\ a_{m1} & a_{m2} & \cdots & a_{mn} \end{bmatrix} = \text{(shorthand) } (a_{ij}).$$

The first index *always* will represent the row position and the second index, the column position.

Vectors in $\mathbf{R}^n$ can also be thought of as matrices. We shall have occasion to write the vector $\mathbf{a} = (a_1, a_2, \ldots, a_n)$ either as a **row vector** (a $1 \times n$ matrix),

$$\mathbf{a} = \begin{bmatrix} a_1 & a_2 & \cdots & a_n \end{bmatrix},$$

or, more typically, as a **column vector** (an $n \times 1$ matrix),

$$\mathbf{a} = \begin{bmatrix} a_1 \\ a_2 \\ \vdots \\ a_n \end{bmatrix}.$$

We did not use double indices since there is only a single row or column present. It will be clear from context (or else indicated explicitly) in which form a vector $\mathbf{a}$ will be viewed. An $m \times n$ matrix A can be thought of as a "vector of vectors" in two ways: (1) as m row vectors in $\mathbf{R}^n$,

$$A = \begin{bmatrix} [\, a_{11} & a_{12} & \cdots & a_{1n} \,] \\ [\, a_{21} & a_{22} & \cdots & a_{2n} \,] \\ & \vdots & & \\ [\, a_{m1} & a_{m2} & \cdots & a_{mn} \,] \end{bmatrix},$$

or (2) as n column vectors in $\mathbf{R}^m$,

$$A = \left[\begin{bmatrix} a_{11} \\ a_{21} \\ \vdots \\ a_{m1} \end{bmatrix} \begin{bmatrix} a_{12} \\ a_{22} \\ \vdots \\ a_{m2} \end{bmatrix} \cdots \begin{bmatrix} a_{1n} \\ a_{2n} \\ \vdots \\ a_{mn} \end{bmatrix} \right].$$

We now define the basic matrix operations. Matrix addition and scalar multiplication are really no different from the corresponding operations on vectors (and, moreover, they satisfy essentially the same properties).

■ **Definition 6.2** (MATRIX ADDITION) Let A and B be two $m \times n$ matrices. Then their **matrix sum** $A + B$ is the $m \times n$ matrix obtained by adding corresponding entries. That is, the entry in the ith row and jth column of $A + B$ is $a_{ij} + b_{ij}$, where a_{ij} and b_{ij} are the ijth entries of A and B, respectively.

EXAMPLE 2 If

$$A = \begin{bmatrix} 1 & 2 & 3 \\ 4 & 5 & 6 \end{bmatrix} \quad \text{and} \quad B = \begin{bmatrix} 7 & 0 & -1 \\ -2 & 5 & 0 \end{bmatrix},$$

then

$$A + B = \begin{bmatrix} 8 & 2 & 2 \\ 2 & 10 & 6 \end{bmatrix}.$$

However, if $B = \begin{bmatrix} 7 & 1 \\ 5 & 3 \end{bmatrix}$, then $A + B$ is not defined, since B does not have the same dimensions as A. ◆

Properties of matrix addition. For all $m \times n$ matrices A, B, and C we have

1. $A + B = B + A$ (commutativity);
2. $A + (B + C) = (A + B) + C$ (associativity);
3. An $m \times n$ matrix O (the **zero matrix**) with the property that $A + O = A$ for all $m \times n$ matrices A.

■ **Definition 6.3** (SCALAR MULTIPLICATION) If A is an $m \times n$ matrix and $k \in \mathbf{R}$ is any scalar, then the **product** kA of the scalar k and the matrix A is obtained by multiplying every entry in A by k. That is, the ijth entry of kA is ka_{ij} (where a_{ij} is the ijth entry of A).

EXAMPLE 3 If $A = \begin{bmatrix} 1 & 2 & 3 \\ 4 & 5 & 6 \end{bmatrix}$, then $3A = \begin{bmatrix} 3 & 6 & 9 \\ 12 & 15 & 18 \end{bmatrix}$. ◆

Properties of scalar multiplication. If A and B are any $m \times n$ matrices and k and l are any scalars, then

1. $(k + l)A = kA + lA$ (distributivity);
2. $k(A + B) = kA + kB$ (distributivity);
3. $k(lA) = (kl)A = l(kA)$.

We leave it to you to supply proofs of these addition and scalar multiplication properties if you wish.

Just as defining products of vectors needed to be "unexpected" in order to be useful, so it is with defining products of matrices. To a degree, matrix multiplication is a generalization of the dot product of two vectors.

■ **Definition 6.4** (MATRIX MULTIPLICATION) Let A be an $m \times n$ matrix and B an $n \times p$ matrix. Then the **matrix product** AB is the $m \times p$ matrix whose ijth entry is the dot product of the ith row of A and the jth column of B (considered as vectors in $\mathbf{R}^n$). That is, the ijth entry of

$$\begin{bmatrix} a_{11} & a_{12} & \cdots & a_{1n} \\ \vdots & & & \vdots \\ [a_{i1} & a_{i2} & \cdots & a_{in}] \\ \vdots & & & \vdots \\ a_{m1} & a_{m2} & \cdots & a_{mn} \end{bmatrix} \begin{bmatrix} b_{11} & \cdots & \begin{bmatrix} b_{1j} \\ b_{2j} \\ \vdots \\ b_{nj} \end{bmatrix} & \cdots & b_{1p} \\ b_{21} & & & & b_{2p} \\ \vdots & & & & \vdots \\ b_{n1} & \cdots & & \cdots & b_{np} \end{bmatrix}$$

is

$$a_{i1}b_{1j} + a_{i2}b_{2j} + \cdots + a_{in}b_{nj} = \text{(more compactly)} \sum_{k=1}^{n} a_{ik}b_{kj}.$$

EXAMPLE 4 If

$$A = \begin{bmatrix} 1 & 2 & 3 \\ 4 & 5 & 6 \end{bmatrix} \quad \text{and} \quad B = \begin{bmatrix} 0 & 1 \\ 7 & 0 \\ 2 & 4 \end{bmatrix},$$

then the (2, 1) entry of AB is the dot product of the second row of A and the first column of B:

$$(2, 1) \text{ entry} = \begin{bmatrix} 4 & 5 & 6 \end{bmatrix} \cdot \begin{bmatrix} 0 \\ 7 \\ 2 \end{bmatrix} = (4)(0) + (5)(7) + (6)(2) = 47.$$

The full product AB is the 2×2 matrix

$$\begin{bmatrix} 20 & 13 \\ 47 & 28 \end{bmatrix}.$$

On the other hand, BA is the 3×3 matrix

$$\begin{bmatrix} 4 & 5 & 6 \\ 7 & 14 & 21 \\ 18 & 24 & 30 \end{bmatrix}.$$

◆

Order matters in matrix multiplication. To multiply two matrices we must have

$$\text{Number of columns of left matrix} = \text{number of rows of right matrix.}$$

In Example 4, the products AB and BA are matrices of different dimensions, hence, they could not possibly be the same. A worse situation occurs when the matrix product is defined in one order and not the other. For example, if A is 2×3 and B is 3×3, then AB is defined (and is a 2×3 matrix), but BA is not. However, even if both products AB and BA are defined and of the same dimensions (as is the case if A and B are both $n \times n$, for example), it is in general still true that

$$AB \neq BA.$$

Despite this negative news, matrix multiplication does behave well in a number of respects, as the following results indicate:

Properties of matrix multiplication. Suppose A, B, and C are matrices of appropriate dimensions (meaning so that the expressions that follow are all defined) and that k is a scalar. Then

1. $A(BC) = (AB)C$;
2. $k(AB) = (kA)B = A(kB)$;
3. $A(B + C) = AB + AC$;
4. $(A + B)C = AC + BC$.

The proofs of these properties involve little more than Definition 6.4, although the notation can become somewhat involved, as in the proof of property 1.

One simple operation on matrices that has no analogue in the real number system is the **transpose**. The transpose of an $m \times n$ matrix A is the $n \times m$ matrix A^T obtained by writing the rows of A as columns. For example, if

$$A = \begin{bmatrix} 1 & 2 & 3 \\ 4 & 5 & 6 \end{bmatrix}, \quad \text{then} \quad A^T = \begin{bmatrix} 1 & 4 \\ 2 & 5 \\ 3 & 6 \end{bmatrix}.$$

More abstractly, the ijth entry of A^T is a_{ji}, the jith entry of A.

The transpose operation turns row vectors into column vectors and vice versa. We also have the following results:

$$(A^T)^T = A, \qquad \text{for any matrix } A. \tag{3}$$

$$(AB)^T = B^T A^T, \qquad \text{where } A \text{ is } m \times n \text{ and } B \text{ is } n \times p. \tag{4}$$

The transpose will largely function as a notational convenience for us. For example, consider $\mathbf{a}, \mathbf{b} \in \mathbf{R}^n$ to be column vectors. Then the dot product $\mathbf{a} \cdot \mathbf{b}$ can be written in matrix form as

$$\mathbf{a} \cdot \mathbf{b} = a_1 b_1 + a_2 b_2 + \cdots + a_n b_n = \begin{bmatrix} a_1 & a_2 & \cdots & a_n \end{bmatrix} \begin{bmatrix} b_1 \\ b_2 \\ \vdots \\ b_n \end{bmatrix} = \mathbf{a}^T \mathbf{b}.$$

EXAMPLE 5 Matrix multiplication is defined the way it is so that, roughly speaking, working with vectors or quantities involving several variables can be made to look as much as possible like working with a single variable. This idea will become clearer throughout the text, but we can provide an important example now. A **linear function** in a single variable is a function of the form $f(x) = ax$ where a is a constant. The natural generalization of this to higher dimensions is a **linear mapping** $\mathbf{F}: \mathbf{R}^n \to \mathbf{R}^m$, $\mathbf{F}(\mathbf{x}) = A\mathbf{x}$, where A is a (constant) $m \times n$ matrix and $\mathbf{x} \in \mathbf{R}^n$. More explicitly, $\mathbf{F}$ is a function that takes a vector in $\mathbf{R}^n$ (written as a column vector) and returns a vector in $\mathbf{R}^m$ (also written as a column). That is,

$$\mathbf{F}(\mathbf{x}) = A\mathbf{x} = \begin{bmatrix} a_{11} & a_{12} & \cdots & a_{1n} \\ a_{21} & a_{22} & \cdots & a_{2n} \\ \vdots & \vdots & \ddots & \vdots \\ a_{m1} & a_{m2} & \cdots & a_{mn} \end{bmatrix} \begin{bmatrix} x_1 \\ x_2 \\ \vdots \\ x_n \end{bmatrix}.$$

The function $\mathbf{F}$ has the properties that $\mathbf{F}(\mathbf{x} + \mathbf{y}) = \mathbf{F}(\mathbf{x}) + \mathbf{F}(\mathbf{y})$ for all $\mathbf{x}, \mathbf{y} \in \mathbf{R}^n$ and $\mathbf{F}(k\mathbf{x}) = k\mathbf{F}(\mathbf{x})$ for all $\mathbf{x} \in \mathbf{R}^n$, $k \in \mathbf{R}$. These properties are also satisfied by $f(x) = ax$, of course. Perhaps more important, however, is the fact that linear mappings behave nicely with respect to composition. Suppose $\mathbf{F}$ is as just defined and $\mathbf{G}: \mathbf{R}^m \to \mathbf{R}^p$ is another linear mapping defined by $\mathbf{G}(\mathbf{x}) = B\mathbf{x}$, where B is a $p \times m$ matrix. Then there is a composite function $\mathbf{G} \circ \mathbf{F}: \mathbf{R}^n \to \mathbf{R}^p$ defined by

$$\mathbf{G} \circ \mathbf{F}(\mathbf{x}) = \mathbf{G}(\mathbf{F}(\mathbf{x})) = \mathbf{G}(A\mathbf{x}) = B(A\mathbf{x}) = (BA)\mathbf{x}$$

by the associativity property of matrix multiplication. Note that BA is defined and is a $p \times n$ matrix. Hence, we see that the composition of two linear mappings is again a linear mapping. Part of the reason we defined matrix multiplication the way we did is so that this is the case. ◆

EXAMPLE 6 We saw that by interpreting equation (1) in §1.2 in n dimensions, we obtain parametric equations of a line in $\mathbf{R}^n$. Equation (2) of §1.5, the equation for a plane in $\mathbf{R}^3$ through a given point (x_0, y_0, z_0) with given normal vector $\mathbf{n} = A\mathbf{i} + B\mathbf{j} + C\mathbf{k}$, can also be generalized to n dimensions:

$$A_1(x_1 - b_1) + A_2(x_2 - b_2) + \cdots + A_n(x_n - b_n) = 0.$$

If we let $\mathbf{A} = (A_1, A_2, \ldots, A_n)$, $\mathbf{b} = (b_1, b_2, \ldots, b_n)$ ("constant" vectors), and $\mathbf{x} = (x_1, x_2, \ldots, x_n)$ (a "variable" vector), then the aforementioned equation can be rewritten as

$$\mathbf{A} \cdot (\mathbf{x} - \mathbf{b}) = 0$$

or, considering $\mathbf{A}$, $\mathbf{b}$, and $\mathbf{x}$ as $n \times 1$ matrices, as

$$\mathbf{A}^T(\mathbf{x} - \mathbf{b}) = 0.$$

This is the equation for a **hyperplane** in $\mathbf{R}^n$ through the point $\mathbf{b}$ with normal vector $\mathbf{A}$. The points $\mathbf{x}$ that satisfy this equation fill out an $(n-1)$-dimensional subset of $\mathbf{R}^n$. ◆

At this point, it is easy to think that matrix arithmetic and the vector geometry of $\mathbf{R}^n$, although elegant, are so abstract and formal as to be of little practical use.

However, the next example, from the field of economics[2], shows that this is not the case.

EXAMPLE 7 Suppose that we have n commodities. If the price per unit of the ith commodity is p_i, then the cost of purchasing x_i (> 0) units of commodity i is $p_i x_i$. If $\mathbf{p} = (p_1, \ldots, p_n)$ is the price vector of all the commodities and $\mathbf{x} = (x_1, \ldots, x_n)$ is the **commodity bundle** vector, then

$$\mathbf{p} \cdot \mathbf{x} = p_1 x_1 + p_2 x_2 + \cdots + p_n x_n$$

represents the total cost of the commodity bundle.

Now suppose that we have an exchange economy, so that we may buy and sell items. If you have an endowment vector $\mathbf{w} = (w_1, \ldots, w_n)$, where w_i is the amount of commodity i that you can sell (trade), then, with prices given by the price vector $\mathbf{p}$, you can afford any commodity bundle $\mathbf{x}$ where

$$\mathbf{p} \cdot \mathbf{x} \leq \mathbf{p} \cdot \mathbf{w}.$$

We may rewrite this last equation as

$$\mathbf{p} \cdot (\mathbf{x} - \mathbf{w}) \leq 0.$$

In other words, you can afford any commodity bundle $\mathbf{x}$ in the budget set $\{\mathbf{x} \mid \mathbf{p} \cdot (\mathbf{x} - \mathbf{w}) \leq 0\}$. The equation $\mathbf{p} \cdot (\mathbf{x} - \mathbf{w}) = 0$ defines a **budget hyperplane** passing through $\mathbf{w}$ with normal vector $\mathbf{p}$. ◆

Determinants

We have already defined determinants of 2×2 and 3×3 matrices. (See §1.4.) Now we define the determinant of any $n \times n$ (square) matrix in terms of determinants of $(n-1) \times (n-1)$ matrices. By "iterating the definition," we can calculate any determinant.

■ **Definition 6.5** Let $A = (a_{ij})$ be an $n \times n$ matrix. The **determinant** of A is the real number given by

$$|A| = (-1)^{1+1} a_{11} |A_{11}| + (-1)^{1+2} a_{12} |A_{12}| + \cdots + (-1)^{1+n} a_{1n} |A_{1n}|,$$

where A_{ij} is the $(n-1) \times (n-1)$ submatrix of A obtained by deleting the ith row and jth column of A.

EXAMPLE 8 If $A = \begin{bmatrix} 1 & 2 & 1 & 3 \\ -2 & 1 & 0 & 5 \\ 4 & 2 & -1 & 0 \\ 3 & -2 & 1 & 1 \end{bmatrix}$, then

$$A_{12} = \begin{bmatrix} 1 & 2 & 1 & 3 \\ -2 & 1 & 0 & 5 \\ 4 & 2 & -1 & 0 \\ 3 & -2 & 1 & 1 \end{bmatrix} = \begin{bmatrix} -2 & 0 & 5 \\ 4 & -1 & 0 \\ 3 & 1 & 1 \end{bmatrix}.$$

[2] See D. Saari, "Mathematical complexity of simple economics," *Notices of the American Mathematical Society* **42** (1995), no. 2, 222–230.

According to Definition 6.5,

$$\det\begin{bmatrix} 1 & 2 & 1 & 3 \\ -2 & 1 & 0 & 5 \\ 4 & 2 & -1 & 0 \\ 3 & -2 & 1 & 1 \end{bmatrix} = (-1)^{1+1}(1)\det\begin{bmatrix} 1 & 0 & 5 \\ 2 & -1 & 0 \\ -2 & 1 & 1 \end{bmatrix}$$

$$+(-1)^{1+2}(2)\det\begin{bmatrix} -2 & 0 & 5 \\ 4 & -1 & 0 \\ 3 & 1 & 1 \end{bmatrix}$$

$$+(-1)^{1+3}(1)\det\begin{bmatrix} -2 & 1 & 5 \\ 4 & 2 & 0 \\ 3 & -2 & 1 \end{bmatrix}$$

$$+(-1)^{1+4}(3)\det\begin{bmatrix} -2 & 1 & 0 \\ 4 & 2 & -1 \\ 3 & -2 & 1 \end{bmatrix}$$

$$= (1)(1)(-1) + (-1)(2)(37) + (1)(1)(-78) + (-1)(3)(-7)$$

$$= -132.$$

◆

The determinant of the submatrix A_{ij} of A is called the ijth **minor** of A, and the quantity $(-1)^{i+j}|A_{ij}|$ is called the ijth **cofactor**. Definition 6.5 is known as **cofactor expansion** of the determinant along the first row, since det A is written as the sum of the products of each entry of the first row and the corresponding cofactor (i.e., the sum of the terms a_{1j} times $(-1)^{i+j}|A_{ij}|$).

It is natural to ask if one can compute determinants by cofactor expansion along other rows or columns of A. Happily, the answer is *yes* (although we shall not prove this).

Convenient Fact. The determinant of A can be computed by cofactor expansion along any row or column. That is,

$$|A| = (-1)^{i+1}a_{i1}|A_{i1}| + (-1)^{i+2}a_{i2}|A_{i2}| + \cdots + (-1)^{i+n}a_{in}|A_{in}|$$

(expansion along the ith row),

$$|A| = (-1)^{1+j}a_{1j}|A_{1j}| + (-1)^{2+j}a_{2j}|A_{2j}| + \cdots + (-1)^{n+j}a_{nj}|A_{nj}|$$

(expansion along the jth column).

EXAMPLE 9 To compute the determinant of

$$\begin{bmatrix} 1 & 2 & 0 & 4 & 5 \\ 2 & 0 & 0 & 9 & 0 \\ 7 & 5 & 1 & -1 & 0 \\ 0 & 2 & 0 & 0 & 2 \\ 3 & 1 & 0 & 0 & 0 \end{bmatrix},$$

expansion along the first row involves more calculation than necessary. In particular, one would need to calculate four 4×4 determinants on the way to finding the desired 5×5 determinant. (To make matters worse, these 4×4 determinants would, in turn, need to be expanded also.) However, if we expand along the third column, we find that

$$\begin{aligned}\det A &= (-1)^{1+3}(0)\det A_{13} + (-1)^{2+3}(0)\det A_{23} + (-1)^{3+3}(1)\det A_{33}\\ &\quad + (-1)^{4+3}(0)\det A_{43} + (-1)^{5+3}(0)\det A_{53}\\ &= \det A_{33}\\ &= \begin{vmatrix} 1 & 2 & 4 & 5\\ 2 & 0 & 9 & 0\\ 0 & 2 & 0 & 2\\ 3 & 1 & 0 & 0 \end{vmatrix}.\end{aligned}$$

There are several good ways to evaluate this 4×4 determinant. We'll expand about the bottom row:

$$\begin{aligned}\begin{vmatrix} 1 & 2 & 4 & 5\\ 2 & 0 & 9 & 0\\ 0 & 2 & 0 & 2\\ 3 & 1 & 0 & 0 \end{vmatrix} &= (-1)^{4+1}(3)\begin{vmatrix} 2 & 4 & 5\\ 0 & 9 & 0\\ 2 & 0 & 2 \end{vmatrix} + (-1)^{4+2}(1)\begin{vmatrix} 1 & 4 & 5\\ 2 & 9 & 0\\ 0 & 0 & 2 \end{vmatrix}\\ &= (-1)(3)(-54) + (1)(1)(2)\\ &= 164.\end{aligned}$$

◆

Of course, not all matrices contain well-distributed zeros as in Example 9, so there is by no means always an obvious choice for an expansion that avoids much calculation. Indeed, one does not compute determinants of large matrices by means of cofactor expansion. Instead, certain properties of determinants are used to make hand computations feasible. Since we shall rarely need to consider determinants larger than 3×3, we leave such properties and their significance to the exercises. (See, in particular, Exercises 26 and 27.)

Exercises

1. Rewrite in terms of the standard basis for $\mathbf{R}^n$:

(a) $(1, 2, 3, \ldots, n)$

(b) $(1, 0, -1, 1, 0, -1, \ldots, 1, 0, -1)$ (Assume that n is a multiple of 3.)

In Exercises 2–4 write the given vectors without *recourse to standard basis notation.*

2. $\mathbf{e}_1 + \mathbf{e}_2 + \cdots + \mathbf{e}_n$

3. $\mathbf{e}_1 - 2\mathbf{e}_2 + 3\mathbf{e}_3 - 4\mathbf{e}_4 + \cdots + (-1)^{n+1}n\mathbf{e}_n$

4. $\mathbf{e}_1 + \mathbf{e}_n$

5. Calculate the following, where $\mathbf{a} = (1, 3, 5, \ldots, 2n-1)$ and $\mathbf{b} = (2, -4, 6, \ldots, (-1)^{n+1}2n)$:

(a) $\mathbf{a} + \mathbf{b}$ (b) $\mathbf{a} - \mathbf{b}$ (c) $-3\mathbf{a}$

(d) $\|\mathbf{a}\|$ (e) $\mathbf{a} \cdot \mathbf{b}$

6. Let n be an even number. Verify the triangle inequality in $\mathbf{R}^n$ for $\mathbf{a} = (1, 0, 1, 0, \ldots, 0)$ and $\mathbf{b} = (0, 1, 0, 1, \ldots, 1)$.

7. Verify that the Cauchy–Schwarz inequality holds for the vectors $\mathbf{a} = (1, 2, \ldots, n)$ and $\mathbf{b} = (1, 1, \ldots, 1)$.

8. If $\mathbf{a} = (1, -1, 7, 3, 2)$ and $\mathbf{b} = (2, 5, 0, 9, -1)$, calculate the projection $\text{proj}_{\mathbf{a}}\mathbf{b}$.

9. Show, for all vectors $\mathbf{a}, \mathbf{b}, \mathbf{c} \in \mathbf{R}^n$, that

$$\|\mathbf{a} - \mathbf{b}\| \le \|\mathbf{a} - \mathbf{c}\| + \|\mathbf{c} - \mathbf{b}\|.$$

10. Prove the Pythagorean theorem. That is, if $\mathbf{a}$, $\mathbf{b}$, and $\mathbf{c}$ are vectors in $\mathbf{R}^n$ such that $\mathbf{a} + \mathbf{b} = \mathbf{c}$ and $\mathbf{a} \cdot \mathbf{b} = 0$, then

$$\|\mathbf{a}\|^2 + \|\mathbf{b}\|^2 = \|\mathbf{c}\|^2.$$

Why is this called the Pythagorean theorem?

11. Let $\mathbf{a}$ and $\mathbf{b}$ be vectors in $\mathbf{R}^n$. Show that if $\|\mathbf{a}+\mathbf{b}\| = \|\mathbf{a}-\mathbf{b}\|$, then $\mathbf{a}$ and $\mathbf{b}$ are orthogonal.

12. Let $\mathbf{a}$ and $\mathbf{b}$ be vectors in $\mathbf{R}^n$. Show that if $\|\mathbf{a}-\mathbf{b}\| > \|\mathbf{a}+\mathbf{b}\|$, then the angle between $\mathbf{a}$ and $\mathbf{b}$ is obtuse (i.e., more than $\pi/2$).

13. Describe "geometrically" the set of points in $\mathbf{R}^5$ satisfying the equation

$$2(x_1-1)+3(x_2+2)-7x_3+x_4-4-5(x_5+1)=0.$$

14. To make some extra money, you decide to print four types of silk-screened T-shirts that you sell at various prices. You have an inventory of 20 shirts that you can sell for \$8 each, 30 shirts that you sell for \$10 each, 24 shirts that you sell for \$12 each, and 20 shirts that you sell for \$15 each. A friend of yours runs a side business selling embroidered baseball caps and has an inventory of 30 caps that can be sold for \$10 each, 16 caps that can be sold for \$10 each, 20 caps that can be sold for \$12 each, and 28 caps that can be sold for \$15 each. You suggest swapping half your inventory of each type of T-shirt for half his inventory of each type of baseball cap. Is your friend likely to accept your offer? Why or why not?

15. Suppose that you run a grain farm that produces six types of grain at prices of \$200, \$250, \$300, \$375, \$450, \$500 per ton.

(a) If $\mathbf{x}=(x_1,\ldots,x_6)$ is the commodity bundle vector (meaning that x_i is the number of tons of grain i to be purchased), express the total cost of the commodity bundle as a dot product of two vectors in $\mathbf{R}^6$.

(b) A customer has a budget of \$100,000 to be used to purchase your grain. Express the set of possible commodity bundle vectors that the customer can afford. Also describe the relevant budget hyperplane in $\mathbf{R}^6$.

In Exercises 16–19, calculate the indicated matrix quantities where

$$A=\begin{bmatrix}1&2&3\\-2&0&1\end{bmatrix},\quad B=\begin{bmatrix}-4&9&5\\0&3&0\end{bmatrix},$$

$$C=\begin{bmatrix}1&-1&0\\2&0&7\\0&3&-2\end{bmatrix},\quad D=\begin{bmatrix}1&0\\2&-3\end{bmatrix}.$$

16. $3A-2B$

17. AC

18. DB

19. B^TD

20. The $n\times n$ **identity matrix**, denoted I or I_n, is the matrix whose iith entry is 1 and whose other entries are all zero. That is,

$$I_n=\begin{bmatrix}1&0&\cdots&0\\0&1&\cdots&0\\\vdots&\vdots&\ddots&\vdots\\0&0&\cdots&1\end{bmatrix}.$$

(a) Explicitly write out I_2, I_3, and I_4.

(b) The reason I is called the identity matrix is that it behaves as follows: Let A be any $m\times n$ matrix. Then

i. $AI_n=A$.

ii. $I_mA=A$.

Prove these results. (Hint: What are the ijth entries of the products in (i) and (ii)?)

Evaluate the determinants given in Exercises 21–23.

21. $\begin{vmatrix}7&0&-1&0\\2&0&1&3\\1&-3&0&2\\0&5&1&-2\end{vmatrix}$

22. $\begin{vmatrix}8&0&0&0\\15&1&0&0\\-7&6&-1&0\\8&1&9&7\end{vmatrix}$

23. $\begin{vmatrix}5&-1&0&8&11\\0&2&1&9&7\\0&0&4&-3&5\\0&0&0&2&1\\0&0&0&0&-3\end{vmatrix}$

24. Prove that a matrix that has a row or a column consisting entirely of zeros has determinant equal to zero.

25. An **upper triangular** matrix is an $n\times n$ matrix whose entries below the main diagonal are all zero. (Note: The **main diagonal** is the diagonal going from upper left to lower right.) For example, the matrix

$$\begin{bmatrix}1&2&-1&2\\0&3&4&3\\0&0&5&6\\0&0&0&7\end{bmatrix}$$

is upper triangular.

(a) Give an analogous definition for a **lower triangular** matrix and also an example of one.

(b) Use cofactor expansion to show that the determinant of any $n\times n$ upper or lower triangular matrix A is the product of the entries on the main diagonal. That is, $\det A = a_{11}a_{22}\cdots a_{nn}$.

26. Some properties of the determinant. Exercises 24 and 25 show that it is not difficult to compute determinants of even large matrices, provided that the matrices have a nice form. The following operations (called **elementary row operations**) can be used to transform an $n \times n$ matrix into one in upper triangular form:

I. Exchange rows i and j.
II. Multiply row i by a nonzero scalar.
III. Add a multiple of row i to row j. (Row i remains unchanged.)

For example, one can transform the matrix

$$\begin{bmatrix} 0 & 2 & 3 \\ 1 & 7 & -2 \\ 1 & 5 & 9 \end{bmatrix}$$

into one in upper triangular form in three steps:

Step 1. Exchange rows 1 and 2 (this puts a nonzero entry in the upper left corner):

$$\begin{bmatrix} 0 & 2 & 3 \\ 1 & 7 & -2 \\ 1 & 5 & 9 \end{bmatrix} \longrightarrow \begin{bmatrix} 1 & 7 & -2 \\ 0 & 2 & 3 \\ 1 & 5 & 9 \end{bmatrix}.$$

Step 2. Add -1 times row 1 to row 3 (this eliminates the nonzero entries below the entry in the upper left corner):

$$\begin{bmatrix} 1 & 7 & -2 \\ 0 & 2 & 3 \\ 1 & 5 & 9 \end{bmatrix} \longrightarrow \begin{bmatrix} 1 & 7 & -2 \\ 0 & 2 & 3 \\ 0 & -2 & 11 \end{bmatrix}.$$

Step 3. Add row 2 to row 3:

$$\begin{bmatrix} 1 & 7 & -2 \\ 0 & 2 & 3 \\ 0 & -2 & 11 \end{bmatrix} \longrightarrow \begin{bmatrix} 1 & 7 & -2 \\ 0 & 2 & 3 \\ 0 & 0 & 14 \end{bmatrix}.$$

The question is, how do these operations affect the determinant?

(a) By means of examples, make a conjecture as to the effect of a row operation of type I on the determinant. (That is, if matrix B results from matrix A by performing a single row operation of type I, how are $\det A$ and $\det B$ related?) You need not prove your results are correct.

(b) Repeat part (a) in the case of a row operation of type III.

(c) Prove that if B results from A by multiplying the entries in the ith row of A by the scalar c (a type II operation), then $\det B = c \cdot \det A$.

27. Calculate the determinant of the matrix

$$A = \begin{bmatrix} 2 & 1 & -2 & 7 & 8 \\ 1 & 0 & 1 & -2 & 4 \\ -1 & 1 & 2 & 3 & -5 \\ 0 & 2 & 3 & 1 & 7 \\ -3 & 2 & -1 & 0 & 1 \end{bmatrix}$$

by using row operations to transform A into a matrix in upper triangular form and by using the results of Exercise 26 to keep track of how the determinant of A and the determinant of your final matrix are related.

28. (a) Is $\det(A+B) = \det A + \det B$? Why or why not?

(b) Calculate

$$\begin{vmatrix} 1 & 2 & 7 \\ 3+2 & 1-1 & 5+1 \\ 0 & -2 & 0 \end{vmatrix}$$

and

$$\begin{vmatrix} 1 & 2 & 7 \\ 3 & 1 & 5 \\ 0 & -2 & 0 \end{vmatrix} + \begin{vmatrix} 1 & 2 & 7 \\ 2 & -1 & 1 \\ 0 & -2 & 0 \end{vmatrix},$$

and compare your results.

(c) Calculate

$$\begin{vmatrix} 1 & 3 & 2+3 \\ 0 & 4 & -1+5 \\ -1 & 0 & 0-2 \end{vmatrix}$$

and

$$\begin{vmatrix} 1 & 3 & 2 \\ 0 & 4 & -1 \\ -1 & 0 & 0 \end{vmatrix} + \begin{vmatrix} 1 & 3 & 3 \\ 0 & 4 & 5 \\ -1 & 0 & -2 \end{vmatrix},$$

and compare your results.

(d) Conjecture and prove a result about sums of determinants. (You may wish to construct further examples such as those in parts (b) and (c).)

29. It is a fact that, if A and B are any $n \times n$ matrices, then

$$\det(AB) = (\det A)(\det B).$$

Use this fact to show that $\det(AB) = \det(BA)$. (Recall that $AB \neq BA$, in general.)

An $n \times n$ matrix A is said to be ***invertible*** *(or* ***nonsingular****) if there is another $n \times n$ matrix B with the property that*

$$AB = BA = I_n,$$

where I_n denotes the $n \times n$ identity matrix. (See Exercise 20.) The matrix B is called an ***inverse*** *to the matrix A. Exercises 30–38 concern various aspects of matrices and their inverses.*

30. (a) Verify that $\begin{bmatrix} 1 & 0 \\ 1 & 1 \end{bmatrix}$ is an inverse of $\begin{bmatrix} 1 & 0 \\ -1 & 1 \end{bmatrix}$.

(b) Verify that $\begin{bmatrix} 1 & 2 & 3 \\ 2 & 5 & 3 \\ 1 & 0 & 8 \end{bmatrix}$ is an inverse of

$$\begin{bmatrix} -40 & 16 & 9 \\ 13 & -5 & -3 \\ 5 & -2 & -1 \end{bmatrix}.$$

31. Using the definition of an inverse matrix, find an inverse to $\begin{bmatrix} 2 & 2 & 1 \\ 0 & 1 & 0 \\ 0 & 0 & -1 \end{bmatrix}$.

32. Try to find an inverse matrix to $\begin{bmatrix} 0 & 2 & 1 \\ 0 & 1 & 0 \\ 0 & 0 & -1 \end{bmatrix}$. What happens?

33. Show that if an $n \times n$ matrix A is invertible, then A can have only one inverse matrix. Thus, we may write A^{-1} to denote the unique inverse of a nonsingular matrix A. (Hint: Suppose A were to have two inverses B and C. Consider $B(AC)$.)

34. Suppose that A and B are $n \times n$ invertible matrices. Show that the product matrix AB is invertible by verifying that its inverse $(AB)^{-1} = B^{-1}A^{-1}$.

35. (a) Show that if A is invertible, then $\det A \neq 0$. (In fact, the converse is also true.)

(b) Show that if A is invertible, then $\det(A^{-1}) = \dfrac{1}{\det A}$.

36. (a) Show that, if $ad - bc \neq 0$, then a general 2×2 matrix $\begin{bmatrix} a & b \\ c & d \end{bmatrix}$ has the matrix

$$\frac{1}{ad-bc}\begin{bmatrix} d & -b \\ -c & a \end{bmatrix} = \begin{bmatrix} \frac{d}{ad-bc} & -\frac{b}{ad-bc} \\ -\frac{c}{ad-bc} & \frac{a}{ad-bc} \end{bmatrix}$$

as inverse.

(b) Use this formula to find an inverse of $\begin{bmatrix} 2 & 4 \\ -1 & 2 \end{bmatrix}$.

37. If A is a 3×3 matrix and $\det A \neq 0$, then there is a (somewhat complicated) formula for A^{-1}. In particular,

$$A^{-1} = \frac{1}{\det A}\begin{bmatrix} |A_{11}| & -|A_{21}| & |A_{31}| \\ -|A_{12}| & |A_{22}| & -|A_{32}| \\ |A_{13}| & -|A_{23}| & |A_{33}| \end{bmatrix},$$

where A_{ij} denotes the submatrix of A obtained by deleting the ith row and jth column (see Definition 6.5). Use this formula to find the inverse of

$$A = \begin{bmatrix} 2 & 1 & 1 \\ 0 & 2 & 4 \\ 1 & 0 & 3 \end{bmatrix}.$$

More generally, if A is any $n \times n$ matrix and $\det A \neq 0$, then

$$A^{-1} = \frac{1}{\det A}\text{adj}A,$$

where adj A is the **adjoint matrix** of A, that is the matrix whose ijth entry is $(-1)^{i+j}|A_{ji}|$. (Note: The formula for the inverse matrix using the adjoint is typically more of theoretical than practical interest, as there are more efficient computational methods to determine the inverse, when it exists.)

38. Repeat Exercise 37 with the matrix

$$A = \begin{bmatrix} 2 & -1 & 3 \\ 1 & 2 & -2 \\ 3 & 0 & 1 \end{bmatrix}.$$

Cross products in $\mathbf{R}^n$. *Although it is not possible to define a cross product of two vectors in $\mathbf{R}^n$ as we did for two vectors in $\mathbf{R}^3$, we can construct a "cross product" of $n-1$ vectors in $\mathbf{R}^n$ that behaves analogously to the three-dimensional cross product. To be specific, if*

$$\mathbf{a}_1 = (a_{11}, a_{12}, \ldots, a_{1n}), \quad \mathbf{a}_2 = (a_{21}, a_{22}, \ldots, a_{2n}), \ldots,$$
$$\mathbf{a}_{n-1} = (a_{n-11}, a_{n-12}, \ldots, a_{n-1n})$$

are $n-1$ vectors in $\mathbf{R}^n$, we define $\mathbf{a}_1 \times \mathbf{a}_2 \times \cdots \times \mathbf{a}_{n-1}$ to be the vector in $\mathbf{R}^n$ given by the symbolic determinant

$$\mathbf{a}_1 \times \mathbf{a}_2 \times \cdots \times \mathbf{a}_{n-1} = \begin{vmatrix} \mathbf{e}_1 & \mathbf{e}_2 & \cdots & \mathbf{e}_n \\ a_{11} & a_{12} & \cdots & a_{1n} \\ a_{21} & a_{22} & \cdots & a_{2n} \\ \vdots & \vdots & \ddots & \vdots \\ a_{n-11} & a_{n-12} & \cdots & a_{n-1n} \end{vmatrix}.$$

(Here $\mathbf{e}_1, \ldots, \mathbf{e}_n$ are the standard basis vectors for $\mathbf{R}^n$.) Exercises 39–42 concern this generalized notion of cross product.

39. Calculate the following cross product in $\mathbf{R}^4$:

$$(1, 2, -1, 3) \times (0, 2, -3, 1) \times (-5, 1, 6, 0).$$

40. Use the results of Exercises 26 and 28 to show that

(a) $\mathbf{a}_1 \times \cdots \times \mathbf{a}_i \times \cdots \times \mathbf{a}_j \times \cdots \times \mathbf{a}_{n-1}$
$= -(\mathbf{a}_1 \times \cdots \times \mathbf{a}_j \times \cdots \times \mathbf{a}_i \times \cdots \times \mathbf{a}_{n-1})$,
$1 \le i \le n-1, 1 \le j \le n-1$

(b) $\mathbf{a}_1 \times \cdots \times k\mathbf{a}_i \times \cdots \times \mathbf{a}_{n-1}$
$= k(\mathbf{a}_1 \times \cdots \times \mathbf{a}_i \times \cdots \times \mathbf{a}_{n-1})$,
$1 \le i \le n-1$.

(c) $\mathbf{a}_1 \times \cdots \times (\mathbf{a}_i + \mathbf{b}) \times \cdots \times \mathbf{a}_{n-1}$
$= \mathbf{a}_1 \times \cdots \times \mathbf{a}_i \times \cdots \times \mathbf{a}_{n-1} +$
$\mathbf{a}_1 \times \cdots \times \mathbf{b} \times \cdots \times \mathbf{a}_{n-1}$,
$1 \le i \le n-1$, all $\mathbf{b} \in \mathbf{R}^n$.

(d) Show that if $\mathbf{b} = (b_1, \ldots, b_n)$ is any vector in $\mathbf{R}^n$, then

$$\mathbf{b} \cdot (\mathbf{a}_1 \times \mathbf{a}_2 \times \cdots \times \mathbf{a}_{n-1})$$

is given by the determinant

$$\begin{vmatrix} b_1 & \cdots & b_n \\ a_{11} & \cdots & a_{1n} \\ \vdots & & \vdots \\ a_{n-11} & \cdots & a_{n-1n} \end{vmatrix}.$$

41. Show that the vector $\mathbf{b} = \mathbf{a}_1 \times \mathbf{a}_2 \times \cdots \times \mathbf{a}_{n-1}$ is orthogonal to $\mathbf{a}_1, \ldots, \mathbf{a}_{n-1}$.

42. Use the generalized notion of cross products to find an equation of the (four-dimensional) hyperplane in $\mathbf{R}^5$ through the five points $P_0(1, 0, 3, 0, 4)$, $P_1(2, -1, 0, 0, 5)$, $P_2(7, 0, 0, 2, 0)$, $P_3(2, 0, 3, 0, 4)$, and $P_4(1, -1, 3, 0, 4)$.

1.7 New Coordinate Systems

We hope that you are comfortable with Cartesian (rectangular) coordinates for $\mathbf{R}^2$ or $\mathbf{R}^3$. The Cartesian coordinate system will continue to be of prime importance to us, but from time to time, we will find it advantageous to use different coordinate systems. In $\mathbf{R}^2$, polar coordinates are useful for describing figures with circular symmetry. In $\mathbf{R}^3$, there are two particularly valuable coordinate systems besides Cartesian coordinates: cylindrical and spherical coordinates. As we shall see, cylindrical and spherical coordinates are each a way of adapting polar coordinates in the plane for use in three dimensions.

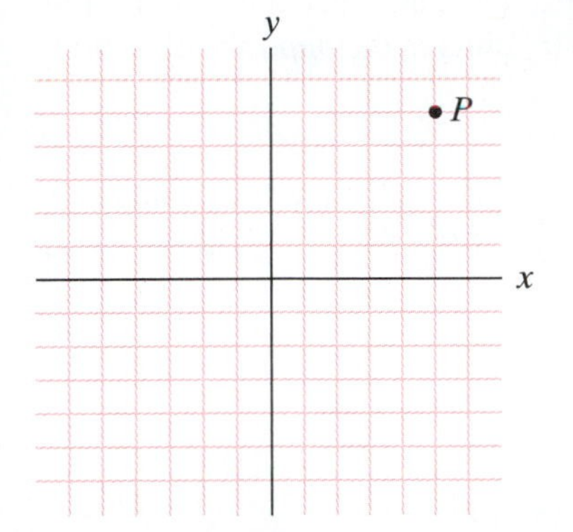

Figure 1.82 The Cartesian coordinate system.

Cartesian and Polar Coordinates on $\mathbf{R}^2$

You can understand the **Cartesian** (or **rectangular**) **coordinates** (x, y) of a point P in $\mathbf{R}^2$ in the following way: Imagine the entire plane filled with horizontal and vertical lines, as in Figure 1.82. Then the point P lies on exactly one vertical line and one horizontal line. The x-coordinate of P is where this vertical line intersects the x-axis, and the y-coordinate is where the horizontal line intersects the y-axis. (See Figure 1.83.) (Of course, we've already assigned coordinates along the axes so that the zero point of each axis is at the point of intersection of the axes. We also normally mark off the same unit distance on each axis.) Note that, because of this geometry, every point in $\mathbf{R}^2$ has a uniquely determined set of Cartesian coordinates.

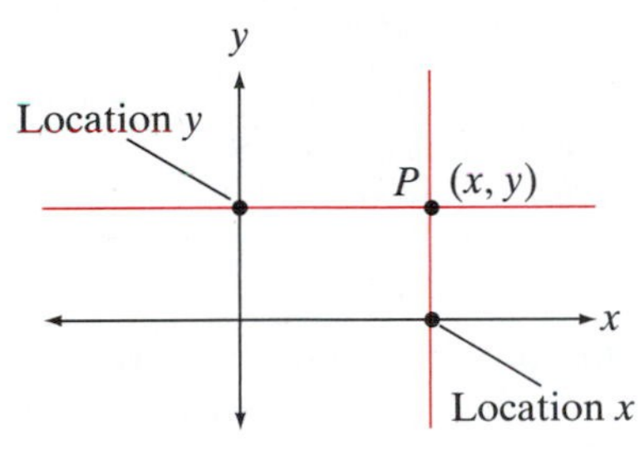

Figure 1.83 Locating a point P, using Cartesian coordinates.

Polar coordinates are defined by considering different geometric information. Now imagine the plane filled with concentric circles centered at the origin and rays emanating from the origin. Then every point except the origin lies on exactly one such circle and one such ray. The origin itself is special: No circle passes through it, and all the rays begin at it. (See Figure 1.84.) For points P other than the origin, we assign to P the polar coordinates (r, θ), where r is the radius of the circle on which P lies and θ is the angle between the positive x-axis and the ray on which P lies. (θ is measured as opening counterclockwise.) The origin is an exception: It is assigned the polar coordinates $(0, \theta)$, where θ can be any angle. (See Figure 1.85.) As we have described polar coordinates, $r \geq 0$ since r is the radius of a circle. It also makes good sense to require $0 \leq \theta < 2\pi$, for then every point in the plane, except the origin, has a uniquely determined pair of polar coordinates. Occasionally, however, it is useful *not* to restrict r to be nonnegative

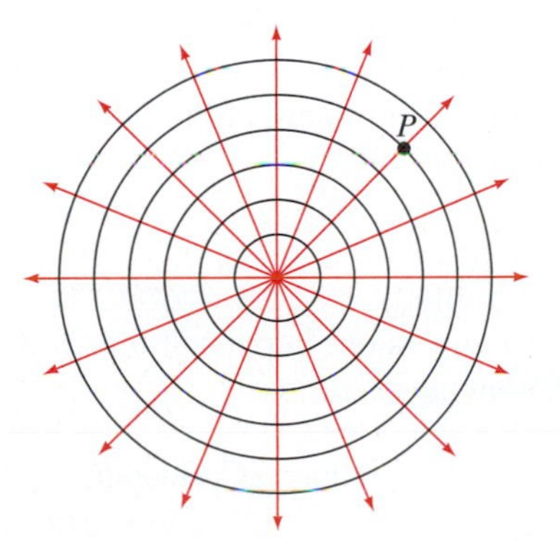

Figure 1.84 The polar coordinate system.

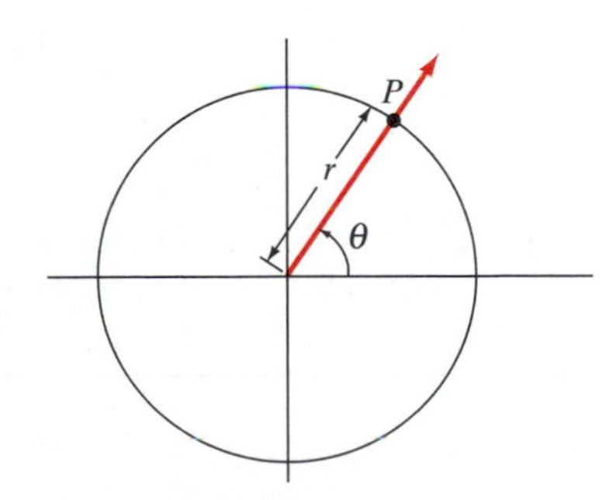

Figure 1.85 Locating a point P, using polar coordinates.

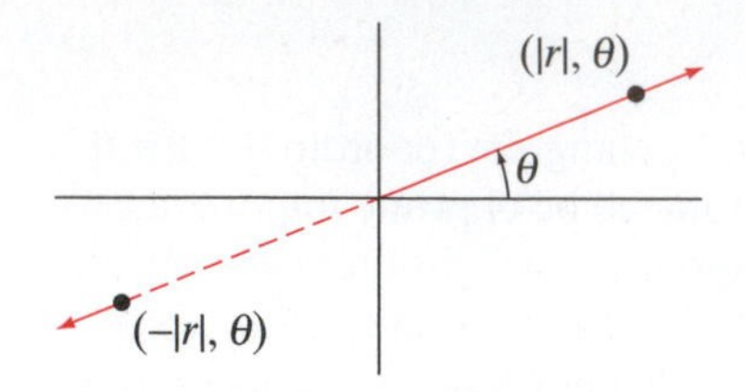

Figure 1.86 Locating the point with polar coordinates (r, θ), where $r < 0$.

and θ to be between 0 and 2π. In such a case, *no* point of $\mathbf{R}^2$ will be described by a unique pair of polar coordinates: If P has polar coordinates (r, θ), then it also has $(r, \theta + 2n\pi)$ and $(-r, \theta + (2n+1)\pi)$ as coordinates, where n can be any integer. (To locate the point having coordinates (r, θ), where $r < 0$, construct the ray making angle θ with respect to the positive x-axis, and instead of marching $|r|$ units away from the origin along this ray, go $|r|$ units in the *opposite* direction, as shown in Figure 1.86.)

EXAMPLE 1 Polar coordinates may already be familiar to you. Nonetheless, make sure you understand that the points pictured in Figure 1.87 have the coordinates indicated. ◆

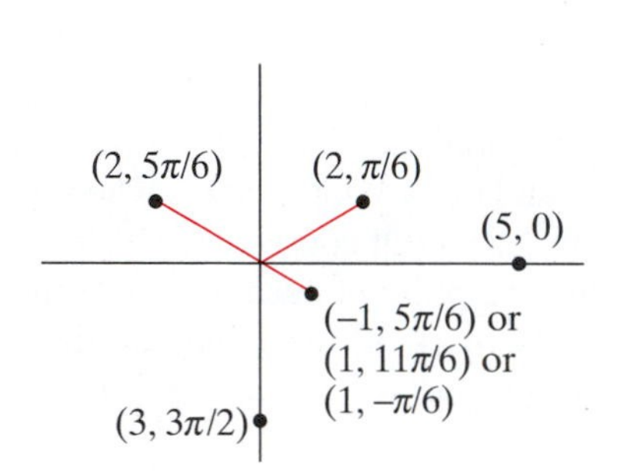

Figure 1.87 Figure for Example 1.

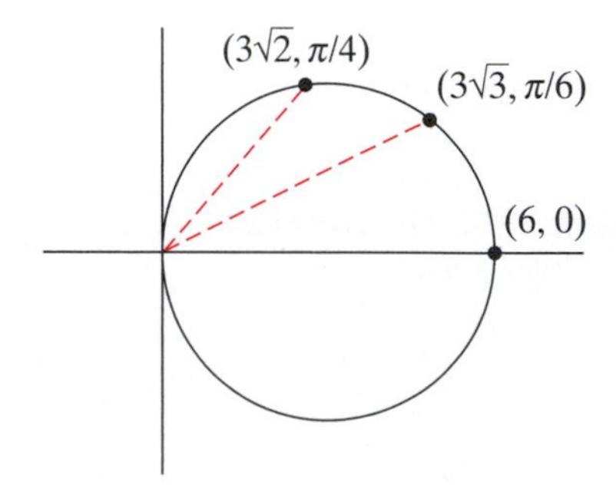

Figure 1.88 The graph of $r = 6\cos\theta$ in Example 2.

θ	$r = 6\cos\theta$
0	6
$\pi/6$	$3\sqrt{3}$
$\pi/4$	$3\sqrt{2}$
$\pi/3$	3
$\pi/2$	0
$2\pi/3$	-3
$3\pi/4$	$-3\sqrt{2}$
$5\pi/6$	$-3\sqrt{3}$
π	-6
$7\pi/6$	$-3\sqrt{3}$
$5\pi/4$	$-3\sqrt{2}$
$4\pi/3$	-3
$3\pi/2$	0
$5\pi/3$	3
$7\pi/4$	$3\sqrt{2}$

EXAMPLE 2 Let's graph the curve given by the polar equation $r = 6\cos\theta$ (Figure 1.88). We can begin to get a feeling for the graph by compiling values, as in the adjacent tabulation.

Thus, r decreases from 6 to 0 as θ increases from 0 to $\pi/2$; r decreases from 0 to -6 (or is not defined, if you take r to be nonnegative) as θ varies from $\pi/2$ to π; r increases from -6 to 0 as θ varies from π to $3\pi/2$; and r increases from 0 to 6 as θ varies from $3\pi/2$ to 2π. To graph the resulting curve, imagine a radar screen: As θ moves counterclockwise from 0 to 2π, the point (r, θ) of the graph is traced as the appropriate "blip" on the radar screen. Note that the curve is actually traced twice: once as θ varies from 0 to π and then again as θ varies from π to 2π. Alternatively, the curve is traced just once if we allow only θ values that yield nonnegative r values. The resulting graph appears to be a circle of radius 3 (not centered at the origin), and, in fact, one can see (as in Example 3) that the graph is indeed such a circle. ◆

The basic conversions between polar and Cartesian coordinates are provided by the following relations:

Polar to Cartesian:
$$\begin{cases} x = r\cos\theta \\ y = r\sin\theta \end{cases}; \tag{1}$$

Cartesian to polar:
$$\begin{cases} r^2 = x^2 + y^2 \\ \tan\theta = y/x \end{cases}. \tag{2}$$

Note that the equations in (2) do not uniquely determine r and θ in terms of x and y. This is quite acceptable, really, since we do not always want to insist that r be nonnegative and θ be between 0 and 2π. If we do restrict r and θ, however, then they are given in terms of x and y by the following formulas:

$$r = \sqrt{x^2 + y^2},$$

$$\theta = \begin{cases} \tan^{-1} y/x & \text{if } x > 0,\ y \geq 0 \\ \tan^{-1} y/x + 2\pi & \text{if } x > 0,\ y < 0 \\ \tan^{-1} y/x + \pi & \text{if } x < 0,\ y \geq 0 \\ \pi/2 & \text{if } x = 0,\ y > 0 \\ 3\pi/2 & \text{if } x = 0,\ y < 0 \\ \text{indeterminate} & \text{if } x = y = 0 \end{cases}.$$

The complicated formula for θ arises because we require $0 \leq \theta < 2\pi$, while the inverse tangent function returns values between $-\pi/2$ and $\pi/2$ only. Now you see why the equations given in (2) are a better bet!

EXAMPLE 3 We can use the formulas in (1) and (2) to prove that the curve in Example 2 really is a circle. The polar equation $r = 6\cos\theta$ that defines the curve requires a little ingenuity to convert to the corresponding Cartesian equation. The trick is to multiply both sides of the equation by r. Doing so, we obtain

$$r^2 = 6r\cos\theta.$$

Now (1) and (2) immediately give

$$x^2 + y^2 = 6x.$$

We complete the square in x to find that this equation can be rewritten as

$$(x-3)^2 + y^2 = 9,$$

which is indeed a circle of radius 3 with center at (3, 0). ◆

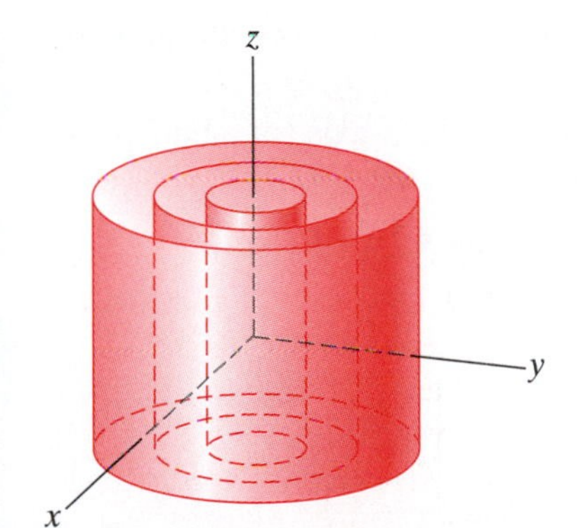

Figure 1.89 The cylindrical coordinate system.

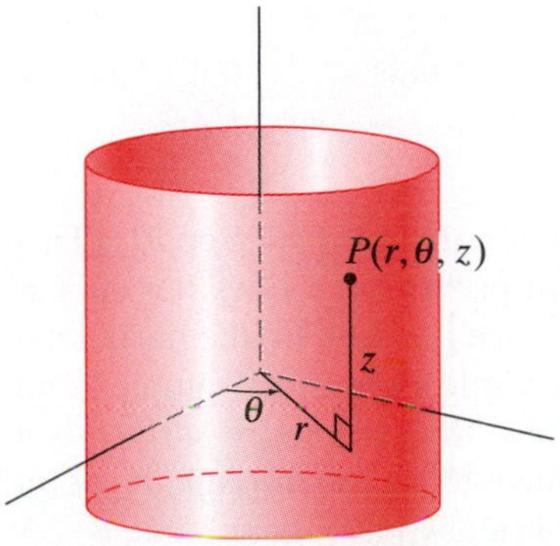

Figure 1.90 Locating a point P, using cylindrical coordinates.

Cylindrical Coordinates

Cylindrical coordinates on $\mathbf{R}^3$ are a "naive" way of generalizing polar coordinates to three dimensions, in the sense that they are nothing more than polar coordinates used in place of the x- and y-coordinates. (The z-coordinate is left unchanged.) The geometry is as follows: Except for the z-axis, fill all of space with infinitely extended circular cylinders with axes along the z-axis as in Figure 1.89. Then, any point P in $\mathbf{R}^3$ not lying on the z-axis lies on exactly one such cylinder. Hence, to locate such a point, it's enough to give the radius of the cylinder, the circumferential angle θ around the cylinder, and the vertical position z along the cylinder. The **cylindrical coordinates** of P are (r, θ, z), as shown in Figure 1.90. Algebraically, the equations in (1) and (2) can be extended to produce the basic conversions between Cartesian and cylindrical coordinates.

The basic conversions between cylindrical and Cartesian coordinates are provided by the following relations:

$$\text{Cylindrical to Cartesian:} \qquad \begin{cases} x = r\cos\theta \\ y = r\sin\theta \\ z = z \end{cases}; \tag{3}$$

$$\text{Cartesian to cylindrical:} \qquad \begin{cases} r^2 = x^2 + y^2 \\ \tan\theta = y/x \\ z = z \end{cases}. \tag{4}$$

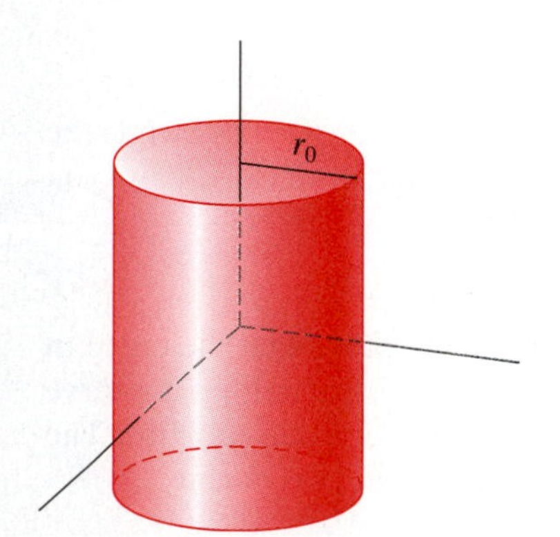

Figure 1.91 The graph of the cylindrical equation $r = r_0$.

As with polar coordinates, if we make the restrictions $r \geq 0, 0 \leq \theta < 2\pi$, then all points of $\mathbf{R}^3$ except the z-axis have a unique set of cylindrical coordinates. A point on the z-axis with Cartesian coordinates $(0, 0, z_0)$ has cylindrical coordinates $(0, \theta, z_0)$, where θ can be any angle.

Cylindrical coordinates are useful for studying objects possessing an axis of symmetry. Before exploring a few examples, let's understand the three "constant coordinate" surfaces.

- The $r = r_0$ surface is, of course, just a cylinder of radius r_0 with axis the z-axis. (See Figure 1.91.)
- The $\theta = \theta_0$ surface is a vertical plane containing the z-axis (or a half-plane with edge the z-axis if we take $r \geq 0$ only). (See Figure 1.92.)
- The $z = z_0$ surface is a horizontal plane. (See Figure 1.93.)

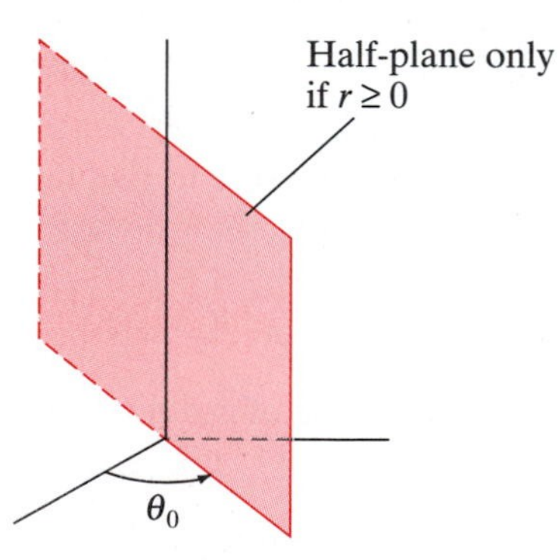

Figure 1.92 The graph of $\theta = \theta_0$.

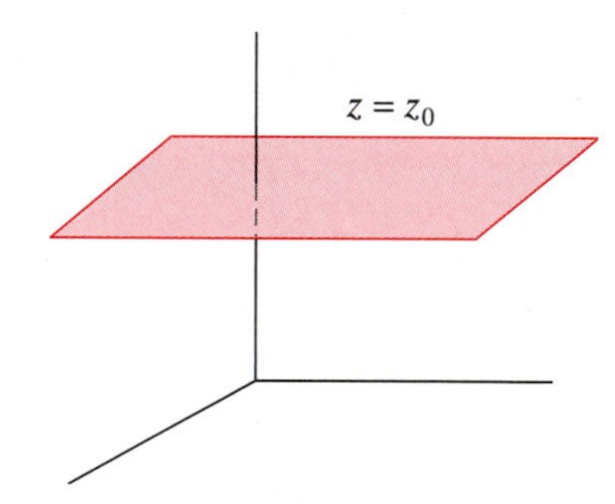

Figure 1.93 The graph of $z = z_0$.

Figure 1.94 The graph of $r = 6\cos\theta$ in cylindrical coordinates.

EXAMPLE 4 Graph the surface having cylindrical equation $r = 6\cos\theta$. (This equation is identical to the one in Example 2.) In particular, z does not appear in this equation. What this means is that if the surface is sliced by the horizontal plane $z = c$ where c is a constant, we will see the circle shown in Example 2, no matter what c is. If we stack these circular sections, then the entire surface is a circular cylinder of radius 3 with axis parallel to the z-axis (and through the point $(3, 0, 0)$ in cylindrical coordinates). This surface is shown in Figure 1.94. ◆

EXAMPLE 5 Graph the surface having equation $z = 2r$ in cylindrical coordinates.

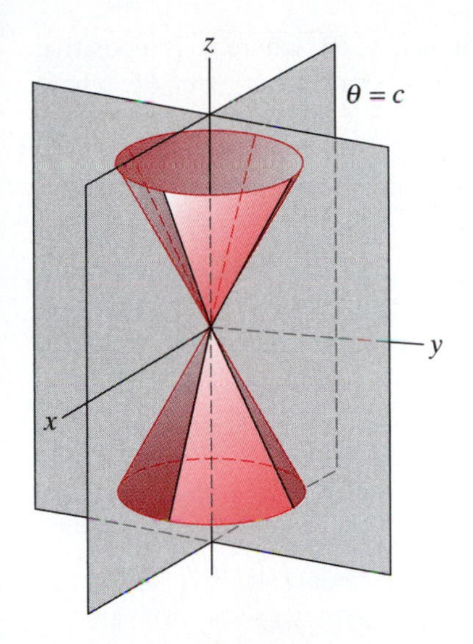

Figure 1.95 The graph of $z = 2r$ in cylindrical coordinates.

Here the variable θ does not appear in the equation, which means that the surface in question will be circularly symmetric about the z-axis. In other words, if we slice the surface by any plane of the form $\theta =$ constant (or half-plane, if we take $r \geq 0$), we see the same curve, namely, a line (respectively, a half-line) of slope 2. As we let the constant-θ plane vary, this line generates a cone, as shown in Figure 1.95. The cone consists only of the top half (**nappe**) when we restrict r to be nonnegative.

The Cartesian equation of this cone is readily determined. Using the formulas in (4), we have

$$z = 2r \quad \Longrightarrow \quad z^2 = 4r^2 \quad \Longleftrightarrow \quad z^2 = 4(x^2 + y^2).$$

Since z can be positive as well as negative, this last Cartesian equation describes the cone with both nappes. If we want the top nappe only, then the equation $z = 2\sqrt{x^2 + y^2}$ describes it. Similarly, $z = -2\sqrt{x^2 + y^2}$ describes the bottom nappe. ◆

Spherical Coordinates

Fill all of space with spheres centered at the origin as in Figure 1.96. Then every point $P \in \mathbf{R}^3$, except the origin, lies on a single such sphere. Roughly speaking, the **spherical coordinates** of P are given by specifying the radius ρ of the sphere containing P and the "latitude and longitude" readings of P along this sphere. More precisely, the spherical coordinates (ρ, φ, θ) of P are defined as follows: ρ is the distance from P to the origin; φ is the angle between the positive z-axis and the ray through the origin and P; and θ is the angle between the positive x-axis and the ray made by dropping a perpendicular from P to the xy-plane. (See Figure 1.97.) The θ-coordinate is exactly the same as the θ-coordinate used in cylindrical coordinates. (**Warning:** Physicists usually prefer to reverse the roles of φ and θ, as do some graphical software packages.)

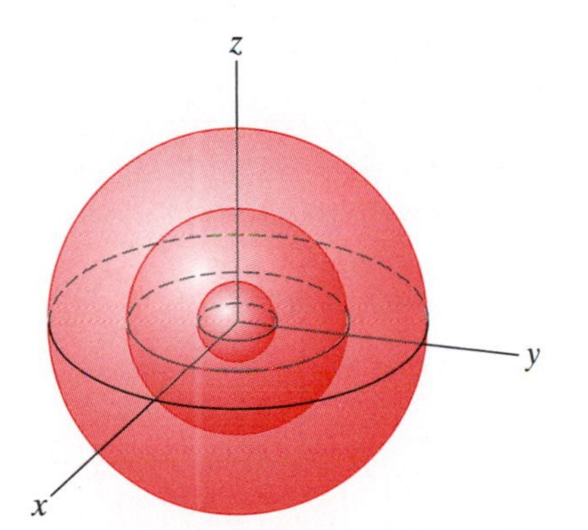

Figure 1.96 The spherical coordinate system.

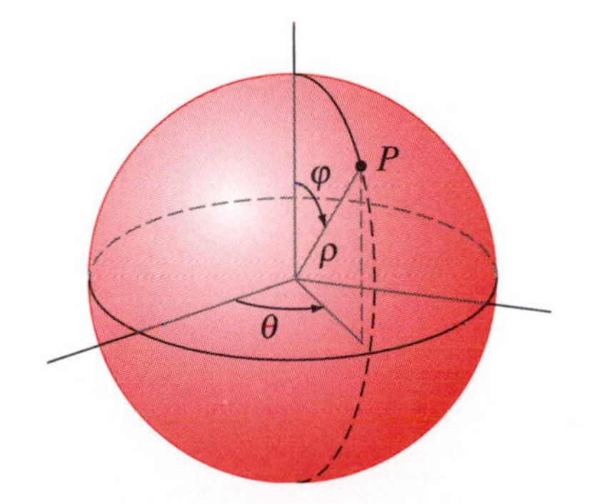

Figure 1.97 Locating the point P, using spherical coordinates.

It is standard practice to impose the following restrictions on the range of values for the individual coordinates:

$$\rho \geq 0, \qquad 0 \leq \varphi \leq \pi, \qquad 0 \leq \theta < 2\pi. \tag{5}$$

With such restrictions, all points of $\mathbf{R}^3$, except those on the z-axis, have a uniquely determined set of spherical coordinates. Points along the z-axis, except for the origin, have coordinates of the form $(\rho_0, 0, \theta)$ or (ρ_0, π, θ), where ρ_0 is a positive constant and θ is arbitrary. The origin has spherical coordinates $(0, \varphi, \theta)$, where both φ and θ are arbitrary.

EXAMPLE 6 Several points and their corresponding spherical coordinates are shown in Figure 1.98. ◆

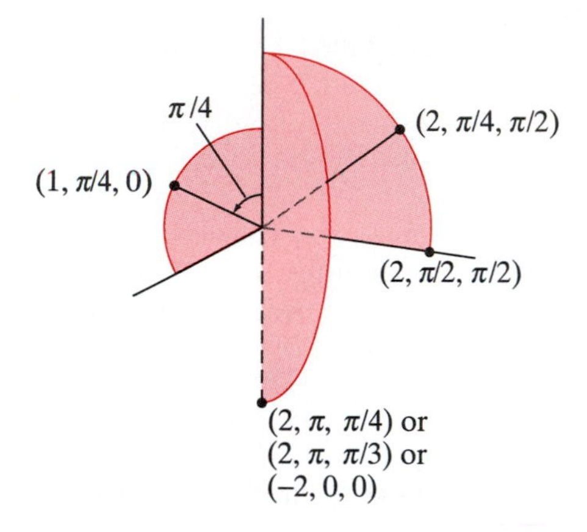

Figure 1.98 Figure for Example 6.

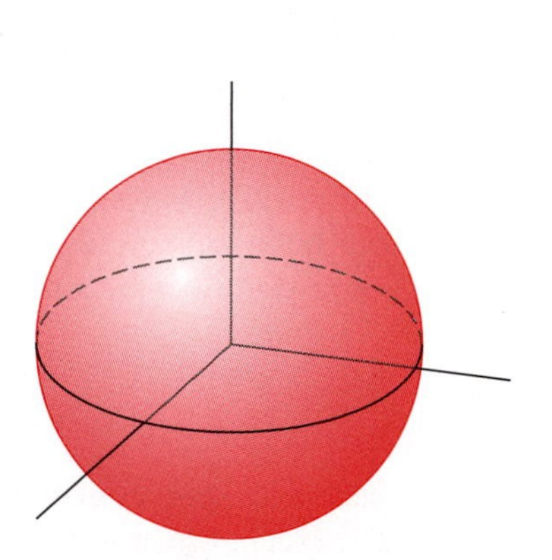

Figure 1.99 The graph of $\rho = \rho_0\ (> 0)$.

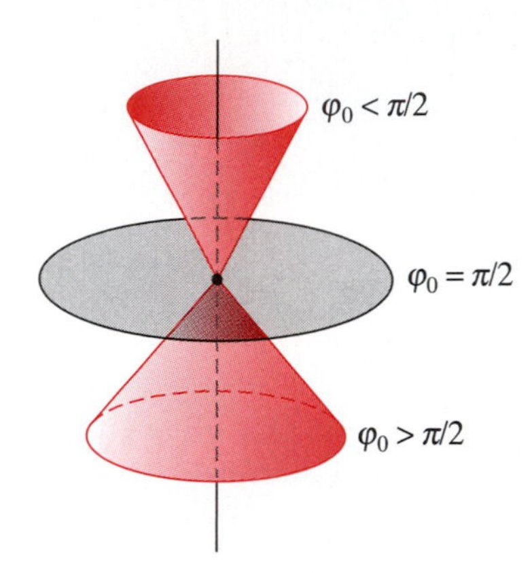

Figure 1.100 The spherical surface $\varphi = \varphi_0$, shown for different values of φ_0.

Spherical coordinates are especially useful for describing objects that have a center of symmetry. With the restrictions given by the inequalities in (5), the constant coordinate surface $\rho = \rho_0$ ($\rho_0 > 0$) is, of course, a sphere of radius ρ_0, as shown in Figure 1.99. The surface given by $\theta = \theta_0$ is a half-plane just as in the cylindrical case. The $\varphi = \varphi_0$ surface is a single-nappe cone if $\varphi_0 \neq \pi/2$ and is the xy-plane if $\varphi_0 = \pi/2$ (and is the positive or negative z-axis if $\varphi_0 = 0$ or π). (See Figure 1.100.) If we do not insist that ρ be nonnegative, then the cones would include both nappes.

The basic equations relating spherical coordinates to both cylindrical and Cartesian coordinates are as follows.

Spherical/cylindrical:

$$\begin{cases} r = \rho\sin\varphi \\ \theta = \theta \\ z = \rho\cos\varphi \end{cases} \qquad \begin{cases} \rho^2 = r^2 + z^2 \\ \tan\varphi = r/z \\ \theta = \theta \end{cases}. \tag{6}$$

Spherical/Cartesian:

$$\begin{cases} x = \rho\sin\varphi\,\cos\theta \\ y = \rho\sin\varphi\,\sin\theta \\ z = \rho\cos\varphi \end{cases} \qquad \begin{cases} \rho^2 = x^2 + y^2 + z^2 \\ \tan\varphi = \sqrt{x^2+y^2}/z \\ \tan\theta = y/x \end{cases}. \tag{7}$$

Using basic trigonometry, it is not difficult to establish the conversions in (6). From the right triangle shown in Figure 1.101, we have

$$\cos\left(\frac{\pi}{2} - \varphi\right) = \frac{r}{\rho}.$$

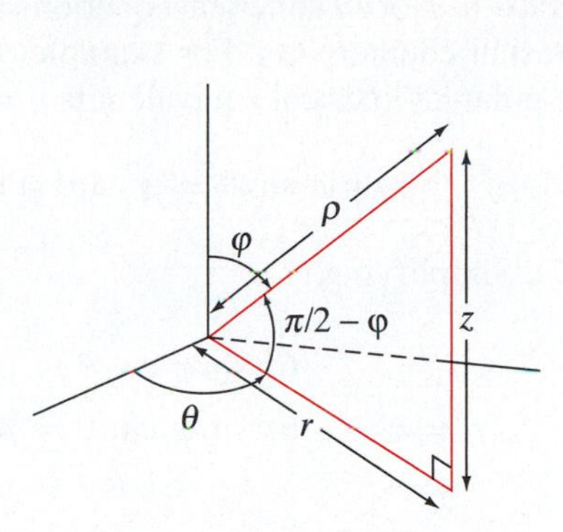

Figure 1.101 Converting spherical to cylindrical coordinates when $0 < \varphi < \frac{\pi}{2}$.

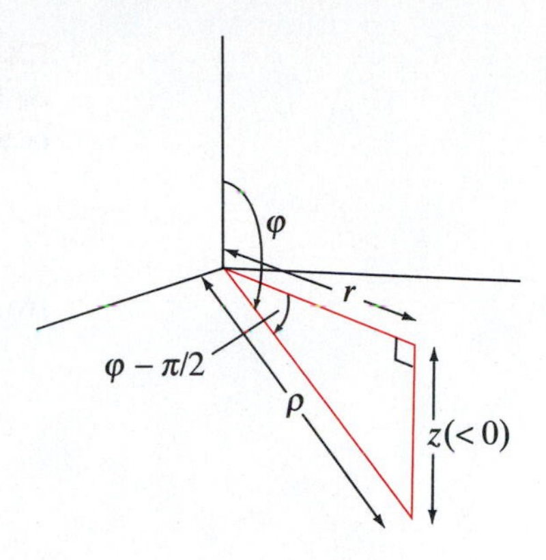

Figure 1.102 Converting spherical to cylindrical coordinates when $\pi/2 < \varphi < \pi$.

Hence,

$$r = \rho \cos\left(\frac{\pi}{2} - \varphi\right) = \rho \sin\varphi.$$

Similarly,

$$\sin\left(\frac{\pi}{2} - \varphi\right) = \frac{z}{\rho},$$

so that

$$z = \rho \sin\left(\frac{\pi}{2} - \varphi\right) = \rho \cos\varphi.$$

Thus, the formulas in (6) follow when $0 \le \varphi \le \pi/2$. If $\pi/2 < \varphi \le \pi$, then we may employ Figure 1.102. So

$$r = \rho \cos\left(\varphi - \frac{\pi}{2}\right) = \rho \sin\varphi,$$

and

$$z = -\rho \sin\left(\varphi - \frac{\pi}{2}\right) = \rho \sin\left(\frac{\pi}{2} - \varphi\right) = \rho \cos\varphi.$$

Hence, the relations in (6) hold in general. The equations in (7) follow by substitution of those in (6) into those of (3) and (4).

EXAMPLE 7 The cylindrical equation $z = 2r$ in Example 5 converts via (6) to the spherical equation

$$\rho \cos\varphi = 2\rho \sin\varphi.$$

Therefore,

$$\tan\varphi = \frac{1}{2} \iff \varphi = \tan^{-1}\frac{1}{2} \approx 26^\circ.$$

Thus, the equation defines a cone (as we just saw). The spherical equation is especially simple in that it involves just a single coordinate. ◆

EXAMPLE 8 Not all spherical equations are improvements over their cylindrical or Cartesian counterparts. For example, the Cartesian equation $6x = x^2 + y^2$ (whose polar–cylindrical equivalent is $r = 6\cos\theta$) becomes

$$6\rho\sin\varphi\,\cos\theta = \rho^2\sin^2\varphi\,\cos^2\theta + \rho^2\sin^2\varphi\,\sin^2\theta$$

from (7). Simplifying,

$$\begin{aligned} & 6\rho\sin\varphi\,\cos\theta = \rho^2\sin^2\varphi\,(\cos^2\theta + \sin^2\theta) \\ \Longleftrightarrow\quad & 6\rho\sin\varphi\,\cos\theta = \rho^2\sin^2\varphi \\ \Longleftrightarrow\quad & 6\cos\theta = \rho\sin\varphi. \end{aligned}$$

This spherical equation is *more* complicated than the original Cartesian equation in that all three spherical coordinates are involved. Therefore, it is not at all obvious that the spherical equation describes a cylinder. ◆

φ	$\rho = 2a\cos\varphi$
0	$2a$
$\pi/6$	$\sqrt{3}a$
$\pi/4$	$\sqrt{2}a$
$\pi/3$	a
$\pi/2$	0
$2\pi/3$	$-a$
$3\pi/4$	$-\sqrt{2}a$
π	$-2a$

EXAMPLE 9 Let's graph the surface with spherical equation $\rho = 2a\cos\varphi$, where $a > 0$. As with the graph of the cone with cylindrical equation $z = 2r$, note that the equation is independent of θ. Thus, all sections of this surface made by slicing with the half-plane $\theta = c$ must be the same. If we compile values as in the adjacent table, then the section of the surface in the half-plane $\theta = 0$ is as shown in Figure 1.103. Since this section must be identical in all other constant-θ half-planes, we see that this surface appears to be a sphere of radius a tangent to the xy-plane, which is shown in Figure 1.104.

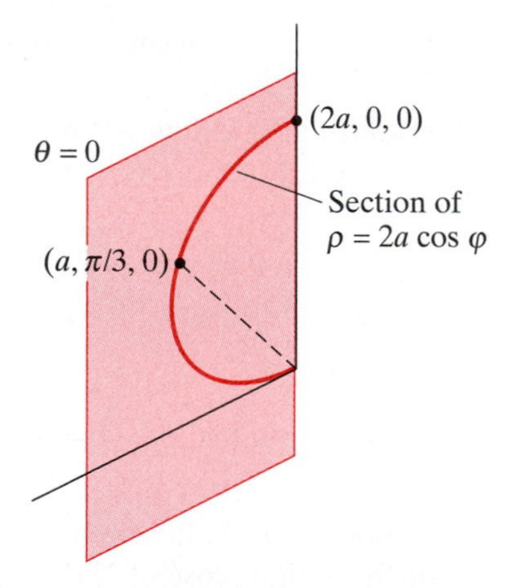

Figure 1.103 The cross-section of $\rho = 2a\cos\varphi$ in the half-plane $\theta = 0$.

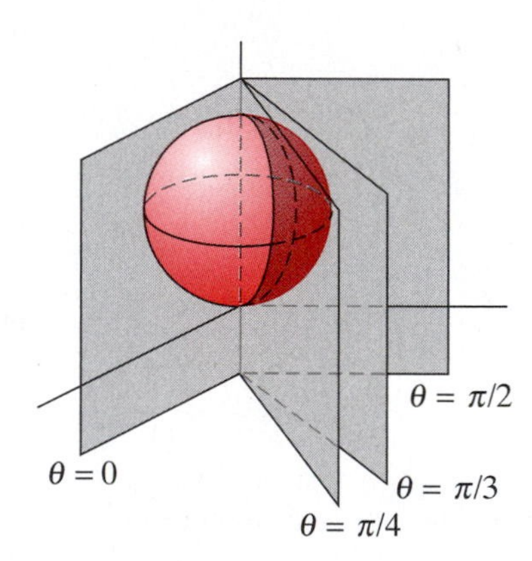

Figure 1.104 The graph of $\rho = 2a\cos\varphi$.

The Cartesian equation of the surface is determined by multiplying both sides of the spherical equation by ρ and using the conversion equations in (7):

$$\begin{aligned} \rho = 2a\cos\varphi \Longrightarrow\ & \rho^2 = 2a\rho\cos\varphi \\ \Longleftrightarrow\ & x^2 + y^2 + z^2 = 2az \\ \Longleftrightarrow\ & x^2 + y^2 + (z - a)^2 = a^2 \end{aligned}$$

by completing the square in z. This last equation can be recognized as that of a sphere of radius a with center at $(0, 0, a)$ in Cartesian coordinates. ◆

EXAMPLE 10 NASA launches a 10-ft-diameter space probe. Unfortunately, a meteor storm pushes the probe off course, and it is partially embedded in the surface of Venus, to a depth of one quarter of its diameter. To attempt to reprogram the probe's on-board computer to remove it from Venus, it is necessary to describe the embedded portion of the probe in spherical coordinates. Let us find the description desired, assuming that the surface of Venus is essentially flat in relation to the probe and that the origin of our coordinate system is at the center of the probe.

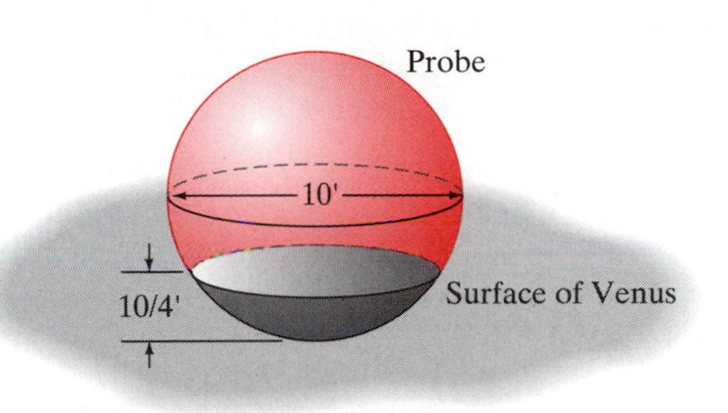

Figure 1.105 The space probe of Example 10.

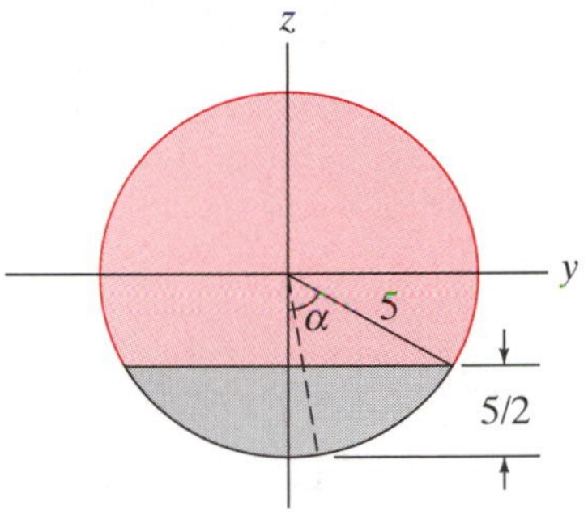

Figure 1.106 A slice of the probe of Example 10.

The situation is illustrated in Figure 1.105. The buried part of the probe clearly has symmetry about the z-axis. That is, any slice by the half-plane $\theta = \text{constant}$ looks the same as any other. Thus, θ can vary between 0 and 2π. A typical slice of the probe is shown in Figure 1.106. Elementary trigonometry indicates that for the angle α in Figure 1.106,

$$\cos\alpha = \frac{\frac{5}{2}}{5} = \frac{1}{2}.$$

Hence, $\alpha = \cos^{-1}\frac{1}{2} = \pi/3$. Thus, the spherical angle φ (which opens from the *positive* z-axis) varies from $\pi - \pi/3 = 2\pi/3$ to π as it generates the buried part of the probe. Finally, note that for a given value of φ between $2\pi/3$ and π, ρ is bounded by the surface of Venus (the plane $z = -\frac{5}{2}$ in Cartesian coordinates) and the spherical surface of the probe (whose equation in spherical coordinates is $\rho = 5$). See Figure 1.107. From the formulas in (7) the equation $z = -\frac{5}{2}$ corresponds to the spherical equation $\rho \cos\varphi = -\frac{5}{2}$ or, equivalently, to $\rho = -\frac{5}{2}\sec\varphi$. Therefore, the embedded part of the probe may be defined by the set

$$\left\{(\rho, \varphi, \theta) \;\middle|\; -\frac{5}{2}\sec\varphi \le \rho \le 5, \frac{2\pi}{3} \le \varphi \le \pi, 0 \le \theta < 2\pi\right\}. \qquad \blacklozenge$$

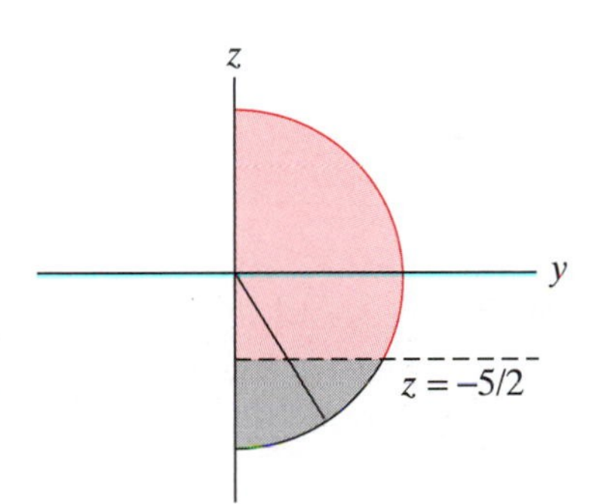

Figure 1.107 Coordinate view of the cross-section of the probe of Example 10.

Standard Bases for Cylindrical and Spherical Coordinates

In Cartesian coordinates, there are three special unit vectors **i**, **j**, and **k** that point in the directions of increasing x-, y-, and z-coordinate, respectively. We close this section by finding corresponding sets of vectors for cylindrical and spherical coordinates. That is, in each set of coordinates, we seek mutually orthogonal unit vectors that point in the directions of increasing coordinate values.

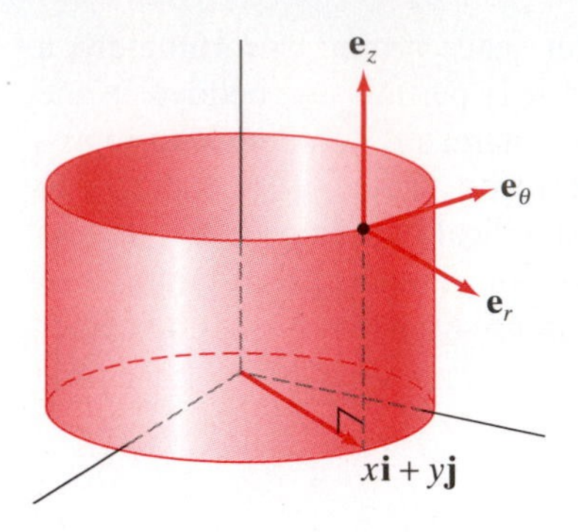

Figure 1.108 The standard basis vectors for the cylindrical coordinate system.

In cylindrical coordinates, the situation is as shown in Figure 1.108. The vectors $\mathbf{e}_r$, $\mathbf{e}_\theta$, and $\mathbf{e}_z$, which form the **standard basis** for cylindrical coordinates, are unit vectors that each point in the direction in which only the coordinate indicated by the subscript increases. There is an important difference between the standard basis vectors in Cartesian and cylindrical coordinates. In the former case, **i**, **j**, and **k** do not vary from point to point. However, the vectors $\mathbf{e}_r$ and $\mathbf{e}_\theta$ *do* change as we move from point to point.

Now we give expressions for $\mathbf{e}_r$, $\mathbf{e}_\theta$, and $\mathbf{e}_z$. Since the cylindrical z-coordinate is the same as the Cartesian z-coordinate, we must have $\mathbf{e}_z = \mathbf{k}$. The vector $\mathbf{e}_r$ must point radially outward from the z-axis with no **k**-component. At a point $(x, y, z) \in \mathbf{R}^3$ (Cartesian coordinates), the vector $x\mathbf{i} + y\mathbf{j}$ has this property. Normalizing it to obtain a unit vector (see Proposition 3.4 of §1.3), we obtain

$$\mathbf{e}_r = \frac{x\mathbf{i} + y\mathbf{j}}{\sqrt{x^2 + y^2}}.$$

With $\mathbf{e}_r$ and $\mathbf{e}_z$ in hand, it's now a simple matter to define $\mathbf{e}_\theta$, since it must be perpendicular to both $\mathbf{e}_r$ and $\mathbf{e}_z$. We take

$$\mathbf{e}_\theta = \mathbf{e}_z \times \mathbf{e}_r = \frac{-y\mathbf{i} + x\mathbf{j}}{\sqrt{x^2 + y^2}}.$$

(The reason for this choice of cross product, as opposed to $\mathbf{e}_r \times \mathbf{e}_z$, is so that $\mathbf{e}_\theta$ points in the direction of *increasing* θ.) To summarize, and using the cylindrical to Cartesian conversions given in (3),

$$\begin{aligned} \mathbf{e}_r &= \frac{x\mathbf{i} + y\mathbf{j}}{\sqrt{x^2 + y^2}} = \cos\theta\,\mathbf{i} + \sin\theta\,\mathbf{j}; \\ \mathbf{e}_\theta &= \frac{-y\mathbf{i} + x\mathbf{j}}{\sqrt{x^2 + y^2}} = -\sin\theta\,\mathbf{i} + \cos\theta\,\mathbf{j}; \\ \mathbf{e}_z &= \mathbf{k}. \end{aligned} \tag{8}$$

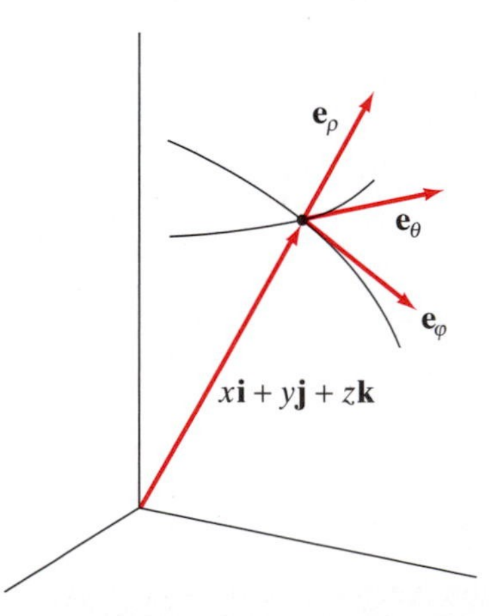

Figure 1.109 The standard basis vectors for the spherical coordinate system.

In spherical coordinates, the situation is shown in Figure 1.109. In particular, there are three unit vectors $\mathbf{e}_\rho$, $\mathbf{e}_\varphi$, and $\mathbf{e}_\theta$ that form the **standard basis** for spherical coordinates. These vectors all change direction as we move from point to point.

We give expressions for $\mathbf{e}_\rho$, $\mathbf{e}_\varphi$, and $\mathbf{e}_\theta$. Since the θ-coordinates in both spherical and cylindrical coordinates mean the same thing, $\mathbf{e}_\theta$ in spherical coordinates is given by the value of $\mathbf{e}_\theta$ in (8). At a point (x, y, z), the vector $\mathbf{e}_\rho$ should point from the origin directly to (x, y, z). Thus, $\mathbf{e}_\rho$ may be obtained by normalizing $x\mathbf{i} + y\mathbf{j} + z\mathbf{k}$. Finally, $\mathbf{e}_\varphi$ is nothing more than $\mathbf{e}_\theta \times \mathbf{e}_\rho$. If we explicitly perform the calculations just described and make use of the conversion formulas in (7), the following are obtained:

$$\begin{aligned}\mathbf{e}_\rho &= \frac{x\mathbf{i}+y\mathbf{j}+z\mathbf{k}}{\sqrt{x^2+y^2+z^2}} = \sin\varphi\,\cos\theta\,\mathbf{i}+\sin\varphi\,\sin\theta\,\mathbf{j}+\cos\varphi\,\mathbf{k};\\ \mathbf{e}_\varphi &= \frac{xz\mathbf{i}+yz\mathbf{j}-(x^2+y^2)\mathbf{k}}{\sqrt{x^2+y^2}\,\sqrt{x^2+y^2+z^2}}\\ &= \cos\varphi\,\cos\theta\,\mathbf{i}+\cos\varphi\,\sin\theta\,\mathbf{j}-\sin\varphi\,\mathbf{k};\\ \mathbf{e}_\theta &= \frac{-y\mathbf{i}+x\mathbf{j}}{\sqrt{x^2+y^2}} = -\sin\theta\,\mathbf{i}+\cos\theta\,\mathbf{j}.\end{aligned} \tag{9}$$

Although the results of (8) and (9) will not be used frequently, they will prove helpful on occasion.

Exercises

In Exercises 1–3, find the Cartesian coordinates of the points whose polar coordinates are given.

1. $(\sqrt{2}, \pi/4)$

2. $(\sqrt{3}, 5\pi/6)$

3. $(3, 0)$

In Exercises 4–6, give a set of polar coordinates for the point whose Cartesian coordinates are given.

4. $(2\sqrt{3}, 2)$

5. $(-2, 2)$

6. $(-1, -2)$

In Exercises 7–9, find the Cartesian coordinates of the points whose cylindrical coordinates are given.

7. $(2, 2, 2)$

8. $(\pi, \pi/2, 1)$

9. $(1, 2\pi/3, -2)$

In Exercises 10–13, find the rectangular coordinates of the points whose spherical coordinates are given.

10. $(4, \pi/2, \pi/3)$

11. $(3, \pi/3, \pi/2)$

12. $(1, 3\pi/4, 2\pi/3)$

13. $(2, \pi, \pi/4)$

In Exercises 14–16, find a set of cylindrical coordinates of the point whose Cartesian coordinates are given.

14. $(-1, 0, 2)$

15. $(-1, \sqrt{3}, 13)$

16. $(5, 6, 3)$

In Exercises 17 and 18, find a set of spherical coordinates of the point whose Cartesian coordinates are given.

17. $(1, -1, \sqrt{6})$

18. $(0, \sqrt{3}, 1)$

19. This problem concerns the surface described by the equation $(r-2)^2 + z^2 = 1$ in cylindrical coordinates. (Assume $r \geq 0$.)

(a) Sketch the intersection of this surface with the half-plane $\theta = \pi/2$.

(b) Sketch the entire surface.

20. (a) Graph the curve in $\mathbf{R}^2$ having polar equation $r = 2a\sin\theta$, where a is a positive constant.

(b) Graph the surface in $\mathbf{R}^3$ having spherical equation $\rho = 2a\sin\varphi$.

21. Graph the surface whose spherical equation is $\rho = 1 - \cos\varphi$.

22. Graph the surface whose spherical equation is $\rho = 1 - \sin\varphi$.

In Exercises 23–25, translate the following equations from the given coordinate system (i.e., Cartesian, cylindrical, or spherical) into equations in each of the other two systems. In addition, identify the surfaces so described by providing appropriate sketches.

23. $\rho\sin\varphi\,\sin\theta = 2$

24. $z^2 = 2x^2 + 2y^2$

25. $r = 0$

In Exercises 26–29, sketch the solid whose cylindrical coordinates (r, θ, z) satisfy the given inequalities.

26. $0 \leq r \leq 3, \quad 0 \leq \theta \leq \pi/2, \quad -1 \leq z \leq 2$

27. $r \leq z \leq 5, \quad 0 \leq \theta \leq \pi$

28. $2r \le z \le 5 - 3r$

29. $r^2 - 1 \le z \le 5 - r^2$

In Exercises 30–35, sketch the solid whose spherical coordinates (ρ, φ, θ) satisfy the given inequalities.

30. $1 \le \rho \le 2$

31. $0 \le \rho \le 1, \quad 0 \le \varphi \le \pi/2$

32. $0 \le \rho \le 1, \quad 0 \le \theta \le \pi/2$

33. $0 \le \varphi \le \pi/4, \quad 0 \le \rho \le 2$

34. $0 \le \rho \le 2/\cos\varphi, \quad 0 \le \varphi \le \pi/4$

35. $2\cos\varphi \le \rho \le 3$

36. (a) Which points P in $\mathbf{R}^2$ have the same rectangular and polar coordinates?

(b) Which points P in $\mathbf{R}^3$ have the same rectangular and cylindrical coordinates?

(c) Which points P in $\mathbf{R}^3$ have the same rectangular and spherical coordinates?

37. (a) How are the graphs of the polar equations $r = f(\theta)$ and $r = -f(\theta)$ related?

(b) How are the graphs of the spherical equations $\rho = f(\varphi, \theta)$ and $\rho = -f(\varphi, \theta)$ related?

(c) Repeat part (a) for the graphs of $r = f(\theta)$ and $r = 3f(\theta)$.

(d) Repeat part (b) for the graphs of $\rho = f(\varphi, \theta)$ and $\rho = 3f(\varphi, \theta)$.

38. Suppose that a surface has an equation in cylindrical coordinates of the form $z = f(r)$. Explain why it must be a surface of revolution.

39. (a) Verify that the basis vectors $\mathbf{e}_r$, $\mathbf{e}_\theta$, and $\mathbf{e}_z$ for cylindrical coordinates are mutually perpendicular unit vectors.

(b) Verify that the basis vectors $\mathbf{e}_\rho$, $\mathbf{e}_\varphi$, and $\mathbf{e}_\theta$ for spherical coordinates are mutually perpendicular unit vectors.

40. Use the formulas in (8) to express $\mathbf{i}, \mathbf{j}, \mathbf{k}$ in terms of $\mathbf{e}_r$, $\mathbf{e}_\theta$, and $\mathbf{e}_z$.

41. Use the formulas in (9) to express $\mathbf{i}, \mathbf{j}, \mathbf{k}$ in terms of $\mathbf{e}_\rho$, $\mathbf{e}_\varphi$, and $\mathbf{e}_\theta$.

42. Consider the solid in $\mathbf{R}^3$ shown in Figure 1.110.

(a) Describe the solid, using spherical coordinates.

(b) Describe the solid, using cylindrical coordinates.

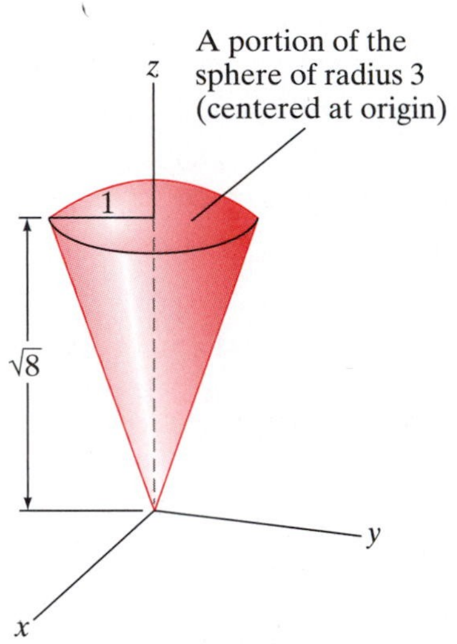

Figure 1.110 The ice-cream-cone–like solid in $\mathbf{R}^3$ in Exercise 42.

1.8 True/False Exercises for Chapter 1

1. If $\mathbf{a} = (1, 7, -9)$ and $\mathbf{b} = (1, -9, 7)$, then $\mathbf{a} = \mathbf{b}$.

2. If $\mathbf{a}$ and $\mathbf{b}$ are two vectors in $\mathbf{R}^3$ and k and l are real numbers, then $(k - l)(\mathbf{a} + \mathbf{b}) = k\mathbf{a} - l\mathbf{a} + k\mathbf{b} - l\mathbf{b}$.

3. The displacement vector from $P_1(1, 0, -1)$ to $P_2(5, 3, 2)$ is $(-4, -3, -3)$.

4. Force and acceleration are vector quantities.

5. Velocity and speed are vector quantities.

6. Displacement and distance are scalar quantities.

7. If a particle is at the point $(2, -1)$ in the plane and moves from that point with velocity vector $\mathbf{v} = (1, 3)$, then after 2 units of time have passed, the particle will be at the point $(5, 1)$.

8. The vector $(2, 3, -2)$ is the same as $2\mathbf{i} + 3\mathbf{j} - 2\mathbf{k}$.

9. A set of parametric equations for the line through $(1, -2, 0)$ that is parallel to $(-2, 4, 7)$ is $x = 1 - 2t$, $y = 4t - 2$, $z = 7$.

10. A set of parametric equations for the line through $(1, 2, 3)$ and $(4, 3, 2)$ is $x = 4 - 3t$, $y = 3 - t$, $z = t + 2$.

11. The line with parametric equations $x = 2 - 3t$, $y = t + 1$, $z = 2t - 3$ has symmetric form $\dfrac{x+2}{-3} = y - 1 = \dfrac{z-3}{2}$.

12. The two sets of parametric equations $x = 3t - 1$, $y = 2 - t$, $z = 2t + 5$ and $x = 2 - 6t$, $y = 2t + 1$, $z = 7 - 4t$ both represent the same line.

13. The parametric equations $x = 2\sin t$, $y = 2\cos t$, where $0 \le t \le \pi$, describe a circle of radius 2.

14. The dot product of two unit vectors is 1.

15. For any vector $\mathbf{a}$ in $\mathbf{R}^n$ and scalar k, we have $\|k\mathbf{a}\| = k\|\mathbf{a}\|$.

16. If $\mathbf{a}, \mathbf{u} \in \mathbf{R}^n$ and $\|\mathbf{u}\| = 1$, then $\text{proj}_{\mathbf{u}}\mathbf{a} = (\mathbf{a} \cdot \mathbf{u})\mathbf{u}$.

17. For any vectors $\mathbf{a}$, $\mathbf{b}$, $\mathbf{c}$ in $\mathbf{R}^3$, we have $\mathbf{a} \times (\mathbf{b} \times \mathbf{c}) = (\mathbf{a} \times \mathbf{b}) \times \mathbf{c}$.

18. The volume of a parallelepiped determined by the vectors $\mathbf{a}, \mathbf{b}, \mathbf{c} \in \mathbf{R}^3$ is $|(\mathbf{a} \times \mathbf{c}) \cdot \mathbf{b}|$.

19. $\|\mathbf{a}\|\mathbf{b} - \|\mathbf{b}\|\mathbf{a}$ is a vector.

20. $(\mathbf{a} \times \mathbf{b}) \cdot \mathbf{c} - (\mathbf{a} \times \mathbf{c}) \cdot \mathbf{b}$ is a scalar.

21. The plane containing the points $(1, 2, 1)$, $(3, -1, 0)$, and $(1, 0, 2)$ has equation $5x + 2y + 4z = 13$.

22. The plane containing the points $(1, 2, 1)$, $(3, -1, 0)$, and $(1, 0, 2)$ is given by the parametric equations $x = 2s$, $y = -3s - 2t$, $z = t - s$.

23. If A is a 5×7 matrix and B is a 7×7 matrix, then BA is a 7×5 matrix.

24. If $A = \begin{bmatrix} 1 & 2 & 0 & 3 \\ -1 & 0 & 2 & 1 \\ 5 & 9 & 2 & 0 \\ 0 & 8 & 0 & -6 \end{bmatrix}$, then

$$\det A = 2\begin{vmatrix} -1 & 2 & 1 \\ 5 & 2 & 0 \\ 0 & 0 & -6 \end{vmatrix} + 9\begin{vmatrix} 1 & 0 & 3 \\ -1 & 2 & 1 \\ 0 & 0 & -6 \end{vmatrix} - 8\begin{vmatrix} 1 & 0 & 3 \\ -1 & 2 & 1 \\ 5 & 2 & 0 \end{vmatrix}.$$

25. If A is an $n \times n$ matrix, then $\det(2A) = 2\det A$.

26. The surface having equation $r = 4\sin\theta$ in cylindrical coordinates is a cylinder of radius 2.

27. The surface having equation $\rho = 4\cos\theta$ in spherical coordinates is a sphere of radius 2.

28. The surface having equation $\rho\cos\theta\sin\varphi = 3$ in spherical coordinates is a plane.

29. The surface having equation $\rho = 3$ in spherical coordinates is the same as the surface whose equation in cylindrical coordinates is $r^2 + z^2 = 9$.

30. The surface whose equation in cylindrical coordinates is $z = 2r$ is the same as the surface whose equation in spherical coordinates is $\varphi = \pi/6$.

1.9 Miscellaneous Exercises for Chapter 1

1. If $P_1, P_2, \ldots, P_n$ are the vertices of a regular polygon having n sides and if O is the center of the polygon, show that $\sum_{i=1}^{n} \overrightarrow{OP_i} = \mathbf{0}$. The case $n = 5$ is shown in Figure 1.111. (Hint: Don't try using coordinates. Use, instead, sketches, geometry, and perhaps translations or rotations.)

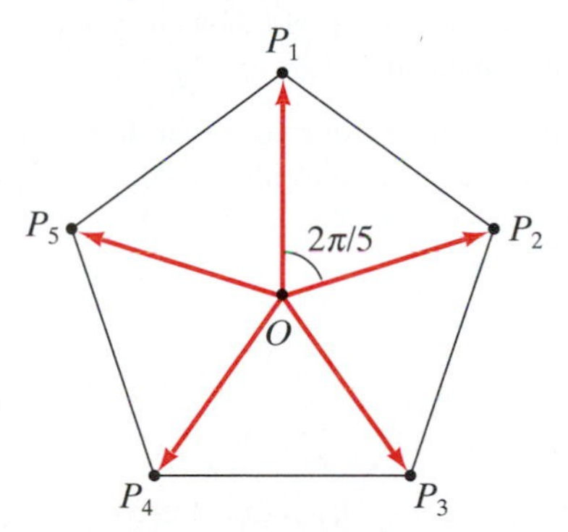

Figure 1.111 The case $n = 5$.

2. Find parametric equations for the line through the point $(1, 0, -2)$ that is parallel to the line $x = 3t + 1$, $y = 5 - 7t$, $z = t + 12$.

3. Find parametric equations for the line through the point $(1, 0, -2)$ that intersects the line $x = 3t + 1$, $y = 5 - 7t$, $z = t + 12$ orthogonally. (Hint: Let $x_0 = 3t_0 + 1$, $y_0 = 5 - 7t_0$, $z_0 = t_0 + 12$ be the point where the desired line intersects the given line.)

4. Given two points $P_0(a_1, a_2, a_3)$ and $P_1(b_1, b_2, b_3)$, we have seen in equations (3) and (4) of §1.2 how to parametrize the line through P_0 and P_1 as $\mathbf{r}(t) = \overrightarrow{OP_0} + t\,\overrightarrow{P_0P_1}$, where t can be any real number. (Recall that $\mathbf{r} = \overrightarrow{OP}$, the position vector of an arbitrary point P on the line.)

(a) For what value of t does $\mathbf{r}(t) = \overrightarrow{OP_0}$? For what value of t does $\mathbf{r}(t) = \overrightarrow{OP_1}$?

(b) Explain how to parametrize the **line segment** joining P_0 and P_1. (See Figure 1.112.)

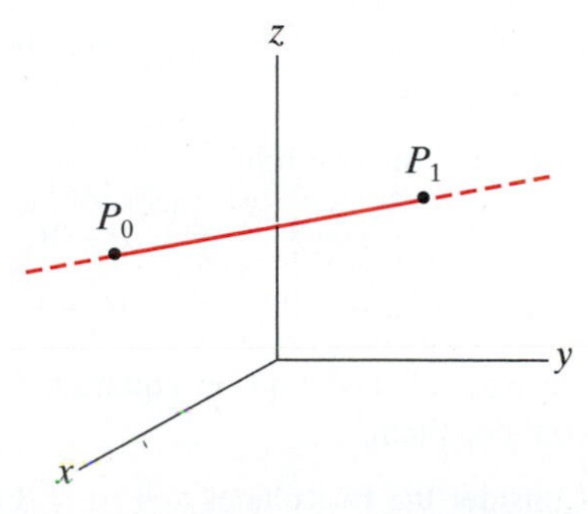

Figure 1.112 The segment joining P_0 and P_1 is a portion of the line containing P_0 and P_1. (See Exercise 4.)

(c) Give a set of parametric equations for the line segment joining the points $(0, 1, 3)$ and $(2, 5, -7)$.

5. Recall that the **perpendicular bisector** of a line segment in $\mathbf{R}^2$ is the line through the midpoint of the segment that is orthogonal to the segment.

(a) Give a set of parametric equations for the perpendicular bisector of the segment joining the points $P_1(-1, 3)$ and $P_2(5, -7)$.

(b) Given general points $P_1(a_1, a_2)$ and $P_2(b_1, b_2)$, provide a set of parametric equations for the perpendicular bisector of the segment joining them.

6. If we want to consider a perpendicular bisector of a line segment in $\mathbf{R}^3$, we will find that the bisector must be a *plane*.

(a) Give an (implicit) equation for the plane that serves as the perpendicular bisector of the segment joining the points $P_1(6, 3, -2)$ and $P_2(-4, 1, 0)$.

(b) Given general points $P_1(a_1, a_2, a_3)$ and $P_2(b_1, b_2, b_3)$, provide an equation for the plane that serves as the perpendicular bisector of the segment joining them.

7. Generalizing Exercises 5 and 6, we may define the perpendicular bisector of a line segment in $\mathbf{R}^n$ to be the hyperplane through the midpoint of the segment that is orthogonal to the segment.

(a) Give an equation for the hyperplane in $\mathbf{R}^5$ that serves as the perpendicular bisector of the segment joining the points $P_1(1, 6, 0, 3, -2)$ and $P_2(-3, -2, 4, 1, 0)$.

(b) Given arbitrary points $P_1(a_1, \ldots, a_n)$ and $P_2(b_1, \ldots, b_n)$ in $\mathbf{R}^n$, provide an equation for the hyperplane that serves as the perpendicular bisector of the segment joining them.

8. If $\mathbf{a}$ and $\mathbf{b}$ are unit vectors in $\mathbf{R}^3$, show that

$$\|\mathbf{a} \times \mathbf{b}\|^2 + (\mathbf{a} \cdot \mathbf{b})^2 = 1.$$

9. (a) If $\mathbf{a} \cdot \mathbf{b} = \mathbf{a} \cdot \mathbf{c}$, does it follow that $\mathbf{b} = \mathbf{c}$? Explain your answer.

(b) If $\mathbf{a} \times \mathbf{b} = \mathbf{a} \times \mathbf{c}$, does it follow that $\mathbf{b} = \mathbf{c}$? Explain.

10. Show that the two lines

$$l_1: \quad x = t - 3, \qquad y = 1 - 2t, \qquad z = 2t + 5$$

$$l_2: \quad x = 4 - 2t, \qquad y = 4t + 3, \qquad z = 6 - 4t$$

are parallel, and find an equation for the plane that contains them.

11. Consider the two planes $x + y = 1$ and $y + z = 1$. These planes intersect in a straight line.

(a) Find the (acute) angle of intersection between these planes.

(b) Give a set of parametric equations for the line of intersection.

12. (a) What is the angle between the diagonal of a cube and one of the edges it meets? (Hint: Locate the cube in space in a convenient way.)

(b) Find the angle between the diagonal of a cube and the diagonal of one of its faces.

13. Mark each of the following statements with a 1 if you agree, -1 if you disagree:

(1) Red is my favorite color.

(2) I consider myself to be a good athlete.

(3) I like cats more than dogs.

(4) I enjoy spicy foods.

(5) Mathematics is my favorite subject.

Your responses to the preceding "questionnaire" may be considered to form a vector in $\mathbf{R}^5$. Suppose that you and a friend calculate your respective "response vectors" for the questionnaire. Explain the significance of the dot product of your two vectors.

14. The **median** of a triangle is the line segment that joins a vertex of a triangle to the midpoint of the opposite side. The purpose of this problem is to use vectors to show that the medians of a triangle all meet at a point.

(a) Using Figure 1.113, write the vectors $\overrightarrow{BM_1}$ and $\overrightarrow{CM_2}$ in terms of $\overrightarrow{AB}$ and $\overrightarrow{AC}$.

(b) Let P be the point of intersection of $\overline{BM_1}$ and $\overline{CM_2}$. Write $\overrightarrow{BP}$ and $\overrightarrow{CP}$ in terms of $\overrightarrow{AB}$ and $\overrightarrow{AC}$.

(c) Use the fact that $\overrightarrow{CB} = \overrightarrow{CP} + \overrightarrow{PB} = \overrightarrow{CA} + \overrightarrow{AB}$ to show that P must lie two-thirds of the way from B to M_1 and two-thirds of the way from C to M_2.

(d) Now use part (c) to show why all three medians must meet at P.

15. Suppose that the four vectors $\mathbf{a}$, $\mathbf{b}$, $\mathbf{c}$, and $\mathbf{d}$ in $\mathbf{R}^3$ are coplanar (i.e., that they all lie in the same plane). Show that then $(\mathbf{a} \times \mathbf{b}) \times (\mathbf{c} \times \mathbf{d}) = \mathbf{0}$.

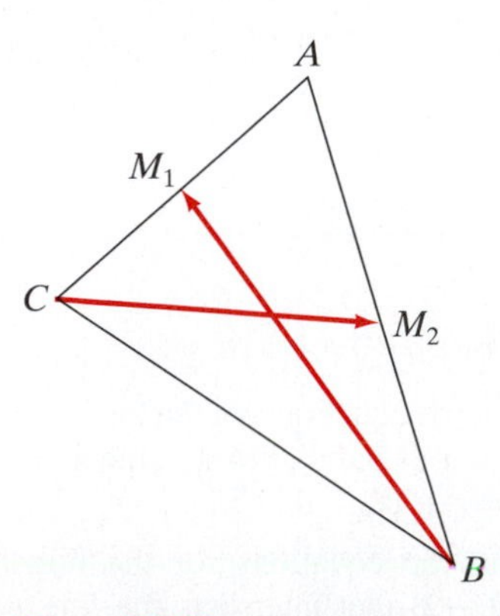

Figure 1.113 Two of the three medians of a triangle in Exercise 14.

16. Show that the area of the triangle, two of whose sides are determined by the vectors **a** and **b** (see Figure 1.114), is given by the formula

$$\text{Area} = \frac{1}{2}\sqrt{\|\mathbf{a}\|^2\|\mathbf{b}\|^2 - (\mathbf{a}\cdot\mathbf{b})^2}.$$

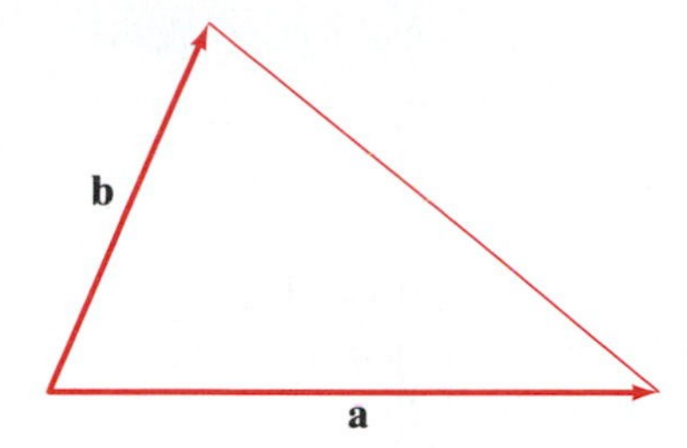

Figure 1.114 The triangle in Exercise 16.

17. Let $A(1, 3, -1)$, $B(4, -1, 3)$, $C(2, 5, 2)$, and $D(5, 1, 6)$ be the vertices of a parallelogram.

(a) Find the area of the parallelogram.

(b) Find the area of the projection of the parallelogram in the xy-plane.

18. (a) For the line l in $\mathbf{R}^2$ given by the equation $ax+by = d$, find a vector **v** that is parallel to l.

(b) Find a vector **n** that is normal to l and has first component equal to a.

(c) If $P_0(x_0, y_0)$ is any point in $\mathbf{R}^2$, use vectors to derive the following formula for the distance from P_0 to l:

$$\text{Distance from } P_0 \text{ to } l = \frac{|ax_0 + by_0 - d|}{\sqrt{a^2 + b^2}}.$$

To do this, you'll find it helpful to use Figure 1.115, where $P_1(x_1, y_1)$ is any point on l.

(d) Find the distance between the point $(3, 5)$ and the line $8x - 5y = 2$.

19. (a) If $P_0(x_0, y_0, z_0)$ is any point in $\mathbf{R}^3$, use vectors to derive the following formula for the distance from P_0 to the plane Π having equation $Ax+By+Cz=D$:

$$\text{Distance from } P_0 \text{ to } \Pi = \frac{|Ax_0 + By_0 + Cz_0 - D|}{\sqrt{A^2 + B^2 + C^2}}.$$

Figure 1.116 should help. ($P_1(x_1, y_1, z_1)$ is any point in Π.)

(b) Find the distance between the point $(1, 5, -3)$ and the plane $x - 2y + 2z + 12 = 0$.

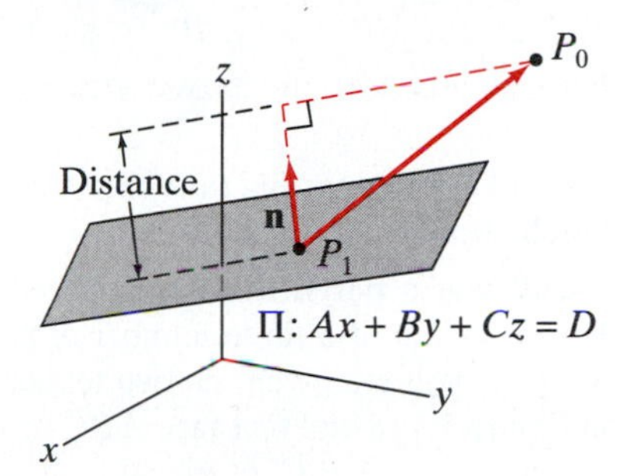

Figure 1.116 Geometric construction for Exercise 19.

20. Let $\mathbf{x} = \overrightarrow{OP}$, the position vector of a point P in $\mathbf{R}^3$. Consider the equation

$$\frac{\mathbf{x}\cdot\mathbf{k}}{\|\mathbf{x}\|} = \frac{1}{\sqrt{2}}.$$

Describe the configuration of points P that satisfy the equation.

21. Let **a** and **b** be two fixed, nonzero vectors in $\mathbf{R}^3$, and let c be a fixed constant. Explain how the pair of equations,

$$\begin{cases} \mathbf{a}\cdot\mathbf{x} = c \\ \mathbf{a}\times\mathbf{x} = \mathbf{b}, \end{cases}$$

completely determine the vector $\mathbf{x} \in \mathbf{R}^3$.

22. (a) Given an arbitrary (i.e., not necessarily regular) tetrahedron, associate to each of its four triangular faces a vector outwardly normal to that face with length equal to the area of that face. (See Figure 1.117.) Show that the sum of these four vectors is zero. (Hint: Describe $\mathbf{v}_1, \ldots, \mathbf{v}_4$ in terms of

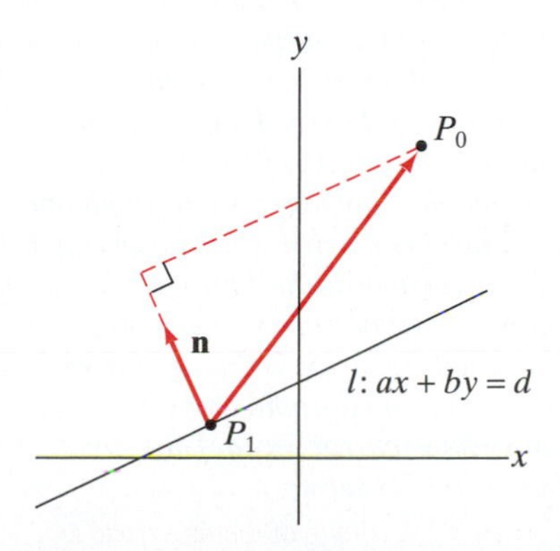

Figure 1.115 Geometric construction for Exercise 18.

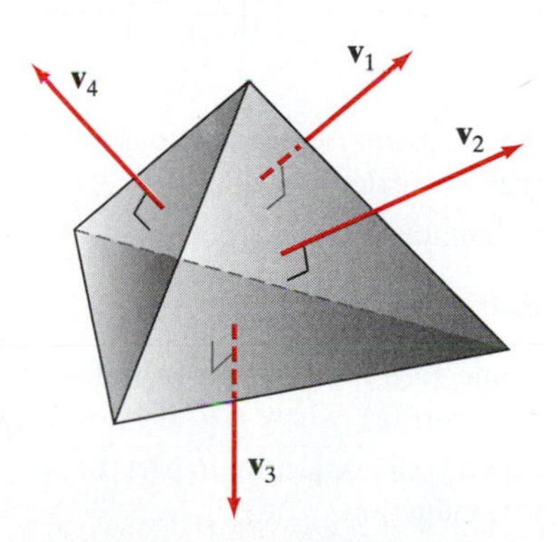

Figure 1.117 The tetrahedron of part (a) of Exercise 22.

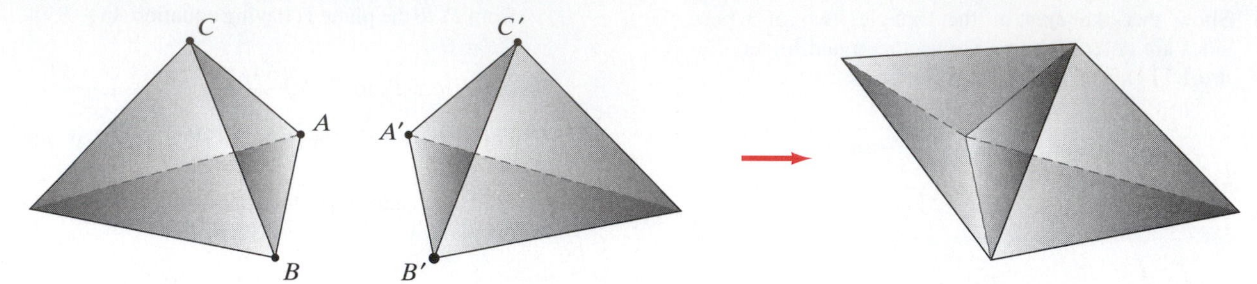

Figure 1.118 In Exercise 22(b), glue the two tetrahedra shown along congruent faces.

some of the vectors that run along the *edges* of the tetrahedron.)

(b) Recall that a **polyhedron** is a closed surface in $\mathbf{R}^3$ consisting of a finite number of planar faces. Suppose you are given the two tetrahedra shown in Figure 1.118 and that face ABC of one is congruent to face $A'B'C'$ of the other. If you glue the tetrahedra together along these congruent faces, then the outer faces give you a six-faced polyhedron. Associate to each face of this polyhedron an outward-pointing normal vector with length equal to the area of that face. Show that the sum of these six vectors is zero.

(c) Outline a proof of the following: Given an n-faced polyhedron, associate to each face an outward-pointing normal vector with length equal to the area of that face. Show that the sum of these n vectors is zero.

23. (a) Use vectors to prove that the sum of the squares of the lengths of the diagonals of a parallelogram equals the sum of the squares of the lengths of the four sides.

(b) Give an algebraic generalization of part (a) for $\mathbf{R}^n$.

24. Show that for any real numbers $a_1, \ldots, a_n, b_1, \ldots, b_n$ we have

$$\left[\sum_{i=1}^{n} a_i b_i\right]^2 \leq \left[\sum_{i=1}^{n} a_i^2\right]\left[\sum_{i=1}^{n} b_i^2\right].$$

25. To raise a square ($n \times n$) matrix A to a positive integer power n, one calculates A^n as $A \cdot A \cdots A$ (n times).

(a) Calculate successive powers A, A^2, A^3, A^4 of the matrix $A = \begin{bmatrix} 1 & 1 \\ 0 & 1 \end{bmatrix}$.

(b) Conjecture the general form of A^n for the matrix A of part (a), where n is any positive integer.

(c) Prove your conjecture in part (b) using mathematical induction.

26. A square matrix A is called **nilpotent** if $A^n = 0$ for some positive power n.

(a) Show that $A = \begin{bmatrix} 0 & 1 & 1 \\ 0 & 0 & 0 \\ 0 & 0 & 0 \end{bmatrix}$ is nilpotent.

(b) Use a calculator or computer to show that $A = \begin{bmatrix} 0 & 0 & 0 & 0 & 0 \\ 1 & 0 & 0 & 0 & 0 \\ 0 & 1 & 0 & 0 & 0 \\ 0 & 0 & 1 & 0 & 0 \\ 0 & 0 & 0 & 1 & 0 \end{bmatrix}$ is nilpotent.

27. The $n \times n$ matrix H_n whose ijth entry is $1/(i + j - 1)$ is called the **Hilbert matrix** of order n.

(a) Write out H_2, H_3, H_4, H_5, and H_6. Use a computer to calculate their determinants exactly. What seems to happen to $\det H_n$ as n gets larger?

(b) Now calculate H_{10} and $\det H_{10}$. If you use exact arithmetic, you should find that $\det H_{10} \neq 0$ and hence that H_{10} is invertible. (See Exercises 30–38 of §1.6 for more about invertible matrices.)

(c) Now give a numerical approximation A for H_{10}. Calculate the inverse matrix B of this approximation, if your computer allows. Then calculate AB and BA. Do you obtain the 10×10 identity matrix I_{10} in both cases?

(d) Explain what parts (b) and (c) suggest about the difficulties in using numerical approximations in matrix arithmetic.

As a child, you may have played with a popular toy called a ***Spirograph***®*. With it one could draw some appealing geometric figures. The Spirograph consists of a small toothed disk with several holes in it and a larger ring with teeth on both inside and outside as shown in Figure 1.119. You can draw pictures by meshing the small disk with either the inside or outside circles of the ring and then poking a pen through one of the holes of the disk while turning the disk. (The large ring is held fixed.)*

An idealized version of the Spirograph can be obtained by taking a large circle (of radius a) and letting a small circle (of radius b) roll either inside or outside it without slipping. A "Spirograph" pattern is produced by tracking a particular point lying anywhere on (or inside) the small circle. Exercises 28–31 concern this set-up.

28. Suppose that the small circle rolls inside the larger circle and that the point P we follow lies on the

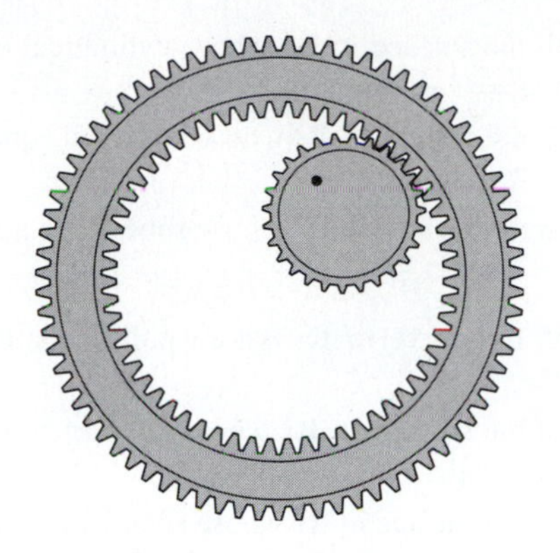

Figure 1.119 The Spirograph.

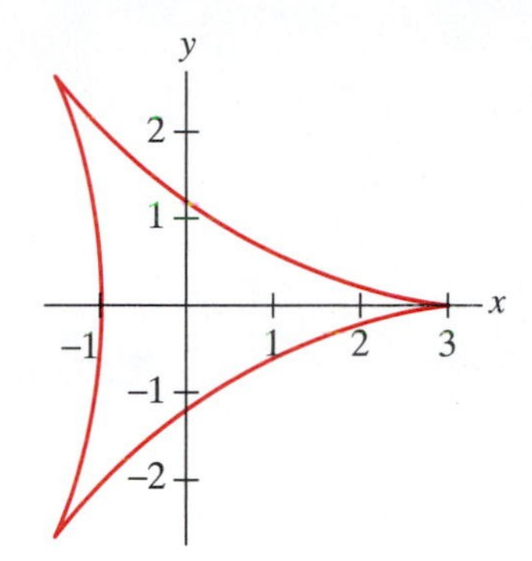

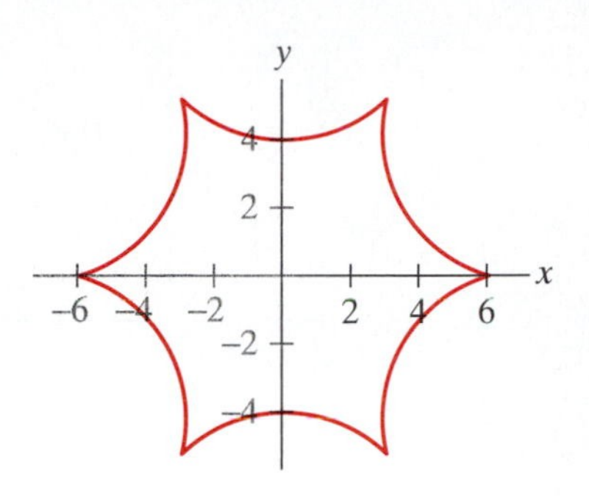

Figure 1.121 Hypocycloids with $a = 3, b = 2$ and $a = 6, b = 5$.

circumference of the small circle. If the initial configuration is such that P is at $(a, 0)$, find parametric equations for the curve traced by P, using angle t from the positive x-axis to the center B of the moving circle. (This configuration is shown in Figure 1.120.) The resulting curve is called a **hypocycloid**. Two examples are shown in Figure 1.121.

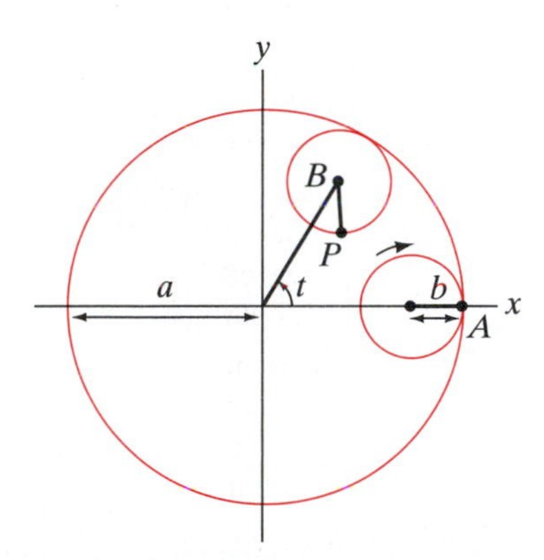

Figure 1.120 The coordinate configuration for finding parametric equations for a hypocycloid.

29. Now suppose that the small circle rolls on the outside of the larger circle. Derive a set of parametric equations for the resulting curve in this case. Such a curve is called an **epicycloid**, shown in Figure 1.122.

30. (a) A **cusp** (or corner) occurs on either the hypocycloid or epicycloid every time the point P on the small circle touches the large circle. Equivalently, this happens whenever the smaller circle rolls through 2π. Assuming that a/b is rational, how many cusps does a hypocycloid or epicycloid have? (Your answer should involve a and b in some way.)

(b) Describe in words and pictures what happens when a/b is not rational.

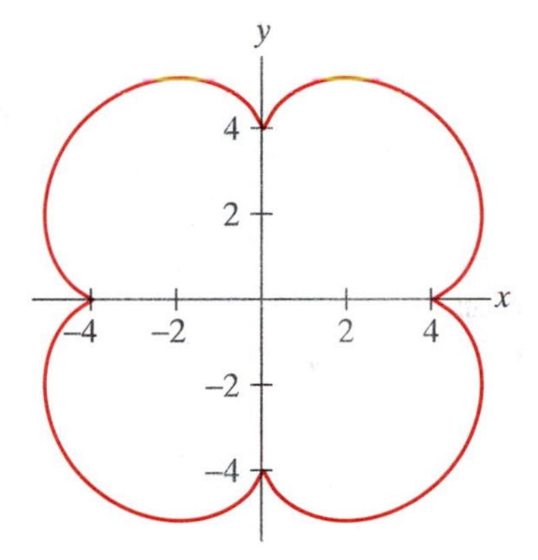

Figure 1.122 An epicycloid with $a = 4, b = 1$.

31. Consider the original Spirograph set-up again. If we now mark a point P at a distance c from the center of the smaller circle, then the curve traced by P is called a **hypotrochoid** (if the smaller circle rolls on the inside of the larger circle) or an **epitrochoid** (if the smaller circle rolls on the outside). Note that we must have $b < a$, but we can have c either larger or smaller than b. (If $c < b$, we get a "true" Spirograph pattern in the sense that the point P will be on the inside of the smaller circle. The situation when $c > b$ is like having P mounted on the end of an elongated spoke on the smaller circle.) Give a set of parametric equations for the curves that result in this way. (See Figure 1.123.)

Exercises 32–37 are made feasible through the use of appropriate software for graphing in polar, cylindrical, and spherical coordinates. (Note: When using software for graphing in spherical coordinates, be sure to check the definitions that are used for the angles φ and θ.)

32. (a) Graph the curve in $\mathbf{R}^2$ whose polar equation is $r = \cos 2\theta$.

(b) Graph the surface in $\mathbf{R}^3$ whose cylindrical equation is $r = \cos 2\theta$.

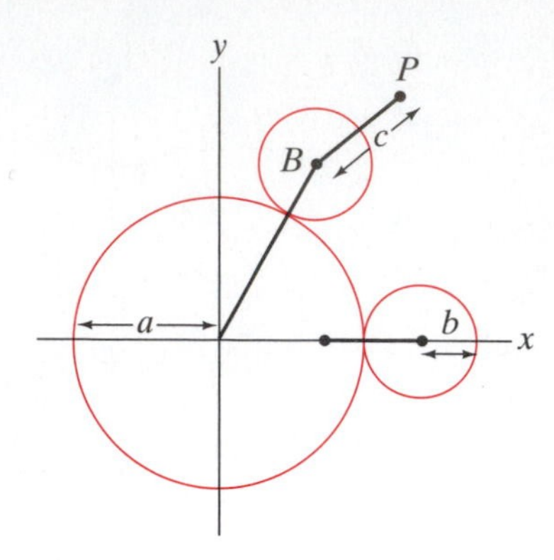

Figure 1.123 The configuration for finding parametric equations for epitrochoids.

(c) Graph the surface in $\mathbf{R}^3$ whose spherical equation is $\rho = \cos 2\varphi$.

(d) Graph the surface in $\mathbf{R}^3$ whose spherical equation is $\rho = \cos 2\theta$.

33. (a) Graph the curve in $\mathbf{R}^2$ whose polar equation is $r = \sin 2\theta$.

(b) Graph the surface in $\mathbf{R}^3$ whose cylindrical equation is $r = \sin 2\theta$.

(c) Graph the surface in $\mathbf{R}^3$ whose spherical equation is $\rho = \sin 2\varphi$.

(d) Graph the surface in $\mathbf{R}^3$ whose spherical equation is $\rho = \sin 2\theta$. Compare the results of this exercise with those of Exercise 32.

34. (a) Graph the curve in $\mathbf{R}^2$ whose polar equation is $r = \cos 3\theta$.

(b) Graph the surface in $\mathbf{R}^3$ whose cylindrical equation is $r = \cos 3\theta$.

(c) Graph the surface in $\mathbf{R}^3$ whose spherical equation is $\rho = \cos 3\varphi$.

(d) Graph the surface in $\mathbf{R}^3$ whose spherical equation is $\rho = \cos 3\theta$.

35. (a) Graph the curve in $\mathbf{R}^2$ whose polar equation is $r = \sin 3\theta$.

(b) Graph the surface in $\mathbf{R}^3$ whose cylindrical equation is $r = \sin 3\theta$.

(c) Graph the surface in $\mathbf{R}^3$ whose spherical equation is $\rho = \sin 3\varphi$.

(d) Graph the surface in $\mathbf{R}^3$ whose spherical equation is $\rho = \sin 3\theta$. Compare the results of this exercise with those of Exercise 34.

36. (a) Graph the curve in $\mathbf{R}^2$ whose polar equation is $r = 1 + \sin\frac{\theta}{2}$. (This curve is known as a **nephroid**, meaning "kidney shaped".)

(b) Graph the surface in $\mathbf{R}^3$ whose cylindrical equation is $r = 1 + \sin\frac{\theta}{2}$.

(c) Graph the surface in $\mathbf{R}^3$ whose spherical equation is $\rho = 1 + \sin\frac{\varphi}{2}$.

(d) Graph the surface in $\mathbf{R}^3$ whose spherical equation is $\rho = 1 + \sin\frac{\theta}{2}$.

37. (a) Graph the curve in $\mathbf{R}^2$ whose polar equation is $r = \theta$.

(b) Graph the surface in $\mathbf{R}^3$ whose cylindrical equation is $r = \theta$.

(c) Graph the surface in $\mathbf{R}^3$ whose spherical equation is $\rho = \varphi$.

(d) Graph the surface in $\mathbf{R}^3$ whose spherical equation is $\rho = \theta$, where $\pi/2 \le \varphi \le \pi$ and $0 \le \theta \le 4\pi$.

38. Consider the solid hemisphere of radius 5 pictured in Figure 1.124.

(a) Describe this solid, using spherical coordinates.

(b) Describe this solid, using cylindrical coordinates.

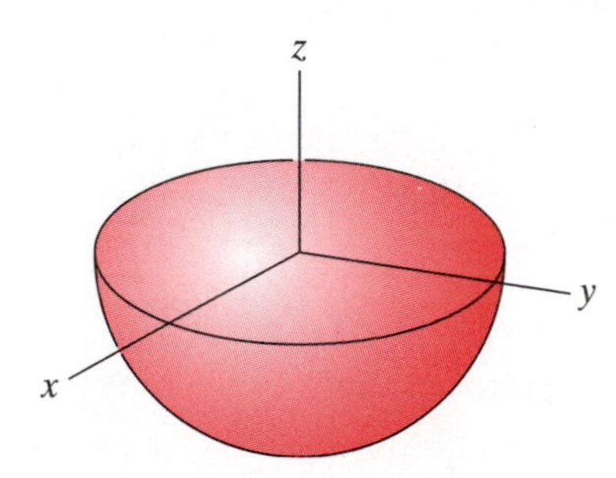

Figure 1.124 The solid hemisphere of Exercise 38.

39. Consider the solid cylinder pictured in Figure 1.125.

(a) Describe this solid, using cylindrical coordinates (position the cylinder conveniently).

(b) Describe this solid, using spherical coordinates.

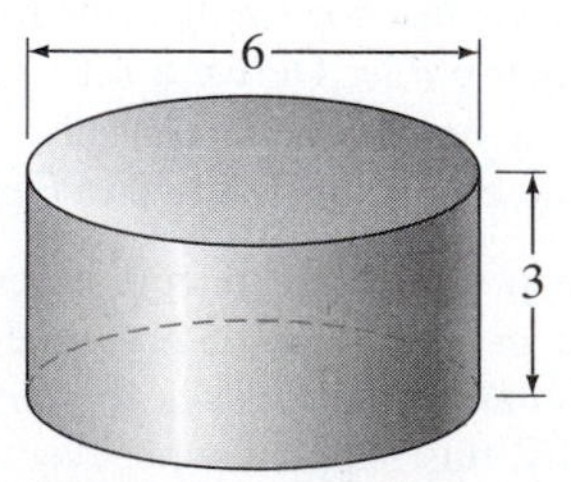

Figure 1.125 The solid cylinder of Exercise 39.

CHAPTER 2

Differentiation in Several Variables

2.1 Functions of Several Variables; Graphing Surfaces

The volume and surface area of a sphere depend on its radius, the formulas describing their relationships being $V = \frac{4}{3}\pi r^3$ and $S = 4\pi r^2$. (Here V and S are, respectively, the volume and surface area of the sphere and r its radius.) These equations define the volume and surface area as **functions** of the radius. The essential characteristic of a function is that the so-called independent variable (in this case the radius) determines a *unique* value of the dependent variable (V or S). No doubt you can think of many quantities that are determined uniquely not by one variable (as the volume of a sphere is determined by its radius), but by several: the area of a rectangle, the volume of a cylinder or cone, the average annual rainfall in Cleveland, or the National Debt. Realistic modeling of the world requires that we understand the concept of a function of more than one variable and how to find meaningful ways to visualize such functions.

Definitions, Notation, and Examples

A function, *any* function, has three features: (1) a **domain** set X, (2) a **codomain** set Y, and (3) a **rule of assignment** that associates to each element x in the domain X a unique element, usually denoted $f(x)$, in the codomain Y. We will frequently use the notation $f: X \to Y$ for a function. Such notation indicates all the ingredients of a particular function, although it does not make the nature of the rule of assignment explicit. This notation also suggests the "mapping" nature of a function, indicated by Figure 2.1.

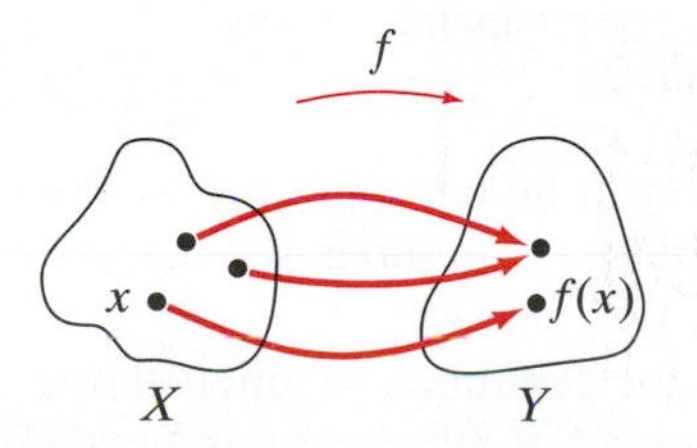

Figure 2.1 The mapping nature of a function.

EXAMPLE 1 Abstract definitions are necessary, but it is just as important that you understand functions as they actually occur. Consider the act of assigning to each U.S. citizen his or her social security number. This pairing defines a function: Each citizen is assigned one social security number. The domain is the set of U.S. citizens and the codomain, the set of all nine-digit strings of numbers.

On the other hand, when a university assigns students to dormitory rooms, it is unlikely that it is creating a function from the set of available rooms to the set of students. This is because some rooms may have more than one student assigned to them, so that a particular room does not necessarily determine a unique student occupant. ◆

■ **Definition 1.1** The **range** of a function $f: X \to Y$ is the set of those elements of Y that are actual values of f. That is, the range of f consists of those y in Y such that $y = f(x)$ for some x in X.

Using set notation, we find that

$$\text{Range } f = \{y \in Y \mid y = f(x) \text{ for some } x \in X\}.$$

In the social security function of Example 1, the range consists of those nine-digit numbers actually used as social security numbers. For example, the number 000-00-0000 is *not* in the range, since no one is actually assigned this number.

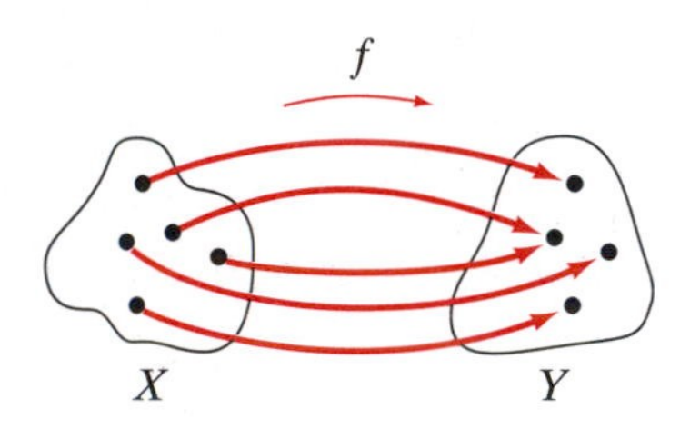

Figure 2.2 Every $y \in Y$ is "hit" by at least one $x \in X$.

■ **Definition 1.2** A function $f: X \to Y$ is said to be **onto** (or **surjective**) if every element of Y is the image of some element of X, that is, if range $f = Y$.

The social security function is *not* onto, since 000-00-0000 is in the codomain, but not in the range. Pictorially, an onto function is suggested by Figure 2.2. A function that is *not* onto looks instead like Figure 2.3. You may find it helpful to think of the codomain of a function f as the set of *possible* (or allowable) values of f, and the range of f as the set of *actual* values attained. Then an onto function is one whose possible and actual values are the same.

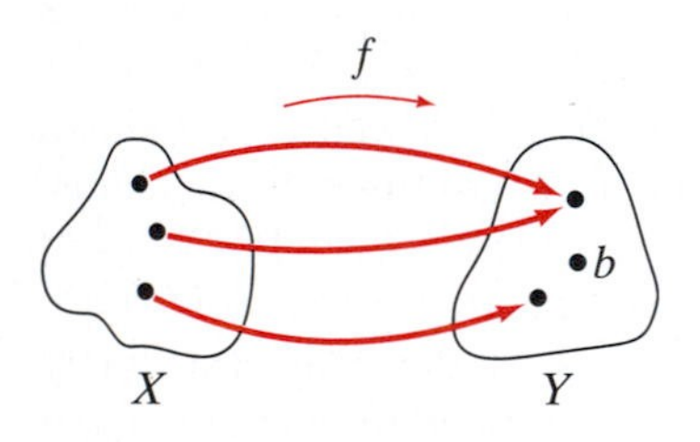

Figure 2.3 The element $b \in Y$ is not the image of any $x \in X$.

■ **Definition 1.3** A function $f: X \to Y$ is called **one-one** (or **injective**) if no two distinct elements of the domain have the same image under f. That is, f is one-one if whenever $x_1, x_2 \in X$ and $x_1 \neq x_2$, then $f(x_1) \neq f(x_2)$. (See Figure 2.4.)

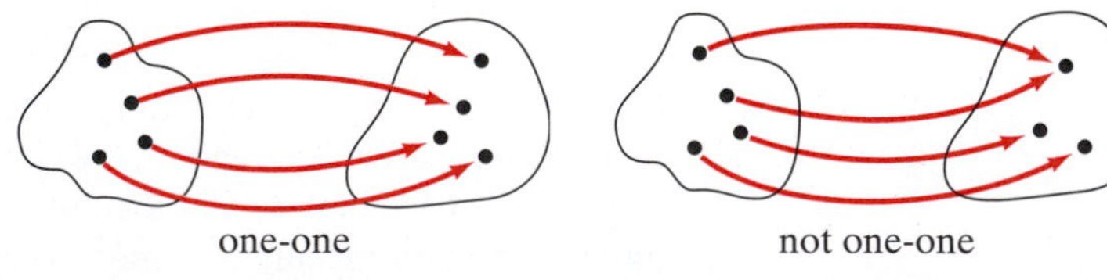

Figure 2.4 The figure on the left depicts a one-one mapping; the one on the right shows a function that is not one-one.

One would expect the social security function to be one-one, but we have heard of cases of two people being assigned the same number so that, alas, apparently it is not.

When you studied single-variable calculus, the functions of interest were those whose domains and codomains were subsets of **R** (the real numbers). It was probably the case that only the rule of assignment was made explicit; it is generally assumed that the domain is the largest possible subset of **R** for which the function makes sense. The codomain is generally taken to be all of **R**.

EXAMPLE 2 Suppose $f: \mathbf{R} \to \mathbf{R}$ is given by $f(x) = x^2$. Then the domain and codomain are, explicitly, all of $\mathbf{R}$, but the range of f is the interval $[0, \infty)$. Thus f is not onto, since the codomain is strictly larger than the range. Note that f is not one-one, since $f(2) = f(-2) = 4$, but $2 \neq -2$. ◆

EXAMPLE 3 Suppose g is a function such that $g(x) = \sqrt{x-1}$. Then if we take the codomain to be all of $\mathbf{R}$, the domain cannot be any larger than $[1, \infty)$. If the domain included any values less than one, the radicand would be negative and, hence, g would not be real-valued. ◆

Now we're ready to think about functions of more than one real variable. In the most general terms, these are the functions whose domains are subsets X of $\mathbf{R}^n$ and whose codomains are subsets of $\mathbf{R}^m$, for some positive integers n and m. (For simplicity of notation, we'll take the codomains to be all of $\mathbf{R}^m$, except when specified otherwise.) That is, such a function is a mapping $\mathbf{f}: X \subseteq \mathbf{R}^n \to \mathbf{R}^m$ that associates to a vector (or point) $\mathbf{x}$ in X a unique vector (point) $\mathbf{f}(\mathbf{x})$ in $\mathbf{R}^m$.

EXAMPLE 4 Let $T: \mathbf{R}^3 \to \mathbf{R}$ be defined by $T(x, y, z) = xy + xz + yz$. We can think of T as a sort of "temperature function." Given a point $\mathbf{x} = (x, y, z)$ in $\mathbf{R}^3$, $T(\mathbf{x})$ calculates the temperature at that point. ◆

EXAMPLE 5 Let $L: \mathbf{R}^n \to \mathbf{R}$ be given by $L(\mathbf{x}) = \|\mathbf{x}\|$. This is a "length function" in that it computes the length of any vector $\mathbf{x}$ in $\mathbf{R}^n$. Note that L is not one-one, since $L(\mathbf{e}_i) = L(\mathbf{e}_j) = 1$, where $\mathbf{e}_i$ and $\mathbf{e}_j$ are any two of the standard basis vectors for $\mathbf{R}^n$. L also fails to be onto, since the length of a vector is always nonnegative. ◆

EXAMPLE 6 Consider the function given by $\mathbf{N}(\mathbf{x}) = \mathbf{x}/\|\mathbf{x}\|$ where $\mathbf{x}$ is a vector in $\mathbf{R}^3$. Note that $\mathbf{N}$ is not defined if $\mathbf{x} = \mathbf{0}$, so the largest possible domain for $\mathbf{N}$ is $\mathbf{R}^3 - \{\mathbf{0}\}$. The range of $\mathbf{N}$ consists of all unit vectors in $\mathbf{R}^3$. The function $\mathbf{N}$ is the "normalization function," that is, the function that takes a nonzero vector in $\mathbf{R}^3$ and returns the unit vector that points in the same direction. ◆

EXAMPLE 7 Sometimes a function may be given numerically by a table. One such example is the notion of **windchill**—the apparent temperature one feels when taking into account both the actual air temperature and the speed of the wind. A standard table of windchill values is shown in Figure 2.5.[1] From it we see that if the air temperature is 20 °F and the windspeed is 25 mph, the windchill temperature ("how cold it feels") is 3 °F. Similarly, if the air temperature is 35 °F and the windspeed is 10 mph, then the windchill is 27 °F. In other words, if s denotes windspeed and t air temperature, then the windchill is a function $W(s, t)$. ◆

The functions described in Examples 4, 5, and 7 are **scalar-valued** functions, that is, functions whose codomains are $\mathbf{R}$ or subsets of $\mathbf{R}$. Scalar-valued functions are our main concern for this chapter. Nonetheless, let's look at a few examples of functions whose codomains are $\mathbf{R}^m$ where $m > 1$.

---

[1]NOAA, National Weather Service, Office of Climate, Water, and Weather Services, "NWS Wind Chill Temperature Index." February 26, 2004. <http://www.nws.noaa.gov/om/ windchill> (June 30, 2004).

Air Temp (deg F)	Windspeed (mph)											
	5	10	15	20	25	30	35	40	45	50	55	60
40	36	34	32	30	29	28	28	27	26	26	25	25
35	31	27	25	24	23	22	21	20	19	19	18	17
30	25	21	19	17	16	15	14	13	12	12	11	10
25	19	15	13	11	9	8	7	6	5	4	4	3
20	13	9	6	4	3	1	0	−1	−2	−3	−3	−4
15	7	3	0	−2	−4	−5	−7	−8	−9	−10	−11	−11
10	1	−4	−7	−9	−11	−12	−14	−15	−16	−17	−18	−19
5	−5	−10	−13	−15	−17	−19	−21	−22	−23	−24	−25	−26
0	−11	−16	−19	−22	−24	−26	−27	−29	−30	−31	−32	−33
−5	−16	−22	−26	−29	−31	−33	−34	−36	−37	−38	−39	−40
−10	−22	−28	−32	−35	−37	−39	−41	−43	−44	−45	−46	−48
−15	−28	−35	−39	−42	−44	−46	−48	−50	−51	−52	−54	−55
−20	−34	−41	−45	−48	−51	−53	−55	−57	−58	−60	−61	−62
−25	−40	−47	−51	−55	−58	−60	−62	−64	−65	−67	−68	−69
−30	−46	−53	−58	−61	−64	−67	−69	−71	−72	−74	−75	−76
−35	−52	−59	−64	−68	−71	−73	−76	−78	−79	−81	−82	−84
−40	−57	−66	−71	−74	−78	−80	−82	−84	−86	−88	−89	−91
−45	−63	−72	−77	−81	−84	−87	−89	−91	−93	−95	−97	−98

Figure 2.5 Table of windchill values in English units.

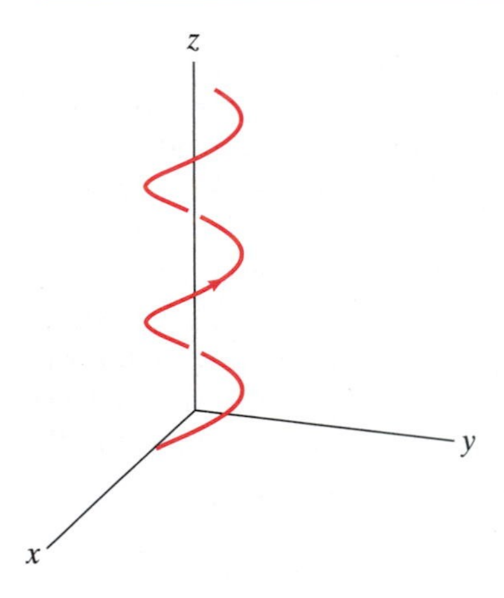

Figure 2.6 The helix of Example 8. The arrow shows the direction of increasing t.

EXAMPLE 8 Define $\mathbf{f}\colon \mathbf{R} \to \mathbf{R}^3$ by $\mathbf{f}(t) = (\cos t, \sin t, t)$. The range of f is the curve in $\mathbf{R}^3$ with parametric equations $x = \cos t$, $y = \sin t$, $z = t$. If we think of t as a time parameter, then this function traces out the corkscrew curve (called a **helix**) shown in Figure 2.6. ◆

EXAMPLE 9 We can think of the velocity of a fluid as a vector in $\mathbf{R}^3$. This vector depends on (at least) the point at which one measures the velocity and also the time at which one makes the measurement. In other words, velocity may be considered to be a function $\mathbf{v}\colon X \subseteq \mathbf{R}^4 \to \mathbf{R}^3$. The domain X is a subset of $\mathbf{R}^4$ because three variables x, y, z are required to describe a point in the fluid and a fourth variable t is needed to keep track of time. (See Figure 2.7.) For instance, such a function $\mathbf{v}$ might be given by the expression

$$\mathbf{v}(x, y, z, t) = xyzt\mathbf{i} + (x^2 - y^2)\mathbf{j} + (3z + t)\mathbf{k}. \quad ◆$$

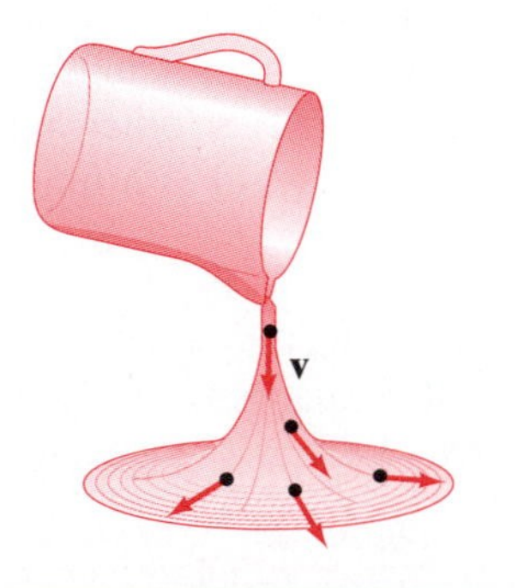

Figure 2.7 A water pitcher. The velocity $\mathbf{v}$ of the water is a function from a subset of $\mathbf{R}^4$ to $\mathbf{R}^3$.

You may have noted that the expression for $\mathbf{v}$ in Example 9 is considerably more complicated than those for the functions given in Examples 4–8. This is because all the variables and vector components have been written out explicitly. In general, if we have a function $\mathbf{f}\colon X \subseteq \mathbf{R}^n \to \mathbf{R}^m$, then $\mathbf{x} \in X$ can be written as $\mathbf{x} = (x_1, x_2, \dots, x_n)$ and $\mathbf{f}$ can be written in terms of its **component functions** $f_1, f_2, \dots, f_m$. The component functions are **scalar-valued** functions of $\mathbf{x} \in X$ that define the components of the vector $\mathbf{f}(\mathbf{x}) \in \mathbf{R}^m$. What results is a morass of symbols:

$$\begin{aligned}
\mathbf{f}(\mathbf{x}) &= \mathbf{f}(x_1, x_2, \dots, x_n) && \text{(emphasizing the variables)}\\
&= (f_1(\mathbf{x}), f_2(\mathbf{x}), \dots, f_m(\mathbf{x})) && \text{(emphasizing the component functions)}\\
&= (f_1(x_1, x_2, \dots, x_n), f_2(x_1, x_2, \dots, x_n), \dots, f_m(x_1, x_2, \dots, x_n)) \\
& && \text{(writing out all components).}
\end{aligned}$$

For example, the function L of Example 5, when expanded, becomes

$$L(\mathbf{x}) = L(x_1, x_2, \ldots, x_n) = \sqrt{x_1^2 + x_2^2 + \cdots + x_n^2}.$$

The function $\mathbf{N}$ of Example 6 becomes

$$\mathbf{N}(\mathbf{x}) = \frac{\mathbf{x}}{\|\mathbf{x}\|} = \frac{(x_1, x_2, x_3)}{\sqrt{x_1^2 + x_2^2 + x_3^2}}$$

$$= \left(\frac{x_1}{\sqrt{x_1^2 + x_2^2 + x_3^2}}, \frac{x_2}{\sqrt{x_1^2 + x_2^2 + x_3^2}}, \frac{x_3}{\sqrt{x_1^2 + x_2^2 + x_3^2}} \right).$$

Although writing a function in terms of all its variables and components has the advantage of being explicit, quite a lot of paper and ink are used in the process. The use of vector notation not only saves space and trees, but helps to make the meaning of a function clear by emphasizing that a function maps points in $\mathbf{R}^n$ to points in $\mathbf{R}^m$. Vector notation makes a function of 300 variables look "just like" a function of one variable. Try to avoid writing out components as much as you can (except when you want to impress your friends).

Visualizing Functions

No doubt you have been graphing scalar-valued functions of one variable for so long that you give the matter little thought. Let's scrutinize what you've been doing, however. A function $f: X \subseteq \mathbf{R} \to \mathbf{R}$ takes a real number and returns another real number as suggested by Figure 2.8. The **graph** of f is something that "lives" in $\mathbf{R}^2$. (See Figure 2.9.) It consists of points (x, y) such that $y = f(x)$. That is,

$$\text{Graph } f = \{(x, f(x)) \mid x \in X\} = \{(x, y) \mid x \in X, y = f(x)\}.$$

The important fact is that, in general, the graph of a scalar-valued function of a single variable is a curve—a one-dimensional object—sitting inside two-dimensional space.

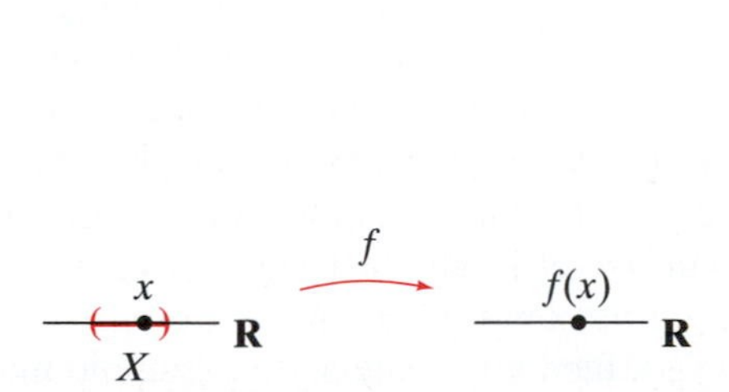

Figure 2.8 A function $f: X \subseteq \mathbf{R} \to \mathbf{R}$.

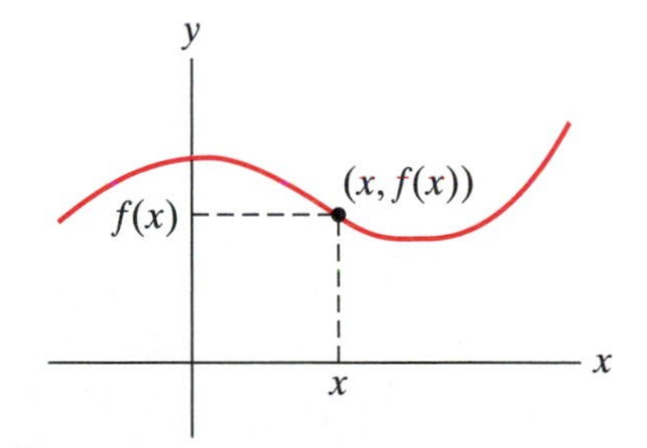

Figure 2.9 The graph of f.

Now suppose we have a function $f: X \subseteq \mathbf{R}^2 \to \mathbf{R}$, that is, a function of two variables. We make essentially the same definition for the graph:

$$\text{Graph } f = \{(\mathbf{x}, f(\mathbf{x})) \mid \mathbf{x} \in X\}. \tag{1}$$

Of course, $\mathbf{x} = (x, y)$ is a point of $\mathbf{R}^2$. Thus, $\{(\mathbf{x}, f(\mathbf{x}))\}$ may also be written as

$$\{(x, y, f(x, y))\}, \quad \text{or as} \quad \{(x, y, z) \mid (x, y) \in X, z = f(x, y)\}.$$

Hence, the graph of a scalar-valued function of two variables is something that sits in $\mathbf{R}^3$. Generally speaking, the graph will be a surface.

EXAMPLE 10 The graph of the function

$$f: \mathbf{R}^2 \to \mathbf{R}, \qquad f(x, y) = \frac{1}{12}y^3 - y - \frac{1}{4}x^2 + \frac{7}{2}$$

is shown in Figure 2.10. For each point $\mathbf{x} = (x, y)$ in $\mathbf{R}^2$, the point in $\mathbf{R}^3$ with coordinates $\left(x, y, \frac{1}{12}y^3 - y - \frac{1}{4}x^2 + \frac{7}{2}\right)$ is graphed. ◆

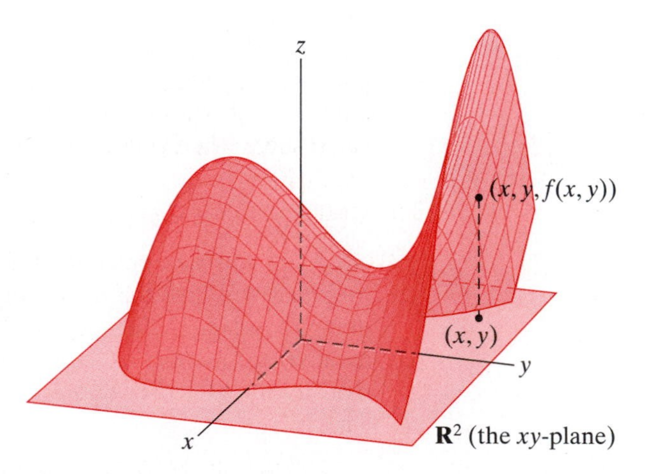

Figure 2.10 The graph of $f(x, y) = \frac{1}{12}y^3 - y - \frac{1}{4}x^2 + \frac{7}{2}$.

Graphing functions of two variables is a much more difficult task than graphing functions of one variable. Of course, one method is to let a computer do the work. Nonetheless, if you want to get a feeling for functions of more than one variable, being able to sketch a rough graph by hand is still a valuable skill. The trick to putting together a reasonable graph is to find a way to cut down on the dimensions involved. One way this can be achieved is by drawing certain special curves that lie on the surface $z = f(x, y)$. These special curves, called **contour curves,** are the ones obtained by intersecting the surface with horizontal planes $z = c$ for various values of the constant c. Some contour curves drawn on the surface of Example 10 are shown in Figure 2.11. If we compress all the contour curves onto the xy-plane (in essence, if we look down along the positive z-axis), then we create a "topographic map" of the surface that is shown in Figure 2.12. These curves in the xy-plane are called the **level curves** of the original function f.

The point of the preceding discussion is that we can reverse the process in order to sketch systematically the graph of a function f of two variables: We first construct a topographic map in $\mathbf{R}^2$ by finding the level curves of f, then situate these curves in $\mathbf{R}^3$ as contour curves at the appropriate heights, and finally complete the graph of the function. Before we give an example, let's restate our terminology with greater precision.

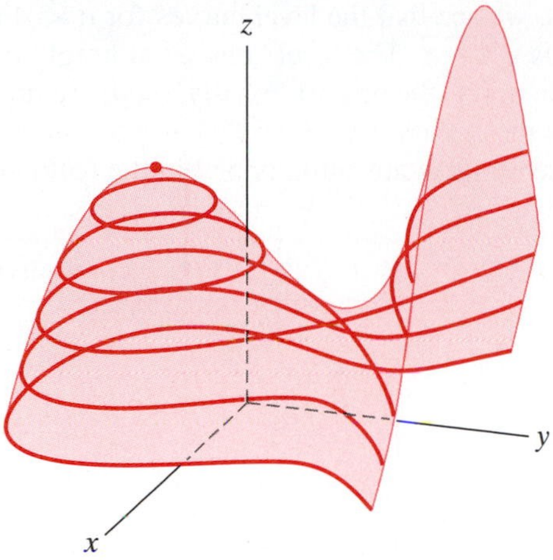

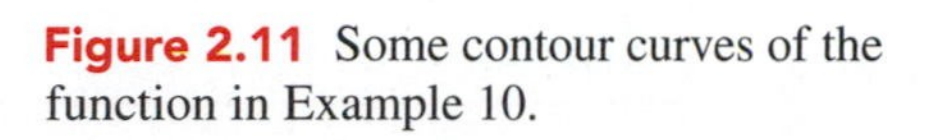

Figure 2.11 Some contour curves of the function in Example 10.

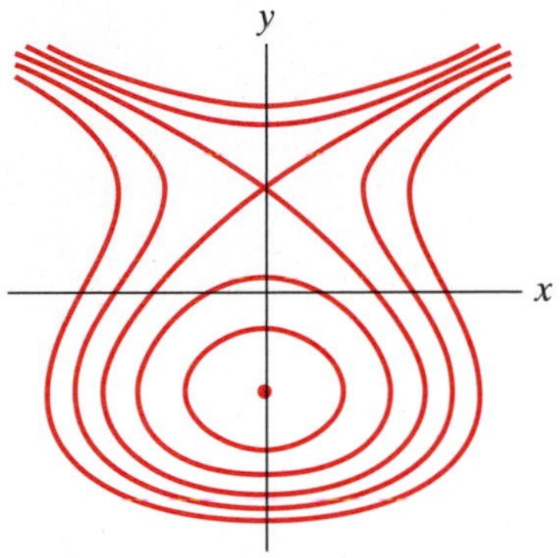

Figure 2.12 Some level curves of the function in Example 10.

■ **Definition 1.4** Let $f: X \subseteq \mathbf{R}^2 \to \mathbf{R}$ be a scalar-valued function of two variables. The **level curve at height** c **of** f is the curve in $\mathbf{R}^2$ defined by the equation $f(x, y) = c$, where c is a constant. In mathematical notation,

$$\text{Level curve at height } c = \left\{(x, y) \in \mathbf{R}^2 \mid f(x, y) = c\right\}.$$

The **contour curve at height** c **of** f is the curve in $\mathbf{R}^3$ defined by the two equations $z = f(x, y)$ and $z = c$. Symbolized,

$$\text{Contour curve at height } c = \left\{(x, y, z) \in \mathbf{R}^3 \mid z = f(x, y) = c\right\}.$$

In addition to level and contour curves, consideration of the **sections** of a surface by the planes where x or y is held constant is also helpful. A **section** of a surface by a plane is just the intersection of the surface with that plane. Formally, we have the following definition:

■ **Definition 1.5** Let $f: X \subseteq \mathbf{R}^2 \to \mathbf{R}$ be a scalar-valued function of two variables. The **section of the graph of** f **by the plane** $x = c$ (where c is a constant) is the set of points (x, y, z), where $z = f(x, y)$ and $x = c$. Symbolized,

$$\text{Section by } x = c \text{ is } \left\{(x, y, z) \in \mathbf{R}^3 \mid z = f(x, y), x = c\right\}.$$

Similarly, the **section of the graph of** f **by the plane** $y = c$ is the set of points described as follows:

$$\text{Section by } y = c \text{ is } \left\{(x, y, z) \in \mathbf{R}^3 \mid z = f(x, y), y = c\right\}.$$

EXAMPLE 11 We'll use level and contour curves to construct the graph of the function

$$f: \mathbf{R}^2 \to \mathbf{R}, \qquad f(x, y) = 4 - x^2 - y^2.$$

By Definition 1.4, the level curve at height c is

$$\left\{(x, y) \in \mathbf{R}^2 \mid 4 - x^2 - y^2 = c\right\} = \left\{(x, y) \mid x^2 + y^2 = 4 - c\right\}.$$

Thus, we see that the level curves for $c < 4$ are circles centered at the origin of radius $\sqrt{4-c}$. The level "curve" at height $c = 4$ is not a curve at all, but just a single point (the origin). Finally, there are no level curves at heights larger than 4 since the equation $x^2 + y^2 = 4 - c$ has no real solutions in x and y. (Why not?) These remarks are summarized in the following table:

c	Level curve $x^2 + y^2 = 4 - c$
-5	$x^2 + y^2 = 9$
-1	$x^2 + y^2 = 5$
0	$x^2 + y^2 = 4$
1	$x^2 + y^2 = 3$
3	$x^2 + y^2 = 1$
4	$x^2 + y^2 = 0 \iff x = y = 0$
c, where $c > 4$	empty

Thus, the family of level curves, the "topographic map" of the surface $z = 4 - x^2 - y^2$, is shown in Figure 2.13. Some contour curves, *which sit in* $\mathbf{R}^3$, are shown in Figure 2.14, where we can get a feeling for the complete graph of $z = 4 - x^2 - y^2$. It is a surface that looks like an inverted dish and is called a **paraboloid**. (See Figure 2.15.) To make the picture clearer, we have also sketched in the sections of the surface by the planes $x = 0$ and $y = 0$. The section by $x = 0$ is given analytically by the set

$$\{(x, y, z) \in \mathbf{R}^3 \mid z = 4 - x^2 - y^2, x = 0\} = \{(0, y, z) \mid z = 4 - y^2\}.$$

Similarly, the section by $y = 0$ is

$$\{(x, y, z) \in \mathbf{R}^3 \mid z = 4 - x^2 - y^2, y = 0\} = \{(x, 0, z) \mid z = 4 - x^2\}.$$

Since these sections are parabolas, it is obvious to see how this surface obtained its name. ◆

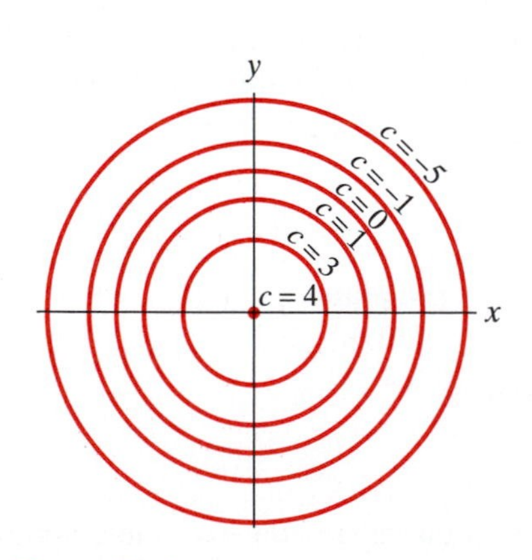

Figure 2.13 The topographic map of $z = 4 - x^2 - y^2$ (i.e., several of its level curves).

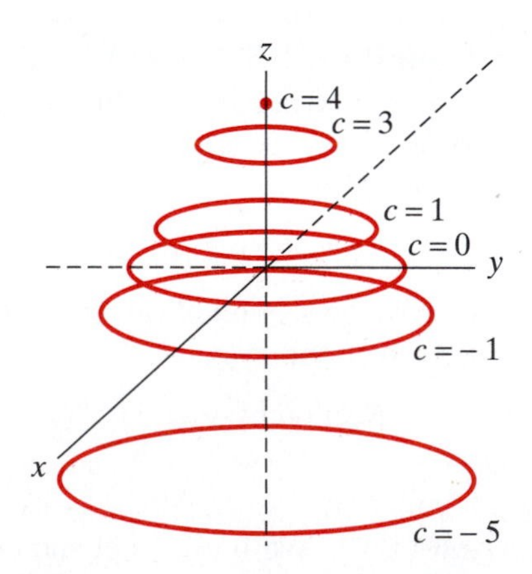

Figure 2.14 Some contour curves of $z = 4 - x^2 - y^2$.

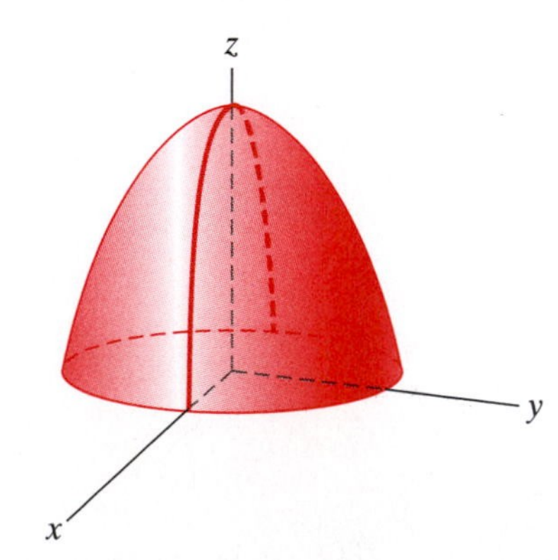

Figure 2.15 The graph of $f(x, y) = 4 - x^2 - y^2$.

EXAMPLE 12 We'll graph the function $g\colon \mathbf{R}^2 \to \mathbf{R}$, $g(x, y) = y^2 - x^2$. The level curves are all hyperbolas, with the exception of the level curve at height 0, which is a pair of intersecting lines.

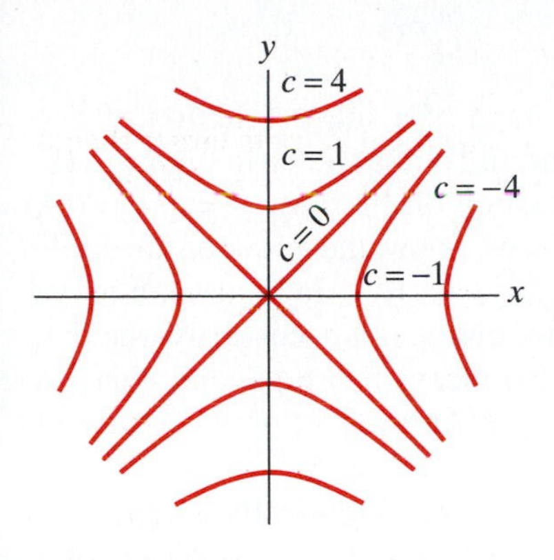

Figure 2.16 Some level curves of $g(x, y) = y^2 - x^2$.

c	Level curve $y^2 - x^2 = c$
-4	$x^2 - y^2 = 4$
-1	$x^2 - y^2 = 1$
0	$y^2 - x^2 = 0 \iff (y - x)(y + x) = 0 \iff y = \pm x$
1	$y^2 - x^2 = 1$
4	$y^2 - x^2 = 4$

The collection of level curves is graphed in Figure 2.16. The sections by $x = c$ are

$$\{(x, y, z) \mid z = y^2 - x^2, x = c\} = \{(c, y, z) \mid z = y^2 - c^2\}.$$

These are clearly parabolas in the planes $x = c$. The sections by $y = c$ are

$$\{(x, y, z) \mid z = y^2 - x^2, y = c\} = \{(c, y, z) \mid z = c^2 - x^2\},$$

which are again parabolas. The level curves and sections generate the contour curves and surface depicted in Figure 2.17. Perhaps understandably, this surface is called a **hyperbolic paraboloid**. ◆

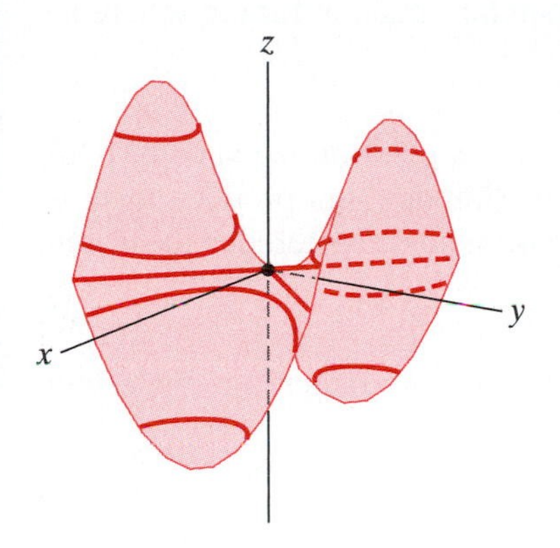

Figure 2.17 The contour curves and graph of $g(x, y) = y^2 - x^2$.

The preceding discussion has been devoted entirely to graphing scalar-valued functions of just two variables. However, all the ideas can be extended to more variables and higher dimensions. If $f: X \subseteq \mathbf{R}^n \to \mathbf{R}$ is a (scalar-valued) function of n variables, then the **graph** of f is the subset of $\mathbf{R}^{n+1}$ given by

$$\begin{aligned}\text{Graph } f &= \{(\mathbf{x}, f(\mathbf{x})) \mid \mathbf{x} \in X\} \\ &= \{(x_1, \ldots, x_n, x_{n+1}) \mid (x_1, \ldots, x_n) \in X, \\ &\qquad x_{n+1} = f(x_1, \ldots, x_n)\}.\end{aligned} \tag{2}$$

(The compactness of vector notation makes the definition of the graph of a function of n variables *exactly* the same as in (1).) The **level set at height** c of such a function is defined by

$$\begin{aligned}\text{Level set at height } c &= \{\mathbf{x} \in \mathbf{R}^n \mid f(\mathbf{x}) = c\} \\ &= \{(x_1, x_2, \ldots, x_n) \mid f(x_1, x_2, \ldots, x_n) = c\}.\end{aligned}$$

While the graph of f is a subset of $\mathbf{R}^{n+1}$, a level set of f is a subset of $\mathbf{R}^n$. This makes it possible to get some geometric insight into graphs of functions of *three* variables, even though we cannot actually visualize them.

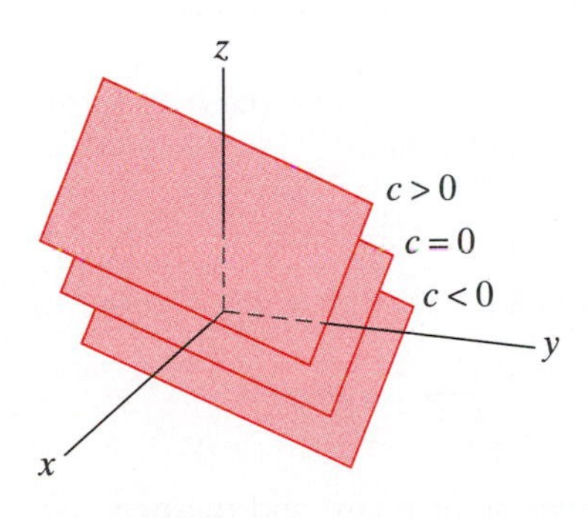

Figure 2.18 The level sets of the function $F(x, y, z) = x + y + z$ are planes in $\mathbf{R}^3$.

EXAMPLE 13 Let $F: \mathbf{R}^3 \to \mathbf{R}$ be given by $F(x, y, z) = x + y + z$. Then the graph of F is the set $\{(x, y, z, w) \mid w = x + y + z\}$ and is a subset (called a **hypersurface**) of $\mathbf{R}^4$, which we cannot depict adequately. Nonetheless, we can look at the level sets of F, which are surfaces in $\mathbf{R}^3$. (See Figure 2.18.) We have

$$\text{Level set at height } c = \{(x, y, z) \mid x + y + z = c\}.$$

Thus, the level sets form a family of parallel planes with normal vector $\mathbf{i} + \mathbf{j} + \mathbf{k}$. ◆

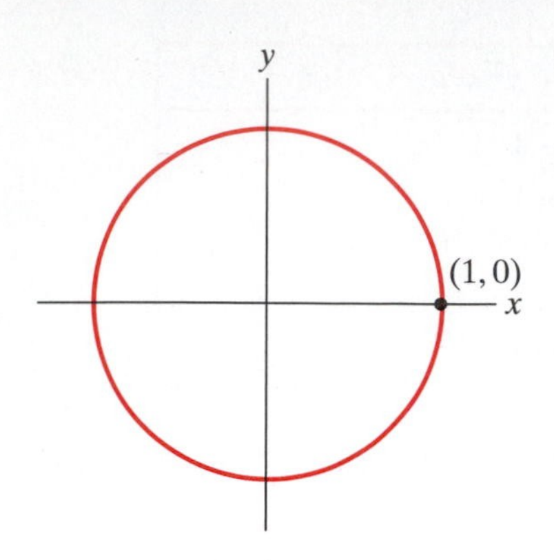

Figure 2.19 The unit circle $x^2 + y^2 = 1$.

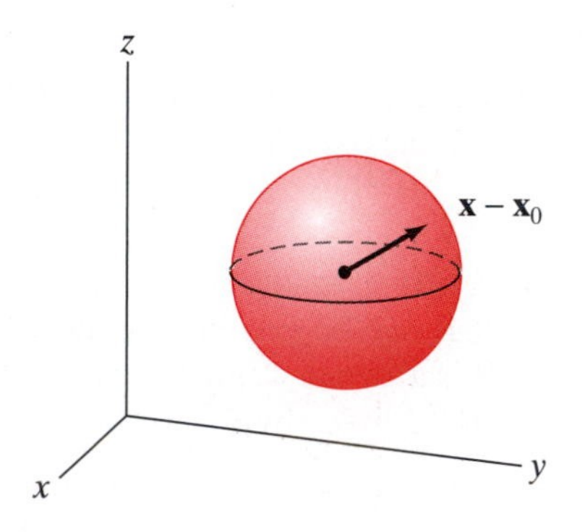

Figure 2.20 The sphere of radius a, centered at (x_0, y_0, z_0).

Surfaces in General

Not all curves in $\mathbf{R}^2$ can be described as the graph of a single function of one variable. Perhaps the most familiar example is the unit circle shown in Figure 2.19. Its graph *cannot* be determined by a single equation of the form $y = f(x)$ (or, for that matter, by one of the form $x = g(y)$). As we know, the graph of the circle may be described analytically by the equation $x^2 + y^2 = 1$. In general, a curve in $\mathbf{R}^2$ is determined by an arbitrary equation in x and y, not necessarily one that isolates y alone on one side. In other words, this means that a general curve is given by an equation of the form $F(x, y) = c$ (i.e., a level set of a function of *two* variables).

The analogous situation occurs with surfaces in $\mathbf{R}^3$. Frequently a surface is determined by an equation of the form $F(x, y, z) = c$ (i.e., as a level set of a function of three variables), *not* necessarily one of the form $z = f(x, y)$.

EXAMPLE 14 A **sphere** is a surface in $\mathbf{R}^3$ whose points are all equidistant from a fixed point. If this fixed point is the origin, then the equation for the sphere is

$$\|\mathbf{x} - \mathbf{0}\| = \|\mathbf{x}\| = a, \tag{3}$$

where a is a positive constant and $\mathbf{x} = (x, y, z)$ is a point on the sphere. If we square both sides of equation (3) and expand the (implicit) dot product, then we obtain the perhaps familiar equation of a sphere of radius a centered at the origin:

$$x^2 + y^2 + z^2 = a^2. \tag{4}$$

If the center of the sphere is at the point $\mathbf{x}_0 = (x_0, y_0, z_0)$, rather than the origin, then equation (3) should be modified to

$$\|\mathbf{x} - \mathbf{x}_0\| = a. \tag{5}$$

(See Figure 2.20.)

When equation (5) is expanded, the following general equation for a sphere is obtained:

$$(x - x_0)^2 + (y - y_0)^2 + (z - z_0)^2 = a^2 \tag{6}$$

In the equation for a sphere, there is no way to solve for z *uniquely* in terms of x and y. Indeed, if we try to isolate z in equation (4), then

$$z^2 = a^2 - x^2 - y^2,$$

so we are forced to make a *choice* of positive or negative square roots in order to solve for z:

$$z = \sqrt{a^2 - x^2 - y^2} \qquad \text{or} \qquad z = -\sqrt{a^2 - x^2 - y^2}.$$

The positive square root corresponds to the upper hemisphere and the negative square root to the lower one. In any case, the entire sphere *cannot* be the graph of a single function of two variables. ◆

Of course, the graph of a function of two variables does describe a surface in the "level set" sense. If a surface happens to be given by an equation of the form

$$z = f(x, y)$$

for some appropriate function $f: X \subseteq \mathbf{R}^2 \to \mathbf{R}$, then we can move z to the opposite side, obtaining

$$f(x, y) - z = 0.$$

If we define a new function F of *three* variables by

$$F(x, y, z) = f(x, y) - z,$$

then the graph of f is precisely the level set at height 0 of F. We reiterate this point since it is all too often forgotten: *The graph of a function of two variables is a surface in* $\mathbf{R}^3$ *and is a level set of a function of three variables. However, not all level sets of functions of three variables are graphs of functions of two variables.* We urge you to understand this distinction.

Quadric Surfaces

Conic sections, those curves obtained from the intersection of a cone with various planes, are among the simplest, yet also the most interesting, of plane curves: They are the circle, the ellipse, the parabola, and the hyperbola. Besides being produced in a similar geometric manner, conic sections have an elegant algebraic connection: Every conic section is described analytically by a polynomial equation of degree two in two variables. That is, every conic can be described by an equation that looks like

$$Ax^2 + Bxy + Cy^2 + Dx + Ey + F = 0$$

for suitable constants $A, \ldots, F$.

In $\mathbf{R}^3$, the analytic analogue of the conic section is called a **quadric surface**. Quadric surfaces are those defined by equations that are polynomials of degree two in three variables:

$$Ax^2 + Bxy + Cxz + Dy^2 + Eyz + Fz^2 + Gx + Hy + Iz + J = 0.$$

To pass from this equation to the appropriate graph is, in general, a cumbersome process without the aid of either a computer or more linear algebra than we currently have at our disposal. So, instead, we offer examples of those quadric surfaces whose axes of symmetry lie along the coordinate axes in $\mathbf{R}^3$ and whose corresponding analytic equations are relatively simple. In the discussion that follows, a, b, and c are constants, which, for convenience, we take to be positive.

Ellipsoid (Figure 2.21.) $x^2/a^2 + y^2/b^2 + z^2/c^2 = 1$.
This is the three-dimensional analogue of an ellipse in the plane. The sections of the ellipsoid by planes perpendicular to the coordinate axes are all ellipses.

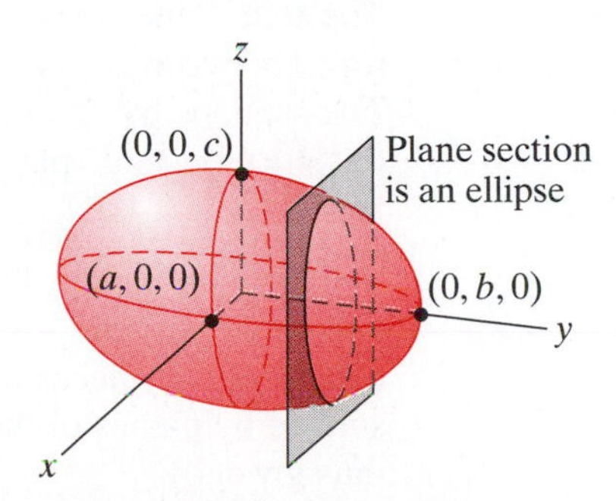

Figure 2.21 The ellipsoid
$$\frac{x^2}{a^2} + \frac{y^2}{b^2} + \frac{z^2}{c^2} = 1.$$

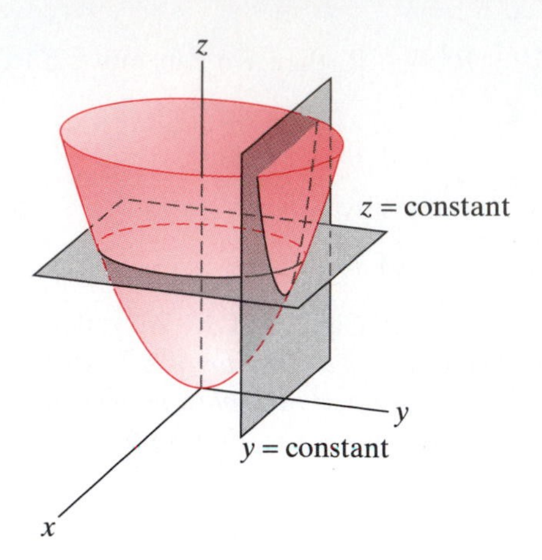

Figure 2.22 The elliptic paraboloid $\frac{z}{c} = \frac{x^2}{a^2} + \frac{y^2}{b^2}$.

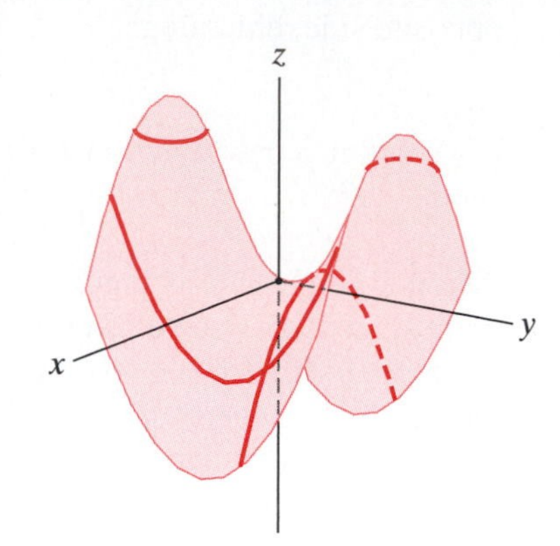

Figure 2.23 The hyperbolic paraboloid $\frac{z}{c} = \frac{y^2}{b^2} - \frac{x^2}{a^2}$.

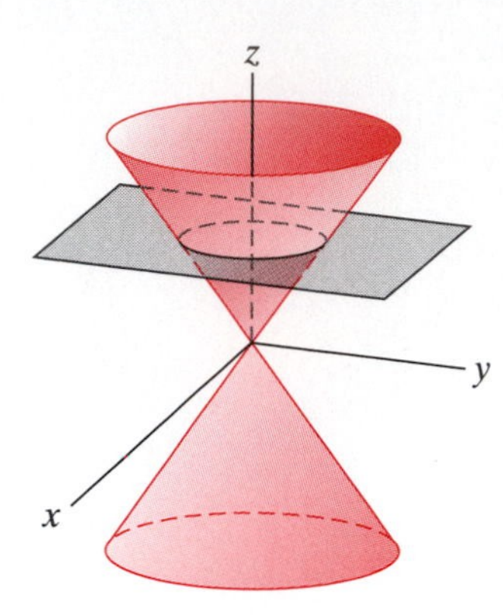

Figure 2.24 The elliptic cone $\frac{z^2}{c^2} = \frac{x^2}{a^2} + \frac{y^2}{b^2}$.

For example, if the ellipsoid is intersected with the plane $z = 0$, one obtains the standard ellipse $x^2/a^2 + y^2/b^2 = 1$, $z = 0$. If $a = b = c$, then the ellipsoid is a sphere of radius a.

Elliptic paraboloid (Figure 2.22.) $z/c = x^2/a^2 + y^2/b^2$.
(The roles of x, y, and z may be interchanged.) This surface is the graph of a function of x and y. The paraboloid has elliptical (or empty) sections by the planes "$z =$ constant" and parabolic sections by "$x =$ constant" or "$y =$ constant" planes. The constants a and b affect the aspect ratio of the elliptical cross-sections, and the constant c affects the steepness of the dish. (Larger values of c produce steeper paraboloids.)

Hyperbolic paraboloid (Figure 2.23.) $z/c = y^2/b^2 - x^2/a^2$.
(Again the roles of x, y, and z may be interchanged.) We saw the graph of this surface earlier in Example 12 of this section. It is shaped like a saddle whose "$x =$ constant" or "$y =$ constant" sections are parabolas and "$z =$ constant" sections are hyperbolas.

Elliptic cone (Figure 2.24.) $z^2/c^2 = x^2/a^2 + y^2/b^2$.
The sections by "$z =$ constant" planes are ellipses. The sections by $x = 0$ or $y = 0$ are each a pair of intersecting lines.

Hyperboloid of one sheet (Figure 2.25.) $x^2/a^2 + y^2/b^2 - z^2/c^2 = 1$.
The term "one sheet" signifies that the surface is *connected* (i.e., that you can travel between any two points on the surface without having to leave the surface). The sections by "$z =$ constant" planes are ellipses and those by "$x =$ constant" or "$y =$ constant" planes are hyperbolas, hence, this surface's name.

Hyperboloid of two sheets (Figure 2.26.) $z^2/c^2 - x^2/a^2 - y^2/b^2 = 1$.
The fact that the left-hand side of the defining equation is the opposite of the left side of the equation for the previous hyperboloid is what causes this surface to consist of two pieces instead of one. More precisely, consider the sections of the surface by planes of the form $z = k$ for different constants k. These sections are thus given by

$$\frac{k^2}{c^2} - \frac{x^2}{a^2} - \frac{y^2}{b^2} = 1, \quad z = k$$

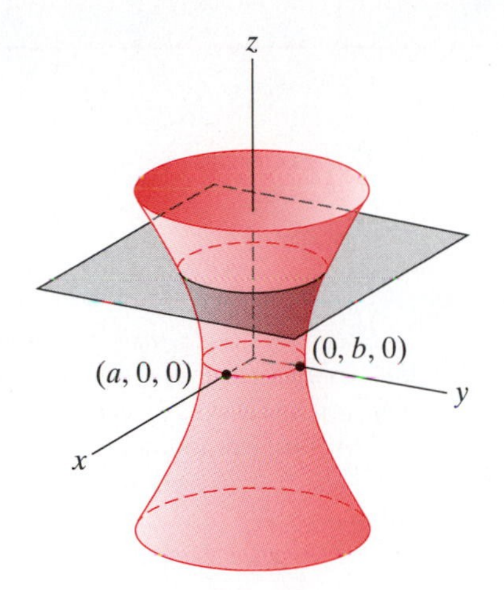

Figure 2.25 The graph of the equation $\frac{x^2}{a^2} + \frac{y^2}{b^2} - \frac{z^2}{c^2} = 1$ is a hyperboloid of one sheet.

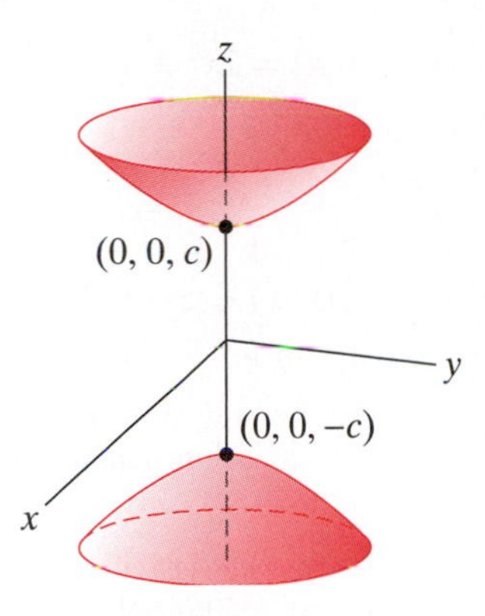

Figure 2.26 The graph of the equation $\frac{z^2}{c^2} - \frac{x^2}{a^2} - \frac{y^2}{b^2} = 1$ is a hyperboloid of two sheets.

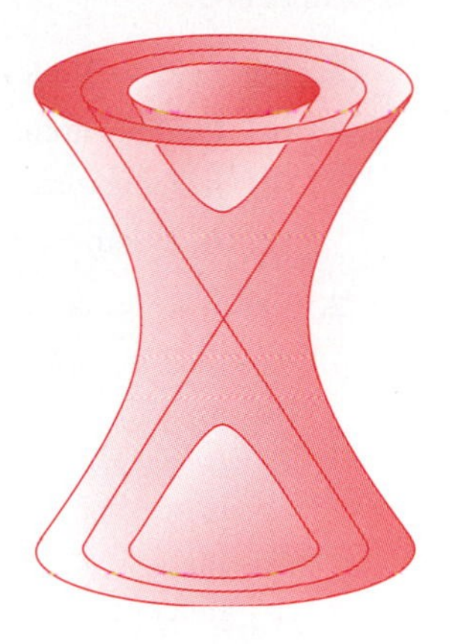

Figure 2.27 The hyperboloids $\frac{z^2}{c^2} = \frac{x^2}{a^2} + \frac{y^2}{b^2} \pm 1$ are asymptotic to the cone $\frac{z^2}{c^2} = \frac{x^2}{a^2} + \frac{y^2}{b^2}$.

or, equivalently, by

$$\frac{x^2}{a^2} + \frac{y^2}{b^2} = \frac{k^2}{c^2} - 1, \quad z = k.$$

If $-c < k < c$, then $0 \le k^2/c^2 < 1$. Thus, $k^2/c^2 - 1 < 0$, and so the preceding equation has no solution in x and y. Hence, the section by $z = k$, where $|k| < c$, is empty. If $|k| > c$, then the section is an ellipse. The sections by "$x =$ constant" or "$y =$ constant" planes are hyperbolas.

In the same way that the hyperbolas

$$\frac{x^2}{a^2} - \frac{y^2}{b^2} = \pm 1$$

are asymptotic to the lines $y = \pm(b/a)x$, the hyperboloids

$$\frac{z^2}{c^2} = \frac{x^2}{a^2} + \frac{y^2}{b^2} \pm 1$$

are asymptotic to the cone

$$\frac{z^2}{c^2} = \frac{x^2}{a^2} + \frac{y^2}{b^2}.$$

This is perhaps intuitively clear from Figure 2.27, but let's see how to prove it rigorously. In our present context, to say that the hyperboloids are asymptotic to the cones means that they look more and more like the cones as $|z|$ becomes (arbitrarily) large. Analytically, this should mean that the equations for the hyperboloids should approximate the equation for the cone for sufficiently large $|z|$. The equations of the hyperboloids can be written as follows:

$$\frac{x^2}{a^2} + \frac{y^2}{b^2} = \frac{z^2}{c^2} \pm 1 = \frac{z^2}{c^2}\left(1 \pm \frac{c^2}{z^2}\right).$$

As $|z| \to \infty$, $c^2/z^2 \to 0$, so the right side of the equation for the hyperboloids approaches z^2/c^2. Hence, the equations for the hyperboloids approximate that of the cone, as desired.

Exercises

1. Let $f: \mathbf{R} \to \mathbf{R}$ be given by $f(x) = 2x^2 + 1$.

(a) Find the domain and range of f.

(b) Is f one-one?

(c) Is f onto?

2. Let $g: \mathbf{R}^2 \to \mathbf{R}$ be given by $g(x, y) = 2x^2 + 3y^2 - 7$.

(a) Find the domain and range of g.

(b) Find a way to restrict the domain to make a new function with the same rule of assignment as g that is one-one.

(c) Find a way to restrict the codomain to make a new function with the same rule of assignment as g that is onto.

Find the domain and range of each of the functions given in Exercises 3–7.

3. $f(x, y) = \dfrac{x}{y}$

4. $f(x, y) = \ln(x + y)$

5. $g(x, y, z) = \sqrt{x^2 + (y - 2)^2 + (z + 1)^2}$

6. $g(x, y, z) = \dfrac{1}{\sqrt{4 - x^2 - y^2 - z^2}}$

7. $\mathbf{f}(x, y) = \left(x + y, \dfrac{1}{y - 1}, x^2 + y^2\right)$

8. Let $\mathbf{f}: \mathbf{R}^3 \to \mathbf{R}^3$ be defined by $\mathbf{f}(\mathbf{x}) = \mathbf{x} + 3\mathbf{j}$. Write out the component functions of $\mathbf{f}$ in terms of the components of the vector $\mathbf{x}$.

9. Consider the mapping that assigns to a nonzero vector $\mathbf{x}$ in $\mathbf{R}^3$ the vector of length 2 that points in the direction opposite to $\mathbf{x}$.

(a) Give an analytic (symbolic) description of this mapping.

(b) If $\mathbf{x} = (x, y, z)$, determine the component functions of this mapping.

In each of Exercises 10–19, (a) determine several level curves of the given function f (make sure to indicate the height c of each curve); (b) use the information obtained in part (a) to sketch the graph of f.

10. $f(x, y) = 3$

11. $f(x, y) = x^2 + y^2$

12. $f(x, y) = x^2 + y^2 - 9$

13. $f(x, y) = \sqrt{x^2 + y^2}$

14. $f(x, y) = 4x^2 + 9y^2$

15. $f(x, y) = xy$

16. $f(x, y) = \dfrac{y}{x}$

17. $f(x, y) = \dfrac{x}{y}$

18. $f(x, y) = 3 - 2x - y$

19. $f(x, y) = |x|$

In Exercises 20–23, use a computer to provide a portrait of the given function $g(x, y)$. To do this, (a) use the computer to help you understand some of the level curves of the function, and (b) use the computer to graph (a portion of) the surface $z = g(x, y)$. In addition, mark on your surface some of the contour curves corresponding to the level curves you obtained in part (a). (See Figures 2.10 and 2.11.)

20. $g(x, y) = ye^x$

21. $g(x, y) = x^2 - xy$

22. $g(x, y) = (x^2 + 3y^2)e^{1-x^2-y^2}$

23. $g(x, y) = \dfrac{\sin(2 - x^2 - y^2)}{x^2 + y^2 + 1}$

24. The **ideal gas law** is the equation $PV = kT$, where P denotes the pressure of the gas, V the volume, T the temperature, and k is a positive constant.

(a) Describe the temperature T of the gas as a function of volume and pressure. Sketch some level curves for this function.

(b) Describe the volume V of the gas as a function of pressure and temperature. Sketch some level curves.

25. (a) Graph the surfaces $z = x^2$ and $z = y^2$.

(b) Explain how one can understand the graph of the *surfaces* $z = f(x)$ and $z = f(y)$ by considering the *curve* in the uv-plane given by $v = f(u)$.

(c) Graph the surface in $\mathbf{R}^3$ with equation $y = x^2$.

26. Use a computer to graph the family of level curves for the functions in Exercises 16 and 17 and compare your results with those obtained by hand sketching. How do you account for any differences?

27. Given a function $f(x, y)$, can two different level curves of f intersect? Why or why not?

In Exercises 28–32, describe the graph of $g(x, y, z)$ by computing some level surfaces. (If you prefer, use a computer to assist you.)

28. $g(x, y, z) = x - 2y + 3z$

29. $g(x, y, z) = x^2 + y^2 - z$

30. $g(x, y, z) = x^2 + y^2 + z^2$

31. $g(x, y, z) = x^2 + 9y^2 + 4z^2$

32. $g(x, y, z) = xy - yz$

33. (a) Describe the graph of $g(x, y, z) = x^2 + y^2$ by computing some level surfaces.

(b) Suppose g is a function such that the expression for $g(x, y, z)$ involves only x and y (i.e., $g(x, y, z) = h(x, y)$). What can you say about the level surfaces of g?

(c) Suppose g is a function such that the expression for $g(x, y, z)$ involves only x and z. What can you say about the level surfaces of g?

(d) Suppose g is a function such that the expression for $g(x, y, z)$ involves only x. What can you say about the level surfaces of g?

34. This problem concerns the surface determined by the graph of the equation $x^2 + xy - xz = 2$.

(a) Find a function $F(x, y, z)$ of three variables so that this surface may be considered to be a level set of F.

(b) Find a function $f(x, y)$ of two variables so that this surface may be considered to be the graph of $z = f(x, y)$.

35. Graph the ellipsoid

$$\frac{x^2}{4} + \frac{y^2}{9} + z^2 = 1.$$

Is it possible to find a function $f(x, y)$ so that this ellipsoid may be considered to be the graph of $z = f(x, y)$? Explain.

Sketch or describe the surfaces in $\mathbf{R}^3$ *determined by the equations in Exercises 36–42.*

36. $z = \dfrac{x^2}{4} - y^2$

37. $z^2 = \dfrac{x^2}{4} - y^2$

38. $x = \dfrac{y^2}{4} - \dfrac{z^2}{9}$

39. $x^2 + \dfrac{y^2}{9} - \dfrac{z^2}{16} = 0$

40. $\dfrac{x^2}{4} - \dfrac{y^2}{16} + \dfrac{z^2}{9} = 1$

41. $\dfrac{x^2}{25} + \dfrac{y^2}{16} = z^2 - 1$

42. $z = y^2 + 2$

We can look at examples of quadric surfaces with centers or vertices at points other than the origin by employing a change of coordinates of the form $\bar{x} = x - x_0$, $\bar{y} = y - y_0$, *and* $\bar{z} = z - z_0$. *This coordinate change simply puts the point* (x_0, y_0, z_0) *of the* xyz*-coordinate system at the origin of the* $\bar{x}\bar{y}\bar{z}$*-coordinate system by a translation of axes. Then, for example, the surface having equation*

$$\frac{(x-1)^2}{4} + \frac{(y+2)^2}{9} + (z-5)^2 = 1$$

can be identified by setting $\bar{x} = x - 1$, $\bar{y} = y + 2$, *and* $\bar{z} = z - 5$, *so that we obtain*

$$\frac{\bar{x}^2}{4} + \frac{\bar{y}^2}{9} + \bar{z}^2 = 1,$$

which is readily seen to be an ellipsoid centered at $(1, -2, 5)$ *of the* xyz*-coordinate system. By completing the square in* x, y, *or* z *as necessary, identify and sketch the quadric surfaces in Exercises 43–48.*

43. $(x-1)^2 + (y+1)^2 = (z+3)^2$

44. $z = 4x^2 + (y+2)^2$

45. $4x^2 + y^2 + z^2 + 8x = 0$

46. $4x^2 + y^2 - 4z^2 + 8x - 4y + 4 = 0$

47. $x^2 + 2y^2 - 6x - z + 10 = 0$

48. $9x^2 + 4y^2 - 36z^2 - 8y - 144z = 104$

2.2 Limits

As you may recall, limit processes are central to the development of calculus. The mathematical and philosophical debate in the 18th and 19th centuries surrounding the meaning and soundness of techniques of taking limits was intense, questioning the very foundations of calculus. By the middle of the 19th century, the infamous "$\epsilon - \delta$" definition of limits had been devised, chiefly by Karl Weierstrass and Augustin Cauchy, much to the chagrin of many 20th (and 21st) century students of calculus. In the ensuing discussion, we study both the intuitive and rigorous meanings of the limit of a function $\mathbf{f}\colon X \subseteq \mathbf{R}^n \to \mathbf{R}^m$ and how limits lead to the notion of a continuous function, our main object of study for the remainder of this text.

The Notion of a Limit

For a scalar-valued function of a single variable, $f\colon X \subseteq \mathbf{R} \to \mathbf{R}$, you have seen the statement

$$\lim_{x \to a} f(x) = L$$

and perhaps have an intuitive understanding of its meaning. In imprecise terms, the preceding equation (read "The limit of $f(x)$ as x approaches a is L.") means that you can make the numerical value of $f(x)$ arbitrarily close to L by keeping x sufficiently close (but not equal) to a. This idea generalizes immediately to functions $\mathbf{f}\colon X \subseteq \mathbf{R}^n \to \mathbf{R}^m$. In particular, by writing the equation

$$\lim_{\mathbf{x}\to\mathbf{a}} \mathbf{f}(\mathbf{x}) = \mathbf{L},$$

where $\mathbf{f}\colon X \subseteq \mathbf{R}^n \to \mathbf{R}^m$, we mean that we can make the vector $\mathbf{f}(\mathbf{x})$ arbitrarily close to the **limit vector L** by keeping the vector $\mathbf{x} \in X$ sufficiently close (but not equal) to $\mathbf{a}$.

The word "close" means that the distance (in the sense of §1.6) between $\mathbf{f}(\mathbf{x})$ and $\mathbf{L}$ is small. Thus, we offer a first definition of limit using the notation for distance.

■ Definition 2.1 (INTUITIVE DEFINITION OF LIMIT) The equation

$$\lim_{\mathbf{x}\to\mathbf{a}} \mathbf{f}(\mathbf{x}) = \mathbf{L},$$

where $\mathbf{f}\colon X \subseteq \mathbf{R}^n \to \mathbf{R}^m$, means that we can make $\|\mathbf{f}(\mathbf{x}) - \mathbf{L}\|$ arbitrarily small (i.e., near zero) by keeping $\|\mathbf{x} - \mathbf{a}\|$ sufficiently small (but nonzero).

In the case of a scalar-valued function $f\colon X \subseteq \mathbf{R}^n \to \mathbf{R}$, the vector length $\|\mathbf{f}(\mathbf{x}) - \mathbf{L}\|$ can be replaced by the absolute value $|f(\mathbf{x}) - L|$. Similarly, if f is a function of just one variable, then $\|\mathbf{x} - \mathbf{a}\|$ can be replaced by $|x - a|$.

EXAMPLE 1 Suppose that $f\colon \mathbf{R} \to \mathbf{R}$ is given by

$$f(x) = \begin{cases} 0 & \text{if } x < 1 \\ 2 & \text{if } x \geq 1 \end{cases}.$$

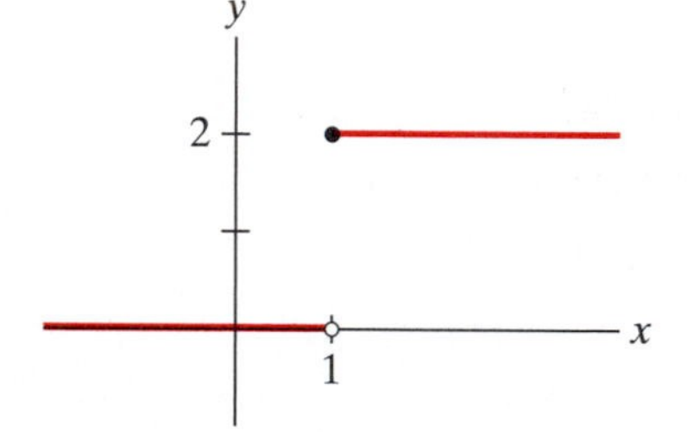

Figure 2.28 The graph of f of Example 1.

The graph of f is shown in Figure 2.28. What should $\lim_{x\to 1} f(x)$ be? The limit can't be 0, because no matter how near we make x to 1 (i.e., no matter how small we take $|x - 1|$), the values of x can be both slightly larger and slightly smaller than 1. The values of f corresponding to those values of x larger than 1 will be 2. Thus, for such values of x, we cannot make $|f(x) - 0|$ arbitrarily small, since, for $x \geq 1$, $|f(x) - 0| = |2 - 0| = 2$. Similarly, the limit can't be 2, since no matter how small we take $|x - 1|$, x can be slightly smaller than 1. For $x < 1$, $f(x) = 0$ and, therefore, we cannot make $|f(x) - 2| = |0 - 2| = 2$ arbitrarily small. Indeed, it should now be clear that the limit can't be L for *any* $L \in \mathbf{R}$. Hence, $\lim_{x\to 1} f(x)$ *does not exist* for this function. ◆

EXAMPLE 2 Let $\mathbf{f}\colon \mathbf{R}^2 \to \mathbf{R}^2$ be defined by $\mathbf{f}(\mathbf{x}) = \mathbf{x}$. (That is, $\mathbf{f}$ is the **identity function**.) Then it should be obvious intuitively that

$$\lim_{\mathbf{x}\to\mathbf{i}+\mathbf{j}} \mathbf{f}(\mathbf{x}) = \lim_{\mathbf{x}\to\mathbf{i}+\mathbf{j}} \mathbf{x} = \mathbf{i} + \mathbf{j}.$$

Indeed, if we write $\mathbf{x} = x\mathbf{i} + y\mathbf{j}$, then

$$\begin{aligned} \|\mathbf{f}(\mathbf{x}) - (\mathbf{i}+\mathbf{j})\| &= \|(x\mathbf{i} + y\mathbf{j}) - (\mathbf{i}+\mathbf{j})\| \\ &= \|(x-1)\mathbf{i} + (y-1)\mathbf{j}\| = \sqrt{(x-1)^2 + (y-1)^2}. \end{aligned}$$

This quantity can be made as small as we wish by keeping

$$\|\mathbf{x} - (\mathbf{i}+\mathbf{j})\| = \sqrt{(x-1)^2 + (y-1)^2}$$

appropriately small. More generally, if $\mathbf{g}: \mathbf{R}^n \to \mathbf{R}^n$ is the identity function (i.e., $\mathbf{g}(\mathbf{x}) = \mathbf{x}$) and $\mathbf{a} \in \mathbf{R}^n$ is *any* vector, then

$$\lim_{\mathbf{x}\to\mathbf{a}} \mathbf{g}(\mathbf{x}) = \lim_{\mathbf{x}\to\mathbf{a}} \mathbf{x} = \mathbf{a}. \qquad \blacklozenge$$

EXAMPLE 3 Now suppose that $\mathbf{h}: \mathbf{R}^n \to \mathbf{R}^n$ is defined by $\mathbf{h}(\mathbf{x}) = 3\mathbf{x}$. We claim that

$$\lim_{\mathbf{x}\to\mathbf{a}} 3\mathbf{x} = 3\mathbf{a}.$$

In other words, we claim that $\|3\mathbf{x} - 3\mathbf{a}\|$ can be made as small as we like by keeping $\|\mathbf{x} - \mathbf{a}\|$ sufficiently small. Note that

$$\|3\mathbf{x} - 3\mathbf{a}\| = \|3(\mathbf{x} - \mathbf{a})\| = 3\|\mathbf{x} - \mathbf{a}\|.$$

This means that if we wish to make $\|3\mathbf{x} - 3\mathbf{a}\|$ no more than, say, 0.003, then we may do so by making sure that $\|\mathbf{x} - \mathbf{a}\|$ is no more than 0.001. If, instead, we want $\|3\mathbf{x} - 3\mathbf{a}\|$ to be no more than 0.0003, we can achieve this by keeping $\|\mathbf{x} - \mathbf{a}\|$ no more than 0.0001. Indeed, if we want $\|3\mathbf{x} - 3\mathbf{a}\|$ to be no more than *any* specified amount (no matter how small), then we can achieve this by making sure that $\|\mathbf{x} - \mathbf{a}\|$ is no more than one third of that amount. ◆

The main difficulty with Definition 2.1 lies in the terms "arbitrarily small" and "sufficiently small." They are simply too vague. We can add some precision to our intuition as follows: Think of applying the function $\mathbf{f}: X \subseteq \mathbf{R}^n \to \mathbf{R}^m$ as performing some sort of scientific experiment. Letting the variable $\mathbf{x}$ take on a particular value in X amounts to making certain measurements of the input variables to the experiment, and the resulting value $\mathbf{f}(\mathbf{x})$ can be considered to be the outcome of the experiment. Experiments are designed to test theories, so suppose that this hypothetical experiment is designed to test the theory that as the input is closer and closer to $\mathbf{a}$, then the outcome gets closer and closer to $\mathbf{L}$. To verify this theory, you should establish some acceptable (absolute) experimental error for the outcome, say 0.05. That is, you want $\|\mathbf{f}(\mathbf{x}) - \mathbf{L}\| < 0.05$, if $\|\mathbf{x} - \mathbf{a}\|$ is sufficiently small. Then just how small does $\|\mathbf{x} - \mathbf{a}\|$ need to be? Perhaps it turns out that you must have $\|\mathbf{x} - \mathbf{a}\| < 0.02$, and that if you do take $\|\mathbf{x} - \mathbf{a}\| < 0.02$, then indeed $\|\mathbf{f}(\mathbf{x}) - \mathbf{L}\| < 0.05$. Does this mean that your theory is correct? Not yet. Now, suppose that you decide to be more exacting and will only accept an experimental error of 0.005 instead of 0.05. In other words, you desire $\|\mathbf{f}(\mathbf{x}) - \mathbf{L}\| < 0.005$. Perhaps you find that if you take $\|\mathbf{x} - \mathbf{a}\| < 0.001$, then this new goal can be achieved. Is your theory correct? Well, there's nothing sacred about the number 0.005, so perhaps you should insist that $\|\mathbf{f}(\mathbf{x}) - \mathbf{L}\| < 0.001$, or that $\|\mathbf{f}(\mathbf{x}) - \mathbf{L}\| < 0.00001$. The point is that if your theory really is correct, then *no matter what* (absolute) experimental error ϵ you choose for your outcome, you should be able to find a "tolerance level" δ for your input $\mathbf{x}$ so that if $\|\mathbf{x} - \mathbf{a}\| < \delta$, then $\|\mathbf{f}(\mathbf{x}) - \mathbf{L}\| < \epsilon$. It is this heuristic approach that motivates the technical definition of the limit.

■ **Definition 2.2** (Rigorous definition of limit) Let $\mathbf{f}: X \subseteq \mathbf{R}^n \to \mathbf{R}^m$ be a function. Then to say

$$\lim_{\mathbf{x}\to\mathbf{a}} \mathbf{f}(\mathbf{x}) = \mathbf{L}$$

means that given any $\epsilon > 0$, you can find a $\delta > 0$ (which will, in general, depend on ϵ) such that if $\mathbf{x} \in X$ and $0 < \|\mathbf{x} - \mathbf{a}\| < \delta$, then $\|\mathbf{f}(\mathbf{x}) - \mathbf{L}\| < \epsilon$.

The condition $0 < \|\mathbf{x} - \mathbf{a}\|$ simply means that we care only about values $\mathbf{f}(\mathbf{x})$ when $\mathbf{x}$ is *near* $\mathbf{a}$, but *not* equal to $\mathbf{a}$. Definition 2.2 is not easy to use in practice (and we will not use it frequently). Moreover, it is of little value insofar as actually *evaluating* limits of functions is concerned. (The evaluation of the limit of a function of more than one variable is, in general, a difficult task.)

EXAMPLE 4 So that you have some feeling for working with Definition 2.2, let's see rigorously that

$$\lim_{(x,y,z)\to(1,-1,2)} (3x - 5y + 2z) = 12$$

(as should be "obvious"). This means that given any number $\epsilon > 0$, we can find a corresponding $\delta > 0$ such that

$$\text{if } 0 < \|(x, y, z) - (1, -1, 2)\| < \delta, \quad \text{then } |3x - 5y + 2z - 12| < \epsilon.$$

(Note the uses of vector lengths and absolute values.) We'll present a formal proof in the next paragraph, but for now we'll do the necessary background calculations in order to provide such a proof. First, we need to rewrite the two inequalities in such a way as to make it more plausible that the ϵ-inequality could arise algebraically from the δ-inequality. From the definition of vector length, the δ-inequality becomes

$$0 < \sqrt{(x-1)^2 + (y+1)^2 + (z-2)^2} < \delta.$$

If this is true, then we certainly have the three inequalities

$$\begin{aligned} \sqrt{(x-1)^2} &= |x - 1| < \delta, \\ \sqrt{(y+1)^2} &= |y + 1| < \delta, \\ \sqrt{(z-2)^2} &= |z - 2| < \delta. \end{aligned}$$

Now, rewrite the left side of the ϵ-inequality and use the triangle inequality (2) of §1.6:

$$\begin{aligned} |3x - 5y + 2z - 12| &= |3(x-1) - 5(y+1) + 2(z-2)| \\ &\le |3(x-1)| + |5(y+1)| + |2(z-2)| \\ &= 3|x-1| + 5|y+1| + 2|z-2|. \end{aligned}$$

Thus, if

$$0 < \|(x, y, z) - (1, -1, 2)\| < \delta,$$

then

$$|x - 1| < \delta, \quad |y + 1| < \delta, \quad \text{and} \quad |z - 2| < \delta,$$

so that

$$\begin{aligned} |3x - 5y + 2z - 12| &\le 3|x-1| + 5|y+1| + 2|z-2| \\ &< 3\delta + 5\delta + 2\delta = 10\delta. \end{aligned}$$

If we think of δ as a positive quantity that we can make as small as desired, then 10δ can also be made small. In fact, it is 10δ that plays the role of ϵ.

Now for a formal, "textbook" proof: Given any $\epsilon > 0$, choose $\delta > 0$ so that $\delta \leq \epsilon/10$. Then, if

$$0 < \|(x, y, z) - (1, -1, 2)\| < \delta,$$

it follows that

$$|x - 1| < \delta, \quad |y + 1| < \delta, \quad \text{and} \quad |z - 2| < \delta,$$

so that

$$\begin{aligned} |3x - 5y + 2z - 12| &\leq 3|x - 1| + 5|y + 1| + 2|z - 2| \\ &< 3\delta + 5\delta + 2\delta \\ &= 10\delta \leq 10\frac{\epsilon}{10} = \epsilon. \end{aligned}$$

Thus, $\lim_{(x,y,z)\to(1,-1,2)}(3x - 5y + 2z) = 12$, as desired. ◆

Using the same methods as in Example 4, you can show that

$$\lim_{\mathbf{x}\to\mathbf{b}}(a_1x_1 + a_2x_2 + \cdots + a_nx_n) = a_1b_1 + a_2b_2 + \cdots + a_nb_n$$

for any a_i, $i = 1, 2, \ldots, n$.

Some Topological Terminology

Before discussing the geometric meaning of the limit of a function, we need to introduce some standard terminology regarding sets of points in $\mathbf{R}^n$. The underlying geometry of point sets of a space is known as the **topology** of that space.

Recall from §2.1 that the vector equation $\|\mathbf{x} - \mathbf{a}\| = r$, where $\mathbf{x}$ and $\mathbf{a}$ are in $\mathbf{R}^3$ and $r > 0$, defines a sphere of radius r centered at $\mathbf{a}$. If we modify this equation so that it becomes the inequality

$$\|\mathbf{x} - \mathbf{a}\| \leq r, \tag{1}$$

then the points $\mathbf{x} \in \mathbf{R}^3$ that satisfy it fill out what is called a **closed ball** shown in Figure 2.29. Similarly, the *strict* inequality

$$\|\mathbf{x} - \mathbf{a}\| < r \tag{2}$$

describes points $\mathbf{x} \in \mathbf{R}^3$ that are a distance of less than r from $\mathbf{a}$. Such points determine an **open ball** of radius r centered at $\mathbf{a}$, that is, a solid ball *without* the boundary sphere.

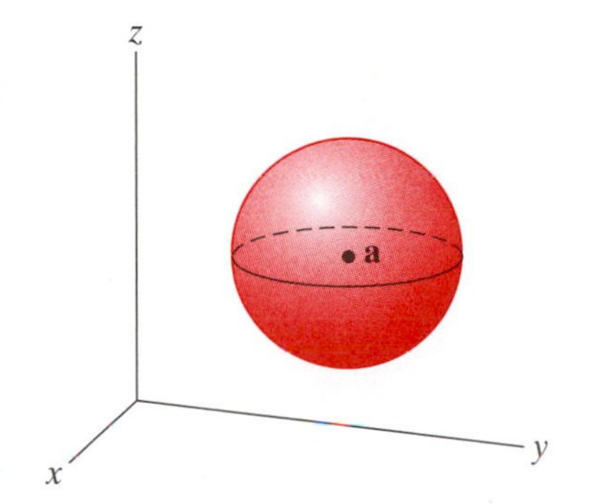

Figure 2.29 A closed ball centered at $\mathbf{a}$.

There is nothing about the inequalities (1) and (2) that tie them to $\mathbf{R}^3$. In fact, if we take $\mathbf{x}$ and $\mathbf{a}$ to be points of $\mathbf{R}^n$, then (1) and (2) define, respectively, closed and open n-dimensional balls of radius r centered at $\mathbf{a}$. While we cannot draw sketches when $n > 3$, we can see what (1) and (2) mean when n is 1 or 2. (See Figures 2.30 and 2.31.)

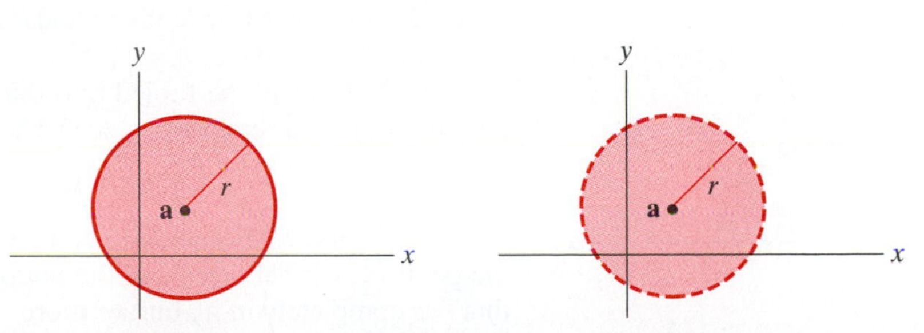

Figure 2.30 The closed and open balls (disks) in $\mathbf{R}^2$ defined by $\|\mathbf{x} - \mathbf{a}\| \leq r$ and $\|\mathbf{x} - \mathbf{a}\| < r$.

Figure 2.31 The closed and open balls (intervals) in $\mathbf{R}$ defined by $|x - a| \leq r$ and $|x - a| < r$.

> ■ **Definition 2.3** A set $X \subseteq \mathbf{R}^n$ is said to be **open** in $\mathbf{R}^n$ if, for each point $\mathbf{x} \in X$, there is some open ball centered at $\mathbf{x}$ that lies entirely within X. A point $\mathbf{x} \in \mathbf{R}^n$ is said to be in the **boundary** of a set $X \subseteq \mathbf{R}^n$ if every open ball centered at $\mathbf{x}$, no matter how small, contains some points that are in X and also some points that are not in X. A set $X \subseteq \mathbf{R}^n$ is said to be **closed** in $\mathbf{R}^n$ if it contains all of its boundary points. Finally, a **neighborhood** of a point $\mathbf{x} \in X$ is an open set containing $\mathbf{x}$ and contained in X.

It is an easy consequence of Definition 2.3 that a set X is closed in $\mathbf{R}^n$ precisely if its complement $\mathbf{R}^n - X$ is open.

EXAMPLE 5 The rectangular region

$$X = \{(x, y) \in \mathbf{R}^2 \mid -1 < x < 1, -1 < y < 2\}$$

is open in $\mathbf{R}^2$. (See Figure 2.32.) Each point in X has an open disk around it contained entirely in the rectangle. The boundary of X consists of the four sides of the rectangle. (See Figure 2.33.) ◆

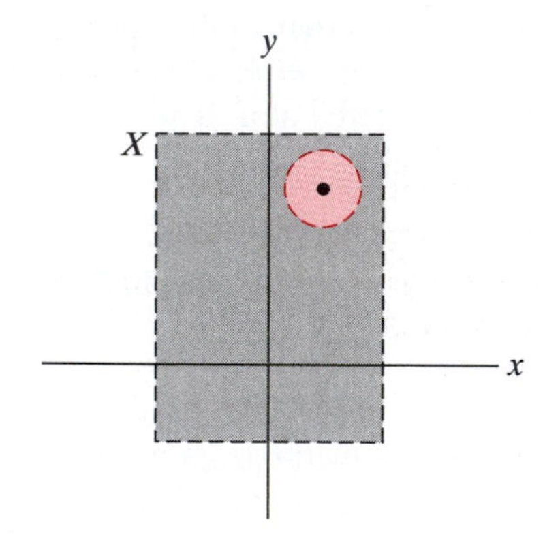

Figure 2.32 The graph of X.

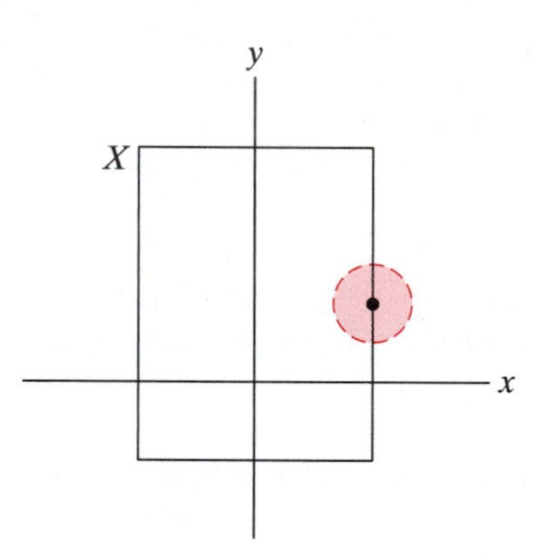

Figure 2.33 Every open disk about a point on a side of rectangle X of Example 5 contains points in both X and $\mathbf{R}^2 - X$.

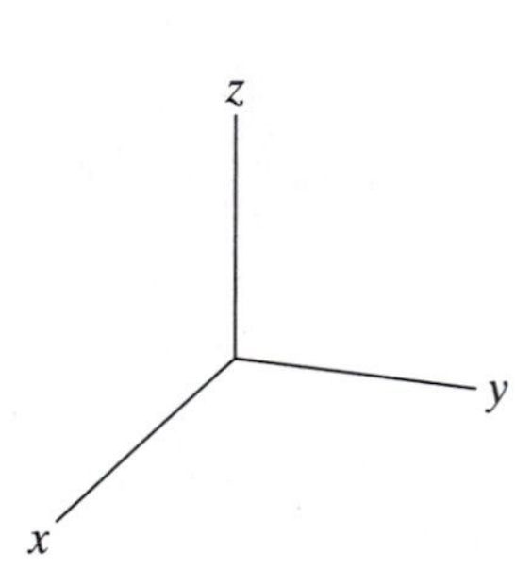

Figure 2.34 The set X of Example 6 consists of the nonnegative coordinate axes.

EXAMPLE 6 The set X consisting of the nonnegative coordinate axes in $\mathbf{R}^3$ in Figure 2.34 is closed since the boundary of X is just X itself. ◆

EXAMPLE 7 Don't be fooled into thinking that sets are always either open or closed. (That is, a set is not a door.) The set

$$X = \{(x, y) \in \mathbf{R}^2 \mid 0 \leq x < 1, 0 \leq y < 1\}$$

shown in Figure 2.35 is neither open nor closed. It's not open since, for example, the point $(\frac{1}{2}, 0)$ that lies along the bottom edge of X has no open disk around it that lies completely in X. Furthermore, X is not closed, since the boundary of X includes points of the form $(x, 1)$ for $0 \leq x \leq 1$ (why?), which are not part of X. ◆

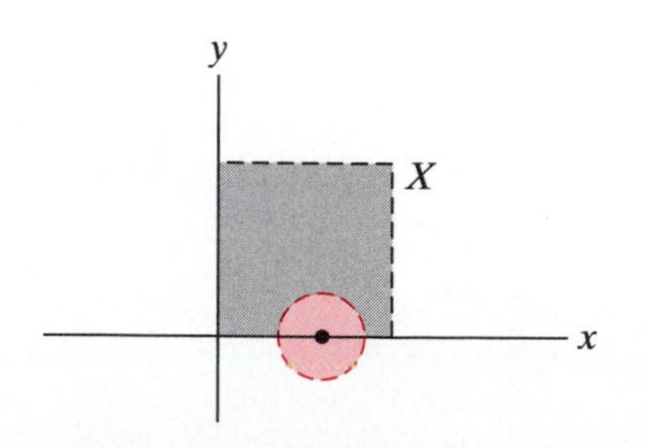

Figure 2.35 The set X of Example 7.

The Geometric Interpretation of a Limit

Suppose that $\mathbf{f}: X \subseteq \mathbf{R}^n \to \mathbf{R}^m$. Then the geometric meaning of the statement

$$\lim_{\mathbf{x}\to\mathbf{a}} \mathbf{f}(\mathbf{x}) = \mathbf{L}$$

is as follows: Given any $\epsilon > 0$, you can find a corresponding $\delta > 0$ such that if points $\mathbf{x} \in X$ are inside an open ball of radius δ centered at $\mathbf{a}$, then the corresponding points $\mathbf{f}(\mathbf{x})$ will remain inside an open ball of radius ϵ centered at $\mathbf{L}$. (See Figure 2.36.)

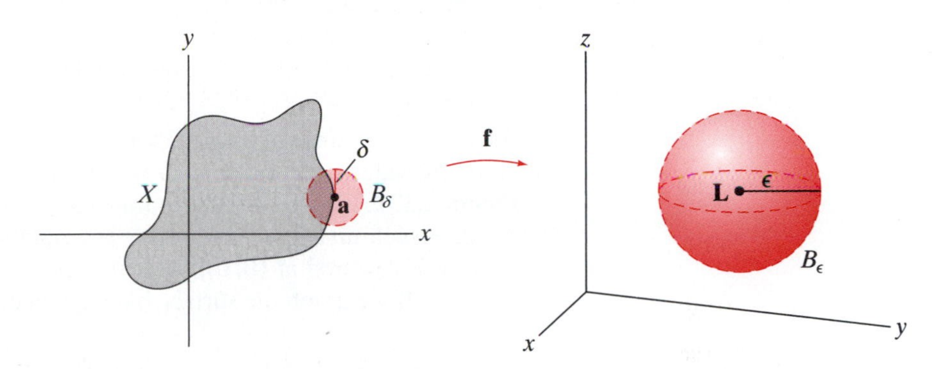

Figure 2.36 Definition of a limit: Given an open ball B_ϵ centered at $\mathbf{L}$ (right), you can always find a corresponding ball B_δ centered at $\mathbf{a}$ (left), so that points in $B_\delta \cap X$ are mapped by $\mathbf{f}$ to points in B_ϵ.

We remark that for this definition to make sense, the point $\mathbf{a}$ cannot be isolated from X. In particular, $\mathbf{a}$ must be such that every neighborhood of it contains points $\mathbf{x} \in X$ distinct from $\mathbf{a}$. Such a point $\mathbf{a}$ is called an **accumulation point** of X. (Technically, this assumption should also be made in Definition 2.2.)

From these considerations, we see that the statement $\lim_{\mathbf{x}\to\mathbf{a}} \mathbf{f}(\mathbf{x}) = \mathbf{L}$ really does mean that as $\mathbf{x}$ moves towards $\mathbf{a}$, $\mathbf{f}(\mathbf{x})$ moves towards $\mathbf{L}$. The significance of the "open ball" geometry is that entirely arbitrary motion is allowed.

EXAMPLE 8 Let $f: \mathbf{R}^2 - \{(0,0)\} \to \mathbf{R}$ be defined by

$$f(x, y) = \frac{x^2 - y^2}{x^2 + y^2}.$$

Let's see what happens to f as $\mathbf{x} = (x, y)$ approaches $\mathbf{0} = (0, 0)$. (Note that f is undefined at the origin, although this is of no consequence insofar as evaluating limits is concerned.) Along the x-axis (i.e., the line $y = 0$), we calculate the value of f to be

$$f(x, 0) = \frac{x^2 - 0}{x^2 + 0} = 1.$$

Thus, as $\mathbf{x}$ approaches $\mathbf{0}$ *along the line* $y = 0$, the values of f remain constant, and so

$$\lim_{\mathbf{x}\to\mathbf{0} \text{ along } y=0} f(\mathbf{x}) = 1.$$

Along the y-axis, however, the value of f is

$$f(0, y) = \frac{0 - y^2}{0 + y^2} = -1.$$

Hence,

$$\lim_{\substack{\mathbf{x}\to\mathbf{0} \text{ along } x=0}} f(\mathbf{x}) = -1.$$

Indeed, the value of f is constant along each line through the origin. Along the line $y = mx$, m constant, we have

$$f(x, mx) = \frac{x^2 - m^2x^2}{x^2 + m^2x^2} = \frac{x^2(1 - m^2)}{x^2(1 + m^2)} = \frac{1 - m^2}{1 + m^2}.$$

Therefore,

$$\lim_{\substack{\mathbf{x}\to\mathbf{0} \text{ along } y=mx}} f(\mathbf{x}) = \frac{1 - m^2}{1 + m^2}.$$

As a result, the limit of f as $\mathbf{x}$ approaches $\mathbf{0}$ *does not exist,* since f has different "limiting values" depending on which direction we approach the origin. (See Figure 2.37). That is, no matter how close we come to the origin, we can find points $\mathbf{x}$ such that $f(\mathbf{x})$ is *not* near any number $L \in \mathbf{R}$. (In other words, every open disk centered at $(0, 0)$, no matter how small, is mapped onto the interval $[-1, 1]$.) If we graph the surface having equation

$$z = \frac{x^2 - y^2}{x^2 + y^2}$$

(Figure 2.38), we can see quite clearly that there is no limiting value as x approaches the origin. ◆

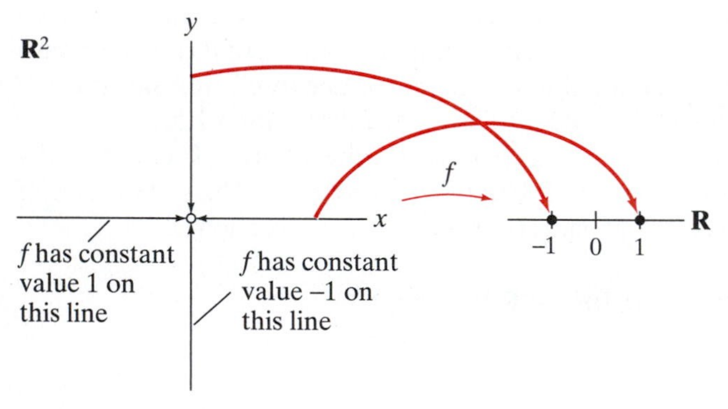

Figure 2.37 The function $f(x, y) = (x^2 - y^2)/(x^2 + y^2)$ of Example 8 has value 1 along the x-axis and value -1 along the y-axis (except at the origin).

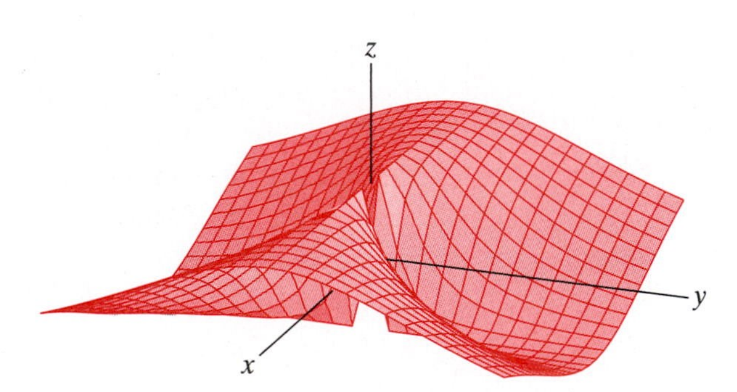

Figure 2.38 The graph of $f(x, y) = (x^2 - y^2)/(x^2 + y^2)$ of Example 8.

WARNING Example 8 might lead you to think you can establish that $\lim_{\mathbf{x}\to\mathbf{a}} \mathbf{f}(\mathbf{x}) = \mathbf{L}$ by showing that the values of $\mathbf{f}$ as $\mathbf{x}$ approaches $\mathbf{a}$ along straight line paths all tend toward the same value $\mathbf{L}$. Although this is certainly good evidence that the limit should be $\mathbf{L}$, it is by no means conclusive. See Exercise 23 for an example that shows what can happen.

Properties of Limits

One of the biggest drawbacks to Definition 2.2 is that it is not at all useful for determining the value of a limit. You must already have a "candidate limit" in mind and must also be prepared to confront some delicate work with inequalities to use Definition 2.2. The results that follow (which we state without proof), plus a little faith, can be quite helpful for establishing limits, as the subsequent examples demonstrate.

Theorem 2.4 (UNIQUENESS OF LIMITS) If a limit exists, it is unique. That is, let $\mathbf{f}: X \subseteq \mathbf{R}^n \to \mathbf{R}^m$. If $\lim_{\mathbf{x}\to\mathbf{a}} \mathbf{f}(\mathbf{x}) = \mathbf{L}$ and $\lim_{\mathbf{x}\to\mathbf{a}} \mathbf{f}(\mathbf{x}) = \mathbf{M}$, then $\mathbf{L} = \mathbf{M}$.

Theorem 2.5 (ALGEBRAIC PROPERTIES) Let $\mathbf{F}, \mathbf{G}: X \subseteq \mathbf{R}^n \to \mathbf{R}^m$ be vector-valued functions, $f, g: X \subseteq \mathbf{R}^n \to \mathbf{R}$ be scalar-valued functions, and let $k \in \mathbf{R}$ be a scalar.

1. If $\lim_{\mathbf{x}\to\mathbf{a}} \mathbf{F}(\mathbf{x}) = \mathbf{L}$ and $\lim_{\mathbf{x}\to\mathbf{a}} \mathbf{G}(\mathbf{x}) = \mathbf{M}$, then $\lim_{\mathbf{x}\to\mathbf{a}}(\mathbf{F}+\mathbf{G})(\mathbf{x}) = \mathbf{L}+\mathbf{M}$.
2. If $\lim_{\mathbf{x}\to\mathbf{a}} \mathbf{F}(\mathbf{x}) = \mathbf{L}$, then $\lim_{\mathbf{x}\to\mathbf{a}} k\mathbf{F}(\mathbf{x}) = k\mathbf{L}$.
3. If $\lim_{\mathbf{x}\to\mathbf{a}} f(\mathbf{x}) = L$ and $\lim_{\mathbf{x}\to\mathbf{a}} g(\mathbf{x}) = M$, then $\lim_{\mathbf{x}\to\mathbf{a}}(fg)(\mathbf{x}) = LM$.
4. If $\lim_{\mathbf{x}\to\mathbf{a}} f(\mathbf{x}) = L$, $g(\mathbf{x}) \neq 0$ for $\mathbf{x} \in X$, and $\lim_{\mathbf{x}\to\mathbf{a}} g(\mathbf{x}) = M \neq 0$, then $\lim_{\mathbf{x}\to\mathbf{a}}(f/g)(\mathbf{x}) = L/M$.

There is nothing surprising about these theorems—they are exactly the same as the corresponding results for scalar-valued functions of a single variable. Moreover, Theorem 2.5 renders the evaluation of many limits relatively straightforward.

EXAMPLE 9 Either from rigorous considerations or blind faith, you should find it plausible that

$$\lim_{(x,y)\to(a,b)} x = a \quad \text{and} \quad \lim_{(x,y)\to(a,b)} y = b.$$

From these facts, it follows from Theorem 2.5 parts 1, 2, and 3 that

$$\lim_{(x,y)\to(a,b)} (x^2 + 2xy - y^3) = a^2 + 2ab - b^3,$$

because, by part 1 of Theorem 2.5,

$$\lim_{(x,y)\to(a,b)} (x^2 + 2xy - y^3) = \lim x^2 + \lim 2xy + \lim(-y^3)$$

and, by parts 2 and 3,

$$\lim_{(x,y)\to(a,b)} (x^2 + 2xy - y^3) = (\lim x)^2 + 2(\lim x)(\lim y) - (\lim y)^3$$

so that, from the facts just cited,

$$\lim_{(x,y)\to(a,b)} (x^2 + 2xy - y^3) = a^2 + 2ab - b^3.$$ ◆

EXAMPLE 10 More generally, a **polynomial** in two variables x and y is any expression of the form

$$p(x, y) = \sum_{k=0}^{d} \sum_{l=0}^{d} c_{kl} x^k y^l,$$

where d is some nonnegative integer and $c_{kl} \in \mathbf{R}$ for $k, l = 0, \ldots, d$. That is, $p(x, y)$ is an expression consisting of a (finite) sum of terms that are real number coefficients times powers of x and y. For instance, the expression $x^2 + 2xy - y^3$ in Example 9 is a polynomial. For any $(a, b) \in \mathbf{R}^2$, we have, by part 1 of

Theorem 2.5,

$$\lim_{(x,y)\to(a,b)} p(x,y) = \sum_{k=0}^{d}\sum_{l=0}^{d} \lim_{(x,y)\to(a,b)} (c_{kl}x^k y^l),$$

so that, from part 2,

$$\lim_{(x,y)\to(a,b)} p(x,y) = \sum_{k=0}^{d}\sum_{l=0}^{d} c_{kl} \lim_{(x,y)\to(a,b)} x^k y^l$$

and, from part 3,

$$\begin{aligned}\lim_{(x,y)\to(a,b)} p(x,y) &= \sum_{k=0}^{d}\sum_{l=0}^{d} c_{kl}(\lim x^k)(\lim y^l)\\ &= \sum_{k=0}^{d}\sum_{l=0}^{d} c_{kl}a^k b^l.\end{aligned}$$

Similarly, a **polynomial** in n variables $x_1, x_2, \ldots, x_n$ is an expression of the form

$$p(x_1, x_2, \ldots, x_n) = \sum_{k_1,\ldots,k_n=0}^{d} c_{k_1\cdots k_n} x_1^{k_1} x_2^{k_2} \cdots x_n^{k_n},$$

where d is some nonnegative integer and $c_{k_1\cdots k_n} \in \mathbf{R}$ for $k_1, \ldots, k_n = 0, \ldots, d$. For example, a polynomial in four variables might look like this:

$$p(x_1, \ldots, x_4) = 3x_1^2 x_2 + x_1x_2x_3x_4 - 7x_3^8 x_4^2.$$

Theorem 2.5 implies readily that

$$\lim_{\mathbf{x}\to\mathbf{a}} \sum c_{k_1\cdots k_n} x_1^{k_1} x_2^{k_2} \cdots x_n^{k_n} = \sum c_{k_1\cdots k_n} a_1^{k_1} a_2^{k_2} \cdots a_n^{k_n}.$$ ◆

EXAMPLE 11 We evaluate $\displaystyle\lim_{(x,y)\to(-1,0)} \frac{x^2+xy+3}{x^2y-5xy+y^2+1}$.

Using Example 10, we see that

$$\lim_{(x,y)\to(-1,0)} x^2+xy+3 = 4,$$

and

$$\lim_{(x,y)\to(-1,0)} x^2y-5xy+y^2+1 = 1(\neq 0).$$

Thus, from part 4 of Theorem 2.5, we conclude that

$$\lim_{(x,y)\to(-1,0)} \frac{x^2+xy+3}{x^2y-5xy+y^2+1} = \frac{4}{1} = 4.$$ ◆

EXAMPLE 12 Of course, not all limits of quotient expressions are as simple to evaluate as that of Example 11. For instance, we cannot use Theorem 2.5 to evaluate

$$\lim_{(x,y)\to(0,0)} \frac{x^2-y^4}{x^2+y^4} \tag{3}$$

since $\lim_{(x,y)\to(0,0)}(x^2+y^4) = 0$. Indeed, since $\lim_{(x,y)\to(0,0)}(x^2-y^4) = 0$ as well, the expression $(x^2-y^4)/(x^2+y^4)$ becomes indeterminate as $(x,y)\to(0,0)$. To see what happens to the expression, we note that

$$\lim_{x\to 0 \text{ along } y=0} \frac{x^2-y^4}{x^2+y^4} = \lim_{x\to 0}\frac{x^2}{x^2} = 1,$$

while

$$\lim_{y\to 0 \text{ along } x=0} \frac{x^2 - y^4}{x^2 + y^4} = \lim_{y\to 0} \frac{-y^4}{y^4} = -1.$$

Thus, the limit in (3) does not exist. (Compare this with Example 8.) ◆

The following result shows that evaluating the limit of a function $\mathbf{f}: X \subseteq \mathbf{R}^n \to \mathbf{R}^m$ is equivalent to evaluating the limits of its (scalar-valued) component functions. First recall from §2.1 that $\mathbf{f}(\mathbf{x})$ may be rewritten as $(f_1(\mathbf{x}), f_2(\mathbf{x}), \ldots, f_m(\mathbf{x}))$.

■ **Theorem 2.6** Suppose $\mathbf{f}: X \subseteq \mathbf{R}^n \to \mathbf{R}^m$ is a vector-valued function. Then $\lim_{\mathbf{x}\to\mathbf{a}} \mathbf{f}(\mathbf{x}) = \mathbf{L}$, where $\mathbf{L} = (L_1, \ldots, L_m)$, if and only if $\lim_{\mathbf{x}\to\mathbf{a}} f_i(\mathbf{x}) = L_i$ for $i = 1, \ldots, m$.

EXAMPLE 13 Consider the linear mapping $\mathbf{f}: \mathbf{R}^n \to \mathbf{R}^m$ defined by $\mathbf{f}(\mathbf{x}) = A\mathbf{x}$, where $A = (a_{ij})$ is an $m \times n$ matrix of real numbers. (See Example 5 of §1.6.) Theorem 2.6 shows us that

$$\lim_{\mathbf{x}\to\mathbf{b}} \mathbf{f}(\mathbf{x}) = A\mathbf{b}$$

for any $\mathbf{b} = (b_1, \ldots, b_n)$ in $\mathbf{R}^n$. If we write out the matrix multiplication, we have

$$\mathbf{f}(\mathbf{x}) = A\mathbf{x} = \begin{bmatrix} a_{11} & \cdots & a_{1n} \\ a_{21} & \cdots & a_{2n} \\ \vdots & \ddots & \vdots \\ a_{m1} & \cdots & a_{mn} \end{bmatrix} \begin{bmatrix} x_1 \\ x_2 \\ \vdots \\ x_n \end{bmatrix}$$

$$= \begin{bmatrix} a_{11}x_1 + a_{12}x_2 + \cdots + a_{1n}x_n \\ a_{21}x_1 + a_{22}x_2 + \cdots + a_{2n}x_n \\ \vdots \\ a_{m1}x_1 + a_{m2}x_2 + \cdots + a_{mn}x_n \end{bmatrix}.$$

Therefore, the ith component function of $\mathbf{f}$ is

$$f_i(x) = a_{i1}x_1 + a_{i2}x_2 + \cdots + a_{in}x_n.$$

From Example 4, we have that

$$\lim_{\mathbf{x}\to\mathbf{b}} f_i(\mathbf{x}) = a_{i1}b_1 + a_{i2}b_2 + \cdots + a_{in}b_n$$

for each i. Hence, Theorem 2.6 tells us that the limits of the component functions fit together to form a limit vector. We can, therefore, conclude that

$$\begin{aligned} \lim_{\mathbf{x}\to\mathbf{b}} \mathbf{f}(\mathbf{x}) &= (\lim_{\mathbf{x}\to\mathbf{b}} f_1(\mathbf{x}), \ldots, \lim_{\mathbf{x}\to\mathbf{b}} f_m(\mathbf{x})) \\ &= (a_{11}b_1 + \cdots + a_{1n}b_n, \ldots, a_{m1}b_1 + \cdots + a_{mn}b_n) \\ &= \begin{bmatrix} a_{11}b_1 + \cdots + a_{1n}b_n \\ a_{21}b_1 + \cdots + a_{2n}b_n \\ \vdots \\ a_{m1}b_1 + \cdots + a_{mn}b_n \end{bmatrix} = A\mathbf{b}, \end{aligned}$$

once we take advantage of matrix notation. ◆

Continuous Functions

For scalar-valued functions of a single variable, one often adopts the following attitude toward the notion of continuity: A function $f: X \subseteq \mathbf{R} \to \mathbf{R}$ is **continuous** if its graph can be drawn without taking the pen off the paper. By this criterion, Figure 2.39 describes a continuous function $y = f(x)$, while Figure 2.40 does not.

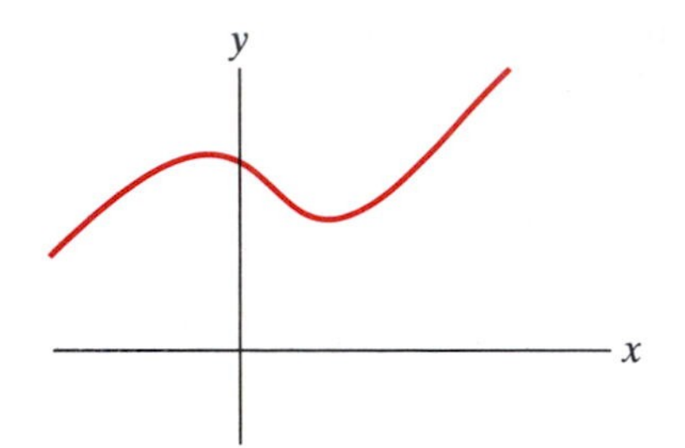

Figure 2.39 The graph of a continuous function.

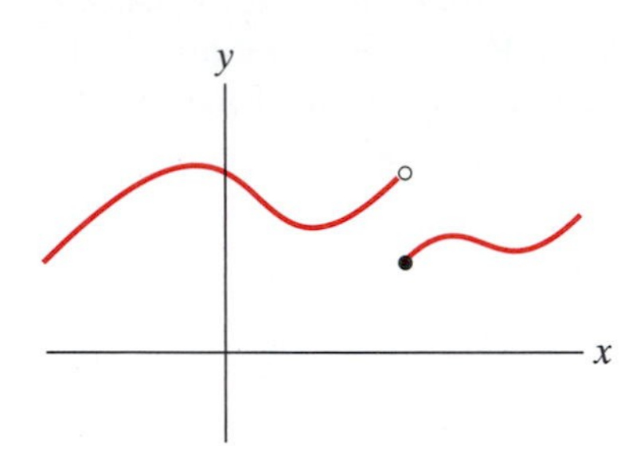

Figure 2.40 The graph of a function that is not continuous.

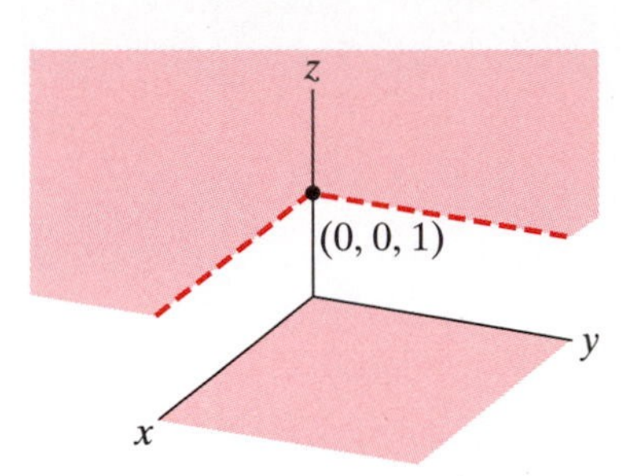

Figure 2.41 The graph of f where $f(x, y) = 0$ if both $x \geq 0$ and $y \geq 0$, and where $f(x, y) = 1$ otherwise.

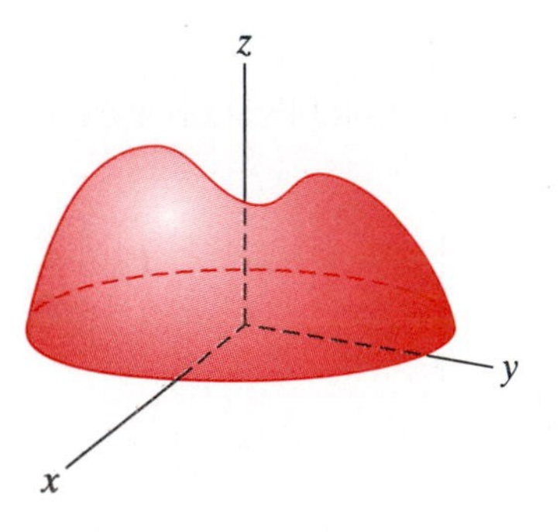

Figure 2.42 The graph of a continuous function $f(x, y)$.

We can try to extend this idea to scalar-valued function of two variables: A function $f: X \subseteq \mathbf{R}^2 \to \mathbf{R}$ is **continuous** if its graph (in $\mathbf{R}^3$) has no breaks in it. Then, the function shown in Figure 2.41 fails to be continuous, but Figure 2.42 depicts a continuous function. Although this graphical approach to continuity is pleasantly geometric and intuitive, it does have real and fatal flaws. For one thing, we can't visualize graphs of functions of more than two variables, so how will we be able to tell in general if a function $f: X \subseteq \mathbf{R}^n \to \mathbf{R}^m$ is continuous? Moreover, it is not always so easy to produce a graph of a function of two variables that is sufficient to make a visual determination of continuity. This said, we now give a rigorous definition of continuity of functions of several variables.

■ **Definition 2.7** Let $\mathbf{f}: X \subseteq \mathbf{R}^n \to \mathbf{R}^m$ and let $\mathbf{a} \in X$. Then, $\mathbf{f}$ is said to be **continuous at a** if

$$\lim_{\mathbf{x}\to\mathbf{a}} \mathbf{f}(\mathbf{x}) = \mathbf{f}(\mathbf{a}).$$

If $\mathbf{f}$ is continuous at all points of its domain X, then we simply say that $\mathbf{f}$ is **continuous.**

EXAMPLE 14 Consider the function $f: \mathbf{R}^2 \to \mathbf{R}$ defined by

$$f(x, y) = \begin{cases} \dfrac{x^2 + xy - 2y^2}{x^2 + y^2} & \text{if } (x, y) \neq (0, 0) \\ 0 & \text{if } (x, y) = (0, 0) \end{cases}.$$

Therefore, $f(0, 0) = 0$, but $\lim_{(x,y)\to(0,0)} f(x, y)$ does not exist. (To see this, check what happens as (x, y) approaches (0,0) first along $y = 0$ and then along $x = 0$.) Hence, f is not continuous at (0,0). ◆

It is worth noting that Definition 2.7 is nothing more than the "vectorized" version of the usual definition of continuity of a (scalar-valued) function of one variable. This definition thus provides another example of the power of our vector notation: Continuity looks the same no matter what the context.

One way of thinking about continuous functions is that they are the ones whose limits are easy to evaluate: When $\mathbf{f}$ is continuous, the *limit* of $\mathbf{f}$ as $\mathbf{x}$ approaches $\mathbf{a}$ is just the *value* of $\mathbf{f}$ at $\mathbf{a}$. It's all too tempting to get into the habit of behaving as if all functions are continuous, especially since the functions that will be of primary interest to us will be continuous. Try to avoid such an impulse.

EXAMPLE 15 Polynomial functions in n variables are continuous. Example 10 gives a sketch of the fact that

$$\lim_{\mathbf{x}\to\mathbf{a}} \sum c_{k_1\cdots k_n} x_1^{k_1} \cdots x_n^{k_n} = \sum c_{k_1\cdots k_n} a_1^{k_1} \cdots a_n^{k_n},$$

where $\mathbf{x} = (x_1, \ldots, x_n)$ and $\mathbf{a} = (a_1, \ldots, a_n)$ are in $\mathbf{R}^n$. If $f\colon \mathbf{R}^n \to \mathbf{R}$ is defined by

$$f(\mathbf{x}) = \sum c_{k_1\cdots k_n} x_1^{k_1} \cdots x_n^{k_n},$$

then the preceding limit statement says precisely that f is continuous at $\mathbf{a}$. ◆

EXAMPLE 16 Linear mappings are continuous. If $\mathbf{f}\colon \mathbf{R}^n \to \mathbf{R}^m$ is defined by $\mathbf{f}(\mathbf{x}) = A\mathbf{x}$, where A is an $m \times n$ matrix, then Example 13 establishes that

$$\lim_{\mathbf{x}\to\mathbf{b}} \mathbf{f}(\mathbf{x}) = A\mathbf{b} = \mathbf{f}(\mathbf{b})$$

for all $\mathbf{b} \in \mathbf{R}^n$. Thus, $\mathbf{f}$ is continuous. ◆

The geometric interpretation of the $\epsilon - \delta$ definition of a limit gives rise to a similar interpretation of continuity at a point: $\mathbf{f}\colon X \subseteq \mathbf{R}^n \to \mathbf{R}^m$ is continuous at a point $\mathbf{a} \in X$ if, for every open ball B_ϵ in $\mathbf{R}^m$ of radius ϵ centered at $\mathbf{f}(\mathbf{a})$, there is a corresponding open ball B_δ in $\mathbf{R}^n$ of radius δ centered at $\mathbf{a}$ such that points $\mathbf{x} \in X$ inside B_δ are mapped by $\mathbf{f}$ to points inside B_ϵ. (See Figure 2.43.) Roughly speaking, continuity of $\mathbf{f}$ means that "close" points in $X \subseteq \mathbf{R}^n$ are mapped to "close" points in $\mathbf{R}^m$.

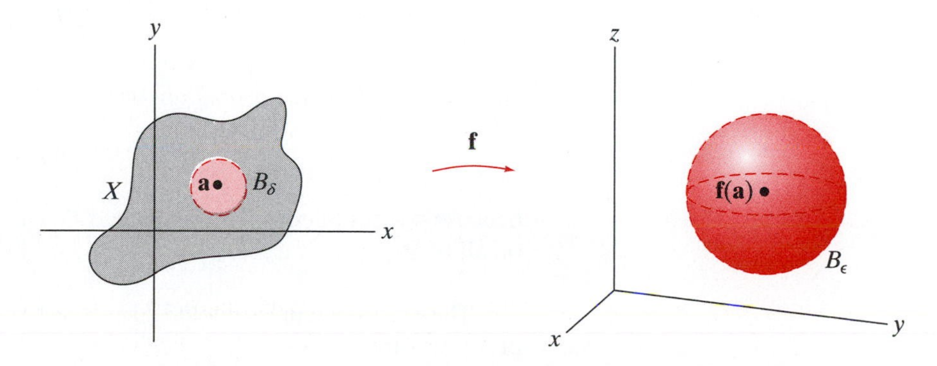

Figure 2.43 Given an open ball B_ϵ about $\mathbf{f}(\mathbf{a})$ (right), you can always find a corresponding open ball B_δ so that points in $B_\delta \cap X$ are mapped to points in B_ϵ.

In practice, we usually establish continuity of a function through the use of Theorems 2.5 and 2.6. These theorems, when interpreted in the context of continuity, tell us the following:

- The sum $\mathbf{F}+\mathbf{G}$ of two functions $\mathbf{F}, \mathbf{G}: X \subseteq \mathbf{R}^n \to \mathbf{R}^m$ that are continuous at $\mathbf{a} \in X$ is continuous at $\mathbf{a}$.
- For all $k \in \mathbf{R}$, the scalar multiple $k\mathbf{F}$ of a function $\mathbf{F}: X \subseteq \mathbf{R}^n \to \mathbf{R}^m$ that is continuous at $\mathbf{a} \in X$ is continuous at $\mathbf{a}$.
- The product fg and the quotient f/g $(g \neq 0)$ of two scalar-valued functions $f, g: X \subseteq \mathbf{R}^n \to \mathbf{R}$ that are continous at $\mathbf{a} \in X$ are continuous at $\mathbf{a}$.
- $\mathbf{F}: X \subseteq \mathbf{R}^n \to \mathbf{R}^m$ is continuous at $\mathbf{a} \in X$ if and only if its component functions $F_i: X \subseteq \mathbf{R}^n \to \mathbf{R}$, $i = 1, \ldots, m$ are all continuous at $\mathbf{a}$.

EXAMPLE 17 The function $\mathbf{f}: \mathbf{R}^2 \to \mathbf{R}^3$ defined by

$$\mathbf{f}(x, y) = (x + y, x^2y, y \sin(xy))$$

is continuous. In view of the remarks above, we can see this by checking that the three component functions

$$f_1(x, y) = x + y, \quad f_2(x, y) = x^2y, \quad \text{and} \quad f_3(x, y) = y \sin(xy)$$

are each continuous (as scalar-valued functions). Now f_1 and f_2 are continuous, since they are polynomials in the two variables x and y. (See Example 15.) The function f_3 is the product of two further functions; that is,

$$f_3(x, y) = g(x, y)h(x, y),$$

where $g(x, y) = y$ and $h(x, y) = \sin(xy)$. The function g is clearly continuous. (It's a polynomial in two variables—one variable doesn't appear explicitly!) The function h is a *composite* of the sine function (which is continuous as a function of one variable) and the continuous function $p(x, y) = xy$. From these remarks, it's not difficult to see that

$$\begin{aligned} \lim_{(x,y)\to(a,b)} h(x, y) &= \lim_{(x,y)\to(a,b)} \sin(p(x, y)) \\ &= \sin\left(\lim_{(x,y)\to(a,b)} p(x, y)\right), \end{aligned}$$

since the sine function is continuous. Thus,

$$\lim_{(x,y)\to(a,b)} h(x, y) = \sin p(a, b) = h(a, b),$$

because p is continuous. Thus, h, hence f_3, and, consequently, $\mathbf{f}$ are all continuous on all of $\mathbf{R}^2$. ◆

The discussion in Example 17 leads us to the following general result, whose proof we omit:

■ **Theorem 2.8** If $\mathbf{f}: X \subseteq \mathbf{R}^n \to \mathbf{R}^m$ and $\mathbf{g}: Y \subseteq \mathbf{R}^m \to \mathbf{R}^p$ are continuous functions such that range $\mathbf{f} \subseteq Y$, then the composite function $\mathbf{g} \circ \mathbf{f}: X \subseteq \mathbf{R}^n \to \mathbf{R}^p$ is defined and is also continuous.

Exercises

In Exercises 1–6, determine whether the given set is open or closed (or neither).

1. $\{(x, y) \in \mathbf{R}^2 \mid 1 < x^2 + y^2 < 4\}$

2. $\{(x, y) \in \mathbf{R}^2 \mid 1 \leq x^2 + y^2 \leq 4\}$

3. $\{(x, y) \in \mathbf{R}^2 \mid 1 \leq x^2 + y^2 < 4\}$

4. $\{(x, y, z) \in \mathbf{R}^3 \mid 1 \leq x^2 + y^2 + z^2 \leq 4\}$

5. $\{(x, y) \in \mathbf{R}^2 \mid -1 < x < 1\} \cup \{(x, y) \in \mathbf{R}^2 \mid x = 2\}$

6. $\{(x, y, z) \in \mathbf{R}^3 \mid 1 < x^2 + y^2 < 4\}$

Evaluate the limits in Exercises 7–21, or explain why the limit fails to exist.

7. $\displaystyle\lim_{(x,y,z)\to(0,0,0)} x^2 + 2xy + yz + z^3 + 2$

8. $\displaystyle\lim_{(x,y)\to(0,0)} \frac{|y|}{\sqrt{x^2 + y^2}}$

9. $\displaystyle\lim_{(x,y)\to(0,0)} \frac{(x + y)^2}{x^2 + y^2}$

10. $\displaystyle\lim_{(x,y)\to(0,0)} \frac{e^x e^y}{x + y + 2}$

11. $\displaystyle\lim_{(x,y)\to(0,0)} \frac{2x^2 + y^2}{x^2 + y^2}$

12. $\displaystyle\lim_{(x,y)\to(-1,2)} \frac{2x^2 + y^2}{x^2 + y^2}$

13. $\displaystyle\lim_{(x,y)\to(0,0)} \frac{x^2 + 2xy + y^2}{x + y}$

14. $\displaystyle\lim_{(x,y)\to(0,0)} \frac{xy}{x^2 + y^2}$

15. $\displaystyle\lim_{(x,y)\to(0,0)} \frac{x^4 - y^4}{x^2 + y^2}$

16. $\displaystyle\lim_{(x,y)\to(0,0)} \frac{x^2}{x^2 + y^2}$

17. $\displaystyle\lim_{(x,y)\to(0,0),x\neq y} \frac{x^2 - xy}{\sqrt{x} - \sqrt{y}}$

18. $\displaystyle\lim_{(x,y)\to(2,0)} \frac{x^2 - y^2 - 4x + 4}{x^2 + y^2 - 4x + 4}$

19. $\displaystyle\lim_{(x,y,z)\to(0,\sqrt{\pi},1)} e^{xz} \cos y^2 - x$

20. $\displaystyle\lim_{(x,y,z)\to(0,0,0)} \frac{2x^2 + 3y^2 + z^2}{x^2 + y^2 + z^2}$

21. $\displaystyle\lim_{(x,y,z)\to(0,0,0)} \frac{xy - xz + yz}{x^2 + y^2 + z^2}$

22. (a) What is $\displaystyle\lim_{\theta\to 0} \frac{\sin\theta}{\theta}$?

(b) What is $\displaystyle\lim_{(x,y)\to(0,0)} \frac{\sin(x + y)}{x + y}$?

(c) What is $\displaystyle\lim_{(x,y)\to(0,0)} \frac{\sin(xy)}{xy}$?

23. Examine the behavior of $f(x, y) = x^4 y^4/(x^2 + y^4)^3$ as (x, y) approaches $(0, 0)$ along various straight lines. From your observations, what might you conjecture $\lim_{(x,y)\to(0,0)} f(x, y)$ to be? Next, consider what happens when (x, y) approaches $(0, 0)$ along the curve $x = y^2$. Does $\lim_{(x,y)\to(0,0)} f(x, y)$ exist? Why or why not?

In Exercises 24–27, (a) use a computer to graph $z = f(x, y)$; (b) use your graph in part (a) to give a geometric discussion as to whether $\lim_{(x,y)\to(0,0)} f(x, y)$ exists; (c) give an analytic (i.e., nongraphical) argument for your answer in part (b).

24. $f(x, y) = \dfrac{4x^2 + 2xy + 5y^2}{3x^2 + 5y^2}$

25. $f(x, y) = \dfrac{x^2 - y}{x^2 + y^2}$

26. $f(x, y) = \dfrac{xy^5}{x^2 + y^{10}}$

27. $f(x, y) = \begin{cases} x \sin \dfrac{1}{y} & \text{if } y \neq 0 \\ 0 & \text{if } y = 0 \end{cases}$

Some limits become easier to identify if we switch to a different coordinate system. In Exercises 28–30 switch from Cartesian to polar coordinates to evaluate the given limits. In Exercises 31–33 switch to spherical coordinates.

28. $\displaystyle\lim_{(x,y)\to(0,0)} \frac{x^2 y}{x^2 + y^2}$

29. $\displaystyle\lim_{(x,y)\to(0,0)} \frac{x^2}{x^2 + y^2}$

30. $\displaystyle\lim_{(x,y)\to(0,0)} \frac{x^2 + xy + y^2}{x^2 + y^2}$

31. $\displaystyle\lim_{(x,y,z)\to(0,0,0)} \frac{xyz}{x^2 + y^2 + z^2}$

32. $\displaystyle\lim_{(x,y,z)\to(0,0,0)} \frac{x^2 + y^2}{\sqrt{x^2 + y^2 + z^2}}$

33. $\displaystyle\lim_{(x,y,z)\to(0,0,0)} \frac{xz}{x^2 + y^2 + z^2}$

In Exercises 34–41, determine whether the functions are continuous throughout their domains:

34. $f(x, y) = x^2 + 2xy - y^7$

35. $f(x, y, z) = x^2 + 3xyz + yz^3 + 2$

36. $g(x, y) = \dfrac{x^2 - y^2}{x^2 + 1}$

37. $h(x, y) = \cos\left(\dfrac{x^2 - y^2}{x^2 + 1}\right)$

38. $f(x, y) = \cos^2 x - 2\sin^2 xy$

39. $f(x, y) = \begin{cases} \dfrac{x^2 - y^2}{x^2 + y^2} & \text{if } (x, y) \neq (0, 0) \\ 0 & \text{if } (x, y) = (0, 0) \end{cases}$

40. $g(x, y) = \begin{cases} \dfrac{x^3 + x^2 + xy^2 + y^2}{x^2 + y^2} & \text{if } (x, y) \neq (0, 0) \\ 2 & \text{if } (x, y) = (0, 0) \end{cases}$

41. $\mathbf{F}(x, y, z) = \left(x^2 + 3xy, \dfrac{e^x e^y}{2x^2 + y^4 + 3}, \sin\left(\dfrac{xy}{y^2 + 1}\right)\right)$

42. Determine the value of the constant c so that

$$g(x, y) = \begin{cases} \dfrac{x^3 + xy^2 + 2x^2 + 2y^2}{x^2 + y^2} & \text{if } (x, y) \neq (0, 0) \\ c & \text{if } (x, y) = (0, 0) \end{cases}$$

is continuous.

43. Show that the function $f: \mathbf{R}^3 \to \mathbf{R}$ given by $f(\mathbf{x}) = (2\mathbf{i} - 3\mathbf{j} + \mathbf{k}) \cdot \mathbf{x}$ is continuous.

44. Show that the function $\mathbf{f}: \mathbf{R}^3 \to \mathbf{R}^3$ given by $\mathbf{f}(\mathbf{x}) = (6\mathbf{i} - 5\mathbf{k}) \times \mathbf{x}$ is continuous.

Exercises 45–49 involve Definition 2.2 of the limit.

45. Consider the function $f(x) = 2x - 3$.

(a) Show that if $|x - 5| < \delta$, then $|f(x) - 7| < 2\delta$.

(b) Use part (a) to prove that $\lim_{x\to 5} f(x) = 7$.

46. Consider the function $f(x, y) = 2x - 10y + 3$.

(a) Show that if $\|(x, y) - (5, 1)\| < \delta$, then $|x - 5| < \delta$ and $|y - 1| < \delta$.

(b) Use part (a) to show that if $\|(x, y) - (5, 1)\| < \delta$, then $|f(x, y) - 3| < 12\delta$.

(c) Show that $\lim_{(x,y)\to(5,1)} f(x, y) = 3$.

47. If A, B, and C are constants and $f(x, y) = Ax + By + C$, show that

$$\lim_{(x,y)\to(x_0,y_0)} f(x, y) = f(x_0, y_0) = Ax_0 + By_0 + C.$$

48. In this problem, you will establish rigorously that

$$\lim_{(x,y)\to(0,0)} \frac{x^3 + y^3}{x^2 + y^2} = 0.$$

(a) Show that $|x| \leq \|(x, y)\|$ and $|y| \leq \|(x, y)\|$.

(b) Show that $|x^3 + y^3| \leq 2(x^2 + y^2)^{3/2}$. (Hint: Begin with the triangle inequality, and then use part (a).)

(c) Show that if $0 < \|(x, y)\| < \delta$, then $|(x^3 + y^3)/(x^2 + y^2)| < 2\delta$.

(d) Now prove that $\lim_{(x,y)\to(0,0)} (x^3 + y^3)/(x^2 + y^2) = 0$.

49. (a) If a and b are any real numbers, show that $2|ab| \leq a^2 + b^2$.

(b) Let

$$f(x, y) = xy\left(\frac{x^2 - y^2}{x^2 + y^2}\right).$$

Use part (a) to show that if $0 < \|(x, y)\| < \delta$, then $|f(x, y)| < \delta^2/2$.

(c) Prove that $\lim_{(x,y)\to(0,0)} f(x, y)$ exists, and find its value.

2.3 The Derivative

Our goal for this section is to define the derivative of a function $\mathbf{f}: X \subseteq \mathbf{R}^n \to \mathbf{R}^m$, where n and m are arbitrary positive integers. Predictably, the derivative of a vector-valued function of several variables is a more complicated object than the derivative of a scalar-valued function of a single variable. In addition, the notion of differentiability is quite subtle in the case of a function of more than one variable.

We first define the basic computational tool of partial derivatives. After doing so, we can begin to understand differentiability via the geometry of tangent planes to surfaces. Finally, we generalize these relatively concrete ideas to higher dimensions.

Partial Derivatives

Recall that if $F: X \subseteq \mathbf{R} \to \mathbf{R}$ is a scalar-valued function of one variable, then the **derivative** of F at a number $a \in X$ is

$$F'(a) = \lim_{h\to 0} \frac{F(a + h) - F(a)}{h}. \tag{1}$$

Moreover, F is said to be **differentiable at** a precisely when the limit in equation (1) exists.

Definition 3.1 Suppose $f: X \subseteq \mathbf{R}^n \to \mathbf{R}$ is a scalar-valued function of n variables. Let $\mathbf{x} = (x_1, x_2, \ldots, x_n)$ denote a point of $\mathbf{R}^n$. A **partial function** F **with respect to the variable** x_i is a one-variable function obtained from f by holding all variables constant except x_i. That is, we set x_j equal to a constant a_j for $j \neq i$. Then, the partial function in x_i is defined by

$$F(x_i) = f(a_1, a_2, \ldots, x_i, \ldots, a_n).$$

EXAMPLE 1 If $f(x, y) = (x^2 - y^2)/(x^2 + y^2)$, then the partial functions with respect to x are given by

$$F(x) = f(x, a_2) = \frac{x^2 - a_2^2}{x^2 + a_2^2},$$

where a_2 may be any constant. If, for example, $a_2 = 0$, then the partial function is

$$F(x) = f(x, 0) = \frac{x^2}{x^2} \equiv 1.$$

Geometrically, this partial function is nothing more than the restriction of f to the horizontal line $y = 0$. Note that since the origin is not in the domain of f, 0 should not be taken to be in the domain of F. (See Figure 2.44.) ◆

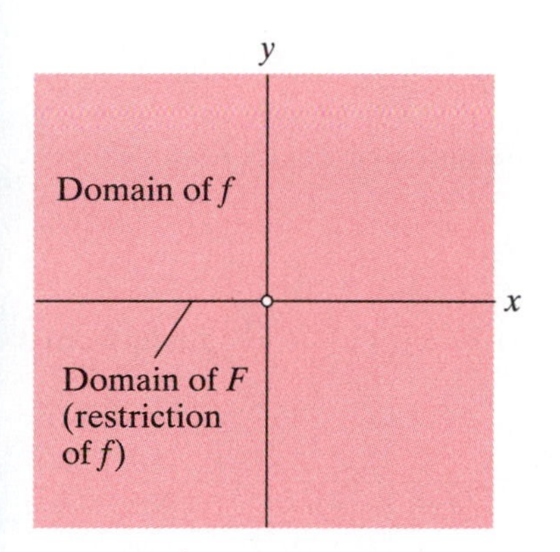

Figure 2.44 The function f of Example 1 is defined on $\mathbf{R}^2 - \{(0,0)\}$, while its partial function F along $y = 0$ is defined on the x-axis minus the origin.

REMARK In practice, we usually do not go to the notational trouble of explicitly replacing the x_j's ($j \neq i$) by constants when working with partial functions. Instead, we make a mental note that the partial function is obtained by allowing only one variable to vary, while all the other variables are held fixed.

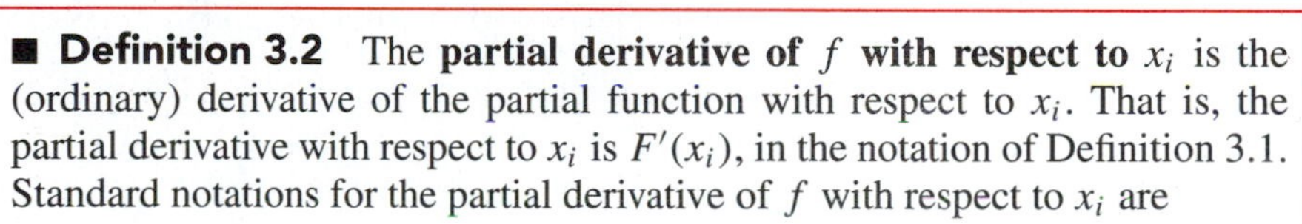

Definition 3.2 The **partial derivative of** f **with respect to** x_i is the (ordinary) derivative of the partial function with respect to x_i. That is, the partial derivative with respect to x_i is $F'(x_i)$, in the notation of Definition 3.1. Standard notations for the partial derivative of f with respect to x_i are

$$\frac{\partial f}{\partial x_i}, \quad D_{x_i} f(x_1, \ldots, x_n), \quad \text{and} \quad f_{x_i}(x_1, \ldots, x_n).$$

Symbolically, we have

$$\frac{\partial f}{\partial x_i} = \lim_{h \to 0} \frac{f(x_1, \ldots, x_i + h, \ldots, x_n) - f(x_1, \ldots, x_n)}{h}. \tag{2}$$

By definition, the partial derivative is the (instantaneous) rate of change of f when all variables, except the specified one, are held fixed. In the case where f is a (scalar-valued) function of two variables, we can understand

$$\frac{\partial f}{\partial x}(a, b)$$

geometrically as the slope at the point $(a, b, f(a, b))$ of the curve obtained by intersecting the surface $z = f(x, y)$ with the plane $y = b$, as shown in Figure 2.45.

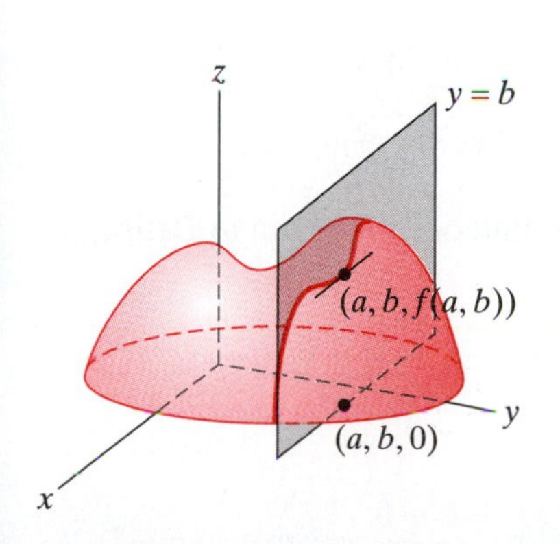

Figure 2.45 Visualizing the partial derivative $\frac{\partial f}{\partial x}(a, b)$.

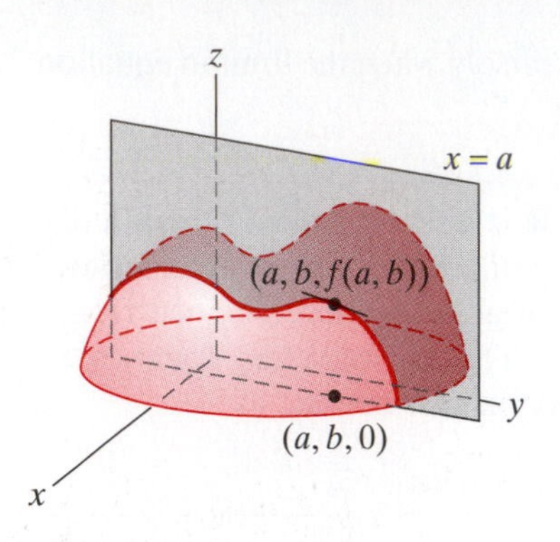

Figure 2.46 Visualizing the partial derivative $\frac{\partial f}{\partial y}(a, b)$.

Similarly,

$$\frac{\partial f}{\partial y}(a, b)$$

is the slope at $(a, b, f(a, b))$ of the curve formed by the intersection of $z = f(x, y)$ and $x = a$, shown in Figure 2.46.

EXAMPLE 2 For the most part, partial derivatives are quite easy to compute, once you become adept at treating variables like constants. If

$$f(x, y) = x^2 y + \cos(x + y),$$

then we have

$$\frac{\partial f}{\partial x} = 2xy - \sin(x + y).$$

(Imagine y to be a constant throughout the differentiation process.) Also,

$$\frac{\partial f}{\partial y} = x^2 - \sin(x + y).$$

(Imagine x to be a constant.) Similarly, if $g(x, y) = xy/(x^2 + y^2)$, then, from the quotient rule of ordinary calculus, we have

$$g_x(x, y) = \frac{(x^2 + y^2)y - xy(2x)}{(x^2 + y^2)^2} = \frac{y(y^2 - x^2)}{(x^2 + y^2)^2},$$

and

$$g_y(x, y) = \frac{(x^2 + y^2)x - xy(2y)}{(x^2 + y^2)^2} = \frac{x(x^2 - y^2)}{(x^2 + y^2)^2}.$$

Note that, of course, neither g nor its partial derivatives are defined at $(0, 0)$. ◆

EXAMPLE 3 Occasionally, it is necessary to appeal explicitly to limits to evaluate partial derivatives. Suppose $f: \mathbf{R}^2 \to \mathbf{R}$ is defined by

$$f(x, y) = \begin{cases} \dfrac{3x^2 y - y^3}{x^2 + y^2} & \text{if } (x, y) \neq (0, 0) \\ 0 & \text{if } (x, y) = (0, 0) \end{cases}.$$

Then, for $(x, y) \neq (0, 0)$, we have

$$\frac{\partial f}{\partial x} = \frac{8xy^3}{(x^2 + y^2)^2} \quad \text{and} \quad \frac{\partial f}{\partial y} = \frac{3x^4 - 6x^2y^2 - y^4}{(x^2 + y^2)^2}.$$

But what should $\frac{\partial f}{\partial x}(0, 0)$ and $\frac{\partial f}{\partial y}(0, 0)$ be? To find out, we return to Definition 3.2 of the partial derivatives:

$$\frac{\partial f}{\partial x}(0, 0) = \lim_{h \to 0} \frac{f(0 + h, 0) - f(0, 0)}{h} = \lim_{h \to 0} \frac{0 - 0}{h} = 0,$$

and

$$\frac{\partial f}{\partial y}(0, 0) = \lim_{h \to 0} \frac{f(0, 0 + h) - f(0, 0)}{h} = \lim_{h \to 0} \frac{-h - 0}{h} = \lim_{h \to 0} -1 = -1.$$

◆

Tangency and Differentiability

If $F: X \subseteq \mathbf{R} \to \mathbf{R}$ is a scalar-valued function of one variable, then to have F differentiable at a number $a \in X$ means precisely that the graph of the curve $y = F(x)$ has a tangent line at the point $(a, F(a))$. (See Figure 2.47.) Moreover, this tangent line is given by the equation

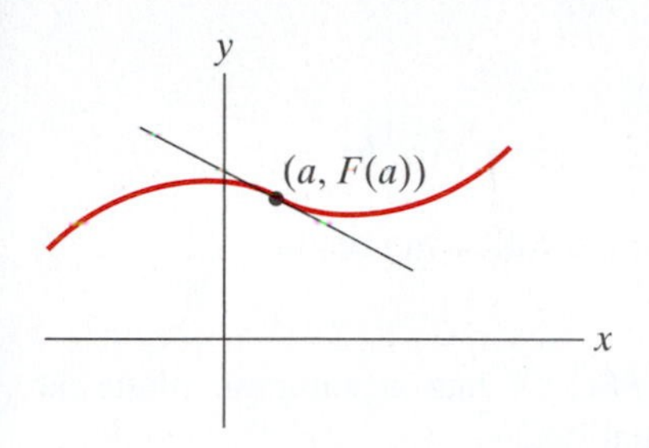

Figure 2.47 The tangent line to $y = F(x)$ at $x = a$ has equation $y = F(a) + F'(a)(x - a)$.

$$y = F(a) + F'(a)(x - a). \tag{3}$$

If we define the function $H(x)$ to be $F(a) + F'(a)(x - a)$ (i.e., $H(x)$ is the right side of equation (3) that gives the equation for the tangent line), then H has two properties:

1. $H(a) = F(a)$
2. $H'(a) = F'(a)$.

In other words, the line defined by $y = H(x)$ passes through the point $(a, F(a))$ and has the same slope at $(a, F(a))$ as the curve defined by $y = F(x)$. (Hence, the term "tangent line.")

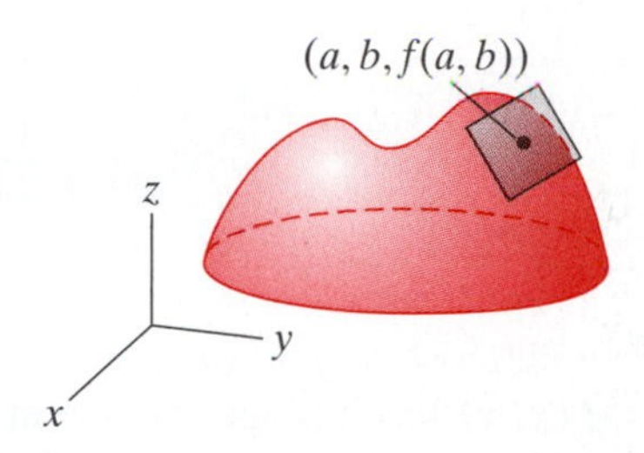

Figure 2.48 The plane tangent to $z = f(x, y)$ at $(a, b, f(a, b))$.

Now suppose $f: X \subseteq \mathbf{R}^2 \to \mathbf{R}$ is a scalar-valued function of two variables, where X is open in $\mathbf{R}^2$. Then the graph of f is a surface. What should the **tangent plane** to the graph of $z = f(x, y)$ at the point $(a, b, f(a, b))$ be? Geometrically, the situation is as depicted in Figure 2.48. From our earlier observations, we know that the partial derivative $f_x(a, b)$ is the slope of the line tangent at the point $(a, b, f(a, b))$ to the curve obtained by intersecting the surface $z = f(x, y)$ with the plane $y = b$. (See Figure 2.49.) This means that if we travel along this tangent line, then for every unit change in the positive x-direction, there's a change of $f_x(a, b)$ units in the z-direction. Hence, by using formula (1) of §1.2, the tangent line is given in vector parametric form as

$$\mathbf{l}_1(t) = (a, b, f(a, b)) + t(1, 0, f_x(a, b)).$$

Thus, a vector parallel to this tangent line is

$$\mathbf{u} = \mathbf{i} + f_x(a, b)\,\mathbf{k}.$$

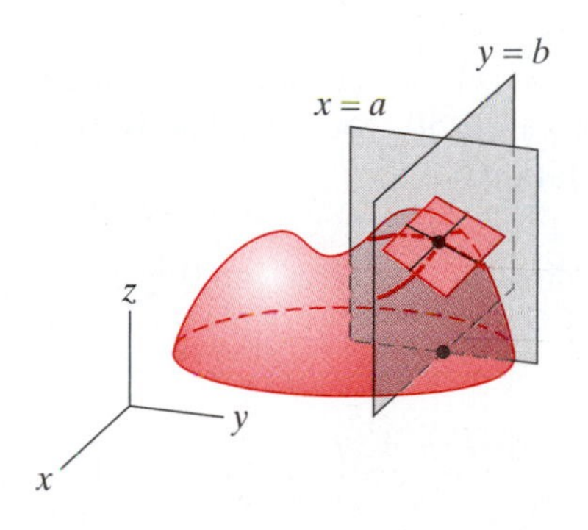

Figure 2.49 The tangent plane at $(a, b, f(a, b))$ contains the lines tangent to the curves formed by intersecting the surface $z = f(x, y)$ by the planes $x = a$ and $y = b$.

Similarly, the partial derivative $f_y(a, b)$ is the slope of the line tangent at the point $(a, b, f(a, b))$ to the curve obtained by intersecting the surface $z = f(x, y)$ with the plane $x = a$. (Again see Figure 2.49.) Consequently, the tangent line is given by

$$\mathbf{l}_2(t) = (a, b, f(a, b)) + t(0, 1, f_y(a, b)),$$

so a vector parallel to this tangent line is

$$\mathbf{v} = \mathbf{j} + f_y(a, b)\,\mathbf{k}.$$

Both of the aforementioned tangent lines must be contained in the plane tangent to $z = f(x, y)$ at $(a, b, f(a, b))$, if one exists. Hence, a vector $\mathbf{n}$ *normal* to the tangent plane must be perpendicular to both $\mathbf{u}$ and $\mathbf{v}$. Therefore, we may take $\mathbf{n}$ to be

$$\mathbf{n} = \mathbf{u} \times \mathbf{v} = -f_x(a, b)\,\mathbf{i} - f_y(a, b)\,\mathbf{j} + \mathbf{k}.$$

Now, use equation (1) of §1.5 to find that the equation for the tangent plane—that is, the plane through $(a, b, f(a, b))$ with normal $\mathbf{n}$—is

$$(-f_x(a, b), -f_y(a, b), 1) \cdot (x - a, y - b, z - f(a, b)) = 0$$

or, equivalently,

$$-f_x(a, b)(x - a) - f_y(a, b)(y - b) + z - f(a, b) = 0.$$

By rewriting this last equation, we have shown the following result:

Theorem 3.3 If the graph of $z = f(x, y)$ has a tangent plane at $(a, b, f(a, b))$, then that tangent plane has equation

$$z = f(a, b) + f_x(a, b)(x - a) + f_y(a, b)(y - b). \tag{4}$$

Note that if we define the function $h(x, y)$ to be equal to $f(a, b) + f_x(a, b)(x - a) + f_y(a, b)(y - b)$ (i.e., $h(x, y)$ is the right side of equation (4)), then h has the following properties:

1. $h(a, b) = f(a, b)$
2. $\dfrac{\partial h}{\partial x}(a, b) = \dfrac{\partial f}{\partial x}(a, b)$ and $\dfrac{\partial h}{\partial y}(a, b) = \dfrac{\partial f}{\partial y}(a, b)$.

In other words, h and its partial derivatives agree with those of f at (a, b).

It is tempting to think that the surface $z = f(x, y)$ has a tangent plane at $(a, b, f(a, b))$ as long as you can make sense of equation (4), that is, as long as the partial derivatives $f_x(a, b)$ and $f_y(a, b)$ exist. Indeed, this would be analogous to the one-variable situation where the existence of the derivative and the existence of the tangent line mean exactly the same thing. However, it is possible for a function of two variables to have well-defined partial derivatives (so that equation (4) makes sense), yet *not* have a tangent plane.

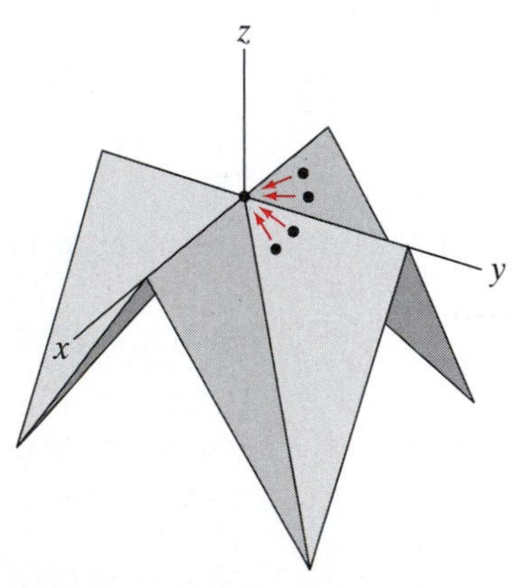

Figure 2.50 If two points approach (0, 0, 0) while remaining on one face of the surface described in Example 4, the limiting plane they and (0, 0, 0) determine is different from the one determined by letting the two points approach (0, 0, 0) while remaining on another face.

EXAMPLE 4 Let $f(x, y) = ||x| - |y|| - |x| - |y|$ and consider the surface defined by the graph of $z = f(x, y)$ shown in Figure 2.50. The partial derivatives of f at the origin may be calculated from Definition 3.2 as

$$f_x(0, 0) = \lim_{h \to 0} \frac{f(0 + h, 0) - f(0, 0)}{h} = \lim_{h \to 0} \frac{||h|| - |h|}{h} = \lim_{h \to 0} 0 = 0$$

and

$$f_y(0, 0) = \lim_{h \to 0} \frac{f(0, 0 + h) - f(0, 0)}{h} = \lim_{h \to 0} \frac{|-|h|| - |h|}{h} = \lim_{h \to 0} 0 = 0.$$

(Indeed, the partial functions $F(x) = f(x, 0)$ and $G(y) = f(0, y)$ are both identically zero and, thus, have zero derivatives.) Consequently, *if* the surface in question has a tangent plane at the origin, then equation (4) tells us that it has equation $z = 0$. But there is no geometric sense in which the surface $z = f(x, y)$ has a tangent plane at the origin. If we think of a tangent plane as the geometric limit of planes that pass through the point of tangency and two other "moving" points on the surface as those two points approach the point of tangency, then Figure 2.50 shows that there is no uniquely determined limiting plane. ◆

Example 4 shows that the existence of a tangent plane to the graph of $z = f(x, y)$ is a stronger condition than the existence of partial derivatives. It turns out that such a stronger condition is more useful in that theorems from the calculus of functions of a single variable carry over to the context of functions of several variables. What we must do now is find a suitable analytic definition of differentiability that captures this idea. We begin by looking at the definition of the one-variable derivative with fresh eyes.

By replacing the quantity $a+h$ by the variable x, the limit equation in formula (1) may be rewritten as

$$F'(a) = \lim_{x \to a} \frac{F(x) - F(a)}{x - a}.$$

This is equivalent to the equation

$$\lim_{x \to a} \left(\frac{F(x) - F(a)}{x - a} \right) - F'(a) = 0.$$

The quantity $F'(a)$ does not depend on x and therefore may be brought inside the limit. We thus obtain the equation

$$\lim_{x \to a} \left\{ \frac{F(x) - F(a)}{x - a} - F'(a) \right\} = 0.$$

Finally, some easy algebra enables us to conclude that the function F is differentiable at a if there is a number $F'(a)$ such that

$$\lim_{x \to a} \frac{F(x) - [F(a) + F'(a)(x - a)]}{x - a} = 0. \tag{5}$$

What have we learned from writing equation (5)? Note that the expression in brackets in the numerator of the limit expression in equation (5) is the function $H(x)$ that was used to define the tangent line to $y = F(x)$ at $(a, F(a))$. Thus, we may rewrite equation (5) as

$$\lim_{x \to a} \frac{F(x) - H(x)}{x - a} = 0.$$

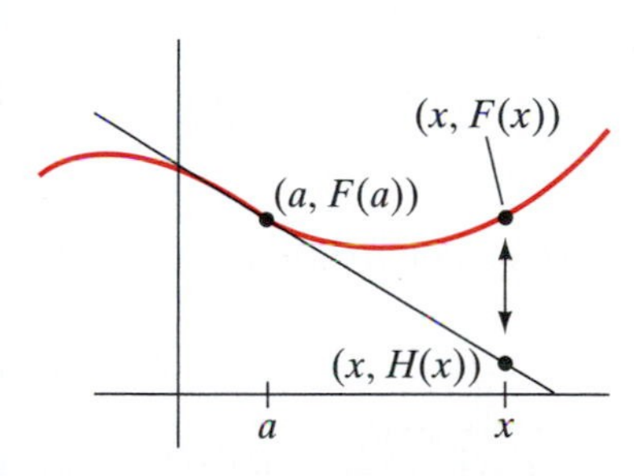

Figure 2.51 If F is differentiable at a, the vertical distance between $F(x)$ and $H(x)$ must approach zero faster than the horizontal distance between x and a does.

For the limit above to be zero, we certainly must have that the limit of the numerator is zero. But since the limit of the denominator is also zero, we can say even more, namely, that the difference between the y-values of the graph of F and of its tangent line must approach zero faster than x approaches a. This is what is meant when we say that "H is a good linear approximation to F near a." (See Figure 2.51.) Geometrically, it means that, near the point of tangency, the graph of $y = F(x)$ is approximately straight like the graph of $y = H(x)$.

If we now pass to the case of a scalar-valued function $f(x, y)$ of two variables, then to say that $z = f(x, y)$ has a tangent plane at $(a, b, f(a, b))$ (i.e., that f is differentiable at (a, b)) should mean that the vertical distance between the graph of f and the "candidate" tangent plane given by

$$z = h(x, y) = f(a, b) + f_x(a, b)(x - a) + f_y(a, b)(y - b)$$

must approach zero faster than the point (x, y) approaches (a, b). (See Figure 2.52.) In other words, near the point of tangency, the graph of $z = f(x, y)$ is approximately flat just like the graph of $z = h(x, y)$. We can capture this geometric idea with the following formal definition of differentiability:

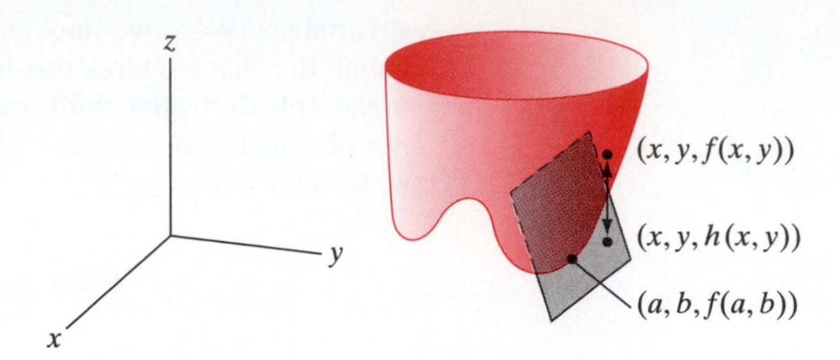

Figure 2.52 If f is differentiable at (a, b), the distance between $f(x, y)$ and $h(x, y)$ must approach zero faster than the distance between (x, y) and (a, b) does.

■ **Definition 3.4** Let X be open in $\mathbf{R}^2$ and $f: X \subseteq \mathbf{R}^2 \to \mathbf{R}$ be a scalar-valued function of two variables. We say that f is **differentiable at** $(a, b) \in X$ if the partial derivatives $f_x(a, b)$ and $f_y(a, b)$ exist and if the function

$$h(x, y) = f(a, b) + f_x(a, b)(x - a) + f_y(a, b)(y - b)$$

is a good linear approximation to f near (a, b)—that is, if

$$\lim_{(x,y)\to(a,b)} \frac{f(x, y) - h(x, y)}{\|(x, y) - (a, b)\|} = 0.$$

Moreover, if f is differentiable at (a, b), then the equation $z = h(x, y)$ defines the **tangent plane** to the graph of f at the point $(a, b, f(a, b))$. If f is differentiable at all points of its domain, then we simply say that f is **differentiable.**

EXAMPLE 5 Let us return to the function $f(x, y) = ||x| - |y|| - |x| - |y|$ of Example 4. We already know that the partial derivatives $f_x(0, 0)$ and $f_y(0, 0)$ exist and equal zero. Thus, the function h of Definition 3.4 is the zero function. Consequently, f will be differentiable at (0,0) just in case

$$\lim_{(x,y)\to(0,0)} \frac{f(x, y) - h(x, y)}{\|(x, y) - (0, 0)\|} = \lim_{(x,y)\to(0,0)} \frac{f(x, y)}{\|(x, y)\|} = \lim_{(x,y)\to(0,0)} \frac{||x| - |y|| - |x| - |y|}{\sqrt{x^2 + y^2}}$$

is zero. However, it is not hard to see that the limit in question fails to exist. Along the line $y = 0$, we have

$$\frac{f(x, y)}{\|(x, y)\|} = \frac{||x| - 0| - |x| - |0|}{\sqrt{x^2}} = \frac{0}{|x|} = 0,$$

but along the line $y = x$, we have

$$\frac{f(x, y)}{\|(x, y)\|} = \frac{||x| - |x|| - |x| - |x|}{\sqrt{x^2 + x^2}} = \frac{-2|x|}{\sqrt{2}|x|} = -\sqrt{2}.$$

Hence, f fails to be differentiable at (0,0) and has no tangent plane at (0,0,0). ◆

The limit condition in Definition 3.4 can be difficult to apply in practice. Fortunately, the following result, which we will not prove, simplifies matters in many instances. Recall from Definition 2.3 that the phrase "a **neighborhood** of a point P in a set X" just means an open set containing P and contained in X.

Theorem 3.5 Suppose X is open in $\mathbf{R}^2$. If $f\colon X \to \mathbf{R}$ has continuous partial derivatives in a neighborhood of (a, b) in X, then f is differentiable at (a, b).

A proof of a more general result (Theorem 3.10) is provided in the addendum to this section.

EXAMPLE 6 Let $f(x, y) = x^2 + 2y^2$. Then $\partial f/\partial x = 2x$ and $\partial f/\partial y = 4y$, both of which are continuous functions on all of $\mathbf{R}^2$. Thus, Theorem 3.5 implies that f is differentiable everywhere. The surface $z = x^2 + 2y^2$ must therefore have a tangent plane at every point. At the point $(2, -1)$, for example, this tangent plane is given by the equation

$$z = 6 + 4(x - 2) - 4(y + 1)$$

(or, equivalently, by $4x - 4y - z = 6$). ◆

While we're on the subject of continuity and differentiability, the next result is the multivariable analogue of a familiar theorem about functions of one variable.

Theorem 3.6 If $f\colon X \subseteq \mathbf{R}^2 \to \mathbf{R}$ is differentiable at (a, b), then it is continuous at (a, b).

EXAMPLE 7 Let the function $f\colon \mathbf{R}^2 \to \mathbf{R}$ be defined by

$$f(x, y) = \begin{cases} \dfrac{x^2y^2}{x^4 + y^4} & \text{if } (x, y) \neq (0, 0) \\ 0 & \text{if } (x, y) = (0, 0) \end{cases}.$$

The function f is not continuous at the origin, since $\lim_{(x,y)\to(0,0)} f(x, y)$ does not exist. (However, f is continuous everywhere else in $\mathbf{R}^2$.) By Theorem 3.6, f therefore cannot be differentiable at the origin. Nonetheless, the partial derivatives of f do exist at the origin, and we have

$$f(x, 0) = \frac{0}{x^4 + 0} \equiv 0 \quad \Longrightarrow \quad \frac{\partial f}{\partial x}(0, 0) = 0,$$

and

$$f(0, y) = \frac{0}{0 + y^4} \equiv 0 \quad \Longrightarrow \quad \frac{\partial f}{\partial y}(0, 0) = 0,$$

since the partial functions are constant. Thus, we see that if we want something like Theorem 3.6 to be true, the existence of partial derivatives alone is not enough. ◆

Differentiability in General

It is not difficult now to see how to generalize Definition 3.4 to three (or more) variables: For a scalar-valued function of three variables to be differentiable at a point (a, b, c), we must have that (i) the three partial derivatives exist at (a, b, c) and (ii) the function $h\colon \mathbf{R}^3 \to \mathbf{R}$ defined by

$$\begin{aligned} h(x, y, z) = f(a, b, c) &+ f_x(a, b, c)(x - a) \\ &+ f_y(a, b, c)(y - b) + f_z(a, b, c)(z - c) \end{aligned}$$

is a good linear approximation to f near (a, b, c). In other words, (ii) means that

$$\lim_{(x,y,z)\to(a,b,c)} \frac{f(x, y, z) - h(x, y, z)}{\|(x, y, z) - (a, b, c)\|} = 0.$$

The passage from three variables to arbitrarily many is now straightforward.

■ **Definition 3.7** Let X be open in $\mathbf{R}^n$ and $f: X \to \mathbf{R}$ be a scalar-valued function; let $\mathbf{a} = (a_1, a_2, \ldots, a_n) \in X$. We say that f is **differentiable at a** if all the partial derivatives $f_{x_i}(\mathbf{a})$, $i = 1, \ldots, n$, exist and if the function $h: \mathbf{R}^n \to \mathbf{R}$ defined by

$$\begin{aligned} h(\mathbf{x}) = f(\mathbf{a}) &+ f_{x_1}(\mathbf{a})(x_1 - a_1) + f_{x_2}(\mathbf{a})(x_2 - a_2) \\ &+ \cdots + f_{x_n}(\mathbf{a})(x_n - a_n) \end{aligned} \tag{6}$$

is a good linear approximation to f near $\mathbf{a}$, meaning that

$$\lim_{\mathbf{x}\to\mathbf{a}} \frac{f(\mathbf{x}) - h(\mathbf{x})}{\|\mathbf{x} - \mathbf{a}\|} = 0.$$

We can use vector and matrix notation to rewrite things a bit. Define the **gradient** of a scalar-valued function $f: X \subseteq \mathbf{R}^n \to \mathbf{R}$ to be the *vector*

$$\nabla f(\mathbf{x}) = \left(\frac{\partial f}{\partial x_1}, \frac{\partial f}{\partial x_2}, \ldots, \frac{\partial f}{\partial x_n}\right).$$

Consequently,

$$\nabla f(\mathbf{a}) = (f_{x_1}(\mathbf{a}), f_{x_2}(\mathbf{a}), \ldots, f_{x_n}(\mathbf{a})).$$

Alternatively, we can use matrix notation and define the **derivative** of f at $\mathbf{a}$, denoted $Df(\mathbf{a})$, to be the row matrix whose entries are the components of $\nabla f(\mathbf{a})$, that is,

$$Df(\mathbf{a}) = \begin{bmatrix} f_{x_1}(\mathbf{a}) & f_{x_2}(\mathbf{a}) & \cdots & f_{x_n}(\mathbf{a}) \end{bmatrix}.$$

Then, by identifying the vector $\mathbf{x} - \mathbf{a}$ with the $n \times 1$ column matrix whose entries are the components of $\mathbf{x} - \mathbf{a}$, we have

$$\begin{aligned} \nabla f(\mathbf{a}) \cdot (\mathbf{x} - \mathbf{a}) = Df(\mathbf{a})(\mathbf{x} - \mathbf{a}) &= \begin{bmatrix} f_{x_1}(\mathbf{a}) & f_{x_2}(\mathbf{a}) & \cdots & f_{x_n}(\mathbf{a}) \end{bmatrix} \begin{bmatrix} x_1 - a_1 \\ x_2 - a_2 \\ \vdots \\ x_n - a_n \end{bmatrix} \\ &= f_{x_1}(\mathbf{a})(x_1 - a_1) + f_{x_2}(\mathbf{a})(x_2 - a_2) \\ &\quad + \cdots + f_{x_n}(\mathbf{a})(x_n - a_n). \end{aligned}$$

Hence, vector notation allows us to rewrite equation (6) quite compactly as

$$h(\mathbf{x}) = f(\mathbf{a}) + \nabla f(\mathbf{a}) \cdot (\mathbf{x} - \mathbf{a}).$$

Thus, to say that h is a good linear approximation to f near $\mathbf{a}$ in equation (6) means that

$$\lim_{\mathbf{x}\to\mathbf{a}} \frac{f(\mathbf{x}) - [f(\mathbf{a}) + \nabla f(\mathbf{a}) \cdot (\mathbf{x} - \mathbf{a})]}{\|\mathbf{x} - \mathbf{a}\|} = 0. \tag{7}$$

Compare equation (7) with equation (5). Differentiability of functions of one and several variables should really look very much the same to you. It is worth noting that the analogues of Theorems 3.5 and 3.6 hold in the case of n variables.

The gradient of a function is an extremely important construction, and we consider it in greater detail in §2.6.

You may be wondering what, if any, geometry is embedded in this general notion of differentiability. Recall that the graph of the function $f: X \subseteq \mathbf{R}^n \to \mathbf{R}$ is the **hypersurface** in $\mathbf{R}^{n+1}$ given by the equation $x_{n+1} = f(x_1, x_2, \ldots, x_n)$. (See equation (2) of §2.1.) If f is differentiable at $\mathbf{a}$, then the hypersurface determined by the graph has a **tangent hyperplane** at $(\mathbf{a}, f(\mathbf{a}))$ given by the equation

$$\begin{aligned} x_{n+1} = h(x_1, x_2, \ldots, x_n) &= f(\mathbf{a}) + \nabla f(\mathbf{a}) \cdot (\mathbf{x} - \mathbf{a}) \\ &= f(\mathbf{a}) + Df(\mathbf{a})(\mathbf{x} - \mathbf{a}). \end{aligned} \tag{8}$$

Compare equation (8) with equation (3) for the tangent line to the curve $y = F(x)$ at $(\mathbf{a}, F(\mathbf{a}))$. Although we cannot visualize the graph of a function of more than two variables, nonetheless, we can use vector notation to lend real meaning to tangency in n dimensions.

EXAMPLE 8 Before we drown in a sea of abstraction and generalization, let's do some concrete computation. An example of an "n-dimensional paraboloid" in $\mathbf{R}^{n+1}$ is given by the equation

$$x_{n+1} = x_1^2 + x_2^2 + \cdots + x_n^2,$$

that is, by the graph of the function $f(x_1, \ldots, x_n) = x_1^2 + x_2^2 + \cdots + x_n^2$. We have

$$\frac{\partial f}{\partial x_i} = 2x_i, \quad i = 1, 2, \ldots, n,$$

so that

$$\nabla f(x_1, \ldots, x_n) = (2x_1, 2x_2, \ldots, 2x_n).$$

Note that the partial derivatives of f are continuous everywhere. Hence, the n-dimensional version of Theorem 3.5 tells us that f is differentiable everywhere. In particular, f is differentiable at the point $(1, 2, \ldots, n)$,

$$\nabla f(1, 2, \ldots, n) = (2, 4, \ldots, 2n),$$

and

$$Df(1, 2, \ldots, n) = \begin{bmatrix} 2 & 4 & \cdots & 2n \end{bmatrix}.$$

Thus, the paraboloid has a tangent hyperplane at the point

$$(1, 2, \ldots, n, 1^2 + 2^2 + \cdots + n^2)$$

whose equation is given by equation (8):

$$\begin{aligned} x_{n+1} &= (1^2 + 2^2 + \cdots + n^2) + \begin{bmatrix} 2 & 4 & \cdots & 2n \end{bmatrix} \begin{bmatrix} x_1 - 1 \\ x_2 - 2 \\ \vdots \\ x_n - n \end{bmatrix} \\ &= (1^2 + 2^2 + \cdots + n^2) + 2(x_1 - 1) + 4(x_2 - 2) + \cdots + 2n(x_n - n) \\ &= (1^2 + 2^2 + \cdots + n^2) + 2x_1 + 4x_2 + \cdots + 2nx_n \\ &\quad - (2 \cdot 1 + 4 \cdot 2 + \cdots + 2n \cdot n) \\ &= 2x_1 + 4x_2 + \cdots + 2nx_n - (1^2 + 2^2 + \cdots + n^2) \\ &= \sum_{i=1}^{n} 2ix_i - \frac{n(n+1)(2n+1)}{6}. \end{aligned}$$

(The formula $1^2+2^2+\cdots+n^2 = n(n+1)(2n+1)/6$ is a well-known identity, encountered when you first learned about the definite integral. It's straightforward to prove using mathematical induction.) ◆

At last we're ready to take a look at differentiability in the most general setting of all. Let X be open in $\mathbf{R}^n$ and let $\mathbf{f}: X \to \mathbf{R}^m$ be a vector-valued function of n variables. We define the **matrix of partial derivatives** of $\mathbf{f}$, denoted $D\mathbf{f}$, to be the $m \times n$ matrix whose ijth entry is $\partial f_i/\partial x_j$, where $f_i: X \subseteq \mathbf{R}^n \to \mathbf{R}$ is the ith component function of $\mathbf{f}$. That is,

$$D\mathbf{f}(x_1, x_2, \ldots, x_n) = \begin{bmatrix} \dfrac{\partial f_1}{\partial x_1} & \dfrac{\partial f_1}{\partial x_2} & \cdots & \dfrac{\partial f_1}{\partial x_n} \\ \dfrac{\partial f_2}{\partial x_1} & \dfrac{\partial f_2}{\partial x_2} & \cdots & \dfrac{\partial f_2}{\partial x_n} \\ \vdots & \vdots & \ddots & \vdots \\ \dfrac{\partial f_m}{\partial x_1} & \dfrac{\partial f_m}{\partial x_2} & \cdots & \dfrac{\partial f_m}{\partial x_n} \end{bmatrix}.$$

The ith row of $D\mathbf{f}$ is nothing more than Df_i—and the entries of Df_i are precisely the components of the gradient vector ∇f_i. (Indeed, in the case where $m = 1$, ∇f and Df mean exactly the same thing.)

EXAMPLE 9 Suppose $\mathbf{f}: \mathbf{R}^3 \to \mathbf{R}^2$ is given by $\mathbf{f}(x, y, z) = (x \cos y + z, xy)$. Then, we have

$$D\mathbf{f}(x, y, z) = \begin{bmatrix} \cos y & -x \sin y & 1 \\ y & x & 0 \end{bmatrix}.$$

◆

We generalize equation (7) and Definition 3.7 in an obvious way to make the following definition:

> ■ **Definition 3.8** (GRAND DEFINITION OF DIFFERENTIABILITY) Let $X \subseteq \mathbf{R}^n$ be open, let $\mathbf{f}: X \to \mathbf{R}^m$, and let $\mathbf{a} \in X$. We say that $\mathbf{f}$ is **differentiable at a** if $D\mathbf{f}(\mathbf{a})$ exists and if the function $\mathbf{h}: \mathbf{R}^n \to \mathbf{R}^m$ defined by
>
> $$\mathbf{h}(\mathbf{x}) = \mathbf{f}(\mathbf{a}) + D\mathbf{f}(\mathbf{a})(\mathbf{x} - \mathbf{a})$$
>
> is a good linear approximation to $\mathbf{f}$ near $\mathbf{a}$. That is, we must have
>
> $$\lim_{\mathbf{x}\to\mathbf{a}} \frac{\|\mathbf{f}(\mathbf{x}) - \mathbf{h}(\mathbf{x})\|}{\|\mathbf{x} - \mathbf{a}\|} = \lim_{\mathbf{x}\to\mathbf{a}} \frac{\|\mathbf{f}(\mathbf{x}) - [\mathbf{f}(\mathbf{a}) + D\mathbf{f}(\mathbf{a})(\mathbf{x} - \mathbf{a})]\|}{\|\mathbf{x} - \mathbf{a}\|} = 0.$$

Some remarks are in order. First, the reason for having the vector length appearing in the numerator in the limit equation in Definition 3.8 is so that there is a quotient of real numbers of which we can take a limit. (Definition 3.7 concerns scalar-valued functions only, so there is automatically a quotient of real numbers.) Second, the term $D\mathbf{f}(\mathbf{a})(\mathbf{x} - \mathbf{a})$ in the definition of $\mathbf{h}$ should be interpreted as the product of the $m \times n$ matrix $\mathbf{Df}(\mathbf{a})$ and the $n \times 1$ column matrix

$$\begin{bmatrix} x_1 - a_1 \\ x_2 - a_2 \\ \vdots \\ x_n - a_n \end{bmatrix}.$$

Because of the consistency of our definitions, the following results should not surprise you:

Theorem 3.9 If $\mathbf{f}: X \subseteq \mathbf{R}^n \to \mathbf{R}^m$ is differentiable at $\mathbf{a}$, then it is continuous at $\mathbf{a}$.

Theorem 3.10 If $\mathbf{f}: X \subseteq \mathbf{R}^n \to \mathbf{R}^m$ is such that, for $i = 1, \ldots, m$ and $j = 1, \ldots, n$, all $\partial f_i / \partial x_j$ exist and are continuous in a neighborhood of $\mathbf{a}$ in X, then $\mathbf{f}$ is differentiable at $\mathbf{a}$.

Theorem 3.11 A function $\mathbf{f}: X \subseteq \mathbf{R}^n \to \mathbf{R}^m$ is differentiable at $\mathbf{a} \in X$ (in the sense of Definition 3.8) if and only if each of its component functions $f_i: X \subseteq \mathbf{R}^n \to \mathbf{R}$, $i = 1, \ldots, m$, is differentiable at $\mathbf{a}$ (in the sense of Definition 3.7).

The proofs of Theorems 3.9, 3.10, and 3.11 are provided in the addendum to this section. Note that Theorems 3.10 and 3.11 frequently make it a straightforward matter to check that a function is differentiable: Just look at the partial derivatives of the component functions and verify that they are continuous. Thus, in many—but not all—circumstances, we can avoid working directly with the limit in Definition 3.8.

EXAMPLE 10 The function $\mathbf{g}: \mathbf{R}^3 - \{(0, 0, 0)\} \to \mathbf{R}^3$ given by

$$\mathbf{g}(x, y, z) = \left(\frac{3}{x^2 + y^2 + z^2}, xy, xz \right)$$

has

$$D\mathbf{g}(x, y, z) = \begin{bmatrix} \dfrac{-6x}{(x^2 + y^2 + z^2)^2} & \dfrac{-6y}{(x^2 + y^2 + z^2)^2} & \dfrac{-6z}{(x^2 + y^2 + z^2)^2} \\ y & x & 0 \\ z & 0 & x \end{bmatrix}.$$

Each of the entries of this matrix is continuous over $\mathbf{R}^3 - \{(0, 0, 0)\}$. Hence, by Theorem 3.10, $\mathbf{g}$ is differentiable over its entire domain. ◆

What Is a Derivative?

Although we have defined quite carefully what it means for a function to be differentiable, the derivative itself has really taken a "backseat" in the preceding discussion. It is time to get some perspective on the concept of the derivative.

In the case of a (differentiable) scalar-valued function of a single variable, $f: X \subseteq \mathbf{R} \to \mathbf{R}$, the derivative $f'(a)$ is simply a real number, the slope of the tangent line to the graph of f at the point $(a, f(a))$. From a more sophisticated (and slightly less geometric) point of view, the derivative $f'(a)$ is the number such that the function

$$h(x) = f(a) + f'(a)(x - a)$$

is a good linear approximation to $f(x)$ for x near a. (And, of course, $y = h(x)$ is the equation of the tangent line.)

If a function $f: X \subseteq \mathbf{R}^n \to \mathbf{R}$ of n variables is differentiable, there must exist n partial derivatives $\partial f/\partial x_1, \ldots, \partial f/\partial x_n$. These partial derivatives form the components of the gradient vector ∇f (or the entries of the $1 \times n$ matrix Df). It is the gradient that should properly be considered to be the derivative of f, but in the following sense: $\nabla f(\mathbf{a})$ is the vector such that the function $h: \mathbf{R}^n \to \mathbf{R}$ given by

$$h(\mathbf{x}) = f(\mathbf{a}) + \nabla f(\mathbf{a}) \cdot (\mathbf{x} - \mathbf{a})$$

is a good linear approximation to $f(\mathbf{x})$ for $\mathbf{x}$ near $\mathbf{a}$. Finally, the derivative of a differentiable vector-valued function $\mathbf{f}: X \subseteq \mathbf{R}^n \to \mathbf{R}^m$ may be taken to be the matrix $D\mathbf{f}$ of partial derivatives, but in the sense that the function $\mathbf{h}: \mathbf{R}^n \to \mathbf{R}^m$ given by

$$\mathbf{h}(\mathbf{x}) = \mathbf{f}(\mathbf{a}) + D\mathbf{f}(\mathbf{a})(\mathbf{x} - \mathbf{a})$$

is a good linear approximation to $\mathbf{f}(\mathbf{x})$ near $\mathbf{a}$. You should view the derivative $D\mathbf{f}(\mathbf{a})$ not as a "static" matrix of numbers, but rather as a matrix that defines a *linear mapping* from $\mathbf{R}^n$ to $\mathbf{R}^m$. (See Example 5 of §1.6.) This is embodied in the limit equation of Definition 3.8 and, though a subtle idea, is truly the heart of differential calculus of several variables.

In fact, we could have approached our discussion of differentiability much more abstractly right from the beginning. We could have defined a function $\mathbf{f}: X \subseteq \mathbf{R}^n \to \mathbf{R}^m$ to be differentiable at a point $\mathbf{a} \in X$ to mean that there exists some linear mapping $\mathbf{L}: \mathbf{R}^n \to \mathbf{R}^m$ such that

$$\lim_{\mathbf{x} \to \mathbf{a}} \frac{\|\mathbf{f}(\mathbf{x}) - [\mathbf{f}(\mathbf{a}) + \mathbf{L}(\mathbf{x} - \mathbf{a})]\|}{\|\mathbf{x} - \mathbf{a}\|} = 0.$$

Recall that any linear mapping $\mathbf{L}: \mathbf{R}^n \to \mathbf{R}^m$ is really nothing more than multiplication by a suitable $m \times n$ matrix A (i.e., that $\mathbf{L}(\mathbf{y}) = A\mathbf{y}$). It is possible to show that if there is a linear mapping that satisfies the aforementioned limit equation, then the matrix A that defines it is both uniquely determined and is precisely the matrix of partial derivatives $D\mathbf{f}(\mathbf{a})$. However, to begin with such a definition, though equivalent to Definition 3.8, strikes us as less well motivated than the approach we have taken. Hence, we have presented the notions of differentiability and the derivative from what we hope is a somewhat more concrete and geometric perspective.

Addendum: Proofs of Theorems 3.9, 3.10, and 3.11

Proof of Theorem 3.9 We begin by claiming the following: Let $\mathbf{x} \in \mathbf{R}^n$ and $B = (b_{ij})$ be an $m \times n$ matrix. If $\mathbf{y} = B\mathbf{x}$, (so $\mathbf{y} \in \mathbf{R}^m$), then

$$\|\mathbf{y}\| \le K\|\mathbf{x}\|, \tag{9}$$

where $K = \left(\sum_{i,j} b_{ij}^2\right)^{1/2}$. We postpone the proof of (9) until we establish the main theorem.

To show that $\mathbf{f}$ is continuous at $\mathbf{a}$, we will show that $\|\mathbf{f}(\mathbf{x}) - \mathbf{f}(\mathbf{a})\| \to 0$ as $\mathbf{x} \to \mathbf{a}$. We do so by using the fact that $\mathbf{f}$ is differentiable at $\mathbf{a}$ (Definition 3.8). We have

$$\begin{aligned}\|\mathbf{f}(\mathbf{x}) - \mathbf{f}(\mathbf{a})\| &= \|\mathbf{f}(\mathbf{x}) - \mathbf{f}(\mathbf{a}) - D\mathbf{f}(\mathbf{a})(\mathbf{x}-\mathbf{a}) + D\mathbf{f}(\mathbf{a})(\mathbf{x}-\mathbf{a})\| \\ &\le \|\mathbf{f}(\mathbf{x}) - \mathbf{f}(\mathbf{a}) - D\mathbf{f}(\mathbf{a})(\mathbf{x}-\mathbf{a})\| + \|D\mathbf{f}(\mathbf{a})(\mathbf{x}-\mathbf{a})\|, \qquad (10)\end{aligned}$$

using the triangle inequality. Note that the first term in the right side of inequality (10) is the numerator of the limit expression in Definition 3.8. Thus, since $\mathbf{f}$ is differentiable at $\mathbf{a}$, we can make $\|\mathbf{f}(\mathbf{x}) - \mathbf{f}(\mathbf{a}) - D\mathbf{f}(\mathbf{a})(\mathbf{x}-\mathbf{a})\|$ as small as we wish by keeping $\|\mathbf{x}-\mathbf{a}\|$ appropriately small. In particular,

$$\|\mathbf{f}(\mathbf{x}) - \mathbf{f}(\mathbf{a}) - D\mathbf{f}(\mathbf{a})(\mathbf{x}-\mathbf{a})\| \le \|\mathbf{x}-\mathbf{a}\|$$

if $\|\mathbf{x}-\mathbf{a}\|$ is sufficiently small. To the second term in the right side of inequality (10), we may apply (9), since $D\mathbf{f}(\mathbf{a})$ is an $m \times n$ matrix. Therefore, we see that if $\|\mathbf{x}-\mathbf{a}\|$ is made sufficiently small,

$$\|\mathbf{f}(\mathbf{x}) - \mathbf{f}(\mathbf{a})\| \le \|\mathbf{x}-\mathbf{a}\| + K\|\mathbf{x}-\mathbf{a}\| = (1+K)\|\mathbf{x}-\mathbf{a}\|.$$

The constant K does not depend on $\mathbf{x}$. Thus, as $\mathbf{x} \to \mathbf{a}$, we have

$$\|\mathbf{f}(\mathbf{x}) - \mathbf{f}(\mathbf{a})\| \to 0,$$

as desired.

To complete the proof, we establish inequality (9). Writing out the matrix multiplication,

$$\mathbf{y} = B\mathbf{x} = \begin{bmatrix} b_{11}x_1 + b_{12}x_2 + \cdots + b_{1n}x_n \\ b_{21}x_1 + b_{22}x_2 + \cdots + b_{2n}x_n \\ \vdots \\ b_{m1}x_1 + b_{m2}x_2 + \cdots + b_{mn}x_n \end{bmatrix} = \begin{bmatrix} \mathbf{b}_1 \cdot \mathbf{x} \\ \mathbf{b}_2 \cdot \mathbf{x} \\ \vdots \\ \mathbf{b}_m \cdot \mathbf{x} \end{bmatrix},$$

where $\mathbf{b}_i$ denotes the ith row of B, considered as a vector in $\mathbf{R}^n$. Therefore, using the Cauchy–Schwarz inequality,

$$\begin{aligned}\|\mathbf{y}\| &= \left((\mathbf{b}_1 \cdot \mathbf{x})^2 + (\mathbf{b}_2 \cdot \mathbf{x})^2 + \cdots + (\mathbf{b}_m \cdot \mathbf{x})^2\right)^{1/2} \\ &\le \left(\|\mathbf{b}_1\|^2\|\mathbf{x}\|^2 + \|\mathbf{b}_2\|^2\|\mathbf{x}\|^2 + \cdots + \|\mathbf{b}_m\|^2\|\mathbf{x}\|^2\right)^{1/2} \\ &= \left(\|\mathbf{b}_1\|^2 + \|\mathbf{b}_2\|^2 + \cdots + \|\mathbf{b}_m\|^2\right)^{1/2}\|\mathbf{x}\|.\end{aligned}$$

Now,

$$\|\mathbf{b}_i\|^2 = b_{i1}^2 + b_{i2}^2 + \cdots + b_{in}^2 = \sum_{j=1}^{n} b_{ij}^2.$$

Consequently,

$$\|\mathbf{b}_1\|^2 + \|\mathbf{b}_2\|^2 + \cdots + \|\mathbf{b}_m\|^2 = \sum_{i=1}^{m}\|\mathbf{b}_i\|^2 = \sum_{i=1}^{m}\sum_{j=1}^{n} b_{ij}^2 = K^2.$$

Thus, $\|\mathbf{y}\| \le K\|\mathbf{x}\|$, and we have completed the proof of Theorem 3.9. ■

Proof of Theorem 3.10 First, we prove Theorem 3.10 for the case where f is a scalar-valued function of two variables. We begin by writing

$$f(x_1, x_2) - f(a_1, a_2) = f(x_1, x_2) - f(a_1, x_2) + f(a_1, x_2) - f(a_1, a_2).$$

By the mean value theorem,[2] there exists a number c_1 between a_1 and x_1 such that

$$f(x_1, x_2) - f(a_1, x_2) = f_{x_1}(c_1, x_2)(x_1 - a_1)$$

and a number c_2 between a_2 and x_2 such that

$$f(a_1, x_2) - f(a_1, a_2) = f_{x_2}(a_1, c_2)(x_2 - a_2).$$

(This works because in each case we hold all the variables in f constant except one, so that the mean value theorem applies.) Hence,

$$\begin{aligned}
&\left|f(x_1, x_2) - f(a_1, a_2) - f_{x_1}(a_1, a_2)(x_1 - a_1) - f_{x_2}(a_1, a_2)(x_2 - a_2)\right| \\
&\quad = \left|f_{x_1}(c_1, x_2)(x_1 - a_1) + f_{x_2}(a_1, c_2)(x_2 - a_2) - f_{x_1}(a_1, a_2)(x_1 - a_1)\right. \\
&\quad\quad \left. - f_{x_2}(a_1, a_2)(x_2 - a_2)\right| \\
&\quad \leq \left|f_{x_1}(c_1, x_2)(x_1 - a_1) - f_{x_1}(a_1, a_2)(x_1 - a_1)\right| \\
&\quad\quad + \left|f_{x_2}(a_1, c_2)(x_2 - a_2) - f_{x_2}(a_1, a_2)(x_2 - a_2)\right|,
\end{aligned}$$

by the triangle inequality. Hence,

$$\begin{aligned}
&\left|f(x_1, x_2) - f(a_1, a_2) - f_{x_1}(a_1, a_2)(x_1 - a_1) - f_{x_2}(a_1, a_2)(x_2 - a_2)\right| \\
&\quad \leq \left|f_{x_1}(c_1, x_2) - f_{x_1}(a_1, a_2)\right| |x_1 - a_1| \\
&\quad\quad + \left|f_{x_2}(a_1, c_2) - f_{x_2}(a_1, a_2)\right| |x_2 - a_2| \\
&\quad \leq \left\{\left|f_{x_1}(c_1, x_2) - f_{x_1}(a_1, a_2)\right| + \left|f_{x_2}(a_1, c_2) - f_{x_2}(a_1, a_2)\right|\right\} \|\mathbf{x} - \mathbf{a}\|,
\end{aligned}$$

since, for $i = 1, 2$,

$$|x_i - a_i| \leq \|\mathbf{x} - \mathbf{a}\| = ((x_1 - a_1)^2 + (x_2 - a_2)^2)^{1/2}.$$

Thus,

$$\begin{aligned}
&\frac{\left|f(x_1, x_2) - f(a_1, a_2) - f_{x_1}(a_1, a_2)(x_1 - a_1) - f_{x_2}(a_1, a_2)(x_2 - a_2)\right|}{\|\mathbf{x} - \mathbf{a}\|} \\
&\quad \leq \left|f_{x_1}(c_1, x_2) - f_{x_1}(a_1, a_2)\right| + \left|f_{x_2}(a_1, c_2) - f_{x_2}(a_1, a_2)\right|. \qquad (11)
\end{aligned}$$

As $\mathbf{x} \to \mathbf{a}$, we must have that $c_i \to a_i$, for $i = 1, 2$, since c_i is between a_i and x_i. Consequently, by the continuity of the partial derivatives, both terms of the right side of (11) approach zero. Therefore,

$$\lim_{\mathbf{x} \to \mathbf{a}} \frac{\left|f(x_1, x_2) - f(a_1, a_2) - f_{x_1}(a_1, a_2)(x_1 - a_1) - f_{x_2}(a_1, a_2)(x_2 - a_2)\right|}{\|\mathbf{x} - \mathbf{a}\|} = 0$$

as desired.

Exactly the same kind of argument may be used in the case that f is a scalar-valued function of n variables—the details are only slightly more involved, so we omit them. Granting this, we consider the case of a vector-valued function

---

[2]Recall that the mean value theorem says that if F is continuous on the closed interval $[a, b]$ and differentiable on the open interval (a, b), then there is a number c in (a, b) such that $F(b) - F(a) = F'(c)(b - a)$.

$\mathbf{f}: \mathbf{R}^n \to \mathbf{R}^m$. According to Definition 3.8, we must show that

$$\lim_{\mathbf{x}\to\mathbf{a}} \frac{\|\mathbf{f}(\mathbf{x}) - \mathbf{f}(\mathbf{a}) - D\mathbf{f}(\mathbf{a})(\mathbf{x}-\mathbf{a})\|}{\|\mathbf{x}-\mathbf{a}\|} = 0. \tag{12}$$

The component functions of the expression appearing in the numerator may be written as

$$G_i = f_i(\mathbf{x}) - f_i(\mathbf{a}) - Df_i(\mathbf{a})(\mathbf{x}-\mathbf{a}), \tag{13}$$

where f_i, $i = 1, \ldots, m$, denotes the ith component function of $\mathbf{f}$. (Note that, by the cases of Theorem 3.10 already established, each scalar-valued function f_i is differentiable.) Now, we consider

$$\begin{aligned}\frac{\|\mathbf{f}(\mathbf{x}) - \mathbf{f}(\mathbf{a}) - D\mathbf{f}(\mathbf{a})(\mathbf{x}-\mathbf{a})\|}{\|\mathbf{x}-\mathbf{a}\|} &= \frac{\|(G_1, G_2, \ldots, G_m)\|}{\|\mathbf{x}-\mathbf{a}\|} \\ &= \frac{\left(G_1^2 + G_2^2 + \cdots + G_m^2\right)^{1/2}}{\|\mathbf{x}-\mathbf{a}\|} \\ &\leq \frac{|G_1| + |G_2| + \cdots + |G_m|}{\|\mathbf{x}-\mathbf{a}\|} \\ &= \frac{|G_1|}{\|\mathbf{x}-\mathbf{a}\|} + \frac{|G_2|}{\|\mathbf{x}-\mathbf{a}\|} + \cdots + \frac{|G_m|}{\|\mathbf{x}-\mathbf{a}\|}.\end{aligned}$$

As $\mathbf{x} \to \mathbf{a}$, each term $|G_i|/\|\mathbf{x}-\mathbf{a}\| \to 0$, by definition of G_i in equation (13) and the differentiability of the component functions f_i of $\mathbf{f}$. Hence, equation (12) holds and $\mathbf{f}$ is differentiable at $\mathbf{a}$. (To see that $(G_1^2 + \cdots + G_m^2)^{1/2} \leq |G_1| + \cdots + |G_m|$, note that

$$\begin{aligned}(|G_1| + \cdots + |G_m|)^2 &= |G_1|^2 + \cdots + |G_m|^2 \\ &\quad + 2|G_1|\,|G_2| + 2|G_1|\,|G_3| + \cdots + 2|G_{m-1}|\,|G_m| \\ &\geq |G_1|^2 + \cdots + |G_m|^2.\end{aligned}$$

Then, taking square roots provides the inequality.) ■

Proof of Theorem 3.11 In the final paragraph of the proof of Theorem 3.10, we showed that

$$\frac{\|\mathbf{f}(\mathbf{x}) - \mathbf{f}(\mathbf{a}) - D\mathbf{f}(\mathbf{a})(\mathbf{x}-\mathbf{a})\|}{\|\mathbf{x}-\mathbf{a}\|} \leq \frac{|G_1|}{\|\mathbf{x}-\mathbf{a}\|} + \frac{|G_2|}{\|\mathbf{x}-\mathbf{a}\|} + \cdots + \frac{|G_m|}{\|\mathbf{x}-\mathbf{a}\|},$$

where $G_i = f_i(\mathbf{x}) - f_i(\mathbf{a}) - Df_i(\mathbf{a})(\mathbf{x}-\mathbf{a})$ as in equation (13). From this, it follows immediately that differentiability of the component functions $f_1, \ldots, f_m$ at $\mathbf{a}$ implies differentiability of $\mathbf{f}$ at $\mathbf{a}$. Conversely, for $i = 1, \ldots, m$,

$$\frac{\|\mathbf{f}(\mathbf{x}) - \mathbf{f}(\mathbf{a}) - D\mathbf{f}(\mathbf{a})(\mathbf{x}-\mathbf{a})\|}{\|\mathbf{x}-\mathbf{a}\|} = \frac{\|(G_1, G_2, \ldots, G_m)\|}{\|\mathbf{x}-\mathbf{a}\|} \geq \frac{|G_i|}{\|\mathbf{x}-\mathbf{a}\|}.$$

Hence, differentiability of $\mathbf{f}$ at $\mathbf{a}$ forces differentiability of each component function. ■

Exercises

In Exercises 1–7, calculate $\partial f/\partial x$ and $\partial f/\partial y$.

1. $f(x, y) = xy^2 + x^2 y$
2. $f(x, y) = e^{x^2+y^2}$
3. $f(x, y) = \sin xy + \cos xy$
4. $f(x, y) = \dfrac{x^3 - y^2}{1 + x^2 + 3y^4}$
5. $f(x, y) = \dfrac{x^2 - y^2}{x^2 + y^2}$
6. $f(x, y) = \ln(x^2 + y^2)$
7. $f(x, y) = \cos x^3 y$

In Exercises 8–13, evaluate the partial derivatives $\partial F/\partial x$, $\partial F/\partial y$, and $\partial F/\partial z$ for the given functions F.

8. $F(x, y, z) = xyz$
9. $F(x, y, z) = \sqrt{x^2 + y^2 + z^2}$
10. $F(x, y, z) = e^{ax} \cos by + e^{az} \sin bx$
11. $F(x, y, z) = \dfrac{x + y + z}{(1 + x^2 + y^2 + z^2)^{3/2}}$
12. $F(x, y, z) = \sin x^2 y^3 z^4$
13. $F(x, y, z) = \dfrac{x^3 + yz}{x^2 + z^2 + 1}$

Find the gradient $\nabla f(\mathbf{a})$, where f and $\mathbf{a}$ are given in Exercises 14–19.

14. $f(x, y) = x^2 y + e^{y/x}, \quad \mathbf{a} = (1, 0)$
15. $f(x, y) = \dfrac{x - y}{x^2 + y^2 + 1}, \quad \mathbf{a} = (2, -1)$
16. $f(x, y, z) = \sin xyz, \quad \mathbf{a} = (\pi, 0, \pi/2)$
17. $f(x, y, z) = xy + y \cos z - x \sin yz$, $\mathbf{a} = (2, -1, \pi)$
18. $f(x, y) = e^{xy} + \ln(x - y), \quad \mathbf{a} = (2, 1)$
19. $f(x, y, z) = \dfrac{x + y}{e^z}, \quad \mathbf{a} = (3, -1, 0)$

In Exercises 20–25, find the matrix $D\mathbf{f}(\mathbf{a})$ of partial derivatives, where $\mathbf{f}$ and $\mathbf{a}$ are as indicated.

20. $f(x, y) = \dfrac{x}{y}, \quad \mathbf{a} = (3, 2)$
21. $\mathbf{f}(x, y, z) = \left(xyz, \sqrt{x^2 + y^2 + z^2}\right)$, $\mathbf{a} = (1, 0, -2)$
22. $\mathbf{f}(t) = (t, \cos 2t, \sin 5t), \quad a = 0$
23. $\mathbf{f}(x, y, z, w) = (3x - 7y + z, 5x + 2z - 8w, y - 17z + 3w), \quad \mathbf{a} = (1, 2, 3, 4)$
24. $\mathbf{f}(x, y) = (x^2 y, x + y^2, \cos \pi xy), \quad \mathbf{a} = (2, -1)$
25. $\mathbf{f}(s, t) = (s^2, st, t^2), \quad \mathbf{a} = (-1, 1)$

Explain why each of the functions given in Exercises 26–28 is differentiable at every point in its domain.

26. $f(x, y) = xy - 7x^8 y^2 + \cos x$
27. $f(x, y, z) = \dfrac{x + y + z}{x^2 + y^2 + z^2}$
28. $\mathbf{f}(x, y) = \left(\dfrac{xy^2}{x^2 + y^4}, \dfrac{x}{y} + \dfrac{y}{x}\right)$
29. (a) Explain why the graph of $z = x^3 - 7xy + e^y$ has a tangent plane at $(-1, 0, 0)$.
    (b) Give an equation for this tangent plane.
30. Find an equation for the plane tangent to the graph of $z = 4 \cos xy$ at the point $(\pi/3, 1, 2)$.
31. Find an equation for the plane tangent to the graph of $z = e^{x+y} \cos xy$ at the point $(0, 1, e)$.
32. Find equations for the planes tangent to $z = x^2 - 6x + y^3$ that are parallel to the plane $4x - 12y + z = 7$.
33. Use formula (8) to find an equation for the hyperplane tangent to the 4-dimensional paraboloid $x_5 = 10 - (x_1^2 + 3x_2^2 + 2x_3^2 + x_4^2)$ at the point $(2, -1, 1, 3, -8)$.
34. Suppose that you have the following information concerning a differentiable function f:

$$f(2, 3) = 12, \quad f(1.98, 3) = 12.1, \quad f(2, 3.01) = 12.2.$$

    (a) Give an approximate equation for the plane tangent to the graph of f at $(2, 3, 12)$.
    (b) Use the result of part (a) to estimate $f(1.98, 2.98)$.

In Exercises 35–37, (a) Use the linear function $h(\mathbf{x})$ in Definition 3.8 to approximate the indicated value of the given function f. (b) How accurate is the approximation determined in part (a)?

35. $f(x, y) = e^{x+y}$, $f(0.1, -0.1)$
36. $f(x, y) = 3 + \cos \pi xy$, $f(0.98, 0.51)$
37. $f(x, y, z) = x^2 + xyz + y^3 z$, $f(1.01, 1.95, 2.2)$
38. Calculate the partial derivatives of

$$f(x_1, x_2, \ldots, x_n) = \frac{x_1 + x_2 + \cdots + x_n}{\sqrt{x_1^2 + x_2^2 + \cdots + x_n^2}}.$$

39. Let

$$f(x, y) = \begin{cases} \dfrac{xy^2 - x^2y + 3x^3 - y^3}{x^2 + y^2} & \text{if } (x, y) \neq (0, 0) \\ 0 & \text{if } (x, y) = (0, 0) \end{cases}.$$

(a) Calculate $\partial f/\partial x$ and $\partial f/\partial y$ for $(x, y) \neq (0, 0)$. (You may wish to use a computer algebra system for this part.)

(b) Find $f_x(0, 0)$ and $f_y(0, 0)$.

As mentioned in the text, if a function $F(x)$ of a single variable is differentiable at a, then, as we zoom in on the point $(a, F(a))$, the graph of $y = F(x)$ will "straighten out" and look like its tangent line at $(a, F(a))$. For the differentiable functions given in Exercises 40–43, (a) calculate the tangent line at the indicated point, and (b) use a computer to graph the function and the tangent line on the same set of axes. Zoom in on the point of tangency to illustrate how the graph of $y = F(x)$ looks like its tangent line near $(a, F(a))$.

40. $F(x) = x^3 - 2x + 3, \quad a = 1$

41. $F(x) = x + \sin x, \quad a = \dfrac{\pi}{4}$

42. $F(x) = \dfrac{x^3 - 3x^2 + x}{x^2 + 1}, \quad a = 0$

43. $F(x) = \ln(x^2 + 1), \quad a = -1$

44. (a) Use a computer to graph the function $F(x) = (x - 2)^{2/3}$.

(b) By zooming in near $x = 2$, offer a geometric discussion concerning the differentiability of F at $x = 2$.

As discussed in the text, a function $f(x, y)$ may have partial derivatives $f_x(a, b)$ and $f_y(a, b)$, yet fail to be differentiable at (a, b). Geometrically, if a function $f(x, y)$ is differentiable at (a, b), then, as we zoom in on the point $(a, b, f(a, b))$, the graph of $z = f(x, y)$ will "flatten out" and look like the plane given by equation (4) in this section. For the functions $f(x, y)$ given in Exercises 45–49, (a) calculate $f_x(a, b)$ and $f_y(a, b)$ at the indicated point (a, b) and write the equation for the plane given by formula (4) of this section, (b) use a computer to graph the equation $z = f(x, y)$ together with the plane calculated in part (a). Zoom in near the point $(a, b, f(a, b))$ and discuss whether or not $f(x, y)$ is differentiable at (a, b). (c) Give an analytic (i.e., nongraphical) argument for your answer in part (b).

45. $f(x, y) = x^3 - xy + y^2, \quad (a, b) = (2, 1)$

46. $f(x, y) = ((x - 1)y)^{2/3}, \quad (a, b) = (1, 0)$

47. $f(x, y) = \dfrac{xy}{x^2 + y^2 + 1}, \quad (a, b) = (0, 0)$

48. $f(x, y) = \sin x \cos y, \quad (a, b) = \left(\dfrac{\pi}{6}, \dfrac{3\pi}{4}\right)$

49. $f(x, y) = x^2 \sin y + y^2 \cos x, \quad (a, b) = \left(\dfrac{\pi}{3}, \dfrac{\pi}{4}\right)$

50. Let $g(x, y) = \sqrt[3]{xy}$.

(a) Is g continuous at $(0, 0)$?

(b) Calculate $\partial g/\partial x$ and $\partial g/\partial y$ when $xy \neq 0$.

(c) Show that $g_x(0, 0)$ and $g_y(0, 0)$ exist by supplying values for them.

(d) Are $\partial g/\partial x$ and $\partial g/\partial y$ continuous at $(0, 0)$?

(e) Does the graph of $z = g(x, y)$ have a tangent plane at $(0, 0)$? You might consider creating a graph of this surface.

(f) Is g differentiable at $(0, 0)$?

51. Suppose $\mathbf{f}: \mathbf{R}^n \to \mathbf{R}^m$ is a linear mapping; that is,

$$\mathbf{f}(\mathbf{x}) = A\mathbf{x}, \quad \text{where } \mathbf{x} = (x_1, x_2, \ldots, x_n) \in \mathbf{R}^n$$

and A is an $m \times n$ matrix. Calculate $D\mathbf{f}(\mathbf{x})$ and relate your result to the derivative of the one-variable linear function $f(x) = ax$.

2.4 Properties; Higher-order Partial Derivatives; Newton's Method

Properties of the Derivative

From our work in the previous section, we know that the derivative of a function $\mathbf{f}: X \subseteq \mathbf{R}^n \to \mathbf{R}^m$ can be identified with its matrix of partial derivatives. We next note several properties that the derivative must satisfy. The proofs of these results involve Definition 3.8 of the derivative, properties of ordinary differentiation, and matrix algebra.

■ **Proposition 4.1** (Linearity of Differentiation) Let $\mathbf{f}, \mathbf{g}: X \subseteq \mathbf{R}^n \to \mathbf{R}^m$ be two functions that are both differentiable at a point $\mathbf{a} \in X$, and let $c \in \mathbf{R}$ be any scalar. Then,

1. The function $\mathbf{h} = \mathbf{f} + \mathbf{g}$ is also differentiable at $\mathbf{a}$, and we have

$$D\mathbf{h}(\mathbf{a}) = D(\mathbf{f} + \mathbf{g})(\mathbf{a}) = D\mathbf{f}(\mathbf{a}) + D\mathbf{g}(\mathbf{a}).$$

2. The function $\mathbf{k} = c\mathbf{f}$ is differentiable at $\mathbf{a}$ and

$$D\mathbf{k}(\mathbf{a}) = D(c\mathbf{f})(\mathbf{a}) = cD\mathbf{f}(\mathbf{a}).$$

EXAMPLE 1 Let $\mathbf{f}$ and $\mathbf{g}$ be defined by $\mathbf{f}(x, y) = (x + y, xy \sin y, y/x)$ and $\mathbf{g}(x, y) = (x^2 + y^2, ye^{xy}, 2x^3 - 7y^5)$. We have

$$D\mathbf{f}(x, y) = \begin{bmatrix} 1 & 1 \\ y \sin y & x \sin y + xy \cos y \\ -y/x^2 & 1/x \end{bmatrix}$$

and

$$D\mathbf{g}(x, y) = \begin{bmatrix} 2x & 2y \\ y^2 e^{xy} & e^{xy} + xye^{xy} \\ 6x^2 & -35y^4 \end{bmatrix}.$$

Thus, by Theorem 3.10, $\mathbf{f}$ is differentiable on $\mathbf{R}^2 - \{y\text{-axis}\}$ and $\mathbf{g}$ is differentiable on all of $\mathbf{R}^2$. If we let $\mathbf{h} = \mathbf{f} + \mathbf{g}$, then part 1 of Proposition 4.1 tells us that $\mathbf{h}$ must be differentiable on all of its domain, and

$$\begin{aligned} D\mathbf{h}(x, y) &= D\mathbf{f}(x, y) + D\mathbf{g}(x, y) \\ &= \begin{bmatrix} 2x + 1 & 2y + 1 \\ y \sin y + y^2 e^{xy} & x \sin y + xy \cos y + e^{xy} + xye^{xy} \\ 6x^2 - y/x^2 & 1/x - 35y^4 \end{bmatrix}. \end{aligned}$$

Note also that the function $\mathbf{k} = 3\mathbf{g}$ must be differentiable everywhere by part 2 of Proposition 4.1. We can readily check that $D\mathbf{k}(x, y) = 3D\mathbf{g}(x, y)$: We have

$$\mathbf{k}(x, y) = (3x^2 + 3y^2, 3ye^{xy}, 6x^3 - 21y^5).$$

Hence,

$$\begin{aligned} D\mathbf{k}(x, y) &= \begin{bmatrix} 6x & 6y \\ 3y^2 e^{xy} & 3e^{xy} + 3xye^{xy} \\ 18x^2 & -105y^4 \end{bmatrix} \\ &= 3 \begin{bmatrix} 2x & 2y \\ y^2 e^{xy} & e^{xy} + xye^{xy} \\ 6x^2 & -35y^4 \end{bmatrix} \\ &= 3D\mathbf{g}(x, y). \end{aligned}$$

◆

Due to the nature of matrix multiplication, general versions of the product and quotient rules do not exist in any particularly simple form. However, for scalar-valued functions, it is possible to prove the following:

■ **Proposition 4.2** Let $f, g: X \subseteq \mathbf{R}^n \to \mathbf{R}$ be differentiable at $\mathbf{a} \in X$. Then,

1. The product function fg is also differentiable at $\mathbf{a}$, and

$$D(fg)(\mathbf{a}) = g(\mathbf{a})Df(\mathbf{a}) + f(\mathbf{a})Dg(\mathbf{a}).$$

2. If $g(\mathbf{a}) \neq 0$, then the quotient function f/g is differentiable at $\mathbf{a}$, and

$$D(f/g)(\mathbf{a}) = \frac{g(\mathbf{a})Df(\mathbf{a}) - f(\mathbf{a})Dg(\mathbf{a})}{g(\mathbf{a})^2}.$$

EXAMPLE 2 If $f(x, y, z) = ze^{xy}$ and $g(x, y, z) = xy + 2yz - xz$, then

$$(fg)(x, y, z) = (xyz + 2yz^2 - xz^2)e^{xy},$$

so that

$$D(fg)(x, y, z) = \begin{bmatrix} (yz - z^2)e^{xy} + (xyz + 2yz^2 - xz^2)ye^{xy} \\ (xz + 2z^2)e^{xy} + (xyz + 2yz^2 - xz^2)xe^{xy} \\ (xy + 4yz - 2xz)e^{xy} \end{bmatrix}^T.$$

Also, we have

$$Df(x, y, z) = \begin{bmatrix} yze^{xy} & xze^{xy} & e^{xy} \end{bmatrix}$$

and

$$Dg(x, y, z) = \begin{bmatrix} y - z & x + 2z & 2y - x \end{bmatrix},$$

so that

$$\begin{aligned} &g(x, y, z)Df(x, y, z) + f(x, y, z)Dg(x, y, z) \\ &= \begin{bmatrix} (xy^2z + 2y^2z^2 - xyz^2)e^{xy} \\ (x^2yz + 2xyz^2 - x^2z^2)e^{xy} \\ (xy + 2yz - xz)e^{xy} \end{bmatrix}^T + \begin{bmatrix} (yz - z^2)e^{xy} \\ (xz + 2z^2)e^{xy} \\ (2yz - xz)e^{xy} \end{bmatrix}^T \\ &= e^{xy} \begin{bmatrix} xy^2z + 2y^2z^2 - xyz^2 + yz - z^2 \\ x^2yz + 2xyz^2 - x^2z^2 + xz + 2z^2 \\ xy + 4yz - 2xz \end{bmatrix}^T, \end{aligned}$$

which checks with part 1 of Proposition 4.2. (Note: The matrix transpose is used simply to conserve space on the page.) ◆

The product rule in part 1 of Proposition 4.2 is not the most general result possible. Indeed, if $f: X \subseteq \mathbf{R}^n \to \mathbf{R}$ is a scalar-valued function and $\mathbf{g}: X \subseteq \mathbf{R}^n \to \mathbf{R}^m$ a vector-valued function, then if f and $\mathbf{g}$ are both differentiable at $\mathbf{a} \in X$, so is $f\mathbf{g}$, and the following formula holds (where we view $\mathbf{g}(\mathbf{a})$ as an $m \times 1$ matrix):

$$D(f\mathbf{g})(\mathbf{a}) = \mathbf{g}(\mathbf{a})Df(\mathbf{a}) + f(\mathbf{a})D\mathbf{g}(\mathbf{a}).$$

Partial Derivatives of Higher Order

Thus far in our study of differentiation, we have been concerned only with partial derivatives of first order. Nonetheless, it is easy to imagine computing second- and third-order partials by iterating the process of differentiating with respect to one variable, while all others are held constant.

EXAMPLE 3 Let $f(x, y, z) = x^2y + y^2z$. Then the first-order partial derivatives are

$$\frac{\partial f}{\partial x} = 2xy, \quad \frac{\partial f}{\partial y} = x^2 + 2yz, \quad \text{and} \quad \frac{\partial f}{\partial z} = y^2.$$

The **second-order partial derivative** with respect to x, denoted by $\partial^2 f/\partial x^2$ or $f_{xx}(x, y, z)$, is

$$\frac{\partial^2 f}{\partial x^2} = \frac{\partial}{\partial x}\left(\frac{\partial f}{\partial x}\right) = \frac{\partial}{\partial x}(2xy) = 2y.$$

Similarly, the second-order partials with respect to y and z are, respectively,

$$\frac{\partial^2 f}{\partial y^2} = \frac{\partial}{\partial y}\left(\frac{\partial f}{\partial y}\right) = \frac{\partial}{\partial y}(x^2 + 2yz) = 2z,$$

and

$$\frac{\partial^2 f}{\partial z^2} = \frac{\partial}{\partial z}\left(\frac{\partial f}{\partial z}\right) = \frac{\partial}{\partial z}(y^2) \equiv 0.$$

There are more second-order partials, however. The **mixed partial derivative** with respect to first x and then y, denoted $\partial^2 f/\partial y\partial x$ or $f_{xy}(x, y, z)$, is

$$\frac{\partial^2 f}{\partial y \partial x} = \frac{\partial}{\partial y}\left(\frac{\partial f}{\partial x}\right) = \frac{\partial}{\partial y}(2xy) = 2x.$$

There are five more mixed partials for this particular function: $\partial^2 f/\partial x\partial y$, $\partial^2 f/\partial z\partial x$, $\partial^2 f/\partial x\partial z$, $\partial^2 f/\partial z\partial y$, and $\partial^2 f/\partial y\partial z$. Compute each of them to get a feeling for the process. ◆

In general, if $f: X \subseteq \mathbf{R}^n \to \mathbf{R}$ is a (scalar-valued) function of n variables, the **kth-order partial derivative** with respect to the variables $x_{i_1}, x_{i_2}, \ldots, x_{i_k}$ (in that order), where $i_1, i_2, \ldots, i_k$ are integers in the set $\{1, 2, \ldots, n\}$ (possibly repeated), is the iterated derivative

$$\frac{\partial^k f}{\partial x_{i_k} \cdots \partial x_{i_2} \partial x_{i_1}} = \frac{\partial}{\partial x_{i_k}} \cdots \frac{\partial}{\partial x_{i_2}} \frac{\partial}{\partial x_{i_1}}(f(x_1, x_2, \ldots, x_n)).$$

Equivalent (and frequently more manageable) notation for this kth-order partial is

$$f_{x_{i_1} x_{i_2} \cdots x_{i_k}}(x_1, x_2, \ldots, x_n).$$

Note that the order in which we write the variables with respect to which we differentiate is different in the two notations: In the subscript notation, we write the differentiation variables from left to right in the order we differentiate, while in the ∂-notation, we write those variables in the *opposite* order (i.e., from right to left).

EXAMPLE 4 Let $f(x, y, z, w) = xyz + xy^2w - \cos(x + zw)$. We then have

$$\begin{aligned} f_{yw}(x, y, z, w) = \frac{\partial^2 f}{\partial w \partial y} &= \frac{\partial}{\partial w}\frac{\partial}{\partial y}(xyz + xy^2w - \cos(x + zw)) \\ &= \frac{\partial}{\partial w}(xz + 2xyw) = 2xy, \end{aligned}$$

and

$$f_{wy}(x, y, z, w) = \frac{\partial^2 f}{\partial y \partial w} = \frac{\partial}{\partial y}\frac{\partial}{\partial w}(xyz + xy^2 w - \cos(x + zw))$$

$$= \frac{\partial}{\partial y}(xy^2 + z\sin(x + zw)) = 2xy. \qquad \blacklozenge$$

Although it is generally ill-advised to formulate conjectures based on a single piece of evidence, Example 4 suggests that there might be an outrageously simple relationship among the mixed second partials. Indeed, such is the case, as the next result, due to the eighteenth-century French mathematician Alexis Clairaut, indicates.

Theorem 4.3 Suppose that X is open in $\mathbf{R}^n$ and $f: X \subseteq \mathbf{R}^n \to \mathbf{R}$ has continuous first- and second-order partial derivatives. Then the order in which we evaluate the mixed second-order partials is immaterial; that is, if i_1 and i_2 are any two integers between 1 and n, then

$$\frac{\partial^2 f}{\partial x_{i_1} \partial x_{i_2}} = \frac{\partial^2 f}{\partial x_{i_2} \partial x_{i_1}}.$$

A proof of Theorem 4.3 is provided in the addendum to this section. We also suggest a second proof (using integrals!) in Exercise 4 of the Miscellaneous Exercises for Chapter 5.

It is natural to speculate about the possibility of an analogue to Theorem 4.3 for kth-order mixed partials. Before we state what should be an easily anticipated result, we need some terminology.

Definition 4.4 Assume X is open in $\mathbf{R}^n$. A scalar-valued function $f: X \subseteq \mathbf{R}^n \to \mathbf{R}$ whose partial derivatives up to (and including) order at least k exist and are continuous on X is said to be **of class** C^k. If f has continuous partial derivatives of all orders on X, then f is said to be **of class** C^∞, or **smooth**. A vector-valued function $\mathbf{f}: X \subseteq \mathbf{R}^n \to \mathbf{R}^m$ is of class C^k (respectively, of class C^∞) if and only if each of its component functions is of class C^k (respectively, C^∞).

Theorem 4.5 Let $f: X \subseteq \mathbf{R}^n \to \mathbf{R}$ be a scalar-valued function of class C^k. Then the order in which we calculate any kth-order partial derivative does not matter: If $(i_1, \ldots, i_k)$ are any k integers (not necessarily distinct) between 1 and n, and if $(j_1, \ldots, j_k)$ is any permutation (rearrangement) of these integers, then

$$\frac{\partial^k f}{\partial x_{i_1} \cdots \partial x_{i_k}} = \frac{\partial^k f}{\partial x_{j_1} \cdots \partial x_{j_k}}.$$

EXAMPLE 5 If $f(x, y, z, w) = x^2 w e^{yz} - z e^{xw} + xyzw$, then you can check that

$$\frac{\partial^5 f}{\partial x \partial w \partial z \partial y \partial x} = 2e^{yz}(yz + 1) = \frac{\partial^5 f}{\partial z \partial y \partial w \partial^2 x},$$

verifying Theorem 4.5 in this case. $\blacklozenge$

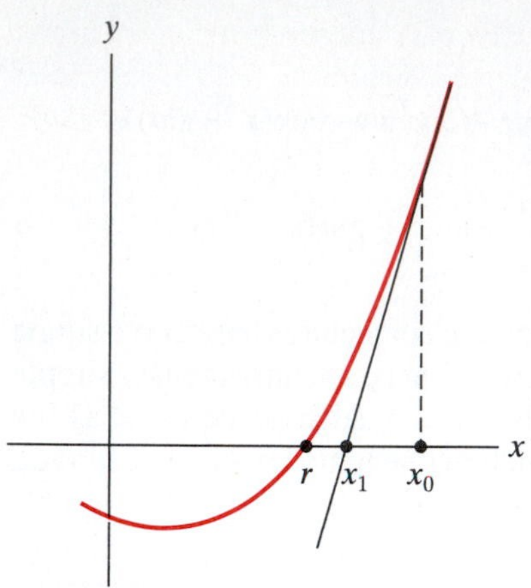

Figure 2.53 The tangent line to $y = f(x)$ at $(x_0, f(x_0))$ crosses the x-axis at $x = x_1$.

Newton's Method (optional)

When you studied single-variable calculus, you may have learned a method, known as **Newton's method** (or the **Newton–Raphson method**), for approximating the solution to an equation of the form $f(x) = 0$, where $f: X \subseteq \mathbf{R} \to \mathbf{R}$ is a differentiable function. Here's a reminder of how the method works.

We wish to find a number r such that $f(r) = 0$. To approximate r, we make an initial guess x_0 for r and, in general, we expect to find that $f(x_0) \neq 0$. So next we look at the tangent line to the graph of f at $(x_0, f(x_0))$. (See Figure 2.53.) Since the tangent line approximates the graph of f near $(x_0, f(x_0))$, we can find where the tangent line crosses the x-axis. The crossing point $(x_1, 0)$ will generally be closer to $(r, 0)$ than $(x_0, 0)$ is, so we take x_1 as a revised and improved approximation to the root r of $f(x) = 0$.

To find x_1, we begin with the equation of the tangent line

$$y = f(x_0) + f'(x_0)(x - x_0),$$

then set $y = 0$ to find where this line crosses the x-axis. Thus we solve the equation

$$f(x_0) + f'(x_0)(x_1 - x_0) = 0$$

for x_1 to find that

$$x_1 = x_0 - \frac{f(x_0)}{f'(x_0)}.$$

Once we have x_1, we can start the process again using x_1 in place of x_0 and produce what we hope will be an even better approximation x_2 via the formula

$$x_2 = x_1 - \frac{f(x_1)}{f'(x_1)}.$$

Indeed, we may iterate this process and define x_k recursively by

$$x_k = x_{k-1} - \frac{f(x_{k-1})}{f'(x_{k-1})} \qquad k = 1, 2, \ldots \tag{1}$$

and thereby produce a sequence of numbers $x_0, x_1, \ldots, x_k, \ldots$.

It is not always the case that the sequence $\{x_k\}$ converges. However, when it does, it must converge to a root of the equation $f(x) = 0$. To see this, let $L = \lim_{k\to\infty} x_k$. Then we also have $\lim_{k\to\infty} x_{k-1} = L$. Taking limits in formula (1), we find

$$L = L - \frac{f(L)}{f'(L)},$$

which immediately implies that $f(L) = 0$. Hence L is a root of the equation.

Now that we have some understanding of derivatives in the multivariable case, we turn to the generalization of Newton's method for solving systems of n equations in n unknowns. We may write such a system as

$$\begin{cases} f_1(x_1, \ldots, x_n) = 0 \\ f_2(x_1, \ldots, x_n) = 0 \\ \quad\vdots \\ f_n(x_1, \ldots, x_n) = 0 \end{cases}. \tag{2}$$

We consider the map $\mathbf{f}: X \subseteq \mathbf{R}^n \to \mathbf{R}^n$ defined as $\mathbf{f}(\mathbf{x}) = (f_1(\mathbf{x}), \ldots, f_n(\mathbf{x}))$ (i.e., $\mathbf{f}$ is the map whose component functions come from the equations in (2).

The domain X of $\mathbf{f}$ may be taken to be the set where all the component functions are defined.) Then to solve system (2) means to find a vector $\mathbf{r} = (r_1, \ldots, r_n)$ such that $\mathbf{f}(\mathbf{r}) = \mathbf{0}$. To approximate such a vector $\mathbf{r}$, we may, as in the single-variable case, make an initial guess $\mathbf{x}_0$ for what $\mathbf{r}$ might be. If $\mathbf{f}$ is differentiable, then we know that $\mathbf{y} = \mathbf{f}(\mathbf{x})$ is approximated by the equation

$$\mathbf{y} = \mathbf{f}(\mathbf{x}_0) + D\mathbf{f}(\mathbf{x}_0)(\mathbf{x} - \mathbf{x}_0).$$

(Here we think of $\mathbf{f}(\mathbf{x}_0)$ and the vectors $\mathbf{x}$ and $\mathbf{x}_0$ as $n \times 1$ matrices.) Then we set $\mathbf{y}$ equal to $\mathbf{0}$ to find where this approximating function is zero. Thus we solve the matrix equation

$$\mathbf{f}(\mathbf{x}_0) + D\mathbf{f}(\mathbf{x}_0)(\mathbf{x}_1 - \mathbf{x}_0) = \mathbf{0} \tag{3}$$

for $\mathbf{x}_1$ to give a revised approximation to the root $\mathbf{r}$. Evidently (3) is equivalent to

$$D\mathbf{f}(\mathbf{x}_0)(\mathbf{x}_1 - \mathbf{x}_0) = -\mathbf{f}(\mathbf{x}_0). \tag{4}$$

To continue our argument, suppose that $D\mathbf{f}(\mathbf{x}_0)$ is an invertible $(n \times n)$ matrix, meaning that there is a second $n \times n$ matrix $[D\mathbf{f}(\mathbf{x}_0)]^{-1}$ with the property that $[D\mathbf{f}(\mathbf{x}_0)]^{-1} D\mathbf{f}(\mathbf{x}_0) = D\mathbf{f}(\mathbf{x}_0)[D\mathbf{f}(\mathbf{x}_0)]^{-1} = I_n$, the $n \times n$ identity matrix. (See Exercises 20 and 30–38 in §1.6.) Then we may multiply equation (4) on the left by $[D\mathbf{f}(\mathbf{x}_0)]^{-1}$ to obtain

$$I_n(\mathbf{x}_1 - \mathbf{x}_0) = -[D\mathbf{f}(\mathbf{x}_0)]^{-1}\mathbf{f}(\mathbf{x}_0).$$

Since $I_n A = A$ for any $n \times k$ matrix A, this last equation implies that

$$\mathbf{x}_1 = \mathbf{x}_0 - [D\mathbf{f}(\mathbf{x}_0)]^{-1}\mathbf{f}(\mathbf{x}_0). \tag{5}$$

As we did in the one-variable case of Newton's method, we may iterate formula (5) to define recursively a sequence $\{\mathbf{x}_k\}$ of *vectors* by

$$\mathbf{x}_k = \mathbf{x}_{k-1} - [D\mathbf{f}(\mathbf{x}_{k-1})]^{-1}\mathbf{f}(\mathbf{x}_{k-1}) \tag{6}$$

Note the similarity between formulas (1) and (6). Moreover, just as in the case of formula (1), although the sequence $\{\mathbf{x}_0, \mathbf{x}_1, \ldots, \mathbf{x}_k, \ldots\}$ may not converge, if it does, it must converge to a root of $\mathbf{f}(\mathbf{x}) = \mathbf{0}$. (See Exercise 26.)

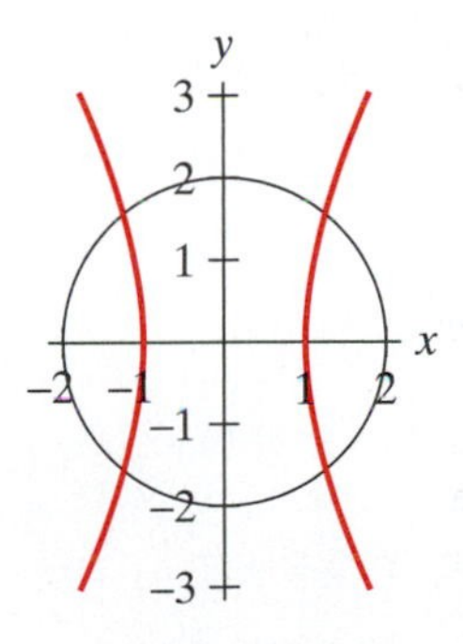

Figure 2.54 Finding the intersection points of the circle $x^2 + y^2 = 4$ and the hyperbola $4x^2 - y^2 = 4$ in Example 6.

EXAMPLE 6 Consider the problem of finding the intersection points of the circle $x^2 + y^2 = 4$ and the hyperbola $4x^2 - y^2 = 4$. (See Figure 2.54.) Analytically, we seek simultaneous solutions to the two equations

$$x^2 + y^2 = 4 \quad \text{and} \quad 4x^2 - y^2 = 4,$$

or, equivalently, solutions to the system

$$\begin{cases} x^2 + y^2 - 4 = 0 \\ 4x^2 - y^2 - 4 = 0 \end{cases}. \tag{7}$$

To use Newton's method, we define a function $\mathbf{f}\colon \mathbf{R}^2 \to \mathbf{R}^2$ by $\mathbf{f}(x, y) = (x^2 + y^2 - 4, 4x^2 - y^2 - 4)$ and try to approximate solutions to the vector equation $\mathbf{f}(x, y) = (0, 0)$. We may begin with any initial guess, say

$$\mathbf{x}_0 = \begin{bmatrix} x_0 \\ y_0 \end{bmatrix} = \begin{bmatrix} 1 \\ 1 \end{bmatrix},$$

and then produce successive approximations $\mathbf{x}_1, \mathbf{x}_2, \ldots$ to a solution using formula (6). In particular, we have

$$D\mathbf{f}(x, y) = \begin{bmatrix} 2x & 2y \\ 8x & -2y \end{bmatrix}.$$

Note that $\det D\mathbf{f}(x, y) = -20xy$. You may verify (see Exercise 36 in §1.6) that

$$[D\mathbf{f}(x, y)]^{-1} = \frac{1}{-20xy}\begin{bmatrix} -2y & -2y \\ -8x & 2x \end{bmatrix} = \begin{bmatrix} \dfrac{1}{10x} & \dfrac{1}{10x} \\ \dfrac{2}{5y} & -\dfrac{1}{10y} \end{bmatrix}.$$

Thus

$$\begin{aligned}
\begin{bmatrix} x_k \\ y_k \end{bmatrix} &= \begin{bmatrix} x_{k-1} \\ y_{k-1} \end{bmatrix} - [D\mathbf{f}(x_{k-1}, y_{k-1})]^{-1}\mathbf{f}(x_{k-1}, y_{k-1}) \\
&= \begin{bmatrix} x_{k-1} \\ y_{k-1} \end{bmatrix} - \begin{bmatrix} \dfrac{1}{10x_{k-1}} & \dfrac{1}{10x_{k-1}} \\ \dfrac{2}{5y_{k-1}} & -\dfrac{1}{10y_{k-1}} \end{bmatrix} \begin{bmatrix} x_{k-1}^2 + y_{k-1}^2 - 4 \\ 4x_{k-1}^2 - y_{k-1}^2 - 4 \end{bmatrix} \\
&= \begin{bmatrix} x_{k-1} \\ y_{k-1} \end{bmatrix} - \begin{bmatrix} \dfrac{5x_{k-1}^2 - 8}{10x_{k-1}} \\ \dfrac{5y_{k-1}^2 - 12}{10y_{k-1}} \end{bmatrix} = \begin{bmatrix} x_{k-1} - \dfrac{5x_{k-1}^2 - 8}{10x_{k-1}} \\ y_{k-1} - \dfrac{5y_{k-1}^2 - 12}{10y_{k-1}} \end{bmatrix}.
\end{aligned}$$

Beginning with $x_0 = y_0 = 1$, we have

$$x_1 = 1 - \frac{5 \cdot 1^2 - 8}{10 \cdot 1} = 1.3 \qquad y_1 = 1 - \frac{5 \cdot 1^2 - 12}{10 \cdot 1} = 1.7$$

$$x_2 = 1.3 - \frac{5(1.3)^2 - 8}{10(1.3)} = 1.265385 \quad y_2 = 1.7 - \frac{5(1.7)^2 - 12}{10(1.7)} = 1.555882, \quad \text{etc.}$$

It is also easy to hand off the details of the computation to a calculator or a computer. One finds the following results:

k	x_k	y_k
0	1	1
1	1.3	1.7
2	1.26538462	1.55588235
3	1.26491115	1.54920772
4	1.26491106	1.54919334
5	1.26491106	1.54919334

Thus it appears that, to eight decimal places, an intersection point of the curves is (1.26491106, 1.54919334).

In this particular example, it is not difficult to find the solutions to (7) exactly. We add the two equations in (7) to obtain

$$5x^2 - 8 = 0 \qquad \Longleftrightarrow \qquad x^2 = \tfrac{8}{5}.$$

Thus $x = \pm\sqrt{8/5}$. If we substitute these values for x into the first equation of (7), we obtain

$$\tfrac{8}{5} + y^2 - 4 = 0 \qquad \Longleftrightarrow \qquad y^2 = \tfrac{12}{5}.$$

Hence $y = \pm\sqrt{12/5}$. Therefore, the four intersection points are

$$\left(\sqrt{\frac{8}{5}}, \sqrt{\frac{12}{5}}\right), \quad \left(-\sqrt{\frac{8}{5}}, \sqrt{\frac{12}{5}}\right), \quad \left(-\sqrt{\frac{8}{5}}, -\sqrt{\frac{12}{5}}\right), \quad \left(\sqrt{\frac{8}{5}}, -\sqrt{\frac{12}{5}}\right).$$

Since $\sqrt{8/5} \approx 1.264911064$ and $\sqrt{12/5} \approx 1.54919334$, we see that Newton's method provided us with an accurate approximate solution very quickly. ◆

EXAMPLE 7 We use Newton's method to find solutions to the system

$$\begin{cases} x^3 - 5x^2 + 2x - y + 13 = 0 \\ x^3 + x^2 - 14x - y - 19 = 0 \end{cases}. \tag{8}$$

As in the previous example, we define $\mathbf{f}\colon \mathbf{R}^2 \to \mathbf{R}^2$ by $\mathbf{f}(x, y) = (x^3 - 5x^2 + 2x - y + 13, x^3 + x^2 - 14x - y - 19)$. Then

$$D\mathbf{f}(x, y) = \begin{bmatrix} 3x^2 - 10x + 2 & -1 \\ 3x^2 + 2x - 14 & -1 \end{bmatrix},$$

so that $\det D\mathbf{f}(x, y) = 12x - 16$ and

$$[D\mathbf{f}(x, y)]^{-1} = \begin{bmatrix} -\dfrac{1}{12x - 16} & \dfrac{1}{12x - 16} \\ \dfrac{-3x^2 - 2x + 14}{12x - 16} & \dfrac{-3x^2 - 10x + 2}{12x - 16} \end{bmatrix}.$$

Thus, formula (6) becomes

$$\begin{bmatrix} x_k \\ y_k \end{bmatrix} = \begin{bmatrix} x_{k-1} \\ y_{k-1} \end{bmatrix} - \begin{bmatrix} -\dfrac{1}{12x_{k-1} - 16} & \dfrac{1}{12x_{k-1} - 16} \\ \dfrac{-3x_{k-1}^2 - 2x_{k-1} + 14}{12x_{k-1} - 16} & \dfrac{-3x_{k-1}^2 - 10x_{k-1} + 2}{12x_{k-1} - 16} \end{bmatrix}$$

$$\times \begin{bmatrix} x_{k-1}^3 - 5x_{k-1}^2 + 2x_{k-1} - y_{k-1} + 13 \\ x_{k-1}^3 + x_{k-1}^2 - 14x_{k-1} - y_{k-1} - 19 \end{bmatrix}$$

$$= \begin{bmatrix} x_{k-1} - \dfrac{6x_{k-1}^2 - 16x_{k-1} - 32}{12x_{k-1} - 16} \\ y_{k-1} - \dfrac{3x_{k-1}^4 - 16x_{k-1}^3 - 14x_{k-1}^2 + 82x_{k-1} - 8y_{k-1} + 6x_{k-1}y_{k-1} + 72}{6x_{k-1} - 8} \end{bmatrix}.$$

This is the formula we iterate to obtain approximate solutions to (8).

If we begin with $\mathbf{x}_0 = (x_0, y_0) = (8, 10)$, then the successive approximations $\mathbf{x}_k$ quickly converge to $(4, 5)$, as demonstrated in the table below.

k	x_k	y_k
0	8	10
1	5.2	−98.2
2	4.1862069	−2.7412414
3	4.00607686	4.82161865
4	4.00000691	4.99981073
5	4.00000000	5.00000000
6	4.00000000	5.00000000

If we begin instead with $\mathbf{x}_0 = (50, 60)$, then convergence is, as you might predict, somewhat slower (although still quite rapid):

k	x_k	y_k
0	50	60
1	25.739726	−57257.438
2	13.682211	−7080.8238
3	7.79569757	−846.58548
4	5.11470969	−86.660453
5	4.1643023	−1.6486813
6	4.00476785	4.86119425
7	4.00000425	4.99988349
8	4.00000000	5.00000000
9	4.00000000	5.00000000

On the other hand, if we begin with $\mathbf{x}_0 = (-2, 12)$, then the sequence of points generated converges to a different solution, namely, $(-4/3, -25/27)$:

k	x_k	y_k
0	−2	12
1	−1.4	1.4
2	−1.3341463	−0.903122
3	−1.3333335	−0.9259225
4	−1.3333333	−0.9259259
5	−1.3333333	−0.9259259

In fact, when a system of equations has multiple solutions, it is not always easy to predict to which solution a given starting vector $\mathbf{x}_0$ will converge under Newton's method (if, indeed, there is convergence at all). ◆

Finally, we make two remarks. First, if at any stage of the iteration process the matrix $D\mathbf{f}(\mathbf{x}_k)$ fails to be invertible (i.e., $[D\mathbf{f}(\mathbf{x}_k)]^{-1}$ does not exist), then formula (6) cannot be used. One way to salvage the situation is to make a different choice of initial vector $\mathbf{x}_0$ in the hope that the sequence $\{\mathbf{x}_k\}$ that it generates will not involve any noninvertible matrices. Second, we note that if, at any stage, $\mathbf{x}_k$ is *exactly* a root of $\mathbf{f}(\mathbf{x}) = \mathbf{0}$, then formula (6) will not change it. (See Exercise 29).

Addendum: Two Technical Proofs

Proof of Part 1 of Proposition 4.1

Step 1. We show that the matrix of partial derivatives of $\mathbf{h}$ is the sum of those of $\mathbf{f}$ and $\mathbf{g}$. If we write $\mathbf{h}(\mathbf{x})$ as $(h_1(\mathbf{x}), h_2(\mathbf{x}), \ldots, h_m(\mathbf{x}))$ (i.e., in terms of its component functions), then the ijth entry of $D\mathbf{h}(\mathbf{a})$ is $\partial h_i/\partial x_j$ evaluated at $\mathbf{a}$. But $h_i(\mathbf{x}) = f_i(\mathbf{x}) + g_i(\mathbf{x})$ by definition of $\mathbf{h}$. Hence,

$$\frac{\partial h_i}{\partial x_j} = \frac{\partial}{\partial x_j}(f_i(\mathbf{x}) + g_i(\mathbf{x})) = \frac{\partial f_i}{\partial x_j} + \frac{\partial g_i}{\partial x_j},$$

by properties of ordinary differentiation (since all variables except x_j are held constant). Thus,

$$\frac{\partial h_i}{\partial x_j}(\mathbf{a}) = \frac{\partial f_i}{\partial x_j}(\mathbf{a}) + \frac{\partial g_i}{\partial x_j}(\mathbf{a}),$$

and, therefore,

$$D\mathbf{h}(\mathbf{a}) = D\mathbf{f}(\mathbf{a}) + D\mathbf{g}(\mathbf{a}).$$

Step 2. Now that we know the desired matrix of partials exists, we must show that $\mathbf{h}$ really is differentiable; that is, we must establish that

$$\lim_{\mathbf{x}\to\mathbf{a}} \frac{\|\mathbf{h}(\mathbf{x}) - [\mathbf{h}(\mathbf{a}) + D\mathbf{h}(\mathbf{a})(\mathbf{x}-\mathbf{a})]\|}{\|\mathbf{x}-\mathbf{a}\|} = 0.$$

As preliminary background, we note that

$$\begin{aligned}
&\frac{\|\mathbf{h}(\mathbf{x}) - [\mathbf{h}(\mathbf{a}) + D\mathbf{h}(\mathbf{a})(\mathbf{x}-\mathbf{a})]\|}{\|\mathbf{x}-\mathbf{a}\|} \\
&\quad = \frac{\|\mathbf{f}(\mathbf{x}) + \mathbf{g}(\mathbf{x}) - [\mathbf{f}(\mathbf{a}) + \mathbf{g}(\mathbf{a}) + D\mathbf{f}(\mathbf{a})(\mathbf{x}-\mathbf{a}) + D\mathbf{g}(\mathbf{a})(\mathbf{x}-\mathbf{a})]\|}{\|\mathbf{x}-\mathbf{a}\|} \\
&\quad = \frac{\|(\mathbf{f}(\mathbf{x}) - [\mathbf{f}(\mathbf{a}) + D\mathbf{f}(\mathbf{a})(\mathbf{x}-\mathbf{a})]) + (\mathbf{g}(\mathbf{x}) - [\mathbf{g}(\mathbf{a}) + D\mathbf{g}(\mathbf{a})(\mathbf{x}-\mathbf{a})])\|}{\|\mathbf{x}-\mathbf{a}\|} \\
&\quad \le \frac{\|\mathbf{f}(\mathbf{x}) - [\mathbf{f}(\mathbf{a}) + D\mathbf{f}(\mathbf{a})(\mathbf{x}-\mathbf{a})]\|}{\|\mathbf{x}-\mathbf{a}\|} + \frac{\|\mathbf{g}(\mathbf{x}) - [\mathbf{g}(\mathbf{a}) + D\mathbf{g}(\mathbf{a})(\mathbf{x}-\mathbf{a})]\|}{\|\mathbf{x}-\mathbf{a}\|},
\end{aligned}$$

by the triangle inequality, formula (2) of §1.6. To show that the desired limit equation for $\mathbf{h}$ follows from the definition of the limit, we must show that given any $\epsilon > 0$, we can find a number $\delta > 0$ such that

$$\text{if } 0 < \|\mathbf{x}-\mathbf{a}\| < \delta, \text{ then } \frac{\|\mathbf{h}(\mathbf{x}) - [\mathbf{h}(\mathbf{a}) + D\mathbf{h}(\mathbf{a})(\mathbf{x}-\mathbf{a})]\|}{\|\mathbf{x}-\mathbf{a}\|} < \epsilon. \qquad (9)$$

Since $\mathbf{f}$ is given to be differentiable at $\mathbf{a}$, this means that given any $\epsilon_1 > 0$, we can find $\delta_1 > 0$ such that

$$\text{if } 0 < \|\mathbf{x}-\mathbf{a}\| < \delta_1, \text{ then } \frac{\|\mathbf{f}(\mathbf{x}) - [\mathbf{f}(\mathbf{a}) + D\mathbf{f}(\mathbf{a})(\mathbf{x}-\mathbf{a})]\|}{\|\mathbf{x}-\mathbf{a}\|} < \epsilon_1. \qquad (10)$$

Similarly, differentiability of $\mathbf{g}$ means that given any $\epsilon_2 > 0$, we can find a $\delta_2 > 0$ such that

$$\text{if } 0 < \|\mathbf{x}-\mathbf{a}\| < \delta_2, \text{ then } \frac{\|\mathbf{g}(\mathbf{x}) - [\mathbf{g}(\mathbf{a}) + D\mathbf{g}(\mathbf{a})(\mathbf{x}-\mathbf{a})]\|}{\|\mathbf{x}-\mathbf{a}\|} < \epsilon_2. \qquad (11)$$

Now we're ready to establish statement (9). Suppose $\epsilon > 0$ is given. Let δ_1 and δ_2 be such that (10) and (11) hold with $\epsilon_1 = \epsilon_2 = \epsilon/2$. Take δ to be the smaller of δ_1 and δ_2. Hence, if $0 < \|\mathbf{x}-\mathbf{a}\| < \delta$, then both statements (10) and (11) hold (with $\epsilon_1 = \epsilon_2 = \epsilon/2$) and, moreover,

$$\begin{aligned}
\frac{\|\mathbf{h}(\mathbf{x}) - [\mathbf{h}(\mathbf{a}) + D\mathbf{h}(\mathbf{a})(\mathbf{x}-\mathbf{a})]\|}{\|\mathbf{x}-\mathbf{a}\|} &\le \frac{\|\mathbf{f}(\mathbf{x}) - [\mathbf{f}(\mathbf{a}) + D\mathbf{f}(\mathbf{a})(\mathbf{x}-\mathbf{a})]\|}{\|\mathbf{x}-\mathbf{a}\|} \\
&\quad + \frac{\|\mathbf{g}(\mathbf{x}) - [\mathbf{g}(\mathbf{a}) + D\mathbf{g}(\mathbf{a})(\mathbf{x}-\mathbf{a})]\|}{\|\mathbf{x}-\mathbf{a}\|} \\
&< \epsilon_1 + \epsilon_2 \\
&= \frac{\epsilon}{2} + \frac{\epsilon}{2} = \epsilon.
\end{aligned}$$

That is, statement (9) holds, as desired. ■

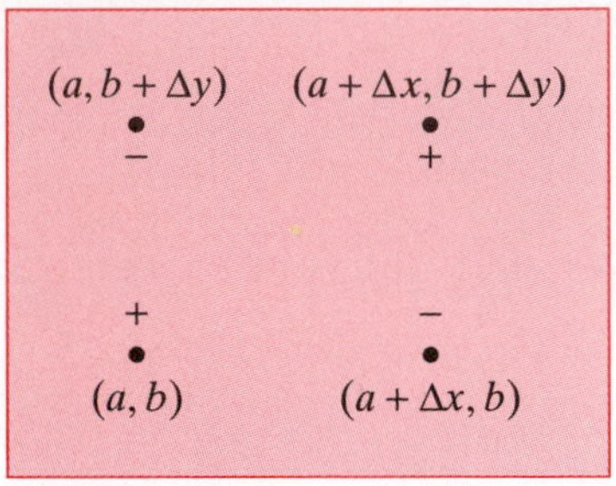

Proof of Theorem 4.3

For simplicity of notation only, we'll assume that f is a function of just two variables (x and y). Let the point $(a, b) \in \mathbf{R}^2$ be in the interior of some rectangle on which $f_x, f_y, f_{xx}, f_{yy}, f_{xy}$, and f_{yx} are all continuous. Consider the following "difference function." (See Figure 2.55.)

$$D(\Delta x, \Delta y) = f(a + \Delta x, b + \Delta y) - f(a + \Delta x, b) \\ - f(a, b + \Delta y) + f(a, b).$$

Figure 2.55 To construct the difference function D used in the proof of Theorem 4.3, evaluate f at the four points shown with the signs as indicated.

Our proof depends upon viewing this function in two ways. We first regard D as a difference of vertical differences in f:

$$\begin{aligned} D(\Delta x, \Delta y) &= [f(a + \Delta x, b + \Delta y) - f(a + \Delta x, b)] \\ &\quad - [f(a, b + \Delta y) - f(a, b)] \\ &= F(a + \Delta x) - F(a). \end{aligned}$$

Here we define the one-variable function $F(x)$ to be $f(x, b + \Delta y) - f(x, b)$. As we will see, the mixed second partial of f can be found from two applications of the mean value theorem of one-variable calculus. Since f has continuous partials, it is differentiable. (See Theorem 3.10.) Hence, F is continuous and differentiable, and, thus, the mean value theorem implies that there is some number c between a and $a + \Delta x$ such that

$$D(\Delta x, \Delta y) = F(a + \Delta x) - F(a) = F'(c)\Delta x. \tag{12}$$

Now $F'(c) = f_x(c, b + \Delta y) - f_x(c, b)$. We again apply the mean value theorem, this time to the function $f_x(c, y)$. (Here, we think of c as constant and y as the variable.) By hypothesis f_x is differentiable since its partial derivatives, f_{xx} and f_{xy}, are assumed to be continuous. Consequently, the mean value theorem applies to give us a number d between b and $b + \Delta y$ such that

$$F'(c) = f_x(c, b + \Delta y) - f_x(c, b) = f_{xy}(c, d)\Delta y. \tag{13}$$

Using equation (13) in equation (12), we have

$$D(\Delta x, \Delta y) = F'(c)\Delta x = f_{xy}(c, d)\Delta y \Delta x.$$

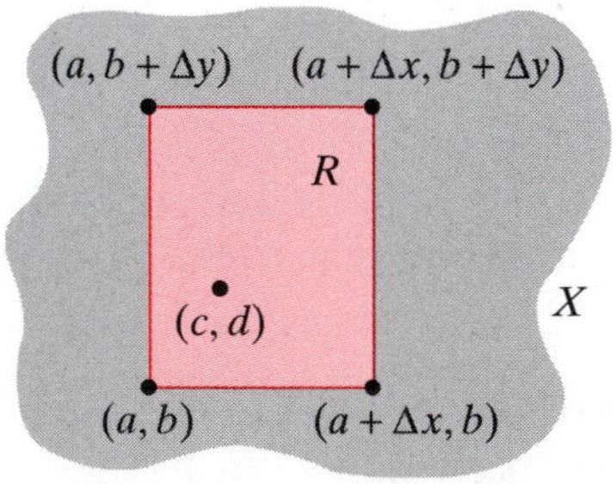

The point (c, d) lies somewhere in the interior of the rectangle R with vertices (a, b), $(a + \Delta x, b)$, $(a, b + \Delta y)$, $(a + \Delta x, b + \Delta y)$, as shown in Figure 2.56. Thus, as $(\Delta x, \Delta y) \to (0, 0)$, we have $(c, d) \to (a, b)$. Hence, it follows that

$$f_{xy}(c, d) \to f_{xy}(a, b) \quad \text{as} \quad (\Delta x, \Delta y) \to (0, 0),$$

since f_{xy} is assumed to be continuous. Therefore,

$$f_{xy}(a, b) = \lim_{(\Delta x, \Delta y) \to (0,0)} f_{xy}(c, d) = \lim_{(\Delta x, \Delta y) \to (0,0)} \frac{D(\Delta x, \Delta y)}{\Delta y \Delta x}.$$

Figure 2.56 Applying the mean value theorem twice.

On the other hand, we could just as well have written D as a difference of horizontal differences in f:

$$\begin{aligned} D(\Delta x, \Delta y) &= [f(a + \Delta x, b + \Delta y) - f(a, b + \Delta y)] \\ &\quad - [f(a + \Delta x, b) - f(a, b)] \\ &= G(b + \Delta y) - G(b). \end{aligned}$$

Here $G(y) = f(a + \Delta x, y) - f(a, y)$. As before, we can apply the mean value theorem twice to find that there must be another point $(\bar{c}, \bar{d})$ in R such that

$$D(\Delta x, \Delta y) = G'(\bar{d})\Delta y = f_{yx}(\bar{c}, \bar{d})\Delta x \Delta y.$$

Therefore,

$$f_{yx}(a, b) = \lim_{(\Delta x, \Delta y)\to(0,0)} f_{yx}(\bar{c}, \bar{d}) = \lim_{(\Delta x, \Delta y)\to(0,0)} \frac{D(\Delta x, \Delta y)}{\Delta x \Delta y}.$$

Because this is the same limit as that for $f_{xy}(a, b)$ just given, we have established the desired result. ■

Exercises

In Exercises 1–4, verify the sum rule for derivative matrices (i.e., part 1 of Proposition 4.1) for each of the given pairs of functions:

1. $f(x, y) = xy + \cos x, \quad g(x, y) = \sin(xy) + y^3$

2. $\mathbf{f}(x, y) = (e^{x+y}, xe^y), \quad \mathbf{g}(x, y) = (\ln(xy), ye^x)$

3. $\mathbf{f}(x, y, z) = (x \sin y + z, ye^z - 3x^2), \quad \mathbf{g}(x, y, z) = (x^3 \cos x, xyz)$

4. $\mathbf{f}(x, y, z) = (xyz^2, xe^{-y}, y \sin xz), \quad \mathbf{g}(x, y, z) = (x - y, x^2 + y^2 + z^2, \ln(xz + 2))$

Verify the product and quotient rules (Proposition 4.2) for the pairs of functions given in Exercises 5–8.

5. $f(x, y) = x^2y + y^3, \quad g(x, y) = \dfrac{x}{y}$

6. $f(x, y) = e^{xy}, \quad g(x, y) = x \sin 2y$

7. $f(x, y) = 3xy + y^5, \quad g(x, y) = x^3 - 2xy^2$

8. $f(x, y, z) = x \cos(yz)$,
$g(x, y, z) = x^2 + x^9y^2 + y^2z^3 + 2$

For the functions given in Exercises 9–17, determine all second-order partial derivatives (including mixed partials).

9. $f(x, y) = x^3y^7 + 3xy^2 - 7xy$

10. $f(x, y) = \cos(xy)$

11. $f(x, y) = e^{y/x} - ye^{-x}$

12. $f(x, y) = \sin\sqrt{x^2 + y^2}$

13. $f(x, y) = \dfrac{1}{\sin^2 x + 2e^y}$

14. $f(x, y) = e^{x^2+y^2}$

15. $f(x, y, z) = x^2yz + xy^2z + xyz^2$

16. $f(x, y, z) = e^{xyz}$

17. $f(x, y, z) = e^{ax} \sin y + e^{bx} \cos z$

18. Consider the function $F(x, y, z) = 2x^3y + xz^2 + y^3z^5 - 7xyz$.

(a) Find F_{xx}, F_{yy}, and F_{zz}.

(b) Calculate the mixed second-order partials F_{xy}, F_{yx}, F_{xz}, F_{zx}, F_{yz}, and F_{zy}, and verify Theorem 4.3.

(c) Is $F_{xyx} = F_{xxy}$? Could you have known this without resorting to calculation?

(d) Is $F_{xyz} = F_{yzx}$?

19. Recall from §2.2 that a polynomial in two variables x and y is an expression of the form

$$p(x, y) = \sum_{k,l=0}^{d} c_{kl}x^k y^l,$$

where c_{kl} can be any real number for $0 \leq k, l \leq d$. The **degree of the term** $c_{kl}x^k y^l$ when $c_{kl} \neq 0$ is $k + l$ and the **degree of the polynomial** p is the largest degree of any nonzero term of the polynomial (i.e., the largest degree of any term for which $c_{kl} \neq 0$). For example, the polynomial

$$p(x, y) = 7x^6y^9 + 2x^2y^3 - 3x^4 - 5xy^3 + 1$$

has five terms of degrees 15, 5, 4, 4, and 0. The degree of p is therefore 15. (Note: The degree of the **zero polynomial** $p(x, y) \equiv 0$ is undefined.)

(a) If $p(x, y) = 8x^7y^{10} - 9x^2y + 2x$, what is the degree of $\partial p/\partial x$? $\partial p/\partial y$? $\partial^2 p/\partial x^2$? $\partial^2 p/\partial y^2$? $\partial^2 p/\partial x \partial y$?

(b) If $p(x, y) = 8x^2y + 2x^3y$, what is the degree of $\partial p/\partial x$? $\partial p/\partial y$? $\partial^2 p/\partial x^2$? $\partial^2 p/\partial y^2$? $\partial^2 p/\partial x \partial y$?

(c) Try to formulate and prove a conjecture relating the degree of a polynomial p to the degree of its partial derivatives.

20. The partial differential equation

$$\frac{\partial^2 f}{\partial x^2} + \frac{\partial^2 f}{\partial y^2} + \frac{\partial^2 f}{\partial z^2} = 0$$

is known as **Laplace's equation,** after Pierre Simon de Laplace (1749–1827). Any function f of class C^2

that satisfies Laplace's equation is called a **harmonic function.**[3]

(a) Is $f(x, y, z) = x^2 + y^2 - 2z^2$ harmonic? What about $f(x, y, z) = x^2 - y^2 + z^2$?

(b) We may generalize Laplace's equation to functions of n variables as

$$\frac{\partial^2 f}{\partial x_1^2} + \frac{\partial^2 f}{\partial x_2^2} + \cdots + \frac{\partial^2 f}{\partial x_n^2} = 0.$$

Give an example of a harmonic function of n variables, and verify that your example is correct.

21. The three-dimensional **heat equation** is the partial differential equation

$$k\left(\frac{\partial^2 T}{\partial x^2} + \frac{\partial^2 T}{\partial y^2} + \frac{\partial^2 T}{\partial z^2}\right) = \frac{\partial T}{\partial t},$$

where k is a constant. It models the temperature $T(x, y, z, t)$ at the point (x, y, z) and time t of a body in space.

(a) First, we examine a simplified version of the heat equation. Consider a straight wire "coordinatized" by x. Then the temperature $T(x, t)$ at time t and position x along the wire is modeled by the one-dimensional heat equation

$$k\frac{\partial^2 T}{\partial x^2} = \frac{\partial T}{\partial t}.$$

Show that the function $T(x, t) = e^{-kt}\cos x$ satisfies this equation. Note that if t is held constant at value t_0, then $T(x, t_0)$ shows how the temperature varies along the wire at time t_0. Graph the curves $z = T(x, t_0)$ for $t_0 = 0, 1, 10$, and use them to understand the graph of the surface $z = T(x, t)$ for $t \geq 0$. Explain what happens to the temperature of the wire after a long period of time.

(b) Show that $T(x, y, t) = e^{-kt}(\cos x + \cos y)$ satisfies the two-dimensional heat equation

$$k\left(\frac{\partial^2 T}{\partial x^2} + \frac{\partial^2 T}{\partial y^2}\right) = \frac{\partial T}{\partial t}.$$

Graph the surfaces given by $z = T(x, y, t_0)$, where $t_0 = 0, 1, 10$. If we view the function $T(x, y, t)$ as modeling the temperature at points (x, y) of a flat plate at time t, then describe what happens to the temperature of the plate after a long period of time.

(c) Now show that $T(x, y, z, t) = e^{-kt}(\cos x + \cos y + \cos z)$ satisfies the three-dimensional heat equation.

22. Let

$$f(x, y) = \begin{cases} xy\left(\dfrac{x^2 - y^2}{x^2 + y^2}\right) & \text{if } (x, y) \neq (0, 0) \\ 0 & \text{if } (x, y) = (0, 0) \end{cases}.$$

(a) Find $f_x(x, y)$ and $f_y(x, y)$ for $(x, y) \neq (0, 0)$. (You will find a computer algebra system helpful.)

(b) Either by hand (using limits) or by means of part (a), find the partial derivatives $f_x(0, y)$ and $f_y(x, 0)$.

(c) Find the values of $f_{xy}(0, 0)$ and $f_{yx}(0, 0)$. Reconcile your answer with Theorem 4.3.

A surface that has the least surface area among all surfaces with a given boundary is called a ***minimal surface.*** *Soap bubbles are naturally occurring examples of minimal surfaces. It is a fact that minimal surfaces having equations of the form $z = f(x, y)$ (where f is of class C^2) satisfy the partial differential equation*

$$\left(1 + z_y^2\right) z_{xx} + \left(1 + z_x^2\right) z_{yy} = 2z_x z_y z_{xy}. \qquad (14)$$

Exercises 23–25 concern minimal surfaces and equation (14).

23. Show that a plane is a minimal surface.

24. **Scherk's surface** is given by the equation $e^z \cos y = \cos x$.

(a) Use a computer to graph a portion of this surface.

(b) Verify that Scherk's surface is a minimal surface.

25. One way to describe the surface known as the **helicoid** is by the equation $x = y \tan z$.

(a) Use a computer to graph a portion of this surface.

(b) Verify that the helicoid is a minimal surface.

26. Consider the sequence of vectors $\mathbf{x}_0, \mathbf{x}_1, \ldots,$ where, for $k \geq 1$, the vector $\mathbf{x}_k$ is defined by the Newton's method recursion formula (6) given an initial "guess" $\mathbf{x}_0$ at a root of the equation $\mathbf{f}(\mathbf{x}) = \mathbf{0}$. (Here we assume that $\mathbf{f}\colon X \subseteq \mathbf{R}^n \to \mathbf{R}^n$ is a differentiable function.) By imitating the argument in the single-variable case, show that if the sequence $\{\mathbf{x}_k\}$ converges to a vector $\mathbf{L}$ and $D\mathbf{f}(\mathbf{L})$ is an invertible matrix, then $\mathbf{L}$ must satisfy $\mathbf{f}(\mathbf{L}) = \mathbf{0}$.

27. This problem concerns the Newton's method iteration in Example 6.

(a) Use initial vector $\mathbf{x}_0 = (-1, 1)$ and calculate the successive approximations $\mathbf{x}_1, \mathbf{x}_2, \mathbf{x}_3$, etc. To what

[3] Laplace did fundamental and far-reaching work in both mathematical physics and probability theory. Laplace's equation and harmonic functions are part of the field of **potential theory,** a subject that Laplace can be credited as having developed. Potential theory has applications to such areas as gravitation, electricity and magnetism, and fluid mechanics, to name a few.

solution of the system of equations (7) do the approximations converge?

(b) Repeat part (a) with $\mathbf{x}_0 = (1, -1)$. Repeat again with $\mathbf{x}_0 = (-1, -1)$.

(c) Comment on the results of parts (a) and (b) and whether you might have predicted them. Describe the results in terms of Figure 2.54.

28. Consider the Newton's method iteration in Example 7.

(a) Use initial vector $\mathbf{x}_0 = (1.4, 10)$ and calculate the successive approximations $\mathbf{x}_1, \mathbf{x}_2, \mathbf{x}_3$, etc. To what solution of the system of equations (8) do the approximations converge?

(b) Repeat part (a) with $\mathbf{x}_0 = (1.3, 10)$.

(c) In Example 7 we saw that (4, 5) was a solution of the given system of equations. Is (1.3, 10) closer to (4, 5) or to the limiting point of the sequence you calculated in part (b)?

(d) Comment on your observations in part (c). What do these observations suggest about how easily you can use the initial vector $\mathbf{x}_0$ to predict the value of $\lim_{k\to\infty} \mathbf{x}_k$ (assuming that the limit exists)?

29. Suppose that at some stage in the Newton's method iteration using formula (6), we obtain a vector $\mathbf{x}_k$ that is an exact solution to the system of equations (2). Show that all the subsequent vectors $\mathbf{x}_{k+1}, \mathbf{x}_{k+2}, \ldots$ are equal to $\mathbf{x}_k$. Hence if we happen to obtain an exact root via Newton's method, we will retain it.

30. Suppose that $\mathbf{f}\colon X \subseteq \mathbf{R}^2 \to \mathbf{R}^2$ is differentiable and that we write $\mathbf{f}(x, y) = (f(x, y), g(x, y))$. Show that formula (6) implies that, for $k \geq 1$,

$$x_k = x_{k-1} - \frac{f(x_{k-1}, y_{k-1})g_y(x_{k-1}, y_{k-1}) - g(x_{k-1}, y_{k-1})f_y(x_{k-1}, y_{k-1})}{f_x(x_{k-1}, y_{k-1})g_y(x_{k-1}, y_{k-1}) - f_y(x_{k-1}, y_{k-1})g_x(x_{k-1}, y_{k-1})}$$

$$y_k = y_{k-1} - \frac{g(x_{k-1}, y_{k-1})f_x(x_{k-1}, y_{k-1}) - f(x_{k-1}, y_{k-1})g_x(x_{k-1}, y_{k-1})}{f_x(x_{k-1}, y_{k-1})g_y(x_{k-1}, y_{k-1}) - f_y(x_{k-1}, y_{k-1})g_x(x_{k-1}, y_{k-1})}.$$

31. As we will see in Chapter 4, when looking for maxima and minima of a differentiable function $F\colon X \subseteq \mathbf{R}^n \to \mathbf{R}$, we need to find the points where $DF(x_1, \ldots, x_n) = [0 \cdots 0]$, called **critical points** of F. Let $F(x, y) = 4\sin(xy) + x^3 + y^3$. Use Newton's method to approximate the critical point that lies near $(x, y) = (-1, -1)$.

32. Consider the problem of finding the intersection points of the sphere $x^2 + y^2 + z^2 = 4$, the circular cylinder $x^2 + y^2 = 1$, and the elliptical cylinder $4y^2 + z^2 = 4$.

(a) Use Newton's method to find one of the intersection points. By choosing a different initial vector $\mathbf{x}_0 = (x_0, y_0, z_0)$, approximate a second intersection point. (Note: You may wish to use a computer algebra system to determine appropriate inverse matrices.)

(b) Find all the intersection points exactly by means of algebra and compare with your results in part (a).

2.5 The Chain Rule

Among the various properties that the derivative satisfies, one that stands alone in both its usefulness and its subtlety is the derivative's behavior with respect to composition of functions. This behavior is described by a formula known as the **chain rule.** In this section, we review the chain rule of one-variable calculus and see how it generalizes to the cases of scalar- and vector-valued functions of several variables.

The Chain Rule for Functions of One Variable: A Review

We begin with a typical example of the use of the chain rule from single-variable calculus.

EXAMPLE 1 Let $f(x) = \sin x$ and $x(t) = t^3 + t$. We may then construct the composite function $f(x(t)) = \sin(t^3 + t)$. The chain rule tells us how to find the derivative of $f \circ x$ with respect to t:

$$(f \circ x)'(t) = \frac{d}{dt}(\sin(t^3 + t)) = (\cos(t^3 + t))(3t^2 + 1).$$

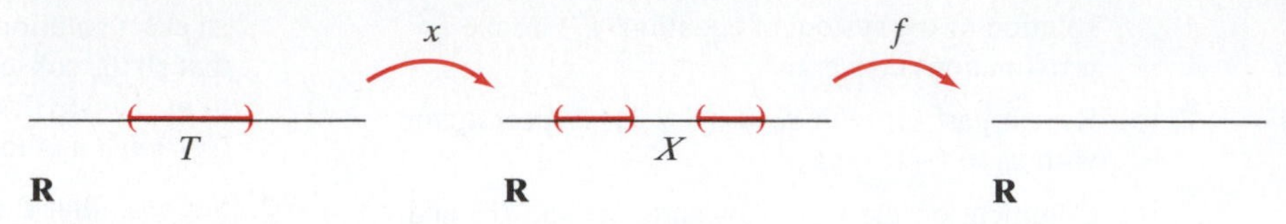

Figure 2.57 The range of the function x must be contained in the domain X of f in order for the composite $f \circ x$ to be defined.

Since $x = t^3 + t$, we have

$$(f \circ x)'(t) = \frac{d}{dx}(\sin x) \cdot \frac{d}{dt}(t^3 + t) = f'(x) \cdot x'(t). \quad \blacklozenge$$

In general, suppose X and T are open subsets of $\mathbf{R}$ and $f: X \subseteq \mathbf{R} \to \mathbf{R}$ and $x: T \subseteq \mathbf{R} \to \mathbf{R}$ are functions defined so that the composite function $f \circ x: T \to \mathbf{R}$ makes sense. (See Figure 2.57.) In particular, this means that the range of the function x must be contained in X, the domain of f. The key result is the following:

Theorem 5.1 (THE CHAIN RULE IN ONE VARIABLE) Under the preceding assumptions, if x is differentiable at $t_0 \in T$ and f is differentiable at $x_0 = x(t_0) \in X$, then the composite $f \circ x$ is differentiable at t_0 and, moreover,

$$(f \circ x)'(t_0) = f'(x_0)x'(t_0). \tag{1}$$

A more common way to write the chain rule formula in Theorem 5.1 is

$$\frac{df}{dt}(t_0) = \frac{df}{dx}(x_0)\,\frac{dx}{dt}(t_0). \tag{2}$$

Although equation (2) is most useful in practice, it does represent an unfortunate abuse of notation in that the symbol f is used to denote both a function of x and one of t. It would be more appropriate to define a new function y by $y(t) = (f \circ x)(t)$ so that $dy/dt = (df/dx)(dx/dt)$. But our original abuse of notation is actually a convenient one, since it avoids the awkwardness of having too many variable names appearing in a single discussion. In the name of simplicity, we will therefore continue to commit such abuses and urge you to do likewise.

The formulas in equations (1) and (2) are so simple that little more needs to be said. We elaborate, nonetheless, because this will prove helpful when we generalize to the case of several variables. The chain rule tells us the following: To understand how f depends on t, we must know how f depends on the "intermediate variable" x and how this intermediate variable depends on the "final" independent variable t. The diagram in Figure 2.58 traces the hierarchy of the variable dependences. The "paths" indicate the derivatives involved in the chain rule formula.

The Chain Rule in Several Variables

Now let's go a step further and assume $f: X \subseteq \mathbf{R}^2 \to \mathbf{R}$ is a C^1 function of two variables and $\mathbf{x}: T \subseteq \mathbf{R} \to \mathbf{R}^2$ is a differentiable vector-valued function of a single variable. If the range of $\mathbf{x}$ is contained in X, then the composite $f \circ \mathbf{x}: T \subseteq \mathbf{R} \to \mathbf{R}$ is defined. (See Figure 2.59.) It's good to think of $\mathbf{x}$ as describing a parametrized curve in $\mathbf{R}^2$ and f as a sort of "temperature function"

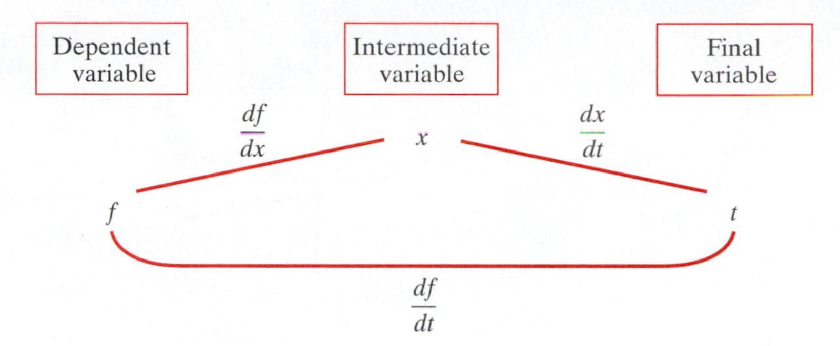

Figure 2.58 The chain rule for functions of a single variable.

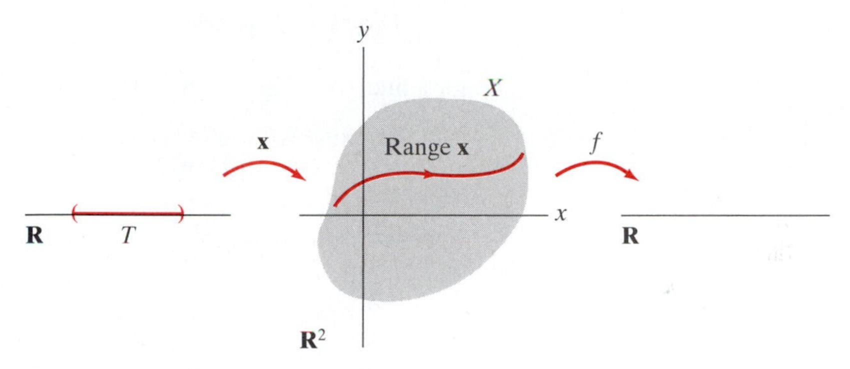

Figure 2.59 The composite function $f \circ \mathbf{x}$.

on X. The composite $f \circ \mathbf{x}$ is then nothing more than the restriction of f to the curve (i.e., the function that measures the temperature along just the curve). The question is, how does f depend on t? We claim the following:

■ **Proposition 5.2** Suppose $\mathbf{x}: T \subseteq \mathbf{R} \to \mathbf{R}^2$ is differentiable at $t_0 \in T$, and $f: X \subseteq \mathbf{R}^2 \to \mathbf{R}$ is differentiable at $\mathbf{x}_0 = \mathbf{x}(t_0) = (x_0, y_0) \in X$, where T and X are open in $\mathbf{R}$ and $\mathbf{R}^2$, respectively, and range $\mathbf{x}$ is contained in X. If, in addition, f is of class C^1, then $f \circ \mathbf{x}: T \to \mathbf{R}$ is differentiable at t_0 and

$$\frac{df}{dt}(t_0) = \frac{\partial f}{\partial x}(\mathbf{x}_0)\frac{dx}{dt}(t_0) + \frac{\partial f}{\partial y}(\mathbf{x}_0)\frac{dy}{dt}(t_0).$$

Before we prove Proposition 5.2, some remarks are in order. First, notice the mixture of ordinary and partial derivatives appearing in the formula for the derivative. These terms make sense if we contruct an appropriate "variable hierarchy" diagram, as shown in Figure 2.60. At the intermediate level, f depends on two variables, x and y (or, equivalently, on the vector variable $\mathbf{x} = (x, y)$), so partial derivatives are in order. On the final or composite level, f depends on just a single independent variable t and hence the use of the ordinary derivative df/dt is warranted. Second, the formula in Proposition 5.2 is a generalization of equation (2): A product term appears for each of the two intermediate variables.

EXAMPLE 2 Suppose $f(x, y) = (x + y^2)/(2x^2 + 1)$ is a temperature function on $\mathbf{R}^2$ and $\mathbf{x}(t) = (2t, t + 1)$. The function $\mathbf{x}$ gives parametric equations

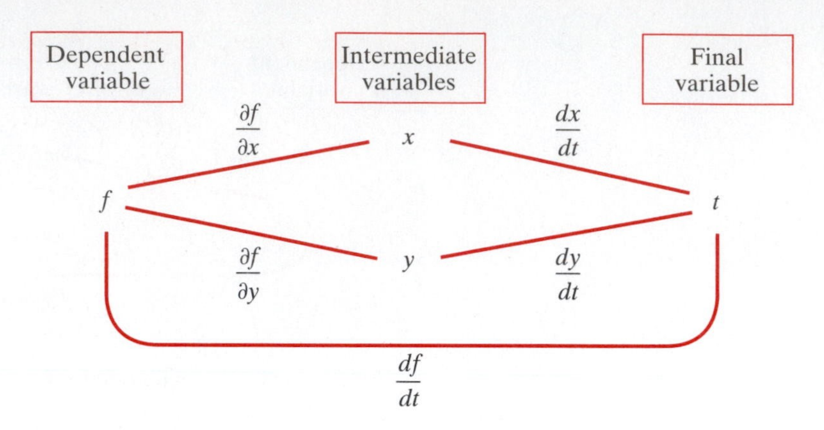

Figure 2.60 The chain rule of Proposition 5.2.

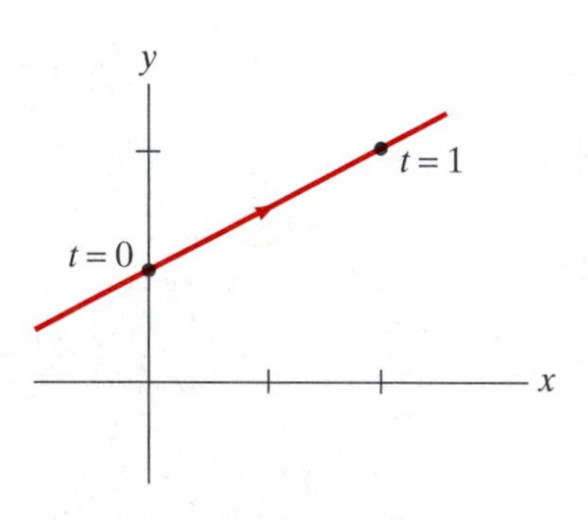

Figure 2.61 The graph of the function $\mathbf{x}$ of Example 2.

for a line. (See Figure 2.61.) Then,

$$(f \circ \mathbf{x})(t) = f(\mathbf{x}(t)) = \frac{2t + (t+1)^2}{8t^2 + 1} = \frac{t^2 + 4t + 1}{8t^2 + 1}$$

is the temperature function along the line, and we have

$$\frac{df}{dt} = \frac{4 - 14t - 32t^2}{(8t^2 + 1)^2},$$

by the quotient rule. Thus, all the hypotheses of Proposition 5.2 are satisfied and so the derivative formula must hold. Indeed, we have

$$\frac{\partial f}{\partial x} = \frac{1 - 2x^2 - 4xy^2}{(2x^2 + 1)^2},$$
$$\frac{\partial f}{\partial y} = \frac{2y}{2x^2 + 1},$$

and

$$\mathbf{x}'(t) = \left(\frac{dx}{dt}, \frac{dy}{dt}\right) = (2, 1).$$

Therefore,

$$\frac{\partial f}{\partial x}\frac{dx}{dt} + \frac{\partial f}{\partial y}\frac{dy}{dt} = \frac{1 - 2x^2 - 4xy^2}{(2x^2 + 1)^2} \cdot 2 + \frac{2y}{2x^2 + 1} \cdot 1$$
$$= \frac{2(1 - 8t^2 - 8t(t+1)^2)}{(8t^2 + 1)^2} + \frac{2(t+1)}{8t^2 + 1},$$

after substitution of $2t$ for x and $t + 1$ for y. Hence,

$$\frac{\partial f}{\partial x}\frac{dx}{dt} + \frac{\partial f}{\partial y}\frac{dy}{dt} = \frac{2(2 - 7t - 16t^2)}{(8t^2 + 1)^2},$$

which checks with our previous result for df/dt. ◆

Proof of Proposition 5.2 Denote the composite function $f \circ \mathbf{x}$ by z. We want to establish a formula for dz/dt at t_0. Since z is just a scalar-valued function of one variable, differentiability and the existence of the derivative mean the same thing. Thus, we consider

$$\frac{dz}{dt}(t_0) = \lim_{t \to t_0} \frac{z(t) - z(t_0)}{t - t_0},$$

and see if this limit exists. We have

$$\frac{dz}{dt}(t_0) = \lim_{t \to t_0} \frac{f(x(t), y(t)) - f(x(t_0), y(t_0))}{t - t_0}.$$

The first step is to rewrite the numerator of the limit expression by subtracting and adding $f(x_0, y)$ and to apply a modicum of algebra. Thus,

$$\begin{aligned}\frac{dz}{dt}(t_0) &= \lim_{t \to t_0} \frac{f(x, y) - f(x_0, y) + f(x_0, y) - f(x_0, y_0)}{t - t_0} \\ &= \lim_{t \to t_0} \frac{f(x, y) - f(x_0, y)}{t - t_0} + \lim_{t \to t_0} \frac{f(x_0, y) - f(x_0, y_0)}{t - t_0}.\end{aligned}$$

(Remember that $\mathbf{x}(t_0) = \mathbf{x}_0 = (x_0, y_0)$.) Now, for the main innovation of the proof. We apply the mean value theorem to the partial functions of f. This tells us that there must be a number c between x_0 and x and another number d between y_0 and y such that

$$f(x, y) - f(x_0, y) = f_x(c, y)(x - x_0)$$

and

$$f(x_0, y) - f(x_0, y_0) = f_y(x_0, d)(y - y_0).$$

Thus,

$$\begin{aligned}\frac{dz}{dt}(t_0) &= \lim_{t \to t_0} f_x(c, y)\frac{x - x_0}{t - t_0} + \lim_{t \to t_0} f_y(x_0, d)\frac{y - y_0}{t - t_0} \\ &= \lim_{t \to t_0} f_x(c, y)\frac{x(t) - x(t_0)}{t - t_0} + \lim_{t \to t_0} f_y(x_0, d)\frac{y(t) - y(t_0)}{t - t_0} \\ &= f_x(x_0, y_0)\frac{dx}{dt}(t_0) + f_y(x_0, y_0)\frac{dy}{dt}(t_0),\end{aligned}$$

by the definition of the derivatives

$$\frac{dx}{dt}(t_0) \quad \text{and} \quad \frac{dy}{dt}(t_0)$$

and the fact that $f_x(c, y)$ and $f_y(x_0, d)$ must approach $f_x(x_0, y_0)$ and $f_y(x_0, y_0)$, respectively, as t approaches t_0, by continuity of the partials. (Recall that f was assumed to be of class C^1.) This completes the proof. ■

Proposition 5.2 and its proof are easy to generalize to the case where f is a function of n variables (i.e., $f: X \subseteq \mathbf{R}^n \to \mathbf{R}$) and $\mathbf{x} : T \subseteq \mathbf{R} \to \mathbf{R}^n$. The appropriate chain rule formula in this case is

$$\frac{df}{dt}(t_0) = \frac{\partial f}{\partial x_1}(\mathbf{x}_0)\frac{dx_1}{dt}(t_0) + \frac{\partial f}{\partial x_2}(\mathbf{x}_0)\frac{dx_2}{dt}(t_0) + \cdots + \frac{\partial f}{\partial x_n}(\mathbf{x}_0)\frac{dx_n}{dt}(t_0). \quad (3)$$

Note that the right side of equation (3) can also be written by using matrix notation so that

$$\frac{df}{dt}(t_0) = \begin{bmatrix} \frac{\partial f}{\partial x_1}(\mathbf{x}_0) & \frac{\partial f}{\partial x_2}(\mathbf{x}_0) & \cdots & \frac{\partial f}{\partial x_n}(\mathbf{x}_0) \end{bmatrix} \begin{bmatrix} \frac{dx_1}{dt}(t_0) \\ \frac{dx_2}{dt}(t_0) \\ \vdots \\ \frac{dx_n}{dt}(t_0) \end{bmatrix}.$$

Thus, we have shown

$$\frac{df}{dt}(t_0) = Df(\mathbf{x}_0)D\mathbf{x}(t_0) = \nabla f(\mathbf{x}_0) \cdot \mathbf{x}'(t_0), \quad (4)$$

where we use $\mathbf{x}'(t_0)$ as a notational alternative to $D\mathbf{x}(t_0)$. The version of the chain rule given in formula (4) is particularly important and will be used a number of times in our subsequent work.

Let us consider further instances of composition of functions of many variables. For example, suppose X is open in $\mathbf{R}^3$, T is open in $\mathbf{R}^2$, and $f: X \subseteq \mathbf{R}^3 \to \mathbf{R}$ and $\mathbf{x}: T \subseteq \mathbf{R}^2 \to \mathbf{R}^3$ are such that the range of $\mathbf{x}$ is contained in X. Then the composite $f \circ \mathbf{x}: T \subseteq \mathbf{R}^2 \to \mathbf{R}$ can be formed, as shown in Figure 2.62. Note that the range of $\mathbf{x}$, that is, $\mathbf{x}(T)$, is just a surface in $\mathbf{R}^3$, so $f \circ \mathbf{x}$ can be thought of as an appropriate "temperature function" restricted to this surface. If we use $\mathbf{x} = (x, y, z)$ to denote the vector variable in $\mathbf{R}^3$ and $\mathbf{t} = (s, t)$ for the vector variable in $\mathbf{R}^2$, then we can write a plausible chain rule formula from an appropriate variable hierarchy diagram. (See Figure 2.63.) Thus, it is

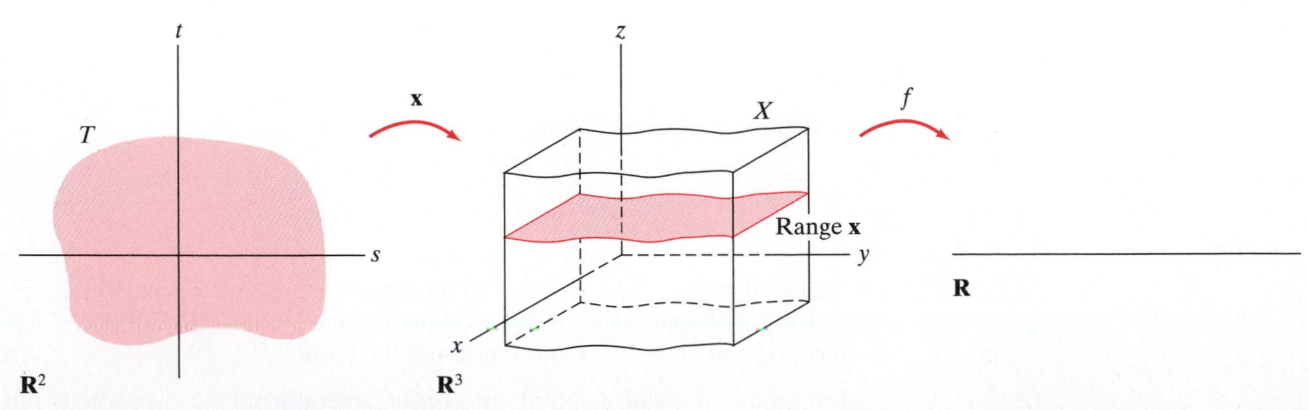

Figure 2.62 The composite $f \circ \mathbf{x}$ where $f: X \subseteq \mathbf{R}^3 \to \mathbf{R}$ and $\mathbf{x}: T \subseteq \mathbf{R}^2 \to \mathbf{R}^3$.

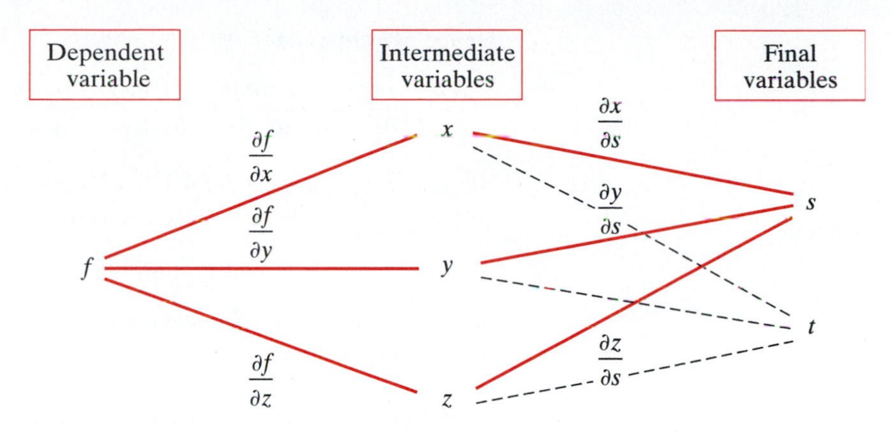

Figure 2.63 The chain rule for $f \circ \mathbf{x}$, where $f: X \subseteq \mathbf{R}^3 \to \mathbf{R}$ and $\mathbf{x}: T \subseteq \mathbf{R}^2 \to \mathbf{R}^3$.

reasonable to expect that the following formulas hold:

$$\frac{\partial f}{\partial s} = \frac{\partial f}{\partial x}\frac{\partial x}{\partial s} + \frac{\partial f}{\partial y}\frac{\partial y}{\partial s} + \frac{\partial f}{\partial z}\frac{\partial z}{\partial s}$$

and (5)

$$\frac{\partial f}{\partial t} = \frac{\partial f}{\partial x}\frac{\partial x}{\partial t} + \frac{\partial f}{\partial y}\frac{\partial y}{\partial t} + \frac{\partial f}{\partial z}\frac{\partial z}{\partial t}.$$

(Again, we abuse notation by writing both $\partial f/\partial s$, $\partial f/\partial t$ and $\partial f/\partial x$, $\partial f/\partial y$, $\partial f/\partial z$.) Indeed, when f is a function of x, y, and z of class C^1, formula (3) with $n = 3$ applies once we realize that $\partial x/\partial s$, $\partial x/\partial t$, etc., represent ordinary differentiation of the partial functions in s or t.

EXAMPLE 3 Suppose

$$f(x, y, z) = x^2 + y^2 + z^2 \quad \text{and} \quad \mathbf{x}(s, t) = (s\cos t, e^{st}, s^2 - t^2).$$

Then, $h(s, t) = f \circ \mathbf{x}(s, t) = s^2\cos^2 t + e^{2st} + (s^2 - t^2)^2$, so that

$$\frac{\partial h}{\partial s} = \frac{\partial(f \circ \mathbf{x})}{\partial s} = 2s\cos^2 t + 2te^{2st} + 4s(s^2 - t^2)$$

$$\frac{\partial h}{\partial t} = \frac{\partial(f \circ \mathbf{x})}{\partial t} = -2s^2\cos t\sin t + 2se^{2st} - 4t(s^2 - t^2).$$

We also have

$$\frac{\partial f}{\partial x} = 2x, \qquad \frac{\partial f}{\partial y} = 2y, \qquad \frac{\partial f}{\partial z} = 2z$$

and

$$\frac{\partial x}{\partial s} = \cos t, \qquad \frac{\partial x}{\partial t} = -s\sin t,$$
$$\frac{\partial y}{\partial s} = te^{st}, \qquad \frac{\partial y}{\partial t} = se^{st},$$
$$\frac{\partial z}{\partial s} = 2s, \qquad \frac{\partial z}{\partial t} = -2t.$$

Hence, we compute

$$\begin{aligned}\frac{\partial f}{\partial s} = \frac{\partial (f \circ \mathbf{x})}{\partial s} &= \frac{\partial f}{\partial x}\frac{\partial x}{\partial s} + \frac{\partial f}{\partial y}\frac{\partial y}{\partial s} + \frac{\partial f}{\partial z}\frac{\partial z}{\partial s} \\ &= 2x(\cos t) + 2y(te^{st}) + 2z(2s) \\ &= 2s\cos t(\cos t) + 2e^{st}(te^{st}) + 2(s^2 - t^2)(2s) \\ &= 2s\cos^2 t + 2te^{2st} + 4s(s^2 - t^2),\end{aligned}$$

just as we saw earlier. We leave it to you to use the chain rule to calculate $\partial f/\partial t$ in a similar manner. ◆

Of course, there is no need for us to stop here. Suppose we have an open set X in $\mathbf{R}^m$, an open set T in $\mathbf{R}^n$, and functions $f\colon X \to \mathbf{R}$ and $\mathbf{x}\colon T \to \mathbf{R}^m$ such that $h = f \circ \mathbf{x}\colon T \to \mathbf{R}$ can be defined. If f is of class C^1 and $\mathbf{x}$ is differentiable, then, from the previous remarks, h must also be differentiable and, moreover,

$$\begin{aligned}\frac{\partial h}{\partial t_j} &= \frac{\partial f}{\partial x_1}\frac{\partial x_1}{\partial t_j} + \frac{\partial f}{\partial x_2}\frac{\partial x_2}{\partial t_j} + \cdots + \frac{\partial f}{\partial x_m}\frac{\partial x_m}{\partial t_j} \\ &= \sum_{k=1}^{m} \frac{\partial f}{\partial x_k}\frac{\partial x_k}{\partial t_j}, \qquad j = 1, 2, \ldots, n.\end{aligned}$$

Since the component functions of a vector-valued function are just scalar-valued functions, we can say even more. Suppose $\mathbf{f}\colon X \subseteq \mathbf{R}^m \to \mathbf{R}^p$ and $\mathbf{x}\colon T \subseteq \mathbf{R}^n \to \mathbf{R}^m$ are such that $\mathbf{h} = \mathbf{f} \circ \mathbf{x}\colon T \subseteq \mathbf{R}^n \to \mathbf{R}^p$ can be defined. (As always, we assume that X is open in $\mathbf{R}^m$ and T is open in $\mathbf{R}^n$.) See Figure 2.64 for a representation of the situation. If $\mathbf{f}$ is of class C^1 and $\mathbf{x}$ is differentiable, then the composite $\mathbf{h} = \mathbf{f} \circ \mathbf{x}$ is differentiable and the following general formula holds:

$$\frac{\partial h_i}{\partial t_j} = \sum_{k=1}^{m} \frac{\partial f_i}{\partial x_k}\frac{\partial x_k}{\partial t_j}, \qquad i = 1, 2, \ldots, p; \quad j = 1, 2, \ldots, n. \tag{6}$$

The plausibility of formula (6) is immediate, given the variable hierarchy diagram shown in Figure 2.65.

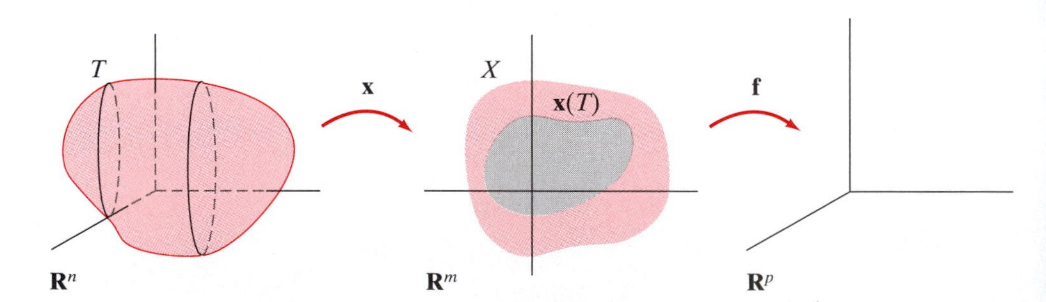

Figure 2.64 The composite $\mathbf{f} \circ \mathbf{x}$ where $\mathbf{f}\colon X \subseteq \mathbf{R}^m \to \mathbf{R}^p$ and $\mathbf{x}\colon T \subseteq \mathbf{R}^n \to \mathbf{R}^m$.

Now comes the real "magic". Recall that if A is a $p \times m$ matrix and B is an $m \times n$ matrix, then the product matrix $C = AB$ is defined and is a $p \times n$ matrix. Moreover, the ijth entry of C is given by

$$c_{ij} = \sum_{k=1}^{m} a_{ik} b_{kj}.$$

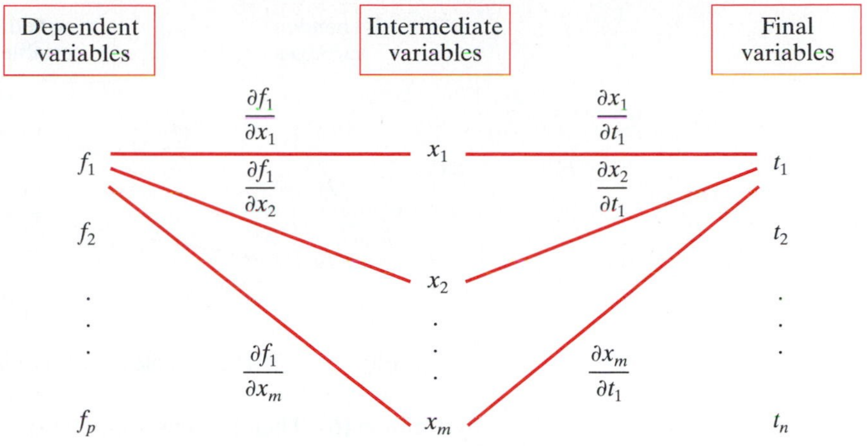

Dependent variables

Intermediate variables

Final variables

f_1, f_2, $\ldots$, f_p — x_1, x_2, $\ldots$, x_m — t_1, t_2, $\ldots$, t_n

$\dfrac{\partial f_1}{\partial x_1}$ $\dfrac{\partial f_1}{\partial x_2}$ $\dfrac{\partial f_1}{\partial x_m}$ $\dfrac{\partial x_1}{\partial t_1}$ $\dfrac{\partial x_2}{\partial t_1}$ $\dfrac{\partial x_m}{\partial t_1}$

Figure 2.65 The chain rule diagram for $\mathbf{f} \circ \mathbf{x}$, where $\mathbf{f}: X \subseteq \mathbf{R}^m \to \mathbf{R}^p$ and $\mathbf{x}: T \subseteq \mathbf{R}^n \to \mathbf{R}^m$.

If we recall that the ijth entry of the matrix $D\mathbf{h}(\mathbf{t})$ is $\partial h_i/\partial t_j$, and similarly for $D\mathbf{f}(\mathbf{x})$ and $D\mathbf{x}(\mathbf{t})$, then we see that formula (6) expresses nothing more than the following equation of matrices:

$$D\mathbf{h}(\mathbf{t}) = D(\mathbf{f} \circ \mathbf{x})(\mathbf{t}) = D\mathbf{f}(\mathbf{x})D\mathbf{x}(\mathbf{t}). \tag{7}$$

The similarity between formulas (7) and (1) is striking. One of the reasons (perhaps the principal reason) for defining matrix multiplication as we have is precisely so that the chain rule in several variables can have the elegant appearance that it has in formula (7).

EXAMPLE 4 Suppose $\mathbf{f}: \mathbf{R}^3 \to \mathbf{R}^2$ is given by $\mathbf{f}(x_1, x_2, x_3) = (x_1 - x_2, x_1x_2x_3)$ and $\mathbf{x}: \mathbf{R}^2 \to \mathbf{R}^3$ is given by $\mathbf{x}(t_1, t_2) = (t_1t_2, t_1^2, t_2^2)$. Then, $\mathbf{f} \circ \mathbf{x}: \mathbf{R}^2 \to \mathbf{R}^2$ is given by $\mathbf{f} \circ \mathbf{x}(t_1, t_2) = (t_1t_2 - t_1^2, t_1^3t_2^3)$, so that

$$D(\mathbf{f} \circ \mathbf{x})(\mathbf{t}) = \begin{bmatrix} t_2 - 2t_1 & t_1 \\ 3t_1^2t_2^3 & 3t_1^3t_2^2 \end{bmatrix}.$$

On the other hand,

$$D\mathbf{f}(\mathbf{x}) = \begin{bmatrix} 1 & -1 & 0 \\ x_2x_3 & x_1x_3 & x_1x_2 \end{bmatrix} \quad \text{and} \quad D\mathbf{x}(t) = \begin{bmatrix} t_2 & t_1 \\ 2t_1 & 0 \\ 0 & 2t_2 \end{bmatrix},$$

so that the product matrix is

$$D\mathbf{f}(\mathbf{x})D\mathbf{x}(\mathbf{t}) = \begin{bmatrix} t_2 - 2t_1 & t_1 \\ x_2x_3t_2 + 2x_1x_3t_1 & x_2x_3t_1 + 2x_1x_2t_2 \end{bmatrix}$$
$$= \begin{bmatrix} t_2 - 2t_1 & t_1 \\ t_1^2t_2^3 + 2t_1^2t_2^3 & t_1^3t_2^2 + 2t_1^3t_2^2 \end{bmatrix},$$

after substituting for x_1, x_2, and x_3. Thus, $D(\mathbf{f} \circ \mathbf{x})(\mathbf{t}) = D\mathbf{f}(\mathbf{x})D\mathbf{x}(\mathbf{t})$, as expected.

Alternatively, we may use the variable hierarchy diagram shown in Figure 2.66 and compute any individual partial derivative we may desire. For example,

$$\frac{\partial f_2}{\partial t_1} = \frac{\partial f_2}{\partial x_1}\frac{\partial x_1}{\partial t_1} + \frac{\partial f_2}{\partial x_2}\frac{\partial x_2}{\partial t_1} + \frac{\partial f_2}{\partial x_3}\frac{\partial x_3}{\partial t_1}$$

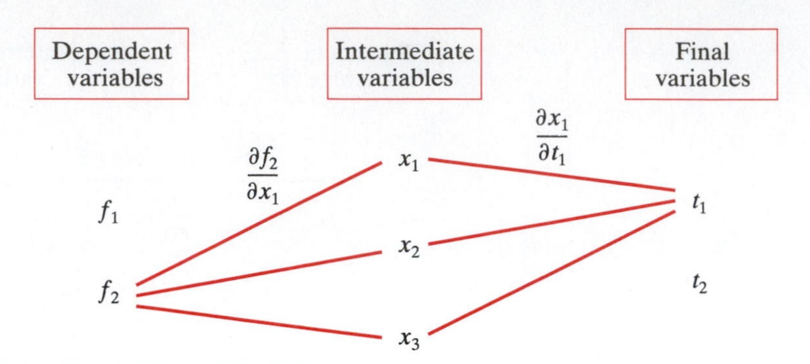

Figure 2.66 The variable hierarchy diagram for Example 4.

by formula (6). Then by abuse of notation,

$$\begin{aligned}\frac{\partial f_2}{\partial t_1} &= (x_2x_3)(t_2) + (x_1x_3)(2t_1) + (x_1x_2)(0) \\ &= (t_1^2t_2^2)(t_2) + (t_1t_2)(t_2^2)(2t_1) \\ &= 3t_1^2t_2^3,\end{aligned}$$

which is indeed the (2, 1) entry of the matrix product. ◆

Although we shall not do so, it is possible to prove the following version of the chain rule, which is somewhat more general from a technical standpoint:

■ **Theorem 5.3** (THE CHAIN RULE) Suppose $X \subseteq \mathbf{R}^m$ and $T \subseteq \mathbf{R}^n$ are open and $\mathbf{f}: X \to \mathbf{R}^p$ and $\mathbf{x}: T \to \mathbf{R}^m$ are defined so that range $\mathbf{x} \subseteq X$. If $\mathbf{x}$ is differentiable at $\mathbf{t}_0 \in T$ and $\mathbf{f}$ is differentiable at $\mathbf{x}_0 = \mathbf{x}(\mathbf{t}_0)$, then the composite $\mathbf{f} \circ \mathbf{x}$ is differentiable at $\mathbf{t}_0$, and we have

$$D(\mathbf{f} \circ \mathbf{x})(\mathbf{t}_0) = D\mathbf{f}(\mathbf{x}_0)D\mathbf{x}(\mathbf{t}_0).$$

The advantage of Theorem 5.3 over the earlier versions of the chain rule we have been discussing is that it requires **f** only to be differentiable at the point in question, not to be of class C^1. Note that, of course, Theorem 5.3 includes all the special cases of the chain rule we have previously discussed. In particular, Theorem 5.3 includes the important case of formula (4).

EXAMPLE 5 (Polar/rectangular conversions) Recall that in §1.7 we provided the basic equations relating polar and rectangular coordinates:

$$\begin{cases} x = r\cos\theta \\ y = r\sin\theta \end{cases}.$$

Now suppose you have an equation defining a quantity w as a function of x and y, that is,

$$w = f(x, y).$$

Then, of course, w may just as well be regarded as a function of r and θ by susbtituting $r\cos\theta$ for x and $r\sin\theta$ for y. That is,

$$w = g(r, \theta) = f(x(r, \theta), y(r, \theta)).$$

Our question is as follows: Assuming all functions involved are differentiable, how are the partial derivatives $\partial w/\partial r$, $\partial w/\partial \theta$ related to $\partial w/\partial x$, $\partial w/\partial y$?

In the situation just described, we have $w = g(r, \theta) = (f \circ \mathbf{x})(r, \theta)$, so that the chain rule implies

$$Dg(r, \theta) = Df(x, y) D\mathbf{x}(r, \theta).$$

Therefore,

$$\begin{aligned} \begin{bmatrix} \dfrac{\partial g}{\partial r} & \dfrac{\partial g}{\partial \theta} \end{bmatrix} &= \begin{bmatrix} \dfrac{\partial f}{\partial x} & \dfrac{\partial f}{\partial y} \end{bmatrix} \begin{bmatrix} \dfrac{\partial x}{\partial r} & \dfrac{\partial x}{\partial \theta} \\ \dfrac{\partial y}{\partial r} & \dfrac{\partial y}{\partial \theta} \end{bmatrix} \\ &= \begin{bmatrix} \dfrac{\partial f}{\partial x} & \dfrac{\partial f}{\partial y} \end{bmatrix} \begin{bmatrix} \cos\theta & -r\sin\theta \\ \sin\theta & r\cos\theta \end{bmatrix}. \end{aligned}$$

By extracting entries, we see that the various partial derivatives of w are related by the following formulas:

$$\begin{cases} \dfrac{\partial w}{\partial r} = \cos\theta \dfrac{\partial w}{\partial x} + \sin\theta \dfrac{\partial w}{\partial y} \\ \dfrac{\partial w}{\partial \theta} = -r\sin\theta \dfrac{\partial w}{\partial x} + r\cos\theta \dfrac{\partial w}{\partial y} \end{cases}. \tag{8}$$

The significance of (8) is that it provides us with a relation of **differential operators:**

$$\begin{cases} \dfrac{\partial}{\partial r} = \cos\theta \dfrac{\partial}{\partial x} + \sin\theta \dfrac{\partial}{\partial y} \\ \dfrac{\partial}{\partial \theta} = -r\sin\theta \dfrac{\partial}{\partial x} + r\cos\theta \dfrac{\partial}{\partial y} \end{cases}. \tag{9}$$

The appropriate interpretation for (9) is the following: Differentiation with respect to the polar coordinate r is the same as a certain combination of differentiation with respect to both Cartesian coordinates x and y (namely, the combination $\cos\theta\, \partial/\partial x + \sin\theta\, \partial/\partial y$). A similar comment applies to differentiation with respect to the polar coordinate θ. Note that, when $r \neq 0$, we can solve algebraically for $\partial/\partial x$ and $\partial/\partial y$ in (9), obtaining

$$\begin{cases} \dfrac{\partial}{\partial x} = \cos\theta \dfrac{\partial}{\partial r} - \dfrac{\sin\theta}{r} \dfrac{\partial}{\partial \theta} \\ \dfrac{\partial}{\partial y} = \sin\theta \dfrac{\partial}{\partial r} + \dfrac{\cos\theta}{r} \dfrac{\partial}{\partial \theta} \end{cases}. \tag{10}$$

We will have occasion to use the relations in (9) and (10), and the method of their derivation, later in this text. ◆

Exercises

1. If $f(x, y, z) = x^2 - y^3 + xyz$, and $x = 6t + 7$, $y = \sin 2t$, $z = t^2$, verify the chain rule by finding df/dt in two different ways.

2. If $f(x, y) = \sin(xy)$ and $x = s + t$, $y = s^2 + t^2$, find $\partial f/\partial s$ and $\partial f/\partial t$ in two ways:

(a) by substitution.

(b) by means of the chain rule.

3. Suppose that a bird flies along the helical curve $x = 2\cos t$, $y = 2\sin t$, $z = 3t$. The bird suddenly encounters a weather front so that the barometric pressure is varying rather wildly from point to point as $P(x, y, z) = 6x^2 z/y$ atm.

(a) Use the chain rule to determine how the pressure is changing at $t = \pi/4$ min.

(b) Check your result in part (a) by direct substitution.

(c) What is the approximate pressure at $t = \pi/4 + 0.01$ min?

4. Suppose that $z = x^2 + y^3$, where $x = st$ and y is a function of s and t. Suppose further that when $(s, t) = (2, 1)$, $\partial y/\partial t = 0$. Determine $\dfrac{\partial z}{\partial t}(2, 1)$.

5. You are the proud new owner of an Acme Deluxe Bread Kneading Machine, which you are using for the first time today. Suppose that at noon the dimensions of your (nearly rectangular) loaf of bread dough are $L = 7$ in (length), $W = 5$ in (width), and $H = 4$ in (height). At that time, you place the loaf in the machine for kneading and the machine begins by stretching the loaf's length at an initial rate of 0.75 in/min, punching down the loaf's height at a rate of 1 in/min, and increasing the loaf's width at a rate of 0.5 in/min. What is the rate of change of the volume of the loaf when the machine starts? Is the dough increasing or decreasing in size at that moment?

6. A rectangular stick of butter is placed in the microwave oven to melt. When the butter's length is 6 in and its square cross-section measures 1.5 in on a side, its length is decreasing at a rate of 0.25 in/min and its cross-sectional edge is decreasing at a rate of 0.125 in/min. How fast is the butter melting (i.e., at what rate is the solid volume of butter turning to liquid) at that instant?

7. Suppose that the following function is used to model the monthly demand for bicycles:

$$P(x, y) = 200 + 20\sqrt{0.1x + 10} - 12\sqrt[3]{y}.$$

In this formula, x represents the price (in dollars per gallon) of automobile gasoline and y represents the selling price (in dollars) of each bicycle. Furthermore, suppose that the price of gasoline t months from now will be

$$x = 1 + 0.1t - \cos\frac{\pi t}{6}$$

and the price of each bicycle will be

$$y = 200 + 2t\sin\frac{\pi t}{6}.$$

At what rate will the monthly demand for bicycles be changing six months from now?

8. The Centers for Disease Control provide information on the **body mass index** (BMI) to give a more meaningful assessment of a person's weight. The BMI is given by the formula

$$\text{BMI} = \frac{10{,}000w}{h^2},$$

where w is an individual's mass in kilograms and h the person's height in centimeters. While monitoring a child's growth, you estimate that at the time he turned 10 years old, his height showed a growth rate of 0.6 cm per month. At the same time, his mass showed a growth rate of 0.4 kg per month. Suppose that he was 140 cm tall and weighed 33 kg on his tenth birthday.

(a) At what rate is his BMI changing on his tenth birthday?

(b) The BMI of a typical 10-year-old male increases at an average rate of 0.04 BMI points per month. Should you be concerned about the child's weight gain?

9. Suppose $z = f(x, y)$ has continuous partial derivatives. Let $x = e^r\cos\theta$, $y = e^r\sin\theta$. Show that then

$$\left(\frac{\partial z}{\partial x}\right)^2 + \left(\frac{\partial z}{\partial y}\right)^2 = e^{-2r}\left[\left(\frac{\partial z}{\partial r}\right)^2 + \left(\frac{\partial z}{\partial \theta}\right)^2\right].$$

10. Suppose that $z = f(x + y, x - y)$ has continuous partial derivatives with respect to $u = x + y$ and $v = x - y$. Show that

$$\frac{\partial z}{\partial x}\frac{\partial z}{\partial y} = \left(\frac{\partial z}{\partial u}\right)^2 - \left(\frac{\partial z}{\partial v}\right)^2.$$

11. If $w = f\left(\dfrac{xy}{x^2 + y^2}\right)$ is a differentiable function of $u = \dfrac{xy}{x^2 + y^2}$, show that

$$x\frac{\partial w}{\partial x} + y\frac{\partial w}{\partial y} = 0.$$

12. If $w = f\left(\dfrac{x^2 - y^2}{x^2 + y^2}\right)$ is a differentiable function of $u = \dfrac{x^2 - y^2}{x^2 + y^2}$, show that then

$$x\frac{\partial w}{\partial x} + y\frac{\partial w}{\partial y} = 0.$$

13. Suppose $w = f\left(\dfrac{y-x}{xy}, \dfrac{z-x}{xz}\right)$ is a differentiable function of $u = \dfrac{y-x}{xy}$ and $v = \dfrac{z-x}{xz}$. Show then that

$$x^2\frac{\partial w}{\partial x} + y^2\frac{\partial w}{\partial y} + z^2\frac{\partial w}{\partial z} = 0.$$

14. Suppose that $w = g\left(\dfrac{x}{y}, \dfrac{z}{y}\right)$ is a differentiable function of $u = x/y$ and $v = z/y$. Show then that

$$x\frac{\partial w}{\partial x} + y\frac{\partial w}{\partial y} + z\frac{\partial w}{\partial z} = 0.$$

In Exercises 15–19, calculate $D(\mathbf{f} \circ \mathbf{g})$ *in two ways: (a) by first evaluating* $\mathbf{f} \circ \mathbf{g}$ *and (b) by using the chain rule and the derivative matrices* $D\mathbf{f}$ *and* $D\mathbf{g}$.

15. $\mathbf{f}(x) = (3x^5, e^{2x})$, $g(s, t) = s - 7t$

16. $f(x, y) = x^2 - 3y^2$, $\mathbf{g}(s, t) = (st, s + t^2)$

17. $\mathbf{f}(x, y) = \left(xy - \dfrac{y}{x}, \dfrac{x}{y} + y^3\right)$, $\mathbf{g}(s, t) = \left(\dfrac{s}{t}, s^2 t\right)$

18. $\mathbf{f}(x, y, z) = (x^2y + y^2z, xyz, e^z)$, $\mathbf{g}(t) = (t - 2, 3t + 7, t^3)$

19. $\mathbf{f}(x, y, z) = (x + y + z, x^3 - e^{yz})$, $\mathbf{g}(s, t, u) = (st, tu, su)$

20. Let $\mathbf{g}\colon \mathbf{R}^3 \to \mathbf{R}^2$ be a differentiable function such that $\mathbf{g}(1, -1, 3) = (2, 5)$ and $D\mathbf{g}(1, -1, 3) = \begin{bmatrix} 1 & -1 & 0 \\ 4 & 0 & 7 \end{bmatrix}$. Suppose that $\mathbf{f}\colon \mathbf{R}^2 \to \mathbf{R}^2$ is defined by $\mathbf{f}(x, y) = (2xy, 3x - y + 5)$. What is $D(\mathbf{f} \circ \mathbf{g})(1, -1, 3)$?

21. Let $\mathbf{g}\colon \mathbf{R}^2 \to \mathbf{R}^2$ and $\mathbf{f}\colon \mathbf{R}^2 \to \mathbf{R}^2$ be differentiable functions such that $\mathbf{g}(0, 0) = (1, 2)$, $\mathbf{g}(1, 2) = (3, 5)$, $\mathbf{f}(0, 0) = (3, 5)$, $\mathbf{f}(4, 1) = (1, 2)$, $D\mathbf{g}(0, 0) = \begin{bmatrix} 1 & 0 \\ -1 & 4 \end{bmatrix}$, $D\mathbf{g}(1, 2) = \begin{bmatrix} 2 & 3 \\ 5 & 7 \end{bmatrix}$, $D\mathbf{f}(3, 5) = \begin{bmatrix} 1 & 1 \\ 3 & 5 \end{bmatrix}$, $D\mathbf{f}(4, 1) = \begin{bmatrix} -1 & 2 \\ 1 & 3 \end{bmatrix}$.

(a) Calculate $D(\mathbf{f} \circ \mathbf{g})(1, 2)$.

(b) Calculate $D(\mathbf{g} \circ \mathbf{f})(4, 1)$.

22. Let $z = f(x, y)$, where f has continuous partial derivatives. If we make the standard polar/rectangular substitution $x = r\cos\theta$, $y = r\sin\theta$, show that

$$\left(\frac{\partial z}{\partial x}\right)^2 + \left(\frac{\partial z}{\partial y}\right)^2 = \left(\frac{\partial z}{\partial r}\right)^2 + \frac{1}{r^2}\left(\frac{\partial z}{\partial \theta}\right)^2.$$

23. (a) Use the methods of Example 5 and formula (10) in this section to determine $\partial^2/\partial x^2$ and $\partial^2/\partial y^2$ in terms of the polar partial differential operators $\partial^2/\partial r^2$, $\partial^2/\partial\theta^2$, $\partial^2/\partial r\,\partial\theta$, $\partial/\partial r$, and $\partial/\partial\theta$. (Hint: You will need to use the product rule.)

(b) Use part (a) to show that the **Laplacian operator** $\partial^2/\partial x^2 + \partial^2/\partial y^2$ is given in polar coordinates by the formula

$$\frac{\partial^2}{\partial x^2} + \frac{\partial^2}{\partial y^2} = \frac{\partial^2}{\partial r^2} + \frac{1}{r}\frac{\partial}{\partial r} + \frac{1}{r^2}\frac{\partial^2}{\partial\theta^2}.$$

24. Show that the Laplacian operator $\partial^2/\partial x^2 + \partial^2/\partial y^2 + \partial^2/\partial z^2$ in three dimensions is given in cylindrical coordinates by the formula

$$\frac{\partial^2}{\partial x^2} + \frac{\partial^2}{\partial y^2} + \frac{\partial^2}{\partial z^2} = \frac{\partial^2}{\partial r^2} + \frac{1}{r}\frac{\partial}{\partial r} + \frac{1}{r^2}\frac{\partial^2}{\partial\theta^2} + \frac{\partial^2}{\partial z^2}.$$

25. In this problem, you will determine the formula for the Laplacian operator in spherical coordinates.

(a) First, note that the cylindrical/spherical conversions given by formula (6) of §1.7 express the cylindrical coordinates z and r in terms of the spherical coordinates ρ and φ by equations of precisely the same form as those that express x and y in terms of the polar coordinates r and θ. Use this fact to write $\partial/\partial r$ in terms of $\partial/\partial\rho$ and $\partial/\partial\varphi$. (Also see formula (10) of this section.)

(b) Use the ideas and result of part (a) to establish the following formula:

$$\frac{\partial^2}{\partial x^2} + \frac{\partial^2}{\partial y^2} + \frac{\partial^2}{\partial z^2} = \frac{\partial^2}{\partial\rho^2} + \frac{1}{\rho^2}\frac{\partial^2}{\partial\varphi^2} + \frac{1}{\rho^2\sin^2\varphi}\frac{\partial^2}{\partial\theta^2} + \frac{2}{\rho}\frac{\partial}{\partial\rho} + \frac{\cot\varphi}{\rho^2}\frac{\partial}{\partial\varphi}.$$

26. Suppose that y is defined implicitly as a function $y(x)$ by an equation of the form

$$F(x, y) = 0.$$

(For example, the equation $x^3 - y^2 = 0$ defines y as two functions of x, namely, $y = x^{3/2}$ and $y = -x^{3/2}$. The equation $\sin(xy) - x^2y^7 + e^y = 0$, on the other hand, cannot readily be solved for y in terms of x. See the end of §2.6 for more about implicit functions.)

(a) Show that if F and $y(x)$ are both assumed to be differentiable functions, then

$$\frac{dy}{dx} = -\frac{F_x(x, y)}{F_y(x, y)}$$

provided $F_y(x, y) \neq 0$.

(b) Use the result of part (a) to find dy/dx when y is defined implicitly in terms of x by the equation $x^3 - y^2 = 0$. Check your result by explicitly solving for y and differentiating.

27. Find dy/dx when y is defined implicitly by the equation $\sin(xy) - x^2y^7 + e^y = 0$. (See Exercise 26.)

28. Suppose that you are given an equation of the form

$$F(x, y, z) = 0,$$

for example, something like $x^3z+y\cos z+(\sin y)/z = 0$. Then we may consider z to be defined implicitly as a function $z(x, y)$.

(a) Use the chain rule to show that if F and $z(x, y)$ are both assumed to be differentiable, then

$$\frac{\partial z}{\partial x} = -\frac{F_x(x, y, z)}{F_z(x, y, z)}, \qquad \frac{\partial z}{\partial y} = -\frac{F_y(x, y, z)}{F_z(x, y, z)}.$$

(b) Use part (a) to find $\partial z/\partial x$ and $\partial z/\partial y$ where z is given by the equation $xyz = 2$. Check your result by explicitly solving for z and then calculating the partial derivatives.

29. Find $\partial z/\partial x$ and $\partial z/\partial y$, where z is given implicitly by the equation

$$x^3z + y\cos z + \frac{\sin y}{z} = 0.$$

(See Exercise 28.)

30. Let

$$f(x, y) = \begin{cases} \dfrac{x^2y}{x^2+y^2} & \text{if } (x, y) \neq (0, 0) \\ 0 & \text{if } (x, y) = (0, 0) \end{cases}.$$

(a) Use the definition of the partial derivative to find $f_x(0, 0)$ and $f_y(0, 0)$.

(b) Let a be a nonzero constant and let $\mathbf{x}(t) = (t, at)$. Show that $f \circ \mathbf{x}$ is differentiable, and find $D(f \circ \mathbf{x})(0)$ directly.

(c) Calculate $Df(0, 0)D\mathbf{x}(0)$. How can you reconcile your answer with your answer in part (b) and the chain rule?

Let $w = f(x, y, z)$ be a differentiable function of x, y, and z. For example, suppose that $w = x + 2y + z$. Regarding the variables x, y, and z as independent, we have $\partial w/\partial x = 1$ and $\partial w/\partial y = 2$. But now suppose that $z = xy$. Then x, y, and z are not all independent and, by substitution, we have that $w = x+2y+xy$ so that $\partial w/\partial x = 1+y$ and $\partial w/\partial y = 2+x$. To overcome the apparent ambiguity in the notation for partial derivatives, it is customary to indicate the complete set of independent variables by writing additional subscripts beside the partial derivative. Thus,

$$\left(\frac{\partial w}{\partial x}\right)_{y,z}$$

would signify the partial derivative of w with respect to x, while holding both y and z constant. Hence, x, y, and z are the complete set of independent variables in this case. On the other hand, we would use $(\partial w/\partial x)_y$ to indicate that x and y alone are the independent variables. In the case that $w = x+2y+z$, this notation gives

$$\left(\frac{\partial w}{\partial x}\right)_{y,z} = 1, \quad \left(\frac{\partial w}{\partial y}\right)_{x,z} = 2, \quad \textit{and} \quad \left(\frac{\partial w}{\partial z}\right)_{x,y} = 1.$$

If $z = xy$, then we also have

$$\left(\frac{\partial w}{\partial x}\right)_{y} = 1 + y, \quad \textit{and} \quad \left(\frac{\partial w}{\partial y}\right)_{x} = 2 + x.$$

In this way, the ambiguity of notation can be avoided. Use this notation in Exercises 31–37.

31. Let $w = x + 7y - 10z$ and $z = x^2 + y^2$.

(a) Find $\left(\frac{\partial w}{\partial x}\right)_{y,z}$, $\left(\frac{\partial w}{\partial y}\right)_{x,z}$, $\left(\frac{\partial w}{\partial z}\right)_{x,y}$, $\left(\frac{\partial w}{\partial x}\right)_{y}$, and $\left(\frac{\partial w}{\partial y}\right)_{x}$.

(b) Relate $(\partial w/\partial x)_{y,z}$ and $(\partial w/\partial x)_y$ by using the chain rule.

32. Repeat Exercise 31 where $w = x^3 + y^3 + z^3$ and $z = 2x - 3y$.

33. Suppose $s = x^2y+xzw-z^2$ and $xyw-y^3z+xz = 0$. Find

$$\left(\frac{\partial s}{\partial z}\right)_{x,y,w} \quad \text{and} \quad \left(\frac{\partial s}{\partial z}\right)_{x,w}.$$

34. Let $U = F(P, V, T)$ denote the internal energy of a gas. Suppose the gas obeys the ideal gas law $PV = kT$, where k is a constant.

(a) Find $\left(\frac{\partial U}{\partial T}\right)_{P}$.

(b) Find $\left(\frac{\partial U}{\partial T}\right)_{V}$.

(c) Find $\left(\frac{\partial U}{\partial P}\right)_{V}$.

35. Show that if x, y, z are related implicitly by an equation of the form $F(x, y, z) = 0$, then

$$\left(\frac{\partial x}{\partial y}\right)_z \left(\frac{\partial y}{\partial z}\right)_x \left(\frac{\partial z}{\partial x}\right)_y = -1.$$

This relation is used in thermodynamics. (Hint: Use Exercise 28.)

36. The ideal gas law $PV = kT$, where k is a constant, relates the pressure P, temperature T, and volume V of a gas. Verify the result of Exercise 35 for the ideal gas law equation.

37. Verify the result of Exercise 35 for the ellipsoid

$$ax^2 + by^2 + cz^2 = d$$

where a, b, c, and d are constants.

2.6 Directional Derivatives and the Gradient

In this section, we will consider some of the key geometric properties of the **gradient vector**

$$\nabla f = \left(\frac{\partial f}{\partial x_1}, \frac{\partial f}{\partial x_2}, \ldots, \frac{\partial f}{\partial x_n}\right)$$

of a scalar-valued function of n variables. In what follows, n will usually be 2 or 3.

The Directional Derivative

Let $f(x, y)$ be a scalar-valued function of two variables. In §2.3, we understood the partial derivative $\frac{\partial f}{\partial x}(a, b)$ as the slope, at the point $(a, b, f(a, b))$, of the curve obtained as the intersection of the surface $z = f(x, y)$ with the plane $y = b$. The other partial derivative $\frac{\partial f}{\partial y}(a, b)$ has a similar geometric interpretation. However, the surface $z = f(x, y)$ contains infinitely many curves passing through $(a, b, f(a, b))$ whose slope we might choose to measure. The directional derivative enables us to do this.

An alternative way to view $\frac{\partial f}{\partial x}(a, b)$ is as the rate of change of f as we move "infinitesimally" from $\mathbf{a} = (a, b)$ in the $\mathbf{i}$-direction, as suggested by Figure 2.67. This is easy to see since, by the definition of the partial derivative,

$$\begin{aligned}\frac{\partial f}{\partial x}(a, b) &= \lim_{h\to 0} \frac{f(a+h, b) - f(a, b)}{h} \\ &= \lim_{h\to 0} \frac{f((a, b) + (h, 0)) - f(a, b)}{h} \\ &= \lim_{h\to 0} \frac{f((a, b) + h(1, 0)) - f(a, b)}{h} \\ &= \lim_{h\to 0} \frac{f(\mathbf{a} + h\mathbf{i}) - f(\mathbf{a})}{h}.\end{aligned}$$

Note that we are identifying the point (a, b) with the vector $\mathbf{a} = (a, b) = a\mathbf{i} + b\mathbf{j}$. Similarly, we have

$$\frac{\partial f}{\partial y}(a, b) = \lim_{h\to 0} \frac{f(\mathbf{a} + h\mathbf{j}) - f(\mathbf{a})}{h}.$$

Writing partial derivatives as we just have enables us to see that they are special cases of a more general type of derivative. Suppose $\mathbf{v}$ is any unit vector in $\mathbf{R}^2$. (The reason for taking a unit vector will be made clear later.) The quantity

$$\lim_{h\to 0} \frac{f(\mathbf{a} + h\mathbf{v}) - f(\mathbf{a})}{h} \tag{1}$$

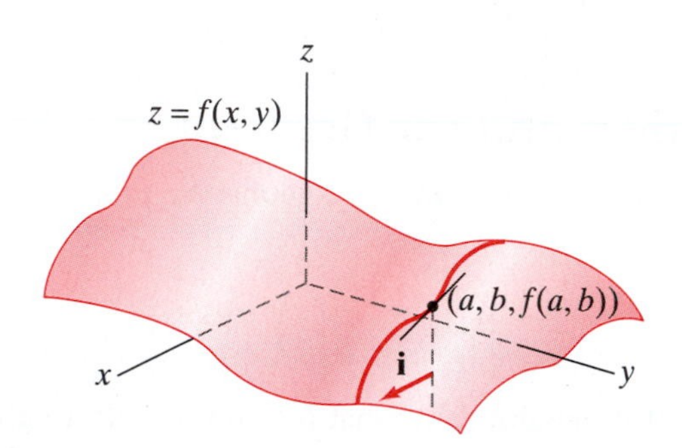

Figure 2.67 Another way to view the partial derivative $\partial f/\partial x$ at a point.

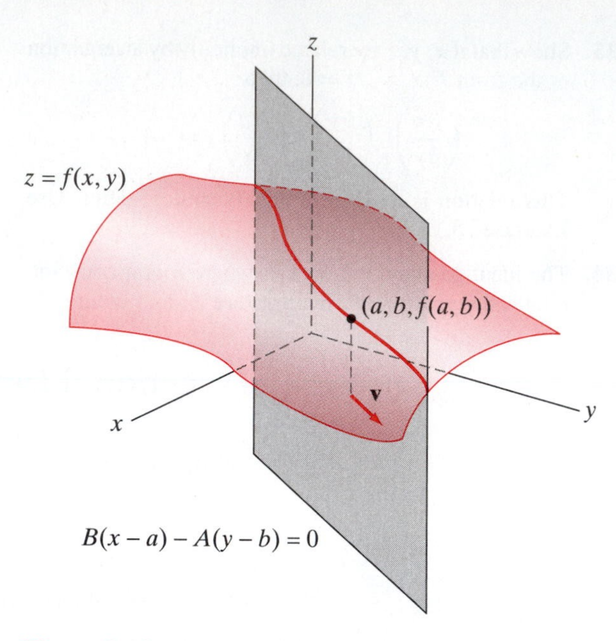

Figure 2.68 The directional derivative.

is nothing more than the rate of change of f as we move (infinitesimally) from $\mathbf{a} = (a, b)$ in the direction specified by $\mathbf{v} = (A, B) = A\mathbf{i} + B\mathbf{j}$. It's also the slope of the curve obtained as the intersection of the surface $z = f(x, y)$ with the vertical plane $B(x - a) - A(y - b) = 0$. (See Figure 2.68.) We can use the limit expression in (1) to define the derivative of any scalar-valued function in a particular direction.

Definition 6.1 Let X be open in $\mathbf{R}^n$, $f: X \subseteq \mathbf{R}^n \to \mathbf{R}$ a scalar-valued function, and $\mathbf{a} \in X$. If $\mathbf{v} \in \mathbf{R}^n$ is any unit vector, then the **directional derivative of f at a in the direction of v**, denoted $D_{\mathbf{v}} f(\mathbf{a})$, is

$$D_{\mathbf{v}} f(\mathbf{a}) = \lim_{h \to 0} \frac{f(\mathbf{a} + h\mathbf{v}) - f(\mathbf{a})}{h}$$

(provided that this limit exists).

EXAMPLE 1 Suppose $f(x, y) = x^2 - 3xy + 2x - 5y$. Then, if $\mathbf{v} = (v, w) \in \mathbf{R}^2$ is any unit vector, it follows that

$$\begin{aligned} D_{\mathbf{v}} f(0, 0) &= \lim_{h \to 0} \frac{f((0, 0) + h(v, w)) - f(0, 0)}{h} \\ &= \lim_{h \to 0} \frac{h^2 v^2 - 3h^2 vw + 2hv - 5hw}{h} \\ &= \lim_{h \to 0} (hv^2 - 3hvw + 2v - 5w) \\ &= 2v - 5w. \end{aligned}$$

Thus, the rate of change of f is $2v - 5w$ if we move from the origin in the direction given by $\mathbf{v}$. The rate of change is zero if $\mathbf{v} = (5/\sqrt{29}, 2/\sqrt{29})$ or $(-5/\sqrt{29}, -2/\sqrt{29})$. ◆

Consequently, we see that the partial derivatives of a function are just the "tip of the iceberg." However, it turns out that when f is differentiable, the partial derivatives actually determine the directional derivatives for all directions $\mathbf{v}$. To see this rather remarkable result, we begin by defining a new function F of a single variable by

$$F(t) = f(\mathbf{a} + t\mathbf{v}).$$

Then, by Definition 6.1, we have

$$D_{\mathbf{v}} f(\mathbf{a}) = \lim_{t\to 0} \frac{f(\mathbf{a}+t\mathbf{v}) - f(\mathbf{a})}{t} = \lim_{t\to 0} \frac{F(t) - F(0)}{t - 0} = F'(0).$$

That is,

$$D_{\mathbf{v}} f(\mathbf{a}) = \frac{d}{dt} f(\mathbf{a} + t\mathbf{v})\Big|_{t=0}. \tag{2}$$

The significance of equation (2) is that, when f is differentiable at $\mathbf{a}$, we can apply the chain rule to the right-hand side. Indeed, let $\mathbf{x}(t) = \mathbf{a} + t\mathbf{v}$. Then, by the chain rule,

$$\frac{d}{dt} f(\mathbf{a} + t\mathbf{v}) = Df(\mathbf{x})D\mathbf{x}(t) = Df(\mathbf{x})\mathbf{v}.$$

Evaluation at $t = 0$ gives

$$D_{\mathbf{v}} f(\mathbf{a}) = Df(\mathbf{a})\mathbf{v} = \nabla f(\mathbf{a}) \cdot \mathbf{v}. \tag{3}$$

The purpose of equation (3) is to emphasize the geometry of the situation. The result above says that the directional derivative is just the dot product of the gradient and the direction vector $\mathbf{v}$. Since the gradient is made up of the partial derivatives, we see that the more general notion of the directional derivative depends entirely on just the direction vector and the partial derivatives. To be more formal, we summarize this discussion with a theorem.

Theorem 6.2 Let $X \subseteq \mathbf{R}^n$ be open and suppose $f: X \to \mathbf{R}$ is differentiable at $\mathbf{a} \in X$. Then the directional derivative $D_{\mathbf{v}} f(\mathbf{a})$ exists for all directions (unit vectors) $\mathbf{v} \in \mathbf{R}^n$ and, moreover, we have

$$D_{\mathbf{v}} f(\mathbf{a}) = \nabla f(\mathbf{a}) \cdot \mathbf{v}.$$

EXAMPLE 2 The function $f(x, y) = x^2 - 3xy + 2x - 5y$ we considered in Example 1 has continuous partials and hence, by Theorem 3.5, is differentiable. Thus, Theorem 6.2 applies to tell us that, for any unit vector $\mathbf{v} = v\mathbf{i} + w\mathbf{j} \in \mathbf{R}^2$,

$$\begin{aligned} D_{\mathbf{v}} f(0,0) &= \nabla f(0,0) \cdot \mathbf{v} = (f_x(0,0)\mathbf{i} + f_y(0,0)\mathbf{j}) \cdot (v\mathbf{i} + w\mathbf{j}) \\ &= (2\mathbf{i} - 5\mathbf{j}) \cdot (v\mathbf{i} + w\mathbf{j}) \\ &= 2v - 5w, \end{aligned}$$

as seen earlier. ◆

EXAMPLE 3 The converse of Theorem 6.2 does not hold. That is, a function may have directional derivatives in all directions at a point yet fail to be differentiable.

To see how this can happen, consider the function $f: \mathbf{R}^2 \to \mathbf{R}$ defined by

$$f(x, y) = \begin{cases} \dfrac{xy^2}{x^2 + y^4} & \text{if } (x, y) \neq (0, 0) \\ 0 & \text{if } (x, y) = (0, 0) \end{cases}.$$

This function is *not* continuous at the origin. (Why?) So, by Theorem 3.6, it fails to be differentiable there; however, we claim that all directional derivatives exist at the origin. To see this, let the direction vector $\mathbf{v}$ be $v\mathbf{i} + w\mathbf{j}$. Hence, by Definition 6.1, we observe that

$$\begin{aligned} D_{\mathbf{v}} f(0, 0) &= \lim_{h \to 0} \frac{f((0, 0) + h(v\mathbf{i} + w\mathbf{j})) - f(0, 0)}{h} \\ &= \lim_{h \to 0} \frac{1}{h} \left[\frac{hv(hw)^2}{(hv)^2 + (hw)^4} - 0 \right] \\ &= \lim_{h \to 0} \frac{h^2 v w^2}{h^2(v^2 + h^2 w^4)} \\ &= \lim_{h \to 0} \frac{v w^2}{v^2 + h^2 w^4} = \frac{v w^2}{v^2} = \frac{w^2}{v}. \end{aligned}$$

Thus, the directional derivative exists whenever $v \neq 0$. When $v = 0$ (in which case $\mathbf{v} = \mathbf{j}$), we, again, must calculate

$$\begin{aligned} D_{\mathbf{j}} f(0, 0) &= \lim_{h \to 0} \frac{f((0, 0) + h\mathbf{j}) - f(0, 0)}{h} \\ &= \lim_{h \to 0} \frac{f(0, h) - f(0, 0)}{h} \\ &= \lim_{h \to 0} \frac{0 - 0}{h} = 0. \end{aligned}$$

Consequently, this directional derivative (which is, in fact, $\partial f / \partial y$) exists as well. ◆

The reason we have restricted the direction vector $\mathbf{v}$ to be of unit length in our discussion of directional derivatives has to do with the meaning of $D_{\mathbf{v}} f(\mathbf{a})$, not with any technicalities pertaining to Definition 6.1 or Theorem 6.2. Indeed, we can certainly define the limit in Definition 6.1 for any vector $\mathbf{v}$, not just one of unit length. So, suppose $\mathbf{w}$ is an arbitrary nonzero vector in $\mathbf{R}^n$ and f is differentiable. Then, the proof of Theorem 6.2 goes through without change to give

$$\lim_{h \to 0} \frac{f(\mathbf{a} + h\mathbf{w}) - f(\mathbf{a})}{h} = \nabla f(\mathbf{a}) \cdot \mathbf{w}.$$

The problem is as follows: If $\mathbf{w} = k\mathbf{v}$ for some (nonzero) scalar k, then

$$\begin{aligned} \lim_{h \to 0} \frac{f(\mathbf{a} + h\mathbf{w}) - f(\mathbf{a})}{h} &= \nabla f(\mathbf{a}) \cdot \mathbf{w} \\ &= \nabla f(\mathbf{a}) \cdot (k\mathbf{v}) \\ &= k(\nabla f(\mathbf{a}) \cdot \mathbf{v}) \\ &= k \left(\lim_{h \to 0} \frac{f(\mathbf{a} + h\mathbf{v}) - f(\mathbf{a})}{h} \right). \end{aligned}$$

That is, the "generalized directional derivative" in the direction of $k\mathbf{v}$ is k times the derivative in the direction of $\mathbf{v}$. But $\mathbf{v}$ and $k\mathbf{v}$ are parallel vectors, and it is undesirable to have this sort of ambiguity of terminology. So we avoid the trouble by insisting upon using unit vectors only (i.e., by allowing k to be ± 1 only) when working with directional derivatives.

Gradients and Steepest Ascent

Suppose you are traveling in space near the planet Nilrebo and that one of your spaceship's instruments measures the external atmospheric pressure on your ship as a function $f(x, y, z)$ of position. Assume, quite reasonably, that this function is differentiable. Then, Theorem 6.2 applies and tells us that if you travel from point $\mathbf{a} = (a, b, c)$ in the direction of the (unit) vector $\mathbf{u} = u\mathbf{i} + v\mathbf{j} + w\mathbf{k}$, the rate of change of pressure is given by

$$D_{\mathbf{u}} f(\mathbf{a}) = \nabla f(\mathbf{a}) \cdot \mathbf{u}.$$

Now, we ask the following: In what direction is the pressure increasing the most? If θ is the angle between $\mathbf{u}$ and the gradient vector $\nabla f(\mathbf{a})$, then we have, by Theorem 3.3 of §1.3, that

$$D_{\mathbf{u}} f(\mathbf{a}) = \|\nabla f(\mathbf{a})\| \, \|\mathbf{u}\| \cos\theta = \|\nabla f(\mathbf{a})\| \cos\theta,$$

since $\mathbf{u}$ is a unit vector. Because $-1 \le \cos\theta \le 1$, we have

$$-\|\nabla f(\mathbf{a})\| \le D_{\mathbf{u}} f(\mathbf{a}) \le \|\nabla f(\mathbf{a})\|.$$

Moreover, $\cos\theta = 1$ when $\theta = 0$ and $\cos\theta = -1$ when $\theta = \pi$. Thus, we have established the following:

Theorem 6.3 The directional derivative $D_{\mathbf{u}} f(\mathbf{a})$ is maximized, with respect to direction, when $\mathbf{u}$ points in the *same* direction as $\nabla f(\mathbf{a})$ and is minimized when $\mathbf{u}$ points in the *opposite* direction. Furthermore, the maximum and minimum values of $D_{\mathbf{u}} f(\mathbf{a})$ are $\|\nabla f(\mathbf{a})\|$ and $-\|\nabla f(\mathbf{a})\|$, respectively.

EXAMPLE 4 If the pressure function on Nilrebo is

$$f(x, y, z) = 5x^2 + 7y^4 + x^2z^2 \text{ atm},$$

where the origin is located at the center of Nilrebo and distance units are measured in thousands of kilometers, then the rate of change of pressure at $(1, -1, 2)$ in the direction of $\mathbf{i} + \mathbf{j} + \mathbf{k}$ may be calculated as $\nabla f(1, -1, 2) \cdot \mathbf{u}$, where $\mathbf{u} = (\mathbf{i} + \mathbf{j} + \mathbf{k})/\sqrt{3}$. (Note that we normalized the vector $\mathbf{i} + \mathbf{j} + \mathbf{k}$ to obtain a unit vector.) Using Theorem 6.2, we compute

$$\begin{aligned} D_{\mathbf{u}} f(1, -1, 2) &= \nabla f(1, -1, 2) \cdot \mathbf{u} \\ &= (18\mathbf{i} - 28\mathbf{j} + 4\mathbf{k}) \cdot \frac{\mathbf{i} + \mathbf{j} + \mathbf{k}}{\sqrt{3}} \\ &= \frac{18 - 28 + 4}{\sqrt{3}} = -2\sqrt{3} \text{ atm/Mm}. \end{aligned}$$

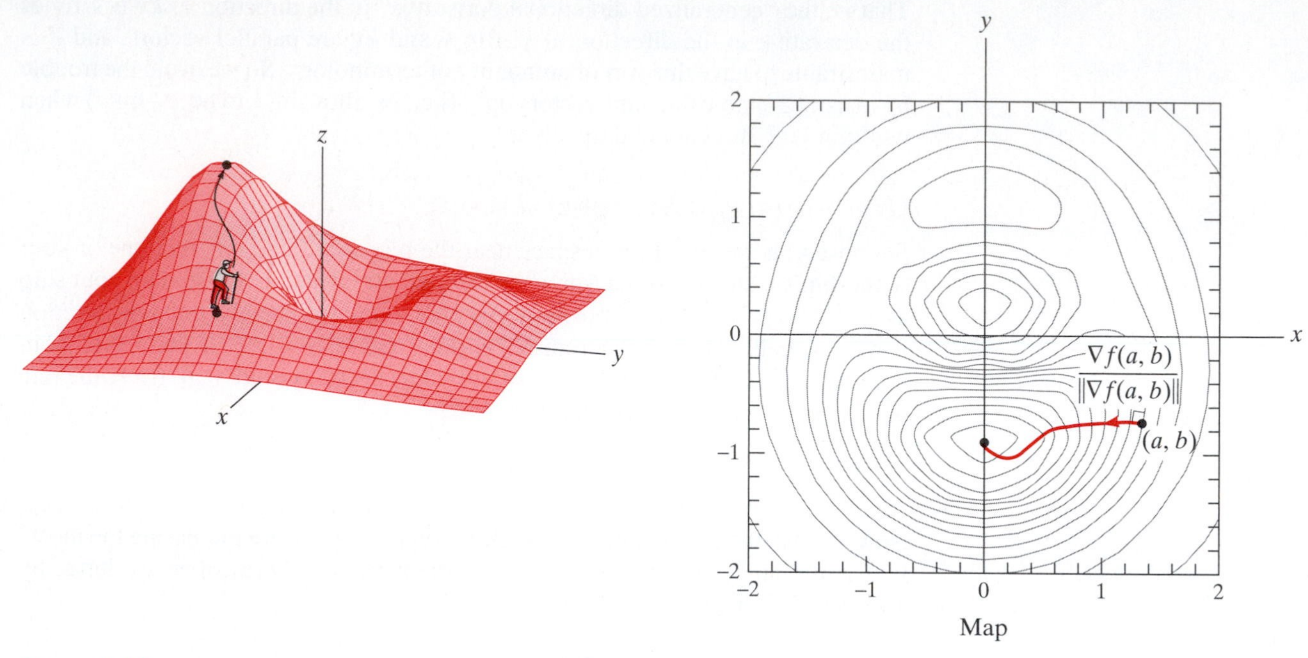

Figure 2.69 Select $\nabla f(a, b)/\|\nabla f(a, b)\|$ for direction of steepest ascent.

Additionally, in view of Theorem 6.3, the pressure will increase most rapidly in the direction of $\nabla f(1, -1, 2)$, that is, in the

$$\frac{18\mathbf{i} - 28\mathbf{j} + 4\mathbf{k}}{\|18\mathbf{i} - 28\mathbf{j} + 4\mathbf{k}\|} = \frac{9\mathbf{i} - 14\mathbf{j} + 2\mathbf{k}}{\sqrt{281}}$$

direction. Moreover, the *rate* of this increase is

$$\|\nabla f(1, -1, 2)\| = 2\sqrt{281} \text{ atm/Mm.}$$ ◆

Theorem 6.3 is stated in a manner which is independent of dimension—that is, so that it applies to functions $f\colon X \subseteq \mathbf{R}^n \to \mathbf{R}$ for any $n \geq 2$. In the case $n = 2$, there is another geometric interpretation of Theorem 6.3: Suppose you are mountain climbing on the surface $z = f(x, y)$. Think of the value of f as the height of the mountain above (or below) sea level. If you are equipped with a map and compass (which supply information in the xy-plane only), then if you are at the point on the mountain with xy-coordinates (map coordinates) (a, b), Theorem 6.3 says that you should move in the direction parallel to the gradient $\nabla f(a, b)$ in order to climb the mountain most rapidly. (See Figure 2.69.) Similarly, you should move in the direction parallel to $-\nabla f(a, b)$ in order to descend most rapidly. Moreover, the slope of your ascent or descent in these cases is $\|\nabla f(a, b)\|$. Be sure that you understand that $\nabla f(a, b)$ *is a vector in* $\mathbf{R}^2$ that gives the optimal north–south, east–west direction of travel.

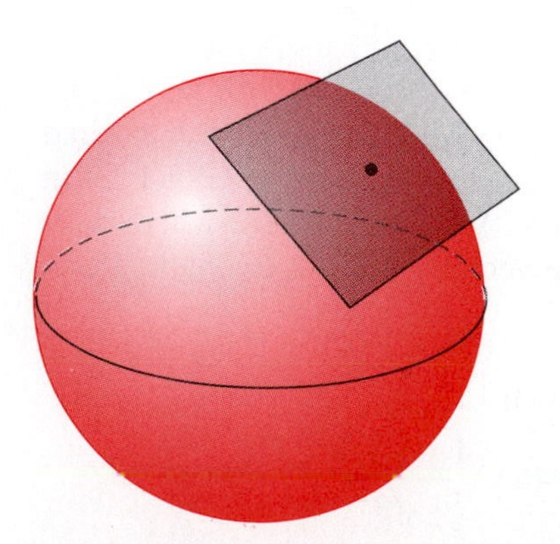

Figure 2.70 A sphere and one of its tangent planes.

Tangent Planes Revisited

In §2.1, we indicated that not all surfaces can be described by equations of the form $z = f(x, y)$. Indeed, a surface as simple and familiar as the sphere is not the graph of any single function of two variables. Yet the sphere is certainly smooth enough for us to see intuitively that it must have a tangent plane at every point. (See Figure 2.70.)

How can we find the equation of the tangent plane? In the case of the unit sphere $x^2 + y^2 + z^2 = 1$, we could proceed as follows: First decide whether the point of tangency is in the top or bottom hemisphere. Then apply equation (4) of §2.3 to the graph of $z = \sqrt{1 - x^2 - y^2}$ or $z = -\sqrt{1 - x^2 - y^2}$, as appropriate. The calculus is tedious, but not conceptually difficult. However, the tangent planes to points on the equator are all vertical and so equation (4) of §2.3 does not apply. (It is possible to modify this approach to accommodate such points, but we will not do so.) In general, given a surface described by an equation of the form $F(x, y, z) = c$ (where c is a constant), it may be entirely impractical to solve for z even as several functions of x and y. Try solving for z in the equation $xyz + ye^{xz} - x^2 + yz^2 = 0$ and you'll see what we mean. We need some other way to get our hands on tangent planes to surfaces described as level sets of functions of three variables.

To get started on our quest, we present the following result, interesting in its own right:

Theorem 6.4 Let $X \subseteq \mathbf{R}^n$ be open and $f: X \to \mathbf{R}$ be a function of class C^1. If $\mathbf{x}_0$ is a point on the level set $S = \{\mathbf{x} \in X \mid f(\mathbf{x}) = c\}$, then the vector $\nabla f(\mathbf{x}_0)$ is perpendicular to S.

Proof We need to establish the following: If $\mathbf{v}$ is any vector tangent to S at $\mathbf{x}_0$, then $\nabla f(\mathbf{x}_0)$ is perpendicular to $\mathbf{v}$ (i.e., $\nabla f(\mathbf{x}_0) \cdot \mathbf{v} = 0$). By a tangent vector to S at $\mathbf{x}_0$, we mean that $\mathbf{v}$ is the velocity vector of a curve C that lies in S and passes through $\mathbf{x}_0$. The situation in $\mathbf{R}^3$ is pictured in Figure 2.71.

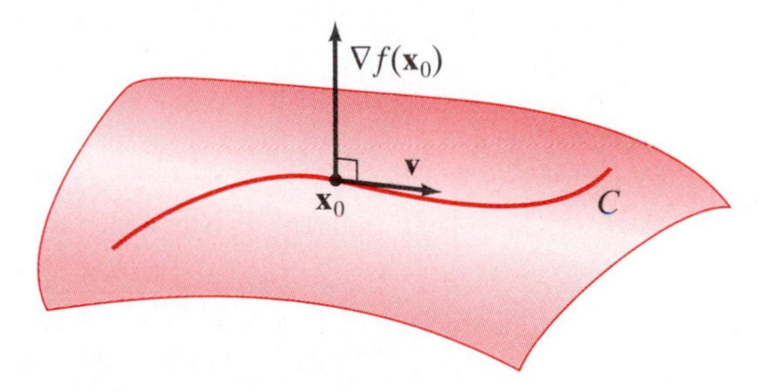

Figure 2.71 The level set surface $S = \{\mathbf{x} \mid f(\mathbf{x}) = c\}$.

Thus, let C be given parametrically by $\mathbf{x}(t) = (x_1(t), x_2(t), \ldots, x_n(t))$, where $a < t < b$ and $\mathbf{x}(t_0) = \mathbf{x}_0$ for some number t_0 in (a, b). (Then, if $\mathbf{v}$ is the velocity vector at $\mathbf{x}_0$, we must have $\mathbf{x}'(t_0) = \mathbf{v}$. See §3.1 for more about velocity vectors.) Since C is contained in S, we have

$$f(\mathbf{x}(t)) = f(x_1(t), x_2(t), \ldots, x_n(t)) = c.$$

Hence,

$$\frac{d}{dt}[f(\mathbf{x}(t))] = \frac{d}{dt}[c] \equiv 0. \tag{4}$$

On the other hand, the chain rule applied to the composite function $f \circ \mathbf{x}: (a, b) \to \mathbf{R}$ tells us

$$\frac{d}{dt}[f(\mathbf{x}(t))] = \nabla f(\mathbf{x}(t)) \cdot \mathbf{x}'(t).$$

Evaluation at t_0 and equation (4) let us conclude that

$$\nabla f(\mathbf{x}(t_0)) \cdot \mathbf{x}'(t_0) = \nabla f(\mathbf{x}_0) \cdot \mathbf{v} = 0,$$

as desired. ■

Here's how we can use the result of Theorem 6.4 to find the plane tangent to the sphere $x^2 + y^2 + z^2 = 1$ at the point $\left(-\frac{1}{\sqrt{2}}, 0, \frac{1}{\sqrt{2}}\right)$. From §1.5, we know that a plane is determined uniquely from two pieces of information: (i) a point in the plane and (ii) a vector perpendicular to the plane. We are given a point in the plane in the form of the point of tangency $\left(-\frac{1}{\sqrt{2}}, 0, \frac{1}{\sqrt{2}}\right)$. As for a vector normal to the plane, Theorem 6.4 tells us that the gradient of the function $f(x, y, z) = x^2 + y^2 + z^2$ that defines the sphere as a level set will do. We have

$$\nabla f(x, y, z) = 2x\mathbf{i} + 2y\mathbf{j} + 2z\mathbf{k},$$

so that

$$\nabla f\left(-\frac{1}{\sqrt{2}}, 0, \frac{1}{\sqrt{2}}\right) = -\sqrt{2}\,\mathbf{i} + \sqrt{2}\,\mathbf{k}.$$

Hence, the equation of the tangent plane is

$$\nabla f\left(-\frac{1}{\sqrt{2}}, 0, \frac{1}{\sqrt{2}}\right) \cdot \left(x + \frac{1}{\sqrt{2}}, y - 0, z - \frac{1}{\sqrt{2}}\right) = 0,$$

$$-\sqrt{2}\left(x + \frac{1}{\sqrt{2}}\right) + \sqrt{2}\left(z - \frac{1}{\sqrt{2}}\right) = 0,$$

or

$$z - x = \sqrt{2}.$$

In general, if S is a surface in $\mathbf{R}^3$ defined by an equation of the form

$$f(x, y, z) = c,$$

then if $\mathbf{x}_0 \in X$, the gradient vector $\nabla f(\mathbf{x}_0)$ is perpendicular to S, and, consequently, if nonzero, is a vector normal to the plane tangent to S at $\mathbf{x}_0$. Thus, the equation

$$\nabla f(\mathbf{x}_0) \cdot (\mathbf{x} - \mathbf{x}_0) = 0 \tag{5}$$

or, equivalently,

$$\begin{aligned} f_x(x_0, y_0, z_0)(x - x_0) + f_y(x_0, y_0, z_0)(y - y_0) \\ + f_z(x_0, y_0, z_0)(z - z_0) = 0 \end{aligned} \tag{6}$$

is an equation for the tangent plane to S at $\mathbf{x}_0$.

Note that formula (5) can be used in $\mathbf{R}^n$ as well as in $\mathbf{R}^3$, in which case it defines the **tangent hyperplane** to the hypersurface $S \subset \mathbf{R}^n$ defined by $f(x_1, x_2, \ldots, x_n) = c$ at the point $\mathbf{x}_0 \in S$.

EXAMPLE 5 Consider the surface S defined by the equation $x^3y - yz^2 + z^5 = 9$. We calculate the plane tangent to S at the point $(3, -1, 2)$.

To do this, we define $f(x, y, z) = x^3y - yz^2 + z^5$. Then

$$\begin{aligned}\nabla f(3, -1, 2) &= \left(3x^2y\mathbf{i} + (x^3 - z^2)\mathbf{j} + (5z^4 - 2yz)\mathbf{k}\right)\big|_{(3,-1,2)} \\ &= -27\mathbf{i} + 23\mathbf{j} + 84\mathbf{k}\end{aligned}$$

is normal to S at $(3, -1, 2)$ by Theorem 6.4. Using formula (6), we see that the tangent plane has equation

$$-27(x - 3) + 23(y + 1) + 84(z - 2) = 0$$

or, equivalently,

$$-27x + 23y + 84z = 64.$$

◆

EXAMPLE 6 Consider the surface defined by $z^4 = x^2 + y^2$. This surface is the level set (at height 0) of the function

$$f(x, y, z) = x^2 + y^2 - z^4.$$

The gradient of f is

$$\nabla f(x, y, z) = 2x\,\mathbf{i} + 2y\,\mathbf{j} - 4z^3\,\mathbf{k}.$$

Note that the point $(0, 0, 0)$ lies on the surface. However, $\nabla f(0, 0, 0) = \mathbf{0}$, which makes the gradient vector unusable as a normal vector to a tangent plane. Thus formula (6) doesn't apply. What we conclude from this example is that the surface fails to have a tangent plane at the origin, a fact that is easy to believe from the graph. (See Figure 2.72.) ◆

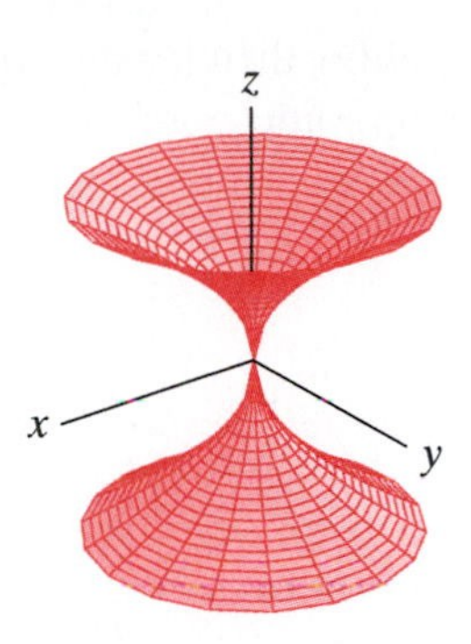

Figure 2.72 The surface of Example 6.

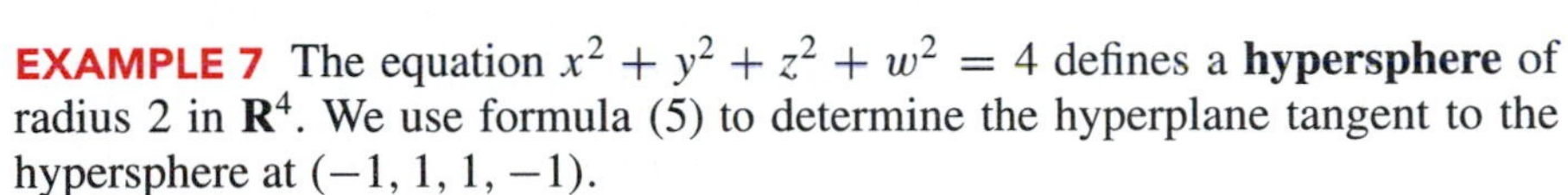

EXAMPLE 7 The equation $x^2 + y^2 + z^2 + w^2 = 4$ defines a **hypersphere** of radius 2 in $\mathbf{R}^4$. We use formula (5) to determine the hyperplane tangent to the hypersphere at $(-1, 1, 1, -1)$.

The hypersphere may be considered to be the level set at height 4 of the function $f(x, y, z, w) = x^2 + y^2 + z^2 + w^2$, so that the gradient vector is

$$\nabla f(x, y, z, w) = (2x, 2y, 2z, 2w),$$

so that

$$\nabla f(-1, 1, 1, -1) = (-2, 2, 2, -2).$$

Using formula (5), we obtain an equation for the tangent hyperplane as

$$(-2, 2, 2, -2) \cdot (x + 1, y - 1, z - 1, w + 1) = 0$$

or

$$-2(x + 1) + 2(y - 1) + 2(z - 1) - 2(w + 1) = 0.$$

Equivalently, we have the equation

$$x - y - z + w + 4 = 0.$$

◆

EXAMPLE 8 We determine the plane tangent to the paraboloid $z = x^2 + 3y^2$ at the point $(-2, 1, 7)$ in two ways: (i) by using formula (4) in §2.3, and (ii) by using our new formula (6).

First, the equation $z = x^2 + 3y^2$ explicitly describes the paraboloid as the graph of the function $f(x, y) = x^2 + 3y^2$; that is, by an equation of the form $z = f(x, y)$. Therefore, formula (4) of §2.3 applies to tell us that the tangent plane at $(-2, 1, 7)$ has equation

$$z = f(-2, 1) + f_x(-2, 1)(x + 2) + f_y(-2, 1)(y - 1)$$

or, equivalently,

$$z = 7 - 4(x + 2) + 6(y - 1). \tag{7}$$

Second, if we write the equation of the paraboloid as $x^2 + 3y^2 - z = 0$, then we see that it describes the paraboloid as the level set of height 0 of the three-variable function $F(x, y, z) = x^2 + 3y^2 - z$. Hence, formula (6) applies and indicates that an equation for the tangent plane at $(-2, 1, 7)$ is

$$F_x(-2, 1, 7)(x + 2) + F_y(-2, 1, 7)(y - 1) + F_z(-2, 1, 7)(z - 7) = 0$$

or

$$-4(x + 2) + 6(y - 1) - 1(z - 7) = 0 \tag{8}$$

As can be seen, equation (7) agrees with equation (8). ◆

Example 8 may be viewed in a more general context. If S is the surface in $\mathbf{R}^3$ given by the equation $z = f(x, y)$ (where f is differentiable), then formula (4) of §2.3 tells us that an equation for the plane tangent to S at the point $(a, b, f(a, b))$ is

$$z = f(a, b) + f_x(a, b)(x - a) + f_y(a, b)(y - b).$$

At the same time, the equation for S may be written as

$$f(x, y) - z = 0.$$

Then, if we let $F(x, y, z) = f(x, y) - z$, we see that S is the level set of F at height 0. Hence, formula (6) tells us that the tangent plane at $(a, b, f(a, b))$ is

$$\begin{aligned} F_x(a, b, f(a, b))(x - a) &+ F_y(a, b, f(a, b))(y - b) \\ &+ F_z(a, b, f(a, b))(z - f(a, b)) = 0. \end{aligned}$$

By construction of F,

$$\frac{\partial F}{\partial x} = \frac{\partial f}{\partial x}, \qquad \frac{\partial F}{\partial y} = \frac{\partial f}{\partial y}, \qquad \frac{\partial F}{\partial z} = -1.$$

Thus, the tangent plane formula becomes

$$f_x(a, b)(x - a) + f_y(a, b)(y - b) - (z - f(a, b)) = 0.$$

The last equation for the tangent plane is the same as the one given above by equation (4) of §2.3.

The result shows that equations (5) and (6) extend the formula (4) of §2.3 to the more general setting of level sets.

The Implicit Function and Inverse Function Theorems (optional)

We have previously noted that not all surfaces that are described by equations of the form $F(x, y, z) = c$ can be described by an equation of the form $z = f(x, y)$. We close this chapter with a brief—but theoretically important—digression about when and how the level set $\{(x, y, z) \mid F(x, y, z) = c\}$ can also be described as the graph of a function of two variables, that is, as the graph of $z = f(x, y)$. We also consider the more general question of when we can solve a system of equations for some of the variables in terms of the others.

We begin with an example.

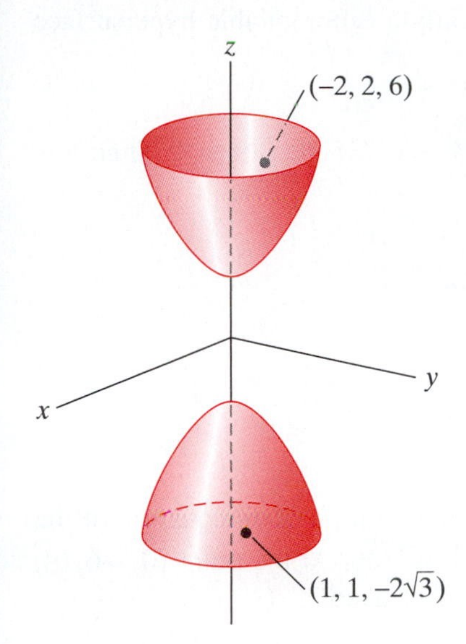

Figure 2.73 The two-sheeted hyperboloid $z^2/4 - x^2 - y^2 = 1$. The point $(-2, 2, 6)$ lies on the sheet given by $z = 2\sqrt{x^2+y^2+1}$, and the point $(1, 1, -2\sqrt{3})$ lies on the sheet given by $z = -2\sqrt{x^2+y^2+1}$.

EXAMPLE 9 Consider the hyperboloid $z^2/4 - x^2 - y^2 = 1$, which may be described as the level set (at height 1) of the function

$$F(x, y, z) = \frac{z^2}{4} - x^2 - y^2.$$

(See Figure 2.73.) This surface *cannot* be described as the graph of an equation of the form $z = f(x, y)$, since particular values for x and y give rise to *two* values for z. Indeed, when we solve for z in terms of x and y, we find that there are two functional solutions:

$$z = 2\sqrt{x^2+y^2+1} \quad \text{and} \quad z = -2\sqrt{x^2+y^2+1}. \tag{9}$$

On the other hand, these two solutions show that, given any particular point (x_0, y_0, z_0) of the hyperboloid, we may solve *locally* for z in terms of x and y. That is, we may identify on which sheet of the hyperboloid the point (x_0, y_0, z_0) lies and then use the appropriate expression in (9) to describe that sheet. ◆

Example 9 prompts us to pose the following question: Given a surface S, described as the level set $\{(x, y, z) \mid F(x, y, z) = c\}$, can we always determine at least a portion of S as the graph of a function $z = f(x, y)$? The result that follows, a special case of what is known as the **implicit function theorem,** provides relatively mild hypotheses under which we can.

Theorem 6.5 (THE IMPLICIT FUNCTION THEOREM) Let $F: X \subseteq \mathbf{R}^n \to \mathbf{R}$ be of class C^1 and let $\mathbf{a}$ be a point of the level set $S = \{\mathbf{x} \in \mathbf{R}^n \mid F(\mathbf{x}) = c\}$. If $F_{x_n}(\mathbf{a}) \neq 0$, then there is a neighborhood U of $(a_1, a_2, \ldots, a_{n-1})$ in $\mathbf{R}^{n-1}$, a neighborhood V of a_n in $\mathbf{R}$, and a function $f: U \subseteq \mathbf{R}^{n-1} \to V$ of class C^1 such that if $(x_1, x_2, \ldots, x_{n-1}) \in U$ and $x_n \in V$ satisfy $F(x_1, x_2, \ldots, x_n) = c$ (i.e., $(x_1, x_2, \ldots, x_n) \in S$), then $x_n = f(x_1, x_2, \ldots, x_{n-1})$.

The significance of Theorem 6.5 is that it tells us that *near* a point $\mathbf{a} \in S$ such that $\partial F/\partial x_n \neq 0$, the level set S given by the equation $F(x_1, \ldots, x_n) = c$ is *locally* also the graph of a function $x_n = f(x_1, \ldots, x_{n-1})$. In other words, we may solve locally for x_n in terms of $x_1, \ldots, x_{n-1}$, so that S is, at least locally, a differentiable hypersurface in $\mathbf{R}^n$.

EXAMPLE 10 Returning to Example 9, we recall that the hyperboloid is the level set (at height 1) of the function $F(x, y, z) = z^2/4 - x^2 - y^2$. We have

$$\frac{\partial F}{\partial z} = \frac{z}{2}.$$

Note that for any point (x_0, y_0, z_0) in the hyperboloid, we have $|z_0| \geq 2$. Hence, $\partial F_z(x_0, y_0, z_0) \neq 0$. Thus, Theorem 6.5 implies that we may describe a portion of the hyperboloid near any point as the graph of a function of two variables. This is consistent with what we observed in Example 9. ◆

Of course, there is nothing special about solving for the particular variable x_n in terms of $x_1, \ldots, x_{n-1}$. Suppose $\mathbf{a}$ is a point on the level set S determined by the equation $F(\mathbf{x}) = c$ and suppose $\nabla F(\mathbf{a}) \neq \mathbf{0}$. Then, $F_{x_i}(\mathbf{a}) \neq 0$ for some i. Hence, we can solve locally near $\mathbf{a}$ for x_i as a differentiable function of $x_1, \ldots, x_{i-1}, x_{i+1}, \ldots, x_n$. Therefore, S is locally a differentiable hypersurface in $\mathbf{R}^n$.

EXAMPLE 11 Let S denote the ellipsoid $x^2/4 + y^2/36 + z^2/9 = 1$. Then S is the level set (at height 1) of the function

$$F(x, y, z) = \frac{x^2}{4} + \frac{y^2}{36} + \frac{z^2}{9}.$$

At the point $(\sqrt{2}, \sqrt{6}, \sqrt{3})$, we have

$$\left.\frac{\partial F}{\partial z}\right|_{(\sqrt{2},\sqrt{6},\sqrt{3})} = \left.\frac{2z}{9}\right|_{(\sqrt{2},\sqrt{6},\sqrt{3})} = \frac{2\sqrt{3}}{9} \neq 0.$$

Thus, S may be realized near $(\sqrt{2}, \sqrt{6}, \sqrt{3})$ as the graph of an equation of the form $z = f(x, y)$, namely, $z = 3\sqrt{1 - x^2/4 - y^2/36}$. At the point $(0, -6, 0)$, however, we see that $\partial F/\partial z$ vanishes. On the other hand,

$$\left.\frac{\partial F}{\partial y}\right|_{(0,-6,0)} = \left.\frac{2y}{36}\right|_{(0,-6,0)} = -\frac{1}{3} \neq 0.$$

Consequently, near $(0, -6, 0)$, the ellipsoid may be described by solving for y as a function of x and z, namely, $y = -6\sqrt{1 - x^2/4 - z^2/9}$. ◆

EXAMPLE 12 Consider the set of points S defined by the equation $x^2z^2 - y = 0$. Then S is the level set at height 0 of the function $F(x, y, z) = x^2z^2 - y$. Note that

$$\nabla F(x, y, z) = (2xz^2, -1, 2x^2z).$$

Since $\partial F/\partial y$ never vanishes, we see that we can always solve for y as a function of x and z. (This is, of course, obvious from the equation.) On the other hand, near points where x and z are nonzero, both $\partial F/\partial x$ and $\partial F/\partial z$ are nonzero. Hence, we can solve for either x or z in this case. For example, near $(1, 1, -1)$, we have

$$x = \sqrt{\frac{y}{z^2}} \quad \text{and} \quad z = -\sqrt{\frac{y}{x^2}}.$$ ◆

As just mentioned, Theorem 6.5 is actually a special case of a more general result. In Theorem 6.5 we are attempting to solve the equation

$$F(x_1, x_2, \ldots, x_n) = c$$

for x_n in terms of $x_1, \ldots, x_{n-1}$. In the general case, we have a *system* of m equations

$$\begin{cases} F_1(x_1, \ldots, x_n, y_1, \ldots, y_m) = c_1 \\ F_2(x_1, \ldots, x_n, y_1, \ldots, y_m) = c_2 \\ \vdots \\ F_m(x_1, \ldots, x_n, y_1, \ldots, y_m) = c_m \end{cases}, \tag{10}$$

and we desire to solve the system for $y_1, \ldots, y_m$ in terms of $x_1, \ldots, x_n$. Using vector notation, we can also write this system as $\mathbf{F}(\mathbf{x}, \mathbf{y}) = \mathbf{c}$, where $\mathbf{x} = (x_1, \ldots, x_n)$,

$\mathbf{y} = (y_1, \ldots, y_m)$, $\mathbf{c} = (c_1, \ldots, c_m)$, and $F_1, \ldots, F_m$ make up the component functions of $\mathbf{F}$. With this notation, the general result is the following:

Theorem 6.6 (THE IMPLICIT FUNCTION THEOREM, GENERAL CASE) Suppose $\mathbf{F}: A \to \mathbf{R}^m$ is of class C^1, where A is open in $\mathbf{R}^{n+m}$. Let $(\mathbf{a}, \mathbf{b}) = (a_1, \ldots, a_n, b_1, \ldots, b_m) \in A$ satisfy $\mathbf{F}(\mathbf{a}, \mathbf{b}) = \mathbf{c}$. If the determinant

$$\Delta(\mathbf{a}, \mathbf{b}) = \det \begin{bmatrix} \dfrac{\partial F_1}{\partial y_1}(\mathbf{a}, \mathbf{b}) & \cdots & \dfrac{\partial F_1}{\partial y_m}(\mathbf{a}, \mathbf{b}) \\ \vdots & \ddots & \vdots \\ \dfrac{\partial F_m}{\partial y_1}(\mathbf{a}, \mathbf{b}) & \cdots & \dfrac{\partial F_m}{\partial y_m}(\mathbf{a}, \mathbf{b}) \end{bmatrix} \neq 0,$$

then there is a neighborhood U of $\mathbf{a}$ in $\mathbf{R}^n$ and a unique function $\mathbf{f}: U \to \mathbf{R}^m$ of class C^1 such that $\mathbf{F}(\mathbf{x}, \mathbf{f}(\mathbf{x})) = \mathbf{c}$ for all $\mathbf{x} \in U$. In other words, we can solve locally for $\mathbf{y}$ as a function $\mathbf{f}(\mathbf{x})$.

EXAMPLE 13 We show that, near the point $(x_1, x_2, x_3, y_1, y_2) = (-1, 1, 1, 2, 1)$, we can solve the system

$$\begin{cases} x_1 y_2 + x_2 y_1 = 1 \\ x_1^2 x_3 y_1 + x_2 y_2^3 = 3 \end{cases} \tag{11}$$

for y_1 and y_2 in terms of x_1, x_2, x_3.

We apply the general implicit function theorem (Theorem 6.6) to the system

$$\begin{cases} F_1(x_1, x_2, x_3, y_1, y_2) = x_1 y_2 + x_2 y_1 = 1 \\ F_2(x_1, x_2, x_3, y_1, y_2) = x_1^2 x_3 y_1 + x_2 y_2^3 = 3 \end{cases}.$$

The relevant determinant is

$$\begin{aligned} \Delta(-1, 1, 1, 2, 1) &= \det \begin{bmatrix} \dfrac{\partial F_1}{\partial y_1} & \dfrac{\partial F_1}{\partial y_2} \\ \dfrac{\partial F_2}{\partial y_1} & \dfrac{\partial F_2}{\partial y_2} \end{bmatrix} \Bigg|_{(x_1,x_2,x_3,y_1,y_2)=(-1,1,1,2,1)} \\ &= \det \begin{bmatrix} x_2 & x_1 \\ x_1^2 x_3 & 3x_2 y_2^2 \end{bmatrix} \Bigg|_{(x_1,x_2,x_3,y_1,y_2)=(-1,1,1,2,1)} \\ &= \det \begin{bmatrix} 1 & -1 \\ 1 & 3 \end{bmatrix} = 4 \neq 0. \end{aligned}$$

Hence, we may solve locally, at least in principle.

We can also use the equations in (11) to determine, for example, $\dfrac{\partial y_2}{\partial x_1}(-1, 1, 1)$, where we treat x_1, x_2, x_3 as independent variables and y_1 and y_2 as functions of them.

Differentiating the equations in (11) implicitly with respect to x_1 and using the chain rule, we obtain

$$\begin{cases} y_2 + x_1\dfrac{\partial y_2}{\partial x_1} + x_2\dfrac{\partial y_1}{\partial x_1} = 0 \\ 2x_1x_3y_1 + x_1^2x_3\dfrac{\partial y_1}{\partial x_1} + 3x_2y_2^2\dfrac{\partial y_2}{\partial x_1} = 0 \end{cases}.$$

Now, let $(x_1, x_2, x_3, y_1, y_2) = (-1, 1, 1, 2, 1)$, so that the system becomes

$$\begin{cases} \dfrac{\partial y_1}{\partial x_1}(-1, 1, 1) - \dfrac{\partial y_2}{\partial x_1}(-1, 1, 1) = -1 \\ \dfrac{\partial y_1}{\partial x_1}(-1, 1, 1) + 3\dfrac{\partial y_2}{\partial x_1}(-1, 1, 1) = 4 \end{cases}.$$

We may easily solve this last system to find that $\dfrac{\partial y_2}{\partial x_1}(-1, 1, 1) = \dfrac{5}{4}$. ◆

Now, suppose we have a system of n equations that defines the variables $y_1, \ldots, y_n$ in terms of the variables $x_1, \ldots, x_n$, that is,

$$\begin{cases} y_1 = f_1(x_1, \ldots, x_n) \\ y_2 = f_2(x_1, \ldots, x_n) \\ \quad\vdots \\ y_n = f_n(x_1, \ldots, x_n) \end{cases}. \tag{12}$$

Note that the system given in (12) can be written in vector form as $\mathbf{y} = \mathbf{f}(\mathbf{x})$. The question we ask is, when can we *invert* this system? In other words, when can we solve for $x_1, \ldots, x_n$ in terms of $y_1, \ldots, y_n$, or, equivalently, when can we find a function $\mathbf{g}$ so that $\mathbf{x} = \mathbf{g}(\mathbf{y})$?

The solution is to apply Theorem 6.6 to the system

$$\begin{cases} F_1(x_1, \ldots, x_n, y_1, \ldots, y_n) = 0 \\ F_2(x_1, \ldots, x_n, y_1, \ldots, y_n) = 0 \\ \qquad\qquad\vdots \\ F_m(x_1, \ldots, x_n, y_1, \ldots, y_n) = 0 \end{cases},$$

where $F_i(x_1, \ldots, x_n, y_1, \ldots, y_n) = f_i(x_1, \ldots, x_n) - y_i$. (In vector form, we are setting $\mathbf{F}(\mathbf{x}, \mathbf{y}) = \mathbf{f}(\mathbf{x}) - \mathbf{y}$.) Then solvability near $\mathbf{x} = \mathbf{a}$, $\mathbf{y} = \mathbf{b}$ is governed by the nonvanishing of the determinant

$$\det D\mathbf{f}(\mathbf{a}) = \det \begin{bmatrix} \dfrac{\partial f_1}{\partial x_1}(\mathbf{a}) & \cdots & \dfrac{\partial f_1}{\partial x_n}(\mathbf{a}) \\ \vdots & \ddots & \vdots \\ \dfrac{\partial f_n}{\partial x_1}(\mathbf{a}) & \cdots & \dfrac{\partial f_n}{\partial x_n}(\mathbf{a}) \end{bmatrix}.$$

This determinant is also denoted by

$$\left.\frac{\partial(f_1, \ldots, f_n)}{\partial(x_1, \ldots, x_n)}\right|_{\mathbf{x}=\mathbf{a}}$$

and is called the **Jacobian** of $\mathbf{f} = (f_1, \ldots, f_n)$. A more precise and complete statement of what we are observing is the following:

Theorem 6.7 (THE INVERSE FUNCTION THEOREM) Suppose $\mathbf{f} = (f_1, \ldots, f_n)$ is of class C^1 on an open set $A \subseteq \mathbf{R}^n$. If

$$\det D\mathbf{f}(\mathbf{a}) = \left.\frac{\partial(f_1, \ldots, f_n)}{\partial(x_1, \ldots, x_n)}\right|_{\mathbf{x}=\mathbf{a}} \neq 0,$$

then there is an open set $U \subseteq \mathbf{R}^n$ containing $\mathbf{a}$ such that $\mathbf{f}$ is one-one on U, the set $V = \mathbf{f}(U)$ is also open, and there is a uniquely determined inverse function $\mathbf{g}\colon V \to U$ to $\mathbf{f}$, which is also of class C^1. In other words, the system of equations $\mathbf{y} = \mathbf{f}(\mathbf{x})$ may be solved uniquely as $\mathbf{x} = \mathbf{g}(\mathbf{y})$ for $\mathbf{x}$ near $\mathbf{a}$ and $\mathbf{y}$ near $\mathbf{b}$.

EXAMPLE 14 Consider the equations that relate polar and Cartesian coordinates:

$$\begin{cases} x = r\cos\theta \\ y = r\sin\theta \end{cases}.$$

These equations define x and y as functions of r and θ. We use Theorem 6.7 to see near which points of the plane we can invert these equations, that is, solve for r and θ in terms of x and y.

To use Theorem 6.7, we compute the Jacobian

$$\frac{\partial(x, y)}{\partial(r, \theta)} = \begin{vmatrix} \cos\theta & -r\sin\theta \\ \sin\theta & r\cos\theta \end{vmatrix} = r.$$

Thus, we see that, away from the origin ($r = 0$), we can solve (locally) for r and θ uniquely in terms of x and y. At the origin, however, the inverse function theorem does not apply. Geometrically, this makes perfect sense, since at the origin, the polar angle θ can have any value. ◆

Exercises

1. Suppose $f(x, y, z)$ is a differentiable function of three variables.

(a) Explain what the quantity $\nabla f(x, y, z) \cdot (-\mathbf{k})$ represents.

(b) How does $\nabla f(x, y, z) \cdot (-\mathbf{k})$ relate to $\partial f/\partial z$?

In Exercises 2–8, calculate the directional derivative of the given function f at the point $\mathbf{a}$ in the direction parallel to the vector $\mathbf{u}$.

2. $f(x, y) = e^y \sin x$, $\mathbf{a} = \left(\frac{\pi}{3}, 0\right)$, $\mathbf{u} = \dfrac{3\mathbf{i} - \mathbf{j}}{\sqrt{10}}$

3. $f(x, y) = x^2 - 2x^3y + 2y^3$, $\mathbf{a} = (2, -1)$, $\mathbf{u} = \dfrac{\mathbf{i} + 2\mathbf{j}}{\sqrt{5}}$

4. $f(x, y) = \dfrac{1}{(x^2 + y^2)}$, $\mathbf{a} = (3, -2)$, $\mathbf{u} = \mathbf{i} - \mathbf{j}$

5. $f(x, y) = e^x - x^2y$, $\mathbf{a} = (1, 2)$, $\mathbf{u} = 2\mathbf{i} + \mathbf{j}$

6. $f(x, y, z) = xyz$, $\mathbf{a} = (-1, 0, 2)$, $\mathbf{u} = \dfrac{2\mathbf{k} - \mathbf{i}}{\sqrt{5}}$

7. $f(x, y, z) = e^{-(x^2+y^2+z^2)}$, $\mathbf{a} = (1, 2, 3)$, $\mathbf{u} = \mathbf{i} + \mathbf{j} + \mathbf{k}$

8. $f(x, y, z) = \dfrac{xe^y}{3z^2 + 1}$, $\mathbf{a} = (2, -1, 0)$, $\mathbf{u} = \mathbf{i} - 2\mathbf{j} + 3\mathbf{k}$

9. For the function

$$f(x, y) = \begin{cases} \dfrac{x|y|}{\sqrt{x^2 + y^2}} & \text{if } (x, y) \neq (0, 0) \\ 0 & \text{if } (x, y) = (0, 0) \end{cases},$$

(a) calculate $f_x(0, 0)$ and $f_y(0, 0)$. (You will need to use the definition of the partial derivative.)

(b) use Definition 6.1 to determine for which unit vectors $\mathbf{v} = v\mathbf{i} + w\mathbf{j}$ the directional derivative $D_{\mathbf{v}} f(0, 0)$ exists.

(c) use a computer to graph the surface $z = f(x, y)$.

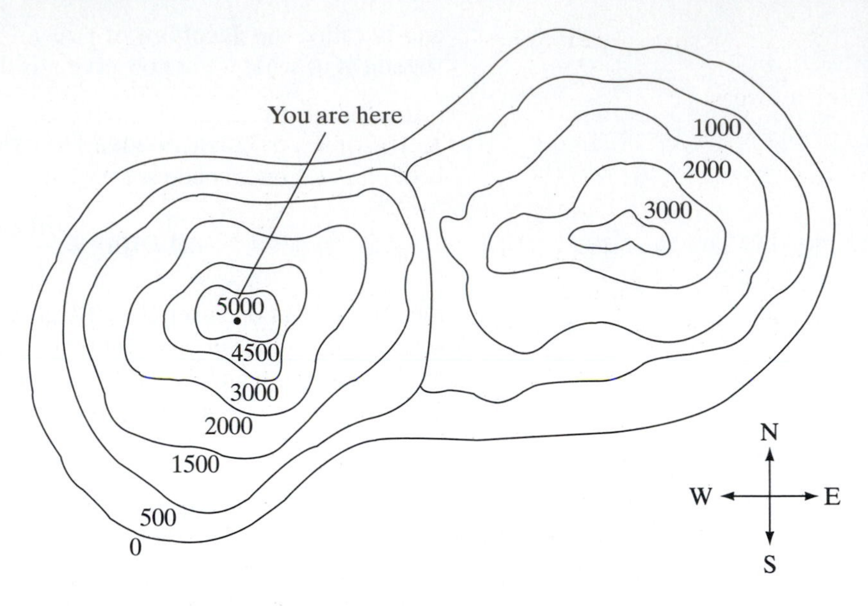

Figure 2.74 The topographic map of Mt. Gradient in Exercise 13.

10. For the function

$$f(x, y) = \begin{cases} \dfrac{xy}{\sqrt{x^2 + y^2}} & \text{if } (x, y) \neq (0, 0) \\ 0 & \text{if } (x, y) = (0, 0) \end{cases},$$

(a) calculate $f_x(0, 0)$ and $f_y(0, 0)$.

(b) use Definition 6.1 to determine for which unit vectors $\mathbf{v} = v\mathbf{i} + w\mathbf{j}$ the directional derivative $D_{\mathbf{v}}f(0, 0)$ exists.

(c) use a computer to graph the surface $z = f(x, y)$.

11. The surface of Lake Erehwon can be represented by a region D in the xy-plane such that the lake's depth (in meters) at the point (x, y) is given by the expression $400 - 3x^2y^2$. If your calculus instructor is in the water at the point $(1, -2)$, in which direction should she swim

(a) so that the depth increases most rapidly (i.e., so that she is most likely to drown)?

(b) so that the depth remains constant?

12. A ladybug (who is very sensitive to temperature) is crawling on graph paper. She is at the point $(3, 7)$ and notices that if she moves in the **i**-direction, the temperature *increases* at a rate of 3 deg/cm. If she moves in the **j**-direction, she finds that her temperature *decreases* at a rate of 2 deg/cm. In what direction should the ladybug move if

(a) she wants to warm up most rapidly?

(b) she wants to cool off most rapidly?

(c) she desires her temperature *not* to change?

13. You are atop Mt. Gradient, 5000 ft above sea level, equipped with the topographic map shown in Figure 2.74. A storm suddenly begins to blow, necessitating your immediate return home. If you begin heading due east from the top of the mountain, sketch the path that will take you down to sea level most rapidly.

14. It is raining and rainwater is running off an ellipsoidal dome with equation $4x^2 + y^2 + 4z^2 = 16$, where $z \geq 0$. Given that gravity will cause the raindrops to slide down the dome as rapidly as possible, describe the curves whose paths the raindrops must follow. (Hint: You will need to solve a simple differential equation.)

15. Igor, the inchworm, is crawling along graph paper in a magnetic field. The intensity of the field at the point (x, y) is given by $M(x, y) = 3x^2 + y^2 + 5000$. If Igor is at the point $(8, 6)$, describe the curve along which he should travel if he wishes to reduce the field intensity as rapidly as possible.

In Exercises 16–19, find an equation for the tangent plane to the surface given by the equation at the indicated point (x_0, y_0, z_0).

16. $x^3 + y^3 + z^3 = 7$, $(x_0, y_0, z_0) = (0, -1, 2)$

17. $ze^y \cos x = 1$, $(x_0, y_0, z_0) = (\pi, 0, -1)$

18. $2xz + yz - x^2y + 10 = 0$, $(x_0, y_0, z_0) = (1, -5, 5)$

19. $2xy^2 = 2z^2 - xyz$, $(x_0, y_0, z_0) = (2, -3, 3)$

20. Calculate the plane tangent to the surface whose equation is $x^2 - 2y^2 + 5xz = 7$ at the point $(-1, 0, -\frac{6}{5})$ in two ways:

(a) by solving for z in terms of x and y and using formula (4) in §2.3

(b) by using formula (6) in this section.

21. Calculate the plane tangent to the surface $x \sin y + xz^2 = 2e^{yz}$ at the point $\left(2, \frac{\pi}{2}, 0\right)$ in two ways:

(a) by solving for x in terms of y and z and using a variant of formula (4) in §2.3

(b) by using formula (6) in this section.

22. Find the point on the surface $x^3 - 2y^2 + z^2 = 27$ where the tangent plane is perpendicular to the line given parametrically as $x = 3t - 5$, $y = 2t + 7$, $z = 1 - \sqrt{2}t$.

23. Find the points on the hyperboloid $9x^2 - 45y^2 + 5z^2 = 45$ where the tangent plane is parallel to the plane $x + 5y - 2z = 7$.

24. Show that the surfaces $z = 7x^2 - 12x - 5y^2$ and $xyz^2 = 2$ intersect orthogonally at the point $(2, 1, -1)$.

25. Suppose that two surfaces are given by the equations

$$F(x, y, z) = c \quad \text{and} \quad G(x, y, z) = k.$$

Moreover, suppose that these surfaces intersect at the point (x_0, y_0, z_0). Show that the surfaces are tangent at (x_0, y_0, z_0) if and only if

$$\nabla F(x_0, y_0, z_0) \times \nabla G(x_0, y_0, z_0) = \mathbf{0}.$$

26. Let S denote the cone $x^2 + 4y^2 = z^2$.

(a) Find an equation for the plane tangent to S at the point $(3, -2, -5)$.

(b) What happens if you try to find an equation for a tangent plane to S at the origin? Discuss how your findings relate to the appearance of S.

27. Consider the surface S defined by the equation $x^3 - x^2y^2 + z^2 = 0$.

(a) Find an equation for the plane tangent to S at the point $(2, -3/2, 1)$.

(b) Does S have a tangent plane at the origin? Why or why not?

If a curve is given by an equation of the form $f(x, y) = 0$, then the tangent line to the curve at a given point (x_0, y_0) on it may be found in two ways: (a) by using the technique of implicit differentiation from single-variable calculus and (b) by using a formula analogous to formula (6). In Exercises 28–30, use both of these methods to find the lines tangent to the given curves at the indicated points.

28. $x^2 + y^2 = 4$, $(x_0, y_0) = (-\sqrt{2}, \sqrt{2})$

29. $y^3 = x^2 + x^3$, $(x_0, y_0) = (1, \sqrt[3]{2})$

30. $x^5 + 2xy + y^3 = 16$, $(x_0, y_0) = (2, -2)$

*Let C be a curve in $\mathbf{R}^2$ given by an equation of the form $f(x, y) = 0$. The **normal line** to C at a point (x_0, y_0) on it is the line that passes through (x_0, y_0) and is perpendicular to C (meaning that it is perpendicular to the tangent line to C at (x_0, y_0)). In Exercises 31–33, find the normal lines to the given curves at the indicated points. Give both a set of parametric equations for the lines and an equation in the form $Ax + By = C$. (Hint: Use gradients.)*

31. $x^2 - y^2 = 9$, $(x_0, y_0) = (5, -4)$

32. $x^2 - x^3 = y^2$, $(x_0, y_0) = (-1, \sqrt{2})$

33. $x^3 - 2xy + y^5 = 11$, $(x_0, y_0) = (2, -1)$

34. This problem concerns the surface defined by the equation

$$x^3z + x^2y^2 + \sin(yz) = -3.$$

(a) Find an equation for the plane tangent to this surface at the point $(-1, 0, 3)$.

(b) The **normal line** to a surface S in $\mathbf{R}^3$ at a point (x_0, y_0, z_0) on it is the line that passes through (x_0, y_0, z_0) and is perpendicular to S. Find a set of parametric equations for the line normal to the surface given above at the point $(-1, 0, 3)$.

35. Give a set of parametric equations for the normal line to the surface defined by the equation $e^{xy} + e^{xz} - 2e^{yz} = 0$ at the point $(-1, -1, -1)$. (See Exercise 34.)

36. Give a general formula for parametric equations for the normal line to a surface given by the equation $F(x, y, z) = 0$ at the point (x_0, y_0, z_0) on the surface. (See Exercise 34.)

37. Generalizing upon the techniques of this section, find an equation for the hyperplane tangent to the hypersurface $\sin x_1 + \cos x_2 + \sin x_3 + \cos x_4 + \sin x_5 = -1$ at the point $(\pi, \pi, 3\pi/2, 2\pi, 2\pi) \in \mathbf{R}^5$.

38. Find an equation for the hyperplane tangent to the $(n-1)$-dimensional ellipsoid

$$x_1^2 + 2x_2^2 + 3x_3^2 + \cdots + nx_n^2 = \frac{n(n+1)}{2}$$

at the point $(-1, -1, \ldots, -1) \in \mathbf{R}^n$.

39. Find an equation for the tangent hyperplane to the $(n-1)$-dimensional sphere $x_1^2 + x_2^2 + \cdots + x_n^2 = 1$ in $\mathbf{R}^n$ at the point $(1/\sqrt{n}, 1/\sqrt{n}, \ldots, 1/\sqrt{n}, -1/\sqrt{n})$.

Exercises 40–49 concern the implicit function theorems and the inverse function theorem (Theorems 6.5, 6.6, and 6.7).

40. Let S be described by $z^2y^3 + x^2y = 2$.

(a) Use the implicit function theorem to determine near which points S can be described locally as the graph of a C^1 function $z = f(x, y)$.

(b) Near which points can S be described (locally) as the graph of a function $x = g(y, z)$?

(c) Near which points can S be described (locally) as the graph of a function $y = h(x, z)$?

41. Let S be the set of points described by the equation $\sin xy + e^{xz} + x^3y = 1$.

(a) Near which points can we describe S as the graph of a C^1 function $z = f(x, y)$? What is $f(x, y)$ in this case?

(b) Describe the set of "bad" points of S, that is, the points $(x_0, y_0, z_0) \in S$ where we *cannot* describe S as the graph of a function $z = f(x, y)$.

(c) Use a computer to help give a *complete* picture of S.

42. Let $F(x, y) = c$ define a curve C in $\mathbf{R}^2$. Suppose (x_0, y_0) is a point of C such that $\nabla F(x_0, y_0) \neq \mathbf{0}$. Show that the curve can be represented near (x_0, y_0) as either the graph of a function $y = f(x)$ or the graph of a function $x = g(y)$.

43. Let $F(x, y) = x^2 - y^3$, and consider the curve C defined by the equation $F(x, y) = 0$.

(a) Show that $(0, 0)$ lies on C and that $F_y(0, 0) = 0$.

(b) Can we describe C as the graph of a function $y = f(x)$? Graph C.

(c) Comment on the results of parts (a) and (b) in light of the implicit function theorem (Theorem 6.5).

44. (a) Consider the family of level sets of the function $F(x, y) = xy + 1$. Use the implicit function theorem to identify which level sets of this family are actually unions of smooth curves in $\mathbf{R}^2$ (i.e., locally graphs of C^1 functions of a single variable).

(b) Now consider the family of level sets of $F(x, y, z) = xyz + 1$. Which level sets of this family are unions of smooth surfaces in $\mathbf{R}^3$?

45. Suppose that $F(u, v)$ is of class C^1 and is such that $F(-2, 1) = 0$ and $F_u(-2, 1) = 7$, $F_v(-2, 1) = 5$. Let $G(x, y, z) = F(x^3 - 2y^2 + z^5, xy - x^2z + 3)$.

(a) Check that $G(-1, 1, 1) = 0$.

(b) Show that we can solve the equation $G(x, y, z) = 0$ for z in terms of x and y (i.e., as $z = g(x, y)$, for (x, y) near $(-1, 1)$ so that $g(-1, 1) = 1$).

46. Can you solve

$$\begin{cases} x_2y_2 - x_1 \cos y_1 = 5 \\ x_2 \sin y_1 + x_1y_2 = 2 \end{cases}$$

for y_1, y_2 as functions of x_1, x_2 near the point $(x_1, x_2, y_1, y_2) = (2, 3, \pi, 1)$? What about near the point $(x_1, x_2, y_1, y_2) = (0, 2, \pi/2, 5/2)$?

47. Consider the system

$$\begin{cases} x_1y_2^2 - 2x_2y_3 = 1 \\ x_1y_1^5 + x_2y_2 - 4y_2y_3 = -9. \\ x_2y_1 + 3x_1y_3^2 = 12 \end{cases}$$

(a) Show that, near the point $(x_1, x_2, y_1, y_2, y_3) = (1, 0, -1, 1, 2)$, it is possible to solve for y_1, y_2, y_3 in terms of x_1, x_2.

(b) From the result of part (a), we may consider y_1, y_2, y_3 to be functions of x_1 and x_2. Use implicit differentiation and the chain rule to evaluate $\dfrac{\partial y_1}{\partial x_1}(1, 0)$, $\dfrac{\partial y_2}{\partial x_1}(1, 0)$, and $\dfrac{\partial y_3}{\partial x_1}(1, 0)$.

48. Consider the equations that relate cylindrical and Cartesian coordinates in $\mathbf{R}^3$:

$$\begin{cases} x = r\cos\theta \\ y = r\sin\theta. \\ z = z \end{cases}$$

(a) Near which points of $\mathbf{R}^3$ can we solve for r, θ, and z in terms of the Cartesian coordinates?

(b) Explain the geometry behind your answer in part (a).

49. Recall that the equations relating spherical and Cartesian coordinates in $\mathbf{R}^3$ are

$$\begin{cases} x = \rho \sin\varphi \cos\theta \\ y = \rho \sin\varphi \sin\theta. \\ z = \rho \cos\varphi \end{cases}$$

(a) Near which points of $\mathbf{R}^3$ can we solve for ρ, φ, and θ in terms of x, y, and z?

(b) Describe the geometry behind your answer in part (a).

2.7 True/False Exercises for Chapter 2

1. The component functions of a vector-valued function are vectors.

2. The domain of $\mathbf{f}(x, y) = \left(x^2 + y^2 + 1, \dfrac{3}{x + y}, \dfrac{x}{y}\right)$ is $\{(x, y) \in \mathbf{R}^2 \mid y \neq 0, x \neq y\}$.

3. The range of $\mathbf{f}(x, y) = \left(x^2 + y^2 + 1, \dfrac{3}{x+y}, \dfrac{x}{y}\right)$ is $\{(u, v, w) \in \mathbf{R}^3 \mid u \geq 1\}$.

4. The function $\mathbf{f}: \mathbf{R}^3 - \{(0,0,0)\} \to \mathbf{R}^3$, $\mathbf{f}(\mathbf{x}) = 2\mathbf{x}/\|\mathbf{x}\|$ is one-one.

5. The graph of $x = 9y^2 + z^2/4$ is a paraboloid.

6. The graph of $z + x^2 = y^2$ is a hyperboloid.

7. The level set of a function $f(x, y, z)$ is either empty or a surface.

8. The graph of any function of two variables is a level set of a function of three variables.

9. The level set of any function of three variables is the graph of a function of two variables.

10. $\displaystyle\lim_{(x,y)\to(0,0)} \frac{x^2 - 2y^2}{x^2 + y^2} = 1.$

11. If $f(x, y) = \begin{cases} \dfrac{y^4 - x^4}{x^2 + y^2} & \text{when } (x, y) \neq (0,0) \\ 2 & \text{when } (x, y) = (0,0) \end{cases}$, then f is continuous.

12. If $f(x, y)$ approaches a number L as $(x, y) \to (a, b)$ along all lines through (a, b), then $\lim_{(x,y)\to(a,b)} f(x, y) = L$.

13. If $\lim_{\mathbf{x}\to\mathbf{a}} \mathbf{f}(\mathbf{x})$ exists and is finite, then $\mathbf{f}$ is continuous at $\mathbf{a}$.

14. $f_x(a, b) = \displaystyle\lim_{x\to a} \frac{f(x, b) - f(a, b)}{x - a}.$

15. If $f(x, y, z) = \sin y$, then $\nabla f(x, y, z) = \cos y$.

16. If $\mathbf{f}: \mathbf{R}^3 \to \mathbf{R}^4$ is differentiable, then $D\mathbf{f}(\mathbf{x})$ is a 3×4 matrix.

17. If $\mathbf{f}$ is differentiable at $\mathbf{a}$, then $\mathbf{f}$ is continuous at $\mathbf{a}$.

18. If $\mathbf{f}$ is continuous at $\mathbf{a}$, then $\mathbf{f}$ is differentiable at $\mathbf{a}$.

19. If all partial derivatives $\partial f/\partial x_1, \ldots, \partial f/\partial x_n$ of a function $f(x_1, \ldots, x_n)$ exist at $\mathbf{a} = (a_1, \ldots, a_n)$, then f is differentiable at $\mathbf{a}$.

20. If $\mathbf{f}: \mathbf{R}^4 \to \mathbf{R}^5$ and $\mathbf{g}: \mathbf{R}^4 \to \mathbf{R}^5$ are both differentiable at $\mathbf{a} \in \mathbf{R}^4$, then $D(\mathbf{f} - \mathbf{g})(\mathbf{a}) = D\mathbf{f}(\mathbf{a}) - D\mathbf{g}(\mathbf{a})$.

21. There's a function f of class C^2 such that $\dfrac{\partial f}{\partial x} = y^3 - 2x$ and $\dfrac{\partial f}{\partial y} = y - 3xy^2$.

22. If the second-order partial derivatives of f exist at (a, b), then $f_{xy}(a, b) = f_{yx}(a, b)$.

23. If $w = F(x, y, z)$ and $z = g(x, y)$ where F and g are differentiable, then

$$\frac{\partial w}{\partial x} = \frac{\partial F}{\partial x} + \frac{\partial F}{\partial z}\frac{\partial g}{\partial x}.$$

24. The tangent plane to $z = x^3/(y+1)$ at the point $(-2, 0, -8)$ has equation $z = 12x + 8y + 16$.

25. The plane tangent to $xy/z = 1$ at $(2, 8, -4)$ has equation $x - y - 2z = 2$.

26. The plane tangent to the surface $x^2 + xye^z + y^3 = 1$ at the point $(2, -1, 0)$ is parallel to the vector $3\mathbf{i} + 5\mathbf{j} - 3\mathbf{k}$.

27. $D_{\mathbf{j}} f(x, y, z) = \dfrac{\partial f}{\partial y}$.

28. $D_{-\mathbf{k}} f(x, y, z) = \dfrac{\partial f}{\partial z}$.

29. If $f(x, y) = \sin x \cos y$ and $\mathbf{v}$ is a unit vector in $\mathbf{R}^2$, then $0 \leq D_{\mathbf{v}} f\left(\dfrac{\pi}{4}, \dfrac{\pi}{3}\right) \leq \dfrac{\sqrt{2}}{2}$.

30. If $\mathbf{v}$ is a unit vector in $\mathbf{R}^3$ and $f(x, y, z) = \sin x - \cos y + \sin z$, then

$$-\sqrt{3} \leq D_{\mathbf{v}} f(x, y, z) \leq \sqrt{3}.$$

2.8 Miscellaneous Exercises for Chapter 2

1. Let $\mathbf{f}(\mathbf{x}) = (\mathbf{i} + \mathbf{k}) \times \mathbf{x}$.
   (a) Write the component functions of $\mathbf{f}$.
   (b) Describe the domain and range of $\mathbf{f}$.

2. Let $\mathbf{f}(\mathbf{x}) = \text{proj}_{3\mathbf{i}-2\mathbf{j}+\mathbf{k}}\mathbf{x}$, where $\mathbf{x} = x\mathbf{i} + y\mathbf{j} + z\mathbf{k}$.
   (a) Describe the domain and range of $\mathbf{f}$.
   (b) Write the component functions of $\mathbf{f}$.

3. Let $f(x, y) = \sqrt{xy}$.
   (a) Find the domain and range of f.
   (b) Is the domain of f open or closed? Why?

4. Let $g(x, y) = \sqrt{\dfrac{x}{y}}$.
   (a) Determine the domain and range of g.
   (b) Is the domain of g open or closed? Why?

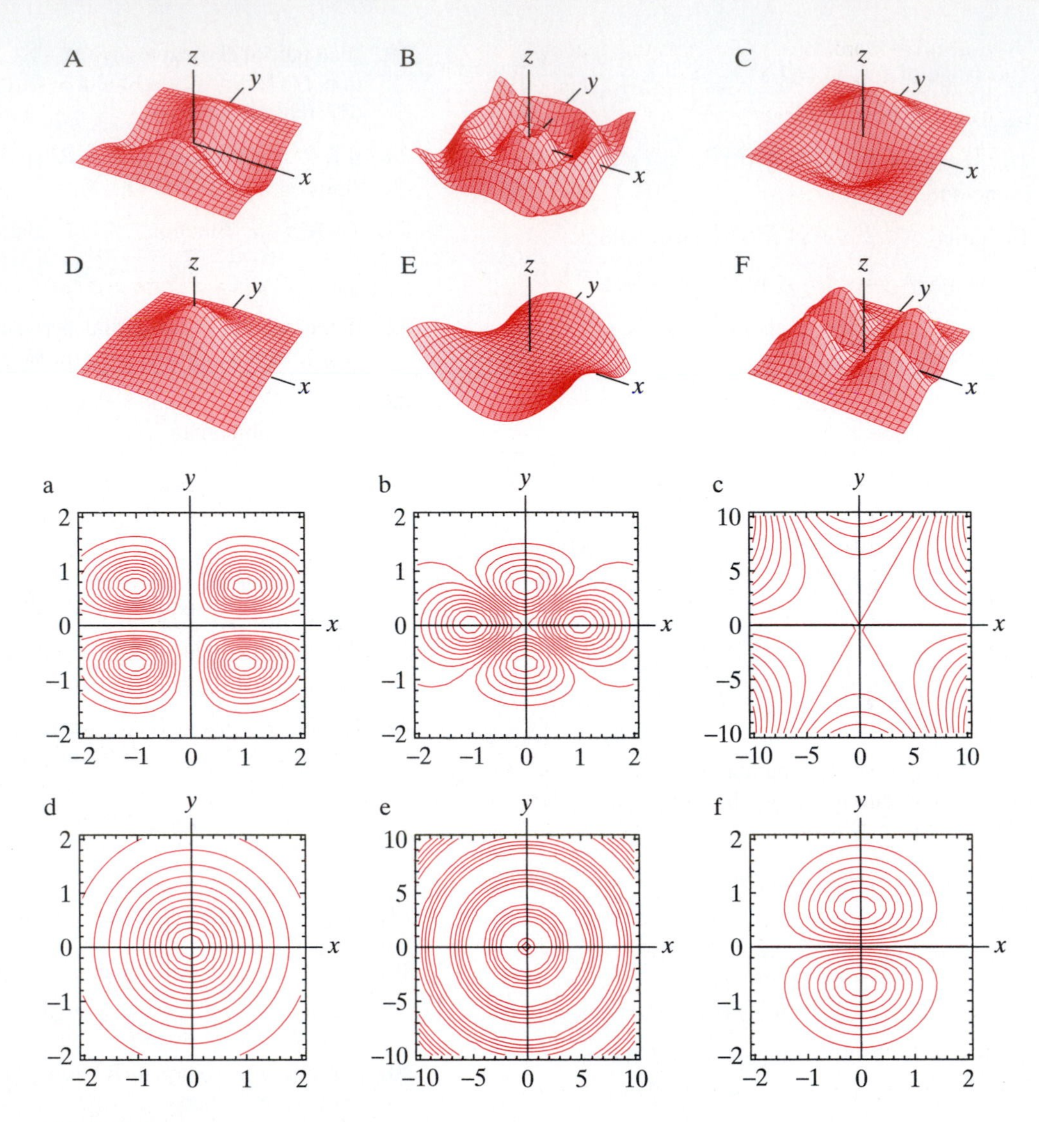

Figure 2.75 Figures for Exercise 5.

5. Figure 2.75 shows the graphs of six functions $f(x, y)$ and plots of the collections of their level curves in some order. Complete the following table by matching each function in the table with its graph and plot of its level curves.

Function $f(x, y)$	Graph (uppercase letter)	Level curves (lowercase letter)
$f(x, y) = \dfrac{1}{x^2 + y^2 + 1}$		
$f(x, y) = \sin\sqrt{x^2 + y^2}$		
$f(x, y) = (3y^2 - 2x^2)e^{-x^2-2y^2}$		
$f(x, y) = y^3 - 3x^2y$		
$f(x, y) = x^2y^2e^{-x^2-2y^2}$		
$f(x, y) = ye^{-x^2-y^2}$		

6. Consider the function $f(x, y) = 2 + \ln(x^2 + y^2)$.

(a) Sketch some level curves of f. Give at least those at heights, 0, 1, and 2. (It will probably help if you give a few more.)

(b) Using part (a) or otherwise, give a rough sketch of the graph of $z = f(x, y)$.

7. Use polar coordinates to evaluate

$$\lim_{(x,y)\to(0,0)} \frac{yx^2 - y^3}{x^2 + y^2}.$$

8. This problem concerns the function

$$f(x, y) = \begin{cases} \dfrac{2xy}{x^2 + y^2} & \text{if } (x, y) \neq (0, 0) \\ 0 & \text{if } (x, y) = (0, 0) \end{cases}.$$

(a) Use polar coordinates to describe this function.

(b) Using the polar coordinate description obtained in part (a), give some level curves for this function.

(c) Prepare a rough sketch of the graph of f.

(d) Determine $\lim_{(x,y)\to(0,0)} f(x, y)$, if it exists.

(e) Is f continuous? Why or why not?

9. Let

$$F(x, y) = \begin{cases} \dfrac{xy(xy + x^2)}{x^4 + y^4} & \text{if } (x, y) \neq (0, 0) \\ 0 & \text{if } (x, y) = (0, 0) \end{cases}.$$

Show that the function $g(x) = F(x, 0)$ is continuous at $x = 0$. Show that the function $h(y) = F(0, y)$ is continuous at $y = 0$. However, show that F fails to be continuous at $(0, 0)$. (Thus, continuity in each variable separately does not necessarily imply continuity of the function.)

10. Suppose $f: U \subseteq \mathbf{R}^n \to \mathbf{R}$ is not defined at a point $\mathbf{a} \in \mathbf{R}^n$, but is defined for all $\mathbf{x}$ near $\mathbf{a}$. In other words, the domain U of f includes, for some $r > 0$, the set $B_r = \{\mathbf{x} \in \mathbf{R}^n \mid 0 < \|\mathbf{x} - \mathbf{a}\| < r\}$. (The set B_r is just an open ball of radius r centered at $\mathbf{a}$ with the point $\mathbf{a}$ deleted.) Then, we say $\lim_{\mathbf{x}\to\mathbf{a}} f(\mathbf{x}) = +\infty$ if $f(\mathbf{x})$ grows without bound as $\mathbf{x} \to \mathbf{a}$. More precisely, this means that given any $N > 0$ (no matter how large), there is some $\delta > 0$ such that if $0 < \|\mathbf{x} - \mathbf{a}\| < \delta$ (i.e., if $\mathbf{x} \in B_r$), then $f(\mathbf{a}) > N$.

(a) Using intuitive arguments or the preceding technical definition, explain why $\lim_{x\to 0} 1/x^2 = \infty$.

(b) Explain why

$$\lim_{(x,y)\to(1,3)} \frac{2}{(x-1)^2 + (y-3)^2} = \infty.$$

(c) Formulate a definition of what it means to say that

$$\lim_{\mathbf{x}\to\mathbf{a}} f(\mathbf{x}) = -\infty.$$

(d) Explain why

$$\lim_{(x,y)\to(0,0)} \frac{1-x}{xy^4 - y^4 + x^3 - x^2} = -\infty.$$

Exercises 11–17 involve the notion of windchill temperature—see Example 7 in §2.1, and refer to the table of windchill values on page 82.

11. (a) Find the windchill temperature when the air temperature is 25 °F and the windspeed is 10 mph.

(b) If the windspeed is 20 mph, what air temperature causes a windchill temperature of −15 °F?

12. (a) If the air temperature is 10 °F, estimate (to the nearest unit) what windspeed would give a windchill temperature of −5 °F.

(b) Do you think your estimate in part (a) is high or low? Why?

13. At a windspeed of 30 mph and air temperature of 35 °F, estimate the rate of change of the windchill temperature with respect to air temperature if the windspeed is held constant.

14. At a windspeed of 15 mph and air temperature of 25 °F, estimate the rate of change of the windchill temperature with respect to windspeed.

15. Windchill tables are constructed from empirically derived formulas for heat loss from an exposed surface. Early experimental work of P. A. Siple and C. F. Passel,[4] resulted in the following formula:

$$W = 91.4 + (t - 91.4)(0.474 + 0.304\sqrt{s} - 0.0203s).$$

Here W denotes windchill temperature (in degrees Fahrenheit), t the air temperature (for $t < 91.4$°F), and s the windspeed in miles per hour (for $s \geq 4$ mph).[5]

(a) Compare your answers in Exercises 11 and 12 with those computed directly from the Siple formula just mentioned.

(b) Discuss any differences you observe between your answers to Exercises 11 and 12 and your answers to part (a).

(c) Why is it necessary to take $t < 91.4$°F and $s \geq 4$ mph in the Siple formula? (Don't look for a purely mathematical reason; think about the model.)

16. Recent research led the United States National Weather Service to employ a new formula for calculating windchill values beginning November 1, 2001. In particular, the table on page 82 was constructed from the formula

$$W = 35.74 + 0.621t - 35.75s^{0.16} + 0.4275ts^{0.16}.$$

Here, as in the Siple formula of Exercise 15, W denotes windchill temperature (in degrees Fahrenheit), t the air temperature (for $t \leq 50$°F), and s the windspeed

[4] "Measurements of dry atmospheric cooling in subfreezing temperatures," *Proc. Amer. Phil. Soc.*, **89** (1945), 177–199.

[5] The aforementioned windchill formula appears in Bob Rilling, Atmospheric Technology Division, National Center for Atmospheric Research, "Calculating windchill values," February 12, 1996. <http://www.atd.ucar.edu/homes/rilling/wc_formula.html> (June 30, 2004).

in miles per hour (for $s \geq 3$ mph).[6] Compare your answers in Exercises 13 and 14 with those computed directly from the National Weather Service formula above.

17. In this problem you will compare graphically the two windchill formulas given in Exercises 15 and 16.

(a) If $W_1(s, t)$ denotes the windchill function given by the Siple formula in Exercise 15 and $W_2(s, t)$ the windchill function given by the National Weather Service formula in Exercise 16, graph the curves $y = W_1(s, 40)$ and $y = W_2(s, 40)$ on the same set of axes. (Let s vary between 3 and 120 mph.) In addition, graph other pairs of curves $y = W_1(s, t_0)$, $y = W_2(s, t_0)$ for other values of t_0. Discuss what your results tell you about the two windchill formulas.

(b) Now graph pairs of curves $y = W_1(s_0, t)$, $y = W_2(s_0, t)$ for various constant values s_0 for windspeed. Discuss your results.

(c) Finally, graph the surfaces $z = W_1(s, t)$ and $z = W_2(s, t)$ and comment.

18. Consider the sphere of radius 3 centered at the origin. The plane tangent to the sphere at $(1, 2, 2)$ intersects the x-axis at a point P. Find the coordinates of P.

19. Show that the plane tangent to a sphere at a point P on the sphere is always perpendicular to the vector $\overrightarrow{OP}$ from the center O of the sphere to P. (Hint: Locate the sphere so its center is at the origin in $\mathbf{R}^3$.)

20. The surface $z = 3x^2 + \frac{1}{6}x^3 - \frac{1}{8}x^4 - 4y^2$ is intersected by the plane $2x - y = 1$. The resulting intersection is a curve on the surface. Find a set of parametric equations for the line tangent to this curve at the point $(1, 1, -\frac{23}{24})$.

21. Consider the cone $z^2 = x^2 + y^2$.

(a) Find an equation of the plane tangent to the cone at the point $(3, -4, 5)$.

(b) Find an equation of the plane tangent to the cone at the point (a, b, c).

(c) Show that every tangent plane to the cone must pass through the origin.

22. Show that the two surfaces

$$S_1\colon z = xy \qquad \text{and} \qquad S_2\colon z = \tfrac{3}{4}x^2 - y^2$$

intersect perpendicularly at the point $(2, 1, 2)$.

23. Consider the surface $z = x^2 + 4y^2$.

(a) Find an equation for the plane that is tangent to the surface at the point $(1, -1, 5)$.

(b) Now suppose that the surface is intersected with the plane $x = 1$. The resulting intersection is a curve on the surface (and is a curve in the plane $x = 1$ as well). Give a set of parametric equations for the line in $\mathbf{R}^3$ that is tangent to this curve at the point $(1, -1, 5)$. A rough sketch may help your thinking.

24. A turtleneck sweater has been washed and is now tumbling in the dryer, along with the rest of the laundry. At a particular moment t_0, the neck of the sweater measures 18 inches in circumference and 3 inches in length. However, the sweater is 100% cotton, so that at t_0 the heat of the dryer is causing the neck circumference to shrink at a rate of 0.2 in/min, while the twisting and tumbling action is causing the length of the neck to stretch at the rate of 0.1 in/min. How is the volume V of the space inside the neck changing at $t = t_0$? Is V increasing or decreasing at that moment?

25. A factory generates air pollution each day according to the formula

$$P(S, T) = 330S^{2/3}T^{4/5},$$

where S denotes the number of machine stations in operation and T denotes the average daily temperature. At the moment, 75 stations are in regular use and the average daily temperature is 15°C. If the average temperature is rising at the rate of 0.2°C/day and the number of stations being used is falling at a rate of 2 per month, at what rate is the amount of pollution changing? (Note: Assume that there are 24 workdays per month.)

26. Economists attempt to quantify how useful or satisfying people find goods or services by means of **utility functions.** Suppose that the utility a particular individual derives from consuming x ounces of soda per week and watching y minutes of television per week is

$$u(x, y) = 1 - e^{-0.001x^2 - 0.00005y^2}.$$

Further suppose that she currently drinks 80 oz of soda per week and watches 240 min of TV each week. If she were to increase her soda consumption by 5 oz/week and cut back on her TV viewing by 15 min/week, is the utility she derives from these changes increasing or decreasing? At what rate?

27. Suppose that $w = x^2 + y^2 + z^2$ and $x = \rho \cos\theta \sin\varphi$, $y = \rho \sin\theta \sin\varphi$, $z = \rho \cos\varphi$. (Note that the equations for x, y, and z in terms of ρ, φ, and θ are just the conversion relations from spherical to rectangular coordinates.)

[6]NOAA, National Weather Service, Office of Climate, Water, and Weather Services, "NWS Wind Chill Temperature Index." February 26, 2004. <http://www.nws.noaa.gov/om/ windchill> (June 30, 2004).

(a) Use the chain rule to compute $\partial w/\partial \rho$, $\partial w/\partial \varphi$, and $\partial w/\partial \theta$. Simplify your answers as much as possible.

(b) Substitute ρ, φ, and θ for x, y, and z in the original expression for w. Can you explain your answer in part (a)?

28. If $w = f\left(\dfrac{x+y}{xy}\right)$, show that

$$x^2\frac{\partial w}{\partial x} - y^2\frac{\partial w}{\partial y} = 0.$$

(You should assume that f is a differentiable function of one variable.)

29. Let $z = g(x, y)$ be a function of class C^2, and let $x = e^r \cos\theta$, $y = e^r \sin\theta$.

(a) Use the chain rule to find $\partial z/\partial r$ and $\partial z/\partial \theta$ in terms of $\partial z/\partial x$ and $\partial z/\partial y$. Use your results to solve for $\partial z/\partial x$ and $\partial z/\partial y$ in terms of $\partial z/\partial r$ and $\partial z/\partial \theta$.

(b) Use part (a) and the product rule to show that

$$\frac{\partial^2 z}{\partial x^2} + \frac{\partial^2 z}{\partial y^2} = e^{-2r}\left(\frac{\partial^2 z}{\partial r^2} + \frac{\partial^2 z}{\partial \theta^2}\right).$$

30. (a) Use the function $f(x, y) = x^y$ $(= e^{y\ln x})$ and the multivariable chain rule to calculate $\dfrac{d}{du}(u^u)$.

(b) Use the multivariable chain rule to calulate $\dfrac{d}{dt}((\sin t)^{\cos t})$.

31. Use the function $f(x, y, z) = x^{y^z}$ and the multivariable chain rule to calculate $\dfrac{d}{du}\left(u^{u^u}\right)$.

32. Suppose that $f\colon \mathbf{R}^n \to \mathbf{R}$ is a function of class C^2. The **Laplacian** of f, denoted $\nabla^2 f$, is defined to be

$$\nabla^2 f = \frac{\partial^2 f}{\partial x_1^2} + \frac{\partial^2 f}{\partial x_2^2} + \cdots + \frac{\partial^2 f}{\partial x_n^2}.$$

When $n = 2$ or 3, this construction is important when studying certain differential equations that model physical phenomena, such as the heat or wave equations. (See Exercises 20 and 21 of §2.4.) Now suppose that f depends only on the distance $\mathbf{x} = (x_1, \ldots, x_n)$ is from the origin in $\mathbf{R}^n$, that is, suppose that $f(\mathbf{x}) = g(r)$ for some function g, where $r = \|\mathbf{x}\|$. Show that for all $\mathbf{x} \neq \mathbf{0}$, the Laplacian is given by

$$\nabla^2 f = \frac{n-1}{r}g'(r) + g''(r).$$

33. (a) Consider a function $f(x, y)$ of class C^4. Show that if we apply the Laplacian operator $\nabla^2 = \partial^2/\partial x^2 + \partial^2/\partial y^2$ twice to f, we obtain

$$\nabla^2(\nabla^2 f) = \frac{\partial^4 f}{\partial x^4} + 2\frac{\partial^4 f}{\partial x^2 \partial y^2} + \frac{\partial^4 f}{\partial y^4}.$$

(b) Now suppose that f is a function of n variables of class C^4. Show that

$$\nabla^2(\nabla^2 f) = \sum_{i,j=1}^{n} \frac{\partial^4 f}{\partial x_i^2 \partial x_j^2}.$$

Functions that satisfy the partial differential equation $\nabla^2(\nabla^2 f) = 0$ are called **biharmonic functions** and arise in the theoretical study of elasticity.

34. Livinia, the housefly, finds herself caught in the oven at the point $(0, 0, 1)$. The temperature at points in the oven is given by the function

$$T(x, y, z) = 10(xe^{-y^2} + ze^{-x^2}),$$

where the units are in degrees Celsius.

(a) If Livinia begins to move towards the point $(2, 3, 1)$, at what rate (in deg/cm) does she find the temperature changing?

(b) In what direction should she move in order to cool off as rapidly as possible?

(c) Suppose that Livinia can fly at a speed of 3 cm/sec. If she moves in the direction of part (b), at what (instantaneous) rate (in deg/sec) will she find the temperature to be changing?

35. Consider the surface given in cylindrical coordinates by the equation $z = r\cos 3\theta$.

(a) Describe this surface in Cartesian coordinates, that is, as $z = f(x, y)$.

(b) Is f continuous at the origin? (Hint: Think cylindrical.)

(c) Find expressions for $\partial f/\partial x$ and $\partial f/\partial y$ at points other than $(0, 0)$. Give values for $\partial f/\partial x$ and $\partial f/\partial y$ at $(0, 0)$ by looking at the partial functions of f through $(x, 0)$ and $(0, y)$ and taking one-variable limits.

(d) Show that the directional derivative $D_{\mathbf{u}} f(0, 0)$ exists for every direction (unit vector) $\mathbf{u}$. (Hint: Think in cylindrical coordinates again and note that you can specify a direction through the origin in the xy-plane by choosing a particular constant value for θ.)

(e) Show directly (by examining the expression for $\partial f/\partial y$ when $(x, y) \neq (0, 0)$ and also using part (c)) that $\partial f/\partial y$ is *not* continuous at $(0, 0)$.

(f) Sketch the graph of the surface, perhaps using a computer to do so.

36. The partial differential equation

$$\frac{\partial^2 u}{\partial x^2} + \frac{\partial^2 u}{\partial y^2} + \frac{\partial^2 u}{\partial z^2} = c\frac{\partial^2 u}{\partial t^2}$$

is known as the **wave equation.** It models the motion of a wave $u(x, y, z, t)$ in $\mathbf{R}^3$ and was originally derived by Johann Bernoulli in 1727. In this equation, c is a positive constant, the variables x, y, and z represent spatial coordinates, and the variable t represents time.

(a) Let $u = \cos(x-t) + \sin(x+t) - 2e^{z+t} - (y-t)^3$. Show that u satisfies the wave equation with $c = 1$.

(b) More generally, show that if f_1, f_2, g_1, g_2, h_1, and h_2 are any twice differentiable functions of a single variable, then

$$\begin{aligned} u(x, y, z, t) = f_1(x-t) + f_2(x+t) \\ + g_1(y-t) + g_2(y+t) \\ + h_1(z-t) + h_2(z+t) \end{aligned}$$

satisfies the wave equation with $c = 1$.

*Let X be an open set in $\mathbf{R}^n$. A function $F: X \to \mathbf{R}$ is said to be **homogeneous of degree** d if, for all $\mathbf{x} = (x_1, x_2, \ldots, x_n) \in X$ and all $t \in \mathbf{R}$ such that $t\mathbf{x} \in X$, we have*

$$F(tx_1, tx_2, \ldots, tx_n) = t^d F(x_1, x_2, \ldots, x_n).$$

Exercises 37–44 concern homogeneous functions.

In Exercises 37–41, which of the given functions are homogeneous? For those that are, indicate the degree d of homogeneity.

37. $F(x, y) = x^3 + xy^2 - 6y^3$

38. $F(x, y, z) = x^3y - x^2z^2 + z^8$

39. $F(x, y, z) = zy^2 - x^3 + x^2z$

40. $F(x, y) = e^{y/x}$

41. $F(x, y, z) = \dfrac{x^3 + x^2y - yz^2}{xyz + 7xz^2}$

42. If $F(x, y, z)$ is a polynomial, characterize what it means to say that F is homogeneous of degree d (i.e., explain what must be true about the polynomial if it is to be homogeneous of degree d).

43. Suppose $F(x_1, x_2, \ldots, x_n)$ is differentiable and homogeneous of degree d. Prove **Euler's formula:**

$$x_1 \frac{\partial F}{\partial x_1} + x_2 \frac{\partial F}{\partial x_2} + \cdots + x_n \frac{\partial F}{\partial x_n} = dF.$$

(Hint: Take the equation $F(tx_1, tx_2, \ldots, tx_n) = t^d F(x_1, x_2, \ldots, x_n)$ that defines homogeneity and differentiate with respect to t.)

44. Generalize Euler's formula as follows: If F is of class C^2 and homogeneous of degree d, then

$$\sum_{i,j=1}^{n} x_i x_j \frac{\partial^2 F}{\partial x_i \partial x_j} = d(d-1)F.$$

Can you conjecture what an analogous formula involving the kth-order partial derivatives should look like?

CHAPTER 3

Vector-Valued Functions

Introduction

The primary focus of Chapter 2 was on scalar-valued functions, although general mappings from $\mathbf{R}^n$ to $\mathbf{R}^m$ were considered occasionally. This chapter concerns vector-valued functions of two special types:

1. Continuous mappings of one variable (i.e., functions $\mathbf{x}: I \subseteq \mathbf{R} \to \mathbf{R}^n$, where I is an interval, called **paths** in $\mathbf{R}^n$).
2. Mappings from (subsets of) $\mathbf{R}^n$ to itself (called **vector fields**).

An understanding of both concepts is required later, when we discuss line and surface integrals.

3.1 Parametrized Curves and Kepler's Laws

Paths in $\mathbf{R}^n$

We begin with a simple definition. Let I denote any interval in $\mathbf{R}$. (So I can be of the form $[a, b]$, (a, b), $[a, b)$, $(a, b]$, $[a, \infty)$, (a, ∞), $(-\infty, b]$, $(-\infty, b)$, or $(-\infty, \infty) = \mathbf{R}$.)

Definition 1.1 A **path in $\mathbf{R}^n$** is a continuous function $\mathbf{x}: I \to \mathbf{R}^n$. If $I = [a, b]$ for some numbers $a < b$, then the points $\mathbf{x}(a)$ and $\mathbf{x}(b)$ are called the **endpoints** of the path $\mathbf{x}$. (Similar definitions apply if $I = [a, b)$, $[a, \infty)$, etc.)

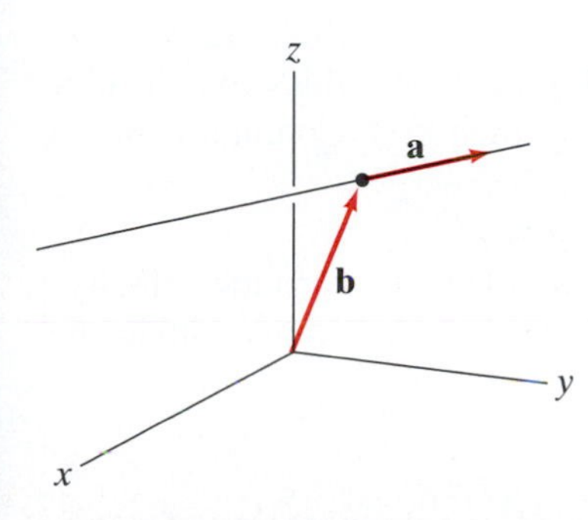

Figure 3.1 The path **x** of Example 1.

EXAMPLE 1 Let **a** and **b** be vectors in $\mathbf{R}^3$ with $\mathbf{a} \neq \mathbf{0}$. Then the function $\mathbf{x}: (-\infty, \infty) \to \mathbf{R}^3$ given by

$$\mathbf{x}(t) = \mathbf{b} + t\mathbf{a}$$

defines the path along the straight line parallel to **a** and passing through the endpoint of the position vector of **b** as in Figure 3.1. (See formula (1) of §1.2.) ◆

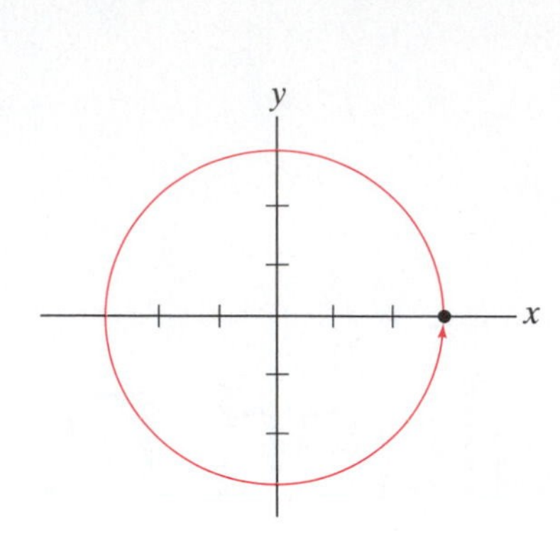

Figure 3.2 The path **y** of Example 2.

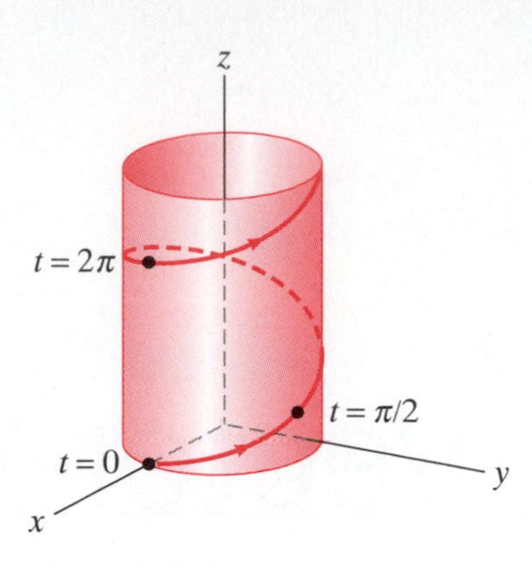

Figure 3.3 The path **z** of Example 3.

EXAMPLE 2 The path $\mathbf{y}\colon [0,\ 2\pi) \to \mathbf{R}^2$ given by

$$\mathbf{y}(t) = (3\cos t,\, 3\sin t)$$

can be thought of as the path of a particle that travels once, counterclockwise, around a circle of radius 3 (Figure 3.2). ◆

EXAMPLE 3 The map $\mathbf{z}\colon \mathbf{R} \to \mathbf{R}^3$ defined by

$$\mathbf{z}(t) = (a\cos t,\, a\sin t,\, bt), \qquad a, b \text{ constants } (a > 0)$$

is called a **circular helix,** so named because its projection in the xy-plane is a circle of radius a. The helix itself lies in the right circular cylinder $x^2 + y^2 = a^2$ (Figure 3.3). The value of b determines how tightly the helix twists. ◆

We distinguish between a path $\mathbf{x}$ and its range or image set $\mathbf{x}(I)$, the latter being a curve in $\mathbf{R}^n$. By definition, a path is a function, a dynamic object (at least when we imagine the independent variable t to represent time), whereas a curve is a static figure in space. With such a point of view, it is natural for us to consider the derivative $D\mathbf{x}(t)$, which we also write as $\mathbf{x}'(t)$ or $\mathbf{v}(t)$, to be the **velocity** vector of the path. We can readily justify such terminology. Since

$$\mathbf{x}(t) = (x_1(t), x_2(t), \ldots, x_n(t))$$

is a function of just one variable,

$$\mathbf{v}(t) = \mathbf{x}'(t) = \lim_{\Delta t \to 0} \frac{\mathbf{x}(t + \Delta t) - \mathbf{x}(t)}{\Delta t}.$$

Thus, $\mathbf{v}(t)$ is the instantaneous rate of change of position $\mathbf{x}(t)$ with respect to t (time), so it can appropriately be called velocity. Figure 3.4 provides an indication as to why we draw $\mathbf{v}(t)$ as a vector tangent to the path at $\mathbf{x}(t)$. Continuing in this vein, we introduce the following terminology:

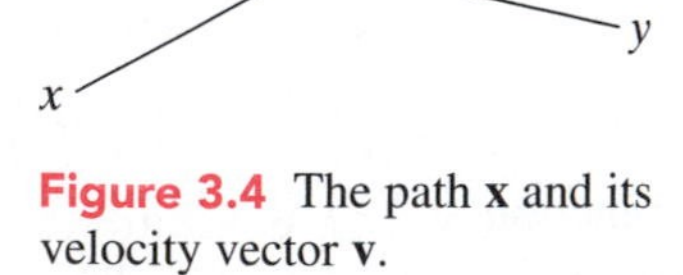

Figure 3.4 The path **x** and its velocity vector **v**.

> ■ **Definition 1.2** Let $\mathbf{x}\colon I \to \mathbf{R}^n$ be a differentiable path. Then the velocity $\mathbf{v}(t) = \mathbf{x}'(t)$ exists, and we define the **speed** of $\mathbf{x}$ to be the magnitude of velocity; that is,
>
> $$\text{Speed} = \|\mathbf{v}(t)\|.$$
>
> If $\mathbf{v}$ is itself differentiable, then we call $\mathbf{v}'(t) = \mathbf{x}''(t)$ the **acceleration** of $\mathbf{x}$ and denote it by $\mathbf{a}(t)$.

EXAMPLE 4 The helix $\mathbf{x}(t) = (a\cos t, a\sin t, bt)$ has

$$\mathbf{v}(t) = -a\sin t\,\mathbf{i} + a\cos t\,\mathbf{j} + b\,\mathbf{k} \quad \text{and} \quad \mathbf{a}(t) = -a\cos t\,\mathbf{i} - a\sin t\,\mathbf{j}.$$

Thus, the acceleration vector is parallel to the xy-plane (i.e., is horizontal). The speed of this helical path is

$$\|\mathbf{v}(t)\| = \sqrt{(-a\sin t)^2 + (a\cos t)^2 + b^2} = \sqrt{a^2 + b^2},$$

which is constant. ◆

The velocity vector $\mathbf{v}$ is important for another reason, namely, for finding equations of tangent lines to paths. The **tangent line** to a differentiable path $\mathbf{x}$, at the point $\mathbf{x}_0 = \mathbf{x}(t_0)$, is the line through $\mathbf{x}_0$ that is parallel to any (nonzero) tangent vector to $\mathbf{x}$ at $\mathbf{x}_0$. Since $\mathbf{v}(t)$, when nonzero, is always tangent to $\mathbf{x}(t)$, we may use equation (1) of §1.2 to obtain the following vector parametric equation for the tangent line:

$$\mathbf{l}(s) = \mathbf{x}_0 + s\mathbf{v}_0. \tag{1}$$

Here $\mathbf{v}_0 = \mathbf{v}(t_0)$ and s may be any real number.

In equation (1), we have $\mathbf{l}(0) = \mathbf{x}_0$. To relate the new parameter s to the original parameter t for the path, we set $s = t - t_0$ and establish the following result:

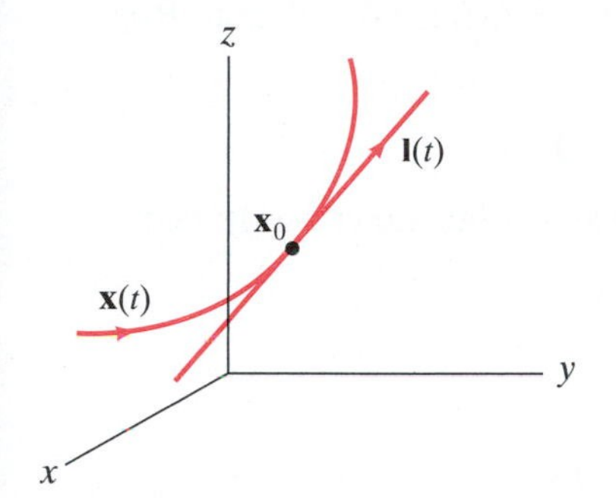

Figure 3.5 The path of the line tangent to $\mathbf{x}(t)$ at the point $\mathbf{x}_0$.

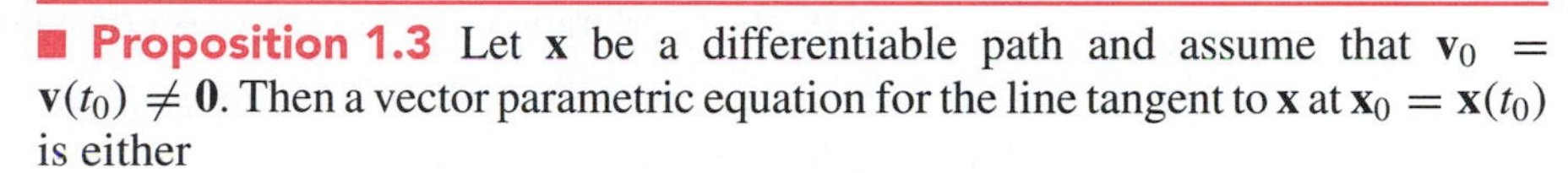

Proposition 1.3 Let $\mathbf{x}$ be a differentiable path and assume that $\mathbf{v}_0 = \mathbf{v}(t_0) \neq \mathbf{0}$. Then a vector parametric equation for the line tangent to $\mathbf{x}$ at $\mathbf{x}_0 = \mathbf{x}(t_0)$ is either

$$\mathbf{l}(s) = \mathbf{x}_0 + s\mathbf{v}_0 \tag{2}$$

or

$$\mathbf{l}(t) = \mathbf{x}_0 + (t - t_0)\mathbf{v}_0. \tag{3}$$

(See Figure 3.5.)

EXAMPLE 5 If $\mathbf{x}(t) = (3t + 2, t^2 - 7, t - t^2)$, we find parametric equations for the line tangent to $\mathbf{x}$ at $(5, -6, 0) = \mathbf{x}(1)$.

For this path, $\mathbf{v}(t) = \mathbf{x}'(t) = 3\mathbf{i} + 2t\mathbf{j} + (1 - 2t)\mathbf{k}$, so that

$$\mathbf{v}_0 = \mathbf{v}(1) = 3\mathbf{i} + 2\mathbf{j} - \mathbf{k}.$$

Thus, by formula (3),

$$\mathbf{l}(t) = (5\mathbf{i} - 6\mathbf{j}) + (t - 1)(3\mathbf{i} + 2\mathbf{j} - \mathbf{k}).$$

Taking components, we read off the parametric equations for the coordinates of the tangent line as

$$\begin{cases} x = 3t + 2 \\ y = 2t - 8. \\ z = 1 - t \end{cases}$$

◆

The physical significance of the tangent line is this: Suppose a particle of mass m travels along a path $\mathbf{x}$. If, suddenly, at $t = t_0$, all forces cease to act on the particle (so that, by Newton's second law of motion $\mathbf{F} = m\mathbf{a}$, we have $\mathbf{a}(t) \equiv \mathbf{0}$ for $t \geq t_0$), then the particle will follow the tangent line path of equation (3).

EXAMPLE 6 If Roger Ramjet is fired from a cannon, then we can use vectors to describe his trajectory. (See Figure 3.6.)

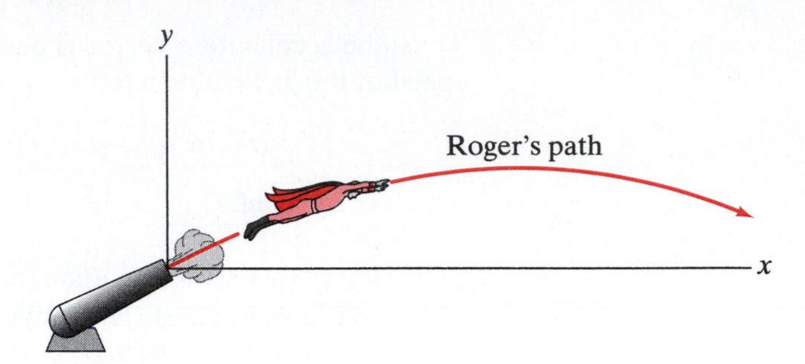

Figure 3.6 Roger Ramjet's path.

We'll assume that Roger is given an initial velocity vector $\mathbf{v}_0$ by virtue of the firing of the cannon, and that thereafter the only force acting on Roger is that due to gravity (so, in particular, we neglect any air resistance). Let us choose coordinates so that Roger is initially at the origin, and throughout our calculations we'll neglect the height of the cannon. Let $\mathbf{x}(t) = (x(t), y(t))$ denote Roger's path. Then the information we have is

$$\mathbf{a}(t) = \mathbf{x}''(t) = -g\,\mathbf{j}$$

(i.e., the acceleration due to gravity is constant and points downward); hence,

$$\mathbf{v}(0) = \mathbf{x}'(0) = \mathbf{v}_0$$

and

$$\mathbf{x}(0) = \mathbf{0}.$$

Since $\mathbf{a}(t) = \mathbf{v}'(t)$, we simply integrate the expression for acceleration componentwise to find the velocity:

$$\mathbf{v}(t) = \int \mathbf{a}(t)\,dt = \int -g\mathbf{j}\,dt = -gt\,\mathbf{j} + \mathbf{c}.$$

Here $\mathbf{c}$ is an arbitrary constant *vector* (the "constant of integration"). Since $\mathbf{v}(0) = \mathbf{v}_0$, we must have $\mathbf{c} = \mathbf{v}_0$, so that

$$\mathbf{v}(t) = -gt\,\mathbf{j} + \mathbf{v}_0.$$

Integrating again to find the path,

$$\mathbf{x}(t) = \int \mathbf{v}(t)\,dt = \int (-gt\,\mathbf{j} + \mathbf{v}_0)\,dt = -\frac{1}{2}gt^2\,\mathbf{j} + t\,\mathbf{v}_0 + \mathbf{d},$$

where $\mathbf{d}$ is another arbitrary constant vector. From the remaining fact that $\mathbf{x}(0) = \mathbf{0}$, we conclude that

$$\mathbf{x}(t) = -\frac{1}{2}gt^2\mathbf{j} + t\mathbf{v}_0 \tag{4}$$

describes Roger's path.

To understand equation (4) better, we write $\mathbf{v}_0$ in terms of its components:

$$\mathbf{v}_0 = v_0 \cos\theta\,\mathbf{i} + v_0 \sin\theta\,\mathbf{j}.$$

Here $v_0 = \|\mathbf{v}_0\|$ is the initial speed. (We're really doing nothing more than expressing the rectangular components of $\mathbf{v}_0$ in terms of polar coordinates.

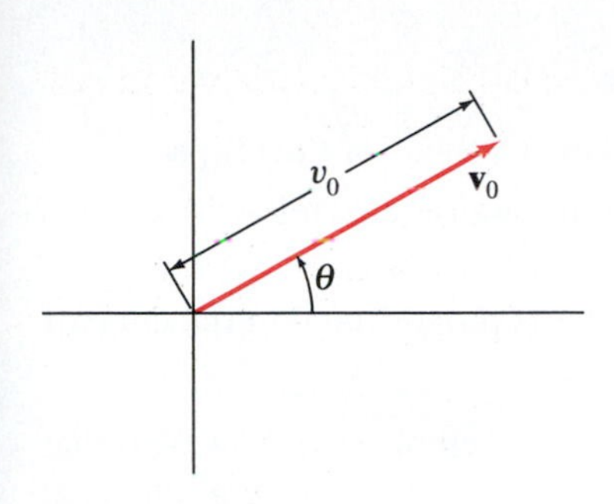

Figure 3.7 Roger's initial velocity.

See Figure 3.7.) Thus,

$$\begin{aligned}\mathbf{x}(t) &= -\tfrac{1}{2}gt^2\mathbf{j} + t(v_0\cos\theta\,\mathbf{i} + v_0\sin\theta\,\mathbf{j})\\ &= (v_0\cos\theta)t\,\mathbf{i} + \left((v_0\sin\theta)t - \tfrac{1}{2}gt^2\right)\mathbf{j}.\end{aligned}$$

From this, we may read off the parametric equations:

$$\begin{cases} x = (v_0\cos\theta)t \\ y = (v_0\sin\theta)t - \tfrac{1}{2}gt^2 \end{cases},$$

from which it is not difficult to check that Roger's path traces a parabola. ◆

Here are two practical questions concerning the set-up of Example 6: First, for a given initial velocity, how far does Roger travel horizontally? Second, for a given initial speed, how should the cannon be aimed so that Roger travels (horizontally) as far as possible? To find the range of the cannon shot and thereby answer the first question, we need to know when $y = 0$ (i.e., when Roger hits the ground). Thus, we solve

$$(v_0\sin\theta)t - \tfrac{1}{2}gt^2 = t(v_0\sin\theta - \tfrac{1}{2}gt) = 0$$

for t. Hence, $y = 0$ when $t = 0$ (which is when Roger blasts off) and when $t = (2v_0\sin\theta)/g$. At this later time,

$$x = (v_0\cos\theta)\cdot\left(\frac{2v_0\sin\theta}{g}\right) = \frac{v_0^2\sin 2\theta}{g}. \tag{5}$$

Formula (5) is Roger's horizontal range for a given initial velocity. To maximize the range for a given initial speed v_0, we must choose θ so that $(v_0^2\sin 2\theta)/g$ is as large as possible. Clearly, this happens when $\sin 2\theta = 1$ (i.e., when $\theta = \pi/4$).

Kepler's Laws of Planetary Motion (optional)

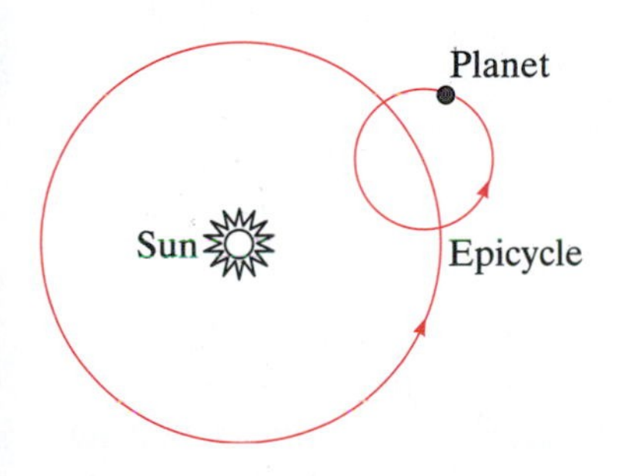

Figure 3.8 An epicycle.

Since classical antiquity, individuals have sought to understand the motions of the planets and stars. The majority of the ancient astronomers, using a combination of crude observation and faith, believed all heavenly bodies revolved around the earth. Fortunately, the heliocentric (or "sun-centered") theory of Nicholas Copernicus (1473–1543) did eventually gain favor as observational techniques improved. However, it was still believed that the planets traveled in circular orbits around the sun. This circular orbit theory did not correctly predict planetary positions, so astronomers postulated the existence of **epicycles**, smaller circular orbits traveling along the major circular arc, an example of which is shown in Figure 3.8. Although positional calculations with epicycles yielded results closer to the observed data, they still were not correct. Attempts at further improvements were made using second- and third-order epicycles, but any gains in predictive power were made at a cost of considerable calculational complexity. A new idea was needed. Such inspiration came from Johannes Kepler (1571–1630), son of a saloonkeeper and assistant to the Danish astronomer Tycho Brahe. The classical astronomers were "stuck on circles" for they believed the circle to be a perfect form, and that God would use only such perfect figures for planetary motion. Kepler, however, considered the other conic sections to be as elegant as the circle and so hypothesized the simple theory that planetary orbits are elliptical. Empirical evidence bore out this theory.

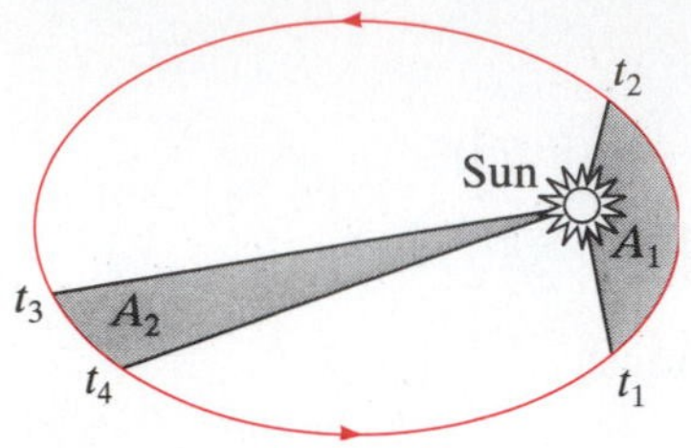

Figure 3.9 Kepler's second law of planetary motion: If $t_2 - t_1 = t_4 - t_3$, then $A_1 = A_2$, where A_1 and A_2 are the areas of the shaded regions.

Kepler's three laws of planetary motion are:

1. The orbit of a planet is elliptical, with the sun at a focus of the ellipse.
2. During equal periods of time, a planet sweeps through equal areas with respect to the sun. (See Figure 3.9.)
3. The square of the period of one elliptical orbit is proportional to the cube of the length of the semimajor axis of the ellipse.

Kepler's laws changed the face of astronomy. We emphasize, however, that they were discovered empirically, not analytically derived from general physical laws. The first analytic derivation is frequently credited to Newton, who claimed to have established Kepler's laws (at least the first and third laws) in Book I of his *Philosophiae Naturalis Principia Mathematica* (1687). However, a number of scientists and historians of science now consider Newton's proof of Kepler's first law to be flawed, and that Johann Bernoulli (1667–1748) offered the first rigorous derivation in 1710.[1] In the discussion that follows, Newton's law of universal gravitation is used to prove all three of Kepler's laws.

In our work below, we assume that the only physical effects are those between the sun and a single planet—the so-called two-body problem. (The n-body problem, where $n \geq 3$ is, by contrast, an important area of current mathematical research.) To set the stage for our calculations, we take the sun to be fixed at the origin O in $\mathbf{R}^3$ and the planet to be at the moving position P. We also need the following two "vector product rules," whose proofs we leave to you:

Proposition 1.4

1. If $\mathbf{x}$ and $\mathbf{y}$ are differentiable paths in $\mathbf{R}^n$, then
$$\frac{d}{dt}(\mathbf{x}\cdot\mathbf{y}) = \mathbf{y}\cdot\frac{d\mathbf{x}}{dt} + \mathbf{x}\cdot\frac{d\mathbf{y}}{dt}.$$
2. If $\mathbf{x}$ and $\mathbf{y}$ are differentiable paths in $\mathbf{R}^3$, then
$$\frac{d}{dt}(\mathbf{x}\times\mathbf{y}) = \frac{d\mathbf{x}}{dt}\times\mathbf{y} + \mathbf{x}\times\frac{d\mathbf{y}}{dt}.$$

First, we establish the following preliminary result:

Proposition 1.5 The motion of the planet is planar, and the sun lies in the planet's plane of motion.

Proof Let $\mathbf{r} = \overrightarrow{OP}$. Then $\mathbf{r}$ is a vector whose representative arrow has its tail fixed at O. (Note that $\mathbf{r} = \mathbf{r}(t)$; that is, $\mathbf{r}$ is a function of time.) If $\mathbf{v} = \mathbf{r}'(t)$, we will show that $\mathbf{r}\times\mathbf{v}$ is a constant vector $\mathbf{c}$. This result, in turn, implies that $\mathbf{r}$ must always be perpendicular to $\mathbf{c}$, and hence that $\mathbf{r}$ always lies in a plane with $\mathbf{c}$ as normal vector.

To show that $\mathbf{r}\times\mathbf{v}$ is constant, we show that its derivative is zero. By part 2 of Proposition 1.4,
$$\frac{d}{dt}(\mathbf{r}\times\mathbf{v}) = \frac{d\mathbf{r}}{dt}\times\mathbf{v} + \mathbf{r}\times\frac{d\mathbf{v}}{dt} = \mathbf{v}\times\mathbf{v} + \mathbf{r}\times\mathbf{a},$$

[1] For an indication of the more recent controversy surrounding Newton's mathematical accomplishments, see R. Weinstock, "Isaac Newton: Credit where credit won't do," *The College Mathematics Journal,* **25** (1994), no. 3, 179–192, and C. Wilson, "Newton's orbit problem: A historian's response," *Ibid.,* 193–200, and related papers.

by the definitions of velocity and acceleration. We know that $\mathbf{v} \times \mathbf{v} = \mathbf{0}$ (why?), so

$$\frac{d}{dt}(\mathbf{r} \times \mathbf{v}) = \mathbf{r} \times \mathbf{a}. \tag{6}$$

Now we use Newton's laws. Newton's law of gravitation tells us that the planet is attracted to the sun with a force

$$\mathbf{F} = -\frac{GMm}{r^2}\mathbf{u}, \tag{7}$$

where G is Newton's gravitational constant ($= 6.6720 \times 10^{-11}\ \mathrm{Nm^2/kg^2}$), M is the mass of the sun, m is the mass of the planet (in kilograms), $r = \|\mathbf{r}\|$, and $\mathbf{u} = \mathbf{r}/\|\mathbf{r}\|$ (distances in meters). On the other hand, Newton's second law of motion states that, for the planet,

$$\mathbf{F} = m\mathbf{a}.$$

Thus,

$$m\mathbf{a} = -\frac{GMm}{r^2}\mathbf{u},$$

or

$$\mathbf{a} = -\frac{GM}{r^3}\mathbf{r}. \tag{8}$$

Therefore, $\mathbf{a}$ is just a scalar multiple of $\mathbf{r}$ and hence is always parallel to $\mathbf{r}$. In view of equations (6) and (8), we conclude that

$$\frac{d}{dt}(\mathbf{r} \times \mathbf{v}) = \mathbf{r} \times \mathbf{a} = \mathbf{0}$$

(i.e., that $\mathbf{r} \times \mathbf{v}$ is constant). ■

■ **Theorem 1.6** (KEPLER'S FIRST LAW) In a two-body system consisting of one sun and one planet, the planet's orbit is an ellipse and the sun lies at one focus of that ellipse.

Proof We will eventually find a polar equation for the planet's orbit and see that this equation defines an ellipse as described. We retain the notation from the proof of Proposition 1.5 and take coordinates for $\mathbf{R}^3$ so that the sun is at the origin, and the path of the planet lies in the xy-plane. Then the constant vector $\mathbf{c} = \mathbf{r} \times \mathbf{v}$ used in the proof of Proposition 1.5 may be written as $c\mathbf{k}$, where c is some nonzero real number. This set-up is shown in Figure 3.10.

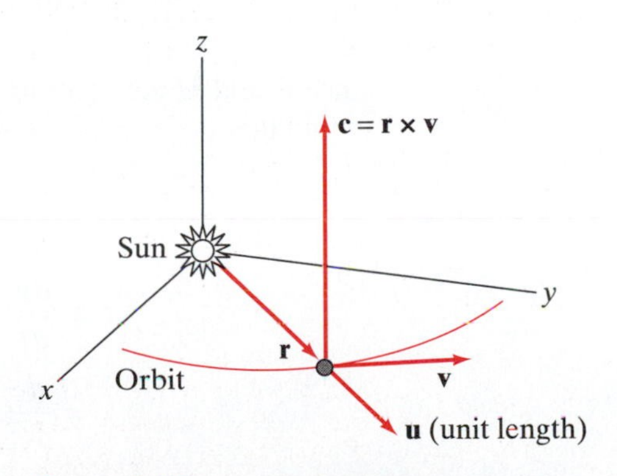

Figure 3.10 Establishing Kepler's laws.

Step 1. We find another expression for $\mathbf{c}$. By definition of $\mathbf{u}$ in formula (7), $\mathbf{r} = r\mathbf{u}$, so that, by the product rule,

$$\mathbf{v} = \frac{d}{dt}(r\mathbf{u}) = r\frac{d\mathbf{u}}{dt} + \frac{dr}{dt}\mathbf{u}.$$

Hence,

$$\mathbf{c} = \mathbf{r} \times \mathbf{v} = (r\mathbf{u}) \times \left(r\frac{d\mathbf{u}}{dt} + \frac{dr}{dt}\mathbf{u}\right) = r^2\left(\mathbf{u} \times \frac{d\mathbf{u}}{dt}\right) + r\frac{dr}{dt}(\mathbf{u} \times \mathbf{u}).$$

Since $\mathbf{u} \times \mathbf{u}$ must be zero, we conclude that

$$\mathbf{c} = r^2\left(\mathbf{u} \times \frac{d\mathbf{u}}{dt}\right). \tag{9}$$

Step 2. We derive the polar equation for the orbit. Before doing so, however, note the following result, whose proof is left to you as an exercise:

Proposition 1.7 If $\mathbf{x}(t)$ has constant length (i.e., $\|\mathbf{x}(t)\|$ is constant for all t), then $\mathbf{x}$ is perpendicular to its derivative $d\mathbf{x}/dt$.

Continuing now with the main argument, note that the vector $\mathbf{r}(t)$ is defined so that its magnitude is precisely the polar coordinate r of the planet's position. Using equations (8) and (9), we find that

$$\begin{aligned}
\mathbf{a} \times \mathbf{c} &= \left(-\frac{GM}{r^2}\mathbf{u}\right) \times r^2\left(\mathbf{u} \times \frac{d\mathbf{u}}{dt}\right) \\
&= -GM\left[\mathbf{u} \times \left(\mathbf{u} \times \frac{d\mathbf{u}}{dt}\right)\right] \\
&= GM\left[\left(\mathbf{u} \times \frac{d\mathbf{u}}{dt}\right) \times \mathbf{u}\right] \\
&= GM\left[(\mathbf{u} \cdot \mathbf{u})\frac{d\mathbf{u}}{dt} - \left(\mathbf{u} \cdot \frac{d\mathbf{u}}{dt}\right)\mathbf{u}\right] && \text{(see Exercise 27 of §1.4)} \\
&= GM\left[1\frac{d\mathbf{u}}{dt} - 0\mathbf{u}\right] && \text{(by Proposition 1.7)} \\
&= \frac{d}{dt}(GM\mathbf{u}),
\end{aligned}$$

since G and M are constant. On the other hand, we can "reverse" the product rule to find that

$$\begin{aligned}
\mathbf{a} \times \mathbf{c} &= \frac{d\mathbf{v}}{dt} \times \mathbf{c} \\
&= \frac{d\mathbf{v}}{dt} \times \mathbf{c} + \mathbf{v} \times \frac{d\mathbf{c}}{dt} && \text{(since } \mathbf{c} \text{ is constant)} \\
&= \frac{d}{dt}(\mathbf{v} \times \mathbf{c}).
\end{aligned}$$

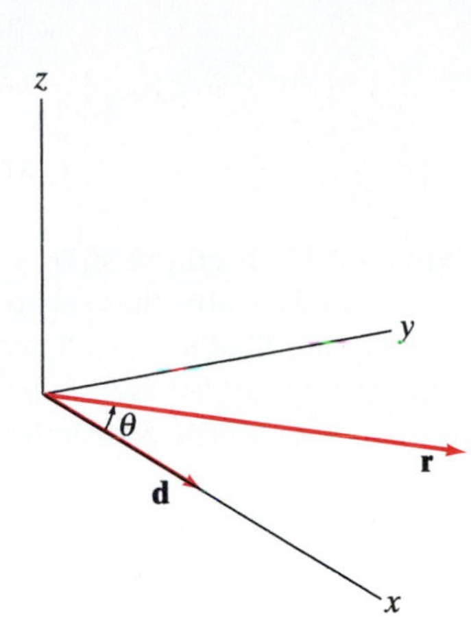

Figure 3.11 The angle θ is the angle between **r** and **d**.

Thus,

$$\mathbf{a} \times \mathbf{c} = \frac{d}{dt}(GM\mathbf{u}) = \frac{d}{dt}(\mathbf{v} \times \mathbf{c}),$$

and hence,

$$\mathbf{v} \times \mathbf{c} = GM\mathbf{u} + \mathbf{d}, \tag{10}$$

where **d** is an arbitrary constant vector. Because both $\mathbf{v} \times \mathbf{c}$ and **u** lie in the xy-plane, so must **d**.

Let us adjust coordinates, if necessary, so that **d** points in the **i**-direction (i.e., so that $\mathbf{d} = d\mathbf{i}$ for some $d \in \mathbf{R}$). This can be accomplished by rotating the whole set-up about the z-axis, which does not lift anything lying in the xy-plane out of that plane. Then the angle between **r** (and hence **u**) and **d** is the polar angle θ as shown in Figure 3.11.

By Theorem 3.3 of Chapter 1,

$$\mathbf{u} \cdot \mathbf{d} = \|\mathbf{u}\| \, \|\mathbf{d}\| \cos\theta = d\cos\theta. \tag{11}$$

Since $c = \|\mathbf{c}\|$,

$$\begin{aligned} c^2 &= \mathbf{c} \cdot \mathbf{c} \\ &= (\mathbf{r} \times \mathbf{v}) \cdot \mathbf{c} \\ &= \mathbf{r} \cdot (\mathbf{v} \times \mathbf{c}) && \text{(Why? See formula (4) of §1.4.)} \\ &= r\mathbf{u} \cdot (GM\mathbf{u} + \mathbf{d}) && \text{by equation (10).} \end{aligned}$$

Hence,

$$c^2 = GMr + rd\cos\theta$$

by equation (11). We can readily solve this equation for r to obtain

$$r = \frac{c^2}{GM + d\cos\theta}, \tag{12}$$

the polar equation for the planet's orbit.

Step 3. We now check that equation (12) really does define an ellipse by converting to Cartesian coordinates. First, we'll rewrite the equation as

$$r = \frac{c^2}{GM + d\cos\theta} = \frac{(c^2/GM)}{1 + (d/GM)\cos\theta},$$

and then let $p = c^2/GM, e = d/GM$ for convenience. (Note that $p > 0$.) Hence, equation (12) becomes

$$r = \frac{p}{1 + e\cos\theta}. \tag{13}$$

A little algebra provides the equivalent equation,

$$r = p - er\cos\theta. \tag{14}$$

Now $r\cos\theta = x$ (x being the usual Cartesian coordinate), so that equation (14) is equivalent to

$$r = p - ex.$$

To complete the conversion, we square both sides and find, by virtue of the fact that $r^2 = x^2 + y^2$,

$$x^2 + y^2 = p^2 - 2pex + e^2x^2.$$

A little more algebra reveals that

$$(1 - e^2)x^2 + 2pex + y^2 = p^2. \tag{15}$$

Therefore, the curve described by the preceding equation is an ellipse if $0 < |e| < 1$, a parabola if $e = \pm 1$, and a hyperbola if $|e| > 1$. Analytically, there is no way to eliminate the last two possibilities. Indeed, "uncaptured" objects such as comets or expendable deep space probes can have hyperbolic or parabolic orbits. However, to have a *closed* orbit (so that the planet repeats its transit across the sky), we are forced to conclude that the orbit must be elliptical.

More can be said about the elliptical orbit. Dividing equation (15) through by $1 - e^2$ and completing the square in x, we have

$$\left(x + \frac{pe}{1 - e^2}\right)^2 + \frac{y^2}{1 - e^2} = \frac{p^2}{(1 - e^2)^2}.$$

This is equivalent to the rather awkward-looking equation

$$\frac{\left(x + pe/(1 - e^2)\right)^2}{p^2/(1 - e^2)^2} + \frac{y^2}{p^2/(1 - e^2)} = 1. \tag{16}$$

From equation (16), we see that the ellipse is centered at the point $(-pe/(1-e^2), 0)$, that its semimajor axis has length $a = p/(1 - e^2)$, and that its semiminor axis has length $b = p/\sqrt{1 - e^2}$. The foci of the ellipse are at a distance

$$\sqrt{a^2 - b^2} = \sqrt{\frac{p^2}{(1 - e^2)^2} - \frac{p^2}{1 - e^2}} = \frac{p|e|}{1 - e^2}$$

from the center. (See Figure 3.12.) Hence, we see that one focus must be at the origin, the location of the sun. Our proof is, therefore, complete. ■

Fortunately, all the toil involved in proving the first law will pay off in proofs of the second and third laws, which are considerably shorter. Again, we retain all the notation we already introduced.

■ **Theorem 1.8** (**KEPLER'S SECOND LAW**) During equal intervals of time, a planet sweeps through equal areas with respect to the sun.

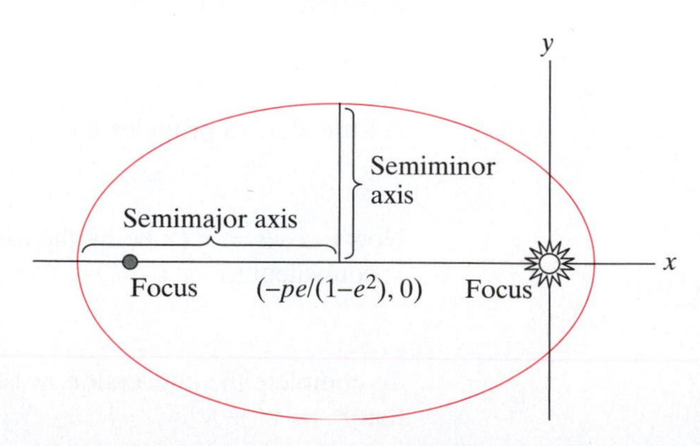

Figure 3.12 The ellipse of equation (16).

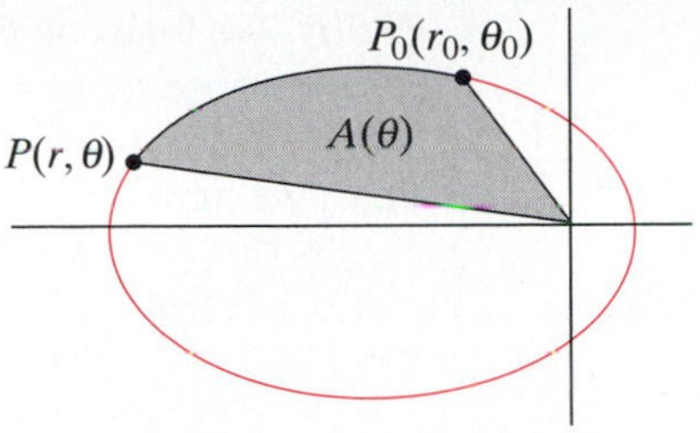

Figure 3.13 The shaded area $A(\theta)$ is given by $\int_{\theta_0}^{\theta} \frac{1}{2} r^2 \, d\varphi$.

Proof Fix one point P_0 on the planet's orbit. Then, the area A swept between P_0 and a second (moving) point P on the orbit is given by the polar area integral

$$A(\theta) = \int_{\theta_0}^{\theta} \frac{1}{2} r^2 \, d\varphi.$$

(See Figure 3.13.) Thus, we may reformulate Kepler's law to say that dA/dt is constant. We establish this reformulation by relating dA/dt to a known constant, namely, the vector $\mathbf{c} = \mathbf{r} \times \mathbf{v}$.

By the chain rule (in one variable),

$$\frac{dA}{dt} = \frac{dA}{d\theta} \frac{d\theta}{dt}.$$

By the fundamental theorem of calculus,

$$\frac{dA}{d\theta} = \frac{d}{d\theta} \int_{\theta_0}^{\theta} \frac{1}{2} r^2 \, d\varphi = \frac{1}{2} [r(\theta)]^2.$$

Hence,

$$\frac{dA}{dt} = \frac{1}{2} r^2 \frac{d\theta}{dt}. \tag{17}$$

Now, we relate $\mathbf{c}$ to $d\theta/dt$ by means of equation (9). Therefore, we compute $\mathbf{u} \times d\mathbf{u}/dt$ in terms of θ. Recall that $\mathbf{u} = \dfrac{1}{r} \mathbf{r}$ and $\mathbf{r} = r \cos\theta \, \mathbf{i} + r \sin\theta \, \mathbf{j}$. Thus,

$$\mathbf{u} = \cos\theta \, \mathbf{i} + \sin\theta \, \mathbf{j}$$

$$\frac{d\mathbf{u}}{dt} = -\sin\theta \frac{d\theta}{dt} \mathbf{i} + \cos\theta \frac{d\theta}{dt} \mathbf{j}.$$

Hence, it follows by direct calculation of the cross product that

$$\mathbf{c} = r^2 \left(\mathbf{u} \times \frac{d\mathbf{u}}{dt} \right) = r^2 \frac{d\theta}{dt} \mathbf{k},$$

so $c = \|\mathbf{c}\| = r^2 d\theta/dt$, and equation (17) implies that

$$\frac{dA}{dt} = \frac{1}{2} c, \tag{18}$$

a constant. ■

■ **Theorem 1.9** (KEPLER'S THIRD LAW) If T is the length of time for one planetary orbit, and a is the length of the semimajor axis of this orbit, then $T^2 = Ka^3$ for some constant K.

Proof We focus on the total area enclosed by the elliptical orbit. The area of an ellipse whose semimajor and semiminor axes have lengths a and b, respectively, is πab. This area must also be that swept by the planet in the time interval $[0, T]$. Thus, we have

$$\begin{aligned} \pi ab &= \int_0^T \frac{dA}{dt}\, dt \\ &= \int_0^T \tfrac{1}{2} c\, dt \qquad \text{by equation (18)} \\ &= \tfrac{1}{2} cT. \end{aligned}$$

Hence,

$$T = \frac{2\pi ab}{c}, \quad \text{so} \quad T^2 = \frac{4\pi^2 a^2 b^2}{c^2}. \tag{19}$$

Now, b and c are related to a, so these quantities must be replaced before we are done. In particular, from equation (16), $b^2 = p^2/(1 - e^2)$, so

$$b^2 = pa.$$

Also,

$$p = \frac{c^2}{GM}.$$

(See equations (12) and (13).) With these substitutions, the result in (19) becomes

$$T^2 = \frac{4\pi^2 a^2 (pa)}{pGM} = \left(\frac{4\pi^2}{GM}\right) a^3.$$

This last equation shows that T^2 is proportional to a^3, but it says even more: The constant of proportionality $4\pi^2/GM$ depends entirely on the mass of the sun—the constant is the same for *any* planet that might revolve around the sun. ■

Exercises

In Exercises 1–6, sketch the images of the following paths, using arrows to indicate the direction in which the parameter increases:

1. $\begin{cases} x = 2t - 1 \\ y = 3 - t \end{cases}, \quad -1 \le t \le 1$

2. $\mathbf{x}(t) = e^t\,\mathbf{i} + e^{-t}\,\mathbf{j}$

3. $\begin{cases} x = t\cos t \\ y = t\sin t \end{cases}, \quad -6\pi \le t \le 6\pi$

4. $\begin{cases} x = 3\cos t \\ y = 2\sin 2t \end{cases}, \quad 0 \le t \le 2\pi$

5. $\mathbf{x}(t) = (t, 3t^2 + 1, 0)$

6. $\mathbf{x}(t) = (t, t^2, t^3)$

Calculate the velocity, speed, and acceleration of the paths given in Exercises 7–10.

7. $\mathbf{x}(t) = (3t - 5)\mathbf{i} + (2t + 7)\mathbf{j}$

8. $\mathbf{x}(t) = 5\cos t\,\mathbf{i} + 3\sin t\,\mathbf{j}$

9. $\mathbf{x}(t) = (t\sin t, t\cos t, t^2)$

10. $\mathbf{x}(t) = (e^t, e^{2t}, 2e^t)$

In Exercises 11–14, (a) use a computer to give a plot of the given path **x** *over the indicated interval for t; identify the direction in which t increases. (b) Show that the path lies on the given surface S.*

11. $\mathbf{x}(t) = (3\cos\pi t, 4\sin\pi t, 2t)$, $-4 \le t \le 4$; S is elliptical cylinder $\dfrac{x^2}{9} + \dfrac{y^2}{16} = 1$.

12. $\mathbf{x}(t) = (t\cos t, t\sin t, t)$, $-20 \le t \le 20$; S is cone $z^2 = x^2 + y^2$.

13. $\mathbf{x}(t) = (t\sin 2t, t\cos 2t, t^2)$, $-6 \le t \le 6$; S is paraboloid $z = x^2 + y^2$.

14. $\mathbf{x}(t) = (2\cos t, 2\sin t, 3\sin 8t)$, $0 \le t \le 2\pi$; S is cylinder $x^2 + y^2 = 4$.

In Exercises 15–18, find an equation for the line tangent to the given path at the indicated value for the parameter.

15. $\mathbf{x}(t) = te^{-t}\,\mathbf{i} + e^{3t}\,\mathbf{j}$, $t = 0$

16. $\mathbf{x}(t) = 4\cos t\,\mathbf{i} - 3\sin t\,\mathbf{j} + 5t\,\mathbf{k}$, $t = \pi/3$

17. $\mathbf{x}(t) = (t^2, t^3, t^5)$, $t = 2$

18. $\mathbf{x}(t) = (\cos(e^t), 3 - t^2, t)$, $t = 1$

19. (a) Sketch the path $\mathbf{x}(t) = (t, t^3 - 2t + 1)$.

(b) Calculate the line tangent to $\mathbf{x}$ when $t = 2$.

(c) Describe the image of $\mathbf{x}$ by an equation of the form $y = f(x)$ by eliminating t.

(d) Verify your answer in part (b) by recalculating the tangent line, using your result in part (c).

Exercises 20–23 concern Roger Ramjet and his trajectory when he is shot from a cannon as in Example 6 of this section.

20. Verify that Roger Ramjet's path in Example 6 is indeed a parabola.

21. Suppose that Roger is fired from the cannon with an angle of inclination θ of $60°$ and an initial speed v_0 of 100 ft/sec. What is the maximum height Roger attains?

22. Suppose that Roger is fired from the cannon with an angle of inclination θ of $60°$ and that he hits the ground $1/2$ mile from the cannon. What, then, was Roger's initial speed?

23. If Roger is fired from the cannon with an initial speed of 250 ft/sec, what angle of inclination θ should be used so that Roger hits the ground 1500 ft from the cannon?

24. Gertrude is aiming a Super Drencher water pistol at Egbert, who is 1.6 m tall and is standing 5 m away. Gertrude holds the water gun 1 m above ground at an angle α of elevation. (See Figure 3.14.)

(a) If the water pistol fires with an initial speed of 7 m/sec and an elevation angle of $45°$, does Egbert get wet?

(b) If the water pistol fires with an initial speed of 8 m/sec, what possible angles of elevation will cause Egbert to get wet? (Note: You will want to use a computer algebra system or a graphics calculator for this part.)

25. A malfunctioning rocket is traveling according to the path $\mathbf{x}(t) = \left(e^{2t}, 3t^3 - 2t, t - \frac{1}{t}\right)$ in the hope of reaching a repair station at the point $(7e^4, 35, 5)$. (Here t represents time in minutes and spatial coordinates are measured in miles.) At $t = 2$, the rocket's engines suddenly cease. Will the rocket coast into the repair station?

26. Two billiard balls are moving on a (coordinatized) pool table according to the respective paths $\mathbf{x}(t) = \left(t^2 - 2, \frac{t^2}{2} - 1\right)$ and $\mathbf{y}(t) = (t, 5 - t^2)$, where t represents time measured in seconds.

(a) When and where do the balls collide?

(b) What is the angle formed by the paths of the balls at the collision point?

27. Establish part 1 of Proposition 1.4 in this section: If $\mathbf{x}$ and $\mathbf{y}$ are differentiable paths in $\mathbf{R}^n$, show that

$$\frac{d}{dt}(\mathbf{x}\cdot\mathbf{y}) = \mathbf{y}\cdot\frac{d\mathbf{x}}{dt} + \mathbf{x}\cdot\frac{d\mathbf{y}}{dt}.$$

28. Establish part 2 of Proposition 1.4 in this section: If $\mathbf{x}$ and $\mathbf{y}$ are differentiable paths in $\mathbf{R}^3$, show that

$$\frac{d}{dt}(\mathbf{x}\times\mathbf{y}) = \frac{d\mathbf{x}}{dt}\times\mathbf{y} + \mathbf{x}\times\frac{d\mathbf{y}}{dt}.$$

29. Prove Proposition 1.7.

30. (a) Show that the path $\mathbf{x}(t) = (\cos t, \cos t\sin t, \sin^2 t)$ lies on a unit sphere.

(b) Verify that $\mathbf{x}(t)$ is always perpendicular to the velocity vector $\mathbf{v}(t)$.

(c) Use Proposition 1.7 to show that if a differentiable path lies on a sphere centered at the origin, then its position vector is always perpendicular to its velocity vector.

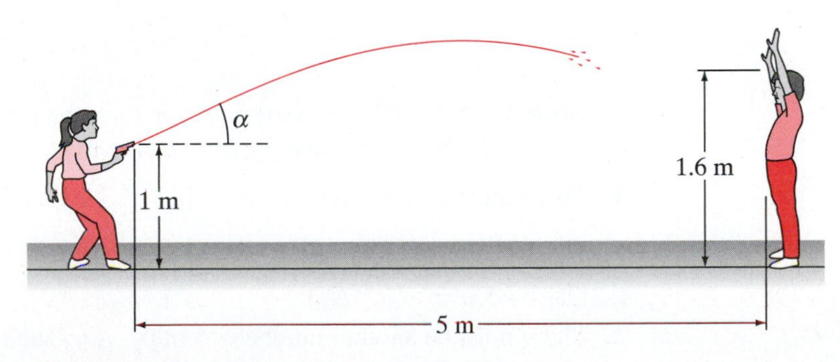

Figure 3.14 Figure for Exercise 24.

31. Consider the path

$$\begin{cases} x = (a + b\cos\omega t)\cos t \\ y = (a + b\cos\omega t)\sin t\,, \\ z = b\sin\omega t \end{cases}$$

where a, b, and ω are positive constants and $a > b$.

(a) Use a computer to plot this path when

i. $a = 3$, $b = 1$, and $\omega = 15$.

ii. $a = 5$, $b = 1$, and $\omega = 15$.

iii. $a = 5$, $b = 1$, and $\omega = 25$.

Comment on how the values of a, b, and ω affect the shapes of the image curves.

(b) Show that the image curve lies on the torus

$$(\sqrt{x^2 + y^2} - a)^2 + z^2 = b^2.$$

(A **torus** is the surface of a doughnut.)

32. Let $\mathbf{x}(t)$ be a path of class C^1 that does not pass through the origin in $\mathbf{R}^3$. If $\mathbf{x}(t_0)$ is the point on the image of $\mathbf{x}$ closest to the origin and $\mathbf{x}'(t_0) \neq \mathbf{0}$, show that the position vector $\mathbf{x}(t_0)$ is orthogonal to the velocity vector $\mathbf{x}'(t_0)$.

3.2 Arclength and Differential Geometry

In this section, we continue our general study of parametrized curves in $\mathbf{R}^3$, considering how to measure such geometric properties as length and curvature. This can be done by defining three mutually perpendicular unit vectors that form the so-called moving frame specially adapted to a path $\mathbf{x}$. Our study takes us briefly into the branch of mathematics called **differential geometry,** an area where calculus and analysis are used to understand the geometry of curves, surfaces, and certain higher-dimensional objects (called **manifolds**).

Length of a Path

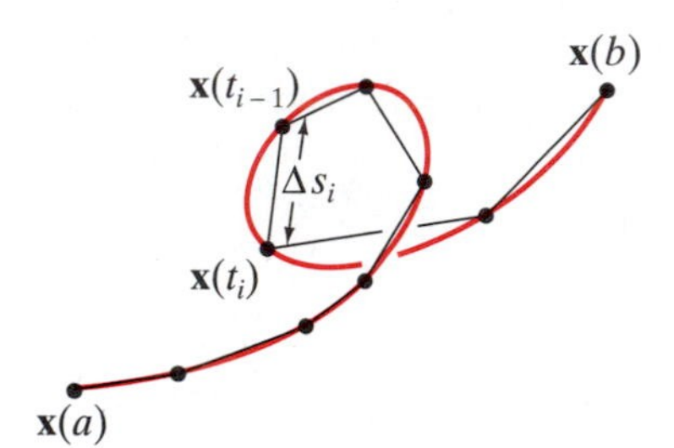

Figure 3.15 Approximating the length of a C^1 path.

For now, let $\mathbf{x}\colon [a, b] \to \mathbf{R}^3$ be a C^1 path in $\mathbf{R}^3$. Then we can approximate the length L of $\mathbf{x}$ as follows: First, partition the interval $[a, b]$ into n subintervals. That is, choose numbers $t_0, t_1, \ldots, t_n$ such that $a = t_0 < t_1 < \cdots < t_n = b$. If, for $i = 1, \ldots, n$, we let Δs_i denote the distance between the points $\mathbf{x}(t_{i-1})$ and $\mathbf{x}(t_i)$ on the path, then

$$L \approx \sum_{i=1}^{n} \Delta s_i. \tag{1}$$

(See Figure 3.15.) We have $\mathbf{x}(t) = (x(t), y(t), z(t))$, so that the distance formula (i.e., the Pythagorean theorem) implies

$$\Delta s_i = \sqrt{\Delta x_i^2 + \Delta y_i^2 + \Delta z_i^2},$$

where $\Delta x_i = x(t_i) - x(t_{i-1})$, $\Delta y_i = y(t_i) - y(t_{i-1})$, and $\Delta z_i = z(t_i) - z(t_{i-1})$. It is entirely reasonable to hope that the approximation in (1) improves as the Δt_i's become closer to zero. Hence, we *define* the length L of $\mathbf{x}$ to be

$$L = \lim_{\max \Delta t_i \to 0} \sum_{i=1}^{n} \sqrt{\Delta x_i{}^2 + \Delta y_i{}^2 + \Delta z_i{}^2}. \tag{2}$$

Now, we find a way to rewrite equation (2) as an integral. On each subinterval $[t_{i-1}, t_i]$, apply the mean value theorem (three times) to conclude the following:

1. There must be some number t_i^* in $[t_{i-1}, t_i]$ such that

$$x(t_i) - x(t_{i-1}) = x'(t_i^*)(t_i - t_{i-1});$$

that is, $\Delta x_i = x'(t_i^*)\Delta t_i$.

2. There must be another number t_i^{**} in $[t_{i-1}, t_i]$ such that

$$\Delta y_i = y'(t_i^{**})\Delta t_i.$$

3. There must be a third number t_i^{***} in $[t_{i-1}, t_i]$ such that

$$\Delta z_i = z'(t_i^{***})\Delta t_i.$$

Therefore, with a little algebra, equation (2) becomes

$$L = \lim_{\max \Delta t_i \to 0} \sum_{i=1}^{n} \sqrt{x'(t_i^{*})^2 + y'(t_i^{**})^2 + z'(t_i^{***})^2}\, \Delta t_i. \tag{3}$$

When the limit appearing in equation (3) is finite, it gives the value of the definite integral

$$\int_a^b \sqrt{x'(t)^2 + y'(t)^2 + z'(t)^2}\, dt.$$

Note that the integrand is precisely $\|\mathbf{x}'(t)\|$, the speed of the path. (This makes perfect sense, of course. Speed measures the rate of distance traveled per unit time, so integrating the speed over the elapsed time interval should give the total distance traveled.) Moreover, it's not hard to see how we should go about defining the length of a path in $\mathbf{R}^n$ for arbitrary n.

■ **Definition 2.1** The **length** $L(\mathbf{x})$ of a C^1 path $\mathbf{x}\colon [a, b] \to \mathbf{R}^n$ is found by integrating its speed:

$$L(\mathbf{x}) = \int_a^b \|\mathbf{x}'(t)\|\, dt.$$

EXAMPLE 1 To check our definition in a well-known situation, we compute the length of the path

$$\mathbf{x}\colon [0, 2\pi] \to \mathbf{R}^2, \quad \mathbf{x}(t) = (a\cos t, a\sin t), \quad a > 0.$$

We have

$$\mathbf{x}'(t) = -a\sin t\,\mathbf{i} + a\cos t\,\mathbf{j},$$

so

$$\|\mathbf{x}'(t)\| = \sqrt{a^2\sin^2 t + a^2\cos^2 t} = a.$$

Thus, Definition 2.1 gives

$$L(\mathbf{x}) = \int_0^{2\pi} a\, dt = 2\pi a.$$

Since the path traces a circle of radius a once, the length integral works out to be the circumference of the circle, as it should. ◆

EXAMPLE 2 For the helix $\mathbf{x}(t) = (a\cos t, a\sin t, bt)$, $0 \le t \le 2\pi$, we have

$$\mathbf{x}'(t) = -a\sin t\,\mathbf{i} + a\cos t\,\mathbf{j} + b\,\mathbf{k},$$

so that $\|\mathbf{x}'(t)\| = \sqrt{a^2 + b^2}$, and

$$L(\mathbf{x}) = \int_0^{2\pi} \sqrt{a^2 + b^2}\, dt = 2\pi\sqrt{a^2 + b^2}.$$

When $b = 0$, the helix reverts to a circle and the length integral agrees with the previous example. ◆

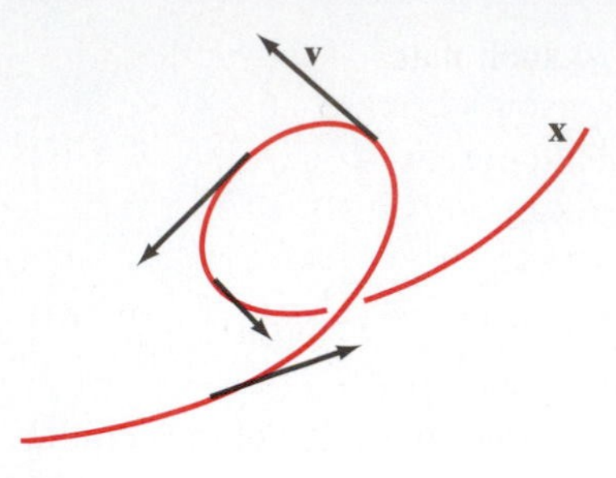

Figure 3.16 A C^1 path.

Although we have defined the length integral only for C^1 (or "smooth") paths, there is no problem with extending our definition to the **piecewise** C^1 case. By definition, a C^1 path is one with a continuously varying velocity vector, and so it looks like Figure 3.16. A piecewise C^1 path is one which may not be itself C^1, but consists of finitely many C^1 chunks. A continuous, piecewise C^1 path that is *not* C^1 looks like Figure 3.17. Each of the three portions of the path defined for (i) $a \le t \le t_1$, (ii) $t_1 \le t \le t_2$, and (iii) $t_2 \le t \le b$ is of class C^1, but the velocity is discontinuous at $t = t_1$ and $t = t_2$. To define the length of a piecewise C^1 path, all we need do is break up the path into its C^1 pieces, calculate the length of each piece, and add to get the total length. For the piecewise C^1 path shown in Figure 3.17, this means we would take

$$\int_a^{t_1} \|\mathbf{x}'(t)\|\, dt + \int_{t_1}^{t_2} \|\mathbf{x}'(t)\|\, dt + \int_{t_2}^{b} \|\mathbf{x}'(t)\|\, dt$$

to be the length.

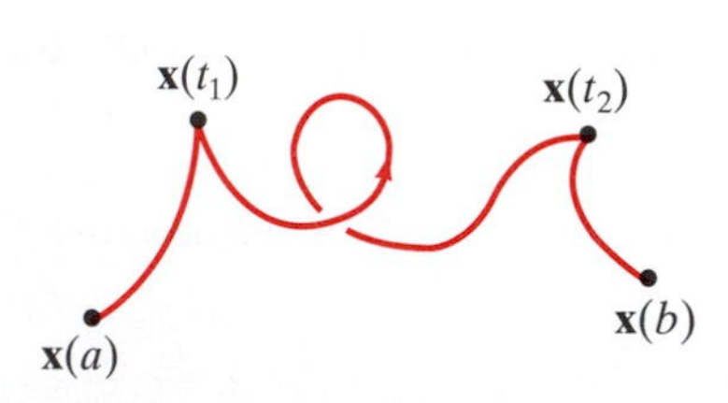

Figure 3.17 A piecewise C^1 path $\mathbf{x}\colon [a, b] \to \mathbf{R}^3$.

WARNING Even if a path is continuous, the definite integral in Definition 2.1 may fail to exist. An example of such an unfortunate situation is furnished by the path $\mathbf{x}\colon [0, 1] \to \mathbf{R}^2$,

$$\mathbf{x}(t) = (t, y(t)), \quad \text{where} \quad y(t) = \begin{cases} t \sin \dfrac{1}{t} & \text{if } t \neq 0 \\ 0 & \text{if } t = 0 \end{cases}.$$

Such a path is called **nonrectifiable.** It is a fact that any C^1 path with endpoints is rectifiable, which is why we made such a condition part of Definition 2.1.

The Arclength Parameter

The calculation of the length of a path is not only useful (and moderately interesting) in itself, but it also provides a way for us to **reparametrize** the path with a parameter that depends solely on the geometry of the curve traced by the path, not on the way in which the curve is traced.

Let $\mathbf{x}$ be any C^1 path and assume that the velocity $\mathbf{x}'$ is never zero. Fix a point P_0 on the path and let a be such that $\mathbf{x}(a) = P_0$. We define a one-variable function s of the given parameter t that measures the length of the path from P_0 to any other (moving) point P by

$$s(t) = \int_a^t \|\mathbf{x}'(\tau)\|\, d\tau. \tag{4}$$

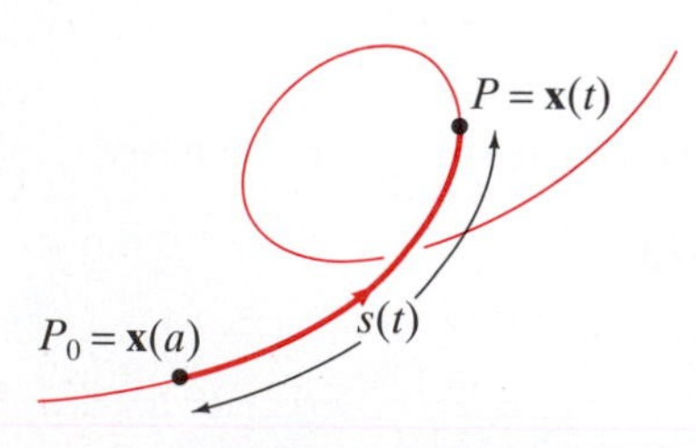

Figure 3.18 The arclength reparametrization.

(See Figure 3.18. The Greek letter tau, τ, is used purely as a dummy variable—the standard convention is never to have the same variable appearing in both the integrand and either of the limits of integration.) If t happens to be less than a, then the value of s in formula (4) will be negative. This is nothing more than a consequence of how the "base point" P_0 is chosen.

Here's how to get the new parameter: From formula (4) and from the fundamental theorem of calculus,

$$\frac{ds}{dt} = \frac{d}{dt} \int_a^t \|\mathbf{x}'(\tau)\|\, d\tau = \|\mathbf{x}'(t)\| = \text{speed}. \tag{5}$$

Since we have assumed that $\mathbf{x}'(t) \neq \mathbf{0}$, it follows that ds/dt is nonzero. Hence, ds/dt is always positive, so s is a strictly increasing function of t. Thus, s is, in fact, an invertible function; that is, it is at least theoretically possible to solve the equation $s = s(t)$ for t in terms of s. If we imagine doing this, then we can reparametrize the path $\mathbf{x}$, using the arclength parameter s as independent variable.

EXAMPLE 3 For the helix $\mathbf{x}(t) = (a\cos t, a\sin t, bt)$, if we choose the "base point" P_0 to be $\mathbf{x}(0) = (a, 0, 0)$, then we have

$$s(t) = \int_0^t \|\mathbf{x}'(\tau)\|\,d\tau = \int_0^t \sqrt{a^2+b^2}\,d\tau = \sqrt{a^2+b^2}\,t,$$

so that

$$s = \sqrt{a^2+b^2}\,t,$$

or

$$t = \frac{s}{\sqrt{a^2+b^2}}.$$

(What the preceding tells us is that this reparametrization just rescales the time variable.) Hence, we can rewrite the helical path as

$$\mathbf{x}(s) = \left(a\cos\left(\frac{s}{\sqrt{a^2+b^2}}\right), a\sin\left(\frac{s}{\sqrt{a^2+b^2}}\right), \frac{bs}{\sqrt{a^2+b^2}}\right).$$ ◆

EXAMPLE 4 The explicit determination of the arclength parameter for a given parametrized path is a delicate matter. Consider the path

$$\mathbf{x}(t) = \left(t, \frac{\sqrt{2}}{2}t^2, \frac{1}{3}t^3\right).$$

Then $\mathbf{x}'(t) = (1, \sqrt{2}t, t^2)$ and, if we take the base point to be $\mathbf{x}(0) = (0, 0, 0)$, then

$$\begin{aligned} s(t) &= \int_0^t \sqrt{1+2\tau^2+\tau^4}\,d\tau \\ &= \int_0^t \sqrt{(1+\tau^2)^2}\,d\tau = \int_0^t (1+\tau^2)\,d\tau = t + \frac{t^3}{3}. \end{aligned}$$

On the other hand, the path $\mathbf{y}(t) = (t, t^2, t^3)$ is quite similar to $\mathbf{x}$, yet it has no readily calculable arclength parameter. In this case, $\mathbf{y}'(t) = (1, 2t, 3t^2)$ and the resulting integral for $s(t)$ is

$$s(t) = \int_0^t \sqrt{1+4\tau^2+9\tau^4}\,d\tau.$$

It can be shown that this integral has no "closed form" formula (i.e., a formula that involves only finitely many algebraic and transcendental functions). ◆

The significance of the arclength parameter s is that it is an **intrinsic** parameter; it depends only on how the curve itself bends, not on how fast (or slowly) the curve is traced. To see more precisely what this means, we resort to the chain rule. Consider s as an intermediate variable and t as a final variable. Then we have

$$\begin{aligned} \mathbf{x}'(t) &= \mathbf{x}'(s)\frac{ds}{dt} && \text{by the chain rule,} \\ &= \mathbf{x}'(s)\|\mathbf{x}'(t)\| && \text{by (5).} \end{aligned}$$

Since $\mathbf{x}'(t) \neq \mathbf{0}$, we can solve for $\mathbf{x}'(s)$ to find

$$\mathbf{x}'(s) = \frac{\mathbf{x}'(t)}{\|\mathbf{x}'(t)\|}. \tag{6}$$

Therefore, $\mathbf{x}'(s)$ is precisely the normalization of the original velocity vector, and so it is a unit vector. Hence, the reparametrized path $\mathbf{x}(s)$ has **unit speed**, regardless of the speed of the original path $\mathbf{x}(t)$. (This result makes good geometric sense, too. If arclength, rather than time, is the parameter, then speed is measured in units of "length per length," which necessarily must be one.)

The only unfortunate note to our story is that the integral in formula (4) is usually impossible to compute exactly, thus making it impossible to compute s as a simple function of t. (The case of the helix is a convenient and rather special exception.) One generally prefers to work indirectly, letting the chain rule come to the rescue. We shall see this indirect approach next.

The Unit Tangent Vector and Curvature

Let $\mathbf{x}\colon I \subseteq \mathbf{R} \to \mathbf{R}^3$ be a C^3 path and assume that $\mathbf{x}'$ is never zero.

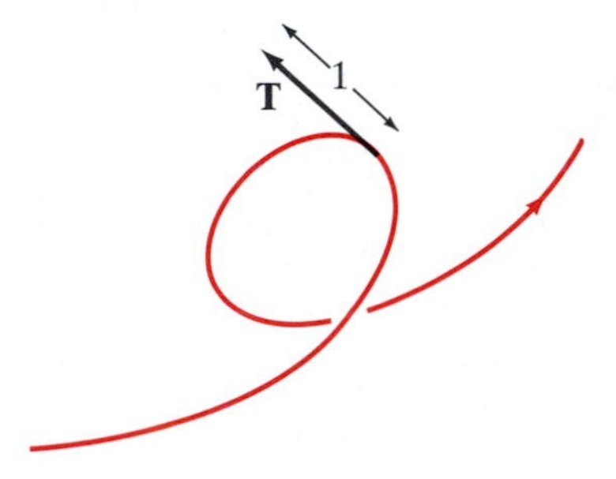

Figure 3.19 A unit tangent vector.

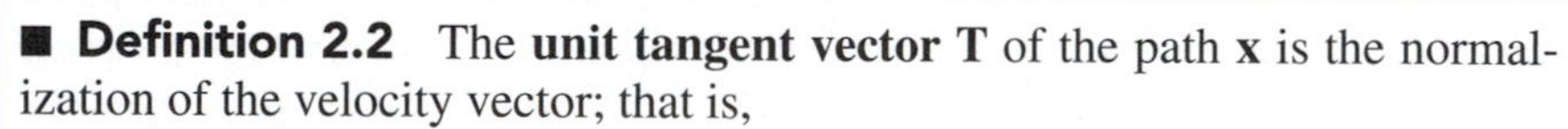

■ **Definition 2.2** The **unit tangent vector T** of the path $\mathbf{x}$ is the normalization of the velocity vector; that is,

$$\mathbf{T} = \frac{\mathbf{v}}{\|\mathbf{v}\|} = \frac{\mathbf{x}'(t)}{\|\mathbf{x}'(t)\|}.$$

We see from Definition 2.2 that the unit tangent vector is undefined when the speed of the path is zero. Also note that, from equation (6), $\mathbf{T}$ is $d\mathbf{x}/ds$, where s is the arclength parameter. Geometrically, $\mathbf{T}$ is the tangent vector of unit length that points in the direction of increasing arclength, as suggested by Figure 3.19.

EXAMPLE 5 For the helix $\mathbf{x}(t) = (a\cos t, a\sin t, bt)$, we have

$$\mathbf{T}(t) = \frac{\mathbf{x}'(t)}{\|\mathbf{x}'(t)\|} = \frac{-a\sin t\,\mathbf{i} + a\cos t\,\mathbf{j} + b\,\mathbf{k}}{\sqrt{a^2+b^2}}.$$

On the other hand, if we parametrize the helix using arclength so that

$$\mathbf{x}(s) = \left(a\cos\left(\frac{s}{\sqrt{a^2+b^2}}\right), a\sin\left(\frac{s}{\sqrt{a^2+b^2}}\right), \frac{bs}{\sqrt{a^2+b^2}}\right),$$

then

$$\mathbf{T}(s) = \mathbf{x}'(s) = \frac{-a}{\sqrt{a^2+b^2}}\sin\left(\frac{s}{\sqrt{a^2+b^2}}\right)\mathbf{i} + \frac{a}{\sqrt{a^2+b^2}}\cos\left(\frac{s}{\sqrt{a^2+b^2}}\right)\mathbf{j} + \frac{b}{\sqrt{a^2+b^2}}\mathbf{k}.$$

This agrees (as it should) with the first expression for $\mathbf{T}$, since $s = \sqrt{a^2+b^2}\,t$, as shown in Example 3. ◆

Using the unit tangent vector, we can define a quantity that measures how much a path bends as we travel along it. To do so, note the following key facts:

■ **Proposition 2.3** Assume that the path $\mathbf{x}$ always has nonzero speed. Then

1. $d\mathbf{T}/dt$ is perpendicular to $\mathbf{T}$ for all t in I (the domain of the path $\mathbf{x}$).

2. $\|d\mathbf{T}/dt\|\,|_{t=t_0}$ equals the angular rate of change (as t increases) of the direction of $\mathbf{T}$ when $t = t_0$.

Proof (You can omit reading this proof for the moment if you are interested in the main flow of ideas.) To prove part 1, we have

$$\mathbf{T}(t) \cdot \mathbf{T}(t) = 1,$$

since $\mathbf{T}$ is a unit vector. Hence,

$$\frac{d}{dt}(\mathbf{T} \cdot \mathbf{T}) = 0,$$

because the derivative of a constant is zero. Also, we have

$$\frac{d}{dt}(\mathbf{T} \cdot \mathbf{T}) = \mathbf{T} \cdot \frac{d\mathbf{T}}{dt} + \frac{d\mathbf{T}}{dt} \cdot \mathbf{T},$$

by the product rule (Proposition 1.4). Thus,

$$2\mathbf{T} \cdot \frac{d\mathbf{T}}{dt} = 0.$$

Therefore, $\mathbf{T}$ is always perpendicular to $d\mathbf{T}/dt$. (See Proposition 1.7.)

Now we prove part 2. Because $\mathbf{T}$ is a unit vector for all t, only its direction can change as t increases. This angular rate of change of $\mathbf{T}$ is precisely

$$\lim_{\Delta t \to 0^+} \frac{\Delta\theta}{\Delta t},$$

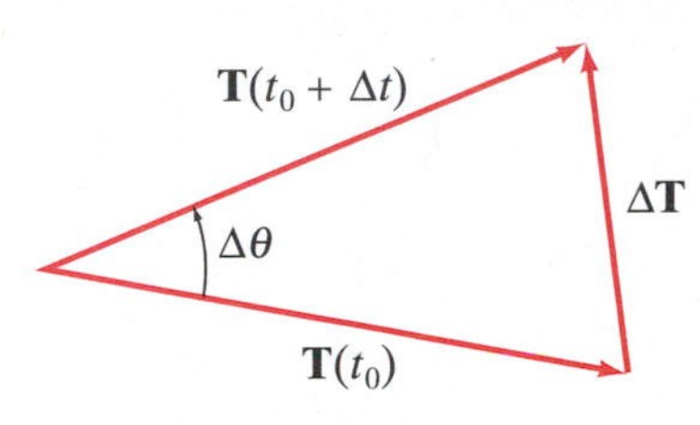

Figure 3.20 The vector triangle used in the proof of Proposition 2.3.

where $\Delta\theta$ comes from the vector triangle shown in Figure 3.20. To make the argument technically simpler, we shall assume that $\Delta\mathbf{T} \neq \mathbf{0}$. We claim that

$$\lim_{\Delta t \to 0^+} \frac{\Delta\theta}{\|\Delta\mathbf{T}\|} = 1. \tag{7}$$

Then, from equation (7),

$$\begin{aligned}
\lim_{\Delta t \to 0^+} \frac{\Delta\theta}{\Delta t} &= \lim_{\Delta t \to 0^+} \frac{\Delta\theta}{\|\Delta\mathbf{T}\|} \frac{\|\Delta\mathbf{T}\|}{\Delta t} = \lim_{\Delta t \to 0^+} \frac{\Delta\theta}{\|\Delta\mathbf{T}\|} \lim_{\Delta t \to 0^+} \frac{\|\Delta\mathbf{T}\|}{\Delta t} \\
&= 1 \cdot \lim_{\Delta t \to 0^+} \frac{\|\Delta\mathbf{T}\|}{\Delta t}.
\end{aligned}$$

Since Δt is assumed to be positive in the limit, we may conclude that

$$\lim_{\Delta t \to 0^+} \frac{\Delta\theta}{\Delta t} = \lim_{\Delta t \to 0^+} \left\| \frac{\Delta\mathbf{T}}{\Delta t} \right\| = \left\| \frac{d\mathbf{T}}{dt} \right\|,$$

as desired.

To establish equation (7), the law of cosines applied to the vector triangle in Figure 3.20 implies

$$\begin{aligned}
\|\Delta\mathbf{T}\|^2 &= \|\mathbf{T}(t + \Delta t)\|^2 + \|\mathbf{T}(t)\|^2 - 2\|\mathbf{T}(t + \Delta t)\|\,\|\mathbf{T}(t)\| \cos\Delta\theta \\
&= 2 - 2\cos\Delta\theta,
\end{aligned}$$

because $\mathbf{T}$ is always a unit vector. Thus,

$$\begin{aligned}
\lim_{\Delta t \to 0^+} \frac{\Delta\theta}{\|\Delta\mathbf{T}\|} &= \lim_{\Delta t \to 0^+} \frac{\Delta\theta}{\sqrt{2 - 2\cos\Delta\theta}} \\
&= \lim_{\Delta t \to 0^+} \frac{\Delta\theta}{\sqrt{2 \cdot 2(\sin^2(\Delta\theta/2))}}
\end{aligned}$$

from the half-angle formula, and so

$$\lim_{\Delta t \to 0^+} \frac{\Delta\theta}{\|\Delta \mathbf{T}\|} = \lim_{\Delta t \to 0^+} \frac{\Delta\theta/2}{\sin(\Delta\theta/2)} = 1,$$

from the well-known trigonometric limit (or from L'Hôpital's rule). ■

Part 2 of Proposition 2.3 provides a precise way of measuring the bending of a path.

■ **Definition 2.4** The **curvature** κ of a path $\mathbf{x}$ in $\mathbf{R}^3$ is the angular rate of change of the direction of $\mathbf{T}$ per unit change in distance along the path.

The reason for taking the rate of change of $\mathbf{T}$ *per unit change in distance* in the definition of κ is so that the curvature is an intrinsic quantity (which we certainly want it to be). Figure 3.21 should help you develop some intuition about κ.

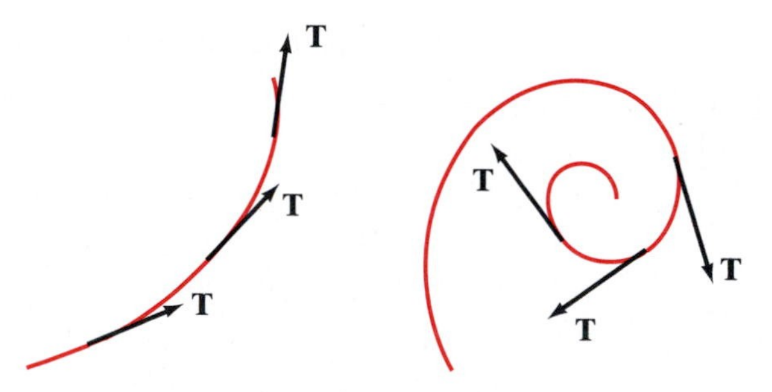

Figure 3.21 In the left figure, κ is not large, since the path's unit tangent vector turns only a small amount per unit change in distance along the path. In the right figure, κ is much larger, because $\mathbf{T}$ turns a great deal relative to distance traveled.

Because $\|d\mathbf{T}/dt\|$ measures the angular rate of change of the direction of $\mathbf{T}$ per unit change in parameter (by part 2 of Proposition 2.3) and ds/dt is the rate of change of distance per unit change in parameter, we see that

$$\kappa(t) = \frac{\|d\mathbf{T}/dt\|}{ds/dt} = \left\| \frac{d\mathbf{T}}{ds} \right\|, \tag{8}$$

where the last equality holds by the chain rule. It is formula (8) that we will use when making calculations.

EXAMPLE 6 For the circle $\mathbf{x}(t) = (a\cos t, a\sin t)$, $0 \le t < 2\pi$,

$$\mathbf{x}'(t) = -a\sin t\,\mathbf{i} + a\cos t\,\mathbf{j}, \qquad \|\mathbf{x}'(t)\| = \frac{ds}{dt} = a,$$

so that

$$\mathbf{T}(t) = \frac{\mathbf{x}'(t)}{\|\mathbf{x}'(t)\|} = -\sin t\,\mathbf{i} + \cos t\,\mathbf{j}.$$

Hence,

$$\kappa = \frac{\|d\mathbf{T}/dt\|}{ds/dt} = \frac{1}{a}\|-\cos t\,\mathbf{i} - \sin t\,\mathbf{j}\| = \frac{1}{a}.$$

Thus, we see that the curvature of a circle is always constant with value equal to the reciprocal of the radius. Therefore, the smaller the circle, the greater the curvature. (Draw a sketch to convince yourself.) ◆

EXAMPLE 7 If $\mathbf{a}$ and $\mathbf{b}$ are constant vectors in $\mathbf{R}^3$ and $\mathbf{a} \neq \mathbf{0}$, the path

$$\mathbf{x}(t) = \mathbf{a}\,t + \mathbf{b}$$

traces a line. We have

$$\mathbf{x}'(t) = \mathbf{a},$$

so

$$\frac{ds}{dt} = \|\mathbf{a}\|.$$

Hence,

$$\mathbf{T}(t) = \frac{\mathbf{a}}{\|\mathbf{a}\|},$$

which is a constant vector. Thus, $\mathbf{T}'(t) \equiv \mathbf{0}$ and formula (8) implies immediately that κ is zero, which agrees with the intuitive fact that a line doesn't curve. ◆

EXAMPLE 8 Returning to our friend the helix

$$\mathbf{x}(t) = (a\cos t, a\sin t, bt),$$

we have already seen that

$$\frac{ds}{dt} = \sqrt{a^2+b^2} \quad \text{and} \quad \mathbf{T}(t) = \frac{-a\sin t\,\mathbf{i} + a\cos t\,\mathbf{j} + b\,\mathbf{k}}{\sqrt{a^2+b^2}}.$$

Thus, formula (8) gives

$$\kappa = \frac{1}{\sqrt{a^2+b^2}}\left\|\frac{-a\cos t\,\mathbf{i} - a\sin t\,\mathbf{j}}{\sqrt{a^2+b^2}}\right\| = \frac{a}{a^2+b^2}.$$

We see that the curvature of the helix is constant, just like the circle. In fact, as b approaches zero, the helix degenerates to a circle, and the resulting curvature is consistent with that of Example 6.

We can also compute the curvature from the parametrization given by arclength. The same helix is also described by

$$\mathbf{x}(s) = \left(a\cos\left(\frac{s}{\sqrt{a^2+b^2}}\right), a\sin\left(\frac{s}{\sqrt{a^2+b^2}}\right), \frac{bs}{\sqrt{a^2+b^2}}\right),$$

and we have

$$\mathbf{T}(s) = \frac{d\mathbf{x}}{ds} = -\frac{a}{\sqrt{a^2+b^2}}\sin\left(\frac{s}{\sqrt{a^2+b^2}}\right)\mathbf{i} + \frac{a}{\sqrt{a^2+b^2}}\cos\left(\frac{s}{\sqrt{a^2+b^2}}\right)\mathbf{j} + \frac{b}{\sqrt{a^2+b^2}}\mathbf{k}.$$

We can, therefore, compute

$$\frac{d\mathbf{T}}{ds} = -\frac{a}{a^2+b^2}\cos\left(\frac{s}{\sqrt{a^2+b^2}}\right)\mathbf{i} - \frac{a}{a^2+b^2}\sin\left(\frac{s}{\sqrt{a^2+b^2}}\right)\mathbf{j},$$

and hence, from formula (8), that

$$\kappa = \left\|\frac{d\mathbf{T}}{ds}\right\| = \frac{a}{a^2+b^2},$$

which checks. ◆

The Moving Frame and Torsion

We now introduce a triple of mutually orthogonal unit vectors that "travel" with a given path $\mathbf{x}: I \to \mathbf{R}^3$, known as the **moving frame** of the path. (Note: In general, the term "frame" means an ordered collection of mutually orthogonal unit vectors in $\mathbf{R}^n$.) These vectors should be thought of as a set of special vector "coordinate axes" that move from point to point along the path.

To begin, assume that (i) $\mathbf{x}'(t) \neq \mathbf{0}$ and (ii) $\mathbf{x}'(t) \times \mathbf{x}''(t) \neq \mathbf{0}$ for all t in I. (The first condition assures us that $\mathbf{x}$ never has zero speed and the second that $\mathbf{x}$ is not a straight-line path.) Then the first vector of the moving frame is just the unit tangent vector:

$$\mathbf{T} = \frac{d\mathbf{x}}{ds} = \frac{\mathbf{x}'(t)}{\|\mathbf{x}'(t)\|}.$$

(Now you see why condition (i) is needed.) For a second vector orthogonal to $\mathbf{T}$, recall that part 1 of Proposition 2.3 says that $d\mathbf{T}/dt$ must be perpendicular to $\mathbf{T}$. Hence, we define

$$\mathbf{N} = \frac{d\mathbf{T}/dt}{\|d\mathbf{T}/dt\|}. \tag{9}$$

(That $d\mathbf{T}/dt$ is not zero follows from assumptions (i) and (ii).) The vector $\mathbf{N}$ is called the **principal normal vector** of $\mathbf{x}$. By the chain rule, $\mathbf{N}$ is also given by

$$\mathbf{N} = \frac{d\mathbf{T}/ds}{\|d\mathbf{T}/ds\|}. \tag{10}$$

Since $\kappa = \|d\mathbf{T}/ds\|$ by formula (8), we also see that

$$\frac{d\mathbf{T}}{ds} = \kappa\mathbf{N}. \tag{11}$$

At a given point P along the path, the vectors $\mathbf{T}$ and $\mathbf{N}$ (and also the vectors $\mathbf{x}'$ and $\mathbf{x}''$) determine what is called the **osculating plane** of the path at P. (See Figure 3.22.) This is the plane that "instantaneously" contains the path at P. (More precisely, it is the plane obtained by taking points P_1 and P_2 on the path near P and finding the limiting position of the plane through P, P_1, and P_2 as P_1 and P_2 approach P along $\mathbf{x}$. The word "osculating" derives from the Latin *osculare,* meaning "to kiss.")

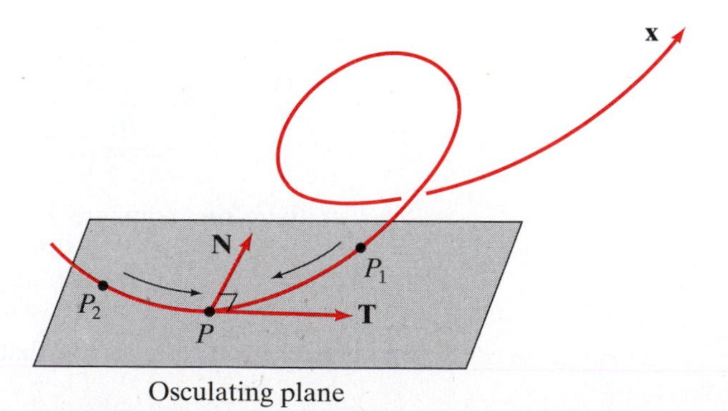

Figure 3.22 The osculating plane of the path $\mathbf{x}$ at the point P.

Now that we have defined two orthogonal unit vectors $\mathbf{T}$ and $\mathbf{N}$, we can produce a third unit vector perpendicular to both:

$$\mathbf{B} = \mathbf{T} \times \mathbf{N}. \tag{12}$$

The vector $\mathbf{B}$, called the **binormal vector**, is defined so that the ordered triple $(\mathbf{T}, \mathbf{N}, \mathbf{B})$ is a right-handed system. Thus, $\mathbf{B}$ is a unit vector since

$$\|\mathbf{B}\| = \|\mathbf{T}\|\,\|\mathbf{N}\| \sin\frac{\pi}{2} = 1 \cdot 1 \cdot 1 = 1.$$

EXAMPLE 9 For the helix $\mathbf{x}(t) = (a\cos t, a\sin t, bt)$, the moving frame vectors are

$$\mathbf{T}(t) = \frac{-a\sin t\,\mathbf{i} + a\cos t\,\mathbf{j} + b\,\mathbf{k}}{\sqrt{a^2+b^2}}$$

(as we have already seen),

$$\mathbf{N}(t) = \frac{\mathbf{T}'(t)}{\|\mathbf{T}'(t)\|} = \frac{(-a\cos t\,\mathbf{i} - a\sin t\,\mathbf{j})/\sqrt{a^2+b^2}}{a/\sqrt{a^2+b^2}} = -\cos t\,\mathbf{i} - \sin t\,\mathbf{j},$$

and

$$\mathbf{B}(t) = \mathbf{T} \times \mathbf{N} = \begin{vmatrix} \mathbf{i} & \mathbf{j} & \mathbf{k} \\ -a\sin t/\sqrt{a^2+b^2} & a\cos t/\sqrt{a^2+b^2} & b/\sqrt{a^2+b^2} \\ -\cos t & -\sin t & 0 \end{vmatrix}$$

$$= \left(\frac{b}{\sqrt{a^2+b^2}}\sin t\right)\mathbf{i} - \left(\frac{b}{\sqrt{a^2+b^2}}\cos t\right)\mathbf{j} + \left(\frac{a}{\sqrt{a^2+b^2}}\right)\mathbf{k}. \qquad ◆$$

Equation (11) says that the derivative of $\mathbf{T}$ (with respect to arclength) is a scalar function (namely, the curvature) multiple of the principal normal $\mathbf{N}$. This is not surprising, since $\mathbf{N}$ is defined to be parallel to the derivative of $\mathbf{T}$. A more remarkable result (see the addendum at the end of this section) is that the derivative of the binormal vector is also always parallel to the principal normal; that is,

$$\frac{d\mathbf{B}}{ds} = (\text{scalar function})\,\mathbf{N}.$$

The standard convention is to write this scalar function with a negative sign, so we have

$$\frac{d\mathbf{B}}{ds} = -\tau\mathbf{N}. \tag{13}$$

The scalar function τ thus defined is called the **torsion** of the path $\mathbf{x}$. Roughly speaking, the torsion measures how much the path twists out of the plane, how "three-dimensional" $\mathbf{x}$ is. Note that, according to our conventions, the curvature κ is always nonnegative (why?), while τ can be positive, negative, or zero.

EXAMPLE 10 Consider again the case of circular motion. Thus, let $\mathbf{x}(t) = (a\cos t, a\sin t)$. Then, as shown in Example 6,

$$\mathbf{T}(t) = \frac{\mathbf{x}'(t)}{\|\mathbf{x}'(t)\|} = -\sin t\,\mathbf{i} + \cos t\,\mathbf{j}, \quad \text{and} \quad \kappa = \left\|\frac{d\mathbf{T}}{ds}\right\| = \frac{1}{a}.$$

Now we calculate

$$\mathbf{N} = \frac{\mathbf{T}'(t)}{\|\mathbf{T}'(t)\|} = -\cos t\,\mathbf{i} - \sin t\,\mathbf{j},$$
$$\mathbf{B} = \mathbf{T}\times\mathbf{N} = \mathbf{k}, \quad \text{a constant vector.}$$

Hence, $d\mathbf{B}/ds \equiv \mathbf{0}$, so there is no torsion. This makes sense, since a circle does not twist out of the plane. ◆

EXAMPLE 11 Let $\mathbf{x}(t) = (e^t\cos t, e^t\sin t, e^t)$. We calculate $\mathbf{T}$, $\mathbf{N}$, and $\mathbf{B}$ and identify the curvature and torsion of $\mathbf{x}$.

To begin, we have

$$\mathbf{T}(t) = \frac{\mathbf{x}'(t)}{\|\mathbf{x}'(t)\|} = \frac{e^t(\cos t - \sin t)\,\mathbf{i} + e^t(\cos t + \sin t)\,\mathbf{j} + e^t\,\mathbf{k}}{\sqrt{3}\,e^t}$$
$$= \frac{1}{\sqrt{3}}\left((\cos t - \sin t)\,\mathbf{i} + (\cos t + \sin t)\,\mathbf{j} + \mathbf{k}\right).$$

From this, we may compute

$$\frac{d\mathbf{T}}{ds} = \frac{d\mathbf{T}/dt}{ds/dt} = \frac{\frac{1}{\sqrt{3}}(-(\sin t + \cos t)\,\mathbf{i} + (\cos t - \sin t)\,\mathbf{j})}{\sqrt{3}\,e^t}$$
$$= \frac{e^{-t}}{3}(-(\sin t + \cos t)\,\mathbf{i} + (\cos t - \sin t)\mathbf{j}),$$

so that the curvature is

$$\kappa = \left\|\frac{d\mathbf{T}}{ds}\right\| = \frac{\sqrt{2}\,e^{-t}}{3}.$$

Now we determine the remainder of the moving frame:

$$\mathbf{N} = \frac{\mathbf{T}'(t)}{\|\mathbf{T}'(t)\|} = \frac{1}{\sqrt{2}}(-(\sin t + \cos t)\,\mathbf{i} + (\cos t - \sin t)\,\mathbf{j}),$$
$$\mathbf{B} = \mathbf{T}\times\mathbf{N} = \frac{1}{\sqrt{6}}((\sin t - \cos t)\,\mathbf{i} - (\sin t + \cos t)\,\mathbf{j} + 2\mathbf{k}).$$

Finally, to find the torsion, we calculate

$$\frac{d\mathbf{B}}{ds} = \frac{d\mathbf{B}/dt}{ds/dt} = \frac{\frac{1}{\sqrt{6}}((\cos t + \sin t)\,\mathbf{i} + (\sin t - \cos t)\,\mathbf{j})}{\sqrt{3}\,e^t}$$
$$= \frac{e^{-t}}{3\sqrt{2}}((\cos t + \sin t)\,\mathbf{i} + (\sin t - \cos t)\,\mathbf{j})$$
$$= -\frac{e^{-t}}{3}\mathbf{N},$$

so

$$\tau = \frac{e^{-t}}{3}.$$ ◆

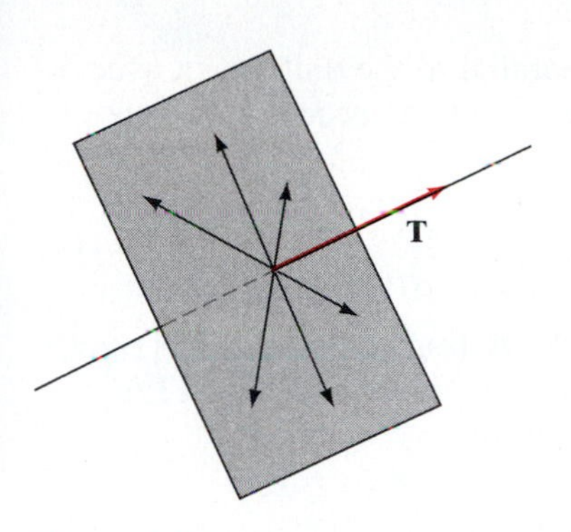

Figure 3.23 Any vector in the plane perpendicular to $\mathbf{T}$ can be used for $\mathbf{N}$.

EXAMPLE 12 If $\mathbf{a}$ and $\mathbf{b}$ are vectors in $\mathbf{R}^3$, then the straight-line path $\mathbf{x}(t) = \mathbf{a}\,t+\mathbf{b}$ has, as we saw in Example 7, $\mathbf{T} = \mathbf{a}/\|\mathbf{a}\|$. Thus, both $d\mathbf{T}/dt$ and $d\mathbf{T}/ds$ are identically zero. Hence, $\kappa \equiv 0$ (as shown in Example 7) and $\mathbf{N}$ cannot be defined using formula (9). From geometric considerations, any unit vector perpendicular to $\mathbf{T}$ can, in principle, be used for $\mathbf{N}$. (See Figure 3.23.) If we choose one such vector, then $\mathbf{B}$ can be calculated from formula (12). Since $\mathbf{T}$, $\mathbf{N}$, and $\mathbf{B}$ are all constant, τ must be zero. This is an example of a moving frame that is *not* uniquely determined by the path $\mathbf{x}$ and serves to illustrate why the assumption $\mathbf{x}' \times \mathbf{x}'' \neq \mathbf{0}$ was made. ◆

It is important to realize that the moving frame, curvature, and torsion are quantities that are *intrinsic* to the curve traced by the path. That is, any parametrized path that traces the same curve (in the same direction) must necessarily have the same $\mathbf{T}$, $\mathbf{N}$, $\mathbf{B}$ vector functions and the same curvature and torsion. This is because all of these quantities can be defined entirely in terms of the intrinsic arclength parameter s. (See Definition 2.2 and formulas (6), (8), (10), (11), (12), and (13).)

Another important fact is that the curvature function κ and the torsion function τ together determine all the geometric information regarding the shape of the curve, except for the curve's particular position in space. To be more precise, we have the following theorem, whose proof we omit:

■ **Theorem 2.5** Let s be the arclength parameter and suppose C_1 and C_2 are two curves of class C^3 in $\mathbf{R}^3$. Assume that the corresponding curvature functions κ_1 and κ_2 are strictly positive. Then if $\kappa_1(s) \equiv \kappa_2(s)$ and $\tau_1(s) \equiv \tau_2(s)$, the two curves must be congruent (in the sense of high school geometry). In fact, given any two continuous functions κ and τ, where $\kappa(s) > 0$ for all s in the closed interval $[0, L]$, there is a unique curve parametrized by arclength on $[0, L]$ (up to position in space) whose curvature and torsion are κ and τ, respectively.

Tangential and Normal Components of Velocity and Acceleration; Other Curvature Formulas

As we have seen, the moving frame provides us with an intrinsic set of vectors, like coordinate axes, that are special to the particular curve traced by a path. In contrast, the velocity and acceleration vectors of a path are definitely *not* intrinsic quantities, but depend on the particular parametrization chosen as well as on the shape of the path. (The speed of a path is entirely independent of the geometry of the curve traced.) We can get some feeling for the relationship between the intrinsic notion of the moving frame and the extrinsic quantities of velocity and acceleration by expressing the latter two vector functions in terms of the moving frame vectors.

Thus, we begin with a C^2 path $\mathbf{x}\colon I \to \mathbf{R}^3$ having $\mathbf{x}' \neq \mathbf{0}$ and $\mathbf{x}' \times \mathbf{x}'' \neq \mathbf{0}$. For notational convenience, let $\dot{s}$ denote ds/dt and $\ddot{s}$ denote d^2s/dt^2. From Definition 2.2, we know that $\mathbf{T} = \mathbf{v}/\|\mathbf{v}\|$ and so, since the speed $\dot{s} = ds/dt = \|\mathbf{v}\|$, we have

$$\mathbf{v}(t) = \dot{s}\mathbf{T}. \tag{14}$$

This formula says that the velocity is always parallel to the unit tangent vector, something we know well. To obtain a similar result for acceleration, we can differentiate (14) and apply the product rule:

$$\mathbf{a}(t) = \mathbf{v}'(t) = \frac{d}{dt}(\dot{s}\mathbf{T}) = \ddot{s}\mathbf{T} + \dot{s}\frac{d\mathbf{T}}{dt}. \tag{15}$$

Next, we express $d\mathbf{T}/dt$ in terms of the $\mathbf{T}$, $\mathbf{N}$, $\mathbf{B}$ frame. Formula (11) gives the derivative of $d\mathbf{T}/ds$ in terms of $\mathbf{N}$. The chain rule says that $d\mathbf{T}/ds = (d\mathbf{T}/dt)/(ds/dt)$. Thus, from formula (11), we have

$$\frac{d\mathbf{T}}{dt} = \dot{s}\frac{d\mathbf{T}}{ds} = \dot{s}\kappa\mathbf{N}.$$

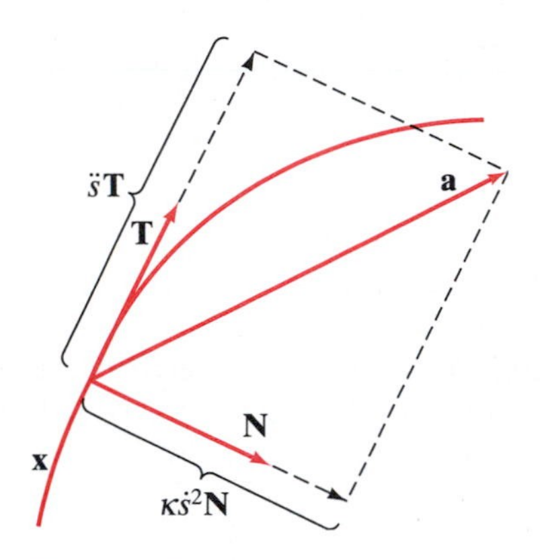

Figure 3.24 Decomposition of acceleration **a** into tangential and normal components.

Hence, we may rewrite equation (15) as

$$\mathbf{a}(t) = \ddot{s}\mathbf{T} + \kappa\dot{s}^2\mathbf{N}. \tag{16}$$

WARNING $\ddot{s} = d^2s/dt^2$ is the derivative of the speed, which is a scalar function. The acceleration **a** is the derivative of *velocity* and so is a vector function.

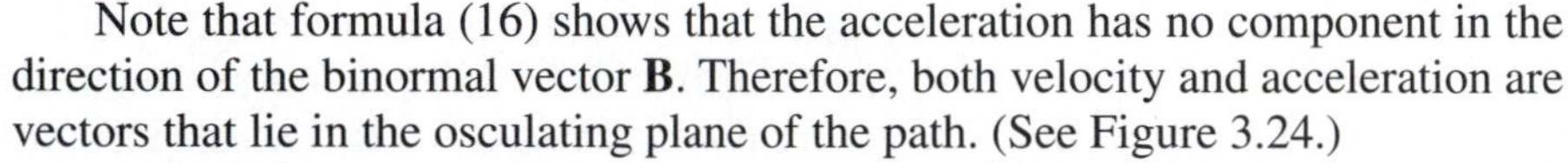

Note that formula (16) shows that the acceleration has no component in the direction of the binormal vector **B**. Therefore, both velocity and acceleration are vectors that lie in the osculating plane of the path. (See Figure 3.24.)

At first glance, it may not appear to be especially easy to use formula (16) to resolve acceleration into its tangential and normal components because of the curvature term. However,

$$\|\mathbf{a}\|^2 = \mathbf{a}\cdot\mathbf{a} = (\ddot{s}\mathbf{T} + \kappa\dot{s}^2\mathbf{N})\cdot(\ddot{s}\mathbf{T} + \kappa\dot{s}^2\mathbf{N}) = \ddot{s}^2 + (\kappa\dot{s}^2)^2,$$

since **T** and **N** are perpendicular vectors. Consequently, we may calculate the components as follows:

Tangential component of acceleration $= a_{\text{tang}} = \ddot{s}$.

Normal component of acceleration $= a_{\text{norm}} = \kappa\dot{s}^2 = \sqrt{\|\mathbf{a}\|^2 - a_{\text{tang}}^2}$.

EXAMPLE 13 Let $\mathbf{x}(t) = (t, 2t, t^2)$. Then $\mathbf{v}(t) = \mathbf{i} + 2\mathbf{j} + 2t\mathbf{k}$ and $\mathbf{a}(t) = 2\mathbf{k}$. We have $\dot{s} = \|\mathbf{v}(t)\| = \sqrt{5 + 4t^2}$. Therefore,

$$a_{\text{tang}} = \ddot{s} = \frac{4t}{\sqrt{5 + 4t^2}}.$$

Since $\|\mathbf{a}\| = 2$, we see that

$$a_{\text{norm}} = \sqrt{\|\mathbf{a}\|^2 - a_{\text{tang}}^2} = \sqrt{4 - \frac{16t^2}{5 + 4t^2}} = \frac{2\sqrt{5}}{\sqrt{5 + 4t^2}}.$$ ◆

Formulas (14) and (16) enable us to find an alternative equation for the curvature of the path. We simply calculate that

$$\mathbf{v}\times\mathbf{a} = (\dot{s}\mathbf{T}) \times (\ddot{s}\mathbf{T} + \kappa\dot{s}^2\mathbf{N}) = \dot{s}\ddot{s}(\mathbf{T}\times\mathbf{T}) + \kappa\dot{s}^3(\mathbf{T}\times\mathbf{N}) = \kappa\dot{s}^3\mathbf{B}.$$

Recalling that $\dot{s} = \|\mathbf{v}\|$, we have, by taking magnitudes,

$$\|\mathbf{v} \times \mathbf{a}\| = \kappa \|\mathbf{v}\|^3 \|\mathbf{B}\| = \kappa \|\mathbf{v}\|^3,$$

since $\mathbf{B}$ is a unit vector. Thus,

$$\kappa = \frac{\|\mathbf{v} \times \mathbf{a}\|}{\|\mathbf{v}\|^3}. \tag{17}$$

This relatively simple formula expresses the curvature (an intrinsic quantity) in terms of the nonintrinsic quantities of velocity and acceleration.

EXAMPLE 14 For the path $\mathbf{x}(t) = (2t^3 + 1, t^4, t^5)$, we have

$$\mathbf{v}(t) = 6t^2\mathbf{i} + 4t^3\mathbf{j} + 5t^4\mathbf{k}$$

and

$$\mathbf{a}(t) = 12t\mathbf{i} + 12t^2\mathbf{j} + 20t^3\mathbf{k}.$$

You can check that

$$\|\mathbf{v}\| = t^2\sqrt{25t^4 + 16t^2 + 36}$$

and

$$\|\mathbf{v} \times \mathbf{a}\| = \left\|4t^4(5t^2\mathbf{i} - 15t\mathbf{j} + 6\mathbf{k})\right\| = 4t^4\sqrt{25t^4 + 225t^2 + 36}.$$

Therefore, formula (17) yields

$$\kappa = \frac{\|\mathbf{v} \times \mathbf{a}\|}{\|\mathbf{v}\|^3} = \frac{4(25t^4 + 225t^2 + 36)^{1/2}}{t^2(25t^4 + 16t^2 + 36)^{3/2}},$$

which is certainly a more convenient way to determine curvature in this case. ◆

Summary

You have seen many formulas in this section, and, at first, it may seem difficult to sort out the primary statements from the secondary results. We list the more fundamental facts here:

For a path $\mathbf{x}\colon I \to \mathbf{R}^3$:

Nonintrinsic quantities:

$$\text{Velocity } \mathbf{v}(t) = \mathbf{x}'(t).$$

$$\text{Speed } \frac{ds}{dt} = \|\mathbf{v}(t)\|.$$

$$\text{Acceleration } \mathbf{a}(t) = \mathbf{x}''(t).$$

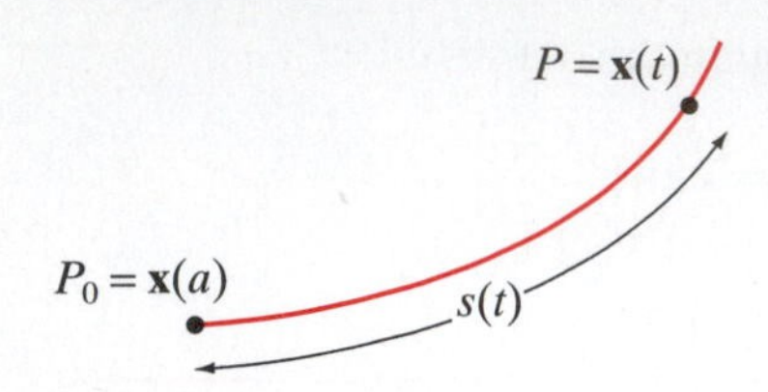

Figure 3.25 The arclength function.

Arclength function: (See Figure 3.25.)

$$s(t) = \int_a^t \|\mathbf{x}'(\tau)\|\, d\tau \quad (\text{basepoint is } P_0 = \mathbf{x}(a))$$

Intrinsic quantities:

The moving frame:

$$\text{Unit tangent vector } \mathbf{T} = \frac{d\mathbf{x}}{ds} = \frac{\mathbf{x}'(t)}{\|\mathbf{x}'(t)\|}.$$

$$\text{Principal normal vector } \mathbf{N} = \frac{d\mathbf{T}/ds}{\|d\mathbf{T}/ds\|} = \frac{d\mathbf{T}/dt}{\|d\mathbf{T}/dt\|}.$$

$$\text{Binormal vector } \mathbf{B} = \mathbf{T} \times \mathbf{N}.$$

$$\text{Curvature } \kappa = \left\| \frac{d\mathbf{T}}{ds} \right\| = \frac{\|d\mathbf{T}/dt\|}{ds/dt}.$$

$$\text{Torsion } \tau \text{ is defined so that } \frac{d\mathbf{B}}{ds} = -\tau \mathbf{N}.$$

Additional formulas:

$$\mathbf{v}(t) = \dot{s}\,\mathbf{T} \quad (\dot{s} \text{ is speed}).$$

$$\mathbf{a}(t) = \ddot{s}\,\mathbf{T} + \kappa \dot{s}^2\,\mathbf{N} \quad (\ddot{s} \text{ is derivative of speed}).$$

$$\kappa = \frac{\|\mathbf{v} \times \mathbf{a}\|}{\|\mathbf{v}\|^3}.$$

Addendum: More About Torsion and the Frenet–Serret Formulas

We now derive formula (13), the basis for the definition of the torsion of a curve. That is, we show that the derivative of the binormal vector $\mathbf{B}$ (with respect to arclength) is always parallel to the principal normal $\mathbf{N}$ (i.e., that $d\mathbf{B}/ds$ is a scalar function times $\mathbf{N}$). The two main ingredients in our derivation are part 1 of Proposition 2.3 and the product rule.

We begin by noting that, since the ordered triple of vectors $(\mathbf{T}, \mathbf{N}, \mathbf{B})$ forms a frame for $\mathbf{R}^3$, any moving vector, including $d\mathbf{B}/ds$, can be expressed as a **linear combination** of these vectors; that is, we must have

$$\frac{d\mathbf{B}}{ds} = a(s)\mathbf{T} + b(s)\mathbf{N} + c(s)\mathbf{B}, \tag{18}$$

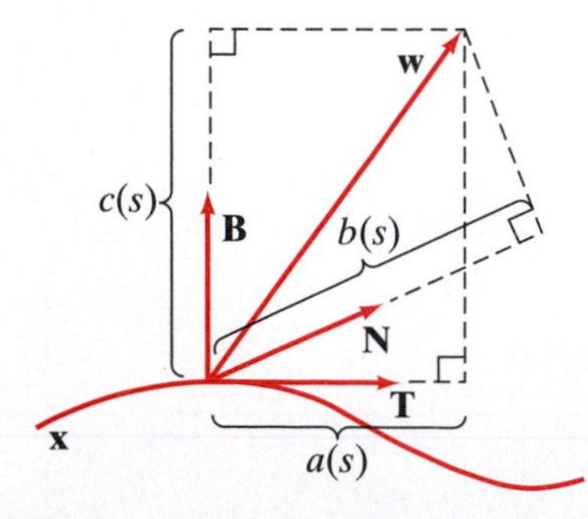

Figure 3.26 $\mathbf{w}(s) = a\mathbf{T} + b\mathbf{N} + c\mathbf{B}$.

where a, b, and c are appropriate scalar-valued functions. (Because $\mathbf{T}$, $\mathbf{N}$, and $\mathbf{B}$ are mutually perpendicular unit vectors, any (moving) vector $\mathbf{w}$ in $\mathbf{R}^3$ can be decomposed into its components with respect to $\mathbf{T}$, $\mathbf{N}$, and $\mathbf{B}$ in much the same way that it can be decomposed into $\mathbf{i}$, $\mathbf{j}$, and $\mathbf{k}$ components—see Figure 3.26.) To find the particular values of the component functions a, b, and c, it turns out that we can solve for each function by applying appropriate dot products to equation (18). Specifically,

$$\begin{aligned}\frac{d\mathbf{B}}{ds}\cdot\mathbf{T} &= a(s)\mathbf{T}\cdot\mathbf{T} + b(s)\mathbf{N}\cdot\mathbf{T} + c(s)\mathbf{B}\cdot\mathbf{T}\\ &= a(s)\cdot 1 + b(s)\cdot 0 + c(s)\cdot 0\\ &= a(s),\end{aligned}$$

and, similarly,

$$\frac{d\mathbf{B}}{ds}\cdot\mathbf{N} = b(s), \qquad \frac{d\mathbf{B}}{ds}\cdot\mathbf{B} = c(s).$$

From Proposition 1.7, $d\mathbf{B}/ds$ is perpendicular to $\mathbf{B}$, and hence, c must be zero. To find a, we use an ingenious trick with the product rule: Because $\mathbf{T}\cdot\mathbf{B} = 0$, it follows that $d/ds(\mathbf{T}\cdot\mathbf{B}) = 0$. Now, by the product rule,

$$\frac{d}{ds}(\mathbf{T}\cdot\mathbf{B}) = \mathbf{T}\cdot\frac{d\mathbf{B}}{ds} + \frac{d\mathbf{T}}{ds}\cdot\mathbf{B}.$$

Consequently, $(d\mathbf{B}/ds)\cdot\mathbf{T} = -(d\mathbf{T}/ds)\cdot\mathbf{B}$. Thus,

$$\begin{aligned}a(s) = \frac{d\mathbf{B}}{ds}\cdot\mathbf{T} &= -\frac{d\mathbf{T}}{ds}\cdot\mathbf{B}\\ &= -\kappa\mathbf{N}\cdot\mathbf{B} \qquad \text{by formula (11),}\\ &= 0,\end{aligned}$$

and equation (18) reduces to

$$\frac{d\mathbf{B}}{ds} = b(s)\mathbf{N}.$$

No further reductions are possible, and we have proved that the derivative of $\mathbf{B}$ is parallel to $\mathbf{N}$. The torsion τ can, therefore, be defined by $\tau(s) = -b(s)$.

Formulas (11) and (13) gave us intrinsic expressions for $d\mathbf{T}/ds$ and $d\mathbf{B}/ds$, respectively. We can complete the set by finding an expression for $d\mathbf{N}/ds$. The method is the same as the one just used. Begin by writing

$$\frac{d\mathbf{N}}{ds} = a(s)\mathbf{T} + b(s)\mathbf{N} + c(s)\mathbf{B}, \tag{19}$$

where a, b, and c are suitable scalar functions. Taking the dot product of equation (19) with, in turn, $\mathbf{T}$, $\mathbf{N}$, and $\mathbf{B}$, yields the following:

$$a(s) = \frac{d\mathbf{N}}{ds}\cdot\mathbf{T}, \qquad b(s) = \frac{d\mathbf{N}}{ds}\cdot\mathbf{N}, \qquad c(s) = \frac{d\mathbf{N}}{ds}\cdot\mathbf{B}.$$

The "product rule trick" used here then reveals that

$$\begin{aligned}a(s) = \frac{d\mathbf{N}}{ds}\cdot\mathbf{T} &= -\mathbf{N}\cdot\frac{d\mathbf{T}}{ds}\\ &= -\mathbf{N}\cdot\kappa\mathbf{N} \qquad \text{by formula (11)}\\ &= -\kappa,\end{aligned}$$

and

$$\begin{aligned}c(s) = \frac{d\mathbf{N}}{ds}\cdot\mathbf{B} &= -\mathbf{N}\cdot\frac{d\mathbf{B}}{ds}\\ &= -\mathbf{N}\cdot(-\tau\mathbf{N}) \qquad \text{by formula (13)}\\ &= \tau.\end{aligned}$$

Moreover,

$$b(s) = \frac{d\mathbf{N}}{ds} \cdot \mathbf{N} = -\mathbf{N} \cdot \frac{d\mathbf{N}}{ds},$$

which implies that $b(s)$ is zero. Hence, equation (19) becomes

$$\frac{d\mathbf{N}}{ds} = -\kappa \mathbf{T} + \tau \mathbf{B}.$$

The formulas for $d\mathbf{T}/ds$, $d\mathbf{N}/ds$, and $d\mathbf{B}/ds$ are usually taken together as

$$\begin{cases} \mathbf{T}'(s) = \kappa \mathbf{N} \\ \mathbf{N}'(s) = -\kappa \mathbf{T} + \tau \mathbf{B} \\ \mathbf{B}'(s) = -\tau \mathbf{N} \end{cases}$$

and are known as the **Frenet–Serret formulas** for a curve in space. They are so named for Frédéric-Jean Frenet and Joseph Alfred Serret, who published them separately in 1852 and 1851, respectively. The Frenet–Serret formulas give a system of differential equations for a curve and are key to proving a result like Theorem 2.5. They are often written in matrix form, in which case, they have an especially appealing appearance, namely,

$$\begin{bmatrix} \mathbf{T}' \\ \mathbf{N}' \\ \mathbf{B}' \end{bmatrix} = \begin{bmatrix} 0 & \kappa & 0 \\ -\kappa & 0 & \tau \\ 0 & -\tau & 0 \end{bmatrix} \begin{bmatrix} \mathbf{T} \\ \mathbf{N} \\ \mathbf{B} \end{bmatrix}.$$

Exercises

Calculate the length of each of the paths given in Exercises 1–6.

1. $\mathbf{x}(t) = (2t + 1, 7 - 3t)$, $-1 \le t \le 2$

2. $\mathbf{x}(t) = t^2\,\mathbf{i} + \frac{2}{3}(2t + 1)^{3/2}\,\mathbf{j}$, $0 \le t \le 4$

3. $\mathbf{x}(t) = (\cos 3t, \sin 3t, 2t^{3/2})$, $0 \le t \le 2$

4. $\mathbf{x}(t) = 7\mathbf{i} + t\,\mathbf{j} + t^2\mathbf{k}$, $1 \le t \le 3$

5. $\mathbf{x}(t) = (\ln t, t^2/2, \sqrt{2}t)$, $1 \le t \le 4$

6. $\mathbf{x}(t) = (2t \cos t, 2t \sin t, 2\sqrt{2}t^2)$, $0 \le t \le 3$

7. The path $\mathbf{x}(t) = (a \cos^3 t, a \sin^3 t)$, where a is a positive constant, traces a curve known as an **astroid** or a **hypocycloid of four cusps.** Sketch this curve and find its total length. (Be careful when you do this.)

8. If f is a continuously differentiable function, show how Definition 2.1 may be used to establish the formula

$$L = \int_a^b \sqrt{1 + (f'(x))^2}\, dx$$

for the length of the curve $y = f(x)$ between $(a, f(a))$ and $(b, f(b))$.

9. Use Exercise 8 or Definition 2.1 (or both) to calculate the length of the line segment $y = mx + b$ between (x_0, y_0) and (x_1, y_1). Explain your result with an appropriate sketch.

10. (a) Calculate the length of the line segment determined by the path

$$\mathbf{x}(t) = (a_1 t + b_1, a_2 t + b_2)$$

as t varies from t_0 to t_1.

(b) Compare your result with that of Exercise 9.

(c) Now calculate the length of the line segment determined by the path $\mathbf{x}(t) = \mathbf{a}t + \mathbf{b}$ as t varies from t_0 to t_1.

11. This problem concerns the path $\mathbf{x} = |t - 1|\,\mathbf{i} + |t|\,\mathbf{j}$, $-2 \le t \le 2$.

(a) Sketch this path.

(b) The path fails to be of class C^1, but is piecewise C^1. Explain.

(c) Calculate the length of the path.

12. (a) Find the arclength parameter $s = s(t)$ for the path

$$\mathbf{x} = e^{at} \cos bt\,\mathbf{i} + e^{at} \sin bt\,\mathbf{j} + e^{at}\,\mathbf{k}.$$

(b) Express the original parameter t in terms of s, and, thereby, reparametrize $\mathbf{x}$ in terms of s.

Determine the moving frame $\{\mathbf{T}, \mathbf{N}, \mathbf{B}\}$, *and compute the curvature and torsion for the paths given in Exercises 13–16.*

13. $\mathbf{x}(t) = 5\cos 3t\,\mathbf{i} + 6t\,\mathbf{j} + 5\sin 3t\,\mathbf{k}$

14. $\mathbf{x}(t) = (\sin t - t\cos t)\,\mathbf{i} + (\cos t + t\sin t)\,\mathbf{j} + 2\mathbf{k},\ t \geq 0$

15. $\mathbf{x}(t) = \left(t, \frac{1}{3}(t+1)^{3/2}, \frac{1}{3}(1-t)^{3/2}\right),\ -1 < t < 1$

16. $\mathbf{x}(t) = (e^{2t}\sin t, e^{2t}\cos t, 1)$

17. (a) Use formula (17) in this section to establish the following well-known formula for the curvature of a plane curve $y = f(x)$:

$$\kappa = \frac{|f''(x)|}{[1 + (f'(x))^2]^{3/2}}.$$

(Assume that f is of class C^2.)

(b) Use your result in (a) to find the curvature of $y = \ln(\sin x)$.

18. (a) Let $\mathbf{x}(s) = (x(s), y(s))$ be a plane curve parametrized by arclength. Show that the curvature is given by the formula

$$\kappa = |x'y'' - x''y'|.$$

(b) Show that $\mathbf{x}(s) = \left(\frac{1}{2}(1-s^2), \frac{1}{2}(\cos^{-1}s - s\sqrt{1-s^2})\right)$ is parametrized by arclength, and compute its curvature.

In Exercises 19–22, (a) use a computer algebra system to calculate the curvature κ *of the indicated path* $\mathbf{x}$ *and (b) plot the path* $\mathbf{x}$ *and, separately, plot the curvature* κ *as a function of* t *over the indicated interval for* t *and value(s) of the constants.*

19. $\mathbf{x}(t) = (a\cos t, b\sin t), \quad 0 \leq t \leq 2\pi; \quad a = 2, b = 1$

20. $\mathbf{x}(t) = (2a(1+\cos t)\cos t, 2a(1+\cos t)\sin t), \quad 0 \leq t \leq 2\pi; \quad a = 1$

21. $\mathbf{x}(t) = (2a\cos t(1+\cos t) - a, 2a\sin t(1+\cos t)), \quad 0 \leq t \leq 2\pi; \quad a = 1$

22. $\mathbf{x}(t) = (a\sin nt, b\sin mt), \quad 0 \leq t \leq 2\pi; \quad a = 3, b = 2, n = 4, m = 3$

Find the tangential and normal components of acceleration for the paths given in Exercises 23–28.

23. $\mathbf{x}(t) = t^2\,\mathbf{i} + t\,\mathbf{j}$

24. $\mathbf{x}(t) = (2t, e^{2t})$

25. $\mathbf{x}(t) = (e^t\cos 2t, e^t\sin 2t)$

26. $\mathbf{x}(t) = (4\cos 5t, 5\sin 4t, 3t)$

27. $\mathbf{x}(t) = (t, t, t^2)$

28. $\mathbf{x}(t) = \frac{3}{5}(1-\cos t)\,\mathbf{i} + \sin t\,\mathbf{j} + \frac{4}{5}\cos t\,\mathbf{k}$

29. (a) Show that the tangential and normal components of acceleration a_{tang} and a_{norm} satisfy the equations

$$a_{\text{tang}} = \frac{\mathbf{x}'\cdot\mathbf{x}''}{\|\mathbf{x}'\|}, \qquad a_{\text{norm}} = \frac{\|\mathbf{x}'\times\mathbf{x}''\|}{\|\mathbf{x}'\|}.$$

(b) Use these formulas to find the tangential and normal components of acceleration for the path $\mathbf{x}(t) = (t+2)\,\mathbf{i} + t^2\,\mathbf{j} + 3t\,\mathbf{k}$.

30. Use Exercise 29 to show that, for the plane curve $y = f(x)$,

$$a_{\text{tang}} = \frac{f'(x)f''(x)}{\sqrt{1+(f'(x))^2}},$$

$$a_{\text{norm}} = \frac{|f''(x)|}{\sqrt{1+(f'(x))^2}}.$$

31. Establish the following formula for the torsion:

$$\tau = \frac{(\mathbf{v}\times\mathbf{a})\cdot\mathbf{a}'}{\|\mathbf{v}\times\mathbf{a}\|^2}.$$

32. Show that $\kappa\tau = -\mathbf{T}'\cdot\mathbf{B}'$, where differentiation is with respect to the arclength parameter s.

33. Show that if $\mathbf{x}$ is a path parametrized by arclength and $\mathbf{x}'\times\mathbf{x}'' \neq \mathbf{0}$, then

$$\kappa^2\tau = (\mathbf{x}'\times\mathbf{x}'')\cdot\mathbf{x}'''.$$

34. Suppose $\mathbf{x}\colon I \to \mathbf{R}^3$ is a path with $\mathbf{x}'(t)\times\mathbf{x}''(t) \neq \mathbf{0}$ for all $t \in I$. The **osculating plane** to the path at $t = t_0$ is the plane containing $\mathbf{x}(t_0)$ and determined by (i.e., parallel to) the tangent and normal vectors $\mathbf{T}(t_0)$ and $\mathbf{N}(t_0)$. The **rectifying plane** at $t = t_0$ is the plane containing $\mathbf{x}(t_0)$ and determined by the tangent and binormal vectors $\mathbf{T}(t_0)$ and $\mathbf{B}(t_0)$. Finally, the **normal plane** at $t = t_0$ is the plane containing $\mathbf{x}(t_0)$ and determined by the normal and binormal vectors $\mathbf{N}(t_0)$ and $\mathbf{B}(t_0)$. Note that both the osculating and rectifying planes may be considered to be tangent planes to the path at t_0 since they are both parallel to $\mathbf{T}(t_0)$.

(a) Show that $\mathbf{B}(t_0)$ is perpendicular to the osculating plane at t_0, that $\mathbf{N}(t_0)$ is perpendicular to the rectifying plane at t_0, and that $\mathbf{T}(t_0)$ is perpendicular to the normal plane at t_0.

(b) Calculate the equations for the osculating, rectifying, and normal planes to the helix $\mathbf{x}(t) = (a\cos t, a\sin t, bt)$ at any t_0. (Hint: To speed your calculations, use the results of Example 9.)

35. Recall that the equation for a sphere of radius $a > 0$ and center $\mathbf{x}_0$ may be written as $\|\mathbf{x} - \mathbf{x}_0\| = a$. (See Example 14 of §2.1.) Explain why the image of a path $\mathbf{x}$ with the property that

$$(\mathbf{x}(t) - \mathbf{x}_0)\cdot(\mathbf{x}(t) - \mathbf{x}_0) = a^2$$

for all t must lie on a sphere of radius a.

36. Let $\mathbf{x}$ be a path with $\mathbf{x}'\times\mathbf{x}'' \neq \mathbf{0}$ and suppose that there is a point $\mathbf{x}_0$ that lies on every normal plane to $\mathbf{x}$. Show that the image of $\mathbf{x}$ lies on a sphere. (See Exercise 34 concerning normal planes to paths.)

37. Use the result of Exercise 36 to show that $\mathbf{x}(t) = (\cos 2t, -\sin 2t, 2\cos t)$ lies on a sphere by showing that $(1, 0, 0)$ lies on every normal plane to $\mathbf{x}$.

38. Use the result of Exercise 27 of §1.4 to show that

$$\mathbf{N} \times \mathbf{B} = \mathbf{T} \quad \text{and} \quad \mathbf{B} \times \mathbf{T} = \mathbf{N}.$$

As a result, we can arrange $\mathbf{T}$, $\mathbf{N}$, and $\mathbf{B}$ in a circle so that they correspond, respectively, to the vectors $\mathbf{i}$, $\mathbf{j}$, $\mathbf{k}$ appearing in Figure 1.54 and so that we may use a mnemonic for identifying cross products that is similar to the one described in Example 1 of §1.4.

Let $\mathbf{x}$ be a path of class C^3, parametrized by arclength s, with $\mathbf{x}' \times \mathbf{x}'' \neq \mathbf{0}$. We define the ***Darboux rotation vector*** *(also called the* ***angular velocity vector****) by*

$$\mathbf{w} = \tau\mathbf{T} + \kappa\mathbf{B}.$$

Note that $\mathbf{w}(s_0)$ is parallel to the rectifying plane to $\mathbf{x}(s_0)$. The direction of the Darboux vector $\mathbf{w}$ gives the axis of the "screw-like" motion of the path $\mathbf{x}$ and its length give the angular velocity of the motion. Exercises 39–41 concern the Darboux vector.

39. Show that $\|\mathbf{w}\| = \sqrt{\kappa^2 + \tau^2}$. (Hint: The vectors $\mathbf{T}$, $\mathbf{N}$, and $\mathbf{B}$ are pairwise orthogonal.)

40. (a) Use the Frenet–Serret formulas to establish the **Darboux formulas:**

$$\mathbf{T}' = \mathbf{w} \times \mathbf{T}$$
$$\mathbf{N}' = \mathbf{w} \times \mathbf{N}$$
$$\mathbf{B}' = \mathbf{w} \times \mathbf{B}.$$

(b) Use the Darboux formulas to establish the Frenet–Serret formulas. Hence the two sets of equations are equivalent. (Hint: Use Exercise 38.)

41. Show that $\mathbf{x}$ is a helix if and only if $\mathbf{w}$ is a constant vector. (Hint: Consider $\mathbf{w}'$ and use Theorem 2.5.)

3.3 Vector Fields: An Introduction

We begin with a simple definition.

■ **Definition 3.1** A **vector field** on $\mathbf{R}^n$ is a mapping

$$\mathbf{F}: X \subseteq \mathbf{R}^n \to \mathbf{R}^n.$$

We are concerned primarily with vector fields on $\mathbf{R}^2$ or $\mathbf{R}^3$. In such cases, we adopt the point of view that a vector field assigns to each *point* $\mathbf{x}$ in X a *vector* $\mathbf{F}(\mathbf{x})$ in $\mathbf{R}^n$, represented by an arrow whose tail is at the point $\mathbf{x}$. This perspective allows us to visualize vector fields in a reasonable way.

EXAMPLE 1 Suppose $\mathbf{F}: \mathbf{R}^2 \to \mathbf{R}^2$ is defined by $\mathbf{F}(\mathbf{x}) = \mathbf{a}$, where $\mathbf{a}$ is a constant vector. Then $\mathbf{F}$ assigns $\mathbf{a}$ to each point of $\mathbf{R}^2$, and so we can picture $\mathbf{F}$ by drawing the same vector (parallel translated, of course) emanating from each point in the plane, as suggested by Figure 3.27. ◆

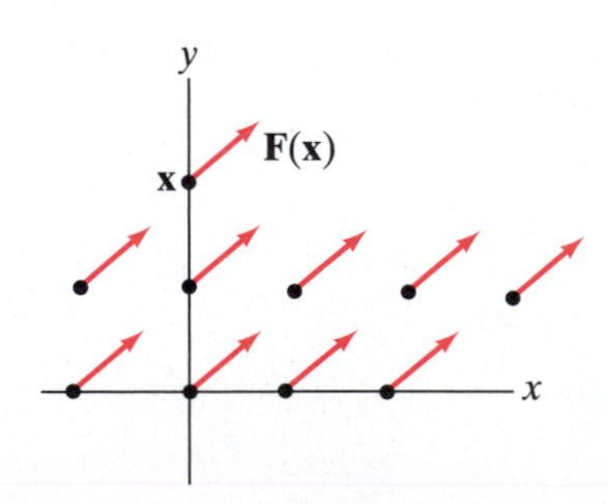

Figure 3.27 The constant vector field $\mathbf{F}(\mathbf{x}) = \mathbf{i} + \mathbf{j}$.

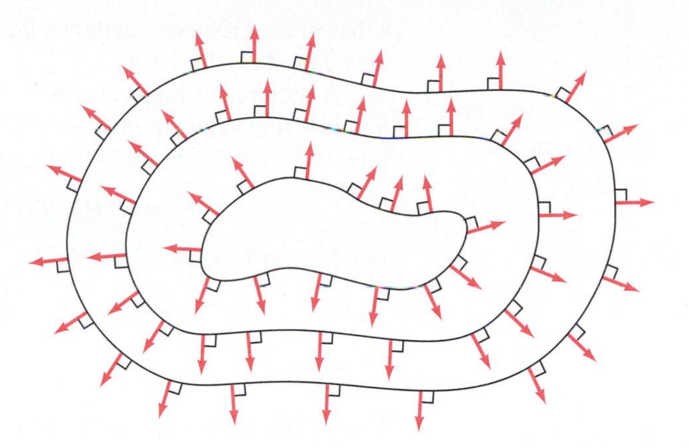

Figure 3.30 A gradient vector field $\mathbf{F} = \nabla f$. Equipotential lines are shown where f is constant.

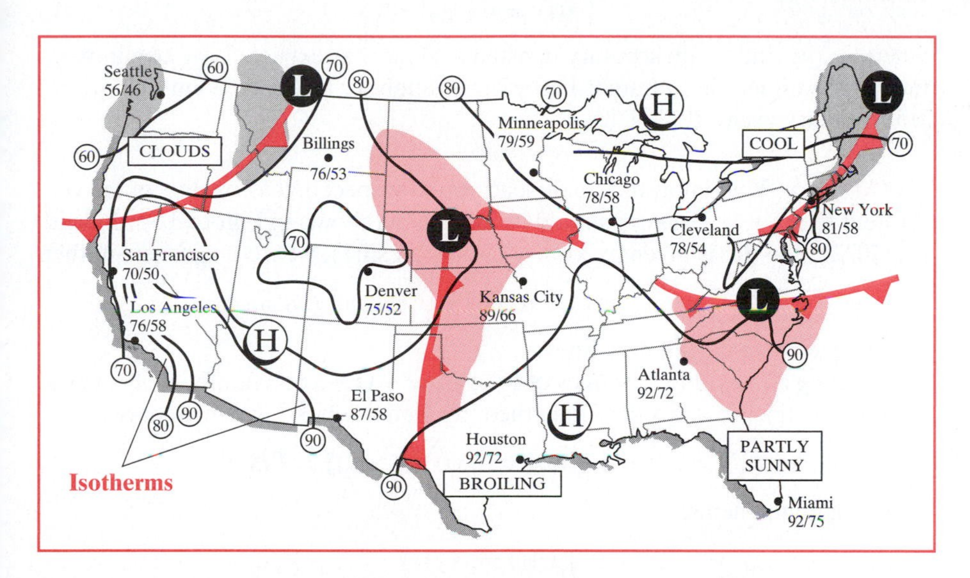

Figure 3.31 A weather map. (Weather graphics courtesy of AccuWeather, Inc., 385 Science Park Road, State College, PA 16803, (814) 237-0309, © 2004.)

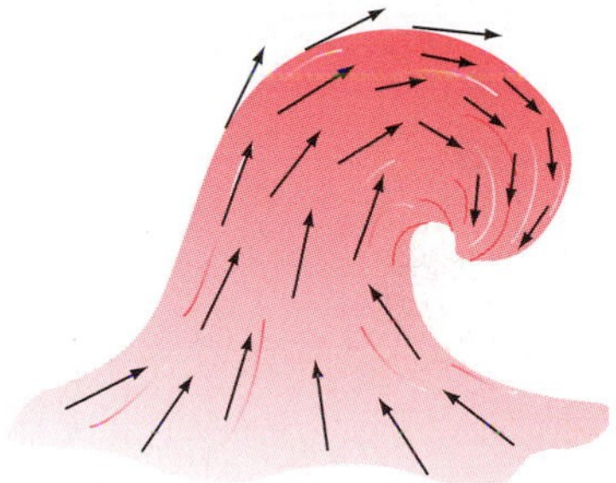

Figure 3.32 A fluid moving through space.

Path $\mathbf{x}(t)$

Vector field $\mathbf{F}$

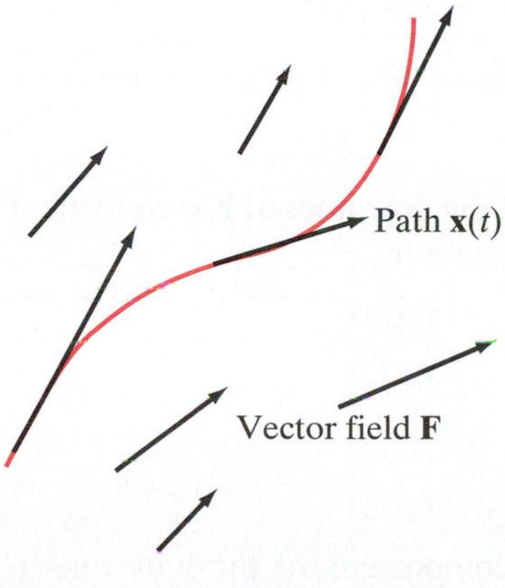

Figure 3.33 A flow line.

Flow Lines of Vector Fields

When you draw a sketch of a vector field on $\mathbf{R}^2$ or $\mathbf{R}^3$, it is easy to imagine that the arrows represent the velocity of some fluid moving through space as in Figure 3.32. It's natural to let the arrows blend into complete curves. What you're doing analytically is drawing paths whose velocity vectors coincide with those of the vector field.

■ **Definition 3.2** A **flow line** of a vector field $\mathbf{F}\colon X \subseteq \mathbf{R}^n \to \mathbf{R}^n$ is a differentiable path $\mathbf{x}\colon I \to \mathbf{R}^n$ such that

$$\mathbf{x}'(t) = \mathbf{F}(\mathbf{x}(t)).$$

That is, the velocity vector of $\mathbf{x}$ at time t is given by the value of the vector field $\mathbf{F}$ at the point on $\mathbf{x}$ at time t. (See Figure 3.33.)

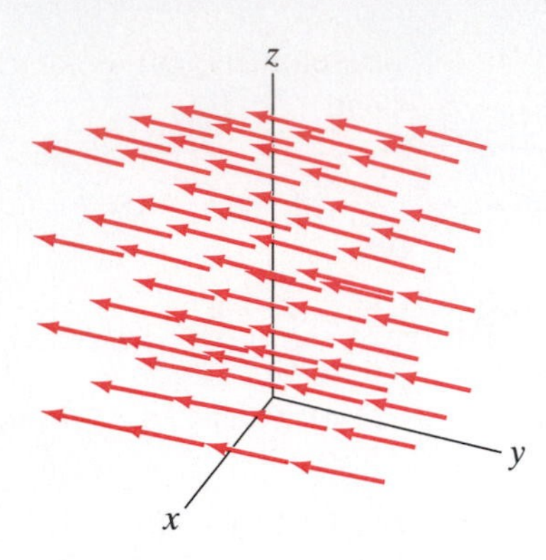

Figure 3.34 The vector field $\mathbf{F}(x, y, z) = 2\mathbf{i} - 3\mathbf{j} + \mathbf{k}$ of Example 4.

EXAMPLE 4 We calculate the flow lines of the constant vector field $\mathbf{F}(x, y, z) = 2\mathbf{i} - 3\mathbf{j} + \mathbf{k}$.

A picture of this vector field (see Figure 3.34) makes it easy to believe that the flow lines are straight-line paths. Indeed, if $\mathbf{x}(t) = (x(t), y(t), z(t))$ is a flow line, then, by Definition 3.2, we must have

$$\mathbf{x}'(t) = (x'(t), y'(t), z'(t)) = (2, -3, 1) = \mathbf{F}(\mathbf{x}(t)).$$

Equating components, we see

$$\begin{cases} x'(t) = 2 \\ y'(t) = -3 \\ z'(t) = 1 \end{cases}.$$

These differential equations are readily solved by direct integration; we obtain

$$\begin{cases} x(t) = 2t + x_0 \\ y(t) = -3t + y_0 \\ z(t) = t + z_0 \end{cases},$$

where x_0, y_0, and z_0 are arbitrary constants. Hence, as expected, we obtain parametric equations for a straight-line path through an arbitrary point (x_0, y_0, z_0) with velocity vector $(2, -3, 1)$. ◆

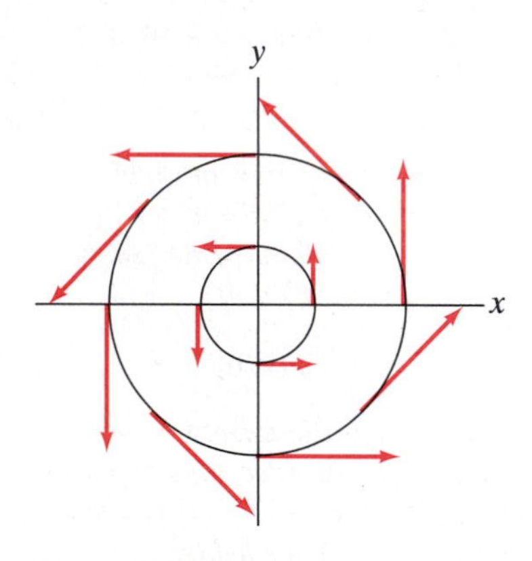

Figure 3.35 Flow lines of $\mathbf{F}(x, y) = -y\mathbf{i} + x\mathbf{j}$ of Example 5.

EXAMPLE 5 Your intuition should lead you to suspect that a flow line of the vector field $\mathbf{F}(x, y) = -y\mathbf{i} + x\mathbf{j}$ should be circular as shown in Figure 3.35. Indeed, if $\mathbf{x}\colon [0, 2\pi) \to \mathbf{R}^2$ is given by $\mathbf{x}(t) = (a\cos t, a\sin t)$, where a is constant, then

$$\mathbf{x}'(t) = -a\sin t\,\mathbf{i} + a\cos t\,\mathbf{j} = \mathbf{F}(a\cos t, a\sin t),$$

so such paths are indeed flow lines.

Finding all possible flow lines of $\mathbf{F}(x, y) = -y\mathbf{i} + x\mathbf{j}$ is a more involved task. If $\mathbf{x}(t) = (x(t), y(t))$ is a flow line, then, by Definition 3.2, we must have

$$\mathbf{x}'(t) = x'(t)\mathbf{i} + y'(t)\mathbf{j} = -y(t)\mathbf{i} + x(t)\mathbf{j} = \mathbf{F}(\mathbf{x}(t)).$$

Equating components,

$$\begin{cases} x'(t) = -y(t) \\ y'(t) = x(t) \end{cases}.$$

This is an example of a first-order system of differential equations. It turns out that *all* solutions to this system are of the form

$$\mathbf{x}(t) = (a\cos t - b\sin t, a\sin t + b\cos t),$$

where a and b are arbitrary constants. It's not difficult to see that such paths trace circles when at least one of a or b is nonzero. ◆

In general, if $\mathbf{F}$ is a vector field on $\mathbf{R}^n$, finding the flow lines of $\mathbf{F}$ is equivalent to solving the first-order system of differential equations

$$\begin{cases} x_1'(t) = F_1(x_1(t), x_2(t), \ldots, x_n(t)) \\ x_2'(t) = F_2(x_1(t), x_2(t), \ldots, x_n(t)) \\ \quad\vdots \\ x_n'(t) = F_n(x_1(t), x_2(t), \ldots, x_n(t)) \end{cases}$$

for the functions $x_1(t), \ldots, x_n(t)$ that are the components of the flow line $\mathbf{x}$. (The function F_i is just the ith component function of the vector field $\mathbf{F}$.) Such a problem takes us squarely into the realm of the theory of differential equations, a fascinating subject, but not of primary concern at the moment.

Exercises

In Exercises 1–6, sketch the given vector fields on $\mathbf{R}^2$.

1. $\mathbf{F} = y\mathbf{i} - x\mathbf{j}$
2. $\mathbf{F} = x\mathbf{i} - y\mathbf{j}$
3. $\mathbf{F} = (-x, y)$
4. $\mathbf{F} = (x, x^2)$
5. $\mathbf{F} = (x^2, x)$
6. $\mathbf{F} = (y^2, y)$

In Exercises 7–12, sketch the given vector field on $\mathbf{R}^3$.

7. $\mathbf{F} = 3\mathbf{i} + 2\mathbf{j} + \mathbf{k}$
8. $\mathbf{F} = (y, -x, 0)$
9. $\mathbf{F} = (0, z, -y)$
10. $\mathbf{F} = (y, -x, 2)$
11. $\mathbf{F} = (y, -x, z)$
12. $\mathbf{F} = \dfrac{y}{\sqrt{x^2+y^2+z^2}}\mathbf{i} - \dfrac{x}{\sqrt{x^2+y^2+z^2}}\mathbf{j} + \dfrac{z}{\sqrt{x^2+y^2+z^2}}\mathbf{k}$

In Exercises 13–16, use a computer to plot the given vector fields over the indicated ranges.

13. $\mathbf{F} = (x - y, x + y);\quad -1 \le x \le 1,\ -1 \le y \le 1$
14. $\mathbf{F} = (y^3 x, x^2 y);\quad -2 \le x \le 2,\ -2 \le y \le 2$
15. $\mathbf{F} = (x \sin y, y \cos x);\quad -2\pi \le x \le 2\pi,\ -2\pi \le y \le 2\pi$
16. $\mathbf{F} = (\cos(x - y), \sin(x + y));\quad -2\pi \le x \le 2\pi,\ -2\pi \le y \le 2\pi$

In Exercises 17–19, verify that the path given is a flow line of the indicated vector field. Justify the result geometrically with an appropriate sketch.

17. $\mathbf{x}(t) = (\sin t, \cos t, 0)$, $\mathbf{F} = (y, -x, 0)$
18. $\mathbf{x}(t) = (\sin t, \cos t, 2t)$, $\mathbf{F} = (y, -x, 2)$
19. $\mathbf{x}(t) = (\sin t, \cos t, e^{2t})$, $\mathbf{F} = (y, -x, 2z)$

In Exercises 20–22, calculate the flow line $\mathbf{x}(t)$ of the given vector field $\mathbf{F}$ that passes through the indicated point at the specified value of t.

20. $\mathbf{F}(x, y) = -x\,\mathbf{i} + y\,\mathbf{j};\quad \mathbf{x}(0) = (2, 1)$
21. $\mathbf{F}(x, y) = (x^2, y);\quad \mathbf{x}(1) = (1, e)$
22. $\mathbf{F}(x, y, z) = 2\,\mathbf{i} - 3y\,\mathbf{j} + z^3\mathbf{k};\quad \mathbf{x}(0) = (3, 5, 7)$

23. Consider the vector field $\mathbf{F} = 3\,\mathbf{i} - 2\,\mathbf{j} + \mathbf{k}$.
   (a) Show that $\mathbf{F}$ is a gradient field.
   (b) Describe the equipotential surfaces of $\mathbf{F}$ in words and with sketches.

24. Consider the vector field $\mathbf{F} = 2x\,\mathbf{i} + 2y\,\mathbf{j} - 3\mathbf{k}$.
   (a) Show that $\mathbf{F}$ is a gradient field.
   (b) Describe the equipotential surfaces of $\mathbf{F}$ in words and with sketches.

25. If $\mathbf{x}$ is a flow line of a gradient vector field $\mathbf{F} = \nabla f$, show that the function $G(t) = f(\mathbf{x}(t))$ is an increasing function of t. (Hint: Show that $G'(t)$ is always nonnegative.) Thus, we see that a particle traveling along a flow line of the gradient field $\mathbf{F} = \nabla f$ will move from lower to higher values of the potential function f. That's why physicists define a potential function of a gradient vector field $\mathbf{F}$ to be a function g such that $\mathbf{F} = -\nabla g$ (i.e., so that particles traveling along flow lines move from higher to lower values of g).

*Let $\mathbf{F}\colon X \subseteq \mathbf{R}^n \to \mathbf{R}^n$ be a continuous vector field. Let (a, b) be an interval in $\mathbf{R}$ that contains 0. (Think of (a, b) as a "time interval.") A **flow** of $\mathbf{F}$ is a differentiable function $\phi\colon X \times (a, b) \to \mathbf{R}^n$ of $n + 1$ variables such that*

$$\frac{\partial}{\partial t}\phi(\mathbf{x}, t) = \mathbf{F}(\phi(\mathbf{x}, t)); \qquad \phi(\mathbf{x}, 0) = \mathbf{x}.$$

Intuitively, we think of $\phi(\mathbf{x}, t)$ as the point at time t on the flow line of $\mathbf{F}$ that passes through $\mathbf{x}$ at time 0. (See Figure 3.36.) Thus, the flow of $\mathbf{F}$ is, in a sense, the collection of all flow lines of $\mathbf{F}$. Exercises 26–31 concern flows of vector fields.

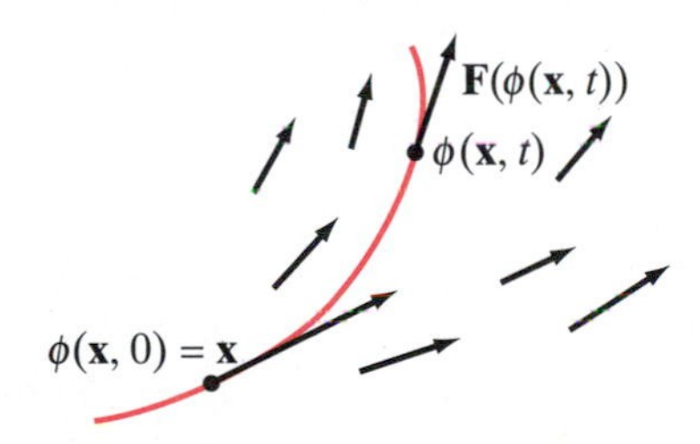

Figure 3.36 The flow of the vector field $\mathbf{F}$.

26. Verify that $\phi\colon \mathbf{R}^2 \times \mathbf{R} \to \mathbf{R}^2$,

$$\phi(x, y, t) = \left(\frac{x+y}{2}e^t + \frac{x-y}{2}e^{-t},\ \frac{x+y}{2}e^t + \frac{y-x}{2}e^{-t}\right)$$

is a flow of the vector field $\mathbf{F}(x, y) = (y, x)$.

27. Verify that

$$\phi: \mathbf{R}^2 \times \mathbf{R} \to \mathbf{R}^2,$$
$$\phi(x, y, t) = (y \sin t + x \cos t, y \cos t - x \sin t)$$

is a flow of the vector field $\mathbf{F}(x, y) = (y, -x)$.

28. Verify that

$$\phi: \mathbf{R}^3 \times \mathbf{R} \to \mathbf{R}^3,$$
$$\phi(x, y, z, t) = (x \cos 2t - y \sin 2t, y \cos 2t + x \sin 2t, ze^{-t})$$

is a flow of the vector field $\mathbf{F}(x, y, z) = -2y\,\mathbf{i} + 2x\,\mathbf{j} - z\,\mathbf{k}$.

29. Show that if $\phi: X \times (a, b) \to \mathbf{R}^n$ is a flow of $\mathbf{F}$, then, for a fixed point $\mathbf{x}_0$ in X, the map $\mathbf{x}: (a, b) \to \mathbf{R}^n$ given by $\mathbf{x}(t) = \phi(\mathbf{x}_0, t)$ is a flow line of $\mathbf{F}$.

30. If ϕ is a flow of the vector field $\mathbf{F}$, explain why $\phi(\phi(\mathbf{x}, t), s) = \phi(\mathbf{x}, s + t)$. (Hint: Relate the value of the flow ϕ at $(\mathbf{x}, t)$ to the flow line of $\mathbf{F}$ through $\mathbf{x}$. You may assume the fact that the flow line of a continuous vector field at a given point and time is determined uniquely.)

31. Derive the **equation of first variation** for a flow of a vector field. That is, if $\mathbf{F}$ is a vector field of class C^1 with flow ϕ of class C^2, show that

$$\frac{\partial}{\partial t} D_{\mathbf{x}}\phi(\mathbf{x}, t) = D\mathbf{F}(\phi(\mathbf{x}, t)) D_{\mathbf{x}}\phi(\mathbf{x}, t).$$

Here the expression "$D_{\mathbf{x}}\phi(\mathbf{x}, t)$" means to differentiate ϕ with respect to the variables $x_1, x_2, \ldots, x_n$, that is, by holding t fixed.

3.4 Gradient, Divergence, Curl, and the Del Operator

In this section, we consider certain types of differentiation operations on vector and scalar fields. These operations are as follows:

1. The gradient, which turns a scalar field into a vector field.
2. The divergence, which turns a vector field into a scalar field.
3. The curl, which turns a vector field into another vector field. (Note: The curl will be defined only for vector fields on $\mathbf{R}^3$.)

We begin by defining these operations from a purely computational point of view. Gradually, we shall come to understand their geometric significance.

The Del Operator

The **del operator,** denoted ∇, is an odd creature. It leads a double life as both differential operator and vector. In Cartesian coordinates on $\mathbf{R}^3$, del is defined by the curious expression

$$\nabla = \mathbf{i}\frac{\partial}{\partial x} + \mathbf{j}\frac{\partial}{\partial y} + \mathbf{k}\frac{\partial}{\partial z}. \tag{1}$$

The "empty" partial derivatives are the components of a vector that awaits suitable scalar and vector fields on which to act. Del operates on (i.e., transforms) fields via "multiplication" of vectors, interpreted by using partial differentiation.

For example, if $f: X \subseteq \mathbf{R}^3 \to \mathbf{R}$ is a differentiable function (scalar field), the gradient of f may be considered to be the result of multiplying the vector ∇ by the scalar f, except that when we "multiply" each component of ∇ by f, we actually compute the appropriate partial derivative:

$$\nabla f(x, y, z) = \left(\mathbf{i}\frac{\partial}{\partial x} + \mathbf{j}\frac{\partial}{\partial y} + \mathbf{k}\frac{\partial}{\partial z}\right) f(x, y, z) = \frac{\partial f}{\partial x}\mathbf{i} + \frac{\partial f}{\partial y}\mathbf{j} + \frac{\partial f}{\partial z}\mathbf{k}.$$

The del operator can also be defined in $\mathbf{R}^n$, for arbitrary n. If we take $x_1, x_2, \ldots, x_n$ to be coordinates for $\mathbf{R}^n$, then del is simply

$$\nabla = \left(\frac{\partial}{\partial x_1}, \frac{\partial}{\partial x_2}, \ldots, \frac{\partial}{\partial x_n}\right) = \mathbf{e}_1 \frac{\partial}{\partial x_1} + \mathbf{e}_2 \frac{\partial}{\partial x_2} + \cdots + \mathbf{e}_n \frac{\partial}{\partial x_n}, \quad (2)$$

where $\mathbf{e}_i = (0, \ldots, 1, \ldots, 0)$, $i = 1, \ldots, n$, is the standard basis vector for $\mathbf{R}^n$.

The Divergence of a Vector Field

Whereas taking the gradient of a scalar field yields a vector field, the process of taking the divergence does just the opposite: It turns a vector field into a scalar field.

■ **Definition 4.1** Let $\mathbf{F}: X \subseteq \mathbf{R}^n \to \mathbf{R}^n$ be a differentiable vector field. Then the **divergence** of $\mathbf{F}$, denoted div $\mathbf{F}$ or $\nabla \cdot \mathbf{F}$ (the latter read "del dot $\mathbf{F}$"), is the scalar field

$$\text{div } \mathbf{F} = \nabla \cdot \mathbf{F} = \frac{\partial F_1}{\partial x_1} + \frac{\partial F_2}{\partial x_2} + \cdots + \frac{\partial F_n}{\partial x_n},$$

where $x_1, \ldots, x_n$ are Cartesian coordinates for $\mathbf{R}^n$ and $F_1, \ldots, F_n$ are the component functions of $\mathbf{F}$.

It is essential that Cartesian coordinates be used in the formula of Definition 4.1. (Later in this section we shall see what div $\mathbf{F}$ looks like in cylindrical and spherical coordinates for $\mathbf{R}^3$.)

EXAMPLE 1 If $\mathbf{F} = x^2y\mathbf{i} + xz\mathbf{j} + xyz\mathbf{k}$, then

$$\text{div } \mathbf{F} = \frac{\partial}{\partial x}(x^2y) + \frac{\partial}{\partial y}(xz) + \frac{\partial}{\partial z}(xyz) = 2xy + 0 + xy = 3xy. \quad ◆$$

The notation for the divergence involving the dot product and the del operator is especially apt: If we write

$$\mathbf{F} = F_1\mathbf{e}_1 + F_2\mathbf{e}_2 + \cdots + F_n\mathbf{e}_n,$$

then,

$$\begin{aligned} \nabla \cdot \mathbf{F} &= \left(\mathbf{e}_1 \frac{\partial}{\partial x_1} + \mathbf{e}_2 \frac{\partial}{\partial x_2} + \cdots + \mathbf{e}_n \frac{\partial}{\partial x_n}\right) \cdot (F_1\mathbf{e}_1 + F_2\mathbf{e}_2 + \cdots + F_n\mathbf{e}_n) \\ &= \frac{\partial F_1}{\partial x_1} + \frac{\partial F_2}{\partial x_2} + \cdots + \frac{\partial F_n}{\partial x_n}, \end{aligned}$$

where, once again, we interpret "multiplying" a function by a partial differential operator as performing that partial differentiation on the given function.

Intuitively, the value of the divergence of a vector field at a particular point gives a measure of the "net mass flow" or "flux density" of the vector field in or out of that point. To understand what such a statement means, imagine that the vector field $\mathbf{F}$ represents velocity of a fluid. If $\nabla \cdot \mathbf{F}$ is zero at a point, then the rate at which fluid is flowing into that point is equal to the rate at which fluid

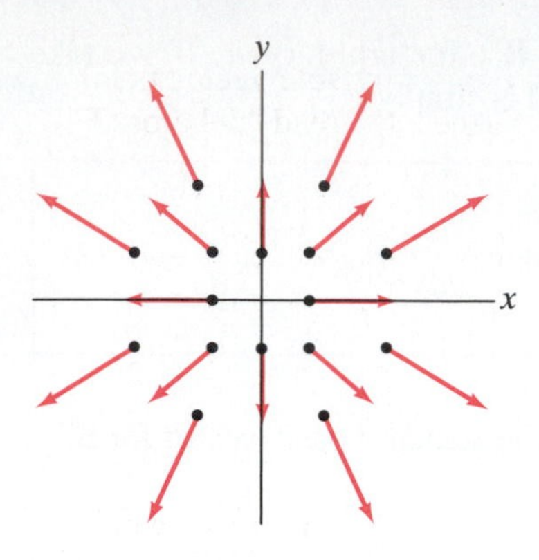

Figure 3.37 The vector field $\mathbf{F} = x\mathbf{i} + y\mathbf{j}$ of Example 2.

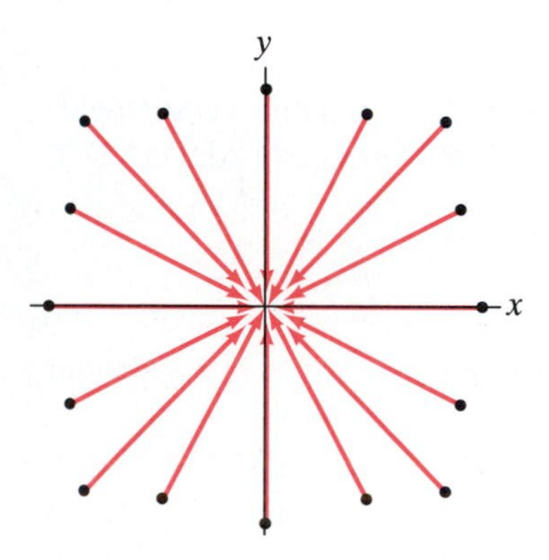

Figure 3.38 The vector field $\mathbf{G} = -x\mathbf{i} - y\mathbf{j}$ of Example 2.

is flowing out. Positive divergence at a point signifies more fluid flowing out than in, while negative divergence signifies just the opposite. We will make these assertions more precise, even prove them, when we have some integral vector calculus at our disposal. For now, however, we remark that a vector field $\mathbf{F}$ such that $\nabla \cdot \mathbf{F} = 0$ everywhere is called **incompressible** or **solenoidal.**

EXAMPLE 2 The vector field $\mathbf{F} = x\mathbf{i} + y\mathbf{j}$ has

$$\nabla \cdot \mathbf{F} = \frac{\partial}{\partial x}(x) + \frac{\partial}{\partial y}(y) = 2.$$

This vector field is shown in Figure 3.37. At any point in $\mathbf{R}^2$, the arrow whose tail is at that point is longer than the arrow whose head is there. Hence, there is greater flow *away* from each point than into it, that is, $\mathbf{F}$ is "diverging" at every point. (Thus, we see the origin of the term "divergence.")

The vector field $\mathbf{G} = -x\mathbf{i} - y\mathbf{j}$ points in the direction opposite to the vector field $\mathbf{F}$ of Figure 3.37 (see Figure 3.38), and it should be clear how $\mathbf{G}$'s divergence of -2 is reflected in the diagram. ◆

EXAMPLE 3 The constant vector field $\mathbf{F}(x, y, z) = \mathbf{a}$ shown in Figure 3.39 is incompressible. Intuitively, we can see that each point of $\mathbf{R}^3$ has an arrow representing $\mathbf{a}$ with its tail at that point and another arrow, also representing $\mathbf{a}$, with its head there.

The vector field $\mathbf{G} = y\mathbf{i} - x\mathbf{j}$ has

$$\nabla \cdot \mathbf{G} = \frac{\partial}{\partial x}(y) + \frac{\partial}{\partial y}(-x) \equiv 0.$$

A sketch of $\mathbf{G}$ reveals that it looks like the velocity field of a rotating fluid, without either a source or a sink. (See Figure 3.40.) ◆

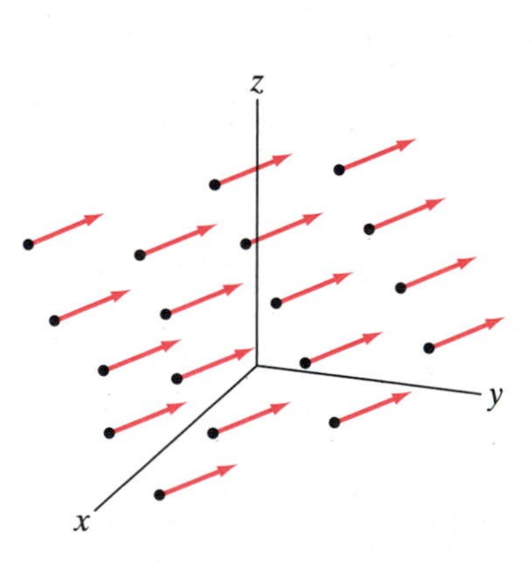

Figure 3.39 The constant vector field $\mathbf{F} = \mathbf{a}$.

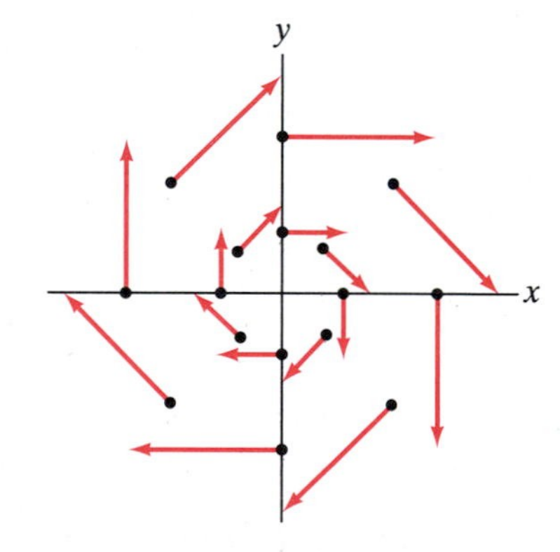

Figure 3.40 The vector field $\mathbf{G} = y\mathbf{i} - x\mathbf{j}$ resembles the velocity field of a rotating fluid.

The Curl of a Vector Field

If the gradient is the result of performing "scalar multiplication" with the del operator and a scalar field, and the divergence is the result of performing the "dot product" of del with a vector field, then there seems to be only one simple differential operation left to be built from del. We call it the **curl** of a vector field and define it as follows:

■ **Definition 4.2** Let $\mathbf{F}\colon X \subseteq \mathbf{R}^3 \to \mathbf{R}^3$ be a differentiable vector field *on* $\mathbf{R}^3$ *only*. The **curl** of $\mathbf{F}$, denoted curl $\mathbf{F}$ or $\nabla \times \mathbf{F}$ (the latter read "del cross $\mathbf{F}$"), is the vector field

$$\begin{aligned}
\operatorname{curl} \mathbf{F} = \nabla \times \mathbf{F} &= \left(\mathbf{i}\frac{\partial}{\partial x} + \mathbf{j}\frac{\partial}{\partial y} + \mathbf{k}\frac{\partial}{\partial z}\right) \times (F_1\mathbf{i} + F_2\mathbf{j} + F_3\mathbf{k}) \\
&= \begin{vmatrix} \mathbf{i} & \mathbf{j} & \mathbf{k} \\ \partial/\partial x & \partial/\partial y & \partial/\partial z \\ F_1 & F_2 & F_3 \end{vmatrix} \\
&= \left(\frac{\partial F_3}{\partial y} - \frac{\partial F_2}{\partial z}\right)\mathbf{i} + \left(\frac{\partial F_1}{\partial z} - \frac{\partial F_3}{\partial x}\right)\mathbf{j} + \left(\frac{\partial F_2}{\partial x} - \frac{\partial F_1}{\partial y}\right)\mathbf{k}.
\end{aligned}$$

There is no good reason to remember the formula for the components of the curl—instead, simply compute the cross product explicitly.

EXAMPLE 4 If $\mathbf{F} = x^2y\mathbf{i} - 2xz\mathbf{j} + (x + y - z)\mathbf{k}$, then

$$\begin{aligned}
\nabla \times \mathbf{F} &= \begin{vmatrix} \mathbf{i} & \mathbf{j} & \mathbf{k} \\ \partial/\partial x & \partial/\partial y & \partial/\partial z \\ x^2y & -2xz & x + y - z \end{vmatrix} \\
&= \left(\frac{\partial}{\partial y}(x + y - z) - \frac{\partial}{\partial z}(-2xz)\right)\mathbf{i} + \left(\frac{\partial}{\partial z}(x^2y) - \frac{\partial}{\partial x}(x + y - z)\right)\mathbf{j} \\
&\quad + \left(\frac{\partial}{\partial x}(-2xz) - \frac{\partial}{\partial y}(x^2y)\right)\mathbf{k} \\
&= (1 + 2x)\mathbf{i} - \mathbf{j} - (x^2 + 2z)\mathbf{k}.
\end{aligned}$$

◆

One would think that, with a name like "curl," $\nabla \times \mathbf{F}$ should measure how much a vector field curls. Indeed, the curl does measure, in a sense, the twisting or circulation of a vector field, but in a subtle way: Imagine that $\mathbf{F}$ represents the velocity of a stream or lake. Drop a small twig in the lake and watch it travel. The twig may perhaps be pushed by the current so that it travels in a large circle, but the curl will not detect this. What curl $\mathbf{F}$ measures is how quickly and in what orientation the twig itself rotates as it moves. (See Figure 3.41.) We prove this

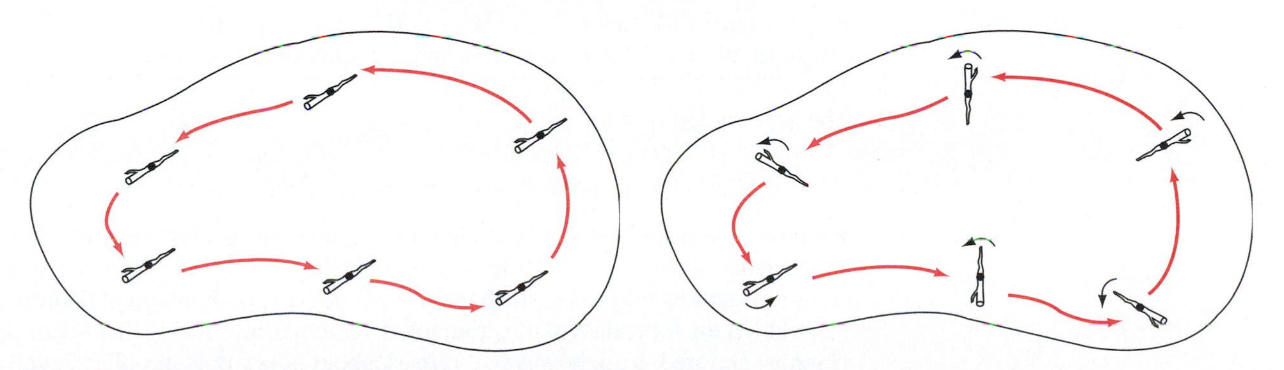

Figure 3.41 A twig in a pond where water moves with velocity given by a vector field $\mathbf{F}$. In the left figure, the twig does not rotate as it travels, so curl $\mathbf{F} = \mathbf{0}$. In the right figure, curl $\mathbf{F} \neq \mathbf{0}$, since the twig rotates.

assertion much later, when we know something about line and surface integrals. For now, we simply point out some terminology: A vector field $\mathbf{F}$ is said to be **irrotational** if $\nabla \times \mathbf{F} = \mathbf{0}$ everywhere.

Two Vector-analytic Results

Without further delay, we state a basic result about scalar-valued functions and the del operator.

Theorem 4.3 Let $f: X \subseteq \mathbf{R}^3 \to \mathbf{R}$ be of class C^2. Then curl (grad f) $= \mathbf{0}$. That is, gradient fields are irrotational.

Proof Using the del operator, we rewrite the conclusion as

$$\nabla \times (\nabla f) = \mathbf{0},$$

which might lead you to think that the proof involves nothing more than noting that ∇f is a "scalar" times ∇, hence, "parallel" to ∇, so that the cross product must be the zero vector. However, ∇ is not an ordinary vector, and the multiplications involved are not the usual ones. A real proof is needed.

Such a proof is not hard to produce: We need only start calculating $\nabla \times (\nabla f)$. We have

$$\nabla f = \frac{\partial f}{\partial x}\mathbf{i} + \frac{\partial f}{\partial y}\mathbf{j} + \frac{\partial f}{\partial z}\mathbf{k}.$$

Therefore,

$$\nabla \times (\nabla f) = \begin{vmatrix} \mathbf{i} & \mathbf{j} & \mathbf{k} \\ \partial/\partial x & \partial/\partial y & \partial/\partial z \\ \partial f/\partial x & \partial f/\partial y & \partial f/\partial z \end{vmatrix}$$

$$= \left(\frac{\partial^2 f}{\partial y \partial z} - \frac{\partial^2 f}{\partial z \partial y}\right)\mathbf{i} + \left(\frac{\partial^2 f}{\partial z \partial x} - \frac{\partial^2 f}{\partial x \partial z}\right)\mathbf{j} + \left(\frac{\partial^2 f}{\partial x \partial y} - \frac{\partial^2 f}{\partial y \partial x}\right)\mathbf{k}.$$

Since f is of class C^2, we know that the mixed second partials don't depend on the order of differentiation. Hence, each component of $\nabla \times (\nabla f)$ is zero, as desired. ■

The following result is similar in nature to that of Theorem 4.3:

Theorem 4.4 Let $\mathbf{F}: X \subseteq \mathbf{R}^3 \to \mathbf{R}^3$ be a vector field of class C^2. Then div (curl $\mathbf{F}$) $= 0$. That is, curl $\mathbf{F}$ is an incompressible vector field.

The proof is left to you.

Other Coordinate Formulations (optional)

We have introduced the gradient, divergence, and curl by formulas in Cartesian coordinates and have, at least briefly, discussed their geometric significance. Since certain situations may necessitate the use of cylindrical or spherical coordinates, we next list the formulas for the gradient, divergence, and curl in these coordinate systems. Before we do, however, a remark about notation is in order. Recall that in cylindrical coordinates, there are three unit vectors $\mathbf{e}_r$, $\mathbf{e}_\theta$, and $\mathbf{e}_z$ that point in the directions of increasing r, θ, and z coordinates, respectively. Thus, a vector field $\mathbf{F}$ on $\mathbf{R}^3$ may be written as

$$\mathbf{F} = F_r\mathbf{e}_r + F_\theta\mathbf{e}_\theta + F_z\mathbf{e}_z.$$

In general, the component functions F_r, F_θ, and F_z are each functions of the three coordinates r, θ, and z; the subscripts serve only to indicate to which of the vectors $\mathbf{e}_r$, $\mathbf{e}_\theta$, and $\mathbf{e}_z$ that particular component function should be attached. Similar comments apply to spherical coordinates, of course: There are three unit vectors $\mathbf{e}_\rho$, $\mathbf{e}_\varphi$, and $\mathbf{e}_\theta$, and any vector field $\mathbf{F}$ can be written as

$$\mathbf{F} = F_\rho\mathbf{e}_\rho + F_\varphi\mathbf{e}_\varphi + F_\theta\mathbf{e}_\theta.$$

Theorem 4.5 Let $f\colon X \subseteq \mathbf{R}^3 \to \mathbf{R}$ and $\mathbf{F}\colon Y \subseteq \mathbf{R}^3 \to \mathbf{R}^3$ be differentiable scalar and vector fields, respectively. Then

$$\nabla f = \frac{\partial f}{\partial r}\mathbf{e}_r + \frac{1}{r}\frac{\partial f}{\partial \theta}\mathbf{e}_\theta + \frac{\partial f}{\partial z}\mathbf{e}_z; \tag{3}$$

$$\operatorname{div}\mathbf{F} = \frac{1}{r}\left[\frac{\partial}{\partial r}(rF_r) + \frac{\partial F_\theta}{\partial \theta} + \frac{\partial}{\partial z}(rF_z)\right]; \tag{4}$$

$$\operatorname{curl}\mathbf{F} = \frac{1}{r}\begin{vmatrix} \mathbf{e}_r & r\mathbf{e}_\theta & \mathbf{e}_z \\ \partial/\partial r & \partial/\partial\theta & \partial/\partial z \\ F_r & rF_\theta & F_z \end{vmatrix}. \tag{5}$$

Proof We'll prove formula (4) only, since the argument should be sufficiently clear so that it can be modified to give proofs of formulas (3) and (5). The idea is simply to rewrite all rectangular symbols in terms of cylindrical ones.

From the equations in (8) of §1.7, we have

$$\begin{cases} \mathbf{e}_r = \cos\theta\,\mathbf{i} + \sin\theta\,\mathbf{j} \\ \mathbf{e}_\theta = -\sin\theta\,\mathbf{i} + \cos\theta\,\mathbf{j}. \\ \mathbf{e}_z = \mathbf{k} \end{cases} \tag{6}$$

From the chain rule, we have the following relations between rectangular and cylindrical differential operators:

$$\begin{cases} \dfrac{\partial}{\partial r} = \cos\theta\dfrac{\partial}{\partial x} + \sin\theta\dfrac{\partial}{\partial y} \\ \dfrac{\partial}{\partial \theta} = -r\sin\theta\dfrac{\partial}{\partial x} + r\cos\theta\dfrac{\partial}{\partial y}. \\ \dfrac{\partial}{\partial z} = \dfrac{\partial}{\partial z} \end{cases}$$

These relations can be solved algebraically for $\partial/\partial x$, $\partial/\partial y$, and $\partial/\partial z$ to yield

$$\begin{cases} \dfrac{\partial}{\partial x} = \cos\theta\dfrac{\partial}{\partial r} - \dfrac{\sin\theta}{r}\dfrac{\partial}{\partial \theta} \\ \dfrac{\partial}{\partial y} = \sin\theta\dfrac{\partial}{\partial r} + \dfrac{\cos\theta}{r}\dfrac{\partial}{\partial \theta}. \\ \dfrac{\partial}{\partial z} = \dfrac{\partial}{\partial z} \end{cases} \tag{7}$$

Hence, we can use (6) and (7) to rewrite the expression for the divergence of a vector field on $\mathbf{R}^3$:

$$\begin{aligned}\nabla\cdot\mathbf{F} &= \left(\frac{\partial}{\partial x}\mathbf{i}+\frac{\partial}{\partial y}\mathbf{j}+\frac{\partial}{\partial z}\mathbf{k}\right)\cdot(F_r\mathbf{e}_r+F_\theta\mathbf{e}_\theta+F_z\mathbf{e}_z)\\ &= \left[\mathbf{i}\left(\cos\theta\frac{\partial}{\partial r}-\frac{\sin\theta}{r}\frac{\partial}{\partial\theta}\right)+\mathbf{j}\left(\sin\theta\frac{\partial}{\partial r}+\frac{\cos\theta}{r}\frac{\partial}{\partial\theta}\right)+\mathbf{k}\frac{\partial}{\partial z}\right]\\ &\quad\cdot\left[(F_r\cos\theta-F_\theta\sin\theta)\,\mathbf{i}+(F_r\sin\theta+F_\theta\cos\theta)\,\mathbf{j}+F_z\,\mathbf{k}\right].\end{aligned}$$

(We used the equations in (7) to rewrite the partial operators $\partial/\partial x$, $\partial/\partial y$, and $\partial/\partial z$ appearing in del and the equations in (6) to replace the cylindrical basis vectors $\mathbf{e}_r$, $\mathbf{e}_\theta$, and $\mathbf{e}_z$ by expressions involving $\mathbf{i}$, $\mathbf{j}$, and $\mathbf{k}$.) Performing the dot product and using the product rule yields

$$\begin{aligned}\nabla\cdot\mathbf{F} &= \left(\cos\theta\frac{\partial}{\partial r}-\frac{\sin\theta}{r}\frac{\partial}{\partial\theta}\right)(F_r\cos\theta-F_\theta\sin\theta)\\ &\quad+\left(\sin\theta\frac{\partial}{\partial r}+\frac{\cos\theta}{r}\frac{\partial}{\partial\theta}\right)(F_r\sin\theta+F_\theta\cos\theta)+\frac{\partial}{\partial z}F_z\\ &= \cos\theta\left(\cos\theta\frac{\partial F_r}{\partial r}+F_r\frac{\partial}{\partial r}(\cos\theta)\right)-\frac{\sin\theta}{r}\left(\cos\theta\frac{\partial F_r}{\partial\theta}+F_r\frac{\partial}{\partial\theta}(\cos\theta)\right)\\ &\quad-\cos\theta\left(\sin\theta\frac{\partial F_\theta}{\partial r}+F_\theta\frac{\partial}{\partial r}(\sin\theta)\right)+\frac{\sin\theta}{r}\left(\sin\theta\frac{\partial F_\theta}{\partial\theta}+F_\theta\frac{\partial}{\partial\theta}(\sin\theta)\right)\\ &\quad+\sin\theta\left(\sin\theta\frac{\partial F_r}{\partial r}+F_r\frac{\partial}{\partial r}(\sin\theta)\right)+\frac{\cos\theta}{r}\left(\sin\theta\frac{\partial F_r}{\partial\theta}+F_r\frac{\partial}{\partial\theta}(\sin\theta)\right)\\ &\quad+\sin\theta\left(\cos\theta\frac{\partial F_\theta}{\partial r}+F_\theta\frac{\partial}{\partial r}(\cos\theta)\right)+\frac{\cos\theta}{r}\left(\cos\theta\frac{\partial F_\theta}{\partial\theta}+F_\theta\frac{\partial}{\partial\theta}(\cos\theta)\right)\\ &\quad+\frac{\partial F_z}{\partial z}.\end{aligned}$$

After some additional algebra, we find that

$$\begin{aligned}\nabla\cdot\mathbf{F} &= (\cos^2\theta+\sin^2\theta)\frac{\partial F_r}{\partial r}+\left(\frac{\sin^2\theta+\cos^2\theta}{r}\right)F_r\\ &\quad+\left(\frac{\sin^2\theta+\cos^2\theta}{r}\right)\frac{\partial F_\theta}{\partial\theta}+\frac{\partial F_z}{\partial z}\\ &= \frac{\partial F_r}{\partial r}+\frac{1}{r}F_r+\frac{1}{r}\frac{\partial F_\theta}{\partial\theta}+\frac{\partial F_z}{\partial z}\\ &= \frac{1}{r}\left(\underbrace{r\frac{\partial F_r}{\partial r}+F_r}_{\frac{\partial}{\partial r}(rF_r)}+\frac{\partial F_\theta}{\partial\theta}+\frac{\partial}{\partial z}(rF_z)\right),\end{aligned}$$

as desired. ■

In spherical coordinates, the story for the gradient, divergence, and curl is more complicated algebraically, although the ideas behind the proof are essentially the same. We state the relevant results and leave to you the rather tedious task of verifying them.

Theorem 4.6 Let $f: X \subseteq \mathbf{R}^3 \to \mathbf{R}$ and $\mathbf{F}: Y \subseteq \mathbf{R}^3 \to \mathbf{R}^3$ be differentiable scalar and vector fields, respectively. Then the following formulas hold:

$$\nabla f = \frac{\partial f}{\partial \rho}\mathbf{e}_\rho + \frac{1}{\rho}\frac{\partial f}{\partial \varphi}\mathbf{e}_\varphi + \frac{1}{\rho \sin\varphi}\frac{\partial f}{\partial \theta}\mathbf{e}_\theta; \tag{8}$$

$$\nabla \cdot \mathbf{F} = \frac{1}{\rho^2}\frac{\partial}{\partial \rho}(\rho^2 F_\rho) + \frac{1}{\rho \sin\varphi}\frac{\partial}{\partial \varphi}(\sin\varphi\, F_\varphi) + \frac{1}{\rho \sin\varphi}\frac{\partial F_\theta}{\partial \theta}; \tag{9}$$

$$\nabla \times \mathbf{F} = \frac{1}{\rho^2 \sin\varphi}\begin{vmatrix} \mathbf{e}_\rho & \rho\,\mathbf{e}_\varphi & \rho \sin\varphi\, \mathbf{e}_\theta \\ \partial/\partial\rho & \partial/\partial\varphi & \partial/\partial\theta \\ F_\rho & \rho F_\varphi & \rho\sin\varphi\, F_\theta \end{vmatrix}. \tag{10}$$

Exercises

Calculate the divergence of the vector fields given in Exercises 1–6.

1. $\mathbf{F} = x^2\mathbf{i} + y^2\mathbf{j}$
2. $\mathbf{F} = y^2\mathbf{i} + x^2\mathbf{j}$
3. $\mathbf{F} = (x+y)\mathbf{i} + (y+z)\mathbf{j} + (x+z)\mathbf{k}$
4. $\mathbf{F} = z\cos(e^{y^2})\,\mathbf{i} + x\sqrt{z^2+1}\,\mathbf{j} + e^{2y}\sin 3x\,\mathbf{k}$
5. $\mathbf{F} = x_1^2\mathbf{e}_1 + 2x_2^2\mathbf{e}_2 + \cdots + nx_n^2\mathbf{e}_n$
6. $\mathbf{F} = x_1\mathbf{e}_1 + 2x_1\mathbf{e}_2 + \cdots + nx_1\mathbf{e}_n$

Find the curl of the vector fields given in Exercises 7–11.

7. $\mathbf{F} = x^2\,\mathbf{i} - xe^y\,\mathbf{j} + 2xyz\,\mathbf{k}$
8. $\mathbf{F} = x\mathbf{i} + y\mathbf{j} + z\mathbf{k}$
9. $\mathbf{F} = (x+yz)\mathbf{i} + (y+xz)\mathbf{j} + (z+xy)\mathbf{k}$
10. $\mathbf{F} = (\cos yz - x)\mathbf{i} + (\cos xz - y)\mathbf{j} + (\cos xy - z)\mathbf{k}$
11. $\mathbf{F} = y^2z\,\mathbf{i} + e^{xyz}\mathbf{j} + x^2y\,\mathbf{k}$
12. (a) Consider again the vector field in Exercise 8 and its curl. Sketch the vector field and use your picture to explain geometrically why the curl is as you calculated.
    (b) Use geometry to determine $\nabla \times \mathbf{F}$, where $\mathbf{F} = \dfrac{(x\mathbf{i} + y\mathbf{j} + z\mathbf{k})}{\sqrt{x^2+y^2+z^2}}$.
    (c) For $\mathbf{F}$ as in part (b), verify your intuition by explicitly computing $\nabla \times \mathbf{F}$.
13. Can you tell in what portions of $\mathbf{R}^2$, the vector fields shown in Figures 3.42–3.45 have positive divergence? Negative divergence?
14. Check that if $f(x, y, z) = x^2 \sin y + y^2 \cos z$, then
$$\nabla \times (\nabla f) = \mathbf{0}.$$

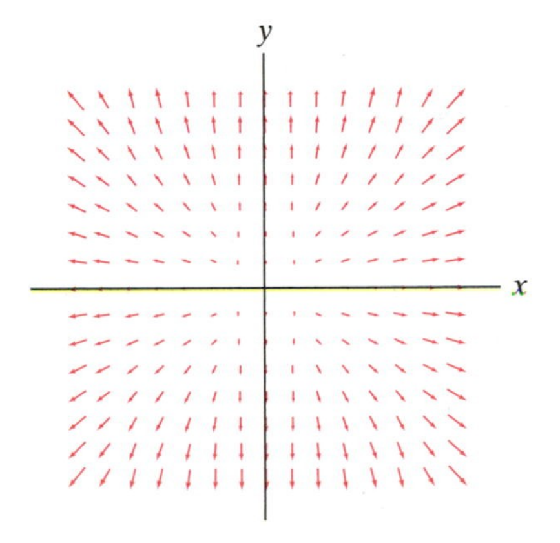

Figure 3.42 Vector field for Exercise 13(a).

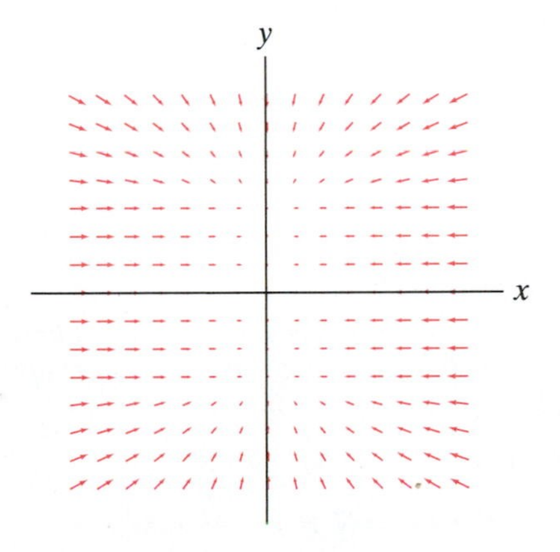

Figure 3.43 Vector field for Exercise 13(b).

15. Check that if $\mathbf{F}(x, y, z) = xyz\mathbf{i} - e^z\cos x\mathbf{j} + xy^2z^3\mathbf{k}$, then
$$\nabla \cdot (\nabla \times \mathbf{F}) = 0.$$

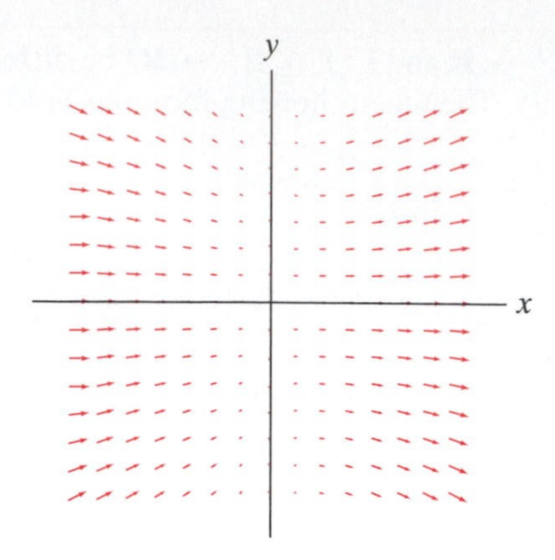

Figure 3.44 Vector field for Exercise 13(c).

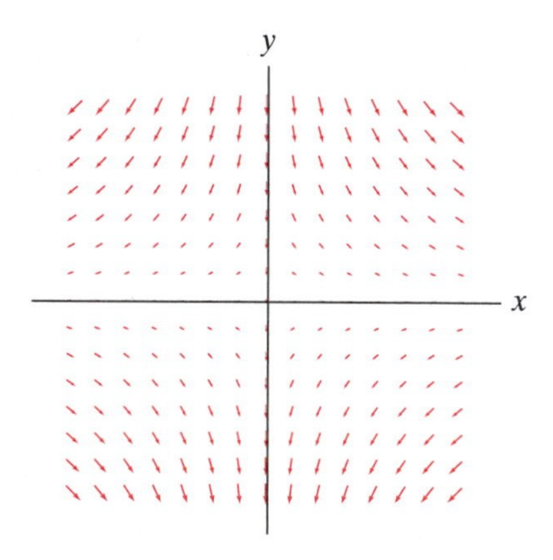

Figure 3.45 Vector field for Exercise 13(d).

16. Prove Theorem 4.4.

In Exercises 17–20, let $\mathbf{r} = x\,\mathbf{i} + y\,\mathbf{j} + z\,\mathbf{k}$ *and let* r *denote* $\|\mathbf{r}\|$. *Verify the following:*

17. $\nabla r^n = nr^{n-2}\mathbf{r}$

18. $\nabla(\ln r) = \dfrac{\mathbf{r}}{r^2}$

19. $\nabla \cdot (r^n \mathbf{r}) = (n+3)r^n$

20. $\nabla \times (r^n \mathbf{r}) = \mathbf{0}$

In Exercises 21–25, establish the given identities. (You may assume that any functions and vector fields are appropriately differentiable.)

21. $\nabla \cdot (\mathbf{F} + \mathbf{G}) = \nabla \cdot \mathbf{F} + \nabla \cdot \mathbf{G}$

22. $\nabla \times (\mathbf{F} + \mathbf{G}) = \nabla \times \mathbf{F} + \nabla \times \mathbf{G}$

23. $\nabla \cdot (f\mathbf{F}) = f\nabla \cdot \mathbf{F} + \mathbf{F} \cdot \nabla f$

24. $\nabla \times (f\mathbf{F}) = f\nabla \times \mathbf{F} + \nabla f \times \mathbf{F}$

25. $\nabla \cdot (\mathbf{F} \times \mathbf{G}) = \mathbf{G} \cdot \nabla \times \mathbf{F} - \mathbf{F} \cdot \nabla \times \mathbf{G}$

26. Prove formulas (3) and (5) of Theorem 4.5.

27. Establish the formula for the gradient of a function in spherical coordinates given in Theorem 4.6.

28. The Laplacian operator, denoted ∇^2, is the second-order partial differential operator defined by

$$\nabla^2 = \frac{\partial^2}{\partial x^2} + \frac{\partial^2}{\partial y^2} + \frac{\partial^2}{\partial z^2}.$$

(a) Explain why it makes sense to think of ∇^2 as $\nabla \cdot \nabla$.

(b) Show that if f and g are functions of class C^2, then

$$\nabla^2(fg) = f\nabla^2 g + g\nabla^2 f + 2(\nabla f \cdot \nabla g).$$

(c) Show that

$$\nabla \cdot (f\nabla g - g\nabla f) = f\nabla^2 g - g\nabla^2 f.$$

29. Show that $\nabla \cdot (f\nabla f) = \|\nabla f\|^2 + f\nabla^2 f$.

30. Show that $\nabla \times (\nabla \times \mathbf{F}) = \nabla(\nabla \cdot \mathbf{F}) - \nabla^2 \mathbf{F}$. (Here $\nabla^2 \mathbf{F}$ means to take the Laplacian of each component function of $\mathbf{F}$.)

Let X *be an open set in* $\mathbf{R}^n$, $\mathbf{F}\colon X \subseteq \mathbf{R}^n \to \mathbf{R}^n$ *a vector field on* X, *and* $\mathbf{a} \in X$. *If* $\mathbf{v}$ *is any unit vector in* $\mathbf{R}^n$, *we define the* ***directional derivative*** *of* $\mathbf{F}$ *at* $\mathbf{a}$ *in the direction of* $\mathbf{v}$, *denoted* $D_{\mathbf{v}}\mathbf{F}(\mathbf{a})$, *by*

$$D_{\mathbf{v}}\mathbf{F}(\mathbf{a}) = \lim_{h \to 0} \frac{1}{h}(\mathbf{F}(\mathbf{a} + h\mathbf{v}) - \mathbf{F}(\mathbf{a})),$$

provided that the limit exists. Exercises 31–34 involve directional derivatives of vector fields.

31. (a) In analogy with the directional derivative of a scalar-valued function defined in §2.6, show that

$$D_{\mathbf{v}}\mathbf{F}(\mathbf{a}) = \left.\frac{d}{dt}\mathbf{F}(\mathbf{a} + t\mathbf{v})\right|_{t=0}.$$

(b) Use the result of part (a) and the chain rule to show that, if $\mathbf{F}$ is differentiable at $\mathbf{a}$, then

$$D_{\mathbf{v}}\mathbf{F}(\mathbf{a}) = D\mathbf{F}(\mathbf{a})\mathbf{v},$$

where $\mathbf{v}$ is interpreted to be an $n \times 1$ matrix. (Note that this result makes it straightforward to calculate directional derivatives of vector fields.)

32. Show that the directional derivative of a vector field $\mathbf{F}$ is the vector whose components are the directional derivatives of the component functions $F_1, \dots, F_n$ of $\mathbf{F}$, that is, that

$$D_{\mathbf{v}}\mathbf{F}(\mathbf{a}) = (D_{\mathbf{v}}F_1(\mathbf{a}), D_{\mathbf{v}}F_2(\mathbf{a}), \dots, D_{\mathbf{v}}F_n(\mathbf{a})).$$

33. Let $\mathbf{F} = yz\,\mathbf{i} + xz\,\mathbf{j} + xy\,\mathbf{k}$. Find $D_{(\mathbf{i}-\mathbf{j}+\mathbf{k})/\sqrt{3}}\mathbf{F}(3, 2, 1)$. (Hint: See Exercise 31.)

34. Let $\mathbf{F} = x\,\mathbf{i} + y\,\mathbf{j} + z\,\mathbf{k}$. Show that $D_{\mathbf{v}}\mathbf{F}(\mathbf{a}) = \mathbf{v}$ for any point $\mathbf{a} \in \mathbf{R}^3$ and any unit vector $\mathbf{v} \in \mathbf{R}^3$. More generally, if $\mathbf{F} = (x_1, x_2, \dots, x_n)$, $\mathbf{a} = (a_1, a_2, \dots, a_n)$, and $\mathbf{v} = (v_1, v_2, \dots, v_n)$, show that $D_{\mathbf{v}}\mathbf{F}(\mathbf{a}) = \mathbf{v}$.

3.5 True/False Exercises for Chapter 3

1. If a path $\mathbf{x}$ remains a constant distance from the origin, then the velocity of $\mathbf{x}$ is perpendicular to $\mathbf{x}$.

2. If a path is parametrized by arclength, then its velocity vector is constant.

3. If a path is parametrized by arclength, then its velocity and acceleration are orthogonal.

4. $\dfrac{d}{dt}\|\mathbf{x}(t)\| = \|\mathbf{x}'(t)\|$.

5. $\dfrac{d}{dt}(\mathbf{x}\times\mathbf{y}) = \mathbf{x}\times\dfrac{d\mathbf{y}}{dt} + \mathbf{y}\times\dfrac{d\mathbf{x}}{dt}$.

6. $\kappa = \left\|\dfrac{d\mathbf{T}}{dt}\right\|$.

7. $|\tau| = \left\|\dfrac{d\mathbf{B}}{ds}\right\|$.

8. The curvature κ is always nonnegative.

9. The torsion τ is always nonnegative.

10. $\mathbf{N} = \dfrac{d\mathbf{T}}{ds}$.

11. If a path $\mathbf{x}$ has zero curvature, then its acceleration is always parallel to its velocity.

12. If a path $\mathbf{x}$ has a constant binormal vector $\mathbf{B}$, then $\tau \equiv 0$.

13. $\left(\dfrac{d^2s}{dt^2}\right)^2 + \kappa^2\left(\dfrac{ds}{dt}\right)^4 = \|\mathbf{a}(t)\|^2$.

14. grad f is a scalar field.

15. div $\mathbf{F}$ is a vector field.

16. curl $\mathbf{F}$ is a vector field.

17. grad(div $\mathbf{F}$) is a vector field.

18. div(curl(grad f)) is a vector field.

19. grad f $\times$ div $\mathbf{F}$ is a vector field.

20. The path $\mathbf{x}(t) = (2\cos t, 4\sin t, t)$ is a flow line of the vector field $\mathbf{F}(x, y, z) = -\dfrac{y}{2}\mathbf{i} + 2x\,\mathbf{j} + z\,\mathbf{k}$.

21. The path $\mathbf{x}(t) = (e^t\cos t, e^t(\cos t + \sin t), e^t\sin t)$ is a flow line of the vector field $\mathbf{F}(x, y, z) = (x - z)\,\mathbf{i} + 2x\,\mathbf{j} + y\,\mathbf{k}$.

22. The vector field $\mathbf{F} = 2xy\cos z\,\mathbf{i} - y^2\cos z\,\mathbf{j} + e^{xy}\,\mathbf{k}$ is incompressible.

23. The vector field $\mathbf{F} = 2xy\cos z\,\mathbf{i} - y^2\cos z\,\mathbf{j} + e^{xy}\,\mathbf{k}$ is irrotational.

24. $\nabla\times(\nabla f) = \mathbf{0}$ for all functions $f\colon \mathbf{R}^3 \to \mathbf{R}$.

25. If $\nabla\cdot\mathbf{F} = 0$ and $\nabla\times\mathbf{F} = \mathbf{0}$, then $\mathbf{F} = \mathbf{0}$.

26. $\nabla\cdot(\mathbf{F}\times\mathbf{G}) = \mathbf{F}\cdot(\nabla\times\mathbf{G}) + \mathbf{G}\cdot(\nabla\times\mathbf{F})$.

27. If $\mathbf{F} = \text{curl}\,\mathbf{G}$, then $\mathbf{F}$ is solenoidal.

28. The vector field $\mathbf{F} = 2x\sin y\cos z\,\mathbf{i} + x^2\cos y\cos z\,\mathbf{j} + x^2\sin y\sin z\,\mathbf{k}$ is the gradient of a function f of class C^2.

29. There is a vector field $\mathbf{F}$ of class C^2 on $\mathbf{R}^3$ such that $\nabla\times\mathbf{F} = x\cos^2 y\,\mathbf{i} + 3y\,\mathbf{j} - xyz^2\,\mathbf{k}$.

30. If $\mathbf{F}$ and $\mathbf{G}$ are gradient fields, then $\mathbf{F}\times\mathbf{G}$ is incompressible.

3.6 Miscellaneous Exercises for Chapter 3

1. Figure 3.46 shows the plots of six paths $\mathbf{x}$ in the plane. Match each parametric description with the correct graph.

(a) $\mathbf{x}(t) = (\sin 2t, \sin 3t)$

(b) $\mathbf{x}(t) = (t + \sin 5t, t^2 + \cos 6t)$

(c) $\mathbf{x}(t) = (t^2 + 1, t^3 - t)$

(d) $\mathbf{x}(t) = (2t + \sin 4t, t - \sin 5t)$

(e) $\mathbf{x}(t) = (t - t^2, t^3 - t)$,

(f) $\mathbf{x}(t) = (\sin(t + \sin 3t), \cos t)$

2. Figure 3.47 shows the plots of six paths $\mathbf{x}$ in $\mathbf{R}^3$. Match each parametric description with the correct graph.

(a) $\mathbf{x}(t) = (t + \cos 3t, t^2 + \sin 5t, \sin 4t)$

(b) $\mathbf{x}(t) = (2\cos^3 t, 3\sin^3 t, \cos 2t)$

(c) $\mathbf{x}(t) = (15\cos t, 23\sin t, 4t)$

(d) $\mathbf{x}(t) = (\cos 3t, \cos 5t, \sin 4t)$

(e) $\mathbf{x}(t) = (2t\cos t, 2t\sin t, 4t)$,

(f) $\mathbf{x}(t) = ((t^2 + 1, t^3 - t, t^4 - t^2)$

3. Suppose that $\mathbf{x}$ is a C^2 path with nonzero velocity. Show that $\mathbf{x}$ has constant speed if and only if its velocity and acceleration vectors are always perpendicular to one another.

4. You are at Vertigo Amusement Park riding the new Vector roller coaster. The path of your car is given by

$$\mathbf{x}(t) = \left(e^{t/60}\cos\frac{\pi t}{30}, e^{t/60}\sin\frac{\pi t}{30}, 80 + \frac{2t(10 - t)(t - 90)^2}{10^6}\right),$$

where $t = 0$ corresponds to the beginning of your three-minute ride, measured in seconds, and spatial dimensions are measured in feet. It is a calm day, but after 90 sec of your ride your glasses suddenly fly off your face.

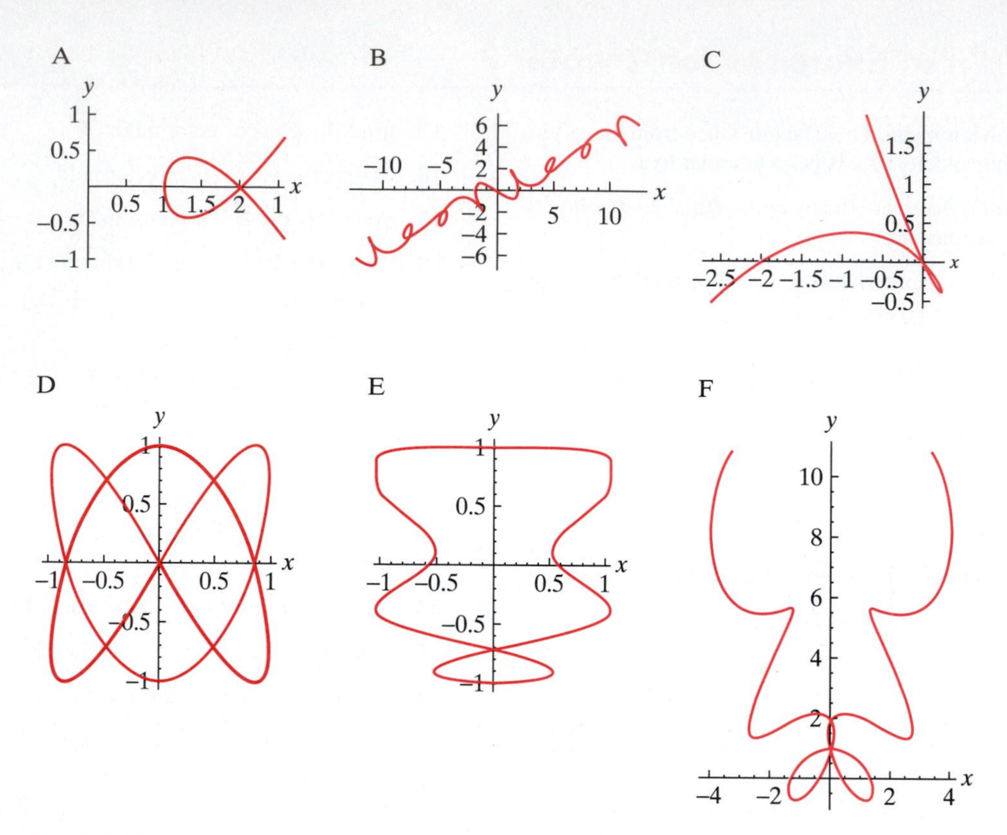

Figure 3.46 Figures for Exercise 1.

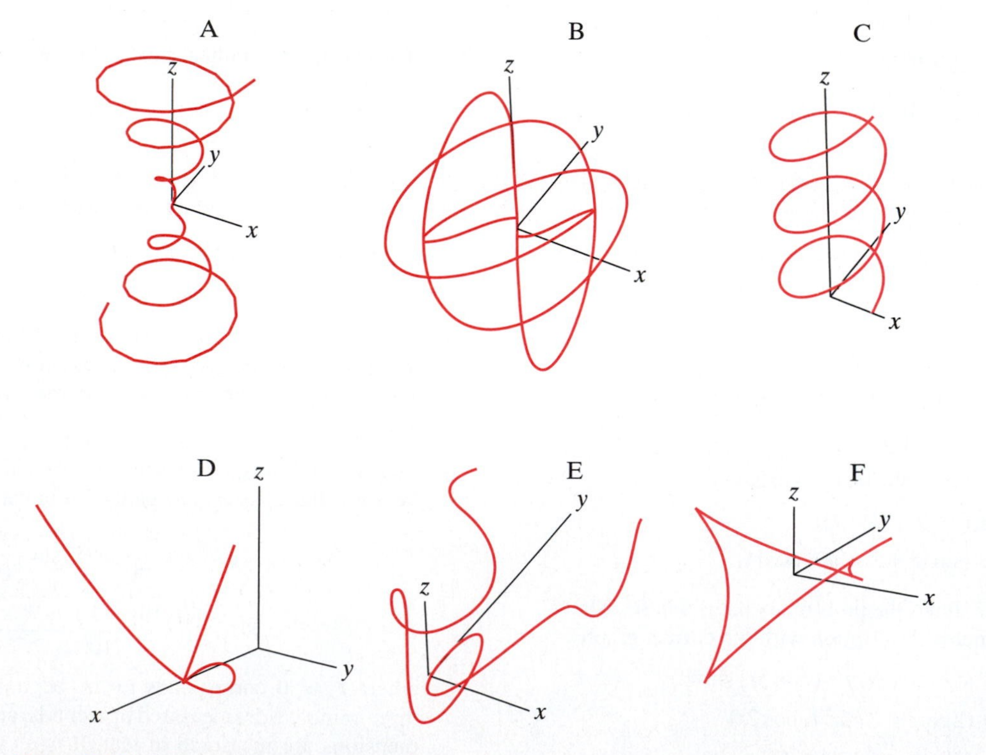

Figure 3.47 Figures for Exercise 2.

(a) Neglecting the effect of gravity, where will your glasses be 2 sec later?

(b) What if gravity is taken into account?

5. Show that the curve traced parametrically by

$$\mathbf{x}(t) = \left(\cos(t-1), t^3 - 1, \frac{1}{t} - 2\right)$$

is tangent to the surface $x^3 + y^3 + z^3 - xyz = 0$ when $t = 1$.

6. Gregor, the cockroach, is on the edge of a Ferris wheel that is rotating at a rate of 2 rev/min (counterclockwise as you observe him). Gregor is crawling along a spoke towards the center of the wheel at a rate of 3 in/min.

(a) Using polar coordinates with the center of the wheel as origin, assume that Gregor starts (at time $t = 0$) at the point $r = 20$ ft, $\theta = 0$. Give parametric equations for Gregor's polar coordinates r and θ at time t (in minutes).

(b) Give parametric equations for Gregor's Cartesian coordinates at time t.

(c) Determine the distance Gregor has traveled once he reaches the center of the wheel. Express your answer as an integral and evaluate it numerically.

If you have used a drawing program on a computer, you have probably worked with a curve known as a ***Bézier curve.***[2] *Such a curve is defined parametrically by using several* ***control points*** *in the plane to shape the curve. In Exercises 7–12, we discuss various aspects of quadratic Bézier curves. These curves are defined by using three fixed control points* (x_1, y_1), (x_2, y_2), *and* (x_3, y_3) *and a nonnegative constant* w. *The Bézier curve defined by this information is given by* $\mathbf{x}\colon [0,1] \to \mathbf{R}^2$, $\mathbf{x}(t) = (x(t), y(t))$, *where*

$$\begin{cases} x(t) = \dfrac{(1-t)^2 x_1 + 2wt(1-t)x_2 + t^2 x_3}{(1-t)^2 + 2wt(1-t) + t^2} \\[2ex] y(t) = \dfrac{(1-t)^2 y_1 + 2wt(1-t)y_2 + t^2 y_3}{(1-t)^2 + 2wt(1-t) + t^2} \end{cases}, \quad 0 \le t \le 1. \tag{1}$$

7. Let the control points be $(1, 0)$, $(0, 1)$, and $(1, 1)$. Use a computer to graph the Bézier curve for $w = 0, 1/2, 1, 2, 5$. What happens as w increases?

8. Repeat Exercise 7 for the control points $(-1, -1)$, $(1, 3)$, and $(4, 1)$.

9. (a) Show that the Bézier curve given by the parametric equations in (1) has (x_1, y_1) as initial point and (x_3, y_3) as terminal point.

(b) Show that $\mathbf{x}(\frac{1}{2})$ lies on the line segment joining (x_2, y_2) to the midpoint of the line segment joining (x_1, y_1) to (x_3, y_3).

10. In general the control points (x_1, y_1), (x_2, y_2), and (x_3, y_3) will form a triangle, known as the **control polygon** for the curve. Assume in this problem that $w > 0$. By calculating $\mathbf{x}'(0)$ and $\mathbf{x}'(1)$, show that the tangent lines to the curve at $\mathbf{x}(0)$ and $\mathbf{x}(1)$ intersect at (x_2, y_2). Hence, the control triangle has two of its sides tangent to the curve.

11. In this problem, you will establish the geometric significance of the constant w appearing in the equations in (1).

(a) Calculate the distance a between $\mathbf{x}(\frac{1}{2})$ and (x_2, y_2).

(b) Calculate the distance b between $\mathbf{x}(\frac{1}{2})$ and the midpoint of the line segment joining (x_1, y_1) and (x_3, y_3).

(c) Show that $w = b/a$. By part (b) of Exercise 9, $\mathbf{x}(\frac{1}{2})$ divides the line segment joining (x_2, y_2) to the midpoint of the line segment joining (x_1, y_1) to (x_3, y_3) into two pieces, and w represents the ratio of the lengths of the two pieces.

12. Determine the Bézier parametrization for the portion of the parabola $y = x^2$ between the points $(-2, 4)$ and $(2, 4)$ as follows:

(a) Two of the three control points must be $(-2, 4)$ and $(2, 4)$. Find the third control point using the result of Exercise 10.

(b) Using part (a) and Exercise 9, we must have that $\mathbf{x}(\frac{1}{2})$ lies on the y-axis, and hence at the point $(0, 0)$. Use the result of Exercise 11 to determine the constant w.

(c) Now write the Bézier parametrization. You should be able to check that your answer is correct.

13. Let $\mathbf{x}\colon (0, \pi) \to \mathbf{R}^2$ be the path given by

$$\mathbf{x}(t) = \left(\sin t, \cos t + \ln \tan \tfrac{t}{2}\right),$$

where t is the angle that the y-axis makes with the vector $\mathbf{x}(t)$. The image of $\mathbf{x}$ is called the **tractrix**. (See Figure 3.48.)

(a) Show that $\mathbf{x}$ has nonzero speed except when $t = \pi/2$.

(b) Show that the length of the segment of the tangent to the tractrix between the point of tangency and the y-axis is always equal to 1. This means that the image curve has the following description: Let a horse pull a heavy load by a rope of length 1.

[2] P. Bézier was an automobile design engineer for Renault. See D. Cox, J. Little, and D. O'Shea, *Ideals, Varieties, and Algorithms: An Introduction to Computational Algebraic Geometry and Commutative Algebra,* 2nd ed. (Springer-Verlag, New York, 1997), pp. 27–29. Exercises 7–11 adapted with permission.

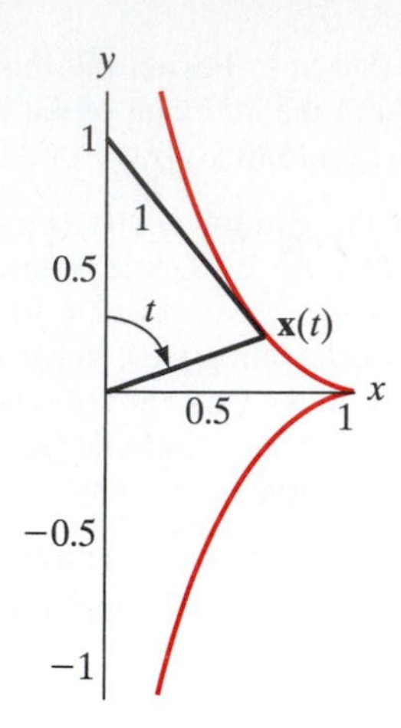

Figure 3.48 The tractrix of Exercise 13.

Suppose that the horse initially is at $(0, 0)$, the load at $(1, 0)$, and let the horse walk along the y-axis. The load follows the image of the tractrix.

14. Another way to parametrize the tractrix path given in Exercise 13 is

$$\mathbf{y}: (-\infty, 0) \to \mathbf{R}^2,$$

$$\text{where} \quad \mathbf{y}(r) = \left(e^r, \int_0^r \sqrt{1 - e^{2\rho}}\, d\rho \right).$$

(a) Show that $\mathbf{y}$ satisfies the property described in part (b) of Exercise 13.

(b) In fact, $\mathbf{y}$ is actually a reparametrization of *part* of the path $\mathbf{x}$ of Exercise 13. Without proving this fact in detail, indicate what portion of the image of $\mathbf{x}$ the image of $\mathbf{y}$ covers.

15. Suppose that a plane curve is given in polar coordinates by the equation $r = f(\theta)$, where f is a function of class C^2. Use equation (17) in §3.2 to derive the curvature formula

$$\kappa(\theta) = \frac{|r^2 - rr'' + 2r'^2|}{(r^2 + r'^2)^{3/2}}.$$

(Hint: First give parametric equations for the curve in Cartesian coordinates using θ as the parameter.)

16. Use the result of Exercise 15 to calculate the curvature of the lemniscate $r^2 = \cos 2\theta$.

Let $\mathbf{x}: I \to \mathbf{R}^2$ be a path of class C^2 that is not a straight line and such that $\mathbf{x}'(t) \neq \mathbf{0}$. Choose some $t_0 \in I$ and let

$$\mathbf{y}(t) = \mathbf{x}(t) - s(t)\mathbf{T}(t),$$

where $s(t) = \int_{t_0}^t \|\mathbf{x}'(\tau)\|\, d\tau$ is the arclength function and $\mathbf{T}$ is the unit tangent vector. The path $\mathbf{y}: I \to \mathbf{R}^2$ is called the ***involute*** *of $\mathbf{x}$. Exercises 17–19 concern involutes of paths.*

17. (a) Calculate the involute of the circular path of radius a, that is, $\mathbf{x}(t) = (a \cos t, a \sin t)$. (Take t_0 to be 0.)

(b) Let $a = 1$ and use a computer to graph the path $\mathbf{x}$ and the involute path $\mathbf{y}$ on the same set of axes.

18. Show that the unit tangent vector to the involute at t is the opposite of the unit normal vector $\mathbf{N}(t)$ to the original path $\mathbf{x}$. (Hint: Use the Frenet–Serret formulas and the fact that a plane curve has torsion equal to zero everywhere.)

19. Show that the involute $\mathbf{y}$ of the path $\mathbf{x}$ is formed by unwinding a taut string that has been wrapped around $\mathbf{x}$ as follows:

(a) Show that the distance in $\mathbf{R}^2$ between a point $\mathbf{x}(t)$ on the original path and the corresponding point $\mathbf{y}(t)$ on the involute is equal to the distance traveled from $\mathbf{x}(t_0)$ to $\mathbf{x}(t)$ along the underlying curve of $\mathbf{x}$.

(b) Show that the distance between a point $\mathbf{x}(t)$ on the path and the corresponding point $\mathbf{y}(t)$ on the involute is equal to the distance from $\mathbf{x}(t)$ to $\mathbf{y}(t)$ measured along the tangent emanating from $\mathbf{x}(t)$. Then finish the argument.

Let $\mathbf{x}: I \to \mathbf{R}^2$ be a path of class C^2 that is not a straight line and such that $\mathbf{x}'(t) \neq \mathbf{0}$. Let

$$\mathbf{e}(t) = \mathbf{x}(t) + \frac{1}{\kappa}\mathbf{N}(t).$$

This is the path traced by the center of the osculating circle of the path $\mathbf{x}$. The quantity $\rho = 1/\kappa$ is the radius of the osculating circle and is called the ***radius of curvature*** *of the path $\mathbf{x}$. The path $\mathbf{e}$ is called the* ***evolute*** *of the path $\mathbf{x}$. Exercises 20–25 involve evolutes of paths.*

20. Let $\mathbf{x}(t) = (t, t^2)$ be a parabolic path. (See Figure 3.49.)

(a) Find the unit tangent vector $\mathbf{T}$, the unit normal vector $\mathbf{N}$, and the curvature κ as functions of t.

(b) Calculate the evolute of $\mathbf{x}$.

(c) Use a computer to plot $\mathbf{x}(t)$ and $\mathbf{e}(t)$ on the same set of axes.

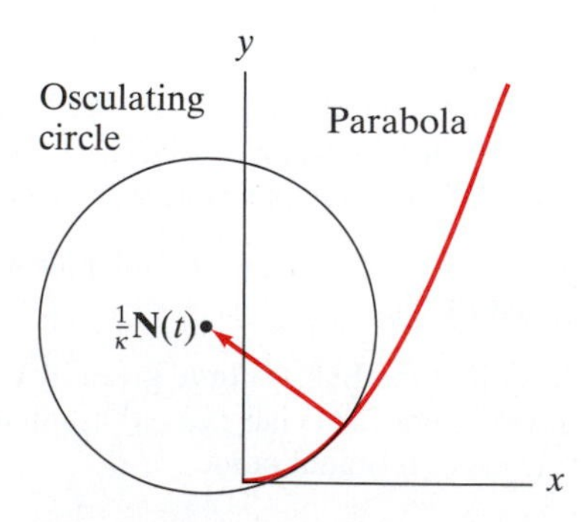

Figure 3.49 The parabola and its osculating circle at a point. The centers of the osculating circles at all points of the parabola trace the evolute of the parabola as described in Exercise 20.

21. Show that the evolute of a circular path is a point.

22. (a) Use a computer algebra system to calculate the formula for the evolute of the elliptical path $\mathbf{x}(t) = (a\cos t, b\sin t)$.

(b) Use a computer to plot $\mathbf{x}(t)$ and the evolute $\mathbf{e}(t)$ on the same set of axes for various values of the constants a and b. What happens to the evolute when a becomes close in value to b?

23. Use a computer algebra system to calculate the formula for the evolute of the cycloid $\mathbf{x}(t) = (at - a\sin t, a - a\cos t)$. What do you find?

24. Use a computer algebra system to calculate the formula for the evolute of the cardioid $\mathbf{x}(t) = (2a\cos t(1 + a\cos t), 2a\sin t(1 + a\cos t))$.

25. Assuming $\kappa'(t) \neq 0$, show that the unit tangent vector to the evolute $\mathbf{e}(t)$ is parallel to the unit normal vector $\mathbf{N}(t)$ to the original path $\mathbf{x}(t)$.

26. Suppose that a C^1 path $\mathbf{x}(t)$ is such that both its velocity and acceleration are unit vectors for all t. Show that $\kappa = 1$ for all t.

27. Consider the plane curve parametrized by

$$x(s) = \int_0^s \cos g(t)\,dt, \quad y(s) = \int_0^s \sin g(t)\,dt,$$

where g is a differentiable function.

(a) Show that the parameter s is the arclength parameter.

(b) Calculate the curvature $\kappa(s)$.

(c) Use part (b) to explain how you can create a parametrized plane curve with any specified (nonnegative) curvature function $\kappa(s)$.

(d) Give a set of parametric equations for a curve whose curvature $\kappa(s) = |s|$. (Your answer should involve integrals.)

(e) Use a computer to graph the curve you found in part (d), known as a **clothoid** or a **spiral of Cornu**. (Note: The integrals involved are known as **Fresnel integrals** and arise in the study of optics. You must evaluate these integrals numerically in order to graph the curve.)

28. Suppose that $\mathbf{x}$ is a C^3 path in $\mathbf{R}^3$ with torsion τ always equal to 0.

(a) Explain why $\mathbf{x}$ must have a constant binormal vector (i.e., one whose direction must remain fixed for all t).

(b) Suppose we have chosen coordinates so that $\mathbf{x}(0) = \mathbf{0}$ and that $\mathbf{v}(0)$ and $\mathbf{a}(0)$ lie in the xy-plane (i.e., have no $\mathbf{k}$-component). Then what must the binormal vector $\mathbf{B}$ be?

(c) Using the coordinate assumptions in part (b), show that $\mathbf{x}(t)$ must lie in the xy-plane for all t. (Hint: Begin by explaining why $\mathbf{v}(t)\cdot\mathbf{k} = \mathbf{a}(t)\cdot\mathbf{k} = 0$ for all t. Then show that if

$$\mathbf{x}(t) = x(t)\mathbf{i} + y(t)\mathbf{j} + z(t)\mathbf{k},$$

we must have $z(t) = 0$ for all t.)

(d) Now explain how we may conclude that curves with zero torsion must lie in a plane.

29. Suppose that $\mathbf{x}$ is a C^3 path in $\mathbf{R}^3$, parametrized by arclength, with $\kappa \neq 0$. Suppose that the image of $\mathbf{x}$ lies in the xy-plane.

(a) Explain why $\mathbf{x}$ must have a constant binormal vector.

(b) Show that the torsion τ must always be zero.

Note that there is really nothing special about the image of $\mathbf{x}$ lying in the xy-plane, so that this exercise, combined with the results of Exercise 28, shows that the image of $\mathbf{x}$ is a plane curve if and only if τ is always zero and if and only if $\mathbf{B}$ is a constant vector.

30. In Example 7 of §3.2 we saw that if $\mathbf{x}$ is a straight-line path, then $\mathbf{x}$ has zero curvature. Demonstrate the converse, that is, if $\mathbf{x}$ is a C^2 path parametrized by arclength s and has zero curvature for all s, then $\mathbf{x}$ traces a straight line.

31. A large piece of cylindrical metal pipe is to be manufactured to include a **strake,** which is a spiraling strip of metal that offers structural support for the pipe. (See Figure 3.50.) The pieces of the strake are to be made from flat pieces of flexible metal whose curved sides are arcs of circles as shown in Figure 3.51. Assume that the pipe has a radius of a ft and that the strake makes one complete revolution around the pipe every h ft.[3]

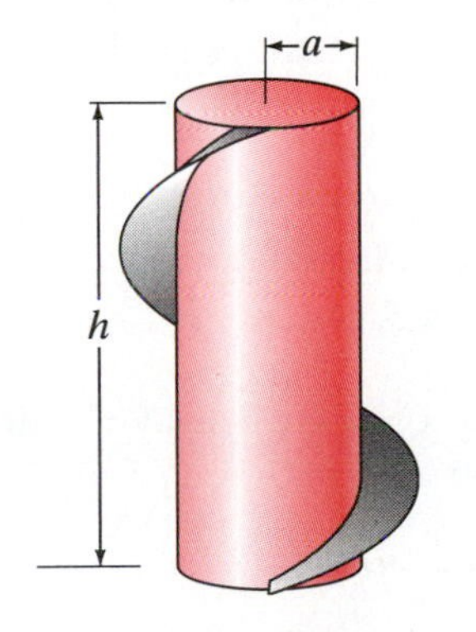

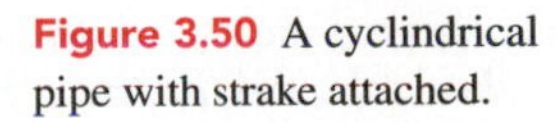
Figure 3.50 A cyclindrical pipe with strake attached.

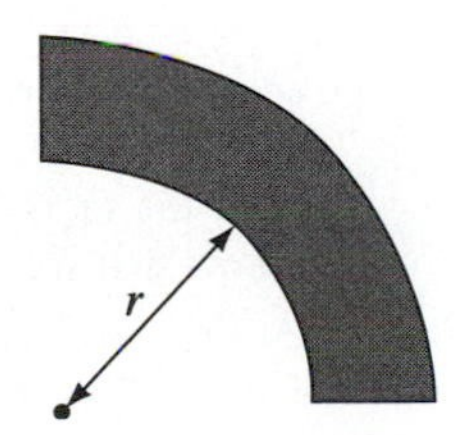

Figure 3.51 A section of the strake. (See Exercise 31.)

[3]See F. Morgan, *Riemannian Geometry: A Beginner's Guide*, 2nd ed. (A K Peters, Wellesley, 1998), pp. 7–10. Figures 3.50 and 3.51 adapted with permission.

(a) In terms of a and h, what should the inner radius r be so that the strake will fit snugly against the pipe?

(b) Suppose $a = 3$ ft and $h = 25$ ft. What is r?

Suppose that $\mathbf{x}: I \to \mathbf{R}^3$ *is a path of class* C^3 *parametrized by arclength. Then the unit tangent vector* $\mathbf{T}(s)$ *defines a vector-valued function* $\mathbf{T}: I \to \mathbf{R}^3$ *that may also be considered to be a path (although not necessarily one parametrized by arclength, nor necessarily one with nonvanishing velocity). Since* $\mathbf{T}$ *is a* unit vector, *the image of the path* $\mathbf{T}$ *must lie on a sphere of radius* 1 *centered at the origin. This image curve is called the* ***tangent spherical image*** *of* $\mathbf{x}$. *Likewise, we may consider the functions defined by the normal and binormal vectors* $\mathbf{N}$ *and* $\mathbf{B}$ *to give paths called, respectively, the* ***normal spherical image*** *and* ***binormal spherical image*** *of* $\mathbf{x}$. *Exercises 32–35 concern these notions.*

32. Find the tangent spherical image, normal spherical image, and binormal spherical image of the circular helix $\mathbf{x}(t) = (a\cos t, a\sin t, bt)$. (Note: The path $\mathbf{x}$ is not parametrized by arclength.)

33. Suppose that $\mathbf{x}$ is parametrized by arclength. Show that $\mathbf{x}$ is a straight-line path if and only if its tangent spherical image is a constant path. (See Example 7 of §3.2 and Exercise 30.)

34. Suppose that $\mathbf{x}$ is parametrized by arclength. Show that the image of $\mathbf{x}$ lies in a plane if and only if its binormal spherical image is constant. (See Exercises 28 and 29.)

35. Suppose that $\mathbf{x}$ is parametrized by arclength. Show that the normal spherical image of $\mathbf{x}$ can never be constant.

36. In this problem, we will find expressions for velocity and acceleration in cylindrical coordinates. We begin with the expression

$$\mathbf{x}(t) = x(t)\mathbf{i} + y(t)\mathbf{j} + z(t)\mathbf{k}$$

for the path in Cartesian coordinates.

(a) Recall that the standard basis vectors for cylindrical coordinates are

$$\begin{aligned}\mathbf{e}_r &= \cos\theta\,\mathbf{i} + \sin\theta\,\mathbf{j},\\ \mathbf{e}_\theta &= -\sin\theta\,\mathbf{i} + \cos\theta\,\mathbf{j},\\ \mathbf{e}_z &= \mathbf{k}.\end{aligned}$$

Use the facts that $x = r\cos\theta$ and $y = r\sin\theta$ to show that we may write $\mathbf{x}(t)$ as

$$\mathbf{x}(t) = r(t)\,\mathbf{e}_r + z(t)\,\mathbf{e}_z.$$

(b) Use the definitions of $\mathbf{e}_r$, $\mathbf{e}_\theta$, and $\mathbf{e}_z$ just given and the chain rule to find $d\mathbf{e}_r/dt$, $d\mathbf{e}_\theta/dt$, and $d\mathbf{e}_z/dt$ in terms of $\mathbf{e}_r$, $\mathbf{e}_\theta$, and $\mathbf{e}_z$.

(c) Now use the product rule to give expressions for $\mathbf{v}$ and $\mathbf{a}$ in terms of the standard basis for cylindrical coordinates.

37. Suppose that the path

$$\mathbf{x}(t) = (\sin 2t, \sqrt{2}\cos 2t, \sin 2t - 2)$$

describes the position of the Starship Inertia at time t.

(a) Lt. Commander Agnes notices that the ship is tracing a closed loop. What is the length of this loop?

(b) Ensign Egbert reports that the Inertia's path is actually a flow line of the Martian vector field $\mathbf{F}(x, y, z) = y\mathbf{i} - 2x\mathbf{j} + y\mathbf{k}$, but he omitted a constant factor when he entered this information in his log. Help him set things right by finding the correct vector field.

38. Suppose that the temperature at points inside a room is given by a differentiable function $T(x, y, z)$. Livinia, the housefly (who is recovering from a head cold), is in the room and desires to warm up as rapidly as possible.

(a) Show that Livinia's path $\mathbf{x}(t)$ must be a flow line of $k\nabla T$, where k is a positive constant.

(b) If $T(x, y, z) = x^2 - 2y^2 + 3z^2$ and Livinia is initially at the point $(2, 3, -1)$, describe her path explicitly.

39. Let $\mathbf{F} = u(x, y)\,\mathbf{i} - v(x, y)\,\mathbf{j}$ be an incompressible, irrotational vector field of class C^2.

(a) Show that the functions u and v (which determine the component functions of $\mathbf{F}$) satisfy the **Cauchy–Riemann equations**

$$\frac{\partial u}{\partial x} = \frac{\partial v}{\partial y}, \quad \text{and} \quad \frac{\partial u}{\partial y} = -\frac{\partial v}{\partial x}.$$

(b) Show that u and v are **harmonic,** that is, that

$$\frac{\partial^2 u}{\partial x^2} + \frac{\partial^2 u}{\partial y^2} = 0 \quad \text{and} \quad \frac{\partial^2 v}{\partial x^2} + \frac{\partial^2 v}{\partial y^2} = 0.$$

40. Suppose that a particle of mass m travels along a path $\mathbf{x}$ according to Newton's second law $\mathbf{F} = m\mathbf{a}$, where $\mathbf{F}$ is a gradient vector field. If the particle is also constrained to lie on an equipotential surface of $\mathbf{F}$, show that then it must have constant speed.

41. Let a particle of mass m travel along a differentiable path $\mathbf{x}$ in a Newtonian vector field $\mathbf{F}$ (i.e., one that satisfies Newton's second law $\mathbf{F} = m\mathbf{a}$, where $\mathbf{a}$ is the acceleration of $\mathbf{x}$). We define the **angular momentum** $\mathbf{l}(t)$ of the particle to be the cross product of the position vector and the linear momentum $m\mathbf{v}$, i.e.,

$$\mathbf{l}(t) = \mathbf{x}(t) \times m\mathbf{v}(t).$$

(Here $\mathbf{v}$ denotes the velocity of $\mathbf{x}$.) The **torque** about the origin of the coordinate system due to the force $\mathbf{F}$ is the cross product of position and force:

$$\mathbf{M}(t) = \mathbf{x}(t) \times \mathbf{F}(t) = \mathbf{x}(t) \times m\mathbf{a}(t).$$

(Also see §1.4 concerning the notion of torque.) Show that

$$\frac{d\mathbf{l}}{dt} = \mathbf{M}.$$

Thus, we see that the rate of change of angular momentum is equal to the torque imparted to the particle by the vector field $\mathbf{F}$.

42. Consider the situation in Exercise 41 and suppose that $\mathbf{F}$ is a central force (i.e., a force that always points directly towards or away from the origin). Show that in this case the angular momentum is *conserved*, that is, that it must remain constant.

43. Can the vector field

$$\mathbf{F} = (e^x \cos y + e^{-x} \sin z)\,\mathbf{i} - e^x \sin y\,\mathbf{j} + e^{-x} \cos z\,\mathbf{k}$$

be the gradient of a function $f(x, y, z)$ of class C^2? Why or why not?

44. Can the vector field

$$\mathbf{F} = x(y^2 + 1)\,\mathbf{i} + (ye^x - e^z)\,\mathbf{j} + x^2 e^z\,\mathbf{k}$$

be the curl of another vector field $\mathbf{G}(x, y, z)$ of class C^2? Why or why not?

CHAPTER 4

Maxima and Minima in Several Variables

4.1 Differentials and Taylor's Theorem

Among all classes of functions of one or several variables, polynomials are without a doubt the nicest in that they are continuous and differentiable everywhere and display intricate and interesting behavior. Our goal in this section is to provide a means of approximating any scalar-valued function by a polynomial of given degree, known as the **Taylor polynomial.** Because of the relative ease with which one can calculate with them, Taylor polynomials are useful for work in computer graphics and computer-aided design, to name just two areas.

Taylor's Theorem in One Variable: a Review

Suppose you have a function $f: X \subseteq \mathbf{R} \to \mathbf{R}$ that is differentiable at a point a in X. Then the equation for the tangent line gives the best linear approximation for f near a. That is, when we define p_1 by

$$p_1(x) = f(a) + f'(a)(x - a), \quad \text{we have} \quad p_1(x) \approx f(x) \text{ if } x \approx a.$$

(See Figure 4.1.) As explained in §2.3, the phrase "best linear approximation" means that if we take $R_1(x, a)$ to be $f(x) - p_1(x)$, then

$$\lim_{x \to a} \frac{R_1(x, a)}{x - a} = 0.$$

Note that, in particular, we have $p_1(a) = f(a)$ and $p_1'(a) = f'(a)$.

Generally, tangent lines approximate graphs of functions only over very small neighborhoods containing the point of tangency. For a better approximation, we might try to fit a parabola that hugs the function's graph more closely as in

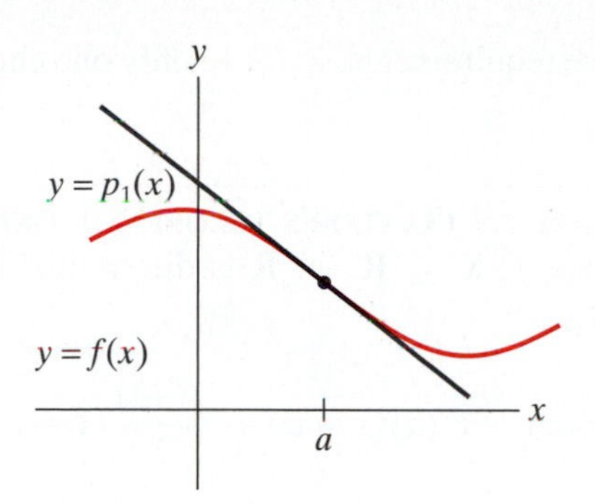

Figure 4.1 The graph of $y = f(x)$ and its tangent line $y = p_1(x)$ at $x = a$.

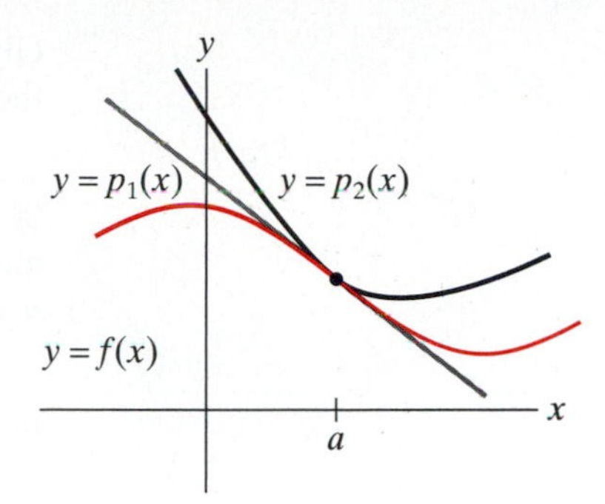

Figure 4.2 The tangent graphs of f, p_1, and p_2.

Figure 4.2. In this case, we want p_2 to be the quadratic function such that

$$p_2(a) = f(a), \quad p_2'(a) = f'(a), \quad \text{and} \quad p_2''(a) = f''(a).$$

The only quadratic polynomial that satisfies these three conditions is

$$p_2(x) = f(a) + f'(a)(x-a) + \frac{f''(a)}{2}(x-a)^2.$$

It can be proved that, if f is of class C^2, then

$$f(x) = p_2(x) + R_2(x, a),$$

where

$$\lim_{x \to a} \frac{R_2(x, a)}{(x-a)^2} = 0.$$

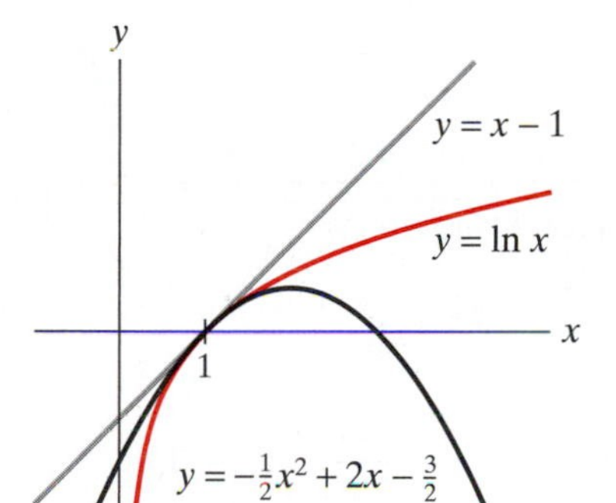

Figure 4.3 Approximations to $f(x) = \ln x$.

EXAMPLE 1 If $f(x) = \ln x$, then, for $a = 1$, we have

$$f(1) = \ln 1 = 0,$$
$$f'(1) = \frac{1}{1} = 1,$$
$$f''(1) = -\frac{1}{1^2} = -1.$$

Hence,

$$p_1(x) = 0 + 1(x-1) = x - 1,$$
$$p_2(x) = 0 + 1(x-1) - \tfrac{1}{2}(x-1)^2 = -\tfrac{1}{2}x^2 + 2x - \tfrac{3}{2}.$$

The approximating polynomials p_1 and p_2 are shown in Figure 4.3. ◆

There is no reason to stop with quadratic polynomials. Suppose we want to approximate f by a polynomial p_k of degree k, where k is a positive integer. Analogous to the work above, we require that p_k and its first k derivatives agree with f and its first k derivatives at the point a. Thus, we demand that

$$p_k(a) = f(a),$$
$$p_k'(a) = f'(a),$$
$$p_k''(a) = f''(a),$$
$$\vdots$$
$$p_k^{(k)}(a) = f^{(k)}(a).$$

Given these requirements, we have only one choice for p_k, stated in the following theorem:

Theorem 1.1 (TAYLOR'S THEOREM IN ONE VARIABLE) Let X be open in $\mathbf{R}$ and suppose $f: X \subseteq \mathbf{R} \to \mathbf{R}$ is differentiable up to (at least) order k. Given $a \in X$, let

$$p_k(x) = f(a) + f'(a)(x-a) + \frac{f''(a)}{2}(x-a)^2 + \cdots + \frac{f^{(k)}(a)}{k!}(x-a)^k. \quad (1)$$

Then

$$f(x) = p_k(x) + R_k(x, a),$$

where the remainder term R_k is such that $R_k(x, a)/(x-a)^k \to 0$ as $x \to a$.

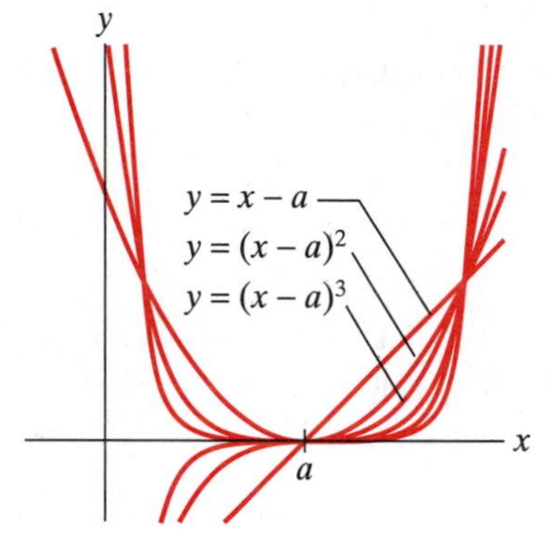

Figure 4.4 The graphs of
(1) $y = x - a$,
(2) $y = (x-a)^2$, and
(3) $y = (x-a)^3$.
Note how much more closely the graph of (3) hugs the x-axis than that of (1) or (2).

The polynomial defined by formula (1) is called the **kth-order Taylor polynomial of f at a**. The essence of Taylor's theorem is this: For x near a, the Taylor polynomial p_k approximates f in the sense that the error R_k involved in making this approximation tends to zero even faster than $(x-a)^k$ does. When k is large, this is very fast indeed, as we see graphically in Figure 4.4.

EXAMPLE 2 Consider $\ln x$ with $a = 1$ again. We calculate

$$\begin{aligned} f(1) &= \ln 1 = 0, \\ f'(1) &= \frac{1}{1} = 1, \\ f''(1) &= -\frac{1}{1^2} = -1, \\ &\vdots \\ f^{(k)}(1) &= \frac{(-1)^{k-1}(k-1)!}{1^k} = (-1)^{k+1}(k-1)!. \end{aligned}$$

Therefore,

$$p_k(x) = (x-1) - \frac{1}{2}(x-1)^2 + \frac{1}{3}(x-1)^3 - \cdots + \frac{(-1)^{k-1}}{k}(x-1)^k. \quad ◆$$

Taylor's theorem as stated in Theorem 1.1 says nothing explicit about the remainder term R_k. However, it is possible to establish the following derivative form for the remainder:

Proposition 1.2 If f is differentiable up to order $k+1$, then there exists some number z between a and x such that

$$R_k(x, a) = \frac{f^{(k+1)}(z)}{(k+1)!}(x-a)^{k+1}. \quad (2)$$

In practice, formula (2) is quite useful for estimating the error involved with a Taylor polynomial approximation.

EXAMPLE 3 The fifth-order Taylor polynomial of $f(x) = \cos x$ about $x = \pi/2$ is

$$p_5(x) = -\left(x - \frac{\pi}{2}\right) + \frac{1}{6}\left(x - \frac{\pi}{2}\right)^3 - \frac{1}{120}\left(x - \frac{\pi}{2}\right)^5.$$

(You should verify this calculation.) According to formula (2), the difference between p_5 and $\cos x$ is

$$R_5\left(x, \frac{\pi}{2}\right) = \frac{f^{(6)}(z)}{6!}\left(x - \frac{\pi}{2}\right)^6 = -\frac{\cos z}{6!}\left(x - \frac{\pi}{2}\right)^6,$$

where z is some number between $\pi/2$ and x. Since $|\cos x|$ is never larger than 1, we have

$$\left|R_5\left(x, \frac{\pi}{2}\right)\right| = \left|\frac{\cos z}{6!}\left(x - \frac{\pi}{2}\right)^6\right| \le \frac{(x - \pi/2)^6}{720}.$$

Thus, for x in the interval $[0, \pi]$, we have

$$\left|R_5\left(x, \frac{\pi}{2}\right)\right| \le \frac{(\pi - \pi/2)^6}{720} = \frac{\pi^6}{46{,}080} \approx 0.0209.$$

In other words, the use of the polynomial p_5 above in place of $\cos x$ will be accurate to at least 0.0209 throughout the interval $[0, \pi]$. ◆

Taylor's Theorem in Several Variables: The First-order Formula

For the moment, suppose that $f: X \subseteq \mathbf{R}^2 \to \mathbf{R}$ is a function of two variables, where X is open in $\mathbf{R}^2$ and of class C^1. Then near the point $(a, b) \in X$, the best linear approximation to f is provided by the equation giving the tangent plane at $(a, b, f(a, b))$. That is,

$$f(x, y) \approx p_1(x, y),$$

where

$$p_1(x, y) = f(a, b) + f_x(a, b)(x - a) + f_y(a, b)(y - b).$$

Note that the linear polynomial p_1 has the property that

$$p_1(a, b) = f(a, b);$$
$$\frac{\partial p_1}{\partial x}(a, b) = \frac{\partial f}{\partial x}(a, b),$$
$$\frac{\partial p_1}{\partial y}(a, b) = \frac{\partial f}{\partial y}(a, b).$$

Such an approximation is shown in Figure 4.5.

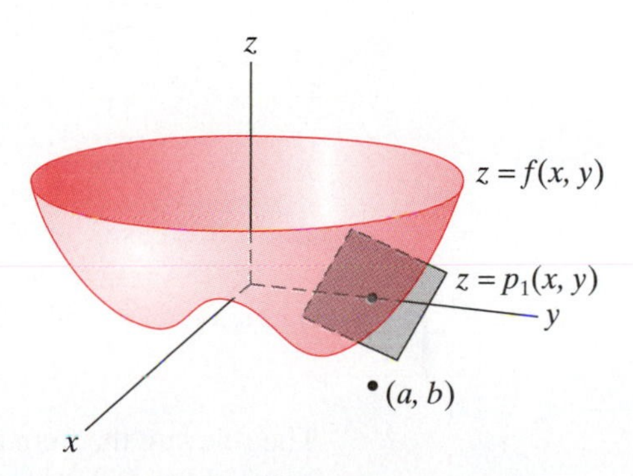

Figure 4.5 The graph of $z = f(x, y)$ and $z = p_1(x, y)$.

To generalize this situation to the case of a function $f: X \subseteq \mathbf{R}^n \to \mathbf{R}$ of class C^1, we naturally use the equation for the *tangent hyperplane*. That is, if $\mathbf{a} = (a_1, a_2, \dots, a_n) \in X$, then

$$f(x_1, x_2, \dots, x_n) \approx p_1(x_1, x_2, \dots, x_n),$$

where

$$p_1(x_1, \dots, x_n) = f(\mathbf{a}) + f_{x_1}(\mathbf{a})(x_1 - a_1) + f_{x_2}(\mathbf{a})(x_2 - a_2) + \cdots + f_{x_n}(\mathbf{a})(x_n - a_n).$$

Of course, the formula for p_1 can be written more compactly using either Σ-notation or, better still, matrices:

$$p_1(x_1, \dots, x_n) = f(\mathbf{a}) + \sum_{i=1}^{n} f_{x_i}(\mathbf{a})(x_i - a_i) = f(\mathbf{a}) + Df(\mathbf{a})(\mathbf{x} - \mathbf{a}). \quad (3)$$

EXAMPLE 4 Let $f(x_1, x_2, x_3, x_4) = x_1 + 2x_2 + 3x_3 + 4x_4 + x_1x_2x_3x_4$. Then

$$\frac{\partial f}{\partial x_1} = 1 + x_2x_3x_4, \qquad \frac{\partial f}{\partial x_2} = 2 + x_1x_3x_4,$$

$$\frac{\partial f}{\partial x_3} = 3 + x_1x_2x_4, \qquad \frac{\partial f}{\partial x_4} = 4 + x_1x_2x_3.$$

At $\mathbf{a} = \mathbf{0} = (0, 0, 0, 0)$, we have

$$\frac{\partial f}{\partial x_1}(\mathbf{0}) = 1, \qquad \frac{\partial f}{\partial x_2}(\mathbf{0}) = 2, \qquad \frac{\partial f}{\partial x_3}(\mathbf{0}) = 3, \qquad \frac{\partial f}{\partial x_4}(\mathbf{0}) = 4.$$

Thus,

$$p_1(x_1, x_2, x_3, x_4) = 0 + 1(x_1 - 0) + 2(x_2 - 0) + 3(x_3 - 0) + 4(x_4 - 0) = x_1 + 2x_2 + 3x_3 + 4x_4.$$

Note that p_1 contains precisely the linear terms of the original function f. On the other hand, if $\mathbf{a} = (1, 2, 3, 4)$, then

$$\frac{\partial f}{\partial x_1}(1, 2, 3, 4) = 25, \qquad \frac{\partial f}{\partial x_2}(1, 2, 3, 4) = 14,$$

$$\frac{\partial f}{\partial x_3}(1, 2, 3, 4) = 11, \qquad \frac{\partial f}{\partial x_4}(1, 2, 3, 4) = 10,$$

so that, in this case,

$$p_1(x_1, x_2, x_3, x_4) = 54 + 25(x_1 - 1) + 14(x_2 - 2) + 11(x_3 - 3) + 10(x_4 - 4).$$

◆

The relevant theorem regarding the first-order Taylor polynomial is just a restatement of the definition of differentiability. However, since we plan to consider higher-order Taylor polynomials, we state the theorem explicitly.

■ Theorem 1.3 (FIRST-ORDER TAYLOR'S FORMULA IN SEVERAL VARIABLES) Let X be open in $\mathbf{R}^n$ and suppose that $f\colon X \subseteq \mathbf{R}^n \to \mathbf{R}$ is differentiable at the point $\mathbf{a}$ in X. Let

$$p_1(\mathbf{x}) = f(\mathbf{a}) + Df(\mathbf{a})(\mathbf{x} - \mathbf{a}). \tag{4}$$

Then

$$f(\mathbf{x}) = p_1(\mathbf{x}) + R_1(\mathbf{x}, \mathbf{a}),$$

where $R_1(\mathbf{x}, \mathbf{a})/\|\mathbf{x} - \mathbf{a}\| \to 0$ as $\mathbf{x} \to \mathbf{a}$.

Differentials

Before we explore higher-order versions of Taylor's theorem in several variables, we consider the linear (or first-order) approximation in further detail.

Let $\mathbf{h} = \mathbf{x} - \mathbf{a}$. Then formula (3) becomes

$$p_1(\mathbf{x}) = f(\mathbf{a}) + Df(\mathbf{a})\mathbf{h} = f(\mathbf{a}) + \sum_{i=1}^{n} \frac{\partial f}{\partial x_i}(\mathbf{a})h_i. \tag{5}$$

We focus on the sum appearing in formula (5) and summarize its salient features as follows:

■ Definition 1.4 Let $f : X \subseteq \mathbf{R}^n \to \mathbf{R}$ and let $\mathbf{a} \in X$. The **incremental change of** f, denoted Δf, is

$$\Delta f = f(\mathbf{a} + \mathbf{h}) - f(\mathbf{a}).$$

The **total differential of** f, denoted $df(\mathbf{a}, \mathbf{h})$, is

$$df(\mathbf{a}, \mathbf{h}) = \frac{\partial f}{\partial x_1}(\mathbf{a})h_1 + \frac{\partial f}{\partial x_2}(\mathbf{a})h_2 + \cdots + \frac{\partial f}{\partial x_n}(\mathbf{a})h_n.$$

The significance of the differential is that for $\mathbf{h} \approx \mathbf{0}$,

$$\Delta f \approx df.$$

(We have abbreviated $df(\mathbf{a}, \mathbf{h})$ by df.)

Sometimes h_i is replaced by the expression Δx_i or dx_i to emphasize that it represents a change in the ith independent variable, in which case we write

$$df = \frac{\partial f}{\partial x_1}\,dx_1 + \frac{\partial f}{\partial x_2}\,dx_2 + \cdots + \frac{\partial f}{\partial x_n}\,dx_n.$$

(We've supressed the evaluation of the partial derivatives at $\mathbf{a}$, as is customary.)

EXAMPLE 5 Suppose $f(x, y, z) = \sin(xyz) + \cos(xyz)$. Then

$$\begin{aligned} df &= \frac{\partial f}{\partial x}\,dx + \frac{\partial f}{\partial y}\,dy + \frac{\partial f}{\partial z}\,dz \\ &= yz[\cos(xyz) - \sin(xyz)]dx + xz[\cos(xyz) - \sin(xyz)]dy \\ &\quad + xy[\cos(xyz) - \sin(xyz)]dz \\ &= (\cos(xyz) - \sin(xyz))(yz\,dx + xz\,dy + xy\,dz). \end{aligned}$$

◆

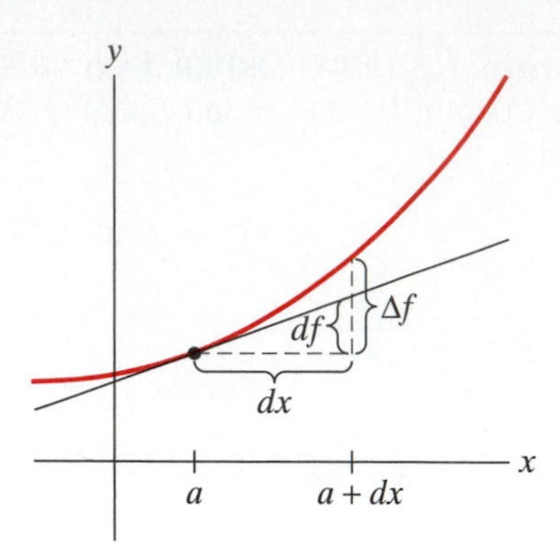

Figure 4.6 The incremental change Δf equals the change in y-coordinate of the graph of $y = f(x)$ as the x-coordinate of a point changes from a to $a + dx$. The differential df equals the change in y-coordinate of the graph of the tangent line at a (i.e., the graph of $y = p_1(x)$).

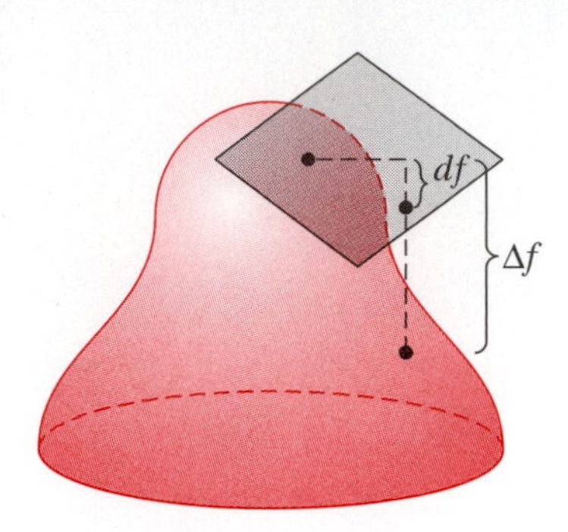

Figure 4.7 The incremental change Δf equals the change in z-coordinate of the graph of $z = f(x, y)$ as a point in $\mathbf{R}^2$ changes from $\mathbf{a} = (a, b)$ to $\mathbf{a} + \mathbf{h} = (a + h, b + k)$. The differential df equals the change in z-coordinate of the graph of the tangent plane at (a, b).

The geometry of the differential arises, naturally enough, from tangent lines and planes. (See Figures 4.6 and 4.7.)

EXAMPLE 6 Let $f(x, y) = x - y + 2x^2 + xy^2$. Then for $(a, b) = (2, -1)$, we have that the increment is

$$\begin{aligned}\Delta f &= f(2 + \Delta x, -1 + \Delta y) - f(2, -1)\\ &= 2 + \Delta x - (-1 + \Delta y) + 2(2 + \Delta x)^2 + (2 + \Delta x)(-1 + \Delta y)^2 - 13\\ &= 10\Delta x - 5\Delta y + 2(\Delta x)^2 - 2\Delta x\Delta y + 2(\Delta y)^2 + \Delta x(\Delta y)^2.\end{aligned}$$

On the other hand,

$$\begin{aligned}df((2, -1), (\Delta x, \Delta y)) &= f_x(2, -1)\Delta x + f_y(2, -1)\Delta y\\ &= (1 + 4x + y^2)|_{(2,-1)}\Delta x + (-1 + 2xy)|_{(2,-1)}\Delta y\\ &= 10\Delta x - 5\Delta y.\end{aligned}$$

We see that df consists of exactly the terms of Δf that are linear in Δx and Δy (i.e., appear to first power only). This will always be the case, of course, since that is the nature of the first-order Taylor approximation. Use of the differential approximation is often sufficient in practice, for when Δx and Δy are small, higher powers of them will be small enough to make virtually negligible contributions to Δf. For example, if Δx and Δy are both 0.01, then

$$df = (0.1 - 0.05) = 0.05$$

and

$$\begin{aligned}\Delta f &= (0.1 - 0.05) + 0.0002 - 0.0002 + 0.0002 + 0.000001\\ &= 0.05 + 0.000201 = 0.050201.\end{aligned}$$

Thus, the values of df and Δf are the same to three decimal places. ◆

EXAMPLE 7 A wooden rectangular block is to be manufactured with dimensions 3 in × 4 in × 6 in. Suppose that the possible error in measuring each dimension of the block is the same. We use differentials to estimate how accurately we must measure the dimensions so that the resulting calculated error in volume is no more than 0.1 in^3.

Let the dimensions of the block be denoted by x (≈ 3 in), y (≈ 4 in), and z (≈ 6 in). Then the volume of the block is

$$V = xyz \quad \text{and} \quad V \approx 3 \cdot 4 \cdot 6 = 72 \text{ in}^3.$$

The error in calculated volume is ΔV, which is approximated by the total differential dV. Thus,

$$\begin{aligned} \Delta V \approx dV &= V_x(3, 4, 6)\Delta x + V_y(3, 4, 6)\Delta y + V_z(3, 4, 6)\Delta z \\ &= 24\Delta x + 18\Delta y + 12\Delta z. \end{aligned}$$

If the error in measuring each dimension is ϵ, then we have $\Delta x = \Delta y = \Delta z = \epsilon$. Therefore,

$$dV = 24\Delta x + 18\Delta y + 12\Delta z = 24\epsilon + 18\epsilon + 12\epsilon = 54\epsilon.$$

To ensure (approximately) that $|\Delta V| \leq 0.1$, we demand

$$|dV| = |54\epsilon| \leq 0.1.$$

Hence,

$$|\epsilon| \leq \frac{0.1}{54} = 0.0019 \text{ in}.$$

So the measurements in each dimension must be accurate to within 0.0019 in. ◆

EXAMPLE 8 The formula for the volume of a cylinder of radius r and height h is $V(r, h) = \pi r^2 h$. If the dimensions are changed by small amounts Δr and Δh, then the resulting change ΔV in volume is approximated by the differential change dV. That is,

$$\Delta V \approx dV = \frac{\partial V}{\partial r}\Delta r + \frac{\partial V}{\partial h}\Delta h = 2\pi r h \Delta r + \pi r^2 \Delta h.$$

Suppose the cylinder is actually a beer can, so that it has approximate dimensions of $r = 1$ in and $h = 5$ in. Then

$$dV = \pi(10\Delta r + \Delta h).$$

This statement shows that, for these particular values of r and h, the volume is approximately 10 times **more sensitive** to changes in radius than changes in height. That is, if the radius is changed by an amount ϵ, then the height must be changed by roughly 10ϵ to keep the volume constant (i.e., to make ΔV zero). We use the word "approximate" because our analysis arises from considering the differential change dV, rather than the actual incremental change ΔV.

This beer can example has real application to product marketing strategies. Because the volume is so much more sensitive to changes in radius than height, it is possible to make a can appear to be larger than standard by decreasing its radius slightly (little enough so as to be hardly noticeable) and increasing the height so no change in volume results. (See Figure 4.8.) This sensitivity analysis shows that even a tiny decrease in radius can force an appreciable compensating increase in height. The result can be quite striking, and these ideas apparently

Figure 4.8 Which would you buy?

have been adopted by at least one brewery. Indeed, this is how the author came to fully appreciate differentials and sensitivity analysis.[1] ◆

Taylor's Theorem in Several Variables: The Second-order Formula

Suppose $f: X \subseteq \mathbf{R}^2 \to \mathbf{R}$ is a C^2 function of two variables. Then we know that the tangent plane gives rise to a linear approximation p_1 of f near a given point (a, b) of X. We can improve on this result by looking for the *quadric surface* that best approximates the graph of $z = f(x, y)$ near $(a, b, f(a, b))$. See Figure 4.9 for an illustration. That is, we search for a degree two polynomial $p_2(x, y) = Ax^2 + Bxy + Cy^2 + Dx + Ey + F$ such that, for $(x, y) \approx (a, b)$,

$$f(x, y) \approx p_2(x, y).$$

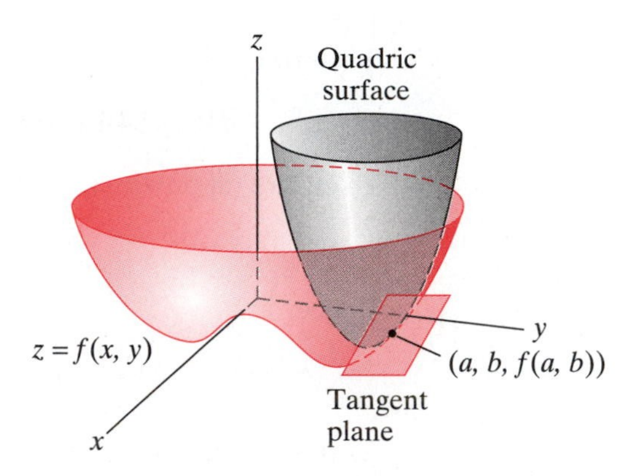

Figure 4.9 The tangent plane and quadric surface.

Analogous to the linear approximation p_1, it is reasonable to require that p_2 and all of its first- and second-order partial derivatives agree with those of f at the point (a, b). That is, we demand

$$\begin{gathered} p_2(a, b) = f(a, b), \\ \frac{\partial p_2}{\partial x}(a, b) = \frac{\partial f}{\partial x}(a, b), \qquad \frac{\partial p_2}{\partial y}(a, b) = \frac{\partial f}{\partial y}(a, b), \\ \frac{\partial^2 p_2}{\partial x^2}(a, b) = \frac{\partial^2 f}{\partial x^2}(a, b), \qquad \frac{\partial^2 p_2}{\partial x \partial y}(a, b) = \frac{\partial^2 f}{\partial x \partial y}(a, b), \\ \frac{\partial^2 p_2}{\partial y^2}(a, b) = \frac{\partial^2 f}{\partial y^2}(a, b). \end{gathered} \tag{6}$$

After some algebra, we see that the only second-degree polynomial meeting these requirements is

$$\begin{aligned} p_2(x, y) = {} & f(a, b) + f_x(a, b)(x - a) + f_y(a, b)(y - b) \\ & + \tfrac{1}{2} f_{xx}(a, b)(x - a)^2 + f_{xy}(a, b)(x - a)(y - b) \\ & + \tfrac{1}{2} f_{yy}(a, b)(y - b)^2. \end{aligned} \tag{7}$$

[1] See S. J. Colley, *The College Mathematics Journal*, **25** (1994), no. 3, 226–227.

How does formula (7) generalize to functions of n variables? We need to begin by demanding conditions analogous to those in (6) for a function $f: X \subseteq \mathbf{R}^n \to \mathbf{R}$. For $\mathbf{a} = (a_1, a_2, \ldots, a_n) \in X$, these conditions are

$$\begin{aligned} p_2(\mathbf{a}) &= f(\mathbf{a}), \\ \frac{\partial p_2}{\partial x_i}(\mathbf{a}) &= \frac{\partial f}{\partial x_i}(\mathbf{a}), \qquad i = 1, 2, \ldots, n, \\ \frac{\partial^2 p_2}{\partial x_i \partial x_j}(\mathbf{a}) &= \frac{\partial^2 f}{\partial x_i \partial x_j}(\mathbf{a}), \quad i, j = 1, 2, \ldots, n. \end{aligned} \tag{8}$$

If you do some algebra (which we omit), you will find that the only polynomial of degree two that satisfies the conditions in (8) is

$$p_2(\mathbf{x}) = f(\mathbf{a}) + \sum_{i=1}^{n} f_{x_i}(\mathbf{a})(x_i - a_i) + \frac{1}{2}\sum_{i,j=1}^{n} f_{x_i x_j}(\mathbf{a})(x_i - a_i)(x_j - a_j). \tag{9}$$

(Note that the second sum appearing in (9) is a double sum consisting of n^2 terms.) To check that everything is consistent when $n = 2$, we have

$$\begin{aligned} p_2(x_1, x_2) = {} & f(a_1, a_2) + f_{x_1}(a_1, a_2)(x_1 - a_1) + f_{x_2}(a_1, a_2)(x_2 - a_2) \\ & + \tfrac{1}{2}\left[f_{x_1 x_1}(a_1, a_2)(x_1 - a_1)^2 + f_{x_1 x_2}(a_1, a_2)(x_1 - a_1)(x_2 - a_2) \right. \\ & \left. + f_{x_2 x_1}(a_1, a_2)(x_2 - a_2)(x_1 - a_1) + f_{x_2 x_2}(a_1, a_2)(x_2 - a_2)^2\right]. \end{aligned}$$

When f is a C^2 function, the two mixed partials are the same, so this formula agrees with formula (7).

EXAMPLE 9 Let $f(x, y, z) = e^{x+y+z}$ and let $\mathbf{a} = (a, b, c) = (0, 0, 0)$. Then

$$\begin{aligned} f(0,0,0) &= e^0 = 1, \\ f_x(0,0,0) &= f_y(0,0,0) = f_z(0,0,0) = e^0 = 1, \\ f_{xx}(0,0,0) &= f_{xy}(0,0,0) = f_{xz}(0,0,0) = f_{yy}(0,0,0) \\ &= f_{yz}(0,0,0) = f_{zz}(0,0,0) = e^0 = 1. \end{aligned}$$

Thus,

$$\begin{aligned} p_2(x, y, z) &= 1 + 1(x-0) + 1(y-0) + 1(z-0) \\ &\quad + \tfrac{1}{2}\left[1(x-0)^2 + 2\cdot 1(x-0)(y-0) + 2\cdot 1(x-0)(z-0)\right. \\ &\quad \left. + 1(y-0)^2 + 2\cdot 1(y-0)(z-0) + 1(z-0)^2\right] \\ &= 1 + x + y + z + \tfrac{1}{2}x^2 + xy + xz + \tfrac{1}{2}y^2 + yz + \tfrac{1}{2}z^2 \\ &= 1 + (x+y+z) + \tfrac{1}{2}(x+y+z)^2. \end{aligned}$$

We have made use of the fact that, since f is of class C^2, a term like

$$f_{xy}(0,0,0)(x-0)(y-0) \quad \text{is equal to} \quad f_{yx}(0,0,0)(y-0)(x-0). \qquad \blacklozenge$$

Now we state the second-order version of Taylor's theorem precisely.

■ **Theorem 1.5** (SECOND-ORDER TAYLOR'S FORMULA) Let X be open in $\mathbf{R}^n$, and suppose that $f: X \subseteq \mathbf{R}^n \to \mathbf{R}$ is of class C^2. Let

$$p_2(\mathbf{x}) = f(\mathbf{a}) + \sum_{i=1}^{n} f_{x_i}(\mathbf{a})(x_i - a_i) + \frac{1}{2}\sum_{i,j=1}^{n} f_{x_i x_j}(\mathbf{a})(x_i - a_i)(x_j - a_j).$$

Then

$$f(\mathbf{x}) = p_2(\mathbf{x}) + R_2(\mathbf{x}, \mathbf{a}),$$

where $|R_2|/\|\mathbf{x} - \mathbf{a}\|^2 \to 0$ as $\mathbf{x} \to \mathbf{a}$.

EXAMPLE 10 Let $f(x, y) = \cos x \cos y$ and $(a, b) = (0, 0)$. Then

$$f(0,0) = 1;$$

$$f_x(0,0) = -\sin x \cos y|_{(0,0)} = 0, \qquad f_y(0,0) = -\cos x \sin y|_{(0,0)} = 0;$$

$$f_{xx}(0,0) = -\cos x \cos y|_{(0,0)} = -1,$$

$$f_{xy}(0,0) = \sin x \sin y|_{(0,0)} = 0,$$

$$f_{yy}(0,0) = -\cos x \cos y|_{(0,0)} = -1.$$

Hence,

$$f(x, y) \approx p_2(x, y) = 1 + \tfrac{1}{2}(-1 \cdot x^2 - 1 \cdot y^2) = 1 - \tfrac{1}{2}x^2 - \tfrac{1}{2}y^2.$$

We can also solve this problem another way since f is a product of two functions. We can multiply the two Taylor polynomials:

$$\begin{aligned} p_2(x, y) &= (\text{Taylor polynomial for } \cos x) \cdot (\text{Taylor polynomial for } \cos y) \\ &= \left(1 - \tfrac{1}{2}x^2\right)\left(1 - \tfrac{1}{2}y^2\right) \\ &= 1 - \tfrac{1}{2}x^2 - \tfrac{1}{2}y^2 + \tfrac{1}{4}x^2y^2 \\ &= 1 - \tfrac{1}{2}x^2 - \tfrac{1}{2}y^2 \qquad \text{up to terms of degree 2.} \end{aligned}$$

This method is justified by noting that if q_2 is the Taylor polynomial for cosine and R_2 is the corresponding remainder term, then

$$\begin{aligned} \cos x \, \cos y &= [q_2(x) + R_2(x, 0)][q_2(y) + R_2(y, 0)] \\ &= q_2(x)q_2(y) + q_2(y)R_2(x, 0) + q_2(x)R_2(y, 0) + R_2(x, 0)R_2(y, 0) \\ &= q_2(x)q_2(y) + \text{other stuff,} \end{aligned}$$

where (other stuff)$/\|(x, y)\|^2 \to 0$ as $(x, y) \to (0, 0)$, since both $R_2(x, 0)$ and $R_2(y, 0)$ do. ◆

The Hessian

Recall that the formula for the first-order Taylor polynomial p_1 was written quite concisely in formula (5) by using vector and matrix notation. It turns out that it is possible to do something similar for the second-order polynomial p_2.

■ **Definition 1.6** The **Hessian** of a function $f\colon X \subseteq \mathbf{R}^n \to \mathbf{R}$ is the matrix whose ijth entry is $\partial^2 f/\partial x_j \partial x_i$. That is,

$$Hf = \begin{bmatrix} f_{x_1x_1} & f_{x_1x_2} & \cdots & f_{x_1x_n} \\ f_{x_2x_1} & f_{x_2x_2} & \cdots & f_{x_2x_n} \\ \vdots & \vdots & \ddots & \vdots \\ f_{x_nx_1} & f_{x_nx_2} & \cdots & f_{x_nx_n} \end{bmatrix}.$$

The term "Hessian" comes from Ludwig Otto Hesse, the mathematician who first introduced it, not from the German mercenaries who fought in the American revolution.

Now let's look again at the formula for p_2 in Theorem 1.5:

$$p_2(\mathbf{x}) = f(\mathbf{a}) + \sum_{i=1}^{n} f_{x_i}(\mathbf{a})h_i + \frac{1}{2}\sum_{i,j=1}^{n} f_{x_i x_j}(\mathbf{a})h_i h_j.$$

(We have let $\mathbf{h} = (h_1, \ldots, h_n) = \mathbf{x} - \mathbf{a}$.) This can be written as

$$p_2(\mathbf{x}) = f(\mathbf{a}) + \begin{bmatrix} f_{x_1}(\mathbf{a}) & f_{x_2}(\mathbf{a}) & \cdots & f_{x_n}(\mathbf{a}) \end{bmatrix} \begin{bmatrix} h_1 \\ h_2 \\ \vdots \\ h_n \end{bmatrix}$$

$$+ \frac{1}{2}\begin{bmatrix} h_1 & h_2 & \cdots & h_n \end{bmatrix} \begin{bmatrix} f_{x_1x_1}(\mathbf{a}) & f_{x_1x_2}(\mathbf{a}) & \cdots & f_{x_1x_n}(\mathbf{a}) \\ f_{x_2x_1}(\mathbf{a}) & f_{x_2x_2}(\mathbf{a}) & \cdots & f_{x_2x_n}(\mathbf{a}) \\ \vdots & \vdots & \ddots & \vdots \\ f_{x_nx_1}(\mathbf{a}) & f_{x_nx_2}(\mathbf{a}) & \cdots & f_{x_nx_n}(\mathbf{a}) \end{bmatrix} \begin{bmatrix} h_1 \\ h_2 \\ \vdots \\ h_n \end{bmatrix}.$$

Thus, we see that

$$p_2(\mathbf{x}) = f(\mathbf{a}) + Df(\mathbf{a})\mathbf{h} + \tfrac{1}{2}\mathbf{h}^T Hf(\mathbf{a})\mathbf{h}. \tag{10}$$

(Remember that $\mathbf{h}^T$ is the *transpose* of the $n \times 1$ matrix $\mathbf{h}$.)

EXAMPLE 11 (Example 10 revisited) For $f(x, y) = \cos x \cos y$, $\mathbf{a} = (0, 0)$, we have

$$Df(x, y) = \begin{bmatrix} -\sin x \cos y & -\cos x \sin y \end{bmatrix}$$

and

$$Hf(x, y) = \begin{bmatrix} -\cos x \cos y & \sin x \sin y \\ \sin x \sin y & -\cos x \cos y \end{bmatrix}.$$

Hence,

$$\begin{aligned} p_2(x, y) &= f(0, 0) + Df(0, 0)\mathbf{h} + \tfrac{1}{2}\mathbf{h}^T Hf(0, 0)\mathbf{h} \\ &= 1 + \begin{bmatrix} 0 & 0 \end{bmatrix}\begin{bmatrix} h_1 \\ h_2 \end{bmatrix} + \tfrac{1}{2}\begin{bmatrix} h_1 & h_2 \end{bmatrix}\begin{bmatrix} -1 & 0 \\ 0 & -1 \end{bmatrix}\begin{bmatrix} h_1 \\ h_2 \end{bmatrix} \\ &= 1 - \tfrac{1}{2}h_1^2 - \tfrac{1}{2}h_2^2. \end{aligned}$$

Once we recall that $\mathbf{h} = (h_1, h_2) = (x - 0, y - 0) = (x, y)$, we see that this result checks with our work in Example 10, just as it should. ◆

Higher-order Taylor Polynomials

So far we have said nothing about Taylor polynomials of degree greater than two in the case of functions of several variables. The main reasons for this are (i) the general formula is quite complicated and has no compact matrix reformulation analogous to (10) and (ii) we will have little need for such formulas in this text. Nonetheless, if your curiosity cannot be denied, here is the third-order Taylor polynomial for a function $f\colon X \subseteq \mathbf{R}^n \to \mathbf{R}$ of class C^3 near $\mathbf{a} \in X$:

$$p_3(\mathbf{x}) = f(\mathbf{a}) + \sum_{i=1}^{n} f_{x_i}(\mathbf{a})(x_i - a_i) + \frac{1}{2}\sum_{i,j=1}^{n} f_{x_i x_j}(\mathbf{a})(x_i - a_i)(x_j - a_j) + \frac{1}{3!}\sum_{i,j,k=1}^{n} f_{x_i x_j x_k}(\mathbf{a})(x_i - a_i)(x_j - a_j)(x_k - a_k).$$

(The relevant theorem regarding p_3 is that $f(\mathbf{x}) = p_3(\mathbf{x}) + R_3(\mathbf{x}, \mathbf{a})$, where $|R_3(\mathbf{x}, \mathbf{a})|/\|\mathbf{x} - \mathbf{a}\|^3 \to 0$ as $\mathbf{x} \to \mathbf{a}$.) If you must know even more, the kth-order Taylor polynomial is

$$p_k(\mathbf{x}) = f(\mathbf{a}) + \sum_{i=1}^{n} f_{x_i}(\mathbf{a})(x_i - a_i) + \frac{1}{2}\sum_{i,j=1}^{n} f_{x_i x_j}(\mathbf{a})(x_i - a_i)(x_j - a_j) + \cdots + \frac{1}{k!}\sum_{i_1,\ldots,i_k=1}^{n} f_{x_{i_1}\cdots x_{i_k}}(\mathbf{a})(x_{i_1} - a_{i_1})\cdots(x_{i_k} - a_{i_k}).$$

Formulas for Remainder Terms (optional)

Under slightly stricter hypotheses than those appearing in Theorems 1.3 and 1.5, integral formulas for the remainder terms may be derived as follows. Set $\mathbf{h} = \mathbf{x} - \mathbf{a}$. If f is of class C^2, then

$$\begin{aligned} R_1(\mathbf{x}, \mathbf{a}) &= \sum_{i,j=1}^{n} \int_0^1 (1 - t) f_{x_i x_j}(\mathbf{a} + t\mathbf{h}) h_i h_j \, dt \\ &= \int_0^1 \left[\mathbf{h}^T Hf(\mathbf{a} + t\mathbf{h})\mathbf{h}\right](1 - t)\, dt. \end{aligned}$$

If f is of class C^3, then

$$R_2(\mathbf{x}, \mathbf{a}) = \sum_{i,j,k=1}^{n} \int_0^1 \frac{(1-t)^2}{2} f_{x_i x_j x_k}(\mathbf{a} + t\mathbf{h}) h_i h_j h_k \, dt,$$

and if f is of class C^{k+1}, then

$$R_k(\mathbf{x}, \mathbf{a}) = \sum_{i_i,\ldots,i_{k+1}=1}^{n} \int_0^1 \frac{(1-t)^k}{k!} f_{x_{i_1} x_{i_2} \cdots x_{i_{k+1}}}(\mathbf{a} + t\mathbf{h}) h_{i_1} h_{i_2} \cdots h_{i_{k+1}} \, dt.$$

Although explicit, these formulas are not very useful in practice. By artful application of Taylor's formula for a single variable, we can arrive at derivative versions of these remainder terms (known as **Lagrange's form of the remainder**) that are similar to those in the one-variable case.

Lagrange's form of the remainder. If f is of class C^2, then in Theorem 1.3 the remainder R_1 is

$$R_1(\mathbf{x}, \mathbf{a}) = \frac{1}{2} \sum_{i,j=1}^{n} f_{x_i x_j}(\mathbf{z}) h_i h_j$$

for a suitable point $\mathbf{z}$ on the line in the domain of f joining $\mathbf{a}$ and $\mathbf{x} = \mathbf{a} + \mathbf{h}$. Similarly, if f is of class C^3, then the remainder R_2 in Theorem 1.5 is

$$R_2(\mathbf{x}, \mathbf{a}) = \frac{1}{3!} \sum_{i,j,k=1}^{n} f_{x_i x_j x_k}(\mathbf{z}) h_i h_j h_k$$

for a suitable point $\mathbf{z}$ on the line joining $\mathbf{a}$ and $\mathbf{x} = \mathbf{a} + \mathbf{h}$. More generally, if f is of class C^{k+1}, then the remainder R_k is

$$R_k(\mathbf{x}, \mathbf{a}) = \frac{1}{(k+1)!} \sum_{i_1, \ldots, i_{k+1}=1}^{n} f_{x_{i_1} x_{i_2} \cdots x_{i_{k+1}}}(\mathbf{z}) h_{i_1} h_{i_2} \cdots h_{i_{k+1}}$$

for a suitable point $\mathbf{z}$ on the line joining $\mathbf{a}$ and $\mathbf{x} = \mathbf{a} + \mathbf{h}$.

EXAMPLE 12 For $f(x, y) = \cos x \cos y$, we have

$$|R_2(x, y, 0, 0)| = \frac{1}{3!} \left| \sum_{i,j,k=1}^{2} f_{x_i x_j x_k}(\mathbf{z}) h_i h_j h_k \right|$$

$$\leq \frac{1}{3!} \sum_{i,j,k=1}^{2} 1 \cdot |h_i h_j h_k|,$$

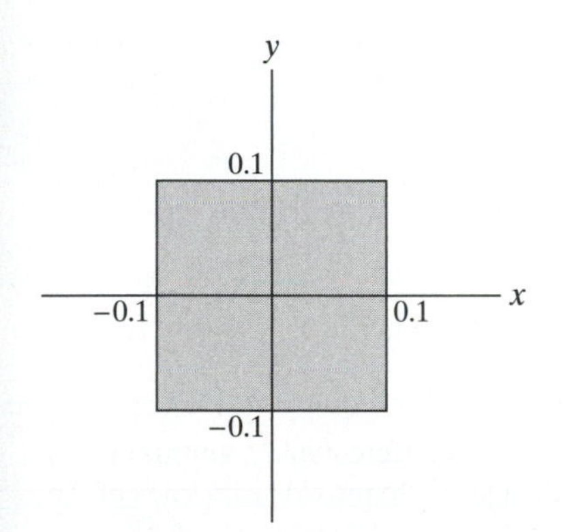

Figure 4.10 The polynomial p_2 approximates f to within $0.001\bar{3}$ on the square shown. (See Example 12.)

since all partial derivatives of f will be a product of sines and cosines and hence no larger than 1 in magnitude. Expanding the sum, we get

$$|R_2(x, y, 0, 0)| \leq \tfrac{1}{6} \left(|h_1|^3 + 3h_1^2 |h_2| + 3|h_1| h_2^2 + |h_2|^3 \right).$$

If both $|h_1|$ and $|h_2|$ are no more than, say, 0.1, then

$$|R_2(x, y, 0, 0)| \leq \tfrac{1}{6} \left(8 \cdot (0.1)^3 \right) = 0.001\bar{3}.$$

So throughout the square of side 0.2 centered at the origin and shown in Figure 4.10, the second-order Taylor polynomial is accurate to at least $0.001\bar{3}$ (i.e., to two decimal places) as an approximation of $f(x, y) = \cos x \; \cos y$. In Figure 4.11,

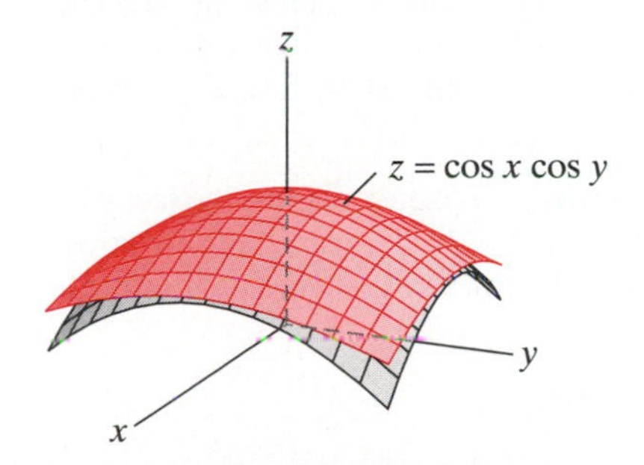

Figure 4.11 The graph of $f(x, y) = \cos x \cos y$ and its Taylor polynomial $p_2(x, y) = 1 - \frac{1}{2}x^2 - \frac{1}{2}y^2$ over the square $\{(x, y) \mid -1 \leq x \leq 1, -1 \leq y \leq 1\}$.

we show the graph of $f(x, y) = \cos x \cos y$ over the square domain $\{(x, y) \mid -1 \leq x \leq 1, -1 \leq y \leq 1\}$ together with the graph of its second-order Taylor polynomial $p_2(x, y) = 1 - \frac{1}{2}x^2 - \frac{1}{2}y^2$ (calculated in Example 10). Note how closely the surfaces coincide near the point $(0, 0, 1)$, just as the analysis above indicates. ◆

Exercises

In Exercises 1–7, find the Taylor polynomials p_k of given order k at the indicated point a.

1. $f(x) = e^{2x}, a = 0, k = 4$

2. $f(x) = \ln(1 + x), a = 0, k = 3$

3. $f(x) = 1/x^2, a = 1, k = 4$

4. $f(x) = \sqrt{x}, a = 1, k = 3$

5. $f(x) = \sqrt{x}, a = 9, k = 3$

6. $f(x) = \sin x, a = 0, k = 5$

7. $f(x) = \sin x, a = \pi/2, k = 5$

In Exercises 8–13, find the first- and second-order Talyor polynomials for the given function f at the given point **a**.

8. $f(x, y) = 1/(x^2 + y^2 + 1), \mathbf{a} = (0, 0)$

9. $f(x, y) = 1/(x^2 + y^2 + 1), \mathbf{a} = (1, -1)$

10. $f(x, y) = e^{2x+y}, \mathbf{a} = (0, 0)$

11. $f(x, y) = e^{2x} \cos 3y, \mathbf{a} = (0, \pi)$

12. $f(x, y, z) = 1/(x^2 + y^2 + z^2 + 1), \mathbf{a} = (0, 0, 0)$

13. $f(x, y, z) = \sin xyz, \mathbf{a} = (0, 0, 0)$

In Exercises 14–16, calculate the Hessian matrix $Hf(\mathbf{a})$ for the indicated function f at the indicated point **a**.

14. $f(x, y) = 1/(x^2 + y^2 + 1), \mathbf{a} = (0, 0)$

15. $f(x, y, z) = x^3 + x^2y - yz^2 + 2z^3, \mathbf{a} = (1, 0, 1)$

16. $f(x, y, z) = e^{2x-3y} \sin 5z, \mathbf{a} = (0, 0, 0)$

17. For f and **a** as given in Exercise 8, express the second-order Taylor polynomial $p_2(x, y)$, using the derivative matrix and the Hessian matrix as in formula (10) of this section.

18. For f and **a** as given in Exercise 15, express the second-order Taylor polynomial $p_2(x, y)$, using the derivative matrix and the Hessian matrix as in formula (10) of this section.

19. Consider the function

$$f(x_1, x_2, \ldots, x_n) = e^{x_1+2x_2+\cdots+nx_n}.$$

(a) Calculate $Df(0, 0, \ldots, 0)$ and $Hf(0, 0, \ldots, 0)$.

(b) Determine the first- and second-order Taylor polynomials of f at $\mathbf{0}$.

(c) Use formulas (3) and (10) to write the Taylor polynomials in terms of the derivative and Hessian matrices.

20. Find the third-order Taylor polynomial $p_3(x, y, z)$ of

$$f(x, y, z) = e^{x+2y+3z}$$

at $(0, 0, 0)$.

21. Find the third-order Taylor polynomial of

$$f(x, y, z) = x^4 + x^3y + 2y^3 - xz^2 + x^2y + 3xy - z + 2$$

(a) at $(0, 0, 0)$.

(b) at $(1, -1, 0)$.

Determine the total differential of the functions given in Exercises 22–26.

22. $f(x, y) = x^2y^3$

23. $f(x, y, z) = x^2 + 3y^2 - 2z^3$

24. $f(x, y, z) = \cos(xyz)$

25. $f(x, y, z) = e^x \cos y + e^y \sin z$

26. $f(x, y, z) = 1/\sqrt{xyz}$

27. Use the fact that the total differential df approximates the incremental change Δf to provide estimates of the following quantities:

(a) $(7.07)^2(1.98)^3$

(b) $1/\sqrt{(4.1)(1.96)(2.05)}$

(c) $(1.1)\cos((\pi - 0.03)(0.12))$

28. Near the point $(1, -2, 1)$, is the function $g(x, y, z) = x^3 - 2xy + x^2z + 7z$ most sensitive to changes in x, y, or z?

29. To which entry in the matrix is the value of the determinant

$$\begin{vmatrix} 2 & 3 \\ -1 & 5 \end{vmatrix}$$

most sensitive?

30. If you measure the radius of a cylinder to be 2 in, with a possible error of ± 0.1 in, and the height to be 3 in, with a possible error of ± 0.05 in, use differentials to determine the approximate error in

(a) the calculated volume of the cylinder.

(b) the calculated surface area.

31. To estimate the volume of a cone of radius approximately 2 m and height approximately 6 m, how accurately should the radius and height be measured so that the error in the calculated volume estimate does not exceed 0.2 m^3? Assume that the possible error in measuring the radius and height are the same.

32. Suppose that you measure the dimensions of a block of tofu to be (approximately) 3 in by 4 in by 2 in. Assuming that the possible error in each of your measurements is the same, about how accurate must your measurements be so that the error in the calculated volume of the tofu is not more than 0.2 in^3? What percentage error in volume does this represent?

33. (a) Calculate the second-order Taylor polynomial for $f(x, y) = \cos x \sin y$ at the point $(0, \pi/2)$.

(b) If $\mathbf{h} = (h_1, h_2) = (x, y) - (0, \pi/2)$ is such that $|h_1|$ and $|h_2|$ are no more than 0.3, estimate how accurate your Taylor approximation is.

34. (a) Determine the second-order Taylor polynomial of $f(x, y) = e^{x+2y}$ at the origin.

(b) Estimate the accuracy of the approximation if $|x|$ and $|y|$ are no more than 0.1.

4.2 Extrema of Functions

The power of calculus resides at least in part in its role in helping to solve a wide variety of optimization problems. With any quantity that changes, it is natural to ask when, if ever, does that quantity reach its largest, its smallest, its fastest or slowest? You have already learned how to find maxima and minima of a function of a single variable, and no doubt you have applied your techniques to a number of situations. However, many phenomena are not appropriately modeled by functions of only one variable. Thus, there is a genuine need to adapt and extend optimization methods to the case of functions of more than one variable. We develop the necessary theory in this and the next section and explore a few applications in §4.4.

Critical Points of Functions

Let X be open in $\mathbf{R}^n$ and $f: X \subseteq \mathbf{R}^n \to \mathbf{R}$ a scalar-valued function.

> ■ **Definition 2.1** We say that f has a **local minimum** at the point $\mathbf{a}$ in X if there is some neighborhood U of $\mathbf{a}$ such that $f(\mathbf{x}) \geq f(\mathbf{a})$ for all $\mathbf{x}$ in U. Similarly, we say that f has a **local maximum** at $\mathbf{a}$ if there is some neighborhood U of $\mathbf{a}$ such that $f(\mathbf{x}) \leq f(\mathbf{a})$ for all $\mathbf{x}$ in U.

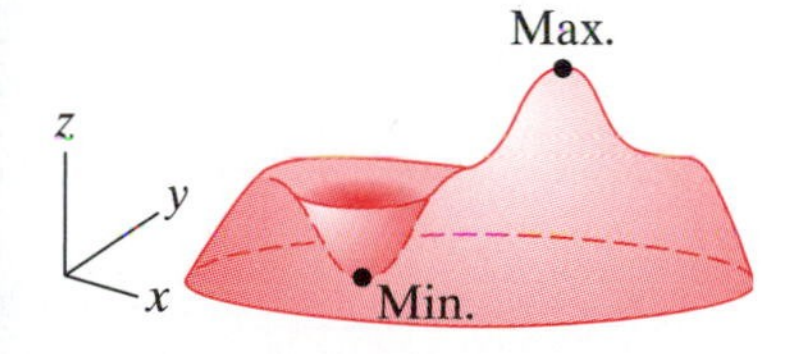

Figure 4.12 The graph of $z = f(x, y)$.

When $n = 2$, local extrema of $f(x, y)$ are precisely the pits and peaks of the surface given by the graph of $z = f(x, y)$, as suggested by Figure 4.12.

We emphasize our use of the adjective "local." When a local maximum of a function f occurs at a point $\mathbf{a}$, this means that the values of f at points *near* $\mathbf{a}$ can be no larger, *not* that *all* values of f are no larger. Indeed, f may have local maxima and no **global** (or **absolute**) maximum. Consider the graphs in Figure 4.13. (Of course, analogous comments apply to local and global minima.)

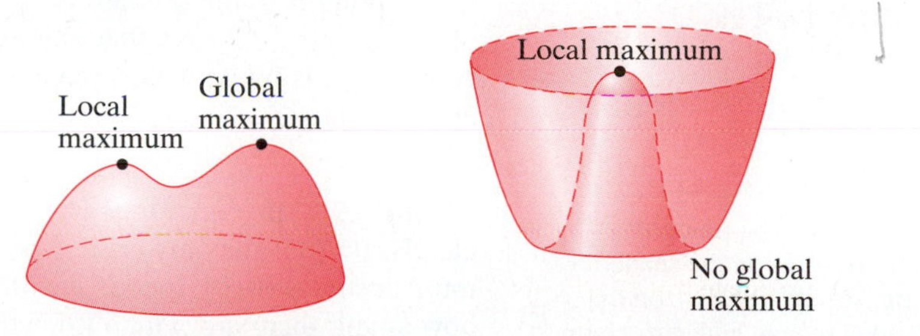

Figure 4.13 Examples of local and global maxima.

Recall that, if a differentiable function of one variable has a local extremum at a point, then the derivative vanishes there (i.e., the tangent line to the graph of the function is horizontal). Figures 4.12 and 4.13 suggest strongly that, if a function of two variables has a local maximum or minimum at a point in the domain, then the tangent plane at the corresponding point of the graph must be horizontal. Such is indeed the case, as the following general result (plus formula (4) of §2.3) implies.

Theorem 2.2 Let X be open in $\mathbf{R}^n$ and let $f: X \subseteq \mathbf{R}^n \to \mathbf{R}$ be differentiable. If f has a local extremum at $\mathbf{a} \in X$, then $Df(\mathbf{a}) = \mathbf{0}$.

Proof Suppose, for argument's sake, that f has a local maximum at $\mathbf{a}$. Then the one-variable function F defined by $F(t) = f(\mathbf{a} + t\mathbf{h})$ must have a local maximum at $t = 0$ for *any* $\mathbf{h}$. (Geometrically, the function F is just the restriction of f to the line through $\mathbf{a}$ parallel to $\mathbf{h}$ as shown in Figure 4.14.) From one-variable calculus, we must therefore have $F'(0) = 0$. By the chain rule

$$F'(t) = \frac{d}{dt}[f(\mathbf{a} + t\mathbf{h})] = Df(\mathbf{a} + t\mathbf{h})\mathbf{h} = \nabla f(\mathbf{a} + t\mathbf{h}) \cdot \mathbf{h}.$$

Hence,

$$0 = F'(0) = Df(\mathbf{a})\mathbf{h} = f_{x_1}(\mathbf{a})h_1 + f_{x_2}(\mathbf{a})h_2 + \cdots + f_{x_n}(\mathbf{a})h_n.$$

Since this last result must hold for all $\mathbf{h} \in \mathbf{R}^n$, we find that by setting $\mathbf{h}$ in turn equal to $(1, 0, \ldots, 0)$, $(0, 1, 0, \ldots, 0)$, $\ldots$, $(0, \ldots, 0, 1)$, we have

$$f_{x_1}(\mathbf{a}) = f_{x_2}(\mathbf{a}) = \cdots = f_{x_n}(\mathbf{a}) = 0.$$

Therefore, $Df(\mathbf{a}) = \mathbf{0}$, as desired. ■

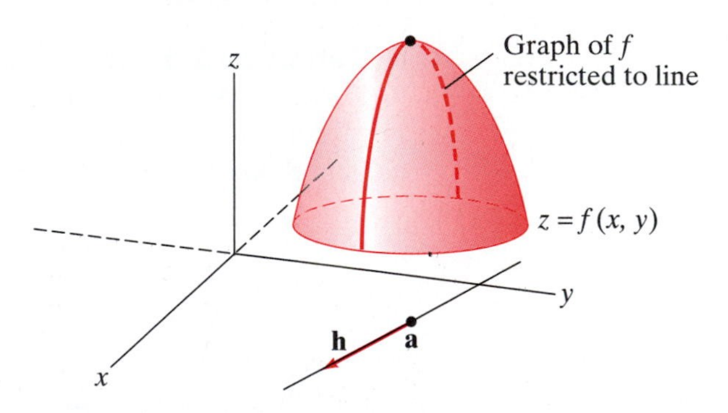

Figure 4.14 The graph of f restricted to a line.

A point $\mathbf{a}$ in the domain of f where $Df(\mathbf{a})$ is $\mathbf{0}$ is called a **critical point** of f. Theorem 2.2 says that any extremum of f must occur at a critical point. However, it is by no means the case that every critical point must be the site of an extremum.

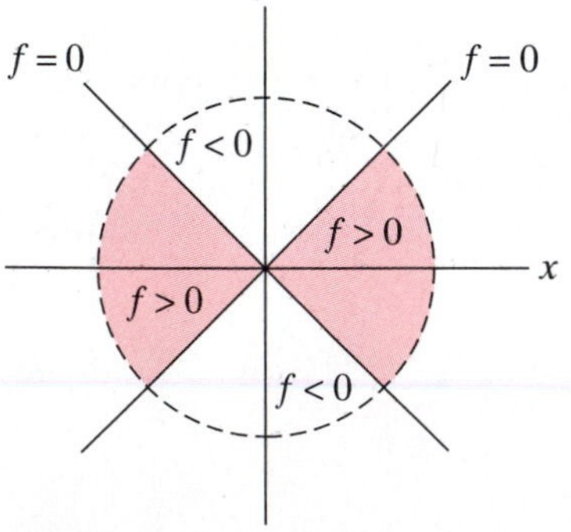

Figure 4.15 The function f is strictly positive on the shaded region, strictly negative on the unshaded region, and zero along the lines $y = \pm x$.

EXAMPLE 1 If $f(x, y) = x^2 - y^2$, then $Df(x, y) = \begin{bmatrix} 2x & -2y \end{bmatrix}$ so that, clearly, $(0, 0)$ is the only critical point. However, neither a maximum nor a minimum occurs at $(0, 0)$. Indeed, inside every open disk centered at $(0, 0)$, no matter how small, there are points for which $f(x, y) > f(0, 0) = 0$ and also points where $f(x, y) < f(0, 0)$. (See Figure 4.15.) ◆

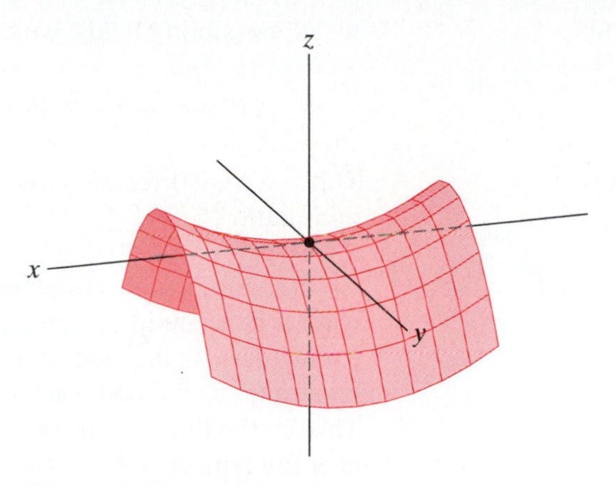

Figure 4.16 A saddle point.

This type of critical point is called a **saddle point**. Its name derives from the fact that the graph of $z = f(x, y)$ looks somewhat like a saddle. (See Figure 4.16.)

The Nature of a Critical Point: The Hessian Criterion

We illustrate our current understanding regarding extrema with the following example:

EXAMPLE 2 We find the extrema of

$$f(x, y) = x^2 + xy + y^2 + 2x - 2y + 5.$$

Since f is a polynomial, it is differentiable everywhere, and Theorem 2.2 implies that any extremum must occur where $\partial f/\partial x$ and $\partial f/\partial y$ vanish simultaneously. Thus, we solve

$$\begin{cases} \dfrac{\partial f}{\partial x} = 2x + y + 2 = 0 \\ \dfrac{\partial f}{\partial y} = x + 2y - 2 = 0 \end{cases},$$

and find that the only solution is $x = -2$, $y = 2$. Consequently, $(-2, 2)$ is the only critical point of this function.

To determine whether $(-2, 2)$ is a maximum or minimum (or neither), we could try graphing the function and drawing what we hope would be an obvious conclusion. Of course, such a technique does not extend to functions of more than two variables, so a graphical method is of limited value at best. Instead, we'll see how f changes as we move away from the critical point:

$$\begin{aligned} \Delta f &= f(-2 + h, 2 + k) - f(-2, 2) \\ &= [(-2 + h)^2 + (-2 + h)(2 + k) + (2 + k)^2 \\ &\quad + 2(-2 + h) - 2(2 + k) + 5] - 1 \\ &= h^2 + hk + k^2. \end{aligned}$$

If the quantity $\Delta f = h^2 + hk + k^2$ is nonnegative for all small values of h and k, then $(-2, 2)$ yields a local minimum. Similarly, if Δf is always nonpositive, then $(-2, 2)$ must yield a local maximum. Finally, if Δf is positive for some values of h and k and negative for others, then $(-2, 2)$ is a saddle point. To determine which possibility holds, we complete the square:

$$\Delta f = h^2 + hk + k^2 = h^2 + hk + \tfrac{1}{4}k^2 + \tfrac{3}{4}k^2 = \left(h + \tfrac{1}{2}k\right)^2 + \tfrac{3}{4}k^2.$$

Thus, $\Delta f \geq 0$ for all values of h and k, so $(-2, 2)$ necessarily yields a local minimum. ◆

Example 2 with its attendant algebra clearly demonstrates the need for a better way of determining when a critical point yields a local maximum or minimum (or neither). In the case of a one-variable function $f\colon X \subseteq \mathbf{R} \to \mathbf{R}$, you already know a quick method, namely, consideration of the sign of the second derivative. This method derives from looking at the second-order Taylor polynomial of f near the critical point a, viz.,

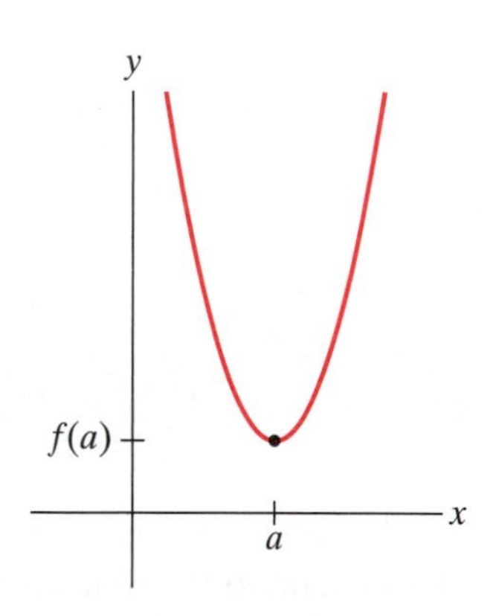

Figure 4.17 An upward-opening parabola.

$$\begin{aligned} f(x) \approx p_2(x) &= f(a) + f'(a)(x-a) + \frac{f''(a)}{2}(x-a)^2 \\ &= f(a) + \frac{f''(a)}{2}(x-a)^2, \end{aligned}$$

since f' is zero at the critical point a of f. If $f''(a) > 0$, the graph of $y = p_2(x)$ is an upward-opening parabola, as in Figure 4.17, whereas if $f''(a) < 0$, then the graph of $y = p_2(x)$ looks like the one shown in Figure 4.18. If $f''(a) = 0$, then the graph of $y = p_2(x)$ is just a horizontal line, and we would need to use a higher-order Taylor polynomial to determine if f has an extremum at a. (You may recall that when $f''(a) = 0$, the second derivative test from single-variable calculus gives no information about the nature of the critical point a.)

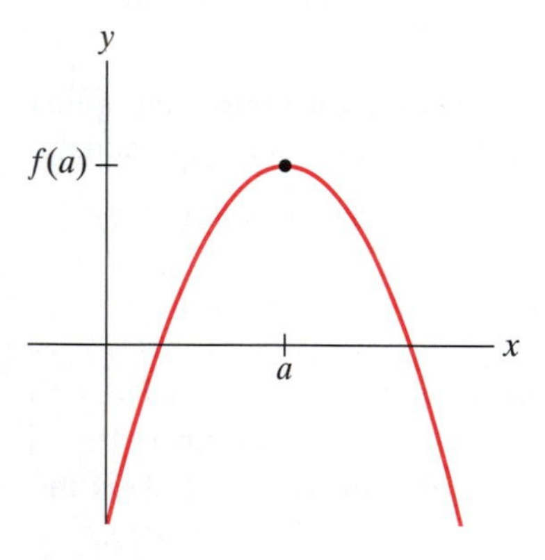

Figure 4.18 A downward-opening parabola.

The concept is similar in the context of n variables. Suppose that

$$f(\mathbf{x}) = f(x_1, x_2, \ldots, x_n)$$

is of class C^2 and that $\mathbf{a} = (a_1, a_2, \ldots, a_n)$ is a critical point of f. Then the second-order Taylor approximation to f gives

$$\begin{aligned} \Delta f = f(\mathbf{x}) - f(\mathbf{a}) &\approx p_2(\mathbf{x}) - f(\mathbf{a}) \\ &= Df(\mathbf{a})(\mathbf{x}-\mathbf{a}) + \tfrac{1}{2}(\mathbf{x}-\mathbf{a})^T Hf(\mathbf{a})(\mathbf{x}-\mathbf{a}) \end{aligned}$$

when $\mathbf{x} \approx \mathbf{a}$. (See Theorem 1.5 and formula (10) in §4.1.) Since $\mathbf{a}$ is a critical point, all the partial derivatives vanish at $\mathbf{a}$, so that we have $Df(\mathbf{a}) = \mathbf{0}$ and hence

$$\Delta f \approx \tfrac{1}{2}(\mathbf{x}-\mathbf{a})^T Hf(\mathbf{a})(\mathbf{x}-\mathbf{a}). \tag{1}$$

The approximation in (1) suggests that we may be able to see whether the increment Δf remains positive (respectively, remains negative) for $\mathbf{x}$ near $\mathbf{a}$ and, hence, whether f has a local minimum (respectively, a local maximum) at $\mathbf{a}$ by seeing what happens to the right side.

Note that the right side of (1), when expanded, is quadratic in the terms $(x_i - a_i)$. More generally, a **quadratic form** in $h_1, h_2, \ldots, h_n$ is a function Q that can be written as

$$Q(h_1, h_2, \ldots, h_n) = \sum_{i,j=1}^{n} b_{ij} h_i h_j,$$

where the b_{ij}'s are constants. The quadratic form Q can also be written in terms of matrices as

$$Q(\mathbf{h}) = \begin{bmatrix} h_1 & h_2 & \cdots & h_n \end{bmatrix} \begin{bmatrix} b_{11} & b_{12} & \cdots & b_{1n} \\ b_{21} & b_{22} & \cdots & b_{2n} \\ \vdots & \vdots & \ddots & \vdots \\ b_{n1} & b_{n2} & \cdots & b_{nn} \end{bmatrix} \begin{bmatrix} h_1 \\ h_2 \\ \vdots \\ h_n \end{bmatrix} = \mathbf{h}^T B\mathbf{h}, \tag{2}$$

where $B = (b_{ij})$. Note that the function Q is unchanged if we replace b_{ij} with $\frac{1}{2}(b_{ij} + b_{ji})$. Hence, we may always assume that the matrix B associated to Q is **symmetric**, that is, that $b_{ij} = b_{ji}$ (or, equivalently, that $B^T = B$). Ignoring the factor of $1/2$, we see that the right side of (1) is the quadratic form in $\mathbf{h} = \mathbf{x} - \mathbf{a}$, corresponding to the matrix $B = Hf(\mathbf{a})$.

A quadratic form Q (respectively, its associated symmetric matrix B) is said to be **positive definite** if $Q(\mathbf{h}) > 0$ for all $\mathbf{h} \neq \mathbf{0}$ and **negative definite** if $Q(\mathbf{h}) < 0$ for all $\mathbf{h} \neq \mathbf{0}$. Note that if Q is positive definite, then Q has a global minimum (of 0) at $\mathbf{h} = \mathbf{0}$. Similarly, if Q is negative definite, then Q has a global maximum at $\mathbf{h} = \mathbf{0}$.

The importance of quadratic forms to us is that we can judge whether f has a local extremum at a critical point $\mathbf{a}$ by seeing if the quadratic form in the right side of (1) has a maximum or minimum at $\mathbf{x} = \mathbf{a}$. The precise result, whose proof is given in the addendum to this section, is the following:

Theorem 2.3 Let $U \subseteq \mathbf{R}^n$ be open and $f: U \to \mathbf{R}$ a function of class C^2. Suppose that $\mathbf{a} \in U$ is a critical point of f.

1. If the Hessian $Hf(\mathbf{a})$ is positive definite, then f has a local minimum at $\mathbf{a}$.
2. If the Hessian $Hf(\mathbf{a})$ is negative definite, then f has a local maximum at $\mathbf{a}$.
3. If $\det Hf(\mathbf{a}) \neq 0$ but $Hf(\mathbf{a})$ is neither positive nor negative definite, then f has a saddle point at $\mathbf{a}$.

In view of Theorem 2.3, the issue thus becomes to determine when the Hessian $Hf(\mathbf{a})$ is positive or negative definite. Fortunately, linear algebra provides an effective means for making such a determination, which we state without proof. Given a symmetric matrix B (which, as we have seen, corresponds to a quadratic form Q), let B_k, for $k = 1, \ldots, n$, denote the upper leftmost $k \times k$ submatrix of B. Calculate the following sequence of determinants:

$$\det B_1 = b_{11}, \quad \det B_2 = \begin{vmatrix} b_{11} & b_{12} \\ b_{21} & b_{22} \end{vmatrix},$$

$$\det B_3 = \begin{vmatrix} b_{11} & b_{12} & b_{13} \\ b_{21} & b_{22} & b_{23} \\ b_{31} & b_{32} & b_{33} \end{vmatrix}, \ldots, \det B_n = \det B.$$

If this sequence consists entirely of positive numbers, then B and Q are positive definite. If this sequence is such that $\det B_k < 0$ for k odd and $\det B_k > 0$ for k even, then B and Q are negative definite. Finally, if $\det B \neq 0$, but the sequence of determinants $\det B_1, \det B_2, \ldots, \det B_n$ is neither of the first two types, then B and Q are neither positive nor negative definite. Combining these remarks with

Theorem 2.3, we can establish the following test for local extrema:

Second derivative test for local extrema. Given a critical point $\mathbf{a}$ of a function f of class C^2, look at the Hessian matrix evaluated at $\mathbf{a}$:

$$Hf(\mathbf{a}) = \begin{bmatrix} f_{x_1x_1}(\mathbf{a}) & f_{x_1x_2}(\mathbf{a}) & \cdots & f_{x_1x_n}(\mathbf{a}) \\ f_{x_2x_1}(\mathbf{a}) & f_{x_2x_2}(\mathbf{a}) & \cdots & f_{x_2x_n}(\mathbf{a}) \\ \vdots & \vdots & \ddots & \vdots \\ f_{x_nx_1}(\mathbf{a}) & f_{x_nx_2}(\mathbf{a}) & \cdots & f_{x_nx_n}(\mathbf{a}) \end{bmatrix}.$$

From the Hessian, calculate the **sequence of principal minors** of $Hf(\mathbf{a})$. This is the sequence of the determinants of the upper leftmost square submatrices of $Hf(\mathbf{a})$. More explicitly, this is the sequence $d_1, d_2, \ldots, d_n$, where $d_k = \det H_k$, and H_k is the upper leftmost $k \times k$ submatrix of $Hf(\mathbf{a})$. That is,

$$d_1 = f_{x_1x_1}(\mathbf{a}),$$

$$d_2 = \begin{vmatrix} f_{x_1x_1}(\mathbf{a}) & f_{x_1x_2}(\mathbf{a}) \\ f_{x_2x_1}(\mathbf{a}) & f_{x_2x_2}(\mathbf{a}) \end{vmatrix},$$

$$d_3 = \begin{vmatrix} f_{x_1x_1}(\mathbf{a}) & f_{x_1x_2}(\mathbf{a}) & f_{x_1x_3}(\mathbf{a}) \\ f_{x_2x_1}(\mathbf{a}) & f_{x_2x_2}(\mathbf{a}) & f_{x_2x_3}(\mathbf{a}) \\ f_{x_3x_1}(\mathbf{a}) & f_{x_3x_2}(\mathbf{a}) & f_{x_3x_3}(\mathbf{a}) \end{vmatrix}, \ldots, d_n = |Hf(\mathbf{a})|.$$

The numerical test is as follows:
Assume that $d_n = \det Hf(\mathbf{a}) \neq 0$.

1. If $d_k > 0$ for $k = 1, 2, \ldots, n$, then f has a local minimum at $\mathbf{a}$.
2. If $d_k < 0$ for k odd and $d_k > 0$ for k even, then f has a local maximum at $\mathbf{a}$.
3. If neither case 1 nor case 2 holds, then f has a saddle point at $\mathbf{a}$.

In the event that $\det Hf(\mathbf{a}) = 0$, we say that the critical point $\mathbf{a}$ is **degenerate** and must use another method to determine whether or not it is the site of an extremum of f.

EXAMPLE 3 Consider the function

$$f(x, y) = x^2 + xy + y^2 + 2x - 2y + 5$$

in Example 2. We have already seen that $(-2, 2)$ is the only critical point. The Hessian is

$$Hf(x, y) = \begin{bmatrix} f_{xx} & f_{xy} \\ f_{yx} & f_{yy} \end{bmatrix} = \begin{bmatrix} 2 & 1 \\ 1 & 2 \end{bmatrix}.$$

The sequence of principal minors is $2\ (> 0)$, $3\ (> 0)$. Hence, f has a minimum at $(-2, 2)$, as we saw before, but this method uses less algebra. ◆

EXAMPLE 4 (Second derivative test for functions of two variables) Let us generalize Example 3. Suppose that $f(x, y)$ is a function of two variables of class C^2 and further suppose that f has a critical point at $\mathbf{a} = (a, b)$. The Hessian matrix of f evaluated at (a, b) is

$$Hf(a,b) = \begin{bmatrix} f_{xx}(a,b) & f_{xy}(a,b) \\ f_{xy}(a,b) & f_{yy}(a,b) \end{bmatrix}.$$

Note that we have used the fact that $f_{xy} = f_{yx}$ (since f is of class C^2) in constructing the Hessian. The sequence of principal minors thus consists of two numbers:

$$d_1 = f_{xx}(a,b) \quad \text{and} \quad d_2 = f_{xx}(a,b)f_{yy}(a,b) - f_{xy}(a,b)^2.$$

Hence, in this case, the second derivative test tells us that:

1. f has a local minimum at (a, b) if
$$f_{xx}(a,b) > 0 \quad \text{and} \quad f_{xx}(a,b)f_{yy}(a,b) - f_{xy}(a,b)^2 > 0.$$
2. f has a local maximum at (a, b) if
$$f_{xx}(a,b) < 0 \quad \text{and} \quad f_{xx}(a,b)f_{yy}(a,b) - f_{xy}(a,b)^2 > 0.$$
3. f has a saddle point at (a, b) if
$$f_{xx}(a,b)f_{yy}(a,b) - f_{xy}(a,b)^2 < 0.$$

Note that if $f_{xx}(a,b)f_{yy}(a,b) - f_{xy}(a,b)^2 = 0$, then f has a *degenerate* critical point at (a, b) and we cannot immediately determine if (a, b) is the site of a local extremum of f. ◆

EXAMPLE 5 Let $f(x, y, z) = x^3 + xy^2 + x^2 + y^2 + 3z^2$. To find any local extrema of f, we must first identify the critical points. Thus, we solve

$$Df(x,y,z) = \begin{bmatrix} 3x^2 + y^2 + 2x & 2xy + 2y & 6z \end{bmatrix} = \begin{bmatrix} 0 & 0 & 0 \end{bmatrix}.$$

From this, it is not hard to see that there are two critical points: $(0, 0, 0)$ and $\left(-\frac{2}{3}, 0, 0\right)$. The Hessian of f is

$$Hf(x,y,z) = \begin{bmatrix} 6x+2 & 2y & 0 \\ 2y & 2x+2 & 0 \\ 0 & 0 & 6 \end{bmatrix}.$$

At the critical point $(0, 0, 0)$, we have

$$Hf(0,0,0) = \begin{bmatrix} 2 & 0 & 0 \\ 0 & 2 & 0 \\ 0 & 0 & 6 \end{bmatrix},$$

and its sequence of principal minors is 2, 4, 24. Since these determinants are all positive, we conclude that f has a local minimum at (0,0,0). At $\left(-\frac{2}{3}, 0, 0\right)$, we calculate that

$$Hf\left(-\frac{2}{3}, 0, 0\right) = \begin{bmatrix} -2 & 0 & 0 \\ 0 & \frac{2}{3} & 0 \\ 0 & 0 & 6 \end{bmatrix}.$$

The sequence of minors is $-2, -\frac{4}{3}, -8$. Hence, f has a saddle point at $\left(-\frac{2}{3}, 0, 0\right)$. ◆

EXAMPLE 6 To get a feeling for what happens in the case of a degenerate critical point (i.e., a critical point $\mathbf{a}$ such that $\det Hf(\mathbf{a}) = 0$), consider the three functions

$$f(x, y) = x^4 + x^2 + y^4,$$
$$g(x, y) = -x^4 - x^2 - y^4,$$

and

$$h(x, y) = x^4 - x^2 + y^4.$$

We leave it to you to check that the origin (0, 0) is a degenerate critical point of each of these functions. (In fact, the Hessians themselves look very similar.) Since f is 0 at (0, 0) and strictly positive at all $(x, y) \neq (0, 0)$, we see that f has a strict minimum at the origin. Similar reasoning shows that g has a strict maximum at the origin. For h, the situation is slightly more complicated. Along the y-axis, we have $h(0, y) = y^4$, which is zero at $y = 0$ (the origin) and strictly positive everywhere else. Along the x-axis,

$$h(x, 0) = x^4 - x^2 = x^2(x - 1)(x + 1).$$

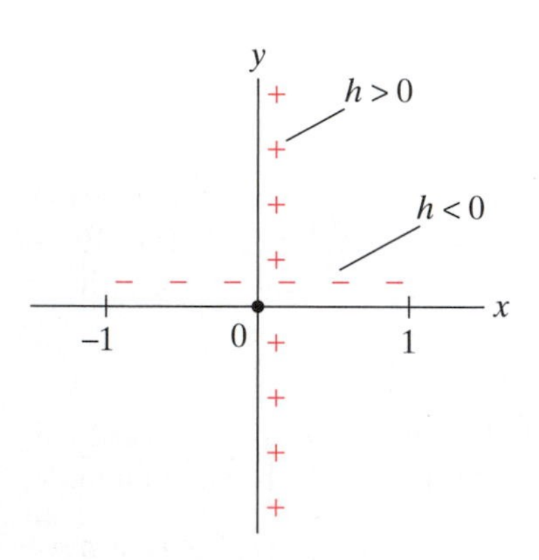

Figure 4.19 Away from the origin, the function h of Example 6 is negative along the x-axis and positive along the y-axis.

For $-1 < x < 1$ and $x \neq 0$, $h(x, 0) < 0$. We have the situation depicted in Figure 4.19. Thus, every neighborhood of (0, 0) contains some points (x, y) where h is positive and also some points where h is negative. Therefore, h has a saddle point at the origin. The "moral of the story" is that a degenerate critical point can exhibit any type of behavior, and more detailed consideration of the function itself, rather than its Hessian, is necessary to understand its nature as a site of an extremum. ◆

Global Extrema on Compact Regions

Thus far our discussion has been limited to consideration of only local extrema. We have said nothing about how to identify global extrema, because there really is no general, effective method for looking at an arbitrary function and determining whether and where it reaches an absolute maximum or minimum value. For the purpose of applications, where finding an absolute maximum or minimum is essential, such a state of affairs is indeed unfortunate. Nonetheless, we can say something about global extrema for functions defined on a certain type of domain.

■ **Definition 2.4** A subset $X \subseteq \mathbf{R}^n$ is said to be **compact** if it is both closed and bounded.

Recall that X is closed if it contains all the points that make up its boundary. (See Definition 2.3 of §2.2.) To say that X is bounded means that there is some (open or closed) ball B that contains it. (That is, X is bounded if there is some positive number M such that $\|\mathbf{x}\| < M$ for all $\mathbf{x} \in X$.) Thus, compact sets contain their boundaries (a consequence of being closed) and have only finite extent (a consequence of being bounded). Some typical compact sets in $\mathbf{R}^2$ and $\mathbf{R}^3$ are shown in Figure 4.20.

For our purposes the notion of compactness is of value because of the next result, which we state without proof.

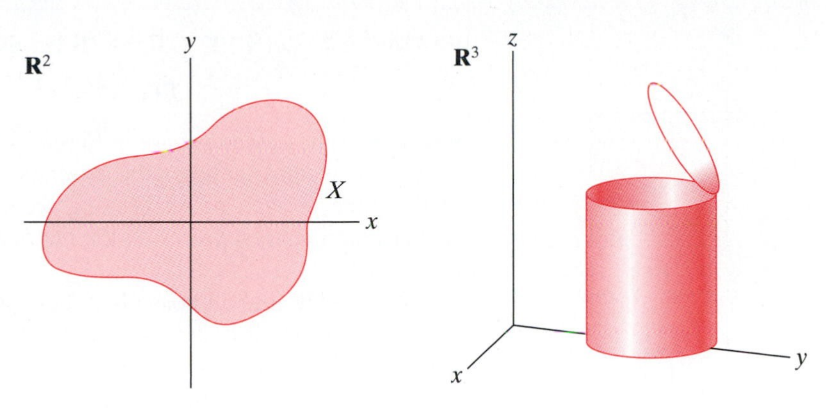

Figure 4.20 Compact regions.

■ **Theorem 2.5** (EXTREME VALUE THEOREM) If $X \subseteq \mathbf{R}^n$ is compact and $f: X \to \mathbf{R}$ is continuous, then f must have both a global maximum and a global minimum somewhere on X. That is, there must exist points $\mathbf{a}_{\text{max}}$ and $\mathbf{a}_{\text{min}}$ in X such that, for all $\mathbf{x} \in X$,

$$f(\mathbf{a}_{\text{min}}) \leq f(\mathbf{x}) \leq f(\mathbf{a}_{\text{max}}).$$

We need the compactness hypothesis since a function defined over a noncompact domain may increase or decrease without bound and hence fail to have any global extremum, as suggested by Figure 4.21. This is analogous to the situation in one variable where a continuous function defined on an open interval may fail to have any extrema, but one defined on a closed interval (which is a compact subset of $\mathbf{R}$) must attain both maximum and minimum values. (See Figure 4.22.) In the one-variable case, extrema can occur either in the interior of the interval or else at the endpoints. Therefore, you must compare the values of f at any interior critical points with those at the endpoints to determine which is largest and smallest. In the case of functions of n variables, we do something similar, namely, compare the values of f at any critical points with values at any restricted critical points that may occur along the boundary of the domain.

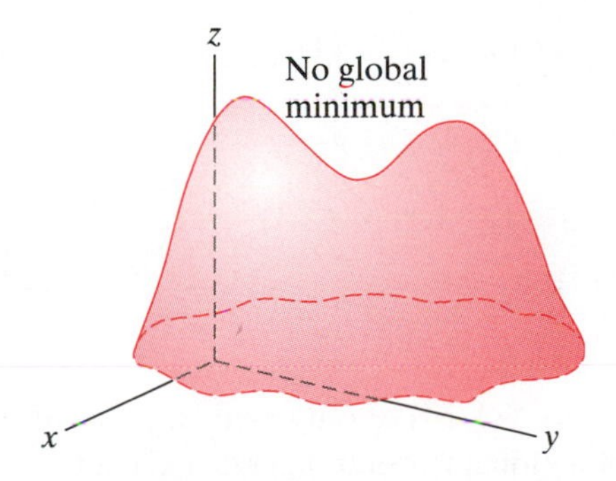

Figure 4.21 A graph that lacks a global minimum.

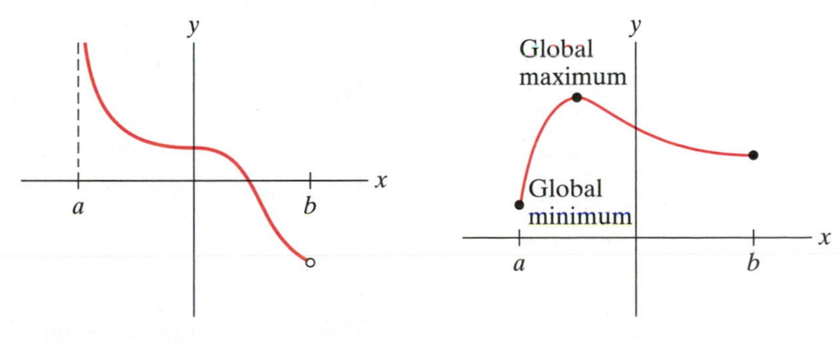

Figure 4.22 The function depicted by the graph on the left has no global extrema—the function is defined on the open interval (a, b). By contrast, the function defined on the closed interval $[a, b]$, and with the graph on the right, has both a global maximum and minimum.

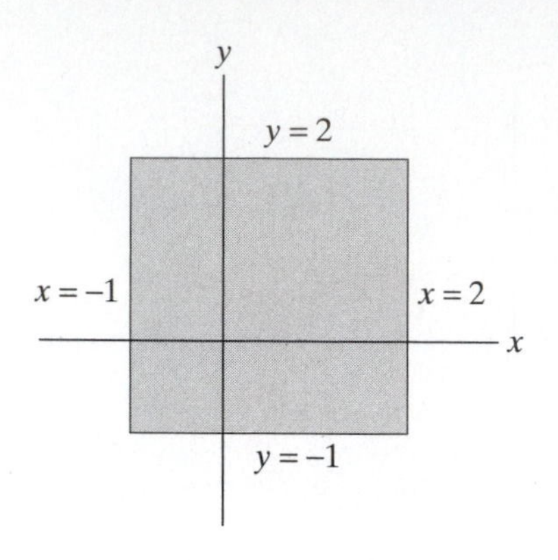

Figure 4.23 The domain of the function T of Example 7.

EXAMPLE 7 Let $T: X \subseteq \mathbf{R}^2 \to \mathbf{R}$ be given by

$$T(x, y) = x^2 - xy + y^2 + 1,$$

where X is the closed square in Figure 4.23. (Note that X is compact.) Think of the square as representing a flat metal plate and the function T as the temperature of the plate at each point. Finding the global extrema amounts to finding the warmest and coldest points on the plate. According to Theorem 2.5, such points must exist.

We need to find all possible critical points of T. Momentarily considering T as a function on all of $\mathbf{R}^2$, we find the usual critical points by setting $DT(x, y)$ equal to $\mathbf{0}$. The result is the system of two equations

$$\begin{cases} 2x - y = 0 \\ -x + 2y = 0 \end{cases},$$

which has $(0, 0)$ as its only solution. Whether it is a local maximum or minimum is not important for now, because we seek global extrema. Because there is only one critical point, at least one global extremum must occur along the boundary of X (which consists of the four edges of the square). We now find all critical points of the *restriction* of T to this boundary:

1. The bottom edge of X is the set

   $$E_1 = \{(x, y) \mid y = -1, -1 \le x \le 2\}.$$

   The restriction of T to E_1 defines a new function $f_1: [-1, 2] \to \mathbf{R}$ given by

   $$f_1(x) = T(x, -1) = x^2 + x + 2.$$

   As $f_1'(x) = 2x + 1$, the function f_1 has a critical point at $x = -\frac{1}{2}$. Thus, we must examine the following points of X for possible extrema: $\left(-\frac{1}{2}, -1\right)$, $(-1, -1)$, and $(2, -1)$. (The first point is the critical point of f_1, and the second two are the vertices of X that lie on E_1.)

2. The top edge of X is given by

   $$E_2 = \{(x, y) \mid y = 2, -1 \le x \le 2\}.$$

   Consequently, we define $f_2: [-1, 2] \to \mathbf{R}$ by

   $$f_2(x) = T(x, 2) = x^2 - 2x + 5.$$

   (f_2 is the restriction of T to E_2.) We calculate $f_2'(x) = 2x - 2$, which implies that $x = 1$ is a critical point of f_2. Hence, we must consider $(1, 2)$, $(-1, 2)$, and $(2, 2)$ as possible sites for global extrema of T. (The points $(-1, 2)$ and $(2, 2)$ are the remaining two vertices of X.)

3. The left edge of X is

   $$E_3 = \{(x, y) \mid x = -1, -1 \le y \le 2\}.$$

   Therefore, we define $f_3 : [-1, 2] \to \mathbf{R}$ by

   $$f_3(y) = T(-1, y) = y^2 + y + 2.$$

   We have $f_3'(y) = 2y + 1$, and so $y = -\frac{1}{2}$ is the only critical point of f_3. Thus $\left(-1, -\frac{1}{2}\right)$ is a potential site of a global extremum. (We need not worry again about the vertices $(-1, -1)$ and $(-1, 2)$.)

4. The right edge of X is

   $$E_4 = \{(x, y) \mid x = 2, -1 \le y \le 2\}.$$

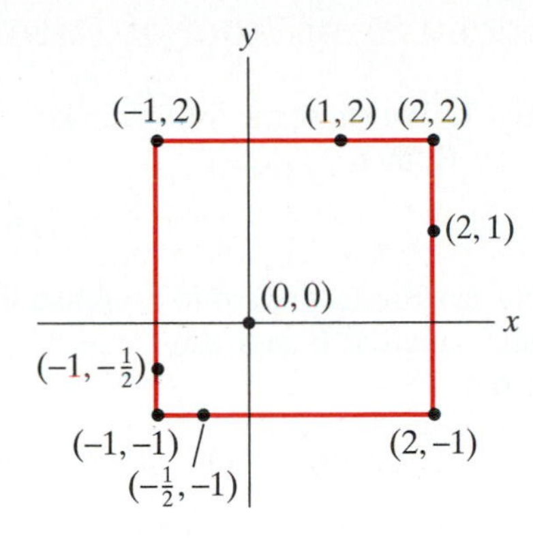

(x,y)	$T(x,y)$
$(0,0)$	1
$(-\frac{1}{2},-1)$	$\frac{7}{4}$
$(-1,-1)$	2
$(2,-1)$	8
$(1,2)$	4
$(-1,2)$	8
$(2,2)$	5
$(-1,-\frac{1}{2})$	$\frac{7}{4}$
$(2,1)$	4

Figure 4.24 Possible global extrema for T.

We define $f_4\colon [-1, 2] \to \mathbf{R}$ by

$$f_4(y) = T(2, y) = y^2 - 2y + 5.$$

We have $f_4'(y) = 2y - 2$, and so $y = 1$ is the only critical point of f_4. Hence, we must include (2, 1) in our consideration.

Consequently, we have nine possible locations for global extrema, shown in Figure 4.24. Now we need only to compare the actual values of T at these points to see that (0, 0) is the coldest point on the plate and both $(2, -1)$ and $(-1, 2)$ are the hottest points. ◆

If a function is defined over a noncompact region, there is no general result like the extreme value theorem (Theorem 2.5) to guarantee existence of any global extrema. However, ad hoc arguments frequently can be used to identify global extrema.

EXAMPLE 8 Consider the function $f(x, y, z) = e^{1-3x^2-y^2}$ defined on all of $\mathbf{R}^2$ (so the domain is certainly not compact). Verifying that f has a unique critical point at (0, 0) is straightforward. We leave it to you to check that the Hessian criterion implies that f has a local maximum there. In any case, for all $(x, y) \in \mathbf{R}^2$, we have

$$1 - 3x^2 - y^2 \leq 1.$$

Therefore, because the exponential function is always increasing (i.e., if $u_1 \leq u_2$, then $e^{u_1} \leq e^{u_2}$),

$$e^{1-3x^2-y^2} \leq e^1 = e.$$

As $f(0, 0) = e$, we see that f has a global maximum at (0, 0). ◆

WARNING It is tempting to assume that if a function has a unique critical point that is a local extremum, then it must be a global extremum as well. Although true for the case of a function of a single variable, it is not true for functions of two or more variables. (See Exercise 46 for an example.)

Addendum: Proof of Theorem 2.3

Step 1. We show the following key property of a quadratic form Q, namely, that if $\lambda \in \mathbf{R}$, then

$$Q(\lambda\mathbf{h}) = \lambda^2 Q(\mathbf{h}). \tag{3}$$

This is straightforward to establish if we write Q in terms of its associated symmetric matrix B and use some of the properties of matrix arithmetic given in §1.6:

$$Q(\lambda\mathbf{h}) = (\lambda\mathbf{h})^T B(\lambda\mathbf{h}) = \lambda\mathbf{h}^T B(\lambda\mathbf{h}) = \lambda^2\mathbf{h}^T B\mathbf{h} = \lambda^2 Q(\mathbf{h}).$$

Step 2. We show that if B is the symmetric matrix associated to a positive definite quadratic form Q, then there is a positive constant M such that

$$Q(\mathbf{h}) \geq M\|\mathbf{h}\|^2$$

for all $\mathbf{h} \in \mathbf{R}^n$.

First note that when $\mathbf{h} = \mathbf{0}$, then $Q(\mathbf{h}) = Q(\mathbf{0}) = 0$ so the conclusion holds trivially in this case.

Next, suppose that $\mathbf{h}$ is a unit vector (i.e., $\|\mathbf{h}\| = 1$). The (endpoints of the) set of all unit vectors in $\mathbf{R}^n$ is an $(n-1)$-dimensional sphere S, which is a compact set. Hence, by the extreme value theorem (Theorem 2.5), the restriction of Q to S must achieve a global minimum value M somewhere on S. Thus, $Q(\mathbf{h}) \geq M$ for all $\mathbf{h} \in S$.

Finally, let $\mathbf{h}$ be any nonzero vector in $\mathbf{R}^n$. Then its normalization $\mathbf{h}/\|\mathbf{h}\|$ is a unit vector and so lies in S. Therefore, by the result of Step 1, we have

$$Q(\mathbf{h}) = Q\left(\|\mathbf{h}\|\frac{\mathbf{h}}{\|\mathbf{h}\|}\right) = \|\mathbf{h}\|^2 Q\left(\frac{\mathbf{h}}{\|\mathbf{h}\|}\right) \geq \|\mathbf{h}\|^2 M,$$

since $\mathbf{h}/\|\mathbf{h}\|$ is in S.

Step 3. Now we prove the theorem. By the second-order Taylor formula Theorem 1.5 and formula (10) of §4.1, we have that, for the critical point $\mathbf{a}$ of f,

$$\Delta f = f(\mathbf{x}) - f(\mathbf{a}) = \tfrac{1}{2}(\mathbf{x}-\mathbf{a})^T Hf(\mathbf{a})(\mathbf{x}-\mathbf{a}) + R_2(\mathbf{x},\mathbf{a}), \tag{4}$$

where $|R_2(\mathbf{x},\mathbf{a})|/\|\mathbf{x}-\mathbf{a}\|^2 \to 0$ as $\mathbf{x} \to \mathbf{a}$.

Suppose first that $Hf(\mathbf{a})$ is positive definite. Then by Step 2 with $\mathbf{h} = \mathbf{x} - \mathbf{a}$, there must exist a constant $M > 0$ such that

$$\tfrac{1}{2}(\mathbf{x}-\mathbf{a})^T Hf(\mathbf{a})(\mathbf{x}-\mathbf{a}) \geq M\|\mathbf{x}-\mathbf{a}\|^2. \tag{5}$$

Because $|R_2(\mathbf{x},\mathbf{a})|/\|\mathbf{x}-\mathbf{a}\|^2 \to 0$ as $\mathbf{x} \to \mathbf{a}$, there must be some $\delta > 0$ so that if $0 < \|\mathbf{x}-\mathbf{a}\| < \delta$, then $|R_2(\mathbf{x},\mathbf{a})|/\|\mathbf{x}-\mathbf{a}\|^2 < M$, or, equivalently,

$$|R_2(\mathbf{x},\mathbf{a})| < M\|\mathbf{x}-\mathbf{a}\|^2. \tag{6}$$

Therefore, (4), (5), and (6) imply that, for $0 < \|\mathbf{x}-\mathbf{a}\| < \delta$,

$$\Delta f > 0$$

so that f has a (strict) local minimum at $\mathbf{a}$.

If $Hf(\mathbf{a})$ is negative definite, then consider $g = -f$. We see that $\mathbf{a}$ is also a critical point of g and that $Hg(\mathbf{a}) = -Hf(\mathbf{a})$, so $Hg(\mathbf{a})$ is positive definite. Hence, the argument in the preceding paragraph shows that g has a local minimum at $\mathbf{a}$, so f has a local maximum at $\mathbf{a}$.

Now suppose $\det Hf(\mathbf{a}) \neq 0$, but that $Hf(\mathbf{a})$ is neither positive nor negative definite. Let $\mathbf{x}_1$ be such that

$$\tfrac{1}{2}(\mathbf{x}_1 - \mathbf{a})^T Hf(\mathbf{a})(\mathbf{x}_1 - \mathbf{a}) > 0$$

and $\mathbf{x}_2$ such that

$$\tfrac{1}{2}(\mathbf{x}_2 - \mathbf{a})^T Hf(\mathbf{a})(\mathbf{x}_2 - \mathbf{a}) < 0.$$

(Since $\det Hf(\mathbf{a}) \neq 0$ such points must exist.) For $i = 1, 2$ let

$$\mathbf{y}_i(t) = t(\mathbf{x}_i - \mathbf{a}) + \mathbf{a},$$

the vector parametric equation for the line though $\mathbf{a}$ and $\mathbf{x}_i$. Applying formula (4) with $\mathbf{x} = \mathbf{y}_i(t)$, we see

$$\begin{aligned}\Delta f = f(\mathbf{y}_i(t)) - f(\mathbf{a}) &= \tfrac{1}{2}(\mathbf{y}_i(t) - \mathbf{a})^T Hf(\mathbf{a})(\mathbf{y}_i(t) - \mathbf{a}) + R_2(\mathbf{y}_i(t), \mathbf{a}) \\ &= \tfrac{1}{2}(\mathbf{y}_i(t) - \mathbf{a})^T Hf(\mathbf{a})(\mathbf{y}_i(t) - \mathbf{a}) + \|\mathbf{y}_i(t) - \mathbf{a}\|^2 \frac{R_2(\mathbf{y}_i(t), \mathbf{a})}{\|\mathbf{y}_i(t) - \mathbf{a}\|^2}.\end{aligned}$$

Note that $\mathbf{y}_i(t) - \mathbf{a} = t(\mathbf{x}_i - \mathbf{a})$. Therefore, using the property of quadratic forms given in Step 1 and the fact that $\|\mathbf{y}_i(t) - \mathbf{a}\|^2 = \|t(\mathbf{x}_i - \mathbf{a})\|^2 = t^2\|\mathbf{x}_i - \mathbf{a}\|^2$, we have

$$\begin{aligned} & f(\mathbf{y}_i(t)) - f(\mathbf{a}) \\ &\quad = t^2 \left[\tfrac{1}{2}(\mathbf{x}_i - \mathbf{a})^T Hf(\mathbf{a})(\mathbf{x}_i - \mathbf{a}) + \|\mathbf{x}_i - \mathbf{a}\|^2 \frac{R_2(\mathbf{y}_i(t), \mathbf{a})}{\|\mathbf{y}_i(t) - \mathbf{a}\|^2} \right]. \end{aligned} \tag{7}$$

Now note that, for $i = 1$, the first term in the brackets in the right side of (7) is a positive number P and, for $i = 2$, it is a negative number N. Set

$$M = \min\left(\frac{P}{\|\mathbf{x}_1 - \mathbf{a}\|^2}, -\frac{N}{\|\mathbf{x}_2 - \mathbf{a}\|^2} \right).$$

Because we know that $|R_2(\mathbf{y}_i(t), \mathbf{a})|/\|\mathbf{y}_i(t) - \mathbf{a}\|^2 \to 0$ as $t \to 0$, we can find some $\delta > 0$ so that if $0 < t < \delta$, then

$$\frac{|R_2(\mathbf{y}_i(t), \mathbf{a})|}{\|\mathbf{y}_i(t) - \mathbf{a}\|^2} < M.$$

But this implies that, for $0 < t < \delta$,

$$\Delta f = f(\mathbf{y}_1(t)) - f(\mathbf{a}) > 0,$$

while

$$\Delta f = f(\mathbf{y}_2(t)) - f(\mathbf{a}) < 0.$$

Thus, f has a saddle point at $\mathbf{x} = \mathbf{a}$. ■

Exercises

1. Concerning the function $f(x, y) = 4x + 6y - 12 - x^2 - y^2$:

(a) There is a unique critical point. Find it.

(b) By considering the increment Δf, determine whether this critical point is a maximum, a minimum, or a saddle point.

(c) Now use the Hessian criterion to determine the nature of the critical point.

2. This problem concerns the function $g(x, y) = x^2 - 2y^2 + 2x + 3$.

(a) Find any critical points of g.

(b) Use the increment Δg to determine the nature of the critical points of g.

(c) Use the Hessian criterion to determine the nature of the critical points.

In Exercises 3–20, identify and determine the nature of the critical points of the given functions.

3. $f(x, y) = 2xy - 2x^2 - 5y^2 + 4y - 3$

4. $f(x, y) = \ln(x^2 + y^2 + 1)$

5. $f(x, y) = x^2 + y^3 - 6xy + 3x + 6y$

6. $f(x, y) = y^4 - 2xy^2 + x^3 - x$

7. $f(x, y) = xy + \dfrac{8}{x} + \dfrac{1}{y}$

8. $f(x, y) = e^x \sin y$

9. $f(x, y) = e^{-y}(x^2 - y^2)$

10. $f(x, y) = (x + y)(1 - xy)$

11. $f(x, y) = x^2 - y^3 - x^2y + y$

12. $f(x, y) = e^{-x}(x^2 + 3y^2)$

13. $f(x, y) = 2x - 3y + \ln xy$

14. $f(x, y) = \cos x \sin y$

15. $f(x, y, z) = x^2 - xy + z^2 - 2xz + 6z$

16. $f(x, y, z) = (x^2 + 2y^2 + 1)\cos z$

17. $f(x, y, z) = x^2 + y^2 + 2z^2 + xz$

18. $f(x, y, z) = x^3 + xz^2 - 3x^2 + y^2 + 2z^2$

19. $f(x, y, z) = xy + xz + 2yz + \dfrac{1}{x}$

20. $f(x, y, z) = e^x(x^2 - y^2 - 2z^2)$

21. (a) Find all critical points of $f(x, y) = \dfrac{2y^3 - 3y^2 - 36y + 2}{1 + 3x^2}$.

(b) Identify any and all extrema of f.

22. (a) Under what conditions on the constant k will the function

$$f(x, y) = kx^2 - 2xy + ky^2$$

have a nondegenerate local minimum at $(0, 0)$? What about a local maximum?

(b) Under what conditions on the constant k will the function

$$g(x, y, z) = kx^2 + kxz - 2yz - y^2 + \frac{k}{2}z^2$$

have a nondegenerate local maximum at $(0, 0, 0)$? What about a nondegenerate local minimum?

23. (a) Consider the function $f(x, y) = ax^2 + by^2$, where a and b are nonzero constants. Show that the origin is the only critical point of f, and determine the nature of that critical point in terms of a and b.

(b) Now consider the function $f(x, y, z) = ax^2 + by^2 + cz^2$, where a, b, and c are all nonzero. Show that the origin in $\mathbf{R}^3$ is the only critical point of f, and determine the nature of that critical point in terms of a, b, and c.

(c) Finally, let $f(x_1, x_2, \ldots, x_n) = a_1x_1^2 + a_2x_2^2 + \cdots + a_nx_n^2$, where a_i is a nonzero constant for $i = 1, 2, \ldots, n$. Show that the origin in $\mathbf{R}^n$ is the only critical point of f, and determine its nature.

Sometimes it can be difficult to determine the critical point of a function f because the system of equations that arises from setting ∇f equal to zero may be very complicated to solve by hand. For the functions given in Exercises 24–27, (a) use a computer to assist you in identifying all the critical points of the given function f, and (b) use a computer to construct the Hessian matrix and determine the nature of the critical points found in part (a).

24. $f(x, y) = y^4 + x^3 - 2xy^2 - x$

25. $f(x, y) = 2x^3y - y^2 - 3xy$

26. $f(x, y, z) = yz - xyz - x^2 - y^2 - 2z^2$

27. $f(x, y, z, w) = yw - xyz - x^2 - 2z^2 + w^2$

28. Show that the largest rectangular box having a fixed surface area must be a cube.

29. What point on the plane $3x - 4y - z = 24$ is closest to the origin?

30. Find the absolute extrema of $f(x, y) = x^2 + xy + y^2 - 6y$ on the rectangle $\{(x, y) \mid -3 \le x \le 3,\ 0 \le y \le 5\}$.

31. Find the absolute maximum and minimum of

$$f(x, y, z) = x^2 + xz - y^2 + 2z^2 + xy + 5x$$

on the block $\{(x, y, z) \mid -5 \le x \le 0,\ 0 \le y \le 3,\ 0 \le z \le 2\}$.

32. A metal plate has the shape of the region $x^2 + y^2 \le 1$. The plate is heated so that the temperature at any point (x, y) on it is indicated by

$$T(x, y) = 2x^2 + y^2 - y + 3.$$

Find the hottest and coldest points on the plate, and the temperature at each of these points. (Hint: Parametrize the boundary of the plate in order to find any critical points there.)

33. Find the (absolute) maximum and minimum values of $f(x, y) = \sin x \cos y$ on the square $R = \{(x, y) \mid 0 \le x \le 2\pi,\ 0 \le y \le 2\pi\}$.

34. Find the absolute extrema of $f(x, y) = 2\cos x + 3\sin y$ on the rectangle $\{(x, y) \mid 0 \le x \le 4,\ 0 \le y \le 3\}$.

35. Find the absolute extrema of $f(x, y, z) = e^{1-x^2-y^2+2y-z^2-4z}$ on the ball $\{(x, y, z) \mid x^2 + y^2 - 2y + z^2 + 4z \le 0\}$.

Each of the functions in Exercises 36–41 has a critical point at the origin. For each function, (a) check that the Hessian fails to provide any information about the nature of the critical point at the origin, and (b) find another way to determine if the function has a maximum, minimum, or neither at the origin.

36. $f(x, y) = x^2y^2$

37. $f(x, y) = 4 - 3x^2y^2$

38. $f(x, y) = x^3y^3$

39. $f(x, y, z) = x^2y^3z^4$

40. $f(x, y, z) = x^2y^2z^4$

41. $f(x, y, z) = 2 - x^4y^4 - z^4$

In Exercises 42–44, (a) find all critical points of the given function f and identify their nature as local extrema and (b) determine, with explanation, any global extrema of f.

42. $f(x, y) = e^{x^2+5y^2}$

43. $f(x, y, z) = e^{2-x^2-2y^2-3z^4}$

44. $f(x, y) = x^3 + y^3 - 3xy + 7$

45. Determine the global extrema, if any, of

$$f(x, y) = xy + 2y - \ln x - 2\ln y,$$

where $x, y > 0$.

46. (a) Suppose $f: \mathbf{R} \to \mathbf{R}$ is a differentiable function of a single variable. Show that if f has a unique critical point at x_0 that is the site of a strict local extremum of f, then f must attain a global extremum at x_0.

(b) Let $f(x, y) = 3ye^x - e^{3x} - y^3$. Verify that f has a unique critical point and that f attains a local maximum there. However, show that f does not have a global maximum by considering how f behaves along the y-axis. Hence, the result of part (a) does not carry over to functions of more than one variable.

47. (a) Let f be a continuous function of one variable. Show that if f has two local maxima, then f must also have a local minimum.

(b) The analogue of part (a) does not necessarily hold for continuous functions of more than one variable, as we now see. Consider the function

$$f(x, y) = 2 - (xy^2 - y - 1)^2 - (y^2 - 1)^2.$$

Show that f has just two critical points—and that both of them are local maxima.

(c) Use a computer to graph the function f in part (b).

4.3 Lagrange Multipliers

Constrained Extrema

Frequently, when working with applications of calculus, you will find that you do not need simply to maximize or minimize a function, but that you must do so subject to one or more additional constraints that depend on the specifics of the situation. The following example is a typical situation:

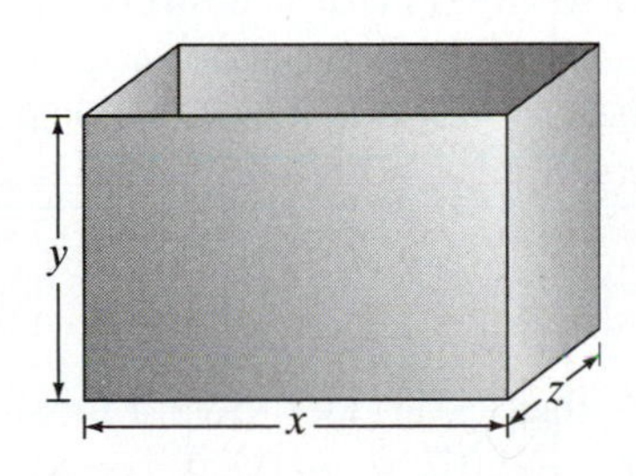

Figure 4.25 The open box of Example 1.

EXAMPLE 1 An open rectangular box is to be manufactured having a (fixed) volume of 4 ft^3. What dimensions should the box have so as to minimize the amount of material used to make it?

We'll let the three dimensions of the box be independent variables x, y, and z, shown in Figure 4.25. To determine how to use as little material as possible, we need to minimize the surface area function A given by

$$A(x, y, z) = \underbrace{2xy}_{\text{front and back}} + \underbrace{2yz}_{\text{sides}} + \underbrace{xz}_{\text{bottom only}}.$$

For $x, y, z > 0$, this function has neither minimum nor maximum. However, we have not yet made use of the fact that the volume is to be maintained at a constant 4 ft^3. This fact provides a **constraint equation,**

$$V(x, y, z) = xyz = 4.$$

The constraint is absolutely essential if we are to solve the problem. In particular, the constraint enables us to solve for z in terms of x and y:

$$z = \frac{4}{xy}.$$

We can thus create a new area function of only two variables:

$$\begin{aligned} a(x, y) &= A\left(x, y, \frac{4}{xy}\right) \\ &= 2xy + 2y\left(\frac{4}{xy}\right) + x\left(\frac{4}{xy}\right) \\ &= 2xy + \frac{8}{x} + \frac{4}{y}. \end{aligned}$$

Now we can find the critical points of a by setting Da equal to $\mathbf{0}$:

$$\begin{cases} \dfrac{\partial a}{\partial x} = 2y - \dfrac{8}{x^2} = 0 \\ \dfrac{\partial a}{\partial y} = 2x - \dfrac{4}{y^2} = 0 \end{cases}.$$

The first equation implies

$$y = \frac{4}{x^2},$$

so that the second equation becomes

$$2x - 4\left(\frac{x^4}{16}\right) = 0$$

or, equivalently,

$$x\left(1 - \frac{1}{8}x^3\right) = 0.$$

The solutions to this equation are $x = 0$ (which we reject) and $x = 2$. Thus, the critical point of a of interest is $(2, 1)$, and the **constrained critical point** of the original function A is $(2, 1, 2)$.

We can use the Hessian criterion to check that $x = 2$, $y = 1$ yields a local minimum of a:

$$Ha(x, y) = \begin{bmatrix} 16/x^3 & 2 \\ 2 & 8/y^3 \end{bmatrix} \quad \text{so} \quad Ha(2, 1) = \begin{bmatrix} 2 & 2 \\ 2 & 8 \end{bmatrix}.$$

The sequence of minors is 2, 12 so we conclude that $(2, 1)$ does yield a local minimum of a. Because $a(x, y) \to \infty$ as either $x \to 0^+$, $y \to 0^+$, $x \to \infty$, or $y \to \infty$, we conclude that the critical point must yield a global minimum as well. Thus, the solution to the original question is to make the box with a square base of side 2 ft and a height of 1 ft. ◆

The abstract setting for the situation discussed in Example 1 is to find maxima or minima of a function $f(x_1, x_2, \ldots, x_n)$ subject to the **constraint** that $g(x_1, x_2, \ldots, x_n) = c$ for some function g and constant c. (In Example 1, the function f is $A(x, y, z)$, and the constraint is $xyz = 4$.) One method for finding constrained critical points is used implicitly in Example 1: Use the constraint equation $g(\mathbf{x}) = c$ to solve for one of the variables in terms of the others. Then substitute for this variable in the expression for $f(\mathbf{x})$, thereby creating a new function of one fewer variables. This new function can then be maximized or minimized, using the techniques of §4.2. In theory, this is an entirely appropriate way to approach such problems, but in practice there is one major drawback: It may be impossible to solve explicitly for any one of the variables in terms of the others. For example, you might wish to maximize

$$f(x, y, z) = x^2 + 3y^2 + y^2z^4$$

subject to

$$g(x, y, z) = e^{xy} - x^5y^2z + \cos\left(\frac{x}{yz}\right) = 2.$$

There is no means of isolating any of x, y, or z on one side of the constraint equation, and so it is impossible for us to proceed any further along the lines of Example 1.

The Lagrange Multiplier

The previous discussion points to the desirability of having another method for solving constrained optimization problems. The key to such an alternative method is the following theorem:

Theorem 3.1 Let X be open in $\mathbf{R}^n$ and $f, g\colon X \to \mathbf{R}$ be functions of class C^1. Let $S = \{\mathbf{x} \in X \mid g(\mathbf{x}) = c\}$ denote the level set of g at height c. Then if $f|_S$ (the restriction of f to S) has an extremum at a point $\mathbf{x}_0 \in S$ such that $\nabla g(\mathbf{x}_0) \neq \mathbf{0}$, there must be some scalar λ such that

$$\nabla f(\mathbf{x}_0) = \lambda \nabla g(\mathbf{x}_0).$$

The conclusion of Theorem 3.1 implies that to find possible sites for extrema of f subject to the constraint that $g(\mathbf{x}) = c$, we can proceed in the following manner:

1. Form the vector equation $\nabla f(\mathbf{x}) = \lambda \nabla g(\mathbf{x})$.
2. Solve the system

$$\begin{cases} \nabla f(\mathbf{x}) = \lambda \nabla g(\mathbf{x}) \\ g(\mathbf{x}) = c \end{cases}$$

for $\mathbf{x}$ and λ. When expanded, this is actually a system of $n+1$ equations in $n+1$ unknowns $x_1, x_2, \ldots, x_n, \lambda$, namely,

$$\begin{cases} f_{x_1}(x_1, x_2, \ldots, x_n) = \lambda g_{x_1}(x_1, x_2, \ldots, x_n) \\ f_{x_2}(x_1, x_2, \ldots, x_n) = \lambda g_{x_2}(x_1, x_2, \ldots, x_n) \\ \qquad \vdots \\ f_{x_n}(x_1, x_2, \ldots, x_n) = \lambda g_{x_n}(x_1, x_2, \ldots, x_n) \\ g(x_1, x_2, \ldots, x_n) = c \end{cases}.$$

The solutions for $\mathbf{x} = (x_1, x_2, \ldots, x_n)$, along with any other points satisfying $g(\mathbf{x}) = c$ and $\nabla g(\mathbf{x}) = \mathbf{0}$, are the candidates for extrema for the problem.

3. Determine the nature of f (as maximum, minimum, or neither) at the critical points found in Step 2.

The scalar λ appearing in Theorem 3.1 is called a **Lagrange multiplier,** after the French mathematician Joseph Louis Lagrange (1736–1813) who first developed this method for solving constrained optimization problems. In practice, Step 2 can involve some algebra, so it is important to keep your work organized. (Alternatively, you can use a computer to solve the system.) In fact, since the Lagrange multiplier λ is usually not of primary interest, you can avoid solving for it explicitly, thereby reducing the algebra and arithmetic somewhat. Determining the nature of a constrained critical point (Step 3) can be a tricky business. We'll have more to say about that issue in the examples and discussions that follow.

EXAMPLE 2 Let us use the method of Lagrange multipliers to identify the critical point found in Example 1. Thus, we wish to find the minimum of

$$A(x, y, z) = 2xy + 2yz + xz$$

subject to the constraint

$$V(x, y, z) = xyz = 4.$$

Theorem 3.1 suggests that we form the equation

$$\nabla A(x, y, z) = \lambda \nabla V(x, y, z).$$

This relation of gradients coupled with the constraint equation gives rise to the system

$$\begin{cases} 2y + z = \lambda yz \\ 2x + 2z = \lambda xz \\ 2y + x = \lambda xy \\ xyz = 4 \end{cases}.$$

Since λ is not essential for our final solution, we can eliminate it by means of any of the first three equations. Hence,

$$\lambda = \frac{2y + z}{yz} = \frac{2x + 2z}{xz} = \frac{2y + x}{xy}.$$

Simplifying, this implies that

$$\frac{2}{z} + \frac{1}{y} = \frac{2}{z} + \frac{2}{x} = \frac{2}{x} + \frac{1}{y}.$$

The first equality yields

$$\frac{1}{y} = \frac{2}{x} \quad \text{or} \quad x = 2y,$$

while the second equality implies that

$$\frac{2}{z} = \frac{1}{y} \quad \text{or} \quad z = 2y.$$

Substituting these relations into the constraint equation $xyz = 4$ yields

$$(2y)(y)(2y) = 4,$$

so that we find that the only solution is $y = 1, x = z = 2$, which agrees with our work in Example 1. (Note that $\nabla V = \mathbf{0}$ only along the coordinate axes, and such points do not satisfy the constraint $V(x, y, z) = 4$.) ◆

An interesting consequence of Theorem 3.1 is this: By Theorem 6.4 of Chapter 2, we know that the gradient ∇g, when nonzero, is perpendicular to the level sets of g. Thus, the equation $\nabla f = \lambda \nabla g$ gives the condition for the normal vector to a level set of f to be parallel to that of a level set of g. Hence, for a point $\mathbf{x}_0$ to be the site of an extremum of f on the level set $S = \{\mathbf{x} \mid g(\mathbf{x}) = c\}$, where $\nabla g(\mathbf{x}_0) \neq \mathbf{0}$, we must have that the level set R of f that contains $\mathbf{x}_0$ is tangent to S at $\mathbf{x}_0$.

EXAMPLE 3 Consider the problem of finding the extrema of $f(x, y) = x^2/4 + y^2$ subject to the condition that $x^2 + y^2 = 1$. We let $g(x, y) = x^2 + y^2$, and so the Lagrange multiplier equation $\nabla f(x, y) = \lambda \nabla g(x, y)$, along with the constraint equation, yields the system

$$\begin{cases} \dfrac{x}{2} = 2\lambda x \\ 2y = 2\lambda y \\ x^2 + y^2 = 1 \end{cases}.$$

(There are no points simultaneously satisfying $g(x, y) = 1$ and $\nabla g(x, y) = (0, 0)$.) The first equation of this system implies that either $x = 0$ or $\lambda = \frac{1}{4}$. If $x = 0$, then the second two equations, taken together, imply that $y = \pm 1$ and $\lambda = 1$. If $\lambda = \frac{1}{4}$, then the second two equations imply $y = 0$ and $x = \pm 1$. Therefore, there are four constrained critical points: $(0, \pm 1)$, corresponding to $\lambda = 1$, and $(\pm 1, 0)$, corresponding to $\lambda = \frac{1}{4}$.

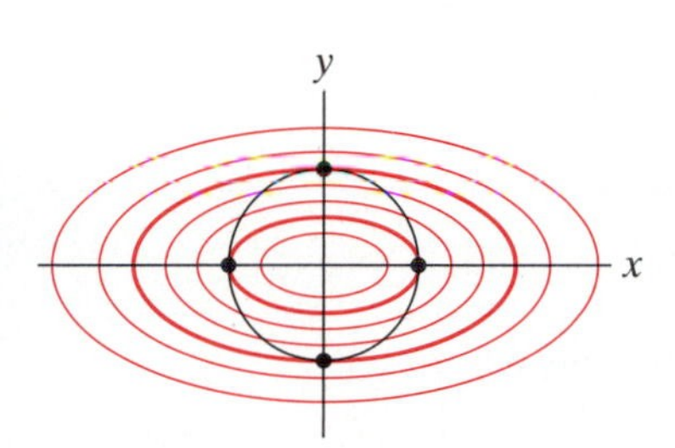

Figure 4.26 The level sets of the function $f(x, y) = x^2/4 + y^2$ define a family of ellipses. The extrema of f subject to the constraint that $x^2 + y^2 = 1$ (i.e., that lie on the unit circle) occur at points where an ellipse of the family is tangent to the unit circle.

We can understand the nature of these critical points by using geometry and the preceding remarks. The collection of level sets of the function f are the family of ellipses $x^2/4 + y^2 = k$ whose major and minor axes lie along the x- and y-axes, respectively. In fact, the value $f(x, y) = x^2/4 + y^2 = k$ is the square of the length of the semiminor axis of the ellipse $x^2/4 + y^2 = k$. The optimization problem then is to find those points on the unit circle $x^2 + y^2 = 1$ that, when considered as points in the family of ellipses, minimize and maximize the length of the minor axis. When we view the problem in this way, we see that such points must occur where the circle is tangent to one of the ellipses in the family. A sketch shows that constrained minima of f occur at $(\pm 1, 0)$ and constrained maxima at $(0, \pm 1)$. In this case, the Lagrange multiplier λ represents the square of the length of the semiminor axis. (See Figure 4.26.) ◆

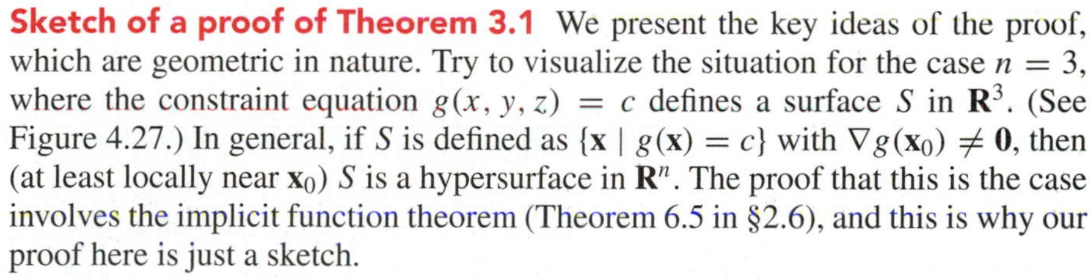

Sketch of a proof of Theorem 3.1 We present the key ideas of the proof, which are geometric in nature. Try to visualize the situation for the case $n = 3$, where the constraint equation $g(x, y, z) = c$ defines a surface S in $\mathbf{R}^3$. (See Figure 4.27.) In general, if S is defined as $\{\mathbf{x} \mid g(\mathbf{x}) = c\}$ with $\nabla g(\mathbf{x}_0) \neq \mathbf{0}$, then (at least locally near $\mathbf{x}_0$) S is a hypersurface in $\mathbf{R}^n$. The proof that this is the case involves the implicit function theorem (Theorem 6.5 in §2.6), and this is why our proof here is just a sketch.

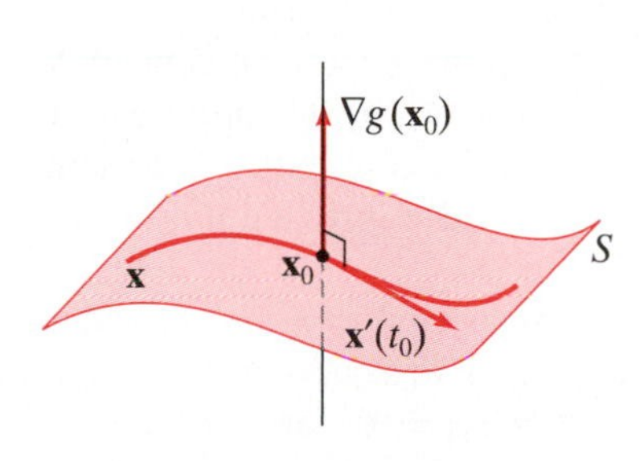

Figure 4.27 The gradient $\nabla g(\mathbf{x}_0)$ is perpendicular to $S = \{\mathbf{x} \mid g(\mathbf{x}) = c\}$, hence to the tangent vector at $\mathbf{x}_0$ to any curve $\mathbf{x}(t)$ lying in S and passing through $\mathbf{x}_0$. If f has an extremum at $\mathbf{x}_0$, then the restriction of f to the curve also has an extremum at $\mathbf{x}_0$.

Thus, suppose that $\mathbf{x}_0$ is an extremum of f restricted to S. We consider a further restriction of f—to a curve lying in S and passing through $\mathbf{x}_0$. This will enable us to use results from one-variable calculus. The notation and analytic particulars are as follows: Let $\mathbf{x}: I \subseteq \mathbf{R} \to S \subset \mathbf{R}^3$ be a C^1 path lying in S with $\mathbf{x}(t_0) = \mathbf{x}_0$ for some $t_0 \in I$. Then the restriction of f to $\mathbf{x}$ is given by the function F, where

$$F(t) = f(\mathbf{x}(t)).$$

Because $\mathbf{x}_0$ is an extremum of f on S, it must also be an extremum on $\mathbf{x}$. Consequently, we must have $F'(t_0) = 0$, and the chain rule implies that

$$0 = F'(t_0) = \frac{d}{dt} f(\mathbf{x}(t))\Big|_{t=t_0} = \nabla f(\mathbf{x}(t_0)) \cdot \mathbf{x}'(t_0) = \nabla f(\mathbf{x}_0) \cdot \mathbf{x}'(t_0).$$

Thus, $\nabla f(\mathbf{x}_0)$ is perpendicular to any curve in S passing through $\mathbf{x}_0$; that is, $\nabla f(\mathbf{x}_0)$ is normal to S at $\mathbf{x}_0$. We've seen previously in §2.6 that the gradient $\nabla g(\mathbf{x}_0)$ is also normal to S at $\mathbf{x}_0$. Since the normal direction to the level set S is uniquely determined and $\nabla g(\mathbf{x}_0) \neq \mathbf{0}$, we must conclude that $\nabla f(\mathbf{x}_0)$ and $\nabla g(\mathbf{x}_0)$ are parallel vectors. Therefore,

$$\nabla f(\mathbf{x}_0) = \lambda \nabla g(\mathbf{x}_0)$$

for some scalar $\lambda \in \mathbf{R}$, as desired. ■

The Case of More than One Constraint

It is natural to generalize the situation of finding extrema of a function f subject to a single constraint equation to that of finding extrema subject to several constraints. In other words, we may wish to maximize or minimize f subject to k simultaneous conditions of the form

$$\begin{cases} g_1(\mathbf{x}) = c_1 \\ g_2(\mathbf{x}) = c_2 \\ \quad\vdots \\ g_k(\mathbf{x}) = c_k \end{cases}.$$

The result that generalizes Theorem 3.1 is as follows:

■ **Theorem 3.2** Let X be open in $\mathbf{R}^n$ and let $f, g_1, \ldots, g_k \colon X \subseteq \mathbf{R}^n \to \mathbf{R}$ be C^1 functions, where $k < n$. Let $S = \{\mathbf{x} \in X \mid g_1(\mathbf{x}) = c_1, \ldots, g_k(\mathbf{x}) = c_k\}$. If $f|_S$ has an extremum at a point $\mathbf{x}_0$, where $\nabla g_1(\mathbf{x}_0), \ldots, \nabla g_k(\mathbf{x}_0)$ are linearly independent vectors, then there must exist scalars $\lambda_1, \ldots, \lambda_k$ such that

$$\nabla f(\mathbf{x}_0) = \lambda_1 \nabla g_1(\mathbf{x}_0) + \lambda_2 \nabla g_2(\mathbf{x}_0) + \cdots + \lambda_k \nabla g_k(\mathbf{x}_0).$$

(Note: k vectors $\mathbf{v}_1, \ldots, \mathbf{v}_k$ in $\mathbf{R}^n$ are said to be **linearly independent** if the only way to satisfy $a_1\mathbf{v}_1 + \cdots + a_k\mathbf{v}_k = \mathbf{0}$ for scalars $a_1, \ldots, a_k$ is if $a_1 = a_2 = \cdots = a_k = 0$.)

Idea of proof First, note that S is the intersection of the k hypersurfaces $S_1, \ldots, S_k$, where $S_j = \{\mathbf{x} \in \mathbf{R}^n \mid g_j(\mathbf{x}) = c_j\}$. Therefore, any vector tangent to S must also be tangent to each of these hypersurfaces, and so, by Theorem 6.4 of Chapter 2, perpendicular to each of the ∇g_j's. Given these remarks, the main ideas of the proof of Theorem 3.1 can be readily adapted to provide a proof of Theorem 3.2.

Therefore, we let $\mathbf{x}_0 \in S$ be an extremum of f restricted to S and consider the one-variable function obtained by further restricting f to a curve in S through $\mathbf{x}_0$. Thus, let $\mathbf{x} \colon I \to S \subset \mathbf{R}^n$ be a C^1 curve in S with $\mathbf{x}(t_0) = \mathbf{x}_0$ for some $t_0 \in I$. Then, as in the proof of Theorem 3.1, we define F by

$$F(t) = f(\mathbf{x}(t)).$$

It follows, since $\mathbf{x}_0$ is assumed to be a constrained extremum, that

$$F'(t_0) = 0.$$

The chain rule then tells us that

$$0 = F'(t_0) = \nabla f(\mathbf{x}(t_0)) \cdot \mathbf{x}'(t_0) = \nabla f(\mathbf{x}_0) \cdot \mathbf{x}'(t_0).$$

That is, $\nabla f(\mathbf{x}_0)$ is perpendicular to all vectors tangent to S at $\mathbf{x}_0$. Therefore, it can be shown that $\nabla f(\mathbf{x}_0)$ is in the k-dimensional plane spanned by the normal vectors to the individual hypersurfaces $S_1, \ldots, S_k$ whose intersection is S. It follows (via

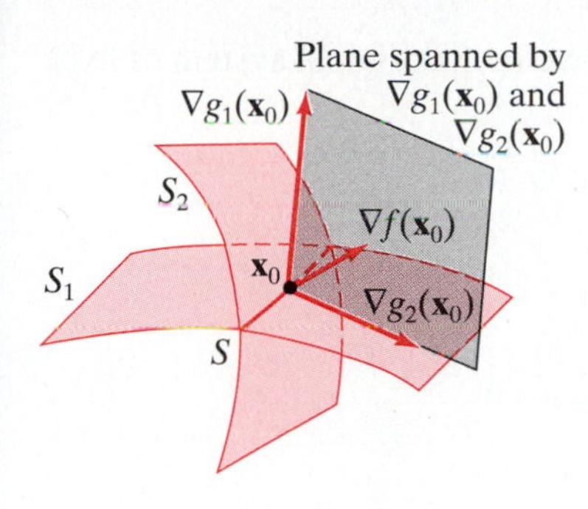

Figure 4.28 Illustration of the proof of Theorem 3.2. The constraints $g_1(\mathbf{x}) = c_1$ and $g_2(\mathbf{x}) = c_2$ are the surfaces S_1 and S_2. Any extremum of f must occur at points where ∇f is in the plane spanned by ∇g_1 and ∇g_2.

a little more linear algebra) that there must be scalars $\lambda_1, \ldots, \lambda_k$ such that

$$\nabla f(\mathbf{x}_0) = \lambda_1 \nabla g_1(\mathbf{x}_0) + \lambda_2 \nabla g_2(\mathbf{x}_0) + \cdots + \lambda_k \nabla g_k(\mathbf{x}_0).$$

A suggestion of the geometry of this proof is provided by Figure 4.28 (where $k = 2$ and $n = 3$). ■

EXAMPLE 4 Suppose the cone $z^2 = x^2 + y^2$ is sliced by the plane $z = x + y + 2$ so that a conic section C is created. We use Lagrange multipliers to find the points on C that are nearest to and farthest from the origin in $\mathbf{R}^3$.

The problem is to find the minimum and maximum distances from $(0, 0, 0)$ of points (x, y, z) on C. For algebraic simplicity, we look at the square of the distance, rather than the actual distance. Thus, we desire to find the extrema of

$$f(x, y, z) = x^2 + y^2 + z^2$$

(the square of the distance from the origin to (x, y, z)) subject to the constraints

$$\begin{cases} g_1(x, y, z) = x^2 + y^2 - z^2 = 0 \\ g_2(x, y, z) = x + y - z = -2 \end{cases}.$$

Note that

$$\nabla g_1(x, y, z) = (2x, 2y, -2z) \quad \text{and} \quad \nabla g_2(x, y, z) = (1, 1, -1).$$

These vectors are linearly dependent only when $x = y = z$. However, no point of the form (x, x, x) simultaneously satisfies $g_1 = 0$ and $g_2 = -2$. Hence, ∇g_1 and ∇g_2 are linearly independent at all points that satisfy the two constraints. Therefore, by Theorem 3.2, we know that any constrained critical points (x_0, y_0, z_0) must satisfy

$$\nabla f(x_0, y_0, z_0) = \lambda_1 \nabla g_1(x_0, y_0, z_0) + \lambda_2 \nabla g_2(x_0, y_0, z_0),$$

as well as the two constraint equations. Thus, we must solve the system

$$\begin{cases} 2x = 2\lambda_1 x + \lambda_2 \\ 2y = 2\lambda_1 y + \lambda_2 \\ 2z = -2\lambda_1 z - \lambda_2 \\ x^2 + y^2 - z^2 = 0 \\ x + y - z = -2 \end{cases}.$$

Eliminating λ_2 from the first two equations yields

$$\lambda_2 = 2x - 2\lambda_1 x = 2y - 2\lambda_1 y,$$

which implies that

$$2(x - y)(1 - \lambda_1) = 0.$$

Therefore, either

$$x = y \quad \text{or} \quad \lambda_1 = 1.$$

The condition $\lambda_1 = 1$ implies immediately $\lambda_2 = 0$, and the third equation of the system becomes $2z = -2z$, so z must equal 0. If $z = 0$, then x and y must be zero by the fourth equation. However, $(0, 0, 0)$ is not a point on the plane $z = x + y + 2$. Thus, the condition $\lambda_1 = 1$ leads to no critical point. On the other hand, if $x = y$, then the constraint equations (the last two in the original system of five) become

$$\begin{cases} 2x^2 - z^2 = 0 \\ 2x - z = -2 \end{cases}.$$

Substituting $z = 2x + 2$ yields

$$2x^2 - (2x + 2)^2 = 0,$$

equivalent to

$$2x^2 + 8x + 4 = 0,$$

whose solutions are $x = -2 \pm \sqrt{2}$. Therefore, there are two constrained critical points

$$\mathbf{a}_1 = \left(-2 + \sqrt{2}, -2 + \sqrt{2}, -2 + 2\sqrt{2}\right)$$

and

$$\mathbf{a}_2 = \left(-2 - \sqrt{2}, -2 - \sqrt{2}, -2 - 2\sqrt{2}\right).$$

We can check that

$$f(\mathbf{a}_1) = 24 - 16\sqrt{2}, \qquad f(\mathbf{a}_2) = 24 + 16\sqrt{2},$$

so it seems that $\mathbf{a}_1$ must be the point on C lying nearest the origin, and $\mathbf{a}_2$ must be the point that lies farthest. However, we don't know a priori if there is a farthest point. If the conic section C is a hyperbola or a parabola, then there is no point that is farthest from the origin. To understand what kind of curve C is, note that $\mathbf{a}_1$ has positive z-coordinate and $\mathbf{a}_2$ has negative z-coordinate. Therefore, the plane $z = x + y + 2$ intersects both nappes of the cone $z^2 = x^2 + y^2$. The only conic section that intersects both nappes of a cone is a hyperbola. Hence, C is a hyperbola, and we see that the point $\mathbf{a}_1$ is indeed the point nearest the origin, but the point $\mathbf{a}_2$ is not the farthest point. Instead, $\mathbf{a}_2$ is the point nearest the origin on the branch of the hyperbola not containing $\mathbf{a}_1$. That is, local constrained minima occur at both $\mathbf{a}_1$ and $\mathbf{a}_2$, but only $\mathbf{a}_1$ is the site of the global minimum. (See Figure 4.29.) ◆

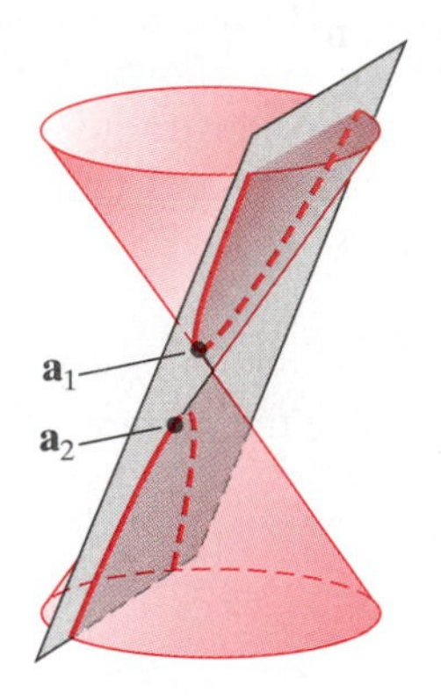

Figure 4.29 The point $\mathbf{a}_1$ is the point on the hyperbola closest to the origin. The point $\mathbf{a}_2$ is the point on the lower branch of the hyperbola closest to the origin.

A Hessian Criterion for Constrained Extrema (optional)

As Example 4 indicates, it is often possible to determine the nature of a critical point (constrained or unconstrained) from considerations particular to the problem at hand. Sometimes this is not difficult to do in practice and can provide useful insight into the problem. Nonetheless, occasionally it is advantageous to have a more automatic means of discerning the nature of a constrained critical point. We therefore present a Hessian criterion for constrained critical points. Like the one in the unconstrained case, this criterion only determines the *local* nature of a critical point. It does not provide information about global constrained extrema.[2]

[2]We invite the reader to consult D. Spring, *Amer. Math. Monthly*, **92** (1985), no. 9, 631–643 for a more complete discussion.

In general, the context for the Hessian criterion is this: We seek extrema of a function $f: X \subseteq \mathbf{R}^n \to \mathbf{R}$ subject to the k constraints

$$\begin{cases} g_1(x_1, x_2, \ldots, x_n) = c_1 \\ g_2(x_1, x_2, \ldots, x_n) = c_2 \\ \vdots \\ g_k(x_1, x_2, \ldots, x_n) = c_k \end{cases}.$$

We assume that $f, g_1, \ldots, g_k$ are all of class C^2, and assume, for simplicity, that f and the g_j's all have the same domain X. Finally, we assume that $\nabla g_1, \ldots, \nabla g_k$ are linearly independent at the constrained critical point $\mathbf{a}$. Then, by Theorem 3.2, any constrained extremum $\mathbf{a}$ must satisfy

$$\nabla f(\mathbf{a}) = \lambda_1 \nabla g_1(\mathbf{a}) + \lambda_2 \nabla g_2(\mathbf{a}) + \cdots + \lambda_k \nabla g_k(\mathbf{a})$$

for some scalars $\lambda_1, \ldots, \lambda_k$. We can consider a constrained critical point to be a pair of vectors

$$(\boldsymbol{\lambda}; \mathbf{a}) = (\lambda_1, \ldots, \lambda_k; a_1, \ldots, a_n)$$

satisfying the aforementioned equation. In fact, we can check that $(\boldsymbol{\lambda}; \mathbf{a})$ is an **unconstrained** critical point of the so-called Lagrangian function L defined by

$$L(l_1, \ldots, l_k; x_1, \ldots, x_n) = f(x_1, \ldots, x_n) - \sum_{i=1}^{k} l_i(g_i(x_1, \ldots, x_n) - c_i).$$

The Hessian criterion comes from considering the Hessian of L at the critical point $(\boldsymbol{\lambda}; \mathbf{a})$. Before we give the criterion, we note the following fact from linear algebra: Since $\nabla g_1(\mathbf{a}), \ldots, \nabla g_k(\mathbf{a})$ are assumed to be linearly independent, the derivative matrix of $\mathbf{g} = (g_1, \ldots, g_k)$ at $\mathbf{a}$,

$$D\mathbf{g}(\mathbf{a}) = \begin{bmatrix} \frac{\partial g_1}{\partial x_1}(\mathbf{a}) & \cdots & \frac{\partial g_1}{\partial x_n}(\mathbf{a}) \\ \vdots & \ddots & \vdots \\ \frac{\partial g_k}{\partial x_1}(\mathbf{a}) & \cdots & \frac{\partial g_k}{\partial x_n}(\mathbf{a}) \end{bmatrix},$$

has a $k \times k$ submatrix (obtained by deleting $n-k$ columns of $D\mathbf{g}(\mathbf{a})$) with nonzero determinant. By relabeling the variables if necessary, we will assume that

$$\det \begin{bmatrix} \frac{\partial g_1}{\partial x_1}(\mathbf{a}) & \cdots & \frac{\partial g_1}{\partial x_k}(\mathbf{a}) \\ \vdots & \ddots & \vdots \\ \frac{\partial g_k}{\partial x_1}(\mathbf{a}) & \cdots & \frac{\partial g_k}{\partial x_k}(\mathbf{a}) \end{bmatrix} \neq 0$$

(i.e., that we may delete the *last* $n - k$ columns).

Second derivative test for constrained local extrema. Given a constrained critical point $\mathbf{a}$ of f subject to the conditions $g_1(\mathbf{x}) = c_1, g_2(\mathbf{x}) = c_2, \ldots, g_k(\mathbf{x}) = c_k$, consider the matrix

$$HL(\boldsymbol{\lambda}; \mathbf{a}) = \begin{bmatrix} 0 & \cdots & 0 & -\frac{\partial g_1}{\partial x_1}(\mathbf{a}) & \cdots & -\frac{\partial g_1}{\partial x_n}(\mathbf{a}) \\ \vdots & \ddots & \vdots & \vdots & \ddots & \vdots \\ 0 & \cdots & 0 & -\frac{\partial g_k}{\partial x_1}(\mathbf{a}) & \cdots & -\frac{\partial g_k}{\partial x_n}(\mathbf{a}) \\ -\frac{\partial g_1}{\partial x_1}(\mathbf{a}) & \cdots & -\frac{\partial g_k}{\partial x_1}(\mathbf{a}) & h_{11} & \cdots & h_{1n} \\ \vdots & \ddots & \vdots & \vdots & \ddots & \vdots \\ -\frac{\partial g_1}{\partial x_n}(\mathbf{a}) & \cdots & -\frac{\partial g_k}{\partial x_n}(\mathbf{a}) & h_{n1} & \cdots & h_{nn} \end{bmatrix},$$

where

$$h_{ij} = \frac{\partial^2 f}{\partial x_j \partial x_i}(\mathbf{a}) - \lambda_1 \frac{\partial^2 g_1}{\partial x_j \partial x_i}(\mathbf{a}) - \lambda_2 \frac{\partial^2 g_2}{\partial x_j \partial x_i}(\mathbf{a}) - \cdots - \lambda_k \frac{\partial^2 g_k}{\partial x_j \partial x_i}(\mathbf{a}).$$

(Note that $HL(\boldsymbol{\lambda}; \mathbf{a})$ is an $(n+k) \times (n+k)$ matrix.) By relabeling the variables as necessary, assume that

$$\det \begin{bmatrix} \frac{\partial g_1}{\partial x_1}(\mathbf{a}) & \cdots & \frac{\partial g_1}{\partial x_k}(\mathbf{a}) \\ \vdots & \ddots & \vdots \\ \frac{\partial g_k}{\partial x_1}(\mathbf{a}) & \cdots & \frac{\partial g_k}{\partial x_k}(\mathbf{a}) \end{bmatrix} \neq 0.$$

As in the unconstrained case, let H_j be the upper leftmost $j \times j$ submatrix of $HL(\boldsymbol{\lambda}, \mathbf{a})$. For $j = 1, 2, \ldots, k+n$, let $d_j = \det H_j$, and calculate the following sequence of $n - k$ numbers:

$$(-1)^k d_{2k+1}, \quad (-1)^k d_{2k+2}, \ldots, \quad (-1)^k d_{k+n}. \tag{1}$$

Note that, if $k \geq 1$, the sequence in (1) is *not* the complete sequence of principal minors of $HL(\boldsymbol{\lambda}, \mathbf{a})$. Assume $d_{k+n} = \det HL(\boldsymbol{\lambda}, \mathbf{a}) \neq 0$. The numerical test is as follows:

1. If the sequence in (1) consists entirely of positive numbers, then f has a local minimum at $\mathbf{a}$ subject to the constraints

$$g_1(\mathbf{x}) = c_1, \quad g_2(\mathbf{x}) = c_2, \ldots, \quad g_k(\mathbf{x}) = c_k.$$

2. If the sequence in (1) begins with a negative number and thereafter alternates in sign, then f has a local maximum at $\mathbf{a}$ subject to the constraints

$$g_1(\mathbf{x}) = c_1, \quad g_2(\mathbf{x}) = c_2, \ldots, \quad g_k(\mathbf{x}) = c_k.$$

3. If neither case 1 nor case 2 holds, then f has a constrained saddle point at $\mathbf{a}$.

In the event that $\det HL(\boldsymbol{\lambda}, \mathbf{a}) = 0$, the constrained critical point $\mathbf{a}$ is **degenerate**, and we must use another method to determine whether or not it is the site of an extremum.

Finally, in the case of no constraint equations $g_i(\mathbf{x}) = c_i$ (i.e., $k = 0$), the preceding criterion becomes the usual Hessian test for a function f of n variables.

EXAMPLE 5 In Example 1, we found the minimum of the area function

$$A(x, y, z) = 2xy + 2yz + xz$$

of an open rectangular box subject to the condition

$$V(x, y, z) = xyz = 4.$$

Using Lagrange multipliers, we found that the only constrained critical point was (2, 1, 2). The value of the multiplier λ corresponding to this point is 2. To use the Hessian criterion to check that (2, 1, 2) really does yield a local minimum, we construct the Lagrangian function

$$\begin{aligned} L(l; x, y, z) &= A(x, y, z) - l(V(x, y, z) - 4) \\ &= 2xy + 2yz + xz - l(xyz - 4). \end{aligned}$$

Then

$$HL(l; x, y, z) = \begin{bmatrix} 0 & -yz & -xz & -xy \\ -yz & 0 & 2 - lz & 1 - ly \\ -xz & 2 - lx & 0 & 2 - lx \\ -xy & 1 - ly & 2 - lx & 0 \end{bmatrix}.$$

At the constrained critical point (2; 2, 1, 2), we have

$$HL(2; 2, 1, 2) = \begin{bmatrix} 0 & -2 & -4 & -2 \\ -2 & 0 & -2 & -1 \\ -4 & -2 & 0 & -2 \\ -2 & -1 & -2 & 0 \end{bmatrix}.$$

The sequence of determinants to consider is

$$(-1)^1 \det H_{2(1)+1} = -\det \begin{bmatrix} 0 & -2 & -4 \\ -2 & 0 & -2 \\ -4 & -2 & 0 \end{bmatrix} = 32,$$

$$(-1)^1 \det H_4 = -\det \begin{bmatrix} 0 & -2 & -4 & -2 \\ -2 & 0 & -2 & -1 \\ -4 & -2 & 0 & -2 \\ -2 & -1 & -2 & 0 \end{bmatrix} = 48.$$

Since these numbers are both positive, we see that (2, 1, 2) indeed minimizes the area of the box subject to the constant volume constraint. ◆

EXAMPLE 6 In Example 4, we found points on the conic section C defined by equations

$$\begin{cases} g_1(x, y, z) = x^2 + y^2 - z^2 = 0 \\ g_2(x, y, z) = x + y - z = -2 \end{cases}$$

that are (constrained) critical points of the "distance" function

$$f(x, y, z) = x^2 + y^2 + z^2.$$

To apply the Hessian criterion in this case, we construct the Lagrangian function

$$L(l, m; x, y, z) = x^2 + y^2 + z^2 - l(x^2 + y^2 - z^2) - m(x + y - z + 2).$$

The critical points of L, found by setting $DL(l, m; x, y, z)$ equal to $\mathbf{0}$, are

$$(\boldsymbol{\lambda}_1; \mathbf{a}_1) = (-3 + 2\sqrt{2}, -24 + 16\sqrt{2}; -2 + \sqrt{2}, -2 + \sqrt{2}, -2 + 2\sqrt{2})$$

and

$$(\boldsymbol{\lambda}_2; \mathbf{a}_2) = (-3 - 2\sqrt{2}, -24 - 16\sqrt{2}; -2 - \sqrt{2}, -2 - \sqrt{2}, -2 - 2\sqrt{2}).$$

The Hessian of L is

$$HL(l, m; x, y, z) = \begin{bmatrix} 0 & 0 & -2x & -2y & 2z \\ 0 & 0 & -1 & -1 & 1 \\ -2x & -1 & 2 - 2l & 0 & 0 \\ -2y & -1 & 0 & 2 - 2l & 0 \\ 2z & 1 & 0 & 0 & 2 + 2l \end{bmatrix}.$$

After we evaluate this matrix at each of the critical points, we need to compute

$$(-1)^2 \det H_{2(2)+1} = \det H_5.$$

We leave it to you to check that for $(\boldsymbol{\lambda}_1; \mathbf{a}_1)$ this determinant is $128 - 64\sqrt{2} \approx 37.49$, and for $(\boldsymbol{\lambda}_2; \mathbf{a}_2)$ it is $128 + 64\sqrt{2} \approx 218.51$. Since both numbers are positive, the points $(-2 \pm \sqrt{2}, -2 \pm \sqrt{2}, -2 \pm 2\sqrt{2})$ are both sites of local minima. By comparing the values of f at these two points, we see that $(-2 + \sqrt{2}, -2 + \sqrt{2}, -2 + 2\sqrt{2})$ must be the global minimum. ◆

Exercises

1. In this problem, find the point on the plane $2x - 3y - z = 4$ that is closest to the origin in two ways:

(a) by using the methods in §4.2 (that is, by finding the minimum value of an appropriate function of two variables);

(b) by using a Lagrange multiplier.

In Exercises 2–8, use Lagrange multipliers to identify the critical points of f subject to the given constraints.

2. $f(x, y) = y, \quad 2x^2 + y^2 = 4$

3. $f(x, y) = 5x + 2y, \quad 5x^2 + 2y^2 = 14$

4. $f(x, y) = xy, \quad 2x - 3y = 6$

5. $f(x, y, z) = xyz, \quad 2x + 3y + z = 6$

6. $f(x, y, z) = x^2 + y^2 + z^2, \quad x + y - z = 1$

7. $f(x, y, z) = 2x + y^2 - z^2, \quad x - 2y = 0, \ x + z = 0$

8. $f(x, y, z) = x + y + z, \quad y^2 - x^2 = 1, \quad x + 2z = 1$

9. (a) Find the critical points of $f(x, y) = x^2 + y$ subject to $x^2 + 2y^2 = 1$.

(b) Use the Hessian criterion to determine the nature of the critical point.

10. (a) Find any critical points of $f(x, y, z, w) = x^2 + y^2 + z^2 + w^2$ subject to $2x + y + z = 1$, $x - 2z - w = -2$, $3x + y + 2w = -1$.

(b) Use the Hessian criterion to determine the nature of the critical point. (Note: You may wish to use a computer algebra system for the calculations.)

Just as sometimes is the case when finding ordinary (i.e., unconstrained) critical points of functions, it can be difficult to solve a Lagrange multiplier problem because the system of equations that results may be prohibitively difficult to solve by hand. In Exercises 11–15, use a computer algebra system to find the critical points of the given function f subject to the constraints indicated. (Note: You may find it helpful to provide numerical approximations in some cases.)

11. $f(x, y, z) = 3xy - 4z$, $3x + y - 2xz = 1$

12. $f(x, y, z) = 3xy - 4yz + 5xz$, $3x + y + 2z = 12$, $2x - 3y + 5z = 0$

13. $f(x, y, z) = y^3 + 2xyz - x^2$, $x^2 + y^2 + z^2 = 1$

14. $f(x, y, z) = x^2 + y^2 - xz^2$, $xy + z^2 = 1$

15. $f(x, y, z, w) = x^2 + y^2 + z^2 + w^2$, $x^2 + y^2 = 1$, $x + y + z + w = 1$, $x - y + z - w = 0$

16. Consider the problem of determining the extreme values of the function $f(x, y) = x^3 + 3y^2$ subject to the constraint that $xy = -4$.

(a) Use a Lagrange multiplier to find the critical points of f that satisfy the constraint.

(b) Give an analytic argument to determine if the critical points you found in part (a) yield (constrained) maxima or minima of f.

(c) Use a computer to plot, on a single set of axes, several level curves of f together with the constraint curve $xy = -4$. Use your plot to give a geometric justification for your answers in parts (a) and (b).

17. Find three positive numbers whose sum is 18 and whose product is as large as possible.

18. Find the maximum and minimum values of $f(x, y, z) = x + y - z$ on the sphere $x^2 + y^2 + z^2 = 81$. Explain how you know that there must be both a maximum and a minimum attained.

19. Find the maximum and minimum values of $f(x, y) = x^2 + xy + y^2$ on the closed disk $D = \{(x, y) \mid x^2 + y^2 \leq 4\}$.

20. You are sending a birthday present to your calculus instructor. Fly-By-Night Delivery Service insists that any package they ship be such that the sum of the length plus the girth be at most 108 in. (The girth is the perimeter of the cross-section perpendicular to the length axis—see Figure 4.30.) What are the dimensions of the largest present you can send?

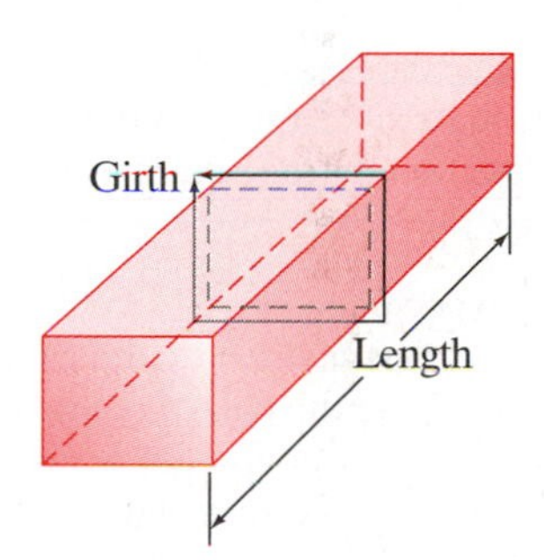

Figure 4.30 Diagram for Exercise 20.

21. A cylindrical metal can is to be manufactured from a fixed amount of sheet metal. Use the method of Lagrange multipliers to determine the ratio between the dimensions of the can with the largest capacity.

22. An industrious farmer is designing a silo to hold her 900π ft^3 supply of grain. The silo is to be cylindrical in shape with a hemispherical roof. (See Figure 4.31.) Suppose that it costs five times as much (per square foot of sheet metal used) to fashion the roof of the silo as it does to make the circular floor and twice as much to make the cylindrical walls as the floor. If you were to act as consultant for this project, what dimensions would you recommend so that the *total* cost would be a minimum? On what do you base your recommendation? (Assume that the entire silo can be filled with grain.)

Figure 4.31 The grain silo of Exercise 22.

23. You are in charge of erecting a space probe on the newly discovered planet Nilrebo. To minimize interference to the probe's sensors, you must place the probe where the magnetic field of the planet is weakest. Nilrebo is perfectly spherical with a radius of 3 (where the units are thousands of miles). Based on a coordinate system whose origin is at the center of Nilrebo, the strength of the magnetic field in space is given by the function $M(x, y, z) = xz - y^2 + 3x + 3$. Where should you locate the probe?

24. Heron's formula for the area of a triangle whose sides have lengths x, y, and z is

$$\text{Area} = \sqrt{s(s-x)(s-y)(s-z)},$$

where $s = \frac{1}{2}(x+y+z)$ is the so-called semi-perimeter of the triangle. Use Heron's formula to show that, for a fixed perimeter P, the triangle with the largest area is equilateral.

25. Use a Lagrange multiplier to find the largest sphere centered at the origin that can be inscribed in the ellipsoid $3x^2 + 2y^2 + z^2 = 6$. (Be careful with this problem; drawing a picture may help.)

26. Find the point closest to the origin and on the line of intersection of the planes $2x + y + 3z = 9$ and $3x + 2y + z = 6$.

27. Find the point closest to the point $(2, 5, -1)$ and on the line of intersection of the planes $x - 2y + 3z = 8$ and $2z - y = 3$.

28. The plane $x + y + z = 4$ intersects the paraboloid $z = x^2 + y^2$ in an ellipse. Find the points on the ellipse nearest to and farthest from the origin.

29. Find the highest and lowest points on the ellipse obtained by intersecting the paraboloid $z = x^2 + y^2$ with the plane $x + y + 2z = 2$.

30. Find the minimum distance between a point on the ellipse $x^2+2y^2=1$ and a point on the line $x+y=4$. (Hint: Consider a point (x, y) on the ellipse and a point (u, v) on the line. Minimize the square of the distance between them as a function of four variables. This problem is difficult to solve without a computer.)

31. (a) Use the method of Lagrange multipliers to find critical points of the function $f(x, y) = x + y$ subject to the constraint $xy = 6$.

(b) Explain geometrically why f has no extrema on the set $\{(x, y) \mid xy = 6\}$.

32. The cylinder $x^2+y^2=4$ and the plane $2x+2y+z=2$ intersect in an ellipse. Find the points on the ellipse that are nearest to and farthest from the origin.

33. Consider the problem of finding extrema of $f(x, y) = x$ subject to the constraint $y^2 - 4x^3 + 4x^4 = 0$.

(a) Use a Lagrange multiplier and solve the system of equations

$$\begin{cases} \nabla f(x, y) = \lambda \nabla g(x, y) \\ g(x, y) = 0 \end{cases},$$

where $g(x, y) = y^2 - 4x^3 + 4x^4$. By doing so, you will identify critical points of f subject to the given constraint.

(b) Graph the curve $y^2 - 4x^3 + 4x^4 = 0$ and use the graph to determine where the extrema of $f(x, y) = x$ occur.

(c) Compare your result in part (a) with what you found in part (b). What accounts for any differences that you observed?

34. Consider the problem of finding extrema of $f(x, y, z) = x^2 + y^2$ subject to the constraint $z = c$, where c is any constant.

(a) Use the method of Lagrange multipliers to identify the critical points of f subject to the constraint given above.

(b) Using the usual alphabetical ordering of variables (i.e., $x_1 = x, x_2 = y, x_3 = z$), construct the Hessian matrix $HL(\lambda; a_1, a_2, a_3)$ (where $L(l; x, y, z) = f(x, y, z) - l(z - c)$) for each critical point you found in part (a). Try to use the second derivative test for constrained extrema to determine the nature of the critical points you found in part (a). What happens?

(c) Repeat part (b), this time using the variable ordering $x_1 = z, x_2 = y, x_3 = x$. What does the second derivative test tell you now?

(d) Without making any detailed calculations, discuss why f must attain its minimum value at the point $(0, 0, c)$. Then try to reconcile your results in parts (b) and (c). This exercise demonstrates that the assumption that

$$\det \begin{bmatrix} \dfrac{\partial g_1}{\partial x_1}(\mathbf{a}) & \cdots & \dfrac{\partial g_1}{\partial x_k}(\mathbf{a}) \\ \vdots & \ddots & \vdots \\ \dfrac{\partial g_k}{\partial x_1}(\mathbf{a}) & \cdots & \dfrac{\partial g_k}{\partial x_k}(\mathbf{a}) \end{bmatrix} \neq 0$$

is important.

35. Consider the problem of finding critical points of the function $f(x_1, \dots, x_n)$ subject to the set of k constraints

$$g_1(x_1, \dots, x_n) = c_1, \quad g_2(x_1, \dots, x_n) = c_2, \dots,$$
$$g_k(x_1, \dots, x_n) = c_k.$$

Assume that $f, g_1, g_2, \dots, g_k$ are all of class C^2.

(a) Show that we can relate the method of Lagrange multipliers for determining constrained critical points to the techniques in §4.2 for finding unconstrained critical points as follows: If

$$(\boldsymbol{\lambda}, \mathbf{a}) = (\lambda_1, \dots, \lambda_k; a_1, \dots, a_n)$$

is a pair consisting of k values for Lagrange multipliers $\lambda_1, \dots, \lambda_k$ and n values $a_1, \dots, a_n$ for the variables $x_1, \dots, x_n$ such that $\mathbf{a}$ is a constrained critical point, then $(\boldsymbol{\lambda}, \mathbf{a})$ is an ordinary (i.e., unconstrained) critical point of the function

$$L(l_1, \dots, l_k; x_1, \dots, x_n) = f(x_1, \dots, x_n) - \sum_{i=1}^{k} l_i (g_i(x_1, \dots, x_n) - c_i).$$

(b) Calculate the Hessian $HL(\boldsymbol{\lambda}, \mathbf{a})$, and verify that it is the matrix used in §4.3 to provide the criterion for determining the nature of constrained critical points.

36. The unit hypersphere in $\mathbf{R}^n$ (centered at the origin $\mathbf{0} = (0, \dots, 0)$) is defined by the equation $x_1^2 + x_2^2 + \cdots + x_n^2 = 1$. Find the pair of points $\mathbf{x} = (x_1, \dots, x_n)$ and $\mathbf{y} = (y_1, \dots, y_n)$, each of which lies on the unit hypersphere, that maximize and minimize the function

$$f(x_1, \dots, x_n, y_1, \dots, y_n) = \sum_{i=1}^{n} x_i y_i.$$

What are the maximum and minimum values of f?

37. Let $\mathbf{x} = (x_1, \dots, x_n)$ and $\mathbf{y} = (y_1, \dots, y_n)$ be any vectors in $\mathbf{R}^n$ and, for $i = 1, \dots, n$, set

$$u_i = \frac{x_i}{\sqrt{\sum_{i=1}^{n} x_i^2}} \quad \text{and} \quad v_i = \frac{y_i}{\sqrt{\sum_{i=1}^{n} y_i^2}}.$$

(a) Show that $\mathbf{u} = (u_1, \dots, u_n)$ and $\mathbf{v} = (v_1, \dots, v_n)$ lie on the unit hypersphere in $\mathbf{R}^n$.

(b) Use the result of Exercise 36 to establish the Cauchy–Schwarz inequality

$$|\mathbf{x} \cdot \mathbf{y}| \leq \|\mathbf{x}\| \, \|\mathbf{y}\|.$$

4.4 Some Applications of Extrema

In this section, we present several applications of the methods for finding both constrained and unconstrained extrema discussed previously.

Least Squares Approximation

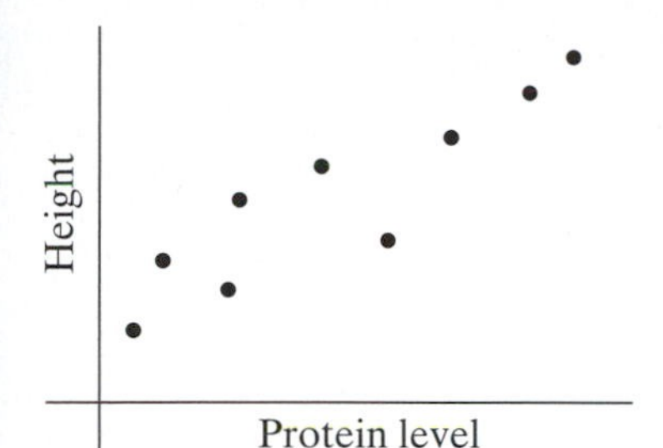

Figure 4.32 Height versus protein level.

The simplest relation between two quantities x and y is, without doubt, a linear one: $y = mx + b$ (where m and b are constants). When a biologist, chemist, psychologist, or economist postulates the most direct connection between two types of observed data, that connection is assumed to be linear. Suppose that Bob Biologist and Carol Chemist have measured certain blood protein levels in an adult population and have graphed these levels versus the heights of the subjects as in Figure 4.32. If Prof. Biologist and Dr. Chemist assume a linear relationship between the protein and height, then they desire to pass a line through the data as closely as possible, as suggested by Figure 4.33.

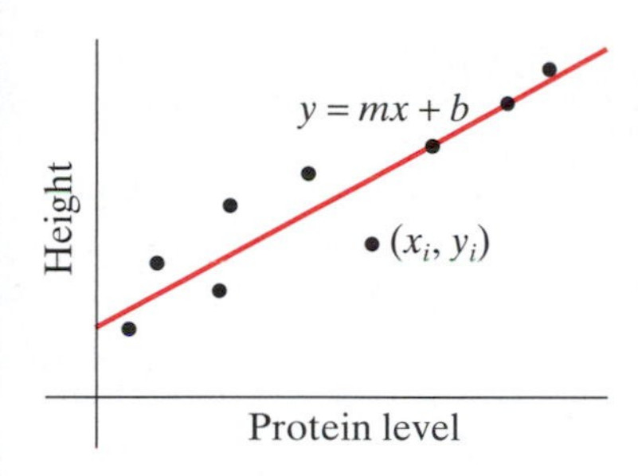

Figure 4.33 Fitting a line to the data.

To make this standard empirical method of **linear regression** precise (instead of merely graphical and intuitive), we first need some notation. Suppose we have collected n pairs of data $(x_1, y_1), (x_2, y_2), \ldots, (x_n, y_n)$. (In the example just described, x_i is the protein level of the ith subject and y_i his or her height.) We assume that there is some underlying relationship of the form $y = mx + b$, and we want to find the constants m and b so that the line fits the data as accurately as possible. Normally, we use the **method of least squares**. The idea is to find the values of m and b that minimize the sum of the squares of the differences between the observed y-values and those predicted by the linear formula. That is, we minimize the quantity

$$\begin{aligned} D(m, b) = [y_1 - (mx_1 + b)]^2 + [y_2 - (mx_2 + b)]^2 \\ + \cdots + [y_n - (mx_n + b)]^2, \end{aligned} \tag{1}$$

where, for $i = 1, \ldots, n$, y_i represents the observed y-value of the data, and $mx_i + b$ represents the y-value predicted by the linear relationship. Hence, each expression in D of the form $y_i - (mx_i + b)$ represents the error between the observed and predicted y-values. (See Figure 4.34.) They are squared in the expression for D in order to avoid the possibility of having large negative and positive terms cancel one another, thereby leaving little or no "net error," which would be misleading. Moreover, $D(m, b)$ is the square of the distance in $\mathbf{R}^n$ between the point $(y_1, y_2, \ldots, y_n)$ and the point $(mx_1 + b, mx_2 + b, \ldots, mx_n + b)$.

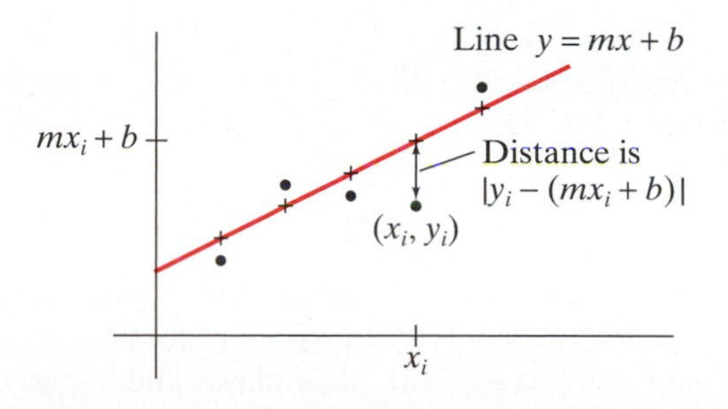

Figure 4.34 The method of least squares.

Thus, we have an ordinary minimization problem at hand. To solve it, we need to find the critical points of D. First, we can rewrite D as

$$\begin{aligned} D(m,b) &= \sum_{i=1}^{n}[y_i - (mx_i + b)]^2 \\ &= \sum_{i=1}^{n} y_i^2 - 2m\sum_{i=1}^{n} x_i y_i - 2b\sum_{i=1}^{n} y_i + \sum_{i=1}^{n}(mx_i + b)^2. \end{aligned}$$

Then

$$\begin{aligned} \frac{\partial D}{\partial m} &= -2\sum_{i=1}^{n} x_i y_i + \sum_{i=1}^{n} 2(mx_i + b)x_i \\ &= -2\sum_{i=1}^{n} x_i y_i + 2m\sum_{i=1}^{n} x_i^2 + 2b\sum_{i=1}^{n} x_i \end{aligned}$$

and

$$\begin{aligned} \frac{\partial D}{\partial b} &= -2\sum_{i=1}^{n} y_i + \sum_{i=1}^{n} 2(mx_i + b) \\ &= -2\sum_{i=1}^{n} y_i + 2m\sum_{i=1}^{n} x_i + 2nb. \end{aligned}$$

When we set both partial derivatives equal to zero, we obtain the following pair of equations, which have been simplified slightly:

$$\begin{cases} \left(\sum x_i^2\right) m + \left(\sum x_i\right) b = \sum x_i y_i \\ \left(\sum x_i\right) m + nb = \sum y_i \end{cases}. \tag{2}$$

(All sums are taken from $i = 1$ to n.) Although (2) may look complicated, it is nothing more than a linear system of two equations in the two unknowns m and b. It is not difficult to see that system (2) has a single solution. Therefore, we have shown the following:

Proposition 4.1 Given n data points $(x_1, y_1), (x_2, y_2), \ldots, (x_n, y_n)$ with not all of $x_1, x_2, \ldots, x_n$ equal, the function

$$D(m,b) = \sum_{i=1}^{n}[y_i - (mx_i + b)]^2$$

has a single critical point (m_0, b_0) given by

$$m_0 = \frac{n\sum x_i y_i - \left(\sum x_i\right)\left(\sum y_i\right)}{n\sum x_i^2 - \left(\sum x_i\right)^2},$$

and

$$b_0 = \frac{\left(\sum x_i^2\right)\left(\sum y_i\right) - \left(\sum x_i\right)\left(\sum x_i y_i\right)}{n\sum x_i^2 - \left(\sum x_i\right)^2}.$$

Since $D(m, b)$ is a quadratic polynomial in m and b, the graph of $z = D(m, b)$ is a quadric surface. (See §2.1.) The only such surfaces that are graphs of functions are paraboloids and hyperbolic paraboloids. We show that, in the present case, the graph is that of a paraboloid, by demonstrating that D has a local minimum at the critical point (m_0, b_0) given in Proposition 4.1.

We can use the Hessian criterion to check that D has a local minimum at (m_0, b_0). We have

$$HD(m,b) = \begin{bmatrix} 2\sum x_i^2 & 2\sum x_i \\ 2\sum x_i & 2n \end{bmatrix}.$$

The principal minors are $2\sum x_i^2$ and $4n\sum x_i^2 - 4\left(\sum x_i\right)^2$. The first minor is obviously positive, but determining the sign of the second requires a bit more algebra. (If you wish, you can omit reading the details of this next calulation and rest assured that the story has a happy ending.) Ignoring the factor of 4, we examine the expression $n\sum x_i^2 - \left(\sum x_i\right)^2$. Expanding the second term yields

$$\begin{aligned} n\sum_{i=1}^n x_i^2 - \left(\sum_{i=1}^n x_i\right)^2 &= n\sum_{i=1}^n x_i^2 - \left(\sum_{i=1}^n x_i^2 + \sum_{i<j} 2x_i x_j\right) \\ &= (n-1)\sum_{i=1}^n x_i^2 - \sum_{i<j} 2x_i x_j. \end{aligned} \qquad (3)$$

On the other hand, we have

$$\sum_{i<j}(x_i - x_j)^2 = \sum_{i<j}\left(x_i^2 - 2x_i x_j + x_j^2\right) = (n-1)\sum_{i=1}^n x_i^2 - \sum_{i<j} 2x_i x_j. \qquad (4)$$

(To see that equation (4) holds, you need to convince yourself that

$$\sum_{i<j}\left(x_i^2 + x_j^2\right) = (n-1)\sum_{i=1}^n x_i^2$$

by counting the number of times a particular term of the form x_k^2 appears in the left-hand sum.) Thus, we have

$$\begin{aligned} \det HD(m,b) &= 4\left(n\sum_{i=1}^n x_i^2 - \left(\sum_{i=1}^n x_i\right)^2\right) \\ &= 4\left((n-1)\sum_{i=1}^n x_i^2 - \sum_{i<j} 2x_i x_j\right) \quad \text{by equation (3),} \\ &= 4\sum_{i<j}(x_i - x_j)^2 \quad \text{by equation (4).} \end{aligned}$$

Because this last expression is a sum of squares, it is nonnegative. Therefore, the Hessian criterion shows that D does indeed have a local minimum at the critical point. Hence, the graph of $z = D(m,b)$ is that of a paraboloid. Since the (unique) local minimum of a paraboloid is in fact a global minimum (consider a typical graph), we see that D is indeed minimized at (m_0, b_0).

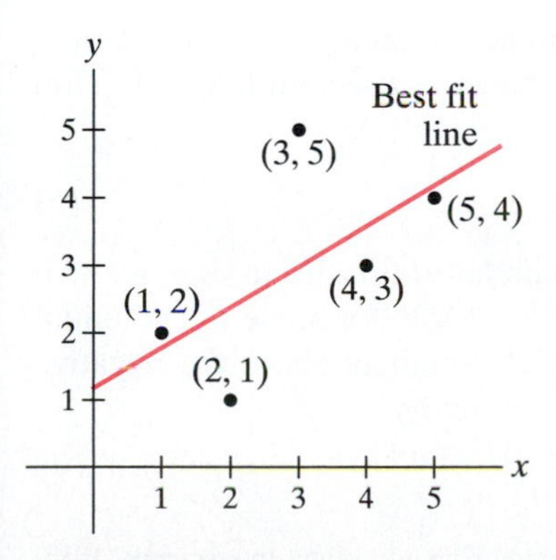

Figure 4.35 Data for the linear regression of Example 1.

EXAMPLE 1 To see how the preceding discussion applies to a specific set of data, consider the situation depicted in Figure 4.35.

We have $n = 5$, and the function D to be minimized is

$$\begin{aligned} D(m,b) = {} & [2 - (m+b)]^2 + [1 - (2m+b)]^2 + [5 - (3m+b)]^2 \\ & + [3 - (4m+b)]^2 + [4 - (5m+b)]^2. \end{aligned}$$

We compute

$$\sum x_i = 15, \qquad \sum x_i^2 = 55, \qquad \sum y_i = 15, \qquad \sum x_i y_i = 51.$$

Thus, using Proposition 4.1,

$$m = \frac{5 \cdot 51 - 15 \cdot 15}{5 \cdot 55 - 15 \cdot 15} = \frac{3}{5}, \qquad b = \frac{55 \cdot 15 - 15 \cdot 51}{5 \cdot 55 - 15 \cdot 15} = \frac{6}{5}$$

The best fit line in terms of least squares approximation is

$$y = \frac{3}{5}x + \frac{6}{5}.$$

◆

Of course, linear regression is not always an appropriate technique. It may not be reasonable to assume that the data points fall nearly on a straight line. Some formula other than $y = mx + b$ may have to be assumed to describe the data with any accuracy. Such a postulated relation might be quadratic,

$$y = ax^2 + bx + c,$$

or x and y might be inversely related,

$$y = \frac{a}{x} + b.$$

You can still apply the method of least squares to construct a function analogous to D in equation (1) to find the relation of a given form that best fits the data.

Another way that least squares arise is if y depends not on one variable, but on several: $x_1, x_2, \ldots, x_n$. For example, perhaps adult height is measured against blood levels of 10 different proteins, instead of just one. **Multiple regression** is the statistical method of finding the linear function

$$y = a_1x_1 + a_2x_2 + \cdots + a_nx_n + b$$

that best fits a data set of $(n+1)$-tuples

$$\left\{(x_1^{(1)}, x_2^{(1)}, \ldots, x_n^{(1)}, y_1), (x_1^{(2)}, x_2^{(2)}, \ldots, x_n^{(2)}, y_2), \ldots, (x_1^{(k)}, x_2^{(k)}, \ldots, x_n^{(k)}, y_k)\right\}.$$

We can find such a "best fit hyperplane" by minimizing the sum of the squares of the differences between the y-values furnished by the data set and those predicted by the linear formula. We leave the details to you.[3]

Physical Equilibria

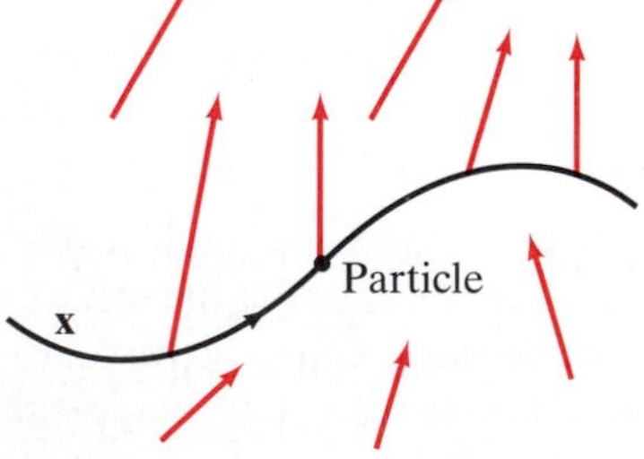

Let $\mathbf{F}\colon X \subseteq \mathbf{R}^3 \to \mathbf{R}^3$ be a continuous force field acting on a particle that moves along a path $\mathbf{x}\colon I \subseteq \mathbf{R} \to \mathbf{R}^3$ as in Figure 4.36. Newton's second law of motion states that

$$\mathbf{F}(\mathbf{x}(t)) = m\mathbf{x}''(t), \tag{5}$$

where m is the mass of the particle. For the remainder of this discussion, we will assume that $\mathbf{F}$ is a gradient field, that is, that $\mathbf{F} = -\nabla V$ for some C^1 potential function $V\colon X \subseteq \mathbf{R}^3 \to \mathbf{R}$. (See §3.3 for a brief comment about the negative sign.) We first establish the law of conservation of energy.

Figure 4.36 A particle traveling in a force field $\mathbf{F}$.

[3]Or you might consult S. Weisberg, *Applied Linear Regression,* 2nd ed., Wiley-Interscience, 1985, Chapter 2. Be forewarned, however, that to treat multiple regression with any elegance requires somewhat more linear algebra than we have presented.

Theorem 4.2 (Conservation of Energy) Given the set-up above, the quantity

$$\tfrac{1}{2}m\|\mathbf{x}'(t)\|^2 + V(\mathbf{x}(t))$$

is constant.

The term $\frac{1}{2}m\|\mathbf{x}'(t)\|^2$ is usually referred to as the **kinetic energy** of the particle and the term $V(\mathbf{x}(t))$ as the **potential energy**. The significance of Theorem 4.2 is that it states that the sum of the kinetic and potential energies of a particle is always fixed (conserved) when the particle travels along a path in a gradient vector field. For this reason, gradient vector fields are also called **conservative** vector fields.

Proof of Theorem 4.2 As usual, we show that the total energy is constant by showing that its derivative is zero. Thus, using the product rule and the chain rule, we calculate

$$\begin{aligned}\frac{d}{dt}\left[\tfrac{1}{2}m\mathbf{x}'(t)\cdot\mathbf{x}'(t) + V(\mathbf{x}(t))\right] &= m\mathbf{x}''(t)\cdot\mathbf{x}'(t) + \nabla V(\mathbf{x}(t))\cdot\mathbf{x}'(t)\\ &= m\mathbf{x}''(t)\cdot\mathbf{x}'(t) - \mathbf{F}(\mathbf{x}(t))\cdot\mathbf{x}'(t)\\ &= m\mathbf{x}''(t)\cdot\mathbf{x}'(t) - m\mathbf{x}''(t)\cdot\mathbf{x}'(t)\\ &= 0,\end{aligned}$$

from the definitions of $\mathbf{F}$ and V and by formula (5). ■

In physical applications it is important to identify those points in space that are "rest positions" for particles moving under the influence of a force field. These positions, known as **equilibrium points**, are such that the force field does not act on the particle so as to move it from that position. Equilibrium points are of two kinds: **stable** equilibria, namely, equilibrium points such that a particle perturbed slightly from these positions tends to remain nearby (for example, a pendulum hanging down at rest) and **unstable** equilibria, such as the act of balancing a ball on your nose. The precise definition is somewhat technical.

Definition 4.3 Let $\mathbf{F}: X \subseteq \mathbf{R}^n \to \mathbf{R}^n$ be any force field. Then $\mathbf{x}_0 \in X$ is called an **equilibrium point** of $\mathbf{F}$ if $\mathbf{F}(\mathbf{x}_0) = \mathbf{0}$. An equilibrium point $\mathbf{x}_0$ is said to be **stable** if, for every $r, \epsilon > 0$, we can find other numbers $r_0, \epsilon_0 > 0$ such that if we place a particle at position $\mathbf{x}$ with $\|\mathbf{x} - \mathbf{x}_0\| < r_0$ and provide it with a kinetic energy less than ϵ_0, then the particle will always remain within distance r of $\mathbf{x}_0$ with kinetic energy less than ϵ.

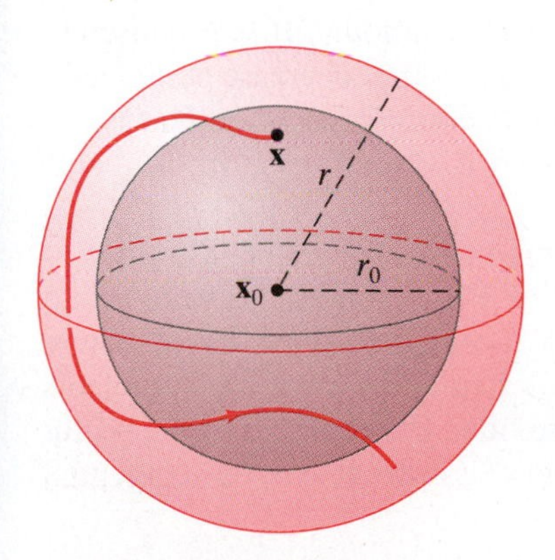

Figure 4.37 For a stable equilibrium point, the path of a nearby particle with a sufficiently small kinetic energy will remain nearby with a bounded kinetic energy.

In other words, a stable equilibrium point $\mathbf{x}_0$ has the following property: You can keep a particle inside a specific ball centered at $\mathbf{x}_0$ with a small kinetic energy by starting the particle inside some other (possibly smaller) ball about $\mathbf{x}_0$ and imparting to it some (possibly smaller) initial kinetic energy. (See Figure 4.37.)

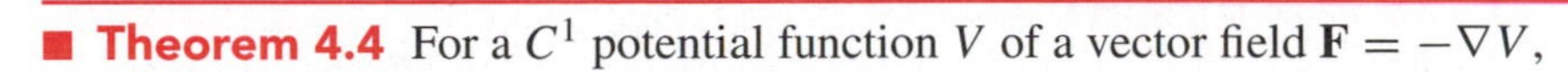

Theorem 4.4 For a C^1 potential function V of a vector field $\mathbf{F} = -\nabla V$,

1. The critical points of the potential function are precisely the equilibrium points of $\mathbf{F}$.
2. If $\mathbf{x}_0$ gives a strict local minimum of V, then $\mathbf{x}_0$ is a stable equilibrium point of $\mathbf{F}$.

EXAMPLE 2 The vector field $\mathbf{F} = (-6x - 2y - 2)\mathbf{i} + (-2x - 4y + 2)\mathbf{j}$ is conservative and has

$$V(x, y) = 3x^2 + 2xy + 2x + 2y^2 - 2y + 4$$

as a potential function (meaning that $\mathbf{F} = -\nabla V$, according to our current sign convention). There is only one equilibrium point, namely, $\left(-\frac{3}{5}, \frac{4}{5}\right)$. To see if it is stable, we look at the Hessian of V :

$$HV\left(-\tfrac{3}{5}, \tfrac{4}{5}\right) = \begin{bmatrix} 6 & 2 \\ 2 & 4 \end{bmatrix}.$$

The sequence of principal minors is 6, 20. By the Hessian criterion, $\left(-\frac{3}{5}, \frac{4}{5}\right)$ is a strict local minimum of V and, by Theorem 4.4, it must be a stable equilibrium point of $\mathbf{F}$. ◆

Proof of Theorem 4.4 The proof of part 1 is straightforward. Since $\mathbf{F} = -\nabla V$, we see that $\mathbf{F}(\mathbf{x}) = \mathbf{0}$ if and only if $\nabla V(\mathbf{x}) = \mathbf{0}$. Thus, equilibrium points of $\mathbf{F}$ are the critical points of V.

To prove part 2, let $\mathbf{x}_0$ be a strict local minimum of V and $\mathbf{x}\colon I \to \mathbf{R}^n$ a C^1 path such that $\mathbf{x}(t_0) = \mathbf{x}_0$ for some $t_0 \in I$. By conservation of energy, we must have, for all $t \in I$, that

$$\tfrac{1}{2}m\|\mathbf{x}'(t)\|^2 + V(\mathbf{x}(t)) = \tfrac{1}{2}m\|\mathbf{x}'(t_0)\|^2 + V(\mathbf{x}(t_0)).$$

To show that $\mathbf{x}_0$ is a stable equilibrium point, we desire to show that we can bound the distance between $\mathbf{x}(t)$ and $\mathbf{x}_0 = \mathbf{x}(t_0)$ by any amount r and the kinetic energy by any amount ϵ. That is, we want to show we can achieve

$$\|\mathbf{x}(t) - \mathbf{x}_0\| < r$$

(i.e., $\mathbf{x}(t) \in B_r(\mathbf{x}_0)$ in the notation of §2.2) and

$$\tfrac{1}{2}m\|\mathbf{x}'(t)\|^2 < \epsilon.$$

As the particle moves along $\mathbf{x}$ away from $\mathbf{x}_0$, the potential energy must increase (since $\mathbf{x}_0$ is assumed to be a strict local minimum of potential energy), so the kinetic energy must decrease by the same amount. For the particle to escape from $B_r(\mathbf{x}_0)$, the potential energy must increase by a certain amount. If ϵ_0 is chosen to be smaller than that amount, then the kinetic energy cannot decrease sufficiently (so that the conservation equation holds) without becoming negative. This being clearly impossible, the particle cannot escape from $B_r(\mathbf{x}_0)$. ■

Often a particle is not only acted on by a force field, but also constrained to lie in a surface in space. The set-up is as follows: $\mathbf{F}$ is a continuous vector field on $\mathbf{R}^3$ acting on a particle that lies in the surface $S = \{\mathbf{x} \in \mathbf{R}^3 \mid g(\mathbf{x}) = c\}$, where g is a C^1 function such that $\nabla g(\mathbf{x}) \neq \mathbf{0}$ for all $\mathbf{x}$ in S. Most of the comments made in the unconstrained case still hold true, provided $\mathbf{F}$ is replaced by the vector component of $\mathbf{F}$ tangent to S. Since, at $\mathbf{x} \in S$, $\nabla g(\mathbf{x})$ is normal to S, this tangential component of $\mathbf{F}$ at $\mathbf{x}$ is

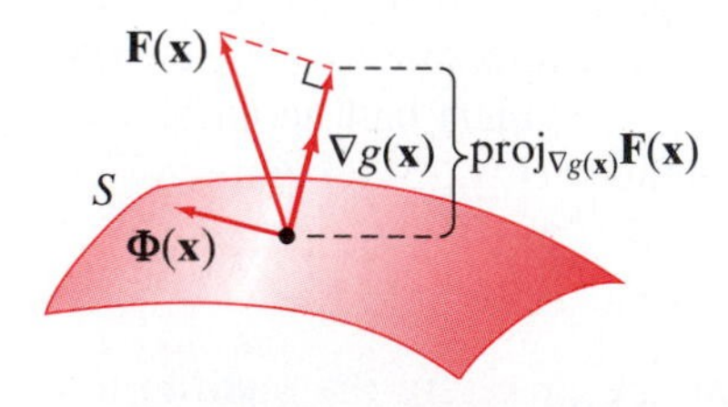

Figure 4.38 On the surface $S = \{\mathbf{x} \mid g(\mathbf{x}) = c\}$, the component of $\mathbf{F}$ that is tangent to S at $\mathbf{x}$ is denoted by $\mathbf{\Phi}(\mathbf{x})$.

$$\mathbf{\Phi}(\mathbf{x}) = \mathbf{F}(\mathbf{x}) - \mathrm{proj}_{\nabla g(\mathbf{x})}\mathbf{F}(\mathbf{x}). \tag{6}$$

(See Figure 4.38.) Then in place of formula (5), we have, for a path $\mathbf{x}\colon I \subseteq \mathbf{R} \to S$,

$$\mathbf{\Phi}(\mathbf{x}(t)) = m\mathbf{x}''(t). \tag{7}$$

We can now state a "constrained version" of Theorem 4.4.

Theorem 4.5 For a C^1 potential function V of a vector field $\mathbf{F} = -\nabla V$,

1. If $V|_S$ has an extremum at $\mathbf{x}_0 \in S$, then $\mathbf{x}_0$ is an equilibrium point in S.
2. If $V|_S$ has a strict local minimum at $\mathbf{x}_0 \in S$, then $\mathbf{x}_0$ is a stable equilibrium point.

Sketch of proof For part 1, if $V|_S$ has an extremum at $\mathbf{x}_0$, then, by Theorem 3.1, we have, for some scalar λ, that

$$\nabla V(\mathbf{x}_0) = \lambda \nabla g(\mathbf{x}_0).$$

Hence, because $\mathbf{F} = -\nabla V$,

$$\mathbf{F}(\mathbf{x}_0) = -\lambda \nabla g(\mathbf{x}_0),$$

implying that $\mathbf{F}$ is normal to S at $\mathbf{x}_0$. Thus, there can be no component of $\mathbf{F}$ tangent to S at $\mathbf{x}_0$ (i.e., $\mathbf{\Phi}(\mathbf{x}_0) = \mathbf{0}$). Since the particle is constrained to lie in S, we see that the particle is in equilibrium in S.

The proof of part 2 is essentially the same as the proof of part 2 of Theorem 4.4. The main modification is that the conservation of energy formula in Theorem 4.2 must be established anew, as its derivation rests on formula (5), which has been replaced by formula (7). Consequently, using the product and chain rules, we check, for $\mathbf{x}\colon I \to S$,

$$\begin{aligned}\frac{d}{dt}\left[\tfrac{1}{2}m\|\mathbf{x}'(t)\|^2 + V(\mathbf{x}(t))\right] &= \frac{d}{dt}\left[\tfrac{1}{2}m\mathbf{x}'(t)\cdot\mathbf{x}'(t) + V(\mathbf{x}(t))\right] \\ &= m\mathbf{x}''(t)\cdot\mathbf{x}'(t) + \nabla V(\mathbf{x}(t))\cdot\mathbf{x}'(t).\end{aligned}$$

Then, using formula (6), we have

$$\begin{aligned}\frac{d}{dt}\left[\tfrac{1}{2}m\|\mathbf{x}'(t)\|^2 + V(\mathbf{x}(t))\right] &= \mathbf{x}'(t)\cdot m\mathbf{x}''(t) - \mathbf{F}(\mathbf{x}(t))\cdot\mathbf{x}'(t) \\ &= \mathbf{x}'(t)\cdot\mathbf{\Phi}(\mathbf{x}(t)) - \mathbf{F}(\mathbf{x}(t))\cdot\mathbf{x}'(t) \\ &= \mathbf{x}'(t)\cdot\left[\mathbf{F}(\mathbf{x}(t)) - \operatorname{proj}_{\nabla g(\mathbf{x}(t))}\mathbf{F}(\mathbf{x}(t))\right] \\ &\quad - \mathbf{F}(\mathbf{x}(t))\cdot\mathbf{x}'(t) \\ &= -\mathbf{x}'(t)\cdot\operatorname{proj}_{\nabla g(\mathbf{x}(t))}\mathbf{F}(\mathbf{x}(t))\end{aligned}$$

after cancellation. Thus, we conclude that

$$\frac{d}{dt}\left[\tfrac{1}{2}m\|\mathbf{x}'(t)\|^2 + V(\mathbf{x}(t))\right] = 0,$$

since $\mathbf{x}'(t)$ is tangent to the path in S and hence tangent to S itself at $\mathbf{x}(t)$, while $\operatorname{proj}_{\nabla g(\mathbf{x}(t))}\mathbf{F}(\mathbf{x}(t))$ is parallel to $\nabla g(\mathbf{x}(t))$ and hence perpendicular to S at $\mathbf{x}(t)$. ■

EXAMPLE 3 Near the surface of the earth, the gravitational field is approximately

$$\mathbf{F} = -mg\mathbf{k}.$$

(We're assuming that, locally, the surface of the earth is represented by the plane $z = 0$.) Note that $\mathbf{F} = -\nabla V$, where

$$V(x, y, z) = mgz.$$

Now suppose a particle of mass m lies on a small sphere with equation

$$h(x, y, z) = x^2 + y^2 + (z - 2r)^2 = r^2.$$

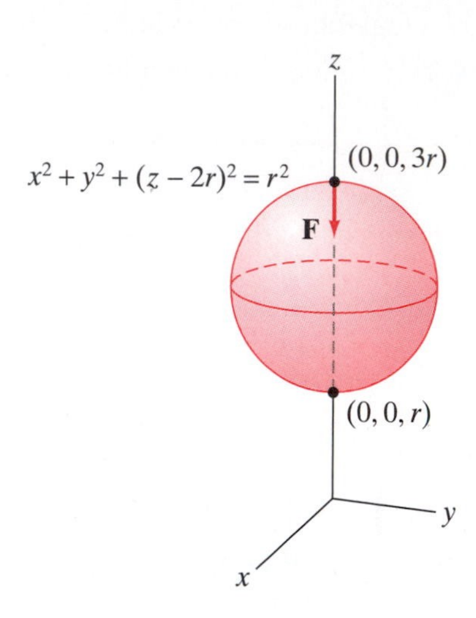

Figure 4.39 On the sphere $x^2 + y^2 + (z - 2r)^2 = r^2$, the points $(0, 0, r)$ and $(0, 0, 3r)$ are equilibrium points for the gravitational force field $\mathbf{F} = -mg\mathbf{k}$.

We can find constrained equilibria for this situation, using a Lagrange multiplier. The gradient equation $\nabla V = \lambda \nabla h$, along with the constraint, yields the system

$$\begin{cases} 0 = 2\lambda x \\ 0 = 2\lambda y \\ mg = 2\lambda(z - 2r) \\ x^2 + y^2 + (z - 2r)^2 = r^2 \end{cases}.$$

Because m and g are nonzero, λ cannot be zero. The first two equations imply $x = y = 0$. Therefore, the last equation becomes

$$(z - 2r)^2 = r^2,$$

which implies

$$z = r, 3r$$

are the solutions. Consequently, the positions of equilibrium are $(0, 0, r)$ and $(0, 0, 3r)$ (corresponding to $\lambda = -mg/2r$ and $+mg/2r$, respectively). From geometric considerations, we see V is strictly minimized at S at $(0, 0, r)$ and maximized at $(0, 0, 3r)$ as shown in Figure 4.39. From physical considerations, $(0, 0, r)$ is a stable equilibrium and $(0, 0, 3r)$ is an unstable one. (Try balancing a marble on top of a ball.) ◆

Applications to Economics

We present two illustrations of how Lagrange multipliers occur in problems involving economic models.

EXAMPLE 4 The usefulness of amounts $x_1, x_2, \ldots, x_n$ of (respectively) different capital goods $G_1, G_2, \ldots, G_n$ can sometimes be measured by a function $U(x_1, x_2, \ldots, x_n)$, called the **utility** of these goods. Perhaps the goods are individual electronic components needed in the manufacture of a stereo or computer, or perhaps U measures an individual consumer's utility for different commodities available at different prices. If item G_i costs a_i per unit and if M is the total amount of money allocated for the purchase of these n goods, then the consumer or the company needs to maximize $U(x_1, x_2, \ldots, x_n)$ subject to

$$a_1x_1 + a_2x_2 + \cdots + a_nx_n = M.$$

This is a standard constrained optimization problem that can readily be approached by using the method of Lagrange multipliers.

For instance, suppose you have a job ordering stationery supplies for an office. The office needs three different types of products a, b, and c, which you will order in amounts x, y, and z, respectively. The usefulness of these products to the smooth operation of the office turns out to be modeled fairly well by the utility function $U(x, y, z) = xy + xyz$. If product a costs \$3 per unit, product b \$2 a unit, and product c \$1 a unit and the budget allows a total expenditure of not more than \$899, what should you do? The answer should be clear: You need to maximize

$$U(x, y, z) = xy + xyz \quad \text{subject to} \quad B(x, y, z) = 3x + 2y + z = 899.$$

The Lagrange multiplier equation, $\nabla U(x, y, z) = \lambda \nabla B(x, y, z)$, and the budget constraint yield the system

$$\begin{cases} y + yz = 3\lambda \\ x + xz = 2\lambda \\ xy = \lambda \\ 3x + 2y + z = 899 \end{cases}.$$

Solving for λ in the first three equations yields

$$\lambda = y\left(\frac{z+1}{3}\right) = x\left(\frac{z+1}{2}\right) = xy.$$

The last equality implies that either $x = 0$ or $y = (z+1)/2$. We can reject the first possibility, since $U(0, y, z) = 0$ and the utility $U(x, y, z) > 0$ whenever x, y, and z are all positive. Thus, we are left with $y = (z+1)/2$. This in turn implies that $\lambda = (z+1)^2/6$. Substituting for y in the constraint equation shows that $x = (898 - 2z)/3$, so that equation $xy = \lambda$ becomes

$$\left(\frac{898 - 2z}{3}\right)\left(\frac{z+1}{2}\right) = \frac{(z+1)^2}{6},$$

which is satisfied by either $z = -1$ (which we reject) or by $z = 299$. The only realistic critical point for this problem is $(100, 150, 299)$. We leave it to you to check that this point is indeed the site of a maximum value for the utility. ◆

EXAMPLE 5 In 1928, C. W. Cobb and P. M. Douglas developed a simple model for the gross output Q of a company or a nation, indicated by the function

$$Q(K, L) = AK^a L^{1-a},$$

where K represents the capital investment (in the form of machinery or other equipment), L the amount of labor used, and A and a positive constants with $0 < a < 1$. (The function Q is known now as the **Cobb–Douglas production function.**) If you are president of a company or nation, you naturally wish to maximize output, but equipment and labor cost money and you have a total amount of M dollars to invest. If the price of capital is p dollars per unit and the cost of labor (in the form of wages) is w dollars per unit, so that you are constrained by

$$B(K, L) = pK + wL \leq M,$$

what do you do?

Again, we have a situation ripe for the use of Lagrange multipliers. Before we consider the technical formalities, however, we consider a graphical solution. Draw the level curves of Q, called **isoquants**, as in Figure 4.40. Note that Q increases as we move away from the origin in the first quadrant. The budget constraint means that you can only consider values of K and L that lie inside or on the shaded triangle. It is clear that the optimum solution occurs at the point (K, L) where the level curve is tangent to the constraint line $pK + wL = M$.

Here is the analytical solution: From the equation $\nabla Q(K, L) = \lambda \nabla B(K, L)$ plus the constraint, we obtain the system

$$\begin{cases} AaK^{a-1}L^{1-a} = \lambda p \\ A(1-a)K^a L^{-a} = \lambda w \\ pK + wL = M \end{cases}.$$

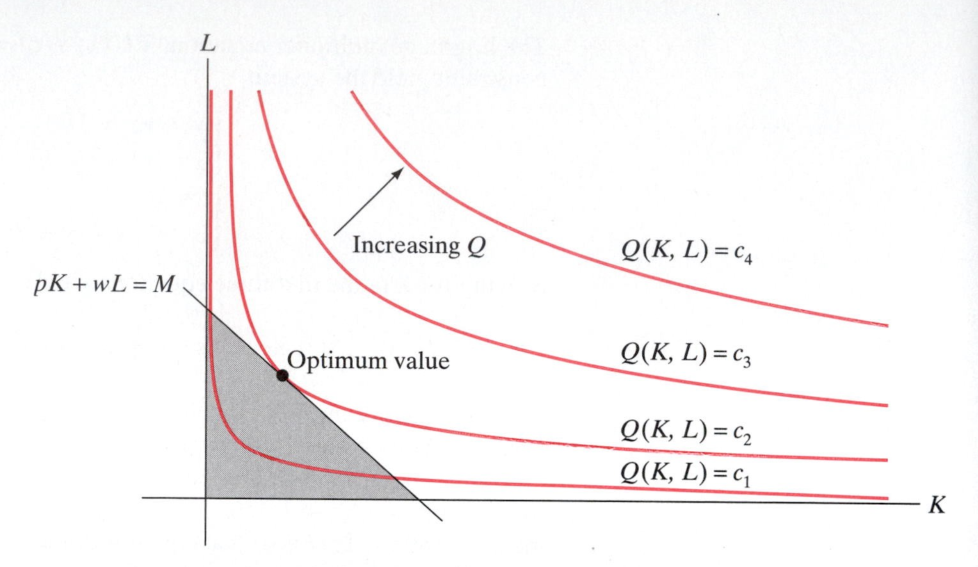

Figure 4.40 A family of isoquants. The optimum value of $Q(K, L)$ subject to the constraint $pK + wL = M$ occurs where a curve of the form $Q = c$ is tangent to the constraint line.

Solving for p and w in the first two equations yields

$$p = \frac{Aa}{\lambda}K^{a-1}L^{1-a} \quad \text{and} \quad w = \frac{A(1-a)}{\lambda}K^{a}L^{-a}.$$

Substitution of these values into the third equation gives

$$\frac{Aa}{\lambda}K^{a}L^{1-a} + \frac{A(1-a)}{\lambda}K^{a}L^{1-a} = M.$$

Thus,

$$\lambda = \frac{A}{M}K^{a}L^{1-a},$$

and the only critical point is

$$(K, L) = \left(\frac{Ma}{p}, \frac{M(1-a)}{w}\right).$$

From this geometric discussion, we know that the critical point must yield the maximum output Q.

From the Lagrange multiplier equation, at the optimum values for L and K, we have

$$\lambda = \frac{1}{p}\frac{\partial Q}{\partial K} = \frac{1}{w}\frac{\partial Q}{\partial L}.$$

This relation says that, at the optimum values, the marginal change in output per dollar's worth of extra capital equals the marginal change per dollar's worth of extra labor. In other words, at the optimum values, exchanging labor for capital (or vice versa) won't change the output. This is by no means the case away from the optimum values.

There is not much that is special about the function Q chosen. Most of our observations remain true for any C^2 function Q that satisfies the conditions

$$\frac{\partial Q}{\partial K}, \frac{\partial Q}{\partial L} \geq 0, \qquad \frac{\partial^2 Q}{\partial K^2}, \frac{\partial^2 Q}{\partial L^2} < 0.$$

If you consider what these relations mean qualitatively about the behavior of the output function with respect to increases in capital and labor, you will see that they are entirely reasonable assumptions.[4] ◆

Exercises

1. Find the line that best fits the following data: $(0, 2)$, $(1, 3)$, $(2, 5)$, $(3, 3)$, $(4, 2)$, $(5, 7)$, $(6, 7)$.

2. Show that if you have only two data points (x_1, y_1) and (x_2, y_2), then the best fit line given by the method of least squares is, in fact, the line through (x_1, y_1) and (x_2, y_2).

3. Suppose that you are given n pairs of data (x_1, y_1), $(x_2, y_2), \ldots, (x_n, y_n)$ and you seek to fit a function of the form $y = a/x + b$ to these data.
   (a) Use the method of least squares as outlined in this section to construct a function $D(a, b)$ that gives the sum of the squares of the distances between observed and predicted y-values of the data.
   (b) Show that the "best fit" curve of the form $y = a/x + b$ should have

$$a = \frac{n\sum y_i/x_i - \left(\sum 1/x_i\right)\left(\sum y_i\right)}{n\sum 1/x_i^2 - \left(\sum 1/x_i\right)^2}$$

and

$$b = \frac{\left(\sum 1/x_i^2\right)\left(\sum y_i\right) - \left(\sum 1/x_i\right)\left(\sum y_i/x_i\right)}{n\sum 1/x_i^2 - \left(\sum 1/x_i\right)^2}.$$

(All sums are from $i = 1$ to n.)

4. Find the curve of the form $y = a/x + b$ that best fits the following data: $(1, 0)$, $(2, -1)$, $\left(\frac{1}{2}, 1\right)$, and $\left(3, -\frac{1}{2}\right)$. (See Exercise 3.)

5. Suppose that you have n pairs of data (x_1, y_1), $(x_2, y_2), \ldots, (x_n, y_n)$ and you desire to fit a quadratic function of the form $y = ax^2+bx+c$ to the data. Show that the "best fit" parabola must have coefficients a, b, and c satisfying

$$\begin{cases} \left(\sum x_i^4\right)a + \left(\sum x_i^3\right)b + \left(\sum x_i^2\right)c = \sum x_i^2 y_i \\ \left(\sum x_i^3\right)a + \left(\sum x_i^2\right)b + \left(\sum x_i\right)c = \sum x_i y_i \\ \left(\sum x_i^2\right)a + \left(\sum x_i\right)b + nc = \sum y_i \end{cases}.$$

(All sums are from $i = 1$ to n.)

6. (Note: This exercise will be facilitated by the use of a spreadsheet or computer algebra system.) Egbert recorded the number of hours he slept the night before a major exam versus the score he earned, as shown in the table below.
   (a) Find the line that best fits these data.
   (b) Find the parabola $y = ax^2 + bx + c$ that best fits these data. (See Exercise 5.)
   (c) Last night Egbert slept 6.8 hr. What do your answers in parts (a) and (b) predict for his score on the calculus final he takes today?

Hours of sleep	Test score
8	85
8.5	72
9	95
7	68
4	52
8.5	75
7.5	90
6	65

7. Let $\mathbf{F} = (-2x - 2y - 1)\mathbf{i} + (-2x - 6y - 2)\mathbf{j}$.
   (a) Show that $\mathbf{F}$ is conservative and has potential function

$$V(x, y) = x^2 + 2xy + 3y^2 + x + 2y$$

   (i.e., $\mathbf{F} = -\nabla V$).
   (b) What are the equilibrium points of $\mathbf{F}$? The stable equilibria?

8. Suppose a particle moves in a vector field $\mathbf{F}$ in $\mathbf{R}^2$ with physical potential

$$V(x, y) = 2x^2 - 8xy - y^2 + 12x - 8y + 12.$$

Find all equilibrium points of $\mathbf{F}$ and indicate which, if any, are stable equilibria.

9. Let a particle move in the vector field $\mathbf{F}$ in $\mathbf{R}^3$ whose physical potential is given by

$$V(x, y, z) = 3x^2 + 2xy + z^2 - 2yz + 3x + 5y - 10.$$

Determine the equilibria of $\mathbf{F}$ and identify those that are stable.

[4] For more about the history and derivation of the Cobb–Douglas function, consult R. Geitz, "The Cobb–Douglas production function," *UMAP Module No. 509*, Birkhäuser, 1981.

10. Suppose that a particle of mass m is constrained to move on the ellipsoid $2x^2 + 3y^2 + z^2 = 1$ subject to both a gravitational force $\mathbf{F} = -mg\mathbf{k}$, as well as to an additional potential $V(x, y, z) = 2x$.

    (a) Find any equilibrium points for this situation.

    (b) Are there any stable equilibria?

11. The Sukolux Vacuum Cleaner Company manufactures and sells three types of vacuum cleaners: the standard, executive, and deluxe models. The annual revenue in dollars as a function of the numbers x, y, and z (respectively) of standard, executive, and deluxe models sold is

$$R(x, y, z) = xyz^2 - 25{,}000x - 25{,}000y - 25{,}000z.$$

    The manufacturing plant can produce 200,000 total units annually. Assuming that everything that is manufactured is sold, how should production be distributed among the models so as to maximize the annual revenue?

12. Some simple electronic devices are to be designed to include three digital component modules, types 1, 2, and 3, which are to be kept in inventory in respective amounts x_1, x_2, and x_3. Suppose that the relative importance of these components to the various devices is modeled by the utility function

$$U(x_1, x_2, x_3) = x_1x_2 + 2x_1x_3 + x_1x_2x_3.$$

    You are authorized to purchase \$90 worth of these parts to make prototype devices. If type 1 costs \$1 per component, type 2 \$4 per component, and type 3 \$2 per component, how should you place your order?

13. The CEO of the Wild Widget Company has decided to invest \$360,000 in his Michigan factory. His economic analysts have noted that the output of this factory is modeled by the function $Q(K, L) = 60K^{1/3}L^{2/3}$, where K represents the amount (in thousands of dollars) spent on capital equipment and L represents the amount (also in thousands of dollars) spent on labor.

    (a) How should the CEO allocate the \$360,000 between labor and equipment?

    (b) Check that $\partial Q/\partial K = \partial Q/\partial L$ at the optimal values for K and L.

14. Let $Q(K, L)$ be a production function for a company where K and L represent the respective amounts spent on capital equipment and labor. Let p denote the price of capital equipment per unit and w the cost of labor per unit. Show that, subject to a fixed production $Q(K, L) = c$, the total cost M of production is minimized when K and L are such that

$$\frac{1}{p}\frac{\partial Q}{\partial K} = \frac{1}{w}\frac{\partial Q}{\partial L}.$$

4.5 True/False Exercises for Chapter 4

1. If f is a function of class C^2 and p_2 denotes the second-order Taylor polynomial of f at $\mathbf{a}$, then $f(\mathbf{x}) \approx p_2(\mathbf{x})$ when $\mathbf{x} \approx \mathbf{a}$.

2. The increment Δf of a function $f(x, y)$ measures the change in the z-coordinate of the tangent plane to the graph of f.

3. The differential df of a function $f(x, y)$ measures the change in the z-coordinate of the tangent plane to the graph of f.

4. The second-order Taylor polynomial of $f(x, y, z) = x^2 + 3xz + y^2$ at $(1, -1, 2)$ is $p_2(x, y, z) = x^2 + 3xz + y^2$.

5. The second-order Taylor polynomial of $f(x, y) = x^3 + 2xy + y$ at $(0, 0)$ is $p_2(x, y) = 2xy + y$.

6. The second-order Taylor polynomial of $f(x, y) = x^3 + 2xy + y$ at $(1, -1)$ is $p_2(x, y) = 2xy + y$.

7. Near the point $(1, 3, 5)$, the function $f(x, y, z) = 3x^4 + 2y^3 + z^2$ is most sensitive to changes in z.

8. The Hessian matrix $Hf(x_1, \ldots, x_n)$ of f has the property that $Hf(x_1, \ldots, x_n)^T = Hf(x_1, \ldots, x_n)$.

9. If $\nabla f(a_1, \ldots, a_n) = \mathbf{0}$, then f has a local extremum at $\mathbf{a} = (a_1, \ldots, a_n)$.

10. If f is differentiable and has a local extremum at $\mathbf{a} = (a_1, \ldots, a_n)$, then $\nabla f(\mathbf{a}) = \mathbf{0}$.

11. The set $\{(x, y, z) \mid 4 \le x^2 + y^2 + z^2 \le 9\}$ is compact.

12. The set $\{(x, y) \mid 2x - 3y = 1\}$ is compact.

13. Any continuous function $f(x, y)$ must attain a global maximum on the disk $\{(x, y) \mid x^2 + y^2 < 1\}$.

14. Any continuous function $f(x, y, z)$ must attain a global maximum on the ball $\{(x, y, z) \mid (x-1)^2 + (y+1)^2 + z^2 \le 4\}$.

15. If $f(x, y)$ is of class C^2, has a critical point at (a, b), and $f_{xx}(a, b)f_{yy}(a, b) - f_{xy}(a, b)^2 < 0$, then f has a saddle point at (a, b).

16. If $\det Hf(\mathbf{a}) = 0$, then f has a saddle point at $\mathbf{a}$.

17. The function $f(x, y, z) = x^3y^2z - x^2(y + z)$ has a saddle point at $(1, -1, 2)$.

18. The function $f(x, y, z) = x^2 + y^2 + z^2 - yz$ has a local maximum at $(0, 0, 0)$.

19. The function $f(x, y, z) = xy^3 - x^2z + z$ has a degenerate critical point at $(-1, 0, 0)$.

20. The function $F(x_1, \ldots, x_n) = 2(x_1 - 1)^2 - 3(x_2 - 2)^2 + \cdots + (-1)^{n+1}(n + 1)(x_n - n)^2$ has a critical point at $(1, 2, \ldots, n)$.

21. The function $F(x_1, \ldots, x_n) = 2(x_1 - 1)^2 - 3(x_2 - 2)^2 + \cdots + (-1)^{n+1}(n + 1)(x_n - n)^2$ has a minimum at $(1, 2, \ldots, n)$.

22. All local extrema of a function of more than one variable occur where all partial derivatives simultaneously vanish.

23. All points $\mathbf{a} = (a_1, \ldots, a_2)$ where the function $f(x_1, \ldots, x_n)$ has an extremum subject to the constraint that $g(x_1, \ldots, x_n) = c$, are solutions to the system of equations

$$\begin{cases} \dfrac{\partial f}{\partial x_1} = \lambda \dfrac{\partial g}{\partial x_1} \\ \quad \vdots \\ \dfrac{\partial f}{\partial x_n} = \lambda \dfrac{\partial g}{\partial x_n} \\ g(x_1, \ldots, x_n) = c \end{cases}$$

24. Any solution $(\lambda_1, \ldots, \lambda_k, x_1, \ldots, x_n)$ to the system of equations

$$\begin{cases} \dfrac{\partial f}{\partial x_1} = \lambda_1 \dfrac{\partial g_1}{\partial x_1} + \cdots + \lambda_k \dfrac{\partial g_k}{\partial x_1} \\ \quad \vdots \\ \dfrac{\partial f}{\partial x_n} = \lambda_1 \dfrac{\partial g_1}{\partial x_n} + \cdots + \lambda_k \dfrac{\partial g_k}{\partial x_n} \\ g_1(x_1, \ldots, x_n) = c_1 \\ \quad \vdots \\ g_1(x_1, \ldots, x_n) = c_k \end{cases}$$

yields a point $(x_1, \ldots, x_n)$ that is an extreme value of f subject to the simultaneous constraints $g_1 = c_1, \ldots, g_k = c_k$.

25. To find the critical points of the function $f(x, y, z, w)$ subject to the simultaneous constraints $g(x, y, z, w) = c$, $h(x, y, z, w) = d$, $k(x, y, z, w) = e$ using the technique of Lagrange multipliers, one will have to solve a system of 4 equations in 4 unknowns.

26. Suppose that $f(x, y, z)$ and $g(x, y, z)$ are of class C^1 and that (x_0, y_0, z_0) is a point where f achieves a maximum value subject to the constraint that $g(x, y, z) = c$ and that $\nabla g(x_0, y_0, z_0)$ is nonzero. Then the level set of f that contains (x_0, y_0, z_0) must be tangent to the level set $S = \{(x, y, z) \mid g(x, y, z) = c\}$.

27. The critical points of $f(x, y, z) = xy + 2xz + 2yz$ subject to the constraint that $xyz = 4$ are the same as the critical points of the function $F(x, y) = xy + \dfrac{8}{x} + \dfrac{8}{y}$.

28. Given data points (3, 1), (4, 10), (5, 8), (6, 12), to find the best fit line by regression, we find the minimum value of the function $D(m, b) = (3m + b - 1)^2 + (4m + b - 10)^2 + (5m + b - 8)^2 + (6m + b - 12)^2$.

29. All equilibrium points of a gradient vector field are minimum points of the vector field's potential function.

30. Given an output function for a company, the marginal change in output per dollar investment in capital is the same as the marginal change in the output per dollar investment in labor.

4.6 Miscellaneous Exercises for Chapter 4

1. Let $V = \pi r^2 h$, where $r \approx r_0$ and $h \approx h_0$. What relationship must hold between r_0 and h_0 for V to be equally sensitive to small changes in r and h?

2. (a) Find the unique critical point of the function

$$f(x_1, x_2, \ldots, x_n) = e^{-x_1^2 - x_2^2 - \cdots - x_n^2}.$$

(b) Use the Hessian criterion to determine the nature of this critical point.

3. The Java Joint Gourmet Coffee House sells top-of-the-line Arabian Mocha and Hawaiian Kona beans. If Mocha beans are priced at x dollars per pound and Kona beans at y dollars per pound, then market research has shown that each week approximately $80 - 100x + 40y$ pounds of Mocha beans will be sold and $20 + 60x - 35y$ pounds of Kona beans will be sold. The wholesale cost to the Java Joint owners is \$2 per pound for Mocha beans and \$4 per pound for Kona beans. How should the owners price the coffee beans in order to maximize their profits?

4. The Crispy Crunchy Cereal Company produces three brands, X, Y, and Z, of breakfast cereal. Each month, x, y, and z (respectively) 1000-box cases of brands X, Y, and Z are sold at a selling price (per box) of each cereal given as follows:

Brand	No. cases sold	Selling price per box
X	x	$4.00 - 0.02x$
Y	y	$4.50 - 0.05y$
Z	z	$5.00 - 0.10z$

(a) What is the total revenue R if x cases of brand X, y cases of brand Y, and z cases of brand Z are sold?

(b) Suppose that during the month of November, brand X sells for \$3.88 per box, brand Y for \$4.25, and brand Z for \$4.60. If the price of each brand is increased by \$0.10, what effect will this have on the total revenue?

(c) What selling prices maximize the total revenue?

5. Find the maximum and minimum values of the function

$$f(x, y, z) = x - \sqrt{3}\, y$$

on the sphere $x^2 + y^2 + z^2 = 4$ in two ways:

(a) by using a Lagrange multiplier;

(b) by substituting spherical coordinates (thereby describing the point (x, y, z) on the sphere as $x = 2 \sin\varphi \cos\theta$, $y = 2 \sin\varphi \sin\theta$, $z = 2\cos\varphi$) and then finding the ordinary (i.e., unconstrained) extrema of $f(x(\varphi, \theta), y(\varphi, \theta), z(\varphi, \theta))$.

6. Suppose that the temperature in a space is given by the function

$$T(x, y, z) = 200xyz^2.$$

Find the hottest point(s) on the unit sphere in two ways:

(a) by using Lagrange multipliers;

(b) by letting $x = \sin\varphi \cos\theta$, $y = \sin\varphi \sin\theta$, $z = \cos\varphi$ and maximizing T as a function of the two independent variables φ and θ. (Note: It will help if you use appropriate trigonometric identities where possible.)

7. Consider the function $f(x, y) = (y - 2x^2)(y - x^2)$.

(a) Show that f has a single critical point at the origin.

(b) Show that this critical point is *degenerate*. Hence, it will require means other than the Hessian criterion to determine the nature of the critical point as a local extremum.

(c) Show that, when restricted to any line that passes through the origin, f has a minimum at $(0, 0)$. (That is, consider the function $F(x) = f(x, mx)$, where m is a constant and the function $G(y) = f(0, y)$.)

(d) However, show that, when restricted to the parabola $y = \frac{3}{2}x^2$, the function f has a global maximum at $(0, 0)$. Thus, the origin must be a saddle point.

(e) Use a computer to graph the surface $z = f(x, y)$.

8. (a) Find all critical points of $f(x, y) = xy$ that satisfy $x^2 + y^2 = 1$.

(b) Draw a collection of level curves of f and, on the same set of axes, the constraint curve $x^2 + y^2 = 1$, and the critical points you found in part (a).

(c) Use the plot you obtained in part (b) and a geometric argument to determine the nature of the critical points found in part (a).

9. (a) Find all critical points of $f(x, y, z) = xy$ that satisfy $x^2 + y^2 + z^2 = 1$.

(b) Give a rough sketch of a collection of level surfaces of f and, on the same set of axes, the constraint surface $x^2 + y^2 + z^2 = 1$, and the critical points you found in part (a).

(c) Use part (b) and a geometric argument to determine the nature of the critical points found in part (a).

10. Find the area A of the largest rectangle so that two squares of total area 1 can be placed snugly inside the rectangle without overlapping, except along their edges. (See Figure 4.41.)

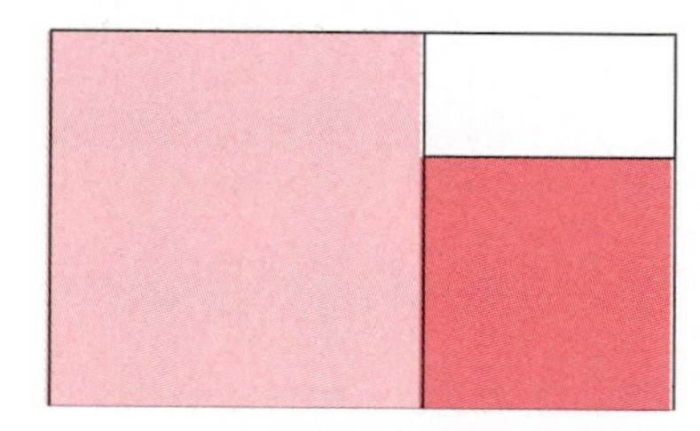

Figure 4.41 Figure for Exercise 10.

11. Find the minimum value of

$$f(x_1, x_2, \ldots, x_n) = x_1^2 + x_2^2 + \cdots + x_n^2$$

subject to the constraint that $a_1x_1 + a_2x_2 + \cdots + a_nx_2 = 1$, assuming that $a_1^2 + a_2^2 + \cdots + a_n^2 > 0$.

12. Find the maximum value of

$$f(x_1, x_2, \ldots, x_n) = (a_1x_1 + a_2x_2 + \cdots + a_nx_n)^2,$$

subject to $x_1^2 + x_2^2 + \cdots + x_n^2 = 1$. Assume that not all of the a_i's are zero.

13. Find the dimensions of the largest rectangular box that can be inscribed in the ellipsoid $x^2 + 2y^2 + 4z^2 = 12$. Assume that the faces of the box are parallel to the coordinate planes.

14. Your company must design a storage tank for Super Suds liquid laundry detergent. The customer's specifications call for a cylindrical tank with hemispherical ends (see Figure 4.42), and the tank is to hold 8000 gal of detergent. Suppose that it costs twice as much (per square foot of sheet metal used) to machine the hemispherical ends of the tank as it does to make the cylindrical part. What radius and height do you recommend for the cylindrical portion so as to minimize the *total* cost of manufacturing the tank?

15. Find the minimum distance from the origin to the surface $x^2 - (y - z)^2 = 1$.

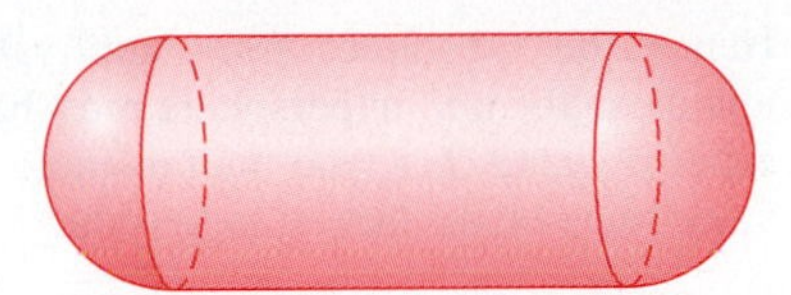

Figure 4.42 The storage tank of Exercise 14.

16. Determine the dimensions of the largest cone that can be inscribed in a sphere of radius a.

17. Find the dimensions of the largest rectangular box (whose faces are parallel to the coordinate planes) that can be inscribed in the tetrahedron having three faces in the coordinate planes and fourth face in the plane with equation $bcx + acy + abz = abc$, where a, b, and c are positive constants. (See Figure 4.43.)

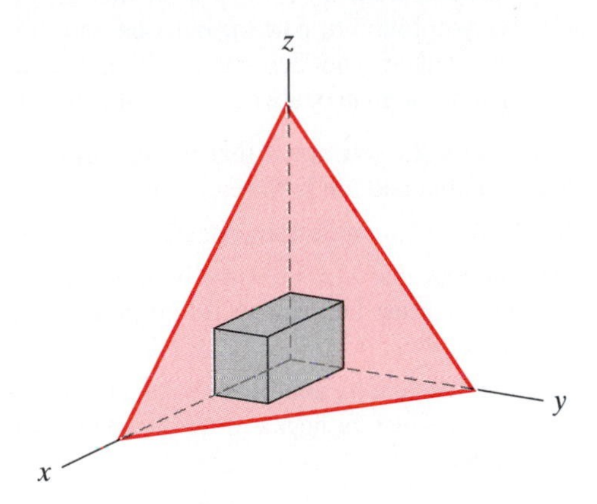

Figure 4.43 Figure for Exercise 17.

18. You seek to mail a poster to your friend as a gift. You roll up the poster and put it in a cylindrical tube of diameter x and length y. The postal regulations demand that the sum of the length of the tube plus its girth (i.e., the circumference of the tube) be at most 108 in.

(a) Use the method of Lagrange multipliers to find the dimensions of the largest volume tube that you can mail.

(b) Use techniques from single-variable calculus to solve this problem in another way.

19. Find the distance between the line $y = 2x + 2$ and the parabola $x = y^2$ by minimizing the distance between a point (x_1, y_1) on the line and a point (x_2, y_2) on the parabola. Draw a sketch indicating that you have found the minimum value.

20. A ray of light travels at a constant speed in a uniform medium, but in different media (such as air and water) light travels at different speeds. For example, if a ray of light passes from air to water, it is bent (or **refracted**) as shown in Figure 4.44. Suppose the speed of light in medium 1 is v_1 and in medium 2 is v_2. Then, by Fermat's principle of least time, the light will strike the boundary between medium 1 and medium 2 at a point P so that the total time the light travels is minimized.

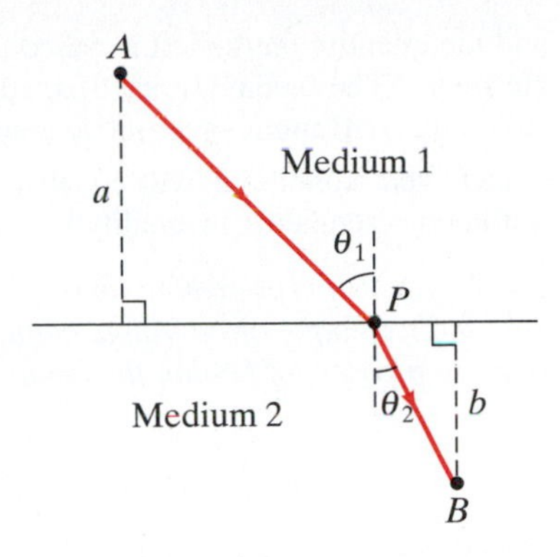

Figure 4.44 Snell's law of refraction.

(a) Determine the total time the light travels in going from point A to point B via point P as shown in Figure 4.44.

(b) Use the method of Lagrange multipliers to establish **Snell's law of refraction:** That the total travel time is minimized when

$$\frac{\sin\theta_1}{\sin\theta_2} = \frac{v_1}{v_2}.$$

(Hint: The horizontal and vertical separation of A and B are constant.)

21. Use Lagrange multipliers to establish the formula

$$D = \frac{|ax_0 + by_0 - d|}{\sqrt{a^2 + b^2}}$$

for the distance D from the point (x_0, y_0) to the line $ax + by = d$.

22. Use Lagrange multipliers to establish the formula

$$D = \frac{|ax_0 + by_0 + cz_0 - d|}{\sqrt{a^2 + b^2 + c^2}}$$

for the distance D from the point (x_0, y_0, z_0) to the plane $ax + by + cz = d$.

23. (a) Show that the maximum value of $f(x, y, z) = x^2y^2z^2$ subject to the constraint that $x^2+y^2+z^2 = a^2$ is

$$\frac{a^6}{27} = \left(\frac{a^2}{3}\right)^3.$$

(b) Use part (a) to show that, for all x, y, and z,

$$(x^2y^2z^2)^{1/3} \le \frac{x^2 + y^2 + z^2}{3}.$$

(c) Show that, for any positive numbers $x_1, x_2, \ldots, x_n$,

$$(x_1x_2\cdots x_n)^{1/n} \le \frac{x_1 + x_2 + \cdots + x_n}{n}.$$

The quantity on the right of the inequality is the **arithmetic mean** of the numbers $x_1, x_2, \ldots, x_n$, and the quantity on the left is called the **geometric mean**. The inequality itself is, appropriately, called the **arithmetic–geometric inequality**.

(d) Under what conditions will equality hold in the arithmetic–geometric inequality?

In Exercises 24–27 you will explore how some ideas from matrix algebra and the technique of Lagrange multipliers come together to treat the problem of finding the points on the unit hypersphere

$$g(x_1, \ldots, x_n) = x_1^2 + x_2^2 + \cdots + x_n^2 = 1$$

that give extreme values of the quadratic form

$$f(x_1, \ldots, x_n) = \sum_{i,j=1}^{n} a_{ij} x_i x_j,$$

where the a_{ij}'s are constants.

24. (a) Use a Lagrange multiplier λ to set up a system of $n+1$ equations in $n+1$ unknowns $x_1, \ldots, x_n, \lambda$ whose solutions provide the appropriate constrained critical points.

(b) Recall that formula (2) in §4.2 shows that the quadratic form f may be written in terms of matrices as

$$f(x_1, \ldots, x_n) = \mathbf{x}^T A\mathbf{x}, \qquad (1)$$

where the vector $\mathbf{x}$ is written as the $n \times 1$ matrix $\begin{bmatrix} x_1 \\ \vdots \\ x_n \end{bmatrix}$ and A is the $n \times n$ matrix whose ijth entry is a_{ij}. Moreover, as noted in the discussion in §4.2, the matrix A may be taken to be symmetric (i.e., so that $A^T = A$), and we will therefore assume that A is symmetric. Show that the gradient equation $\nabla f = \lambda \nabla g$ is equivalent to the matrix equation

$$A\mathbf{x} = \lambda\mathbf{x}. \qquad (2)$$

Since the point $(x_1, \ldots, x_n)$ satisfies the constraint $x_1^2 + \cdots + x_n^2 = 1$, the vector $\mathbf{x}$ is nonzero. If you have studied some linear algebra, you will recognize that you have shown that a constrained critical point $(x_1, \ldots, x_n)$ for this problem corresponds precisely to an **eigenvector** of the matrix A associated with the **eigenvalue** λ.

(c) Now suppose that $\mathbf{x} = \begin{bmatrix} x_1 \\ \vdots \\ x_n \end{bmatrix}$ is one of the eigenvectors of the symmetric matrix A, with associated eigenvalue λ. Use equations (1) and (2) to show, if $\mathbf{x}$ is a unit vector, that

$$f(x_1, \ldots, x_n) = \lambda.$$

Hence the (absolute) minimum value that f attains on the unit hypersphere must be the smallest eigenvalue of A and the (absolute) maximum value must be the largest eigenvalue.

25. Let $n = 2$ in the situation of Exercise 24, so that we are considering the problem of finding points on the circle $x^2 + y^2 = 1$ that give extreme values of the function

$$f(x, y) = ax^2 + 2bxy + cy^2 = \begin{bmatrix} x & y \end{bmatrix} \begin{bmatrix} a & b \\ b & c \end{bmatrix} \begin{bmatrix} x \\ y \end{bmatrix}.$$

(a) Find the eigenvalues of $A = \begin{bmatrix} a & b \\ b & c \end{bmatrix}$ by identifying the constrained critical points of the optimization problem described above.

(b) Now use some algebra to show that the eigenvalues you found in part (a) must be real. It is a fact (that you need not demonstrate here) that any $n \times n$ symmetric matrix always has real eigenvalues.

26. In Exercise 25 you noted that the eigenvalues λ_1, λ_2 that you obtained are both real.

(a) Under what conditions does $\lambda_1 = \lambda_2$?

(b) Suppose that λ_1 and λ_2 are both positive. Explain why f must be positive on all points of the unit circle.

(c) Suppose that λ_1 and λ_2 are both negative. Explain why f must be negative on all points of the unit circle.

27. Let f be a general quadratic form in n variables determined by an $n \times n$ symmetric matrix A, that is, $f(x_1, \ldots, x_n) = \sum_{i,j=1}^{n} a_{ij} x_i x_j = \mathbf{x}^T A\mathbf{x}$.

(a) Show, for any real number k, that $f(kx_1, \ldots, kx_n) = k^2 f(x_1, \ldots, x_n)$. (This means that a quadratic form is a homogeneous polynomial of degree 2—see Exercises 37–44 of the Miscellaneous Exercises for Chapter 2 for more about homogeneous functions.)

(b) Use part (a) to show that if f has a positive minimum on the unit hypersphere, then f must be positive for *all* nonzero $\mathbf{x} \in \mathbf{R}^n$ and that if f has a negative maximum on the unit hypersphere, then f must be negative for all nonzero $\mathbf{x} \in \mathbf{R}^n$. (Hint: For $\mathbf{x} \neq \mathbf{0}$, let $\mathbf{u} = \mathbf{x}/\|\mathbf{x}\|$, so that $\mathbf{x} = k\mathbf{u}$, where $k = \|\mathbf{x}\|$.)

(c) Recall from §4.2 that a quadratic form f is said to be **positive definite** if $f(\mathbf{x}) > 0$ for all nonzero $\mathbf{x} \in \mathbf{R}^n$ and **negative definite** if $f(\mathbf{x}) < 0$ for all nonzero $\mathbf{x} \in \mathbf{R}^n$. Use part (b) and Exercise 24 to show that the quadratic form f is positive definite if and only if all eigenvalues of A are positive, and negative definite if and only if all eigenvalues of A are negative. (Note: As remarked in part (b) of Exercise 25, all the eigenvalues of A will be real.)

Multiple Integration

5.1 Introduction: Areas and Volumes

Our purpose in this chapter is to find ways to generalize the notion of the definite integral of a function of a single variable to the cases of functions of two or three variables. We also explore how these multiple integrals may be used to meaningfully represent various physical quantities.

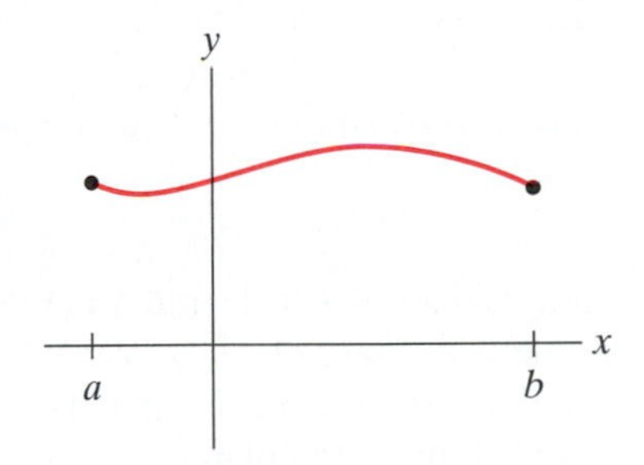

Figure 5.1 The graph of $y = f(x)$.

Let f be a continuous function of one variable defined on the closed interval $[a, b]$ and suppose that f has only nonnegative values. Then the graph of f looks like Figure 5.1. That f is continuous is reflected in the fact that the graph consists of an unbroken curve. That f is nonnegative-valued means that this curve does not dip below the x-axis. We know from one-variable calculus that the definite integral $\int_a^b f(x)\,dx$ exists and gives the area under the curve, as shown in Figure 5.2.

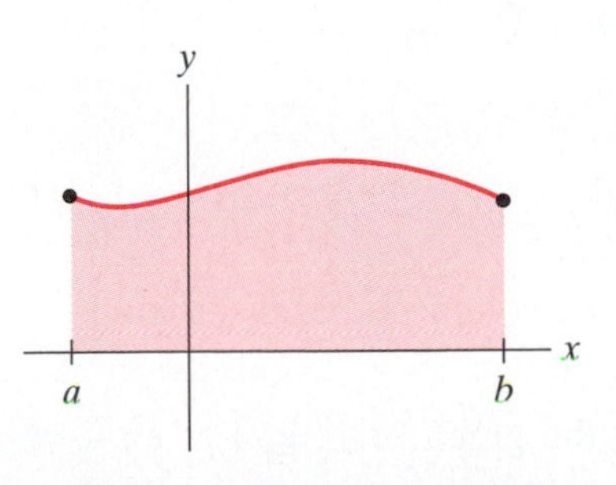

Figure 5.2 The shaded region has area $\int_a^b f(x)\,dx$.

Now suppose that f is a continuous, nonnegative-valued function of *two* variables defined on the **closed rectangle**

$$R = \{(x, y) \in \mathbf{R}^2 \mid a \le x \le b, c \le y \le d\}$$

in $\mathbf{R}^2$. Then the graph of f over R looks like an unbroken surface that never dips below the xy-plane, as shown in Figure 5.3. In analogy with the single variable case, there should be some sort of integral that represents the **volume** under the part of the graph that lies over R. (See Figure 5.4.) We can find such an integral by using **Cavalieri's Principle,** which is nothing more than a fancy term for the method of slicing. Suppose we slice by the vertical plane $x = x_0$, where x_0 is a constant between a and b. Let $A(x_0)$ denote the cross-sectional area of such a slice. Then, roughly, one can think of the quantity $A(x_0)\,dx$ as giving the volume of an "infinitely thin" slab of thickness dx and cross-sectional area $A(x_0)$. (See Figure 5.5.) Hence, the definite integral

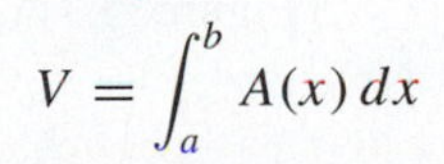

$$V = \int_a^b A(x)\,dx$$

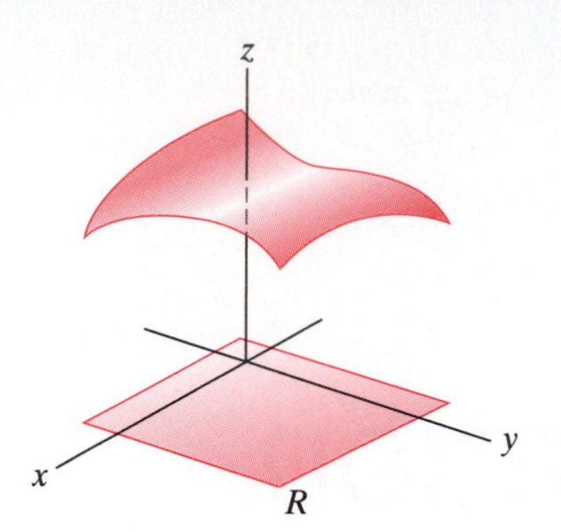

Figure 5.3 The graph of $z = f(x, y)$.

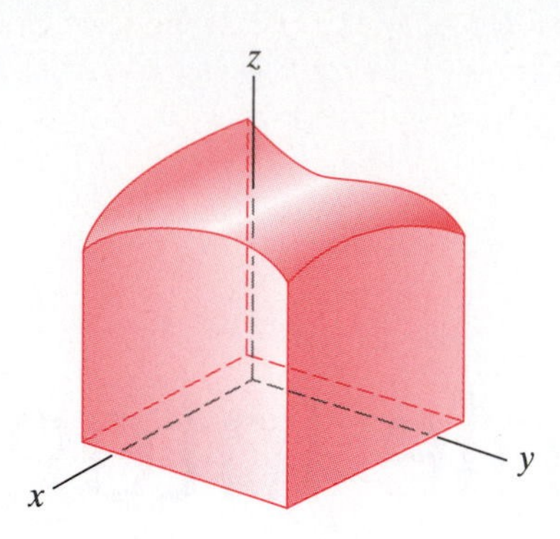

Figure 5.4 The region under the portion of the graph of f lying over R has volume that is given by an integral.

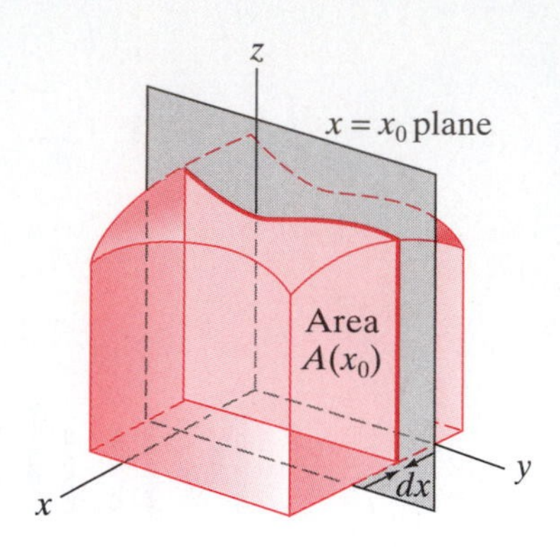

Figure 5.5 A slab of "volume" $dV = A(x_0)\,dx$.

gives a "sum" of the volumes of such slabs and can be considered to provide a reasonable definition of the total volume of the solid.

But what about the value of $A(x_0)$? Note that $A(x_0)$ is nothing more than the area under the curve $z = f(x_0, y)$, obtained by slicing the surface $z = f(x, y)$ with the plane $x = x_0$. Therefore,

$$A(x_0) = \int_c^d f(x_0, y)\,dy$$

(remember x_0 is a constant), and so we find that

$$V = \int_a^b A(x)\,dx = \int_a^b \left[\int_c^d f(x, y)\,dy\right] dx. \tag{1}$$

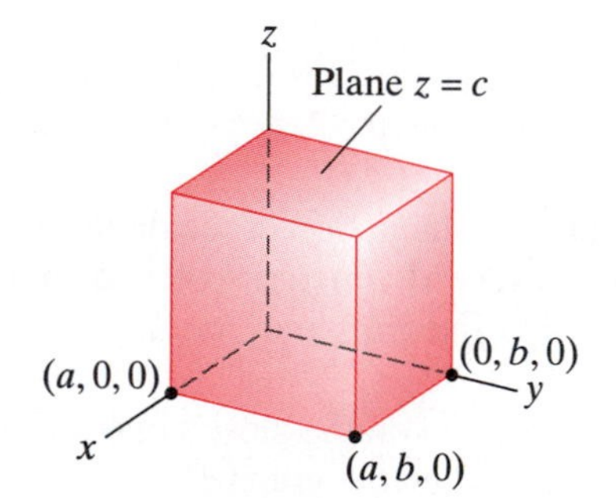

Figure 5.6 Calculating the volume of the box of Example 1.

The right-hand side of formula (1) is called an **iterated integral.** To calculate it, first find an "antiderivative" of $f(x, y)$ with respect to y (by treating x as a constant), evaluate at the integration limits $y = c$ and $y = d$, and then repeat the process with respect to x.

EXAMPLE 1 Let's make sure that the iterated integral defined in formula (1) gives the correct answer in a case we know well, namely, the case of a box. We'll picture the box as in Figure 5.6. That is, the box is bounded on top and bottom by the planes $z = c$ (where $c > 0$) and $z = 0$, on left and right by the planes $y = 0$ and $y = b$ (where $b > 0$), and on back and front by the planes $x = 0$ and $x = a$ ($a > 0$). Hence, the volume of the box may be found by computing the volume under the graph of $z = c$ over the rectangle

$$R = \{(x, y) \mid 0 \le x \le a, 0 \le y \le b\}.$$

Using formula (1), we obtain

$$V = \int_0^a \int_0^b c\,dy\,dx = \int_0^a \left(cy\big|_{y=0}^{y=b}\right) dx = \int_0^a cb\,dx = cbx\big|_{x=0}^{x=a} = cba.$$

This result checks with what we already know the volume to be, as it should. ◆

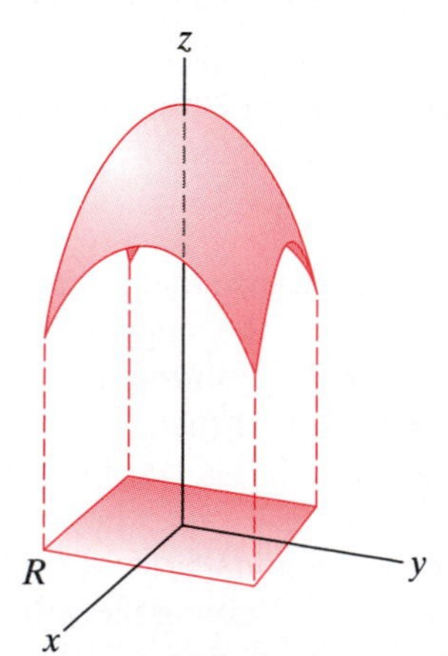

Figure 5.7 The graph of $z = 4 - x^2 - y^2$ of Example 2.

EXAMPLE 2 We calculate the volume under the graph of $z = 4 - x^2 - y^2$ (Figure 5.7) over the square

$$R = \{(x, y) \mid -1 \le x \le 1,\ -1 \le y \le 1\}.$$

Using formula (1) once again, we calculate the volume by first integrating with respect to y (i.e., by treating x as a constant in the inside integral), and then by integrating with respect to x. The details are as follows:

$$V = \int_{-1}^{1}\int_{-1}^{1}(4 - x^2 - y^2)\,dy\,dx = \int_{-1}^{1}\left(4y - x^2y - \frac{1}{3}y^3\right)\Big|_{y=-1}^{y=1} dx$$

$$= \int_{-1}^{1}\left(\left(4 - x^2 - \frac{1}{3}\right) - \left(-4 + x^2 + \frac{1}{3}\right)\right) dx$$

$$= \int_{-1}^{1}\left(8 - 2x^2 - \frac{2}{3}\right) dx$$

$$= \left(\frac{22}{3}x - \frac{2}{3}x^3\right)\Big|_{-1}^{1} = \left(\frac{22}{3} - \frac{2}{3}\right) - \left(-\frac{22}{3} + \frac{2}{3}\right) = \frac{40}{3}. \quad ◆$$

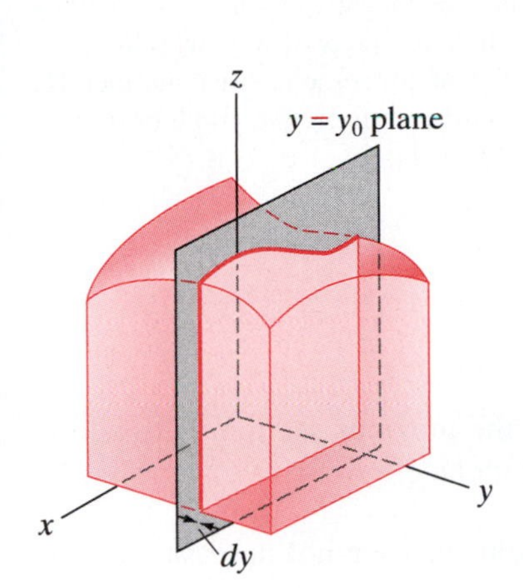

Figure 5.8 Slicing by $y = y_0$ first.

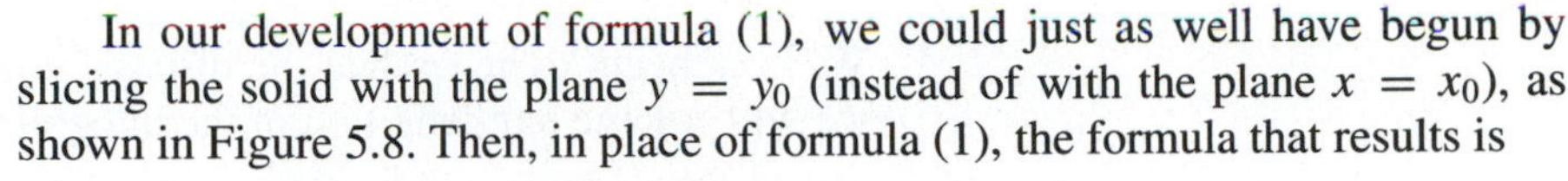

In our development of formula (1), we could just as well have begun by slicing the solid with the plane $y = y_0$ (instead of with the plane $x = x_0$), as shown in Figure 5.8. Then, in place of formula (1), the formula that results is

$$V = \int_c^d \int_a^b f(x, y)\,dx\,dy. \tag{2}$$

Since the iterated integrals in formulas (1) and (2) both represent the volume of the same geometric object, we can summarize the preceding discussion as follows.

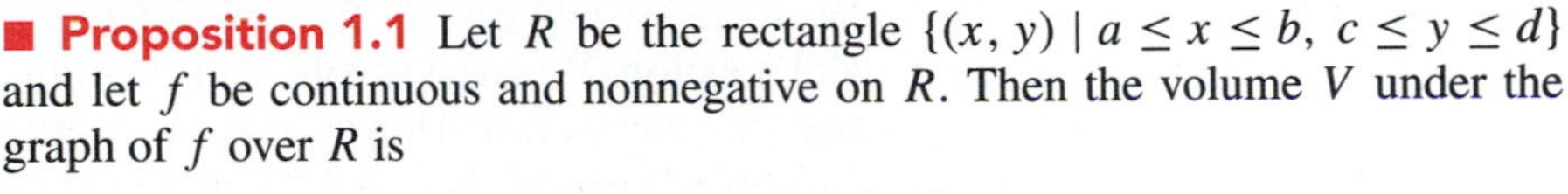

■ **Proposition 1.1** Let R be the rectangle $\{(x, y) \mid a \le x \le b,\ c \le y \le d\}$ and let f be continuous and nonnegative on R. Then the volume V under the graph of f over R is

$$\int_a^b \int_c^d f(x, y)\,dy\,dx = \int_c^d \int_a^b f(x, y)\,dx\,dy.$$

Exercises

Evaluate the iterated integrals given in Exercises 1–6.

1. $\int_0^2 \int_1^3 (x^2 + y)\,dy\,dx$

2. $\int_0^\pi \int_1^2 y \sin x\,dy\,dx$

3. $\int_{-2}^4 \int_0^1 x e^y\,dy\,dx$

4. $\int_0^{\pi/2} \int_0^1 e^x \cos y\,dx\,dy$

5. $\int_1^2 \int_0^1 (e^{x+y} + x^2 + \ln y)\,dx\,dy$

6. $\int_1^9 \int_1^e \frac{\ln\sqrt{x}}{xy}\,dx\,dy$

7. Find the volume of the region that lies under the graph of the paraboloid $z = x^2 + y^2 + 2$ and over the rectangle $R = \{(x, y) \mid -1 \le x \le 2, 0 \le y \le 2\}$ in two ways:

(a) by using Cavalieri's principle to write the volume as an iterated integral that results from slicing the region by parallel planes of the form $x =$ constant;

(b) by using Cavalieri's principle to write the volume as an iterated integral that results from slicing the region by parallel planes of the form $y =$ constant.

8. Find the volume of the region bounded on top by the plane $z = x + 3y + 1$, on the bottom by the xy-plane, and on the sides by the planes $x = 0$, $x = 3$, $y = 1$, $y = 2$.

9. Find the volume of the region bounded by the graph of $f(x, y) = 2x^2 + y^4 \sin \pi x$, the xy-plane, and the planes $x = 0$, $x = 1$, $y = -1$, $y = 2$.

In Exercises 10–15, calculate the given iterated integrals and indicate of what regions in $\mathbf{R}^3$ they may be considered to represent the volumes.

10. $\int_0^2 \int_1^3 2\,dx\,dy$

11. $\int_1^3 \int_{-2}^2 (16 - x^2 - y^2)\,dy\,dx$

12. $\int_{-\pi/2}^{\pi/2} \int_0^{\pi} \sin x \cos y\,dx\,dy$

13. $\int_0^5 \int_{-2}^2 (4 - x^2)\,dx\,dy$

14. $\int_{-2}^3 \int_0^1 |x| \sin \pi y\,dy\,dx$

15. $\int_{-5}^5 \int_{-1}^2 (5 - |y|)\,dx\,dy$

16. Suppose that f is a nonnegative-valued, continuous function defined on $R = \{(x, y) \mid a \le x \le b,\ c \le y \le d\}$. If $f(x, y) \le M$ for some positive number M, explain why the volume V under the graph of f over R is at most $M(b - a)(d - c)$.

5.2 Double Integrals

We now begin to generalize the discussion of the previous section. Ultimately, we will enlarge our definition of the integral to include

1. integrals of arbitrary functions (i.e., functions that are not necessarily nonnegative or continuous) and
2. integrals over arbitrary regions in the plane (i.e., rather than integrals over rectangles only).

We focus first on case 1. To do this, we start anew with some careful definitions and notation. The ideas involved in Definitions 2.1–2.3 are different from those in the previous section. However, we will see that there is a key connection (called **Fubini's theorem**) between the notion of an **iterated integral** discussed in §5.1 and that of a **double integral,** which will be described in Definition 2.3.

The Integral over a Rectangle

We also denote a (closed) rectangle

$$R = \{(x, y) \in \mathbf{R}^2 \mid a \le x \le b, c \le y \le d\}$$

by $[a, b] \times [c, d]$. This notation is intended to be analogous to the notation for a closed interval.

■ Definition 2.1 Given a closed rectangle $R = [a, b] \times [c, d]$, a **partition of R of order n** consists of two collections of **partition points** that break up R into a union of n^2 subrectangles. More specifically, for $i, j = 0, \ldots, n$, we introduce the collections $\{x_i\}$ and $\{y_j\}$, so that

$$a = x_0 < x_1 < \cdots < x_{i-1} < x_i < \cdots < x_n = b,$$

and

$$c = y_0 < y_1 < \cdots < y_{j-1} < y_j < \cdots < y_n = d.$$

Let $\Delta x_i = x_i - x_{i-1}$ (for $i = 1, \ldots, n$) and $\Delta y_j = y_j - y_{j-1}$ (for $j = 1, \ldots, n$). Note that Δx_i and Δy_j are just the width and height (respectively) of the ijth subrectangle (reading left to right and bottom to top) of the partition.

An example of a partitioned rectangle is shown in Figure 5.9. We do not assume that the partition is **regular** (i.e., that all the subrectangles have the same dimensions).

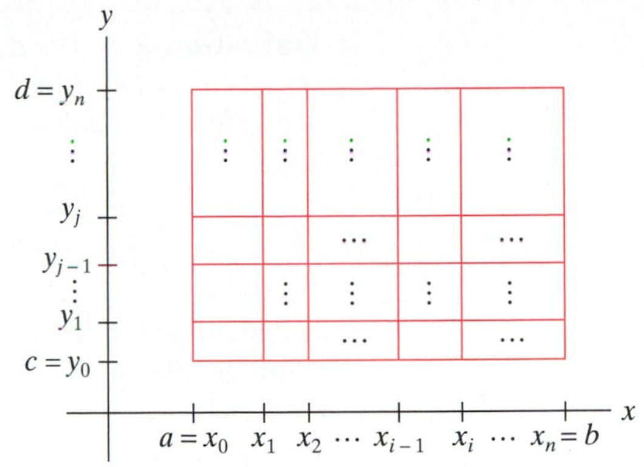

Figure 5.9 A partition of the rectangle $[a, b] \times [c, d]$.

■ **Definition 2.2** Suppose that f is any function defined on $R = [a, b] \times [c, d]$ and partition R in some way. Let $\mathbf{c}_{ij}$ be any point in the subrectangle

$$R_{ij} = [x_{i-1}, x_i] \times [y_{j-1}, y_j] \quad (i, j = 1, \ldots, n).$$

Then the quantity

$$S = \sum_{i,j=1}^{n} f(\mathbf{c}_{ij}) \Delta A_{ij},$$

where $\Delta A_{ij} = \Delta x_i \Delta y_j$ is the area of R_{ij}, is called a **Riemann sum** of f on R corresponding to the partition.

The Riemann sum

$$S = \sum_{i,j} f(\mathbf{c}_{ij}) \, \Delta A_{ij}$$

depends on the function f, the choice of partition, and the choice of the "test point" $\mathbf{c}_{ij}$ in each subrectangle R_{ij} of the partition. The Riemann sum itself is just a *weighted sum of areas* ΔA_{ij} of subrectangles of the original rectangle R, the weighting being given by the value $f(\mathbf{c}_{ij})$.

If f happens to be nonnegative on R, then, for $i, j = 1, \ldots, n$, the individual terms $f(\mathbf{c}_{ij}) \, \Delta A_{ij}$ in S may be considered to be volumes of boxes having base area $\Delta x_i \Delta y_j$ and height $f(\mathbf{c}_{ij})$. Therefore, S can be considered to be an approximation to the volume under the graph of f over R, as suggested by Figure 5.10. If f is not necessarily nonnegative, then the Riemann sum S is a *signed* sum of such volumes (because, with $f(\mathbf{c}_{ij}) < 0$, the term $f(\mathbf{c}_{ij})\Delta A_{ij}$ is the negative of the volume of the appropriate box—see Figure 5.11).

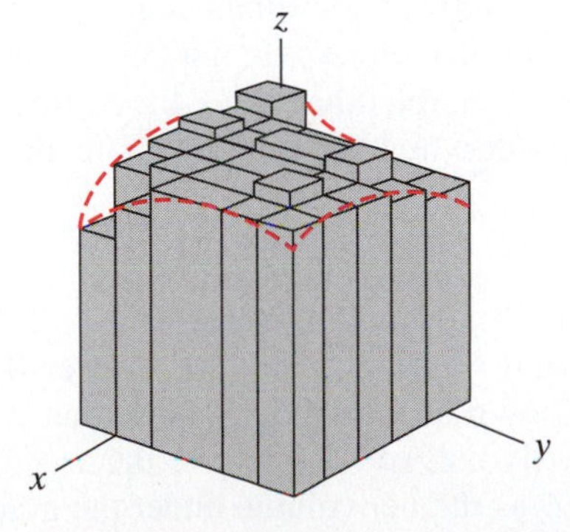

Figure 5.10 The volume under the graph of f is approximated by the Riemann sum.

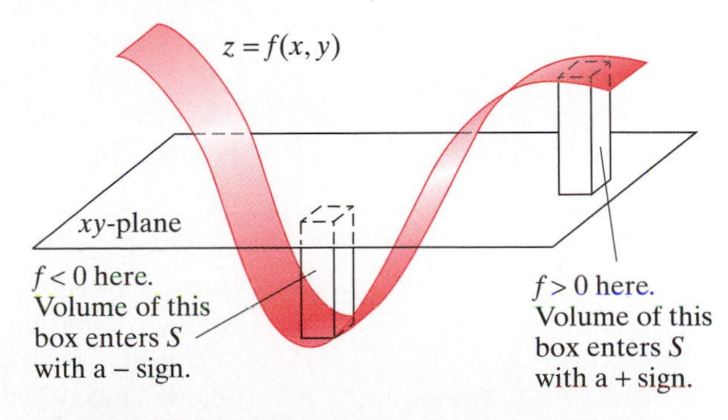

Figure 5.11 The Riemann sum as a signed sum of volumes of boxes.

> ■ **Definition 2.3** The **double integral** of f on R, denoted by $\iint_R f\,dA$ (or by $\iint_R f(x, y)\,dA$ or by $\iint_R f(x, y)\,dx\,dy$), is the limit of the Riemann sum S as the dimensions Δx_i and Δy_j of the subrectangles R_{ij} all approach zero, that is,
>
> $$\iint_R f\,dA = \lim_{\text{all } \Delta x_i, \Delta y_j \to 0} \sum_{i,j=1}^{n} f(\mathbf{c}_{ij})\Delta x_i \Delta y_j,$$
>
> provided, of course, that this limit exists. When $\iint_R f\,dA$ exists, we say that f is **integrable** on R.

Figure 5.12 If A_1, A_2, A_3 represent the values of the shaded areas, then $\int_a^b f(x)\,dx = A_1 - A_2 + A_3$.

The crucial idea to remember—indeed, the defining idea—is that the integral $\iint_R f\,dA$ is a limit of Riemann sums S, for this concept is what is needed to properly apply double integrals to physical situations.

From a geometric point of view, just as the single-variable definite integral $\int_a^b f(x)\,dx$ can be used to compute the "net area" under the graph of the curve $y = f(x)$ (as in Figure 5.12), the double integral $\iint_R f\,dA$ can be used to compute the "net volume" under the graph of $z = f(x, y)$ (as in Figure 5.13).

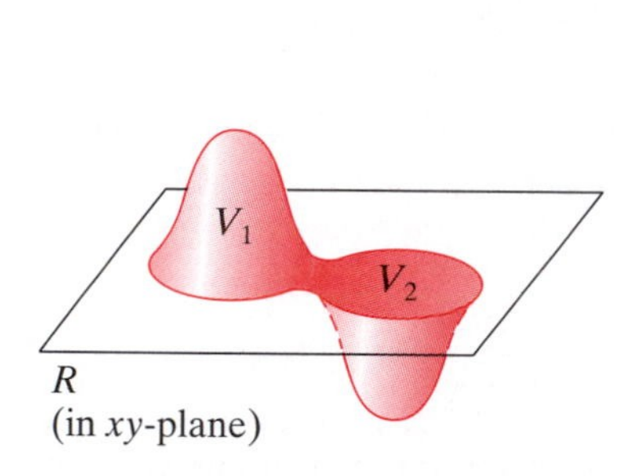

Figure 5.13 If V_1, V_2 represent the volumes of the shaded regions, then $\iint_R f(x, y)\,dA = V_1 - V_2$.

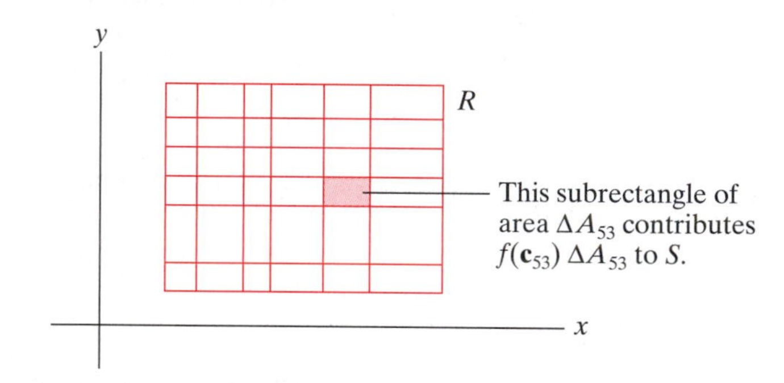

Figure 5.14 $S = \sum_{i,j} f(\mathbf{c}_{ij})\Delta A_{ij}$.

Another way to view the double integral $\iint_R f\,dA$ is somewhat less geometric, but is more in keeping with the notion of the integral as the limit of Riemann sums and provides a perspective that generalizes to triple integrals of functions of three variables. Instead of visualizing the graph of $z = f(x, y)$ as a surface and $S = \sum_{i,j=1}^{n} f(\mathbf{c}_{ij})\Delta x_i \Delta y_j$ as a (signed) sum of volumes of boxes related to the graph, consider S as a weighted sum of *areas* and the integral $\iint_R f\,dA$ as the limiting value of such weighted sums as the dimensions of all the subrectangles approach zero. With this point of view, we do not depict the integrand f when we try to visualize the integral. On the other hand, the distinction between the integrand and the rectangle R over which we integrate is made clearer. (See Figure 5.14.)

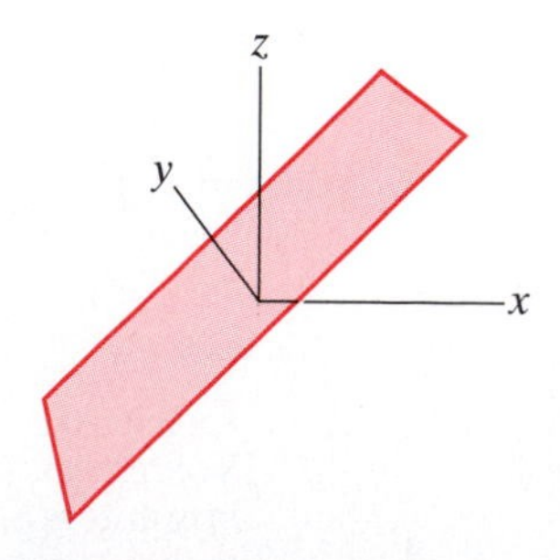

Figure 5.15 The graph of $z = x$ of Example 1.

EXAMPLE 1 We determine the value of $\iint_R x\,dA$, where $R = [-2, 2] \times [-1, 3]$. Here the integrand $f(x, y) = x$ and if we graph $z = f(x, y)$ over R, we see that we have a portion of a plane, as shown in Figure 5.15. Note that the portion of the plane is positioned so that exactly half of it lies above the xy-plane and half below. Thus, if we regard $\iint_R x\,dA$ as the net volume under the graph of $z = x$, then we conclude that $\iint_R x\,dA$ (if it exists) must be zero.

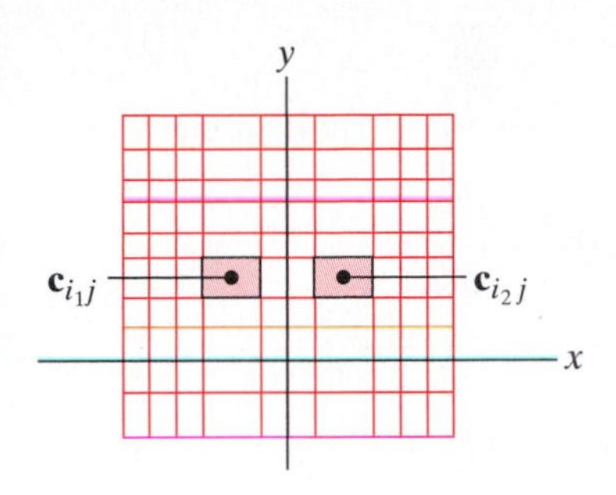

Figure 5.16 The two subrectangles $R_{i_1 j}$ and $R_{i_2 j}$ are symmetrically placed with respect to the y-axis. The corresponding test points $\mathbf{c}_{i_1 j}$ and $\mathbf{c}_{i_2 j}$ are chosen so that they have the same y-coordinates and opposite x-coordinates.

On the other hand, we need not resort to visualization in three dimensions. Consider a Riemann sum corresponding to $\iint_R x\, dA$ obtained by partitioning $R = [-2, 2] \times [-1, 3]$ symmetrically with respect to the y-axis and by choosing the "test points" $\mathbf{c}_{ij}$ symmetrically also. (See Figure 5.16.) It follows that the value of

$$S = \sum f(\mathbf{c}_{ij})\Delta A_{ij} = \sum x_{ij}\Delta A_{ij}$$

(where x_{ij} denotes the x-coordinate of $\mathbf{c}_{ij}$) must be zero since the terms of the sum cancel in pairs. Furthermore, we can arrange things so that, as we shrink the dimensions of the subrectangles to zero (as we must do to get at the integral itself), we preserve all the symmetry just described. Hence, the limit under these restrictions will be zero, and thus, the overall limit (where we do not impose such symmetry restrictions on the Riemann sum), if it exists at all, must be zero as well. ◆

Example 1 points out fundamental difficulties with Definition 2.3, namely, that we never did determine whether $\iint_R f\, dA$ really exists. To do this, we would have to be able to calculate the limit of Riemann sums of f over *all possible* partitions of R by using *all possible* choices for the test points $\mathbf{c}_{ij}$, a practically impossible task. Fortunately, the following result (which we will not prove) provides an easy criterion for integrability:

■ Theorem 2.4 If f is continuous on the closed rectangle R, then $\iint_R f\, dA$ exists.

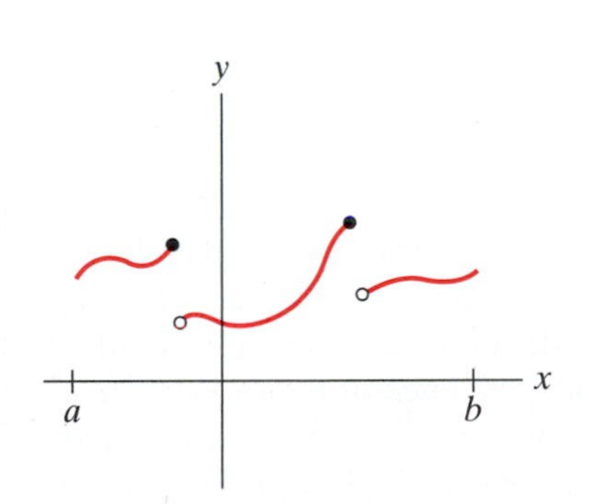

Figure 5.17 The graph of a piecewise continuous function.

In Example 1, $f(x, y) = x$ is a continuous function and hence integrable by Theorem 2.4. The symmetry arguments used in the example then show that $\iint_R x\, dA = 0$.

Continuous functions are not the only examples of integrable functions. In the case of a function of a single variable, piecewise continuous functions are also integrable. (Recall that a function $f(x)$ is **piecewise continuous** on the closed interval $[a, b]$ if f is bounded on $[a, b]$ and has at most finitely many points of discontinuity on the interior of $[a, b]$. Its graph, therefore, consists of finitely many continuous "chunks" as shown in Figure 5.17.) For a function of two variables, there is the following result, which generalizes Theorem 2.4.

■ Theorem 2.5 If f is bounded on R and if the set of discontinuities of f on R has zero area, then $\iint_R f\, dA$ exists.

A function f satisfying the hypotheses of Theorem 2.5 has a graph that looks roughly like the one in Figure 5.18. Theorem 2.5 is the most general sufficient condition for integrability that we will consider. It is of particular use to us when we define the double integral of a function over an arbitrary region in the plane.

Although Theorems 2.4 and 2.5 make it relatively straightforward to check that a given integral exists, they do little to help provide the numerical value of the integral. To mechanize the evaluation of double integrals, we will use the following result:

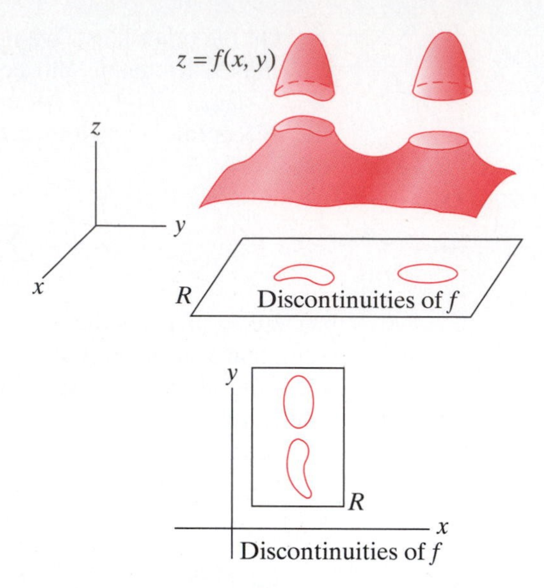

Figure 5.18 The graph of an integrable function.

■ **Theorem 2.6** (FUBINI'S THEOREM) Let f be bounded on $R = [a, b] \times [c, d]$ and assume that the set S of discontinuities of f on R has zero area. If every line parallel to the coordinate axes meets S in at most finitely many points, then

$$\iint_R f\, dA = \int_a^b \int_c^d f(x, y)\, dy\, dx = \int_c^d \int_a^b f(x, y)\, dx\, dy.$$

Fubini's theorem demonstrates that under certain assumptions the *double integral* over a rectangle (i.e., the limit of Riemann sums) can be calculated by using *iterated integrals* and, moreover, that the order of integration for the iterated integral does not matter. We remark that the independence of the order of integration depends strongly on the fact that the region of integration is rectangular; it will not generalize to more arbitrary regions in such a simple way. (A proof of Theorem 2.6 is given in the addendum to this section.)

EXAMPLE 2 We revisit $\iint_R x\, dA$ in Example 1, where $R = [-2, 2] \times [-1, 3]$. By Theorem 2.6, we know that $\iint_R x\, dA$ exists and by Fubini's theorem, we calculate

$$\iint_R x\, dA = \int_{-2}^{2} \int_{-1}^{3} x\, dy\, dx = \int_{-2}^{2} \left(xy \Big|_{y=-1}^{y=3} \right) dx$$

$$= \int_{-2}^{2} x(3 - (-1))\, dx = \int_{-2}^{2} 4x\, dx = 2x^2\Big|_{-2}^{2} = 8 - 8 = 0,$$

which checks. Furthermore, we also have

$$\iint_R x\, dA = \int_{-1}^{3} \int_{-2}^{2} x\, dx\, dy = \int_{-1}^{3} \tfrac{1}{2}x^2 \Big|_{x=-2}^{x=2} dy = \textstyle\int_{-1}^{3} (2 - 2)\, dy = 0.$$ ◆

Proposition 2.7 (PROPERTIES OF THE INTEGRAL) Suppose that f and g are both integrable on the closed rectangle R. Then the following properties hold:

1. $f+g$ is also integrable on R and
$$\iint_R (f+g)\,dA = \iint_R f\,dA + \iint_R g\,dA.$$
2. cf is also integrable on R, where $c \in \mathbf{R}$ is any constant, and
$$\iint_R cf\,dA = c\iint_R f\,dA.$$
3. If $f(x,y) \le g(x,y)$ for all $(x,y) \in R$, then
$$\iint_R f(x,y)\,dA \le \iint_R g(x,y)\,dA.$$
4. $|f|$ is also integrable on R and
$$\left|\iint_R f\,dA\right| \le \iint_R |f|\,dA.$$

Properties 1 and 2 are called the **linearity** properties of the double integral. They can be proved by considering the appropriate Riemann sums and taking limits. For example, to prove property 1, note that the Riemann sum whose limit is $\iint_R (f+g)\,dA$ is

$$\begin{aligned}\sum_{i,j=1}^{n}(f+g)(\mathbf{c}_{ij})\Delta A_{ij} &= \sum_{i,j=1}^{n}\left(f(\mathbf{c}_{ij})+g(\mathbf{c}_{ij})\right)\Delta A_{ij}\\ &= \sum_{i,j=1}^{n} f(\mathbf{c}_{ij})\Delta A_{ij} + \sum_{i,j=1}^{n} g(\mathbf{c}_{ij})\Delta A_{ij}\\ &\to \iint_R f\,dA + \iint_R g\,dA.\end{aligned}$$

Property 3 (known as **monotonicity**) and property 4 can also be proved using Riemann sums. For property 4, one needs to use the fact that

$$\left|\sum_{k=1}^{n} a_k\right| \le \sum_{k=1}^{n}|a_k|.$$

Double Integrals over General Regions in the Plane

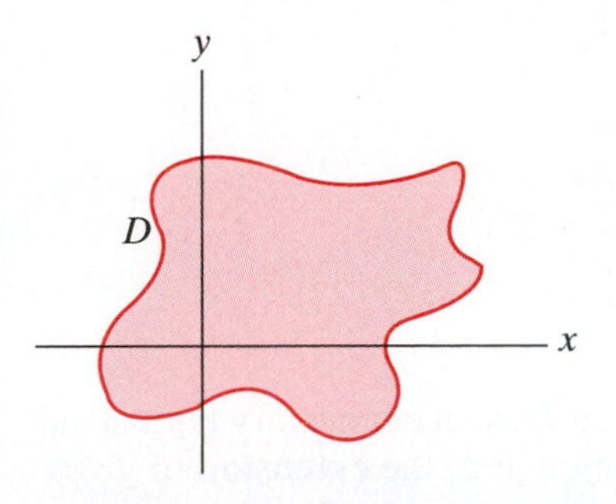

Figure 5.19 A bounded region D in the plane.

Our next step is to understand how to define the integral of a function over an arbitrary bounded region D in the plane. Ideally, we would like to give a precise definition of $\iint_D f\,dA$, where D is the amoeba-shaped blob shown in Figure 5.19 and where f is bounded on D. In keeping with the definition of the integral over a rectangle, $\iint_D f\,dA$ should be a limit of some type of Riemann sum and should represent the net volume under the graph of f over D. Unfortunately, the technicalities involved in making such a direct approach work are prohibitive. Instead, we shall consider only certain special regions (rather than entirely arbitrary ones), and we shall assume that the integrand f is continuous over the region of integration (which will allow us to use what we already know about integrals over rectangles). Although this approach will not provide us with a completely general definition, it is sufficient for essentially all the practical situations we will encounter.

To begin, we define the types of **elementary regions** we wish to consider.

■ **Definition 2.8** We say that D is an **elementary region** in the plane if it can be described as a subset of $\mathbf{R}^2$ of one of the following three types:

Type 1 (see Figure 5.20):

$$D = \{(x, y) \mid \gamma(x) \leq y \leq \delta(x), a \leq x \leq b\},$$

where γ and δ are continuous on $[a, b]$.

Type 2 (see Figure 5.21):

$$D = \{(x, y) \mid \alpha(y) \leq x \leq \beta(y), c \leq y \leq d\},$$

where α and β are continuous on $[c, d]$.

Type 3 D is of both type 1 and type 2.

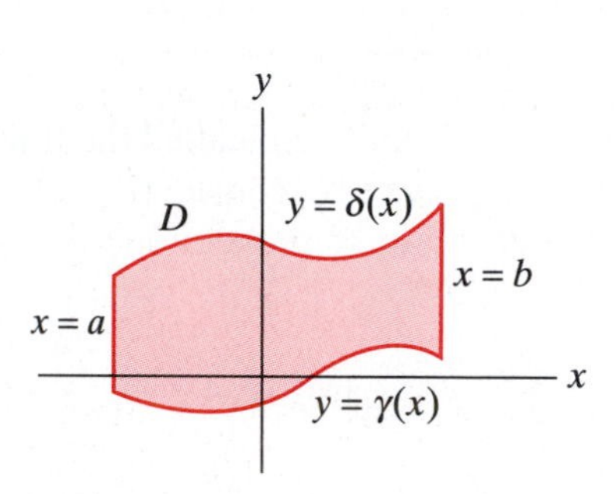

Figure 5.20 A type 1 elementary region.

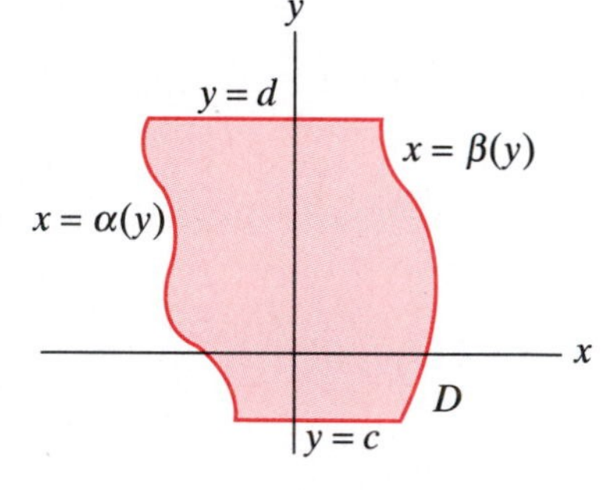

Figure 5.21 A type 2 elementary region.

Thus, a type 1 elementary region D has a boundary (denoted ∂D) consisting of straight segments (possibly single points) on the left and on the right, and graphs of continuous functions of x on the top and on the bottom. A type 2 elementary region has a boundary that is straight on the top and bottom and consists of graphs of continuous functions of y on the left and right.

EXAMPLE 3 The unit disk, shown in Figure 5.22, is an example of a type 3 elementary region. It is a type 1 region since

$$D = \left\{(x, y) \mid -\sqrt{1 - x^2} \leq y \leq \sqrt{1 - x^2}, -1 \leq x \leq 1\right\}.$$

(See Figure 5.23.) It is also a type 2 region since

$$D = \left\{(x, y) \mid -\sqrt{1 - y^2} \leq x \leq \sqrt{1 - y^2}, -1 \leq y \leq 1\right\}.$$

(See Figure 5.24.) ◆

Now we are ready to define $\iint_D f\, dA$, where D is an elementary region and f is continuous on D. We construct a new function f^{ext}, the **extension** of f, by

$$f^{\text{ext}}(x, y) = \begin{cases} f(x, y) & \text{if } (x, y) \in D \\ 0 & \text{if } (x, y) \notin D \end{cases}.$$

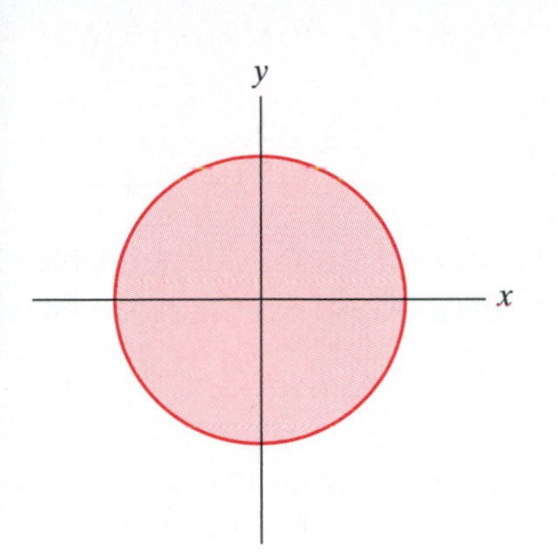

Figure 5.22 The unit disk $D = \{(x, y) \mid x^2 + y^2 \leq 1\}$ is a type 3 region.

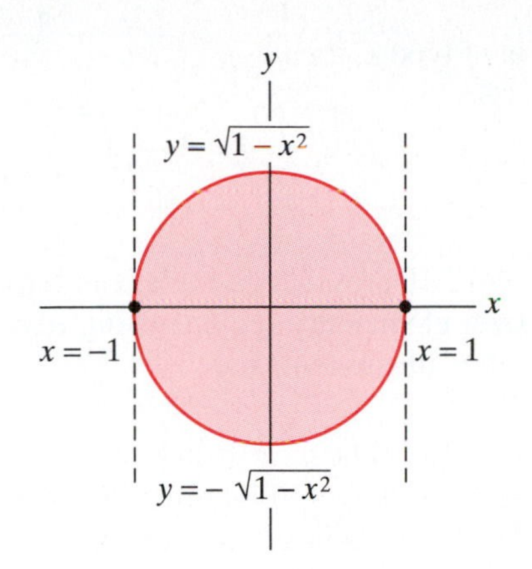

Figure 5.23 The unit disk D as a type 1 region.

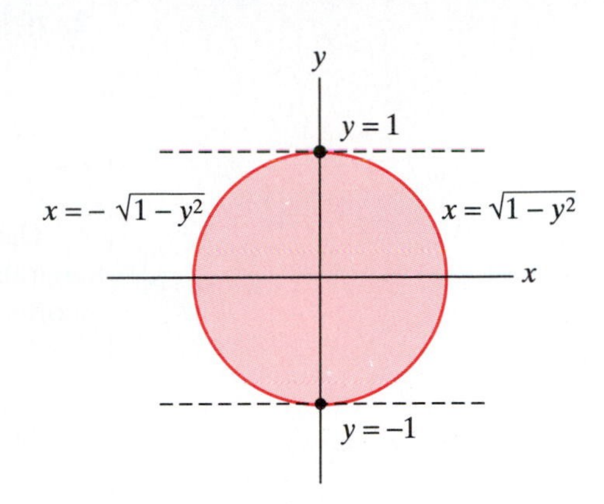

Figure 5.24 The unit disk D as a type 2 region.

Note that, in general, f^{ext} will not be continuous, but the discontinuities of f^{ext} will all be contained in ∂D, which has no area. Hence, by Theorem 2.5, f^{ext} is integrable on any closed rectangle R that contains D. (See Figure 5.25.)

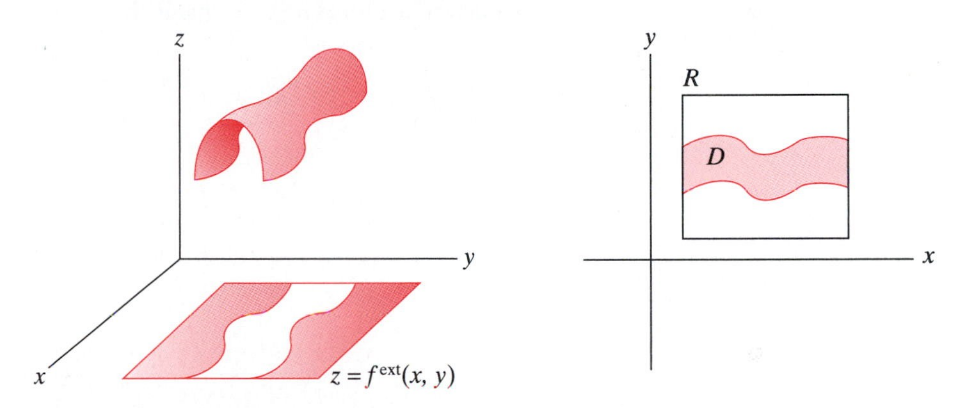

Figure 5.25 The graph of $z = f^{\text{ext}}(x, y)$.

Definition 2.9 Under the previous assumptions and notation, if R is any rectangle that contains D, we define

$$\iint_D f\, dA \quad \text{to be} \quad \iint_R f^{\text{ext}}\, dA.$$

Note that Definition 2.9 implicitly assumes that the choice of the rectangle R that contains D does not affect the value of $\iint_R f^{\text{ext}}\, dA$. This is almost obvious, but still should be proved. We shall not do so directly, but instead establish the following key result:

Theorem 2.10 Let D be an elementary region in $\mathbf{R}^2$ and f a continuous function on D.

1. If D is of type 1 (as described in Definition 2.8), then

$$\iint_D f\, dA = \int_a^b \int_{\gamma(x)}^{\delta(x)} f(x, y)\, dy\, dx.$$

2. If D is of type 2, then

$$\iint_D f\,dA = \int_c^d \int_{\alpha(y)}^{\beta(y)} f(x, y)\,dx\,dy.$$

Theorem 2.10 provides an explicit and straightforward way to evaluate double integrals over elementary regions using iterated integrals. Before we prove the theorem, let us illustrate its use.

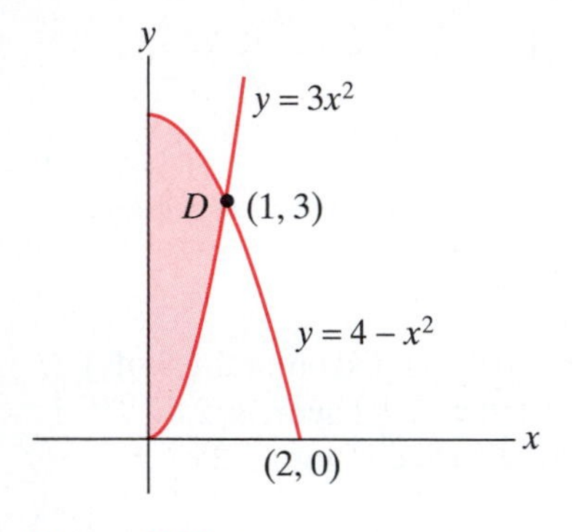

Figure 5.26 The domain of f of Example 4.

EXAMPLE 4 Let D be the region bounded by the parabolas $y = 3x^2$, $y = 4 - x^2$ and the y-axis as shown in Figure 5.26. (Note that the parabolas intersect at the point $(1, 3)$.) Since D is a type 1 elementary region, we may use Theorem 2.10 with $f(x, y) = x^2 y$ to find that

$$\iint_D x^2 y\,dA = \int_0^1 \int_{3x^2}^{4-x^2} x^2 y\,dy\,dx.$$

The limits for the first (inside) integration come from the y-values of the top and bottom boundary curves of D. The limits for second (outside) integration are the constant x-values that correspond to the straight left and right sides of D. The evaluation itself is fairly mechanical:

$$\begin{aligned}
\int_0^1 \int_{3x^2}^{4-x^2} x^2 y\,dy\,dx &= \int_0^1 \left(\frac{x^2 y^2}{2}\right)\Bigg|_{y=3x^2}^{y=4-x^2} dx \\
&= \int_0^1 \frac{x^2}{2}\left((4 - x^2)^2 - (3x^2)^2\right) dx \\
&= \frac{1}{2}\int_0^1 x^2\,(16 - 8x^2 + x^4 - 9x^4)\,dx \\
&= \int_0^1 (8x^2 - 4x^4 - 4x^6)\,dx = \tfrac{8}{3} - \tfrac{4}{5} - \tfrac{4}{7} = \tfrac{136}{105}.
\end{aligned}$$

Note that after the y-integration and evaluation, what remains is a single definite integral in x. The result of calculating this x-integral is, of course, a number. Such a situation where the number of variables appearing in the integral decreases with each integration should always be the case. ◆

Proof of Theorem 2.10 For part 1, we may take D to be described as

$$D = \{(x, y) \mid \gamma(x) \le y \le \delta(x),\ a \le x \le b\}.$$

We have, by Definition 2.9, that

$$\iint_D f\,dA = \iint_R f^{\text{ext}}\,dA,$$

where R is any rectangle containing D. Let $R = [a', b'] \times [c', d']$, where $a' \le a$, $b' \ge b$, and $c' \le \gamma(x)$, $d' \ge \delta(x)$ for all x in $[a, b]$. That is, we have the situation depicted in Figure 5.27. Since f^{ext} is zero outside of the subrectangle $R_2 = [a, b] \times [c', d']$,

$$\iint_R f^{\text{ext}}\,dA = \iint_{R_2} f^{\text{ext}}\,dA = \int_a^b \int_{c'}^{d'} f^{\text{ext}}(x, y)\,dy\,dx$$

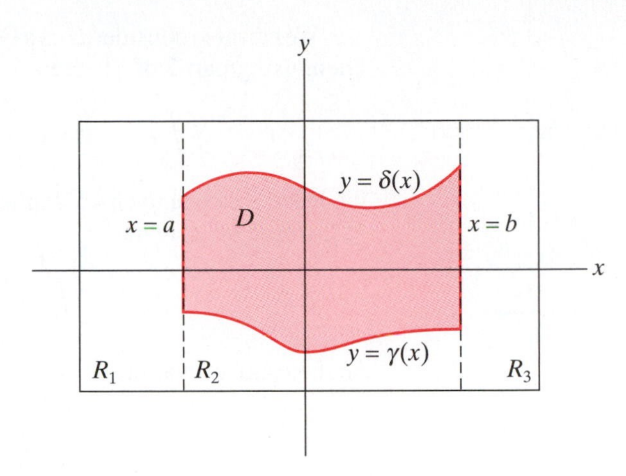

Figure 5.27 The region R is the union of R_1, R_2, and R_3.

by Fubini's theorem. For a fixed value of x between a and b, consider the y-integral $\int_{c'}^{d'} f^{\text{ext}}(x, y)\,dy$. Since $f^{\text{ext}}(x, y) = 0$ unless $\gamma(x) \leq y \leq \delta(x)$ (in which case $f^{\text{ext}}(x, y) = f(x, y)$),

$$\int_{c'}^{d'} f^{\text{ext}}(x, y)\,dy = \int_{\gamma(x)}^{\delta(x)} f(x, y)\,dy,$$

and so

$$\iint_D f(x, y)\,dA = \iint_R f^{\text{ext}}\,dA = \int_a^b \int_{c'}^{d'} f^{\text{ext}}(x, y)\,dy\,dx$$
$$= \int_a^b \int_{\gamma(x)}^{\delta(x)} f(x, y)\,dy\,dx,$$

as desired.

The proof of part 2 is very similar. ■

We continue analyzing examples of double integral calculations.

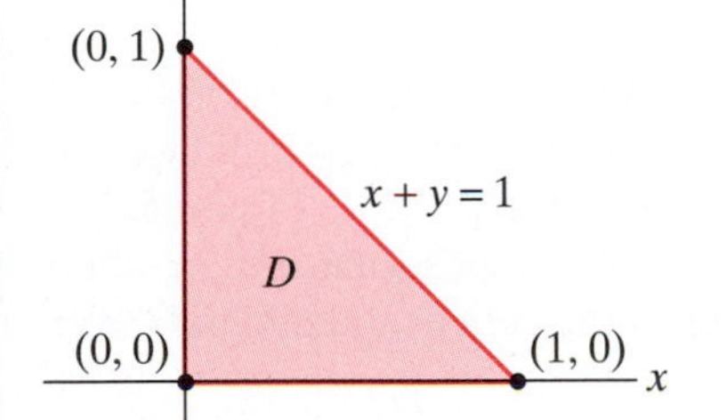

Figure 5.28 The region D of Example 5.

EXAMPLE 5 Let D be the region shown in Figure 5.28 having a triangular border. Consider $\iint_D (1 - x - y)\,dA$. Note that D is a type 3 elementary region, so there should be two ways to evaluate the double integral.

Considering D as a type 1 elementary region (see Figure 5.29), we may apply part 1 of Theorem 2.10 so that

$$\iint_D (1 - x - y)\,dA = \int_0^1 \int_0^{1-x} (1 - x - y)\,dy\,dx$$
$$= \int_0^1 \left(y - xy - \frac{y^2}{2} \right)\Big|_{y=0}^{y=1-x} dx$$
$$= \int_0^1 \left((1 - x) - x(1 - x) - \frac{(1 - x)^2}{2} \right) dx$$
$$= \int_0^1 \frac{(1 - x)^2}{2}\,dx = -\tfrac{1}{6}(1 - x)^3\big|_0^1 = \tfrac{1}{6}.$$

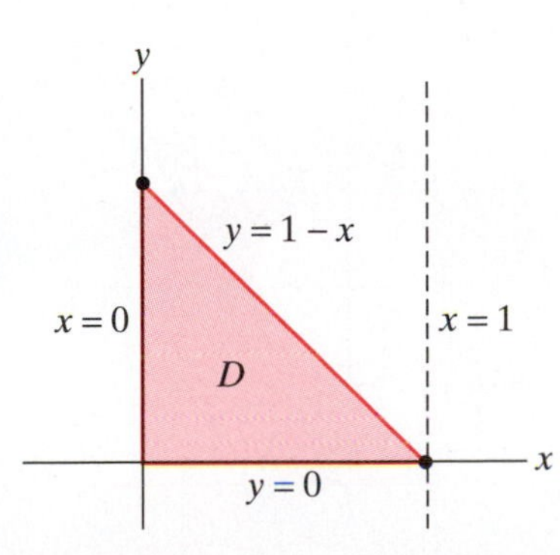

Figure 5.29 The region D of Example 5 as a type 1 region.

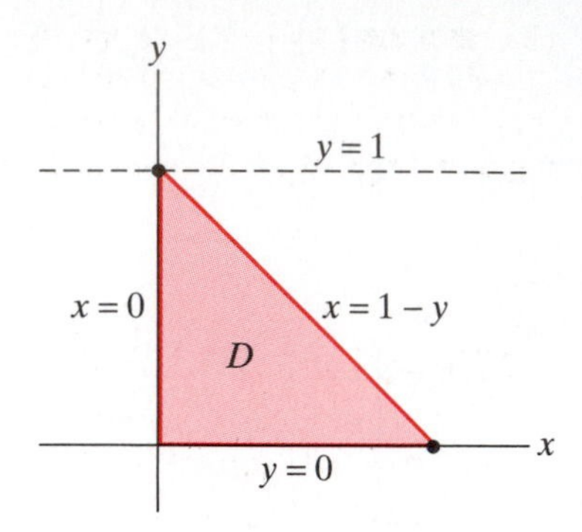

Figure 5.30 The region D of Example 5 as a type 2 region.

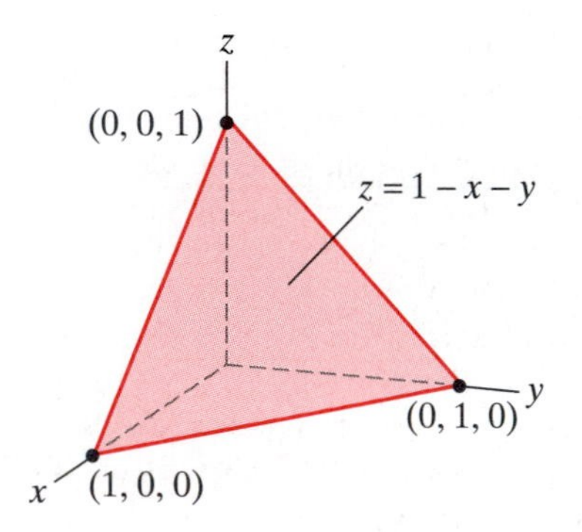

Figure 5.31 The double integral of Example 5 represents the volume of the tetrahedron.

We can also consider D as a type 2 elementary region, as shown in Figure 5.30. Then, using part 2 of Theorem 2.10, we obtain

$$\iint_D (1 - x - y)\, dA = \int_0^1 \int_0^{1-y} (1 - x - y)\, dx\, dy.$$

We leave it to you to check explicitly that this iterated integral also has a value of $\frac{1}{6}$. Instead, we note that

$$\int_0^1 \int_0^{1-x} (1 - x - y)\, dy\, dx$$

can be transformed into

$$\int_0^1 \int_0^{1-y} (1 - x - y)\, dx\, dy$$

by exchanging the roles of x and y. Hence, the two integrals must have the same value. In any case, the double integral

$$\iint_D (1 - x - y)\, dA$$

represents the volume under the graph of $z = 1 - x - y$ over the triangular region D. If we picture the situation in $\mathbf{R}^3$, as in Figure 5.31, we see that the double integral represents the volume of a tetrahedron. ◆

Of course not all regions in the plane are elementary, including even some relatively simple ones. To integrate continuous functions over such regions, the best advice is to attempt to subdivide the region into finitely many of elementary type.

EXAMPLE 6 Let D be the annular region between the two concentric circles of radii 1 and 2 shown in Figure 5.32. Then D is *not* an elementary region, but we can break D up into four subregions that are of elementary type. (See Figure 5.33.) If $f(x, y)$ is any function of two variables that is continuous (hence integrable) on D, then we may compute the double integral as the sum of the integrals over the subregions. That is,

$$\iint_D f\, dA = \iint_{D_1} f\, dA + \iint_{D_2} f\, dA + \iint_{D_3} f\, dA + \iint_{D_4} f\, dA.$$

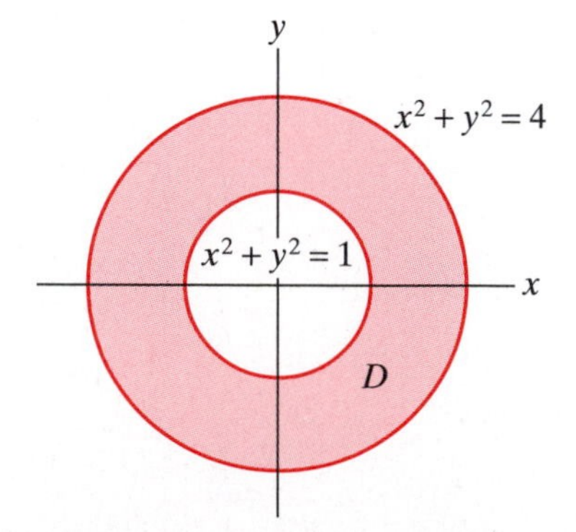

Figure 5.32 The region D of Example 6.

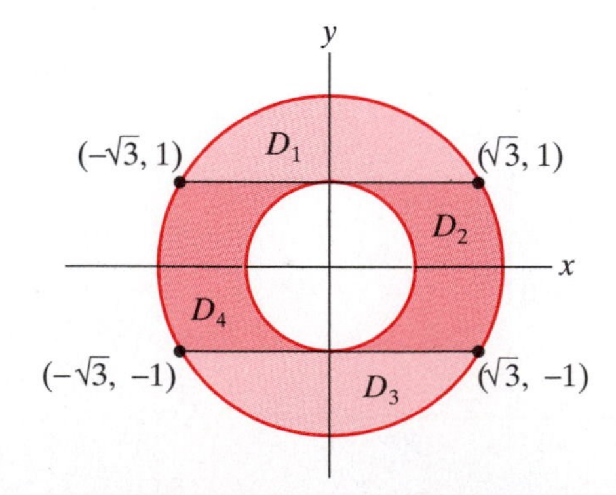

Figure 5.33 The region D of Example 6 subdivided into four elementary regions.

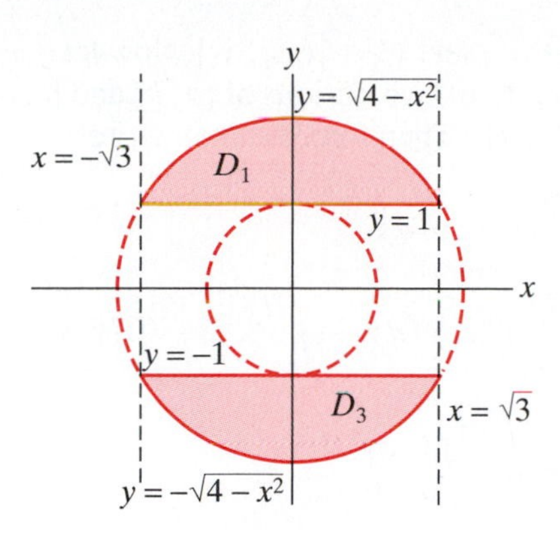

Figure 5.34 The subregions D_1 and D_3 of Example 6 are of type 1.

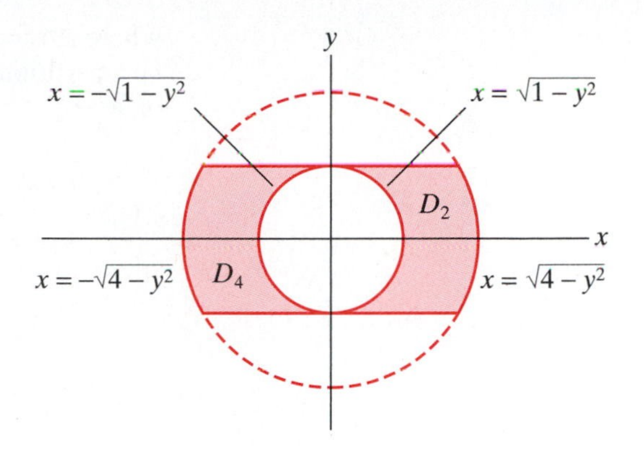

Figure 5.35 The subregions D_2 and D_4 of Example 6 are of type 2.

For the type 1 subregions, we have the set-up shown in Figure 5.34:

$$\iint_{D_1} f\,dA = \int_{-\sqrt{3}}^{\sqrt{3}} \int_{1}^{\sqrt{4-x^2}} f(x, y)\,dy\,dx$$

and

$$\iint_{D_3} f\,dA = \int_{-\sqrt{3}}^{\sqrt{3}} \int_{-\sqrt{4-x^2}}^{-1} f(x, y)\,dy\,dx.$$

For the type 2 subregions, we use the set-up shown in Figure 5.35:

$$\iint_{D_2} f\,dA = \int_{-1}^{1} \int_{\sqrt{1-y^2}}^{\sqrt{4-y^2}} f(x, y)\,dx\,dy$$

and

$$\iint_{D_4} f\,dA = \int_{-1}^{1} \int_{-\sqrt{4-y^2}}^{-\sqrt{1-y^2}} f(x, y)\,dx\,dy.$$

The difficulty of evaluating each of the preceding four iterated integrals then depends on the complexity of the integrand. ◆

Addendum: Proof of Theorem 2.6

Step 1. First we establish Theorem 2.6 in the case where f is continuous on $R = [a, b] \times [c, d]$. By Theorem 2.4, we know that $\iint_R f\,dA$ exists. Let F be the single-variable function defined by

$$F(x) = \int_c^d f(x, y)\,dy.$$

(Note: Since f is continuous on R, the partial function in y is continuous on $[c, d]$. Hence, $\int_c^d f(x, y)\,dy$ exists for every x in $[a, b]$.) We show that

$$\int_a^b F(x)\,dx = \int_a^b \left[\int_c^d f(x, y)\,dy\right] dx = \iint_R f\,dA.$$

Let $a = x_0 < x_1 < \cdots < x_n = b$ be any partition of $[a, b]$. Then a general Riemann sum that approximates $\int_a^b F(x)\,dx$ is

$$\sum_{i=1}^{n} F(x_i^*)\Delta x_i, \tag{1}$$

where $\Delta x_i = x_i - x_{i-1}$ and $x_i^* \in [x_{i-1}, x_i]$. Now let $c = y_0 < y_1 < \cdots < y_n = d$ be a partition of $[c, d]$. (The partitions of $[a, b]$ and $[c, d]$ together give a partition of $R = [a, b] \times [c, d]$.) Therefore, we may write

$$\begin{aligned} F(x) &= \int_c^d f(x, y)\,dy \\ &= \int_c^{y_1} f(x, y)\,dy + \int_{y_1}^{y_2} f(x, y)\,dy + \cdots + \int_{y_{n-1}}^{d} f(x, y)\,dy \\ &= \sum_{j=1}^{n} \int_{y_{j-1}}^{y_j} f(x, y)\,dy. \end{aligned}$$

By the mean value theorem for integrals[1], on each subinterval $[y_{j-1}, y_j]$ there exists a number y_j^* such that

$$\int_{y_{j-1}}^{y_j} f(x, y)\,dy = (y_j - y_{j-1}) f(x, y_j^*) = f(x, y_j^*)\Delta y_j.$$

The choice of y_j^* in general depends on x, so henceforth, we will write $y_j^*(x)$ for y_j^*. Consequently,

$$F(x) = \sum_{j=1}^{n} f(x, y_j^*(x))\Delta y_j,$$

and the Riemann sum (1) may be written as

$$\sum_{i=1}^{n} F(x_i^*)\Delta x_i = \sum_{i=1}^{n}\left\{\sum_{j=1}^{n} f(x_i^*, y_j^*(x_i^*))\Delta y_j\right\}\Delta x_i = \sum_{i,j=1}^{n} f(\mathbf{c}_{ij})\Delta x_i \Delta y_j,$$

where $\mathbf{c}_{ij} = (x_i^*, y_j^*(x_i^*))$. Note that $\mathbf{c}_{ij} \in [x_{i-1}, x_i] \times [y_{j-1}, y_j]$. (See Figure 5.36.)

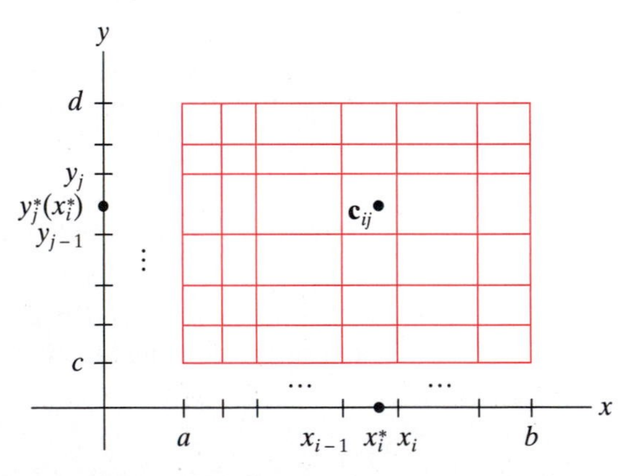

Figure 5.36 The point $\mathbf{c}_{ij} = (x_i^*, y_j^*(x_i^*))$ used in the proof of Theorem 2.6.

We have thus shown that given any partition of $[a, b]$, we can associate a suitable partition of $R = [a, b] \times [c, d]$ such that the Riemann sum (1) that approximates $\int_a^b F(x)\,dx$ is equal to a Riemann sum (namely, $\sum_{i,j} f(\mathbf{c}_{ij})\Delta x_i \Delta y_j$)

[1]The mean value theorem for integrals says that, if g is continuous on $[a, b]$, then there is some number c with $a \le c \le b$ such that $\int_a^b g(x)\,dx = (b - a)g(c)$.

that approximates $\iint_R f\,dA$. Since f is continuous, we know that

$$\sum_{i,j} f(\mathbf{c}_{ij})\Delta x_i \Delta y_j \quad \text{approaches} \quad \iint_R f\,dA$$

as Δx_i and Δy_j tend to zero. Hence,

$$\int_a^b F(x)\,dx = \iint_R f\,dA.$$

By exchanging the roles of x and y in the foregoing argument, we can show that

$$\iint_R f\,dA = \int_c^d \int_a^b f(x, y)\,dx\,dy.$$

Step 2. Now we prove the general case of Theorem 2.6 (i.e., the case that f has discontinuities in $R = [a, b] \times [c, d]$). By hypothesis, the set S of discontinuities of f in R are such that every vertical line meets S in at most finitely many points. Thus, the partial function in y of $f(x, y)$ is continuous throughout $[c, d]$, except possibly at finitely many points. (In other words, the partial function is piecewise continuous.) Then, because f is bounded,

$$F(x) = \int_c^d f(x, y)\,dy$$

exists.

Now we proceed as in Step 1. That is, we begin with a partition of $[a, b]$ into n subintervals and a corresponding Riemann sum

$$\sum_{i=1}^n F(x_i^*)\Delta x_i.$$

Next, we partition $[c, d]$ into n subintervals. Hence,

$$F(x_i^*) = \int_c^d f(x_i^*, y)\,dy = \sum_{j=1}^n \int_{y_{j-1}}^{y_j} f(x_i^*, y)\,dy. \tag{2}$$

As in Step 1, the partitions of $[a, b]$ and $[c, d]$ combine to give a partition of R.

Write R as $R_1 \cup R_2$, where R_1 is the union of all subrectangles

$$R_{ij} = [x_{i-1}, x_i] \times [y_{j-1}, y_j]$$

that intersect S and R_2 is the union of the remaining subrectangles. Then we may apply the mean value theorem for integrals to those intervals $[y_{j-1}, y_j]$ on which $f(x_i^*, y)$ is continuous in y, thus replacing the integral

$$\int_{y_{j-1}}^{y_j} f(x_i^*, y)\,dy$$

by

$$f(x_i^*, y_j^*(x_i^*))\Delta y_j = f(\mathbf{c}_{ij})\Delta y_j.$$

Since f is bounded, we know that

$$|f(x, y)| \le M$$

for some M and all $(x, y) \in R$. Therefore, on the intervals $[y_{j-1}, y_j]$ where $f(x_i^*, y)$ fails to be continuous, we have

$$\begin{aligned} \left| \int_{y_{j-1}}^{y_j} f(x_i^*, y)\,dy \right| &\le \int_{y_{j-1}}^{y_j} |f(x_i^*, y)|\,dy \\ &\le \int_{y_{j-1}}^{y_j} M\,dy = M(y_j - y_{j-1}) = M\Delta y_j. \end{aligned} \tag{3}$$

From equation (2), we know that

$$\begin{aligned}\sum_{i=1}^{n} F(x_i^*)\Delta x_i &= \sum_{i,j=1}^{n} \left\{ \int_{y_{j-1}}^{y_j} f(x_i^*, y)\,dy \right\} \Delta x_i \\ &= \sum_{R_{ij}\subset R_1\cup R_2} \left\{ \int_{y_{j-1}}^{y_j} f(x_i^*, y)\,dy \right\} \Delta x_i \\ &= \sum_{R_{ij}\subset R_1} \left\{ \int_{y_{j-1}}^{y_j} f(x_i^*, y)\,dy \right\} \Delta x_i \\ &\quad + \sum_{R_{ij}\subset R_2} \left\{ \int_{y_{j-1}}^{y_j} f(x_i^*, y)\,dy \right\} \Delta x_i.\end{aligned}$$

Therefore,

$$\begin{aligned}&\left| \sum_{i=1}^{n} F(x_i^*)\Delta x_i - \sum_{R_{ij}\subset R_2} \left\{ \int_{y_{j-1}}^{y_j} f(x_i^*, y)\,dy \right\} \Delta x_i \right| \\ &\qquad = \left| \sum_{R_{ij}\subset R_1} \left\{ \int_{y_{j-1}}^{y_j} f(x_i^*, y)\,dy \right\} \Delta x_i \right|. \qquad (4)\end{aligned}$$

Applying the mean value theorem for integrals to the left side of equation (4) and inequality (3) to the right side, we obtain

$$\begin{aligned}\left| \sum_{i=1}^{n} F(x_i^*)\Delta x_i - \sum_{R_{ij}\subset R_2} f(\mathbf{c}_{ij})\Delta x_i \Delta y_j \right| &\leq \sum_{R_{ij}\subset R_1} M\Delta x_i \Delta y_j \\ &= M \cdot \text{area of } R_1.\end{aligned}$$

Now S has zero area (by hypothesis) and is contained in R_1. By letting the partition of R become sufficiently fine (i.e., by making Δx_i, Δy_j small), the term $M \cdot$ area of R_1 can be made arbitrarily small. (See Figure 5.37.)

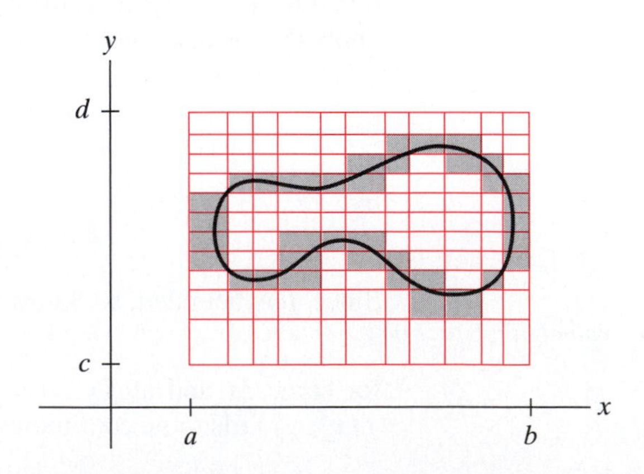

Figure 5.37 The set R_1 (shaded area) consists of the subrectangles of the partition of R that meet S, the set of discontinuities of f on R. As the partition becomes finer, the area of R_1 tends toward zero.

Therefore, as all Δx_i and Δy_j tend to zero, we have that the sums

$$\sum_i F(x_i^*)\Delta x_i \quad \text{and} \quad \sum_{R_{ij}\subset R_2} f(\mathbf{c}_{ij})\Delta x_i \Delta y_j,$$

and the term $M \cdot$ area of R_1 converge (respectively) to

$$\int_a^b F(x)\,dx, \quad \iint_R f\,dA, \quad \text{and} \quad 0.$$

We conclude that

$$\int_a^b F(x)\,dx - \iint_R f\,dA = 0,$$

that is,

$$\iint_R f\,dA = \int_a^b \int_c^d f(x, y)\,dy\,dx.$$

Again, by exchanging the roles of x and y, we can show that

$$\iint_R f\,dA = \int_c^d \int_a^b f(x, y)\,dx\,dy$$

as well. ■

Exercises

1. This problem concerns the double integral $\iint_D x^3\,dA$, where D is the region pictured in Figure 5.38.
   (a) Determine $\iint_D x^3\,dA$ by "brute force" (i.e., by setting up and explicitly evaluating an appropriate iterated integral).
   (b) Now argue what the value of $\iint_D x^3\,dA$ must be by *inspection,* that is, *without* resorting to explicit calculation.

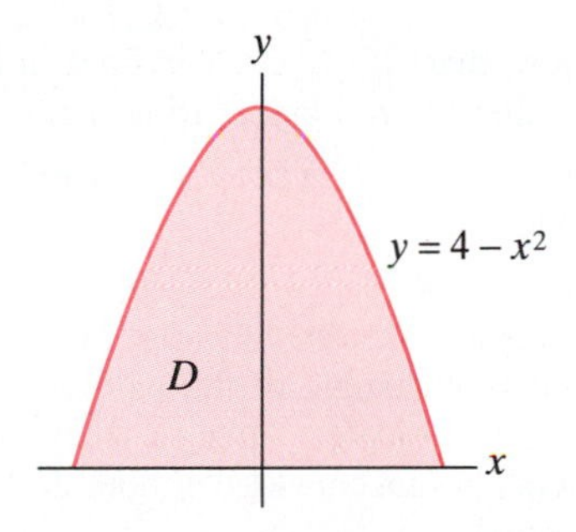

Figure 5.38 The region D of Exercise 1.

In Exercises 2–9, evaluate the given iterated integrals. In addition, sketch the regions D that are determined by the limits of integration.

2. $\int_0^1 \int_0^{x^3} 3\,dy\,dx$

3. $\int_0^2 \int_0^{y^2} y\,dx\,dy$

4. $\int_0^2 \int_0^{x^2} y\,dy\,dx$

5. $\int_{-1}^3 \int_x^{2x+1} xy\,dy\,dx$

6. $\int_0^\pi \int_0^{\sin x} y\cos x\,dy\,dx$

7. $\int_0^1 \int_{-\sqrt{1-x^2}}^{\sqrt{1-x^2}} 3\,dy\,dx$

8. $\int_{-1}^1 \int_0^{\sqrt{1-y^2}} 3\,dx\,dy$

9. $\int_0^1 \int_{-e^x}^{e^x} y^3\,dy\,dx$

10. Figure 5.39 shows the level curves indicating the varying depth (in feet) of a 25 ft by 50 ft swimming pool. Use a Riemann sum to estimate, to the nearest 100 ft^3, the volume of water that the pool contains.

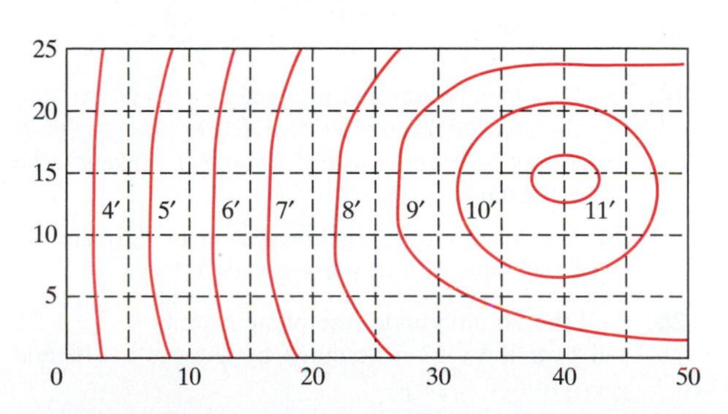

Figure 5.39

11. Integrate the function $f(x, y) = 1 - xy$ over the triangular region whose vertices are $(0, 0)$, $(2, 0)$, $(0, 2)$.

12. Integrate the function $f(x, y) = 3xy$ over the region bounded by $y = 32x^3$ and $y = \sqrt{x}$.

13. Integrate the function $f(x, y) = x + y$ over the region bounded by $x + y = 2$ and $y^2 - 2y - x = 0$.

14. Evaluate $\iint_D 3y\, dA$, where D is the region bounded by $xy^2 = 1$, $y = x$, $x = 0$, and $y = 3$.

15. Evaluate $\iint_D (x - 2y)\, dA$, where D is the region bounded by $y = x^2 + 2$ and $y = 2x^2 - 2$.

16. Evaluate $\iint_D (x^2 + y^2)\, dA$, where D is the region in the first quadrant bounded by $y = x$, $y = 3x$, and $xy = 3$.

17. Prove property 2 of Proposition 2.7.

18. Prove property 3 of Proposition 2.7.

19. Prove property 4 of Proposition 2.7.

20. (a) Let D be an elementary region in $\mathbf{R}^2$. Use the definition of the double integral to explain why $\iint_D 1\, dA$ gives the area of D.

(b) Use part (a) to show that the area inside a circle of radius a is πa^2.

21. Use double integrals to find the area of the region bounded by $y = x^2$ and $y = x^3$.

22. Use double integrals to calculate the area of the region bounded by $y = 2x$, $x = 0$, and $y = 1 - 2x - x^2$.

23. Use double integrals to calculate the area inside the ellipse whose semiaxes have lengths a and b. (See Figure 5.40.)

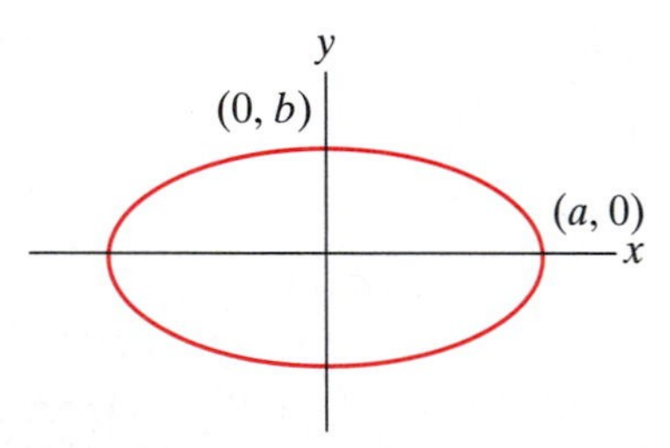

Figure 5.40 The ellipse of Exercise 23.

24. (a) Set up an appropriate iterated integral to find the area of the region bounded by the graphs of $y = x^3 - x$ and $y = ax^2$ for $x \geq 0$. (Take a to be a constant.)

(b) Use a computer algebra system to estimate for what value of a this area equals 1.

25. Find the volume under the plane $z = 4x + 2y + 25$ and over the region bounded by $y = x^2 - 10$ and $y = 31 - (x - 1)^2$.

26. (a) Set up an iterated integral to compute the volume under the hyperbolic paraboloid $z = x^2 - y^2 + 5$ and over the disk

$$D = \{(x, y) \mid x^2 + y^2 \leq 4\}$$

in the xy-plane.

(b) Use a computer algebra system to evaluate the integral.

27. Find the volume of the region under the graph of

$$f(x, y) = 2 - |x| - |y|$$

and above the xy-plane.

28. (a) Show that if $R = [a, b] \times [c, d]$, f is continuous on $[a, b]$, and g is continuous on $[c, d]$, then

$$\iint_R f(x)g(y)\, dA = \left(\int_a^b f(x)\, dx\right)\left(\int_c^d g(y)\, dy\right).$$

(b) What can you say about

$$\iint_D f(x)g(y)\, dA$$

if D is not a rectangle? More specifically, what if D is an elementary region of type 1?

29. Let

$$f(x, y) = \begin{cases} 1 & \text{if } x \text{ is rational} \\ 0 & \text{if } x \text{ is irrational and } y \leq 1 \\ 2 & \text{if } x \text{ is irrational and } y > 1 \end{cases}.$$

(a) Show that $\int_0^2 f(x, y)\, dy$ does not depend on whether x is rational or irrational.

(b) Show that $\int_0^1 \int_0^2 f(x, y)\, dy\, dx$ exists and find its value.

(c) Partition $R = [0, 1] \times [0, 2]$ and construct a Riemann sum by choosing "test points" $\mathbf{c}_{ij}$ in each subrectangle of the partition to have rational x-coordinates. Then to what value must this Riemann sum converge as both Δx_i and Δy_j tend to zero?

(d) Partition R and construct a Riemann sum by choosing test points $\mathbf{c}_{ij} = (x_i^*, y_j^*)$ such that x_i^* is rational if $y_j^* \leq 1$ and x_i^* is irrational if $y_j^* > 1$. What happens to this Riemann sum as both Δx_i and Δy_j tend to zero?

(e) Show that f fails to be integrable on R by using Definition 2.3. Thus, we see that double integrals and iterated integrals are actually different notions.

5.3 Changing the Order of Integration

Frequently, it is useful to think about the evaluation of double integrals over elementary regions essentially as the determination of an appropriate order of integration. When the region of integration is a rectangle, Fubini's theorem (Theorem 2.6) says the order in which we integrate has no significance, that is,

$$\iint_R f\,dA = \int_a^b \int_c^d f(x, y)\,dy\,dx = \int_c^d \int_a^b f(x, y)\,dx\,dy.$$

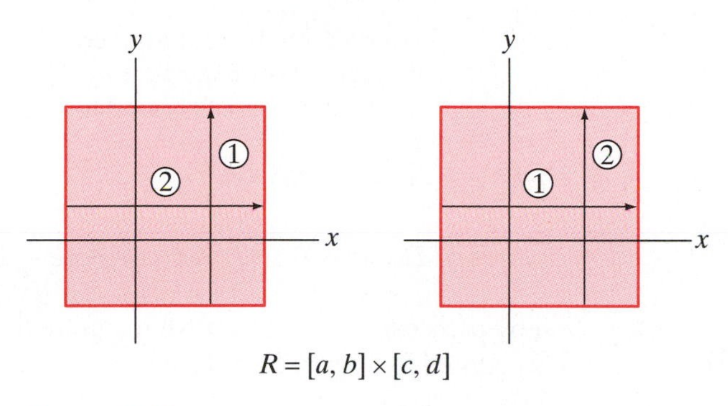

Figure 5.41 Changing the order of integration over a rectangle.

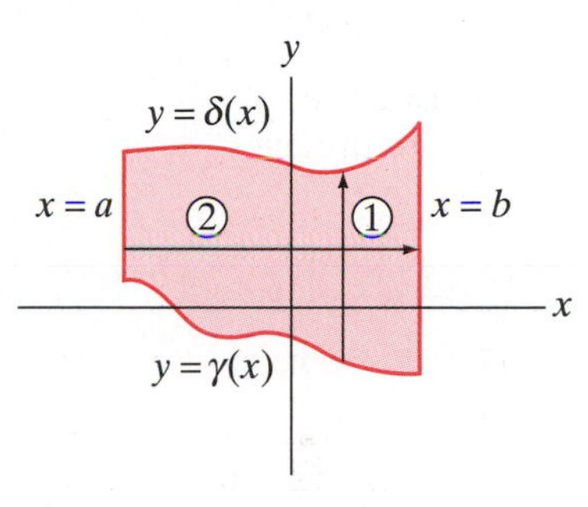

Figure 5.42 A type 1 region forces us to integrate with respect to y first.

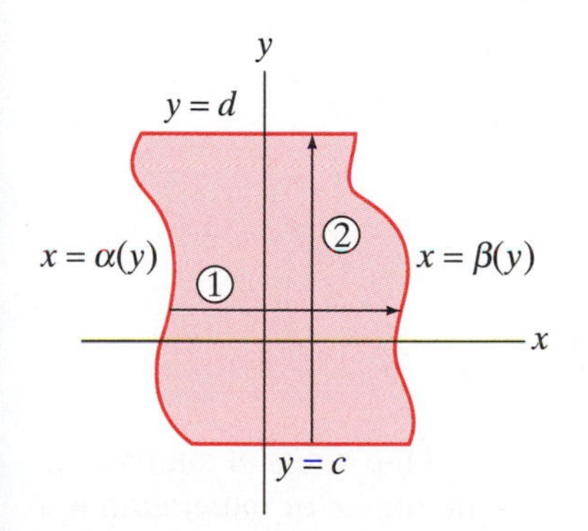

Figure 5.43 A type 2 region requires integration with respect to x first.

(See Figure 5.41.) When the region is elementary of type 1 only, one *must* integrate first with respect to y, then with respect to x, as shown in Figure 5.42. Then

$$\iint_D f\,dA = \int_a^b \int_{\gamma(x)}^{\delta(x)} f(x, y)\,dy\,dx.$$

In the same way, when the region is elementary of type 2 only, one must integrate first with respect to x, so that

$$\iint_D f\,dA = \int_c^d \int_{\alpha(y)}^{\beta(y)} f(x, y)\,dx\,dy.$$

(See Figure 5.43.) When the region is elementary of type 3, however, we can choose either order of integration, at least in principle. Often, this flexibility can be used to advantage, as the following examples illustrate:

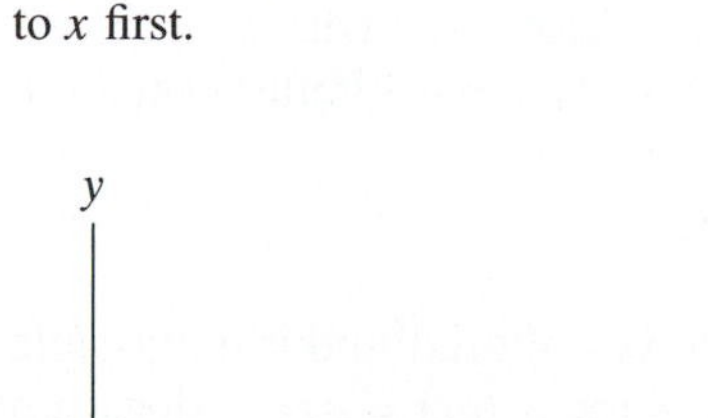

Figure 5.44 The region D of Example 1.

EXAMPLE 1 We calculate the area of the region shown in Figure 5.44. Considering D as a type 1 region, we obtain

$$\text{Area of } D = \iint_D 1\,dA \text{ (Why?)} = \int_1^e \int_0^{\ln x} 1\,dy\,dx$$

$$= \int_1^e y\Big|_0^{\ln x} dx = \int_1^e \ln x\,dx.$$

The single definite integral that results gives the area under the graph of $y = \ln x$ over the x-interval $[1, e]$, just as it should. To evaluate this integral, we need to use integration by parts: Let $u = \ln x$ (so $du = 1/x\,dx$) and $dv = dx$ (so $v = x$). Then

$$\text{Area of } D = \int_1^e \ln x\,dx = \ln x \cdot x\Big|_1^e - \int_1^e x \cdot \frac{1}{x}\,dx$$

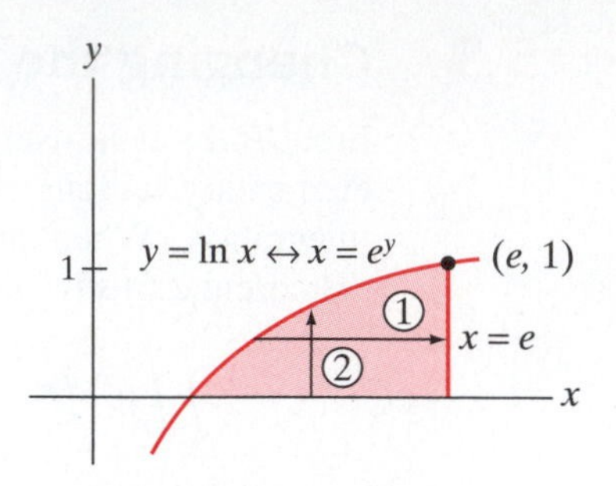

Figure 5.45 Integrating over the region D of Example 1 by integrating first with respect to x.

(remember $\int u\,dv = u \cdot v - \int v\,du$), so

$$\text{Area of } D = e - 0 - \int_1^e dx = e - (e - 1) = 1.$$

Integration by parts can be avoided if we integrate first with respect to x, as schematically suggested by Figure 5.45. Hence,

$$\text{Area of } D = \iint_D 1\,dA = \int_0^1 \int_{e^y}^e 1\,dx\,dy = \int_0^1 x\big|_{e^y}^e dy = \int_0^1 (e - e^y)\,dy$$

$$= (ey - e^y)\big|_0^1 = (e - e) - (0 - e^0) = 1,$$

which checks (just as it should). ◆

Note that the two iterated integrals we used to calculate the area in Example 1, namely,

$$\int_1^e \int_0^{\ln x} dy\,dx \quad \text{and} \quad \int_0^1 \int_{e^y}^e dx\,dy,$$

are *not* obtained from each other by a simple exchange of the limits of integration. The only time such an exchange is justified is when the region of integration is a rectangle of the form $[a, b] \times [c, d]$ so that *all* limits of integration are constants.

EXAMPLE 2 Sometimes changing the order of integration can make an impossible calculation possible. Consider the evaluation of the following iterated integral:

$$\int_0^2 \int_{y^2}^4 y\cos(x^2)\,dx\,dy.$$

After some effort (and maybe some scratchwork), you should find it impossible even to begin this calculation. In fact it can be shown that $\cos(x^2)$ does not have an antiderivative that can be expressed in terms of elementary functions. Consequently, we appear to be stuck.

On the other hand, it is easy to integrate $y\cos(x^2)$ with respect to y. This suggests finding a way to change the order of integration. We do so in two steps:

1. Use the limits of integration in the original iterated integral to identify the region D in $\mathbf{R}^2$ over which the integration takes place. (While doing this, you should make a wish that D turns out to be a type 3 region.)
2. Assuming that the region D in Step 1 is of type 3, change the order of integration.

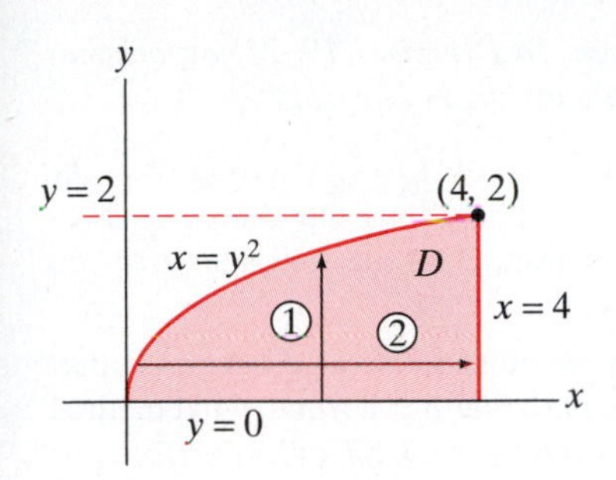

Figure 5.46 Note that $x = y^2$ corresponds to $y = \sqrt{x}$ over the region shown.

The limits of integration in the preceding example imply that D can be described as

$$D = \{(x, y) \mid y^2 \le x \le 4,\ 0 \le y \le 2\},$$

as suggested by Figure 5.46. Now Figure 5.46 can be used to change the order of integration. We have

$$\int_0^2 \int_{y^2}^4 y\cos(x^2)\, dx\, dy = \int_0^4 \int_0^{\sqrt{x}} y\cos(x^2)\, dy\, dx.$$

It is now possible to complete the calculation; that is,

$$\int_0^4 \int_0^{\sqrt{x}} y\cos(x^2)\, dy\, dx = \int_0^4 \left(\frac{y^2}{2}\cos(x^2)\right)\Bigg|_{y=0}^{y=\sqrt{x}} dx$$

$$= \int_0^4 \frac{x}{2}\cos(x^2)\, dx$$

$$= \frac{1}{4}\int_0^{16} \cos u\, du,$$

where $u = x^2$ and $du = 2x\, dx$, so that, finally,

$$\int_0^2 \int_{y^2}^4 y\cos(x^2)\, dx\, dy = \tfrac{1}{4}\sin u\big|_0^{16} = \tfrac{1}{4}\sin 16. \quad \blacklozenge$$

The technique of changing the order of integration is a very powerful one, but it is by no means a panacea for all cumbersome (or impossible) integrals. It relies on an appropriate interaction between the integrals and the region of integration that often fails to occur in practice.

Exercises

1. Consider the integral

$$\int_0^2 \int_{x^2}^{2x} (2x + 1)\, dy\, dx.$$

(a) Evaluate this integral.

(b) Sketch the region of integration.

(c) Write an equivalent iterated integral with the order of integration reversed. Evaluate this new integral and check that your answer agrees with part (a).

In Exercises 2–9, sketch the region of integration, reverse the order of integration, and evaluate both iterated integrals.

2. $\displaystyle\int_0^1 \int_0^x (2 - x - y)\, dy\, dx$

3. $\displaystyle\int_0^2 \int_0^{4-2x} y\, dy\, dx$

4. $\displaystyle\int_0^2 \int_0^{4-y^2} x\, dx\, dy$

5. $\displaystyle\int_0^9 \int_{\sqrt{y}}^3 (x + y)\, dx\, dy$

6. $\displaystyle\int_0^3 \int_1^{e^x} 2\, dy\, dx$

7. $\displaystyle\int_0^1 \int_y^{2y} e^x\, dx\, dy$

8. $\displaystyle\int_0^{\pi/2} \int_0^{\cos x} \sin x\, dy\, dx$

9. $\displaystyle\int_0^2 \int_{-\sqrt{4-y^2}}^{\sqrt{4-y^2}} y\, dx\, dy$

When you reverse the order of integration in Exercises 10 and 11, you should obtain a sum of iterated integrals. Make the reversals and evaluate.

10. $\displaystyle\int_{-2}^1 \int_{x^2-2}^{-x} (x - y)\, dy\, dx$

11. $\displaystyle\int_{-1}^4 \int_{y-4}^{4y-y^2} (y + 1)\, dx\, dy$

In Exercises 12 and 13, rewrite the given sum of iterated integrals as a single iterated integral by reversing the order of integration, and evaluate.

12. $\displaystyle\int_0^1\int_0^x \sin x\,dy\,dx + \int_1^2\int_0^{2-x} \sin x\,dy\,dx.$

13. $\displaystyle\int_0^8\int_0^{\sqrt{y/3}} y\,dx\,dy + \int_8^{12}\int_{\sqrt{y-8}}^{\sqrt{y/3}} y\,dx\,dy.$

In Exercises 14–18, evaluate the given iterated integral.

14. $\displaystyle\int_0^1\int_{3y}^3 \cos(x^2)\,dx\,dy$

15. $\displaystyle\int_0^1\int_y^1 x^2 \sin xy\,dx\,dy$

16. $\displaystyle\int_0^\pi\int_y^\pi \frac{\sin x}{x}\,dx\,dy$

17. $\displaystyle\int_0^3\int_0^{9-x^2} \frac{xe^{3y}}{9-y}\,dy\,dx$

18. $\displaystyle\int_0^2\int_{y/2}^1 e^{-x^2}\,dx\,dy$

It is interesting to see what a computer algebra system does with iterated integrals that are difficult or impossible to integrate in the order given. In Exercises 19–21, experiment with a computer to evaluate the given integrals.

19. (a) Determine the value of $\int_0^2\int_{x/2}^1 y^2\cos(xy)\,dy\,dx$ via computer. Note how long the computer takes to deliver the answer. Does the computer give you a useful answer?

(b) If you were to calculate the iterated integral in part (a) by hand, *in the order it is written*, what method of integration would you use? (Don't actually carry out the evaluation, just think about how you would accomplish it.)

(c) Now reverse the order of integration and let your computer evaluate this iterated integral. Does your computer supply the answer more quickly than in part (a)?

20. (a) See if your computer can calculate $\int_0^3\int_{x^2}^9 x\sin(y^2)\,dy\,dx$ as it is written.

(b) Now reverse the order of integration and have your computer evaluate your new iterated integral. Which of the computations in parts (a) or (b) is easier for your computer?

21. (a) Can your computer evaluate $\int_0^1\int_{\sin^{-1}y}^{\pi/2} e^{\cos x}\,dx\,dy$?

(b) Reverse the order of integration and have it try again. What happens?

5.4 Triple Integrals

Let $f(x, y, z)$ be a function of three variables. Analogous to the double integral, we define the triple integral of f over a solid region in space to be the limit of appropriate Riemann sums. We begin by defining this integral over box-shaped regions and then proceed to define the integral over more general solid regions.

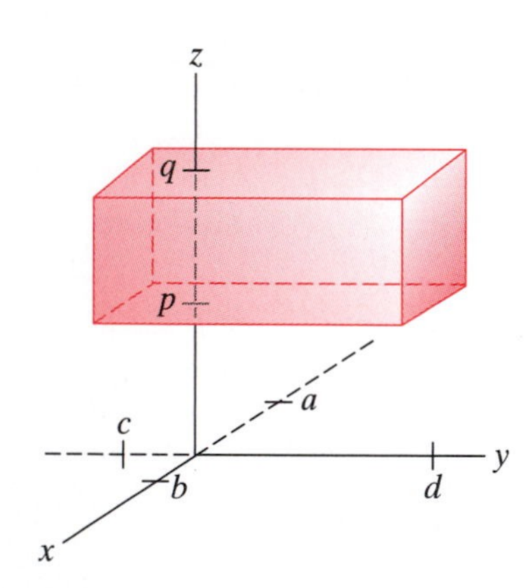

Figure 5.47 The box $B = [a, b] \times [c, d] \times [p, q]$.

The Integral over a Box

Let B be a **closed box** in $\mathbf{R}^3$ whose faces are parallel to the coordinate planes. That is,

$$B = \{(x, y, z) \in \mathbf{R}^3 \mid a \le x \le b,\ c \le y \le d,\ p \le z \le q\}.$$

(See Figure 5.47.) We also use the following shorthand notation for B:

$$B = [a, b] \times [c, d] \times [p, q].$$

■ **Definition 4.1** A **partition** of B of order n consists of three collections of **partition points** that break up B into a union of n^3 subboxes. That is, for $i, j, k = 0, \dots, n$, we introduce the collections $\{x_i\}$, $\{y_j\}$, and $\{z_k\}$, such that

$$a = x_0 < x_1 < \cdots < x_{i-1} < x_i < \cdots < x_n = b,$$
$$c = y_0 < y_1 < \cdots < y_{j-1} < y_j < \cdots < y_n = d,$$
$$p = z_0 < z_1 < \cdots < z_{k-1} < z_k < \cdots < z_n = q.$$

(See Figure 5.48.) In addition, for $i, j, k = 1, \dots, n$, let

$$\Delta x_i = x_i - x_{i-1}, \quad \Delta y_j = y_j - y_{j-1}, \quad \text{and} \quad \Delta z_k = z_k - z_{k-1}.$$

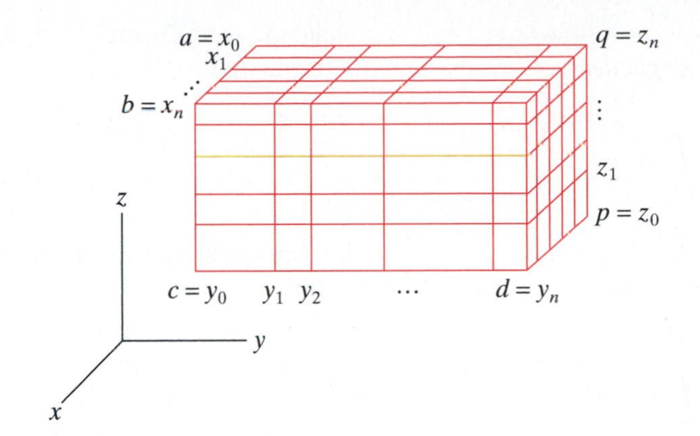

Figure 5.48 A partitioned box.

■ **Definition 4.2** Let f be any function defined on $B = [a, b] \times [c, d] \times [p, q]$. Partition B in some way. Let $\mathbf{c}_{ijk}$ be any point in the subbox

$$B_{ijk} = [x_{i-1}, x_i] \times [y_{j-1}, y_j] \times [z_{k-1}, z_k] \quad (i, j, k = 1, \ldots, n).$$

Then the quantity

$$S = \sum_{i,j,k=1}^{n} f(\mathbf{c}_{ijk}) \Delta V_{ijk},$$

where $\Delta V_{ijk} = \Delta x_i \Delta y_j \Delta z_k$ is the volume of B_{ijk}, is called the **Riemann sum** of f on B corresponding to the partition.

You can think of the Riemann sum $\sum f(\mathbf{c}_{ijk}) \Delta V_{ijk}$ as a weighted sum of volumes of subboxes of B, the weighting given by the value of the function f at particular "test points" $\mathbf{c}_{ijk}$ in each subbox.

■ **Definition 4.3** The **triple integral** of f on B, denoted by

$$\iiint_B f\, dV,$$

by $\displaystyle\iiint_B f(x, y, z)\, dV$, or by $\displaystyle\iiint_B f(x, y, z)\, dx\, dy\, dz$,

is the limit of the Riemann sum S as the dimensions Δx_i, Δy_j, and Δz_k of the subboxes B_{ijk} all approach zero, that is,

$$\iiint_B f\, dV = \lim_{\text{all } \Delta x_i, \Delta y_j, \Delta z_k \to 0} \sum_{i,j,k=1}^{n} f(\mathbf{c}_{ijk}) \Delta x_i \Delta y_j \Delta z_k,$$

provided that this limit exists. When $\iiint_B f\, dV$ exists, we say that f is **integrable** on B.

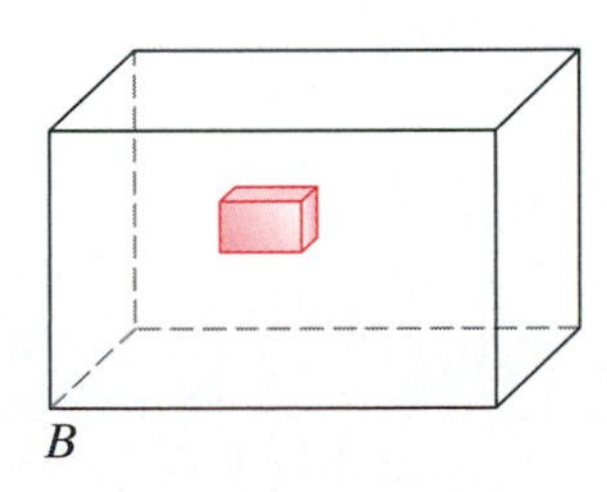

Figure 5.49 The subbox contributes $f(\mathbf{c}_{ijk}) \Delta V_{ijk}$ to the Riemann sum S. If we think of f as representing a density function, then the total mass of the entire box B is $\iiint_B f\, dV$.

The key point to remember is that the triple integral is the limit of Riemann sums. It is this notion that enables useful and important applications of integrals. For example, if we view the integrand f as a type of generalized density function ("generalized" because we allow negative density!), then the Riemann sum S is a sum of approximate masses (densities times volumes) of subboxes of B. These approximations should improve as the subboxes become smaller and smaller. Hence, we can use the triple integral $\iiint_B f\, dV$, when it exists, to compute the total mass of a solid box B whose density varies according to f, as suggested by Figure 5.49.

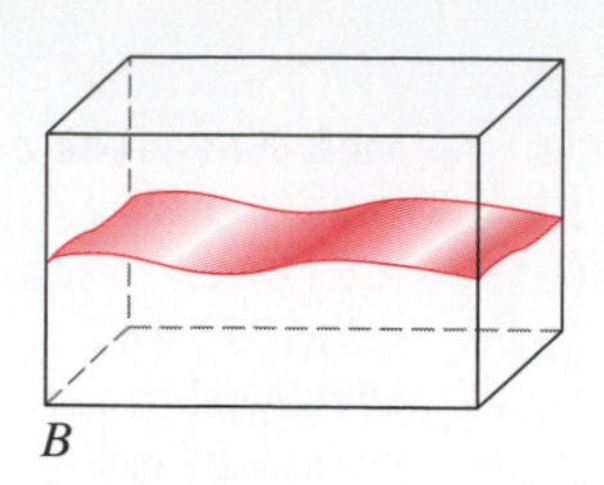

Figure 5.50 In Theorem 4.4 the discontinuities of f on B (shown shaded) must have zero volume.

Analogous to Theorem 2.5, we have the following result regarding integrability of functions:

Theorem 4.4 If f is bounded on B and the set of discontinuities of f on B has zero volume, then $\iiint_B f\,dV$ exists. (See Figure 5.50.)

To evaluate a triple integral over a box, we can use a three-dimensional version of Fubini's theorem.

Theorem 4.5 (FUBINI'S THEOREM) Let f be bounded on $B = [a, b] \times [c, d] \times [p, q]$ and assume that the set S of discontinuities of f has zero volume. If every line parallel to the coordinate axes meets S in at most finitely many points, then

$$\iiint_B f\,dV$$

$$= \int_a^b \int_c^d \int_p^q f(x, y, z)\,dz\,dy\,dx = \int_a^b \int_p^q \int_c^d f(x, y, z)\,dy\,dz\,dx$$

$$= \int_c^d \int_a^b \int_p^q f(x, y, z)\,dz\,dx\,dy = \int_c^d \int_p^q \int_a^b f(x, y, z)\,dx\,dz\,dy$$

$$= \int_p^q \int_a^b \int_c^d f(x, y, z)\,dy\,dx\,dz = \int_p^q \int_c^d \int_a^b f(x, y, z)\,dx\,dy\,dz.$$

EXAMPLE 1 Let

$$B = [-2, 3] \times [0, 1] \times [0, 5], \quad \text{and let} \quad f(x, y, z) = x^2 e^y + xyz.$$

Thus, f is continuous and hence certainly satisfies the hypotheses of Fubini's theorem. Therefore,

$$\begin{aligned}
\iiint_B (x^2 e^y + xyz)\,dV &= \int_{-2}^3 \int_0^1 \int_0^5 (x^2 e^y + xyz)\,dz\,dy\,dx \\
&= \int_{-2}^3 \int_0^1 \left(x^2 e^y z + \tfrac{1}{2} xyz^2\right)\Big|_{z=0}^{z=5} dy\,dx \\
&= \int_{-2}^3 \int_0^1 \left(5x^2 e^y + \tfrac{25}{2} xy\right) dy\,dx \\
&= \int_{-2}^3 \left(5x^2 e^y + \tfrac{25}{4} xy^2\right)\Big|_{y=0}^{y=1} dx \\
&= \int_{-2}^3 \left(5(e-1)x^2 + \tfrac{25}{4} x\right) dx \\
&= \left(\tfrac{5}{3}(e-1)x^3 + \tfrac{25}{8} x^2\right)\Big|_{-2}^3 \\
&= \left(45(e-1) + \tfrac{225}{8}\right) - \left(-\tfrac{40}{3}(e-1) + \tfrac{25}{2}\right) \\
&= \tfrac{175}{3}(e-1) + \tfrac{125}{8}.
\end{aligned}$$

You can check that integrating in any of the other five possible orders produces the same result. ◆

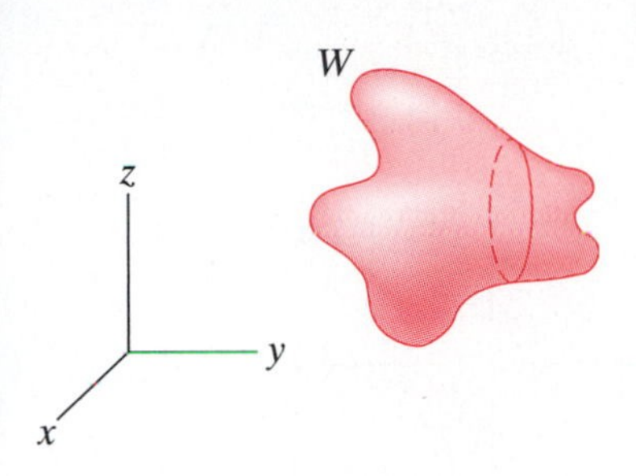

Figure 5.51 The function f is continuous on W.

Elementary Regions in Space

Now suppose W denotes a fairly arbitrary solid region in space, like a rock or a slab of tofu. Suppose f is a continuous function defined on W, such as a mass density function. (See Figure 5.51.) Then the triple integral of f over W should give the total mass of W. As was the case with general double integrals, we need to find a way to properly define $\iiint_W f\,dV$ and to calculate it in practical situations. The course of action is much like before: We see how to calculate integrals over certain types of elementary regions and treat integrals over more general regions by subdividing them into regions of elementary type.

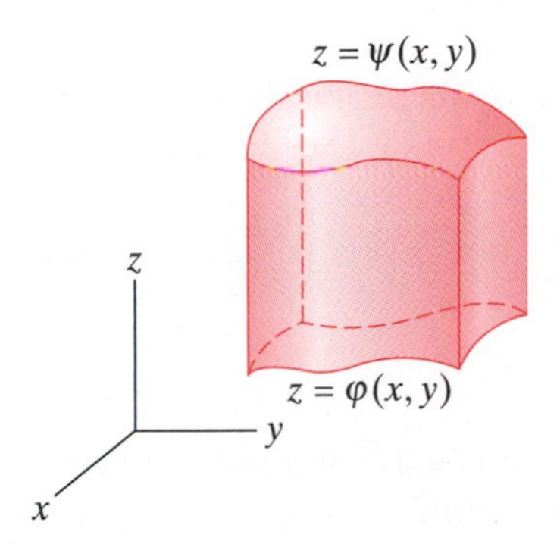

Figure 5.52 An elementary region of type 1.

Definition 4.6 We say that W is an **elementary region** in space if it can be described as a subset of $\mathbf{R}^3$ of one of the following four types:

Type 1 (see Figures 5.52 and 5.53)

(a) $W = \{(x, y, z) \mid \varphi(x, y) \le z \le \psi(x, y),\ \gamma(x) \le y \le \delta(x),\ a \le x \le b\}$,

or

(b) $W = \{(x, y, z) \mid \varphi(x, y) \le z \le \psi(x, y),\ \alpha(y) \le x \le \beta(y),\ c \le y \le d\}$.

Type 2 (see Figure 5.54)

(a) $W = \{(x, y, z) \mid \alpha(y, z) \le x \le \beta(y, z),\ \gamma(z) \le y \le \delta(z),\ p \le z \le q\}$,

or

(b) $W = \{(x, y, z) \mid \alpha(y, z) \le x \le \beta(y, z),\ \varphi(y) \le z \le \psi(y),\ c \le y \le d\}$.

Type 3 (see Figure 5.55)

(a) $W = \{(x, y, z) \mid \gamma(x, z) \le y \le \delta(x, z),\ \alpha(z) \le x \le \beta(z),\ p \le z \le q\}$,

or

(b) $W = \{(x, y, z) \mid \gamma(x, z) \le y \le \delta(x, z),\ \varphi(x) \le z \le \psi(x),\ a \le x \le b\}$.

Type 4

W is of all three previously described types.

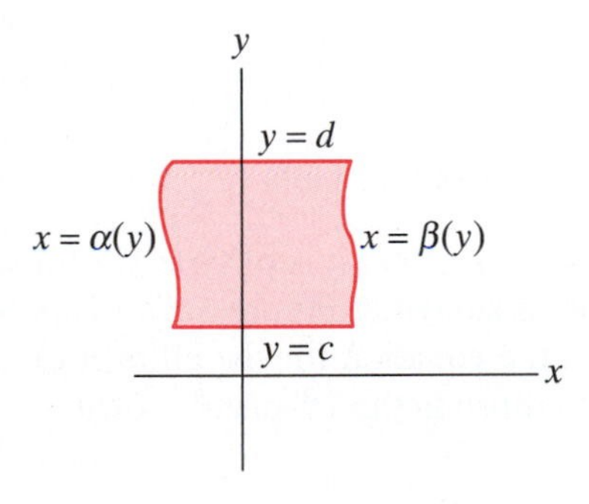

Figure 5.53 The "shadow" (projection) of W into the xy-plane should be an elementary region in the plane.

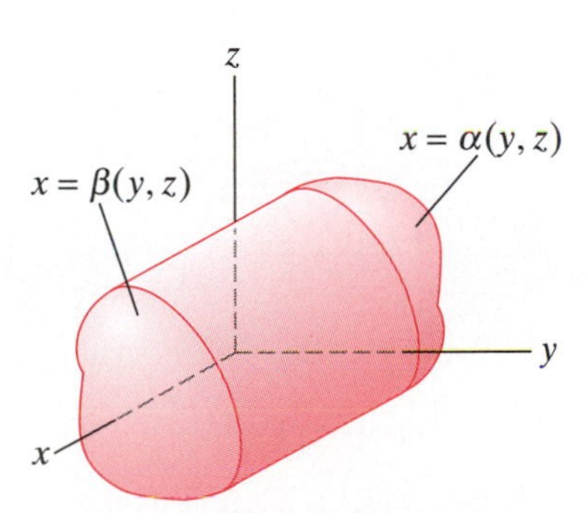

Figure 5.54 For an elementary region of type 2, the shadow in the yz-plane should be an elementary region in the plane.

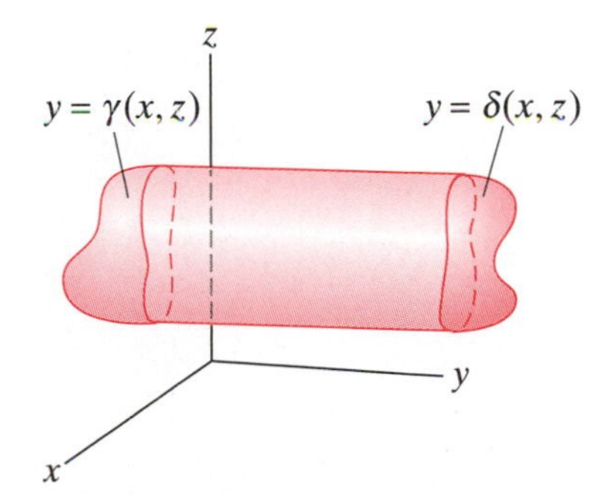

Figure 5.55 For an elementary region of type 3, the shadow in the xz-plane should be an elementary region in the plane.

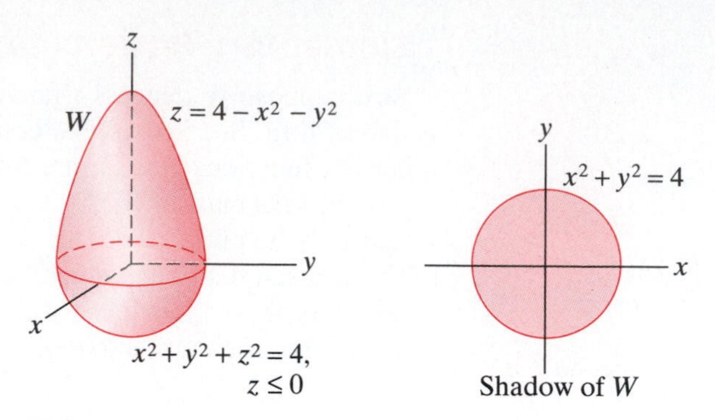

Figure 5.56 The solid region W of Example 2.

EXAMPLE 2 Let W be the solid region bounded by the hemisphere $x^2 + y^2 + z^2 = 4$, where $z \leq 0$, and the paraboloid $z = 4 - x^2 - y^2$. (See Figure 5.56.) It is an elementary region of type 1 since we may describe it as

$$W = \Big\{(x, y, z) \mid -\sqrt{4 - x^2 - y^2} \leq z \leq 4 - x^2 - y^2,$$
$$-\sqrt{4 - x^2} \leq y \leq \sqrt{4 - x^2},\ -2 \leq x \leq 2\Big\}.$$

This description was obtained by noting that W is bounded on top and bottom by a pair of surfaces, each of which is the graph of a function of the form $z = g(x, y)$ and the shadow of W in the xy-plane is a disk D of radius 2, which we have chosen to describe as

$$D = \Big\{(x, y) \mid -\sqrt{4 - x^2} \leq y \leq \sqrt{4 - x^2}, -2 \leq x \leq 2\Big\}.$$ ◆

EXAMPLE 3 The solid bounded by the ellipsoid

$$E : \frac{x^2}{a^2} + \frac{y^2}{b^2} + \frac{z^2}{c^2} = 1, \quad a, b, c \text{ positive constants}$$

can be seen to be an elementary region of type 4. To see that it is of type 1, split the boundary surface in half via the $z = 0$ plane as shown in Figure 5.57. (This is accomplished analytically by solving for z in the equation for the ellipsoid.) Then the shadow D of E is the region inside the ellipse in the xy-plane shown in Figure 5.58.

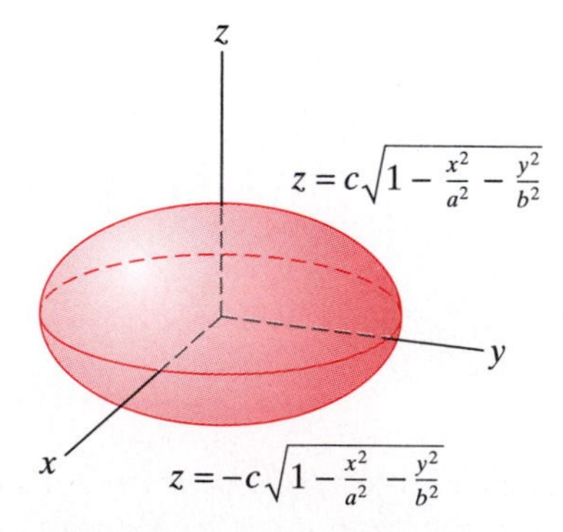

Figure 5.57 The ellipsoid of Example 3 as an elementary region of type 1.

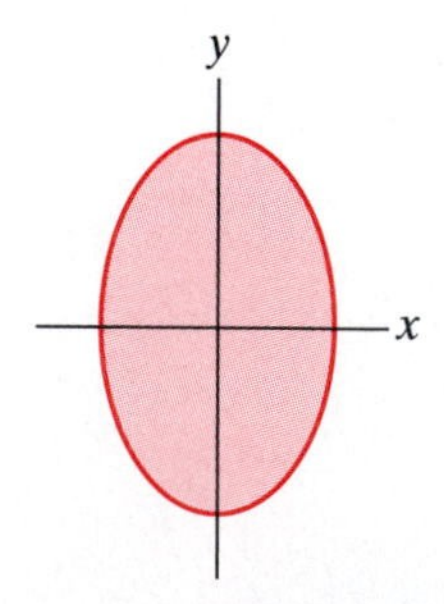

Figure 5.58 The shadow of the type 1 ellipsoid in Figure 5.57 is the region inside the ellipse $x^2/a^2 + y^2/b^2 = 1$ in the xy-plane.

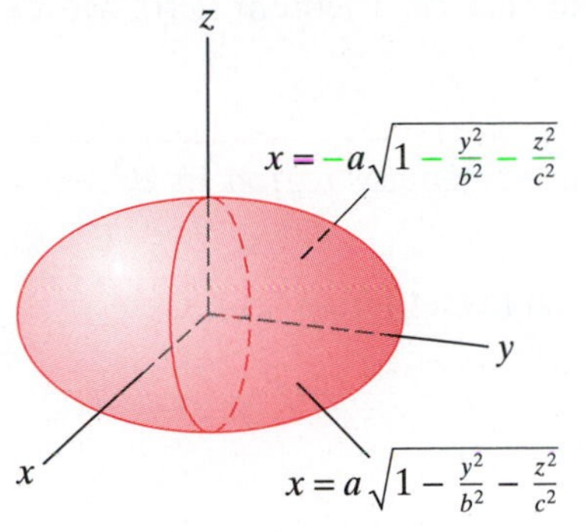

z

$x = -a\sqrt{1 - \frac{y^2}{b^2} - \frac{z^2}{c^2}}$

y

x

$x = a\sqrt{1 - \frac{y^2}{b^2} - \frac{z^2}{c^2}}$

Figure 5.59 The ellipsoid of Example 3 as a type 2 elementary region.

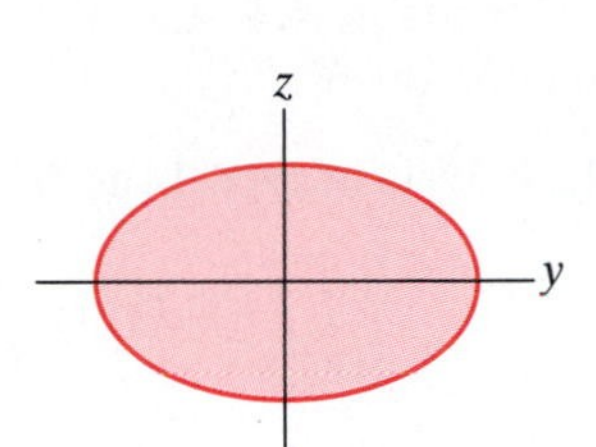

Figure 5.60 The shadow of the ellipsoid in Figure 5.59 is the region inside the ellipse $y^2/b^2 + z^2/c^2 = 1$ in the yz-plane.

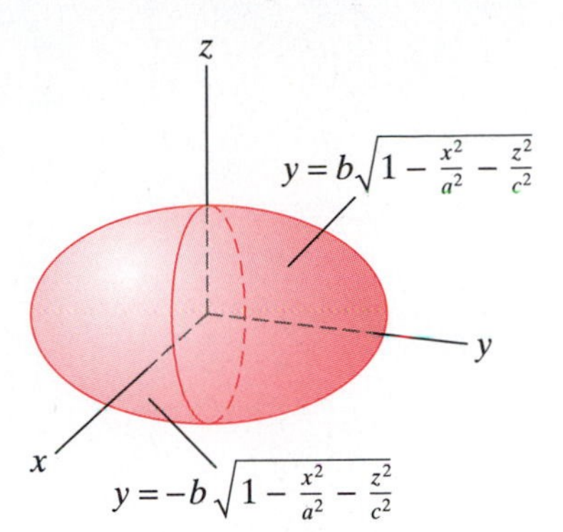

Figure 5.61 The ellipsoid of Example 3 as a type 3 region.

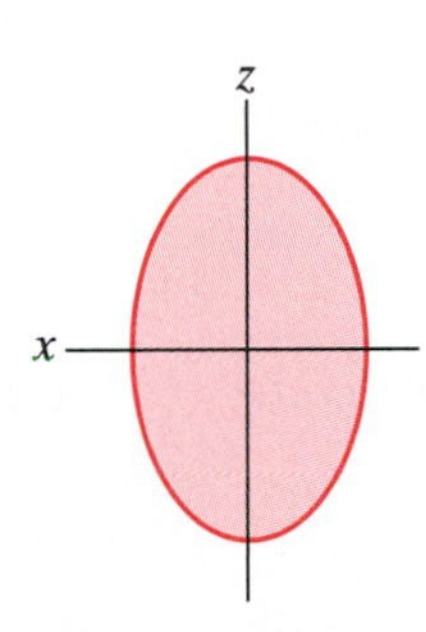

Figure 5.62 The shadow of the ellipsoid in Figure 5.61 in the xz-plane.

We have

$$D = \left\{ (x, y) \;\middle|\; -b\sqrt{1 - \frac{x^2}{a^2}} \leq y \leq b\sqrt{1 - \frac{x^2}{a^2}},\ -a \leq x \leq a \right\}$$

$$= \left\{ (x, y) \;\middle|\; -a\sqrt{1 - \frac{y^2}{b^2}} \leq x \leq a\sqrt{1 - \frac{y^2}{b^2}},\ -b \leq y \leq b \right\},$$

so D is in fact a type 3 elementary region in $\mathbf{R}^2$.

To see that E is of type 2, split the boundary at the $x = 0$ plane as in Figure 5.59. The shadow in the yz-plane is again the region inside an ellipse. (See Figure 5.60.) Finally, to see that E is of type 3, split along $y = 0$. (See Figures 5.61 and 5.62.) ◆

Triple Integrals in General

Suppose W is an elementary region in $\mathbf{R}^3$ and f is a continuous function on W. Then, just as in the case of double integrals, we define the **extension** of f by

$$f^{\text{ext}}(x, y, z) = \begin{cases} f(x, y, z) & \text{if } (x, y, z) \in W \\ 0 & \text{if } (x, y, z) \notin W \end{cases}.$$

By Theorem 4.4, f^{ext} is integrable on any box B that contains W. Thus, we can make the following definition:

> ■ **Definition 4.7** Under the assumptions that W is an elementary region and f is continuous on W, we define the triple integral
>
> $$\iiint_W f\, dV \quad \text{to be} \quad \iiint_B f^{\text{ext}} dV,$$
>
> where B is any box containing W.

Using a proof analogous to that of Theorem 2.10, we can establish the following:

Theorem 4.8 Let W be an elementary region in $\mathbf{R}^3$ and f a continuous function on W.

1. If W is of type 1 (as described in Definition 4.6), then

$$\iiint_W f\,dV = \int_a^b \int_{\gamma(x)}^{\delta(x)} \int_{\varphi(x,y)}^{\psi(x,y)} f(x, y, z)\,dz\,dy\,dx, \qquad \text{(type 1a)}$$

or

$$\iiint_W f\,dV = \int_c^d \int_{\alpha(y)}^{\beta(y)} \int_{\varphi(x,y)}^{\psi(x,y)} f(x, y, z)\,dz\,dx\,dy. \qquad \text{(type 1b)}$$

2. If W is of type 2, then

$$\iiint_W f\,dV = \int_p^q \int_{\gamma(z)}^{\delta(z)} \int_{\alpha(y,z)}^{\beta(y,z)} f(x, y, z)\,dx\,dy\,dz, \qquad \text{(type 2a)}$$

or

$$\iiint_W f\,dV = \int_c^d \int_{\varphi(y)}^{\psi(y)} \int_{\alpha(y,z)}^{\beta(y,z)} f(x, y, z)\,dx\,dz\,dy. \qquad \text{(type 2b)}$$

3. If W is of type 3, then

$$\iiint_W f\,dV = \int_p^q \int_{\alpha(z)}^{\beta(z)} \int_{\gamma(x,z)}^{\delta(x,z)} f(x, y, z)\,dy\,dx\,dz, \qquad \text{(type 3a)}$$

or

$$\iiint_W f\,dV = \int_a^b \int_{\varphi(x)}^{\psi(x)} \int_{\gamma(x,z)}^{\delta(x,z)} f(x, y, z)\,dy\,dz\,dx. \qquad \text{(type 3b)}$$

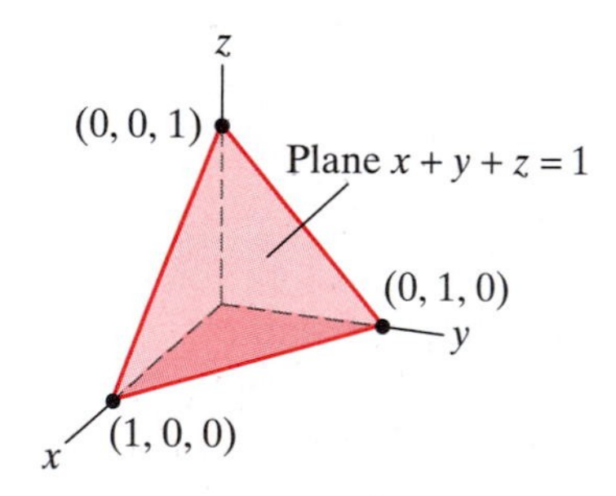

Figure 5.63 The tetrahedron of Example 4.

EXAMPLE 4 Let W denote the (solid) tetrahedron with vertices at $(0, 0, 0)$, $(1, 0, 0)$, $(0, 1, 0)$, and $(0, 0, 1)$ as shown in Figure 5.63. Suppose that the mass density at a point (x, y, z) inside the tetrahedron varies as $f(x, y, z) = 1 + xy$. We will use a triple integral to find the total mass of the tetrahedron.

The total mass M is

$$\iiint_W f\,dV = \iiint_W (1 + xy)\,dV.$$

(See the remark before Theorem 4.4.) To evaluate this triple integral using iterated integrals, note that we can view the tetrahedron as a type 1 elementary region. (Actually, it is a type 4 region, but that will not matter.) The slanted face is given by the equation $x + y + z = 1$, which describes the plane that contains the three points $(1, 0, 0)$, $(0, 1, 0)$, and $(0, 0, 1)$. Hence, by first integrating with respect to z and holding x and y constant,

$$\begin{aligned} M &= \iint_{\text{shadow}} \left(\int_0^{1-x-y} (1 + xy)\,dz \right) dA \\ &= \iint_{\text{shadow}} (1 + xy)(1 - x - y)\,dA \\ &= \iint_{\text{shadow}} (1 - x - y + xy - x^2y - xy^2)\,dA. \end{aligned}$$

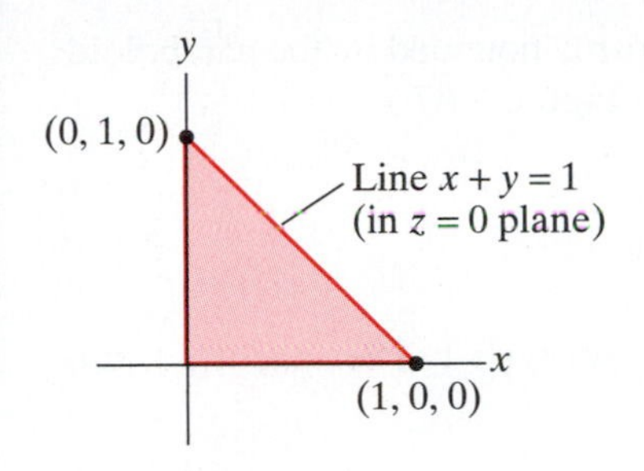

Figure 5.64 The shadow in the xy-plane of the tetrahedron of Example 4 is a triangular region.

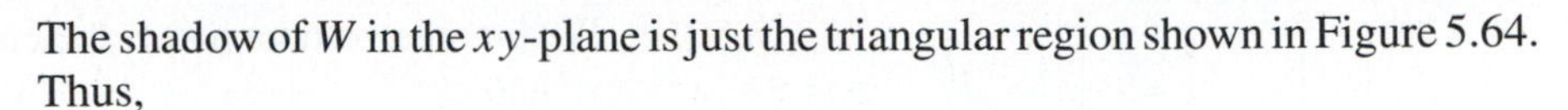

The shadow of W in the xy-plane is just the triangular region shown in Figure 5.64. Thus,

$$\begin{aligned}
M &= \iint_{\text{shadow}} (1 - x - y + xy - x^2y - xy^2)\, dA \\
&= \int_0^1 \int_0^{1-x} (1 - x - y + xy - x^2y - xy^2)\, dy\, dx \\
&= \int_0^1 \Big((1-x) - x(1-x) - \tfrac{1}{2}(1-x)^2 + \tfrac{1}{2}x(1-x)^2 \\
&\qquad - \tfrac{1}{2}x^2(1-x)^2 - \tfrac{1}{3}x(1-x)^3\Big)\, dx \\
&= \int_0^1 \left(\tfrac{1}{2} - \tfrac{5}{6}x + \tfrac{1}{2}x^3 - \tfrac{1}{6}x^4\right) dx = \left(\tfrac{1}{2}x - \tfrac{5}{12}x^2 + \tfrac{1}{8}x^4 - \tfrac{1}{30}x^5\right)\Big|_0^1 = \tfrac{7}{40}.
\end{aligned}$$

Note that M can also be written as a single iterated integral, namely,

$$M = \int_0^1 \int_0^{1-x} \int_0^{1-x-y} (1 + xy)\, dz\, dy\, dx.$$

◆

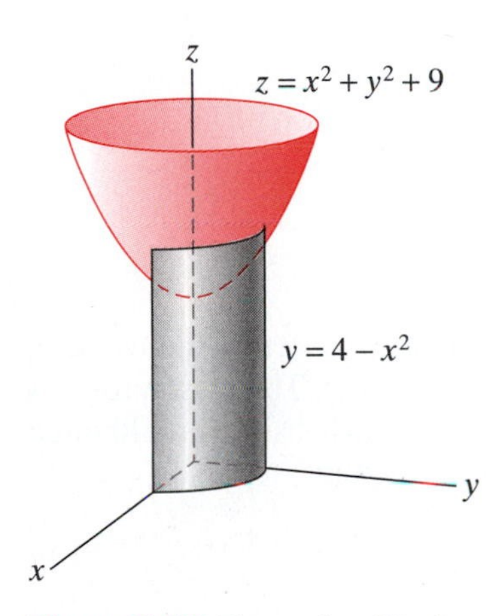

Figure 5.65 The region W of Example 5.

EXAMPLE 5 We calculate the volume of the solid W sitting in the first octant and bounded by the coordinate planes, the paraboloid $z = x^2 + y^2 + 9$, and the parabolic cylinder $y = 4 - x^2$. (See Figure 5.65.)

By definition, the triple integral is a limit of a weighted sum of volumes of tiny subboxes that fill out the region of integration. If the weights in the sum are all taken to be 1, then we obtain an approximation to the volume:

$$V \approx \sum_{i,j,k} 1 \cdot \Delta V_{ijk}.$$

Therefore, taking the limit as the dimensions of the subboxes all approach zero, it makes sense to define

$$V = \iiint_W 1\, dV.$$

In our situation, W is a type 1 region whose shadow in the xy-plane looks like the region shown in Figure 5.66. Thus, by Theorem 4.4,

$$\begin{aligned}
V = \iiint_W dV &= \int_0^2 \int_0^{4-x^2} \int_0^{x^2+y^2+9} dz\, dy\, dx \\
&= \int_0^2 \int_0^{4-x^2} (x^2 + y^2 + 9)\, dy\, dx \\
&= \int_0^2 \left(x^2y + \tfrac{1}{3}y^3 + 9y\Big|_{y=0}^{y=4-x^2}\right) dx \\
&= \int_0^2 \left(x^2(4-x^2) + \tfrac{1}{3}(4-x^2)^3 + 9(4-x^2)\right) dx \\
&= \int_0^2 \left(\tfrac{172}{3} - 21x^2 + 3x^4 - \tfrac{1}{3}x^6\right) dx \\
&= \left(\tfrac{172}{3}x - 7x^3 + \tfrac{3}{5}x^5 - \tfrac{1}{21}x^7\right)\Big|_0^2 = \tfrac{2512}{35}.
\end{aligned}$$

◆

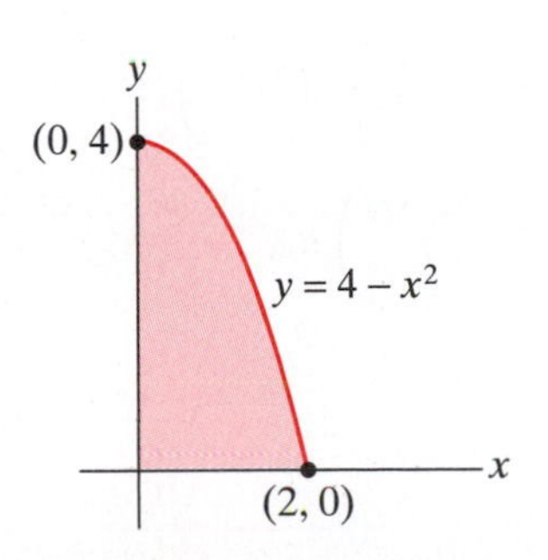

Figure 5.66 The shadow in the xy-plane of the region in Figure 5.65.

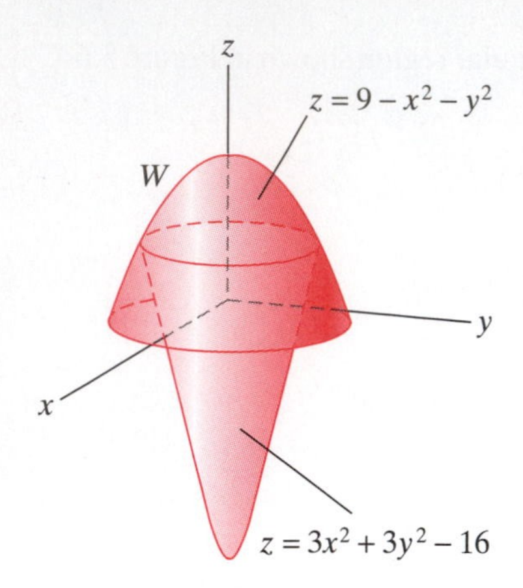

Figure 5.67 The capsule-shaped region of Example 6.

EXAMPLE 6 We find the volume inside the capsule bounded by the paraboloids $z = 9 - x^2 - y^2$ and $z = 3x^2 + 3y^2 - 16$. (See Figure 5.67.)

Once again, we have

$$V = \iiint_W 1\, dV,$$

and again the region W of interest is elementary of type 1. The shadow, or projection, of W in the xy-plane is determined by

$$\{(x, y) \in \mathbf{R}^2 \mid \text{there is some } z \text{ such that } (x, y, z) \in W\}.$$

Physically, one can also imagine the shadow as the hole produced by allowing W to "fall through" the xy-plane. In other words, the shadow is the widest part of W perpendicular to the z-axis. From Figure 5.67, one can see that it is determined by the intersection of the two boundary paraboloids. The shadow itself is shown in Figure 5.68. The intersection may be obtained by equating the z-coordinates of the boundary paraboloids. Therefore,

$$\begin{aligned} 9 - x^2 - y^2 = 3x^2 + 3y^2 - 16 \quad &\Longleftrightarrow \quad 4x^2 + 4y^2 = 25 \\ &\Longleftrightarrow \quad x^2 + y^2 = \tfrac{25}{4} = \left(\tfrac{5}{2}\right)^2. \end{aligned}$$

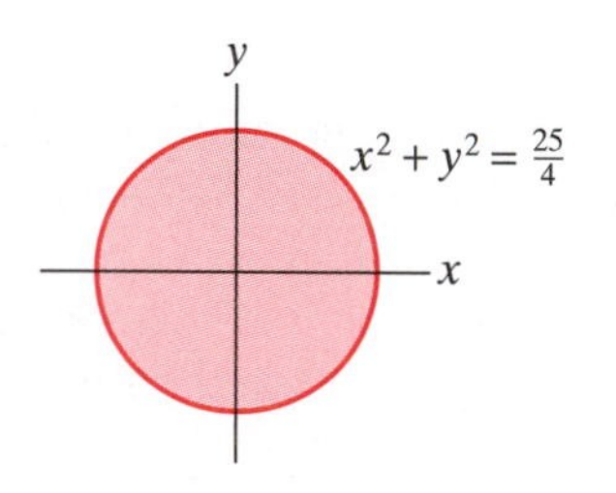

Figure 5.68 The shadow of the region W of Example 6, obtained by projecting the intersection curves of the defining paraboloids onto the xy-plane.

Thus, by Theorem 4.4,

$$\begin{aligned} V = \iiint_W dV &= \int_{-5/2}^{5/2} \int_{-\sqrt{25/4-x^2}}^{\sqrt{25/4-x^2}} \int_{3x^2+3y^2-16}^{9-x^2-y^2} dz\, dy\, dx \\ &= 4 \int_0^{5/2} \int_0^{\sqrt{25/4-x^2}} \int_{3x^2+3y^2-16}^{9-x^2-y^2} dz\, dy\, dx. \end{aligned}$$

This last iterated integral represents the volume of one quarter of the capsule. Hence, we multiply its value by 4 to obtain the total volume. The reason for this manipulation is to make the subsequent calculations somewhat simpler (although the computation that follows is clearly best left to a computer).

We compute

$$\begin{aligned} V &= 4 \int_0^{5/2} \int_0^{\sqrt{25/4-x^2}} \int_{3x^2+3y^2-16}^{9-x^2-y^2} dz\, dy\, dx \\ &= 4 \int_0^{5/2} \int_0^{\sqrt{25/4-x^2}} (25 - 4x^2 - 4y^2)\, dy\, dx \\ &= 4 \int_0^{5/2} \left(25\sqrt{\tfrac{25}{4} - x^2} - 4x^2\sqrt{\tfrac{25}{4} - x^2} - \tfrac{4}{3}\left(\tfrac{25}{4} - x^2\right)^{3/2} \right) dx \\ &= 4 \int_0^{5/2} \left((25 - 4x^2)\sqrt{\tfrac{25}{4} - x^2} - \tfrac{4}{3}\left(\tfrac{25}{4} - x^2\right)^{3/2} \right) dx \\ &= 4 \int_0^{5/2} \left(4\left(\tfrac{25}{4} - x^2\right)\sqrt{\tfrac{25}{4} - x^2} - \tfrac{4}{3}\left(\tfrac{25}{4} - x^2\right)^{3/2} \right) dx \\ &= 4 \int_0^{5/2} \left(4 - \tfrac{4}{3}\right)\left(\tfrac{25}{4} - x^2\right)^{3/2} dx \\ &= \tfrac{32}{3} \int_0^{5/2} \left(\tfrac{25}{4} - x^2\right)^{3/2} dx. \end{aligned}$$

Now let $x = \frac{5}{2}\sin\theta$, so $dx = \frac{5}{2}\cos\theta\, d\theta$. Then

$$\begin{aligned} V &= \frac{32}{3}\int_0^{\pi/2}\left(\frac{5}{2}\cos\theta\right)^3 \frac{5}{2}\cos\theta\, d\theta = \frac{1250}{3}\int_0^{\pi/2} \cos^4\theta d\theta \\ &= \frac{1250}{3}\int_0^{\pi/2}\left(\frac{1}{2}(1+\cos 2\theta)\right)^2 d\theta \\ &= \frac{625}{6}\int_0^{\pi/2}(1 + 2\cos 2\theta + \cos^2 2\theta)\, d\theta \\ &= \frac{625}{6}(\theta + \sin 2\theta)\Big|_0^{\pi/2} + \frac{625}{6}\int_0^{\pi/2}\frac{1}{2}(1+\cos 4\theta)\, d\theta \\ &= \frac{625}{12}\pi + \frac{625}{12}\left(\frac{\pi}{2}+0\right) = \frac{625\pi}{8}. \end{aligned}$$

◆

Exercises

Evaluate the triple integrals given in Exercises 1–3.

1. $\iiint_{[-1,1]\times[0,2]\times[1,3]} xyz\, dV$

2. $\iiint_{[0,1]\times[0,2]\times[0,3]} (x^2+y^2+z^2)\, dV$

3. $\iiint_{[1,e]\times[1,e]\times[1,e]} \frac{1}{xyz}\, dV$

4. Find the value of $\iiint_W z\, dV$, where $W = [-1, 2] \times [2, 5] \times [-3, 3]$, without resorting to explicit calculation.

Evaluate the iterated integrals given in Exercises 5–7.

5. $\int_{-1}^{2}\int_{1}^{z^2}\int_{0}^{y+z} 3yz^2\, dx\, dy\, dz$

6. $\int_{1}^{3}\int_{0}^{z}\int_{1}^{xz} (x+2y+z)\, dy\, dx\, dz$

7. $\int_{0}^{1}\int_{1+y}^{2y}\int_{z}^{y+z} z\, dx\, dz\, dy$

8. (a) Let W be an elementary region in $\mathbf{R}^3$. Use the definition of the triple integral to explain why $\iiint_W 1\, dV$ gives the volume of W.
   (b) Use part (a) to find the volume of the region W bounded by the surfaces $z = x^2 + y^2$ and $z = 9 - x^2 - y^2$.

9. Use triple integrals to verify that the volume of a ball of radius a is $4\pi a^3/3$.

10. Use triple integrals to calculate the volume of a cone of radius r and height h. (You may wish to use a computer algebra system for the evaluation.)

In Exercises 11–18, integrate the given function over the indicated region W.

11. $f(x, y, z) = 2x - y + z$; W is the region bounded by the cylinder $z = y^2$, the xy-plane, and the planes $x = 0, x = 1, y = -2, y = 2$.

12. $f(x, y, z) = y$; W is the region bounded by the plane $x + y + z = 2$, the cylinder $x^2 + z^2 = 1$, and $y = 0$.

13. $f(x, y, z) = 8xyz$; W is the region bounded by the cylinder $y = x^2$, the plane $y + z = 9$, and the xy-plane.

14. $f(x, y, z) = z$; W is the region in the first octant bounded by the cylinder $y^2 + z^2 = 9$ and the planes $y = x$, $x = 0$, and $z = 0$.

15. $f(x, y, z) = 1 - z^2$; W is the tetrahedron with vertices $(0, 0, 0)$, $(1, 0, 0)$, $(0, 2, 0)$, and $(0, 0, 3)$.

16. $f(x, y, z) = 3x$; W is the region in the first octant bounded by $z = x^2 + y^2$, $x = 0$, $y = 0$, and $z = 4$.

17. $f(x, y, z) = x + y$; W is the region bounded by the cylinder $x^2 + 3z^2 = 9$ and the planes $y = 0, x + y = 3$.

18. $f(x, y, z) = z$; W is the region bounded by $z = 0$, $x^2 + 4y^2 = 4$, and $z = x + 2$.

19. Find the volume of the solid bounded by the paraboloid $z = 4x^2 + y^2$ and the cylinder $y^2 + z = 2$.

20. Find the volume of the region inside both of the cylinders $x^2 + y^2 = a^2$ and $x^2 + z^2 = a^2$.

21. Consider the iterated integral

$$\int_{-1}^{1}\int_{y^2}^{1}\int_{0}^{1-x} f(x, y, z)\, dz\, dx\, dy.$$

Sketch the region of integration and rewrite the integral as an equivalent iterated integral in each of the five other orders of integration.

22. Change the order of integration of

$$\int_0^1 \int_0^1 \int_0^{x^2} f(x, y, z)\, dz\, dx\, dy$$

to give five other equivalent iterated integrals.

23. Change the order of integration of

$$\int_0^2 \int_0^x \int_0^y f(x, y, z)\, dz\, dy\, dx$$

to give five other equivalent iterated integrals.

24. Consider the iterated integral

$$\int_0^2 \int_0^{\frac{1}{2}\sqrt{36-9x^2}} \int_{5x^2}^{36-4x^2-4y^2} 2\, dz\, dy\, dx.$$

(a) This integral is equal to a triple integral over a solid region W in $\mathbf{R}^3$. Describe W.

(b) Set up an equivalent iterated integral by integrating first with respect to z, then with respect to x, then with respect to y. Do not evaluate your answer.

(c) Set up an equivalent iterated integral by integrating first with respect to y, then with respect to z, then with respect to x. Do not evaluate your answer.

(d) Now consider integrating first with respect to y, then x, then z. Set up a *sum* of iterated integrals that, when evaluated, give the same result. Do not evaluate your answer.

(e) Repeat part (d) for integration first with respect to x, then z, then y.

25. Consider the iterated integral

$$\int_{-2}^2 \int_0^{\frac{1}{2}\sqrt{4-x^2}} \int_{x^2+3y^2}^{4-y^2} (x^3 + y^3)\, dz\, dy\, dx.$$

(a) This integral is equal to a triple integral over a solid region W in $\mathbf{R}^3$. Describe W.

(b) Set up an equivalent iterated integral by integrating first with respect to z, then with respect to x, then with respect to y. Do not evaluate your answer.

(c) Set up an equivalent iterated integral by integrating first with respect to x, then with respect to z, then with respect to y. Do not evaluate your answer.

(d) Now consider integrating first with respect to x, then y, then z. Set up a *sum* of iterated integrals that, when evaluated, give the same result. Do not evaluate your answer.

(e) Repeat part (d) for integration first with respect to y, then z, then x.

5.5 Change of Variables

As some of the examples in the previous sections suggest, the evaluation of a multiple integral by means of iterated integrals can be a complicated process. Both the integrand and the region of integration can contribute computational difficulties. Our goal for this section is to see ways in which changes in coordinates can be used to transform iterated integrals into ones that are relatively straightforward to calculate. We begin by studying the coordinate transformations themselves and how such transformations affect the relevant integrals.

Coordinate Transformations

Let $\mathbf{T}: \mathbf{R}^2 \to \mathbf{R}^2$ be a map of class C^1 that transforms the uv-plane into the xy-plane. We are interested particularly in how certain subsets D^* of the uv-plane are distorted under $\mathbf{T}$ into subsets D of the xy-plane. (See Figure 5.69.)

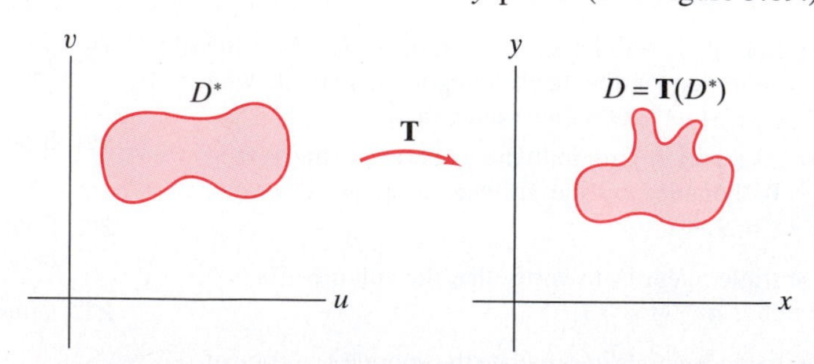

Figure 5.69 The transformation $\mathbf{T}(u, v) = (x(u, v), y(u, v))$ takes the subset D^* in the uv-plane to the subset $D = \{(x, y) \mid (x, y) = \mathbf{T}(u, v) \text{ for some } (u, v) \in D^*\}$ of the xy-plane.

EXAMPLE 1 Let $\mathbf{T}(u, v) = (u+1, v+2)$; that is, let $x = u+1$, $y = v+2$. This transformation translates the origin in the uv-plane to the point $(1, 2)$ in the xy-plane and shifts all other points accordingly. The unit square $D^* = [0, 1] \times [0, 1]$, for example, is shifted one unit to the right and two units up, but is otherwise unchanged, as shown in Figure 5.70. Thus, the image of D^* is $D = [1, 2] \times [2, 3]$. ◆

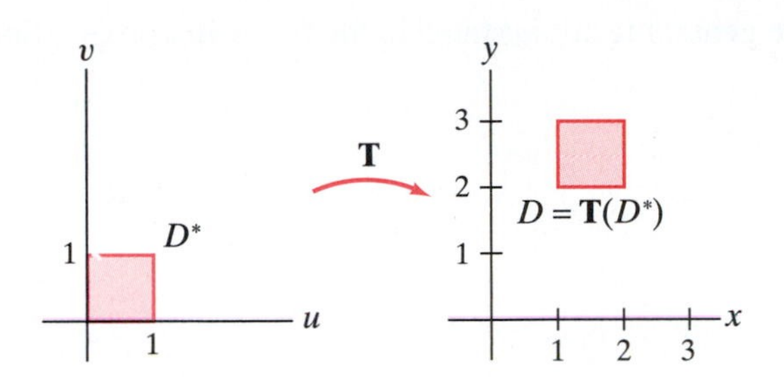

Figure 5.70 The image of $D^* = [0, 1] \times [0, 1]$ is $D = [1, 2] \times [2, 3]$ under the translation $\mathbf{T}(u, v) = (u + 1, v + 2)$ of Example 1.

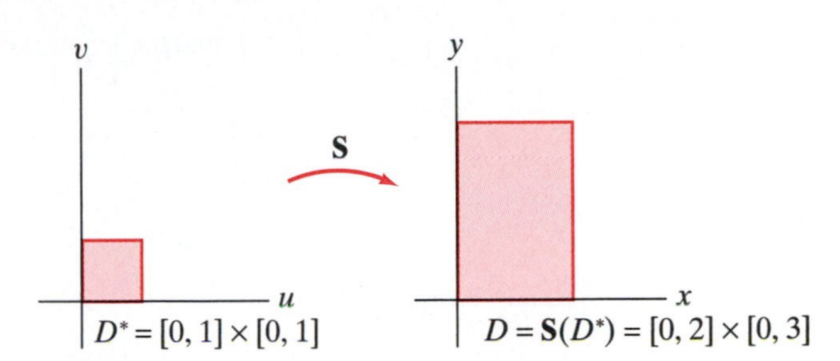

Figure 5.71 The transformation **S** of Example 2 is a scaling by a factor of 2 in the horizontal direction and 3 in the vertical direction.

EXAMPLE 2 Let $\mathbf{S}(u, v) = (2u, 3v)$. The origin is left fixed, but **S** stretches all other points by a factor of two in the horizontal direction and by a factor of three in the vertical direction. (See Figure 5.71.) ◆

EXAMPLE 3 Composing the transformations in Examples 1 and 2, we obtain

$$(\mathbf{T} \circ \mathbf{S})(u, v) = \mathbf{T}(2u, 3v) = (2u + 1, 3v + 2).$$

Such a transformation must both stretch and translate as shown in Figure 5.72. ◆

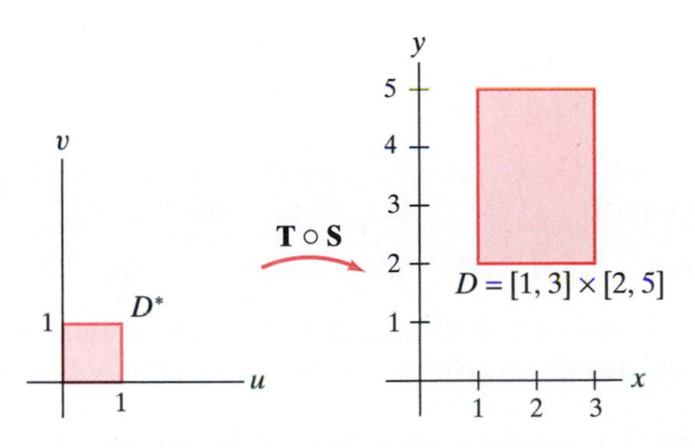

Figure 5.72 Composition of the transformations of Examples 1 and 2.

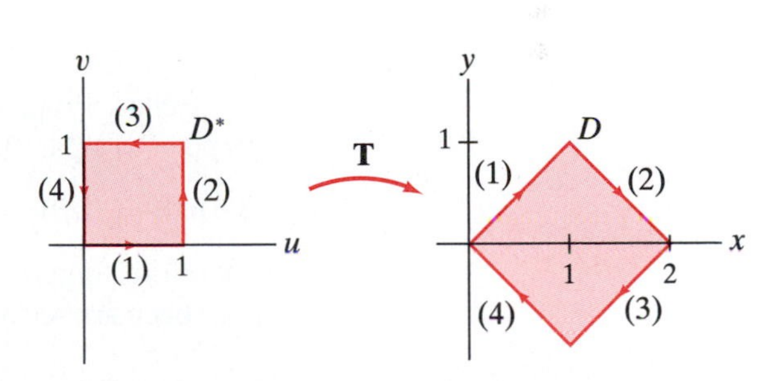

Figure 5.73 The transformation **T** of Example 4.

EXAMPLE 4 Let $\mathbf{T}(u, v) = (u + v, u - v)$. Because each of the component functions of **T** involves both variables u and v, it is less obvious how the unit square $D^* = [0, 1] \times [0, 1]$ transforms. We can begin to get some idea of the geometry by seeing how **T** maps the edges of D^*:

$$\begin{aligned}
\text{Bottom edge:} &\quad (u, 0), \quad 0 \le u \le 1; \quad \mathbf{T}(u, 0) = (u, u);\\
\text{Top edge:} &\quad (u, 1), \quad 0 \le u \le 1; \quad \mathbf{T}(u, 1) = (u + 1, u - 1);\\
\text{Left edge:} &\quad (0, v), \quad 0 \le v \le 1; \quad \mathbf{T}(0, v) = (v, -v);\\
\text{Right edge:} &\quad (1, v), \quad 0 \le v \le 1; \quad \mathbf{T}(1, v) = (v + 1, 1 - v).
\end{aligned}$$

By sketching the images of the edges, it is now plausible that the image of D^* under **T** is as shown in Figure 5.73. ◆

More generally, we consider **linear transformations** $\mathbf{T}: \mathbf{R}^2 \to \mathbf{R}^2$ defined by

$$\mathbf{T}(u, v) = (au + bv, cu + dv) = \begin{bmatrix} a & b \\ c & d \end{bmatrix} \begin{bmatrix} u \\ v \end{bmatrix},$$

where a, b, c, and d are constants. (Note: The vector (u, v) is identified with the 2×1 matrix $\begin{bmatrix} u \\ v \end{bmatrix}$.) One general result is stated in the following proposition:

Proposition 5.1 Let $A = \begin{bmatrix} a & b \\ c & d \end{bmatrix}$, where $\det A \neq 0$. If $\mathbf{T}: \mathbf{R}^2 \to \mathbf{R}^2$ is defined by

$$\mathbf{T}(u, v) = A \begin{bmatrix} u \\ v \end{bmatrix},$$

then $\mathbf{T}$ is one-one, onto, takes parallelograms to parallelograms and the vertices of parallelograms to vertices. (See §2.1 to review the notions of one-one and onto functions.) Moreover, if D^* is a parallelogram in the uv-plane that is mapped onto the parallelogram $D = \mathbf{T}(D^*)$ in the xy-plane, then

$$\text{Area of } D = |\det A| \cdot (\text{Area of } D^*).$$

EXAMPLE 5 We may write the transformation $\mathbf{T}(u, v) = (u + v, u - v)$ in Example 4 as

$$\mathbf{T}(u, v) = \begin{bmatrix} 1 & 1 \\ 1 & -1 \end{bmatrix} \begin{bmatrix} u \\ v \end{bmatrix}.$$

Note that

$$\det \begin{bmatrix} 1 & 1 \\ 1 & -1 \end{bmatrix} = -2 \neq 0.$$

Hence, Proposition 5.1 tells us that the square $D^* = [0, 1] \times [0, 1]$ must be mapped to a parallelogram $D = \mathbf{T}(D^*)$ whose vertices are

$$\mathbf{T}(0, 0) = (0, 0), \quad \mathbf{T}(0, 1) = (1, -1), \quad \mathbf{T}(1, 0) = (1, 1), \quad \mathbf{T}(1, 1) = (2, 0).$$

Therefore, Figure 5.73 is indeed correct and, in view of Proposition 5.1, could have been arrived at quite quickly. Also note that the area of D is $|-2| \cdot 1 = 2$. ◆

Proof of Proposition 5.1 First we show that $\mathbf{T}$ is one-one. So suppose $\mathbf{T}(u, v) = \mathbf{T}(u', v')$. We show that then $u = u'$, $v = v'$. We have

$$\mathbf{T}(u, v) = \mathbf{T}(u', v')$$

if and only if

$$(au + bu, cu + dv) = (au' + bv', cu' + du').$$

By equating components and manipulating, we see this is equivalent to the system

$$\begin{cases} a(u - u') + b(v - v') = 0 \\ c(u - u') + d(v - v') = 0 \end{cases}. \tag{1}$$

If $a \neq 0$, then we may use the first equation to solve for $u - u'$:

$$u - u' = -\frac{b}{a}(v - v') \tag{2}$$

Hence, the second equation in (1) becomes

$$-\frac{bc}{a}(v - v') + d(v - v') = 0$$

or, equivalently,

$$\frac{-bc + ad}{a}(v - v') = 0.$$

By hypothesis, $\det A = ad - bc \neq 0$. Thus, we must have $v - v' = 0$ and, therefore, $u - u' = 0$ by equation (2). If $a = 0$, then we must have both $b \neq 0$ and $c \neq 0$, since $\det A \neq 0$. Consequently, the system (1) becomes

$$\begin{cases} b(v - v') & = 0 \\ c(u - u') + d(v - v') & = 0 \end{cases}.$$

The first equation implies $v - v' = 0$ and hence the second becomes $c(u - u') = 0$, which in turn implies $u - u' = 0$, as desired.

To see that $\mathbf{T}$ is onto, we must show that, given any point $(x, y) \in \mathbf{R}^2$, we can find $(u, v) \in \mathbf{R}^2$ such that $\mathbf{T}(u, v) = (x, y)$. This is equivalent to solving the pair of equations

$$\begin{cases} au + bv = x \\ cu + dv = y \end{cases}$$

for u and v. We leave it to you to check that

$$u = \frac{dx - by}{ad - bc} \quad \text{and} \quad v = \frac{ay - cx}{ad - bc}$$

will work.

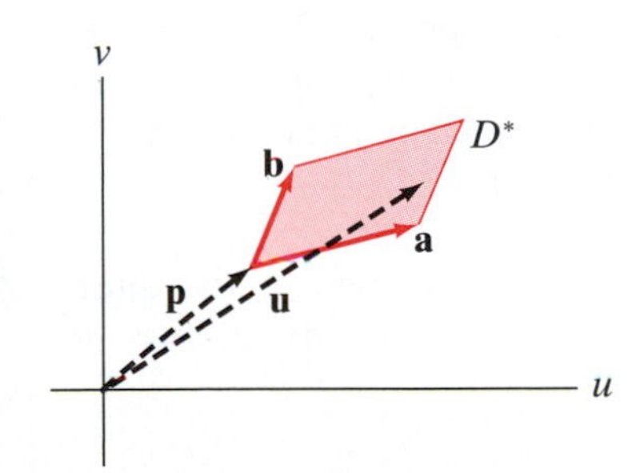

Figure 5.74 The vertices of $D^* = \{\mathbf{p} + s\mathbf{a} + t\mathbf{b} \mid 0 \leq s, t \leq 1\}$ are at $\mathbf{p}$, $\mathbf{p} + \mathbf{a}$, $\mathbf{p} + \mathbf{b}$, $\mathbf{p} + \mathbf{a} + \mathbf{b}$ (i.e., where s and t take on the values 0 or 1).

Now, let D^* be a parallelogram in the uv-plane. (See Figure 5.74.) Then D^* may be described as

$$D^* = \{\mathbf{u} \mid \mathbf{u} = \mathbf{p} + s\mathbf{a} + t\mathbf{b},\ 0 \leq s \leq 1,\ 0 \leq t \leq 1\}.$$

Hence,

$$\begin{aligned} D = \mathbf{T}(D^*) &= \{A\mathbf{u} \mid \mathbf{u} \in D\} \\ &= \{A(\mathbf{p} + s\mathbf{a} + t\mathbf{b}) \mid 0 \leq s \leq 1,\ 0 \leq t \leq 1\} \\ &= \{A\mathbf{p} + sA\mathbf{a} + tA\mathbf{b} \mid 0 \leq s \leq 1,\ 0 \leq t \leq 1\}. \end{aligned}$$

If we let $\mathbf{p}' = A\mathbf{p}$, $\mathbf{a}' = A\mathbf{a}$, and $\mathbf{b}' = A\mathbf{b}$, then

$$D = \{\mathbf{p}' + s\mathbf{a}' + t\mathbf{b}' \mid 0 \leq s \leq 1, 0 \leq t \leq 1\}.$$

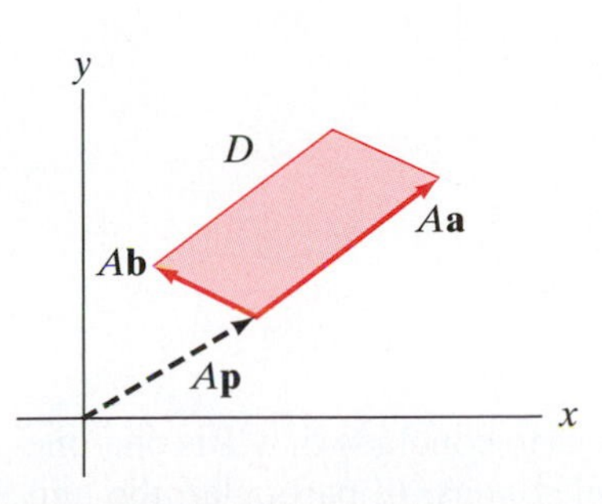

Figure 5.75 The image D of the parallelogram D^* under the linear transformation $\mathbf{T}(\mathbf{u}) = A\mathbf{u}$.

Thus, D is also a parallelogram and, moreover, the vertices of D correspond to those of D^*. (See Figure 5.75.)

Finally, note that the area of the parallelogram D^* whose sides are parallel to

$$\mathbf{a} = \begin{bmatrix} a_1 \\ a_2 \end{bmatrix} \quad \text{and} \quad \mathbf{b} = \begin{bmatrix} b_1 \\ b_2 \end{bmatrix}$$

may be computed as follows:

$$\text{Area of } D^* = \|\mathbf{a} \times \mathbf{b}\| = \left\| \det \begin{bmatrix} \mathbf{i} & \mathbf{j} & \mathbf{k} \\ a_1 & a_2 & 0 \\ b_1 & b_2 & 0 \end{bmatrix} \right\| = |a_1 b_2 - a_2 b_1|.$$

Similarly, the area of $D = \mathbf{T}(D^*)$ whose sides are parallel to

$$\mathbf{a}' = \begin{bmatrix} a_1' \\ a_2' \end{bmatrix} \quad \text{and} \quad \mathbf{b}' = \begin{bmatrix} b_1' \\ b_2' \end{bmatrix}$$

is

$$\text{Area of } D = \|\mathbf{a}' \times \mathbf{b}'\| = |a_1' b_2' - a_2' b_1'|.$$

Now, $\mathbf{a}' = A\mathbf{a}$ and $\mathbf{b}' = A\mathbf{b}$. Therefore,

$$\begin{bmatrix} a_1' \\ a_2' \end{bmatrix} = \begin{bmatrix} a & b \\ c & d \end{bmatrix} \begin{bmatrix} a_1 \\ a_2 \end{bmatrix} = \begin{bmatrix} aa_1 + ba_2 \\ ca_1 + da_2 \end{bmatrix},$$

and

$$\begin{bmatrix} b_1' \\ b_2' \end{bmatrix} = \begin{bmatrix} a & b \\ c & d \end{bmatrix} \begin{bmatrix} b_1 \\ b_2 \end{bmatrix} = \begin{bmatrix} ab_1 + bb_2 \\ cb_1 + db_2 \end{bmatrix}.$$

Hence, by appropriate substitution and algebra,

$$\begin{aligned} \text{Area of } D &= |(aa_1 + ba_2)(cb_1 + db_2) - (ca_1 + da_2)(ab_1 + bb_2)| \\ &= |(ad - bc)(a_1 b_2 - a_2 b_1)| \\ &= |\det A| \cdot \text{area of } D^*. \end{aligned}$$

Note that we have not precluded the possibility of D^*'s being a "degenerate" parallelogram, that is, such that the adjacent sides are represented by vectors $\mathbf{a}$ and $\mathbf{b}$, where $\mathbf{b}$ is a scalar multiple of $\mathbf{a}$. When this happens, D will also be a degenerate parallelogram. The assumption that $\det A \neq 0$ guarantees that a nondegenerate parallelogram D^* will be transformed into another nondegenerate parallelogram, although we have not proved this fact. ■

Essentially all of the preceding comments can be adapted to the three-dimensional case. We omit the formalism and, instead, briefly discuss an example.

EXAMPLE 6 Let $\mathbf{T}: \mathbf{R}^3 \to \mathbf{R}^3$ be given by

$$\mathbf{T}(u, v, w) = (2u, 2u + 3v + w, 3w).$$

Then we rewrite $\mathbf{T}$ by using matrix multiplication:

$$\mathbf{T}(u, v, w) = \begin{bmatrix} 2 & 0 & 0 \\ 2 & 3 & 1 \\ 0 & 0 & 3 \end{bmatrix} \begin{bmatrix} u \\ v \\ w \end{bmatrix}.$$

Note that if

$$A = \begin{bmatrix} 2 & 0 & 0 \\ 2 & 3 & 1 \\ 0 & 0 & 3 \end{bmatrix},$$

then $\det A = 18 \neq 0$.

A result analogous to Proposition 5.1 allows us to conclude that $\mathbf{T}$ is one-one and onto, and $\mathbf{T}$ maps parallelepipeds to parallelepipeds. In particular, the unit cube

$$D^* = [0, 1] \times [0, 1] \times [0, 1]$$

is mapped onto some parallelepiped $D = \mathbf{T}(D^*)$ and, moreover, the volume of D must be

$$|\det A| \cdot \text{volume of } D^* = 18 \cdot 1 = 18.$$

To determine D, we need only determine the images of the vertices of the cube:

$$T(0, 0, 0) = (0, 0, 0); \quad T(1, 0, 0) = (2, 2, 0); \quad T(0, 1, 0) = (0, 3, 0);$$
$$T(0, 0, 1) = (0, 1, 3); \quad T(1, 1, 0) = (2, 5, 0); \quad T(1, 0, 1) = (2, 3, 3);$$
$$T(0, 1, 1) = (0, 4, 3); \quad T(1, 1, 1) = (2, 6, 3).$$

Both D^* and its image D are shown in Figure 5.76. ◆

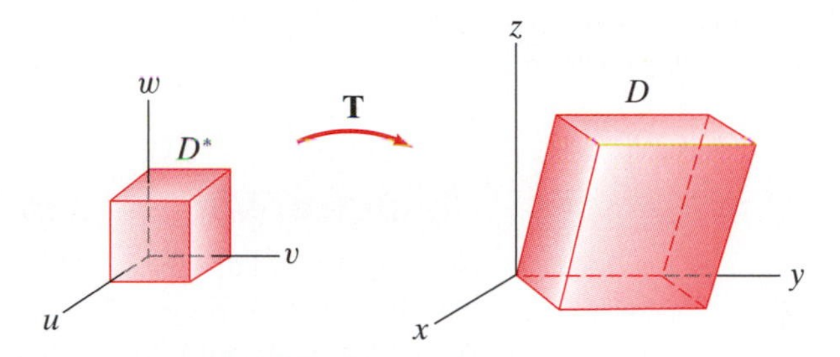

Figure 5.76 The cube D^* and its image D under the linear transformation of Example 6.

EXAMPLE 7 Of course, not all transformations are linear ones. Consider

$$(x, y) = \mathbf{T}(r, \theta) = (r\cos\theta, r\sin\theta).$$

Note that $\mathbf{T}$ is not one-one since $\mathbf{T}(0, 0) = (0, 0) = \mathbf{T}(0, \pi)$. (Indeed $\mathbf{T}(0, \theta) = (0, 0)$ for all real numbers θ.) Note that vertical lines in the $r\theta$-plane given by $r = a$, where a is constant, are mapped to the points $(x, y) = (a\cos\theta, a\sin\theta)$ on a circle of radius a. Horizontal rays $\{(r, \theta) \mid \theta = \alpha, r \geq 0\}$ are mapped to rays emanating from the origin. (See Figure 5.77.) It follows that the rectangle $D^* = [\frac{1}{2}, 1] \times [0, \pi]$ in the $r\theta$-plane is mapped not to a parallelogram, but bent into a region D that is part of the annular region between circles of radii $\frac{1}{2}$ and 1, as shown in Figure 5.78.

Analogously, the transformation $\mathbf{T}: \mathbf{R}^3 \to \mathbf{R}^3$ given by

$$(x, y, z) = \mathbf{T}(r, \theta, z) = (r\cos\theta, r\sin\theta, z)$$

bends the solid box $B^* = [\frac{1}{2}, 1] \times [0, \pi] \times [0, 1]$ into a horseshoe-shaped solid. (See Figure 5.79.) ◆

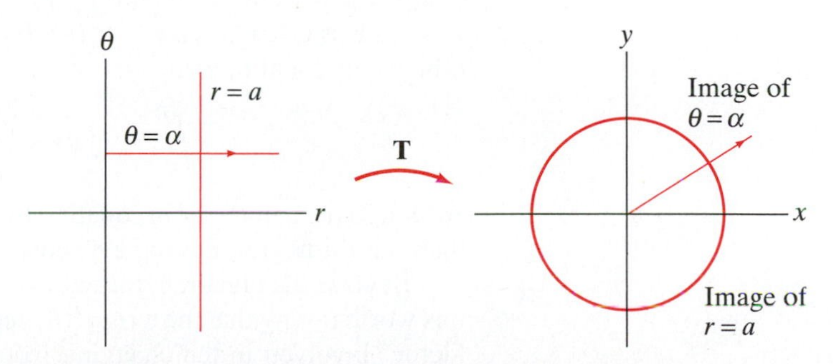

Figure 5.77 The images of lines in the $r\theta$-plane under the transformation $\mathbf{T}(r, \theta) = (r\cos\theta, r\sin\theta)$.

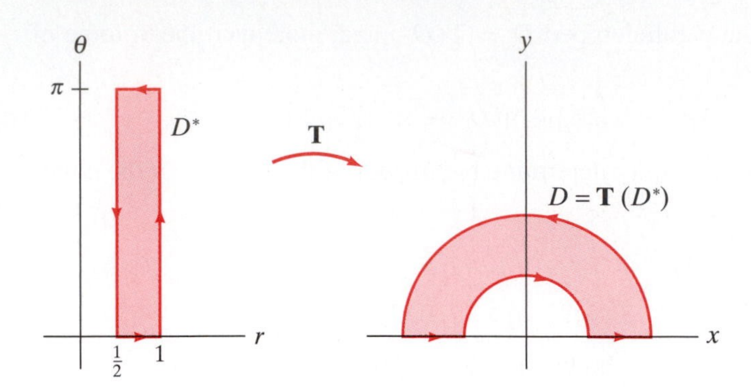

Figure 5.78 The image of the rectangle $D^* = [\frac{1}{2}, 1] \times [0, \pi]$ under $\mathbf{T}(r, \theta) = (r\cos\theta, r\sin\theta)$.

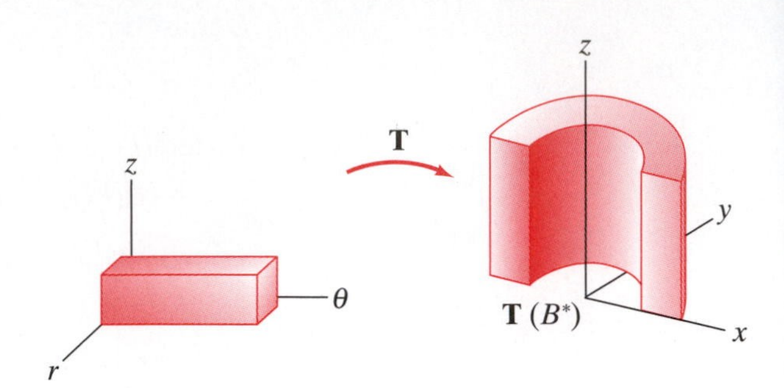

Figure 5.79 The image of $B^* = [\frac{1}{2}, 1] \times [0, \pi] \times [0, 1]$ under $\mathbf{T}(r, \theta, z) = (r\cos\theta, r\sin\theta, z)$.

Change of Variables in Definite Integrals

Now we see what effect a coordinate transformation can have on integrals and how to take advantage of such an effect. To begin, consider a case with which you are already familiar, namely, the method of substitution in single-variable integrals.

EXAMPLE 8 Consider the definite integral $\int_0^2 2x\cos(x^2)\,dx$. To evaluate, one typically makes the substitution $u = x^2$ (so $du = 2x\,dx$). Doing so, we have

$$\int_0^2 2x\cos(x^2)\,dx = \int_0^4 \cos u\,du = \sin u\Big|_{u=0}^{u=4} = \sin 4.$$

Let's dissect this example more carefully. First of all, the substitution $u = x^2$ may be rewritten (restricting x to nonnegative values only) as $x = \sqrt{u}$. Then $dx = du/(2\sqrt{u})$ and

$$\int_0^2 2x\cos(x^2)dx = \int_0^4 2\sqrt{u}\cos(\sqrt{u})^2\frac{du}{2\sqrt{u}} = \int_0^4 \cos u\,du = \sin 4.$$

In other words, the substitution is such that the $2x = 2\sqrt{u}$ factor in the integrand is canceled by the functional part of the differential $dx = du/(2\sqrt{u})$. Hence, a simple integral results. ◆

In general, the method of substitution works as follows: Given a (perhaps complicated) definite integral $\int_A^B f(x)\,dx$, make the substitution $x = x(u)$, where x is of class C^1. Thus, $dx = x'(u)\,du$. If $A = x(a)$, $B = x(b)$, and $x'(u) \neq 0$ for u between a and b, then

$$\int_A^B f(x)\,dx = \int_a^b f(x(u))x'(u)\,du. \tag{3}$$

Although the u-integral in equation (3) may at first appear to be more complicated than the x-integral, Example 8 shows that in fact just the opposite can be true.

Beyond the algebraic formalism of one-variable substitution in equation (3), it is worth noting that the term $x'(u)$ represents the "infinitesimal length distortion factor" involved in the changing from measurement in u to measurement in x. (See Figure 5.80.) We next attempt to understand how these ideas may be adapted to the case of multiple integrals.

Figure 5.80 As $\Delta u = du \to 0$, $\Delta x \to dx = x'(u)\,\Delta u$. Thus, the factor $x'(u)$ measures how length in the u-direction relates to length in the x-direction.

The Change of Variables Theorem for Double Integrals

Suppose we have a differentiable coordinate transformation from the uv-plane to the xy-plane. That is,

$$\mathbf{T}\colon \mathbf{R}^2 \to \mathbf{R}^2, \qquad \mathbf{T}(u, v) = (x(u, v), y(u, v)).$$

■ **Definition 5.2** The **Jacobian** of the transformation $\mathbf{T}$, denoted

$$\frac{\partial(x, y)}{\partial(u, v)},$$

is the determinant of the derivative matrix $D\mathbf{T}(u, v)$. That is,

$$\frac{\partial(x, y)}{\partial(u, v)} = \det D\mathbf{T}(u, v) = \det \begin{bmatrix} \dfrac{\partial x}{\partial u} & \dfrac{\partial x}{\partial v} \\[2mm] \dfrac{\partial y}{\partial u} & \dfrac{\partial y}{\partial v} \end{bmatrix} = \frac{\partial x}{\partial u}\frac{\partial y}{\partial v} - \frac{\partial x}{\partial v}\frac{\partial y}{\partial u}.$$

The notation $\partial(x, y)/\partial(u, v)$ for the Jacobian is a historical convenience. The Jacobian is *not* a partial derivative, but rather the *determinant* of the *matrix* of partial derivatives. It plays the role of an "infinitesimal area distortion factor" when changing variables in double integrals, as in the following key result:

■ **Theorem 5.3** (Change of Variables in Double Integrals) Let D and D^* be elementary regions in (respectively) the xy-plane and the uv-plane. Suppose $\mathbf{T}\colon \mathbf{R}^2 \to \mathbf{R}^2$ is a coordinate transformation of class C^1 that maps D^* onto D in a one-one fashion. If $f\colon D \to \mathbf{R}$ is any integrable function and we use the transformation $\mathbf{T}$ to make the substitution $x = x(u, v)$, $y = y(u, v)$, then

$$\iint_D f(x, y)\, dx\, dy = \iint_{D^*} f(x(u, v), y(u, v)) \left| \frac{\partial(x, y)}{\partial(u, v)} \right| du\, dv.$$

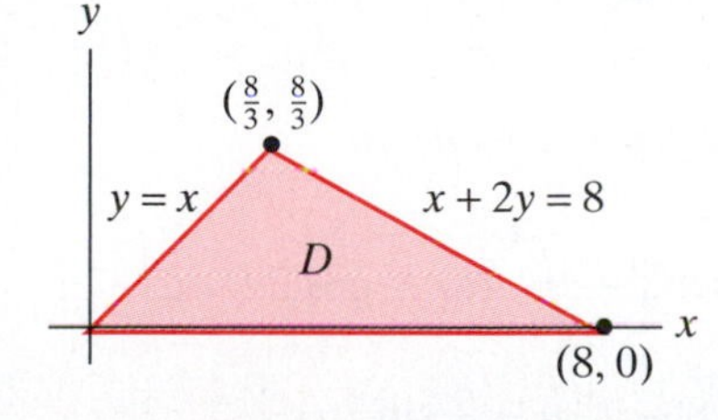

Figure 5.81 The triangular region D of Example 9.

EXAMPLE 9 We use Theorem 5.3 to calculate the integral

$$\iint_D \cos(x + 2y)\sin(x - y)\, dx\, dy$$

over the triangular region D bounded by the lines $y = 0$, $y = x$, and $x + 2y = 8$ as shown in Figure 5.81. It is possible to evaluate this integral by using the relatively straightforward methods of §5.2. However, this would prove to be cumbersome, so, instead, we find a suitable transformation of variables, motivated in this case by the nature of the integrand. In particular, we let $u = x + 2y$, $v = x - y$. Solving for x and y, we obtain

$$x = \frac{u + 2v}{3} \quad \text{and} \quad y = \frac{u - v}{3}.$$

Therefore,

$$\frac{\partial(x, y)}{\partial(u, v)} = \det \begin{bmatrix} x_u & x_v \\ y_u & y_v \end{bmatrix} = \det \begin{bmatrix} \frac{1}{3} & \frac{2}{3} \\ \frac{1}{3} & -\frac{1}{3} \end{bmatrix} = -\tfrac{1}{3}.$$

Considering the coordinate transformation as a mapping $\mathbf{T}(u, v) = (x, y)$ of the plane, we need to identify a region D^* that $\mathbf{T}$ maps in a one-one fashion onto D. To do this, essentially all we need do is to consider the boundaries of D:

$$\begin{aligned} y = x \quad &\Longleftrightarrow \quad x - y = 0 \quad \Longleftrightarrow \quad v = 0; \\ x + 2y = 8 \quad &\Longleftrightarrow \quad u = 8; \\ y = 0 \quad &\Longleftrightarrow \quad \frac{u - v}{3} = 0 \quad \Longleftrightarrow \quad v = u. \end{aligned}$$

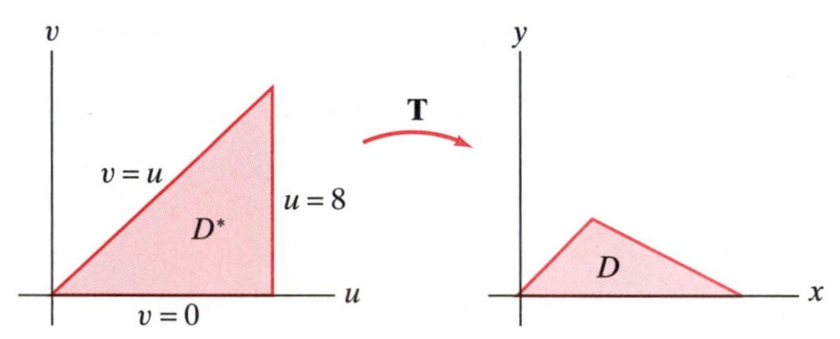

Figure 5.82 The effect of the transformation **T** of Example 9.

Hence, one can see that $\mathbf{T}$ transforms the region D^* shown in Figure 5.82 onto D. Therefore, applying Theorem 5.3,

$$\begin{aligned} \iint_D \cos(x + 2y) \sin(x - y)\, dx\, dy &= \iint_{D^*} \cos u \sin v \left| \frac{\partial(x, y)}{\partial(u, v)} \right| du\, dv \\ &= \iint_{D^*} \cos u \sin v \left| -\tfrac{1}{3} \right| du\, dv \\ &= \int_0^8 \int_0^u \tfrac{1}{3} \cos u \sin v\, dv\, du \\ &= \int_0^8 \tfrac{1}{3} \cos u\, (-\cos v)\Big|_{v=0}^{v=u} du \\ &= \tfrac{1}{3} \int_0^8 \cos u(-\cos u + 1)\, du \\ &= \tfrac{1}{3} \int_0^8 (\cos u - \cos^2 u)\, du \\ &= \tfrac{1}{3} \left[\sin u\Big|_0^8 - \int_0^8 \tfrac{1}{2}(1 + \cos 2u)\, du \right] \\ &= \tfrac{1}{3} \left[\sin 8 - \left(\tfrac{1}{2}u + \tfrac{1}{4} \sin 2u\right)\Big|_0^8 \right] \\ &= \tfrac{1}{3} \left[\sin 8 - 4 - \tfrac{1}{4} \sin 16 \right]. \end{aligned}$$

There is another, faster way to calculate the Jacobian, namely, to calculate $\partial(u, v)/\partial(x, y)$ directly from the variable transformation, and then to take reciprocals. That is, from the equations $u = x + 2y$, $v = x - y$, we have

$$\frac{\partial(u, v)}{\partial(x, y)} = \det \begin{bmatrix} u_x & u_y \\ v_x & v_y \end{bmatrix} = \det \begin{bmatrix} 1 & 2 \\ 1 & -1 \end{bmatrix} = -3.$$

Consequently, $\partial(x, y)/\partial(u, v) = -\frac{1}{3}$, which checks with our previous result. This method works because if $\mathbf{T}(u, v) = (x, y)$, then, under the assumptions of Theorem 5.3, $(u, v) = \mathbf{T}^{-1}(x, y)$. It follows from the chain rule that

$$D\mathbf{T}^{-1}(x, y) = [D\mathbf{T}(u, v)]^{-1}.$$

(That is, $D\mathbf{T}^{-1}$ is the inverse matrix of $D\mathbf{T}$. See Exercises 30–38 in §1.6 for more about inverse matrices.) Hence,

$$\frac{\partial(x, y)}{\partial(u, v)} = \det\left[D\mathbf{T}^{-1}\right] = \det\left[(D\mathbf{T})^{-1}\right] = \frac{1}{\det D\mathbf{T}}.$$

◆

EXAMPLE 10 (Double integrals in polar coordinates) In the previous example, a coordinate transformation was chosen primarily to simplify the integrand of the double integral. Another equally important reason for changing variables is to use a coordinate system better suited to the geometry of the region of integration.

For example, suppose that the region D is a disk of radius a:

$$\begin{aligned} D &= \{(x, y) \mid x^2 + y^2 \le a\} \\ &= \left\{(x, y) \mid -\sqrt{a^2 - x^2} \le y \le \sqrt{a^2 - x^2}, -a \le x \le a\right\}. \end{aligned}$$

Then, to integrate any (integrable) function f over D in Cartesian coordinates, one would write

$$\iint_D f(x, y)\, dx\, dy = \int_{-a}^{a} \int_{-\sqrt{a^2-x^2}}^{\sqrt{a^2-x^2}} f(x, y)\, dy\, dx.$$

Even if it is easy initially to find a partial antiderivative of the integrand, the limits in the preceding double integral may complicate matters considerably. This is because the disk is described rather awkwardly by Cartesian coordinates. We know, however, that it has a much more convenient description in polar coordinates as

$$\{(r, \theta) \mid 0 \le r \le a, 0 \le \theta < 2\pi\}.$$

This suggests that we make the change of variables

$$(x, y) = \mathbf{T}(r, \theta) = (r \cos\theta, r \sin\theta),$$

which is shown in Figure 5.83. (Note that $\mathbf{T}$ maps all points of the form $(0, \theta)$ to the origin in the xy-plane and, thus, cannot map D^* in a one-one fashion onto D. Nonetheless, the points of D^* on which $\mathbf{T}$ fails to be one-one fill out a portion of a line—a one-dimensional locus—and it turns out that it will not affect the double integral transformation.) The Jacobian for this change of variables is

$$\frac{\partial(x, y)}{\partial(r, \theta)} = \det \begin{bmatrix} \cos\theta & -r \sin\theta \\ \sin\theta & r \cos\theta \end{bmatrix} = r \cos^2\theta + r \sin^2\theta = r.$$

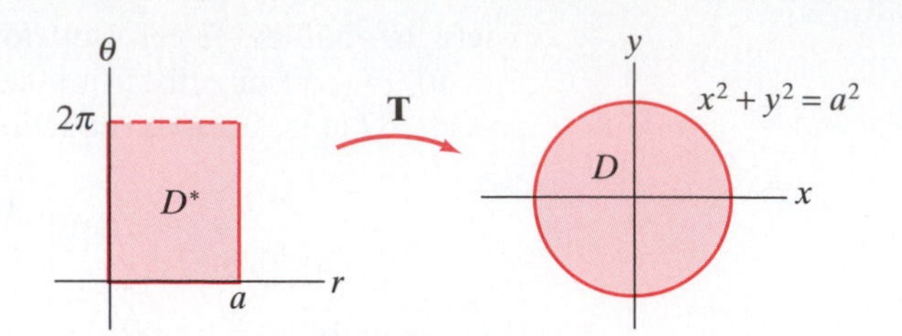

Figure 5.83 **T** maps the (nonclosed) rectangle D^* to the disk D of radius a.

(Note that $r \geq 0$ on D, so $|r| = r$.) Thus, using Theorem 5.3, the double integral can be evaluated by using polar coordinates as follows:

$$\iint_D f(x, y)\, dx\, dy = \int_{-a}^{a} \int_{-\sqrt{a^2-x^2}}^{\sqrt{a^2-x^2}} f(x, y)\, dy\, dx$$

$$= \int_0^{2\pi} \int_0^a f(r\cos\theta, r\sin\theta)\, r\, dr\, d\theta.$$

It is evident that the limits of integration of the $r\theta$-integrals are substantially simpler than those in the xy-integral. Of course, the substitution in the integrand may result in a more complicated expression, but in many situations this will not be the case. Polar coordinate transformations will prove to be especially convenient when dealing with regions whose boundaries are parts of circles. ◆

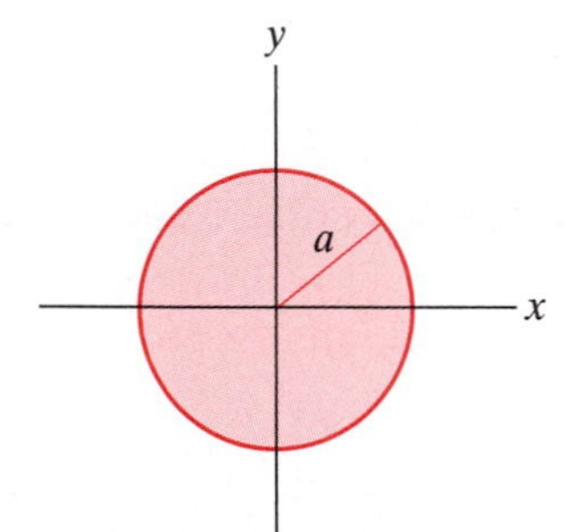

Figure 5.84 The disk of radius a centered at the origin.

EXAMPLE 11 To see polar coordinates "in action," we calculate the area of a circle, using double integrals. Once more, let D be the disk of radius a, centered at the origin as in Figure 5.84. Then we have

$$\text{Area} = \iint_D 1\, dA = \int_{-a}^{a} \int_{-\sqrt{a^2-x^2}}^{\sqrt{a^2-x^2}} dy\, dx = \int_0^{2\pi} \int_0^a r\, dr\, d\theta,$$

following the discussion in Example 10. The last iterated integral is readily evaluated as

$$\int_0^{2\pi} \int_0^a r\, dr\, d\theta = \int_0^{2\pi} \left(\tfrac{1}{2}r^2\big|_0^a\right) d\theta = \textstyle\int_0^{2\pi} \tfrac{1}{2}a^2\, d\theta = \tfrac{1}{2}a^2(2\pi - 0) = \pi a^2,$$

which indeed agrees with what we already know. If you feel so inclined, compare this calculation with the evaluation of the iterated integral in Cartesian coordinates. No doubt you'll agree that the use of polar coordinates offers clear advantages. ◆

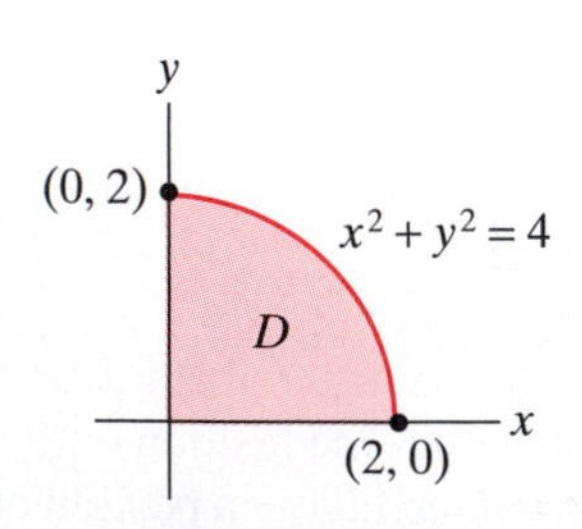

Figure 5.85 The region D of Example 12.

EXAMPLE 12 We evaluate the double integral $\iint_D \sqrt{x^2 + y^2 + 1}\, dx\, dy$, where D is the quarter disk shown in Figure 5.85, using polar coordinates. The region D of integration is given in Cartesian coordinates by

$$D = \{(x, y) \mid 0 \leq y \leq \sqrt{4 - x^2},\ 0 \leq x \leq 2\},$$

so that

$$\iint_D \sqrt{x^2 + y^2 + 1}\, dx\, dy = \int_0^2 \int_0^{\sqrt{4-x^2}} \sqrt{x^2 + y^2 + 1}\, dy\, dx.$$

This iterated integral is extremely difficult to evaluate. However, D corresponds to the polar region

$$D^* = \{(r, \theta) \mid 0 \leq r \leq 2,\ 0 \leq \theta \leq \pi/2\}.$$

Therefore, using Theorem 5.3, we have

$$\begin{aligned}\iint_D \sqrt{x^2+y^2+1}\,dx\,dy &= \iint_{D^*} \sqrt{r^2\cos^2\theta + r^2\sin^2\theta + 1}\cdot r\,dr\,d\theta \\ &= \int_0^{\pi/2}\int_0^2 \sqrt{r^2+1}\,r\,dr\,d\theta \\ &= \int_0^{\pi/2} \tfrac{1}{3}(r^2+1)^{3/2}\Big|_{r=0}^{2}\,d\theta \\ &= \int_0^{\pi/2} \tfrac{1}{3}(5^{3/2}-1)\,d\theta \\ &= \frac{\pi}{6}(5^{3/2}-1).\end{aligned}$$

◆

Sketch of a proof of Theorem 5.3 Let (u_0, v_0) be any point in D^* and let $\Delta u = u - u_0$, $\Delta v = v - v_0$. The coordinate transformation $\mathbf{T}$ maps the rectangle R^* inside D^* (shown in Figure 5.86) onto the region R inside D in the xy-plane. (In general, R will not be a rectangle.) Since $\mathbf{T}$ is of class C^1, the differentiability of $\mathbf{T}$ (see Definition 3.8 of Chapter 2) implies that the linear approximation

$$\begin{aligned}\mathbf{h}(u, v) &= \mathbf{T}(u_0, v_0) + D\mathbf{T}(u_0, v_0)\begin{bmatrix} u - u_0 \\ v - v_0 \end{bmatrix} \\ &= \mathbf{T}(u_0, v_0) + D\mathbf{T}(u_0, v_0)\begin{bmatrix} \Delta u \\ \Delta v \end{bmatrix}\end{aligned}$$

is a good approximation to $\mathbf{T}$ near the point (u_0, v_0). In particular, $\mathbf{h}$ takes the rectangle R^* onto some parallelogram P that approximates R as shown in Figure 5.87. We compare the area of R^* to that of P.

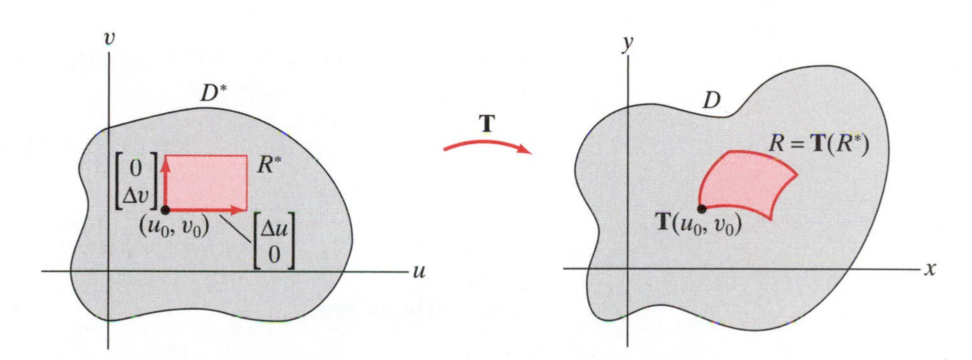

Figure 5.86 $\mathbf{T}$ takes a rectangle R^* inside D^* to a region R inside D.

From Figure 5.87, we see that the rectangle R^* is spanned by

$$\mathbf{a} = \begin{bmatrix} \Delta u \\ 0 \end{bmatrix} \quad \text{and} \quad \mathbf{b} = \begin{bmatrix} 0 \\ \Delta v \end{bmatrix},$$

and the parallelogram P is spanned by the vectors $\mathbf{c} = D\mathbf{T}(u_0, v_0)\mathbf{a}$ and $\mathbf{d} = D\mathbf{T}(u_0, v_0)\mathbf{b}$. Hence,

$$\text{Area of } R^* = \|\mathbf{a}\times\mathbf{b}\| = \Delta u\,\Delta v,$$

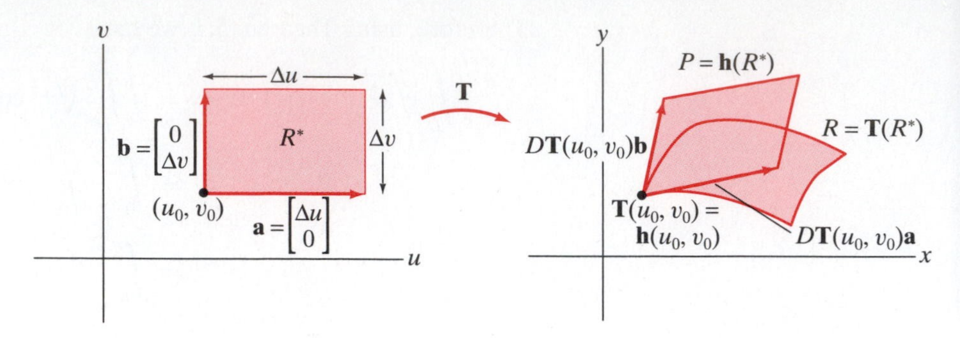

Figure 5.87 The linear approximation $\mathbf{h}$ takes the rectangle R^* onto a parallelogram P that approximates $R = \mathbf{T}(R^*)$.

and thus, by Proposition 5.1,

$$\text{Area of } P = \|\mathbf{c} \times \mathbf{d}\| = |\det D\mathbf{T}(u_0, v_0)|\Delta u \, \Delta v = \left| \frac{\partial(x, y)}{\partial(u, v)}(u_0, v_0) \right| \Delta u \, \Delta v.$$

This result gives us some idea how the Jacobian factor arises.

To complete the sketch of the proof, we need a partitioning argument. Partition D^* by subrectangles R^*_{ij}. Then we obtain a corresponding partition of D into (not necessarily rectangular) subregions $R_{ij} = \mathbf{T}(R^*_{ij})$. Let ΔA_{ij} denote the area of R_{ij}. Let $\mathbf{c}_{ij}$ denote the lower left corner of R^*_{ij} and let $\mathbf{d}_{ij} = \mathbf{T}(\mathbf{c}_{ij})$. (See Figure 5.88.) Then, since f is integrable on D,

$$\iint_D f(x, y) \, dx \, dy = \lim_{\text{all } R_{ij} \to 0} \sum_{i,j} f(\mathbf{d}_{ij}) \Delta A_{ij}.$$

From the remarks in the preceding paragraph, we know that

$$\Delta A_{ij} \approx \text{area of parallelogram } \mathbf{h}(R^*_{ij}) = \left| \frac{\partial(x, y)}{\partial(u, v)}(\mathbf{c}_{ij}) \right| \Delta u_i \, \Delta v_j.$$

Taking limits as all the R_{ij} tend to zero (i.e., as Δu_i and Δv_j approach zero), we find that

$$\begin{aligned} \iint_D f(x, y) \, dx \, dy &= \lim_{\Delta u_i, \Delta v_j \to 0} \sum_{i,j} f\left(\mathbf{T}(\mathbf{c}_{ij})\right) \left| \frac{\partial(x, y)}{\partial(u, v)}(\mathbf{c}_{ij}) \right| \Delta u_i \, \Delta v_j \\ &= \iint_{D^*} f(x(u, v), y(u, v)) \left| \frac{\partial(x, y)}{\partial(u, v)} \right| du \, dv, \end{aligned}$$

as was to be shown. ■

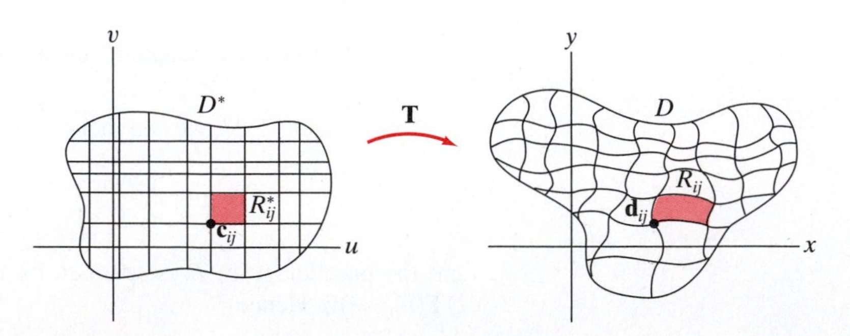

Figure 5.88 A partition of D^* gives rise to a partition of D.

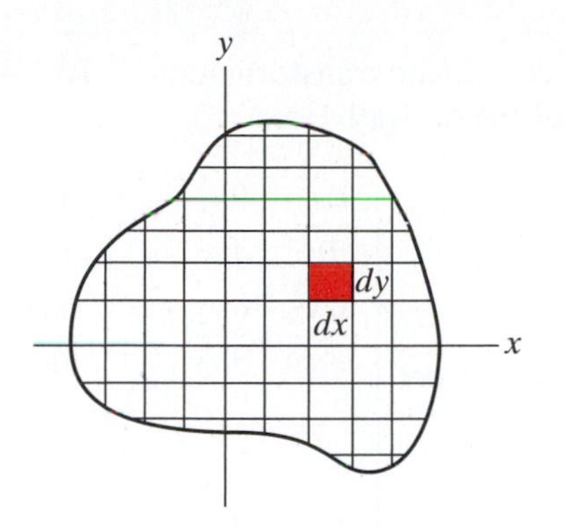

Figure 5.89 The "area element" dA in rectangular coordinates is $dx\,dy$.

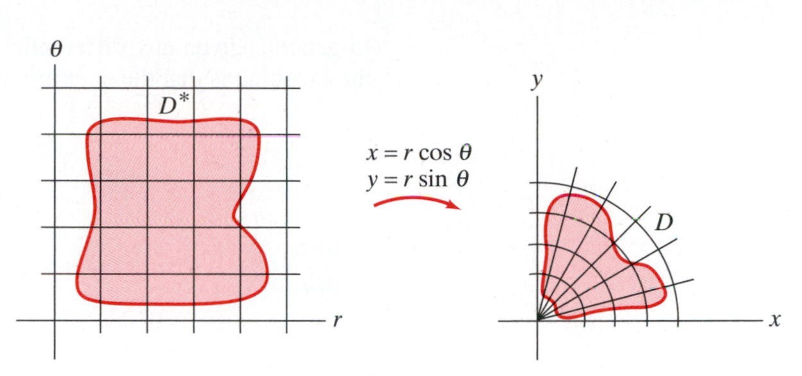

Figure 5.90 The polar-rectangular transformation takes rectangles in the $r\theta$-plane to wedges of disks in the xy-plane.

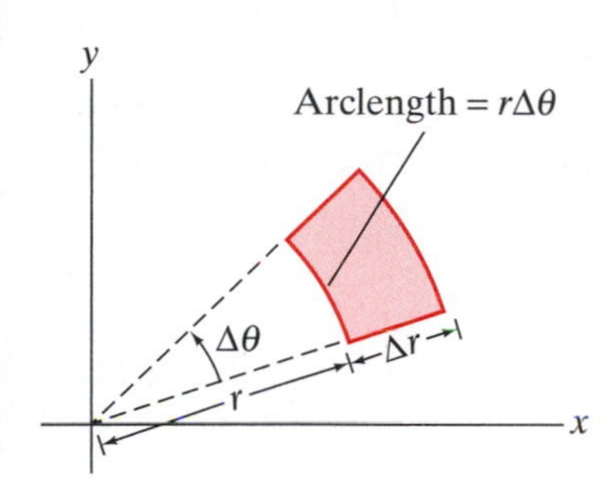

Figure 5.91 An infinitesimal polar wedge.

Consider again the polar-rectangular coordinate tranformation. When we use Cartesian (rectangular) coordinates to calculate a double integral over a region D in the plane, then we are subdividing D into "infinitesimal" rectangles having "area" equal to $dx\,dy$. (See Figure 5.89.) On the other hand, when we use polar coordinates to describe this same region, we are subdividing D into infinitesimal pieces of disks instead. (See Figure 5.90.) These disk wedges arise from transformed rectangles in the $r\theta$-plane. One such infinitesimal wedge in the xy-plane is suggested by Figure 5.91. When $\Delta\theta$ and Δr are very small, the shape is nearly rectangular with approximate area $(r\,\Delta\theta)\,\Delta r$. Thus, in the limit, we frequently say

$$
\begin{aligned}
dA &= dx\,dy && \text{(Cartesian area element)}\\
&= r\,dr\,d\theta && \text{(polar area element).}
\end{aligned}
$$

Change of Variables in Triple Integrals

It is not difficult to adapt the previous reasoning to the case of triple integrals. We omit the details, stating only the main results instead.

■ **Definition 5.4** Let $\mathbf{T}\colon \mathbf{R}^3 \to \mathbf{R}^3$ be a differentiable coordinate transformation

$$\mathbf{T}(u, v, w) = (x(u, v, w), y(u, v, w), z(u, v, w))$$

from uvw-space to xyz-space. The **Jacobian** of $\mathbf{T}$, denoted

$$\frac{\partial(x, y, z)}{\partial(u, v, w)},$$

is $\det(D\mathbf{T}(u, v, w))$. That is,

$$\frac{\partial(x, y, z)}{\partial(u, v, w)} = \det \begin{bmatrix} \dfrac{\partial x}{\partial u} & \dfrac{\partial x}{\partial v} & \dfrac{\partial x}{\partial w} \\[2ex] \dfrac{\partial y}{\partial u} & \dfrac{\partial y}{\partial v} & \dfrac{\partial y}{\partial w} \\[2ex] \dfrac{\partial z}{\partial u} & \dfrac{\partial z}{\partial v} & \dfrac{\partial z}{\partial w} \end{bmatrix}.$$

In general, given any differentiable coordinate transformation $\mathbf{T}: \mathbf{R}^n \to \mathbf{R}^n$, the Jacobian is just the determinant of the derivative matrix:

$$\frac{\partial(x_1, \ldots, x_n)}{\partial(u_1, \cdots, u_n)} = \det D\mathbf{T}(u_1, \ldots, u_n) = \det \begin{bmatrix} \dfrac{\partial x_1}{\partial u_1} & \dfrac{\partial x_1}{\partial u_2} & \cdots & \dfrac{\partial x_1}{\partial u_n} \\ \dfrac{\partial x_2}{\partial u_1} & \dfrac{\partial x_2}{\partial u_2} & \cdots & \dfrac{\partial x_2}{\partial u_n} \\ \vdots & \vdots & \ddots & \vdots \\ \dfrac{\partial x_n}{\partial u_1} & \dfrac{\partial x_n}{\partial u_2} & \cdots & \dfrac{\partial x_n}{\partial u_n} \end{bmatrix}.$$

Theorem 5.5 (CHANGE OF VARIABLES IN TRIPLE INTEGRALS) Let W and W^* be elementary regions in (respectively) xyz-space and uvw-space, and let $\mathbf{T}: \mathbf{R}^3 \to \mathbf{R}^3$ be a coordinate transformation of class C^1 that maps W^* onto W in a one-one fashion. If $f: W \to \mathbf{R}$ is integrable and we use the transformation $\mathbf{T}$ to make the substitution $x = x(u, v, w)$, $y = y(u, v, w)$, $z = z(u, v, w)$, then

$$\iiint_W f(x, y, z)\, dx\, dy\, dz$$
$$= \iiint_{W^*} f(x(u, v, w), y(u, v, w), z(u, v, w)) \left| \frac{\partial(x, y, z)}{\partial(u, v, w)} \right| du\, dv\, dw.$$

(See Figure 5.92.)

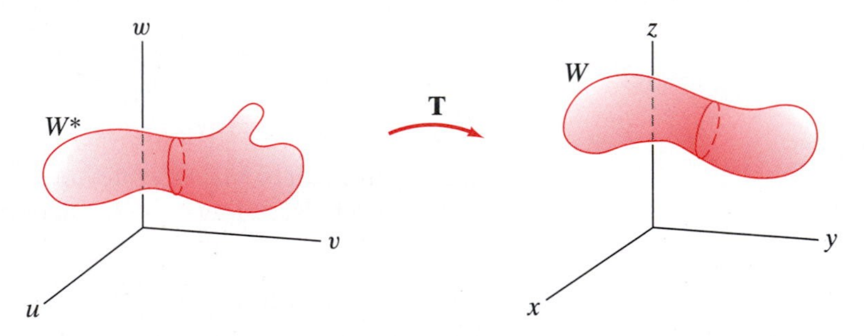

Figure 5.92 A three-dimensional tranformation **T** that takes the solid region W^* in uvw-space to the region W in xyz-space.

In the integral formula of the change of variables theorem (Theorem 5.5), the Jacobian represents the "volume distortion factor" that occurs when the three-dimensional region W is subdivided into pieces that are transformed boxes in uvw-space. (See Figure 5.93.) In other words, the differential **volume elements** (i.e., "infinitesimal" pieces of volumes) in xyz- and uvw-coordinates are related by the formula

$$dV = dx\, dy\, dz = \left| \frac{\partial(x, y, z)}{\partial(u, v, w)} \right| du\, dv\, dw.$$

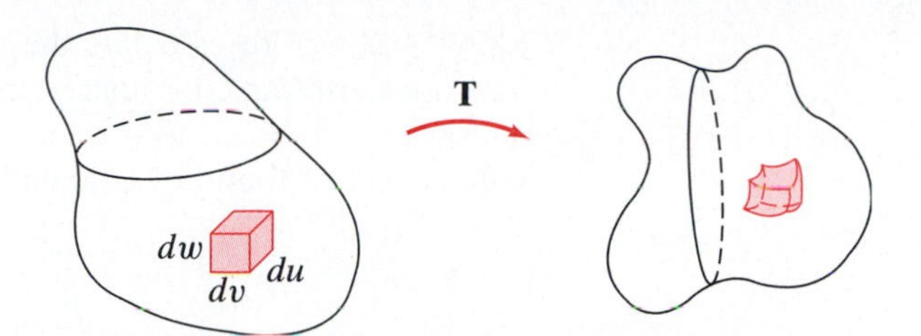

Figure 5.93 The volume of the "infinitesimal box" in uvw-space is $du\,dv\,dw$. The image of this box under $\mathbf{T}$ has volume $|\partial(x, y, z)/\partial(u, v, w)|\,du\,dv\,dw$.

EXAMPLE 13 (Triple integrals in cylindrical coordinates) When integrating over solid objects possessing an axis of rotational symmetry, cylindrical coordinates can be especially helpful. The cylindrical-rectangular coordinate transformation

$$\begin{cases} x = r\cos\theta \\ y = r\sin\theta \\ z = z \end{cases}$$

has Jacobian

$$\frac{\partial(x, y, z)}{\partial(r, \theta, z)} = \det\begin{bmatrix} \cos\theta & -r\sin\theta & 0 \\ \sin\theta & r\cos\theta & 0 \\ 0 & 0 & 1 \end{bmatrix} = r\cos^2\theta + r\sin^2\theta = r.$$

Hence, the formula in Theorem 5.5 becomes

$$\iiint_W f(x, y, z)\,dx\,dy\,dz = \iiint_{W*} f(r\cos\theta, r\sin\theta, z)\,r\,dr\,d\theta\,dz.$$

In particular, we see that the volume element in cylindrical coordinates is

$$dV = r\,dr\,d\theta\,dz.$$

(Recall that the cylindrical coordinate r is usually taken to be nonnegative. Given this convention, we may omit the absolute value sign in the change of variables formula.) The geometry behind this volume element is quite plausible: A "differential box" in $r\theta z$-space is transformed to a portion of a solid cylinder that is nearly a box itself. (See Figure 5.94.) ◆

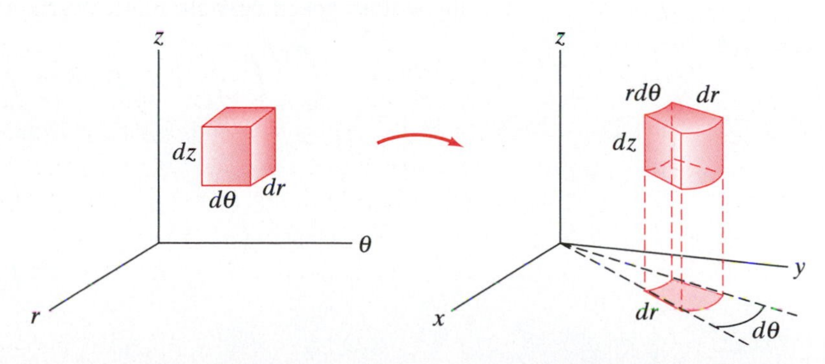

Figure 5.94 A "differential box" in $r\theta z$-space is mapped to a portion of a solid cylinder in xyz-space by the cylindrical–rectangular transformation.

EXAMPLE 14 To calculate the volume of a cone of height h and radius a, we may use Cartesian coordinates, in which case the cone is the solid W bounded by the surface $az = h\sqrt{x^2 + y^2}$ and the plane $z = h$, as shown in Figure 5.95. The volume can be found by calculating the iterated triple integral

$$\int_{-a}^{a} \int_{-\sqrt{a^2-x^2}}^{\sqrt{a^2-x^2}} \int_{\frac{h}{a}\sqrt{x^2+y^2}}^{h} dz\, dy\, dx.$$

We will forgo the details of the evaluation, noting only that trigonometric substitutions are necessary and that they make the resulting computation quite tedious.

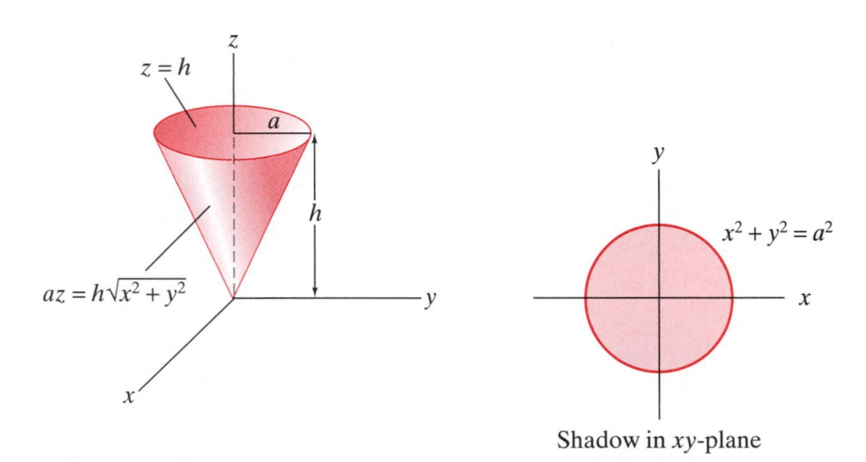

Figure 5.95 The solid cone W of Example 14.

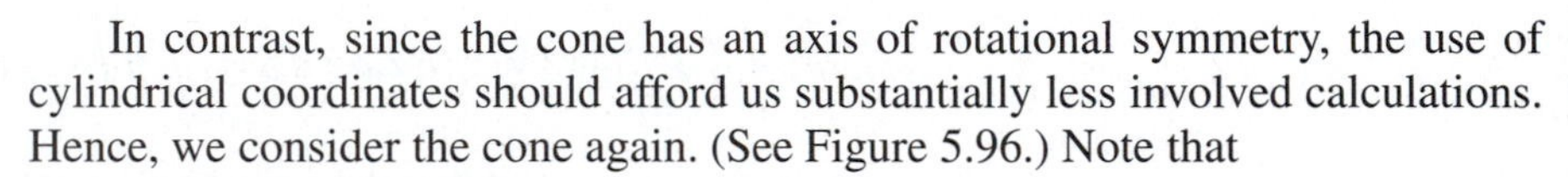

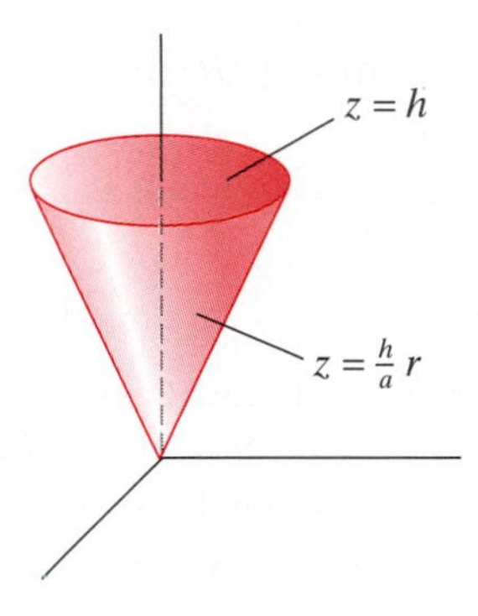

Figure 5.96 The cone of Example 14 described in cylindrical coordinates.

In contrast, since the cone has an axis of rotational symmetry, the use of cylindrical coordinates should afford us substantially less involved calculations. Hence, we consider the cone again. (See Figure 5.96.) Note that

$$W = \left\{ (r, \theta, z) \;\middle|\; \frac{h}{a} r \le z \le h,\ 0 \le r \le a,\ 0 \le \theta < 2\pi \right\}.$$

Thus, the volume is given by

$$\iiint_W dV = \int_0^{2\pi} \int_0^{a} \int_{\frac{h}{a}r}^{h} r\, dz\, dr\, d\theta.$$

(Note the order of integration that we chose.) The evaluation of this iterated integral is exceedingly straightforward; we have

$$\begin{aligned}
\int_0^{2\pi} \int_0^{a} \int_{\frac{h}{a}r}^{h} r\, dz\, dr\, d\theta &= \int_0^{2\pi} \int_0^{a} r\left(h - \frac{h}{a}r\right) dr\, d\theta \\
&= \int_0^{2\pi} \left(\frac{h}{2}r^2 - \frac{h}{3a}r^3\right)\Bigg|_{r=0}^{r=a} d\theta \\
&= \int_0^{2\pi} \left(\frac{h}{2}a^2 - \frac{h}{3}a^2\right) d\theta \\
&= 2\pi\left(\frac{h}{6}a^2\right) = \frac{\pi}{3}a^2 h,
\end{aligned}$$

which agrees with what we already know. ◆

EXAMPLE 15 **(Triple integrals in spherical coordinates)** If a solid object has a center of symmetry, then spherical coordinates can make integration over such an object more convenient. The spherical–rectangular coordinate transformation

$$\begin{cases} x = \rho \sin\varphi \cos\theta \\ y = \rho \sin\varphi \sin\theta \\ z = \rho \cos\varphi \end{cases}$$

has Jacobian

$$\frac{\partial(x, y, z)}{\partial(\rho, \varphi, \theta)} = \det \begin{bmatrix} \sin\varphi\ \cos\theta & \rho\cos\varphi\cos\theta & -\rho\sin\varphi\ \sin\theta \\ \sin\varphi\ \sin\theta & \rho\cos\varphi\ \sin\theta & \rho\sin\varphi\ \cos\theta \\ \cos\varphi & -\rho\sin\varphi & 0 \end{bmatrix}.$$

Using cofactor expansion about the last row, this determinant is equal to

$$\begin{aligned} &\cos\varphi\,(\rho^2\cos^2\theta\sin\varphi\cos\varphi + \rho^2\sin^2\theta\sin\varphi\cos\varphi) \\ &\quad + \rho\sin\varphi\,(\rho\cos^2\theta\sin^2\varphi + \rho\sin^2\theta\sin^2\varphi) \\ &\quad = \rho^2\cos\varphi(\sin\varphi\cos\varphi) + \rho^2\sin^3\varphi \\ &\quad = \rho^2\sin\varphi\,(\cos^2\varphi + \sin^2\varphi) \\ &\quad = \rho^2\sin\varphi. \end{aligned}$$

(Under the restriction that $0 \le \varphi \le \pi$, $\sin\varphi$ will always be nonnegative. Hence, the Jacobian will also be nonnegative.) Therefore, the volume element in spherical coordinates is

$$dV = \rho^2 \sin\varphi\, d\rho\, d\varphi\, d\theta,$$

and the change of variables formula in Theorem 5.5 becomes

$$\begin{aligned} &\iiint_W f(x, y, z)\, dx\, dy\, dz \\ &\quad = \iiint_{W*} f(x(\rho,\varphi,\theta), y(\rho,\varphi,\theta), z(\rho,\varphi,\theta))\rho^2\sin\varphi\, d\rho\, d\varphi\, d\theta. \end{aligned}$$

The volume element in spherical coordinates makes sense geometrically, because a differential box in $\rho\varphi\theta$-space is transformed to a portion of a solid ball that is approximated by a box having volume $\rho^2\sin\varphi\, d\rho\, d\varphi\, d\theta$. (See Figure 5.97.) ◆

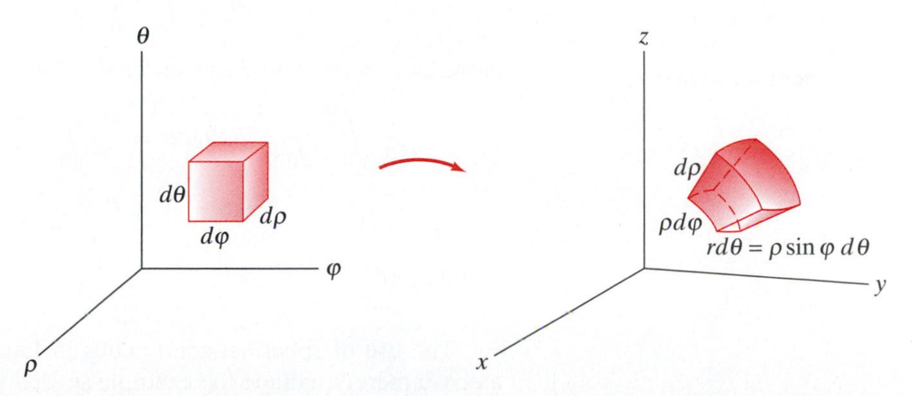

Figure 5.97 A differential box in $\rho\varphi\theta$-space is mapped to a portion of a solid ball in xyz-space by the spherical–rectangular tranformation.

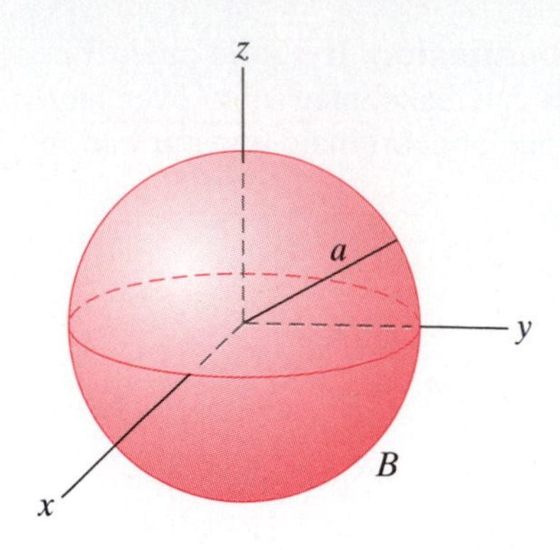

Figure 5.98 The ball B of radius a of Example 16.

EXAMPLE 16 The volume of a ball is easy to calculate in spherical coordinates. A solid ball of radius a may be described as

$$B = \{(\rho, \varphi, \theta) \mid 0 \leq \rho \leq a, 0 \leq \varphi \leq \pi, 0 \leq \theta < 2\pi\}.$$

(See Figure 5.98.) Hence, we may compute the volume by using the triple integral

$$\begin{aligned}\iiint_B dV &= \int_0^{2\pi} \int_0^{\pi} \int_0^{a} \rho^2 \sin\varphi \, d\rho \, d\varphi \, d\theta = \int_0^{2\pi} \int_0^{\pi} \frac{a^3}{3} \sin\varphi \, d\varphi \, d\theta \\ &= \frac{a^3}{3} \int_0^{2\pi} \left(-\cos\varphi|_0^{\pi}\right) d\theta = \frac{a^3}{3} \int_0^{2\pi} (-(-1) + 1) \, d\theta \\ &= \frac{2a^3}{3} \int_0^{2\pi} d\theta = \frac{4\pi a^3}{3},\end{aligned}$$

as expected. ◆

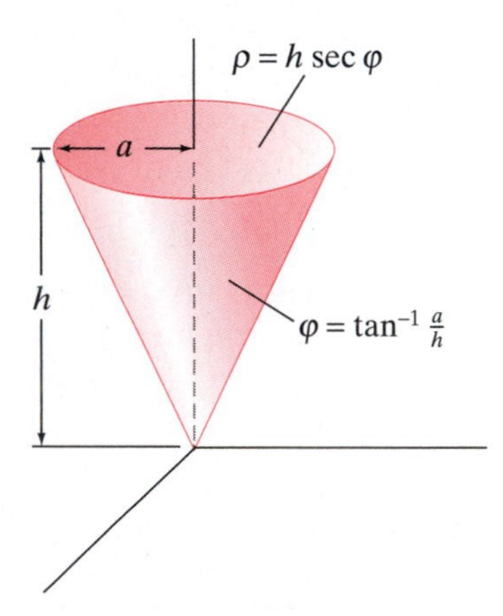

Figure 5.99 The cone of Example 17 described in spherical coordinates.

EXAMPLE 17 We return to the example of the cone of radius a and height h and, this time, use spherical coordinates to calculate its volume. First, note two things: (i) that the cone's lateral surface has the equation $\varphi = \tan^{-1}(a/h)$ in spherical coordinates and (ii) that the planar top having Cartesian equation $z = h$ has spherical equation $\rho \cos\varphi = h$ or, equivalently, $\rho = h \sec\varphi$. (See Figure 5.99.)

For fixed values of the spherical angles φ and θ, the values of ρ that give points inside the cone vary from 0 to $h \sec\varphi$. Any points inside the cone must have spherical angle φ between 0 and $\tan^{-1}(a/h)$. Finally, by symmetry, θ can assume any value between 0 and 2π. Hence, the cone may be described as the set

$$\left\{(\rho, \varphi, \theta) \,\middle|\, 0 \leq \rho \leq h \sec\varphi,\ 0 \leq \varphi \leq \tan^{-1}\frac{a}{h},\ 0 \leq \theta < 2\pi\right\}.$$

Therefore, we calculate the volume as

$$\begin{aligned}&\int_0^{2\pi} \int_0^{\tan^{-1}(a/h)} \int_0^{h \sec\varphi} \rho^2 \sin\varphi \, d\rho \, d\varphi \, d\theta \\ &\quad = \int_0^{2\pi} \int_0^{\tan^{-1}(a/h)} \frac{(h \sec\varphi)^3}{3} \sin\varphi \, d\varphi \, d\theta \\ &\quad = \frac{h^3}{3} \int_0^{2\pi} \int_0^{\tan^{-1}(a/h)} \sec^3\varphi \sin\varphi \, d\varphi \, d\theta \\ &\quad = \frac{h^3}{3} \int_0^{2\pi} \int_0^{\tan^{-1}(a/h)} \tan\varphi \sec^2\varphi \, d\varphi \, d\theta.\end{aligned}$$

Now, let $u = \tan\varphi$ so $du = \sec^2\varphi \, d\varphi$. Then the last integral becomes

$$\begin{aligned}\frac{h^3}{3} \int_0^{2\pi} \int_0^{a/h} u \, du \, d\theta &= \frac{h^3}{3} \int_0^{2\pi} \frac{1}{2} \left(\frac{a}{h}\right)^2 d\theta = \frac{h^3 a^2}{6h^2} \int_0^{2\pi} d\theta \\ &= \frac{a^2 h}{6}(2\pi) = \frac{\pi}{3} a^2 h,\end{aligned}$$

as expected. ◆

The use of spherical coordinates in Example 17 is not the most appropriate. We merely include the example so that you can develop some facility with "thinking spherically." Further practice can be obtained by considering some of the applications in the next section, as well as, of course, some of the exercises.

Summary: Change of Variables Formulas

Change of variables in double integrals:

$$\iint_D f(x, y)\,dx\,dy = \iint_{D^*} f(x(u, v), y(u, v)) \left|\frac{\partial(x, y)}{\partial(u, v)}\right| du\,dv$$

Area elements:

$$\begin{aligned} dA &= dx\,dy && \text{(Cartesian)} \\ &= r\,dr\,d\theta && \text{(polar)} \\ &= \left|\frac{\partial(x, y)}{\partial(u, v)}\right| du\,dv && \text{(general)} \end{aligned}$$

Change of variables in triple integrals:

$$\iiint_W f(x, y, z)\,dx\,dy\,dz$$
$$= \iiint_{W^*} f(x(u, v, w), y(u, v, w), z(u, v, w)) \left|\frac{\partial(x, y, z)}{\partial(u, v, w)}\right| du\,dv\,dw$$

Volume elements:

$$\begin{aligned} dV &= dx\,dy\,dz && \text{(Cartesian)} \\ &= r\,dr\,d\theta\,dz && \text{(cylindrical)} \\ &= \rho^2 \sin\varphi\,d\rho\,d\varphi\,d\theta && \text{(spherical)} \\ &= \left|\frac{\partial(x, y, z)}{\partial(u, v, w)}\right| du\,dv\,dw && \text{(general)} \end{aligned}$$

Exercises

1. Let $\mathbf{T}(u, v) = (3u, -v)$.

(a) Write $\mathbf{T}(u, v)$ as $A \begin{bmatrix} u \\ v \end{bmatrix}$ for a suitable matrix A.

(b) Describe the image $D = \mathbf{T}(D^*)$, where D^* is the unit square $[0, 1] \times [0, 1]$.

2. (a) Let

$$\mathbf{T}(u, v) = \left(\frac{u - v}{\sqrt{2}}, \frac{u + v}{\sqrt{2}}\right).$$

How does $\mathbf{T}$ transform the unit square $D^* = [0, 1] \times [0, 1]$?

(b) Now suppose

$$\mathbf{T}(u, v) = \left(\frac{u + v}{\sqrt{2}}, \frac{u - v}{\sqrt{2}}\right).$$

Describe how $\mathbf{T}$ transforms D^*.

3. If

$$\mathbf{T}(u, v) = \begin{bmatrix} 2 & 3 \\ -1 & 1 \end{bmatrix} \begin{bmatrix} u \\ v \end{bmatrix}$$

and D^* is the parallelogram whose vertices are $(0, 0)$, $(1, 3)$, $(-1, 2)$, and $(0, 5)$, determine $D = \mathbf{T}(D^*)$.

4. If D^* is the parallelogram whose vertices are $(0, 0)$, $(-1, 3)$, $(1, 2)$, and $(0, 5)$ and D is the parallelogram whose vertices are $(0, 0)$, $(3, 2)$, $(1, -1)$, and $(4, 1)$, find a transformation $\mathbf{T}$ such that $\mathbf{T}(D^*) = D$.

5. If $\mathbf{T}(u, v, w) = (3u - v, u - v + 2w, 5u + 3v - w)$, describe how $\mathbf{T}$ transforms the unit cube $W^* = [0, 1] \times [0, 1] \times [0, 1]$.

6. Suppose $\mathbf{T}(u, v) = (u, uv)$. Explain (perhaps by using pictures) how $\mathbf{T}$ transforms the unit square $D^* = [0, 1] \times [0, 1]$. Is $\mathbf{T}$ one-one on D^*?

7. Let $\mathbf{T}: \mathbf{R}^3 \to \mathbf{R}^3$ be the transformation given by

$$\mathbf{T}(\rho, \varphi, \theta) = (\rho \sin\varphi \cos\theta, \rho \sin\varphi \sin\theta, \rho \cos\varphi).$$

(a) Determine $D = \mathbf{T}(D^*)$, where $D^* = [0, 1] \times [0, \pi] \times [0, 2\pi]$.

(b) Determine $D = \mathbf{T}(D^*)$, where $D^* = [0, 1] \times [0, \pi/2] \times [0, \pi/2]$.

(c) Determine $D = \mathbf{T}(D^*)$, where $D^* = [1/2, 1] \times [0, \pi/2] \times [0, \pi/2]$.

8. This problem concerns the iterated integral

$$\int_0^1 \int_{y/2}^{(y/2)+2} (2x - y)\, dx\, dy.$$

(a) Evaluate this integral and sketch the region D of integration in the xy-plane.

(b) Let $u = 2x - y$ and $v = y$. Find the region D^* in the uv-plane that corresponds to D.

(c) Use the change of variables theorem (Theorem 5.3) to evaluate the integral by using the substitution $u = 2x - y$, $v = y$.

9. Evaluate the integral

$$\int_0^2 \int_{x/2}^{(x/2)+1} x^5 (2y - x) e^{(2y-x)^2}\, dy\, dx$$

by making the substitution $u = x$, $v = 2y - x$.

10. Determine the value of

$$\iint_D \sqrt{\frac{x+y}{x-2y}}\, dA,$$

where D is the region in $\mathbf{R}^2$ enclosed by the lines $y = x/2$, $y = 0$, and $x + y = 1$.

11. Evaluate $\iint_D (2x + y)^2 e^{x-y}\, dA$, where D is the region enclosed by $2x + y = 1$, $2x + y = 4$, $x - y = -1$, and $x - y = 1$.

12. Evaluate

$$\iint_D \frac{(2x + y - 3)^2}{(2y - x + 6)^2}\, dx\, dy,$$

where D is the square with vertices $(0, 0)$, $(2, 1)$, $(3, -1)$, and $(1, -2)$. (Hint: First sketch D and find the equations of its sides.)

In Exercises 13–17, transform the given integral in Cartesian coordinates to one in polar coordinates and evaluate the polar integral.

13. $\displaystyle\int_{-1}^{1} \int_{-\sqrt{1-x^2}}^{\sqrt{1-x^2}} 3\, dy\, dx$

14. $\displaystyle\int_0^2 \int_0^{\sqrt{4-x^2}} dy\, dx$

15. $\displaystyle\iint_D (x^2 + y^2)^{3/2}\, dA$, where D is the disk $x^2 + y^2 \le 9$

16. $\displaystyle\int_{-a}^{a} \int_0^{\sqrt{a^2-y^2}} e^{x^2+y^2}\, dx\, dy$

17. $\displaystyle\int_0^3 \int_0^x \frac{dy\, dx}{\sqrt{x^2 + y^2}}$

18. Evaluate

$$\iint_D \frac{1}{\sqrt{4 - x^2 - y^2}}\, dA,$$

where D is the disk of radius 1 with center at $(0, 1)$. (Be careful when you describe D.)

19. Let D be the region between the square with vertices $(1, 1)$, $(-1, 1)$, $(-1, -1)$, $(1, -1)$ and the unit disk centered at the origin. Evaluate $\displaystyle\iint_D y^2\, dA$.

20. Find the total area enclosed inside the rose $r = \sin 2\theta$. (Hint: Sketch the curve and find the area inside a single leaf.)

21. Find the area of the region inside the cardioid $r = 1 - \cos\theta$ and outside the circle $r = 1$.

22. Find the area of the region bounded by the positive x-axis and the spiral $r = 3\theta$, $0 \le \theta \le 2\pi$.

23. Evaluate

$$\iint_D \cos(x^2 + y^2)\, dA,$$

where D is the shaded region in Figure 5.100.

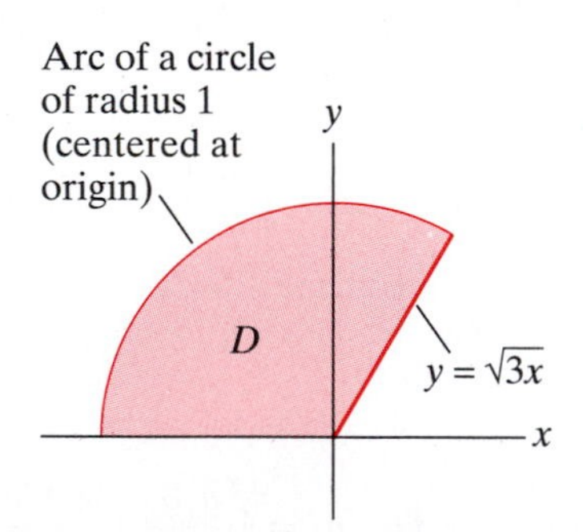

Figure 5.100 The region D of Exercise 23.

24. Evaluate

$$\iiint_B \frac{dV}{\sqrt{x^2 + y^2 + z^2 + 3}},$$

where B is the ball of radius 2 centered at the origin.

25. Determine

$$\iiint_W (x^2 + y^2 + 2z^2)\,dV,$$

where W is the solid cylinder defined by the inequalities $x^2 + y^2 \leq 4$, $-1 \leq z \leq 2$.

In Exercises 26 and 27, determine the values of the given integrals, where W is the region bounded by the two spheres $x^2 + y^2 + z^2 = a^2$ and $x^2 + y^2 + z^2 = b^2$, for $0 < a < b$.

26. $\displaystyle\iiint_W \frac{dV}{\sqrt{x^2 + y^2 + z^2}}$

27. $\displaystyle\iiint_W \sqrt{x^2 + y^2 + z^2}\, e^{x^2+y^2+z^2}\,dV$

28. Determine

$$\iiint_W \left(2 + \sqrt{x^2 + y^2}\right) dV,$$

where $W = \left\{(x, y, z) \mid \sqrt{x^2 + y^2} \leq z/2 \leq 3\right\}$.

29. Find the volume of the region W that represents the intersection of the solid cylinder $x^2 + y^2 \leq 1$ and the solid ellipsoid $2(x^2 + y^2) + z^2 \leq 10$.

30. Find the volume of the solid W that is bounded by the paraboloid $z = 9 - x^2 - y^2$, the xy-plane, and the cylinder $x^2 + y^2 = 4$.

31. Find

$$\iiint_W (2 + x^2 + y^2)\,dV,$$

where W is the region inside the sphere $x^2 + y^2 + z^2 = 25$ and above the plane $z = 3$.

32. Find the volume of the intersection of the three solid cylinders

$$x^2 + y^2 \leq a^2, \quad x^2 + z^2 \leq a^2, \quad \text{and} \quad y^2 + z^2 \leq a^2.$$

(Hint: First draw a careful sketch, then note that, by symmetry, it suffices to calculate the volume of a portion of the intersection.)

5.6 Applications of Integration

In this section, we explore a variety of settings where double and triple integrals arise naturally.

Average value of a function

Day	°F
Monday	65
Tuesday	63
Wednesday	52
Thursday	51
Friday	45
Saturday	43
Sunday	47

Suppose temperatures (shown in the adjacent table) are recorded in Oberlin, Ohio during a particular week. From these data, we calculate the **average** (or **mean**) temperature:

$$\text{Average temperature} = \frac{65 + 63 + 52 + 51 + 45 + 43 + 47}{7} \approx 52.3\,°\text{F}.$$

Of course, this calculation only represents an approximation of the true average value, since the temperature will vary during each day. To determine the true average temperature, we need to know the temperature as a function of time for all instants of time during that one-week period; that is, we consider

$$\text{Temperature} = T(x), \quad x = \text{elapsed time (in days)}, \quad \text{for} \quad 0 \leq x \leq 7.$$

Then, a more accurate determination of the average temperature is as an integral:

$$\text{Average temperature} = \tfrac{1}{7}\int_0^7 T(x)\,dx. \tag{1}$$

Since an integral is nothing more than the limit of a sum, it's not hard to see that the preceding formula is a generalization of the original discrete sum calculation to the continuous case. (See Figure 5.101.)

Note that

$$7 = \int_0^7 dx = \text{length of time interval}.$$

Hence, we may rewrite formula (1) as

$$\text{Average temperature} = \frac{\int_0^7 T(x)\,dx}{\int_0^7 dx}.$$

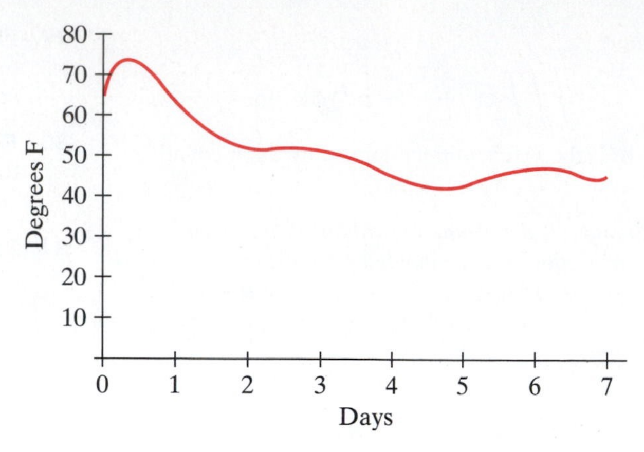

Figure 5.101 A continuous temperature function $T(x)$ over the interval $[0, 7]$. The average temperature for the week is $\frac{1}{7}\int_0^7 T(x)\,dx$.

This observation leads us to make the following definitions concerning average values of functions.

Definition 6.1 (a) Let $f: [a, b] \to \mathbf{R}$ be an integrable function of one variable. The **average (mean) value** of f on $[a, b]$ is

$$[f]_{\text{avg}} = \frac{1}{b-a}\int_a^b f(x)\,dx = \frac{\int_a^b f(x)\,dx}{\int_a^b dx} = \frac{\int_a^b f(x)\,dx}{\text{length of interval } [a, b]}.$$

(b) Let $f: D \subseteq \mathbf{R}^2 \to \mathbf{R}$ be an integrable function of two variables. The **average value** of f on D is

$$[f]_{\text{avg}} = \frac{\iint_D f\,dA}{\iint_D dA} = \frac{\iint_D f\,dA}{\text{area of } D}.$$

(c) Let $f: W \subset \mathbf{R}^3 \to \mathbf{R}$ be an integrable function of three variables. The **average value** of f on W is

$$[f]_{\text{avg}} = \frac{\iiint_W f\,dV}{\iiint_W dV} = \frac{\iiint_W f\,dV}{\text{volume of } W}.$$

EXAMPLE 1 Suppose that the "temperature function" for Oberlin during a week in April is

$$T(x) = \tfrac{113}{5040}x^7 - \tfrac{107}{180}x^6 + \tfrac{1127}{180}x^5 - \tfrac{2393}{72}x^4 + \tfrac{66821}{720}x^3 - \tfrac{45781}{360}x^2 + \tfrac{12581}{210}x + 65,$$

where $0 \le x \le 7$. Then the mean temperature for that week would be

$$\begin{aligned}
[T]_{\text{avg}} &= \tfrac{1}{7-0}\int_0^7 \left(\tfrac{113}{5040}x^7 - \tfrac{107}{180}x^6 + \tfrac{1127}{180}x^5 - \tfrac{2393}{72}x^4\right. \\
&\quad \left. + \tfrac{66821}{720}x^3 - \tfrac{45781}{360}x^2 + \tfrac{12581}{210}x + 65\right)dx \\
&= \tfrac{1}{7}\left(\tfrac{113}{40320}x^8 - \tfrac{107}{1260}x^7 + \tfrac{1127}{1080}x^6 - \tfrac{2393}{360}x^5\right. \\
&\quad \left. + \tfrac{66821}{2880}x^4 - \tfrac{45781}{1080}x^3 + \tfrac{12581}{420}x^2 + 65x\right)\Big|_0^7 \\
&= \tfrac{888709}{17280} \approx 51.43\,^\circ\text{F}.
\end{aligned}$$

◆

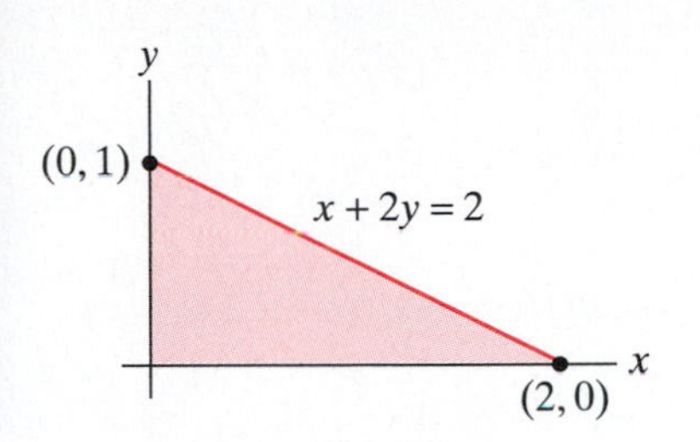

Figure 5.102 The triangular metal plate of Example 2.

EXAMPLE 2 Suppose that the thickness of the triangular metal plate, shown in Figure 5.102, varies as $f(x, y) = xy + 1$, where (x, y) are the coordinates of a point in the plate. The average thickness of the plate is, therefore,

$$\text{Average thickness} = \frac{\int_0^1 \int_0^{2-2y} (xy + 1)\, dx\, dy}{\int_0^1 \int_0^{2-2y} dx\, dy}.$$

Note that

$$\int_0^1 \int_0^{2-2y} dx\, dy = \text{area of triangular plate} = \tfrac{1}{2}(2 \cdot 1) = 1$$

from elementary geometry. Hence, the average thickness is

$$\begin{aligned}
\frac{\int_0^1 \int_0^{2-2y} (xy+1)\, dx\, dy}{1} &= \int_0^1 \left(\tfrac{1}{2}x^2 y + x\right)\Big|_{x=0}^{x=2-2y} dy \\
&= \int_0^1 \left(\tfrac{1}{2}(2-2y)^2 y + (2-2y)\right) dy \\
&= 2\int_0^1 (y^3 - 2y^2 + 1)\, dy = 2\left(\tfrac{y^4}{4} - \tfrac{2}{3}y^3 + y\right)\Big|_0^1 = \tfrac{7}{6}.
\end{aligned}$$ ◆

EXAMPLE 3 (See also Example 6 of §5.4.) Suppose the temperature inside the capsule bounded by the paraboloids $z = 9 - x^2 - y^2$ and $z = 3x^2 + 3y^2 - 16$ varies from point to point as

$$T(x, y, z) = z(x^2 + y^2).$$

We calculate the mean temperature of the capsule.

From Definition 6.1,

$$[T]_{\text{avg}} = \frac{\iiint_W T\, dV}{\iiint_W dV}.$$

The particular iterated integrals we can use for the computation are then

$$[T]_{\text{avg}} = \frac{\displaystyle\int_{-5/2}^{5/2} \int_{-\sqrt{25/4 - x^2}}^{\sqrt{25/4 - x^2}} \int_{3x^2+3y^2-16}^{9-x^2-y^2} z(x^2 + y^2)\, dz\, dy\, dx}{\displaystyle\int_{-5/2}^{5/2} \int_{-\sqrt{25/4 - x^2}}^{\sqrt{25/4 - x^2}} \int_{3x^2+3y^2-16}^{9-x^2-y^2} dz\, dy\, dx}.$$

Unfortunately, the calculations involved in evaluating these integrals are rather tedious.

On the other hand, since the capsule has an axis of rotational symmetry, cylindrical coordinates can be used to simplify the computations. Note that the boundary paraboloids have cylindrical equations of $z = 9 - r^2$ and $z = 3r^2 - 16$ and that the shadow of the capsule in the $z = 0$ plane can be described in polar coordinates as

$$\left\{(r, \theta) \mid 0 \le r \le \tfrac{5}{2}, 0 \le \theta < 2\pi\right\}.$$

(See Figures 5.103 and 5.104.)

In addition, the temperature function may be described in cylindrical coordinates as

$$T(x, y, z) = z(x^2 + y^2) = zr^2.$$

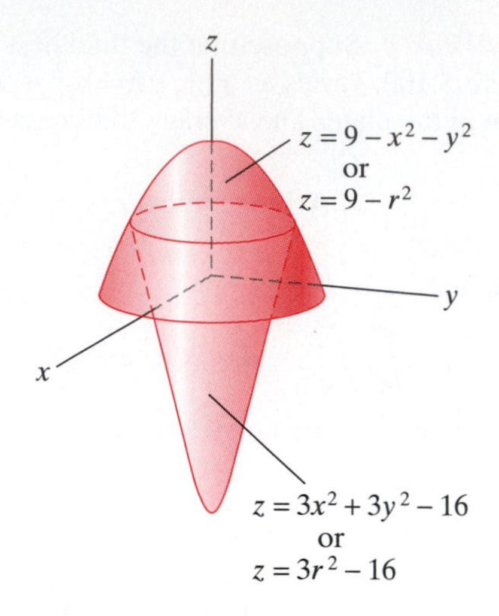

Figure 5.103 The capsule of Example 3.

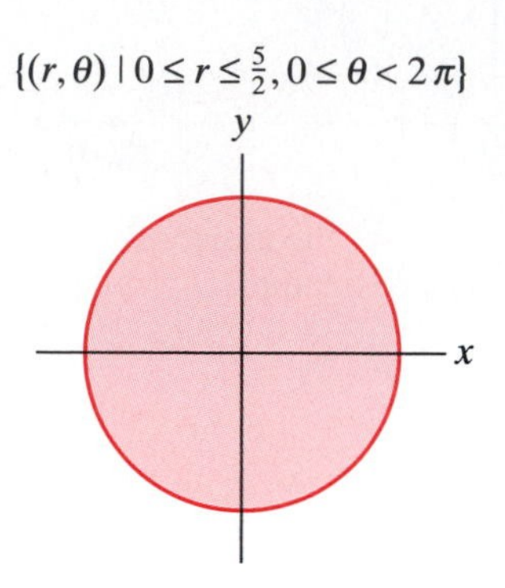

Figure 5.104 The shadow of the capsule in Figure 5.103 in the $z = 0$ plane.

Hence, we may calculate

$$[T]_{\text{avg}} = \frac{\int_0^{2\pi} \int_0^{5/2} \int_{3r^2-16}^{9-r^2} zr^2 \cdot r \, dz \, dr \, d\theta}{\int_0^{2\pi} \int_0^{5/2} \int_{3r^2-16}^{9-r^2} r \, dz \, dr \, d\theta}.$$

For the denominator integral,

$$\begin{aligned}
\int_0^{2\pi} \int_0^{5/2} \int_{3r^2-16}^{9-r^2} r \, dz \, dr \, d\theta &= \int_0^{2\pi} \int_0^{5/2} r \left((9 - r^2) - (3r^2 - 16) \right) dr \, d\theta \\
&= \int_0^{2\pi} \int_0^{5/2} \left(25r - 4r^3 \right) dr \, d\theta \\
&= \int_0^{2\pi} \left(\frac{25}{2} r^2 - r^4 \right) \Big|_0^{5/2} d\theta \\
&= \int_0^{2\pi} \left(\frac{625}{8} - \frac{625}{16} \right) d\theta = \frac{625}{16} \cdot 2\pi = \frac{625\pi}{8}.
\end{aligned}$$

This result agrees with the volume calculation in Example 6 of §5.4, as it should.

For the numerator integral, we compute

$$\begin{aligned}
\int_0^{2\pi} \int_0^{5/2} \int_{3r^2-16}^{9-r^2} zr^3 \, dz \, dr \, d\theta &= \int_0^{2\pi} \int_0^{5/2} \left(\frac{z^2}{2} r^3 \right) \Big|_{z=3r^2-16}^{z=9-r^2} dr \, d\theta \\
&= \int_0^{2\pi} \int_0^{5/2} \frac{r^3}{2} \left((9 - r^2)^2 - (3r^2 - 16)^2 \right) dr \, d\theta \\
&= \int_0^{2\pi} \int_0^{5/2} \frac{r^3}{2} (-8r^4 + 78r^2 - 175) \, dr \, d\theta \\
&= \frac{1}{2} \int_0^{2\pi} \int_0^{5/2} (-8r^7 + 78r^5 - 175r^3) \, dr \, d\theta \\
&= \frac{1}{2} \int_0^{2\pi} \left(-r^8 + 13r^6 - \frac{175}{4} r^4 \right) \Big|_0^{5/2} d\theta \\
&= \frac{1}{2} \int_0^{2\pi} -\frac{15625}{256} \, d\theta = -\frac{15625}{256} \pi.
\end{aligned}$$

Thus,

$$[T]_{\text{avg}} = \frac{-15625\pi/256}{625\pi/8} = -\frac{25}{32}.$$ ◆

Center of Mass: The Discrete Case

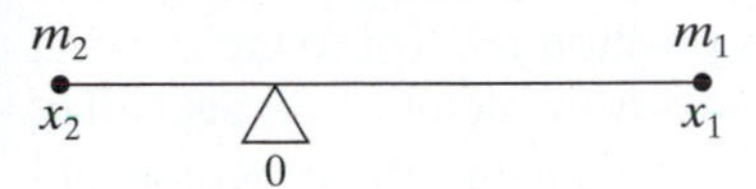

Figure 5.105 This seesaw balances if $m_1x_1 + m_2x_2 = 0$.

Consider a uniform seesaw with two masses m_1 and m_2 placed on either end. If we introduce a coordinate system so that the fulcrum of the seesaw is placed at the origin, then the situation looks something like that shown in Figure 5.105. Note that $x_2 < 0 < x_1$. The seesaw balances if

$$m_1x_1 + m_2x_2 = 0.$$

In this case, the **center of mass** (or "balance point") of the system is at the origin.

But now suppose $m_1x_1 + m_2x_2 \neq 0$. Then, where is the balance point? Let us denote the coordinate of the balance point by $\bar{x}$. Before we find it, we'll introduce a little terminology. The product m_ix_i (in this case, for $i = 1, 2$) of mass and position is called the **moment** of the ith body with respect to the origin of the coordinate system. The sum $m_1x_1 + m_2x_2$ is called the **total moment** with respect to the origin. To find the center of mass, we use the following physical principle, which tells us that a system of several point masses is physically equivalent (in terms of moments) to a system with a single point mass.

> **Guiding physical principle.** The center of mass is the point such that, if all the mass of the system were concentrated there, the total moment of the new system would be the same as that for the original system.

Putting this principle into practice in our situation, we see that total mass M of our system is $m_1 + m_2$. If $\bar{x}$ is the center of mass, then the guiding principle tells that

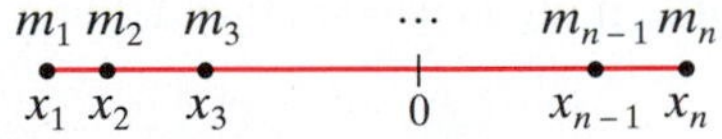

$$M\bar{x} = m_1x_1 + m_2x_2.$$

Figure 5.106 A system of n masses distributed on a line.

That is, the total moment of the new (concentrated) system is the same as the total moment of the original system. Hence,

$$\bar{x} = \frac{m_1x_1 + m_2x_2}{m_1 + m_2}.$$

If we have a system of n masses distributed along a (coordinatized) line, then the same reasoning may be applied. (See Figure 5.106.) We have

$$\bar{x} = \frac{\text{total moment}}{\text{total mass}} = \frac{m_1x_1 + m_2x_2 + \cdots + m_nx_n}{m_1 + m_2 + \cdots + m_n} = \frac{\sum_{i=1}^n m_ix_i}{\sum_{i=1}^n m_i}. \tag{2}$$

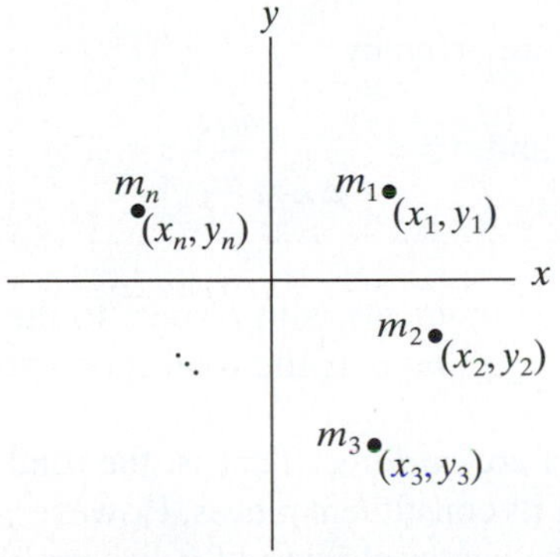

Figure 5.107 A system of n masses in $\mathbf{R}^2$.

Now we move to two and three dimensions. Suppose, first, that we have n particles (or bodies) arranged in the plane as in Figure 5.107. Then there are two moments to consider:

$$\text{Total moment with respect to the } y\text{-axis} = \sum_{i=1}^n m_ix_i,$$

and

$$\text{total moment with respect to the } x\text{-axis} = \sum_{i=1}^{n} m_i y_i.$$

(Admittedly, this terminology may seem confusing at first. The idea is that the moment measures how the system balances with respect to the coordinate axes. It is the x-coordinate—not the y-coordinate—that measures position relative to the y-axis. Similarly, the y-coordinate measures position relative to the x-axis.) The guiding principle tells us that the center of mass is the point $(\bar{x}, \bar{y})$ such that, if all the mass of the system were concentrated there, then the new system would have the same total moments as the original system. That is, if $M = \sum m_i$, then

$$M\bar{x} = \sum_{i=1}^{n} m_i x_i$$

(i.e., the moment with respect to the y-axis of the new system equals the moment with respect to the y-axis of the original system) and

$$M\bar{y} = \sum_{i=1}^{n} m_i y_i.$$

Thus, we have shown the following:

Discrete center of mass in $\mathbf{R}^2$. Given a system of n point masses m_1, $m_2, \ldots, m_n$ at positions

$$(x_1, y_1), \quad (x_2, y_2), \ldots, \quad (x_n, y_n) \quad \text{in } \mathbf{R}^2,$$

the coordinates $(\bar{x}, \bar{y})$ of the center of mass are

$$\bar{x} = \frac{\sum_{i=1}^{n} m_i x_i}{\sum_{i=1}^{n} m_i} \quad \text{and} \quad \bar{y} = \frac{\sum_{i=1}^{n} m_i y_i}{\sum_{i=1}^{n} m_i}. \tag{3}$$

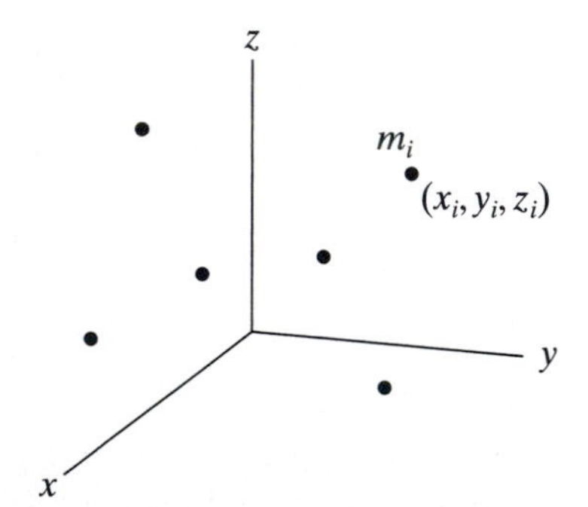

Figure 5.108 A discrete system of masses in $\mathbf{R}^3$.

For particles arranged in three dimensions, little more is needed than adding an additional coordinate. (See Figure 5.108.)

Discrete center of mass in $\mathbf{R}^3$. Given a system of n point masses m_1, $m_2, \ldots, m_n$ at positions

$$(x_1, y_1, z_1), \quad (x_2, y_2, z_2), \ldots, \quad (x_n, y_n, z_n) \quad \text{in } \mathbf{R}^3,$$

the coordinates $(\bar{x}, \bar{y}, \bar{z})$ of the center of mass are given by

$$\bar{x} = \frac{\sum_{i=1}^{n} m_i x_i}{\sum_{i=1}^{n} m_i}, \quad \bar{y} = \frac{\sum_{i=1}^{n} m_i y_i}{\sum_{i=1}^{n} m_i}, \quad \text{and} \quad \bar{z} = \frac{\sum_{i=1}^{n} m_i z_i}{\sum_{i=1}^{n} m_i}. \tag{4}$$

The numerators of the fractions in (4) are the moments with respect to the *coordinate planes*. Thus, for example, the sum $\sum_{i=1}^{n} m_i x_i$ is the total moment with respect to the yz-plane.

By definition, moments of physical systems are *additive*. That is, the total moment of a system is the sum of the moments of its constituent pieces. However, it is by no means the case that a coordinate of the center of mass of a system is the sum of the coordinates of the centers of mass of its pieces. This additivity property makes the study of moments important in its own right.

Center of Mass: The Continuous Case

Now, we turn our attention to physical systems where mass is distributed in a continuous fashion throughout the system, rather than at only finitely many isolated points.

To begin with the one-dimensional case, suppose we have a straight wire placed on a coordinate axis between points $x = a$ and $x = b$ as shown in Figure 5.109. Moreover, suppose that the mass of this wire is distributed according to some continuous density function $\delta(x)$. We seek the coordinate $\bar{x}$ that represents the center of mass, or "balance point," of the wire.

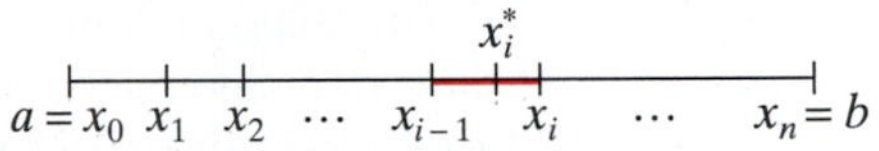

$$a = x_0 \quad x_1 \quad x_2 \quad \cdots \quad x_{i-1} \quad x_i^* \quad x_i \quad \cdots \quad x_n = b$$

Figure 5.109 A "coordinatized" wire. The mass of the segment between x_{i-1} and x_i is approximately $\delta(x_i^*)\Delta x_i$.

Imagine breaking the wire into n small pieces. Since the density is continuous, it will be nearly constant on each small piece. Thus, for $i = 1, \ldots, n$, the mass m_i of each piece is approximately $\delta(x_i^*)\Delta x_i$, where $\Delta x_i = x_i - x_{i-1}$ is the length of each segment of wire, and x_i^* is any number in the subinterval $[x_{i-1}, x_i]$. Hence, the total mass is

$$M = \sum_{i=1}^{n} m_i \approx \sum_{i=1}^{n} \delta(x_i^*)\Delta x_i,$$

and the total moment with respect to the origin is approximately

$$\sum_{i=1}^{n} \underbrace{x_i^*}_{\text{approx. position}} \; \underbrace{\delta(x_i^*)\Delta x_i}_{\text{approx. mass}} .$$

Of course, these results can be used to provide an approximation of the coordinate $\bar{x}$ of the center of mass. For an exact result, however, we let all the pieces of wire become "infinitesimally small"; that is, we take limits of the foregoing approximating sums as all the Δx_i's tend to zero. Such limits give us integrals, and we may reasonably define our terms as follows:

Continuous center of mass in R. For a wire located along the x-axis between $x = a$ and $x = b$ with continuous density per unit length $\delta(x)$:

$$\text{Total mass} = \int_a^b \delta(x)\, dx.$$

$$\text{Total moment} = \int_a^b x\, \delta(x)\, dx. \tag{5}$$

$$\text{Center of mass } \bar{x} = \frac{\text{total moment}}{\text{total mass}} = \frac{\int_a^b x\delta(x)\, dx}{\int_a^b \delta(x)\, dx}.$$

Compare the formulas in (3) with those in (5). Instead of a sum of masses and a sum of products of mass and position, we have an integral of "infinitesimal mass" (the $\delta(x)\, dx$ term) and an integral of infinitesimal mass times position.

EXAMPLE 4 Suppose that a wire is located between $x = -1$ and $x = 1$ along a coordinate line and has density $\delta(x) = x^2 + 1$. Using the formulas in (5), we compute that the center of mass has coordinate

$$\bar{x} = \frac{\int_{-1}^{1} x(x^2+1)\,dx}{\int_{-1}^{1}(x^2+1)\,dx} = \frac{\left(\frac{1}{4}x^4 + \frac{1}{2}x^2\right)\Big|_{-1}^{1}}{\left(\frac{1}{3}x^3 + x\right)\Big|_{-1}^{1}} = \frac{0}{\frac{8}{3}} = 0.$$

This makes sense, since this wire has a symmetric density pattern with respect to the origin (i.e., $\delta(x) = \delta(-x)$). ◆

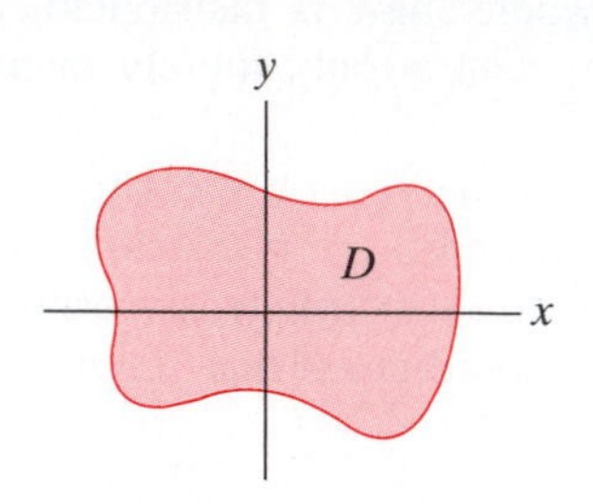

Figure 5.110 A lamina depicted as a region D in the xy-plane with density function δ.

The analogous situation in two dimensions is that of a **lamina** or flat plate of finite extent and continuously varying density $\delta(x, y)$. (See Figure 5.110.) Using reasoning similar to that used to obtain the formulas in (5), we make the following definition for the coordinates $(\bar{x}, \bar{y})$ of the center of mass of the lamina:

Continuous center of mass in $\mathbf{R}^2$. For a lamina represented by the region D in the xy-plane with continuous density per unit area $\delta(x, y)$:

$$\bar{x} = \frac{\text{total moment with respect to } y\text{-axis}}{\text{total mass}} = \frac{\iint_D x\,\delta(x,y)\,dA}{\iint_D \delta(x,y)\,dA};$$

$$\bar{y} = \frac{\text{total moment with respect to } x\text{-axis}}{\text{total mass}} = \frac{\iint_D y\,\delta(x,y)\,dA}{\iint_D \delta(x,y)\,dA}. \qquad (6)$$

Roughly, the term $\delta(x, y)\,dA$ represents the mass of an "infinitesimal two-dimensional" piece of the lamina and the various double integrals the limiting sums of such masses or their corresponding moments.

EXAMPLE 5 We wish to find the center of mass of a lamina represented by the region D in $\mathbf{R}^2$ whose boundary consists of portions of the parabola $y = x^2$ and the line $y = 9$ and whose density varies as $\delta(x, y) = x^2 + y$. (See Figure 5.111.)

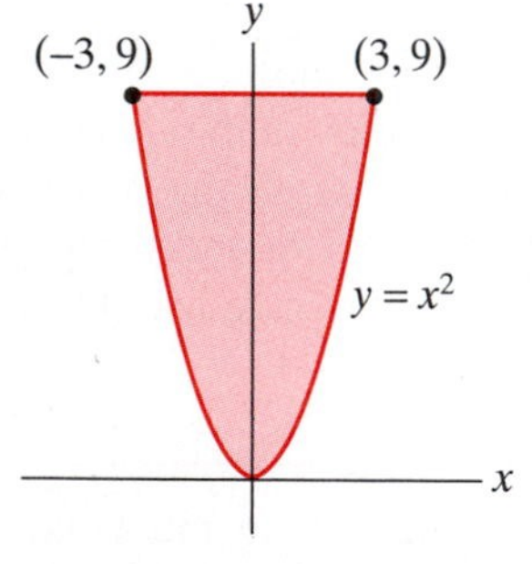

Figure 5.111 The region D representing the lamina of Example 5.

First, note that this lamina is symmetric with respect to the y-axis and that, in addition, the density function has a similar symmetry because $\delta(x, y) = \delta(-x, y)$. We may conclude from these two observations that the center of mass must occur along the y-axis (i.e., that $\bar{x} = 0$). Using the formulas in (6) and noting that the lamina is represented by an elementary region of type 1,

$$\bar{y} = \frac{\iint_D y\,\delta(x,y)\,dA}{\iint_D \delta(x,y)\,dA} = \frac{\int_{-3}^{3}\int_{x^2}^{9} y(x^2+y)\,dy\,dx}{\int_{-3}^{3}\int_{x^2}^{9}(x^2+y)\,dy\,dx}.$$

For the denominator integral, we compute

$$\begin{aligned}\int_{-3}^{3}\int_{x^2}^{9}(x^2+y)\,dy\,dx &= \int_{-3}^{3}\left(x^2y + \frac{1}{2}y^2\right)\Big|_{y=x^2}^{y=9} dx \\ &= \int_{-3}^{3}\left[\left(9x^2 + \frac{81}{2}\right) - \left(x^4 + \frac{1}{2}x^4\right)\right] dx \\ &= \int_{-3}^{3}\left(9x^2 - \frac{3}{2}x^4 + \frac{81}{2}\right) dx \\ &= \left(3x^3 - \frac{3}{10}x^5 + \frac{81}{2}x\right)\Big|_{-3}^{3} = \frac{1296}{5}.\end{aligned}$$

For the numerator,

$$\int_{-3}^{3}\int_{x^2}^{9} y(x^2+y)\,dy\,dx = \int_{-3}^{3}\left(\frac{x^2y^2}{2}+\frac{y^3}{3}\right)\Bigg|_{y=x^2}^{y=9} dx$$

$$= \int_{-3}^{3}\left[\left(\frac{81}{2}x^2+243\right)-\left(\frac{x^6}{2}+\frac{x^6}{3}\right)\right] dx$$

$$= \left(\frac{27}{2}x^3+243x-\frac{5}{42}x^7\right)\Bigg|_{-3}^{3} = \frac{11664}{7}.$$

Hence,

$$\bar{y} = \frac{11664/7}{1296/5} = \frac{45}{7} \approx 6.43.$$

This answer is quite plausible, since the density of the lamina increases with y, and so we should expect the center of mass to be closer to $y = 9$ than to $y = 0$. ◆

We may modify the two-dimensional formulas to produce three-dimensional ones.

Continuous center of mass in $\mathbf{R}^3$. Given a solid W whose density per unit volume varies continuously as $\delta(x, y, z)$, we compute the coordinates $(\bar{x}, \bar{y}, \bar{z})$ of the center of mass of W using the following quotients of triple integrals:

$$\bar{x} = \frac{\text{total moment with respect to } yz\text{-plane}}{\text{total mass}} = \frac{\iiint_W x\,\delta(x,y,z)\,dV}{\iiint_W \delta(x,y,z)\,dV};$$

$$\bar{y} = \frac{\text{total moment with respect to } xz\text{-plane}}{\text{total mass}} = \frac{\iiint_W y\,\delta(x,y,z)\,dV}{\iiint_W \delta(x,y,z)\,dV}; \quad (7)$$

$$\bar{z} = \frac{\text{total moment with respect to } xy\text{-plane}}{\text{total mass}} = \frac{\iiint_W z\,\delta(x,y,z)\,dV}{\iiint_W \delta(x,y,z)\,dV}.$$

In (7) we may think of the term $\delta(x, y, z)\,dV$ as representing the mass of an "infinitesimal three-dimensional" piece of W. Then the triple integrals are the limiting sums of masses or moments of such pieces.

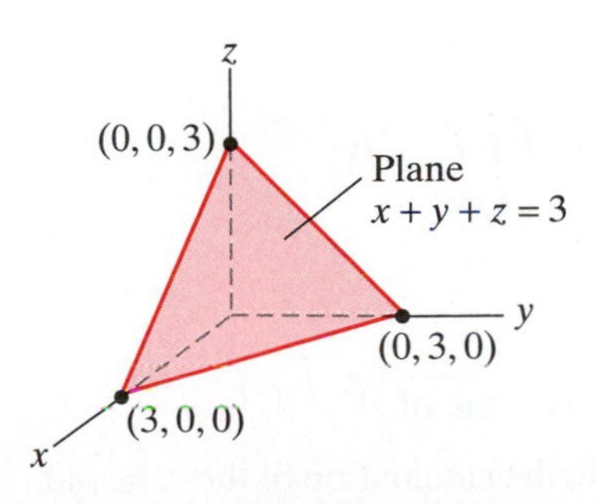

Figure 5.112 The tetrahedron of Example 6.

EXAMPLE 6 Consider the solid tetrahedron W with vertices at (0, 0, 0), (3, 0, 0), (0, 3, 0), and (0, 0, 3). Suppose the mass density at the point (x, y, z) inside the tetrahedron is $\delta(x, y, z) = x + y + z + 1$. We calculate the resulting center of mass. (See Figure 5.112.)

First, note that the position of the tetrahedron in space and the density function are both such that the roles of x, y, and z may be interchanged freely. Hence, the coordinates $(\bar{x}, \bar{y}, \bar{z})$ of the center of mass must satisfy $\bar{x} = \bar{y} = \bar{z}$. Therefore, we may reduce the number of calculations required.

The tetrahedron is a type 4 elementary region in space. Thus, we may calculate the total mass M of W, using the following iterated integral:

$$\begin{aligned} M &= \int_0^3 \int_0^{3-x} \int_0^{3-x-y} (x+y+z+1)\,dz\,dy\,dx \\ &= \int_0^3 \int_0^{3-x} \left((x+y+1)z + \frac{z^2}{2} \right) \Bigg|_{z=0}^{z=3-x-y} dy\,dx \\ &= \int_0^3 \int_0^{3-x} \left(\tfrac{15}{2} - x - \tfrac{1}{2}x^2 - y - xy - \tfrac{1}{2}y^2\right) dy\,dx \\ &= \int_0^3 \left[\left(\tfrac{15}{2} - x - \tfrac{1}{2}x^2\right) y - \tfrac{1}{2}(1+x)y^2 - \tfrac{1}{6}y^3\right]\Big|_{y=0}^{y=3-x} dx \\ &= \int_0^3 \left(\tfrac{27}{2} - \tfrac{15}{2}x + \tfrac{x^2}{2} + \tfrac{x^3}{6}\right) dx = \tfrac{117}{8}. \end{aligned}$$

The total moment with respect to the xy-plane is given by

$$\begin{aligned} &\int_0^3 \int_0^{3-x} \int_0^{3-x-y} z(x+y+z+1)dz\,dy\,dx \\ &\quad = \int_0^3 \int_0^{3-x} \left((x+y+1)\frac{z^2}{2} + \frac{z^3}{3} \right) \Bigg|_{z=0}^{z=3-x-y} dy\,dx \\ &\quad = \int_0^3 \int_0^{3-x} \left(\tfrac{27}{2} - \tfrac{15}{2}x + \tfrac{1}{2}x^2 + \tfrac{1}{6}x^3 - \tfrac{15}{2}y \right. \\ &\qquad\qquad \left. + xy + \tfrac{1}{2}x^2y + \tfrac{1}{2}y^2 + \tfrac{1}{2}xy^2 + \tfrac{1}{6}y^3\right) dy\,dx \\ &\quad = \int_0^3 \left(\tfrac{117}{8} - \tfrac{27}{2}x + \tfrac{15}{4}x^2 - \tfrac{1}{6}x^3 - \tfrac{1}{24}x^4\right) dx = \tfrac{459}{40}. \end{aligned}$$

Hence,

$$\bar{x} = \bar{y} = \bar{z} = \frac{\frac{459}{40}}{\frac{117}{8}} = \frac{51}{65} \approx 0.7846. \quad \blacklozenge$$

If an object is uniform, in the sense that it has *constant* density, then one uses the term **centroid** to refer to the center of mass of that object. Suppose the object is a solid region W in $\mathbf{R}^3$. Then, if the density δ is a constant k, the equations for the coordinates $(\bar{x}, \bar{y}, \bar{z})$ may be deduced from those in (7). For the x-coordinate, we have

$$\begin{aligned} \bar{x} &= \frac{\iiint_W x\delta(x,y,z)\,dV}{\iiint_W \delta(x,y,z)\,dV} = \frac{\iiint_W kx\,dV}{\iiint_W k\,dV} \\ &= \frac{\iiint_W x\,dV}{\iiint_W dV} = \frac{1}{\text{volume of } W} \iiint_W x\,dV. \end{aligned}$$

Similarly,

$$\bar{y} = \frac{1}{\text{volume of } W} \iiint_W y\,dV \quad \text{and} \quad \bar{z} = \frac{1}{\text{volume of } W} \iiint_W z\,dV.$$

In particular, the constant density δ plays no role in the calculation of the centroid, only the geometry of W. (Note: Completely analogous statements can be made in the case of centroids of laminas in $\mathbf{R}^2$.)

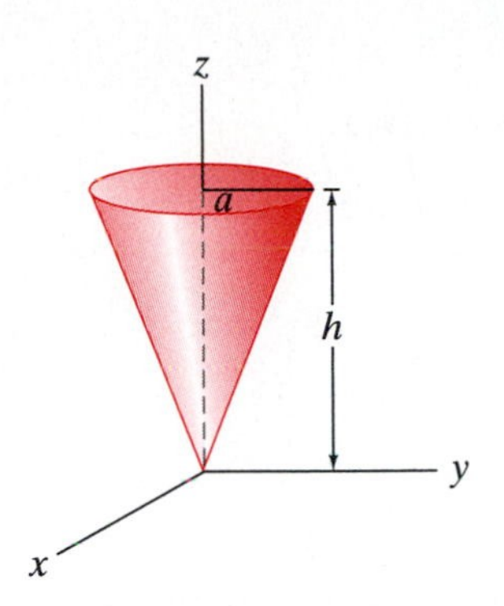

Figure 5.113 The cone of Example 7.

EXAMPLE 7 We compute the centroid of a cone of radius a and height h. (See Figure 5.113.)

By symmetry, $\bar{x} = \bar{y} = 0$. Moreover, we know that the volume of the cone is $(\pi/3)a^2h$. Thus, the z-coordinate of the centroid is

$$\bar{z} = \frac{3}{\pi a^2 h} \iiint_W z\, dV.$$

This triple integral is most readily evaluated by using cylindrical coordinates. (See Example 14 of §5.5.) The lateral surface of the cone is given by $z = \frac{h}{a}r$, so we calculate

$$\bar{z} = \frac{3}{\pi a^2 h} \int_0^{2\pi} \int_0^a \int_{\frac{h}{a}r}^h zr\, dz\, dr\, d\theta = \left(\frac{3}{\pi a^2 h}\right)\left(\frac{\pi a^2 h^2}{4}\right) = \frac{3}{4}h$$

after a straightforward evaluation. Hence, the centroid of the cone is located at $\left(0, 0, \frac{3}{4}h\right)$. ◆

Moments of Inertia

Let W be a rigid solid body in space. As we have seen, the moment integral with respect to the xy-plane is $M_{xy} = \iiint_W z\,\delta(x, y, z)\, dV$—that is, the integral of the product of the position relative to a reference plane (in this case the xy-plane) and the density of the solid. This integral can be considered to measure the ease with which W can be displaced perpendicularly from the reference plane.

Now, consider spinning W about a fixed axis (which may or may not pass through W). The **moment of inertia** I (or **second moment**—the moment integral mentioned in the preceding paragraph is sometimes called the **first moment**) is a measure of the ease with which W can be made to spin about the given axis. Specifically, I is the integral of the product of the density at a point in W and the square of the distance from that point to a fixed axis, that is,

$$I = \iiint_W d^2\, \delta(x, y, z)\, dV, \tag{8}$$

where d is the distance from $(x, y, z) \in W$ to the specified axis.

When the axes of rotation are the coordinate axes in $\mathbf{R}^3$, we have

$$I_x = \text{Moment of inertia about the } x\text{-axis} = \iiint_W (y^2 + z^2)\, \delta(x, y, z)\, dV;$$

$$I_y = \text{Moment of inertia about the } y\text{-axis} = \iiint_W (x^2 + z^2)\, \delta(x, y, z)\, dV;$$

$$I_z = \text{Moment of inertia about the } z\text{-axis} = \iiint_W (x^2 + y^2)\, \delta(x, y, z)\, dV.$$

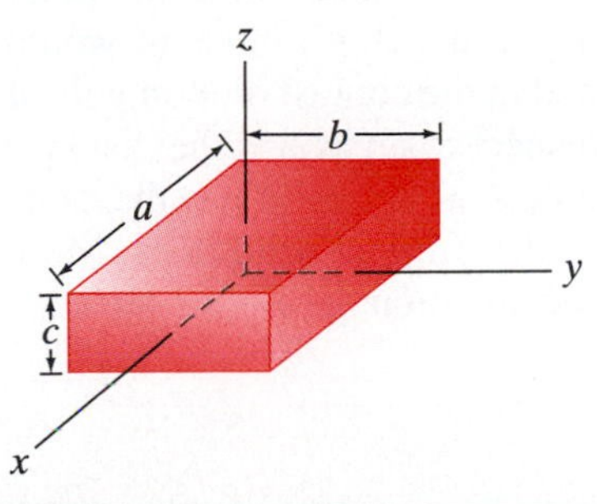

Figure 5.114 The box of Example 8.

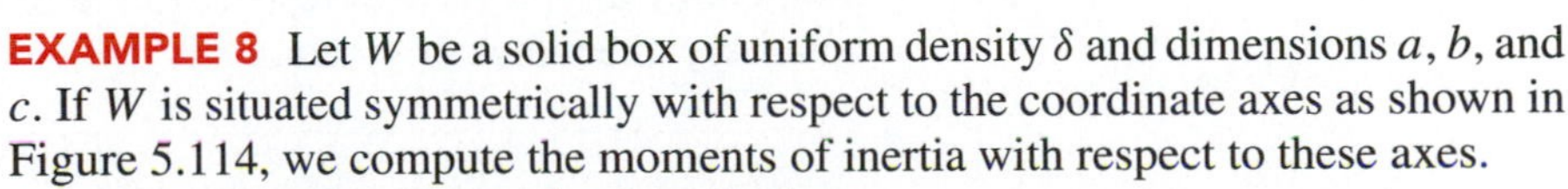

EXAMPLE 8 Let W be a solid box of uniform density δ and dimensions a, b, and c. If W is situated symmetrically with respect to the coordinate axes as shown in Figure 5.114, we compute the moments of inertia with respect to these axes.

Note, first, that W may be described as

$$W = \left\{(x, y, z) \,\middle|\, -\frac{a}{2} \le x \le \frac{a}{2},\ -\frac{b}{2} \le y \le \frac{b}{2},\ -\frac{c}{2} \le z \le \frac{c}{2}\right\}.$$

Hence, the moment of inertia about the x-axis is

$$I_x = \int_{-c/2}^{c/2}\int_{-b/2}^{b/2}\int_{-a/2}^{a/2} (y^2+z^2)\,\delta\,dx\,dy\,dz = \int_{-c/2}^{c/2}\int_{-b/2}^{b/2} (y^2+z^2)\,\delta\,a\,dy\,dz$$

$$= \delta a \int_{-c/2}^{c/2} \left(\frac{y^3}{3}+z^2y\right)\bigg|_{y=-b/2}^{y=b/2} dz = \delta a \int_{-c/2}^{c/2} \left(\frac{b^3}{12}+bz^2\right) dz$$

$$= \delta a \left(\frac{b^3c}{12}+\frac{bc^3}{12}\right) = \frac{\delta abc}{12}(b^2+c^2).$$

By permuting the roles of x, y, and z (and the corresponding constants a, b, and c), we see that

$$I_y = \frac{\delta abc}{12}(a^2+c^2) \quad \text{and} \quad I_z = \frac{\delta abc}{12}(a^2+b^2).$$

Therefore, if $a > b > c$ (as in Figure 5.114), it follows that $I_x < I_y < I_z$. This result may be confirmed by the observation that rotations about the axis parallel to the longest side of the box are easiest to effect in that the same torque applied about each axis will cause the most rapid rotation to occur about the axis through the longest dimension. A related fact is regularly exploited by figure skaters who pull their arms in close to their bodies, thereby reducing their moments of inertia and speeding up their spins. ◆

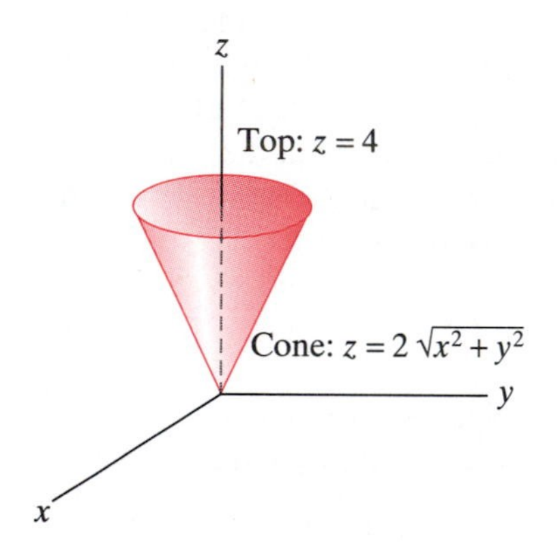

Figure 5.115 The solid W of Example 9.

EXAMPLE 9 Let W be the solid bounded by the cone $z = 2\sqrt{x^2+y^2}$ and the plane $z = 4$ shown in Figure 5.115. Assume that the density of material inside W varies as $\delta(x, y, z) = 5 - z$. Let us calculate the moment of inertia I_z about the z-axis.

Given the geometry of the situation, it is easiest to work in cylindrical coordinates, in which case the cone is given by the cylindrical equation $z = 2r$. Thus, we have

$$I_z = \iiint_W (x^2+y^2)\,\delta(x,y,z)\,dV = \int_0^{2\pi}\int_0^2\int_{2r}^4 r^2(5-z)\,r\,dz\,dr\,d\theta$$

$$= \int_0^{2\pi}\int_0^2 r^3\left(5z-\tfrac{1}{2}z^2\right)\Big|_{z=2r}^{z=4} dr\,d\theta = \int_0^{2\pi}\int_0^2 (12r^3-10r^4+2r^5)\,dr\,d\theta$$

$$= \int_0^{2\pi} \left(3r^4-2r^5+\tfrac{1}{3}r^6\right)\Big|_{r=0}^{r=2} d\theta = \int_0^{2\pi} \frac{16}{3}\,d\theta = \frac{32\pi}{3}.$$ ◆

Recall that the center of mass of a solid object of total mass M is the point such that if all the mass M were concentrated there, the (first) moment would remain the same. An analogous idea may be defined in the context of moments of inertia. The **radius of gyration** of a solid with respect to an axis is the distance r from that axis that we should locate a point of mass M so that it has the same moment of inertia I (with respect to the axis) as the original solid does. More concisely, the radius of gyration r is defined by the equation

$$r^2M = I \quad \text{or} \quad r = \sqrt{\frac{I}{M}}. \tag{9}$$

EXAMPLE 10 We determine the radius of gyration with respect to the z-axis of the cone described in Example 9. Hence, we compute

$$r_z = \sqrt{\frac{I_z}{M}}.$$

From Example 9, $I_z = 32\pi/3$. We determine the total mass M of the cone as follows:

$$M = \int_0^{2\pi}\int_0^2\int_{2r}^4 (5-z)r\,dz\,dr\,d\theta = \int_0^{2\pi}\int_0^2 (12 - 10r + 2r^2)r\,dr\,d\theta$$

$$= \int_0^{2\pi}\left(\left(6r^2 - \tfrac{10}{3}r^3 + \tfrac{1}{2}r^4\right)\Big|_{r=0}^{r=2}\right)d\theta = \int_0^{2\pi}\frac{16}{3}\,d\theta = \frac{32\pi}{3}.$$

Thus,

$$r_z = \sqrt{\frac{32\pi/3}{32\pi/3}} = 1.$$

◆

Exercises

1. The local grocery store receives a shipment of 75 cases of cat food every month. The inventory of cat food (i.e., the number of cases of cat food on hand as a function of days) is given by $I(x) = 75\cos(\pi x/15) + 80$.
   (a) What is the average daily inventory over a month?
   (b) If the cost of storing a case is 2 cents per day, determine the average daily holding cost over the month.

2. Find the average value of $f(x, y) = \sin^2 x \cos^2 y$ over $R = [0, 2\pi] \times [0, 4\pi]$.

3. Find the average value of $f(x, y) = e^{2x+y}$ over the triangular region whose vertices are $(0, 0)$, $(1, 0)$, and $(0, 1)$.

4. Find the average value of $g(x, y, z) = e^z$ over the unit ball given by
$$B = \{(x, y, z) \mid x^2 + y^2 + z^2 \le 1\}.$$

5. Suppose that the temperature at a point in the cube
$$W = [-1, 1] \times [-1, 1] \times [-1, 1]$$
varies in proportion to the square of the point's distance from the origin.
   (a) What is the average temperature of the cube?
   (b) Describe the set of points in the cube where the temperature is equal to the average temperature.

6. Let D be the region between the square with vertices $(1, 1)$, $(-1, 1)$, $(-1, -1)$, $(1, -1)$ and the unit disk centered at the origin. Find the average value of $f(x, y) = x^2 + y^2$ on D.

7. Let W be the region in $\mathbf{R}^3$ between the cube with vertices $(1, 1, 1)$, $(-1, 1, 1)$, $(-1, -1, 1)$, $(1, -1, 1)$, $(1, 1, -1)$, $(-1, 1, -1)$, $(-1, -1, -1)$, $(1, -1, -1)$ and the unit ball centered at the origin. Find the average value of $f(x, y, z) = x^2 + y^2 + z^2$ on W.

8. Suppose that you commute every day to work by subway. You walk to the same subway station, which is served by two subway lines, both stopping near where you work. During rush hour, each subway line sends trains to arrive at the stop every 6 minutes, but the dispatchers begin the schedules at random times. What is the average time you expect to wait for a subway train? (Hint: Model the waiting time for the two subway lines by using a point (x, y) in the square $[0, 6] \times [0, 6]$.)

9. Repeat Exercise 8 in the case that the subway stop is serviced by three subway lines (each with trains arriving every 6 minutes), rather than two.

10. Find the center of mass of the region bounded by the parabola $y = 8 - 2x^2$ and the x-axis
    (a) if the density δ is constant;
    (b) if the density $\delta = 3y$.

11. Find the centroid of a semicircular plate. (Hint: Judicious use of a suitable coordinate system might help.)

12. Find the center of mass of a plate that is shaped like the region between $y = x^2$ and $y = 2x$, where the density varies as $1 + x + y$.

13. Find the center of mass of a lamina shaped like the region
$$\{(x, y) \mid 0 \le y \le \sqrt{x},\ 0 \le x \le 9\},$$
where the density varies as xy.

14. Find the centroid of the region bounded by the cardioid given in polar coordinates by the equation $r = 1 - \sin\theta$. (Hint: Think carefully.)

15. Find the centroid of the lamina described in polar coordinates as

$$\{(r, \theta) \mid 0 \leq r \leq 4\cos\theta,\ 0 \leq \theta \leq \pi/3\}.$$

16. Find the center of mass of the lamina described in polar coordinates as

$$\{(r, \theta) \mid 0 \leq r \leq 3,\ 0 \leq \theta \leq \pi/4\},$$

where the density of the lamina varies as $\delta(r, \theta) = 4 - r$.

17. Find the center of mass of the region inside the cardioid given in polar coordinates as $r = 1 + \cos\theta$, and whose density varies as $\delta(r, \theta) = r$.

18. Find the centroid of the tetrahedron whose vertices are at $(0, 0, 0)$, $(1, 0, 0)$, $(0, 2, 0)$, and $(0, 0, 3)$.

19. A solid is bounded below by $z = 3y^2$, above by the plane $z = 3$, and on the ends by the planes $x = -1$ and $x = 2$.

(a) Find the centroid of this solid.

(b) Now assume that the density of the solid is given by $\delta = z + x^2$. Find the center of mass of the solid.

20. Determine the centroid of the region bounded above by the sphere $x^2 + y^2 + z^2 = 18$ and below by the paraboloid $3z = x^2 + y^2$.

21. Find the centroid of the solid, capsule-shaped region bounded by the paraboloids $z = 3x^2 + 3y^2 - 16$ and $z = 9 - x^2 - y^2$.

22. Find the centroid of the "ice cream cone" shown in Figure 5.116.

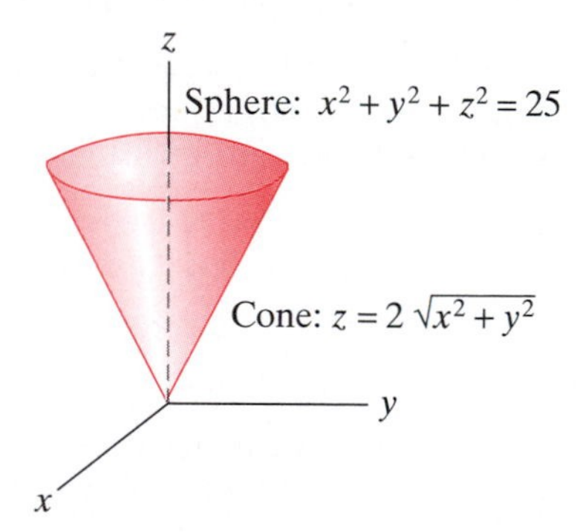

Figure 5.116 The ice cream cone solid of Exercise 22.

23. Find the centroid of the solid shaped as one-eighth of a solid ball of radius a. (Hint: Model the solid as the first octant portion of a ball of radius a with center at the origin.)

24. Find the center of mass of a solid cylindrical peg of radius a and height h whose mass density at a point in the peg varies as the square of the distance of that point from the top of the cylinder.

25. (a) Find the moment of inertia about the coordinate axes of a solid, homogeneous tetrahedron whose vertices are located at $(0,0,0)$, $(1, 0, 0)$, $(0, 1, 0)$, and $(0, 0, 1)$.

(b) What are the radii of gyration about the coordinate axes?

26. Consider the solid cube $W = [0, 2] \times [0, 2] \times [0, 2]$. Find the moments of inertia and the radii of gyration about the coordinate axes if the density of the cube is $\delta(x, y, z) = x + y + z + 1$.

27. A solid is bounded by the paraboloid $z = x^2 + y^2$ and the plane $z = 9$. Find the moment of inertia and radius of gyration about the z-axis if

(a) the density is $\delta(x, y, z) = 2z$;

(b) the density is $\delta(x, y, z) = \sqrt{x^2 + y^2}$.

28. Find the moment of inertia and radius of gyration about the z-axis of a solid ball of radius a, centered at the origin, if

(a) the density δ is constant;

(b) $\delta(x, y, z) = x^2 + y^2 + z^2$;

(c) $\delta(x, y, z) = x^2 + y^2$.

We can find the moment of inertia of a lamina in the plane with density $\delta(x, y)$ by considering the lamina to be a flat plate sitting in the xy-plane in $\mathbf{R}^3$. Then, for example, the distance of a point (x, y) in the lamina to the x-axis is given by $|y|$, the distance to the y-axis is given by $|x|$, and the distance to the z-axis (or the origin) is given by $\sqrt{x^2 + y^2}$. (See Figure 5.117.) Using these ideas, find the specified moments of inertia of the laminas given in Exercises 29–31.

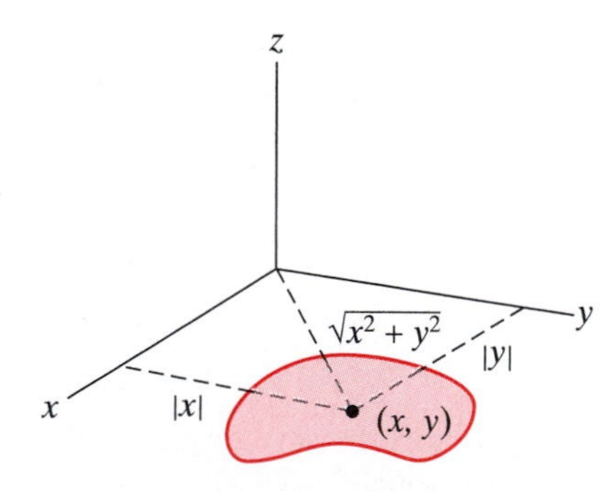

Figure 5.117 A lamina situated in the xy-plane in $\mathbf{R}^3$.

29. The moment of inertia I_x about the x-axis of the lamina that has the shape bounded by the graph of $y = x^2 + 2$ and the line $y = 3$, and whose density varies as $\delta(x, y) = x^2 + 1$.

30. The moment of inertia I_z about the z-axis of the lamina shaped as the rectangle $[0, 2] \times [0, 1]$, and whose density varies as $\delta(x, y) = 1 + y$.

31. The moment of inertia about the line $y = 3$ of the lamina shaped as the disk

$$\{(x, y) \mid x^2 + y^2 \leq 4\},$$

and whose density varies as $\delta(x, y) = x^2$.

The gravitational field between a mass M concentrated at the point (x, y, z) and a mass m concentrated at the point (x_0, y_0, z_0) is

$$\mathbf{F} = -\frac{GMm[(x - x_0)\mathbf{i} + (y - y_0)\mathbf{j} + (z - z_0)\mathbf{k}]}{[(x - x_0)^2 + (y - y_0)^2 + (z - z_0)^2]^{3/2}}.$$

The ***gravitational potential*** *V of $\mathbf{F}$ is*

$$V = -\frac{GMm}{\sqrt{(x - x_0)^2 + (y - y_0)^2 + (z - z_0)^2}}.$$

(We have seen in §3.3 that $\mathbf{F} = -\nabla V$.) Now suppose that, instead of a point mass M, we have a solid region W of density $\delta(x, y, z)$ and total mass M. Then the gravitational potential of W acting on the point mass m may be found by looking at "infinitesimal" point masses $dm = \delta(x, y, z)\, dV$ and adding (via integration) their individual potentials. That is, the potential of W is

$$V(x_0, y_0, z_0) = -\iiint_W \frac{Gm\delta(x, y, z)\, dV}{\sqrt{(x - x_0)^2 + (y - y_0)^2 + (z - z_0)^2}}.$$

In Exercises 32–34, Let W be the region between two concentric spheres of radii $a < b$, centered at the origin. (See Figure 5.118.) Assume that W has total mass M and constant density δ. The object of the following exercises is to compute the gravitational potential $V(x_0, y_0, z_0)$ of W on a mass m concentrated at (x_0, y_0, z_0). Note that, by the spherical symmetry, there is no loss of generality in taking (x_0, y_0, z_0) equal to $(0, 0, r)$. So, in particular, r is the distance from the point mass m to the center of W.

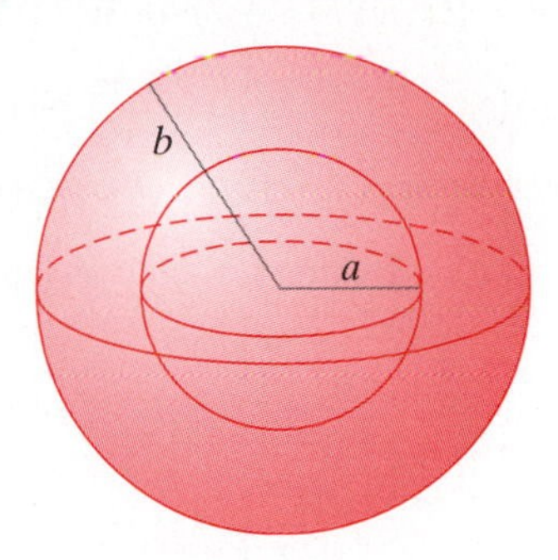

Figure 5.118 The spherical shell of Exercises 32–34.

32. Show that if $r \geq b$, then $V(0, 0, r) = -GMm/r$. This is exactly the same gravitational potential as if all the mass M of W were concentrated at the origin. This is a key result of Newtonian mechanics. (Hint: Use spherical coordinates and integrate with respect to φ before integrating with respect to ρ.)

33. Show that if $r \leq a$, then there is *no* gravitational force. (Hint: Show that $V(0, 0, r)$ is actually independent of r. Then relate the gravitational potential to gravitational force. As in Exercise 32, use spherical coordinates and integrate with respect to φ before integrating with respect to ρ.)

34. (a) Find $V(0, 0, r)$ if $a < r < b$.

(b) Relate your answer in part (a) to the results of Exercises 32 and 33.

5.7 True/False Exercises for Chapter 5

1. Every rectangle in $\mathbf{R}^2$ may be denoted $[a, b] \times [c, d]$.

2. If f is a continuous function and $f(x, y) \geq 0$ on a region D in $\mathbf{R}^2$, then the volume of the solid in $\mathbf{R}^3$ under the graph of the surface $z = f(x, y)$ and above the region D in the xy-plane is $\iint_D f(x, y)\, dA$.

3. $\int_0^{1/2} \int_{-1}^{2} y^3 \sin(\pi x^2)\, dy\, dx = \int_{-1}^{2} \int_0^{1/2} y^3 \sin(\pi x^2)\, dx\, dy.$

4. $\int_0^2 \int_0^x 3\, dy\, dx = \int_0^2 \int_0^y 3\, dx\, dy.$

5. $\int_0^2 \int_0^y f(x, y)\, dx\, dy = \int_0^2 \int_0^x f(x, y)\, dy\, dx$ for all continuous functions f.

6. $0 \leq \iint_D \sin(x^4 + y^4)\, dx\, dy \leq \pi$, where D denotes the unit disk $\{(x, y) \mid x^2 + y^2 \leq 1\}$.

7. $\int_0^2 \int_0^x \sqrt{x^3 + y}\, dy\, dx = \int_0^x \int_0^2 \sqrt{x^3 + y}\, dx\, dy.$

8. $\int_{-1}^{1} \int_0^3 x^2 e^{x+y}\, dy\, dx = \left(\int_{-1}^{1} x^2 e^x\, dx\right)\left(\int_0^3 e^y\, dy\right).$

9. The region in $\mathbf{R}^2$ bounded by the graphs of $y = x^3$ and $y = \sqrt{x}$ is a type 3 elementary region in the plane.

10. The region in $\mathbf{R}^2$ bounded by the graphs of $y = \sin x$, $y = \cos x$, $x = \pi/4$, and $x = 5\pi/4$ is a type 2 elementary region in the plane.

11. The region in $\mathbf{R}^3$ bounded by the graphs of $y^2 - x^2 - z^2 = 1$ and $9x^2 + 4z^2 = 36$ is a type 3 elementary region in space.

12. $\iint_D (y^3+1)\,dx\,dy$ gives the area of the region D, where $D = \{(x, y) \mid (x-2)^2 + 3y^2 \le 5\}$.

13. $\iint_D (y^2 \sin(x^3) + 3)\,dx\,dy = 3/2$, where D is the triangle with vertices $(-1, 0)$, $(1, 0)$, $(0, 1)$.

14. $\iiint_{[-2,2]\times[-1,1]\times[-3,3]} (x + y^3 + z^5)\,dV = 0.$

15. $\iiint_{[-2,2]\times[0,1]\times[-1,1]} (x + z)\,dV = 0.$

16. $\iiint_{[-2,2]\times[0,1]\times[-1,1]} (y + z)\,dV = 0.$

17. $\int_0^4 \int_{y/4}^{\sqrt{y}/2} \int_{-2}^{2} \sin yz\,dz\,dx\,dy = 0.$

18. $\int_{-1}^{1} \int_{-\sqrt{1-x^2}}^{\sqrt{1-x^2}} \int_{-\sqrt{1-x^2-y^2}}^{\sqrt{1-x^2-y^2}} (y - x^2)\,dz\,dy\,dx = 0.$

19. If $\mathbf{T}(u, v) = (2u - v, u + 3v)$, then the area of the image $D = \mathbf{T}(D^*)$ of the unit square $D^* = [0, 1] \times [0, 1]$ is 7 square units.

20. If $\mathbf{T}(u, v) = (v - u, 3u + 2v)$, then the area of the image $D = \mathbf{T}(D^*)$ of the rectangle $D^* = [0, 3] \times [0, 2]$ is 5 square units.

21.

$$\int_0^2 \int_{y/2}^{5-2y} e^{2x-y} \cos(x - 3y)\,dx\,dy = \int_0^{10} \int_{u-5}^{u/2} -5e^u \cos v\,dv\,du.$$

22. If D is the disk $\{(x, y) \mid x^2 + y^2 \le 9\}$, then $\iint_D \sqrt{9 - x^2 - y^2}\,dA = 36\pi.$

23. The iterated integral $\int_0^{2\pi} \int_0^2 \int_{2r}^4 dz\,dr\,d\theta$ represents the volume enclosed by the cone of height 4 and radius 2.

24. The iterated integral

$$\int_{-2}^{2} \int_{-\sqrt{4-x^2}}^{\sqrt{4-x^2}} \int_0^{\sqrt{9-x^2-y^2}} (2 + \sqrt{x^2+y^2})\,dz\,dy\,dx$$

is given by an equivalent integral in cylindrical coordinates as

$$\int_0^{2\pi} \int_0^2 \int_0^{\sqrt{9-r^2}} (r^2 + 2r)\,dz\,dr\,d\theta.$$

25. The iterated integral

$$\int_{-\sqrt{2}}^{\sqrt{2}} \int_0^{\sqrt{2-x^2}} \int_{\sqrt{x^2+y^2}}^{\sqrt{4-x^2-y^2}} \sqrt{x^2+y^2+z^2+5}\,dz\,dy\,dx$$

is given by an equivalent integral in spherical coordinates as

$$\int_0^{\pi} \int_0^{\pi/4} \int_0^2 \rho\sqrt{\rho^2+5}\,\sin\varphi\,d\rho\,d\varphi\,d\theta.$$

26. The average value of the function $f(x, y) = x^2 y$ over the semicircular region $D = \{(x, y) \mid 0 \le y \le \sqrt{4 - x^2}\}$ is given by

$$\frac{1}{\pi} \int_0^{\pi/2} \int_0^2 r^4 \sin\theta \cos^2\theta\,dr\,d\theta$$

or by

$$\frac{1}{\pi} \int_{\pi/2}^{\pi} \int_0^2 r^4 \sin\theta \cos^2\theta\,dr\,d\theta.$$

27. The center of mass of a lamina represented by the triangle with vertices $(2, 0)$, $(0, 1)$, $(0, -1)$, and whose density varies as $\delta(x, y) = (x^2 + 1)\cos y$, has coordinates given by

$$\overline{x} = \frac{\int_0^2 \int_{(x-2)/2}^{(2-x)/2} (x^3 + x)\cos y\,dy\,dx}{\int_0^2 \int_{(x-2)/2}^{(2-x)/2} (x^2 + 1)\cos y\,dy\,dx}, \qquad \overline{y} = 0.$$

28. The centroid of a cone of radius a, height h, with axis the z-axis and vertex at $(0, 0, h)$ is $\left(0, 0, \frac{3}{4}h\right)$.

29. The center of mass of the solid cylinder of radius a, height h, with axis the z-axis whose density at any point varies as e^d, where d is the square of the distance from the point to the z-axis is $(0, 0, \overline{z})$, where

$$\overline{z} = \frac{\int_0^{2\pi} \int_0^a \int_0^h zre^{r^2}\,dz\,dr\,d\theta}{\int_0^{2\pi} \int_0^a \int_0^h re^{r^2}\,dz\,dr\,d\theta} = \tfrac{1}{2}h.$$

30. The integral $\iiint_W \rho^5 \sin 2\varphi \sin\varphi\,d\rho\,d\varphi\,d\theta$ represents the moment of inertia about the z-axis of a solid W with density z, expressed in spherical coordinates.

5.8 Miscellaneous Exercises for Chapter 5

1. Let B be the ball of radius 3; that is,

$$B = \{(x, y, z) \mid x^2 + y^2 + z^2 \le 9\}.$$

Without resorting to any explicit calculation of an iterated integral, determine the value of the triple integral $\iiint_B (z^3 + 2)\,dV$ by using geometry and symmetry considerations.

2. Let W denote half of the solid ball of radius 2; that is,

$$W = \{(x, y, z) \mid x^2 + y^2 + z^2 \le 4,\ z \ge 0\}.$$

Without resorting to explicit calculation of an iterated integral, determine the value of the triple integral

$$\iiint_W (x^3 + y - 3)\, dV.$$

(Hint: Use geometry and symmetry.)

3. Let W be the solid region in $\mathbf{R}^3$ with $x \geq 0$ that is bounded by the three surfaces $z = 9 - x^2$, $z = 2x^2 + y^2$, and $x = 0$.

(a) Set up, but do not evaluate, two different (but equivalent) iterated integrals that both give the value of $\iiint_W 3\, dV$.

(b) Use a computer algebra system to find the value of $\iiint_W 3\, dV$ and to check for consistency in your answers in part (a).

4. Suppose that f is continuous on the rectangle $R = [a, b] \times [c, d]$. For $(x, y) \in (a, b) \times (c, d)$, we define

$$F(x, y) = \int_a^x \int_c^y f(x', y')\, dy'\, dx'.$$

Use Fubini's theorem to show that $\partial^2 F/\partial x \partial y = \partial^2 F/\partial y \partial x$. This provides an alternative proof of the equality of mixed partials. (Hint: Write

$$F(x, y) = \int_a^x g(x', y)\, dx'$$

where

$$g(x', y) = \int_c^y f(x', y')\, dy'.$$

Then, $\partial F/\partial x$ and $\partial^2 F/\partial y \partial x$ may be calculated using the fundamental theorem of calculus. Then, use Fubini's theorem to find $\partial F/\partial y$ and $\partial^2 F/\partial x \partial y$.)

5. Convert the following cylindrical integral to equivalent iterated integrals in (a) Cartesian coordinates and (b) spherical coordinates:

$$\int_0^{2\pi} \int_0^1 \int_0^{\sqrt{9-r^2}} r\, dz\, dr\, d\theta.$$

Evaluate the easiest of the three iterated integrals.

6. The volume of a solid is given by the iterated integral

$$\int_0^4 \int_0^{\sqrt{4y-y^2}} \int_{-\sqrt{16-x^2-y^2}}^{\sqrt{16-x^2-y^2}} dz\, dx\, dy.$$

(a) Sketch the solid and also describe it by giving equations for the surfaces that form its boundary.

(b) Express the volume as an iterated integral in cylindrical coordinates. Determine the volume.

7. Calculate the volume of a cube having edge length a by integrating in cylindrical coordinates. (Hint: Put the center of the cube at the origin.)

8. Calculate the volume of a cube having edge length a by integrating in spherical coordinates.

9. Determine

$$\iint_D \cos\left(\frac{x - 2y}{x + y}\right) dA,$$

where D is the triangular region bounded by the coordinate axes and the line $x + y = 1$.

10. Evaluate

$$\int_0^6 \int_{-2y}^{1-2y} y^3 (x + 2y)^2 e^{(x+2y)^3}\, dx\, dy$$

by making a suitable change of variables.

11. Find the area enclosed by the ellipse E given by the equation

$$\frac{x^2}{a^2} + \frac{y^2}{b^2} = 1$$

in the following way:

(a) First, write the area as the value of an appropriate iterated integral in Cartesian coordinates. Do not evaluate this integral.

(b) Next, scale the variables by letting $x = a\bar{x}$, $y = b\bar{y}$. To what region E^* in the $\bar{x}\bar{y}$-plane does the ellipse E correspond? Rewrite the xy-integral in part (a) as an $\bar{x}\bar{y}$-integral.

(c) Finally, use polar coordinates to transform the $\bar{x}\bar{y}$-integral and thereby show that the area inside the original ellipse is πab.

12. This problem concerns the rotated ellipse E with equation $13x^2 + 14xy + 10y^2 = 9$.

(a) Let $u = 2x - y$, $v = x + y$ and rewrite the equation for E in the form

$$\frac{u^2}{a^2} + \frac{v^2}{b^2} = 1,$$

where a and b are positive constants to be determined.

(b) Use an appropriate change of variables and the result of part (c) of Exercise 11 to find the area enclosed by E.

13. Consider the ellipse E with equation $5x^2 + 6xy + 5y^2 = 4$. Let $u = x - y$, $v = x + y$ and follows the steps of Exercise 12.

14. Imitate the techniques of Exercise 11 to find the volume enclosed by the ellipsoid E given by the equation

$$\frac{x^2}{a^2} + \frac{y^2}{b^2} + \frac{z^2}{c^2} = 1.$$

15. Evaluate

$$\iint_D \frac{xy}{y^2 - x^2}\, dA,$$

where D is the region in the first quadrant bounded by the hyperbolas $x^2 - y^2 = 1$, $x^2 - y^2 = 4$ and the ellipses $x^2/4 + y^2 = 1$, $x^2/16 + y^2/4 = 1$. (Hint: Sketch the region D, and use it to make an appropriate change of variables.)

16. Evaluate

$$\iint_D (x^2 + y^2)e^{x^2-y^2}\, dA,$$

where D is the region in the first quadrant bounded by the hyperbolas $x^2 - y^2 = 1$, $x^2 - y^2 = 9$, $xy = 1$, $xy = 4$.

17. Evaluate

$$\iint_D \frac{1}{x^2y^2 + 1}\, dA,$$

where D is the region bounded by $xy = 1$, $xy = 4$, $y = 1$, $y = 2$.

18. (a) Generalizing the notions of double and triple integrals, develop a definition of the "quadruple integral" $\iiiint_W f\, dV$ of a function $f(x, y, z, w)$ over a four-dimensional region W in $\mathbf{R}^4$.

(b) Use your definition in part (a) to calculate

$$\iiiint_W (x + 2y + 3z - 4w)\, dV,$$

where W is the four-dimensional box

$$W = \{(x, y, z, w) \mid 0 \le x \le 2,\ -1 \le y \le 3,\ 0 \le z \le 4,\ -2 \le w \le 2\}.$$

19. (a) Set up, but do not evaluate, a quadruple iterated integral that computes the four-dimensional volume of the four-dimensional ball of radius a:

$$B = \{(x, y, z, w) \mid x^2 + y^2 + z^2 + w^2 \le a^2\}.$$

(b) Use a computer algebra system to give a formula for the volume.

(c) Use a computer algebra system to give formulas for the n-dimensional volume of the n-dimensional ball

$$B = \{(x_1, x_2, \ldots, x_n) \mid x_1^2 + x_2^2 + \cdots + x_n^2 \le a^2\}$$

in the cases where $n = 5, 6$. Is there any pattern to your answers?

20. A spherical shell with inner radius 3 cm and outer radius 4 cm has a mass density that varies as $0.12d^2$ g/cm^3, where d denotes the distance (in centimeters) from a point in the shell to the center of the shell.

(a) Determine the total mass of the shell.

(b) Will the shell float in water? (Note: The density of water is 1 g/cm^3. To answer this question, you need to determine the average density of the shell.)

(c) Suppose that the shell has a small hole so that the core of the shell fills with water. Now will it float?

21. A dome is shaped as a hemisphere. If a pole whose length is the average height of the dome is to be installed inside the dome in an upright position, where on the floor can it be located?

22. Let f be continuous on $R = [a, b] \times [c, d]$. In this problem, you will establish **Leibniz's rule** for "differentiating under the integral sign":

$$\frac{d}{dy}\int_a^b f(x, y)\, dx = \int_a^b f_y(x, y)\, dx.$$

(a) Let $G(y') = \int_a^b f_y(x, y')\, dx$. For $c \le y \le d$, use the fundamental theorem of calculus to compute $d/dy \int_c^y G(y')\, dy'$ and, therefore,

$$\frac{d}{dy}\int_c^y \int_a^b f_y(x, y')\, dx\, dy'.$$

(b) Use Fubini's theorem and part (a) to establish Leibniz's rule.

23. The function $f(x, y) = 1/\sqrt{xy}$ is unbounded when either x or y is zero. Thus if $D = [0, 1] \times [0, 1]$, we say that $\iint_D \frac{1}{\sqrt{xy}}\, dA$ is an **improper double integral,** analogous to the one-variable improper integral $\int_0^1 \frac{1}{\sqrt{x}}\, dx$. Improper multiple integrals of this type may be evaluated using an appropriate limiting process. In this problem, you will determine the value of $\iint_D \frac{1}{\sqrt{xy}}\, dA$ in the following manner:

(a) For $0 < \epsilon < 1/2$, $0 < \delta < 1/2$, let $D_{\epsilon,\delta} = [\epsilon, 1 - \epsilon] \times [\delta, 1 - \delta]$. (Note that $D_{\epsilon,\delta} \subset D$.) Calculate

$$I(\epsilon, \delta) = \iint_{D_{\epsilon,\delta}} \frac{1}{\sqrt{xy}}\, dA.$$

(b) Evaluate $\lim_{(\epsilon,\delta)\to(0,0)} I(\epsilon, \delta)$. You should obtain a finite value, which may be taken to be the value of $\iint_D \frac{1}{\sqrt{xy}}\, dA$, since $D_{\epsilon,\delta}$ "fills out" D as $(\epsilon, \delta) \to (0, 0)$. (We say that in this case the improper integral **converges**.)

24. Imitate the techniques of Exercise 23 to determine if the improper double integral

$$\iint_{[0,1]\times[0,1]} \frac{1}{x + y}\, dA$$

converges and, if it does, find its value. (Hint: You will need to determine $\lim_{u\to 0^+} u \ln u$.)

25. Imitate the techniques of Exercise 23 to determine if the improper double integral

$$\iint_{[0,1]\times[0,1]} \frac{x}{y}\, dA$$

converges and, if it does, find its value.

26. Calculate $\iint_D \ln\sqrt{x^2+y^2}\,dA$, where D is the unit disk $x^2+y^2 \le 1$. Note that the integrand is not defined at the origin, so this is an example of an improper double integral. Nonetheless, you can find its value by integrating over the annular region

$$D_\epsilon = \{(x, y) \mid \epsilon \le x^2 + y^2 \le 1\}$$

and taking appropriate limits.

27. Find the value of the improper triple integral

$$\iiint_B \ln\sqrt{x^2+y^2+z^2}\,dV,$$

where B is the solid ball $x^2+y^2+z^2 \le 1$. (See Exercise 26.)

28. If D is an unbounded region in $\mathbf{R}^2$, the integral $\iint_D f(x, y)\,dA$ is another type of **improper double integral,** analogous to one-variable improper integrals such as $\int_a^\infty f(x)\,dx$, $\int_{-\infty}^b f(x)\,dx$, or $\int_{-\infty}^\infty f(x)\,dx$. In this problem, you will determine the value of

$$\iint_D \frac{1}{x^2y^3}\,dA,$$

where $D = \{(x, y) \mid x \ge 1, y \ge 1\}$ using a limiting process.

(a) For $a > 1$, $b > 1$, let $D_{a,b} = [1, a] \times [1, b]$. Compute

$$I(a, b) = \iint_{D_{a,b}} \frac{1}{x^2y^3}\,dA.$$

(b) Evaluate $\lim_{(a,b)\to(\infty,\infty)} I(a, b)$. You should obtain a finite value, which may be taken to be the value of $\iint_D \frac{1}{x^2y^3}\,dA$. (In such a case, we say that the improper integral **converges**.)

29. Let $D = \{(x, y) \mid x \ge 1, y \ge 1\}$. For what values of p and q does the improper integral

$$\iint_D \frac{1}{x^p y^q}\,dA$$

converge? For those values of p and q for which the integral converges, what is the value of the integral?

30. This problem concerns the improper integral

$$\iint_{\mathbf{R}^2} (1+x^2+y^2)^p\,dA,$$

where p is a constant.

(a) Determine if the integral converges when $p = -2$ by integrating over the disk $D_a = \{(x, y) \mid x^2+y^2 \le a^2\}$ and then letting $a \to \infty$.

(b) Determine for what values of p the integral $\iint_{\mathbf{R}^2}(1+x^2+y^2)^p\,dA$ converges. What is the value of the integral when it converges?

31. Determine if

$$\iiint_{\mathbf{R}^3} \frac{1}{(1+x^2+y^2+z^2)^{3/2}}\,dV$$

converges by integrating over the ball $B_a = \{(x, y, z) \mid x^2+y^2+z^2 \le a^2\}$ and then letting $a \to \infty$.

32. Show that

$$\iiint_{\mathbf{R}^3} e^{-\sqrt{x^2+y^2+z^2}}\,dV$$

converges by determining its value.

33. In this problem, you will find the value of the one-variable improper integral $\int_{-\infty}^\infty e^{-x^2}\,dx$ by using two-variable improper integrals.

(a) First argue that $\int_{-\infty}^\infty e^{-x^2}\,dx$ converges. (Hint: Note that $e^{-x^2} \le 1/x^2$ for all x; compare integrals.)

(b) Let I denote the value of $\int_{-\infty}^\infty e^{-x^2}\,dx$. Show that

$$I^2 = \int_{-\infty}^\infty \int_{-\infty}^\infty e^{-x^2-y^2}\,dx\,dy = \iint_{\mathbf{R}^2} e^{-x^2-y^2}\,dA.$$

(c) Let D_a denote the disk $x^2+y^2 \le a^2$. Evaluate $\iint_{D_a} e^{-x^2-y^2}\,dA$.

(d) Compute I^2 as $\lim_{a\to\infty}\iint_{D_a} e^{-x^2-y^2}\,dA$.

(e) Now find $\int_{-\infty}^\infty e^{-x^2}\,dx$.

Exercises 34–42 involve the notion of probability densities. A ***probability density*** *function of a single variable is any function $f(x)$ such that $f(x) \ge 0$ for all $x \in \mathbf{R}$, and $\int_{-\infty}^\infty f(x) = 1$. Given such a density function, the probability that a randomly selected number x falls between the values a and b is*

$$\text{Prob}(a \le x \le b) = \int_a^b f(x)\,dx.$$

34. (a) Check that $f(x) = e^{-2|x|}$ is a probability density function.

(b) Egbert turns on the stove in a random manner to heat cooking oil to fry chicken. If the probability density of the temperature x of the oil is given by $f(x) = \frac{1}{2}e^{-|x-300|}$, what is the probability that the oil has a temperature between 250°F and 350°F?

A ***joint probability density*** *function for two random variables x and y is a function $f(x, y)$ such that*

(i) *$f(x, y) \ge 0$ for all $(x, y) \in \mathbf{R}^2$, and*

(ii) *$\iint_{\mathbf{R}^2} f(x, y)\,dA = \int_{-\infty}^\infty\int_{-\infty}^\infty f(x, y)\,dx\,dy = 1$.*

If f is such a probability density and D is a region in $\mathbf{R}^2$, then the probability that a randomly chosen point (x, y) lies in D is

$$\text{Prob}((x, y) \in D) = \iint_D f(x, y)\,dA.$$

35. (a) Show that the function

$$f(x, y) = \begin{cases} \dfrac{2x + y}{140} & \text{if } 0 \le x \le 5, 0 \le y \le 4 \\ 0 & \text{otherwise} \end{cases}$$

is a joint probability density function.

(b) Find the probability that $x \le 1$ and $y \le 1$.

36. (a) Show that the function

$$f(x, y) = \begin{cases} ye^{-x-y} & \text{if } x \ge 0, y \ge 0 \\ 0 & \text{otherwise} \end{cases}$$

is a joint probability density function.

(b) What is the probability that $x + y \le 2$?

37. If a and b are fixed positive constants, what value of C will make the function $f(x, y) = Ce^{-a|x|-b|y|}$ a joint probability density function?

38. Let a and b be fixed nonnegative constants, not both zero. For what value of C is

$$f(x, y) = \begin{cases} C(ax + by) & \text{if } 0 \le x \le 1, 0 \le y \le 1 \\ 0 & \text{otherwise} \end{cases}$$

a joint probability density function?

39. Let a and b be fixed positive constants and, for a given constant C, consider the function

$$f(x, y) = \begin{cases} Cxy & \text{if } 0 \le x \le a, 0 \le y \le b \\ 0 & \text{otherwise} \end{cases}.$$

(a) For what value of C is f a joint probability density function?

(b) Using the value of C that you found in part (a), what is the probability that $bx - ay \ge 0$?

40. The research team for Vertigo Amusement Park determines that the length of time x (in minutes) a customer spends waiting to participate in the new Drown Town water ride, and the length of time y actually spent in the ride, are jointly distributed according to the probability density function

$$f(x, y) = \begin{cases} \frac{1}{250}e^{-x/50-y/5} & \text{if } x \ge 0, y \ge 0 \\ 0 & \text{otherwise} \end{cases}.$$

Find the probability that a customer spends at most an hour involved with the ride (both waiting and participating).

41. Suppose that you randomly shoot arrows at a circular target so that the distribution of your arrows is given by the probability density function

$$f(x, y) = \frac{1}{\pi}e^{-x^2-y^2},$$

where x and y are measured in feet. In the center of the target, there is a bull's-eye that measures 1 ft in diameter. (See Figure 5.119.) What is the probability of your hitting the bull's-eye?

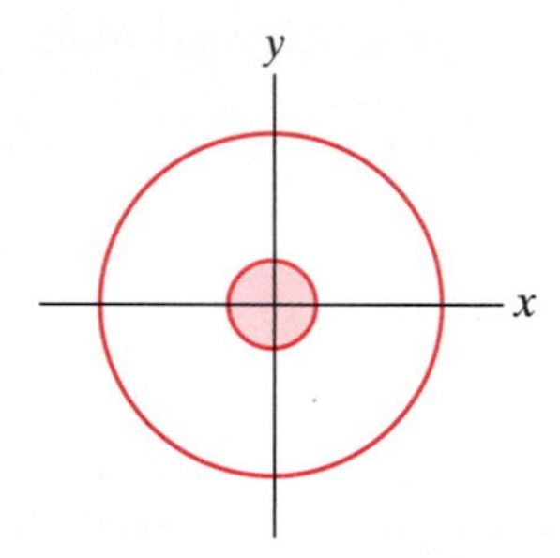

Figure 5.119 The circular target of Exercise 41. The shaded region is the bull's-eye.

42. If x is a random variable with probability density function $f(x)$ and y is a random variable with probability density function $g(y)$, then we say that x and y are **independent random variables** if their joint density function is the product of their individual density functions, that is, if

$$F(x, y) = f(x)g(y).$$

Suppose that an electrical circuit is designed with two identical components whose time x to failure (measured in hours) is given by an exponential probability density function

$$f(x) = \begin{cases} 0 & \text{if } x < 0 \\ \frac{1}{2000}e^{-x/2000} & \text{if } x \ge 0 \end{cases}.$$

Assuming that the components fail independently, what is the probability that they *both* fail within 2000 hours?

43. Let I_L denote the moment of inertia of a solid W (of density $\delta(x, y, z)$) about the line L. (See formula (8) in §5.6.) Let $\bar{L}$ be the line parallel to L that passes through the center of mass of W. Then, if M denotes the mass of W and h the distance between L and $\bar{L}$, prove the **parallel axis theorem**:

$$I_L - I_{\bar{L}} = Mh^2.$$

(Hint: Without loss of generality, you can arrange things so that $\bar{L}$ is the z-axis.)

C H A P T E R 6

Line Integrals

6.1 Scalar and Vector Line Integrals

In this section, we describe two methods of integrating along a curve in space and explore the meaning and significance of our constructions. The main definitions are stated first for parametrized paths. Ultimately, we show that the integrals defined are largely independent of the parametrization, that instead they reflect essentially only the geometry of the underlying curve.

Scalar Line Integrals

To begin, we find a way to integrate a function (a scalar field) along a path. Let $\mathbf{x}\colon [a, b] \to \mathbf{R}^3$ be a path of class C^1. Let $f\colon X \subseteq \mathbf{R}^3 \to \mathbf{R}$ be a continuous function whose domain X contains the image of $\mathbf{x}$ (so that the composite $f(\mathbf{x}(t))$ is defined). As has been the case with every other integral, the scalar line integral is a limit of appropriate Riemann sums.

Let

$$a = t_0 < t_1 < \cdots < t_k < \cdots < t_n = b$$

be a partition of $[a, b]$. Let t_k^* be an arbitrary point in the kth subinterval $[t_{k-1}, t_k]$ of the partition. Then we consider the sum

$$\sum_{k=1}^{n} f(\mathbf{x}(t_k^*))\Delta s_k, \tag{1}$$

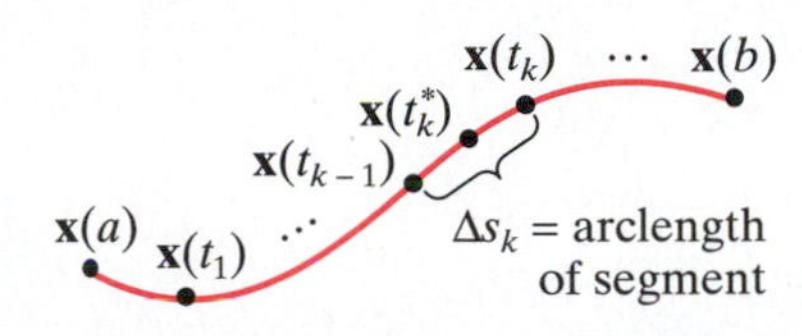

Figure 6.1 The sum $\sum_{k=1}^{n} f(\mathbf{x}(t_k^*))\Delta s_k$ approximates the total charge along an idealized wire described by the path $\mathbf{x}$.

where $\Delta s_k = \int_{t_{k-1}}^{t_k} \|\mathbf{x}'(t)\|\,dt$ is the length of the kth segment of $\mathbf{x}$ (i.e., the portion of $\mathbf{x}$ defined for $t_{k-1} \le t \le t_k$). If we think of the image of the path $\mathbf{x}$ as representing an idealized wire in space and $f(\mathbf{x}(t_k^*))$ as the electrical charge density of the wire at a "test point" $\mathbf{x}(t_k^*)$ in the kth segment, then the product $f(\mathbf{x}(t_k^*))\Delta s_k$ approximates the charge contributed by the segment of curve, and the sum in (1) approximates the total charge of the wire. (See Figure 6.1.) To find the actual charge on the wire, it is reasonable to take a limit as the curve segments become smaller, that is,

$$\begin{aligned}\text{Total charge} &= \lim_{\text{all } \Delta s_k \to 0} \sum_{k=1}^{n} f(\mathbf{x}(t_k^*))\Delta s_k \\ &= \lim_{\text{all } \Delta t_k \to 0} \sum_{k=1}^{n} f(\mathbf{x}(t_k^*))\Delta s_k, \end{aligned} \tag{2}$$

since $\mathbf{x}$ is of class C^1.

The mean value theorem for integrals[1] tells us that there is some number t_k^{**} in $[t_{k-1}, t_k]$ such that

$$\Delta s_k = \int_{t_{k-1}}^{t_k} \|\mathbf{x}'(t)\|\, dt = (t_k - t_{k-1}) \|\mathbf{x}'(t_k^{**})\| = \|\mathbf{x}'(t_k^{**})\|\, \Delta t_k.$$

(Here Δt_k denotes $t_k - t_{k-1}$.) Since t_k^* is an arbitrary point in $[t_{k-1}, t_k]$, we may take it to be equal to t_k^{**}. Therefore, by substituting for Δs_k in equation (2), and letting t_k^* equal t_k^{**}, we have

$$\begin{aligned}\text{Total charge} &= \lim_{\text{all } \Delta t_k \to 0} \sum_{k=1}^{n} f(\mathbf{x}(t_k^{**})) \|\mathbf{x}'(t_k^{**})\|\, \Delta t_k \\ &= \int_a^b f(\mathbf{x}(t)) \|\mathbf{x}'(t)\|\, dt.\end{aligned}$$

This last result prompts the following definition:

■ **Definition 1.1** The **scalar line integral** of f along the C^1 path $\mathbf{x}$ is

$$\int_a^b f(\mathbf{x}(t)) \|\mathbf{x}'(t)\|\, dt.$$

We denote this integral $\int_{\mathbf{x}} f\, ds$.

EXAMPLE 1 Let $\mathbf{x}\colon [0, 2\pi] \to \mathbf{R}^3$ be the helix $\mathbf{x}(t) = (\cos t, \sin t, t)$ and let $f(x, y, z) = xy + z$. We compute

$$\int_{\mathbf{x}} f\, ds = \int_0^{2\pi} f(\mathbf{x}(t)) \|\mathbf{x}'(t)\|\, dt.$$

First,

$$\mathbf{x}'(t) = (-\sin t, \cos t, 1),$$

so that

$$\|\mathbf{x}'(t)\| = \sqrt{\sin^2 t + \cos^2 t + 1} = \sqrt{2}.$$

We also have

$$f(\mathbf{x}(t)) = \cos t \sin t + t = \tfrac{1}{2} \sin 2t + t$$

[1]Recall that this theorem says that if F is a continuous function, then there is some number c with $a \le c \le b$ such that $\int_a^b F(x)\, dx = (b - a)F(c)$.

from the double-angle formula. Thus,

$$\begin{aligned}\int_{\mathbf{x}} f\,ds &= \int_0^{2\pi} \left(\tfrac{1}{2}\sin 2t + t\right)\sqrt{2}\,dt = \sqrt{2}\int_0^{2\pi}\left(\tfrac{1}{2}\sin 2t + t\right)\,dt\\ &= \sqrt{2}\left(-\tfrac{1}{4}\cos 2t + \tfrac{1}{2}t^2\right)\Big|_0^{2\pi} = \sqrt{2}\left(\left(-\tfrac{1}{4} + 2\pi^2\right) - \left(-\tfrac{1}{4} + 0\right)\right)\\ &= 2\sqrt{2}\,\pi^2.\end{aligned}$$

◆

Given the discussion preceding the formal definition of the scalar line integral, it is both convenient and appropriate to view the notation $\int_{\mathbf{x}} f\,ds$ as suggesting that the line integral represents a sum of values of f along $\mathbf{x}$ times "infinitesimal" pieces of arclength of $\mathbf{x}$.

Definition 1.1 is made only for paths $\mathbf{x}$ in $\mathbf{R}^3$ and functions f defined on domains in $\mathbf{R}^3$. Nonetheless, for arbitrary n, we may certainly use the definite integral

$$\int_a^b f(\mathbf{x}(t))\,\|\mathbf{x}'(t)\|\,dt,$$

where $\mathbf{x}$ is a C^1 path in $\mathbf{R}^n$ and f is an appropriate function of n variables. We call this definite integral the scalar line integral as well (and maintain the notation $\int_{\mathbf{x}} f\,ds$) and rely on the context to make clear the dimensionality of the situation. Also, if $\mathbf{x}$ is not of class C^1, but only "piecewise C^1" (meaning that $\mathbf{x}$ can be broken into a finite number of segments that are individually of class C^1), then we may still define the scalar line integral $\int_{\mathbf{x}} f\,ds$ by breaking it up in a suitable manner. A similar technique must be used if $f(\mathbf{x}(t))$ is only piecewise continuous.

EXAMPLE 2 Let $f(x, y) = y - x$ and let $\mathbf{x}\colon [0, 3] \to \mathbf{R}^2$ be the planar path

$$\mathbf{x}(t) = \begin{cases}(2t, t) & \text{if } 0 \le t \le 1\\ (t+1, 5-4t) & \text{if } 1 < t \le 3\end{cases}.$$

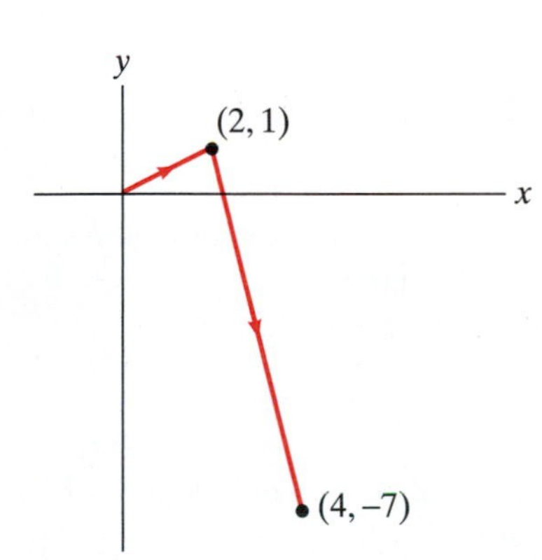

Figure 6.2 A piecewise C^1 path $\mathbf{x}$.

Hence, $\mathbf{x}$ is piecewise C^1 (see Figure 6.2); the two path segments defined for t in $[0, 1]$ and for t in $[1, 3]$ are each of class C^1. Thus,

$$\int_{\mathbf{x}} f\,ds = \int_{\mathbf{x}_1} f\,ds + \int_{\mathbf{x}_2} f\,ds,$$

where $\mathbf{x}_1(t) = (2t, t)$ for $0 \le t \le 1$ and $\mathbf{x}_2(t) = (t+1, 5-4t)$ for $1 \le t \le 3$. Note that

$$\|\mathbf{x}_1'(t)\| = \sqrt{5} \quad\text{and}\quad \|\mathbf{x}_2'(t)\| = \sqrt{17}.$$

Consequently,

$$\int_{\mathbf{x}_1} f\,ds = \int_0^1 f(\mathbf{x}(t))\,\|\mathbf{x}'(t)\|\,dt = \int_0^1 (t - 2t)\cdot\sqrt{5}\,dt = -\frac{\sqrt{5}}{2}t^2\Big|_0^1 = -\frac{\sqrt{5}}{2}.$$

Also,

$$\begin{aligned}\int_{\mathbf{x}_2} f\,ds &= \int_1^3 f(\mathbf{x}_2(t))\,\|\mathbf{x}_2'(t)\|\,dt = \int_1^3 ((5-4t) - (t+1))\sqrt{17}\,dt\\ &= \sqrt{17}\left(4t - \tfrac{5}{2}t^2\right)\Big|_1^3 = -12\sqrt{17}.\end{aligned}$$

Hence,

$$\int_{\mathbf{x}} f\,ds = -\frac{\sqrt{5}}{2} - 12\sqrt{17}.$$

◆

Vector Line Integrals

Now we see how to integrate a vector field along a path. Again, let $\mathbf{x}\colon [a, b] \to \mathbf{R}^n$ be of class C^1. (n will be 2 or 3 in the examples that follow.) Let $\mathbf{F}$ be a vector field defined on a subset X of $\mathbf{R}^n$ such that X contains the image of $\mathbf{x}$. Assume that $\mathbf{F}$ varies continuously along $\mathbf{x}$.

Definition 1.2 The **vector line integral** of $\mathbf{F}$ along $\mathbf{x}\colon [a, b] \to \mathbf{R}^n$, denoted $\int_{\mathbf{x}} \mathbf{F} \cdot d\mathbf{s}$, is

$$\int_{\mathbf{x}} \mathbf{F} \cdot d\mathbf{s} = \int_a^b \mathbf{F}(\mathbf{x}(t)) \cdot \mathbf{x}'(t)\, dt.$$

We caution you to be clear about notation. In the *vector* line integral $\int_{\mathbf{x}} \mathbf{F} \cdot d\mathbf{s}$, the differential term $d\mathbf{s}$ should be thought of as a vector quantity (namely, the "differential" of position along the path), whereas in the *scalar* line integral $\int_{\mathbf{x}} f\, ds$, the differential term ds is a scalar quantity (namely, the differential of arclength).

EXAMPLE 3 Let $\mathbf{F}$ be the radial vector field on $\mathbf{R}^3$ given by $\mathbf{F} = x\mathbf{i} + y\mathbf{j} + z\mathbf{k}$ and let $\mathbf{x}\colon [0, 1] \to \mathbf{R}^3$ be the path $\mathbf{x}(t) = (t, 3t^2, 2t^3)$. Then $\mathbf{x}'(t) = (1, 6t, 6t^2)$, and so

$$\begin{aligned}\int_{\mathbf{x}} \mathbf{F} \cdot d\mathbf{s} &= \int_0^1 \mathbf{F}(\mathbf{x}(t)) \cdot \mathbf{x}'(t)\, dt \\ &= \int_0^1 (t\mathbf{i} + 3t^2\mathbf{j} + 2t^3\mathbf{k}) \cdot (\mathbf{i} + 6t\mathbf{j} + 6t^2\mathbf{k})\, dt \\ &= \int_0^1 (t + 18t^3 + 12t^5)\, dt = \left(\tfrac{1}{2}t^2 + \tfrac{9}{2}t^4 + 2t^6\right)\Big|_0^1 = 7.\end{aligned}$$

◆

As with scalar line integrals, we may define $\int_{\mathbf{x}} \mathbf{F} \cdot d\mathbf{s}$ when $\mathbf{x}$ is only a piecewise C^1 path by breaking up the integral in a suitable manner.

The reason Definition 1.2 is so important is as follows:

Physical interpretation of vector line integrals. Consider $\mathbf{F}$ to be a force field in space. Then $\int_{\mathbf{x}} \mathbf{F} \cdot d\mathbf{s}$ may be taken to represent the **work** done by $\mathbf{F}$ on a particle as the particle moves along the path $\mathbf{x}$.

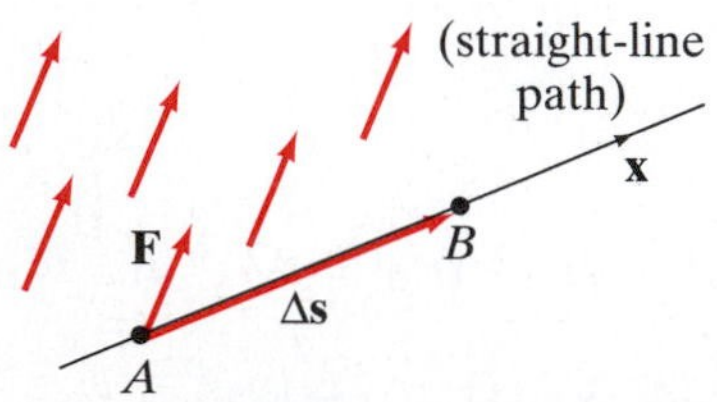

To justify this interpretation, first recall that, if $\mathbf{F}$ is a constant vector field and $\mathbf{x}$ is a straight-line path, then the work done by $\mathbf{F}$ in moving a particle from one point along $\mathbf{x}$ to another is given by

$$\text{Work} = \mathbf{F} \cdot \Delta\mathbf{s},$$

where $\Delta\mathbf{s}$ is the displacement vector from the initial to the final position. (See Figure 6.3.)

Figure 6.3 The work done by $\mathbf{F}$ in moving a particle from A to B in a straight line is $\mathbf{F} \cdot \Delta\mathbf{s}$.

In general, the path $\mathbf{x}$ need not be straight and the force field $\mathbf{F}$ may not be constant along $\mathbf{x}$. Nonetheless, along a short segment of path, $\mathbf{x}$ is nearly straight and $\mathbf{F}$ is roughly constant, by continuity. Partition $[a, b]$ as usual (i.e., take $a = t_0 < \cdots < t_k < \cdots < t_n = b$) and focus on the kth segment.

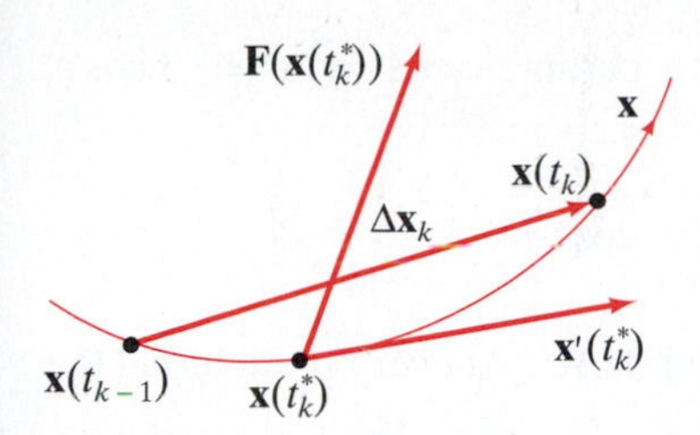

Figure 6.4 Approximating the work along a segment of the path **x**.

(See Figure 6.4.) Then,

$$\text{Work done along } k\text{th segment} \approx \mathbf{F}(\mathbf{x}(t_k^*)) \cdot \Delta\mathbf{x}_k.$$

Since

$$\mathbf{x}'(t) = \lim_{\Delta t \to 0} \frac{\mathbf{x}(t + \Delta t) - \mathbf{x}(t)}{\Delta t} = \lim_{\Delta t \to 0} \frac{\Delta \mathbf{x}}{\Delta t},$$

we must have, for $\Delta t_k = t_k - t_{k-1} \approx 0$ and $t_{k-1} \leq t_k^* \leq t_k$, that

$$\mathbf{x}'(t_k^*) \approx \frac{\mathbf{x}(t_k) - \mathbf{x}(t_{k-1})}{t_k - t_{k-1}} = \frac{\Delta \mathbf{x}_k}{\Delta t_k}.$$

Hence,

$$\text{Total work} \approx \sum_{k=1}^{n} \mathbf{F}(\mathbf{x}(t_k^*)) \cdot \Delta\mathbf{x}_k \approx \sum_{k=1}^{n} \mathbf{F}(\mathbf{x}(t_k^*)) \cdot \mathbf{x}'(t_k^*)\Delta t_k.$$

Therefore, it makes sense to take the limit as all the Δt_k's tend to zero and *define* the total work by

$$\begin{aligned}\text{Work} &= \lim_{\text{all } \Delta t_k \to 0} \sum_{k=1}^{n} \mathbf{F}(\mathbf{x}(t)) \cdot \mathbf{x}'(t_k^*)\Delta t_k \\ &= \int_a^b \mathbf{F}(\mathbf{x}(t)) \cdot \mathbf{x}'(t)\, dt \\ &= \int_{\mathbf{x}} \mathbf{F} \cdot d\mathbf{s}.\end{aligned}$$

Other Interpretations and Formulations

Suppose $\mathbf{x}\colon [a, b] \to \mathbf{R}^n$ is a C^1 path with $\mathbf{x}'(t) \neq \mathbf{0}$ for $a \leq t \leq b$. Recall from §3.2 that we define the **unit tangent vector T** to **x** by normalizing the velocity:

$$\mathbf{T}(t) = \frac{\mathbf{x}'(t)}{\|\mathbf{x}'(t)\|}.$$

We may insinuate this tangent vector into the vector line integral as follows: From Definition 1.2, we have

$$\int_{\mathbf{x}} \mathbf{F} \cdot d\mathbf{s} = \int_a^b \mathbf{F}(\mathbf{x}(t)) \cdot \mathbf{x}'(t)\, dt.$$

Thus,

$$\begin{aligned}\int_{\mathbf{x}} \mathbf{F} \cdot d\mathbf{s} &= \int_a^b \mathbf{F}(\mathbf{x}(t)) \cdot \frac{\mathbf{x}'(t)}{\|\mathbf{x}'(t)\|}\, \|\mathbf{x}'(t)\|\, dt \\ &= \int_a^b (\mathbf{F}(\mathbf{x}(t)) \cdot \mathbf{T}(t))\, \|\mathbf{x}'(t)\|\, dt \\ &= \int_{\mathbf{x}} (\mathbf{F} \cdot \mathbf{T})\, ds. \qquad (3)\end{aligned}$$

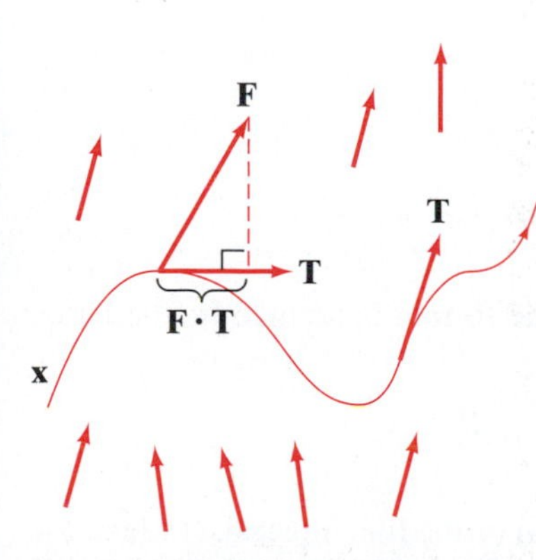

Figure 6.5 The vector line integral $\int_{\mathbf{x}} \mathbf{F} \cdot d\mathbf{s}$ equals $\int_{\mathbf{x}} (\mathbf{F} \cdot \mathbf{T})\, ds$, the scalar line integral of the tangential component of **F** along the path.

Since the dot product $\mathbf{F} \cdot \mathbf{T}$ is a scalar quantity, we have written the original vector line integral as a scalar line integral. We also see that $\int_{\mathbf{x}} \mathbf{F} \cdot d\mathbf{s}$ *represents the (scalar) line integral of the tangential component of* **F** *along the path*—that is, how much of the vector field the underlying curve actually "sees." When **x** is a closed path, the quantity $\int_{\mathbf{x}} \mathbf{F} \cdot d\mathbf{s}$ is called the **circulation** of **F** along **x**. (See Figure 6.5.)

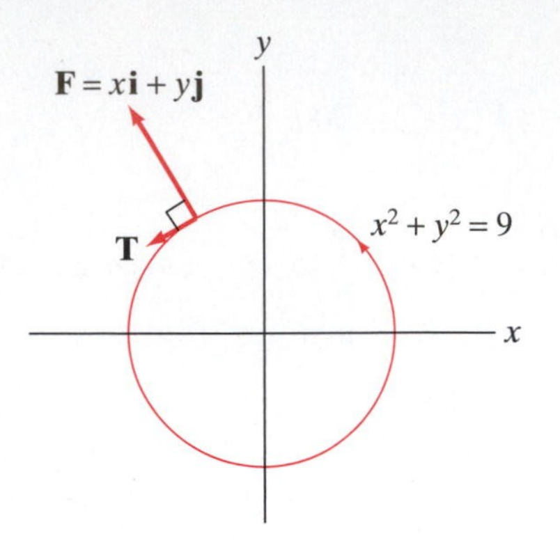

Figure 6.6 At every point along the circle, $\mathbf{F} = x\,\mathbf{i} + y\,\mathbf{j}$ has no tangential component.

EXAMPLE 4 The circle $x^2 + y^2 = 9$ may be parametrized by $x = 3\cos t$, $y = 3\sin t$. Hence, a unit tangent vector is

$$\mathbf{T} = \frac{-3\sin t\,\mathbf{i} + 3\cos t\,\mathbf{j}}{\sqrt{9\sin^2 t + 9\cos^2 t}} = -\sin t\,\mathbf{i} + \cos t\,\mathbf{j} = \frac{-y\mathbf{i} + x\mathbf{j}}{3}.$$

Now consider the radial vector field $\mathbf{F} = x\,\mathbf{i} + y\,\mathbf{j}$ on $\mathbf{R}^2$. At every point along the circle we have

$$\mathbf{F} \cdot \mathbf{T} = (x\,\mathbf{i} + y\,\mathbf{j}) \cdot \left(\frac{-y\,\mathbf{i} + x\,\mathbf{j}}{3}\right) = 0.$$

Thus, $\mathbf{F}$ is always perpendicular to the curve. Therefore,

$$\int_{\mathbf{x}} \mathbf{F} \cdot d\mathbf{s} = \int_{\mathbf{x}} (\mathbf{F} \cdot \mathbf{T})\, ds = \int_{\mathbf{x}} 0\, ds = 0$$

and, considering $\mathbf{F}$ as a force, no work is done. (See Figure 6.6.) ◆

Next, suppose that $\mathbf{x}(t) = (x(t), y(t), z(t))$, $a \le t \le b$, is a C^1 path and

$$\mathbf{F}(x, y, z) = M(x, y, z)\,\mathbf{i} + N(x, y, z)\,\mathbf{j} + P(x, y, z)\,\mathbf{k}$$

is a continuous vector field. Then, from Definition 1.2 of the vector line integral, we have

$$\begin{aligned}\int_{\mathbf{x}} \mathbf{F} \cdot d\mathbf{s} &= \int_a^b (M(x, y, z)\,\mathbf{i} + N(x, y, z)\,\mathbf{j} + P(x, y, z)\,\mathbf{k}) \\ &\qquad \cdot (x'(t)\,\mathbf{i} + y'(t)\,\mathbf{j} + z'(t)\,\mathbf{k})\, dt \\ &= \int_a^b \left[M(x, y, z)x'(t) + N(x, y, z)y'(t) + P(x, y, z)z'(t)\right] dt.\end{aligned}$$

Recall that the differentials of x, y, and z are

$$dx = x'(t)\, dt, \qquad dy = y'(t)\, dt, \qquad dz = z'(t)\, dt.$$

Hence,

$$\int_{\mathbf{x}} \mathbf{F} \cdot d\mathbf{s} = \int_{\mathbf{x}} M(x, y, z)\, dx + N(x, y, z)\, dy + P(x, y, z)\, dz.$$

The integral

$$\int_{\mathbf{x}} M\, dx + N\, dy + P\, dz$$

is a notational alternative to $\int_{\mathbf{x}} \mathbf{F} \cdot d\mathbf{s}$. Indeed, the former is defined by the latter. The integral

$$\int_{\mathbf{x}} M\, dx + N\, dy + P\, dz$$

is commonly referred to as the **differential form** of the line integral. (In fact, the expression $M\,dx + N\,dy + P\,dz$ is itself called a differential form.) It emphasizes the component functions of the vector field $\mathbf{F}$ and arises regularly in applications. Be sure to interpret $M\,dx + N\,dy + P\,dz$ carefully: It should be evaluated using the parametric equations for x, y, and z that come from the path $\mathbf{x}$.

EXAMPLE 5 We compute

$$\int_{\mathbf{x}} (y+z)\,dx + (x+z)\,dy + (x+y)\,dz,$$

where $\mathbf{x}$ is the path $\mathbf{x}(t) = (t, t^2, t^3)$ for $0 \le t \le 1$.

Along the path, we have $x = t$, $y = t^2$, and $z = t^3$ so that $dx = dt$, $dy = 2t\,dt$, and $dz = 3t^2\,dt$. Therefore,

$$\begin{aligned}\int_{\mathbf{x}} (y+z)\,dx + (x+z)\,dy + (x+y)\,dz \\ &= \int_0^1 (t^2+t^3)\,dt + (t+t^3)2t\,dt + (t+t^2)3t^2\,dt \\ &= \int_0^1 (5t^4 + 4t^3 + 3t^2)\,dt = (t^5+t^4+t^3)\Big|_0^1 = 3.\end{aligned}$$

◆

Line Integrals Along Curves: The Effect of Reparametrization

Since the unit tangent vector to a path depends on the geometry of the underlying curve and not on the particular parametrization, we might expect the line integral likewise to depend only on the image curve. We shall see precisely to what degree this observation is true generally for both vector and scalar line integrals.

We begin with an example. Consider the following two paths in the plane:

$$\mathbf{x}\colon [0, 2\pi] \to \mathbf{R}^2, \quad \mathbf{x}(t) = (\cos t, \sin t)$$

and

$$\mathbf{y}\colon [0, \pi] \to \mathbf{R}^2, \quad \mathbf{y}(t) = (\cos 2t, \sin 2t).$$

It is not difficult to see that both $\mathbf{x}$ and $\mathbf{y}$ trace out a circle once in a counterclockwise sense. In fact, if we let $u(t) = 2t$, then we see that $\mathbf{y}(t) = \mathbf{x}(u(t))$. That is, the path $\mathbf{y}$ is nothing more than the path $\mathbf{x}$ together with a change of variables, which suggests the following general definition:

> ■ **Definition 1.3** Let $\mathbf{x}\colon [a, b] \to \mathbf{R}^n$ be a piecewise C^1 path. We say that another C^1 path $\mathbf{y}\colon [c, d] \to \mathbf{R}^n$ is a **reparametrization** of $\mathbf{x}$ if there is a one-one and onto function $u\colon [c, d] \to [a, b]$ of class C^1, with inverse $u^{-1}\colon [a, b] \to [c, d]$ that is also of class C^1, such that $\mathbf{y}(t) = \mathbf{x}(u(t))$; that is, $\mathbf{y} = \mathbf{x} \circ u$.

Reflecting on Definition 1.3 should convince you that any reparametrization of a path must have the same underlying image curve as the original path.

EXAMPLE 6 The path

$$\mathbf{x}(t) = (1+2t, 2-t, 3+5t), \quad 0 \le t \le 1,$$

traces the line segment from the point $(1, 2, 3)$ to the point $(3, 1, 8)$. So does the path

$$\mathbf{y}(t) = (1+2t^2, 2-t^2, 3+5t^2), \quad 0 \le t \le 1.$$

We have that $\mathbf{y}$ is a reparametrization of $\mathbf{x}$ via the change of variable $u(t) = t^2$. However, the path $\mathbf{z}\colon [-1, 1] \to \mathbf{R}^3$ given by

$$\mathbf{z}(t) = (1+2t^2, 2-t^2, 3+5t^2)$$

is *not* a reparametrization of $\mathbf{x}$. We have $\mathbf{z}(t) = \mathbf{x}(u(t))$, where $u(t) = t^2$, but in this case u maps $[-1, 1]$ onto $[0, 1]$ in a way that is not one-one.

On the other hand,

$$\mathbf{w}(t) = (3 - 2t, 1 + t, 8 - 5t), \quad 0 \leq t \leq 1,$$

is a reparametrization of $\mathbf{x}$. We have $\mathbf{w}(t) = \mathbf{x}(1-t)$, so the function $u\colon [0, 1] \to [0, 1]$ given by $u(t) = 1 - t$ provides the change of variable for the reparametrization. Geometrically, $\mathbf{w}$ traces the line segment between $(1, 2, 3)$ and $(3, 1, 8)$ in the opposite direction to that of $\mathbf{x}$. ◆

If $\mathbf{y}\colon [c, d] \to \mathbf{R}^n$ is a reparametrization of $\mathbf{x}\colon [a, b] \to \mathbf{R}^n$ via the change of variable $u\colon [a, b] \to [c, d]$, then, since u is one-one, onto, and continuous, we must have either

(i) $u(a) = c$ and $u(b) = d$, or

(ii) $u(a) = d$ and $u(b) = c$.

In the first case, we say that $\mathbf{y}$ (or u) is **orientation-preserving** and, in the second case, that $\mathbf{y}$ (or u) is **orientation-reversing**. The idea is that if $\mathbf{y}$ is an orientation-preserving reparametrization, then $\mathbf{y}$ traces out the same image curve in the same direction that $\mathbf{x}$ does, and if $\mathbf{y}$ is orientation-reversing, it traces the image in the opposite direction.

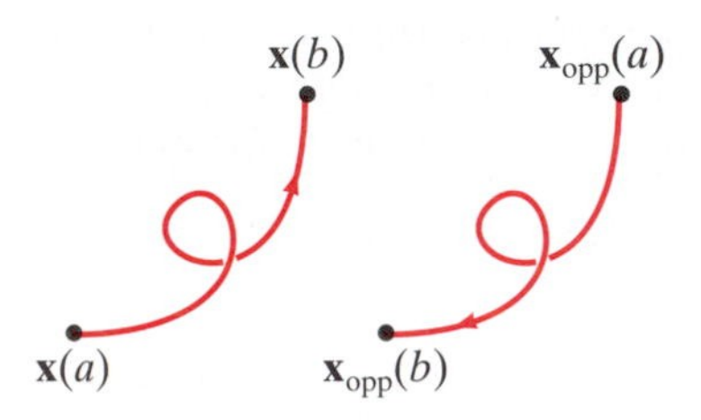

Figure 6.7 A path and its opposite.

EXAMPLE 7 If $\mathbf{x}\colon [a, b] \to \mathbf{R}^n$ is any C^1 path, then we may define the **opposite path** $\mathbf{x}_{\text{opp}}\colon [a, b] \to \mathbf{R}^n$ by

$$\mathbf{x}_{\text{opp}}(t) = \mathbf{x}(a + b - t).$$

(See Figure 6.7.) That is, $\mathbf{x}_{\text{opp}}(t) = \mathbf{x}(u(t))$, where $u\colon [a, b] \to [a, b]$ is given by $u(t) = a + b - t$. Clearly, then, $\mathbf{x}_{\text{opp}}$ is an orientation-reversing reparametrization of $\mathbf{x}$. ◆

In addition to reversing orientation, a reparametrization of a path can change the speed. This follows readily from the chain rule: If $\mathbf{y} = \mathbf{x} \circ u$, then

$$\mathbf{y}'(t) = \frac{d}{dt}(\mathbf{x}(u(t))) = \mathbf{x}'(u(t))\, u'(t). \tag{4}$$

So the velocity vector of the reparametrization $\mathbf{y}$ is just a scalar multiple (namely, $u'(t)$) of the velocity vector of $\mathbf{x}$. In particular, we have

$$\begin{aligned} \text{Speed of } \mathbf{y} &= \|\mathbf{y}'(t)\| = \|u'(t)\,\mathbf{x}'(u(t))\| \\ &= |u'(t)|\, \|\mathbf{x}'(u(t))\| = |u'(t)| \cdot (\text{speed of } \mathbf{x}). \end{aligned} \tag{5}$$

Since u is one-one, it follows that either $u'(t) \geq 0$ for all $t \in [a, b]$ or $u'(t) \leq 0$ for all $t \in [a, b]$. The first case occurs precisely when $\mathbf{y}$ is orientation-preserving and the second when $\mathbf{y}$ is orientation-reversing.

How does the line integral of a function or a vector field along a path differ from the line integral (of the same function or vector field) along a reparametrization of a path? Not much at all. The precise results are stated in Theorems 1.4 and 1.5.

■ **Theorem 1.4** Let $\mathbf{x}\colon [a, b] \to \mathbf{R}^n$ be a piecewise C^1 path and let $f\colon X \subseteq \mathbf{R}^n \to \mathbf{R}$ be a continuous function whose domain X contains the image of $\mathbf{x}$. If $\mathbf{y}\colon [c, d] \to \mathbf{R}^n$ is any reparametrization of $\mathbf{x}$, then

$$\int_{\mathbf{y}} f\, ds = \int_{\mathbf{x}} f\, ds.$$

Proof We will explicitly prove the result in the case where $\mathbf{x}$ (and, therefore, $\mathbf{y}$) is of class C^1. (When $\mathbf{x}$ is only piecewise C^1, we can always break up the integral appropriately.) In the C^1 case, we have, by Definition 1.1 and the observations in equation (5), that

$$\int_{\mathbf{y}} f\,ds = \int_c^d f(\mathbf{y}(t))\,\|\mathbf{y}'(t)\|\,dt = \int_c^d f(\mathbf{x}(u(t)))\,\|\mathbf{x}'(u(t))\|\,|u'(t)|\,dt.$$

If $\mathbf{y}$ is orientation-preserving, then $u(c) = a$, $u(d) = b$, and $|u'(t)| = u'(t)$. Thus, using substitution of variables,

$$\begin{aligned}\int_c^d f(\mathbf{x}(u(t)))\,\|\mathbf{x}'(u(t))\|\,|u'(t)|\,dt &= \int_c^d f(\mathbf{x}(u(t)))\,\|\mathbf{x}'(u(t))\|\,u'(t)\,dt \\ &= \int_a^b f(\mathbf{x}(u))\,\|\mathbf{x}'(u)\|\,du = \int_{\mathbf{x}} f\,ds.\end{aligned}$$

If, on the other hand, $\mathbf{y}$ is orientation-reversing, then $u(c) = b$, $u(d) = a$, and $|u'(t)| = -u'(t)$, since $u'(t) \leq 0$. Therefore, in this case,

$$\begin{aligned}\int_c^d f(\mathbf{x}(u(t)))\,\|\mathbf{x}'(u(t))\|\,|u'(t)|\,dt &= \int_c^d f(\mathbf{x}(u(t)))\,\|\mathbf{x}'(u(t))\|\,(-u'(t))\,dt \\ &= -\int_b^a f(\mathbf{x}(u))\,\|\mathbf{x}'(u)\|\,du \\ &= \int_a^b f(\mathbf{x}(u))\,\|\mathbf{x}'(u)\|\,du = \int_{\mathbf{x}} f\,ds.\end{aligned}$$

Hence, $\int_{\mathbf{y}} f\,ds = \int_{\mathbf{x}} f\,ds$ in either case, as desired. ■

■ **Theorem 1.5** Let $\mathbf{x}\colon [a, b] \to \mathbf{R}^n$ be a piecewise C^1 path and let $\mathbf{F}\colon X \subseteq \mathbf{R}^n \to \mathbf{R}^n$ be a continuous vector field whose domain X contains the image of $\mathbf{x}$. Let $\mathbf{y}\colon [c, d] \to \mathbf{R}^n$ be any reparametrization of $\mathbf{x}$. Then,

1. If $\mathbf{y}$ is orientation-preserving, then $\int_{\mathbf{y}} \mathbf{F}\cdot d\mathbf{s} = \int_{\mathbf{x}} \mathbf{F}\cdot d\mathbf{s}$.
2. If $\mathbf{y}$ is orientation-reversing, then $\int_{\mathbf{y}} \mathbf{F}\cdot d\mathbf{s} = -\int_{\mathbf{x}} \mathbf{F}\cdot d\mathbf{s}$.

Proof As in the proof of Theorem 1.4, we consider only the C^1 case in detail. Using Definition 1.2 for the vector line integral, equation (4), and substitution of variables, we have

$$\int_{\mathbf{y}} \mathbf{F}\cdot d\mathbf{s} = \int_c^d \mathbf{F}(\mathbf{y}(t))\cdot\mathbf{y}'(t)\,dt = \int_c^d \mathbf{F}(\mathbf{x}(u(t)))\cdot\mathbf{x}'(u(t))\,u'(t)\,dt.$$

If $\mathbf{y}$ is orientation-preserving, then $u(c) = a$, $u(d) = b$, so we have

$$\int_c^d \mathbf{F}(\mathbf{x}(u(t)))\cdot\mathbf{x}'(u(t))\,u'(t)\,dt = \int_a^b \mathbf{F}(\mathbf{x}(u))\cdot\mathbf{x}'(u)\,du = \int_{\mathbf{x}} \mathbf{F}\cdot d\mathbf{s}.$$

This proves part 1. If $\mathbf{y}$ is orientation-reversing, then $u(c) = b$, $u(d) = a$, so, instead, we have

$$\int_c^d \mathbf{F}(\mathbf{x}(u(t)))\cdot\mathbf{x}'(u(t))\,u'(t)\,dt = \int_b^a \mathbf{F}(\mathbf{x}(u))\cdot\mathbf{x}'(u)\,du = -\int_{\mathbf{x}} \mathbf{F}\cdot d\mathbf{s},$$

which establishes part 2. ■

Simply put, Theorems 1.4 and 1.5 tell us that scalar line integrals are entirely independent of the way we might choose to reparametrize a path. Vector line integrals are independent of reparametrization up to a sign that depends only on whether the reparametrization preserves or reverses orientation.

EXAMPLE 8 Let $\mathbf{F} = x\,\mathbf{i} + y\,\mathbf{j}$, and consider the following three paths between $(0, 0)$ and $(1, 1)$:

$$\mathbf{x}(t) = (t, t) \qquad 0 \le t \le 1,$$
$$\mathbf{y}(t) = (2t, 2t) \qquad 0 \le t \le \tfrac{1}{2},$$

and

$$\mathbf{z}(t) = (1 - t, 1 - t) \qquad 0 \le t \le 1.$$

The three paths are all reparametrizations of one another; **x**, **y**, and **z** all trace the line segment between $(0, 0)$ and $(1, 1)$—**x** and **y** from $(0, 0)$ to $(1, 1)$ and **z** from $(1, 1)$ to $(0, 0)$.

We can compare the values of the line integrals of **F** along these paths:

$$\begin{aligned}
\int_{\mathbf{x}} \mathbf{F}\cdot d\mathbf{s} &= \int_0^1 \mathbf{F}(\mathbf{x}(t))\cdot \mathbf{x}'(t)\,dt \\
&= \int_0^1 (t\,\mathbf{i} + t\,\mathbf{j})\cdot(\mathbf{i}+\mathbf{j})\,dt = \int_0^1 2t\,dt = t^2\Big|_0^1 = 1;\\
\int_{\mathbf{y}} \mathbf{F}\cdot d\mathbf{s} &= \int_0^{1/2} (2t\,\mathbf{i} + 2t\,\mathbf{j})\cdot(2\mathbf{i}+2\mathbf{j})\,dt = \int_0^{1/2} 8t\,dt = 4t^2\Big|_0^{1/2} = 1;\\
\int_{\mathbf{z}} \mathbf{F}\cdot d\mathbf{s} &= \int_0^1 ((1-t)\mathbf{i} + (1-t)\mathbf{j})\cdot(-\mathbf{i}-\mathbf{j})\,dt \\
&= \int_0^1 2(t-1)\,dt = (t-1)^2\Big|_0^1 = -1.
\end{aligned}$$

The results of these calculations are just what Theorem 1.5 predicts, since **y** is an orientation-preserving reparametrization of **x** and **z** is an orientation-reversing one. ◆

The significance of Theorems 1.4 and 1.5 is not merely that they allow us occasionally to predict the results of line integral computations, but rather that they enable us to *define* line integrals over curves, rather than over parametrized paths. More explicitly, by a **curve** C we now mean the image of a piecewise C^1 map $\mathbf{x}\colon [a, b] \to \mathbf{R}^n$. Such a curve C is **simple** if it has no self-intersections, that is, if the path **x** may be taken to be one-one on $[a, b]$, except possibly that $\mathbf{x}(a)$ may equal $\mathbf{x}(b)$. If $\mathbf{x}(a) = \mathbf{x}(b)$, we say that the path **x** and its image C are **closed**. (See Figure 6.8.) The (nearly) one-one path **x** whose image is C is called a **parametrization** of C. It is a fact (whose proof we omit) that if **x** and **y** are both parametrizations of the same simple curve C, then they must be reparametrizations of each other.

EXAMPLE 9 The ellipse $x^2/25 + y^2/9 = 1$ shown in Figure 6.9 is a simple, closed curve that may be parametrized by either

$$\mathbf{x}\colon [0, 2\pi] \to \mathbf{R}^2, \quad \mathbf{x}(t) = (5\cos t, 3\sin t)$$

or by

$$\mathbf{y}\colon [0, \pi] \to \mathbf{R}^2, \quad \mathbf{y}(t) = (5\cos 2(\pi - t), 3\sin 2(\pi - t)),$$

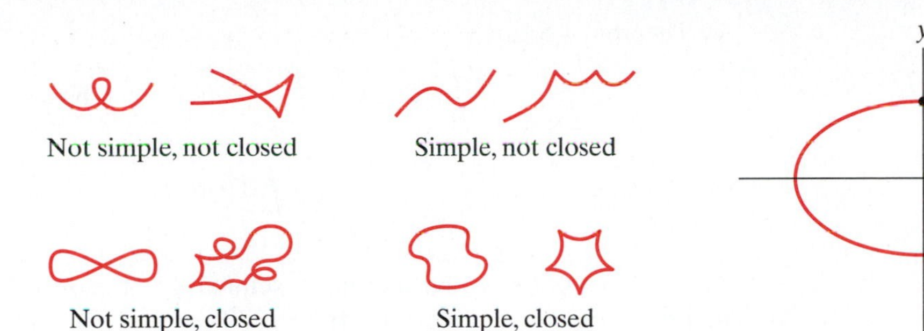

Figure 6.8 Examples of curves.

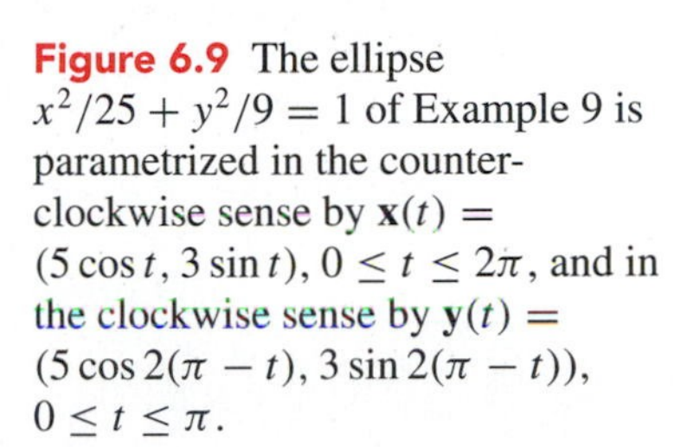

Figure 6.9 The ellipse $x^2/25 + y^2/9 = 1$ of Example 9 is parametrized in the counterclockwise sense by $\mathbf{x}(t) = (5\cos t, 3\sin t)$, $0 \le t \le 2\pi$, and in the clockwise sense by $\mathbf{y}(t) = (5\cos 2(\pi - t), 3\sin 2(\pi - t))$, $0 \le t \le \pi$.

since both paths have the ellipse as image and each map is one-one, except at the endpoints of their respective domain intervals. Note that $\mathbf{y}$ is a reparametrization of $\mathbf{x}$.

However, the path

$$\mathbf{z}\colon [0, 6\pi] \to \mathbf{R}^2, \quad \mathbf{z}(t) = (5\cos t, 3\sin t)$$

is *not* a parametrization of the ellipse, since it traces the ellipse three times as t increases from 0 to 6π. That is, $\mathbf{z}$ is *not* one-one on $(0, 6\pi)$. ◆

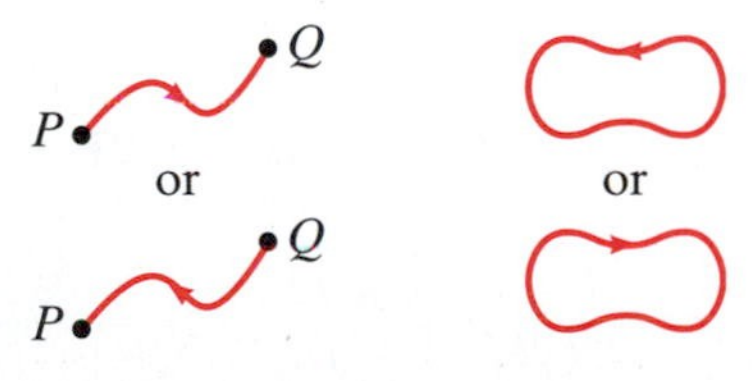

Figure 6.10 The possible orientations for a nonclosed and a closed simple curve.

Simple curves, whether closed or not, always have two **orientations** that correspond to the two possible directions of travel along any parametrizing path. (See Figure 6.10.) We say that a simple curve C is **oriented** if a choice of orientation is made.

The reason for the preceding terminology is that if C is a (piecewise C^1) simple curve, we may unambiguously define the scalar line integral of a continuous function f over C by

$$\int_C f\, ds = \int_{\mathbf{x}} f\, ds,$$

where $\mathbf{x}$ is *any* parametrization of C. Theorem 1.4 guarantees that the choice of how to parametrize C will not matter.

On the other hand, we can define the vector line integral only for *oriented* simple curves. If an orientation for C is chosen and $\mathbf{x}\colon [a, b] \to \mathbf{R}^n$ is a parametrization of C that is consistent with this orientation, then we define the vector line integral of a continuous vector field $\mathbf{F}$ over C by

$$\int_C \mathbf{F} \cdot d\mathbf{s} = \int_{\mathbf{x}} \mathbf{F} \cdot d\mathbf{s}.$$

Theorem 1.5 shows that this definition is independent of the choice of parametrization of C, as long as it is made consistently with the given orientation. Indeed, if C^- denotes the same curve as C, only oriented in the opposite way, then C^- is parametrized by $\mathbf{x}_{\text{opp}}$ (where $\mathbf{x}$ parametrizes C) and we have, by Theorem 1.5,

$$\int_{C-} \mathbf{F}\cdot d\mathbf{s} = \int_{\mathbf{x}_{\text{opp}}} \mathbf{F}\cdot d\mathbf{s} = -\int_{\mathbf{x}} \mathbf{F}\cdot d\mathbf{s}$$

$$= -\int_{C} \mathbf{F}\cdot d\mathbf{s}.$$

EXAMPLE 10 Let C be the upper semicircle of radius 2, centered at $(0, 0)$ and oriented counterclockwise from $(2, 0)$ to $(-2, 0)$. Then we may calculate $\int_C (x^2 - y^2 + 1)\, ds$ by choosing any parametrization for C. For instance, we may parametrize C by

$$\mathbf{x}(t) = (2\cos t, 2\sin t), \quad 0 \le t \le \pi$$

or by

$$\mathbf{y}(t) = (-2\cos 2t, -2\sin 2t), \quad -\pi/2 \le t \le 0.$$

(Note that $\mathbf{y}(t) = \mathbf{x}(2t + \pi)$.) Then

$$\int_C (x^2 - y^2 + 1)\, ds = \int_{\mathbf{x}} (x^2 - y^2 + 1)\, ds$$

$$= \int_0^{\pi} (4\cos^2 t - 4\sin^2 t + 1)\sqrt{4\sin^2 t + 4\cos^2 t}\, dt$$

$$= \int_0^{\pi} (4\cos 2t + 1) 2\, dt = 2\,(2\sin 2t + t)\big|_0^{\pi} = 2\pi,$$

by the double-angle formula. Similarly, we check

$$\int_C (x^2 - y^2 + 1)\, ds = \int_{\mathbf{y}} (x^2 - y^2 + 1)\, ds$$

$$= \int_{-\pi/2}^{0} (4\cos^2 2t - 4\sin^2 2t + 1)\sqrt{16\sin^2 2t + 16\cos^2 2t}\, dt$$

$$= \int_{-\pi/2}^{0} (4\cos 4t + 1)\cdot 4\, dt = 4\,(\sin 4t + t)\big|_{-\pi/2}^{0} = 2\pi.$$

◆

EXAMPLE 11 We calculate the work done by the force

$$\mathbf{F} = x\mathbf{i} - y\mathbf{j} + (x + y + z)\mathbf{k}$$

on a particle that moves along the parabola $y = 3x^2$, $z = 0$ from the origin to the point $(2, 12, 0)$. (See Figure 6.11.)

We parametrize the parabola by $x = t$, $y = 3t^2$, $z = 0$ for $0 \le t \le 2$. Then, by Definition 1.2,

$$\text{Work} = \int_C \mathbf{F}\cdot d\mathbf{s} = \int_{\mathbf{x}} \mathbf{F}\cdot d\mathbf{s} = \int_0^2 \mathbf{F}(\mathbf{x}(t))\cdot \mathbf{x}'(t)\, dt$$

$$= \int_0^2 (t, -3t^2, t + 3t^2)\cdot (1, 6t, 0)\, dt = \int_0^2 (t - 18t^3)\, dt$$

$$= \left(\tfrac{1}{2}t^2 - \tfrac{9}{2}t^4\right)\big|_0^2 = 2 - 72 = -70.$$

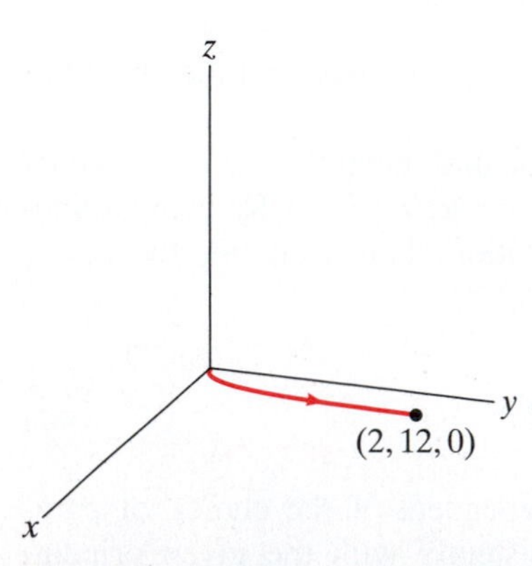

Figure 6.11 The oriented curve of Example 11.

The meaning of the negative sign is that by moving along the curve in the indicated direction, work is done *against* the force. If we orient the curve the opposite way, however, then the work done in moving from (2, 12, 0) to (0, 0, 0) would be 70. ◆

Numerical Evaluation of Line Integrals (optional)

If we have a scalar line integral $\int_C f\, ds$, or a vector line integral $\int_C \mathbf{F} \cdot d\mathbf{s}$, and a suitable parametrization $\mathbf{x}$ of C, then Definitions 1.1 and 1.2 enable the evaluation of the line integrals by means of definite integrals in the parameter variable. These definite integrals may be difficult—or even impossible—to evaluate exactly, so we might need to resort to a numerical method (such as the trapezoidal rule or Simpson's rule) to approximate the value of the integral.

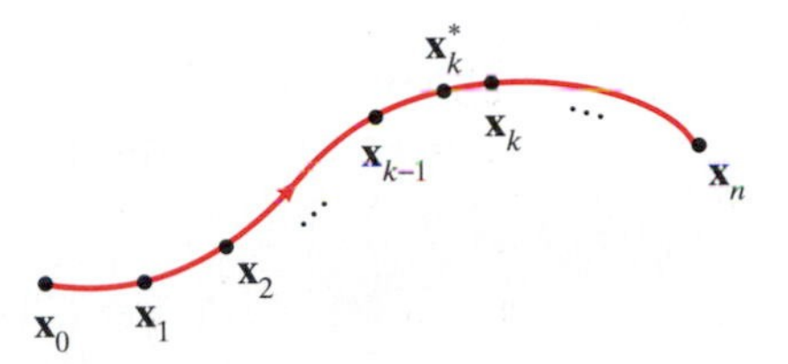

Figure 6.12 The curve C with n points chosen on it.

If a (vector) line integral is given in differential form as $\int_C M\, dx + N\, dy + P\, dz$ (or as $\int_C M\, dx + N\, dy$ if we are working in $\mathbf{R}^2$), then we can give numerical approximations *without* resorting to any parametrization as follows. First, choose points $\mathbf{x}_0, \mathbf{x}_1, \ldots \mathbf{x}_n$ along C, where $\mathbf{x}_0$ is the initial point and $\mathbf{x}_n$ is the terminal point. (See Figure 6.12.) For $k = 0, \ldots, n$, let us write

$$\mathbf{x}_k = (x_k, y_k, z_k)$$

and, for $k = 1, \ldots, n$, let

$$\Delta x_k = x_k - x_{k-1}, \ \Delta y_k = y_k - y_{k-1}, \ \Delta z_k = z_k - z_{k-1}.$$

Finally, let $\mathbf{x}_k^* = (x_k^*, y_k^*, z_k^*)$ denote any point on the arc of C between $\mathbf{x}_{k-1}$ and $\mathbf{x}_k$. Then we may approximate the line integral as

$$\int_C M\, dx + N\, dy + P\, dz \approx \sum_{k=1}^{n} \left[M(\mathbf{x}_k^*)\Delta x_k + N(\mathbf{x}_k^*)\Delta y_k + P(\mathbf{x}_k^*)\Delta z_k \right]. \tag{6}$$

Besides the approximation given in (6), we may also use a version of the trapezoidal rule adapated to the case of line integrals. In particular, the trapezoidal rule approximation T_n to the line integral $\int_C M\, dx + N\, dy + P\, dz$ is

$$\begin{aligned} T_n &= \sum_{k=1}^{n} \left[(M(\mathbf{x}_{k-1}) + M(\mathbf{x}_k)) \frac{\Delta x_k}{2} + (N(\mathbf{x}_{k-1}) + N(\mathbf{x}_k)) \frac{\Delta y_k}{2} \right. \\ &\qquad \left. + (P(\mathbf{x}_{k-1}) + P(\mathbf{x}_k)) \frac{\Delta z_k}{2} \right] \\ &= \sum_{k=1}^{n} (M(\mathbf{x}_{k-1}) + M(\mathbf{x}_k)) \frac{\Delta x_k}{2} + \sum_{k=1}^{n} (N(\mathbf{x}_{k-1}) + N(\mathbf{x}_k)) \frac{\Delta y_k}{2} \\ &\quad + \sum_{k=1}^{n} (P(\mathbf{x}_{k-1}) + P(\mathbf{x}_k)) \frac{\Delta z_k}{2}. \end{aligned} \tag{7}$$

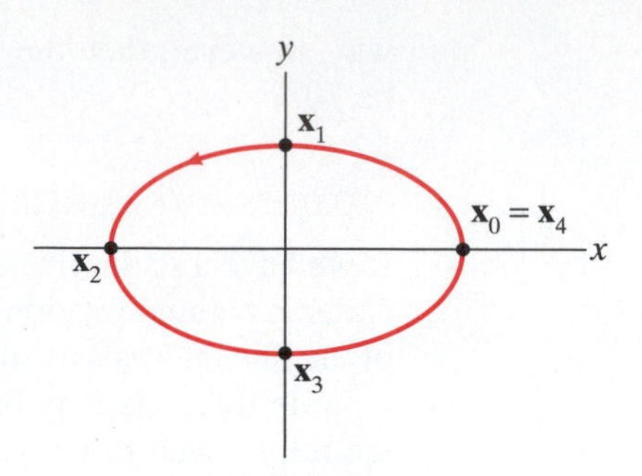

Figure 6.13 The ellipse C in Example 12.

Similar (and shorter) formulas, analogous to (6) and (7) may be given to approximate the two-dimensional line integral $\int_C M\,dx + N\,dy$; we will not state them explicitly.

EXAMPLE 12 Let C be the ellipse $x^2 + 4y^2 = 4$, oriented counterclockwise. (See Figure 6.13.) We approximate $\int_C y^2\,dx + x\,dy$ using the (two-dimensional version of) the trapezoidal rule (7).

To make this approximation, let

$$\mathbf{x}_0 = (2, 0),\ \mathbf{x}_1 = (0, 1),\ \mathbf{x}_2 = (-2, 0),\ \mathbf{x}_3 = (0, -1),\ \mathbf{x}_4 = (2, 0).$$

Then we have

$$\Delta x_1 = \Delta x_2 = -2,\ \Delta x_3 = \Delta x_4 = 2;$$

$$\Delta y_1 = 1,\ \Delta y_2 = \Delta y_3 = -1,\ \Delta y_4 = 1.$$

Hence, from (7), we compute

$$\begin{aligned} T_4 &= (0^2 + 1^2)\tfrac{-2}{2} + (1^2 + 0^2)\tfrac{-2}{2} + (0^2 + (-1)^2)\tfrac{2}{2} + ((-1)^2 + 0^2)\tfrac{2}{2} \\ &\quad + (2 + 0)\tfrac{1}{2} + (0 + (-2))\tfrac{-1}{2} + ((-2) + 0)\tfrac{-1}{2} + (0 + 2)\tfrac{1}{2} \\ &= -1 - 1 + 1 + 1 + 1 + 1 + 1 + 1 = 4. \end{aligned}$$

The exact answer may be found using the parametrization

$$\begin{cases} x = 2\cos t \\ y = \sin t \end{cases} \quad 0 \le t \le 2\pi.$$

We calculate that

$$\begin{aligned} \int_C y^2\,dx + x\,dy &= \int_0^{2\pi} \left(\sin^2 t(-2\sin t) + 2\cos^2 t\right) dt \\ &= \int_0^{2\pi} \left(2(1 - \cos^2 t)(-\sin t) + 1 + \cos 2t\right) dt \\ &= \left(2\cos t - \tfrac{2}{3}\cos^3 t + t + \tfrac{1}{2}\sin 2t\right)\Big|_0^{2\pi} = 2\pi. \end{aligned}$$

From this we see that our approximation is quite rough. It can be improved by increasing n while taking smaller values of Δx_k and Δy_k. For instance (see Figure 6.14), if we let

$$\mathbf{x}_0 = (2, 0),\ \mathbf{x}_1 = \left(\sqrt{3}, \tfrac{1}{2}\right),\ \mathbf{x}_2 = (0, 1),\ \mathbf{x}_3 = \left(-\sqrt{3}, \tfrac{1}{2}\right),\ \mathbf{x}_4 = (-2, 0),$$

$$\mathbf{x}_5 = \left(-\sqrt{3}, -\tfrac{1}{2}\right),\ \mathbf{x}_6 = (0, -1),\ \mathbf{x}_7 = \left(\sqrt{3}, -\tfrac{1}{2}\right),\ \mathbf{x}_8 = (2, 0),$$

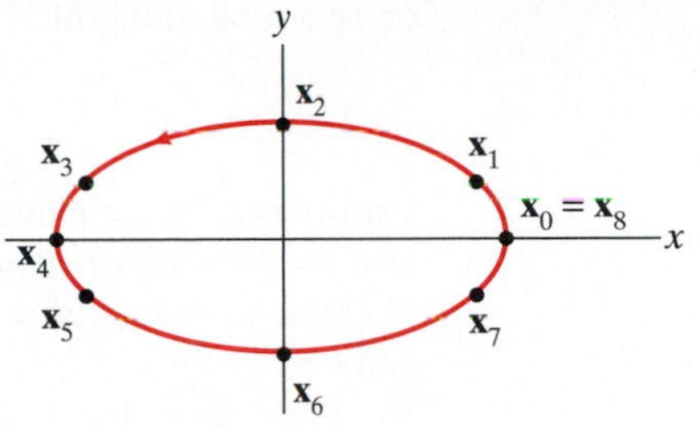

Figure 6.14 The ellipse C in Example 12 with eight points marked.

then

$$\Delta x_1 = \sqrt{3} - 2,\ \Delta x_2 = \Delta x_3 = -\sqrt{3},\ \Delta x_4 = \sqrt{3} - 2,\ \Delta x_5 = 2 - \sqrt{3},$$
$$\Delta x_6 = \Delta x_7 = \sqrt{3},\ \Delta x_8 = 2 - \sqrt{3};$$
$$\Delta y_1 = \Delta y_2 = \tfrac{1}{2},\ \Delta y_3 = \Delta y_4 = \Delta y_5 = \Delta y_6 = -\tfrac{1}{2},\ \Delta y_7 = \Delta y_8 = \tfrac{1}{2}.$$

Therefore, $\int_C y^2\,dx + x\,dy$ is also approximated by

$$\begin{aligned}
T_8 = {} & \left(0^2 + \left(\tfrac{1}{2}\right)^2\right)\frac{\sqrt{3}-2}{2} + \left(\left(\tfrac{1}{2}\right)^2 + 1^2\right)\frac{-\sqrt{3}}{2} + \left(\left(1^2 + \tfrac{1}{2}\right)^2\right)\frac{-\sqrt{3}}{2} \\
& + \left(\left(\tfrac{1}{2}\right)^2 + 0^2\right)\frac{\sqrt{3}-2}{2} + \left(0^2 + \left(-\tfrac{1}{2}\right)^2\right)\frac{2-\sqrt{3}}{2} \\
& + \left(\left(-\tfrac{1}{2}\right)^2 + (-1)^2\right)\frac{\sqrt{3}}{2} + \left((-1)^2 + \left(-\tfrac{1}{2}\right)^2\right)\frac{\sqrt{3}}{2} \\
& + \left(\left(-\tfrac{1}{2}\right)^2 + 0^2\right)\frac{2-\sqrt{3}}{2} + (2+\sqrt{3})\frac{1/2}{2} + (\sqrt{3}+0)\frac{1/2}{2} \\
& + (0 + (-\sqrt{3}))\frac{-1/2}{2} + ((-\sqrt{3}) + (-2))\frac{-1/2}{2} \\
& + ((-2) + (-\sqrt{3}))\frac{-1/2}{2} + ((-\sqrt{3}) + 0)\frac{-1/2}{2} \\
& + (0 + \sqrt{3})\frac{1/2}{2} + (\sqrt{3} + 2)\frac{1/2}{2} \\
= {} & 2 + 2\sqrt{3} \approx 5.4641.
\end{aligned}$$

Although still rough, this represents a better approximation. ◆

EXAMPLE 13 Let C be the line segment from $(0, 0, 0)$ to $(2, 2, 2)$, and consider the line integral $\int_C y\sqrt{1+x^3}\,dx + e^{-xz}\,dy + \cos(z^2)\,dz$. The standard parametrization of the line segment C by

$$\begin{cases} x = t \\ y = t \qquad 0 \le t \le 2 \\ z = t \end{cases}$$

enables us, in theory, to evaluate the line integral above by evaluating the one-variable integral

$$\int_0^2 \left(t\sqrt{1+t^3} + e^{-t^2} + \cos(t^2) \right) dt.$$

Unfortunately, we cannot do so exactly (i.e., by means of the fundamental theorem of calculus, since an antiderivative of the integrand cannot be found in terms of elementary functions). Instead, we provide approximations using formulas (6) and (7).

Let $n = 4$. If we choose points along C that are regularly spaced, then we have

$$\mathbf{x}_0 = (0,0,0),\ \mathbf{x}_1 = \left(\tfrac{1}{2}, \tfrac{1}{2}, \tfrac{1}{2}\right),\ \mathbf{x}_2 = (1,1,1),\ \mathbf{x}_3 = \left(\tfrac{3}{2}, \tfrac{3}{2}, \tfrac{3}{2}\right),\ \mathbf{x}_4 = (2,2,2),$$

so that

$$\Delta x_k = \Delta y_k = \Delta z_k = \tfrac{1}{2}.$$

To calculate an approximation using formula (6), we can, for example, let

$$\mathbf{x}_k^* = \tfrac{1}{2}(\mathbf{x}_{k-1} + \mathbf{x}_k),$$

which is the midpoint of the line segment joining $\mathbf{x}_{k-1}$ and $\mathbf{x}_k$. Then

$$\mathbf{x}_1^* = \left(\tfrac{1}{4}, \tfrac{1}{4}, \tfrac{1}{4}\right),\ \mathbf{x}_2^* = \left(\tfrac{3}{4}, \tfrac{3}{4}, \tfrac{3}{4}\right),\ \mathbf{x}_3^* = \left(\tfrac{5}{4}, \tfrac{5}{4}, \tfrac{5}{4}\right),\ \mathbf{x}_4^* = \left(\tfrac{7}{4}, \tfrac{7}{4}, \tfrac{7}{4}\right),$$

and so formula (6) tells us that

$$\begin{aligned}
&\int_C y\ \sqrt{1+x^3}\,dx + e^{-xz}\,dy + \cos(z^2)\,dz \\
&\approx \left[\tfrac{1}{4}\sqrt{1+\left(\tfrac{1}{4}\right)^3}\,\tfrac{1}{2} + e^{-1/16}\tfrac{1}{2} + \cos\tfrac{1}{16}\cdot\tfrac{1}{2}\right] \\
&\quad + \left[\tfrac{3}{4}\sqrt{1+\left(\tfrac{3}{4}\right)^3}\,\tfrac{1}{2} + e^{-9/16}\tfrac{1}{2} + \cos\tfrac{9}{16}\cdot\tfrac{1}{2}\right] \\
&\quad + \left[\tfrac{5}{4}\sqrt{1+\left(\tfrac{5}{4}\right)^3}\,\tfrac{1}{2} + e^{-25/16}\tfrac{1}{2} + \cos\tfrac{25}{16}\cdot\tfrac{1}{2}\right] \\
&\quad + \left[\tfrac{7}{4}\sqrt{1+\left(\tfrac{7}{4}\right)^3}\,\tfrac{1}{2} + e^{-49/16}\tfrac{1}{2} + \cos\tfrac{49}{16}\cdot\tfrac{1}{2}\right] \\
&\approx 5.16422.
\end{aligned}$$

Using the trapezoidal rule formula (7), we may also approximate the line integral by

$$\begin{aligned}
T_4 &= \left(0 + \tfrac{1}{2}\sqrt{1+\left(\tfrac{1}{2}\right)^3}\right)\tfrac{1/2}{2} + \left(\tfrac{1}{2}\sqrt{1+\left(\tfrac{1}{2}\right)^3} + 1\sqrt{1+1^3}\right)\tfrac{1/2}{2} \\
&\quad + \left(1\sqrt{1+1^3} + \tfrac{3}{2}\sqrt{1+\left(\tfrac{3}{2}\right)^3}\right)\tfrac{1/2}{2} + \left(\tfrac{3}{2}\sqrt{1+\left(\tfrac{3}{2}\right)^3} + 2\sqrt{1+2^3}\right)\tfrac{1/2}{2} \\
&\quad + (e^0 + e^{-1/4})\tfrac{1/2}{2} + (e^{-1/4} + e^{-1})\tfrac{1/2}{2} + (e^{-1} + e^{-9/4})\tfrac{1/2}{2} \\
&\quad + (e^{-9/4} + e^{-4})\tfrac{1/2}{2} + (\cos 0 + \cos\tfrac{1}{4})\tfrac{1/2}{2} + (\cos\tfrac{1}{4} + \cos 1)\tfrac{1/2}{2} \\
&\quad + (\cos 1 + \cos\tfrac{9}{4})\tfrac{1/2}{2} + (\cos\tfrac{9}{4} + \cos 4)\tfrac{1/2}{2} \\
&\approx 5.44874.
\end{aligned}$$

As was the case in Example 12, because of the small number of sampling points used, formulas (6) and (7) provide relatively crude approximations. ◆

Exercises

1. Let $f(x, y) = x + 2y$. Evaluate the scalar line integral $\int_{\mathbf{x}} f\, ds$ over the given path $\mathbf{x}$.
   (a) $\mathbf{x}(t) = (2 - 3t, 4t - 1), 0 \le t \le 2$
   (b) $\mathbf{x}(t) = (\cos t, \sin t), 0 \le t \le \pi$

In Exercises 2–5, calculate $\int_{\mathbf{x}} f\, ds$, where f and $\mathbf{x}$ are as indicated.

2. $f(x, y, z) = xyz, \mathbf{x}(t) = (t, 2t, 3t), 0 \le t \le 2$
3. $f(x, y, z) = \dfrac{x + z}{y + z}, \mathbf{x}(t) = (t, t, t^{3/2}), 1 \le t \le 3$
4. $f(x, y, z) = 3x + xy + z^3, \mathbf{x}(t) = (\cos 4t, \sin 4t, 3t), 0 \le t \le 2\pi$
5. $f(x, y, z) = 2x - y^{1/2} + 2z^2$, $\mathbf{x}(t) = \begin{cases} (t, t^2, 0) & \text{if } 0 \le t \le 1 \\ (1, 1, t - 1) & \text{if } 1 \le t \le 3 \end{cases}$

In Exercises 6–10, find $\int_{\mathbf{x}} \mathbf{F} \cdot d\mathbf{s}$, where the vector field $\mathbf{F}$ and the path $\mathbf{x}$ are given.

6. $\mathbf{F} = x\,\mathbf{i} + y\,\mathbf{j} + z\,\mathbf{k}, \mathbf{x}(t) = (2t + 1, t, 3t - 1), 0 \le t \le 1$
7. $\mathbf{F} = (y + 2)\,\mathbf{i} + x\,\mathbf{j}, \mathbf{x}(t) = (\sin t, -\cos t), 0 \le t \le \pi/2$
8. $\mathbf{F} = x\,\mathbf{i} + y\,\mathbf{j} - z\,\mathbf{k}, \mathbf{x}(t) = (t, 3t^2, 2t^3), -1 \le t \le 1$
9. $\mathbf{F} = 3z\,\mathbf{i} + y^2\,\mathbf{j} + 6z\,\mathbf{k}, \mathbf{x}(t) = (\cos t, \sin t, t/3), 0 \le t \le 4\pi$
10. $\mathbf{F} = y \cos z\,\mathbf{i} + x \sin z\,\mathbf{j} + xy \sin z^2\,\mathbf{k}, \mathbf{x}(t) = (t, t^2, t^3), 0 \le t \le 1$
11. Determine the value of $\int_{\mathbf{x}} x\, dy - y\, dx$, where $\mathbf{x}(t) = (\cos 3t, \sin 3t), 0 \le t \le \pi$.
12. Find the work done by the force field $\mathbf{F} = 2x\,\mathbf{i} + \mathbf{j}$ when a particle moves along the path $\mathbf{x}(t) = (t, 3t^2, 2)$, $0 \le t \le 2$.
13. If $\mathbf{x} = (e^{2t} \cos 3t, e^{2t} \sin 3t), 0 \le t \le 2\pi$, find $$\int_{\mathbf{x}} \frac{x\, dx + y\, dy}{(x^2 + y^2)^{3/2}}.$$
14. Let C be the portion of the curve $y = 2\sqrt{x}$ between $(1, 2)$ and $(9, 6)$. Find $\int_C 3y\, ds$.
15. Let $\mathbf{F} = (x^2 + y)\mathbf{i} + (y - x)\mathbf{j}$ and consider the two paths
$$\mathbf{x}(t) = (t, t^2),\ 0 \le t \le 1 \quad \text{and}$$
$$\mathbf{y}(t) = (1 - 2t, 4t^2 - 4t + 1),\ 0 \le t \le \tfrac{1}{2}.$$
   (a) Calculate $\int_{\mathbf{x}} \mathbf{F} \cdot d\mathbf{s}$ and $\int_{\mathbf{y}} \mathbf{F} \cdot d\mathbf{s}$.
   (b) By considering the image curves of the paths $\mathbf{x}$ and $\mathbf{y}$, discuss your answers in part (a).
16. Find the work done by the force field $\mathbf{F} = x^2 y\,\mathbf{i} + z\,\mathbf{j} + (2x - y)\,\mathbf{k}$ on a particle as the particle moves along a straight line from $(1, 1, 1)$ to $(2, -3, 3)$.
17. Let $\mathbf{F} = (2z^5 - 3xy)\,\mathbf{i} - x^2\,\mathbf{j} + x^2 z\,\mathbf{k}$. Calculate the line integral of $\mathbf{F}$ around the perimeter of the square with vertices $(1, 1, 3)$, $(-1, 1, 3)$, $(-1, -1, 3)$, $(1, -1, 3)$, oriented counterclockwise about the z-axis.
18. Evaluate $\int_C (x^2 - y)\, dx + (x - y^2)\, dy$, where C is the line segment from $(1, 1)$ to $(3, 5)$.
19. Find $\int_C x^2 y\, dx - (x + y)\, dy$, where C is the trapezoid with vertices $(0, 0)$, $(3, 0)$, $(3, 1)$, and $(1, 1)$, oriented counterclockwise.
20. Evaluate $\int_C x^2 y\, dx - xy\, dy$, where C is the curve with equation $y^2 = x^3$, from $(1, -1)$ to $(1, 1)$.
21. Evaluate $\int_C yz\, dx - xz\, dy + xy\, dz$, where C is the line segment from $(1, 1, 2)$ to $(5, 3, 1)$.
22. Calculate $\int_C z\, dx + x\, dy + y\, dz$, where C is the curve obtained by intersecting the surfaces $z = x^2$ and $x^2 + y^2 = 4$ and oriented counterclockwise around the z-axis (as seen from the positive z-axis).
23. Show that $\int_{\mathbf{x}} \mathbf{T} \cdot d\mathbf{s}$ equals the length of the path $\mathbf{x}$, where $\mathbf{T}$ denotes the unit tangent vector of the path.
24. Tom Sawyer is whitewashing a picket fence. The bases of the fenceposts are arranged in the xy-plane as the quarter circle $x^2 + y^2 = 25$, $x, y \ge 0$, and the height of the fencepost at point (x, y) is given by $h(x, y) = 10 - x - y$ (units are feet). Use a scalar line integral to find the area of one side of the fence. (See Figure 6.15.)

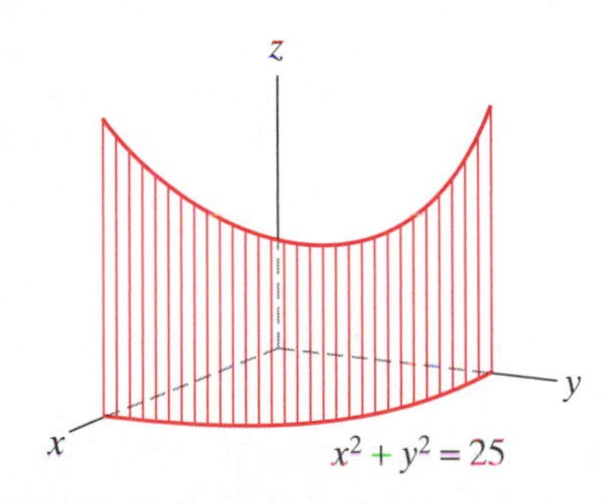

Figure 6.15 The picket fence of Exercise 24. The base of the fence is the quarter circle $x^2 + y^2 = 25$, $x, y \ge 0$.

25. Sisyphus is pushing a boulder up a 100-ft-tall spiral staircase surrounding a cylindrical castle tower. (See Figure 6.16.)
   (a) Suppose Sisyphus's path is described parametrically as
$$\mathbf{x}(t) = (5 \cos 3t, 5 \sin 3t, 10t), \quad 0 \le t \le 10.$$

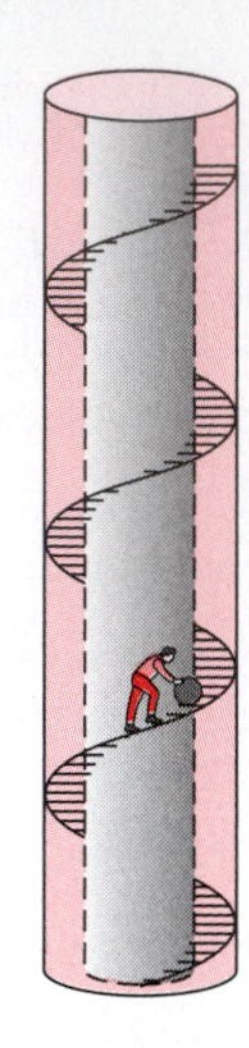

Figure 6.16 Sisyphus's path up the spiral staircase of Exercise 25.

If he exerts a force with a constant magnitude of 50 lb tangent to his path, find the work Sisyphus does in pushing the boulder up to the top of the tower.

(b) Just as Sisyphus reaches the top of the tower, he sneezes and the boulder slides all the way to the bottom. If the boulder weighs 75 lb, how much work is done by gravity when the boulder reaches the bottom?

26. A ship is pulled through a 14-ft-long straight channel by a line that passes from the ship around a pulley. Assume that a coordinate system is set up so that the pulley is at the point $(24, 32)$, and the ship is pulled along the y-axis from the origin to the point $(0, 14)$. If the tension on the line is kept at a constant 25 lb, find the work done in tugging the ship through the channel.

27. Suppose C is the curve $y = f(x)$, oriented from $(a, f(a))$ to $(b, f(b))$ where $a < b$ and where f is positive and continuous on $[a, b]$. If $\mathbf{F} = y\,\mathbf{i}$, show that the value of $\int_C \mathbf{F} \cdot d\mathbf{s}$ is the area under the graph of f between $x = a$ and $x = b$.

28. Let $\mathbf{F}$ be the radial vector field $\mathbf{F} = x\,\mathbf{i} + y\,\mathbf{j} + z\,\mathbf{k}$. Show that if $\mathbf{x}(t)$, $a \le t \le b$, is any path that lies on the sphere $x^2 + y^2 + z^2 = c^2$, then $\int_{\mathbf{x}} \mathbf{F} \cdot d\mathbf{s} = 0$. (Hint: If $\mathbf{x}(t) = (x(t), y(t), z(t))$ lies on the sphere, then $[x(t)]^2 + [y(t)]^2 + [z(t)]^2 = c^2$. Differentiate this last equation with respect to t.)

29. Let C be a level set of the function $f(x, y)$. Show that $\int_C \nabla f \cdot d\mathbf{s} = 0$.

30. You are traveling through Cleveland, famous for its lake-effect snow in winter that makes driving quite treacherous. Suppose that you are currently located 20 miles due east of Cleveland and are attempting to drive to a point 20 miles due west of Cleveland. Further suppose that if you are s miles from the center of Cleveland, where the weather is the worst, you can drive at a rate of at most $v(s) = 2s + 20$ miles per hour.

(a) How long will the trip take if you drive on a straight-line path directly through Cleveland? (Assume that you always drive at the maximum speed possible.)

(b) How long will the trip take if you avoid the middle of the city by driving along a semicircular path with Cleveland at the center? (Again, assume that you drive at the maximum speed possible.)

(c) Repeat parts (a) and (b), this time using $v(s) = (s^2/16) + 25$ miles per hour as the maximum speed that you can drive.

31. Consider a particle of mass m that carries a charge q. Suppose that the particle is under the influence of both an electric field $\mathbf{E}$ and a magnetic field $\mathbf{B}$ so that the particle's trajectory is described by the path $\mathbf{x}(t)$ for $a \le t \le b$. Then the total force acting on the particle is given in mks units by the **Lorentz force**

$$\mathbf{F} = q(\mathbf{E} + \mathbf{v} \times \mathbf{B}),$$

where $\mathbf{v}$ denotes the velocity of the trajectory.

(a) Use Newton's second law of motion (i.e., $\mathbf{F} = m\mathbf{a}$, where $\mathbf{a}$ denotes the acceleration of the particle) to show that

$$m\mathbf{a} \cdot \mathbf{v} = q\mathbf{E} \cdot \mathbf{v}.$$

(b) If the particle moves with constant speed, use part (a) to show that $\mathbf{E}$ does no work along the path of the particle. (Hint: Apply Proposition 1.7 of Chapter 3 to $\mathbf{v}$.)

32. Let C be the segement of the parabola $y = x^2$ between $(0, 0)$ and $(1, 1)$ and consider the line integral $\int_C y^3\,dx - x^2\,dy$.

(a) Let $\mathbf{x}_0 = (0, 0)$, $\mathbf{x}_1 = \left(\frac{1}{4}, \frac{1}{16}\right)$, $\mathbf{x}_2 = \left(\frac{1}{2}, \frac{1}{4}\right)$, $\mathbf{x}_3 = \left(\frac{3}{4}, \frac{9}{16}\right)$, $\mathbf{x}_4 = (1, 1)$. Use the trapezoidal rule formula (7) and these points to approximate the line integral.

(b) Now calculate the exact value of the line integral and compare your results.

33. Let C be the line segement from $(0, 0, 0)$ to $(1, 2, 3)$ and consider the line integral $\int_C yz\,dx + (x + z)\,dy + x^2 y\,dz$.

(a) Divide the segment into five regularly spaced points $\mathbf{x}_0, \mathbf{x}_1, \ldots, \mathbf{x}_4$ and use the trapezoidal rule formula (7) to approximate this line integral.

(b) Compare your approximation with the exact value of the line integral.

34. Suppose that magnetic field measurements are made along a wire shaped as a curve C and the results are tabulated as follows:

k	Point $\mathbf{x}_k =$ (x_k, y_k, z_k)	$\mathbf{B}(x_k, y_k, z_k) =$ $M\,\mathbf{i} + N\,\mathbf{j} + P\,\mathbf{k}$
0	$(-1, -2, -1)$	$\mathbf{k}$
1	$(0, 1, -1)$	$\mathbf{j} + 2\mathbf{k}$
2	$(0, 2, 0)$	$\mathbf{i} + \mathbf{j} + 2\mathbf{k}$
3	$(1, 2, 1)$	$2\mathbf{i} + \mathbf{j} + 2\mathbf{k}$
4	$(1, 2, 2)$	$2\mathbf{i} + 2\mathbf{j} + 2\mathbf{k}$
5	$(1, 1, 2)$	$2\mathbf{i} + 3\mathbf{j} + 3\mathbf{k}$
6	$(1, 1, 1)$	$3\mathbf{i} + 3\mathbf{j} + 3\mathbf{k}$
7	$(1, 0, 0)$	$4\mathbf{i} + 3\mathbf{j} + 3\mathbf{k}$
8	$(0, 0, 0)$	$4\mathbf{i} + 3\mathbf{j} + 4\mathbf{k}$

By writing $\int_C \mathbf{B} \cdot d\mathbf{s}$ as $\int_C M\,dx + N\,dy + P\,dz$, estimate the work done by $\mathbf{B}$ along C using

(a) a trapezoidal rule approximation;

(b) a trapezoidal rule approximation using only the points $\mathbf{x}_0, \mathbf{x}_2, \mathbf{x}_4, \mathbf{x}_6, \mathbf{x}_8$.

6.2 Green's Theorem

Green's theorem relates the *vector line integral* around a closed curve C in $\mathbf{R}^2$ to an appropriate *double integral* over the plane region D bounded by C. The fact that there is such an elegant connection between one- and two-dimensional integrals is at once surprising, satisfying, and powerful. Green's theorem, stated generally, is as follows:

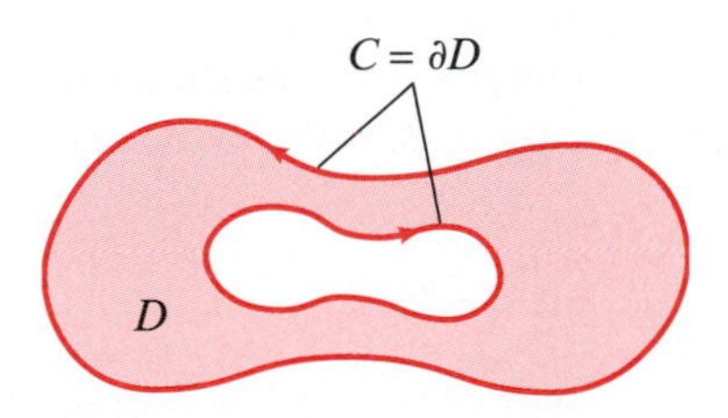

Figure 6.17 The shaded region D has a boundary consisting of two simple, closed curves, each of class C^1, whose union we call C.

Theorem 2.1 (Green's Theorem) Let D be a closed, bounded region in $\mathbf{R}^2$ whose boundary $C = \partial D$ consists of finitely many simple, closed curves. Orient the curves of C so that D is on the left as one traverses C. (See Figure 6.17.) Let $\mathbf{F}(x, y) = M(x, y)\,\mathbf{i} + N(x, y)\,\mathbf{j}$ be a vector field of class C^1 throughout D. Then

$$\oint_C M dx + N dy = \iint_D \left(\frac{\partial N}{\partial x} - \frac{\partial M}{\partial y} \right) dx\,dy.$$

(The symbol $\oint_C$ indicates that the line integral is taken over one or more *closed* curves.)

EXAMPLE 1 Let $\mathbf{F} = xy\,\mathbf{i} + y^2\,\mathbf{j}$ and let D be the first quadrant region bounded by the line $y = x$ and the parabola $y = x^2$. We verify Green's theorem in this case.

The region D and its boundary are shown in Figure 6.18. ∂D is oriented counterclockwise, the orientation stipulated by the statement of Green's theorem. To calculate

$$\oint_{\partial D} \mathbf{F} \cdot d\mathbf{s} = \oint_{\partial D} xy\,dx + y^2 dy,$$

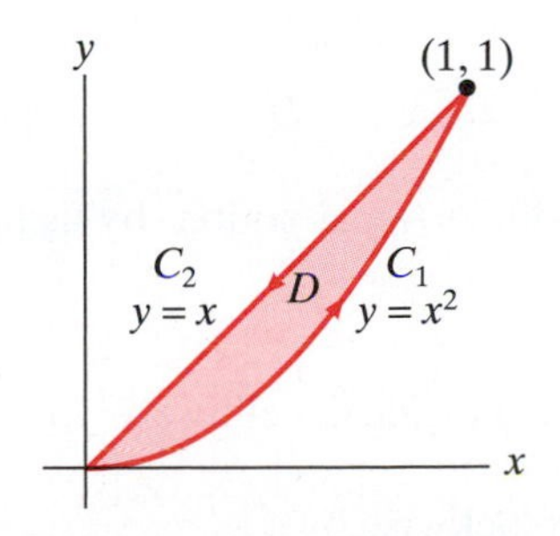

Figure 6.18 The region D of Example 1.

we need to parametrize the two C^1 pieces of ∂D separately:

$$C_1 : \begin{cases} x = t \\ y = t^2 \end{cases} \quad 0 \le t \le 1 \quad \text{and} \quad C_2 : \begin{cases} x = 1 - t \\ y = 1 - t \end{cases} \quad 0 \le t \le 1.$$

(Note the orientations of C_1 and C_2.) Hence,

$$\oint_{\partial D} xy\,dx + y^2dy = \int_{C_1} xy\,dx + y^2dy + \int_{C_2} xy\,dx + y^2dy$$
$$= \int_0^1 (t \cdot t^2 + t^4 \cdot 2t)\,dt + \int_0^1 ((1-t)^2 + (1-t)^2)(-dt)$$
$$= \int_0^1 (t^3 + 2t^5)\,dt + \int_0^1 2(1-t)^2(-dt)$$
$$= \left(\tfrac{1}{4}t^4 + \tfrac{2}{6}t^6\right)\Big|_0^1 + \left(\tfrac{2}{3}(1-t)^3\right)\Big|_0^1$$
$$= \tfrac{1}{4} + \tfrac{2}{6} - \tfrac{2}{3} = -\tfrac{1}{12}.$$

On the other hand,

$$\iint_D \left[\frac{\partial}{\partial x}(y^2) - \frac{\partial}{\partial y}(xy)\right] dx\,dy = \int_0^1 \int_{x^2}^x -x\,dy\,dx = \int_0^1 -x(x - x^2)\,dx$$
$$= \int_0^1 (x^3 - x^2)\,dx = \left(\tfrac{1}{4}x^4 - \tfrac{1}{3}x^3\right)\Big|_0^1$$
$$= \tfrac{1}{4} - \tfrac{1}{3} = -\tfrac{1}{12}.$$

The line integral and the double integral agree, just as Green's theorem says they must. ◆

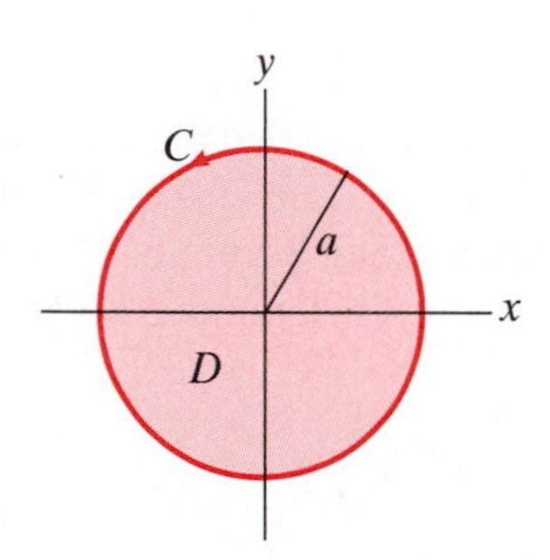

Figure 6.19 The disk of radius a with boundary oriented so that Green's theorem applies.

EXAMPLE 2 Consider $\oint_C -y\,dx + x\,dy$, where C is the circle of radius a (i.e., the boundary of the disk D of radius a), oriented counterclockwise as shown in Figure 6.19. Although we can readily parametrize C and thus evaluate the line integral, let us employ Green's theorem instead:

$$\oint_C -y\,dx + x\,dy = \iint_D \left[\frac{\partial}{\partial x}(x) - \frac{\partial}{\partial y}(-y)\right] dx\,dy$$
$$= \iint_D 2\,dx\,dy = 2(\text{area of } D) = 2\pi a^2.$$

The rightmost expression is twice the area of a disk of radius a. In this case, the double integral is much easier to consider than the line integral. ◆

The use of Green's theorem in Example 2 can be put in a much more general setting: Indeed, if D is *any* region to which Green's theorem can be applied, then, orienting ∂D appropriately, we have

$$\frac{1}{2}\oint_{\partial D} -y\,dx + x\,dy = \frac{1}{2}\iint_D 2\,dx\,dy = \text{area of } D. \tag{1}$$

Thus, we can calculate the area of a region (a two-dimensional notion) by using line integrals (a one-dimensional construction)!

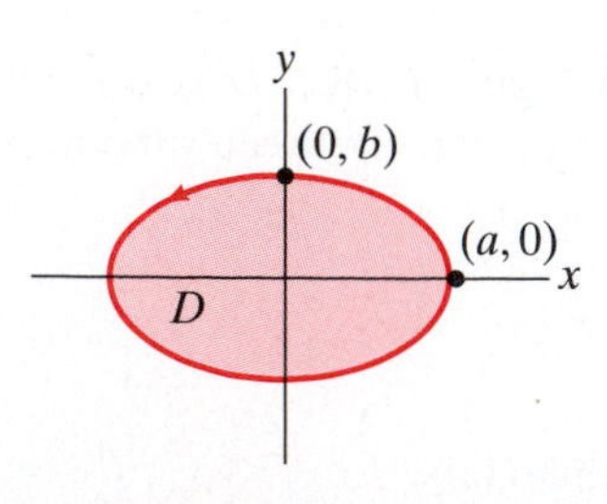

Figure 6.20 The region inside the ellipse $x^2/a^2 + y^2/b^2 = 1$.

EXAMPLE 3 Using formula (1), we compute the area inside the ellipse $x^2/a^2 + y^2/b^2 = 1$ (Figure 6.20).

The ellipse itself may be parametrized counterclockwise by

$$\begin{cases} x = a\cos t \\ y = b\sin t \end{cases} \quad 0 \le t \le 2\pi.$$

Once again, using formula (1), we find that the area inside the ellipse is

$$\begin{aligned}\tfrac{1}{2}\oint_{\partial D} -y\,dx + x\,dy &= \tfrac{1}{2}\int_0^{2\pi} -b\sin t(-a\sin t\,dt) + a\cos t(b\cos t\,dt)\\ &= \tfrac{1}{2}\int_0^{2\pi} (ab\sin^2 t + ab\cos^2 t)\,dt\\ &= \tfrac{1}{2}\int_0^{2\pi} ab\,dt = \pi ab.\end{aligned}$$

◆

Alternative Formulations

We rewrite the line integral–double integral formula appearing in the statement of Theorem 2.1 (Green's theorem) in two ways. These reformulations generalize to higher dimensions and provide some additional insight in interpreting the geometric significance of Green's theorem.

To begin, consider a C^1 vector field

$$\mathbf{F} = M(x, y)\,\mathbf{i} + N(x, y)\,\mathbf{j}$$

to be defined on $\mathbf{R}^3$ by taking its $\mathbf{k}$-component to be zero. Then, if we compute the curl of $\mathbf{F}$, we find

$$\nabla \times \mathbf{F} = \begin{vmatrix} \mathbf{i} & \mathbf{j} & \mathbf{k} \\ \partial/\partial x & \partial/\partial y & \partial/\partial z \\ M & N & 0 \end{vmatrix} = \left(\frac{\partial N}{\partial x} - \frac{\partial M}{\partial y}\right)\mathbf{k}.$$

Therefore, because $\mathbf{k}\cdot\mathbf{k} = 1$, we obtain

$$\iint_D \left(\frac{\partial N}{\partial x} - \frac{\partial M}{\partial y}\right) dA = \iint_D (\nabla \times \mathbf{F})\cdot\mathbf{k}\,dA.$$

Since $\oint_{\partial D} \mathbf{F}\cdot d\mathbf{s} = \oint_{\partial D} M\,dx + N\,dy$, Green's theorem may be rewritten as follows:

■ **Proposition 2.2** (A VECTOR REFORMULATION OF GREEN'S THEOREM) If D is a region to which Green's theorem applies and

$$\mathbf{F} = M(x, y)\,\mathbf{i} + N(x, y)\,\mathbf{j}$$

is a vector field of class C^1 on D, then, orienting ∂D appropriately,

$$\oint_{\partial D} \mathbf{F}\cdot d\mathbf{s} = \iint_D (\nabla \times \mathbf{F})\cdot\mathbf{k}\,dA.$$

To understand this result, visualize the plane region D as sitting in the xy-plane in $\mathbf{R}^3$. (See Figure 6.21.) The vector $\mathbf{k}$ is a unit vector normal to D, and $\iint_D (\nabla \times \mathbf{F})\cdot\mathbf{k}\,dA$ is the double integral of the component of the curl of $\mathbf{F}$ normal to D. Since the line integral $\oint_{\partial D} \mathbf{F}\cdot d\mathbf{s}$ is the circulation of $\mathbf{F}$ along ∂D (see §6.1), the equation of Proposition 2.2 tells us that, under suitable hypotheses, the circulation of a vector field $\mathbf{F}$ along the boundary of a plane region is equal to the total (or net) "infinitesimal rotation" of $\mathbf{F}$ over the entire region. (See also §3.4 where the curl of a vector field is given an intuitive interpretation in terms of rotation—or wait until §7.3 when the notion of curl measuring rotation of a vector field is explained more precisely.) This result generalizes to Stokes's theorem

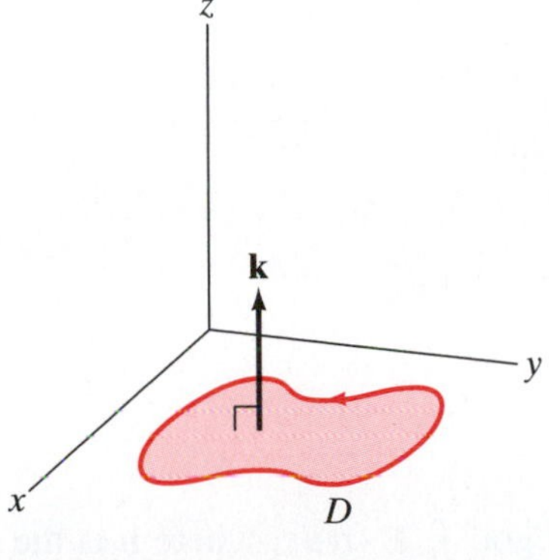

Figure 6.21 A plane region D in $\mathbf{R}^3$. The vector $\mathbf{k}$ is normal to D.

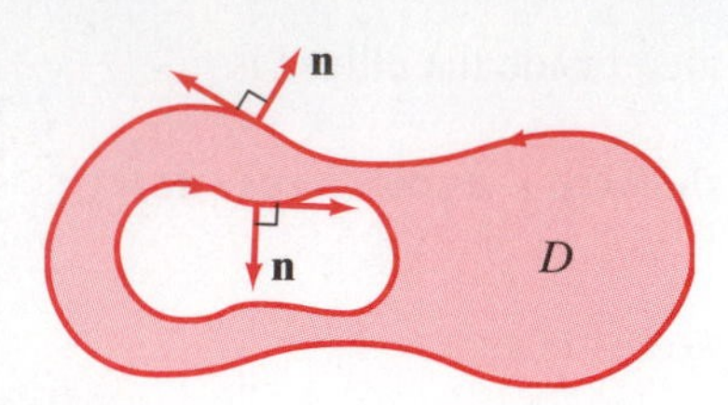

Figure 6.22 The outward unit normal $\mathbf{n}$ to the region D.

in $\mathbf{R}^3$. Stokes's theorem relates the integral of the component of the curl of a three-dimensional vector field $\mathbf{F}$ that is normal to a *surface* in $\mathbf{R}^3$ to the line integral of $\mathbf{F}$ over the boundary curves of the surface.

Next, we reformulate Green's theorem in another way. Once again, assume that D is a region in $\mathbf{R}^2$ to which Green's theorem applies and that its boundary curves are oriented appropriately. At each point along the C^1 segments of ∂D, let $\mathbf{n}$ denote the unit vector that is perpendicular to ∂D and points *away from* the region D. (We call $\mathbf{n}$ the **outward unit normal vector** to D. See Figure 6.22.) Then we can demonstrate the following:

Theorem 2.3 (DIVERGENCE THEOREM IN THE PLANE) If D is a region to which Green's theorem applies, $\mathbf{n}$ is the outward unit normal vector to D, and

$$\mathbf{F} = M(x, y)\,\mathbf{i} + N(x, y)\,\mathbf{j}$$

is a C^1 vector field on D, then

$$\oint_{\partial D} \mathbf{F}\cdot\mathbf{n}\,ds = \iint_D \nabla\cdot\mathbf{F}\,dA.$$

Proof If $\mathbf{x}(t) = (x(t), y(t))$, $a \le t \le b$, parametrizes a C^1 segment of ∂D, then along this segment the unit vector $\mathbf{n}$ may be obtained geometrically by rotating the unit tangent vector $\mathbf{x}'(t)/\|\mathbf{x}'(t)\|$ clockwise by 90°. In particular, along such a parametrized C^1 segment,

$$\mathbf{n} = \frac{y'(t)\,\mathbf{i} - x'(t)\,\mathbf{j}}{\sqrt{x'(t)^2 + y'(t)^2}} = \frac{y'(t)\,\mathbf{i} - x'(t)\,\mathbf{j}}{\|\mathbf{x}'(t)\|}.$$

We calculate the line integral $\oint_{\partial D} \mathbf{F}\cdot\mathbf{n}\,ds$. Along each C^1 segment of ∂D, the contribution to the line integral may be evaluated as

$$\begin{aligned}
&\int_a^b (\mathbf{F}(\mathbf{x}(t))\cdot\mathbf{n}(t))\,\|\mathbf{x}'(t)\|\,dt \\
&= \int_a^b (M(x(t), y(t))\,\mathbf{i} + N(x(t), y(t))\,\mathbf{j})\cdot\frac{y'(t)\,\mathbf{i} - x'(t)\,\mathbf{j}}{\|\mathbf{x}'(t)\|}\,\|\mathbf{x}'(t)\|\,dt \\
&= \int_a^b \big(M(x(t), y(t))y'(t) - N(x(t), y(t))x'(t)\big)\,dt \\
&= \int_{\mathbf{x}} -N\,dx + M\,dy.
\end{aligned}$$

Thus, by Green's theorem,

$$\begin{aligned}
\oint_{\partial D} \mathbf{F}\cdot\mathbf{n}\,ds = \oint_{\partial D} -N dx + M dy &= \iint_D \left(\frac{\partial M}{\partial x} - \frac{\partial(-N)}{\partial y}\right) dA \\
&= \iint_D \left(\frac{\partial M}{\partial x} + \frac{\partial N}{\partial y}\right) dA \\
&= \iint_D \nabla\cdot\mathbf{F}\,dA,
\end{aligned}$$

by the definition of the divergence of $\mathbf{F}$. ■

If C is a simple, oriented curve, the line integral $\int_C \mathbf{F}\cdot\mathbf{n}\,ds$, where $\mathbf{n}$ is the unit normal to C as defined in Theorem 2.3, is known as the **flux** of $\mathbf{F}$ across C. For example, if $\mathbf{F}$ represents the velocity vector field of a planar fluid, then the

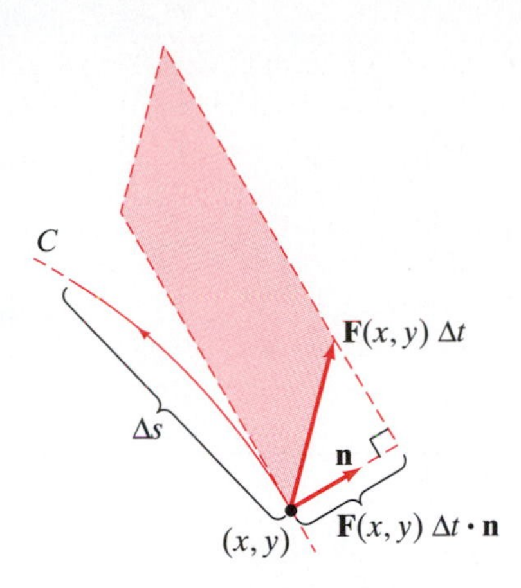

Figure 6.23 The shaded parallelogram has area $\mathbf{F}(x, y)\Delta t \cdot \mathbf{n}\,\Delta s$, the approximate amount of fluid transported across the segment of C.

flux measures the rate of fluid transported across C per unit time. (We assume that $\mathbf{F}$ does not vary with time t.)

To see this, consider the amount of fluid transported across a small segment of C during a brief time interval Δt. As suggested by Figure 6.23, we have

$$\begin{aligned}\text{Amount of fluid transported} &\approx \text{area of parallelogram}\\ &\approx (\mathbf{F}(x, y)\Delta t \cdot \mathbf{n})\Delta s, \end{aligned} \tag{2}$$

where Δs is the length of the segment of the curve C. Formula (2) is only approximate, because the segment of curve need not be completely straight (so the parallelogram geometry is only approximate), and because the vector field $\mathbf{F}$ need not be constant (so the term $\mathbf{F}(x, y)\Delta t$ only approximates a flow line of $\mathbf{F}$ over the time interval). If we divide formula (2) by Δt, then the average *rate* of transport across the segment during the time interval Δt is $(\mathbf{F}(x, y) \cdot \mathbf{n})\,\Delta s$. If we now break up C into finitely many such small segments, sum the contributions of the form $(\mathbf{F}(x, y) \cdot \mathbf{n})\,\Delta s$ for each segment, and let all the lengths Δs tend to zero, we find that the average rate of fluid transport $\Delta M/\Delta t$ is given approximately by

$$\frac{\Delta M}{\Delta t} \approx \int_C \mathbf{F} \cdot \mathbf{n}\, ds.$$

Finally, letting $\Delta t \to 0$, we *define* the (instantaneous) rate of transport dM/dt to be

$$\frac{dM}{dt} = \int_C \mathbf{F} \cdot \mathbf{n}\, ds,$$

which is the flux.

In view of the remarks above, Theorem 2.3 tells us that the flux of $\mathbf{F}$ across the boundary of plane region D (i.e., what of $\mathbf{F}$ flows across ∂D) is equal to the total (or net) divergence of $\mathbf{F}$ over all of D. We revisit the notion of flux in Chapter 7 in the case of a three-dimensional vector field $\mathbf{F}$. In that setting, we are interested in the flux of $\mathbf{F}$ across a *surface* in $\mathbf{R}^3$; defining such a concept requires a surface integral. Then Theorem 2.3 generalizes to three dimensions as Gauss's theorem (also called the **divergence theorem**). Gauss's theorem relates the flux of a three-dimensional vector field $\mathbf{F}$ across a closed surface to the triple integral of the divergence of $\mathbf{F}$ over the solid region enclosed by the surface.

Proof of Green's Theorem

We establish Green's theorem (Theorem 2.1) in three major steps. The first two steps consist of proofs of special cases of Theorem 2.1. The third step is an outline of how the special cases may be used to provide a full proof of the general case. As a result, we fall short of a complete, rigorous proof of the very general version of Green's theorem stated in Theorem 2.1. However, what we do prove makes use of the important geometric ideas of multiple integration and line integrals, and what we do not prove is rather technical.

Step 1. We establish Green's theorem when D is an elementary region in $\mathbf{R}^2$ of type 3. Thus, D can be described in two ways (see Figure 6.24):

$$\begin{aligned} D &= \{(x, y) \in \mathbf{R}^2 \mid \gamma(x) \le y \le \delta(x), a \le x \le b\}\\ &= \{(x, y) \in \mathbf{R}^2 \mid \alpha(y) \le x \le \beta(y), c \le y \le d\}. \end{aligned}$$

The first description of D is as a type 1 elementary region, the second is as a type 2 region. (Recall that a type 3 region is one that is of *both* type 1 and type 2.) We assume that the functions α, β, γ, and δ are all continuous and piecewise C^1.

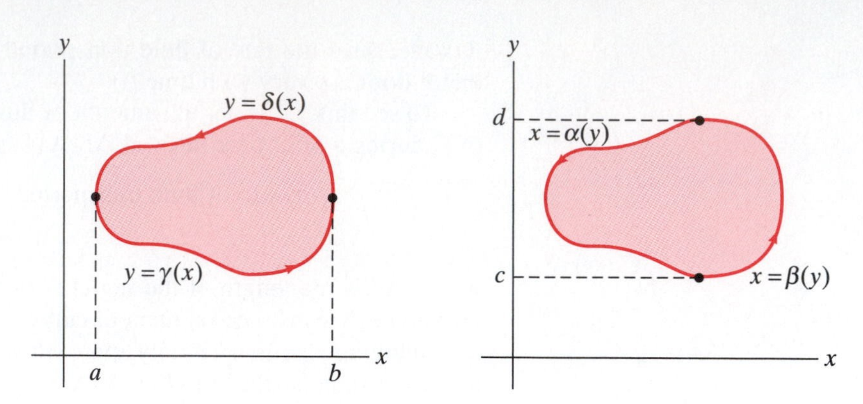

Figure 6.24 A type 3 elementary region D analyzed as both a type 1 and type 2 region. Note the orientations of the boundary curves.

Viewing D as a type 1 elementary region, we evaluate part of $\oint_{\partial D} M\,dx + N\,dy$, namely, $\oint_{\partial D} M\,dx$. Note that ∂D consists of a lower curve C_1 and an upper curve C_2. If we parametrize these curves as follows:

$$C_1 : \begin{cases} x = t \\ y = \gamma(t) \end{cases} \quad a \le t \le b \quad \text{and} \quad C_2 : \begin{cases} x = t \\ y = \delta(t) \end{cases} \quad a \le t \le b,$$

then C_2 is oriented opposite to the desired orientation shown in Figure 6.24. Bearing this in mind, we compute

$$\oint_{\partial D} M(x, y)\,dx = \int_{C_1} M(x, y)\,dx - \int_{C_2} M(x, y)\,dx.$$

(Note the minus sign!) Then

$$\begin{aligned} \oint_{\partial D} M(x, y)\,dx &= \int_a^b M(t, \gamma(t))\,dt - \int_a^b M(t, \delta(t))\,dt \\ &= \int_a^b [M(t, \gamma(t)) - M(t, \delta(t))]\,dt. \end{aligned}$$

Now we compare the calculation of $\oint_{\partial D} M\,dx$ with that of $\iint_D -(\partial M/\partial y)\,dA$. We have

$$\begin{aligned} \iint_D -\frac{\partial M}{\partial y}\,dA &= \int_a^b \int_{\gamma(x)}^{\delta(x)} -\frac{\partial M}{\partial y}\,dy\,dx \\ &= \int_a^b [-M(x, \delta(x)) + M(x, \gamma(x))]\,dx \\ &= \int_a^b [M(x, \gamma(x)) - M(x, \delta(x))]\,dx. \end{aligned}$$

Thus, we see that, in this case,

$$\oint_{\partial D} M\,dx = \iint_D -\frac{\partial M}{\partial y}\,dA.$$

In an analogous manner, we can show

$$\oint_{\partial D} N\,dy = \iint_D \frac{\partial N}{\partial x}\,dA$$

by viewing D as a type 2 elementary region. We omit the details, except to say that both the line integral and the double integral can be shown to be equal to

$$\int_c^d [N(\beta(y), y) - N(\alpha(y), y)]\, dy.$$

Finally, since D is simultaneously of type 1 and type 2,

$$\begin{aligned}\oint_{\partial D} M(x, y)\, dx + N(x, y)\, dy &= \oint_{\partial D} M\, dx + \oint_{\partial D} N\, dy \\ &= \iint_D -\frac{\partial M}{\partial y}\, dA + \iint_D \frac{\partial N}{\partial x}\, dA \\ &= \iint_D \left(\frac{\partial N}{\partial x} - \frac{\partial M}{\partial y}\right) dA.\end{aligned}$$

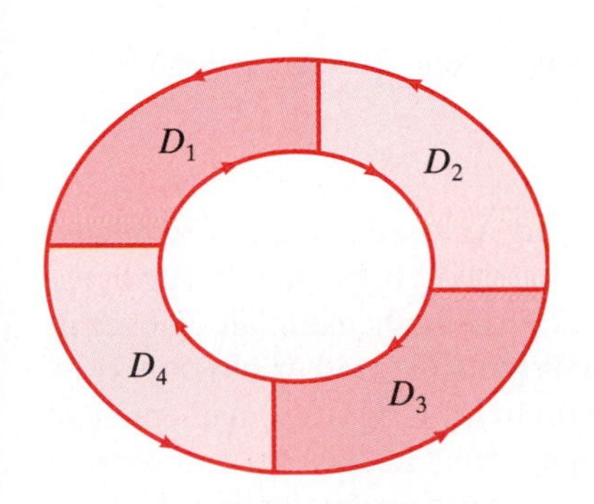

Figure 6.25 The region $D = D_1 \cup D_2 \cup D_3 \cup D_4$, where each subregion D_i, $i = 1, 2, 3, 4$, is elementary of type 3.

Step 2. Now suppose that D is not an elementary region of type 3, but that it can be subdivided into finitely many type 3 regions $D_1, D_2, \ldots, D_n$ in such a way that these subregions overlap at most two at a time and only along common boundaries. Such a region D would look something like the one shown in Figure 6.25. By Step 1, Green's theorem holds for each subregion. Hence, we have

$$\begin{aligned}\iint_D (N_x - M_y)\, dA &= \iint_{D_1} (N_x - M_y)\, dA + \iint_{D_2} (N_x - M_y)\, dA \\ &\quad + \cdots + \iint_{D_n} (N_x - M_y)\, dA \\ &= \oint_{\partial D_1} M\, dx + N\, dy + \oint_{\partial D_2} M\, dx + N\, dy \\ &\quad + \cdots + \oint_{\partial D_n} M\, dx + N\, dy. \qquad (3)\end{aligned}$$

At this point, it is tempting to conclude immediately that the sum of the line integrals in equation (3) is $\oint_{\partial D} M dx + N dy$. However, ∂D_i may contain *more* than only portions of ∂D. The trick is to note that any portion of ∂D_i that is *not* part of ∂D is part of exactly one other ∂D_j. Moreover, this overlapping portion is given one orientation by ∂D_i and the *opposite* orientation by ∂D_j. When we take the sum of the line integrals in equation (3), any contributions arising from the components of the ∂D_i's that are in the interior of D will cancel in pairs. (See Figure 6.26.) Therefore,

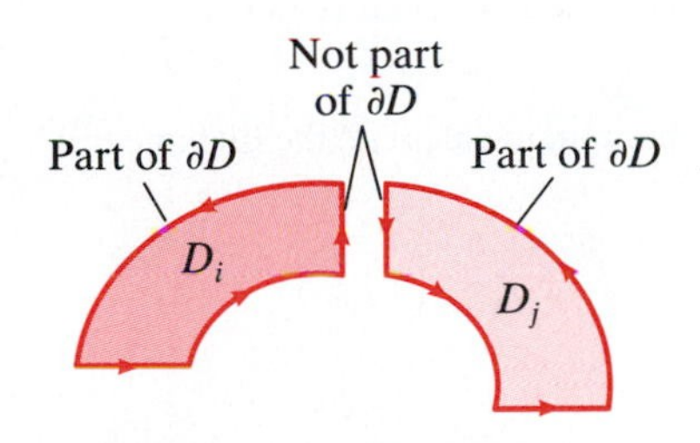

Figure 6.26 The common boundary of subregions D_i and D_j is oriented one way as part of ∂D_i and the opposite way as part of ∂D_j. Hence, $\oint_{\partial D_i} M dx + N dy + \oint_{\partial D_j} M dx + N dy$ will cancel over this common boundary.

$$\begin{aligned}\iint_D (N_x - M_y) dA &= \oint_{\partial D_1} M dx + N dy + \oint_{\partial D_2} M dx + N dy \\ &\quad + \cdots + \oint_{\partial D_n} M dx + N dy \\ &= \oint_{\partial D} M dx + N dy;\end{aligned}$$

Green's theorem is established in this case.

Step 3. Unfortunately, not all regions described in the statement of Theorem 2.1 can be subdivided into finitely many elementary regions of type 3. Here is an outline of what we might do to prove Green's theorem in such generality.

First, we claim (without proof) that for regions D described in the statement of Theorem 2.1 we can produce a sequence of regions $D_1, D_2, \ldots, D_n, \ldots$ whose "limit" as $n \to \infty$ is D and such that each D_n can be subdivided into finitely many type 3 elementary regions. Next, we claim that $\partial D_n \to \partial D$ as $n \to \infty$. Finally, we need to prove that, as $n \to \infty$,

$$\iint_{D_n} (N_x - M_y)\, dA \quad \to \quad \iint_D (N_x - M_y)\, dA$$

and

$$\oint_{\partial D_n} M\, dx + N\, dy \quad \to \quad \oint_{\partial D} M\, dx + N\, dy.$$

Since Green's theorem holds for each D_n (by Steps 1 and 2), we are done.[2] ■

Historical Note[3]

The idea that the line integral of a vector field along a closed curve can be related to a double integral over the region bounded by the curve is frequently attributed to George Green (1793–1841), a self-educated English mathematician. The result we have been calling Green's theorem had its origins in a rather obscure 1828 pamphlet published by Green, in which he sought to lay a rigorous mathematical foundation for the physics of electricity and magnetism. Green's ideas arose from work in partial differential equations concerning gravitational potentials. Green's pamphlet subsequently came to the attention of Lord Kelvin (1824–1907), who had it republished so that, fortunately, Green's results received greater recognition.

Coincidentally, a result similar to Green's theorem was established independently (and also in 1828!) by the Russian mathematician Mikhail Ostrogradsky (1801–1861). Ostrogradsky's name is sometimes associated to what we call Green's theorem.

Exercises

In Exercises 1–4, verify Green's theorem for the given vector field

$$\mathbf{F} = M(x, y)\,\mathbf{i} + N(x, y)\,\mathbf{j}$$

and region D by calculating both

$$\oint_{\partial D} M\, dx + N\, dy \quad \textit{and} \quad \iint_D (N_x - M_y)\, dA.$$

1. $\mathbf{F} = -x^2y\,\mathbf{i} + xy^2\,\mathbf{j}$, D is the disk $x^2 + y^2 \leq 4$.

2. $\mathbf{F} = (x^2 - y)\,\mathbf{i} + (x + y^2)\,\mathbf{j}$, D is the rectangle bounded by $x = 0$, $x = 2$, $y = 0$, and $y = 1$.

3. $\mathbf{F} = y\,\mathbf{i} + x^2\,\mathbf{j}$, D is the square with vertices $(1, 1)$, $(-1, 1)$, $(-1, -1)$, and $(1, -1)$.

4. $\mathbf{F} = 2y\,\mathbf{i} + x\,\mathbf{j}$, D is the semicircular region $x^2 + y^2 \leq a^2$, $y \geq 0$.

5. (a) Use Green's theorem to calculate the line integral

$$\oint_C y^2 dx + x^2 dy,$$

where C is the path formed by the square with vertices $(0, 0)$, $(1, 0)$, $(0, 1)$, and $(1, 1)$, oriented counterclockwise.

(b) Verify your answer for part (a) by calculating the line integral directly.

---

[2] For details of the type of limit argument we have in mind, see O. D. Kellogg, *Foundations of Potential Theory* (Springer, Berlin, 1929; reprinted by Dover Publications, New York, 1954), pp. 113–119, where a limit argument is given in the case of Gauss's theorem, which we explore in §7.3. For a proof of Green's theorem that avoids the limit argument, see D. V. Widder, *Advanced Calculus*, 2nd ed., (Prentice-Hall, Englewood Cliffs, 1961; reprinted by Dover Publications, New York, 1989), pp. 223–225.

[3] See also M. Kline, *Mathematical Thought from Ancient to Modern Times* (Oxford Press, New York, 1972), p. 683.

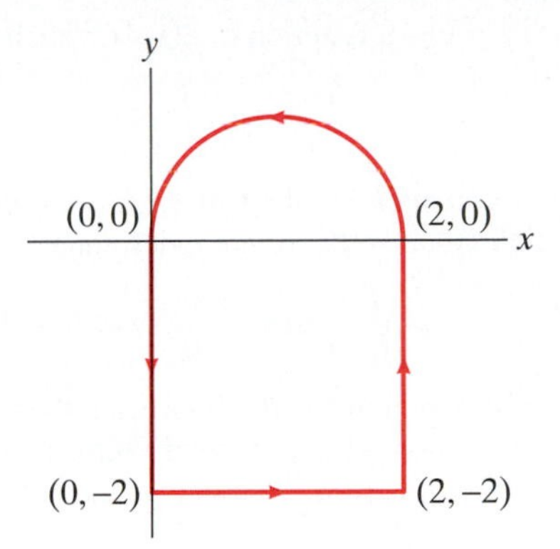

Figure 6.27 The oriented curve C of Exercise 6 consists of three sides of a square plus a semicircular arc.

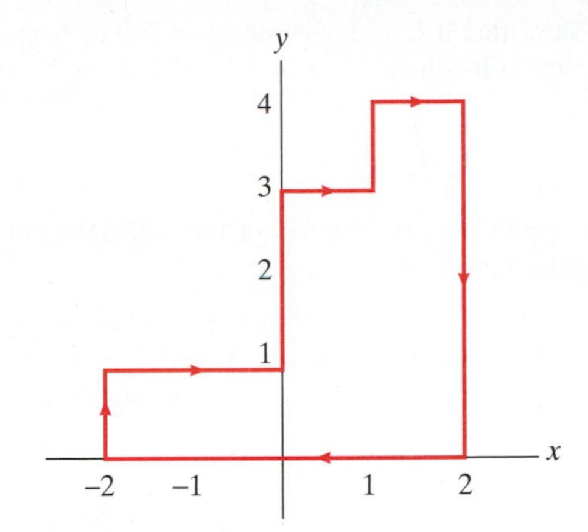

Figure 6.28 The oriented curve C of Exercise 11.

6. Let $\mathbf{F} = 3xy\,\mathbf{i} + 2x^2\,\mathbf{j}$ and suppose C is the oriented curve shown in Figure 6.27. Evaluate

$$\oint_C \mathbf{F} \cdot d\mathbf{s}$$

both directly and also by means of Green's theorem.

7. Evaluate

$$\oint_C (x^2 - y^2)\,dx + (x^2 + y^2)\,dy,$$

where C is the boundary of the square with vertices $(0, 0)$, $(1, 0)$, $(0, 1)$, and $(1, 1)$, oriented *clockwise*. Use whatever method of evaluation seems appropriate.

8. Use Green's theorem to find the work done by the vector field

$$\mathbf{F} = (4y - 3x)\,\mathbf{i} + (x - 4y)\,\mathbf{j}$$

on a particle as the particle moves counterclockwise once around the ellipse $x^2 + 4y^2 = 4$.

9. Verify that the area of the rectangle $R = [0, a] \times [0, b]$ is ab, by calculating an appropriate line integral.

10. Let a be a positive constant. Use Green's theorem to calculate the area under one arch of the cycloid

$$x = a(t - \sin t), \qquad y = a(1 - \cos t).$$

11. Evaluate $\oint_C (x^4y^5 - 2y)\,dx + (3x + x^5y^4)\,dy$, where C is the oriented curve pictured in Figure 6.28.

12. Use Green's theorem to find the area enclosed by the hypocycloid

$$\mathbf{x}(t) = (a\cos^3 t, a\sin^3 t), \quad 0 \le t \le 2\pi.$$

13. (a) Sketch the curve given parametrically by $\mathbf{x}(t) = (1 - t^2, t^3 - t)$.

(b) Find the area inside the closed loop of the curve.

14. Use Green's theorem to find the area between the ellipse $x^2/9 + y^2/4 = 1$ and the circle $x^2 + y^2 = 25$.

15. Show that if D is a region to which Green's theorem applies, and ∂D is oriented so that D is always on the left as we travel along ∂D, then the area of D is given by either of the following two line integrals:

$$\text{Area of } D = \oint_{\partial D} x\,dy = -\oint_{\partial D} y\,dx.$$

16. Find the area inside the quadrilateral with vertices $(2, 0)$, $(1, 2)$, $(1, 1)$, and $(-1, 1)$.

17. (a) Use the divergence theorem (Theorem 2.3) to show that $\oint_C \mathbf{F} \cdot \mathbf{n}\,ds = 0$, where $\mathbf{F} = 2y\,\mathbf{i} - 3x\,\mathbf{j}$ and C is the circle $x^2 + y^2 = 1$.

(b) Now show $\oint_C \mathbf{F} \cdot \mathbf{n}\,ds = 0$ by direct computation of the line integral.

18. Let $\mathbf{F} = M(x, y)\,\mathbf{i} + N(x, y)\,\mathbf{j}$. The divergence theorem shows that the flux of $\mathbf{F}$ across a closed curve C (i.e., $\oint_C \mathbf{F} \cdot \mathbf{n}\,ds$) is equal to $\iint_D (M_x + N_y)\,dA$, where D is the region bounded by C. Use Green's theorem to establish a similar result involving $\oint_C \mathbf{F} \cdot \mathbf{T}\,ds$, the circulation of $\mathbf{F}$ along C. (See also §6.1.)

19. Let C be any simple, closed curve in the plane. Show that

$$\oint_C 3x^2y\,dx + x^3\,dy = 0.$$

20. Show that

$$\oint_C -y^3\,dx + (x^3 + 2x + y)\,dy$$

is positive for any closed curve C to which Green's theorem applies.

21. Show that if C is the boundary of any rectangular region in $\mathbf{R}^2$, then

$$\oint_C (x^2y^3 - 3y)\,dx + x^3y^2\,dy$$

depends only on the area of the rectangle, not on its placement in $\mathbf{R}^2$.

22. Let $\mathbf{r} = x\,\mathbf{i} + y\,\mathbf{j}$ be the position vector of any point in the plane. Show that the flux of $\mathbf{F} = \mathbf{r}$ across any simple closed curve C in $\mathbf{R}^2$ is twice the area inside C.

23. Let D be a region to which Green's theorem applies and suppose that $u(x, y)$ and $v(x, y)$ are two functions of class C^2 whose domains include D. Show that

$$\iint_D \frac{\partial(u, v)}{\partial(x, y)}\,dA = \oint_C (u\nabla v) \cdot d\mathbf{s},$$

where $C = \partial D$ is oriented as in Green's theorem.

24. Let $f(x, y)$ be a function of class C^2 such that

$$\frac{\partial^2 f}{\partial x^2} + \frac{\partial^2 f}{\partial y^2} = 0$$

(i.e., f is *harmonic*). Show that if C is any closed curve to which Green's theorem applies, then

$$\oint_C \frac{\partial f}{\partial y}dx - \frac{\partial f}{\partial x}dy = 0.$$

25. Let D be a region to which Green's theorem applies and $\mathbf{n}$ the outward unit normal vector to D. Suppose $f(x, y)$ is a function of class C^2. Show that

$$\iint_D \nabla^2 f\,dA = \oint_{\partial D} \frac{\partial f}{\partial n}\,ds,$$

where $\nabla^2 f$ denotes the Laplacian of f (namely, $\nabla^2 f = \partial^2 f/\partial x^2 + \partial^2 f/\partial y^2$) and $\partial f/\partial n$ denotes $\nabla f \cdot \mathbf{n}$. (See the proof of Theorem 2.3 for more information about $\mathbf{n}$.)

6.3 Conservative Vector Fields

As seen in §6.1, line integrals of a given vector field depend only on the underlying curve and its orientation, not on the particular parametrization of the curve. In some special instances, however, even the curve itself doesn't matter, only the initial and terminal points. A vector field having the property that line integrals of it depend only on the initial and terminal points of the oriented curve over which the line integral is taken is said to have **path-independent line integrals**. We next state a more careful definition, characterize such vector fields, and explore their significance.

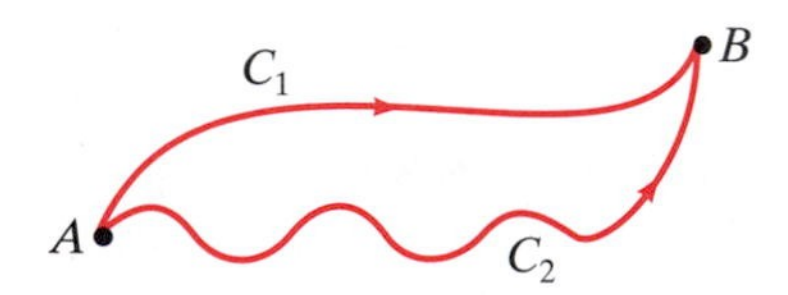

Figure 6.29 If $\mathbf{F}$ has path-independent line integrals, then $\int_{C_1} \mathbf{F} \cdot d\mathbf{s} = \int_{C_2} \mathbf{F} \cdot d\mathbf{s}$ for any two piecewise C^1 oriented curves from A to B.

Path Independence

■ **Definition 3.1** A continuous vector field $\mathbf{F}$ has **path-independent line integrals** if

$$\int_{C_1} \mathbf{F} \cdot d\mathbf{s} = \int_{C_2} \mathbf{F} \cdot d\mathbf{s}$$

for any two simple, piecewise C^1, oriented curves lying in the domain of $\mathbf{F}$ and having the same initial and terminal points. (See Figure 6.29.)

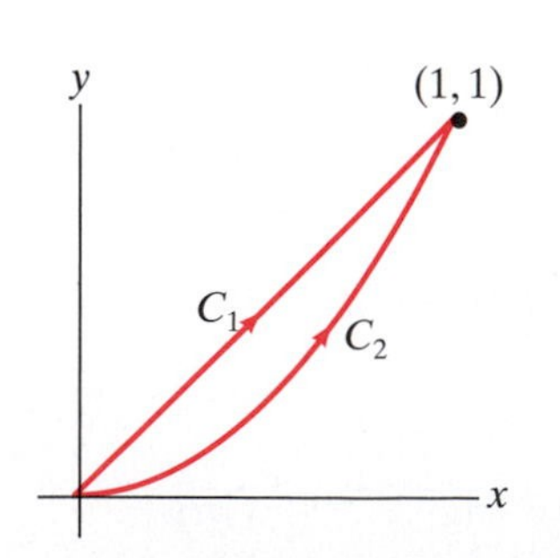

Figure 6.30 The curves C_1 and C_2 of Example 1.

EXAMPLE 1 Let $\mathbf{F} = y\,\mathbf{i} - x\,\mathbf{j}$ and consider the following two curves in $\mathbf{R}^2$ from the origin to $(1, 1)$: C_1, the line segment from $(0, 0)$ to $(1, 1)$, and C_2, the portion of the parabola $y = x^2$ (the curves are shown in Figure 6.30). These curves may be parametrized as

$$C_1 : \begin{cases} x = t \\ y = t \end{cases} \quad 0 \le t \le 1 \quad \text{and} \quad C_2 : \begin{cases} x = t \\ y = t^2 \end{cases} \quad 0 \le t \le 1.$$

Then we calculate

$$\int_{C_1} \mathbf{F} \cdot d\mathbf{s} = \int_0^1 (t\,\mathbf{i} - t\,\mathbf{j}) \cdot (\mathbf{i} + \mathbf{j})\,dt = \int_0^1 0\,dt = 0,$$

while

$$\int_{C_2} \mathbf{F} \cdot d\mathbf{s} = \int_0^1 (t^2\,\mathbf{i} - t\,\mathbf{j}) \cdot (\mathbf{i} + 2t\,\mathbf{j})\,dt$$

$$= \int_0^1 (t^2 - 2t^2)\,dt = -\tfrac{1}{3}t^3\Big|_0^1 = -\tfrac{1}{3}.$$

We see that

$$\int_{C_1} \mathbf{F} \cdot d\mathbf{s} \neq \int_{C_2} \mathbf{F} \cdot d\mathbf{s},$$

and so **F** does not have path-independent line integrals. ◆

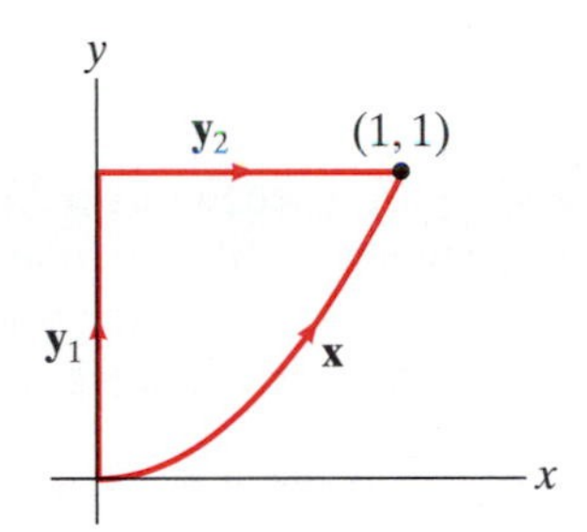

Figure 6.31 The paths **x** and **y** (consisting of the paths $\mathbf{y}_1$ and $\mathbf{y}_2$) of Example 2.

EXAMPLE 2 Let $\mathbf{F} = x\,\mathbf{i} + y\,\mathbf{j}$. This vector field has path-independent line integrals—we will see why presently. For the moment, however, we illustrate (not prove) this fact by considering the parabolic path $\mathbf{x}\colon [0, 1] \to \mathbf{R}^2$, $\mathbf{x}(t) = (t, t^2)$, as well as the path $\mathbf{y}\colon [0, 2] \to \mathbf{R}^2$ made up of the two straight segments

$$\mathbf{y}_1\colon [0, 1] \to \mathbf{R}^2,\ \mathbf{y}_1(t) = (0, t) \quad \text{and} \quad \mathbf{y}_2\colon [1, 2] \to \mathbf{R}^2,\ \mathbf{y}_2(t) = (t - 1, 1).$$

Both **x** and **y** are paths from (0, 0) to (1, 1) and are shown in Figure 6.31. We have

$$\int_{\mathbf{x}} \mathbf{F} \cdot d\mathbf{s} = \int_0^1 (t\,\mathbf{i} + t^2\,\mathbf{j}) \cdot (\mathbf{i} + 2t\,\mathbf{j})\,dt$$

$$= \int_0^1 (t + 2t^3)\,dt = \left(\tfrac{1}{2}t^2 + \tfrac{1}{2}t^4\right)\Big|_0^1 = 1,$$

and

$$\int_{\mathbf{y}} \mathbf{F} \cdot d\mathbf{s} = \int_{\mathbf{y}_1} \mathbf{F} \cdot d\mathbf{s} + \int_{\mathbf{y}_2} \mathbf{F} \cdot d\mathbf{s}$$

$$= \int_0^1 t\,\mathbf{j} \cdot \mathbf{j}\,dt + \int_1^2 ((t - 1)\,\mathbf{i} + \mathbf{j}) \cdot \mathbf{i}\,dt$$

$$= \int_0^1 t\,dt + \int_1^2 (t - 1)\,dt$$

$$= \tfrac{1}{2}t^2\Big|_0^1 + \tfrac{1}{2}(t - 1)^2\Big|_1^2 = \tfrac{1}{2} + \tfrac{1}{2} = 1.$$

To establish that **F** has path-independent line integrals, we would need to check that the value of the line integral of **F** along *any* choice of path between *any* two points is the same as any other—a prohibitive task. ◆

The following result is a reformulation of the path-independence property:

■ **Theorem 3.2** Let **F** be a continuous vector field. Then **F** has path-independent line integrals if and only if $\oint_C \mathbf{F} \cdot d\mathbf{s} = 0$ for all piecewise C^1, simple, closed curves C in the domain of **F**.

Proof First, assume that **F** satisfies the path-independence property. Suppose C is parametrized by $\mathbf{x}(t)$, $a \le t \le b$, where $\mathbf{x}(a) = \mathbf{x}(b) = A$. Let B be another point on C, and break C into two oriented curves C_1 and C_2 from A to B. One of these curves—say, C_1—will be oriented the same way as C and the other the

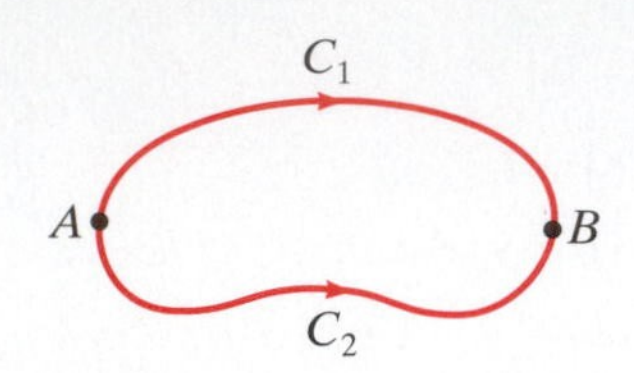

Figure 6.32 The simple, closed curve C consists of two oriented curves from A to B.

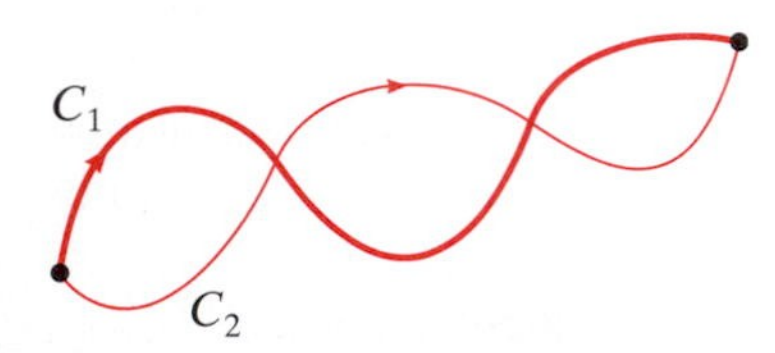

Figure 6.33 The closed curve C constructed from C_1 and the reverse of C_2 need not be simple.

opposite way. (See Figure 6.32.) Thus,

$$\oint_C \mathbf{F}\cdot d\mathbf{s} = \int_{C_1} \mathbf{F}\cdot d\mathbf{s} - \int_{C_2} \mathbf{F}\cdot d\mathbf{s} = 0,$$

since $\mathbf{F}$ has path-independent line integrals.

Conversely, suppose that all line integrals of $\mathbf{F}$ around simple closed curves are zero. Then given two piecewise C^1, oriented, simple curves C_1 and C_2 with the same initial and terminal points, let C be the *closed* curve consisting of C_1 and $-C_2$ (i.e., C_2 with its direction reversed). Then, we have

$$\oint_C \mathbf{F}\cdot d\mathbf{s} = \int_{C_1} \mathbf{F}\cdot d\mathbf{s} + \int_{-C_2} \mathbf{F}\cdot d\mathbf{s} = \int_{C_1} \mathbf{F}\cdot d\mathbf{s} - \int_{C_2} \mathbf{F}\cdot d\mathbf{s}.$$

If C happens to be simple, then $\oint_C \mathbf{F}\cdot d\mathbf{s} = 0$ by assumption, so

$$\int_{C_1} \mathbf{F}\cdot d\mathbf{s} = \int_{C_2} \mathbf{F}\cdot d\mathbf{s},$$

as desired. However, C need not be simple even if C_1 and C_2 are. (See Figure 6.33.) If it is possible to break C_1 and C_2 into finitely many segments so that a segment C_1' of C_1 and a segment C_2' of C_2 either (i) completely coincide or (ii) together form a simple closed curve C', then it is not too difficult to modify the preceding argument to conclude that

$$\int_{C_1} \mathbf{F}\cdot d\mathbf{s} = \int_{C_2} \mathbf{F}\cdot d\mathbf{s}.$$

However, it is not always possible to do this, and further technical arguments (which we omit) are required. ■

Gradient Fields and Line Integrals

We describe next a class of vector fields that satisfy the path-independence property, namely, gradient fields. Suppose that $\mathbf{F}$ is a continuous vector field such that $\mathbf{F} = \nabla f$, where f is some scalar-valued function of class C^1. (Recall that we refer to f as a scalar **potential** of $\mathbf{F}$. We also call $\mathbf{F}$ a **conservative** vector field as well as a gradient field). Then, along any path $\mathbf{x}$ from $A = \mathbf{x}(a)$ to $B = \mathbf{x}(b)$ whose image lies in the domain of $\mathbf{F}$, we have

$$\int_{\mathbf{x}} \mathbf{F}\cdot d\mathbf{s} = \int_{\mathbf{x}} \nabla f \cdot d\mathbf{s} = \int_a^b \nabla f(\mathbf{x}(t))\cdot \mathbf{x}'(t)\,dt.$$

It follows from the chain rule that $d/dt[f(\mathbf{x}(t))] = \nabla f(\mathbf{x}(t))\cdot\mathbf{x}'(t)$. Hence,

$$\begin{aligned}\int_{\mathbf{x}} \nabla f\cdot d\mathbf{s} &= \int_a^b \nabla f(\mathbf{x}(t))\cdot\mathbf{x}'(t)\,dt = \int_a^b \frac{d}{dt}[f(\mathbf{x}(t))]\,dt \\ &= f(\mathbf{x}(t))\big|_a^b = f(\mathbf{x}(b)) - f(\mathbf{x}(a)) = f(B) - f(A).\end{aligned}$$

Therefore, when $\mathbf{F}$ is a gradient field, the line integral of $\mathbf{F}$ depends only on the value of the potential function at the endpoints of the path. Hence, gradient fields have path-independent line integrals. The converse holds as well, as we prove at the end of this section. Stated formally, we have the following theorem:

■ **Theorem 3.3** Let $\mathbf{F}$ be defined and continuous on a connected, open region R of $\mathbf{R}^n$. Then $\mathbf{F} = \nabla f$ (where f is a function of class C^1 on R) if and only if $\mathbf{F}$ has path-independent line integrals over curves in R. Moreover, if C is any piecewise C^1, oriented curve lying in R with initial point A and terminal point B, then

$$\int_C \mathbf{F} \cdot d\mathbf{s} = f(B) - f(A).$$

Note: A region $R \subseteq \mathbf{R}^n$ is **connected** if any two points in R can be joined by a path whose image lies in R.

EXAMPLE 3 Consider the vector field $\mathbf{F} = x\,\mathbf{i} + y\,\mathbf{j}$ of Example 2 again. You can readily check that $\mathbf{F} = \nabla f$, where $f(x, y) = \frac{1}{2}(x^2 + y^2)$. By Theorem 3.3, line integrals of $\mathbf{F}$ will be path independent; this fact was illustrated, but not proved, in Example 2 when we calculated the vector line integral of $\mathbf{F}$ along two paths from $(0, 0)$ to $(1, 1)$. By Theorem 3.3, we see now that for *any* directed piecewise C^1 curve C from $(0, 0)$ to $(1, 1)$, we have

$$\int_C \mathbf{F} \cdot d\mathbf{s} = f(1, 1) - f(0, 0) = \tfrac{1}{2}(1^2 + 1^2) - \tfrac{1}{2}(0^2 + 0^2) = 1,$$

which agrees with our earlier computations. ◆

A Criterion for Conservative Vector Fields

Theorem 3.3 tells us that a vector field $\mathbf{F}$ has path-independent line integrals precisely when it is a conservative (gradient) vector field, and, moreover, that the line integral of $\mathbf{F}$ along any path is determined by the values of the potential function f at the endpoints of the path. Two questions arise naturally:

1. How can we determine whether a given vector field $\mathbf{F}$ is conservative?
2. Assuming that $\mathbf{F}$ is conservative, is there a procedure for finding a scalar potential function f such that $\mathbf{F} = \nabla f$?

We answer the first question by providing a simple and effective test that can be performed on $\mathbf{F}$. Should $\mathbf{F}$ pass this test (i.e., if $\mathbf{F}$ is conservative), then we illustrate via examples how to produce a scalar potential for $\mathbf{F}$, thereby answering the second question.

First, we need additional terminology.

■ **Definition 3.4** A region R in $\mathbf{R}^2$ or $\mathbf{R}^3$ is **simply-connected** if it consists of a single connected piece and if every simple closed curve C in R can be continuously shrunk to a point while remaining in R throughout the deformation.

If R is a region in the plane, then R is simply-connected just in case it is connected and every simple closed curve C lying in R has the property that all the points enclosed by C also lie in R. Loosely speaking, a simply-connected region (in either $\mathbf{R}^2$ or $\mathbf{R}^3$) can have no "essential holes." Illustrative examples are shown in Figures 6.34 and 6.35. The notion of continuously shrinking a curve to a point can be made fully precise, although we shall not take the trouble to do so here.

Now we state our criterion for a vector field to be conservative.

■ **Theorem 3.5** Let $\mathbf{F}$ be a vector field of class C^1 whose domain is a simply-connected region R in either $\mathbf{R}^2$ or $\mathbf{R}^3$. Then $\mathbf{F} = \nabla f$ for some scalar-valued function f of class C^2 on R if and only if $\nabla \times \mathbf{F} = \mathbf{0}$ at all points of R.

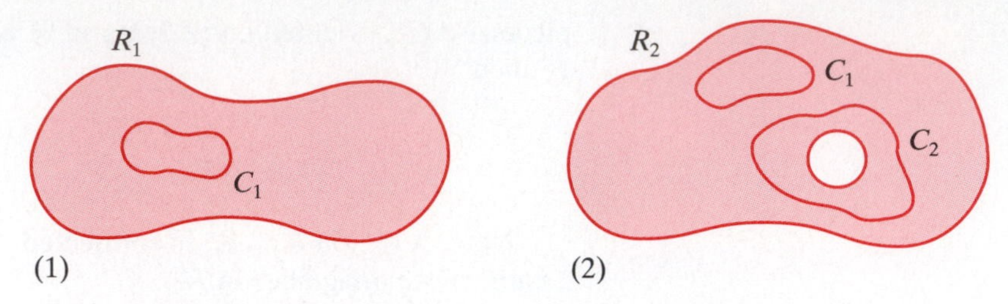

Figure 6.34 (1) The region $R_1 \subset \mathbf{R}^2$ is simply-connected: All points surrounded by any simple closed curve in R_1 lie in R_1. (2) In contrast, R_2 is not simply-connected: Although the curve C_1 encloses points that lie in R_2, the curve C_2 surrounds a hole. Hence, C_2 cannot be continuously shrunk to a point while remaining in R_2.

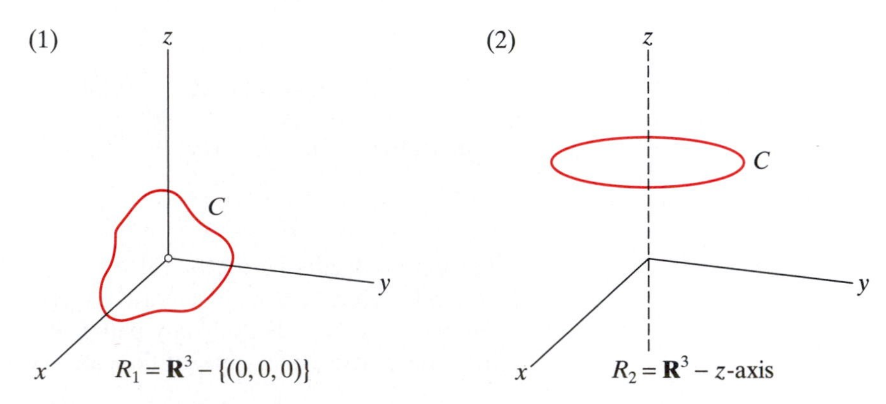

Figure 6.35 (1) The region $R_1 \subset \mathbf{R}^3$ is simply-connected. (2) The region R_2 is *not* simply-connected: The curve C cannot be shrunk continuously to a point without becoming "stuck" on the "missing" z-axis.

Before proving Theorem 3.5, some remarks and examples are appropriate. First, note that Theorem 3.5 provides a straightforward way to determine if a vector field $\mathbf{F}$ is conservative: Check that the domain of $\mathbf{F}$ is simply-connected, and then test if $\nabla \times \mathbf{F} = \mathbf{0}$. If the curl vanishes, it follows that $\mathbf{F}$ has path-independent line integrals. This "curl criterion" can be helpful in practice.

In the case where $\mathbf{F} = M(x, y)\,\mathbf{i} + N(x, y)\,\mathbf{j}$ is a two-dimensional vector field, the condition that the curl of $\mathbf{F}$ vanishes means

$$\nabla \times \mathbf{F} = \begin{vmatrix} \mathbf{i} & \mathbf{j} & \mathbf{k} \\ \partial/\partial x & \partial/\partial y & \partial/\partial z \\ M(x, y) & N(x, y) & 0 \end{vmatrix} = \left(\frac{\partial N}{\partial x} - \frac{\partial M}{\partial y}\right)\mathbf{k} = \mathbf{0}.$$

This is equivalent to the condition

$$\frac{\partial N}{\partial x} = \frac{\partial M}{\partial y}. \tag{1}$$

Equation (1) is a simpler condition to use in this situation.

EXAMPLE 4 Let $\mathbf{F} = x^2y\,\mathbf{i} - 2xy\,\mathbf{j}$. Then

$$\frac{\partial}{\partial x}(-2xy) = -2y \quad \text{and} \quad \frac{\partial}{\partial y}(x^2y) = x^2.$$

Since these partial derivatives are not equal, we conclude that $\mathbf{F}$ is not conservative, by Theorem 3.5. ◆

EXAMPLE 5 Let $\mathbf{F} = (2xy + \cos 2y)\mathbf{i} + (x^2 - 2x\sin 2y)\mathbf{j}$. The vector field $\mathbf{F}$ is defined and of class C^1 on all of $\mathbf{R}^2$ (a simply-connected region). Moreover,

$$\frac{\partial}{\partial x}(x^2 - 2x\sin 2y) = 2x - 2\sin 2y \quad \text{and} \quad \frac{\partial}{\partial y}(2xy + \cos 2y) = 2x - 2\sin 2y.$$

We may conclude that $\mathbf{F}$ is conservative. In addition, if C is the ellipse $x^2/4 + y^2 = 1$ (a simple, closed curve), then, by Theorems 3.2 and 3.3, we conclude, *without any explicit calculation,* that $\oint_C \mathbf{F}\cdot d\mathbf{s} = 0$. ◆

EXAMPLE 6 Let

$$\mathbf{F} = \left(\frac{x}{x^2+y^2+z^2} - 6x\right)\mathbf{i} + \frac{y}{x^2+y^2+z^2}\mathbf{j} + \frac{z}{x^2+y^2+z^2}\mathbf{k}.$$

$\mathbf{F}$ is of class C^1 on all of $\mathbf{R}^3$ except for the origin. Note that $\mathbf{R}^3 - \{(0,0,0)\}$ is simply-connected. We leave it to you to check that $\nabla\times\mathbf{F} = \mathbf{0}$ for all (x, y, z) in the domain of $\mathbf{F}$. Therefore, by Theorem 3.5, $\mathbf{F}$ is conservative.

Now suppose $\mathbf{x}\colon [0,1] \to \mathbf{R}^3$ is the path given by $\mathbf{x}(t) = (1-t, \sin\pi t, t)$. To evaluate $\int_{\mathbf{x}} \mathbf{F}\cdot d\mathbf{s}$ directly, we must calculate

$$\begin{aligned}\int_{\mathbf{x}} \mathbf{F}\cdot d\mathbf{s} &= \int_0^1 \left(\frac{1-t}{(1-t)^2+\sin^2\pi t+t^2} - 6(1-t),\ \frac{\sin\pi t}{(1-t)^2+\sin^2\pi t+t^2},\right.\\ &\qquad \left.\frac{t}{(1-t)^2+\sin^2\pi t+t^2}\right)\cdot(-1, \pi\cos\pi t, 1)\,dt\\ &= \int_0^1 \left(\frac{2t-1+\pi\sin\pi t\cos\pi t}{(1-t)^2+\sin^2\pi t+t^2} + 6(1-t)\right)dt.\end{aligned}$$

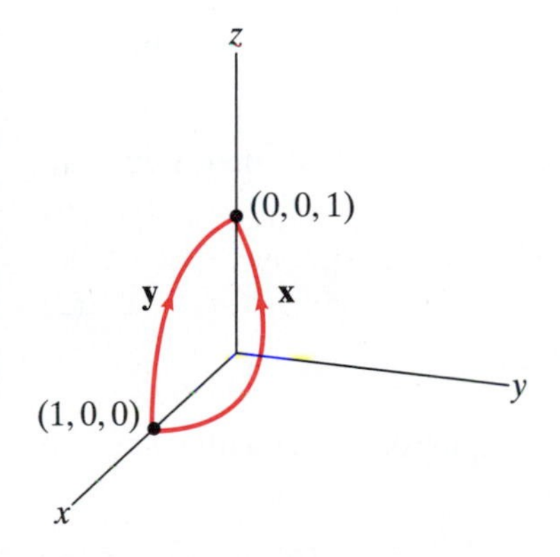

Figure 6.36 The paths $\mathbf{x}(t) = (1-t, \sin\pi t, t), 0 \le t \le 1$ and $\mathbf{y}(t) = (\cos t, 0, \sin t)$, $0 \le t \le \pi/2$.

This last integral is tricky to evaluate. However, since $\mathbf{F}$ is conservative, we may evaluate $\mathbf{F}$ by calculating $\int_{\mathbf{y}} \mathbf{F}\cdot d\mathbf{s}$, where $\mathbf{y}$ is any other path with the same endpoints as $\mathbf{x}$. A good choice is $\mathbf{y}(t) = (\cos t, 0, \sin t)$, $0 \le t \le \pi/2$, because the image of this path lies on the sphere $x^2+y^2+z^2 = 1$, a fact that will enable us to work with a simple integral. (See Figure 6.36 for a graph of the two paths $\mathbf{x}$ and $\mathbf{y}$.) Since $\mathbf{F}$ is conservative (and hence has path-independent line integrals),

$$\begin{aligned}\int_{\mathbf{x}} \mathbf{F}\cdot d\mathbf{s} = \int_{\mathbf{y}} \mathbf{F}\cdot d\mathbf{s} &= \int_0^{\pi/2} \left(\frac{\cos t}{1} - 6\cos t, \frac{0}{1}, \frac{\sin t}{1}\right)\cdot(-\sin t, 0, \cos t)\,dt\\ &= \int_0^{\pi/2} 6\cos t\sin t\,dt = \int_0^{\pi/2} 3\sin 2t\,dt\\ &= -\tfrac{3}{2}\cos 2t\Big|_0^{\pi/2} = -\tfrac{3}{2}(-1-1) = 3.\end{aligned}$$ ◆

Sketch of a proof of Theorem 3.5 By Theorem 4.3 of Chapter 3, note that, if $\mathbf{F} = \nabla f$ for some function f of class C^2, then $\nabla\times\mathbf{F} = \nabla\times(\nabla f) = \mathbf{0}$.

Conversely, suppose that $\nabla\times\mathbf{F} = \mathbf{0}$. We show that if C is any piecewise C^1, simple, closed curve in R, then $\oint_C \mathbf{F}\cdot d\mathbf{s} = 0$. By Theorem 3.2, this implies that $\mathbf{F}$ has path-independent line integrals, which, by Theorem 3.3, is equivalent to $\mathbf{F}$'s being the gradient of some scalar-valued function f. Moreover, since $\mathbf{F}$ is assumed to be of class C^1, it follows that f must be of class C^2.

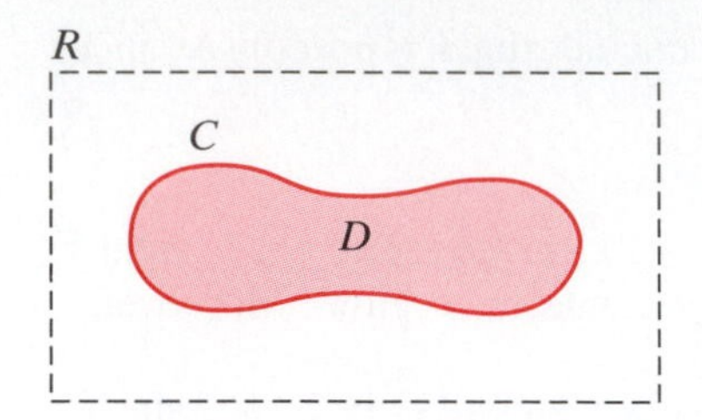

Figure 6.37 Since R is simply-connected, any region D enclosed by C must lie in R.

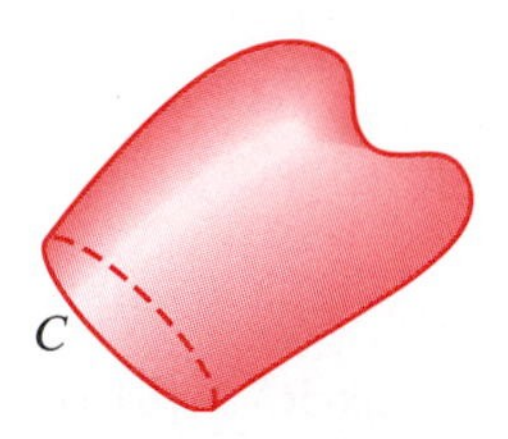

Figure 6.38 A surface in space, bounded by the simple, closed curve C.

To see that $\oint_C \mathbf{F}\cdot d\mathbf{s} = 0$, suppose, first, that $\mathbf{F}$ is defined on a simply-connected region R in $\mathbf{R}^2$. Since R is simply-connected, the closed curve C bounds a region D that is entirely contained in R. (See Figure 6.37.) By Proposition 2.2, which is equivalent to Green's theorem, we have

$$\oint_C \mathbf{F}\cdot d\mathbf{s} = \pm \iint_D (\nabla \times \mathbf{F}\cdot \mathbf{k})\, dA = \pm \iint_D 0\, dA = 0.$$

We use the "$\pm$" sign in the event that C is oriented opposite to the orientation stipulated by Green's theorem. If $\mathbf{F}$ is defined on a simply-connected region R in $\mathbf{R}^3$, then we must apply Stokes's theorem, rather than Green's theorem. This gets us a little ahead of ourselves, although the principle remains the same: If C is a simple, closed curve in $R \subset \mathbf{R}^3$, then $\oint_C \mathbf{F}\cdot d\mathbf{s}$ is equal to a suitable *surface integral* of $\nabla \times \mathbf{F}$ over a surface in R bounded by C. (See Figure 6.38.) Because the curl is assumed to be zero, any integral of it will be zero, and so $\oint_C \mathbf{F}\cdot d\mathbf{s}$ is zero as well. ■

Finding Scalar Potentials

Now that we have a practical test to determine whether a given vector field $\mathbf{F}$ is conservative, we illustrate in Examples 7 and 8 a straightforward method for producing a scalar potential function for $\mathbf{F}$. This technique is a direct consequence of the definition of a gradient field.

EXAMPLE 7 Consider the vector field

$$\mathbf{F} = (2xy + \cos 2y)\,\mathbf{i} + (x^2 - 2x\sin 2y)\,\mathbf{j}$$

of Example 5. We have already seen that $\mathbf{F}$ is conservative in Example 5. To find a scalar potential for $\mathbf{F}$, we seek a suitable function $f(x, y)$ such that

$$\nabla f(x, y) = \frac{\partial f}{\partial x}\mathbf{i} + \frac{\partial f}{\partial y}\mathbf{j} = \mathbf{F}.$$

The components of the gradient of f must agree with those of $\mathbf{F}$; therefore,

$$\begin{cases} \dfrac{\partial f}{\partial x} = 2xy + \cos 2y \\[2ex] \dfrac{\partial f}{\partial y} = x^2 - 2x\sin 2y \end{cases}. \tag{2}$$

We may begin to recover f by integrating the first equation of (2) with respect to x. Thus,

$$f(x, y) = \int \frac{\partial f}{\partial x}\, dx = \int (2xy + \cos 2y)\, dx = x^2 y + x\cos 2y + g(y), \tag{3}$$

where $g(y)$ is an arbitrary function of y. (The function $g(y)$ plays the role of the arbitrary "constant of integration" in the indefinite integral of $\partial f/\partial x$.) Differentiating equation (3) with respect to y yields

$$\frac{\partial f}{\partial y} = x^2 - 2x\sin 2y + g'(y). \tag{4}$$

If we compare equation (4) with the second equation of (2), we see that $g'(y) \equiv 0$, and so g must be a constant function. Therefore, our scalar potential must be of the form

$$f(x, y) = x^2 y + x\cos 2y + C,$$

where C is an arbitrary constant. You may, if you wish, double-check that $\nabla f = \mathbf{F}$. ◆

EXAMPLE 8 Let $\mathbf{F} = (e^x \sin y - yz)\,\mathbf{i} + (e^x \cos y - xz)\,\mathbf{j} + (z - xy)\,\mathbf{k}$. Note that $\mathbf{F}$ is of class C^1 on all of $\mathbf{R}^3$. We calculate

$$\begin{aligned}\nabla \times \mathbf{F} &= \begin{vmatrix} \mathbf{i} & \mathbf{j} & \mathbf{k} \\ \partial/\partial x & \partial/\partial y & \partial/\partial z \\ e^x \sin y - yz & e^x \cos y - xz & z - xy \end{vmatrix} \\ &= \left(\frac{\partial}{\partial y}(z - xy) - \frac{\partial}{\partial z}(e^x \cos y - xz)\right)\mathbf{i} \\ &\quad + \left(\frac{\partial}{\partial z}(e^x \sin y - yz) - \frac{\partial}{\partial x}(z - xy)\right)\mathbf{j} \\ &\quad + \left(\frac{\partial}{\partial x}(e^x \cos y - xz) - \frac{\partial}{\partial y}(e^x \sin y - yz)\right)\mathbf{k} \\ &= \mathbf{0}.\end{aligned}$$

Therefore, by Theorem 3.5, $\mathbf{F}$ is conservative.

Any scalar potential $f(x, y, z)$ for $\mathbf{F}$ must satisfy

$$\begin{cases} \dfrac{\partial f}{\partial x} = e^x \sin y - yz \\ \dfrac{\partial f}{\partial y} = e^x \cos y - xz\,. \\ \dfrac{\partial f}{\partial z} = z - xy \end{cases} \tag{5}$$

Integrating $\partial f/\partial x$ with respect to x, we find that

$$\begin{aligned} f(x, y, z) &= \int \frac{\partial f}{\partial x}\,dx \\ &= \int (e^x \sin y - yz)\,dx \\ &= e^x \sin y - xyz + g(y, z), \end{aligned} \tag{6}$$

where $g(y, z)$ may be any function of y and z. Differentiating equation (6) with respect to y and comparing with the second equation in (5), we see that

$$\frac{\partial f}{\partial y} = e^x \cos y - xz + \frac{\partial g}{\partial y} = e^x \cos y - xz.$$

Hence, $\partial g/\partial y = 0$, so g must be independent of y; that is, $g(y, z) = h(z)$, a function of z alone. So

$$f(x, y, z) = e^x \sin y - xyz + h(z). \tag{7}$$

Finally, we differentiate equation (7) with respect to z and compare with the third equation of (5):

$$\frac{\partial f}{\partial z} = -xy + h'(z) = z - xy.$$

Therefore, $h'(z) = z$, so $h(z) = \frac{1}{2}z^2 + C$, where C is an arbitrary constant. Thus, a scalar potential for the original vector field $\mathbf{F}$ is given by

$$f(x, y, z) = e^x \sin y - xyz + \tfrac{1}{2}z^2 + C.$$

◆

Addendum: Proof of Theorem 3.3

Recall that we have already shown that if a vector field $\mathbf{F}$ is a gradient field, then $\mathbf{F}$ has path-independent line integrals. So now we need only establish the converse. We do this explicitly in the case where $\mathbf{F}$ is defined on a (connected) subset R of $\mathbf{R}^3$, although our proof requires only notational modification in the n-dimensional setting.

Assume that $\mathbf{F}$ has path-independent line integrals. Then we may unambiguously write $\int_A^B \mathbf{F} \cdot d\mathbf{s}$ to denote the vector line integral of $\mathbf{F}$ from the point A to the point B along any path whose image lies in R. In what follows, consider $A(x_0, y_0, z_0)$ to be a fixed point in R, and $B(x, y, z)$ a "variable point." Then we define

$$f(B) = \int_A^B \mathbf{F} \cdot d\mathbf{s}$$

and show that f is a scalar potential for $\mathbf{F}$.

Write $\mathbf{F}$ explicitly as

$$\mathbf{F} = M(x, y, z)\,\mathbf{i} + N(x, y, z)\,\mathbf{j} + P(x, y, z)\,\mathbf{k}.$$

Therefore, we need to verify that the components of ∇f agree with those of $\mathbf{F}$, that is,

$$\frac{\partial f}{\partial x} = M(x, y, z), \quad \frac{\partial f}{\partial y} = N(x, y, z), \quad \text{and} \quad \frac{\partial f}{\partial z} = P(x, y, z).$$

Actually, we check only the first of these equations; the others may be verified in a similar manner.

At the point B, we have, by the definition of the partial derivative, that

$$\frac{\partial f}{\partial x} = \lim_{h \to 0} \frac{f(x+h, y, z) - f(x, y, z)}{h} = \lim_{h \to 0} \frac{f(B') - f(B)}{h}, \tag{8}$$

where B' denotes the point $(x+h, y, z)$. Note that $B' \to B$ as $h \to 0$. The numerator of the difference quotient in equation (8) is

$$f(B') - f(B) = \int_A^{B'} \mathbf{F} \cdot d\mathbf{s} - \int_A^{B} \mathbf{F} \cdot d\mathbf{s} = \int_B^{B'} \mathbf{F} \cdot d\mathbf{s}. \tag{9}$$

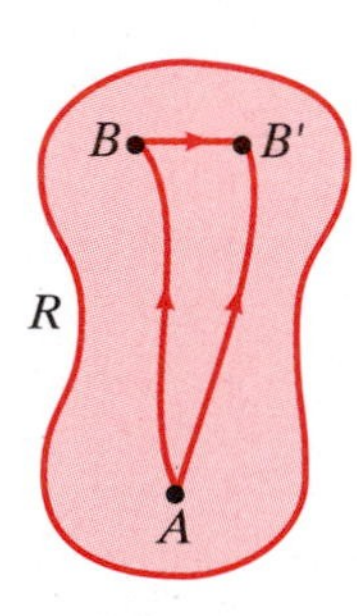

Figure 6.39 If B' is sufficiently close to B, then the straight-line path from B to B' will lie inside R.

If h is sufficiently small, we may evaluate the line integral in equation (9) by using the straight-line path between B and B'. (See Figure 6.39.) Explicitly, this path is

$$\mathbf{x}(t) = (x + th, y, z), \quad 0 \le t \le 1.$$

Then

$$\int_B^{B'} \mathbf{F} \cdot d\mathbf{s} = \int_0^1 \mathbf{F}(x + th, y, z) \cdot (h, 0, 0)\, dt = \int_0^1 hM(x + th, y, z)\, dt.$$

Since t is between 0 and 1, $|th| < |h|$. By the continuity of $\mathbf{F}$ (and therefore M) we have, for $h \approx 0$,

$$M(x + th, y, z) \approx M(x, y, z).$$

This approximation improves as $h \to 0$. Hence,

$$f(B') - f(B) = \int_B^{B'} \mathbf{F} \cdot d\mathbf{s} \approx \int_0^1 hM(x, y, z)\, dt = hM(x, y, z).$$

Using this result in equation (8), we see that

$$\frac{\partial f}{\partial x} = \lim_{h \to 0} \frac{1}{h}[hM(x, y, z)] = M(x, y, z),$$

as desired. ■

Exercises

1. Consider the line integral $\int_C z^2\,dx + 2y\,dy + xz\,dz$.
   (a) Evaluate this integral, where C is the line segment from $(0, 0, 0)$ to $(1, 1, 1)$.
   (b) Evaluate this integral, where C is the path from $(0, 0, 0)$ to $(1, 1, 1)$ parametrized by $\mathbf{x}(t) = (t, t^2, t^3)$, $0 \le t \le 1$.
   (c) Is the vector field $\mathbf{F} = z^2\,\mathbf{i} + 2y\,\mathbf{j} + xz\,\mathbf{k}$ conservative? Why or why not?

2. Let $\mathbf{F} = 2xy\,\mathbf{i} + (x^2 + z^2)\,\mathbf{j} + 2yz\,\mathbf{k}$.
   (a) Calculate $\int_C \mathbf{F}\cdot d\mathbf{s}$, where C is the path parametrized by $\mathbf{x}(t) = (t^2, t^3, t^5)$, $0 \le t \le 1$.
   (b) Calculate $\int_C \mathbf{F}\cdot d\mathbf{s}$, where C is the straight-line path from $(0, 0, 0)$ to $(1, 0, 0)$, followed by the straight-line path from $(1, 0, 0)$ to $(1, 1, 1)$.
   (c) Does $\mathbf{F}$ have path-independent line integrals? Explain your answer.

In Exercises 3–13, determine whether the given vector field $\mathbf{F}$ *is conservative. If it is, find a scalar potential function for* $\mathbf{F}$.

3. $\mathbf{F} = e^{x+y}\,\mathbf{i} + e^{xy}\,\mathbf{j}$

4. $\mathbf{F} = 2x\sin y\,\mathbf{i} + x^2\cos y\,\mathbf{j}$

5. $$\mathbf{F} = \left(3x^2\cos y + \frac{y}{1+x^2y^2}\right)\mathbf{i} + \left(x^3\sin y + \frac{x}{1+x^2y^2}\right)\mathbf{j}$$

6. $\mathbf{F} = \dfrac{xy^2}{(1+x^2)^2}\,\mathbf{i} + \dfrac{x^2y}{1+x^2}\,\mathbf{j}$

7. $\mathbf{F} = (e^{-y} - y\sin xy)\,\mathbf{i} - (xe^{-y} + x\sin xy)\,\mathbf{j}$

8. $\mathbf{F} = (2xz - y^2 + yze^{xyz})\,\mathbf{i} - (2xy + xze^{xyz})\,\mathbf{j} + (x^2 + xye^{xyz})\,\mathbf{k}$

9. $\mathbf{F} = (2x + y)\,\mathbf{i} + (z\cos yz + x)\,\mathbf{j} + (y\cos yz)\,\mathbf{k}$

10. $\mathbf{F} = (y + z)\,\mathbf{i} + 2z\,\mathbf{j} + (x + y)\,\mathbf{k}$

11. $\mathbf{F} = e^x\sin y\,\mathbf{i} + e^x\cos y\,\mathbf{j} + (3z^2 + 2)\,\mathbf{k}$

12. $\mathbf{F} = 3x^2\,\mathbf{i} + \dfrac{z^2}{y}\,\mathbf{j} + 2z\ln y\,\mathbf{k}$

13. $\mathbf{F} = (e^{-yz} - yze^{xyz})\,\mathbf{i} + xz(e^{-yz} + e^{xyz})\,\mathbf{j} + xy(e^{-yz} - e^{xyz})\,\mathbf{k}$

14. Of the two vector fields
$$\mathbf{F} = xy^2z^3\,\mathbf{i} + 2x^2y\,\mathbf{j} + 3x^2y^2z^2\,\mathbf{k}$$
and
$$\mathbf{G} = 2xy\,\mathbf{i} + (x^2 + 2yz)\,\mathbf{j} + y^2\,\mathbf{k},$$
one is conservative and one is not. Determine which is which, and, for the conservative field, find a scalar potential function.

15. Find all functions $N(x, y)$ such that the vector field
$$\mathbf{F} = (ye^{2x} + 3x^2e^y)\,\mathbf{i} + N(x, y)\,\mathbf{j}$$
is conservative.

16. For what values of the constants a and b will the vector field
$$\mathbf{F} = (3x^2 + 3y^2z\sin xz)\,\mathbf{i} + (ay\cos xz + bz)\,\mathbf{j} + (3xy^2\sin xz + 5y)\,\mathbf{k}$$
be conservative?

17. Let $\mathbf{F} = x^2\,\mathbf{i} + \cos y\sin z\,\mathbf{j} + \sin y\cos z\,\mathbf{k}$.
    (a) Show that $\mathbf{F}$ is conservative and find a scalar potential function f for $\mathbf{F}$.
    (b) Evaluate $\int_{\mathbf{x}} \mathbf{F}\cdot d\mathbf{s}$ along the path $\mathbf{x}\colon [0, 1] \to \mathbf{R}^3$, $\mathbf{x}(t) = (t^2 + 1, e^t, e^{2t})$.

Show that the line integrals in Exercises 18–20 are path independent, and evaluate them along the given oriented curve and also by means of Theorem 3.3.

18. $\int_C (3x - 5y)\,dx + (7y - 5x)\,dy$; C is the line segment from $(1, 3)$ to $(5, 2)$.

19. $\int_C \dfrac{x\,dx + y\,dy}{\sqrt{x^2 + y^2}}$; C is the semicircular arc of $x^2 + y^2 = 4$ from $(2, 0)$ to $(-2, 0)$.

20. $\int_C (2y - 3z)\,dx + (2x + z)\,dy + (y - 3x)\,dz$; C is the line segment from the point $(0, 0, 0)$ to $(0, 1, 1)$ followed by the line segment from the point $(0, 1, 1)$ to $(1, 2, 3)$.

21. (a) Determine where the vector field
$$\mathbf{F} = \frac{x + xy^2}{y^2}\,\mathbf{i} - \frac{x^2 + 1}{y^3}\,\mathbf{j}$$
is conservative.
    (b) Determine a scalar potential for $\mathbf{F}$.
    (c) Find the work done by $\mathbf{F}$ in moving a particle along the parabolic curve $y = 1 + x - x^2$ from $(0, 1)$ to $(1, 1)$.

22. Let f, g, and h be functions of class C^1 of a single variable.
    (a) Show that $\mathbf{F} = (f(x) + y + z)\,\mathbf{i} + (x + g(y) + z)\,\mathbf{j} + (x + y + h(z))\,\mathbf{k}$ is conservative.
    (b) Determine a scalar potential for $\mathbf{F}$. (Your answer will involve integrals of f, g, and h.)
    (c) Find $\int_C \mathbf{F}\cdot d\mathbf{s}$, where C is any path from (x_0, y_0, z_0) to (x_1, y_1, z_1).

23. Consider the vector field

$$\mathbf{F} = (2x + z)\cos(x^2 + xz)\,\mathbf{i} - (z + 1)\sin(y + yz)\,\mathbf{j} + (x\cos(x^2 + xz) - y\sin(y + yz))\,\mathbf{k}.$$

(a) Determine if $\mathbf{F}$ is conservative.

(b) If $\mathbf{x}(t) = \left(t^3, t^2, \pi t - \sin\frac{\pi t}{2}\right)$, $0 \le t \le 1$, evaluate $\int_{\mathbf{x}} \mathbf{F} \cdot d\mathbf{s}$.

24. Consider the vector field

$$\mathbf{G} = (2x + z)\cos(x^2 + xz)\,\mathbf{i} + (x - (z + 1)\sin(y + yz))\,\mathbf{j} + (x\cos(x^2 + xz) - y\sin(y + yz))\,\mathbf{k}.$$

(a) How is $\mathbf{G}$ different from the vector field $\mathbf{F}$ in Exercise 23? Is $\mathbf{G}$ conservative?

(b) If $\mathbf{x}(t) = \left(t^3, t^2, \pi t - \sin\frac{\pi t}{2}\right)$, $0 \le t \le 1$, evaluate $\int_{\mathbf{x}} \mathbf{G} \cdot d\mathbf{s}$.

25. Let $\mathbf{F}$ be the gravitational force field of a mass M on a particle of mass m:

$$\mathbf{F} = -\frac{GMm}{(x^2 + y^2 + z^2)^{3/2}}(x\,\mathbf{i} + y\,\mathbf{j} + z\,\mathbf{k}).$$

(This is the force field of Example 3 in §3.3.) Given that G, M, and m are all constants, show that the work done by $\mathbf{F}$ as a particle of mass m moves from $\mathbf{x}_0 = (x_0, y_0, z_0)$ to $\mathbf{x}_1 = (x_1, y_1, z_1)$ depends only on $\|\mathbf{x}_0\|$ and $\|\mathbf{x}_1\|$.

6.4 True/False Exercises for Chapter 6

1. If C is the parabola $y = 4 - x^2$ with $-2 \le x \le 2$, then $\int_C y \sin x \, ds = 0$.

2. If $\mathbf{F} = -\mathbf{i} + \mathbf{j} + \mathbf{k}$ and C is the straight line from the origin to $(2, 2, 2)$, then $\int_C \mathbf{F} \cdot d\mathbf{s} = 2\sqrt{3}$.

3. If $\mathbf{F} = x\mathbf{i} + y\mathbf{j} + z\mathbf{k}$ and C is the straight line from $(3, 3, 3)$ to the origin, then $\int_C \mathbf{F} \cdot d\mathbf{s}$ is positive.

4. Suppose that $f(x) > 0$ for all x. Let $\mathbf{F} = f(x)\mathbf{i}$. If C is the horizontal line segment from $(1, 1)$ to $(2, 1)$, then $\int_C \mathbf{F} \cdot d\mathbf{s} > 0$.

5. Suppose that $f(x) > 0$ for all x. Let $\mathbf{F} = f(x)\mathbf{i}$. If C is the vertical line segment from $(0, 0)$ to $(0, 3)$, then $\int_C \mathbf{F} \cdot d\mathbf{s} > 0$.

6. If $\mathbf{x}$ is a unit-speed path, then $\int_{\mathbf{x}} \mathbf{F} \cdot d\mathbf{s} = \int_a^b (\mathbf{F} \cdot \mathbf{v})\, ds$, where $\mathbf{v}$ denotes the velocity of the path.

7. If $\mathbf{x}$ and $\mathbf{y}$ are two one-one parametrizations of the same curve and $\mathbf{F}$ is a continuous vector field, then $\int_{\mathbf{x}} \mathbf{F} \cdot d\mathbf{s} = \int_{\mathbf{y}} \mathbf{F} \cdot d\mathbf{s}$.

8. If a nonvanishing, continuous vector field $\mathbf{F}$ is everywhere tangent to a smooth curve C, then $\mathbf{F}$ does no work along the curve.

9. If a nonvanishing, continuous vector field $\mathbf{F}$ is everywhere normal to a smooth curve C, then $\mathbf{F}$ does no work along the curve.

10. If the curve C is the level set at height c of a function $f(x, y)$, then $\int_C f(x, y)\, ds$ is c times the length of C.

11. If $f(x, y, z)$ is a continuous function and $\int_C f(x, y, z)\, ds = 0$ for all curves C in $\mathbf{R}^3$, then $f(x, y, z) = 0$ for all $(x, y, z) \in \mathbf{R}^3$.

12. If a closed curve C is a level set of a function $f(x, y)$ of class C^1 and $\nabla f \ne \mathbf{0}$, then the flux of ∇f across C is always zero.

13. If a closed curve C is a level set of a function $f(x, y)$ of class C^1 and $\nabla f \ne \mathbf{0}$, then the circulation of ∇f along C is always zero.

14. If a vector field $\mathbf{F}$ has constant magnitude 3 and makes a constant angle with a curve C, then the work done by $\mathbf{F}$ along C is 3 times the length of C.

15. If $\mathbf{F}$ is a continuous vector field everywhere tangent to an oriented C^1 curve C, then $\int_C \mathbf{F} \cdot d\mathbf{s} = \int_C \|\mathbf{F}\|\, ds$.

16. If $\mathbf{F}$ is a constant vector field on $\mathbf{R}^2$, then $\oint_C \mathbf{F} \cdot d\mathbf{s} = 0$, where C is any simple closed curve.

17. If $\mathbf{F}$ is an incompressible (i.e., divergenceless) C^1 vector field on $\mathbf{R}^2$ and C is a simple closed curve, then the circulation of $\mathbf{F}$ along C is always zero.

18. If $\mathbf{F}$ is an incompressible C^1 vector field on $\mathbf{R}^2$, then the flux across any simple closed curve C in $\mathbf{R}^2$ is always zero.

19. If C is a simple curve in $\mathbf{R}^2$, then $\int_C \nabla f \cdot d\mathbf{s} = 0$.

20. If C is a simple closed curve in $\mathbf{R}^2$ and f is of class C^1, then $\oint_C \nabla f \cdot d\mathbf{s} = 0$.

21. $\mathbf{F} = (e^x \cos y + 3)\mathbf{i} - e^x \sin y\,\mathbf{j}$ is a conservative vector field on $\mathbf{R}^2$.

22. If f and g are functions of class C^1 defined on a region D in $\mathbf{R}^2$, then

$$\oint_{\partial D} f\nabla g \cdot d\mathbf{s} = \oint_{\partial D} g\nabla f \cdot d\mathbf{s}.$$

23. If C is a closed curve in $\mathbf{R}^3$ such that $\oint_C \mathbf{F} \cdot d\mathbf{s} = 0$, then $\mathbf{F}$ is conservative.

24. $\oint_C x\,dx + y\,dy + z\,dz = 0$ for all simple closed curves C in $\mathbf{R}^3$.

25. $\oint_C e^x(\cos y \sin z\,dx + \sin y \sin z\,dy + \cos y \cos z)\,dz = 0$ for all simple closed curves C in $\mathbf{R}^3$.

26. If $\nabla \times \mathbf{F} = \mathbf{0}$, then $\mathbf{F}$ is conservative.

27. Let $M(x, y)$ and $N(x, y)$ be C^1 functions with domain $\mathbf{R}^2 - \{(0, 0)\}$. If $\partial M/\partial y = \partial N/\partial x$, then $\oint_C M\,dx + N\,dy = 0$ for any closed curve C in $\mathbf{R}^2$.

28. Let $M(x, y, z)$, $N(x, y, z)$, and $P(x, y, z)$ be C^1 functions with domain $\mathbf{R}^3 - \{(0, 0, 0)\}$. If $\partial M/\partial y = \partial N/\partial x$, $\partial M/\partial z = \partial P/\partial x$, and $\partial N/\partial z = \partial P/\partial y$, then $\oint_C M\,dx + N\,dy + P\,dz = 0$ for any closed curve C in $\mathbf{R}^3$.

29. If $\mathbf{F}: \mathbf{R}^n \to \mathbf{R}^n$, then there is at most one function $f: \mathbf{R} \to \mathbf{R}$ such that $\nabla f = \mathbf{F}$.

30. If $\mathbf{F}$ is a differentiable vector field and $\mathbf{F} = \nabla \times \mathbf{G}$, then $\nabla \cdot \mathbf{F} = 0$.

6.5 Miscellaneous Exercises for Chapter 6

Let $C \subseteq \mathbf{R}^n$ be a piecewise C^1 curve and $f: X \subseteq \mathbf{R}^n \to \mathbf{R}$ a continuous function whose domain X includes C. Then we define the quantity

$$[f]_{\text{avg}} = \frac{\int_C f\,ds}{\int_C ds} = \frac{\int_C f\,ds}{\text{length of } C}$$

to be the ***average value*** *of f along C. Exercises 1–5 concern the notion of average value along a curve.*

1. Explain why it makes sense to use the preceding integral formula to represent the average value. (A careful explanation involves the use of Riemann sums.)

2. Suppose that a thin wire is shaped as a helical curve parametrized by

$$\mathbf{x}(t) = (\cos t, \sin t, t), \quad 0 \le t \le 3\pi.$$

If $f(x, y, z) = x^2 + y^2 + 2z^2 + 1$ represents the temperature at points along the wire, find the average temperature.

3. Find the average y-coordinate of points on the upper semicircle $y = \sqrt{a^2 - x^2}$.

4. Find the average z-coordinate of points on the broken-line curve pictured in Figure 6.40.

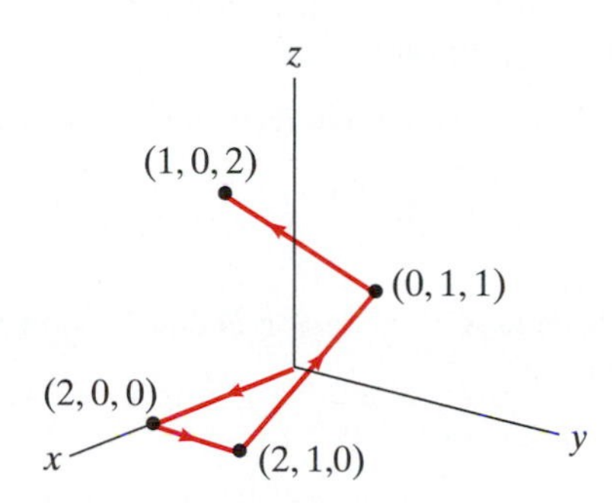

Figure 6.40 The broken-line curve of Exercise 4.

5. Find the average value of $f(x, y, z) = z^2 + xe^y$ on the curve C obtained by intersecting the (elliptic) cylinder $x^2/5 + y^2 = 1$ by the plane $z = 2y$.

6. A metal wire is bent in the shape of the semicircle $x^2 + y^2 = 4$, $y \ge 0$, lying in the xy-plane. Suppose that the mass density at each point (x, y, z) of the wire is $\delta(x, y, z) = 3 - y$.

(a) Find the total mass of the wire.

(b) Using formulas analogous to those in §5.6, we define the (first) moments of the wire to be

$$\int_C x\delta(x, y, z)\,ds, \qquad \int_C y\delta(x, y, z)\,ds,$$

$$\int_C z\delta(x, y, z)\,ds.$$

Using these definitions, find the coordinates of the center of mass of the wire.

7. Suppose that a wire is bent in the shape of a quarter circle of radius a. Find the center of mass of the wire if the density at points on the wire vary as the square of the distance from the center of the wire.

8. (a) Find the centroid of the helical wire $x = 3\cos t$, $y = 3\sin t$, $z = 4t$, where $0 \le t \le 4\pi$. (Hint: No calculation should be necessary.)

(b) Find the center of mass of the same wire if the density at each point of the wire is equal to the square of the point's distance from the origin.

If a thin wire is bent in the shape of a curve C in the xy-plane and has mass density at each point along the curve given by a continuous function $\delta(x, y)$, then we may define the ***moments of inertia*** *of C about the x- and y-axes, respectively, by*

$$I_x = \int_C y^2\delta(x, y)\,ds, \qquad I_y = \int_C x^2\delta(x, y)\,ds,$$

and the corresponding ***radii of gyration*** *of C as*

$$r_x = \sqrt{\frac{I_x}{M}}, \qquad r_y = \sqrt{\frac{I_y}{M}},$$

where M denotes the total mass $M = \int_C \delta(x, y)\,ds$ of the wire. Additionally, the ***moment of inertia*** *of C about the origin (or, equivalently, about the z-axis, if we think of the xy-plane as* *embedded in* $\mathbf{R}^3$*) is*

$$I_z = \int_C (x^2 + y^2)\delta(x, y)\, ds,$$

with corresponding radius of gyration $r_z = \sqrt{I_z/M}$. *Exercises 9–13 concern moments of inertia of curves.*

9. (a) Consider the wire of Exercise 6 again. Find its moment of inertia about the y-axis.

(b) What is the radius of gyration r_z for the wire about the z-axis?

10. Find the moment of inertia I_x and the radius of gyration r_x about the x-axis of a straight wire between $(-2, 1)$ to $(2, 3)$ whose density varies along the wire as $\delta(x, y) = y$.

11. Find the moment of inertia I_x and the radius of gyration r_x about the x-axis of a wire shaped as the curve $y = x^2$ between $(0, 0)$ and $(2, 4)$ and whose density varies as $\delta(x, y) = x$.

12. (a) Suppose that a thin metal wire is bent into a curve C in $\mathbf{R}^3$ and has mass density at each point (x, y, z) along the curve given by a continuous function $\delta(x, y, z)$. Give general formulas analogous to those in §5.6 for the moments of inertia of the wire about the three coordinate axes.

(b) Find the moments of inertia about the coordinate axes of a homogeneous (i.e., constant density) wire shaped like the helix $x = 3\cos t$, $y = 3\sin t$, $z = 4t$, where $0 \le t \le 4\pi$. What are the radii of gyration?

13. Find the moment of inertia I_z about the z-axis of a wire in the shape of the line segment between $(-1, 1, 2)$ and $(2, 2, 3)$ if the density along the segment varies as $\delta(x, y, z) = 1 + z^2$. What is r_z?

14. Let $r = f(\theta)$ be the polar equation of a curve in the plane.

(a) Use scalar line integrals to show that the arclength of the curve between $(f(a), a)$ and $(f(b), b)$ is

$$\int_a^b \sqrt{(f(\theta))^2 + (f'(\theta))^2}\, d\theta.$$

(b) Sketch the curve $r = \sin^2(\theta/2)$ and find its length.

15. (a) Give a formula in polar coordinates for the scalar line integral of a function $g(x, y)$ along the curve $r = f(\theta)$, $a \le \theta \le b$.

(b) Compute $\int_C g\, ds$, where $g(x, y) = x^2 + y^2 - 2x$ and C is the segment of the spiral $r = e^{3\theta}$, $0 \le \theta \le 2\pi$.

Let C be a piecewise C^1*, simple curve in* $\mathbf{R}^3$*. The* ***total curvature*** *K of C is*

$$K = \int_C \kappa\, ds,$$

where κ *denotes the curvature of C. (See §3.2 to review the notion of curvature.) Exercises 16–20 involve the notion of total curvature.*

16. Show that if C is a simple curve of class C^1 parametrized by $\mathbf{x}(t)$, $a \le t \le b$, then

$$K = \int_a^b \frac{\|\mathbf{v} \times \mathbf{a}\|}{\|\mathbf{v}\|^2}\, dt.$$

(Recall that $\mathbf{v}$ and $\mathbf{a}$ denote, respectively, the velocity and acceleration of the path $\mathbf{x}$.)

17. Find the total curvature of the helix

$$\mathbf{x}(t) = (3\cos t, 3\sin t, 4t), \quad 0 \le t \le 10\pi.$$

18. Find the total curvature of the parabola $y = Ax^2$, $A > 0$, for $a \le x \le b$.

19. **Fenchel's theorem** states that if C is a simple, closed C^1 curve in $\mathbf{R}^3$, then $K \ge 2\pi$ and, moreover, $K = 2\pi$ if and only if C is a plane convex curve. (A simple, closed curve C is **convex** if the line segment joining any two points of C lies entirely in the region enclosed by C.) Verify Fenchel's theorem for the ellipse $x^2/a^2 + y^2/b^2 = 1$.

20. Let C be a simple, closed C^1 curve in $\mathbf{R}^3$. Suppose that the curvature κ of C is bounded (i.e., $0 \le \kappa \le 1/a$ for some $a > 0$).

(a) Show that if L denotes the length of C, then $L \ge 2\pi a$.

(b) What can you say about C if $L = 2\pi a$?

21. Calculate the work done by the vector field $\mathbf{F} = \sin x\,\mathbf{i} + \cos y\,\mathbf{j} + xz\,\mathbf{k}$ on a particle moving along the path $\mathbf{x}(t) = (t^3, -t^2, t)$, where $0 \le t \le 1$.

22. Use Green's theorem to find the work done by the vector field $\mathbf{F} = x^2 y\,\mathbf{i} + (x + y)y\,\mathbf{j}$ in moving a particle from the origin along the y-axis to the point $(0, 1)$, then along the line segment from $(0, 1)$ to $(1, 0)$, and then from $(1, 0)$ back to the origin along the x-axis. (**Warning:** Be careful.)

23. Use Green's theorem to recover the formula

$$A = \frac{1}{2}\int_a^b (f(\theta))^2\, d\theta$$

for the area A of the region D described in polar coordinates by

$$D = \{(r, \theta) \mid 0 \le r \le f(\theta), a \le \theta \le b\}.$$

24. Let C be a piecewise C^1, simple, closed curve in $\mathbf{R}^2$. Show that

$$\oint_C f(x)\, dx + g(y)\, dy = 0,$$

where f and g are any single-variable functions of class C^1.

25. Let D be a region in $\mathbf{R}^2$ whose boundary ∂D consists of finitely many piecewise C^1, simple, closed curves oriented so that D is on the left as you travel along any segment of ∂D. If $(\bar{x}, \bar{y})$ denotes the coordinates of the centroid of D, show that

$$\bar{x} = \frac{1}{2 \cdot \text{area of } D} \oint_{\partial D} x^2\, dy$$

and

$$\bar{y} = \frac{1}{\text{area of } D} \oint_{\partial D} xy\, dy.$$

Also show that

$$\bar{x} = -\frac{1}{\text{area of } D} \oint_{\partial D} xy\, dx$$

and

$$\bar{y} = -\frac{1}{2 \cdot \text{area of } D} \oint_{\partial D} y^2\, dx.$$

26. Use the results of Exercise 25 to find the centroid of the triangular region with vertices $(0, 0)$, $(1, 0)$, and $(0, 2)$.

27. Use the results of Exercise 25 to find the centroid of the region in $\mathbf{R}^2$ that lies inside the circle of radius 6 centered at the origin and outside the two circles of radius 1 centered at $(4, 0)$ and $(-2, 2)$, respectively.

28. Let C be a piecewise C^1, simple, closed curve, oriented counterclockwise, enclosing a region D in the plane. Let $\mathbf{n}$ be the outward unit normal vector to D. If $f(x, y)$ and $g(x, y)$ are functions of class C^2 on D, establish **Green's first identity:**

$$\iint_D (f\nabla^2 g + \nabla f \cdot \nabla g)\, dA = \oint_C f\nabla g \cdot \mathbf{n}\, ds.$$

29. Under the hypotheses of Exercise 28, prove **Green's second identity:**

$$\iint_D (f\nabla^2 g - g\nabla^2 f)\, dA = \oint_C (f\nabla g - g\nabla f) \cdot \mathbf{n}\, ds.$$

*A function $g(x, y)$ is said to be **harmonic** at a point (x_0, y_0) if g is of class C^2 and satisfies Laplace's equation*

$$\nabla^2 g = \frac{\partial^2 g}{\partial x^2} + \frac{\partial^2 g}{\partial y^2} = 0$$

on some neighborhood of (x_0, y_0). We say that g is harmonic on a closed region $D \subseteq \mathbf{R}^2$ if it is harmonic at all interior points of D (i.e., not necessarily along ∂D). Exercises 30–33 concern some elementary results about harmonic functions in $\mathbf{R}^2$.

30. Suppose that D is compact (i.e., closed and bounded) and that ∂D is piecewise C^1 and oriented as in Green's theorem. Let $\mathbf{n}$ denote the outward unit normal vector to ∂D and let $\partial g/\partial n$ denote $\nabla g \cdot \mathbf{n}$. (The term $\partial g/\partial n$ is called the **normal derivative** of g.) Use Green's first identity (see Exercise 28) with $f(x, y) \equiv 1$ to show that, if g is harmonic on D, then

$$\oint_{\partial D} \frac{\partial g}{\partial n}\, ds = 0.$$

31. Let f be harmonic on a region D that satisfies the assumptions of Exercise 30. Show that

$$\iint_D \nabla f \cdot \nabla f\, dA = \oint_{\partial D} f \frac{\partial f}{\partial n}\, ds.$$

32. Suppose that $f(x, y) = 0$ for all $(x, y) \in \partial D$. Use Exercise 31 to show that $f(x, y) = 0$ throughout all of D. (Hint: Consider the sign of $\nabla f \cdot \nabla f$.)

33. Let D be a region that satisfies the assumptions of Exercise 30. Use the result of Exercise 31 to show that if f_1 and f_2 are harmonic on D and $f_1(x, y) = f_2(x, y)$ on ∂D, then, in fact, $f_1 = f_2$ on *all* of D. Thus we see that harmonic functions are determined by their values on the boundary of a region. (Hint: Consider $f_1 - f_2$.)

34. We call a vector field $\mathbf{F}$ on $\mathbf{R}^3$ **radially symmetric** if it can be written in spherical coordinates in the form $\mathbf{F} = f(\rho)\mathbf{e}_\rho$, where $\mathbf{e}_\rho$ is the unit vector that points in the direction of increasing ρ-coordinate. (See §1.7.)

(a) Give an example of a (nontrivial) radially symmetric vector field, written in both Cartesian and spherical coordinates.

(b) Show that if f is of class C^1 for all $\rho \geq 0$, then the radially symmetric vector field $\mathbf{F} = f(\rho)\mathbf{e}_\rho$ is conservative.

35. Let $\mathbf{F} = \dfrac{-y\,\mathbf{i} + x\,\mathbf{j}}{x^2 + y^2}$.

(a) Verify Green's theorem over the annular region

$$D = \{(x, y) \mid a^2 \leq x^2 + y^2 \leq 1\}.$$

(See Figure 6.41.)

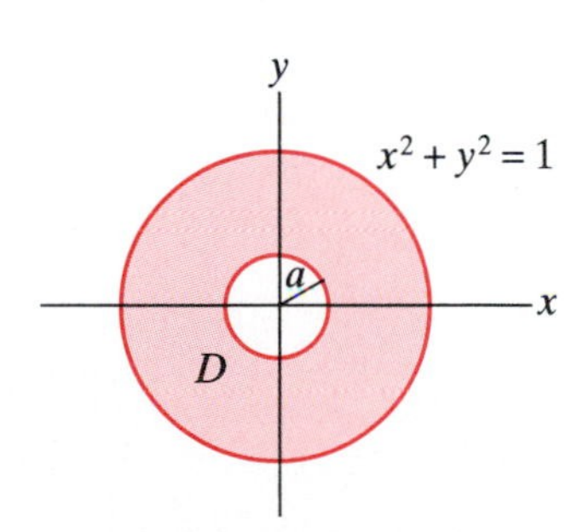

Figure 6.41 The annular region of Exercise 35(a).

(b) Now let D be the unit disk. Does the formula of Green's theorem hold for D? Can you explain why?

(c) Suppose C is any simple closed curve lying outside the circle

$$C_a = \{(x, y) \mid x^2 + y^2 = a^2\}.$$

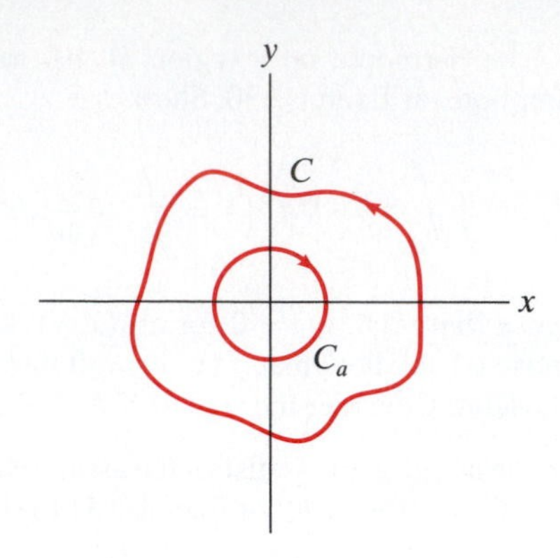

Figure 6.42 The curves C and C_a of Exercise 35(c).

(See Figure 6.42.) Argue that if C is oriented counterclockwise, then

$$\oint_C \frac{x\,dy - y\,dx}{x^2 + y^2} = 2\pi.$$

36. Consider the vector field $\mathbf{F} = -\dfrac{y}{x^2+y^2}\,\mathbf{i} + \dfrac{x}{x^2+y^2}\,\mathbf{j}$.

(a) Calculate $\nabla \times \mathbf{F}$.

(b) Evaluate $\oint_C \mathbf{F} \cdot d\mathbf{s}$, where C is the unit circle $x^2 + y^2 = 1$.

(c) Is $\mathbf{F}$ conservative?

(d) How can you reconcile parts (a) and (b) with Theorem 3.5?

37. (a) Let $\mathbf{F} = e^y\,\mathbf{i} + x^4\,\mathbf{j}$. Calculate the flux $\int_C \mathbf{F} \cdot \mathbf{n}\,ds$ of $\mathbf{F}$ across the boundary of the rectangle $R = [0, 1] \times [0, 5]$.

(b) Let f and g be of class C^1 and let $\mathbf{F} = f(y)\,\mathbf{i} + g(x)\,\mathbf{j}$. Show that the flux of $\mathbf{F}$ across any piecewise C^1 simple, closed curve is zero.

38. Use Newton's second law of motion $\mathbf{F} = m\mathbf{a}$ to show that the work done by a force field $\mathbf{F}$ in moving a particle of mass m along a path $\mathbf{x}(t)$ from $\mathbf{x}(a)$ to $\mathbf{x}(b)$ is equal to the change in kinetic energy of the particle. In other words,

$$\int_{\mathbf{x}} \mathbf{F} \cdot d\mathbf{s} = \tfrac{1}{2}m(v(b))^2 - \tfrac{1}{2}m(v(a))^2,$$

where $v(t) = \|\mathbf{v}(t)\|$, the speed at time t. (Use the product rule for dot products of vector-valued functions.)

39. Let $\mathbf{F}$ be a conservative vector field on $\mathbf{R}^3$ with $\mathbf{F} = -\nabla V$. If a particle travels along a path $\mathbf{x}$, recall that its **potential energy** at time t is defined to be $V(\mathbf{x}(t))$. Use line integrals to prove the **law of conservation of energy:** As a particle of mass m moves between any two points A and B in a conservative force field, the sum of the potential and kinetic energies of the particle remains constant. (Use Exercise 38 and Theorem 3.3.) The use of line integrals provides an alternative proof of Theorem 4.2 in §4.4.

CHAPTER 7

Surface Integrals and Vector Analysis

7.1 Parametrized Surfaces

Introduction

Surfaces in $\mathbf{R}^3$ may be presented analytically in different ways. Here are two familiar descriptions:

1. as a graph of a function of two variables; that is, as points (x, y, z) in $\mathbf{R}^3$ satisfying $z = f(x, y)$ (e.g., $z = x^2 + 4y^2$);
2. as a level set of a function of three variables; that is, as points (x, y, z) such that $F(x, y, z) = c$ for some suitable function F and constant c (e.g., $x^2y - z^2y^5 + x = 1$).

Both of these descriptions are problematical. As noted in §2.1, many common surfaces cannot be described as graphs of functions of two variables. Recall, for example, that the full sphere $x^2 + y^2 + z^2 = 1$ is not the graph of a function of two variables. Therefore, description 1 is not sufficiently general.

There are also problems with description 2. Not all equations of the form $F(x, y, z) = c$ have solutions that fill out surfaces. Indeed, although the level set of $F(x, y, z) = x^2 + y^2 + z^2$ at height 1 is a sphere, at height 0 it is a single point, and at height -1 completely empty. In addition, it is somewhat tricky to describe surfaces (i.e., two-dimensional objects) in $\mathbf{R}^n$ by using level sets when n is larger than 3. Another approach is desirable for presenting surfaces analytically, in order to avoid the problems just mentioned, to emphasize clearly the two-dimensional nature of a surface, and to facilitate subsequent calculations. With this discussion in mind, we state the following definition:

■ **Definition 1.1** Let D be a region in $\mathbf{R}^2$ that consists of a connected open set, possibly together with some or all of its boundary points. A **parametrized surface** in $\mathbf{R}^3$ is a continuous function $\mathbf{X}: D \subseteq \mathbf{R}^2 \to \mathbf{R}^3$ that is one-one on D, except possibly along ∂D. We refer to the image $\mathbf{X}(D)$ as the **underlying surface** of $\mathbf{X}$ (or the surface **parametrized** by $\mathbf{X}$) and denote it by S. (See Figure 7.1.)

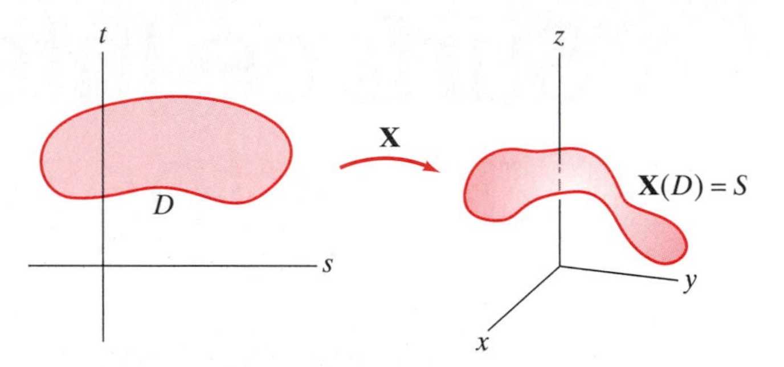

Figure 7.1 A parametrized surface.

The restrictions on the region D and the map $\mathbf{X}$ of Definition 1.1 are meant to ensure that D is a two-dimensional subset of $\mathbf{R}^2$ with a two-dimensional image. If we write the component functions of $\mathbf{X}$, then, for $(s, t) \in D$,

$$\mathbf{X}(s, t) = (x(s, t), y(s, t), z(s, t)),$$

and the underlying surface S can be described by the parametric equations

$$\begin{cases} x = x(s, t) \\ y = y(s, t) \\ z = z(s, t) \end{cases} \quad (s, t) \in D. \tag{1}$$

EXAMPLE 1 Consider the parametrized surface $\mathbf{X}: \mathbf{R}^2 \to \mathbf{R}^3$ described by

$$\mathbf{X}(s, t) = s(\mathbf{i} - \mathbf{j}) + t(\mathbf{i} + 2\mathbf{k}) + 3\mathbf{j}.$$

The image of $\mathbf{X}$, shown in Figure 7.2, is the plane through the point $(0, 3, 0)$, determined by the vectors $\mathbf{a} = \mathbf{i} - \mathbf{j}$ and $\mathbf{b} = \mathbf{i} + 2\mathbf{k}$. (See Proposition 5.1 of §1.5.) ◆

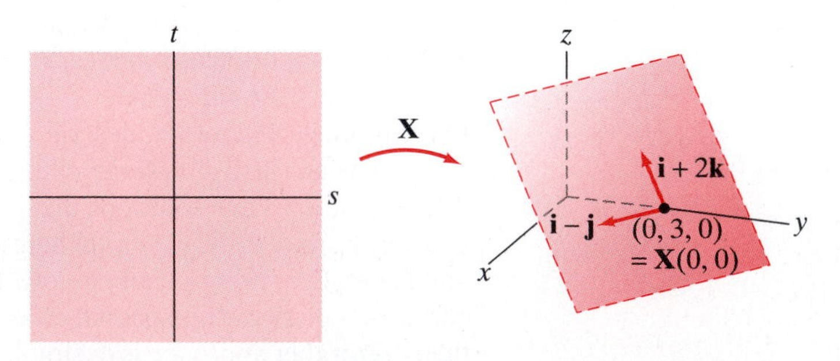

Figure 7.2 The parametrized plane of Example 1.

EXAMPLE 2 Let $D = [0, 2\pi) \times [0, \pi]$ and consider $\mathbf{X}: D \to \mathbf{R}^3$ given by

$$\mathbf{X}(s, t) = (a \cos s \sin t, a \sin s \sin t, a \cos t).$$

The corresponding parametric equations are

$$\begin{cases} x = a\cos s\,\sin t \\ y = a\sin s\,\sin t \\ z = a\cos t \end{cases} \qquad 0 \le s < 2\pi,\ 0 \le t \le \pi.$$

The parametric equations imply that $x^2 + y^2 + z^2 = a^2$, meaning all the points of $S = \mathbf{X}(D)$ lie on a sphere of radius a centered at the origin. The parametric equations are precisely the spherical–rectangular coordinate conversions (see §1.7) with the ρ-coordinate held constant at a and with s and t used instead of θ and φ. Hence, the image of $\mathbf{X}$ is indeed all of the sphere. (See Figure 7.3.) ◆

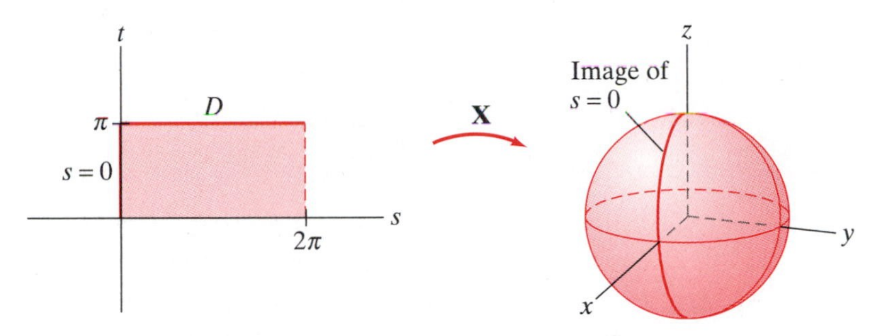

Figure 7.3 A sphere rendered as a parametrized surface.

EXAMPLE 3 The points of the surface parametrized by

$$\begin{cases} x = a\cos s \\ y = a\sin s \\ z = t \end{cases} \qquad 0 \le s \le 2\pi$$

satisfy the equation $x^2 + y^2 = a^2$ and so can be seen to form an infinite cylinder of radius a. Figure 7.4 shows how the function $\mathbf{X}(s, t) = (a\cos s, a\sin s, t)$, maps the infinite strip $D = \{(s, t) \mid 0 \le s \le 2\pi\}$ onto the cylinder by "gluing" the line $s = 0$ to $s = 2\pi$. ◆

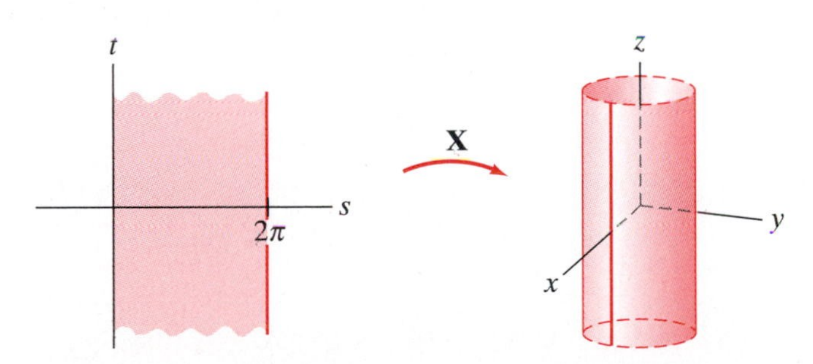

Figure 7.4 The map $\mathbf{X}$ glues together the edges of D to form a cylinder.

EXAMPLE 4 Let $D \subseteq \mathbf{R}^2$ be an open set, possibly together with some or all of its boundary points. If $f: D \to \mathbf{R}$ is a continuous scalar-valued function of two variables, then it is not difficult to parametrize the graph of f: We let

$$\mathbf{X}: D \to \mathbf{R}^3 \quad \text{be} \quad \mathbf{X}(s, t) = (s, t, f(s, t)).$$

That is, the parametric equations

$$\begin{cases} x = s \\ y = t \\ z = f(s, t) \end{cases} \qquad (s, t) \in D$$

describe the points of the graph of f. (See Figure 7.5.) ◆

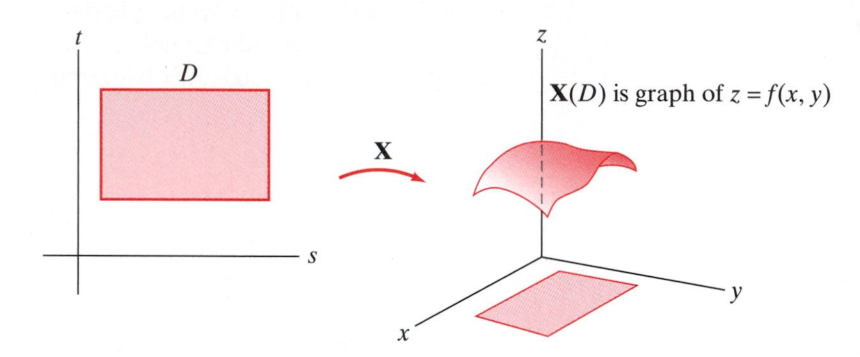

Figure 7.5 The graph of $z = f(x, y)$ as a parametrized surface.

Coordinate Curves, Normal Vectors, and Tangent Planes

Let S be a surface parametrized by $\mathbf{X}: D \to \mathbf{R}^3$. If we fix $t = t_0$ and let only s vary, we obtain a continuous map

$$s \longmapsto \mathbf{X}(s, t_0),$$

whose image is a curve lying in S. We call this curve the s**-coordinate curve** at $t = t_0$. Similarly, we may fix $s = s_0$ and obtain a map

$$t \longmapsto \mathbf{X}(s_0, t),$$

whose image is the t**-coordinate curve** at $s = s_0$. Figure 7.6 suggests the appearance of the coordinate curves.

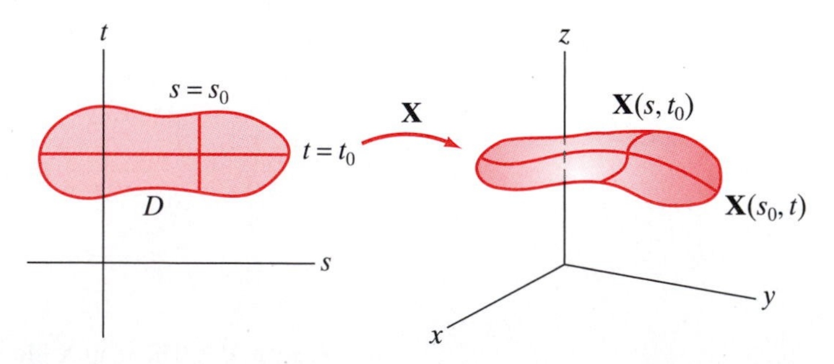

Figure 7.6 The coordinate curves of a parametrized surface.

EXAMPLE 5 The parametrized surface T defined by

$$\begin{cases} x = (a + b\cos t)\cos s \\ y = (a + b\cos t)\sin s \\ z = b\sin t \end{cases} \qquad \begin{array}{l} 0 \le s, t \le 2\pi; \\ a, b \text{ positive constants with } a > b, \end{array}$$

satisfies the equation

$$\left(\sqrt{x^2 + y^2} - a\right)^2 + z^2 = b^2.$$

The s-coordinate curve at $t = 0$ is

$$\begin{cases} x = (a + b)\cos s \\ y = (a + b)\sin s \\ z = 0 \end{cases}$$

and is readily seen to be the circle of radius $a + b$, centered at the origin and lying in the xy-plane. In general, you may check that the s-coordinate curve at $t = t_0$ is a circle of radius $a + b\cos t_0$ (which varies between $a - b$ and $a + b$) in the horizontal plane $z = b\sin t_0$. (See Figure 7.7.)

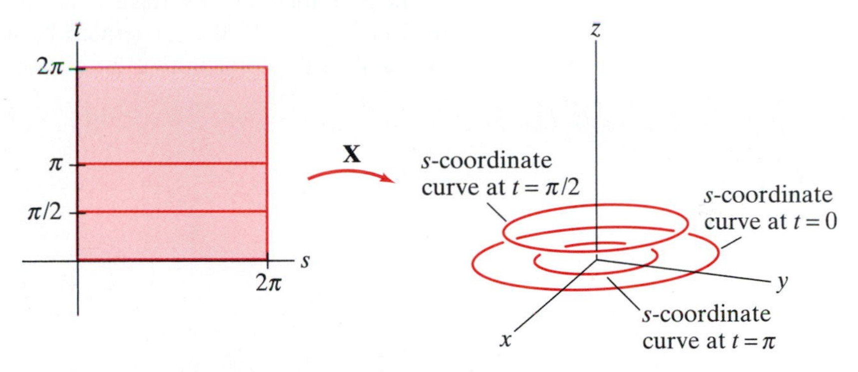

Figure 7.7 Some s-coordinate curves for the torus T of Example 5.

The t-coordinate curve at $s = 0$ is

$$\begin{cases} x = a + b\cos t \\ y = 0 \\ z = b\sin t \end{cases}.$$

The image is a circle of radius b centered at $(a, 0, 0)$ in the xz-plane (i.e., the plane $y = 0$). At $s = s_0$, the t-coordinate curve is

$$\begin{cases} x = \cos s_0\,(a + b\cos t) \\ y = \sin s_0\,(a + b\cos t) \\ z = b\sin t \end{cases}.$$

Along this curve, we have $y/x = \tan s_0$, a constant. The curve lies in the vertical plane $(\sin s_0)x - (\cos s_0)y = 0$. Moreover, it is not hard to see that the distance from any point on this curve to the point $P(a\cos s_0, a\sin s_0, 0)$ is b, and the image is a circle of radius b centered at P. See Figure 7.8 for examples of t-coordinate curves.

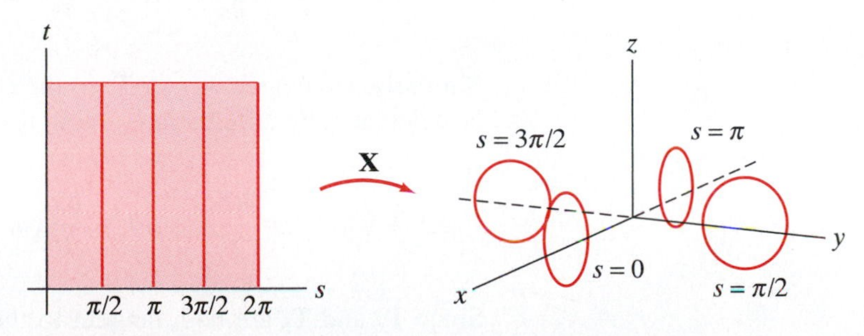

Figure 7.8 Some t-coordinate curves for the torus T of Example 5.

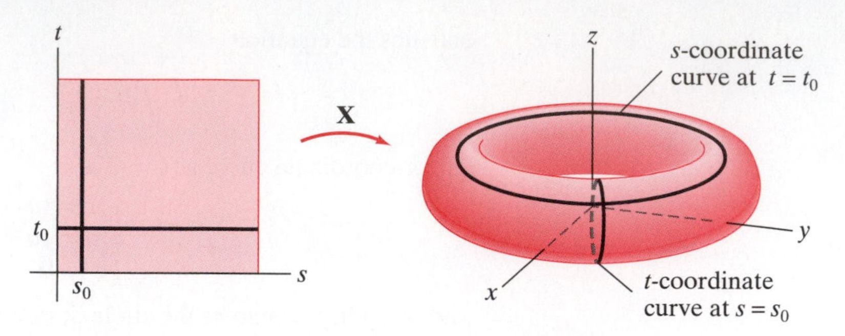

Figure 7.9 The parametrized torus T.

The aforementioned surface T is called a **torus**, a doughnut-shaped surface shown in Figure 7.9. It is generated both by the collection of the s-coordinate curves and by the collection of the t-coordinate curves. ◆

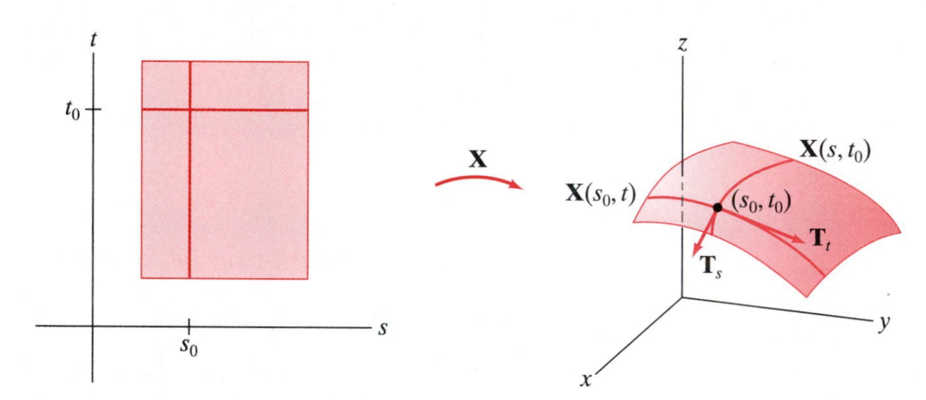

Figure 7.10 The tangent vectors $\mathbf{T}_s$ and $\mathbf{T}_t$ to the coordinate curves.

Suppose that $\mathbf{X}(s, t) = (x(s, t), y(s, t), z(s, t))$, where $(s, t) \in D$, is a differentiable (or C^1) map, in which case we say that the parametrized surface $S = \mathbf{X}(D)$ is **differentiable** (or C^1) as well. Then the coordinate curves $\mathbf{X}(s, t_0)$ and $\mathbf{X}(s_0, t)$ have well-defined tangent vectors at points (s_0, t_0) in D. (See Figure 7.10.) To find the tangent vector $\mathbf{T}_s$ to the s-coordinate curve $\mathbf{X}(s, t_0)$ at (s_0, t_0), we differentiate the component functions of $\mathbf{X}$ with respect to s and evaluate at (s_0, t_0):

$$\mathbf{T}_s(s_0, t_0) = \frac{\partial \mathbf{X}}{\partial s}(s_0, t_0) = \frac{\partial x}{\partial s}(s_0, t_0)\,\mathbf{i} + \frac{\partial y}{\partial s}(s_0, t_0)\,\mathbf{j} + \frac{\partial z}{\partial s}(s_0, t_0)\,\mathbf{k}. \qquad (2)$$

Similarly, the tangent vector $\mathbf{T}_t$ to the t-coordinate curve $\mathbf{X}(s_0, t)$ at (s_0, t_0) may be calculated by differentiating with respect to t:

$$\mathbf{T}_t(s_0, t_0) = \frac{\partial \mathbf{X}}{\partial t}(s_0, t_0) = \frac{\partial x}{\partial t}(s_0, t_0)\,\mathbf{i} + \frac{\partial y}{\partial t}(s_0, t_0)\,\mathbf{j} + \frac{\partial z}{\partial t}(s_0, t_0)\,\mathbf{k}. \qquad (3)$$

Since $\mathbf{T}_s$ and $\mathbf{T}_t$ are both tangent to the surface S at (s_0, t_0), the cross product $\mathbf{T}_s(s_0, t_0) \times \mathbf{T}_t(s_0, t_0)$ will be normal to S at (s_0, t_0), provided it is nonzero.

■ **Definition 1.2** The parametrized surface $S = \mathbf{X}(D)$ is **smooth** at $\mathbf{X}(s_0, t_0)$ if the map $\mathbf{X}$ is of class C^1 in a neighborhood of (s_0, t_0) and if the vector

$$\mathbf{N}(s_0, t_0) = \mathbf{T}_s(s_0, t_0) \times \mathbf{T}_t(s_0, t_0) \neq \mathbf{0}.$$

If S is smooth at every point $\mathbf{X}(s_0, t_0) \in S$, then we simply refer to S as a **smooth** parametrized surface. If S is a smooth parametrized surface, we call the (nonzero) vector $\mathbf{N} = \mathbf{T}_s \times \mathbf{T}_t$ the **standard normal vector** arising from the parametrization $\mathbf{X}$.

EXAMPLE 6 We claim that the torus T of Example 5 is smooth. Recall that T is given as the image of the map

$$\mathbf{X}\colon [0, 2\pi] \times [0, 2\pi] \to \mathbf{R}^3,$$
$$\mathbf{X}(s, t) = ((a + b\cos t)\cos s, (a + b\cos t)\sin s, b\sin t),$$

where $a > b > 0$. Then from formulas (2) and (3), we have

$$\mathbf{T}_s(s_0, t_0) = -(a + b\cos t_0)\sin s_0\,\mathbf{i} + (a + b\cos t_0)\cos s_0\,\mathbf{j}$$

and

$$\mathbf{T}_t(s_0, t_0) = -b\sin t_0\cos s_0\,\mathbf{i} - b\sin t_0\sin s_0\,\mathbf{j} + b\cos t_0\,\mathbf{k},$$

so that

$$\begin{aligned}\mathbf{T}_s \times \mathbf{T}_t &= (a + b\cos t_0)b\cos t_0\cos s_0\,\mathbf{i} + (a + b\cos t_0)b\cos t_0\sin s_0\,\mathbf{j} \\ &\quad + (a + b\cos t_0)b\sin t_0\,\mathbf{k} \\ &= b(a + b\cos t_0)(\cos t_0\cos s_0\,\mathbf{i} + \cos t_0\sin s_0\,\mathbf{j} + \sin t_0\,\mathbf{k}).\end{aligned}$$

Since $a > b > 0$, the factor $b(a + b\cos t_0)$ is never zero. Furthermore, since the sine and cosine functions are never simultaneously zero, at least one component of $\mathbf{T}_s \times \mathbf{T}_t$ is never zero. Hence, the torus is a smooth parametrized surface. ◆

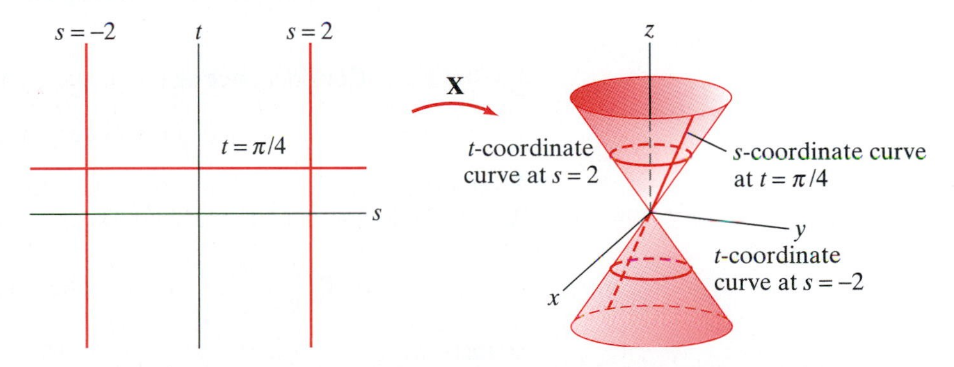

Figure 7.11 The cone $z^2 = x^2 + y^2$ as a parametrized surface.

EXAMPLE 7 The equation $z^2 = x^2 + y^2$ defines a cone in $\mathbf{R}^3$. (See Figure 7.11.) If z is held constant (which corresponds to slicing the surface by a horizontal plane), then the expression $x^2 + y^2$ is constant. Hence, the constant-z cross-sections are circles of radius $|z|$, or a single point in the case of the vertex. This suggests that we can parametrize the cone by using one parameter variable for z and another for the angle around the z-axis. Thus, we obtain the following equations:

$$\begin{cases} x = s\cos t \\ y = s\sin t \quad 0 \leq t \leq 2\pi. \\ z = s \end{cases}$$

Then we have

$$\mathbf{T}_s = \cos t\,\mathbf{i} + \sin t\,\mathbf{j} + \mathbf{k} \quad \text{and} \quad \mathbf{T}_t = -s\sin t\,\mathbf{i} + s\cos t\,\mathbf{j}.$$

Therefore,

$$\mathbf{T}_s \times \mathbf{T}_t = \begin{bmatrix} \mathbf{i} & \mathbf{j} & \mathbf{k} \\ \cos t & \sin t & 1 \\ -s\sin t & s\cos t & 0 \end{bmatrix} = -s\cos t\,\mathbf{i} - s\sin t\,\mathbf{j} + s\,\mathbf{k}.$$

Note that $\mathbf{T}_s \times \mathbf{T}_t = \mathbf{0}$ when (and only when) $s = 0$. The cone fails to be smooth just at its vertex (the single point of the underlying surface corresponding to $s = 0$). ◆

Examples 6 and 7 suggest why the terminology "smooth" is used: Intuitively, a parametrized surface is smooth at a point if it has no sharp "cusps" or "corners" there. This is true of the torus, but not of the cone, which has a singularity at its vertex.

If a parametrized surface is smooth at a point $\mathbf{X}(s_0, t_0)$, then we define the **tangent plane** to $S = \mathbf{X}(D)$ at the point $\mathbf{X}(s_0, t_0)$ to be the plane that passes through $\mathbf{X}(s_0, t_0)$ and has

$$\mathbf{N}(s_0, t_0) = \mathbf{T}_s(s_0, t_0) \times \mathbf{T}_t(s_0, t_0)$$

as normal vector. To write an equation for this plane, we denote (x, y, z) by the (variable) vector $\mathbf{x}$. Then the tangent plane equation is

$$\mathbf{N}(s_0, t_0) \cdot (\mathbf{x} - \mathbf{X}(s_0, t_0)) = 0. \tag{4}$$

If we write the components of $\mathbf{N}(s_0, t_0)$ as $A\,\mathbf{i} + B\,\mathbf{j} + C\,\mathbf{k}$ and $\mathbf{X}(s_0, t_0)$ as $(x(s_0, t_0), y(s_0, t_0), z(s_0, t_0))$, then we may expand equation (4) to obtain

$$A(x - x(s_0, t_0)) + B(y - y(s_0, t_0)) + C(z - z(s_0, t_0)) = 0. \tag{5}$$

EXAMPLE 8 Consider once again the parametrized cone of Example 7 as

$$\mathbf{X}(s, t) = (s\cos t, s\sin t, s).$$

From the calculations in Example 7, the cone is smooth at the point $(0, 1, 1) = \mathbf{X}(1, \pi/2)$, and so the tangent plane exists at that point. We have

$$\mathbf{T}_s\left(1, \frac{\pi}{2}\right) = \mathbf{j} + \mathbf{k} \quad \text{and} \quad \mathbf{T}_t\left(1, \frac{\pi}{2}\right) = -\mathbf{i},$$

so that

$$\mathbf{N} = \mathbf{T}_s\left(1, \frac{\pi}{2}\right) \times \mathbf{T}_t\left(1, \frac{\pi}{2}\right) = \begin{vmatrix} \mathbf{i} & \mathbf{j} & \mathbf{k} \\ 0 & 1 & 1 \\ -1 & 0 & 0 \end{vmatrix} = -\mathbf{j} + \mathbf{k}.$$

Hence, equation (5) can be applied to verify that an equation for the tangent plane is

$$0(x - 0) - 1(y - 1) + 1(z - 1) = 0$$

or, more simply,

$$z = y.$$ ◆

EXAMPLE 9 If $f(x, y)$ is of class C^1 in a neighborhood of a point (x_0, y_0) in its domain D, then the graph of f is a smooth parametrized surface at $(x_0, y_0, f(x_0, y_0))$. Recall from Example 4 that the graph of f is parametrized by $\mathbf{X}: D \to \mathbf{R}^3$, $\mathbf{X}(s, t) = (s, t, f(s, t))$. Then

$$\mathbf{T}_s = \mathbf{i} + \frac{\partial f}{\partial s}\mathbf{k} \quad \text{and} \quad \mathbf{T}_t = \mathbf{j} + \frac{\partial f}{\partial t}\mathbf{k},$$

so that

$$\mathbf{N} = \mathbf{T}_s \times \mathbf{T}_t = -\frac{\partial f}{\partial s}\mathbf{i} - \frac{\partial f}{\partial t}\mathbf{j} + \mathbf{k}.$$

Note that $\mathbf{N}$ is nonzero at any point $(s, t, f(s, t)) = (x, y, f(x, y))$.

Next, consider the surface defined as the level set

$$S = \{(x, y, z) \mid F(x, y, z) = c, c \text{ constant}\}.$$

If F is of class C^1 in a neighborhood of a point $(x_0, y_0, z_0) \in S$ and $\nabla F(x_0, y_0, z_0) \neq \mathbf{0}$, then the implicit function theorem (Theorem 6.5 of §2.6) implies that, in principle at least, the defining equation $F(x, y, z) = c$ of S always can be solved locally for (at least) one of the variables x, y, or z in terms of the other two. In other words, under the given assumptions on F, the level set S is locally the graph of a C^1 function of two variables. It is important to remember that the idea of "solving locally" does not mean that *all* points of S can be described as the graph of a single function (as we already know quite well in the case of the sphere $x^2 + y^2 + z^2 = 1$, for example), but rather that, near points $(x_0, y_0, z_0) \in S$ where $\nabla F(x_0, y_0, z_0) \neq \mathbf{0}$, a *portion* of S may be described as a graph. Hence, graphs of C^1 functions of two variables and level sets of C^1 functions of three variables, under certain smoothness hypotheses, are locally equivalent descriptions for surfaces. Moreover, since the graph of a C^1 function is a smooth parametrized surface, we may shift relatively freely among our three descriptions for surfaces. ◆

Smooth parametrized surfaces are of primary importance because of the ease with which we may adapt techniques of calculus (particularly integral calculus) to them. But we are also interested in piecewise smooth parametrized surfaces.

■ **Definition 1.3** A **piecewise smooth** parametrized surface is the union of images of finitely many parametrized surfaces $\mathbf{X}_i: D_i \to \mathbf{R}^3$, $i = 1, \ldots, m$, where

- Each D_i is a region in $\mathbf{R}^2$ consisting of a connected open set, possibly together with some of its boundary points (for the most part, we want D_i to be an elementary region);
- Each $\mathbf{X}_i$ is of class C^1 and one-one on all of D_i, except possibly along ∂D_i;
- Each $S_i = \mathbf{X}_i(D_i)$ is smooth, except possibly at finitely many points.

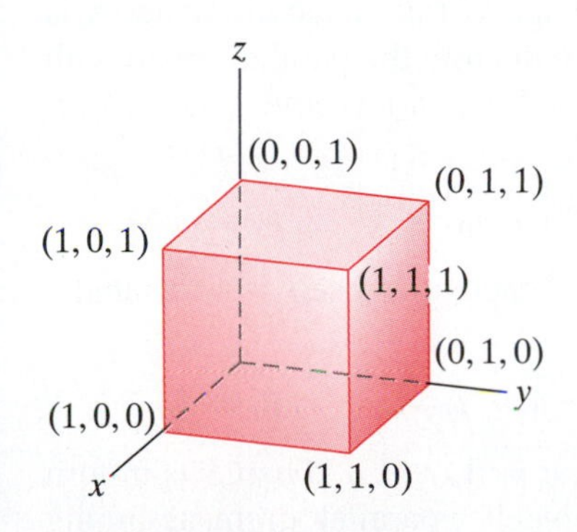

Figure 7.12 A cube.

EXAMPLE 10 The surface of a cube is a piecewise smooth parametrized surface. It is the union of its six faces, each one of which is a smooth parametrized surface, namely, a portion of a plane.

More explicitly, suppose that a cube's faces are portions of the planes $x = 0$, $x = 1$, $y = 0$, $y = 1$, $z = 0$, and $z = 1$ as in Figure 7.12. Then we may parametrize the cube's faces by $\mathbf{X}_i: [0, 1] \times [0, 1] \to \mathbf{R}^3$, $i = 1, \ldots, 6$, where

$$\mathbf{X}_1(s, t) = (0, s, t); \qquad \mathbf{X}_2(s, t) = (1, s, t); \qquad \mathbf{X}_3(s, t) = (s, 0, t);$$

$$\mathbf{X}_4(s, t) = (s, 1, t); \qquad \mathbf{X}_5(s, t) = (s, t, 0); \qquad \mathbf{X}_6(s, t) = (s, t, 1).$$

Each map $\mathbf{X}_i$ is clearly of class C^1 and one-one. In addition, the faces have well-defined nonzero normal vectors. For example, for both $\mathbf{X}_1$ and $\mathbf{X}_2$,

$$\mathbf{N}_1 = \mathbf{N}_2 = \mathbf{T}_s \times \mathbf{T}_t = \mathbf{j} \times \mathbf{k} = \mathbf{i}.$$

Similarly,

$$\mathbf{N}_3 = \mathbf{N}_4 = \mathbf{i} \times \mathbf{k} = -\mathbf{j} \quad \text{and} \quad \mathbf{N}_5 = \mathbf{N}_6 = \mathbf{i} \times \mathbf{j} = \mathbf{k}.$$

None of these vectors vanishes. There is no consistent way to define normal vectors along the edges of the cube (where two faces meet). That is why the cube is only *piecewise* smooth. ◆

Area of a Smooth Parametrized Surface

Now, we use the notion of a parametrized surface to calculate the surface area of a smooth surface. In the discussion that follows, we take $S = \mathbf{X}(D)$ to be a smooth parametrized surface, where D is the union of finitely many elementary regions in $\mathbf{R}^2$ and $\mathbf{X}: D \to \mathbf{R}^3$ is of class C^1 and one-one except possibly along ∂D.

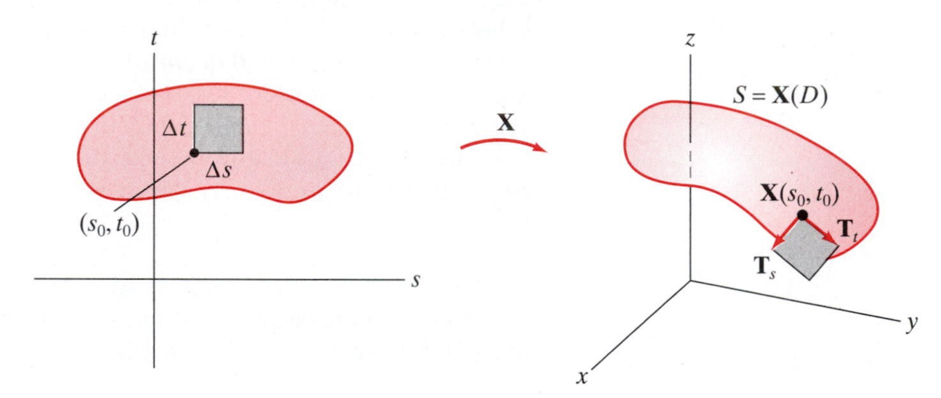

Figure 7.13 The image of the $\Delta s \times \Delta t$ rectangle in D is approximately a parallelogram spanned by $\mathbf{T}_s(s_0, t_0)\Delta s$ and $\mathbf{T}_t(s_0, t_0)\Delta t$.

The key geometric observation is as follows: Consider a small rectangular subset of D whose lower left corner is at the point $(s_0, t_0) \in D$ and whose width and height are Δs and Δt, respectively. The image of this rectangle under $\mathbf{X}$ is a piece of the underlying surface S that is approximately the parallelogram with a corner at $\mathbf{X}(s_0, t_0)$ and spanned by the vectors $\mathbf{T}_s(s_0, t_0)\Delta s$ and $\mathbf{T}_t(s_0, t_0)\Delta t$. (See Figure 7.13.) The area ΔA of this piece is

$$\Delta A \approx \|\mathbf{T}_s(s_0, t_0)\Delta s \times \mathbf{T}_t(s_0, t_0)\Delta t\| = \|\mathbf{T}_s(s_0, t_0) \times \mathbf{T}_t(s_0, t_0)\|\Delta s \Delta t.$$

Now, suppose $D = [a, b] \times [c, d]$; that is, suppose D itself is a rectangle. Partition D into n^2 subrectangles via

$$a = s_0 < s_1 < \cdots < s_n = b \quad \text{and} \quad c = t_0 < t_1 < \cdots < t_n = d.$$

Let $\Delta s_i = s_i - s_{i-1}$ and $\Delta t_j = t_j - t_{j-1}$ for $i, j = 1, \ldots, n$. Then S is in turn partitioned into pieces, each of which is approximately a parallelogram, assuming Δs_i and Δt_j are small for $i, j = 1, \ldots, n$. If ΔA_{ij} denotes the area of the piece of S that is the image of the ijth subrectangle of D, then

$$\begin{aligned}\text{Surface area of } S &= \sum_{i,j=1}^{n} \Delta A_{ij} \\ &\approx \sum_{i,j=1}^{n} \|\mathbf{T}_s(s_{i-1}, t_{j-1}) \times \mathbf{T}_t(s_{i-1}, t_{j-1})\| \Delta s_i \Delta t_j.\end{aligned}$$

Therefore, it makes sense to define

$$\text{Surface area of } S = \lim_{\text{all } \Delta s_i, \Delta t_j \to 0} \sum_{i,j=1}^{n} \Delta A_{ij} = \int_c^d \int_a^b \|\mathbf{T}_s \times \mathbf{T}_t\| \, ds\, dt$$

and, in general, where D is an arbitrary region (that is, not necessarily a rectangle),

$$\text{Surface area of } S = \iint_D \|\mathbf{T}_s \times \mathbf{T}_t\| \, ds\, dt. \tag{6}$$

Formula (6) can be extended readily to the case where S is a piecewise smooth parametrized surface by breaking up the integral in an appropriate manner.

EXAMPLE 11 We use formula (6) to calculate the surface area of a sphere of radius a. Recall from Example 2 that the map

$$\mathbf{X}(s, t) = (a\cos s \sin t, a \sin s \sin t, a \cos t), \quad 0 \le s < 2\pi, \quad 0 \le t \le \pi$$

parametrizes the sphere in a one-one fashion. Then

$$\mathbf{T}_s = -a \sin s \; \sin t\, \mathbf{i} + a \cos s \; \sin t\, \mathbf{j}$$

and

$$\mathbf{T}_t = a \cos s \; \cos t\, \mathbf{i} + a \sin s \; \cos t\, \mathbf{j} - a \sin t\, \mathbf{k}.$$

Hence,

$$\mathbf{T}_s \times \mathbf{T}_t = -a^2 \sin t\, (\cos s \; \sin t\, \mathbf{i} + \sin s \; \sin t\, \mathbf{j} + \cos t\, \mathbf{k}),$$

so that

$$\|\mathbf{T}_s \times \mathbf{T}_t\| = a^2 \sin t.$$

Therefore, formula (6) becomes

$$\begin{aligned}\text{Surface area} &= \int_0^\pi \int_0^{2\pi} a^2 \sin t \, ds\, dt = \int_0^\pi 2\pi a^2 \sin t \, dt \\ &= 2\pi a^2 (-\cos t)\big|_0^\pi = 2\pi a^2(1 + 1) = 4\pi a^2.\end{aligned}$$

This result checks with the well-known formula for the surface area of a sphere. Note, however, that if we let $0 \le s \le 4\pi$, $0 \le t \le \pi$, then the image of $\mathbf{X}$ is the same sphere, but formula (6) would yield

$$\int_0^\pi \int_0^{4\pi} a^2 \sin t \, ds\, dt = 8\pi a^2.$$

This "overcount" is why we must assume that the map $\mathbf{X}$ is (nearly) one-one. (Note: The aforementioned parametrization fails to be smooth at $t = 0$ and at $t = \pi$, but this is along ∂D and so does not affect the surface area integral.) ◆

If we write the component functions of $\mathbf{X}$ as

$$\mathbf{X}(s, t) = (x(s, t), y(s, t), z(s, t)),$$

we find that

$$\begin{aligned}\mathbf{T}_s \times \mathbf{T}_t &= \begin{vmatrix} \mathbf{i} & \mathbf{j} & \mathbf{k} \\ \dfrac{\partial x}{\partial s} & \dfrac{\partial y}{\partial s} & \dfrac{\partial z}{\partial s} \\ \dfrac{\partial x}{\partial t} & \dfrac{\partial y}{\partial t} & \dfrac{\partial z}{\partial t} \end{vmatrix} \\ &= \left(\frac{\partial y}{\partial s}\frac{\partial z}{\partial t} - \frac{\partial y}{\partial t}\frac{\partial z}{\partial s}\right)\mathbf{i} + \left(\frac{\partial x}{\partial t}\frac{\partial z}{\partial s} - \frac{\partial x}{\partial s}\frac{\partial z}{\partial t}\right)\mathbf{j} + \left(\frac{\partial x}{\partial s}\frac{\partial y}{\partial t} - \frac{\partial x}{\partial t}\frac{\partial y}{\partial s}\right)\mathbf{k}.\end{aligned}$$

Using the notation of the Jacobian, we obtain

$$\mathbf{N}(s, t) = \mathbf{T}_s \times \mathbf{T}_t = \frac{\partial(y, z)}{\partial(s, t)}\mathbf{i} - \frac{\partial(x, z)}{\partial(s, t)}\mathbf{j} + \frac{\partial(x, y)}{\partial(s, t)}\mathbf{k}. \tag{7}$$

This alternative formula for the normal vector to a smooth parametrized surface will prove useful to us on occasion. For the moment, we take its magnitude:

$$\|\mathbf{N}(s, t)\| = \sqrt{\left(\frac{\partial(x, y)}{\partial(s, t)}\right)^2 + \left(\frac{\partial(x, z)}{\partial(s, t)}\right)^2 + \left(\frac{\partial(y, z)}{\partial(s, t)}\right)^2}.$$

Hence, formula (6) may also be written as

Surface area of S

$$= \iint_D \sqrt{\left(\frac{\partial(x, y)}{\partial(s, t)}\right)^2 + \left(\frac{\partial(x, z)}{\partial(s, t)}\right)^2 + \left(\frac{\partial(y, z)}{\partial(s, t)}\right)^2}\, ds\, dt. \tag{8}$$

EXAMPLE 12 Find the surface area of the torus described in Example 5. Recall that the torus is parametrized as

$$\begin{cases} x = (a + b\cos t)\cos s \\ y = (a + b\cos t)\sin s \\ z = b\sin t \end{cases} \quad 0 \le s, t \le 2\pi, \quad a > b > 0.$$

Thus,

$$\begin{aligned}\frac{\partial(x, y)}{\partial(s, t)} &= \begin{vmatrix} -(a + b\cos t)\sin s & -b\sin t\, \cos s \\ (a + b\cos t)\cos s & -b\sin t\, \sin s \end{vmatrix} \\ &= (a + b\cos t)(b\sin t\sin^2 s + b\sin t\, \cos^2 s) \\ &= (a + b\cos t)b\sin t,\end{aligned}$$

$$\begin{aligned}\frac{\partial(x, z)}{\partial(s, t)} &= \begin{vmatrix} -(a + b\cos t)\sin s & -b\sin t\, \cos s \\ 0 & b\cos t \end{vmatrix} \\ &= -(a + b\cos t)b\cos t\, \sin s,\end{aligned}$$

and

$$\frac{\partial(y,z)}{\partial(s,t)} = \begin{vmatrix} (a+b\cos t)\cos s & -b\sin t\ \sin s \\ 0 & b\cos t \end{vmatrix}$$
$$= (a+b\cos t)b\cos t\ \cos s.$$

By formula (8), we have

Surface area

$$= \int_0^{2\pi}\int_0^{2\pi} \sqrt{(a+b\cos t)^2[b^2\sin^2 t + b^2\cos^2 t\sin^2 s + b^2\cos^2 t\cos^2 s]}\, ds\, dt.$$

Using the trigonometric identity $\cos^2\theta + \sin^2\theta = 1$ twice, we simplify the integral to

$$\int_0^{2\pi}\int_0^{2\pi} (a+b\cos t)b\, ds\, dt = \int_0^{2\pi} 2\pi b(a+b\cos t)\, dt$$
$$= 2\pi b(at + b\sin t)\Big|_0^{2\pi}$$
$$= 4\pi^2 ab. \quad ◆$$

EXAMPLE 13 Suppose that a smooth surface is described as the graph of a C^1 function $f(x, y)$, that is, by the equation $z = f(x, y)$, where (x, y) varies through a plane region D. Then the standard parametrization $\mathbf{X}(s, t) = (s, t, f(s, t))$ implies

$$\mathbf{T}_s \times \mathbf{T}_t = -f_s\,\mathbf{i} - f_t\,\mathbf{j} + \mathbf{k}.$$

(See Example 9.) Formula (6) yields

$$\text{Surface area} = \iint_D \|\mathbf{T}_s \times \mathbf{T}_t\|\, ds\, dt = \iint_D \sqrt{f_s^2 + f_t^2 + 1}\, ds\, dt.$$

Since $x = s$, $y = t$ in this parametrization of the graph, we conclude that

Surface area of the graph of $f(x, y)$ over D

$$= \iint_D \sqrt{f_x^2 + f_y^2 + 1}\, dx\, dy. \tag{9}$$

◆

One final note: It is not at all clear that either formula (6) or formula (8) depends only on the underlying surface $S = \mathbf{X}(D)$ and not on the particular parametrization $\mathbf{X}$. These formulas are independent of the parametrization, as we shall observe in the following section, in the context of general surface integrals.

Exercises

1. Let $\mathbf{X}\colon \mathbf{R}^2 \to \mathbf{R}^3$ be the parametrized surface given by

$$\mathbf{X}(s, t) = (s^2 - t^2, s + t, s^2 + 3t).$$

(a) Determine a normal vector to this surface at the point

$$(3, 1, 1) = \mathbf{X}(2, -1).$$

(b) Find an equation for the plane tangent to this surface at the point $(3, 1, 1)$.

2. Find an equation for the plane tangent to the torus

$$\mathbf{X}(s, t) = ((5+2\cos t)\cos s, (5+2\cos t)\sin s, 2\sin t)$$

at the point $\left((5-\sqrt{3})/\sqrt{2}, (5-\sqrt{3})/\sqrt{2}, 1\right)$.

3. Find an equation for the plane tangent to the surface

$$x = e^s, \qquad y = t^2 e^{2s}, \qquad z = 2e^{-s} + t$$

at the point $(1, 4, 0)$.

4. Let $\mathbf{X}(s, t) = (s^2 \cos t, s^2 \sin t, s)$, $-3 \le s \le 3$, $0 \le t \le 2\pi$.

(a) Find a normal vector at $(s, t) = (-1, 0)$.

(b) Determine the tangent plane at the point $(1, 0, -1)$.

(c) Find an equation for the image of $\mathbf{X}$ in the form $F(x, y, z) = 0$.

5. Consider the parametrized surface $\mathbf{X}(s, t) = (s, s^2 + t, t^2)$.

(a) Graph this surface for $-2 \le s \le 2$, $-2 \le t \le 2$. (Using a computer may help.)

(b) Is the surface smooth?

(c) Find an equation for the tangent plane at the point $(1, 0, 1)$.

6. Describe the parametrized surface of Exercise 1 by an equation of the form $z = f(x, y)$.

7. Let S be the surface parametrized by $x = s \cos t$, $y = s \sin t$, $z = s^2$, where $s \ge 0$, $0 \le t \le 2\pi$.

(a) At what points is S smooth? Find an equation for the tangent plane at the point $(1, \sqrt{3}, 4)$.

(b) Sketch the graph of S. Can you recognize S as a familiar surface?

(c) Describe S by an equation of the form $z = f(x, y)$.

(d) Using your answer in part (c), discuss whether S has a tangent plane at every point.

8. Verify that the image of the parametrized surface

$$\mathbf{X}(s, t) = (2 \sin s \cos t, 3 \sin s \sin t, \cos s), \qquad 0 \le s \le \pi, \quad 0 \le t \le 2\pi,$$

is an ellipsoid.

9. Verify that, for the torus of Example 5, the s-coordinate curve, when $t = t_0$, is a circle of radius $a + b \cos t_0$.

10. The surface in $\mathbf{R}^3$ parametrized by

$$\mathbf{X}(r, \theta) = (r \cos \theta, r \sin \theta, \theta), \quad r \ge 0, \quad -\infty < \theta < \infty,$$

is called a **helicoid**.

(a) Describe the r-coordinate curve when $\theta = \pi/3$. Give a general description of the r-coordinate curves.

(b) Describe the θ-coordinate curve when $r = 1$. Give a general description of the θ-coordinate curves.

(c) Sketch the graph of the helicoid (perhaps using a computer) for $0 \le r \le 1$, $0 \le \theta \le 4\pi$. Can you see why the surface is called a helicoid?

11. Given the sphere of radius 2 centered at $(2, -1, 0)$, find an equation for the plane tangent to it at the point $(1, 0, \sqrt{2})$ in three ways:

(a) by considering the sphere as the graph of the function

$$f(x, y) = \sqrt{4 - (x-2)^2 - (y+1)^2};$$

(b) by considering the sphere as a level surface of the function

$$F(x, y, z) = (x-2)^2 + (y+1)^2 + z^2;$$

(c) by considering the sphere as the surface parametrized by

$$\mathbf{X}(s, t) = (2 \sin s \cos t + 2, 2 \sin s \sin t - 1, 2 \cos s).$$

12. This problem concerns the parametrized surface $\mathbf{X}(s, t) = (s^3, t^3, st)$.

(a) Find an equation of the plane tangent to this surface at the point $(1, -1, -1)$.

(b) Is this surface smooth? Why or why not?

(c) Use a computer to graph this surface for $-1 \le s \le 1$, $-1 \le t \le 1$.

(d) Verify that this surface may also be described by the xyz-coordinate equation $z = \sqrt[3]{xy}$. Try using a computer to graph the surface when described in this form. Many software systems will have trouble, or will provide an incomplete graph, which is one reason why parametric descriptions of surfaces are desirable.

13. The parametrized surface $\mathbf{X}(s, t) = (st, t, s^2)$ is known as the **Whitney umbrella.**

(a) Verify that this surface may also be described by the xyz-coordinate equation $y^2 z = x^2$.

(b) Is $\mathbf{X}$ smooth?

(c) Use a computer to graph this surface for $-2 \le s \le 2$, $-2 \le t \le 2$.

(d) Some points (x, y, z) of the surface do not correspond to a single parameter point (s, t). Which ones? Explain how this relates to the graph.

(e) Give an equation of the plane tangent to this surface at the point $(2, 1, 4)$.

(f) Show that at the point $(0, 0, 1)$ on the image of $\mathbf{X}$ it's reasonable to conclude that there are *two* tangent planes. Give equations for them.

14. Let S be the surface defined as the graph of a function $f(x, y)$ of class C^1. Then Example 4 shows that S is also a parametrized surface. Show that formula (5) for the tangent plane to S at $(a, b, f(a, b))$ agrees with that of formula (4) in §2.3.

15. (a) Write a formula for the tangent plane to a surface described by the equation $y = g(x, z)$.

(b) Repeat part (a) for a surface described by the equation $x = h(y, z)$.

16. Suppose $\mathbf{X}\colon D \to \mathbf{R}^3$ is a parametrized surface that is smooth at $\mathbf{X}(s_0, t_0)$. Show how the definition of the derivative $D\mathbf{X}(s_0, t_0)$ (see Definition 3.8 of Chapter 2) can be used to give vector parametric equations for the plane tangent to $S = \mathbf{X}(D)$ at the point $\mathbf{X}(s_0, t_0)$.

17. Use the result of Exercise 16 to provide parametric equations for the plane tangent to the surface $\mathbf{X}(s, t) = (s, s^2 + t, t^2)$ at the point $(1, 0, 1)$. Verify that your answer is consistent with that of Exercise 5(c).

18. Use the parametrization in Example 3 to verify that the surface area of a cylinder of radius a and height h is $2\pi ah$.

19. Let D denote the unit disk in the st-plane. Let $\mathbf{X}\colon D \to \mathbf{R}^3$ be defined by $(s + t, s - t, s)$. Find the surface area of $\mathbf{X}(D)$.

20. Find the surface area of the helicoid

$$\mathbf{X}\colon D \to \mathbf{R}^3, \quad \mathbf{X}(r, \theta) = (r\cos\theta, r\sin\theta, \theta)$$

for $0 \le r \le 1$, $0 \le \theta \le 2\pi n$, where n is a positive integer.

21. A cylindrical hole of radius b is bored through a ball of radius a $(> b)$ to form a ring. Find the outer surface area of the ring.

22. Find the area of the portion of the paraboloid $z = 9 - x^2 - y^2$ that lies over the xy-plane.

23. Find the area of the surface cut from the paraboloid $z = 2x^2 + 2y^2$ by the planes $z = 2$ and $z = 8$.

24. Calculate the surface area of the portion of the plane $x + y + z = a$ cut out by the cylinder $x^2 + y^2 = a^2$ in two ways:

(a) by using formula (6);

(b) by using formula (9).

25. Let S be the surface defined by the equation $z = f(x, y)$. If $f_x^2 + f_y^2 = a$, where a is a positive constant, determine the surface area of the portion of S that lies over a region D in the xy-plane in terms of the area of D.

26. Let S be the surface defined by

$$z = \frac{1}{\sqrt{x^2 + y^2}} \quad \text{for } z \ge 1.$$

(a) Sketch the graph of this surface.

(b) Show that the volume of the region bounded by S and the plane $z = 1$ is finite. (You will need to use an improper integral.)

(c) Show that the surface area of S is infinite.

27. Find the surface area of the intersection of the cylinders $x^2 + y^2 = a^2$ and $y^2 + z^2 = a^2$.

28. Suppose that a surface is given in cylindrical coordinates by the equation $z = f(r, \theta)$, where (r, θ) varies through a region D in the $r\theta$-plane where r is nonnegative. Show that the surface area of the surface is given by

$$\iint_D \sqrt{1 + \left(\frac{\partial f}{\partial r}\right)^2 + \frac{1}{r^2}\left(\frac{\partial f}{\partial \theta}\right)^2}\, r\, dr\, d\theta.$$

29. Suppose that a surface is given in spherical coordinates by the equation $\rho = f(\varphi, \theta)$, where (φ, θ) varies through a region D in the $\varphi\theta$-plane and $f(\varphi, \theta)$ is nonnegative. Show that the surface area of the surface is given by

$$\iint_D f(\varphi, \theta) \times \sqrt{\left(f(\varphi, \theta)^2 + f_\varphi(\varphi, \theta)^2\right)\sin^2\varphi + f_\theta(\varphi, \theta)^2}\, d\varphi\, d\theta.$$

7.2 Surface Integrals

In this section, we will learn how to integrate both scalar-valued functions and vector fields along surfaces in $\mathbf{R}^3$. We proceed in a manner that is largely analogous to our explorations of line integrals in §6.1: We begin by defining suitable integrals over parametrized surfaces and then establish that the particular choice of parametrization doesn't much matter—that, really, only the underlying surface is important, and possibly the orientation.

Scalar Surface Integrals

Suppose S is a bounded surface in $\mathbf{R}^3$ and $f(x, y, z)$ is a continuous, scalar-valued function whose domain includes S. Then we want the surface integral $\iint_S f\, dS$ to be a limit of some kind of Riemann sum. So suppose S is partitioned into finitely many small pieces and that the area of the kth piece is ΔA_k. Let $\mathbf{c}_k$ denote an

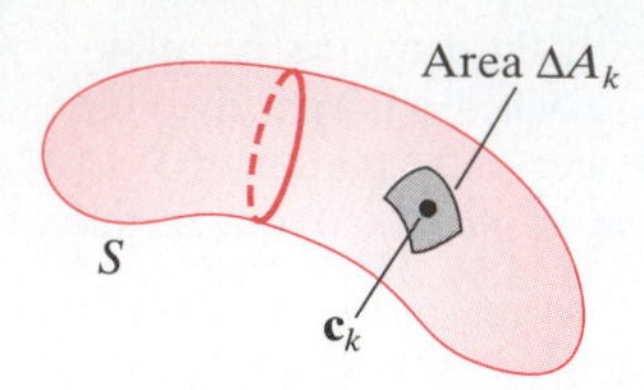

Figure 7.14 A small piece of the surface S has area ΔA_k. The point $\mathbf{c}_k$ is located in this surface piece.

arbitrary "test point" in the kth piece. (See Figure 7.14.) Then the surface integral of f over S should be

$$\iint_S f\,dS = \lim_{\text{all } \Delta A_k \to 0} \sum_k f(\mathbf{c}_k)\Delta A_k, \tag{1}$$

provided, of course, that this limit exists.

Now, we add some formalism to provide a proper definition. Suppose that S is a smooth parametrized surface; that is, suppose that S is the image of the C^1 map $\mathbf{X}: D \to \mathbf{R}^3$, where D is a connected, bounded region in $\mathbf{R}^2$. Let f be a continuous function defined on $S = \mathbf{X}(D)$. As seen in §7.1, the small rectangle in D, having dimensions Δs and Δt with lower left corner at the point (s_0, t_0), is mapped by $\mathbf{X}$ to a piece of surface that is approximately a parallelogram of area

$$\Delta A \approx \|\mathbf{T}_s(s_0, t_0) \times \mathbf{T}_t(s_0, t_0)\|\Delta s \Delta t.$$

(See Figure 7.13, page 414.) Suppose D is the rectangle $[a, b] \times [c, d]$ and that we partition D by

$$a = s_0 < s_1 < \cdots < s_n = b \quad \text{and} \quad c = t_0 < t_1 < \cdots < t_n = d.$$

Then the limiting sum that formula (1) represents is

$$\lim_{\text{all } \Delta s_i, \Delta t_j \to 0} \sum_{i,j=1}^{n} f(\mathbf{X}(s_i^*, t_i^*))\, \|\mathbf{T}_s(s_{i-1}, t_{j-1}) \times \mathbf{T}_t(s_{i-1}, t_{j-1})\|\, \Delta s_i\, \Delta t_j, \tag{2}$$

where

$$\Delta s_i = s_i - s_{i-1}, \quad \Delta t_j = t_j - t_{j-1}, \quad s_{i-1} \le s_i^* \le s_i, \quad \text{and} \quad t_{j-1} \le t_j^* \le t_j.$$

(Thus, $\mathbf{X}(s_i^*, t_j^*)$ is an arbitrary point in the image of the subrectangle $[s_{i-1}, s_i] \times [t_{j-1}, t_j]$, and hence a "test point" in the corresponding small surface piece.) But then the limit in formula (2) is

$$\int_c^d \int_a^b f(\mathbf{X}(s, t))\|\mathbf{T}_s \times \mathbf{T}_t\|\, ds\, dt.$$

When D is a more arbitrary region than a rectangle, it makes sense to use the following definition for the surface integral of a function over a parametrized surface:

■ **Definition 2.1** Let $\mathbf{X}: D \to \mathbf{R}^3$ be a smooth parametrized surface, where $D \subset \mathbf{R}^2$ is a bounded region. Let f be a continuous function whose domain includes $S = \mathbf{X}(D)$. Then the **scalar surface integral** of f along $\mathbf{X}$, denoted $\iint_{\mathbf{X}} f\,dS$, is

$$\begin{aligned}\iint_{\mathbf{X}} f\,dS &= \iint_D f(\mathbf{X}(s, t))\, \|\mathbf{T}_s \times \mathbf{T}_t\|\, ds\, dt \\ &= \iint_D f(\mathbf{X}(s, t))\, \|\mathbf{N}(s, t)\|\, ds\, dt.\end{aligned}$$

Although we need not assume that the map $\mathbf{X}$ is one-one on D in order to work with the integral in Definition 2.1, in practice we usually find it useful to take $\mathbf{X}$ to be one-one, except perhaps along ∂D. If this is the case, and if f is identically 1 on all of $\mathbf{X}(D)$, then

$$\iint_{\mathbf{X}} f\,dS = \iint_{\mathbf{X}} 1\,dS = \iint_D \|\mathbf{T}_s \times \mathbf{T}_t\|\, ds\, dt = \text{surface area of } \mathbf{X}(D),$$

as stated by formula (6) in §7.1. The scalar surface integral in Definition 2.1 is thus a generalization of the integral we use to calculate surface area. We can think of $\iint_{\mathbf{X}} f\, dS$ as the limit of a "weighted sum" of surface area pieces, the weightings given by f. If f represents mass or electrical charge density, then $\iint_{\mathbf{X}} f\, dS$ yields the total mass or total charge on $\mathbf{X}(D)$ (assuming $\mathbf{X}$ is one-one, except perhaps along ∂D).

For computational purposes, recall that if we write the components of $\mathbf{X}$ as

$$\mathbf{X}(s,t) = (x(s,t), y(s,t), z(s,t)),$$

then

$$\mathbf{N}(s,t) = \mathbf{T}_s \times \mathbf{T}_t = \frac{\partial(y,z)}{\partial(s,t)}\mathbf{i} - \frac{\partial(x,z)}{\partial(s,t)}\mathbf{j} + \frac{\partial(x,y)}{\partial(s,t)}\mathbf{k}.$$

We obtain

$$\iint_{\mathbf{X}} f\, dS = \iint_D f(x(s,t), y(s,t), z(s,t)) \times \sqrt{\left(\frac{\partial(x,y)}{\partial(s,t)}\right)^2 + \left(\frac{\partial(x,z)}{\partial(s,t)}\right)^2 + \left(\frac{\partial(y,z)}{\partial(s,t)}\right)^2}\, ds\, dt. \quad (3)$$

EXAMPLE 1 We evaluate $\iint_{\mathbf{X}} z^3\, dS$, where $\mathbf{X}\colon [0, 2\pi] \times [0, \pi] \to \mathbf{R}^3$ is the parametrized sphere of radius a:

$$\mathbf{X}(s,t) = (a\cos s\,\sin t, a\sin s\,\sin t, a\cos t).$$

Using Definition 2.1 or its reformulation in formula (3), we find that

$$\begin{aligned}
\|\mathbf{N}(s,t)\| &= \sqrt{\left(\frac{\partial(x,y)}{\partial(s,t)}\right)^2 + \left(\frac{\partial(x,z)}{\partial(s,t)}\right)^2 + \left(\frac{\partial(y,z)}{\partial(s,t)}\right)^2} \\
&= \sqrt{a^4(\sin^2 t\,\cos^2 t + \sin^2 s\,\sin^4 t + \cos^2 s\,\sin^4 t)} \\
&= a^2\sqrt{\sin^2 t\,\cos^2 t + \sin^4 t} \\
&= a^2\sqrt{\sin^2 t\,(\cos^2 t + \sin^2 t)} \\
&= a^2 \sin t.
\end{aligned}$$

(See also Example 11 in §7.1.) Hence,

$$\begin{aligned}
\iint_{\mathbf{X}} z^3\, dS &= \int_0^{\pi}\int_0^{2\pi} (a\cos t)^3 a^2 \sin t\, ds\, dt = a^5 \int_0^{\pi} 2\pi \cos^3 t\,\sin t\, dt \\
&= 2\pi a^5 \left(-\tfrac{1}{4}\cos^4 t\right)\Big|_0^{\pi} = 2\pi a^5 \left(-\tfrac{1}{4} - \left(-\tfrac{1}{4}\right)\right) = 0.
\end{aligned}$$

◆

To define and evaluate scalar surface integrals over *piecewise* smooth parametrized surfaces, simply calculate the surface integral over each smooth piece and add the results.

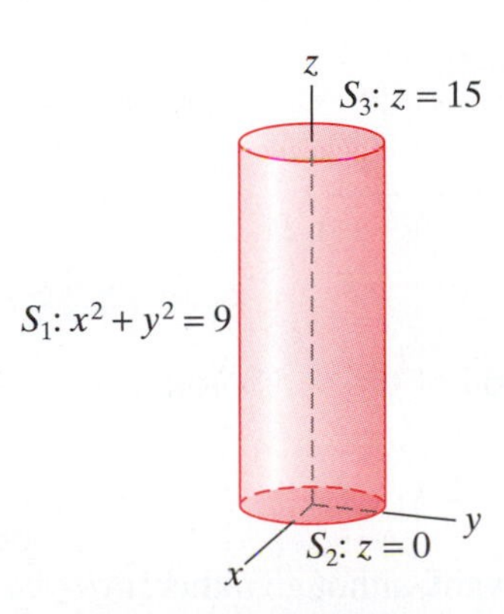

Figure 7.15 The closed cylinder of radius 3 and height 15 of Example 2.

EXAMPLE 2 Let S be the closed cylinder of radius 3 with axis along the z-axis, top face at $z = 15$ and bottom face at $z = 0$, as shown in Figure 7.15. Then S is a piecewise smooth surface; it is the union of the three smooth parametrized surfaces S_1, S_2, and S_3 described next. We calculate $\iint_S z\, dS$.

The three smooth pieces may be parametrized as follows:

$$S_1 \text{ (lateral cylindrical surface):} \quad \begin{cases} x = 3\cos s \\ y = 3\sin s \\ z = t \end{cases} \quad 0 \le s \le 2\pi, \quad 0 \le t \le 15,$$

$$S_2 \text{ (bottom disk):} \quad \begin{cases} x = s\cos t \\ y = s\sin t \\ z = 0 \end{cases} \quad 0 \le s \le 3, \quad 0 \le t \le 2\pi,$$

and

$$S_3 \text{ (top disk):} \quad \begin{cases} x = s\cos t \\ y = s\sin t \\ z = 15 \end{cases} \quad 0 \le s \le 3, \quad 0 \le t \le 2\pi.$$

Using Definition 2.1, we have

$$\begin{aligned} \iint_{S_1} z\,dS &= \int_0^{15}\int_0^{2\pi} t\,\|(-3\sin s\,\mathbf{i} + 3\cos s\,\mathbf{j}) \times \mathbf{k}\|\,ds\,dt \\ &= \int_0^{15}\int_0^{2\pi} t\,\|3\sin s\,\mathbf{j} + 3\cos s\,\mathbf{i}\|\,ds\,dt \\ &= \int_0^{15}\int_0^{2\pi} 3t\,ds\,dt = \int_0^{15} 6\pi t\,dt = 3\pi t^2\Big|_0^{15} = 675\pi. \end{aligned}$$

Now, $\iint_{S_2} z\,dS = 0$, since z vanishes along the bottom of S. For S_3, we have

$$\begin{aligned} \iint_{S_3} z\,dS &= \iint_{S_3} 15\,dS = 15 \cdot \iint_{S_3} 1\,dS \\ &= 15 \cdot \text{area of disk} = 15 \cdot (9\pi) = 135\pi. \end{aligned}$$

Therefore,

$$\begin{aligned} \iint_S z\,dS &= \iint_{S_1} z\,dS + \iint_{S_2} z\,dS + \iint_{S_3} z\,dS \\ &= 675\pi + 0 + 135\pi = 810\pi. \end{aligned}$$

◆

If a surface S is given by the graph of $z = g(x, y)$, where g is of class C^1 on some region D in $\mathbf{R}^2$, then S is parametrized by $\mathbf{X}(x, y) = (x, y, g(x, y))$ with $(x, y) \in D$. (See Example 4 of §7.1.) Then, from Example 13 in §7.1,

$$\mathbf{N}(x, y) = -g_x\,\mathbf{i} - g_y\,\mathbf{j} + \mathbf{k},$$

so that

$$\iint_{\mathbf{X}} f\,dS = \iint_D f(x, y, g(x, y))\sqrt{g_x^2 + g_y^2 + 1}\,dx\,dy. \tag{4}$$

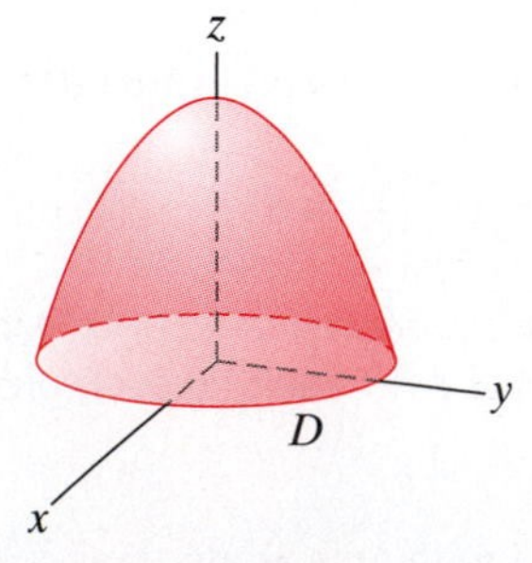

Figure 7.16 The graph of $4 - x^2 - y^2$ over the disk D of radius 2.

EXAMPLE 3 Suppose S is the graph of the portion of the paraboloid $z = 4 - x^2 - y^2$, where (x, y) varies throughout the disk

$$D = \{(x, y) \in \mathbf{R}^2 \mid x^2 + y^2 \le 4\}.$$

(See Figure 7.16.) Formula (4) makes it straightforward, although rather involved, to calculate

$$\iint_{\mathbf{X}} (4 - z)\,dS, \quad \text{where } \mathbf{X}(x, y) = (x, y, 4 - x^2 - y^2).$$

In particular, we have

$$\iint_{\mathbf{X}} (4 - z)\, dS = \iint_D (4 - (4 - x^2 - y^2))\sqrt{4x^2 + 4y^2 + 1}\, dx\, dy$$
$$= \iint_D (x^2 + y^2)\sqrt{4x^2 + 4y^2 + 1}\, dx\, dy.$$

To integrate, we switch to polar coordinates; that is, we let $x = r\cos\theta$ and $y = r\sin\theta$, where $0 \le r \le 2$, $0 \le \theta \le 2\pi$. The desired integral becomes

$$\int_0^{2\pi}\int_0^2 r^2\sqrt{4r^2 + 1}\, r\, dr\, d\theta = \int_0^2\int_0^{2\pi} r^3\sqrt{4r^2 + 1}\, d\theta\, dr$$
$$= 2\pi \int_0^2 r^3\sqrt{4r^2 + 1}\, dr,$$

by Fubini's theorem. Now let $2r = \tan u$; that is, let $r = \frac{1}{2}\tan u$ so that $dr = \frac{1}{2}\sec^2 u\, du$. The previous integral transforms into

$$2\pi \int_0^{\tan^{-1} 4} \frac{1}{8}\tan^3 u \cdot \sqrt{\tan^2 u + 1} \cdot \frac{1}{2}\sec^2 u\, du$$
$$= \frac{\pi}{8}\int_0^{\tan^{-1} 4} \tan^3 u \sec^3 u\, du$$
$$= \frac{\pi}{8}\int_0^{\tan^{-1} 4} \tan^2 u \sec^2 u \cdot (\sec u \tan u\, du)$$
$$= \frac{\pi}{8}\int_0^{\tan^{-1} 4} (\sec^2 u - 1)\sec^2 u \sec u \tan u\, du.$$

Now, let $w = \sec u$ so $dw = \sec u \tan u\, du$. Hence, when $u = 0$, $w = 1$ and when $u = \tan^{-1} 4$, $w = \sqrt{17}$. (See Figure 7.17.) Thus, the u-integral becomes

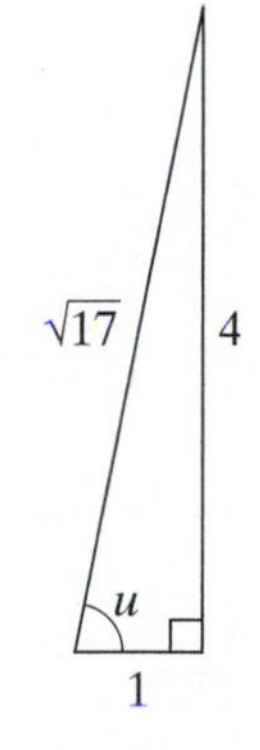

Figure 7.17 If $u = \tan^{-1} 4$, then $\sec u = \sqrt{17}$.

$$\frac{\pi}{8}\int_1^{\sqrt{17}} (w^2 - 1)w^2\, dw = \frac{\pi}{8}\int_1^{\sqrt{17}} (w^4 - w^2)\, dw$$
$$= \frac{\pi}{8}\left(\frac{1}{5}w^5 - \frac{1}{3}w^3\right)\Big|_1^{\sqrt{17}}$$
$$= \frac{\pi}{8}\left(\left(\frac{17^2}{5}\sqrt{17} - \frac{17}{3}\sqrt{17}\right) - \left(\frac{1}{5} - \frac{1}{3}\right)\right)$$
$$= \left(\frac{391\sqrt{17} + 1}{60}\right)\pi.$$

Alternatively, we could calculate the integral $\int_0^2 r^3\sqrt{4r^2 + 1}\, dr$ using integration by parts with $u = r^2$ (so $du = 2r\, dr$) and $dv = r\sqrt{4r^2 + 1}\, dr$ (so $v = \frac{1}{12}(4r^2 + 1)^{3/2}$). ◆

Vector Surface Integrals

Now we develop a means to integrate vector fields along surfaces, beginning with a definition.

■ **Definition 2.2** Let $\mathbf{X}: D \to \mathbf{R}^3$ be a smooth parametrized surface, where D is a bounded region in the plane, and let $\mathbf{F}(x, y, z)$ be a continuous vector field whose domain includes $S = \mathbf{X}(D)$. Then the **vector surface integral** of $\mathbf{F}$ along $\mathbf{X}$, denoted $\iint_{\mathbf{X}} \mathbf{F} \cdot d\mathbf{S}$, is

$$\iint_{\mathbf{X}} \mathbf{F} \cdot d\mathbf{S} = \iint_D \mathbf{F}(\mathbf{X}(s, t)) \cdot \mathbf{N}(s, t)\, ds\, dt,$$

where $\mathbf{N}(s, t) = \mathbf{T}_s \times \mathbf{T}_t$.

As with line integrals, you are cautioned to be careful about notation for surface integrals. In the vector surface integral $\iint_{\mathbf{X}} \mathbf{F} \cdot d\mathbf{S}$, the differential term should be considered to be a vector quantity, whereas in the scalar surface integral $\iint_{\mathbf{X}} f\, dS$, the differential term is a scalar quantity (namely, the differential of surface area).

EXAMPLE 4 Let $\mathbf{F} = x\,\mathbf{i} + y\,\mathbf{j} + (z - 2y)\,\mathbf{k}$. We evaluate $\iint_{\mathbf{X}} \mathbf{F} \cdot d\mathbf{S}$, where $\mathbf{X}$ is the helicoid

$$\mathbf{X}(s, t) = (s \cos t, s \sin t, t), \quad 0 \le s \le 1, \quad 0 \le t \le 2\pi.$$

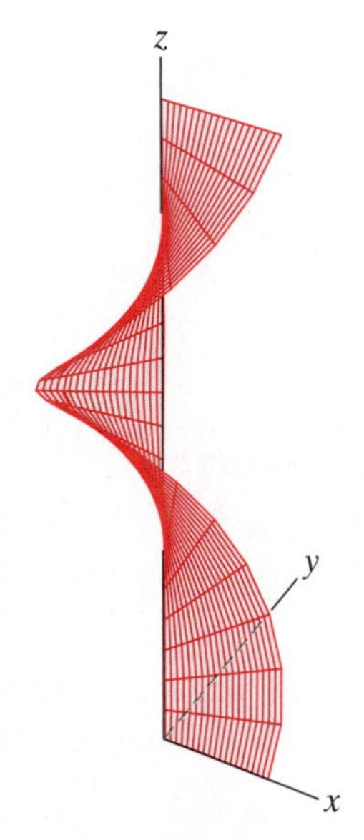

Figure 7.18 The helicoid of Example 4.

The helicoid is shown in Figure 7.18.

We have

$$\begin{aligned}
\mathbf{N}(s, t) &= \frac{\partial(y, z)}{\partial(s, t)}\mathbf{i} - \frac{\partial(x, z)}{\partial(s, t)}\mathbf{j} + \frac{\partial(x, y)}{\partial(s, t)}\mathbf{k} \\
&= \begin{vmatrix} \sin t & s \cos t \\ 0 & 1 \end{vmatrix}\mathbf{i} - \begin{vmatrix} \cos t & -s \sin t \\ 0 & 1 \end{vmatrix}\mathbf{j} + \begin{vmatrix} \cos t & -s \sin t \\ \sin t & s \cos t \end{vmatrix}\mathbf{k} \\
&= \sin t\,\mathbf{i} - \cos t\,\mathbf{j} + s\,\mathbf{k}.
\end{aligned}$$

Using Definition 2.2, we obtain

$$\begin{aligned}
\iint_{\mathbf{X}} \mathbf{F} \cdot d\mathbf{S} &= \int_0^{2\pi} \int_0^1 \mathbf{F}(\mathbf{X}(s, t)) \cdot \mathbf{N}(s, t)\, ds\, dt \\
&= \int_0^{2\pi} \int_0^1 (s \cos t\,\mathbf{i} + s \sin t\,\mathbf{j} \\
&\qquad + (t - 2s \sin t)\,\mathbf{k}) \cdot (\sin t\,\mathbf{i} - \cos t\,\mathbf{j} + s\,\mathbf{k})\, ds\, dt \\
&= \int_0^{2\pi} \int_0^1 (st - 2s^2 \sin t)\, ds\, dt = \int_0^{2\pi} \left(\tfrac{1}{2}s^2 t - \tfrac{2}{3}s^3 \sin t\right)\Big|_{s=0}^{1} dt \\
&= \int_0^{2\pi} \left(\tfrac{1}{2}t - \tfrac{2}{3}\sin t\right) dt = \left(\tfrac{1}{4}t^2 + \tfrac{2}{3}\cos t\right)\Big|_0^{2\pi} = \pi^2.
\end{aligned}$$

◆

EXAMPLE 5 Let $f(x, y)$ be a scalar-valued function of class C^1 on a bounded domain $D \subset \mathbf{R}^2$. Suppose S is the surface described as the graph of $z = f(x, y)$; that is, $S = \mathbf{X}(D)$, where $\mathbf{X}(x, y) = (x, y, f(x, y))$. Then

$$\mathbf{N}(x, y) = -f_x\,\mathbf{i} - f_y\,\mathbf{j} + \mathbf{k},$$

so that Definition 2.2 becomes

$$\iint_{\mathbf{X}} \mathbf{F} \cdot d\mathbf{S} = \iint_D \mathbf{F}(x, y, f(x, y)) \cdot \left(-f_x\,\mathbf{i} - f_y\,\mathbf{j} + \mathbf{k}\right) dx\, dy. \tag{5}$$

Formula (5) will prove to be quite useful. ◆

Further Interpretations

As is the case for vector and scalar *line* integrals, there is a connection between vector and scalar *surface* integrals. Suppose $\mathbf{X}: D \to \mathbf{R}^3$ is a smooth parametrized surface and $\mathbf{F}$ is continuous on $S = \mathbf{X}(D)$. Let $\mathbf{N}(s,t) = \mathbf{T}_s \times \mathbf{T}_t$ be the usual normal vector and let

$$\mathbf{n}(s,t) = \frac{\mathbf{N}(s,t)}{\|\mathbf{N}(s,t)\|}.$$

That is, $\mathbf{n}$ is the unit vector pointing in the same direction as $\mathbf{N}$. In particular,

$$\mathbf{N}(s,t) = \|\mathbf{N}(s,t)\|\,\mathbf{n}(s,t).$$

Using Definition 2.2, we have that the vector surface integral is

$$\begin{aligned}
\iint_{\mathbf{X}} \mathbf{F}\cdot d\mathbf{S} &= \iint_D \mathbf{F}(\mathbf{X}(s,t))\cdot \mathbf{N}(s,t)\,ds\,dt \\
&= \iint_D \mathbf{F}(\mathbf{X}(s,t))\cdot (\|\mathbf{N}(s,t)\|\mathbf{n}(s,t))\,ds\,dt \\
&= \iint_D \mathbf{F}(\mathbf{X}(s,t))\cdot \mathbf{n}(s,t)\|\mathbf{N}(s,t)\|\,ds\,dt \\
&= \iint_{\mathbf{X}} (\mathbf{F}\cdot\mathbf{n})\,dS. \qquad (6)
\end{aligned}$$

Since $\mathbf{n}$ is a unit vector, the quantity $\mathbf{F}\cdot\mathbf{n}$ is precisely the component of $\mathbf{F}$ in the direction of $\mathbf{n}$. In other words, formula (6) says that the *vector* surface integral of $\mathbf{F}$ along $\mathbf{X}$ is the *scalar* surface integral of the component of $\mathbf{F}$ normal to $S = \mathbf{X}(D)$. It is the surface integral analogue of formula (3) of §6.1, which states that the vector line integral of $\mathbf{F}$ along a path $\mathbf{x}$ is the scalar line integral of the component of $\mathbf{F}$ tangent to the image curve. To summarize, we have the following results:

Line integrals:

$$\int_{\mathbf{x}} \mathbf{F}\cdot d\mathbf{s} = \int_{\mathbf{x}} (\mathbf{F}\cdot\mathbf{T})\,ds. \qquad (7)$$

Surface integrals:

$$\iint_{\mathbf{X}} \mathbf{F}\cdot d\mathbf{S} = \iint_{\mathbf{X}} (\mathbf{F}\cdot\mathbf{n})\,dS. \qquad (8)$$

As noted in §6.1, when $\mathbf{x}$ is a closed path, the quantity $\int_{\mathbf{x}} (\mathbf{F}\cdot\mathbf{T})\,ds$ in equation (7) is called the **circulation of F along x**. It measures the tangential flow of $\mathbf{F}$ along the path. On the other hand, the quantity $\iint_{\mathbf{X}} (\mathbf{F}\cdot\mathbf{n})\,dS$ in equation (8) is known as the **flux of F across** $S = \mathbf{X}(D)$. If we think of $\mathbf{F}$ as the velocity vector field of a three-dimensional fluid, then the flux may be thought of as representing the rate of fluid transported across S per unit time, as we now see. (You may wish to compare the following discussion with the one in §6.2 concerning the two-dimensional flux across a curve.)

To avoid notational confusion, we use u and v to denote the parameter variables for $\mathbf{X}$ and t as the variable representing time. Consider a small piece of S, having area ΔS, and the amount of fluid transported across it during a brief time interval Δt. This amount is the volume determined by $\mathbf{F}$ during Δt. Figure 7.19 suggests that if both ΔS and Δt are sufficiently small, then this volume can be

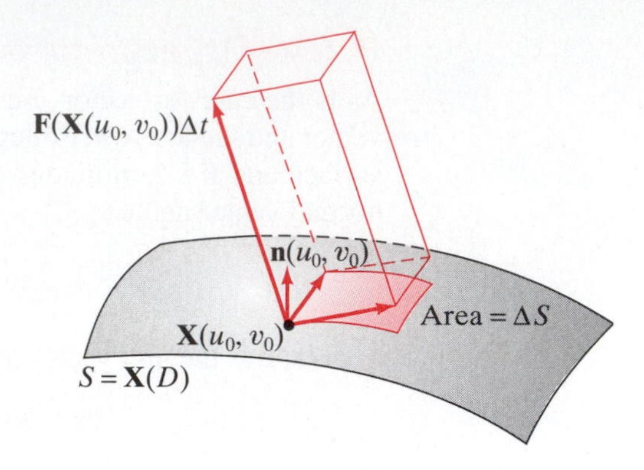

Figure 7.19 The amount of fluid transported across a small piece of S during a brief time interval Δt may be approximated by the volume of a parallelepiped.

approximated by the volume of an appropriate parallelepiped. Therefore,

$$\begin{aligned}\text{Amount of fluid transported} &\approx \text{volume of parallelepiped}\\ &= \text{(height) (area of base)}\\ &= \mathbf{F}(\mathbf{X}(u_0, v_0))\Delta t \cdot \mathbf{n}(u_0, v_0)\Delta S, \end{aligned} \tag{9}$$

since the height of the parallelepiped is the normal component of $\mathbf{F}\Delta t$. We obtain the average rate of transport across the surface piece during the time interval Δt by dividing (9) by Δt:

$$\text{Average rate of transport} \approx \mathbf{F}(\mathbf{X}(u_0, v_0)) \cdot \mathbf{n}(u_0, v_0)\Delta S. \tag{10}$$

Now, break up the entire surface $S = \mathbf{X}(D)$ into many such small pieces and sum the corresponding contributions to the rate of transport in the form given in (10). If we let all the pieces shrink, then, in the limit as all $\Delta S \to 0$, we have that the total average rate $\Delta M/\Delta t$ of fluid transported during Δt is approximately

$$\frac{\Delta M}{\Delta t} \approx \iint_{\mathbf{X}} (\mathbf{F} \cdot \mathbf{n})\, dS.$$

Finally, let $\Delta t \to 0$ and define the (instantaneous) rate of fluid transport to be

$$\frac{dM}{dt} = \iint_{\mathbf{X}} (\mathbf{F} \cdot \mathbf{n})\, dS.$$

Reparametrization of Surfaces

As seen in §6.1, scalar and vector line integrals over curves depend on the geometry of the curve (and possibly its direction), rather than on the particular way in which the curve may be parametrized. Much the same is true for surface integrals. We begin with a definition, analogous to Definition 1.3 of Chapter 6.

■ **Definition 2.3** Let $\mathbf{X}: D_1 \subseteq \mathbf{R}^2 \to \mathbf{R}^3$ and $\mathbf{Y}: D_2 \subseteq \mathbf{R}^2 \to \mathbf{R}^3$ be parametrized surfaces. We say that $\mathbf{Y}$ is a **reparametrization** of $\mathbf{X}$ if there is a one-one and onto function $\mathbf{H}: D_2 \to D_1$ such that $\mathbf{Y}(s, t) = \mathbf{X}(\mathbf{H}(s, t))$; that is, such that $\mathbf{Y} = \mathbf{X} \circ \mathbf{H}$. If $\mathbf{X}$ and $\mathbf{Y}$ are smooth and $\mathbf{H}$ is of class C^1, then we say that $\mathbf{Y}$ is a **smooth** reparametrization of $\mathbf{X}$.

EXAMPLE 6 The helicoid parametrized by

$$\begin{cases} x = s\cos t \\ y = s\sin t \qquad 0 \le s \le 1, \quad 0 \le t \le 2\pi \\ z = t \end{cases}$$

may also be described as

$$\begin{cases} x = \dfrac{s}{2}\cos 2t \\ y = \dfrac{s}{2}\sin 2t \qquad 0 \le s \le 2, \quad 0 \le t \le \pi. \\ z = 2t \end{cases}$$

The first description corresponds to a map $\mathbf{X}\colon [0, 1] \times [0, 2\pi] \to \mathbf{R}^3$, and the second to a map $\mathbf{Y}\colon [0, 2] \times [0, \pi] \to \mathbf{R}^3$. It is not difficult to see that if we make the change of variables by letting

$$u = \frac{s}{2} \quad \text{and} \quad v = 2t,$$

then

$$\mathbf{Y}(s, t) = \mathbf{X}(u, v).$$

Equivalently, we can define a function $\mathbf{H}\colon [0, 2] \times [0, \pi] \to [0, 1] \times [0, 2\pi]$ with $\mathbf{H}(s, t) = (s/2, 2t)$. Then $\mathbf{H}$ is one-one and onto and $\mathbf{Y} = \mathbf{X} \circ \mathbf{H}$. Therefore, $\mathbf{Y}$ is a reparametrization of $\mathbf{X}$. ◆

EXAMPLE 7 Suppose $\mathbf{X}$ is a smooth parametrized surface. Let $\mathbf{Y}(s, t) = \mathbf{X}(u, v)$, where $u = t$, $v = s$. That is, $\mathbf{Y} = \mathbf{X} \circ \mathbf{H}$, where $\mathbf{H}(s, t) = (t, s)$. Then $\mathbf{Y}$ is a (smooth) reparametrization that appears to accomplish little. However, if we let $\mathbf{N_Y}$ denote the usual normal vector $\mathbf{T}_s \times \mathbf{T}_t = \partial\mathbf{Y}/\partial s \times \partial\mathbf{Y}/\partial t$, then we have

$$\frac{\partial \mathbf{Y}}{\partial s} = \frac{\partial \mathbf{X}}{\partial v} \quad \text{and} \quad \frac{\partial \mathbf{Y}}{\partial t} = \frac{\partial \mathbf{X}}{\partial u},$$

so that

$$\mathbf{N_Y} = \frac{\partial \mathbf{Y}}{\partial s} \times \frac{\partial \mathbf{Y}}{\partial t} = \frac{\partial \mathbf{X}}{\partial v} \times \frac{\partial \mathbf{X}}{\partial u} = -\frac{\partial \mathbf{X}}{\partial u} \times \frac{\partial \mathbf{X}}{\partial v} = -\mathbf{N_X}.$$

The parametrized surface $\mathbf{Y}$ is the same as $\mathbf{X}$, except that the standard normal vector arising from $\mathbf{Y}$ points in the opposite direction to the one arising from $\mathbf{X}$. ◆

The calculation in Example 7 generalizes thus: Suppose $\mathbf{X}$ is a smooth parametrized surface and $\mathbf{Y}$ is a smooth reparametrization of $\mathbf{X}$ via $\mathbf{H}$, meaning that

$$\mathbf{Y}(s, t) = \mathbf{X}(u, v) = \mathbf{X}(\mathbf{H}(s, t)).$$

Assume $\mathbf{H}$ is of class C^1. We can show from the chain rule that the standard normal vectors are related by the equation

$$\mathbf{N_Y}(s, t) = \frac{\partial(u, v)}{\partial(s, t)} \mathbf{N_X}(u, v). \tag{11}$$

(See the addendum at the end of this section for a derivation of formula (11).) Formula (11) shows that $\mathbf{N_Y}$ is a scalar multiple of $\mathbf{N_X}$. In addition, since $\mathbf{H}$ is one-one, it follows that the Jacobian of $\mathbf{H}$ is either always nonnegative or always nonpositive. (We assume this fact without proof.) Hence, the standard normal $\mathbf{N_Y}$ either always points in the same direction as $\mathbf{N_X}$, or else always points in the

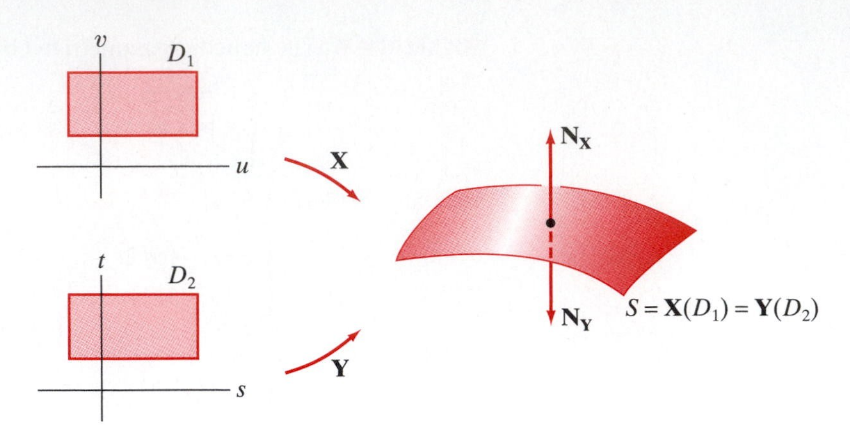

Figure 7.20 If $\mathbf{Y}$ is an orientation-reversing reparametrization of $\mathbf{X}$, then $\mathbf{N_Y}$ points opposite to $\mathbf{N_X}$.

opposite direction (Figure 7.20). Under these assumptions, we say that both $\mathbf{H}$ and $\mathbf{Y}$ are **orientation-preserving** if the Jacobian $\partial(u, v)/\partial(s, t)$ is nonnegative, **orientation-reversing** if $\partial(u, v)/\partial(s, t)$ is nonpositive.

The following result, a close analogue of Theorem 1.4, Chapter 6, shows that smooth reparametrization has no effect on the value of a scalar line integral.

Theorem 2.4 Let $\mathbf{X}: D_1 \to \mathbf{R}^3$ be a smooth parametrized surface and f any continuous function whose domain includes $\mathbf{X}(D_1)$. If $\mathbf{Y}: D_2 \to \mathbf{R}^3$ is any smooth reparametrization of $\mathbf{X}$, then

$$\iint_{\mathbf{Y}} f\,dS = \iint_{\mathbf{X}} f\,dS.$$

The proof of Theorem 2.4 appears in the addendum to this section. With this result, we can define the scalar surface integral over a smooth surface S by taking a smooth parametrization $\mathbf{X}: D \to \mathbf{R}^3$ with $S = \mathbf{X}(D)$ that is one-one, except possibly on ∂D. Then we define the scalar surface integral of f on S by

$$\iint_{S} f\,dS = \iint_{\mathbf{X}} f\,dS.$$

It is a fact (which we shall not prove) that any two smooth parametrizations of S must be reparametrizations of each other, so Theorem 2.4 tells us that any particular choice of parametrization we might make does not matter. We need to assume that $\mathbf{X}$ is (nearly) one-one to ensure that the integral is taken only once over the underlying surface $S = \mathbf{X}(D)$. It is also a straightforward matter to extend these comments to give a definition of a scalar surface integral of a function over a piecewise smooth surface.

Analogous to Theorem 1.5 of Chapter 6, the following result (whose proof is in the addendum) tells us that smooth reparametrizations only affect vector surface integrals by a possible sign change.

Theorem 2.5 Let $\mathbf{X}: D_1 \to \mathbf{R}^3$ be a smooth parametrized surface and $\mathbf{F}$ any continuous vector field whose domain includes $\mathbf{X}(D_1)$. If $\mathbf{Y}: D_2 \to \mathbf{R}^3$ is any smooth reparametrization of $\mathbf{X}$, then either

$$\iint_{\mathbf{Y}} \mathbf{F} \cdot d\mathbf{S} = \iint_{\mathbf{X}} \mathbf{F} \cdot d\mathbf{S},$$

if **Y** is orientation-preserving, or

$$\iint_{\mathbf{Y}} \mathbf{F} \cdot d\mathbf{S} = -\iint_{\mathbf{X}} \mathbf{F} \cdot d\mathbf{S},$$

if **Y** is orientation-reversing.

Because of Theorem 2.5, it is a more subtle and involved matter to define a vector surface integral over a smooth surface than to define a scalar surface integral. Given a smooth, connected surface, we need to choose an **orientation** for it. This is akin to orienting a curve but, perhaps surprisingly, is not always possible, even for a well-behaved, smooth parametrized surface, as Example 8 illustrates.

Here is a formal definition of orientability of a smooth surface.

Definition 2.6 A smooth, connected surface S is **orientable** (or **two-sided**) if it is possible to define a single unit normal vector at each point of S so that the collection of these normal vectors varies continuously over S. (In particular, this means that nearby unit normal vectors must point to the same side of S.) Otherwise, S is called **nonorientable** (or **one-sided**).

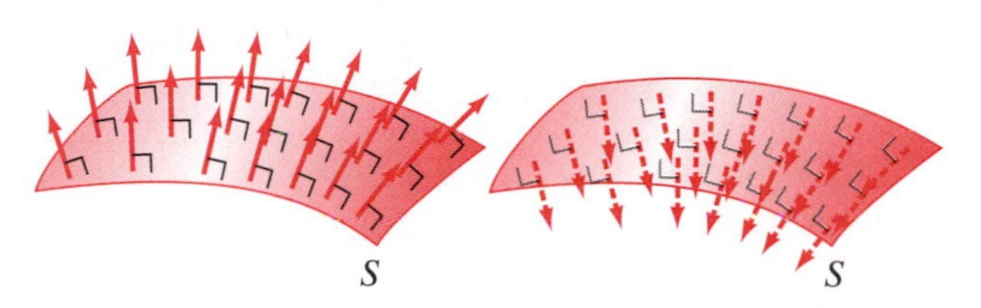

Figure 7.21 The connected orientable surface S shown with its two possible orientations.

It is a fact (clearly suggested by Figure 7.21) that a smooth, connected, orientable surface S has exactly two orientations.

If S happens to be the image $\mathbf{X}(D)$ of a smooth parametrized surface $\mathbf{X}\colon D \to \mathbf{R}^3$, then the normal vectors

$$\mathbf{N}_1 = \mathbf{T}_s \times \mathbf{T}_t \quad \text{and} \quad \mathbf{N}_2 = \mathbf{T}_t \times \mathbf{T}_s = -\mathbf{N}_1$$

can be used to give unit vector normal vectors $\mathbf{n}_1 = \mathbf{N}_1/\|\mathbf{N}_1\|$ and $\mathbf{n}_2 = \mathbf{N}_2/\|\mathbf{N}_2\|$ that point in opposite directions. It is tempting to think that $\mathbf{n}_1$ and $\mathbf{n}_2$ always provide two orientations for S. However, even though both $\mathbf{n}_1$ and $\mathbf{n}_2$ may vary continuously with respect to the parameters s and t, it is not clear that they must vary continuously and consistently with respect to the points on the underlying surface S. Example 8 is a famous instance of a nonorientable surface.

EXAMPLE 8 The surface parametrized by

$$\begin{cases} x = \left(1 + t\cos\dfrac{s}{2}\right)\cos s \\ y = \left(1 + t\cos\dfrac{s}{2}\right)\sin s \qquad 0 \le s \le 2\pi,\ -\tfrac{1}{2} \le t \le \tfrac{1}{2} \\ z = t\sin\dfrac{s}{2} \end{cases}$$

is called a **Möbius strip.** It may be visualized as follows: The t-coordinate curve at $s = s_0$ is

$$\begin{cases} x = \left(\cos s_0 \cos \dfrac{s_0}{2}\right) t + \cos s_0 \\ y = \left(\sin s_0 \cos \dfrac{s_0}{2}\right) t + \sin s_0 \qquad -\frac{1}{2} \le t \le \frac{1}{2}. \\ z = \left(\sin \dfrac{s_0}{2}\right) t \end{cases}$$

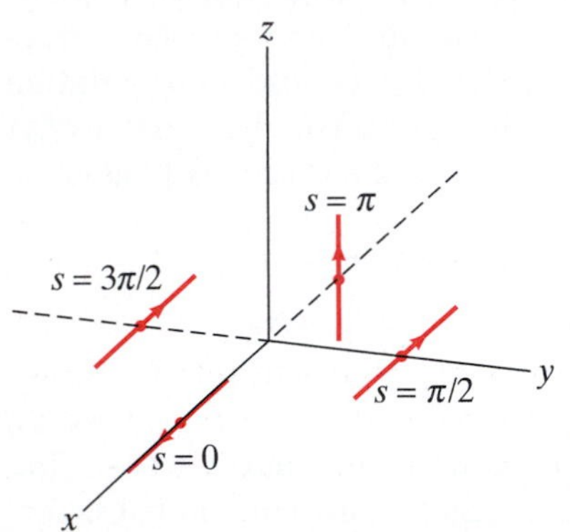

Figure 7.22 Some t-coordinate curves of the parametrized Möbius strip.

This is a line segment through the point $(\cos s_0, \sin s_0, 0)$ and parallel to the vector

$$\mathbf{a} = \cos s_0 \cos\left(\frac{s_0}{2}\right)\mathbf{i} + \sin s_0 \cos\left(\frac{s_0}{2}\right)\mathbf{j} + \sin\left(\frac{s_0}{2}\right)\mathbf{k}.$$

Several such coordinate curves, marked with the direction of increasing t, are shown in Figure 7.22. We see that the Möbius strip is generated by a moving line segment that begins (at $s = 0$) lying along the positive x-axis, rises to a vertical position with center at $(-1, 0, 0)$ when $s = \pi$, and then falls back to horizontal, but with *direction reversed* at $s = 2\pi$. The s-coordinate curve at $t = 0$ is parametrized by

$$\begin{cases} x = \cos s \\ y = \sin s \qquad 0 \le s \le 2\pi, \\ z = 0 \end{cases}$$

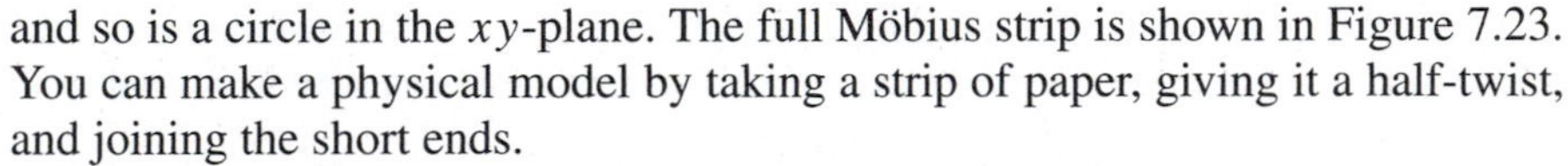

and so is a circle in the xy-plane. The full Möbius strip is shown in Figure 7.23. You can make a physical model by taking a strip of paper, giving it a half-twist, and joining the short ends.

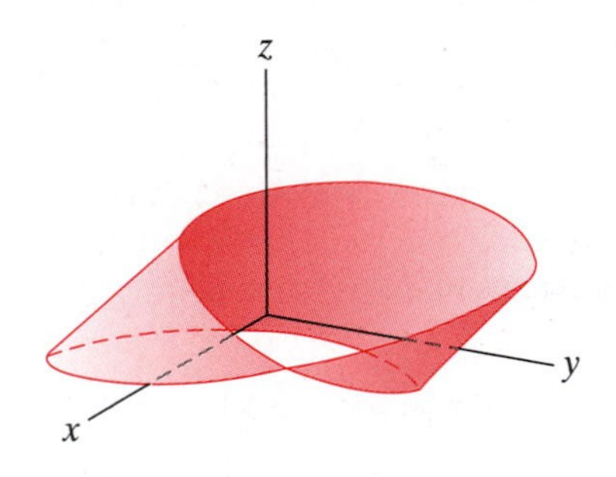

Figure 7.23 The Möbius strip of Example 8.

You can analytically understand the gluing process by noting that the map

$$\mathbf{X}\colon [0, 2\pi] \times \left[-\tfrac{1}{2}, \tfrac{1}{2}\right] \to \mathbf{R}^3$$

defining the Möbius strip as a parametrized surface has the property that $\mathbf{X}(0, t) = \mathbf{X}(2\pi, -t)$, but is otherwise one-one. Therefore, every point $(0, t)$ on the left edge of the domain rectangle $[0, 2\pi] \times \left[-\frac{1}{2}, \frac{1}{2}\right]$ is mapped to the point $(1 + t, 0, 0)$ of the Möbius strip, as is the point $(2\pi, -t)$ on the right edge of the rectangle. (See Figure 7.24.)

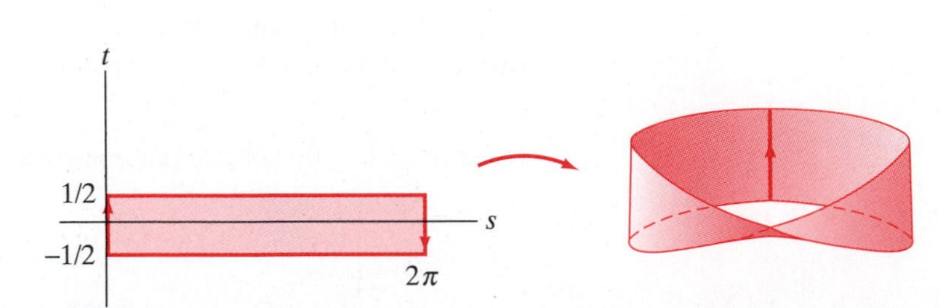

Figure 7.24 Gluing the ends of a strip of paper so that the arrows align provides a model of the Möbius strip.

Now, let's investigate the orientability of the Möbius strip. The standard normal vector is

$$\begin{aligned}\mathbf{N}(s,t) = \mathbf{T}_s \times \mathbf{T}_t &= \frac{\partial(y,z)}{\partial(s,t)}\mathbf{i} - \frac{\partial(x,z)}{\partial(s,t)}\mathbf{j} + \frac{\partial(x,y)}{\partial(s,t)}\mathbf{k} \\ &= \sin\frac{s}{2}\left(\cos s + 2t\left(\cos^3\frac{s}{2} - \cos\frac{s}{2}\right)\right)\mathbf{i} \\ &\quad + \frac{1}{2}\left(4\cos\frac{s}{2} - 4\cos^3\frac{s}{2} + t\left(1 + \cos s - \cos^2 s\right)\right)\mathbf{j} \\ &\quad - \cos\frac{s}{2}\left(1 + t\cos\frac{s}{2}\right)\mathbf{k}.\end{aligned}$$

We have

$$\mathbf{N}(0,t) = \frac{t}{2}\mathbf{j} - (1+t)\,\mathbf{k},$$

and

$$\mathbf{N}(2\pi,-t) = -\frac{t}{2}\mathbf{j} + (1+t)\,\mathbf{k} = -\mathbf{N}(0,t).$$

Therefore, a uniquely determined normal vector has not been defined. More vividly, imagine traveling along the Möbius strip via the s-coordinate path at $t = 0$, that is, along the circular path

$$\mathbf{x}(s) = \mathbf{X}(s,0) = (\cos s, \sin s, 0), \quad 0 \le s \le 2\pi.$$

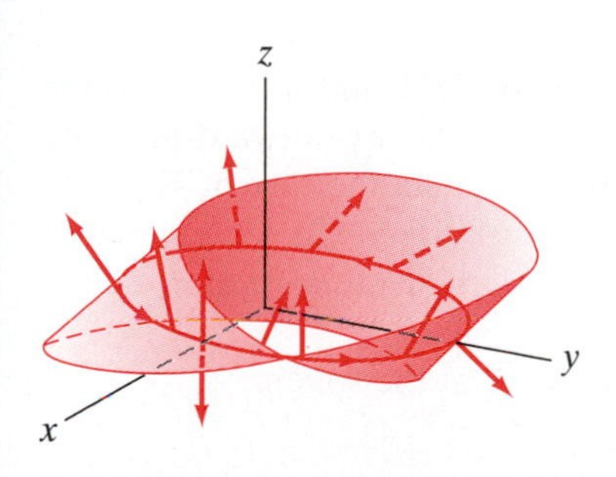

Figure 7.25 Traveling once around the circular path on the Möbius strip forces the normal vector to reverse direction.

Follow the standard normal $\mathbf{N}$. At $s = 0$, it is $\mathbf{N}(0,0) = -\mathbf{k}$, but by the time we close the loop, it is $\mathbf{N}(2\pi, 0) = \mathbf{k}$. This apparent reversal of the normal vector means that the strip is not orientable at all—it is **one-sided**. (See Figure 7.25.) ◆

A smooth, orientable surface together with an explicit choice of orientation for it is called an **oriented** surface. If S is such a smooth oriented surface, then we define the vector surface integral of $\mathbf{F}$ along S by finding a suitable smooth parametrization $\mathbf{X}$ of S such that the unit normal vector $\mathbf{N}(s,t)/\|\mathbf{N}(s,t)\|$ arising from the parametrization agrees with the choice of orientation normal. We take the surface integral to be

$$\iint_S \mathbf{F}\cdot d\mathbf{S} = \iint_{\mathbf{X}} \mathbf{F}\cdot d\mathbf{S}.$$

By Theorem 2.5, if $\mathbf{Y}$ is any orientation-preserving reparametrization of $\mathbf{X}$, the value of $\iint_{\mathbf{Y}} \mathbf{F}\cdot d\mathbf{S}$ is the same as $\iint_{\mathbf{X}} \mathbf{F}\cdot d\mathbf{S}$, so this notion of a surface integral over the underlying oriented surface S is well-defined. Even though we may perfectly well calculate $\iint_{\mathbf{X}} \mathbf{F}\cdot d\mathbf{S}$, where $\mathbf{X}$ is the parametrized Möbius strip of Example 8, it does not make sense to consider the surface integral over the underlying Möbius strip, since there is no way to orient it. Similarly, the interpretation of the vector surface integral as the flux of $\mathbf{F}$ across the surface only makes sense once an orientation of the surface is chosen. Then the flux measures the flow rate, positive or negative, depending on the choice of orientation. (See Figure 7.26.)

Another reason for de-emphasizing the role of parametrization in surface integrals is that we can often exploit the geometry of the underlying surface and vector field when making calculations. If S is a smooth, orientable surface and $\mathbf{n}$ a unit normal that gives an orientation of S (so, in particular, $\mathbf{n}$ is understood

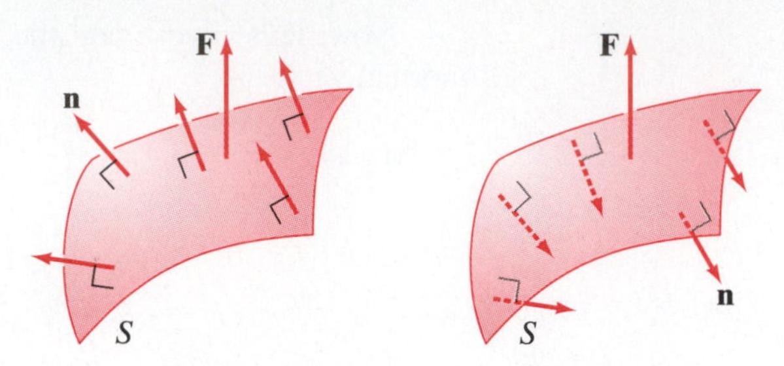

Figure 7.26 How flux depends on orientation. On the left, the surface S is oriented by unit normal vectors so that $\mathbf{F} \cdot \mathbf{n}$ is positive at every point. Hence, $\iint_S \mathbf{F} \cdot d\mathbf{S} = \iint_S \mathbf{F} \cdot \mathbf{n}\, dS$ is positive. On the right, S is given the opposite orientation so that the flux $\iint_S \mathbf{F} \cdot d\mathbf{S} < 0$.

to vary with the points of S), then, for a continuous vector field $\mathbf{F}$ defined on S, we have

$$\iint_S \mathbf{F} \cdot d\mathbf{S} = \iint_S \mathbf{F} \cdot \mathbf{n}\, dS.$$

If we can determine a continuously varying, unit normal vector at each point of S (for example, if S is the graph of a function $f(x, y)$ of two variables or the graph of a level set $f(x, y, z) = c$ of a function of three variables), then there is a good chance that the surface integral can be evaluated readily.

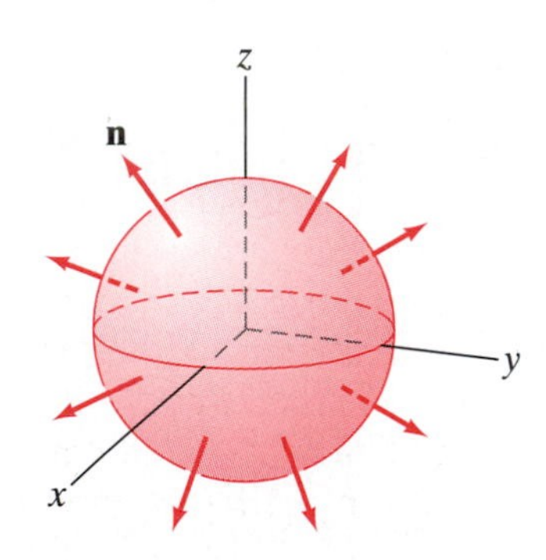

Figure 7.27 The sphere $x^2 + y^2 + z^2 = a^2$ oriented by outward-pointing unit normal vectors.

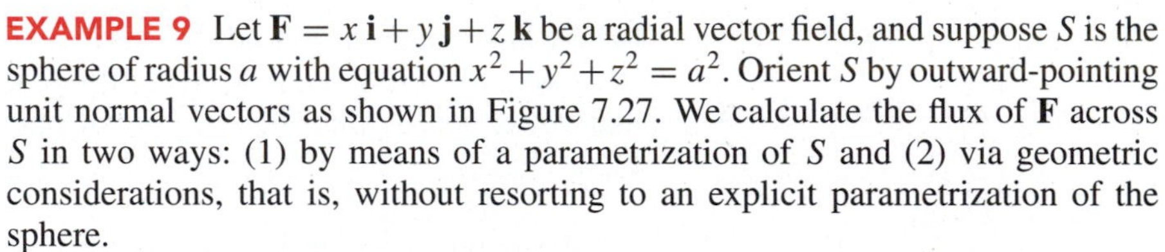

EXAMPLE 9 Let $\mathbf{F} = x\,\mathbf{i} + y\,\mathbf{j} + z\,\mathbf{k}$ be a radial vector field, and suppose S is the sphere of radius a with equation $x^2 + y^2 + z^2 = a^2$. Orient S by outward-pointing unit normal vectors as shown in Figure 7.27. We calculate the flux of $\mathbf{F}$ across S in two ways: (1) by means of a parametrization of S and (2) via geometric considerations, that is, without resorting to an explicit parametrization of the sphere.

For approach (1), use the usual parametrization $\mathbf{X}$ of the sphere:

$$\begin{cases} x = a\cos s\, \sin t \\ y = a\sin s\, \sin t \\ z = a\cos t \end{cases} \quad 0 \le s \le 2\pi,\ 0 \le t \le \pi.$$

The standard normal vector for this parametrization is

$$\mathbf{N}(s, t) = -a^2 \sin t\,(\cos s\, \sin t\,\mathbf{i} + \sin s\, \sin t\,\mathbf{j} + \cos t\,\mathbf{k}).$$

(This normal vector is calculated in Example 11 of §7.1.) If we normalize $\mathbf{N}$, we find that

$$\mathbf{n}(s, t) = \frac{\mathbf{N}(s, t)}{\|\mathbf{N}(s, t)\|} = -(\cos s\, \sin t\,\mathbf{i} + \sin s\, \sin t\,\mathbf{j} + \cos t\,\mathbf{k}).$$

Thus, $\mathbf{n}$ is inward-pointing at every point on the sphere. Therefore, we must make a sign change when we evaluate the vector surface integral, if we use the parametrization just given. Hence, we have

$$\begin{aligned} \iint_S \mathbf{F} \cdot d\mathbf{S} &= -\iint_{\mathbf{X}} \mathbf{F} \cdot d\mathbf{S} = -\int_0^{\pi} \int_0^{2\pi} \mathbf{F}(\mathbf{X}(s, t)) \cdot \mathbf{N}(s, t)\, ds\, dt \\ &= -\int_0^{\pi} \int_0^{2\pi} (a\cos s\, \sin t\,\mathbf{i} + a\sin s\, \sin t\,\mathbf{j} + a\cos t\,\mathbf{k}) \\ &\qquad \cdot \left(-a^2 \sin t\,(\cos s\, \sin t\,\mathbf{i} + \sin s\, \sin t\,\mathbf{j} + \cos t\,\mathbf{k}\right)\, ds\, dt \end{aligned}$$

$$= a^3 \int_0^{\pi} \int_0^{2\pi} \sin t\,(\cos^2 s\,\sin^2 t + \sin^2 s\,\sin^2 t + \cos^2 t)\,ds\,dt$$

$$= a^3 \int_0^{\pi} \int_0^{2\pi} \sin t\,ds\,dt = 2\pi a^3 \int_0^{\pi} \sin t\,dt = 4\pi a^3.$$

Now, reconsider this calculation along the lines of approach (2). Since S is defined as a level set of the function $f(x, y, z) = x^2 + y^2 + z^2$, normal vectors can be obtained from the gradient:

$$\nabla(x^2 + y^2 + z^2) = 2x\,\mathbf{i} + 2y\,\mathbf{j} + 2z\,\mathbf{k}.$$

If we normalize the gradient, then we have unit normal vectors. Thus,

$$\mathbf{n} = \frac{2x\,\mathbf{i} + 2y\,\mathbf{j} + 2z\,\mathbf{k}}{\sqrt{4x^2 + 4y^2 + 4z^2}} = \frac{2x\,\mathbf{i} + 2y\,\mathbf{j} + 2z\,\mathbf{k}}{2\sqrt{x^2 + y^2 + z^2}} = \frac{x\,\mathbf{i} + y\,\mathbf{j} + z\,\mathbf{k}}{a},$$

because $x^2+y^2+z^2 = a^2$ at points on S. (Note that $\mathbf{n}$ is always outward-pointing.) Therefore,

$$\iint_S \mathbf{F} \cdot d\mathbf{S} = \iint_S \mathbf{F} \cdot \mathbf{n}\,dS$$

$$= \iint_S (x\mathbf{i} + y\mathbf{j} + z\mathbf{k}) \cdot \left(\frac{x\mathbf{i} + y\mathbf{j} + z\mathbf{k}}{a}\right) dS$$

$$= \iint_S \frac{x^2 + y^2 + z^2}{a}\,dS = \iint_S \frac{a^2}{a}\,dS$$

$$= a \iint_S dS = a \cdot \text{area of } S = a(4\pi a^2) = 4\pi a^3. \quad ◆$$

All of the preceding remarks concerning scalar and vector surface integrals can be adapted to define integrals over piecewise smooth, connected surfaces. Simply add the contributions of the surface integrals over the various smooth pieces. The only issue is that of orientation, but assuming that each of the smooth pieces is orientable, then it is possible to provide an orientation to the surface as a whole. Here's how: Suppose S_1 and S_2 are two smooth surface pieces that meet along a common edge curve C. Let $\mathbf{n}_1$ and $\mathbf{n}_2$ be the respective unit normal vectors that give the orientations of S_1 and S_2. Then $\mathbf{n}_1$ and $\mathbf{n}_2$ each give rise to an orientation of C via a right-hand rule. (To see this, for $j = 1, 2$, point the thumb of your right hand along $\mathbf{n}_j$; the direction of your fingers will indicate the orientation of C.) If C receives *opposite* orientations from $\mathbf{n}_1$ and $\mathbf{n}_2$, then S_1 and S_2 are oriented consistently; if C receives the same orientation, then S_1 and S_2 are oriented inconsistently. (See Figure 7.28.)

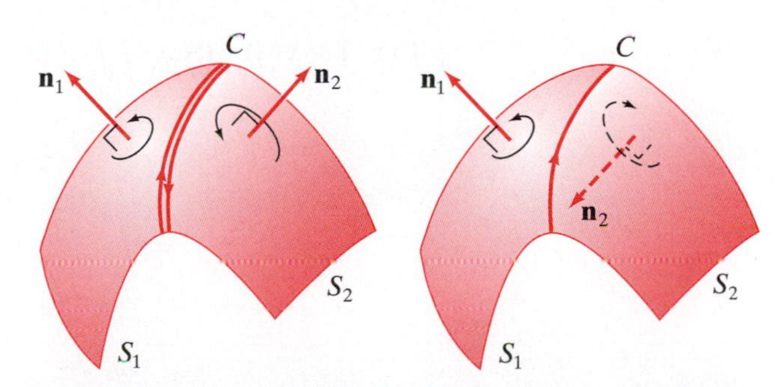

Figure 7.28 The piecewise smooth surface $S = S_1 \cup S_2$ oriented consistently on the left and inconsistently on the right.

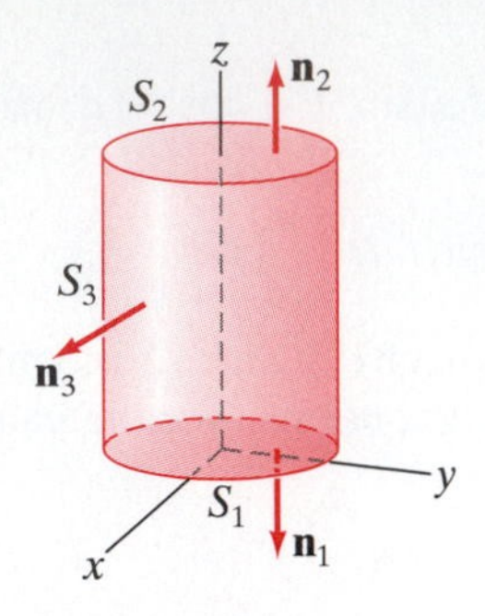

Figure 7.29 The piecewise smooth cylindrical surface S of Example 10 shown with orientation normals.

EXAMPLE 10 We evaluate $\iint_S (x^3\,\mathbf{i} + y^3\,\mathbf{j}) \cdot d\mathbf{S}$, where S is the closed cylinder bounded laterally by $x^2 + y^2 = 4$, and on bottom and top by the planes $z = 0$ and $z = 5$, oriented by outward normal vectors.

Evidently, S is the union of three smooth oriented pieces: (1) the bottom surface S_1, which is a portion of the plane $z = 0$, oriented by $\mathbf{n}_1 = -\mathbf{k}$; (2) the top surface S_2, which is a portion of the plane $z = 5$, oriented by $\mathbf{n}_2 = \mathbf{k}$; and (3) the lateral cylindrical surface S_3 given by the equation $x^2 + y^2 = 4$ and oriented by normalizing the gradient of $x^2 + y^2$ along S_3, namely,

$$\mathbf{n}_3 = \frac{2x\,\mathbf{i} + 2y\,\mathbf{j}}{\sqrt{4x^2 + 4y^2}} = \frac{x\,\mathbf{i} + y\,\mathbf{j}}{\sqrt{x^2 + y^2}} = \frac{x\,\mathbf{i} + y\,\mathbf{j}}{2}.$$

See Figure 7.29 for a depiction of S.

Now we calculate

$$\begin{aligned}
\iint_S (x^3\,\mathbf{i} + y^3\,\mathbf{j}) \cdot d\mathbf{S} &= \iint_{S_1} (x^3\,\mathbf{i} + y^3\,\mathbf{j}) \cdot d\mathbf{S} + \iint_{S_2} (x^3\,\mathbf{i} + y^3\,\mathbf{j}) \cdot d\mathbf{S} \\
&\quad + \iint_{S_3} (x^3\,\mathbf{i} + y^3\,\mathbf{j}) \cdot d\mathbf{S} \\
&= \iint_{S_1} (x^3\,\mathbf{i} + y^3\,\mathbf{j}) \cdot (-\mathbf{k})\, dS + \iint_{S_2} (x^3\,\mathbf{i} + y^3\,\mathbf{j}) \cdot \mathbf{k}\, dS \\
&\quad + \iint_{S_3} (x^3\,\mathbf{i} + y^3\,\mathbf{j}) \cdot \left(\frac{x\,\mathbf{i} + y\,\mathbf{j}}{2}\right) dS \\
&= 0 + 0 + \iint_{S_3} \tfrac{1}{2}(x^4 + y^4)\, dS.
\end{aligned}$$

To finish the evaluation, we may parametrize S_3 by

$$\begin{cases} x = 2\cos s \\ y = 2\sin s \\ z = t \end{cases} \qquad 0 \le s \le 2\pi, \quad 0 \le t \le 5.$$

Then

$$\begin{aligned}
\iint_S (x^3\,\mathbf{i} + y^3\,\mathbf{j}) \cdot d\mathbf{S} &= \iint_{S_3} \tfrac{1}{2}(x^4 + y^4)\, dS \\
&= \int_0^5 \int_0^{2\pi} \tfrac{1}{2}(16\cos^4 s + 16\sin^4 s)\, 2\, ds\, dt \\
&= \int_0^5 \int_0^{2\pi} 16(\cos^4 s + \sin^4 s)\, ds\, dt \\
&= \int_0^5 \int_0^{2\pi} 16((\cos^2 s)^2 + (\sin^2 s)^2)\, ds\, dt \\
&= \int_0^5 \int_0^{2\pi} 16\left(\left(\frac{1 + \cos 2s}{2}\right)^2 + \left(\frac{1 - \cos 2s}{2}\right)^2\right) ds\, dt,
\end{aligned}$$

from the half-angle substitution. Thus,

$$\iint_S (x^3\,\mathbf{i} + y^3\,\mathbf{j}) \cdot d\mathbf{S} = \int_0^5 \int_0^{2\pi} \tfrac{16}{4}(2 + 2\cos^2 2s)\, ds\, dt$$

$$= \int_0^5 \int_0^{2\pi} (8 + 4(1 + \cos 4s))\, ds\, dt.$$

By once again using the half-angle substitution, we get

$$\iint_S (x^3\,\mathbf{i} + y^3\,\mathbf{j}) \cdot d\mathbf{S} = \int_0^5 (12s + \sin 4s)\Big|_{s=0}^{2\pi} dt = \int_0^5 24\pi\, dt = 120\pi. \quad \blacklozenge$$

Summary: Surface Integral Formulas

Scalar surface integrals:

For a surface S parametrized by $\mathbf{X}\colon D \subseteq \mathbf{R}^2 \to \mathbf{R}^3$,

$$\iint_S f\, dS = \iint_{\mathbf{X}} f\, dS = \iint_D f(\mathbf{X}(s,t))\, \|\mathbf{T}_s \times \mathbf{T}_t\|\, ds\, dt.$$

Surface area element is $dS = \|\mathbf{T}_s \times \mathbf{T}_t\|\, ds\, dt$.

For a surface S described as a graph of a function $z = g(x, y)$, where $g\colon D \subseteq \mathbf{R}^2 \to \mathbf{R}$,

$$\iint_S f\, dS = \iint_D f(x, y, g(x, y))\sqrt{g_x(x,y)^2 + g_y(x,y)^2 + 1}\, dx\, dy.$$

Surface area element is $dS = \sqrt{g_x(x,y)^2 + g_y(x,y)^2 + 1}\, dx\, dy$.

Vector surface integrals:

For a surface S parametrized by $\mathbf{X}\colon D \subseteq \mathbf{R}^2 \to \mathbf{R}^3$,

$$\iint_S \mathbf{F} \cdot d\mathbf{S} = \iint_S (\mathbf{F} \cdot \mathbf{n})\, dS = \iint_D \mathbf{F}(\mathbf{X}(s,t)) \cdot \mathbf{N}(s,t)\, ds\, dt,$$

where $\mathbf{N} = \mathbf{T}_s \times \mathbf{T}_t$ and $\mathbf{n} = \mathbf{N}/\|\mathbf{N}\|$.

Vector surface integral element is $d\mathbf{S} = \mathbf{N}(s,t)\, ds\, dt$.

For a surface S described as a graph of a function $z = g(x, y)$, where $g\colon D \subseteq \mathbf{R}^2 \to \mathbf{R}$,

$$\iint_S \mathbf{F} \cdot d\mathbf{S} = \iint_S (\mathbf{F} \cdot \mathbf{n})\, dS$$

$$= \iint_D \mathbf{F}(x, y, g(x, y)) \cdot (-g_x(x,y)\,\mathbf{i} - g_y(x,y)\,\mathbf{j} + \mathbf{k})\, dx\, dy.$$

Here $\mathbf{n} = (-g_x\,\mathbf{i} - g_y\,\mathbf{j} + \mathbf{k})/\sqrt{g_x^2 + g_y^2 + 1}$.

Vector surface integral element is $\mathbf{S} = (-g_x\,\mathbf{i} - g_y\,\mathbf{j} + \mathbf{k})\, dx\, dy$.

Addendum: Proofs of Theorems 2.4 and 2.5

We begin by establishing formula (11) of this section.

■ **Lemma** Suppose $\mathbf{X}: D_1 \to \mathbf{R}^3$ is a smooth parametrized surface and $\mathbf{Y}: D_2 \to \mathbf{R}^3$ is a smooth reparametrization of $\mathbf{X}$ via $\mathbf{H}: D_2 \to D_1$, where we denote $\mathbf{H}(s, t)$ by (u, v). Then the standard normal vectors $\mathbf{N_X}$ and $\mathbf{N_Y}$ are related by the equation

$$\mathbf{N_Y}(s, t) = \frac{\partial(u, v)}{\partial(s, t)} \mathbf{N_X}(u, v).$$

Proof First, we set some notation. Since $\mathbf{Y}$ is a reparametrization of $\mathbf{X}$ via $\mathbf{H}$, we have, from Definition 2.3, that

$$\mathbf{Y}(s, t) = \mathbf{X}(\mathbf{H}(s, t)) = \mathbf{X}(u, v). \tag{12}$$

Write $(x(s, t), y(s, t), z(s, t))$ to denote $\mathbf{Y}(s, t)$ and $(x(u, v), y(u, v), z(u, v))$ to denote $\mathbf{X}(u, v)$, even though this is a small abuse of notation.

By formula (7) of §7.1, we have

$$\mathbf{N_Y}(s, t) = \frac{\partial(y, z)}{\partial(s, t)}\mathbf{i} - \frac{\partial(x, z)}{\partial(s, t)}\mathbf{j} + \frac{\partial(x, y)}{\partial(s, t)}\mathbf{k}.$$

If we apply the chain rule to equation (12), we obtain

$$D\mathbf{Y}(s, t) = D\mathbf{X}(u, v)D\mathbf{H}(s, t).$$

Writing out this matrix equation, we get

$$\begin{bmatrix} x_s & x_t \\ y_s & y_t \\ z_s & z_t \end{bmatrix} = \begin{bmatrix} x_u & x_v \\ y_u & y_v \\ z_u & z_v \end{bmatrix} \begin{bmatrix} u_s & u_t \\ v_s & v_t \end{bmatrix},$$

where x_s, x_t, etc. denote partial derivatives of the component functions of $\mathbf{Y}$ and x_u, x_v, etc. are the partial derivatives of the component functions of $\mathbf{X}$. It is a matter of performing the matrix multiplication to check that

$$\begin{aligned} \begin{bmatrix} x_s & x_t \end{bmatrix} &= \text{first row of } D\mathbf{Y}(s, t) \\ &= \text{first row of the product } D\mathbf{X}(u, v)D\mathbf{H}(s, t) \\ &= (\text{first row of } D\mathbf{X}(u, v)) \cdot D\mathbf{H}(s, t) \\ &= \begin{bmatrix} x_u & x_v \end{bmatrix} \begin{bmatrix} u_s & u_t \\ v_s & v_t \end{bmatrix}. \end{aligned}$$

Similar results hold for the second and third rows of $D\mathbf{Y}(s, t)$. We may recombine these results about rows and establish the following matrix equations:

$$\begin{bmatrix} x_s & x_t \\ y_s & y_t \end{bmatrix} = \begin{bmatrix} x_u & x_v \\ y_u & y_v \end{bmatrix} \begin{bmatrix} u_s & u_t \\ v_s & v_t \end{bmatrix},$$

$$\begin{bmatrix} x_s & x_t \\ z_s & z_t \end{bmatrix} = \begin{bmatrix} x_u & x_v \\ z_u & z_v \end{bmatrix} \begin{bmatrix} u_s & u_t \\ v_s & v_t \end{bmatrix},$$

and

$$\begin{bmatrix} y_s & y_t \\ z_s & z_t \end{bmatrix} = \begin{bmatrix} y_u & y_v \\ z_u & z_v \end{bmatrix} \begin{bmatrix} u_s & u_t \\ v_s & v_t \end{bmatrix}.$$

Taking determinants, we find that

$$\frac{\partial(x, y)}{\partial(s, t)} = \frac{\partial(x, y)}{\partial(u, v)}\frac{\partial(u, v)}{\partial(s, t)},$$

$$\frac{\partial(x, z)}{\partial(s, t)} = \frac{\partial(x, z)}{\partial(u, v)}\frac{\partial(u, v)}{\partial(s, t)},$$

and

$$\frac{\partial(y, z)}{\partial(s, t)} = \frac{\partial(y, z)}{\partial(u, v)}\frac{\partial(u, v)}{\partial(s, t)}.$$

Thus, returning to the original formula for $\mathbf{N_Y}$, we find that

$$\begin{aligned}\mathbf{N_Y}(s, t) &= \frac{\partial(y, z)}{\partial(u, v)}\frac{\partial(u, v)}{\partial(s, t)}\mathbf{i} - \frac{\partial(x, z)}{\partial(u, v)}\frac{\partial(u, v)}{\partial(s, t)}\mathbf{j} + \frac{\partial(x, y)}{\partial(u, v)}\frac{\partial(u, v)}{\partial(s, t)}\mathbf{k} \\ &= \frac{\partial(u, v)}{\partial(s, t)}\mathbf{N_X}(u, v),\end{aligned}$$

as desired. ■

Proof of Theorem 2.4 We use Definition 2.1 and the change of variables theorem for double integrals. Thus, by Definition 2.1 and the lemma just proved,

$$\begin{aligned}\iint_{\mathbf{Y}} f\, dS &= \iint_{D_2} f(\mathbf{Y}(s, t))\, \|\mathbf{N_Y}(s, t)\|\, ds\, dt \\ &= \iint_{D_2} f(\mathbf{X}(\mathbf{H}(s, t))) \left|\frac{\partial(u, v)}{\partial(s, t)}\right| \|\mathbf{N_X}(u(s, t), v(s, t))\|\, ds\, dt.\end{aligned}$$

From the change of variables theorem, it follows that

$$\iint_{\mathbf{Y}} f\, dS = \iint_{D_1} f(\mathbf{X}(u, v))\, \|\mathbf{N_X}(u, v)\|\, du\, dv = \iint_{\mathbf{X}} f\, dS,$$

by Definition 2.1. ■

Proof of Theorem 2.5 This result can be established along the lines of the previous proof. Beginning with Definition 2.2 and using the lemma just established, we have

$$\begin{aligned}\iint_{\mathbf{Y}} \mathbf{F}\cdot d\mathbf{S} &= \iint_{D_2} \mathbf{F}(\mathbf{Y}(s, t))\cdot \mathbf{N_Y}(s, t)\, ds\, dt \\ &= \iint_{D_2} \mathbf{F}(\mathbf{X}(\mathbf{H}(s, t)))\cdot \frac{\partial(u, v)}{\partial(s, t)}\mathbf{N_X}(u(s, t), v(s, t))\, ds\, dt.\end{aligned}$$

Therefore,

$$\iint_{\mathbf{Y}} \mathbf{F}\cdot d\mathbf{S} = \pm \iint_{D_2} \mathbf{F}(\mathbf{X}(\mathbf{H}(s, t)))\cdot \mathbf{N_X}(u(s, t), v(s, t)) \left|\frac{\partial(u, v)}{\partial(s, t)}\right| ds\, dt,$$

where we take the "+" sign if $\mathbf{Y}$ is an orientation-preserving reparametrization of $\mathbf{X}$ (since the Jacobian $\partial(u, v)/\partial(s, t)$ is nonnegative and hence equal to its absolute value) and the "−" sign if $\mathbf{Y}$ is orientation-reversing. By the change of variables theorem, this last expression is equal to

$$\pm \iint_{D_1} \mathbf{F}(\mathbf{X}(u, v))\cdot \mathbf{N_X}(u, v)\, du\, dv = \pm \iint_{\mathbf{X}} \mathbf{F}\cdot d\mathbf{S},$$

by Definition 2.2. ■

Exercises

1. Let $\mathbf{X}(s,t) = (s, s+t, t)$, $0 \le s \le 1$, $0 \le t \le 2$. Find
$$\iint_{\mathbf{X}} (x^2 + y^2 + z^2)\, dS.$$

2. Let $D = \{(s,t) \mid s^2 + t^2 \le 1,\ s \ge 0,\ t \ge 0\}$ and let $\mathbf{X}\colon D \to \mathbf{R}^3$ be defined by $\mathbf{X}(s,t) = (s+t, s-t, st)$.
   (a) Determine $\iint_{\mathbf{X}} f\, dS$, where $f(x,y,z) = 4$.
   (b) Find the value of $\iint_{\mathbf{X}} \mathbf{F} \cdot d\mathbf{S}$, where $\mathbf{F} = x\,\mathbf{i} + y\,\mathbf{j} + z\,\mathbf{k}$.

3. Find the flux of $\mathbf{F} = x\,\mathbf{i} + y\,\mathbf{j} + z\,\mathbf{k}$ across the surface S consisting of the triangular region of the plane $2x - 2y + z = 2$ that is cut out by the coordinate planes. Use an upward-pointing normal to orient S.

4. This problem concerns the two surfaces given parametrically as
$$\mathbf{X}(s,t) = (s\cos t, s\sin t, 3s^2),\quad 0 \le s \le 2,\ 0 \le t \le 2\pi.$$
and
$$\mathbf{Y}(s,t) = (2s\cos t, 2s\sin t, 12s^2),\quad 0 \le s \le 1,\ 0 \le t \le 4\pi.$$
   (a) Show that the images of $\mathbf{X}$ and $\mathbf{Y}$ are the same. (Hint: Give equations in x, y, and z for the surfaces in $\mathbf{R}^3$ parametrized by $\mathbf{X}$ and $\mathbf{Y}$.)
   (b) Calculate $\iint_{\mathbf{X}} (y\,\mathbf{i} - x\,\mathbf{j} + z^2\,\mathbf{k}) \cdot d\mathbf{S}$ and $\iint_{\mathbf{Y}} (y\,\mathbf{i} - x\,\mathbf{j} + z^2\,\mathbf{k}) \cdot d\mathbf{S}$. Reconcile your answers.

5. Find $\iint_S x^2\, dS$, where S is the surface of the cube $[-2,2] \times [-2,2] \times [-2,2]$.

6. Find $\iint_S (x^2 + y^2)\, dS$, where S is the lateral surface of the cylinder of radius a and height h whose axis is the z-axis.

7. Let S be a sphere of radius a.
   (a) Find $\iint_S (x^2 + y^2 + z^2)\, dS$.
   (b) Use symmetry and part (a) to easily find $\iint_S y^2\, dS$.

8. Let S denote the sphere $x^2 + y^2 + z^2 = a^2$.
   (a) Use symmetry considerations to evaluate $\iint_S x\, dS$ *without* resorting to parametrizing the sphere.
   (b) Let $\mathbf{F} = \mathbf{i} + \mathbf{j} + \mathbf{k}$. Use symmetry to determine $\iint_S \mathbf{F} \cdot d\mathbf{S}$ without parametrizing the sphere.

9. Let S denote the surface of the cylinder $x^2 + y^2 = 4$, $-2 \le z \le 2$, and consider the surface integral
$$\iint_S (z - x^2 - y^2)\, dS.$$
   (a) Use an appropriate parametrization of S to calculate the value of the integral.
   (b) Now use geometry and symmetry to evaluate the integral without resorting to a parametrization of the surface.

In Exercises 10–18, let S denote the closed cylinder with bottom given by $z = 0$, top given by $z = 4$, and lateral surface given by the equation $x^2 + y^2 = 9$. Orient S with outward normals. Determine the indicated scalar and vector surface integrals.

10. $\displaystyle\iint_S z\, dS$

11. $\displaystyle\iint_S y\, dS$

12. $\displaystyle\iint_S xyz\, dS$

13. $\displaystyle\iint_S x^2 dS$

14. $\displaystyle\iint_S (x\,\mathbf{i} + y\,\mathbf{j}) \cdot d\mathbf{S}$

15. $\displaystyle\iint_S z\,\mathbf{k} \cdot d\mathbf{S}$

16. $\displaystyle\iint_S y^3\,\mathbf{i} \cdot d\mathbf{S}$

17. $\displaystyle\iint_S (-y\,\mathbf{i} + x\,\mathbf{j}) \cdot d\mathbf{S}$

18. $\displaystyle\iint_S x^2\,\mathbf{i} \cdot d\mathbf{S}$

In Exercises 19–22, find the flux of the given vector field $\mathbf{F}$ across the upper hemisphere $x^2 + y^2 + z^2 = a^2$, $z \ge 0$. Orient the hemisphere with an upward-pointing normal.

19. $\mathbf{F} = y\,\mathbf{j}$

20. $\mathbf{F} = y\,\mathbf{i} - x\,\mathbf{j}$

21. $\mathbf{F} = -y\,\mathbf{i} + x\,\mathbf{j} - \mathbf{k}$

22. $\mathbf{F} = x^2\mathbf{i} + xy\,\mathbf{j} + xz\,\mathbf{k}$

23. Let S be the funnel-shaped surface defined by $x^2 + y^2 = z^2$ for $1 \le z \le 9$ and $x^2 + y^2 = 1$ for $0 \le z \le 1$.
   (a) Sketch S.
   (b) Determine outward-pointing unit normal vectors to S.
   (c) Evaluate $\iint_S \mathbf{F} \cdot d\mathbf{S}$, where $\mathbf{F} = -y\,\mathbf{i} + x\,\mathbf{j} + z\,\mathbf{k}$ and S is oriented by outward normals.

24. The glass dome of a futuristic greenhouse is shaped like the surface $z = 8 - 2x^2 - 2y^2$. The greenhouse has a flat dirt floor at $z = 0$. Suppose that the temperature T, at points in and around the greenhouse, varies as
$$T(x,y,z) = x^2 + y^2 + 3(z-2)^2.$$
Then the temperature gives rise to a **heat flux density field H** given by $\mathbf{H} = -k\nabla T$. (Here k is a positive constant that depends on the insulating properties of the particular medium.) Find the total heat flux outward across the dome and the surface of the ground if $k = 1$ on the glass and $k = 3$ on the ground.

25. The surface given by $\mathbf{X}(s,t) = (x(s,t), y(s,t), z(s,t))$, where

$$\begin{cases} x = \left(a + \cos\dfrac{s}{2}\sin t - \sin\dfrac{s}{2}\sin 2t\right)\cos s \\ y = \left(a + \cos\dfrac{s}{2}\sin t - \sin\dfrac{s}{2}\sin 2t\right)\sin s\,, \\ z = \sin\dfrac{s}{2}\sin t + \cos\dfrac{s}{2}\sin 2t \end{cases}$$

a is a positive constant, and $0 \le s \le 2\pi$, $0 \le t \le 2\pi$, is known as a **Klein bottle.**

(a) Use a computer to plot this surface for $a = 2$.

(b) Determine (and describe) the s-coordinate curve at $t = 0$.

(c) Calculate the standard normal vector $\mathbf{N}$ along the s-coordinate curve at $t = 0$ (i.e., find $\mathbf{N}(s, 0)$). Note that $\mathbf{X}(0,0) = \mathbf{X}(2\pi, 0)$. By comparing $\mathbf{N}(0,0)$ and $\mathbf{N}(2\pi, 0)$, comment regarding the orientability of the Klein bottle. (See Example 8.)

7.3 Stokes's and Gauss's Theorems

Here we contemplate two important results: Stokes's theorem, which relates surface integrals to line integrals, and Gauss's theorem, which relates surface integrals to triple integrals. Along with Green's theorem, Stokes's and Gauss's theorems form the core of integral vector analysis and, as explained in the next section, can be used to establish further results in both mathematics and physics.

Stokes's Theorem

Stokes's theorem equates the surface integral of the curl of a C^1 vector field over a piecewise smooth, orientable surface with the line integral of the vector field along the boundary curve(s) of the surface. Since both vector line and surface integrals are examples of **oriented** integrals (i.e., they depend on the particular orientations chosen), we must comment on the way in which orientations need to be taken.

Definition 3.1 Let S be a bounded, piecewise smooth, oriented surface in $\mathbf{R}^3$. Let C' be any simple, closed curve lying in S. Consider the unit normal vector $\mathbf{n}$ that indicates the orientation of S at any point inside C'. Use $\mathbf{n}$ to orient C' by a right-hand rule, so that if the thumb of your right hand points along $\mathbf{n}$, then the fingers curl in the direction of the orientation of C'. (Equivalently, if you look down the tip of $\mathbf{n}$, the direction of C' should be such that the portion of S bounded by C' is on the *left*.) We say that C' with the orientation just described is **oriented consistently** with S or that the orientation is the one **induced** from that of S. Now suppose the boundary ∂S of S consists of finitely many piecewise C^1, simple, closed curves. Then we say that ∂S is **oriented consistently** (or that ∂S has its **orientation induced** from that of S) if each of its simple, closed pieces is oriented consistently with S.

Some examples of oriented surfaces with consistently oriented boundaries are shown in Figure 7.30. If the orientation of S is reversed, then the orientation of ∂S must also be reversed if it is to remain consistent with the new orientation of S.

Now we state a rather general version of Stokes's theorem, a proof of which is outlined in the addendum to this section.

Theorem 3.2 (STOKES'S THEOREM) Let S be a bounded, piecewise smooth, oriented surface in $\mathbf{R}^3$. Suppose that ∂S consists of finitely many piecewise C^1, simple, closed curves each of which is oriented consistently with S. Let $\mathbf{F}$ be a vector field of class C^1 whose domain includes S. Then

$$\iint_S \nabla \times \mathbf{F} \cdot d\mathbf{S} = \oint_{\partial S} \mathbf{F} \cdot d\mathbf{s}.$$

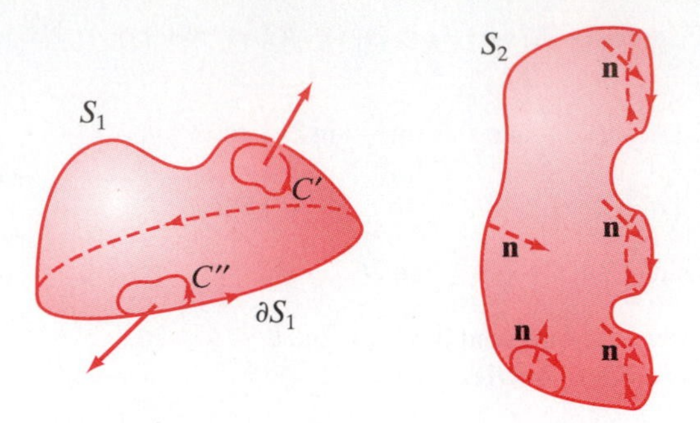

Figure 7.30 Examples of oriented surfaces and curves lying in them having consistent orientations. On the right, the boundary of S_2 consists of three simple closed curves.

Theorem 3.2 says that the total (net) "infinitesimal rotation," or swirling, of a vector field $\mathbf{F}$ over a surface S is equal to the circulation of $\mathbf{F}$ along just the boundary of S.

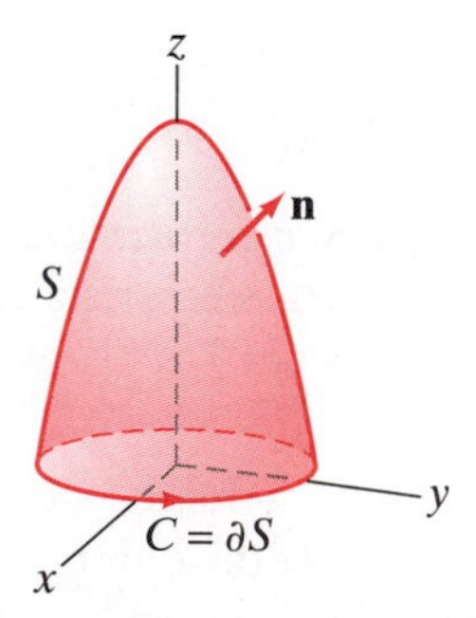

Figure 7.31 The paraboloid $z = 9 - x^2 - y^2$ oriented with upward normal $\mathbf{n}$. Note that the boundary circle C is oriented consistently with S.

EXAMPLE 1 Let S be the paraboloid $z = 9 - x^2 - y^2$ defined over the disk in the xy-plane of radius 3 (i.e., S is defined for $z \geq 0$ only). Then ∂S consists of the circle

$$C = \{(x, y, z) \mid x^2 + y^2 = 9,\ z = 0\}.$$

Orient S with the upward-pointing unit normal vector $\mathbf{n}$. (See Figure 7.31.) We verify Stokes's theorem for the vector field

$$\mathbf{F} = (2z - y)\,\mathbf{i} + (x + z)\,\mathbf{j} + (3x - 2y)\,\mathbf{k}.$$

We calculate

$$\nabla \times \mathbf{F} = \begin{vmatrix} \mathbf{i} & \mathbf{j} & \mathbf{k} \\ \partial/\partial x & \partial/\partial y & \partial/\partial z \\ 2z - y & x + z & 3x - 2y \end{vmatrix}$$
$$= (-2 - 1)\,\mathbf{i} + (2 - 3)\,\mathbf{j} + (1 - (-1))\,\mathbf{k} = -3\mathbf{i} - \mathbf{j} + 2\mathbf{k}.$$

An upward-pointing normal vector $\mathbf{N}$ is given by

$$\mathbf{N} = 2x\,\mathbf{i} + 2y\,\mathbf{j} + \mathbf{k}.$$

(This vector may, of course, be normalized to give an "orientation normal" $\mathbf{n}$.) Therefore, using formula (5) of §7.2 we have, where $D = \{(x, y) \mid x^2 + y^2 \leq 9\}$,

$$\iint_S \nabla \times \mathbf{F} \cdot d\mathbf{S} = \iint_D (-3\mathbf{i} - \mathbf{j} + 2\mathbf{k}) \cdot (2x\,\mathbf{i} + 2y\,\mathbf{j} + \mathbf{k})\,dx\,dy$$
$$= \iint_D (-6x - 2y + 2)\,dx\,dy$$
$$= \iint_D -6x\,dx\,dy - \iint_D 2y\,dx\,dy + \iint_D 2\,dx\,dy.$$

By the symmetry of D and the fact that $-6x$ and $2y$ are odd functions, we have that the first two double integrals are zero. The last double integral gives twice the area of D. Thus,

$$\iint_S \nabla \times \mathbf{F} \cdot d\mathbf{S} = 2 \cdot \pi(3^2) = 18\pi.$$

On the other hand, we may parametrize the boundary of S as

$$\begin{cases} x = 3\cos t \\ y = 3\sin t \\ z = 0 \end{cases} \quad 0 \le t \le 2\pi.$$

(This parametrization yields the orientation desired for ∂S.) Then

$$\oint_{\partial S} \mathbf{F} \cdot d\mathbf{s} = \int_0^{2\pi} \mathbf{F}(\mathbf{x}(t)) \cdot \mathbf{x}'(t)\, dt$$

$$= \int_0^{2\pi} (0 - 3\sin t, 3\cos t + 0, 9\cos t - 6\sin t) \cdot (-3\sin t, 3\cos t, 0)\, dt$$

$$= \int_0^{2\pi} (9\sin^2 t + 9\cos^2 t)\, dt = \int_0^{2\pi} 9\, dt = 18\pi,$$

which checks. ◆

EXAMPLE 2 Consider the surface S defined by the equation $z = e^{-(x^2+y^2)}$ for $z \ge 1/e$ (i.e., S is the graph of $f(x, y) = e^{-(x^2+y^2)}$ defined over $D = \{(x, y) \mid x^2 + y^2 \le 1\}$). Let

$$\mathbf{F} = (e^{y+z} - 2y)\,\mathbf{i} + (xe^{y+z} + y)\,\mathbf{j} + e^{x+y}\,\mathbf{k}.$$

Then, no matter which way we orient S, we can see that $\iint_S \nabla \times \mathbf{F} \cdot d\mathbf{S}$ looks impossible to calculate. Indeed, suppose we take the upward-pointing normal vector

$$\mathbf{N} = 2xe^{-(x^2+y^2)}\,\mathbf{i} + 2ye^{-(x^2+y^2)}\,\mathbf{j} + \mathbf{k}.$$

Then, because

$$\nabla \times \mathbf{F} = (e^{x+y} - xe^{y+z})\,\mathbf{i} + (e^{y+z} - e^{x+y})\,\mathbf{j} + 2\mathbf{k}$$

(you may wish to check this), using formula (5) of §7.2, we find that

$$\iint_S \nabla \times \mathbf{F} \cdot d\mathbf{S}$$

$$= \iint_D \left[2xe^{-(x^2+y^2)}(e^{x+y} - xe^{y+z}) + 2ye^{-(x^2+y^2)}(e^{y+z} - e^{x+y}) + 2 \right] dx\, dy.$$

We will not attempt to proceed any further with this calculation.

It is tempting to use Stokes's theorem at this point, since the boundary of S is the circle $x^2 + y^2 = 1$, $z = 1/e$. (See Figure 7.32.) If we parametrize this circle by

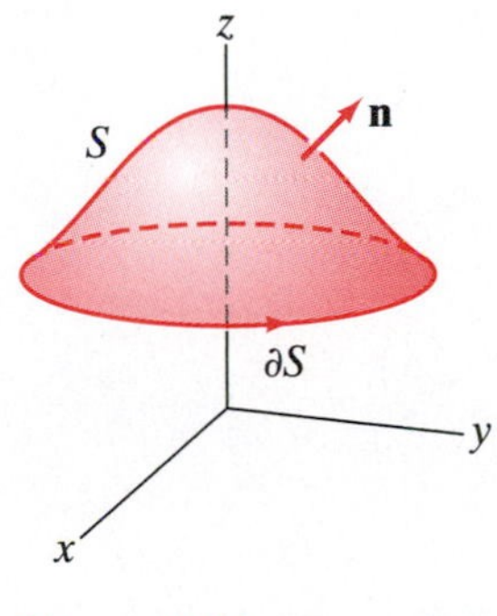

Figure 7.32 The surface $S = \{(x, y, z) \mid z = e^{-(x^2+y^2)}, z \ge 1/e\}$ has boundary $\partial S = \{(x, y, z) \mid x^2 + y^2 = 1, z = 1/e\}$.

$$\begin{cases} x = \cos t \\ y = \sin t \\ z = \dfrac{1}{e} \end{cases} \quad 0 \le t \le 2\pi,$$

then

$$\oint_{\partial S} \mathbf{F} \cdot d\mathbf{s}$$

$$= \int_0^{2\pi} (e^{\sin t + 1/e} - 2\sin t, \cos t\, e^{\sin t + 1/e} + \sin t, e^{\cos t + \sin t}) \cdot (-\sin t, \cos t, 0)\, dt$$

$$= \int_0^{2\pi} (2\sin^2 t - \sin t\, e^{\sin t + 1/e} + \cos^2 t e^{\sin t + 1/e} + \cos t\, \sin t)\, dt.$$

Again, we have difficulties.

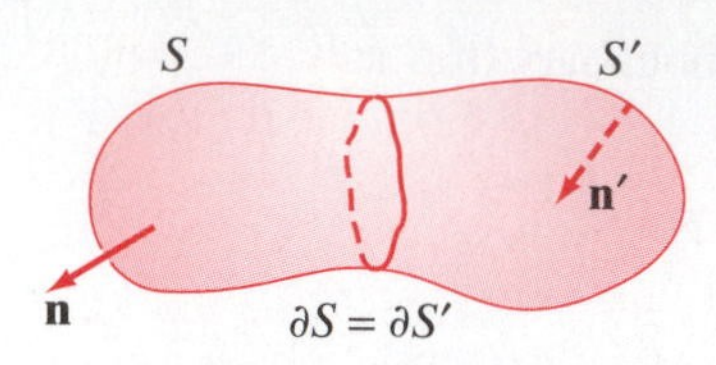

Figure 7.33 Both S and S' have the same boundary and are oriented as indicated. Therefore, by Stokes's theorem, $\iint_S \nabla \times \mathbf{F} \cdot d\mathbf{S} = \iint_{S'} \nabla \times \mathbf{F} \cdot d\mathbf{S}$.

However, the power of Stokes's theorem is that if S' is *any* orientable piecewise smooth surface whose boundary $\partial S'$ is the same as ∂S then, subject to orienting S' appropriately,

$$\iint_S \nabla \times \mathbf{F} \cdot d\mathbf{S} = \oint_{\partial S} \mathbf{F} \cdot d\mathbf{s} = \oint_{\partial S'} \mathbf{F} \cdot d\mathbf{s} = \iint_{S'} \nabla \times \mathbf{F} \cdot d\mathbf{S}.$$

Hence, we may evaluate $\iint_S \nabla \times \mathbf{F} \cdot d\mathbf{S}$ by using a different surface! (See Figure 7.33.)

To use this fact to our advantage, note that $\nabla \times \mathbf{F}$ has a particularly simple $\mathbf{k}$-component. Thus, we let S' be the unit disk at $z = 1/e$:

$$S' = \{(x, y, z) \mid x^2 + y^2 \le 1, z = 1/e\}.$$

Consequently, if we orient S' by the unit normal vector $\mathbf{n} = +\mathbf{k}$, we have

$$\begin{aligned}\iint_S \nabla \times \mathbf{F} \cdot d\mathbf{S} &= \iint_{S'} \nabla \times \mathbf{F} \cdot d\mathbf{S} \\ &= \iint_{S'} (\nabla \times \mathbf{F} \cdot \mathbf{n})\, dS \\ &= \iint_{S'} 2\, dS = 2 \cdot \text{area of } S' = 2\pi.\end{aligned}$$

◆

Gauss's Theorem

Also known as the **divergence theorem, Gauss's theorem** relates the vector surface integral over a closed surface to a triple integral over the solid region enclosed by the surface. Like Stokes's theorem, Gauss's theorem can assist with computational issues, although the significance of the result extends well beyond matters of calculation.

■ **Theorem 3.3** (GAUSS'S THEOREM) Let D be a bounded solid region in $\mathbf{R}^3$ whose boundary ∂D consists of finitely many piecewise smooth, closed orientable surfaces, each of which is oriented by unit normals that point *away from* D. (See Figure 7.34.) Let $\mathbf{F}$ be a vector field of class C^1 whose domain includes D. Then

$$\oiint_{\partial D} \mathbf{F} \cdot d\mathbf{S} = \iiint_D \nabla \cdot \mathbf{F}\, dV.$$

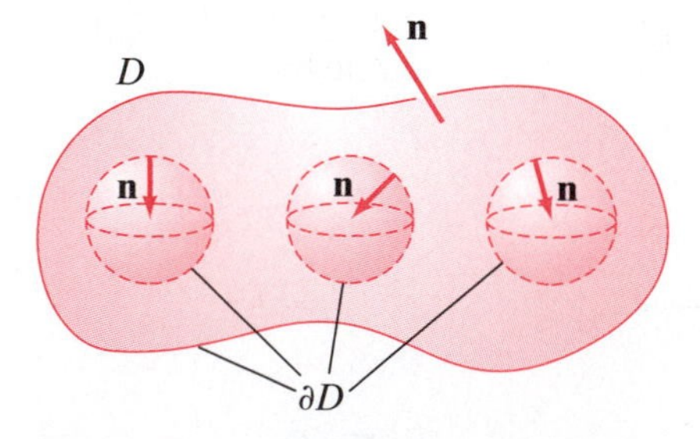

Figure 7.34 A solid region D whose boundary surfaces are oriented so that Gauss's theorem applies.

By a *closed* surface, we mean one without any boundary curves, like a sphere or a cube. The symbol $\oiint$ is used to indicate a surface integral taken over a closed surface or surfaces.

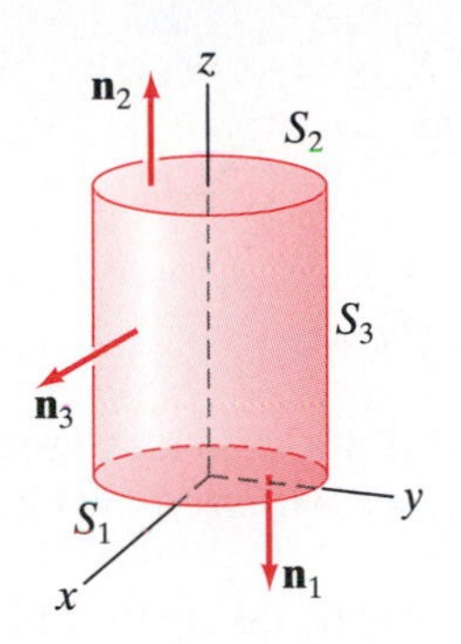

Figure 7.35 The solid cylinder D of Example 3.

Gauss's theorem says that the "total divergence" of a vector field in a bounded region in space is equal to the flux of the vector field away from the region (i.e., the flux across the boundary surface(s)).

EXAMPLE 3 Let $\mathbf{F}$ be the radial vector field $x\,\mathbf{i}+y\,\mathbf{j}+z\,\mathbf{k}$ and let D be the solid cylinder of radius a and height b, located so that axis of the cylinder is the z-axis and the top and bottom of the cylinder are at $z=b$ and $z=0$. (See Figure 7.35.) We verify Gauss's theorem for this vector field and solid region.

The boundary of D consists of three smooth pieces: (1) the bottom surface S_1 that is a portion of the plane $z=0$ and oriented by the normal $\mathbf{n}_1=-\mathbf{k}$, (2) the top surface S_2 that is a portion of the plane $z=b$ and is oriented by the normal vector $\mathbf{n}_2=\mathbf{k}$, and (3) a portion of the lateral cylinder S_3 given by the equation $x^2+y^2=a^2$ and oriented by the unit vector $\mathbf{n}_3=(x\,\mathbf{i}+y\,\mathbf{j})/a$. (The vector $\mathbf{n}_3$ may be obtained by normalizing the gradient of $f(x,y,z)=x^2+y^2$ that defines S_3 as a level set.) Then

$$\begin{aligned}\oint\!\!\!\oint_{\partial D}\mathbf{F}\cdot d\mathbf{S} &= \iint_{S_1}\mathbf{F}\cdot d\mathbf{S}+\iint_{S_2}\mathbf{F}\cdot d\mathbf{S}+\iint_{S_3}\mathbf{F}\cdot d\mathbf{S}\\ &= \iint_{S_1}(x\,\mathbf{i}+y\,\mathbf{j}+z\,\mathbf{k})\cdot(-\mathbf{k})\,dS+\iint_{S_2}(x\,\mathbf{i}+y\,\mathbf{j}+z\,\mathbf{k})\cdot\mathbf{k}\,dS\\ &\quad+\iint_{S_3}(x\,\mathbf{i}+y\,\mathbf{j}+z\,\mathbf{k})\cdot\left(\frac{x\,\mathbf{i}+y\,\mathbf{j}}{a}\right)dS\\ &= \iint_{S_1}-z\,dS+\iint_{S_2}z\,dS+\iint_{S_3}\frac{x^2+y^2}{a}\,dS\\ &= 0+\iint_{S_2}b\,dS+\iint_{S_3}\frac{a^2}{a}\,dS,\end{aligned}$$

since along S_1, z is 0; along S_2, z is equal to b; and along S_3, $x^2+y^2=a^2$. Thus,

$$\oint\!\!\!\oint_{\partial D}\mathbf{F}\cdot d\mathbf{S}=b\cdot\text{area of }S_2+a\cdot\text{area of }S_3=b\pi a^2+a(2\pi ab)=3\pi a^2 b,$$

from familiar geometric formulas.

On the other hand,

$$\nabla\cdot\mathbf{F}=\frac{\partial}{\partial x}(x)+\frac{\partial}{\partial y}(y)+\frac{\partial}{\partial z}(z)=3,$$

so that

$$\iiint_D\nabla\cdot\mathbf{F}\,dV=\iiint_D 3\,dV=3\cdot\text{volume of }D=3\pi a^2 b,$$

which can be checked readily. ◆

In general, if $\mathbf{F}=x\,\mathbf{i}+y\,\mathbf{j}+z\,\mathbf{k}$ and D is a region to which Gauss's theorem applies, then

$$\oint\!\!\!\oint_{\partial D}\mathbf{F}\cdot d\mathbf{S}=\iiint_D\nabla\cdot\mathbf{F}\,dV=\iiint_D 3\,dV=3\cdot\text{volume of }D.$$

Hence,

$$\tfrac{1}{3}\oint\!\!\!\oint_{\partial D}(x\,\mathbf{i}+y\,\mathbf{j}+z\,\mathbf{k})\cdot d\mathbf{S}=\text{volume of }D.$$

Therefore, we may use surface integrals to calculate volumes in much the same way that we used Green's theorem to calculate areas of plane regions by means of suitable line integrals. (See §6.2, especially Examples 2 and 3.)

EXAMPLE 4 Let

$$\mathbf{F} = e^y \cos z\,\mathbf{i} + \sqrt{x^3+1}\sin z\,\mathbf{j} + (x^2+y^2+3)\,\mathbf{k},$$

and let S be the graph of

$$z = (1 - x^2 - y^2)e^{1-x^2-3y^2} \quad \text{for } z \geq 0,$$

oriented by the upward-pointing unit normal vector. It is not difficult to see that $\iint_S \mathbf{F}\cdot d\mathbf{S}$ is impossible to evaluate directly. However, we will see how Gauss's theorem provides us with elegant indirect means.

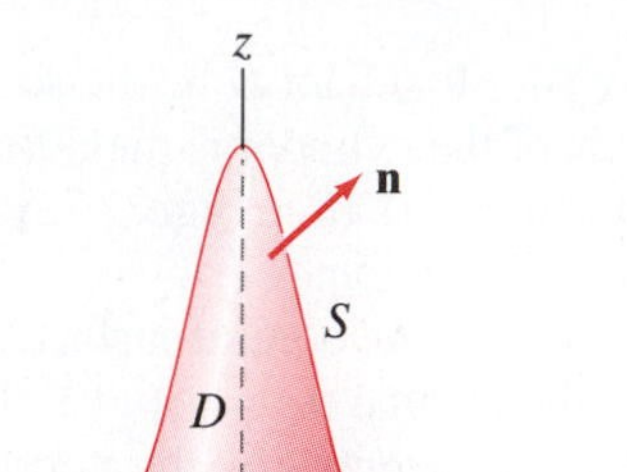

Figure 7.36 The union of the surfaces S and S' enclose a solid region D to which we may apply Gauss's theorem.

Consider the piecewise smooth, *closed* surface created by taking the union of S and S', where S' is the portion of the plane $z = 0$ enclosed by ∂S (i.e., the disk $x^2 + y^2 \leq 1$, $z = 0$). Orient S' by the downward-pointing unit normal $z = -\mathbf{k}$ as shown in Figure 7.36. Note that $S \cup S'$ forms the boundary of a solid region D and, furthermore, that the orientations chosen enable us to apply Gauss's theorem. Doing so, we have

$$\iint_S \mathbf{F}\cdot d\mathbf{S} + \iint_{S'} \mathbf{F}\cdot d\mathbf{S} = \oiint_{\partial D} \mathbf{F}\cdot d\mathbf{S} = \iiint_D \nabla\cdot\mathbf{F}\,dV.$$

Now, it is a simple matter to check that $\nabla\cdot\mathbf{F} = 0$ for all (x, y, z). Therefore, the triple integral is zero, and we find that

$$\iint_S \mathbf{F}\cdot d\mathbf{S} + \iint_{S'} \mathbf{F}\cdot d\mathbf{S} = 0,$$

so that

$$\iint_S \mathbf{F}\cdot d\mathbf{S} = -\iint_{S'} \mathbf{F}\cdot d\mathbf{S}.$$

In other words, because $\mathbf{F}$ is divergenceless, Gauss's theorem allows us to replace the original surface integral by one that is considerably easier to evaluate. Indeed, we have

$$\iint_S \mathbf{F}\cdot d\mathbf{S} = -\iint_{S'} \mathbf{F}\cdot(-\mathbf{k})\,dS = \iint_R (x^2+y^2+3)\,dx\,dy,$$

where R is the unit disk $\{(x, y) \mid x^2 + y^2 \leq 1\}$ in the plane. Now, we switch to polar coordinates to find

$$\begin{aligned}\iint_R (x^2+y^2+3)\,dx\,dy &= \int_0^{2\pi}\int_0^1 (r^2+3)r\,dr\,d\theta \\ &= \int_0^{2\pi} \left(\tfrac{1}{4}r^4 + \tfrac{3}{2}r^2\right)\Big|_{r=0}^{1}\,d\theta \\ &= \int_0^{2\pi} \tfrac{7}{4}\,d\theta = \tfrac{7}{2}\pi.\end{aligned}$$

◆

The Meaning of Divergence and Curl

Part of the significance of Stokes's theorem and Gauss's theorem is that they provide a way to understand the meaning of the divergence and curl of a vector field apart from the coordinate-based definitions of §3.4. By way of explanation, we offer the following two results:

■ **Proposition 3.4** Let $\mathbf{F}$ be a vector field of class C^1 in some neighborhood of the point P in $\mathbf{R}^3$. Let S_a denote the sphere of radius a centered at P, oriented

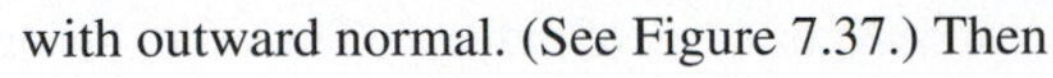

with outward normal. (See Figure 7.37.) Then

$$\operatorname{div} \mathbf{F}(P) = \lim_{a \to 0^+} \frac{3}{4\pi a^3} \oiint_{S_a} \mathbf{F} \cdot d\mathbf{S}.$$

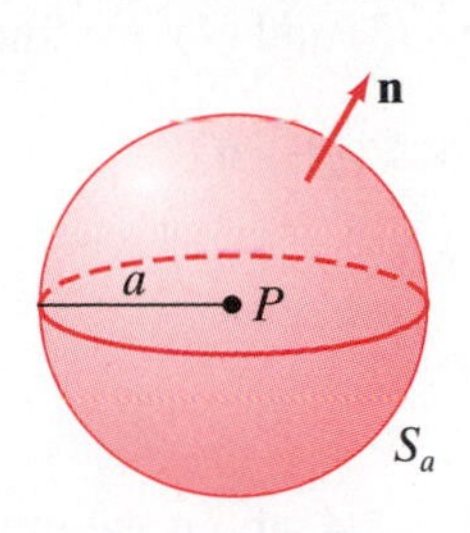

Figure 7.37 A sphere of radius a used to understand the divergence of a vector field.

Proof We have, by Gauss's theorem, that

$$\lim_{a \to 0^+} \frac{3}{4\pi a^3} \oiint_{S_a} \mathbf{F} \cdot d\mathbf{S} = \lim_{a \to 0^+} \frac{3}{4\pi a^3} \iiint_{D_a} \operatorname{div} \mathbf{F} \, dV, \tag{1}$$

where D_a is the solid ball of radius a enclosed by S_a. Next, we use a result known as the **mean value theorem for triple integrals,** which states that if f is a continuous function of three variables and D is a bounded solid region in space, then there is some point $Q \in D$ such that

$$\iiint_D f(x, y, z) \, dV = f(Q) \cdot \text{volume of } D.$$

In our present situation, this result implies that there must be some point $Q \in D_a$ such that

$$\iiint_{D_a} \operatorname{div} \mathbf{F} \, dV = \operatorname{div} \mathbf{F}(Q) \cdot (\text{volume of } D_a) = \left(\tfrac{4}{3}\pi a^3\right) \operatorname{div} \mathbf{F}(Q). \tag{2}$$

Applying formula (2) to formula (1), we have

$$\lim_{a \to 0^+} \frac{3}{4\pi a^3} \oiint_{S_a} \mathbf{F} \cdot d\mathbf{S} = \lim_{a \to 0^+} \operatorname{div} \mathbf{F}(Q) = \operatorname{div} \mathbf{F}(P),$$

since, as $a \to 0^+$, the ball D_a becomes smaller, "crushing" Q onto P. ■

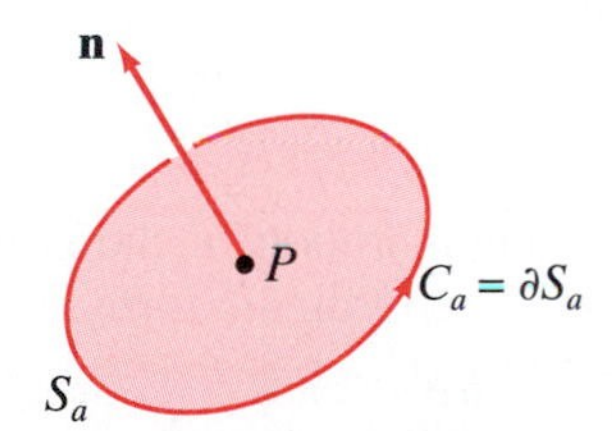

Figure 7.38 A circle of radius a centered at P.

■ **Proposition 3.5** Let $\mathbf{F}$ be a vector field of class C^1 in a neighborhood of the point P in $\mathbf{R}^3$. Let C_a be the circle of radius a centered at P situated in the plane containing P that is perpendicular to the unit vector $\mathbf{n}$. (See Figure 7.38.) Then the component of curl $\mathbf{F}(P)$ in the $\mathbf{n}$-direction is

$$\mathbf{n} \cdot \operatorname{curl} \mathbf{F}(P) = \lim_{a \to 0^+} \frac{1}{\pi a^2} \oint_{C_a} \mathbf{F} \cdot d\mathbf{s},$$

where C_a is oriented by a right-hand rule with respect to $\mathbf{n}$.

Proof Let S_a denote the disk of radius a in the plane of C_a enclosed by C_a. By Stokes's theorem,

$$\begin{aligned} \lim_{a \to 0^+} \frac{1}{\pi a^2} \oint_{C_a} \mathbf{F} \cdot d\mathbf{s} &= \lim_{a \to 0^+} \frac{1}{\pi a^2} \iint_{S_a} \nabla \times \mathbf{F} \cdot d\mathbf{S} \\ &= \lim_{a \to 0^+} \frac{1}{\pi a^2} \iint_{S_a} (\nabla \times \mathbf{F} \cdot \mathbf{n}) \, dS \end{aligned} \tag{3}$$

There is a mean value theorem for surface integrals (similar to the mean value theorem for triple integrals used in the proof of Proposition 3.4) enabling us to conclude that there must be some point Q in S_a for which

$$\begin{aligned} \iint_{S_a} (\nabla \times \mathbf{F} \cdot \mathbf{n}) \, dS &= (\nabla \times \mathbf{F}(Q) \cdot \mathbf{n})(\text{area of } S_a) \\ &= \pi a^2 (\nabla \times \mathbf{F}(Q) \cdot \mathbf{n}), \end{aligned} \tag{4}$$

since S_a is a disk of radius a. Therefore, using equations (3) and (4), we find that

$$\begin{aligned}\lim_{a\to 0^+}\frac{1}{\pi a^2}\oint_{C_a}\mathbf{F}\cdot d\mathbf{s} &= \lim_{a\to 0^+}\frac{1}{\pi a^2}(\pi a^2\nabla\times\mathbf{F}(Q)\cdot\mathbf{n})\\ &= \lim_{a\to 0^+}\nabla\times\mathbf{F}(Q)\cdot\mathbf{n}\\ &= \nabla\times\mathbf{F}(P)\cdot\mathbf{n},\end{aligned}$$

since $Q\to P$ as $a\to 0^+$. ■

Propositions 3.4 and 3.5 justify our claims, made in §3.4, about the intuitive meanings of the divergence and curl of a vector field. The quantity $\oiint_{S_a}\mathbf{F}\cdot d\mathbf{S}$ used in Proposition 3.4 is the flux of $\mathbf{F}$ across the sphere S_a, and so

$$\lim_{a\to 0^+}\frac{3}{4\pi a^2}\oiint_{S_a}\mathbf{F}\cdot d\mathbf{S}$$

is precisely the limit of the flux per unit volume, or the **flux density** of $\mathbf{F}$ at P. Similarly,

$$\lim_{a\to 0^+}\frac{1}{\pi a^2}\oint_{C_a}\mathbf{F}\cdot d\mathbf{s}$$

is the limit of the circulation of $\mathbf{F}$ along C_a per unit area, or the **circulation density** of $\mathbf{F}$ at P around $\mathbf{n}$. In particular, Proposition 3.5 shows that curl $\mathbf{F}(P)$ is the vector whose direction maximizes the circulation density of $\mathbf{F}$ at P and whose magnitude is equal to the circulation density around that direction (or else curl $\mathbf{F}(P)$ is $\mathbf{0}$ if the circulation density is zero).

In fact, we can turn our approach to divergence and curl completely around and, instead of defining the divergence and curl by means of coordinates and the del operator and proving Proposition 3.4 and 3.5, use the surface and line integral formulas of Propositions 3.4 and 3.5 to *define* divergence and curl and derive the coordinate formulations from the limiting integral formulas.

Write $\mathbf{F}$ as

$$M(x,y,z)\,\mathbf{i}+N(x,y,z)\,\mathbf{j}+P(x,y,z)\,\mathbf{k},$$

where M, N, and P are functions of class C^1 in a neighborhood of the point $\mathbf{x}_0=(x_0,y_0,z_0)$. We will demonstrate how to recover the coordinate formula

$$\nabla\times\mathbf{F}=\left(\frac{\partial P}{\partial y}-\frac{\partial N}{\partial z}\right)\mathbf{i}+\left(\frac{\partial M}{\partial z}-\frac{\partial P}{\partial x}\right)+\left(\frac{\partial N}{\partial x}-\frac{\partial M}{\partial y}\right)\mathbf{k}$$

from the formula in Proposition 3.5. (A similar argument can be made to derive the coordinate formula for the divergence, the details of which we leave to you.) The idea is to let the unit vector $\mathbf{n}$ equal, in turn, $\mathbf{i}$, $\mathbf{j}$, and $\mathbf{k}$ in the formula in Proposition 3.5 and thereby to determine the components of the curl.

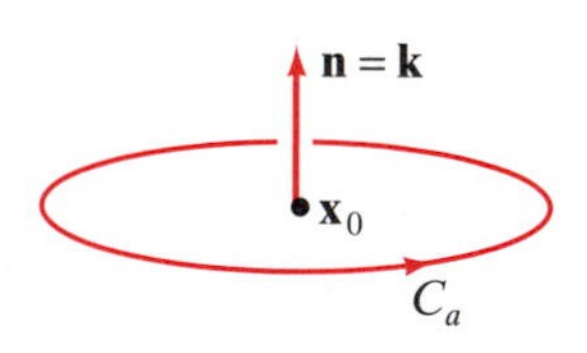

Figure 7.39 The configuration needed to calculate the $\mathbf{k}$-component of curl $\mathbf{F}(\mathbf{x}_0)$, using Proposition 3.5.

First, let $\mathbf{n}=\mathbf{k}$, so that C_a is the circle of radius a in the horizontal plane $z=z_0$, oriented counterclockwise around $\mathbf{k}$. (See Figure 7.39.) Then C_a may be parametrized by

$$\begin{cases} x = x_0 + a\cos t\\ y = y_0 + a\sin t \qquad 0\le t\le 2\pi.\\ z = z_0\end{cases}$$

Therefore,

$$\begin{aligned}\oint_{C_a}\mathbf{F}\cdot d\mathbf{s} &= \int_0^{2\pi}\mathbf{F}(\mathbf{x}(t))\cdot\mathbf{x}'(t)\,dt\\ &= \int_0^{2\pi}(M(\mathbf{x}(t))\,\mathbf{i}+N(\mathbf{x}(t))\,\mathbf{j}+P(\mathbf{x}(t))\,\mathbf{k})\cdot(-a\sin t\,\mathbf{i}+a\cos t\,\mathbf{j})\,dt\\ &= a\int_0^{2\pi}(-\mathrm{sin}t\,M(\mathbf{x}(t))+\cos t\,N(\mathbf{x}(t)))\,dt.\end{aligned}$$

Next, use Taylor's first-order formula (see §4.1) on M and N, which yields, near $\mathbf{x}_0$ (i.e., for $(x, y, z) \approx (x_0, y_0, z_0)$),

$$M(x, y, z) \approx M(\mathbf{x}_0) + M_x(\mathbf{x}_0)(x - x_0) + M_y(\mathbf{x}_0)(y - y_0) + M_z(\mathbf{x}_0)(z - z_0);$$
$$N(x, y, z) \approx N(\mathbf{x}_0) + N_x(\mathbf{x}_0)(x - x_0) + N_y(\mathbf{x}_0)(y - y_0) + N_z(\mathbf{x}_0)(z - z_0).$$

Along the small circle C_a, we have $x - x_0 = a\cos t$, $y - y_0 = a\sin t$, and $z - z_0 = 0$, so that, using the approximations for M and N, we have

$$\begin{aligned}\oint_{C_a} \mathbf{F}\cdot d\mathbf{s} &\approx a\int_0^{2\pi} -\sin t\,[M(\mathbf{x}_0) + M_x(\mathbf{x}_0)a\cos t + M_y(\mathbf{x}_0)a\sin t\,]\,dt \\ &\quad + a\int_0^{2\pi} \cos t[N(\mathbf{x}_0) + N_x(\mathbf{x}_0)a\cos t + N_y(\mathbf{x}_0)a\sin t\,]\,dt \\ &= -aM(\mathbf{x}_0)\int_0^{2\pi}\sin t\,dt - a^2M_x(\mathbf{x}_0)\int_0^{2\pi}\sin t\,\cos t\,dt \\ &\quad - a^2M_y(\mathbf{x}_0)\int_0^{2\pi}\sin^2 t\,dt + aN(\mathbf{x}_0)\int_0^{2\pi}\cos t\,dt \\ &\quad + a^2N_x(\mathbf{x}_0)\int_0^{2\pi}\cos^2 t\,dt + a^2N_y(\mathbf{x}_0)\int_0^{2\pi}\sin t\,\cos t\,dt. \qquad (5)\end{aligned}$$

This last equality holds because $M(\mathbf{x}_0)$, $M_x(\mathbf{x}_0)$, etc., do not involve t and so may be pulled out of the appropriate integrals. You can check that

$$\int_0^{2\pi}\sin t\,dt = \int_0^{2\pi}\cos t\,dt = \int_0^{2\pi}\sin t\,\cos t\,dt = 0,$$
$$\int_0^{2\pi}\sin^2 t\,dt = \int_0^{2\pi}\cos^2 t\,dt = \pi.$$

Therefore, the approximation in (5) simplifies to

$$\oint_{C_a}\mathbf{F}\cdot d\mathbf{s} \approx -\pi a^2 M_y(\mathbf{x}_0) + \pi a^2 N_x(\mathbf{x}_0).$$

Now, the error involved in the approximation for $\oint_{C_a}\mathbf{F}\cdot d\mathbf{s}$ tends to zero as C_a becomes smaller and smaller. Thus,

$$\begin{aligned}\mathbf{k}\cdot\operatorname{curl}\mathbf{F}(\mathbf{x}_0) &= \lim_{a\to 0^+}\frac{1}{\pi a^2}\oint_{C_a}\mathbf{F}\cdot d\mathbf{s} \\ &= \lim_{a\to 0^+}\big(-M_y(\mathbf{x}_0) + N_x(\mathbf{x}_0)\big) \\ &= N_x(\mathbf{x}_0) - M_y(\mathbf{x}_0).\end{aligned}$$

If we let $\mathbf{n} = \mathbf{j}$, so that C_a is the circle parametrized by

$$\begin{cases} x = x_0 + a\sin t \\ y = y_0 \\ z = z_0 + a\cos t \end{cases} \quad 0 \le t \le 2\pi,$$

then a very similar argument to the one just given shows that

$$\mathbf{j}\cdot\operatorname{curl}\mathbf{F}(\mathbf{x}_0) = M_z(\mathbf{x}_0) - P_x(\mathbf{x}_0).$$

Finally, let $\mathbf{n} = \mathbf{i}$, so that C_a is parametrized by

$$\begin{cases} x = x_0 \\ y = y_0 + a\cos t \\ z = z_0 + a\sin t \end{cases} \quad 0 \le t \le 2\pi,$$

to find

$$\mathbf{i} \cdot \operatorname{curl} \mathbf{F}(\mathbf{x}_0) = P_y(\mathbf{x}_0) - N_z(\mathbf{x}_0).$$

We see that the **i**-, **j**-, and **k**-components of the curl of **F** are as stated in §3.4.

Addendum: Proofs of Stokes's and Gauss's Theorems

Proof of Stokes's theorem

Step 1. We begin by establishing a very special case of the theorem, namely, the case where the vector field $\mathbf{F} = M(x, y, z)\mathbf{i}$ (i.e., **F** has an **i**-component only) and where the surface S is the graph of $z = f(x, y)$, where f is of class C^1 on a domain D in the plane that is a type 1 elementary region. (See Figure 7.40.) To be explicit, the region D is

$$\{(x, y) \mid \gamma(x) \le y \le \delta(x), a \le x \le b\},$$

where γ and δ are continuous functions. We assume that S is oriented by the upward-pointing unit normal.

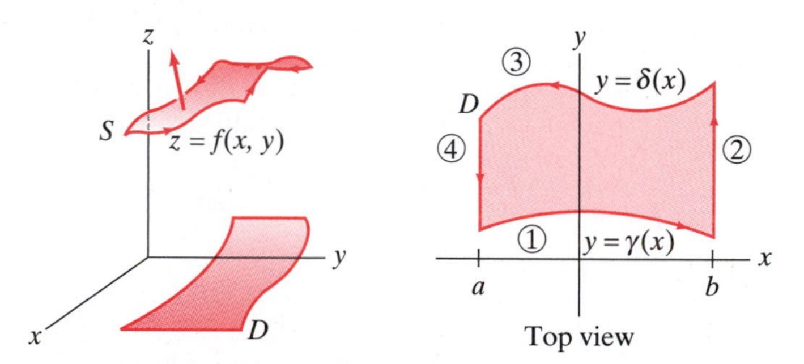

Figure 7.40 The surface S is the graph of $f(x, y)$ for (x, y) in the type 1 region D shown at the right.

First, evaluate $\oint_{\partial S} \mathbf{F} \cdot d\mathbf{s}$. The boundary ∂S consists of (at most) four smooth pieces parametrized as follows. (The subscripted curves correspond to the encircled numbers in Figure 7.40.)

$$C_1 : \begin{cases} x = t \\ y = \gamma(t) \\ z = f(t, \gamma(t)) \end{cases} \quad a \le t \le b; \qquad C_2 : \begin{cases} x = b \\ y = t \\ z = f(b, t) \end{cases} \quad \gamma(b) \le t \le \delta(b);$$

$$C_3 : \begin{cases} x = t \\ y = \delta(t) \\ z = f(t, \delta(t)) \end{cases} \quad a \le t \le b; \qquad C_4 : \begin{cases} x = a \\ y = t \\ z = f(a, t) \end{cases} \quad \gamma(a) \le t \le \delta(a).$$

The parametrizations shown for C_3 and C_4 induce the *opposite* orientations to those indicated in Figure 7.40. Therefore,

$$\oint_{\partial S} \mathbf{F} \cdot d\mathbf{s} = \int_{C_1} \mathbf{F} \cdot d\mathbf{s} + \int_{C_2} \mathbf{F} \cdot d\mathbf{s} - \int_{C_3} \mathbf{F} \cdot d\mathbf{s} - \int_{C_4} \mathbf{F} \cdot d\mathbf{s}.$$

Consider the integral over C_1. Since **F** has only an **i**-component,

$$\int_{C_1} \mathbf{F} \cdot d\mathbf{s} = \int_{C_1} M\, dx + 0\, dy + 0\, dz = \int_a^b M(t, \gamma(t), f(t, \gamma(t)))\, dt.$$

The line integral $\int_{C_3} \mathbf{F} \cdot d\mathbf{s}$ may be calculated in a manner similar to that for $\int_{C_1} \mathbf{F} \cdot d\mathbf{s}$. In particular, we obtain

$$\int_{C_3} \mathbf{F} \cdot d\mathbf{s} = \int_a^b M(t, \delta(t), f(t, \delta(t)))\, dt.$$

For the integral over C_2, note that x is held constant. Thus,

$$\int_{C_2} \mathbf{F} \cdot d\mathbf{s} = \int_{C_2} M\, dx = 0.$$

Likewise, $\int_{C_4} \mathbf{F} \cdot d\mathbf{s} = 0$. The result is

$$\begin{aligned} \oint_{\partial S} \mathbf{F} \cdot d\mathbf{s} &= \int_a^b M(t, \gamma(t), f(t, \gamma(t)))\, dt - \int_a^b M(t, \delta(t), f(t, \delta(t)))\, dt \\ &= \int_a^b [M(x, \gamma(x), f(x, \gamma(x))) - M(x, \delta(x), f(x, \delta(x)))]\, dx. \qquad (6) \end{aligned}$$

(In this last equality we've made a change in the variable of integration.)

Now we compare the line integral to the surface integral $\iint_S \nabla \times \mathbf{F} \cdot d\mathbf{S}$. For $\mathbf{F} = M(x, y, z)\,\mathbf{i}$, we have $\nabla \times \mathbf{F} = M_z\,\mathbf{j} - M_y\,\mathbf{k}$, so formula (5) of §7.2 yields

$$\begin{aligned} \iint_S \nabla \times \mathbf{F} \cdot d\mathbf{S} &= \iint_D (M_z\,\mathbf{j} - M_y\,\mathbf{k}) \cdot (-f_x\,\mathbf{i} - f_y\,\mathbf{j} + \mathbf{k})\, dx\, dy \\ &= \int_a^b \int_{\gamma(x)}^{\delta(x)} \left(-\frac{\partial M}{\partial z}\frac{\partial f}{\partial y} - \frac{\partial M}{\partial y} \right) dy\, dx. \end{aligned}$$

The chain rule implies

$$\frac{\partial}{\partial y}(M(x, y, f(x, y)) = \frac{\partial M}{\partial y} + \frac{\partial M}{\partial z}\frac{\partial f}{\partial y}.$$

Thus,

$$\begin{aligned} \iint_S \nabla \times \mathbf{F} \cdot d\mathbf{S} &= \int_a^b \int_{\gamma(x)}^{\delta(x)} -\frac{\partial}{\partial y}(M(x, y, f(x, y)))\, dy\, dx \\ &= \int_a^b -M(x, y, f(x, y))\Big|_{y=\gamma(x)}^{y=\delta(x)}\, dx \\ &= \int_a^b [-M(x, \delta(x), f(x, \delta(x))) + M(x, \gamma(x), f(x, \gamma(x)))]\, dx, \end{aligned}$$

which agrees with equation (6).

We may readily extend this result to surfaces that are graphs of $z = f(x, y)$, where the point (x, y) varies through an arbitrary region D in the xy-plane via a two-stage process: First, establish the result for regions D that may be subdivided into finitely many elementary regions of the type 1 and then apply a limiting argument similar to the one outlined in the proof of Green's theorem in §6.2.

Step 2. Still keeping $\mathbf{F} = M\mathbf{i}$, note that the argument given in Step 1 works equally well for surfaces of the form $y = f(x, z)$—simply exchange the roles of y and z throughout Step 1.

It is also not difficult to see that if S is a portion of the plane $x = c$ (where c is a constant), then Stokes's theorem for $\mathbf{F} = M\mathbf{i}$ holds for this case, too. On the one hand, $\mathbf{n} = \pm\mathbf{i}$ for such a plane (depending on the orientation chosen) and

$\nabla \times \mathbf{F} = M_z\,\mathbf{j} - M_y\,\mathbf{k}$, as we have seen. Hence,

$$\iint_S \nabla \times \mathbf{F} \cdot d\mathbf{S} = \iint_S (\nabla \times \mathbf{F} \cdot \mathbf{n})\, dS = \iint_S 0\, dS = 0.$$

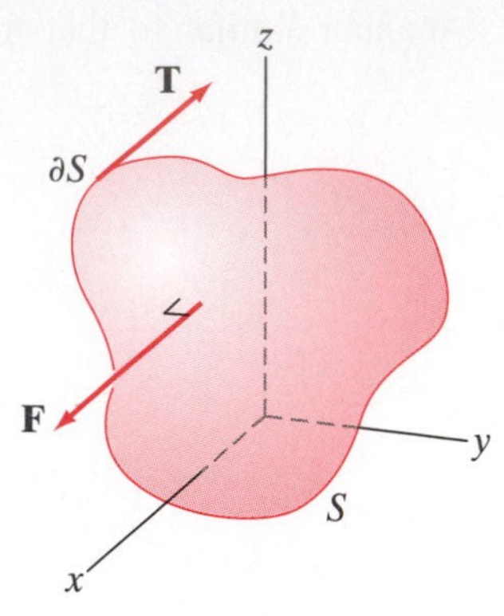

Figure 7.41 The surface S is a portion of the plane $x = c$. If $\mathbf{F} = M(x, y, z)\,\mathbf{i}$, then $\mathbf{F}$ is always perpendicular to S, in particular, to any vector $\mathbf{T}$ tangent to ∂S.

On the other hand, since $\mathbf{F}$ has only an $\mathbf{i}$-component, $\mathbf{F}$ is always parallel to $\mathbf{n}$ and thus perpendicular to any tangent vector to S, including vectors tangent to any boundary curves of S. Therefore, $\oint_{\partial S} \mathbf{F} \cdot d\mathbf{s} = 0$ also. (See Figure 7.41.) Now, suppose that $S = S_1 \cup S_2$, where S_1 and S_2 are each one of the graph of $z = f(x, y)$, the graph of $y = f(x, z)$, or a portion of the plane $x = c$. Assume that S_1 and S_2 coincide along part of their boundaries, as shown in Figure 7.42. The surfaces S_1 and S_2 inherit compatible orientations from S. If C denotes the common part of ∂S_1 and ∂S_2, then we may write ∂S_1 as $C_1 \cup C$ and ∂S_2 as $C_2 \cup C$, where C_1 and C_2 are disjoint (except at their endpoints). If ∂S_i is oriented consistently with S_i for $i = 1, 2$, then note that C will be oriented one way as part of ∂S_1 and the *opposite* way as part of ∂S_2. From this point, let us agree that C denotes the curve oriented so as to agree with the orientation of ∂S_1.

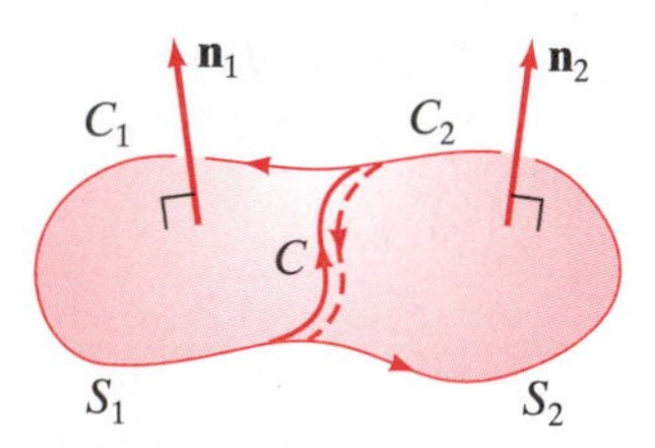

Figure 7.42 The surfaces S_1 and S_2 share the curve C as part of their boundaries. C receives one orientation as part of ∂S_1, and the opposite orientation as part of ∂S_2.

Now Stokes's theorem with $\mathbf{F} = M\mathbf{i}$ holds on both S_1 and S_2; on S_1 we have

$$\iint_{S_1} \nabla \times \mathbf{F} \cdot d\mathbf{S} = \oint_{\partial S_1} \mathbf{F} \cdot d\mathbf{s} = \int_{C_1} \mathbf{F} \cdot d\mathbf{s} + \int_C \mathbf{F} \cdot d\mathbf{s}, \tag{7}$$

whereas on S_2 we have

$$\iint_{S_2} \nabla \times \mathbf{F} \cdot d\mathbf{S} = \oint_{\partial S_2} \mathbf{F} \cdot d\mathbf{s} = \int_{C_2} \mathbf{F} \cdot d\mathbf{s} - \int_C \mathbf{F} \cdot d\mathbf{s}, \tag{8}$$

in view of the remarks made regarding the orientation of C. Now, we consider $S = S_1 \cup S_2$, noting that C is *not* part of ∂S. We see that

$$\iint_S \nabla \times \mathbf{F} \cdot d\mathbf{S} = \iint_{S_1} \nabla \times \mathbf{F} \cdot d\mathbf{S} + \iint_{S_2} \nabla \times \mathbf{F} \cdot d\mathbf{S}.$$

Using equations (7) and (8) and canceling, we find that

$$\int_{C_1} \mathbf{F} \cdot d\mathbf{s} + \int_{C_2} \mathbf{F} \cdot d\mathbf{s} = \oint_{\partial S} \mathbf{F} \cdot d\mathbf{s}.$$

Thus, Stokes's theorem holds in this case, or, indeed, in the case where S can be written as a finite union $S_1 \cup S_2 \cup \cdots \cup S_n$ of the special surfaces just described.

From a practical point of view, any particular surfaces you are likely to encounter will be decomposable as finite unions of the special types of surfaces previously described. However, not all piecewise smooth surfaces are, in fact, of this form. So to finish a truly general proof of Stokes's theorem when $\mathbf{F} = M\mathbf{i}$, some further limit arguments are needed, which we omit.

Step 3. Finally, by permuting variables, we essentially repeat Steps 1 and 2 in the cases where $\mathbf{F} = N(x, y, z)\,\mathbf{j}$ or $\mathbf{F} = P(x, y, z)\,\mathbf{k}$. In general, by the additivity of the curl, we have that

$$\begin{aligned}\iint_S \nabla \times \mathbf{F} \cdot d\mathbf{S} &= \iint_S \nabla \times (M\,\mathbf{i} + N\,\mathbf{j} + P\,\mathbf{k}) \cdot d\mathbf{S} \\ &= \iint_S \nabla \times (M\,\mathbf{i}) \cdot d\mathbf{S} + \iint_S \nabla \times (N\,\mathbf{j}) \cdot d\mathbf{S} \\ &\quad + \iint_S \nabla \times (P\mathbf{k}) \cdot d\mathbf{S}.\end{aligned}$$

Using the versions of Stokes's theorem just established, we see that

$$\iint_S \nabla \times \mathbf{F} \cdot d\mathbf{S} = \oint_{\partial S} M\,\mathbf{i} \cdot d\mathbf{s} + \oint_{\partial S} N\,\mathbf{j} \cdot d\mathbf{s} + \oint_{\partial S} P\,\mathbf{k} \cdot d\mathbf{s},$$

$$= \oint_{\partial S} (M\,\mathbf{i} + N\,\mathbf{j} + P\,\mathbf{k}) \cdot d\mathbf{s} = \oint_{\partial S} \mathbf{F} \cdot d\mathbf{s},$$

as desired. ■

Proof of Gauss's theorem

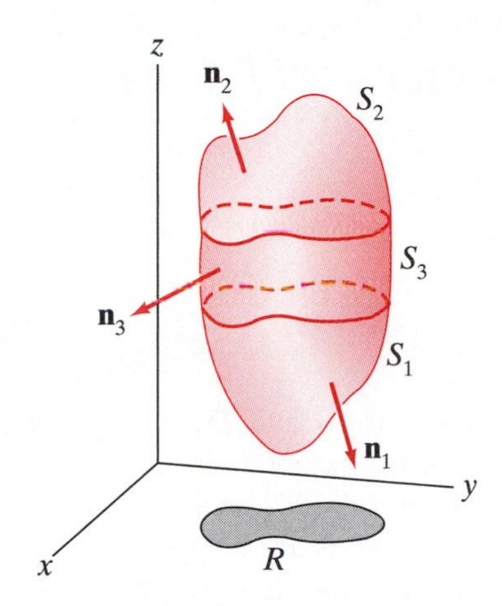

Figure 7.43 The type 1 region D and its shadow region R in the xy-plane.

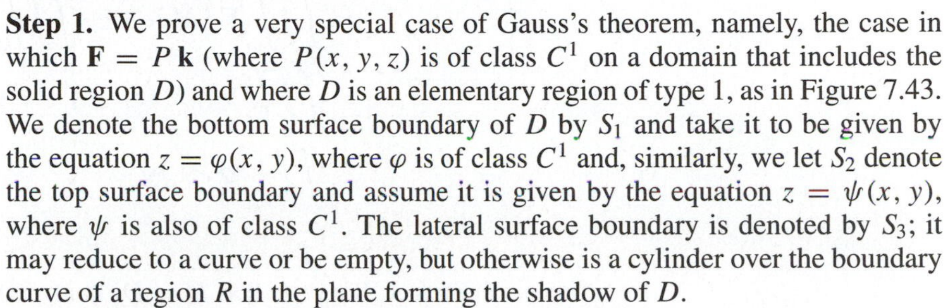

Step 1. We prove a very special case of Gauss's theorem, namely, the case in which $\mathbf{F} = P\,\mathbf{k}$ (where $P(x, y, z)$ is of class C^1 on a domain that includes the solid region D) and where D is an elementary region of type 1, as in Figure 7.43. We denote the bottom surface boundary of D by S_1 and take it to be given by the equation $z = \varphi(x, y)$, where φ is of class C^1 and, similarly, we let S_2 denote the top surface boundary and assume it is given by the equation $z = \psi(x, y)$, where ψ is also of class C^1. The lateral surface boundary is denoted by S_3; it may reduce to a curve or be empty, but otherwise is a cylinder over the boundary curve of a region R in the plane forming the shadow of D.

Orienting $\partial D = S_1 \cup S_2 \cup S_3$ with outward-pointing normal vectors, we have

$$\oiint_{\partial D} \mathbf{F} \cdot d\mathbf{S} = \iint_{S_1} \mathbf{F} \cdot d\mathbf{S} + \iint_{S_2} \mathbf{F} \cdot d\mathbf{S} + \iint_{S_3} \mathbf{F} \cdot d\mathbf{S}.$$

The orientation normal to S_1 should be downward-pointing, hence parallel to $\varphi_x\mathbf{i} + \varphi_y\mathbf{j} - \mathbf{k}$, the opposite of the normal vector obtained from the standard parametrization of S_1. Therefore, using formula (5) of §7.2, we have

$$\iint_{S_1} \mathbf{F} \cdot d\mathbf{S} = \iint_R P(x, y, \varphi(x, y))\,\mathbf{k} \cdot (\varphi_x\,\mathbf{i} + \varphi_y\,\mathbf{j} - \mathbf{k})\,dx\,dy$$

$$= -\iint_R P(x, y, \varphi(x, y))\,dx\,dy.$$

Similarly, the orientation normal to S_2 should be upward-pointing, and so

$$\iint_{S_2} \mathbf{F} \cdot d\mathbf{S} = \iint_R P(x, y, \psi(x, y)) \cdot (-\psi_x\,\mathbf{i} - \psi_y\,\mathbf{j} + \mathbf{k})\,dx\,dy$$

$$= \iint_R P(x, y, \psi(x, y))\,dx\,dy.$$

Now the lateral surface S_3, if it is nonempty, is a cylinder over a curve in the xy-plane. Hence, S_3 is defined by one or more equations of the form $g(x, y) = c$. It follows that any normal vector to S_3 can have *no* $\mathbf{k}$-component. Thus,

$$\iint_{S_3} \mathbf{F} \cdot d\mathbf{S} = \iint_{S_3} (P\,\mathbf{k} \cdot \mathbf{n}_3)\,dS = \iint_{S_3} 0\,dS = 0.$$

Putting these results together, we find that

$$\oiint_{\partial D} \mathbf{F} \cdot d\mathbf{S} = \iint_R [P(x, y, \psi(x, y)) - P(x, y, \varphi(x, y))]\,dx\,dy.$$

On the other hand, if $\mathbf{F} = P\mathbf{k}$, then $\nabla \cdot \mathbf{F} = \partial P/\partial z$, so

$$\iiint_D \nabla \cdot \mathbf{F}\,dV = \iint_R \int_{\varphi(x,y)}^{\psi(x,y)} \frac{\partial P}{\partial z}\,dz\,dx\,dy$$

$$= \iint_R [P(x, y, \psi(x, y)) - P(x, y, \varphi(x, y))]\,dx\,dy.$$

Therefore, Gauss's theorem holds in this special case.

Step 2. We may repeat Step 1 for $\mathbf{F} = M\mathbf{i}$ and D an elementary region of type 2, and for $\mathbf{F} = N\mathbf{j}$ and D of type 3. If D is an elementary region of type 4 (meaning D is simultaneously of types 1, 2, and 3), then

$$\begin{aligned}
\oiint_{\partial D} \mathbf{F}\cdot d\mathbf{S} &= \oiint_{\partial D} (M\,\mathbf{i} + N\,\mathbf{j} + P\,\mathbf{k})\cdot d\mathbf{S} \\
&= \oiint_{\partial D} M\,\mathbf{i}\cdot d\mathbf{S} + \oiint_{\partial D} N\,\mathbf{j}\cdot d\mathbf{S} + \oiint_{\partial D} P\,\mathbf{k}\cdot d\mathbf{S} \\
&= \iiint_D \nabla\cdot M\,\mathbf{i}\,dV + \iiint_D \nabla\cdot N\,\mathbf{j}\,dV + \iiint_D \nabla\cdot P\,\mathbf{k}\,dV \\
&= \iiint_D \nabla\cdot(M\,\mathbf{i} + N\,\mathbf{j} + P\,\mathbf{k})\,dV \\
&= \iiint_D \nabla\cdot\mathbf{F}\,dV.
\end{aligned}$$

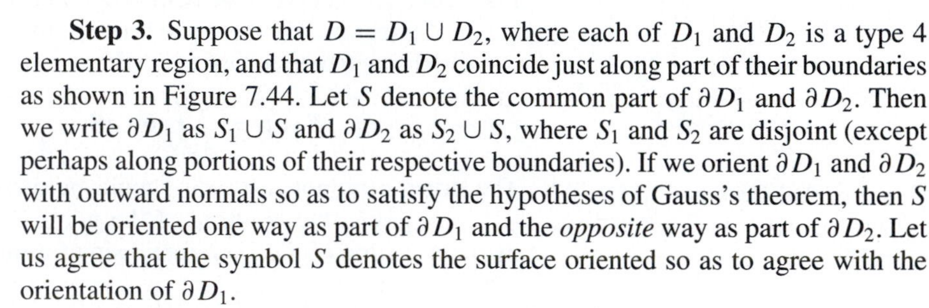

Step 3. Suppose that $D = D_1 \cup D_2$, where each of D_1 and D_2 is a type 4 elementary region, and that D_1 and D_2 coincide just along part of their boundaries as shown in Figure 7.44. Let S denote the common part of ∂D_1 and ∂D_2. Then we write ∂D_1 as $S_1 \cup S$ and ∂D_2 as $S_2 \cup S$, where S_1 and S_2 are disjoint (except perhaps along portions of their respective boundaries). If we orient ∂D_1 and ∂D_2 with outward normals so as to satisfy the hypotheses of Gauss's theorem, then S will be oriented one way as part of ∂D_1 and the *opposite* way as part of ∂D_2. Let us agree that the symbol S denotes the surface oriented so as to agree with the orientation of ∂D_1.

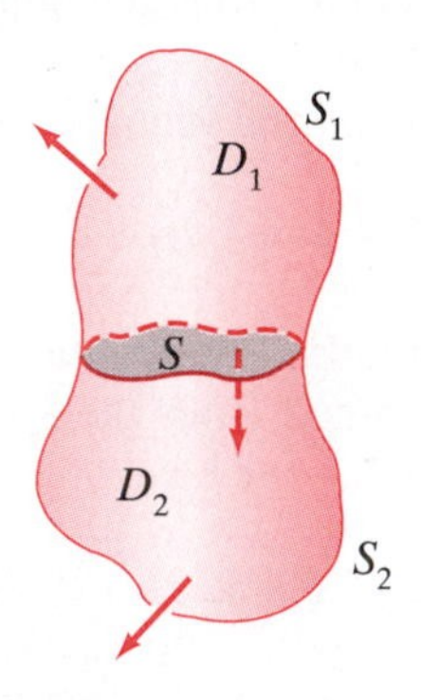

Figure 7.44 The regions D_1 and D_2 share the surface S as part of their boundaries. This common surface S inherits one orientation as part of ∂D_1 and the opposite orientation as part of ∂D_2.

Applying Gauss's theorem to both D_1 and D_2, we obtain

$$\begin{aligned}
\iiint_{D_1} \nabla\cdot\mathbf{F}\,dV &= \oiint_{\partial D_1} \mathbf{F}\cdot d\mathbf{S} = \iint_{S_1} \mathbf{F}\cdot d\mathbf{S} + \iint_S \mathbf{F}\cdot d\mathbf{S} \\
\iiint_{D_2} \nabla\cdot\mathbf{F}\,dV &= \oiint_{\partial D_2} \mathbf{F}\cdot d\mathbf{S} = \iint_{S_2} \mathbf{F}\cdot d\mathbf{S} - \iint_S \mathbf{F}\cdot d\mathbf{S}.
\end{aligned}$$

Combining these last two equations, we find that

$$\begin{aligned}
\iiint_D \nabla\cdot\mathbf{F}\,dV &= \iiint_{D_1} \nabla\cdot\mathbf{F}\,dV + \iiint_{D_2} \nabla\cdot\mathbf{F}\,dV \\
&= \iint_{S_1} \mathbf{F}\cdot d\mathbf{S} + \iint_{S_2} \mathbf{F}\cdot d\mathbf{S} \\
&= \oiint_{\partial D} \mathbf{F}\cdot d\mathbf{S},
\end{aligned}$$

since $\partial D = S_1 \cup S_2$.

Step 4. The result of Step 3 may be extended to regions D that can be decomposed as a union of finitely many type 4 regions. However, not all regions to which Gauss's theorem applies meet this criterion. Consequently, to finish a truly general proof, we once again need an argument using suitable limits of regions and their integrals, which we omit. ■

Exercises

In Exercises 1–4, verify Stokes's theorem for the given surface and vector field.

1. S is defined by $x^2 + y^2 + 5z = 1$, $z \geq 0$, oriented by upward normal;

$$\mathbf{F} = xz\,\mathbf{i} + yz\,\mathbf{j} + (x^2 + y^2)\,\mathbf{k}.$$

2. S is parametrized by $\mathbf{X}(s, t) = (s\cos t, s\sin t, t)$, $0 \leq s \leq 1, 0 \leq t \leq \pi/2$;

$$\mathbf{F} = z\,\mathbf{i} + x\,\mathbf{j} + y\,\mathbf{k}.$$

3. S is defined by $x = \sqrt{16 - y^2 - z^2}$;

$$\mathbf{F} = x\,\mathbf{i} + y\,\mathbf{j} + z\,\mathbf{k}.$$

4. S is defined by $x^2 + y^2 + z^2 = 4$, $z \leq 0$, oriented by downward normal;

$$\mathbf{F} = (2y - z)\,\mathbf{i} + (x + y^2 - z)\,\mathbf{j} + (4y - 3x)\,\mathbf{k}.$$

5. Let S be the "silo surface," that is, S is the union of two smooth surfaces S_1 and S_2, where S_1 is defined by

$$x^2 + y^2 = 9, \quad 0 \leq z \leq 8$$

and S_2 is defined by

$$x^2 + y^2 + (z - 8)^2 = 9, \quad z \geq 8.$$

Find $\iint_S \nabla \times \mathbf{F} \cdot d\mathbf{S}$, where

$$\mathbf{F} = (x^3 + xz + yz^2)\,\mathbf{i} + (xyz^3 + y^7)\,\mathbf{j} + x^2z^5\,\mathbf{k}.$$

In Exercises 6–9, verify Gauss's theorem for the given three-dimensional region D and vector field **F**.

6. $\mathbf{F} = x\,\mathbf{i} + y\,\mathbf{j} + z\,\mathbf{k}$, $D = \{(x, y, z) \mid 0 \leq z \leq 9 - x^2 - y^2\}$

7. $\mathbf{F} = (y - x)\,\mathbf{i} + (y - z)\,\mathbf{j} + (x - y)\,\mathbf{k}$, D is the unit cube $[0, 1] \times [0, 1] \times [0, 1]$

8. $\mathbf{F} = x^2\,\mathbf{i} + y\,\mathbf{j} + z\,\mathbf{k}$, $D = \{(x, y, z) \mid x^2 + y^2 + 1 \leq z \leq 5\}$

9. $\mathbf{F} = \dfrac{x\,\mathbf{i} + y\,\mathbf{j} + z\,\mathbf{k}}{\sqrt{x^2 + y^2 + z^2}}$, $D = \{(x, y, z) \mid a^2 \leq x^2 + y^2 + z^2 \leq b^2\}$

10. Verify that Stokes's theorem implies Green's theorem. (Hint: In Stokes's theorem take $\mathbf{F}(x, y, z) = M(x, y)\,\mathbf{i} + N(x, y)\,\mathbf{j}$; that is, assume $\mathbf{F}$ is independent of z and that its $\mathbf{k}$-component is identically zero.)

11. Let S be the surface defined by $y = 10 - x^2 - z^2$ with $y \geq 1$, oriented with rightward-pointing normal. Let

$$\mathbf{F} = (2xyz + 5z)\,\mathbf{i} + e^x \cos yz\,\mathbf{j} + x^2y\,\mathbf{k}.$$

Determine

$$\iint_S \nabla \times \mathbf{F} \cdot d\mathbf{S}.$$

(Hint: You will need an indirect approach.)

12. Let S be the surface defined as $z = 4 - 4x^2 - y^2$ with $z \geq 0$ and oriented by a normal with nonnegative $\mathbf{k}$-component. Let $\mathbf{F}(x, y, z) = x^3\mathbf{i} + e^{y^2}\mathbf{j} + ze^{xy}\mathbf{k}$. Find $\iint_S \nabla \times \mathbf{F} \cdot d\mathbf{S}$. (Hint: Argue that you can integrate over a different surface.)

13. (a) Show that the path $\mathbf{x}(t) = (\cos t, \sin t, \sin 2t)$ lies on the surface $z = 2xy$.

    (b) Evaluate

$$\oint_C (y^3 + \cos x)\,dx + (\sin y + z^2)\,dy + x\,dz,$$

where C is the closed curve parametrized and oriented by the path $\mathbf{x}$ in part (a).

14. Let S be defined by $z = e^{1-x^2-y^2}$, $z \geq 1$, oriented by upward normal, and let $\mathbf{F} = x\,\mathbf{i} + y\,\mathbf{j} + (2 - 2z)\,\mathbf{k}$. Use Gauss's theorem to calculate

$$\iint_S \mathbf{F} \cdot d\mathbf{S}.$$

15. Give a proof of Stokes's theorem for smooth, parametrized surfaces $S = \mathbf{X}(D)$, where $\mathbf{X}: D \subseteq \mathbf{R}^2 \to \mathbf{R}^3$. To make the proof easier, assume that $\mathbf{X}$ is of class C^2 and that it is one-one on D (in which case $\partial S = \mathbf{X}(\partial D)$).

16. Use Gauss's theorem to evaluate

$$\iint_S \mathbf{F} \cdot d\mathbf{S},$$

where $\mathbf{F} = ze^{x^2}\,\mathbf{i} + 3y\,\mathbf{j} + (2 - yz^7)\,\mathbf{k}$ and S is the union of the five "upper" faces of the unit cube $[0, 1] \times [0, 1] \times [0, 1]$. That is, the $z = 0$ face is *not* part of S. (Hint: Note that S is *not* closed, so to apply Gauss's theorem you will have to close it up.)

17. Write a careful proof of the three-dimensional case of Theorem 3.5 of Chapter 6: If $\mathbf{F}$ is a vector field of class C^1 whose domain is a simply-connected region R in $\mathbf{R}^3$, then $\mathbf{F} = \nabla f$ for some (scalar-valued) function f on R if and only if $\nabla \times \mathbf{F} = \mathbf{0}$ at all points of R.

18. Let S_r denote the sphere of radius r with center at the origin, oriented with outward normal. Suppose $\mathbf{F}$ is of class C^1 on all of $\mathbf{R}^3$ and is such that

$$\oiint_{S_r} \mathbf{F} \cdot d\mathbf{S} = ar + b$$

for some fixed constants a and b.

    (a) Compute

$$\iiint_D \nabla \cdot \mathbf{F}\,dV,$$

where $D = \{(x, y, z) \mid 25 \leq x^2 + y^2 + z^2 \leq 49\}$. (Your answer should be in terms of a and b.)

(b) Suppose, in the situation just described, that $\mathbf{F} = \nabla \times \mathbf{G}$ for some vector field $\mathbf{G}$ of class C^1. What conditions does this place on the constants a and b?

19. Let $\mathbf{n}(x, y, z)$ be a unit normal to a surface S. The directional derivative of a differentiable function $f(x, y, z)$ in the direction of $\mathbf{n}$ is called a **normal derivative** of f, denoted $\partial f/\partial n$. From Theorem 6.2 of Chapter 2, we have

$$\frac{\partial f}{\partial n} = \nabla f \cdot \mathbf{n}.$$

(a) Let S denote the portion of the sphere $x^2 + y^2 + z^2 = a^2$ in the first octant (i.e., where $x \geq 0$, $y \geq 0$, $z \geq 0$), oriented by the unit normal that points away from the origin. Let $f(x, y, z) = \ln(x^2 + y^2 + z^2)$. Evaluate

$$\iint_S \frac{\partial f}{\partial n}\, dS.$$

(b) Let D denote the piece of the solid ball $x^2 + y^2 + z^2 \leq a^2$ in the first octant; that is,

$$D = \{(x, y, z) \mid x^2 + y^2 + z^2 \leq a^2,\ x \geq 0,\ y \geq 0,\ z \geq 0\}.$$

Compute $\iiint_D \nabla \cdot (\nabla f)\, dV$, where f is as in part (a).

(c) Apply Gauss's theorem to the integral in part (b), and reconcile your result with your answer in part (a).

20. Suppose that f is such that for any closed, oriented surface S,

$$\oiint_S \frac{\partial f}{\partial n}\, dS = 0.$$

(See Exercise 19 for the definition of the normal derivative $\partial f/\partial n$.) Show that then

$$\frac{\partial^2 f}{\partial x^2} + \frac{\partial^2 f}{\partial y^2} + \frac{\partial^2 f}{\partial z^2} = 0$$

(i.e., that f is harmonic).

21. Use the integral formula in Proposition 3.4 to establish the formula for the divergence of a C^1 vector field

$$\mathbf{F} = F_1(x, y, z)\,\mathbf{i} + F_2(x, y, z)\,\mathbf{j} + F_3(x, y, z)\,\mathbf{k}.$$

That is, show that

$$\operatorname{div} \mathbf{F} = \frac{\partial F_1}{\partial x} + \frac{\partial F_2}{\partial y} + \frac{\partial F_3}{\partial z}.$$

22. Following Proposition 3.4, show that

$$\operatorname{div} \mathbf{F}(P) = \lim_{V \to 0} \frac{1}{V} \oiint_S \mathbf{F} \cdot d\mathbf{S},$$

where S is a piecewise smooth, orientable, closed surface S enclosing a region D of volume V. (Take S to be oriented by outward normal.) The limiting process should be assumed to be such that D shrinks down to the point P.

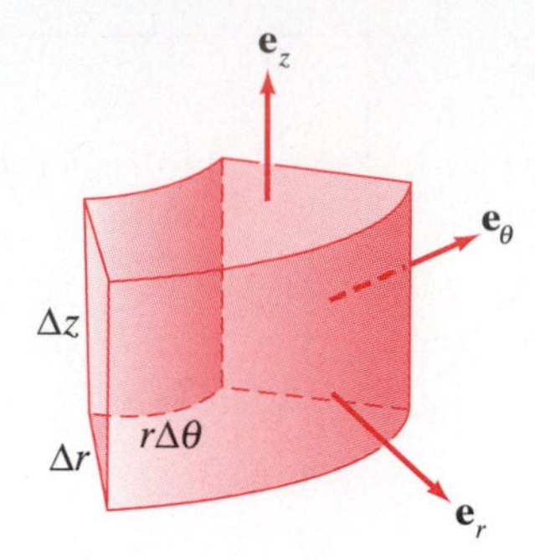

Figure 7.45 The cylindrical coordinate cuboid of Exercise 23.

23. In this problem, you will use the result of Exercise 22 to find an expression for $\nabla \cdot \mathbf{F}$ in cylindrical coordinates. (See Theorem 4.5 of Chapter 3.) Begin by writing

$$\mathbf{F} = F_r\,\mathbf{e}_r + F_\theta\,\mathbf{e}_\theta + F_z\,\mathbf{e}_z,$$

where $F_r(r, \theta, z)$, $F_\theta(r, \theta, z)$, and $F_z(r, \theta, z)$ denote the components of $\mathbf{F}$ in the $\mathbf{e}_r$-, $\mathbf{e}_\theta$-, and $\mathbf{e}_z$-directions (respectively). Let P have cylindrical coordinates (r, θ, z). Consider the small "cylindrical coordinate cuboid" S shown in Figure 7.45. The pairs of opposite faces correspond to values

$$\begin{array}{ccc} r - \Delta r/2 & \text{and} & r + \Delta r/2; \\ \theta - \Delta\theta/2 & \text{and} & \theta + \Delta\theta/2; \\ z - \Delta z/2 & \text{and} & z + \Delta z/2. \end{array}$$

Note that the volume of the cuboid is approximately $r\,\Delta\theta\,\Delta r\,\Delta z$.

(a) Approximate $\oiint_S \mathbf{F} \cdot d\mathbf{S}$ (where S is oriented by outward unit normal) by noting that each face of S is roughly flat with an "obvious" unit normal vector and that $\mathbf{F}$ is approximately constant on each face.

(b) Use your answer in part (a) to calculate the divergence in cylindrical coordinates as

$$\operatorname{div} \mathbf{F} = \frac{\partial F_z}{\partial z} + \frac{1}{r}\frac{\partial}{\partial r}(r F_r) + \frac{1}{r}\frac{\partial F_\theta}{\partial \theta}.$$

(This agrees with formula (4) of §3.4.) Note that if $f(x, y, z)$ is differentiable, then

$$\begin{aligned} \frac{\partial f}{\partial x} &= \lim_{\Delta x \to 0} \frac{f(x + \Delta x, y, z) - f(x, y, z)}{\Delta x} \\ &= \lim_{\Delta x \to 0} \frac{f\left(x + \frac{\Delta x}{2}, y, z\right) - f\left(x - \frac{\Delta x}{2}, y, z\right)}{\Delta x}. \end{aligned}$$

24. Use the ideas of Exercises 22 and 23 to calculate the divergence in spherical coordinates. (See Theorem 4.6 of Chapter 3.) You will want to make use of the small "spherical coordinate cuboid" S shown in Figure 7.46.

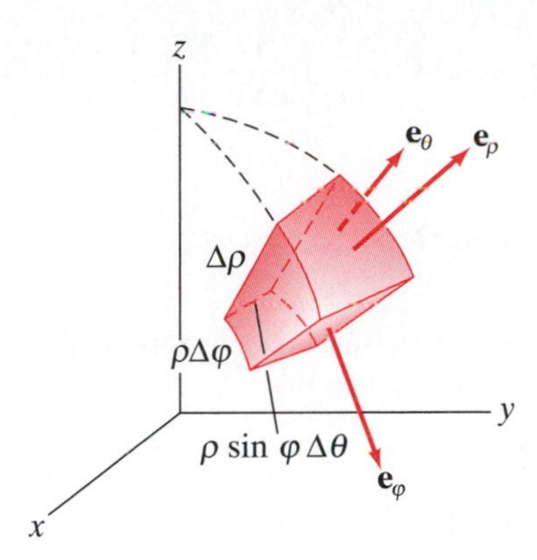

Figure 7.46 The spherical coordinate cuboid of Exercise 24. The volume of the cuboid is approximately $\rho^2 \sin\varphi\, \Delta\theta\, \Delta\varphi\, \Delta\rho$.

The pairs of opposite faces of S correspond to values

$$\begin{aligned} \rho - \Delta\rho/2 \quad &\text{and} \quad \rho + \Delta\rho/2; \\ \varphi - \Delta\varphi/2 \quad &\text{and} \quad \varphi + \Delta\varphi/2; \\ \theta - \Delta\theta/2 \quad &\text{and} \quad \theta + \Delta\theta/2. \end{aligned}$$

25. Let $\mathbf{F}$ be a vector field of class C^1 in a neighborhood of the point P in $\mathbf{R}^3$, and let $\mathbf{n}$ be a unit vector drawn with its tail at P. Let C be a simple, closed curve such that there is an orientable surface S bounded by C that contains P and such that $\mathbf{n}$ is normal to S at P. Orient S by using $\mathbf{n}$, and orient C consistently with S. Following Proposition 3.5, show that, if A denotes the area of S, then

$$\mathbf{n} \cdot \operatorname{curl} \mathbf{F}(P) = \lim_{A \to 0} \frac{1}{A} \oint_C \mathbf{F} \cdot d\mathbf{s}.$$

Here the limiting process is assumed to be such that C shrinks down to the point P. (See Figure 7.47.)

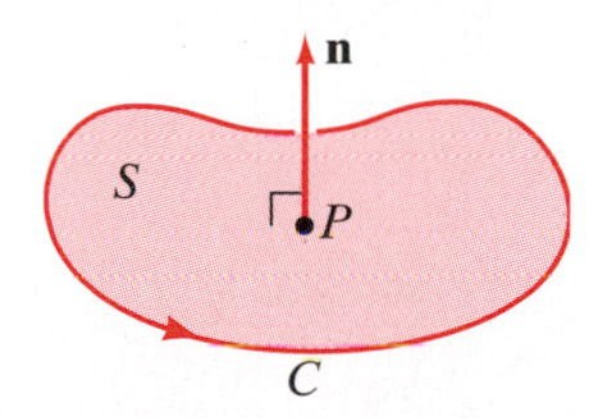

Figure 7.47 Figure for Exercise 25.

26. In this problem, you will use the result of Exercise 25 to determine an expression for curl $\mathbf{F}$ in cylindrical coordinates. Begin by writing

$$\mathbf{F} = F_r\, \mathbf{e}_r + F_\theta\, \mathbf{e}_\theta + F_z\, \mathbf{e}_z.$$

(a) Find the $\mathbf{e}_z$-component of curl $\mathbf{F}$ by considering the planar path shown in Figure 7.48. The pairs of opposite "edges" of the approximately rectangular

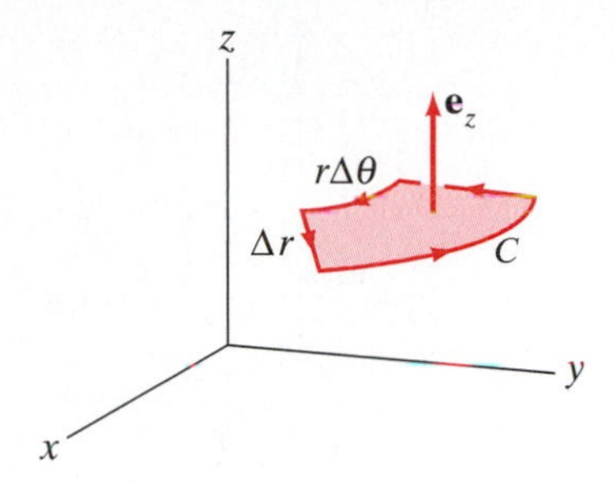

Figure 7.48 The path C of Exercise 26(a).

path C correspond to the values

$$r - \Delta r/2 \quad \text{and} \quad r + \Delta r/2,$$

and

$$\theta - \Delta\theta/2 \quad \text{and} \quad \theta + \Delta\theta/2$$

(all with constant z-coordinate). Note that the area enclosed by C is approximately $r\,\Delta\theta\,\Delta r$. Approximate the line integral $\oint_C \mathbf{F} \cdot d\mathbf{s}$ by using the fact that, for small $\Delta\theta$ and Δr, each edge of C is roughly straight. Show that

$$\mathbf{e}_z \cdot \operatorname{curl} \mathbf{F} = -\frac{1}{r}\frac{\partial F_r}{\partial \theta} + \frac{1}{r}\frac{\partial}{\partial r}(r F_\theta).$$

(b) Use the path in Figure 7.49 to show that

$$\mathbf{e}_r \cdot \operatorname{curl} \mathbf{F} = \frac{1}{r}\frac{\partial F_z}{\partial \theta} - \frac{\partial F_\theta}{\partial z}.$$

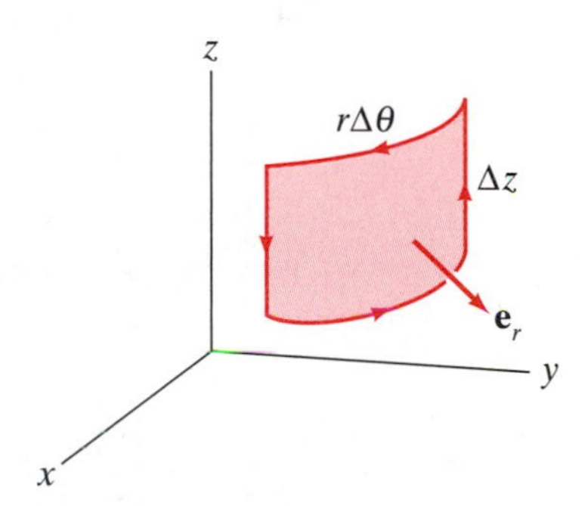

Figure 7.49 The path C of Exercise 26(b).

(c) Use the path in Figure 7.50 to show that

$$\mathbf{e}_\theta \cdot \operatorname{curl} \mathbf{F} = \frac{\partial F_r}{\partial z} - \frac{\partial F_z}{\partial r}.$$

Combine this with the results of parts (a) and (b) to obtain

$$\operatorname{curl} \mathbf{F} = \frac{1}{r}\begin{vmatrix} \mathbf{e}_r & r\mathbf{e}_\theta & \mathbf{e}_z \\ \partial/\partial r & \partial/\partial\theta & \partial/\partial z \\ F_r & r F_\theta & F_z \end{vmatrix}.$$

(See Theorem 4.5 of Chapter 3.)

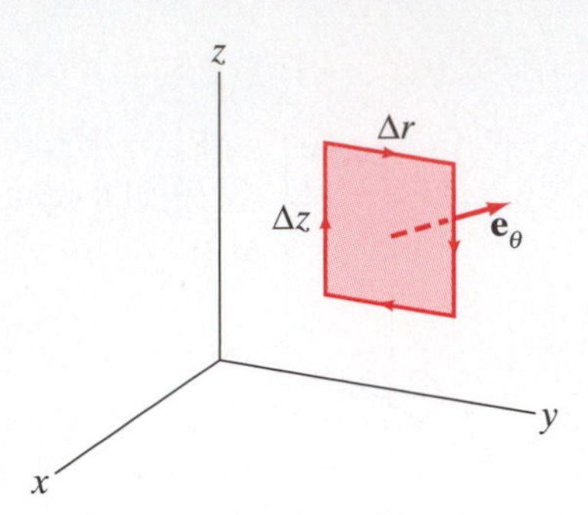

Figure 7.50 The path C of Exercise 26(c).

27. In this problem, you will determine an expression for $\operatorname{curl}\mathbf{F}$ in spherical coordinates. Let $\mathbf{F}$ be a vector field of class C^1, and write

$$\mathbf{F} = F_\rho\,\mathbf{e}_\rho + F_\varphi\,\mathbf{e}_\varphi + F_\theta\,\mathbf{e}_\theta.$$

Show that

$$\mathbf{e}_\rho \cdot \operatorname{curl}\mathbf{F} = \frac{1}{\rho\sin\varphi}\left[\frac{\partial}{\partial\varphi}(\sin\varphi\,F_\theta) - \frac{\partial F_\varphi}{\partial\theta}\right],$$

$$\mathbf{e}_\varphi \cdot \operatorname{curl}\mathbf{F} = \frac{1}{\rho}\left[\frac{1}{\sin\varphi}\frac{\partial F_\rho}{\partial\theta} - \frac{\partial}{\partial\rho}(\rho F_\theta)\right],$$

and

$$\mathbf{e}_\theta \cdot \operatorname{curl}\mathbf{F} = \frac{1}{\rho}\left[\frac{\partial}{\partial\rho}(\rho F_\varphi) - \frac{\partial F_\rho}{\partial\varphi}\right]$$

by using Exercise 23 and the three paths shown in Figure 7.51. Conclude that

$$\operatorname{curl}\mathbf{F} = \frac{1}{\rho^2\sin\varphi}\begin{vmatrix} \mathbf{e}_\rho & \rho\mathbf{e}_\varphi & \rho\sin\varphi\,\mathbf{e}_\theta \\ \partial/\partial\rho & \partial/\partial\varphi & \partial/\partial\theta \\ F_\rho & \rho F_\varphi & \rho\sin\varphi\,F_\theta \end{vmatrix}.$$

(See Theorem 4.6 of Chapter 3.)

28. Of the six planar vector fields shown in Figure 7.52, four have zero divergence in the regions indicated and three have zero curl. By considering appropriate integrals and using the results of Exercises 22 and 25, categorize each vector field.

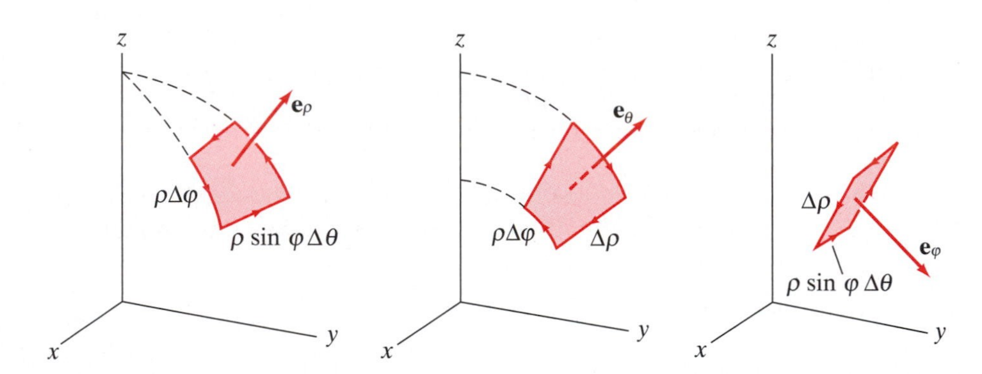

Figure 7.51 The paths of Exercise 27.

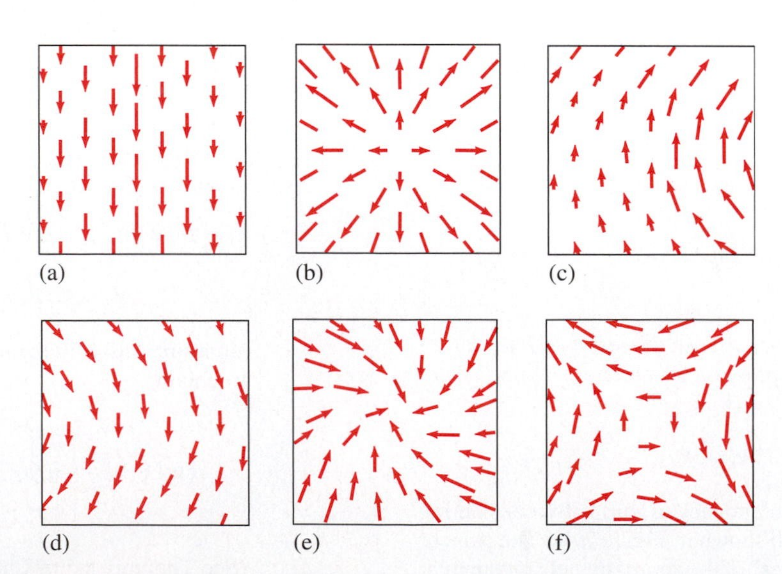

Figure 7.52 Figures for Exercise 28.

7.4 Further Vector Analysis; Maxwell's Equations

In this section, we use Gauss's theorem and Stokes's theorem first to prove some abstract results in vector analysis and then to study Maxwell's equations of electricity and magnetism.

Green's formulas

Our purpose in §7.4 is to establish a few fundamental results of vector analysis. Throughout the discussion all scalar and vector fields are defined on subsets of $\mathbf{R}^3$.

The following pair of results is established readily:

■ **Theorem 4.1** (GREEN'S FIRST AND SECOND FORMULAS) Let f and g be scalar fields of class C^2, and let D be a solid region in space, bounded by a piecewise smooth surface $S = \partial D$, oriented as in Gauss's theorem. Then we have

Green's first formula:

$$\iiint_D \nabla f \cdot \nabla g \, dV + \iiint_D f \, \nabla^2 g \, dV = \oiint_S f \, \nabla g \cdot d\mathbf{S}.$$

Green's second formula:

$$\iiint_D (f \, \nabla^2 g - g \, \nabla^2 f) \, dV = \oiint_S (f \, \nabla g - g \, \nabla f) \cdot d\mathbf{S}.$$

Proof The first formula follows from Gauss's theorem applied to the vector field $\mathbf{F} = f \nabla g$. (We leave the details to you.) The second formula follows from writing the first formula twice:

$$\iiint_D \nabla f \cdot \nabla g \, dV + \iiint_D f \nabla^2 g \, dV = \oiint_S f \nabla g \cdot d\mathbf{S}; \qquad (1)$$

$$\iiint_D \nabla g \cdot \nabla f \, dV + \iiint_D g \nabla^2 f \, dV = \oiint_S g \nabla f \cdot d\mathbf{S}. \qquad (2)$$

Now, subtract equation (2) from equation (1). ■

The third of Green's formulas requires considerably more work to prove.

■ **Theorem 4.2** (GREEN'S THIRD FORMULA) Assume f is a function of class C^2. Then, for $\partial D = S$ oriented as in Gauss's theorem and points $\mathbf{r}$ in the interior of D,

$$f(\mathbf{r}) = -\frac{1}{4\pi} \iiint_D \frac{\nabla^2 f(\mathbf{x})}{\|\mathbf{r} - \mathbf{x}\|} \, dV + \frac{1}{4\pi} \oiint_S \left(-f(\mathbf{x}) \nabla \left(\frac{1}{\|\mathbf{r} - \mathbf{x}\|} \right) + \frac{\nabla f(\mathbf{x})}{\|\mathbf{r} - \mathbf{x}\|} \right) \cdot d\mathbf{S}.$$

In this formula, dV denotes integration with respect to the variables in $\mathbf{x} = (x, y, z)$ (i.e., $\mathbf{r} = (r_1, r_2, r_3)$ is fixed in the integration), and the symbol ∇ means $\nabla_{\mathbf{x}}$, differentiation with respect to x, y, and z.

A proof of Theorem 4.2 appears in the addendum to this section.

An Inversion Formula for the Laplacian

Green's third formula is a type of **inversion formula**—a formula that enables us to recover the values of a function f by knowing certain integrals involving its gradient and Laplacian. Green's third formula is mainly of technical interest in proving further results. We use it here to obtain an inversion formula for the Laplacian operator.

We begin by applying the Laplacian $\nabla_{\mathbf{r}}^2$ to Green's third formula:

$$\nabla_{\mathbf{r}}^2 f(\mathbf{r}) = \nabla_{\mathbf{r}}^2 \left[-\frac{1}{4\pi} \iiint_D \frac{\nabla_{\mathbf{x}}^2 f(\mathbf{x})}{\|\mathbf{r}-\mathbf{x}\|}\, dV + \frac{1}{4\pi} \oiint_S \left(\frac{\nabla_{\mathbf{x}} f(\mathbf{x})}{\|\mathbf{r}-\mathbf{x}\|} - f(\mathbf{x}) \nabla_{\mathbf{x}} \left(\frac{1}{\|\mathbf{r}-\mathbf{x}\|} \right) \right) \cdot d\mathbf{S} \right] \tag{3}$$

The trick is to move $\nabla_{\mathbf{r}}^2$ inside the surface integral, which is justified since $\mathbf{x} \neq \mathbf{r}$ when $\mathbf{x}$ varies over S:

$$\begin{aligned} \nabla_{\mathbf{r}}^2 \oiint_S \left(\frac{\nabla_{\mathbf{x}} f(\mathbf{x})}{\|\mathbf{r}-\mathbf{x}\|} - f(\mathbf{x}) \nabla_{\mathbf{x}} \left(\frac{1}{\|\mathbf{r}-\mathbf{x}\|} \right) \right) \cdot d\mathbf{S} \\ = \oiint_S \nabla_{\mathbf{r}}^2 \left(\frac{\nabla_{\mathbf{x}} f(\mathbf{x})}{\|\mathbf{r}-\mathbf{x}\|} - f(\mathbf{x}) \nabla_{\mathbf{x}} \left(\frac{1}{\|\mathbf{r}-\mathbf{x}\|} \right) \right) \cdot d\mathbf{S}. \end{aligned}$$

By direct calculation, $\nabla_{\mathbf{r}}^2 (1/\|\mathbf{r}-\mathbf{x}\|) = 0$ for $\mathbf{x} \neq \mathbf{r}$, so

$$\nabla_{\mathbf{r}}^2 \left(\frac{\nabla_{\mathbf{x}} f(\mathbf{x})}{\|\mathbf{r}-\mathbf{x}\|} \right) = 0.$$

Similarly, since $f(\mathbf{x})$ does not involve $\mathbf{r}$,

$$\begin{aligned} \nabla_{\mathbf{r}}^2 \left(f(\mathbf{x}) \nabla_{\mathbf{x}} \left(\frac{1}{\|\mathbf{r}-\mathbf{x}\|} \right) \right) &= f(\mathbf{x}) \nabla_{\mathbf{r}}^2 \nabla_{\mathbf{x}} \left(\frac{1}{\|\mathbf{r}-\mathbf{x}\|} \right) \\ &= f(\mathbf{x}) \nabla_{\mathbf{x}} \nabla_{\mathbf{r}}^2 \left(\frac{1}{\|\mathbf{r}-\mathbf{x}\|} \right) \\ &= 0. \end{aligned}$$

Therefore, the Laplacian of the original surface integral is 0. We may conclude from equation (3) that, for $\mathbf{r}$ in the interior of D,

$$\nabla_{\mathbf{r}}^2 f(\mathbf{r}) = -\frac{1}{4\pi} \nabla_{\mathbf{r}}^2 \iiint_D \frac{\nabla_{\mathbf{x}}^2 f(\mathbf{x})}{\|\mathbf{r}-\mathbf{x}\|}\, dV,$$

and we have shown the following:

Theorem 4.3 If $\varphi = \nabla^2 f$ for some function f of class C^2, then for $\mathbf{r}$ in the interior of D,

$$\varphi(\mathbf{r}) = -\frac{1}{4\pi} \nabla_{\mathbf{r}}^2 \iiint_D \frac{\varphi(\mathbf{x})}{\|\mathbf{r}-\mathbf{x}\|}\, dV.$$

Theorem 4.3 provides an inversion formula for the Laplacian in the following sense: If $\nabla^2 f = \varphi$, then

$$f(\mathbf{r}) = -\frac{1}{4\pi} \iiint_D \frac{\varphi(\mathbf{x})}{\|\mathbf{r}-\mathbf{x}\|}\, dV + g(\mathbf{r}),$$

where g is any harmonic function (i.e., g is such that $\nabla^2 g = 0$). That is, if the Laplacian of f is known, we can recover the function f itself, up to addition of a harmonic function.

Finally, it can be shown that the result of Theorem 4.3 holds under considerably less stringent hypotheses than having φ be the Laplacian of another function.[1]

Maxwell's Equations

Maxwell's equations are fundamental results that govern the behavior of—and interactions between—electric and magnetic fields. We see how Maxwell's equations arise from a few simple physical principles coupled with the vector analysis discussed previously.

Gauss's law for electric fields If $\mathbf{E}$ is an electric field, then the flux of $\mathbf{E}$ across a closed surface S is

$$\text{Flux of } \mathbf{E} = \oiint_S \mathbf{E} \cdot d\mathbf{S}. \tag{4}$$

Applying Gauss's theorem to formula (4), we find that

$$\text{Flux of } \mathbf{E} = \iiint_D \nabla \cdot \mathbf{E}\, dV, \tag{5}$$

where D is the region enclosed by S.

If the electric field $\mathbf{E}$ is determined by a single point charge of q coulombs located at the origin, then $\mathbf{E}$ is given by

$$\mathbf{E}(\mathbf{x}) = \frac{q}{4\pi\epsilon_0} \frac{\mathbf{x}}{\|\mathbf{x}\|^3}, \tag{6}$$

where $\mathbf{x} = x\,\mathbf{i} + y\,\mathbf{j} + z\,\mathbf{k}$. In mks units, $\mathbf{E}$ is measured in volts/meter. The constant ϵ_0 is known as the **permittivity of free space;** its value (in mks units) is

$$8.854 \times 10^{-12} \text{ coulomb}^2/\text{newton-meter}^2.$$

For the electric field in equation (6), we can readily verify that $\nabla \cdot \mathbf{E} = 0$ wherever $\mathbf{E}$ is defined. From formulas (4) and (5), if S is any surface that does *not* enclose the origin, then the flux of $\mathbf{E}$ across S is zero.

But now a question arises: How do we calculate the flux of the electric field in equation (6) across surfaces that *do* enclose the origin? The trick is to find an appropriate way to exclude the origin from consideration. To that end, first suppose that S_b denotes a sphere of radius b centered at the origin (i.e., S_b has equation $x^2 + y^2 + z^2 = b^2$). Then the outward unit normal to S_b is

$$\mathbf{n} = \frac{x\,\mathbf{i} + y\,\mathbf{j} + z\,\mathbf{k}}{b} = \frac{1}{b}\mathbf{x}.$$

(See Figure 7.53.) From equation (6),

$$\oiint_{S_b} \mathbf{E} \cdot d\mathbf{S} = \frac{q}{4\pi\epsilon_0} \oiint_{S_b} \frac{\mathbf{x}}{\|\mathbf{x}\|^3} \cdot \frac{1}{b}\mathbf{x}\, dS.$$

On S_b, we have $\|\mathbf{x}\| = b$, so that

$$\begin{aligned}
\oiint_{S_b} \mathbf{E} \cdot d\mathbf{S} &= \frac{q}{4\pi\epsilon_0} \oiint_{S_b} \frac{\mathbf{x}}{b^3} \cdot \frac{\mathbf{x}}{b}\, dS = \frac{q}{4\pi\epsilon_0} \oiint_{S_b} \frac{\|\mathbf{x}\|^2}{b^4}\, dS \\
&= \frac{q}{4\pi\epsilon_0} \oiint_{S_b} \frac{b^2}{b^4}\, dS = \frac{q}{4\pi\epsilon_0 b^2} \oiint_{S_b} dS \\
&= \frac{q}{4\pi\epsilon_0 b^2} (\text{surface area of } S_b) \\
&= \frac{q}{4\pi\epsilon_0 b^2} (4\pi b^2) = \frac{q}{\epsilon_0}.
\end{aligned}$$

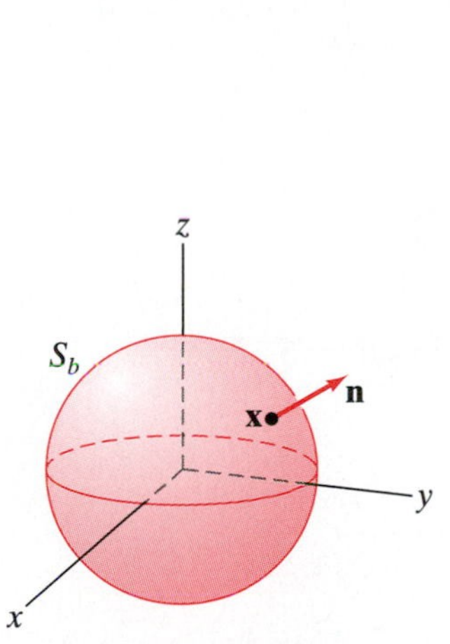

Figure 7.53 The sphere S_b of radius b, centered at the origin, together with an outward unit normal vector.

[1]See, for example, O. D. Kellogg, *Foundations of Potential Theory* (Springer, 1928; reprinted by Dover Publications, 1954), p. 220.

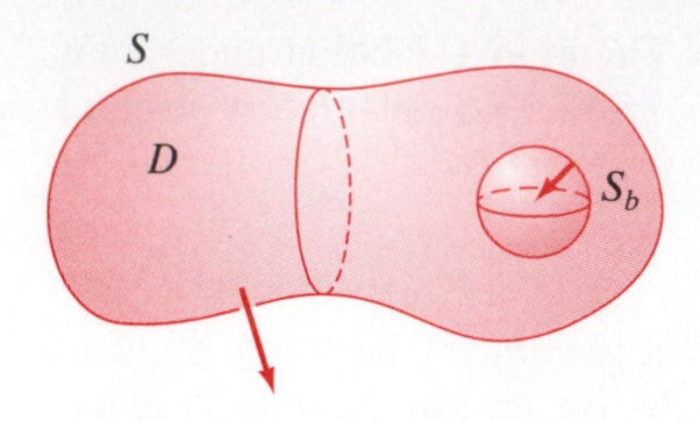

Figure 7.54 The solid region D is the region inside S and outside S_b.

Now, suppose S is *any* surface enclosing the origin. Let S_b be a small sphere centered at the origin and contained inside S. Let D be the solid region in $\mathbf{R}^3$ between S_b and S. (See Figure 7.54.) Note that $\nabla \cdot \mathbf{E} = 0$ throughout D, since D does not contain the origin. ($\mathbf{E}$ is still defined as in equation (6).) Orienting S and S_b with normals that point away from D, we obtain

$$\oiint_S \mathbf{E} \cdot d\mathbf{S} - \oiint_{S_b} \mathbf{E} \cdot d\mathbf{S} = \oiint_{\partial D} \mathbf{E} \cdot d\mathbf{S} = \iiint_D \nabla \cdot \mathbf{E}\, dV = 0,$$

using Gauss's theorem. We conclude that

$$\oiint_S \mathbf{E} \cdot d\mathbf{S} = \frac{q}{\epsilon_0} \tag{7}$$

for any surface that encloses the origin. By modifying equation (6) for $\mathbf{E}$ appropriately, we can show that formula (7) also holds for any closed surface containing a single point charge of q coulombs located anywhere. (Note: The arguments just given hold for *any* inverse square law vector field $\mathbf{F}(\mathbf{x}) = k\mathbf{x}/\|\mathbf{x}\|^3$, where k is a constant. See Exercise 13 in this section for details.)

We can adapt the arguments just given to accommodate the case of n discrete point charges. For $i = 1, \ldots, n$, suppose a point charge of q_i coulombs is located at position $\mathbf{r}_i$. The electric field $\mathbf{E}$ for this configuration is

$$\mathbf{E}(\mathbf{x}) = \frac{1}{4\pi\epsilon_0} \sum_{i=1}^{n} q_i \frac{\mathbf{x} - \mathbf{r}_i}{\|\mathbf{x} - \mathbf{r}_i\|^3}. \tag{8}$$

For $\mathbf{E}$ as given in equation (8), we can calculate that $\nabla \cdot \mathbf{E} = 0$, except at $\mathbf{x} = \mathbf{r}_i$. If S is any piecewise smooth, oriented surface containing the charges, then we may use Gauss's theorem to find the flux of $\mathbf{E}$ across S by taking n small spheres $S_1, S_2, \ldots, S_n$, each enclosing a single point charge. (See Figure 7.55.) If D is the region inside S, but outside all the spheres, we have, by choosing appropriate orientations and using Gauss's theorem,

$$\oiint_S \mathbf{E} \cdot d\mathbf{S} - \sum_{i=1}^{n} \oiint_{S_i} \mathbf{E} \cdot d\mathbf{S} = \oiint_{\partial D} \mathbf{E} \cdot d\mathbf{S} = \iiint_D \nabla \cdot \mathbf{E}\, dV = 0,$$

since $\nabla \cdot \mathbf{E} = 0$ on D. Hence,

$$\begin{aligned} \oiint_S \mathbf{E} \cdot d\mathbf{S} &= \sum_{i=1}^{n} \oiint_{S_i} \mathbf{E} \cdot d\mathbf{S} = \frac{1}{\epsilon_0} \sum_{i=1}^{n} q_i \\ &= \frac{1}{\epsilon_0} (\text{total charge enclosed by } S). \end{aligned} \tag{9}$$

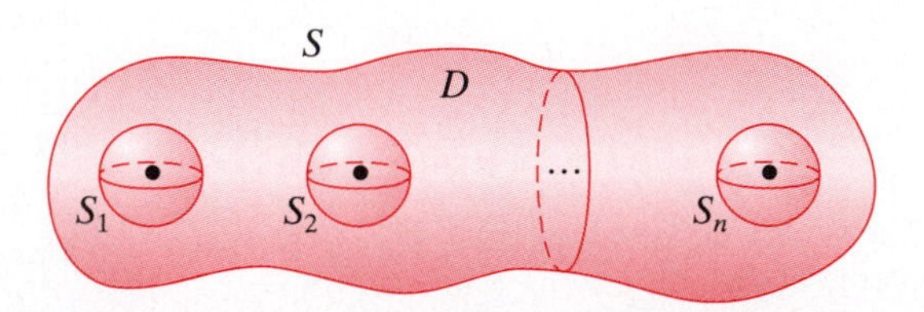

Figure 7.55 D is the solid region inside the surface S and outside the small spheres $S_1, S_2, \ldots, S_n$, each enclosing a point charge.

To establish Gauss's law, consider the case not of an electric field determined by discrete point charges, but rather of one determined by a **continuous charge distribution** given by a **charge density** $\rho(\mathbf{x})$. The total charge over a region D in space is

$$\iiint_D \rho(\mathbf{x})\, dV,$$

so that, in place of formula (8), we have an electric field,

$$\mathbf{E}(\mathbf{r}) = \frac{1}{4\pi\epsilon_0}\iiint_D \rho(\mathbf{x})\frac{\mathbf{r}-\mathbf{x}}{\|\mathbf{r}-\mathbf{x}\|^3}\, dV. \tag{10}$$

In equation (10) the integration occurs with respect to the variables in $\mathbf{x}$. (Note: It is not at all obvious that the integral used to define $\mathbf{E}(\mathbf{r})$ converges at points $\mathbf{r} \in D$, where $\rho(\mathbf{r}) \neq 0$, because at such points the triple integral in equation (10) is improper. See Exercise 20 in this section for an indication of how to deal with this issue.)

The integral form of Gauss's law, analogous to that of formula (9), is

$$\oiint_S \mathbf{E}\cdot d\mathbf{S} = \frac{1}{\epsilon_0}\iiint_D \rho\, dV, \tag{11}$$

where $S = \partial D$. If we apply Gauss's theorem to the left side of formula (11), we find that

$$\iiint_D \nabla\cdot\mathbf{E}\, dV = \frac{1}{\epsilon_0}\iiint_D \rho\, dV.$$

Since the region D is arbitrary, it may be "shrunk to a point." From this, we conclude that

$$\nabla\cdot\mathbf{E} = \frac{\rho}{\epsilon_0}. \tag{12}$$

Equation (12) is the differential form of Gauss's law.

Magnetic fields A moving charged particle generates a magnetic field. To be specific, if a point charge of q coulombs is at position $\mathbf{r}_0$ and is moving with velocity $\mathbf{v}$, then the magnetic field it induces is

$$\mathbf{B}(\mathbf{r}) = \left(\frac{\mu_0 q}{4\pi}\right)\left(\frac{\mathbf{v}\times(\mathbf{r}-\mathbf{r}_0)}{\|\mathbf{r}-\mathbf{r}_0\|^3}\right). \tag{13}$$

In mks units, $\mathbf{B}$ is measured in teslas. The constant μ_0 is known as the **permeability of free space;** in mks units

$$\mu_0 = 4\pi \times 10^{-7}\ \text{N/amp}^2 \approx 1.257 \times 10^{-6}\ \text{N/amp}^2.$$

In the case of a magnetic field that arises from a continuous charged medium (such as electric current moving through a wire), rather than from a single moving charge, we replace q by a suitable charge density function ρ and the velocity of a single particle by the **velocity vector field v** of the charges. Then we define the **current density field J** by

$$\mathbf{J}(\mathbf{x}) = \rho(\mathbf{x})\mathbf{v}(\mathbf{x}). \tag{14}$$

In place of formula (13), we use the following definition for the magnetic field resulting from moving charges in a region D in space:

$$\begin{aligned}\mathbf{B}(\mathbf{r}) &= \frac{\mu_0}{4\pi}\iiint_D \rho(\mathbf{x})\mathbf{v}(\mathbf{x}) \times \frac{\mathbf{r}-\mathbf{x}}{\|\mathbf{r}-\mathbf{x}\|^3}\, dV \\ &= \frac{\mu_0}{4\pi}\iiint_D \mathbf{J}(\mathbf{x}) \times \frac{\mathbf{r}-\mathbf{x}}{\|\mathbf{r}-\mathbf{x}\|^3}\, dV. \end{aligned} \tag{15}$$

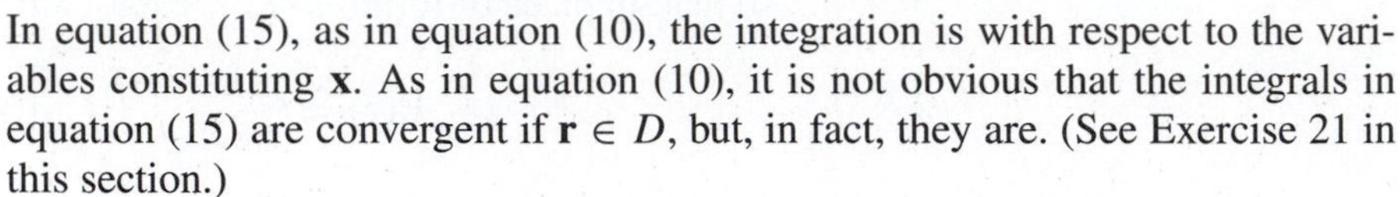

In equation (15), as in equation (10), the integration is with respect to the variables constituting $\mathbf{x}$. As in equation (10), it is not obvious that the integrals in equation (15) are convergent if $\mathbf{r} \in D$, but, in fact, they are. (See Exercise 21 in this section.)

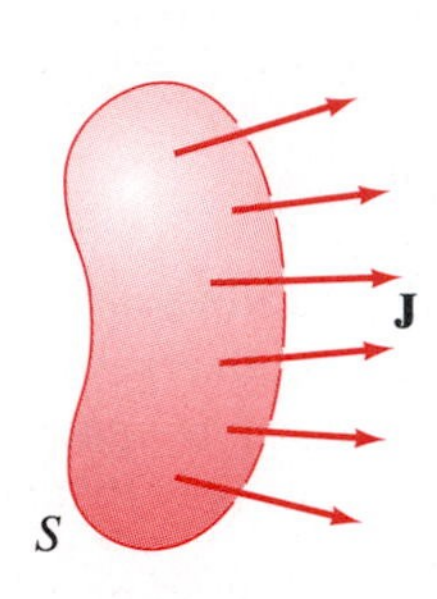

Figure 7.56 The total current I across S is the flux of the current density $\mathbf{J}$ across S.

Before continuing our calculations, we comment further regarding the current density field $\mathbf{J}$. The vector field $\mathbf{J}$ at a point is such that its magnitude is the current per unit area at that point, and its direction is that of the current flow. It is not hard to see then that the total current I across an oriented surface S is given by the flux of $\mathbf{J}$; that is,

$$I = \iint_S \mathbf{J}\cdot d\mathbf{S}. \tag{16}$$

(See Figure 7.56.)

Returning to the magnetic field $\mathbf{B}$ in equation (15), we show that it can be identified as the curl of another vector field $\mathbf{A}$ (to be determined). First, by direct calculation,

$$\nabla_{\mathbf{r}}\left(\frac{1}{\|\mathbf{r}-\mathbf{x}\|}\right) = -\frac{\mathbf{r}-\mathbf{x}}{\|\mathbf{r}-\mathbf{x}\|^3}.$$

Therefore, equation (15) becomes

$$\mathbf{B}(\mathbf{r}) = -\frac{\mu_0}{4\pi}\iiint_D \mathbf{J}(\mathbf{x}) \times \nabla_{\mathbf{r}}\left(\frac{1}{\|\mathbf{r}-\mathbf{x}\|}\right) dV. \tag{17}$$

We rewrite equation (17) using the following standard (and readily verified) identity, where f is a scalar field and $\mathbf{F}$ a vector field (both of class C^2):

$$\nabla \times (f\mathbf{F}) = (\nabla \times \mathbf{F})f - \mathbf{F} \times \nabla f. \tag{18}$$

Formula (18) is equivalent to

$$\mathbf{F} \times \nabla f = (\nabla \times \mathbf{F})f - \nabla \times (f\mathbf{F}).$$

Therefore,

$$\begin{aligned}\mathbf{J}(\mathbf{x}) \times \nabla_{\mathbf{r}}\left(\frac{1}{\|\mathbf{r}-\mathbf{x}\|}\right) &= (\nabla_{\mathbf{r}} \times \mathbf{J}(\mathbf{x}))\frac{1}{\|\mathbf{r}-\mathbf{x}\|} - \nabla_{\mathbf{r}} \times \frac{\mathbf{J}(\mathbf{x})}{\|\mathbf{r}-\mathbf{x}\|} \\ &= -\nabla_{\mathbf{r}} \times \frac{\mathbf{J}(\mathbf{x})}{\|\mathbf{r}-\mathbf{x}\|},\end{aligned}$$

since $\mathbf{J}(\mathbf{x})$ is independent of $\mathbf{r}$. Hence,

$$\mathbf{B}(\mathbf{r}) = \frac{\mu_0}{4\pi}\iiint_D \nabla_{\mathbf{r}} \times \frac{\mathbf{J}(\mathbf{x})}{\|\mathbf{r}-\mathbf{x}\|}\, dV = \frac{\mu_0}{4\pi}\nabla_{\mathbf{r}} \times \iiint_D \frac{\mathbf{J}(\mathbf{x})}{\|\mathbf{r}-\mathbf{x}\|}\, dV,$$

as $\mathbf{r}$ does not contain any of the variables of integration. Consequently,

$$\mathbf{B}(\mathbf{r}) = \nabla \times \mathbf{A}(\mathbf{r}),$$

where

$$\mathbf{A}(\mathbf{r}) = \frac{\mu_0}{4\pi} \iiint_D \frac{\mathbf{J}(\mathbf{x})}{\|\mathbf{r} - \mathbf{x}\|}\, dV. \tag{19}$$

Thus, $\nabla \cdot \mathbf{B} = \nabla \cdot (\nabla \times \mathbf{A})$ and so, by Theorem 4.4 in Chapter 3, we conclude that

$$\nabla \cdot \mathbf{B} = 0. \tag{20}$$

The intuitive content of equation (20) is often expressed by saying that "magnetic monopoles" do not exist.

The vector field $\mathbf{A}$ in equation (19) furnishes an example of a **vector potential** of the field $\mathbf{B}$. (See problems 33–38 in the Miscellaneous Exercises for more about vector potentials.)

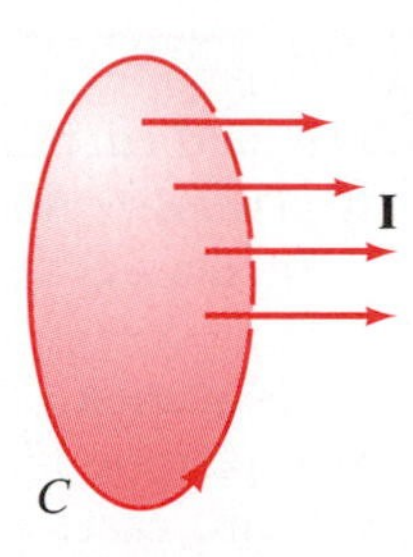

Figure 7.57 The closed loop C is oriented so that it has a right-hand relationship with the direction of current flow it encloses.

Ampère's circuital law If C is a closed loop enclosing a current I, then Ampère's law says that, up to a constant, the current through the loop is equal to the **circulation** of the magnetic field around C. To be precise,

$$\oint_C \mathbf{B} \cdot d\mathbf{s} = \mu_0 I. \tag{21}$$

In equation (21), we assume that C is oriented so that C and I are related by a right-hand rule; that is, that they are related in the same way that the orientation of C and the normal to any surface S that C bounds are related in Stokes's theorem. (See Figure 7.57.)

From equation (16) for the total current, equation (21) may be rewritten as

$$\oint_C \mathbf{B} \cdot d\mathbf{s} = \mu_0 \iint_S \mathbf{J} \cdot d\mathbf{S},$$

where S is any (piecewise smooth, oriented) surface bounded by C. Applying Stokes's theorem to the line integral, we obtain

$$\iint_S \nabla \times \mathbf{B} \cdot d\mathbf{S} = \mu_0 \iint_S \mathbf{J} \cdot d\mathbf{S}.$$

Since the loop C and surface S are arbitrary, we conclude that

$$\nabla \times \mathbf{B} = \mu_0 \mathbf{J}. \tag{22}$$

Equation (22) is the **differential form of Ampère's law in the static case** (i.e., in the case where $\mathbf{B}$ and $\mathbf{E}$ are constant in time). In the event that the magnetic and electric fields are time varying, we need to make some modifications. The so-called **equation of continuity**, established in Exercise 5 in this section, states that

$$\nabla \cdot \mathbf{J} = -\frac{\partial \rho}{\partial t}. \tag{23}$$

The difficulty is that if $\nabla \times \mathbf{B} = \mu_0 \mathbf{J}$ as in equation (22), then equation (23) implies that

$$\nabla \cdot (\nabla \times \mathbf{B}) = \nabla \cdot (\mu_0 \mathbf{J}) = -\mu_0 \frac{\partial \rho}{\partial t}.$$

However, assuming **B** is of class C^2, we must have $\nabla \cdot (\nabla \times \mathbf{B}) = 0$, even in the case where ρ is not constant in time.

The simplest solution to this difficulty is to modify equation (22) by adding an extra term. From Gauss's law, equation (12), we must have

$$\frac{\partial \rho}{\partial t} = \epsilon_0 \nabla \cdot \frac{\partial \mathbf{E}}{\partial t}.$$

Thus, if we replace **J** by $\mathbf{J} + \epsilon_0(\partial \mathbf{E}/\partial t)$ in equation (22), then you can verify that $\nabla \cdot (\nabla \times \mathbf{B}) = 0$. (See Exercise 16 in this section.) Hence, Ampère's law can be generalized as

$$\nabla \times \mathbf{B} = \mu_0 \mathbf{J} + \mu_0 \epsilon_0 \frac{\partial \mathbf{E}}{\partial t}. \tag{24}$$

The term $\epsilon_0(\partial \mathbf{E}/\partial t)$ in equation (24), known as the **displacement current density,** was first postulated by James Clerk Maxwell in order to generalize Ampère's law to the nonstatic case. (In this context, the original current density field **J** is known as the **conduction current density.**)

Equation (24) is not the only possible generalization of equation (22), but it is the simplest one and is consistent with observation. See Exercise 17 in this section for other ways to generalize equation (22) to the nonstatic case.

Faraday's law of induction Michael Faraday observed empirically that the change in magnetic flux across a surface S equals the electromotive force around the boundary C of the surface. This relation can be written as

$$\frac{d\Phi}{dt} = -\oint_C \mathbf{E} \cdot d\mathbf{s}, \tag{25}$$

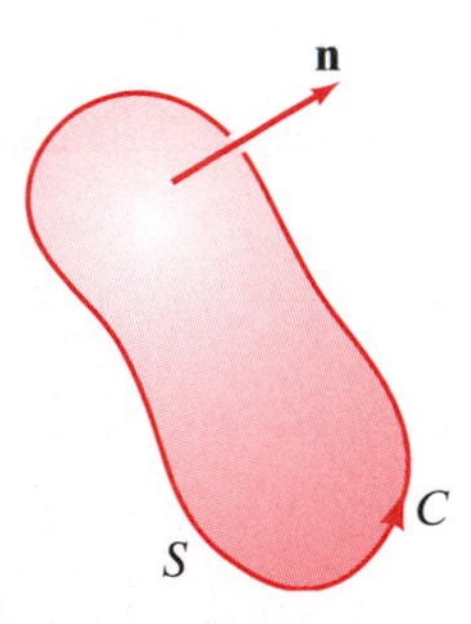

Figure 7.58 The rate of change of magnetic flux across S determines the electromotive force around the boundary C.

where $\Phi(t) = \iint_S \mathbf{B} \cdot d\mathbf{S}$, and C and S are oriented consistently. (See Figure 7.58.) If we apply Stokes's theorem to the line integral, we find that

$$\oint_C \mathbf{E} \cdot d\mathbf{s} = \iint_S \nabla \times \mathbf{E} \cdot d\mathbf{S}.$$

Since

$$\frac{d\Phi}{dt} = \frac{d}{dt} \iint_S \mathbf{B} \cdot d\mathbf{S} = \iint_S \frac{\partial \mathbf{B}}{\partial t} \cdot d\mathbf{S},$$

we have

$$-\iint_S \frac{\partial \mathbf{B}}{\partial t} \cdot d\mathbf{S} = \iint_S \nabla \times \mathbf{E} \cdot d\mathbf{S}. \tag{26}$$

Because equation (26) holds for arbitrary surfaces, we conclude that

$$\nabla \times \mathbf{E} = -\frac{\partial \mathbf{B}}{\partial t}. \tag{27}$$

Summary Equations (12), (20), (24), and (27) together are known as **Maxwell's equations:**

$$\nabla \cdot \mathbf{E} = \frac{\rho}{\epsilon_0} \qquad \text{(Gauss's law);}$$
$$\nabla \cdot \mathbf{B} = 0 \qquad \text{(No magnetic monopoles);}$$
$$\nabla \times \mathbf{E} = -\frac{\partial \mathbf{B}}{\partial t} \qquad \text{(Faraday's law);}$$
$$\nabla \times \mathbf{B} = \mu_0 \mathbf{J} + \mu_0 \epsilon_0 \frac{\partial \mathbf{E}}{\partial t} \qquad \text{(Ampère's law).}$$

Maxwell's equations allow one to reconstruct the electric and magnetic fields from the charge and current densities. They are fundamental to the subject of electricity and magnetism and provide a fitting tribute to the power of the theorems of Stokes and Gauss.

Addendum: Proof of Theorem 4.2

The most obvious idea is to use Green's second formula with

$$g(\mathbf{x}) = \frac{1}{\|\mathbf{r} - \mathbf{x}\|}.$$

However, this function fails to be continuous when $\mathbf{x} = \mathbf{r}$, so Gauss's theorem (and hence Green's formula) cannot be applied so readily. Instead, we need to examine the integrals more carefully.

Throughout the discussion that follows, let S_b denote the sphere of radius b centered at $\mathbf{r}$. First, we establish some subsidiary results.

Lemma 1 If h is a continuous function, then

$$\lim_{b \to 0} \iint_{S_b} \frac{h(\mathbf{x})}{\|\mathbf{r} - \mathbf{x}\|}\, dS = 0.$$

Proof The average value of h on S_b is

$$[h]_{\text{avg}} = \frac{\iint_{S_b} h(\mathbf{x})\, dS}{\text{surface area of } S_b} = \frac{1}{4\pi b^2} \iint_{S_b} h(\mathbf{x})\, dS.$$

(See Exercise 9 of the Miscellaneous Exercises.) Thus,

$$\iint_{S_b} h(\mathbf{x})\, dS = 4\pi b^2 [h]_{\text{avg}}.$$

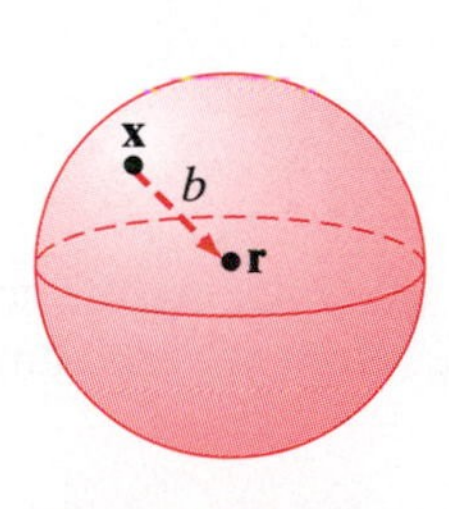

Figure 7.59 If $\mathbf{x}$ is any point on the surface of the sphere of radius b centered at $\mathbf{r}$, then $\|\mathbf{r} - \mathbf{x}\| = b$.

As $\mathbf{x}$ varies over the surface S_b, we have $\|\mathbf{r} - \mathbf{x}\| = b$. (See Figure 7.59.) Hence,

$$\lim_{b \to 0} \iint_{S_b} \frac{h(\mathbf{x})}{\|\mathbf{r} - \mathbf{x}\|}\, dS = \lim_{b \to 0} \iint_{S_b} \frac{1}{b} h(\mathbf{x})\, dS = \lim_{b \to 0} 4\pi b\, [h]_{\text{avg}} = 0. \qquad \blacksquare$$

To clarify the variables with respect to which we differentiate, let $\nabla_{\mathbf{x}}$ denote the del operator with respect to x, y, and z, and $\nabla_{\mathbf{r}}$ the del operator with respect to $\mathbf{r} = (r_1, r_2, r_3)$.

■ **Lemma 2** With h and S_b as in Lemma 1,

$$\lim_{b\to 0} \oint\!\!\!\oint_{S_b} h(\mathbf{x})\nabla_{\mathbf{x}}\left(\frac{1}{\|\mathbf{r}-\mathbf{x}\|}\right)\cdot d\mathbf{S} = -4\pi h(\mathbf{r}).$$

Proof Let $\mathbf{n} = (\mathbf{x}-\mathbf{r})/\|\mathbf{r}-\mathbf{x}\|$, the normalization of $\mathbf{x}-\mathbf{r}$. Straightforward calculations yield

$$\nabla_{\mathbf{x}}\left(\frac{1}{\|\mathbf{r}-\mathbf{x}\|}\right) = -\frac{\mathbf{x}-\mathbf{r}}{\|\mathbf{r}-\mathbf{x}\|^3},$$

and

$$\mathbf{n}\cdot\nabla_{\mathbf{x}}\left(\frac{1}{\|\mathbf{r}-\mathbf{x}\|}\right) = -\frac{(\mathbf{x}-\mathbf{r})\cdot(\mathbf{x}-\mathbf{r})}{\|\mathbf{r}-\mathbf{x}\|^4} = -\frac{\|\mathbf{x}-\mathbf{r}\|^2}{\|\mathbf{r}-\mathbf{x}\|^4} = -\frac{1}{\|\mathbf{r}-\mathbf{x}\|^2} = -\frac{1}{b^2},$$

for $\mathbf{x}$ on S_b. Then

$$\begin{aligned}
\lim_{b\to 0} \oint\!\!\!\oint_{S_b} h(\mathbf{x})\nabla_{\mathbf{x}}\left(\frac{1}{\|\mathbf{r}-\mathbf{x}\|}\right) d\mathbf{S} &= \lim_{b\to 0} \oint\!\!\!\oint_{S_b} \left(h(\mathbf{x})\nabla_{\mathbf{x}}\left(\frac{1}{\|\mathbf{r}-\mathbf{x}\|}\right)\cdot\mathbf{n}\right) dS \\
&= \lim_{b\to 0} - \oint\!\!\!\oint_{S_b} \frac{1}{b^2} h(\mathbf{x})\, dS \\
&= \lim_{b\to 0} -\frac{1}{b^2}\left(4\pi b^2 \,[h]_{\text{avg}}\right) \\
&= -4\pi h(\mathbf{r}).
\end{aligned}$$

(See the proof of Lemma 1.) ■

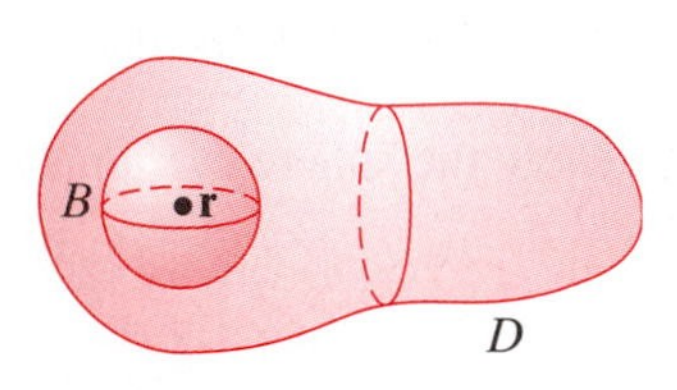

Figure 7.60 The region $D - B$ denotes the solid D with a small ball centered at $\mathbf{r}$ removed.

Returning to the proof of Green's third formula, we look at a region to which we *can* apply Green's second formula, namely, the region $D - B$, where B is a small ball of radius b centered at $\mathbf{r}$. (See Figure 7.60.) By Green's second formula (since $1/\|\mathbf{r}-\mathbf{x}\|$ is not singular on $D - B$), we have

$$\begin{aligned}
&\iiint_{D-B} \left(f(\mathbf{x})\nabla^2_{\mathbf{x}}\left(\frac{1}{\|\mathbf{r}-\mathbf{x}\|}\right) - \frac{\nabla^2_{\mathbf{x}} f(\mathbf{x})}{\|\mathbf{r}-\mathbf{x}\|}\right) dV \\
&\qquad = \iint_{S-S_b} \left(f(\mathbf{x})\nabla_{\mathbf{x}}\left(\frac{1}{\|\mathbf{r}-\mathbf{x}\|}\right) - \frac{\nabla_{\mathbf{x}} f(\mathbf{x})}{\|\mathbf{r}-\mathbf{x}\|}\right)\cdot d\mathbf{S}. \qquad (28)
\end{aligned}$$

By direct calculation $\nabla^2_{\mathbf{x}}(1/\|\mathbf{r}-\mathbf{x}\|) = 0$, so equation (28) becomes

$$\begin{aligned}
&\iiint_{D-B} -\frac{\nabla^2_{\mathbf{x}} f(\mathbf{x})}{\|\mathbf{r}-\mathbf{x}\|}\, dV \\
&\qquad = \oint\!\!\!\oint_{S-S_b} \left(f(\mathbf{x})\nabla_{\mathbf{x}}\left(\frac{1}{\|\mathbf{r}-\mathbf{x}\|}\right) - \frac{\nabla_{\mathbf{x}} f(\mathbf{x})}{\|\mathbf{r}-\mathbf{x}\|}\right)\cdot d\mathbf{S}. \qquad (29)
\end{aligned}$$

We may evaluate the right-hand side of equation (29) by replacing the surface integral over $S - S_b$ by separate integrals over S and S_b. Now, we take limits as $b \to 0$. By Lemma 1 with $h(\mathbf{x}) = \nabla_{\mathbf{x}} f(\mathbf{x})\cdot\mathbf{n}$,

$$\oint\!\!\!\oint_{S_b} \frac{\nabla_{\mathbf{x}} f(\mathbf{x})}{\|\mathbf{r}-\mathbf{x}\|}\cdot d\mathbf{S} = \oint\!\!\!\oint_{S_b} \frac{\nabla_{\mathbf{x}} f(\mathbf{x})\cdot\mathbf{n}}{\|\mathbf{r}-\mathbf{x}\|}\, dS \to 0.$$

By Lemma 2,

$$\oint\!\!\!\oint_{S_b} f(\mathbf{x})\nabla_{\mathbf{x}}\left(\frac{1}{\|\mathbf{r}-\mathbf{x}\|}\right)\cdot d\mathbf{S} \to -4\pi f(\mathbf{r}).$$

Since B shrinks to a point as $b \to 0$, we see that equation (29) becomes

$$\iiint_D -\frac{\nabla_{\mathbf{x}}^2 f(\mathbf{x})}{\|\mathbf{r}-\mathbf{x}\|}\, dV = \oiint_S \left(f(\mathbf{x})\nabla_{\mathbf{x}}\left(\frac{1}{\|\mathbf{r}-\mathbf{x}\|}\right) - \frac{\nabla_{\mathbf{x}} f(\mathbf{x})}{\|\mathbf{r}-\mathbf{x}\|}\right) \cdot d\mathbf{S} + 4\pi f(\mathbf{r}),$$

from which Green's third formula follows immediately. ■

Exercises

1. Prove Green's first formula, stated in Theorem 4.1.

*A function $g(x, y, z)$ is said to be **harmonic** at a point (x_0, y_0, z_0) if g is of class C^2 and satisfies Laplace's equation*

$$\nabla^2 g = \frac{\partial^2 g}{\partial x^2} + \frac{\partial^2 g}{\partial y^2} + \frac{\partial^2 g}{\partial z^2} = 0$$

on some neighborhood of (x_0, y_0, z_0). We say that g is harmonic on a closed region $D \subseteq \mathbf{R}^3$ if it is harmonic at all interior points of D (i.e., not necessarily on the boundary of D). Exercises 2–4 concern some elementary vector analysis of harmonic functions.

2. Assume that D is closed and bounded and that ∂D is a piecewise smooth surface oriented by outward unit normal field $\mathbf{n}$. Let $\partial g/\partial n$ denote $\nabla g \cdot \mathbf{n}$. (The term $\partial g/\partial n$ is called the **normal derivative** of g.) Use Green's first formula with $f(x, y, z) \equiv 1$ to show that, if g is harmonic on D, then

$$\oiint_{\partial D} \frac{\partial g}{\partial n}\, dS = 0.$$

3. Let f be harmonic on a region D that satisfies the assumptions of Exercise 2.

(a) Show that

$$\iiint_D \nabla f \cdot \nabla f\, dV = \oiint_{\partial D} f \frac{\partial f}{\partial n}\, dS.$$

(b) Suppose $f(x, y, z) = 0$ for all $(x, y, z) \in \partial D$. Use part (a) to show that then we must have $f(x, y, z) = 0$ throughout all of D. (Hint: Think about the sign of $\nabla f \cdot \nabla f$.)

4. Let D be a region that satisfies the assumptions of Exercise 2. Use the result of Exercise 3(b) to show that if f_1 and f_2 are harmonic on D and $f_1(x, y, z) = f_2(x, y, z)$ on ∂D, then, in fact, $f_1 = f_2$ on *all* of D. Thus we see that harmonic functions are determined by their boundary values on a region. (Hint: Consider $f_1 - f_2$.)

5. (a) Suppose a fluid of density $\rho(x, y, z, t)$ flows with velocity field $\mathbf{F}(x, y, z, t)$ in a solid region W in space enclosed by a smooth surface S. Use Gauss's theorem to show that, if there are no sources or sinks,

$$\nabla \cdot (\rho \mathbf{F}) = -\frac{\partial \rho}{\partial t}.$$

This equation is called the **continuity equation** in fluid dynamics. (Hint: The triple integral $\iiint_W \frac{\partial \rho}{\partial t}\, dV$ is the rate of fluid flowing into W, and the flux of $\rho\mathbf{F}$ across S gives the rate of fluid flowing out of W.)

(b) Use the argument in part (a) to establish the equation of continuity for current densities given in equation (23):

$$\nabla \cdot \mathbf{J} = -\frac{\partial \rho}{\partial t}.$$

*Let $T(x, y, z, t)$ denote the temperature at the point (x, y, z) of a solid object D at time t. We define the **heat flux density** $\mathbf{H}$ by $\mathbf{H} = -k\nabla T$. (The constant k is the **thermal conductivity.** Note that the symbol ∇ denotes differentiation with respect to x, y, z, not with respect to t.) The vector field $\mathbf{H}$ represents the velocity of heat flow in D. It is a fact from physics that the total heat contained in a solid body D having density ρ and specific heat σ is*

$$\iiint_D \sigma\rho T\, dV.$$

Hence, the total amount of heat leaving D per unit time is

$$-\iiint_D \sigma\rho \frac{\partial T}{\partial t}\, dV.$$

(Here we assume that σ and ρ do not depend on t.) We also know that the heat flux may be calculated as

$$\iint_{\partial D} \mathbf{H} \cdot d\mathbf{S}.$$

Exercises 6–10 concern these notions of temperature, heat, and heat flux density.

6. Use Gauss's theorem to derive the **heat equation,**

$$\sigma\rho \frac{\partial T}{\partial t} = k\nabla^2 T.$$

7. In Exercise 6, suppose that k varies with the points in D; that is, $k = k(x, y, z)$. Show that then we have

$$\sigma\rho \frac{\partial T}{\partial t} = k\nabla^2 T + \nabla k \cdot \nabla T.$$

8. In the heat equation of Exercise 6, suppose that σ, ρ, and k are all constant and the temperature T of the solid D does not vary with time. Show that then T must be harmonic, i.e., that $\nabla^2 T = 0$ at all points in the interior of D.

9. (a) If σ, ρ, and k are constant and the temperature T of the solid D is independent of time, show that the (net) heat flux of $\mathbf{H}$ across the boundary of D must be zero.

(b) Let D be the solid region between two concentric spheres of radii 1 and 2. Suppose that the inner sphere is heated to 120° C and the outer sphere to 20° C. Use the result of part (a) to describe the rate of heat flow across the spheres.

10. Consider the three-dimensional heat equation

$$\nabla^2 u = \frac{\partial u}{\partial t} \tag{30}$$

for functions $u(x, y, z, t)$. (Here $\nabla^2 u$ denotes the Laplacian $\partial^2 u/\partial x^2 + \partial^2 u/\partial y^2 + \partial^2 u/\partial z^2$.) In this exercise, show that any solution $T(x, y, z, t)$ to the heat equation is unique in the following sense: Let D be a bounded solid region in $\mathbf{R}^3$ and suppose that the functions $\alpha(x, y, z)$ and $\phi(x, y, z, t)$ are given. Then there exists a unique solution $T(x, y, z, t)$ to equation (30) that satisifies the conditions

$$T(x, y, z, 0) = \alpha(x, y, z) \quad \text{for } (x, y, z) \in D,$$

and (31)

$$T(x, y, z, t) = \phi(x, y, z, t) \quad \text{for } (x, y, z) \in \partial D \text{ and } t \geq 0.$$

To establish uniqueness, let T_1 and T_2 be two solutions to equation (30) satisfying the conditions in (31) and set $w = T_1 - T_2$.

(a) Show that w must also satisfy equation (30), plus the conditions that

$$w(x, y, z, 0) = 0 \quad \text{for all } (x, y, z) \in D,$$

and

$$w(x, y, z, t) = 0 \quad \text{for all } (x, y, z) \in \partial D \text{ and } t \geq 0.$$

(b) For $t \geq 0$, define the "energy function"

$$E(t) = \frac{1}{2} \iiint_D [w(x, y, z, t)]^2 \, dV.$$

Use Green's first formula in Theorem 4.1 to show that $E'(t) \leq 0$ (i.e., that E does not increase with time).

(c) Show that $E(t) = 0$ for all $t \geq 0$. (Hint: Show that $E(0) = 0$ and use part (b).)

(d) Show that $w(x, y, z, t) = 0$ for all $t \geq 0$ and $(x, y, z) \in D$, and thereby conclude the uniqueness of solutions to equation (30) that satisfy the conditions in (31).

11. Show that Ampère's law and Gauss's law imply the continuity equation for $\mathbf{J}$. (Note: In the text, we use the continuity equation to derive Ampère's law.)

12. Suppose that $\mathbf{E}$ is an electric field, in particular, a vector field that satisfies the equation $\nabla \cdot \mathbf{E} = \rho/\epsilon_0$. A region D in space is said to be **charge-free** if ρ is zero at all points of D. Describe the charge-free regions of $\mathbf{E} = (x^3 - x)\,\mathbf{i} + \frac{1}{4}y^3\,\mathbf{j} + \left(\frac{1}{9}z^3 - 2z\right)\,\mathbf{k}$.

13. By considering the derivation of Gauss's law for electric fields, show that, for any inverse square vector field $\mathbf{F}(\mathbf{x}) = k\mathbf{x}/\|\mathbf{x}\|^3$, the flux of $\mathbf{F}$ across a piecewise smooth, closed, oriented surface S is

$$\oiint_S \mathbf{F} \cdot d\mathbf{S} = \begin{cases} 0 & \text{if } S \text{ does not enclose the origin,} \\ 4\pi k & \text{if } S \text{ encloses the origin.} \end{cases}$$

14. Let a point charge of q coulombs be placed at the origin. Recover the formula

$$\mathbf{E} = \frac{q}{4\pi\epsilon_0} \frac{\mathbf{x}}{\|\mathbf{x}\|^3}$$

by using Gauss's law in the following way:

(a) First, explain that in spherical coordinates, $\mathbf{E}(\mathbf{x}) = E(\mathbf{x})\mathbf{e}_\rho$, that is, that $\mathbf{E}$ has no components in either the $\mathbf{e}_\varphi$- or $\mathbf{e}_\theta$-direction. Next, note that $E(\mathbf{x})$ may be written as $E(\rho)$—that is, that $\|\mathbf{E}\|$ has the same magnitude at all points on a sphere centered at the origin.

(b) Show, using Gauss's law and Gauss's theorem, that

$$\oiint_S E(\rho)\mathbf{e}_\rho \cdot d\mathbf{S} = \frac{q}{\epsilon_0},$$

where S is any smooth, closed surface enclosing the origin.

(c) Now let S be the sphere of radius a centered at the origin. Then the outward unit normal $\mathbf{n}$ to S at (ρ, φ, θ) is $\mathbf{e}_\rho$. Show that

$$\oiint_S E(\rho)\, dS = \frac{q}{\epsilon_0}.$$

(d) Use part (c) to show that $E(\rho) = q/(4\pi\epsilon_0\rho^2)$. Conclude the result desired.

15. (a) Establish the following identity for vector fields $\mathbf{F}$ of class C^2:

$$\nabla \times (\nabla \times \mathbf{F}) = \nabla(\nabla \cdot \mathbf{F}) - \nabla^2\mathbf{F}.$$

(Note: $\nabla^2\mathbf{F} = (\nabla \cdot \nabla)\mathbf{F}$.)

(b) In free space (i.e., in the absence of all charges and currents), use Maxwell's equations to show that $\mathbf{E}$ and $\mathbf{B}$ satisfy the wave equation

$$\nabla^2\mathbf{F} = k\frac{\partial^2\mathbf{F}}{\partial t^2},$$

where k is a constant. What is k in each case?

(c) Use part (a), Faraday's law, and Ampère's law to show that

$$\nabla(\nabla \cdot \mathbf{E}) - (\nabla \cdot \nabla)\mathbf{E} = -\mu_0\frac{\partial}{\partial t}\left[\mathbf{J} + \epsilon_0\frac{\partial \mathbf{E}}{\partial t}\right].$$

(d) Show that, in the absence of any charges (i.e., if $\rho = 0$),

$$\nabla^2 \mathbf{E} = \mu_0 \frac{\partial \mathbf{J}}{\partial t} + \mu_0 \epsilon_0 \frac{\partial^2 \mathbf{E}}{\partial t^2}.$$

16. Verify that if the nonstatic version of Ampère's law (equation (24)) holds, then $\nabla \cdot (\nabla \times \mathbf{B}) = 0$

17. When Maxwell postulated the existence of displacement currents to arrive at a nonstatic version of Ampère's law, he was simply choosing the simplest way to correct equation (22) so that it would be consistent with the continuity equation (23). However, other possibilities are also consistent with the continuity equation.

(a) Show that in order to have equation (22) valid in the static case, then, in general, we must have

$$\nabla \times \mathbf{B} = \mu_0 \mathbf{J} + \frac{\partial \mathbf{F}_1}{\partial t}$$

for some (time-varying) vector field $\mathbf{F}_1$ of class C^2.

(b) By taking the divergence of both sides of the equation in part (a), show that

$$\nabla \cdot \frac{\partial \mathbf{F}_1}{\partial t} = \mu_0 \epsilon_0 \nabla \cdot \frac{\partial \mathbf{E}}{\partial t}.$$

(c) Use part (b) to argue that, from an entirely *mathematical* perspective, Ampère's law can also be generalized as

$$\nabla \times \mathbf{B} = \mu_0 \mathbf{J} + \mu_0 \epsilon_0 \frac{\partial \mathbf{E}}{\partial t} + \mathbf{F}_2,$$

where $\mathbf{F}_2$ is any divergence-free vector field. Since no one has observed any physical evidence for $\mathbf{F}_2$'s being nonzero, it is assumed to be zero, as in equation (24).

18. Suppose that $\mathbf{J} = \sigma \mathbf{E}$. (This is a version of Ohm's law that obtains in some electric conductors—here σ is a positive constant known as the **conductivity**.) If $\rho = 0$, show that $\mathbf{E}$ and $\mathbf{B}$ satisfy the so-called **telegrapher's equation,**

$$\nabla^2 \mathbf{F} = \mu_0 \sigma \frac{\partial \mathbf{F}}{\partial t} + \mu_0 \epsilon_0 \frac{\partial^2 \mathbf{F}}{\partial t^2}.$$

19. Let $\mathbf{E}$ and $\mathbf{B}$ be steady-state electric and magnetic fields (i.e., $\mathbf{E}$ and $\mathbf{B}$ are constant in time). The **Poynting vector field** $\mathbf{P} = \mathbf{E} \times \mathbf{B}$ represents radiation flux density. Use Maxwell's equations to show that, for a smooth, orientable, closed surface S bounding a solid region D,

$$\oint\!\!\oint_S \mathbf{P} \cdot d\mathbf{S} = -\mu_0 \iiint_D \mathbf{E} \cdot \mathbf{J}\, dV.$$

20. Consider the electric field $\mathbf{E}(\mathbf{r})$ defined by equation (10). Note that the integrals in equation (10) are improper in the sense that they become infinite at points $\mathbf{r} \in D$, where $\rho(\mathbf{r})$ is nonzero. In this exercise, you will show that, nonetheless, the integrals in equation (10) converge when D is a bounded region in $\mathbf{R}^3$ and ρ is a continuous charge density function on D.

(a) Write $\mathbf{E}(\mathbf{r})$ in terms of triple integrals for the individual components. Let $\mathbf{r} = (r_1, r_2, r_3)$ and $\mathbf{x} = (x, y, z)$.

(b) Show that if each component of $\mathbf{E}$ is written in the form $\iiint_D f(\mathbf{x})\, dV$, then $|f(\mathbf{x})| \le K/\|\mathbf{r} - \mathbf{x}\|^2$, where K is a positive constant.

(c) It follows from part (b) that if

$$\iiint_D \frac{K}{\|\mathbf{r} - \mathbf{x}\|^2}\, dV$$

converges, so must $\iiint_D f(\mathbf{x})\, dV$. Show that

$$\iiint_D \frac{K}{\|\mathbf{r} - \mathbf{x}\|^2}\, dV$$

converges by considering an iterated integral in spherical coordinates with origin at $\mathbf{r}$. (Hint: Look carefully at the integrand in spherical coordinates.)

21. Consider the magnetic field $\mathbf{B}(\mathbf{r})$ defined by equation (15). As was the case with the electric field in equation (10), it is not obvious that the integrals in (15) converge at all $\mathbf{r} \in D$. Follow the ideas of Exercise 20 to show that $\mathbf{B}(\mathbf{r})$ is, in fact, well-defined at all $\mathbf{r}$, assuming a continuous current density field $\mathbf{J}$ and bounded region D in $\mathbf{R}^3$.

7.5 True/False Exercises for Chapter 7

1. The function $\mathbf{X}: \mathbf{R}^2 \to \mathbf{R}^3$ given by $\mathbf{X}(s, t) = (2s + 3t + 1, 4s - t, s + 2t - 7)$ parametrizes the plane $9x - y - 14z = 107$.

2. The function $\mathbf{X}: \mathbf{R}^2 \to \mathbf{R}^3$ given by $\mathbf{X}(s, t) = (s^2 + 3t - 1, s^2 + 3, -2s^2 + t)$ parametrizes the plane $x - 7y - 3z + 22 = 0$.

3. The function $\mathbf{X}: (-\infty, \infty) \times (-\frac{\pi}{2}, \frac{\pi}{2}) \to \mathbf{R}^3$ given by $\mathbf{X}(s, t) = (s^3 + 3\tan t - 1, s^3 + 3, -2s^3 + \tan t)$ parametrizes the plane $x - 7y - 3z + 22 = 0$.

4. The surface $\mathbf{X}(s, t) = (s^2 t, st^2, st)$ is smooth.

5. The area of the portion of the surface $z = xe^{xy}$ lying over the disk of radius 2 centered at the origin is given by

$$\int_0^2 \int_0^{\sqrt{4-x^2}} \sqrt{1 + e^{2xy}(x^4 + y^2 + 2y + 1)}\, dy\, dx.$$

6. If S is the unit sphere centered at the origin, then $\iint_S x^3\, dS = 0$.

7. If S is the cube with the eight vertices $(\pm 1, \pm 1, \pm 1)$, then $\iint_S (1 + x^3 y)\, dS = 0$.

8. If S denotes the rectangular box with faces given by the planes $x = \pm 1$, $y = \pm 2$, $z = \pm 3$, then $\iint_S xyz\, dS = 0$.

9. If S denotes the sphere of radius a centered at the origin, then

$$\iint_S (z^3 - z + 2)\, dS = \iint_S (x - y^5 + 2)\, dS.$$

10. $\iint_S (-y\mathbf{i} + x\mathbf{j}) \cdot d\mathbf{S} = 0$, where S is the cylinder $x^2 + y^2 = 9$, $0 \le z \le 5$.

11. Let S denote the closed cylinder with lateral surface given by $y^2 + z^2 = 4$, front by $x = 7$, and back by $x = -1$, and oriented by outward normals. Then $\int_S x\mathbf{i} \cdot d\mathbf{S} = 24\pi$.

12. If S is the portion of the cylinder $x^2 + y^2 = 16$, $-2 \le z \le 7$, then $\iint_S \nabla \times (y\mathbf{i}) \cdot d\mathbf{S} = 0$.

13. $\iint_S \mathbf{F} \cdot d\mathbf{S} = 6\pi$, where S is the closed hemisphere $x^2 + y^2 + z^2 = 1$, $z \ge 0$, together with the surface $x^2 + y^2 \le 1$, $z = 0$ and $\mathbf{F} = yz\,\mathbf{i} - xz\,\mathbf{i} + 3\,\mathbf{k}$.

14. If S is the level set of a function $f(x, y, z)$ and $\nabla f \ne \mathbf{0}$, then the flux of ∇f across S is never zero.

15. A smooth surface has at most two orientations.

16. A smooth, connected surface is always orientable.

17. If $\mathbf{F}$ is a vector field of class C^1 and S is the ellipsoid $x^2 + 4y^2 + 9z^2 = 36$, then $\iint_S \nabla \times \mathbf{F} \cdot d\mathbf{S} = 0$.

18. $\iint_S \nabla \times \mathbf{F} \cdot d\mathbf{S}$ has the same value for all piecewise smooth, oriented surfaces S that have the same boundary curve C.

19. If $\mathbf{F}$ is a constant vector field, then $\oiint_S \mathbf{F} \cdot d\mathbf{S} = 0$, where S is any piecewise smooth, closed surface.

20. $\oiint_S \nabla \times \mathbf{F} \cdot d\mathbf{S} = 0$, where S is any closed, orientable, smooth surface in $\mathbf{R}^3$ and $\mathbf{F}$ is of class C^1.

21. Suppose that $\mathbf{F}$ is a vector field of class C^1 whose domain contains the solid region D in $\mathbf{R}^3$ and is such that $\|\mathbf{F}(x, y, z)\| \le 2$ at all points on the boundary surface S of D. Then $\iiint_D \nabla \cdot \mathbf{F}\, dV$ is twice the surface area of S.

22. If S is an orientable, piecewise smooth surface and $\mathbf{F}$ is a vector field of class C^1 that is everywhere tangent to the boundary of S, then $\iint_S \nabla \times \mathbf{F} \cdot d\mathbf{S} = 0$.

23. If S is an orientable, piecewise smooth surface and $\mathbf{F}$ is a vector field of class C^1 that is everywhere perpendicular to the boundary of S, then $\iint_S \nabla \times \mathbf{F} \cdot d\mathbf{S} = 0$.

24. If $\mathbf{F}$ is tangent to a closed surface S that bounds a solid region D in $\mathbf{R}^3$, then $\iiint_D \nabla \cdot \mathbf{F}\, dV = 0$.

25. Let S be a piecewise smooth, orientable surface and $\mathbf{F}$ a vector field of class C^1. Then the flux of $\mathbf{F}$ across S is equal to the circulation of $\mathbf{F}$ around the boundary of S.

26. Let D be a solid region in $\mathbf{R}^3$ and $\mathbf{F}$ a vector field of class C^1. Then the flux of $\mathbf{F}$ across the boundary of D is equal to the integral of the divergence of $\mathbf{F}$ over D.

27. Suppose that f and g are of class C^2 and D is a solid region in $\mathbf{R}^3$ with piecewise smooth boundary surface S that is oriented away from D. If g is harmonic, then $\iiint_D \nabla f \cdot \nabla g\, dV = \oiint_S f \nabla g \cdot d\mathbf{S}$.

28. Suppose that f and g are of class C^2 and D is a solid region in $\mathbf{R}^3$ with piecewise smooth boundary surface S that is oriented away from D. If f and g are harmonic, then $\oiint_S f \nabla g \cdot d\mathbf{S} = -\oiint_S g \nabla f \cdot d\mathbf{S}$.

29. If $\nabla^2 f$ is known, then f is uniquely determined up to a constant.

30. If S is a closed, orientable surface, then

$$\oiint_S \frac{\mathbf{x}}{\|\mathbf{x}\|^3} \cdot d\mathbf{S} = 0.$$

7.6 Miscellaneous Exercises for Chapter 7

1. Figure 7.61 shows the plots of six parametrized surfaces $\mathbf{X}$. Match each parametric description with the correct graph.

(a) $\mathbf{X}(s, t) = (t(s^2 - t^2), s, s^2 - t^2)$

(b) $\mathbf{X}(s, t) = (s \cos t, s \sin t, s)$

(c) $\mathbf{X}(s, t) = ((2 + \cos s) \cos t, (2 + \cos s) \sin t, t + \sin s)$

(d) $\mathbf{X}(s, t) = (\sin s \cos t, s, \sin s \sin t)$

(e) $\mathbf{X}(s, t) = \Big(\cos t + 0.1t\Big(\cos t \cos s + \dfrac{1}{\sqrt{5}} \sin t \sin s\Big), \sin t + 0.1t\Big(\sin t \cos s - \dfrac{1}{\sqrt{5}} \cos t \sin s\Big), \dfrac{t}{2} + \dfrac{0.2t}{\sqrt{5}} \sin s\Big)$

(f) $\mathbf{X}(s, t) = (s \cos t, s \sin t, t)$

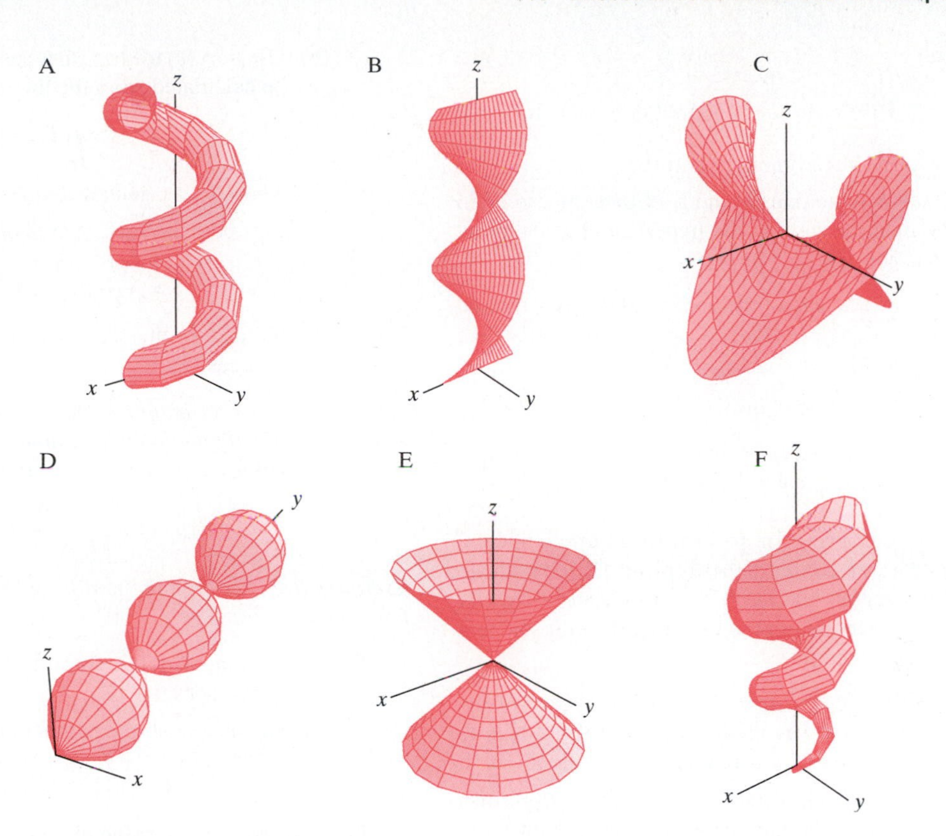

Figure 7.61 Figures for Exercise 1.

2. Consider the unit sphere S with equation $x^2+y^2+z^2 = 1$. In this problem, you will provide a parametrization for (almost all of) the sphere that is different from the one given in Example 2 of §7.1.

(a) First consider the parametrized plane in $\mathbf{R}^3$:

$$D = \{(s, t, 0) \mid (s, t) \in \mathbf{R}^2\}.$$

Note that D is just a copy of $\mathbf{R}^2$ sitting in $\mathbf{R}^3$ as the xy-plane. For any point $(s, t, 0) \in D$, argue geometrically that the line through $(s, t, 0)$ and $(0, 0, 1)$ intersects S in a point other than $(0, 0, 1)$. (See Figure 7.62.)

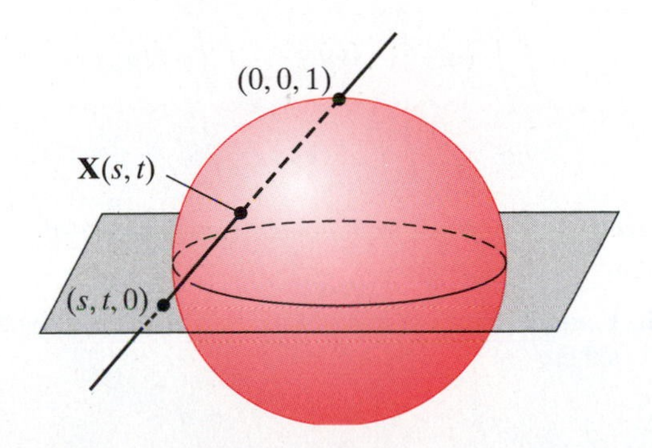

Figure 7.62 Figure for Exercise 2.

(b) Now define $\mathbf{X}\colon \mathbf{R}^2 \to \mathbf{R}^3$ by letting $\mathbf{X}(s, t)$ be the point of intersection of S and the line joining $(s, t, 0)$ and $(0, 0, 1)$. Write a set of parametric equations for the line joining $(s, t, 0)$ and $(0, 0, 1)$ and use it to give a formula for $\mathbf{X}(s, t)$.

(c) Show that $\mathbf{X}(s, t)$ is a smooth parametrization of almost all of S. What points of S are not included in the image of $\mathbf{X}$? For evident geometric reasons, the map $\mathbf{X}$ defined in this problem has an inverse map from (almost all of) the sphere to the xy-plane, called **stereographic projection** of the sphere onto the plane.

3. (a) Provide a parametrization for the hyperboloid $x^2 + y^2 - z^2 = 1$. (Hint: Use the cylindrical coordinates z and θ for parameters.)

(b) Modify your answer in part (a) to give a parametrization of the hyperboloid

$$\frac{x^2}{a^2} + \frac{y^2}{b^2} - \frac{z^2}{c^2} = 1.$$

(c) Let x_0 and y_0 be such that $x_0^2 + y_0^2 = 1$. Show that the lines

$$\mathbf{l}_1(t) = (a(x_0 - y_0 t), b(x_0 t + y_0), ct)$$

and

$$\mathbf{l}_2(t) = (a(y_0 t + x_0), b(y_0 - x_0 t), ct)$$

lie in the hyperboloid of part (b).

(d) Show that the lines $\mathbf{l}_1$ and $\mathbf{l}_2$ of part (c) also lie in the plane tangent to the hyperboloid at the point $(ax_0, by_0, 0)$.

4. Find the surface area of the portion of the hyperboloid $x^2 + y^2 - z^2 = 1$ between $z = -a$ and $z = a$. (See Exercise 3(a).)

5. (a) Parametrize the ellipsoid

$$\frac{x^2}{a^2} + \frac{y^2}{b^2} + \frac{z^2}{c^2} = 1.$$

(b) Use your answer in part (a) to set up an integral for the surface area of the ellipsoid. Do not evaluate this integral, but verify that it indicates correctly the surface area in the case that the ellipsoid is a sphere of radius a.

6. Let $y = f(x)$, $a \le x \le b$, be a curve in the xy-plane. Suppose this curve is revolved around the x-axis to generate a surface of revolution.

(a) Explain why $\mathbf{X}(s, t) = (s, f(s)\cos t, f(s)\sin t)$ parametrizes the surface so described. (Hint: Consider the t-coordinate curve.)

(b) Verify that the area of the surface is

$$2\pi \int_a^b |f(x)|\sqrt{1 + (f'(x))^2}\,dx.$$

7. Let the curve $y = f(x)$, $0 \le a \le x \le b$, be revolved around the y-axis to generate a surface.

(a) Find a parametrization for the surface.

(b) Verify that the area of the surface is

$$2\pi \int_a^b x\sqrt{1 + (f'(x))^2}\,dx.$$

8. Let S denote the *surface* defined by the equation $z = f(x)$, $a \le x \le b$. (The surface is a generalized cylinder over the curve $z = f(x)$ in the xz-plane.) Let C denote a piecewise C^1, simple, closed curve in the xy-plane. Let D denote the region in the xy-plane bounded by C and assume that every point of D has x-coordinate between a and b. Let S_1 denote the portion of S lying over D.

(a) Show that the portion of S lying over D is

$$\iint_D s'(x)\,dA,$$

where $s(x) = \int_a^x \sqrt{1 + (f'(t))^2}\,dt$, that is, s is the arclength function of the curve $z = f(x)$.

(b) Use part (a) to show that the surface area may also be calculated from the line integral

$$\oint_C s(x)\,dy,$$

where C is oriented counterclockwise.

(c) Compute the surface area of the portion of

$$z = \frac{x^3}{3} + \frac{1}{4x}, \qquad 1 \le x \le 3,$$

lying over the rectangle in the xy-plane having vertices $(1, \pm 2)$, $(3, \pm 2)$.

*Let f be a piecewise smooth surface in $\mathbf{R}^3$ and $f\colon X \subseteq \mathbf{R}^3 \to \mathbf{R}$ a scalar-valued function whose domain X contains S. Then the **average value** of f on S is the quantity*

$$[f]_{\text{avg}} = \frac{\iint_S f\,dS}{\iint_S dS} = \frac{\iint_S f\,dS}{\textit{area of } S}.$$

Exercises 9–11 involve the notion of the average value of a function on a surface

9. (a) Explain why the definition of the average value makes sense.

(b) Suppose that the temperature at points on the sphere $x^2 + y^2 + z^2 = 49$ is given by $T(x, y, z) = x^2 + y^2 - 3z$. Find the average temperature.

10. Find the average value of $f(x, y, z) = x^2 e^z - y^2 z$ on the cylinder $x^2 + y^2 = 4$, $0 \le z \le 3$.

11. Find the average value of $f(x, y, z) = x^2 + y^2 - 3$ on the portion of the cone $z^2 = 4x^2 + 4y^2$, $-2 \le z \le 6$.

12. A thin film is made in the shape of the helicoid

$$\mathbf{X}(s, t) = (s\cos t, s\sin t, t), \quad 0 \le s \le 1, 0 \le t \le 4\pi.$$

Suppose that the mass density (per unit area) at each point (x, y, z) of the film varies as

$$\delta(x, y, z) = \sqrt{x^2 + y^2}.$$

Find the total mass of the film.

*Let S be a piecewise smooth surface in $\mathbf{R}^3$. Suppose that the mass density at points (x, y, z) of S is $\delta(x, y, z)$. Using formulas analogous to those in §5.6, we define **the (first) moments** of S to be*

$$\iint_S x\delta(x, y, z)\,dS, \qquad \iint_S y\delta(x, y, z)\,dS,$$

and $$\iint_S z\delta(x, y, z)\,dS.$$

Exercises 13–16 involve first moments and centers of mass of surfaces.

13. Find the center of mass of the first octant portion of the sphere

$$x^2 + y^2 + z^2 = a^2,$$

assuming constant density.

14. Find the center of mass of the piece of the cylinder

$$x^2 + z^2 = a^2, \quad 0 \le y \le a, z \ge 0.$$

Assume that the density of the surface is constant.

15. Find the center of mass of a sphere of radius a, where the density δ varies as the distance from the "south pole" of the sphere.

16. Find the center of mass of the cylinder $x^2 + z^2 = a^2$, between $y = 0$ and $y = 2$, if the density varies as $\delta = x^2 + y$.

Given a piecewise smooth surface S, the ***moment of inertia*** I_z *of S about the z-axis is defined by the surface integral*

$$I_z = \iint_S (x^2 + y^2)\delta\, dS,$$

where $\delta(x, y, z)$ *is mass density. The corresponding* ***radius of gyration*** *about the z-axis is given by*

$$r_z = \sqrt{\frac{I_z}{M}},$$

where $M = \iint_S \delta\, dS$. *Likewise, the* ***moments of inertia*** *of S about the x- and y-axes are given by, respectively, the surface integrals*

$$I_x = \iint_S (y^2+z^2)\delta\, dS \qquad \text{and} \qquad I_y = \iint_S (x^2+z^2)\delta\, dS.$$

(Compare these formulas with the ones in §5.6.) Exercises 17–19 concern moments of inertia and radii of gyration of surfaces.

17. (a) Calculate I_z for the surface S cut from the cone

$$z^2 = 4x^2 + 4y^2$$

by the planes $z = 2$ and $z = 4$. (This surface is known as a **frustum** of the cone.) Assume density is equal to 1.

(b) Find the radius of gyration r_z of the frustum.

(c) Repeat parts (a) and (b), assuming that the density at a point is proportional to the distance from that point to the axis of the cone.

18. Let S denote the cylindrical surface with equation $x^2 + y^2 = a^2$, where $-b \le z \le b$ (a, b positive constants). Assume that the density δ is constant along S.

(a) Find the moment of inertia of S about the z-axis.

(b) Find the radius of gyration of S about the z-axis.

19. Let S be as in Exercise 18.

(a) Find the moments of inertia I_x and I_y of S about the x- and y-axes.

(b) Find the radii of gyration r_x and r_y.

20. (a) Prove the following mean value theorem for double integrals: If f and g are continuous on a compact region $D \subseteq \mathbf{R}^2$, then there is some point $P \in D$ such that

$$\iint_D fg\, dA = f(P) \iint_D g\, dA.$$

(Hint: Consider the ratio $\iint_D fg\, dA / \iint_D g\, dA$ and use the intermediate value theorem.)

(b) Use the result of part (a) to prove the following: Let S be a smooth surface oriented by unit normal $\mathbf{n}$ and let $\mathbf{F}$ be a continuous vector field on S. Assume that S may be parametrized by a single map $\mathbf{X}\colon D \to \mathbf{R}^3$. Then there is some point $P \in S$ such that

$$\iint_S \mathbf{F} \cdot d\mathbf{S} = [\mathbf{F}(P) \cdot \mathbf{n}(P)](\text{area of } S).$$

21. Let $\mathbf{a}$ be a constant vector and C a smooth, simple, closed curve. Show that $\oint_C \mathbf{a} \cdot d\mathbf{s} = 0$ in two ways:

(a) directly;

(b) by assuming that C is the boundary of a smooth surface S.

22. Evaluate $\oint_C (x^2 + z^2)\, dx + y\, dy + z\, dz$, where C is the closed curve parametrized by the path $\mathbf{x}(t) = (\cos t, \sin t, \cos^2 t - \sin^2 t)$.

23. Let f and g be functions of class C^2, and let S be a piecewise smooth, orientable surface. Show that

$$\oint_{\partial S} (f\nabla g) \cdot d\mathbf{s} = \iint_S (\nabla f \times \nabla g) \cdot d\mathbf{S}.$$

24. Let f and g be functions of class C^2, and let S be a piecewise smooth, orientable surface. Show that

$$\oint_{\partial S} (f\nabla g + g\nabla f) \cdot d\mathbf{s} = 0.$$

(Hint: Use Exercise 23.)

25. Let f be of class C^2, and let S be a piecewise smooth, orientable surface. Show that

$$\oint_{\partial S} (f\nabla f) \cdot d\mathbf{s} = 0.$$

26. Let C be a simple, closed, piecewise C^1 *planar* curve in $\mathbf{R}^3$. That is, C is contained in some plane in $\mathbf{R}^3$. Let $\mathbf{n} = a\,\mathbf{i} + b\,\mathbf{j} + c\,\mathbf{k}$ denote a unit vector normal to the plane containing C, and let C be oriented by a right-hand rule with respect to $\mathbf{n}$.

(a) Show that

$$\frac{1}{2}\oint_C (bz - cy)dx + (cx - az)dy + (ay - bx)dz$$
$$= \text{area enclosed by } C.$$

(b) Now show that the result of part (a) reduces to something familiar in the case that C is a curve in the xy-plane. (See Example 2 of §6.2.)

27. Suppose S is a piecewise smooth, orientable surface with boundary ∂S. Use Faraday's law to show that if $\mathbf{E}$

is everywhere perpendicular to ∂S, then the magnetic flux induced by $\mathbf{E}$ does not vary with time.

28. Let $\mathbf{G}$ be a vector field of class C^2. Then by Theorem 4.4 in Chapter 3, $\nabla \cdot (\nabla \times \mathbf{G}) = 0$. Therefore, Gauss's theorem implies that

$$\iint_S (\nabla \times \mathbf{G}) \cdot d\mathbf{S} = \iiint_D \nabla \cdot (\nabla \times \mathbf{G})\, dV = 0.$$

Then Stokes's theorem yields

$$0 = \iint_S (\nabla \times \mathbf{G}) \cdot d\mathbf{S} = \oint_{\partial S} \mathbf{G} \cdot d\mathbf{s}.$$

Hence, all vector fields $\mathbf{G}$ of class C^2 are conservative. How can this be?

Recall that we measure an angle in radians as follows: Place the vertex of the angle at the center O of a circle of radius a, so that the angle subtends an arc of the circle of length s. Then the measure of the angle in radians is

$$\theta = \frac{s}{a}.$$

We can do something similar in the three-dimensional case by defining a ***solid angle*** *as the set of rays beginning at the center O of a sphere of radius a, so that the rays cut out a portion of the sphere having surface area A. Then the measure of the solid angle in* ***steradians*** *is*

$$\Omega = \frac{A}{a^2}.$$

(See Figure 7.63.) Thus the solid angle of the entire sphere of radius a is 4π.

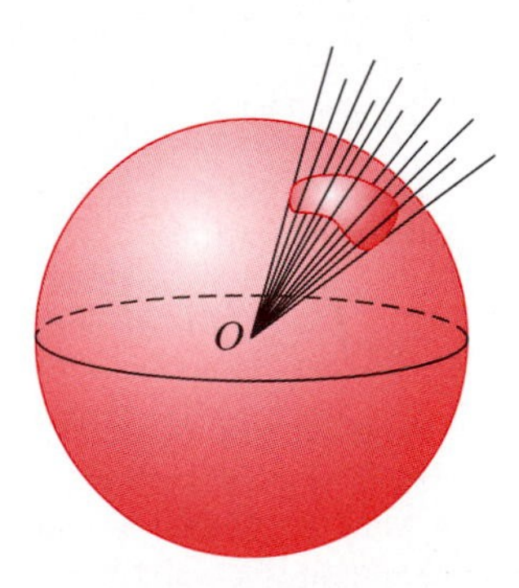

Figure 7.63 A solid angle measured on a sphere.

Now suppose that S is a smooth, oriented surface and that a point O in $\mathbf{R}^3$ is chosen that is not in S and so that every line through O intersects S at most once. In this case, we define the solid angle relative to O subtended by S as the set of rays beginning at O that pass through a point of S. We measure this solid angle by calculating

$$\Omega(S, O) = \iint_S \frac{\mathbf{x}}{\|\mathbf{x}\|^3} \cdot d\mathbf{S}, \qquad (1)$$

where $\mathbf{x}$ denotes the (varying) vector from O to a point P in S and S is oriented by a normal that points away from O.

Exercises 29–32 develop some ideas regarding solid angles.

29. In this problem we'll see how the measure of the solid angle given in equation (1) can be related to the more geometric notion of measuring solid angles using spheres. To that end, in the situation described above, construct a sphere S_a of radius a centered at O. Let $\tilde{S}_a$ denote the intersection of S_a and the solid angle of S relative to O. (See Figure 7.64.) By applying Gauss's theorem to the solid region W between S and $\tilde{S}_a$, show that

$$\Omega(S, O) = \frac{\text{surface area of } \tilde{S}_a}{a^2}.$$

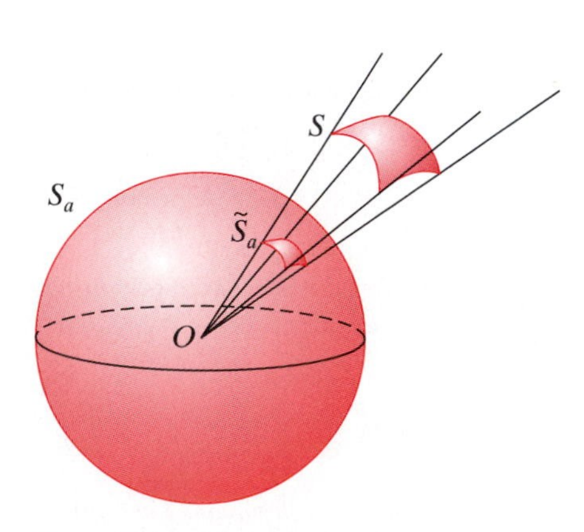

Figure 7.64 A solid angle subtended by a surface S.

30. If S is parametrized by $\mathbf{X}: D \subseteq \mathbf{R}^2 \to \mathbf{R}^3$, and O denotes the origin in $\mathbf{R}^3$, we may use equation (1) to define the measure of the solid angle subtended by S for very general surfaces, including ones that may have some self-intersections. Show that for such a parametrized surface $\Omega(S, O)$ may be calculated as

$$\Omega(S, O) = \iint_D \frac{1}{(x^2 + y^2 + z^2)^{3/2}} \times \det \begin{bmatrix} x & y & z \\ \dfrac{\partial x}{\partial s} & \dfrac{\partial y}{\partial s} & \dfrac{\partial z}{\partial s} \\ \dfrac{\partial x}{\partial t} & \dfrac{\partial y}{\partial t} & \dfrac{\partial z}{\partial t} \end{bmatrix} ds\, dt.$$

31. Suppose that S is closed, oriented, and forms the boundary of a solid, bounded, simply-connected region W in $\mathbf{R}^3$. Using equation (1) to define the measure of the solid angle, show that

$$\Omega(S, O) = \begin{cases} \pm 4\pi & \text{if } S \text{ encloses } O \\ 0 & \text{if } S \text{ does not enclose } O. \end{cases}$$

32. Suppose that S is the circular disk of radius a in the xy-plane and centered at the origin. Let O be a moving point along the z-axis and denote it by $(0, 0, z)$. Use equation (1) to show that $-2\pi \leq \Omega(S, O) \leq 2\pi$. In addition, show that $\Omega(S, O)$ jumps by 4π as O passes through S.

33. Prove the following: Let $\mathbf{F}$ be a vector field of class C^1 defined on $\mathbf{R}^3$. If $\nabla \cdot \mathbf{F} = 0$, then there is some vector field $\mathbf{G}$ of class C^1 such that $\mathbf{F} = \nabla \times \mathbf{G}$. This result provides a converse to Theorem 4.4 of Chapter 3. (Hint: Let

$$\mathbf{G}(x, y, z) = \int_0^1 t\mathbf{F}(xt, yt, zt) \times \mathbf{r}\, dt,$$

where $\mathbf{r} = x\,\mathbf{i} + y\,\mathbf{j} + z\,\mathbf{k}$. Show $\nabla \times \mathbf{G} = \mathbf{F}$. The identities

$$\nabla \times (\mathbf{A} \times \mathbf{B}) = \mathbf{A}\nabla \cdot \mathbf{B} - \mathbf{B}\nabla \cdot \mathbf{A} + (\mathbf{B} \cdot \nabla)\mathbf{A} - (\mathbf{A} \cdot \nabla)\mathbf{B},$$

$$\frac{d}{dt}[t\mathbf{F}(xt, yt, zt)] = (\mathbf{r} \cdot \nabla)\mathbf{F}(xt, yt, zt) + \mathbf{F}(xt, yt, zt),$$

and

$$\frac{d}{dt}\left[t^2\mathbf{F}(xt, yt, zt)\right] = t\left\{\frac{d}{dt}[t\mathbf{F}(xt, yt, zt)] + \mathbf{F}(xt, yt, zt)\right\}$$

will prove useful. Also, let $X = xt$, $Y = yt$, $Z = zt$ and note that, by the chain rule,

$$\frac{\partial \mathbf{F}}{\partial x} = \frac{\partial \mathbf{F}}{\partial X}\frac{\partial X}{\partial x} = t\frac{\partial \mathbf{F}}{\partial X},$$

etc.)

The vector field $\mathbf{G}$ *defined in Exercise 33 is called a* ***vector potential*** *for the vector field* $\mathbf{F}$. *In Exercises 34–36, determine a vector potential for the given vector field or explain why such a potential fails to exist.*

34. $\mathbf{F} = 2x\,\mathbf{i} - y\,\mathbf{j} - z\,\mathbf{k}$

35. $\mathbf{F} = x\,\mathbf{i} + y\,\mathbf{j} + z\,\mathbf{k}$

36. $\mathbf{F} = 3y\,\mathbf{i} + 2xz\,\mathbf{j} - 7x^2y\,\mathbf{k}$

37. The vector potential $\mathbf{G}$ identified in Exercise 33 is not unique. In fact, show that if $\mathbf{G}$ is a vector potential for $\mathbf{F}$, then so is $\mathbf{G} + \nabla\phi$, where ϕ is any scalar-valued function of class C^2. (This is known as the **gauge freedom** in choosing the vector potential.)

38. Suppose that

$$\mathbf{F} = -\frac{GMm}{\|\mathbf{r}\|^3}\mathbf{r}$$

is the gravitational force field defined on $\mathbf{R}^3 - \{(0, 0, 0)\}$.

(a) Show that $\nabla \cdot \mathbf{F} = 0$ by direct calculation.

(b) Show that $\mathbf{F} \neq \nabla \times \mathbf{G}$ for any C^1 vector field $\mathbf{G}$ defined on $\mathbf{R}^3 - \{(0, 0, 0)\}$ by using Stokes's theorem. (Hint: Take a sphere S enclosing the origin and break it up into the upper and lower hemispheres. Consider $\iint_S \mathbf{F} \cdot d\mathbf{S}$ as the sum of the surface integrals over the two hemispheres.)

(c) Why do parts (a) and (b) not contradict the result of Exercise 33?

In Exercises 39–44 below, you will derive a type of wave equation and see how Maxwell's equations can be reduced to this wave equation. Assume that the electric and magnetic fields $\mathbf{E}$ *and* $\mathbf{B}$ *are defined on a simply-connected region in* $\mathbf{R}^3$*; also let the symbol* ∇ *denote differentiation with respect to* x, y, z *(i.e., not with respect to* t*).*

39. Recall that equation (20) in §7.4 was derived by showing that the magnetic field $\mathbf{B}$ had a vector potential $\mathbf{A}$ of class C^1 (see Exercise 33 above), i.e., that $\mathbf{B} = \nabla \times \mathbf{A}$. Use Faraday's law (equation (27)) to show that the vector field $\mathbf{E} + \partial\mathbf{A}/\partial t$ is conservative. Hence we may write

$$\mathbf{E} + \frac{\partial \mathbf{A}}{\partial t} = \nabla f$$

for an appropriate scalar-valued function $f(x, y, z, t)$.

40. Use the vector potential $\mathbf{A}$ for $\mathbf{B}$ in Ampère's law (equation (24)) to conclude that

$$\nabla^2\mathbf{A} - \mu_0\epsilon_0\frac{\partial^2 \mathbf{A}}{\partial t^2} = -\mu_0\mathbf{J} + \nabla\left(\nabla \cdot \mathbf{A} - \mu_0\epsilon_0\frac{\partial f}{\partial t}\right). \tag{2}$$

(Hint: See part (a) of Exercise 15 of §7.4.)

41. Use Gauss's law (equation (12) of §7.4) to show that

$$\nabla^2 f = \frac{\rho}{\epsilon_0} + \frac{\partial}{\partial t}(\nabla \cdot \mathbf{A}). \tag{3}$$

42. As noted in Exercise 37, the vector potential $\mathbf{A}$ is only unique up to addition of $\nabla\phi$, where $\phi(x, y, z, t)$ is a scalar-valued function of class C^2. That is, any vector field $\tilde{\mathbf{A}} = \mathbf{A} + \nabla\phi$ also works as a vector potential for $\mathbf{B}$. However, the function f that arises in Exercises 39–41 will change.

(a) Show that the function $\tilde{f}$ associated to $\tilde{\mathbf{A}}$ is related to f as

$$\tilde{f} = f + \frac{\partial \phi}{\partial t}.$$

(b) Show that the condition

$$\nabla \cdot \tilde{\mathbf{A}} = \mu_0\epsilon_0\frac{\partial \tilde{f}}{\partial t},$$

where $\tilde{\mathbf{A}} = \mathbf{A} + \nabla\phi$, is equivalent to the existence of solutions to the (inhomogeneous) wave equation

$$\nabla^2\phi - \mu_0\epsilon_0\frac{\partial^2 \phi}{\partial t^2} = -\left(\nabla \cdot \mathbf{A} - \mu_0\epsilon_0\frac{\partial f}{\partial t}\right). \tag{4}$$

Given $\mathbf{A}$ and f it is possible to solve equation (4) for ϕ, so we may assume that the condition $\nabla \cdot \mathbf{A} = \mu_0 \epsilon_0 \partial f/\partial t$ holds.

43. Given the condition $\nabla \cdot \mathbf{A} = \mu_0 \epsilon_0 \partial f/\partial t$, show that equations (2) and (3) become

$$\nabla^2 \mathbf{A} - \mu_0 \epsilon_0 \frac{\partial^2 \mathbf{A}}{\partial t^2} = -\mu_0 \mathbf{J}; \tag{5}$$

$$\nabla^2 f - \mu_0 \epsilon_0 \frac{\partial^2 f}{\partial t^2} = \frac{\rho}{\epsilon_0}. \tag{6}$$

44. Conversely, suppose that $\mathbf{A}$ and f satisfy the condition $\nabla \cdot \mathbf{A} = \mu_0 \epsilon_0 \partial f/\partial t$ and equations (5) and (6). Show that then $\mathbf{E} = -\partial \mathbf{A}/\partial t + \nabla f$ and $\mathbf{B} = \nabla \times \mathbf{A}$ must satisfy Maxwell's equations. Hence solutions to (4) enable us to define a vector field $\mathbf{A}$ and a scalar field f from which we may construct $\mathbf{E}$ and $\mathbf{B}$ that satisfy Maxwell's equations.

CHAPTER 8

Vector Analysis in Higher Dimensions

Introduction

In this concluding chapter, our goal is to find a way to unify and extend the three main theorems of vector analysis (namely, the theorems of Green, Gauss, and Stokes). To accomplish such a task, we need to develop the notion of a **differential form** whose integral embraces and generalizes line, surface, and volume integrals.

8.1 An Introduction to Differential Forms

Throughout this section, U will denote an open set in $\mathbf{R}^n$, where $\mathbf{R}^n$ has coordinates $(x_1, x_2, \ldots, x_n)$, as usual. Any functions that appear are assumed to be appropriately differentiable.

Differential Forms

We begin by giving a new name to an old friend. If $f: U \subseteq \mathbf{R}^n \to \mathbf{R}$ is a scalar-valued function (of class C^k), we will also refer to f as a **differential 0-form,** or just a **0-form** for short. 0-forms can be added to one another, and multiplied together, as well we know.

The next step is to describe differential 1-forms. Ultimately, we will see that a differential 1-form is a generalization of $f(x)\,dx$—that is, of something that can be integrated with respect to a single variable, such as with a line integral. More precisely, in $\mathbf{R}^n$, the **basic differential 1-forms** are denoted $dx_1, dx_2, \ldots, dx_n$. A general **(differential) 1-form** ω is an expression that is built from the basic 1-forms as

$$\omega = F_1(x_1, \ldots, x_n)\,dx_1 + F_2(x_1, \ldots, x_n)\,dx_2 + \ldots + F_n(x_1, \ldots, x_n)\,dx_n,$$

where, for $j = 1, \ldots, n$, F_j is a scalar-valued function (of class C^k) on $U \subseteq \mathbf{R}^n$. Differential 1-forms can be added to one another, and we can multiply a 0-form f and a 1-form ω (both defined on $U \subseteq \mathbf{R}^n$) in the obvious way: If

$$\omega = F_1\, dx_1 + F_2\, dx_2 + \cdots + F_n\, dx_n,$$

then

$$f\omega = fF_1\, dx_1 + fF_2\, dx_2 + \cdots + fF_n\, dx_n.$$

EXAMPLE 1 In $\mathbf{R}^3$, let

$$\omega = xyz\, dx + z^2 \cos y\, dy + ze^x\, dz \quad \text{and} \quad \eta = (y - z)dx + z^2 \sin y\, dy - 2dz.$$

Then

$$\omega + \eta = (xyz + y - z)dx + z^2(\cos y + \sin y)dy + (ze^x - 2)dz.$$

If $f(x, y, z) = xe^y - z$, then

$$f\omega = (xe^y - z)xyz\, dx + (xe^y - z)z^2 \cos y\, dy + (xe^y - z)ze^x dz.$$ ◆

Thus far, we have described 1-forms merely as formal expressions in certain symbols. But 1-forms can also be thought of as functions. The basic 1-forms $dx_1, \ldots, dx_n$ take as argument a vector $\mathbf{a} = (a_1, a_2, \ldots, a_n)$ in $\mathbf{R}^n$; the value of dx_i on $\mathbf{a}$ is

$$dx_i(\mathbf{a}) = a_i.$$

In others words, dx_i extracts the ith component of the vector $\mathbf{a}$.

More generally, for each $\mathbf{x}_0 \in U$, the 1-form ω gives rise to a combination $\omega_{\mathbf{x}_0}$ of basic 1-forms

$$\omega_{\mathbf{x}_0} = F_1(\mathbf{x}_0)\, dx_1 + \cdots + F_n(\mathbf{x}_0)\, dx_n;$$

$\omega_{\mathbf{x}_0}$ acts on the vector $\mathbf{a} \in \mathbf{R}^n$ as

$$\omega_{\mathbf{x}_0}(\mathbf{a}) = F_1(\mathbf{x}_0)\, dx_1(\mathbf{a}) + F_2(\mathbf{x}_0)\, dx_2(\mathbf{a}) + \cdots + F_n(\mathbf{x}_0)\, dx_n(\mathbf{a}).$$

EXAMPLE 2 Suppose ω is the 1-form defined on $\mathbf{R}^3$ by

$$\omega = x^2yz\, dx + y^2z\, dy - 3xyz\, dz.$$

If $\mathbf{x}_0 = (1, -2, 5)$ and $\mathbf{a} = (a_1, a_2, a_3)$, then

$$\begin{aligned} \omega_{(1,-2,5)}(\mathbf{a}) &= -10\, dx(\mathbf{a}) + 20\, dy(\mathbf{a}) + 30\, dz(\mathbf{a}) \\ &= -10a_1 + 20a_2 + 30a_3, \end{aligned}$$

and, if $\mathbf{x}_0 = (3, 4, 6)$, then

$$\begin{aligned} \omega_{(3,4,6)}(\mathbf{a}) &= 216\, dx(\mathbf{a}) + 96\, dy(\mathbf{a}) - 216\, dz(\mathbf{a}) \\ &= 216a_1 + 96a_2 - 216a_3. \end{aligned}$$

The notation suggests that a 1-form is a function of the vector $\mathbf{a}$, but that this function "varies" from point to point. Indeed, 1-forms are actually functions on vector fields. ◆

A **basic (differential) 2-form** on $\mathbf{R}^n$ is an expression of the form

$$dx_i \wedge dx_j, \quad i, j = 1, \ldots, n.$$

It is also a function that requires *two* vector arguments $\mathbf{a}$ and $\mathbf{b}$, and we evaluate this function as

$$dx_i \wedge dx_j(\mathbf{a}, \mathbf{b}) = \begin{vmatrix} dx_i(\mathbf{a}) & dx_i(\mathbf{b}) \\ dx_j(\mathbf{a}) & dx_j(\mathbf{b}) \end{vmatrix}.$$

(The determinant represents, up to sign, the area of the parallelogram spanned by the projections of $\mathbf{a}$ and $\mathbf{b}$ in the $x_i x_j$-plane.) It is not difficult to see that, for $i, j = 1, \ldots, n$,

$$dx_i \wedge dx_j = -dx_j \wedge dx_i \tag{1}$$

and

$$dx_i \wedge dx_i = 0. \tag{2}$$

Formula (1) can be established by comparing $dx_i \wedge dx_j(\mathbf{a}, \mathbf{b})$ with $dx_j \wedge dx_i(\mathbf{a}, \mathbf{b})$. Formula (2) follows from formula (1). Given formulas (1) and (2), we see that there must be $n(n-1)/2$ linearly independent, nontrivial basic 2-forms on $\mathbf{R}^n$, namely,

$$\begin{array}{c} dx_1 \wedge dx_2,\ dx_1 \wedge dx_3, \ldots, dx_1 \wedge dx_n, \\ dx_2 \wedge dx_3, \ldots, dx_2 \wedge dx_n, \\ \vdots \\ dx_{n-1} \wedge dx_n. \end{array}$$

Let $\mathbf{x} = (x_1, x_2, \ldots, x_n)$. A general **(differential) 2-form** on $\mathbf{R}^n$ is an expression

$$\omega = F_{12}(\mathbf{x})\, dx_1 \wedge dx_2 + F_{13}(\mathbf{x})\, dx_1 \wedge dx_3 + \cdots + F_{n-1n}(\mathbf{x})\, dx_{n-1} \wedge dx_n,$$

where each F_{ij} is a real-valued function $F_{ij}\colon U \subseteq \mathbf{R}^n \to \mathbf{R}$. The idea here is to generalize something that can be integrated with respect to two variables—such as with a surface integral.

EXAMPLE 3 In $\mathbf{R}^3$, a general 2-form may be written as

$$F_1(x, y, z)\, dy \wedge dz + F_2(x, y, z)\, dz \wedge dx + F_3(x, y, z)\, dx \wedge dy.$$

The reason for using this somewhat curious ordering of the terms in the sum will, we hope, become clear later in the chapter. ◆

Given a point $\mathbf{x}_0 \in U \subseteq \mathbf{R}^n$, to evaluate a general 2-form on the ordered pair $(\mathbf{a}, \mathbf{b})$ of vectors, we have

$$\begin{aligned} \omega_{\mathbf{x}_0}(\mathbf{a}, \mathbf{b}) = {} & F_{12}(\mathbf{x}_0)\, dx_1 \wedge dx_2(\mathbf{a}, \mathbf{b}) + F_{13}(\mathbf{x}_0)\, dx_1 \wedge dx_3(\mathbf{a}, \mathbf{b}) \\ & + \cdots + F_{n-1n}(\mathbf{x}_0)\, dx_{n-1} \wedge dx_n(\mathbf{a}, \mathbf{b}). \end{aligned}$$

EXAMPLE 4 In $\mathbf{R}^3$, let $\omega = 3xy\, dy \wedge dz + (2y + z)\, dz \wedge dx + (x - z)\, dx \wedge dy$. Then

$$\begin{aligned} \omega_{(1,2,-3)}(\mathbf{a}, \mathbf{b}) &= 6\, dy \wedge dz(\mathbf{a}, \mathbf{b}) + dz \wedge dx(\mathbf{a}, \mathbf{b}) + 4\, dx \wedge dy(\mathbf{a}, \mathbf{b}) \\ &= 6\begin{vmatrix} a_2 & b_2 \\ a_3 & b_3 \end{vmatrix} + \begin{vmatrix} a_3 & b_3 \\ a_1 & b_1 \end{vmatrix} + 4\begin{vmatrix} a_1 & b_1 \\ a_2 & b_2 \end{vmatrix} \\ &= 6(a_2 b_3 - a_3 b_2) + (a_3 b_1 - a_1 b_3) + 4(a_1 b_2 - a_2 b_1). \end{aligned}$$ ◆

Finally, we generalize the notions of 1-forms and 2-forms to provide a definition of a ***k*-form.**

■ **Definition 1.1** Let k be a positive integer. A **basic (differential) *k*-form** on $\mathbf{R}^n$ is an expression of the form

$$dx_{i_1} \wedge dx_{i_2} \wedge \cdots \wedge dx_{i_k},$$

where $1 \le i_j \le n$ for $j = 1, \ldots, k$. The basic k-forms are also functions that require k vector arguments $\mathbf{a}_1, \mathbf{a}_2, \ldots, \mathbf{a}_k$ and are evaluated as

$$dx_{i_1} \wedge \cdots \wedge dx_{i_k}(\mathbf{a}_1, \ldots, \mathbf{a}_k) = \det \begin{bmatrix} dx_{i_1}(\mathbf{a}_1) & dx_{i_1}(\mathbf{a}_2) & \cdots & dx_{i_1}(\mathbf{a}_k) \\ dx_{i_2}(\mathbf{a}_1) & dx_{i_2}(\mathbf{a}_2) & \cdots & dx_{i_2}(\mathbf{a}_k) \\ \vdots & \vdots & \ddots & \vdots \\ dx_{i_k}(\mathbf{a}_1) & dx_{i_k}(\mathbf{a}_2) & \cdots & dx_{i_k}(\mathbf{a}_k) \end{bmatrix}.$$

EXAMPLE 5 Let

$$\mathbf{a}_1 = (1, 2, -1, 3, 0), \quad \mathbf{a}_2 = (5, 4, 3, 2, 1), \quad \text{and} \quad \mathbf{a}_3 = (0, 1, 3, -2, 0)$$

be three vectors in $\mathbf{R}^5$. Then we have

$$dx_1 \wedge dx_3 \wedge dx_5(\mathbf{a}_1, \mathbf{a}_2, \mathbf{a}_3) = \det \begin{bmatrix} 1 & 5 & 0 \\ -1 & 3 & 3 \\ 0 & 1 & 0 \end{bmatrix} = -3.$$

◆

Using properties of determinants, we can show that

$$dx_{i_1} \wedge \cdots \wedge dx_{i_j} \wedge \cdots \wedge dx_{i_l} \wedge \cdots \wedge dx_{i_k}$$
$$= -dx_{i_1} \wedge \cdots \wedge dx_{i_l} \wedge \cdots \wedge dx_{i_j} \wedge \cdots \wedge dx_{i_k} \qquad (3)$$

and

$$dx_{i_1} \wedge \cdots \wedge dx_{i_j} \wedge \cdots \wedge dx_{i_j} \wedge \ldots \wedge dx_{i_k} = 0. \qquad (4)$$

Formula (3) says that switching two terms (namely, dx_{i_j} and dx_{i_l}) in the basic k-form $dx_{i_1} \wedge \cdots \wedge dx_{i_k}$ causes a sign change, and formula (4) says that a basic k-form containing two identical terms is zero. Formulas (3) and (4) generalize formulas (1) and (2).

■ **Definition 1.2** A general **(differential) *k*-form** on $U \subseteq \mathbf{R}^n$ is an expression of the form

$$\omega = \sum_{i_1,\ldots,i_k=1}^{n} F_{i_1\ldots i_k}(\mathbf{x})\, dx_{i_1} \wedge \cdots \wedge dx_{i_k},$$

where each $F_{i_1\ldots i_k}$ is a real-valued function $F_{i_1\ldots i_k}\colon U \to \mathbf{R}$. Given a point $\mathbf{x}_0 \in U$, we evaluate ω on an ordered k-tuple $(\mathbf{a}_1, \ldots, \mathbf{a}_k)$ of vectors as

$$\omega_{\mathbf{x}_0}(\mathbf{a}_1, \ldots, \mathbf{a}_k) = \sum_{i_1,\ldots,i_k=1}^{n} F_{i_1\ldots i_k}(\mathbf{x}_0)\, dx_{i_1} \wedge \cdots \wedge dx_{i_k}(\mathbf{a}_1, \ldots, \mathbf{a}_k).$$

Note that a 0-form is so named because, in order to be consistent with a 1-form or 2-form, it must take zero vector arguments!

In view of formulas (3) and (4), we write a general k-form as

$$\omega = \sum_{1 \le i_1 < \cdots < i_k \le n} F_{i_1 \ldots i_k} dx_{i_1} \wedge \cdots \wedge dx_{i_k}.$$

(That is, the sum may be taken over strictly increasing indices $i_1, \ldots, i_k$.) For example, the 4-form

$$\begin{aligned}\omega = x_2\, dx_1 \wedge dx_3 \wedge dx_4 \wedge dx_5 + (x_3 - x_5^2)\, dx_1 \wedge dx_2 \wedge dx_5 \wedge dx_3 \\ + x_1 x_3\, dx_5 \wedge dx_3 \wedge dx_4 \wedge dx_1\end{aligned}$$

may be written in the "standard form" with increasing indices as

$$\omega = (x_2 - x_1 x_3)\, dx_1 \wedge dx_3 \wedge dx_4 \wedge dx_5 + (x_5^2 - x_3)\, dx_1 \wedge dx_2 \wedge dx_3 \wedge dx_5.$$

Two k-forms may be added in the obvious way, and the product of a 0-form f and a k-form ω is analogous to the product of a 0-form and a 1-form.

Exterior Product

The symbol $\wedge$ that we have been using does, in fact, denote a type of multiplication called the **exterior** (or **wedge**) **product.** The exterior product can be extended to general differential forms in the following manner:

■ **Definition 1.3** Let $U \subseteq \mathbf{R}^n$ be open. Let f denote a 0-form on U. Let $\omega = \sum F_{i_1 \ldots i_k} dx_{i_1} \wedge \cdots \wedge dx_{i_k}$ denote a k-form on U and $\eta = \sum G_{j_1 \ldots j_l} dx_{j_1} \wedge \cdots \wedge dx_{j_l}$ an l-form. Then we define

$$\begin{aligned} f \wedge \omega &= f\omega = \sum f F_{i_1 \ldots i_k} dx_{i_1} \wedge \cdots \wedge dx_{i_k}, \\ \omega \wedge \eta &= \sum F_{i_1 \ldots i_k} G_{j_1 \ldots j_l} dx_{i_1} \wedge \cdots \wedge dx_{i_k} \wedge dx_{j_1} \wedge \cdots \wedge dx_{j_l}. \end{aligned}$$

Thus, the wedge product of a k-form and an l-form is a $(k + l)$-form.

EXAMPLE 6 Let

$$\omega = x_1^2\, dx_1 \wedge dx_2 + (2x_3 - x_2)\, dx_1 \wedge dx_3 + e^{x_3}\, dx_3 \wedge dx_4$$

and

$$\eta = x_4\, dx_1 \wedge dx_3 \wedge dx_5 + x_6\, dx_2 \wedge dx_4 \wedge dx_6$$

be, respectively, a 2-form and a 3-form on $\mathbf{R}^6$. Then Definition 1.3 yields

$$\begin{aligned} \omega \wedge \eta = x_1^2 x_4\, & dx_1 \wedge dx_2 \wedge dx_1 \wedge dx_3 \wedge dx_5 \\ &+ (2x_3 - x_2)x_4\, dx_1 \wedge dx_3 \wedge dx_1 \wedge dx_3 \wedge dx_5 \\ &+ e^{x_3} x_4\, dx_3 \wedge dx_4 \wedge dx_1 \wedge dx_3 \wedge dx_5 \\ &+ x_1^2 x_6\, dx_1 \wedge dx_2 \wedge dx_2 \wedge dx_4 \wedge dx_6 \\ &+ (2x_3 - x_2)x_6\, dx_1 \wedge dx_3 \wedge dx_2 \wedge dx_4 \wedge dx_6 \\ &+ e^{x_3} x_6\, dx_3 \wedge dx_4 \wedge dx_2 \wedge dx_4 \wedge dx_6. \end{aligned}$$

Because of formula (4), most of the terms in this sum are zero. In fact,

$$\begin{aligned} \omega \wedge \eta &= (2x_3 - x_2)x_6\, dx_1 \wedge dx_3 \wedge dx_2 \wedge dx_4 \wedge dx_6 \\ &= (x_2 - 2x_3)x_6\, dx_1 \wedge dx_2 \wedge dx_3 \wedge dx_4 \wedge dx_6, \end{aligned}$$

using formula (3). ◆

From the various definitions and observations made so far, we can establish the following results, which are useful when computing with differential forms:

Proposition 1.4 (PROPERTIES OF THE EXTERIOR PRODUCT) Assume that all the differential forms that follow are defined on $U \subseteq \mathbf{R}^n$:

1. **Distributivity.** If ω_1 and ω_2 are k-forms and η is an l-form, then
$$(\omega_1 + \omega_2) \wedge \eta = \omega_1 \wedge \eta + \omega_2 \wedge \eta.$$
2. **Anticommutativity.** If ω is a k-form and η an l-form, then
$$\omega \wedge \eta = (-1)^{kl} \eta \wedge \omega.$$
3. **Associativity.** If ω is a k-form, η an l-form, and τ a p-form, then
$$(\omega \wedge \eta) \wedge \tau = \omega \wedge (\eta \wedge \tau).$$
4. **Homogeneity.** If ω is a k-form, η an l-form, and f a 0-form, then
$$(f\omega) \wedge \eta = f(\omega \wedge \eta) = \omega \wedge (f\eta).$$

Exercises

Determine the values of the following differential forms on the ordered sets of vectors indicated in Exercises 1–5.

1. $dx_1 - 3\,dx_2$; $\mathbf{a} = (7, 3)$
2. $2\,dx + 6\,dy - 5\,dz$; $\mathbf{a} = (1, -1, -2)$
3. $3\,dx_1 \wedge dx_2$; $\mathbf{a} = (4, -1)$, $\mathbf{b} = (2, 0)$
4. $4\,dx \wedge dy - 7\,dy \wedge dz$; $\mathbf{a} = (0, 1, -1)$, $\mathbf{b} = (1, 3, 2)$
5. $2\,dx_1 \wedge dx_3 \wedge dx_4 + dx_2 \wedge dx_3 \wedge dx_5$; $\mathbf{a} = (1, 0, -1, 4, 2)$, $\mathbf{b} = (0, 0, 9, 1, -1)$, $\mathbf{c} = (5, 0, 0, 0, -2)$
6. Let ω be the 1-form on $\mathbf{R}^3$ defined by
$$\omega = x^2 y\,dx + y^2 z\,dy + z^3 x\,dz.$$
Find $\omega_{(3,-1,4)}(\mathbf{a})$, where $\mathbf{a} = (a_1, a_2, a_3)$.
7. Let ω be the 2-form on $\mathbf{R}^4$ given by
$$\omega = x_1 x_3\,dx_1 \wedge dx_3 - x_2 x_4\,dx_2 \wedge dx_4.$$
Find $\omega_{(2,-1,-3,1)}(\mathbf{a}, \mathbf{b})$.
8. Let ω be the 2-form on $\mathbf{R}^3$ given by
$$\omega = \cos x\,dx \wedge dy - \sin z\,dy \wedge dz + (y^2 + 3)\,dx \wedge dz.$$
Find $\omega_{(0,-1,\pi/2)}(\mathbf{a}, \mathbf{b})$, where $\mathbf{a} = (a_1, a_2, a_3)$ and $\mathbf{b} = (b_1, b_2, b_3)$.
9. Let ω be as in Exercise 8. Find $\omega_{(x,y,z)}((2, 0, -1), (1, 7, 5))$.

In Exercises 10–13, determine $\omega \wedge \eta$.

10. On $\mathbf{R}^3$: $\omega = 3\,dx + 2\,dy - x\,dz$; $\eta = x^2 dx - \cos y\,dy + 7\,dz$.
11. On $\mathbf{R}^3$: $\omega = y\,dx - x\,dy$; $\eta = z\,dx \wedge dy + y\,dx \wedge dz + x\,dy \wedge dz$.
12. On $\mathbf{R}^4$: $\omega = 2\,dx_1 \wedge dx_2 - x_3\,dx_2 \wedge dx_4$; $\eta = 2x_4\,dx_1 \wedge dx_3 + (x_3 - x_2) dx_3 \wedge dx_4$
13. On $\mathbf{R}^5$: $\omega = x_1\,dx_2 \wedge dx_3 - x_2 x_3\,dx_1 \wedge dx_5$; $\eta = e^{x_4 x_5} dx_1 \wedge dx_4 \wedge dx_5 - x_1 \cos x_5\,dx_2 \wedge dx_3 \wedge dx_4$.
14. Prove formula (3) by evaluating $dx_{i_1} \wedge dx_{i_2} \wedge \cdots \wedge dx_{i_k}$ on k vectors $\mathbf{a}_1, \ldots, \mathbf{a}_k$ in $\mathbf{R}^n$.
15. Prove formula (4). (Hint: Use formula (3).)
16. Explain why a k-form on $\mathbf{R}^n$ with $k > n$ must be identically zero.
17. Prove property 1 of Proposition 1.4.
18. Prove property 2 of Proposition 1.4. (Hint: Use formula (3).)
19. Prove property 3 of Proposition 1.4.
20. Prove property 4 of Proposition 1.4.

8.2 Manifolds and Integrals of k-forms

In this section, we investigate how to integrate k-forms over k-dimensional objects (i.e., curves, surfaces, and higher-dimensional analogues) in $\mathbf{R}^n$.

Integrals over Curves and Surfaces

We begin by considering integrals of 1-forms and 2-forms over parametrized curves and surfaces.

Definition 2.1 Let $\mathbf{x}: [a, b] \to \mathbf{R}^n$ be a C^1 path in $\mathbf{R}^n$. If ω is a 1-form defined on an open set $U \subseteq \mathbf{R}^n$ that contains the image of $\mathbf{x}$, then the **integral of ω over $\mathbf{x}$**, denoted $\int_{\mathbf{x}} \omega$, is

$$\int_{\mathbf{x}} \omega = \int_a^b \omega_{\mathbf{x}(t)}(\mathbf{x}'(t))\, dt.$$

EXAMPLE 1 Let $\omega = (x^2 + y)\, dx + yz\, dy + (x + y - z)\, dz$. We integrate ω over the path $\mathbf{x}: [0, 1] \to \mathbf{R}^3$, $\mathbf{x}(t) = (2t + 3, 3t, 7 - t)$.

We have $\mathbf{x}'(t) = (2, 3, -1)$ so that, using Definition 2.1, we find that

$$\begin{aligned}
\int_{\mathbf{x}} \omega &= \int_0^1 \omega_{(2t+3,3t,7-t)}(2, 3, -1)\, dt \\
&= \int_0^1 [((2t + 3)^2 + 3t)\, dx(2, 3, -1) + 3t(7 - t)\, dy(2, 3, -1) \\
&\qquad + (2t + 3 + 3t - (7 - t))\, dz(2, 3, -1)]\, dt \\
&= \int_0^1 [(4t^2 + 15t + 9) \cdot 2 + (21t - 3t^2) \cdot 3 + (6t - 4) \cdot (-1)]\, dt \\
&= \int_0^1 (-t^2 + 87t + 22)\, dt = \tfrac{391}{6}.
\end{aligned}$$

◆

In general, if

$$\omega = F_1\, dx_1 + F_2\, dx_2 + \cdots + F_n\, dx_n$$

is a 1-form on $\mathbf{R}^n$ and $\mathbf{x}: [a, b] \to \mathbf{R}^n$ is any path, then

$$\begin{aligned}
\omega_{\mathbf{x}(t)}(\mathbf{x}'(t)) &= F_1(\mathbf{x}(t))\, dx_1(\mathbf{x}'(t)) + F_2(\mathbf{x}(t))\, dx_2(\mathbf{x}'(t)) \\
&\qquad + \cdots + F_n(\mathbf{x}(t))\, dx_n(\mathbf{x}'(t)) \\
&= F_1(\mathbf{x}(t))x_1'(t) + F_2(\mathbf{x}(t))x_2'(t) + \cdots + F_n(\mathbf{x}(t))x_n'(t) \\
&= (F_1(\mathbf{x}(t)), F_2(\mathbf{x}(t)), \ldots, F_n(\mathbf{x}(t))) \cdot \mathbf{x}'(t).
\end{aligned}$$

From this we conclude the following:

Proposition 2.2 If $\mathbf{F}$ denotes the vector field $(F_1, F_2, \ldots, F_n)$ and

$$\omega = F_1\, dx_1 + F_2\, dx_2 + \cdots + F_n\, dx_n$$

and if $\mathbf{x}: [a, b] \to \mathbf{R}^n$ is a C^1 path, then

$$\int_{\mathbf{x}} \omega = \int_a^b \omega_{\mathbf{x}(t)}(\mathbf{x}'(t))\, dt = \int_a^b \mathbf{F}(\mathbf{x}(t)) \cdot \mathbf{x}'(t)\, dt = \int_{\mathbf{x}} \mathbf{F} \cdot d\mathbf{s}.$$

That is, integrating a 1-form over a path (or, indeed, over a simple, piecewise C^1 curve) is exactly the same as computing a vector line integral.

Now we see how to integrate 2-forms over parametrized surfaces in $\mathbf{R}^3$.

■ **Definition 2.3** Let D be a bounded, connected region in $\mathbf{R}^2$ and let $\mathbf{X}\colon D \to \mathbf{R}^3$ be a smooth parametrized surface in $\mathbf{R}^3$. If ω is a 2-form defined on an open set in $\mathbf{R}^3$ that contains $\mathbf{X}(D)$, then we define $\int_{\mathbf{X}} \omega$, **the integral of ω over $\mathbf{X}$**, as

$$\int_{\mathbf{X}} \omega = \iint_D \omega_{\mathbf{X}(s,t)}(\mathbf{T}_s, \mathbf{T}_t)\, ds\, dt.$$

(Recall that $\mathbf{T}_s = \partial\mathbf{X}/\partial s$ and $\mathbf{T}_t = \partial\mathbf{X}/\partial t$.)

Let's work out the integral in Definition 2.3. We write ω as

$$F_1\, dy \wedge dz + F_2\, dz \wedge dx + F_3\, dx \wedge dy$$

and $\mathbf{X}(s,t)$ as $(x(s,t), y(s,t), z(s,t))$. Therefore,

$$\begin{aligned}\int_{\mathbf{X}} \omega &= \iint_D \omega_{\mathbf{X}(s,t)}(\mathbf{T}_s, \mathbf{T}_t)\, ds\, dt \\ &= \iint_D [F_1(\mathbf{X}(s,t))\, dy \wedge dz(\mathbf{T}_s, \mathbf{T}_t) + F_2(\mathbf{X}(s,t))\, dz \wedge dx(\mathbf{T}_s, \mathbf{T}_t) \\ &\qquad + F_3(\mathbf{X}(s,t))\, dx \wedge dy(\mathbf{T}_s, \mathbf{T}_t)]\, ds\, dt.\end{aligned}$$

By definition of the basic 2-forms,

$$\begin{aligned}dy \wedge dz(\mathbf{T}_s, \mathbf{T}_t) &= \det \begin{bmatrix} dy(\mathbf{T}_s) & dy(\mathbf{T}_t) \\ dz(\mathbf{T}_s) & dz(\mathbf{T}_t) \end{bmatrix} \\ &= \det \begin{bmatrix} \partial y/\partial s & \partial y/\partial t \\ \partial z/\partial s & \partial z/\partial t \end{bmatrix} = \frac{\partial(y,z)}{\partial(s,t)}.\end{aligned}$$

Similarly, we have

$$dz \wedge dx(\mathbf{T}_s, \mathbf{T}_t) = \frac{\partial(z,x)}{\partial(s,t)} \quad \text{and} \quad dx \wedge dy(\mathbf{T}_s, \mathbf{T}_t) = \frac{\partial(x,y)}{\partial(s,t)}.$$

Hence, if $\mathbf{F} = (F_1, F_2, F_3)$, then

$$\begin{aligned}\int_{\mathbf{X}} \omega &= \iint_D \left[F_1(\mathbf{X}(s,t)) \frac{\partial(y,z)}{\partial(s,t)} + F_2(\mathbf{X}(s,t)) \frac{\partial(z,x)}{\partial(s,t)} \right. \\ &\qquad \left. + F_3(\mathbf{X}(s,t)) \frac{\partial(x,y)}{\partial(s,t)} \right] ds\, dt \\ &= \iint_D \mathbf{F}(\mathbf{X}(s,t)) \cdot \left(\frac{\partial(y,z)}{\partial(s,t)}, \frac{\partial(z,x)}{\partial(s,t)}, \frac{\partial(x,y)}{\partial(s,t)} \right) ds\, dt.\end{aligned}$$

Recall from formula (7) in §7.1 that

$$\left(\frac{\partial(y,z)}{\partial(s,t)}, \frac{\partial(z,x)}{\partial(s,t)}, \frac{\partial(x,y)}{\partial(s,t)} \right) = \mathbf{N}(s,t),$$

the normal to $\mathbf{X}(D)$ at the point $\mathbf{X}(s,t)$. Therefore, we have established the following Proposition (see also Definition 2.2 of Chapter 7):

■ **Proposition 2.4** If $\mathbf{F}$ denotes the vector field $\mathbf{F} = F_1\,\mathbf{i} + F_2\,\mathbf{j} + F_3\,\mathbf{k}$ and

$$\omega = F_1\,dy \wedge dz + F_2\,dz \wedge dx + F_3\,dx \wedge dy$$

and if $\mathbf{X}\colon D \to \mathbf{R}^3$ is a smooth parametrized surface such that ω (or $\mathbf{F}$) is defined on an open set containing $\mathbf{X}(D)$, then

$$\int_{\mathbf{X}} \omega = \iint_D \omega_{\mathbf{X}(s,t)}(\mathbf{T}_s, \mathbf{T}_t)\,ds\,dt = \iint_D \mathbf{F}(\mathbf{X}_{(s,t)}) \cdot \mathbf{N}(s,t)\,ds\,dt$$
$$= \iint_{\mathbf{X}} \mathbf{F} \cdot d\mathbf{S}.$$

Parametrized Manifolds

Next, we generalize the notions of parametrized curves and surfaces to higher-dimensional objects in $\mathbf{R}^n$. To set notation, let $\mathbf{R}^k$ have coordinates $(u_1, u_2, \ldots, u_k)$.

■ **Definition 2.5** Let D be a region in $\mathbf{R}^k$ that consists of an open, connected set, possibly including some or all of its boundary points. A **parametrized k-manifold** in $\mathbf{R}^n$ is a continuous map $\mathbf{X}\colon D \to \mathbf{R}^n$ that is one-one except, possibly, along ∂D. We refer to the image $M = \mathbf{X}(D)$ as the **underlying manifold of X** (or the manifold **parametrized by X**).

Such a k-manifold possesses k **coordinate curves** defined from $\mathbf{X}$ by holding all the variables $u_1, \ldots, u_k$ fixed except one; namely, the jth coordinate curve is the curve parametrized by

$$u_j \longmapsto \mathbf{X}(a_1, \ldots, a_{j-1}, u_j, a_{j+1}, \ldots, a_k),$$

where the a_i's $(i \neq j)$ are fixed constants. If $\mathbf{X}$ is differentiable and $x_1, x_2, \ldots, x_n$ denote the component functions of $\mathbf{X}$, then the tangent vector to the jth coordinate curve, denoted $\mathbf{T}_{u_j}$, is

$$\mathbf{T}_{u_j} = \frac{\partial \mathbf{X}}{\partial u_j} = \left(\frac{\partial x_1}{\partial u_j}, \frac{\partial x_2}{\partial u_j}, \ldots, \frac{\partial x_n}{\partial u_j} \right).$$

A parametrized k-manifold is said to be **smooth** at a point $\mathbf{X}(\mathbf{u}_0)$ if the mapping $\mathbf{X}$ is of class C^1 in a neighborhood of $\mathbf{u}_0$ and if the k tangent vectors $\mathbf{T}_{u_1}, \ldots, \mathbf{T}_{u_k}$ are linearly independent at $\mathbf{X}(\mathbf{u}_0)$. (Recall that k vectors $\mathbf{v}_1, \ldots, \mathbf{v}_k$ in $\mathbf{R}^n$ are **linearly independent** if the equation $c_1\mathbf{v}_1 + \cdots + c_k\mathbf{v}_k = \mathbf{0}$ holds if and only if $c_1 = c_2 = \cdots = c_k = 0$.) A parametrized k-manifold is said to be **smooth** if it is smooth at every point of $\mathbf{X}(\mathbf{u}_0)$ with $\mathbf{u}_0$ in the interior of D.

Sometimes we will refer to the underlying manifold $M = \mathbf{X}(D)$ of a parametrized manifold $\mathbf{X}\colon D \to \mathbf{R}^n$ as a parametrized manifold; we do not expect any confusion will result from this abuse of terminology.

EXAMPLE 2 Let $D = [0,1] \times [1,2] \times [-1,1]$ and $\mathbf{X}\colon D \to \mathbf{R}^5$ be given by

$$\mathbf{X}(u_1, u_2, u_3) = (u_1 + u_2, 3u_2, u_2u_3^2, u_2 - u_3, 5u_3).$$

We show that $M = \mathbf{X}(D)$ is a smooth parametrized 3-manifold in $\mathbf{R}^5$.

Note first that **X** is continuous (in fact, of class C^∞) since its component functions are polynomials. To see that **X** is one-one, consider the equation

$$\mathbf{X}(\mathbf{u}) = \mathbf{X}(\tilde{\mathbf{u}}); \tag{1}$$

we show that $\mathbf{u} = \tilde{\mathbf{u}}$. Equation (1) is equivalent to a system of five equations:

$$\begin{cases} u_1 + u_2 = \tilde{u}_1 + \tilde{u}_2 \\ 3u_2 = 3\tilde{u}_2 \\ u_2 u_3^2 = \tilde{u}_2 \tilde{u}_3^2 \\ u_2 - u_3 = \tilde{u}_2 - \tilde{u}_3 \\ 5u_3 = 5\tilde{u}_3 \end{cases}.$$

The second equation implies $u_2 = \tilde{u}_2$, and the last equation implies $u_3 = \tilde{u}_3$. Hence, the first equation becomes

$$u_1 + u_2 = \tilde{u}_1 + u_2 \quad \Longleftrightarrow \quad u_1 = \tilde{u}_1.$$

Thus,

$$\mathbf{u} = (u_1, u_2, u_3) = (\tilde{u}_1, \tilde{u}_2, \tilde{u}_3) = \tilde{\mathbf{u}}.$$

To check the smoothness of M, note that the tangent vectors to the three coordinate curves are

$$\mathbf{T}_{u_1} = \frac{\partial \mathbf{X}}{\partial u_1} = (1, 0, 0, 0, 0);$$

$$\mathbf{T}_{u_2} = \frac{\partial \mathbf{X}}{\partial u_2} = (1, 3, u_3^2, 1, 0);$$

$$\mathbf{T}_{u_3} = \frac{\partial \mathbf{X}}{\partial u_3} = (0, 0, 2u_2u_3, -1, 5).$$

Therefore, to have $c_1\mathbf{T}_1 + c_2\mathbf{T}_2 + c_3\mathbf{T}_3 = \mathbf{0}$, we must have

$$(c_1 + c_2, 3c_2, u_3^2 c_2 + 2u_2u_3c_3, c_2 - c_3, 5c_3) = (0, 0, 0, 0, 0).$$

It readily follows that $c_1 = c_2 = c_3 = 0$ is the only possibility for a solution. Hence $\mathbf{T}_{u_1}, \mathbf{T}_{u_2}, \mathbf{T}_{u_3}$ are linearly independent at all $\mathbf{u} \in D$ and so M is smooth at all points. ◆

Parametrized k-manifolds, although seemingly abstract mathematical notions when k is larger than 3, are actually very useful for describing a variety of situations, one of which is illustrated in the next example.

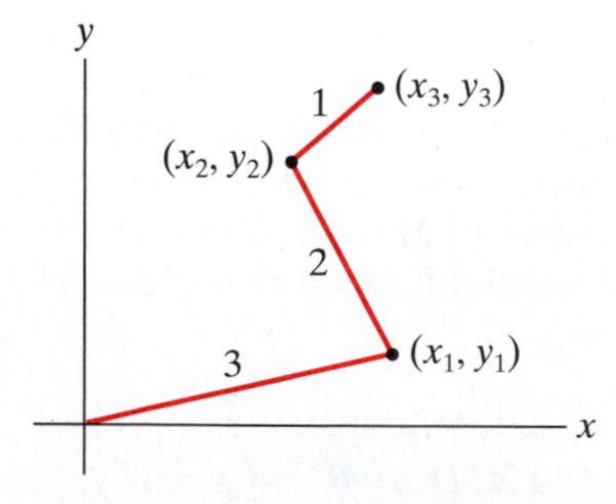

Figure 8.1 The planar robot arm of Example 3. Each rod is free to pivot about the appropriate linkage points.

EXAMPLE 3 A planar robot arm is constructed consisting of three linked rods of lengths 1, 2, and 3. (See Figure 8.1.) The rod of length 3 is anchored at the origin of $\mathbf{R}^2$, but free to rotate about the origin. The rod of length 2 is attached to the free end of the rod of length 3, and the rod of length 1 is, in turn, attached to the free end of the rod of length 2. We describe the set of positions that the arm can take as a parametrized manifold.

Clearly, each state of the robot arm is determined by the coordinates (x_1, y_1), (x_2, y_2), and (x_3, y_3) of the linkage points, which we may consider to form a vector $\mathbf{x} = (x_1, y_1, x_2, y_2, x_3, y_3)$ in $\mathbf{R}^6$. However, not all vectors in $\mathbf{R}^6$ represent a state of the robot arm. In particular, the point (x_1, y_1) must lie on the circle of radius 3, centered at the origin, the point (x_2, y_2) must lie on the circle of radius 2, centered at (x_1, y_1), and the point (x_3, y_3) must lie on the circle of radius 1, centered at (x_2, y_2). Thus, for $\mathbf{x} = (x_1, y_1, x_2, y_2, x_3, y_3)$ to represent a state of the robot arm, we require

$$\begin{cases} x_1^2 + y_1^2 = 9 \\ (x_2 - x_1)^2 + (y_2 - y_1)^2 = 4. \\ (x_3 - x_2)^2 + (y_3 - y_2)^2 = 1 \end{cases} \tag{2}$$

We may parametrize each of the circles in the system (2) in a one-one fashion by using three different angles θ_1, θ_2, and θ_3. Hence, we find

$$\begin{aligned} (x_1, y_1) &= (3\cos\theta_1, 3\sin\theta_1), \\ (x_2, y_2) &= (x_1 + 2\cos\theta_2, y_1 + 2\sin\theta_2) \\ &= (3\cos\theta_1 + 2\cos\theta_2, 3\sin\theta_1 + 2\sin\theta_2), \end{aligned} \tag{3}$$

and

$$\begin{aligned} (x_3, y_3) &= (x_2 + \cos\theta_3, y_2 + \sin\theta_3) \\ &= (3\cos\theta_1 + 2\cos\theta_2 + \cos\theta_3, 3\sin\theta_1 + 2\sin\theta_2 + \sin\theta_3), \end{aligned}$$

where $0 \leq \theta_1, \theta_2, \theta_3 < 2\pi$. Therefore, the map $\mathbf{X}\colon [0, 2\pi) \times [0, 2\pi) \times [0, 2\pi) \to \mathbf{R}^6$ given by

$$\mathbf{X}(\theta_1, \theta_2, \theta_3) = (x_1, y_1, x_2, y_2, x_3, y_3),$$

where $(x_1, y_1, x_2, y_2, x_3, y_3)$ are given in terms of θ_1, θ_2, and θ_3 by means of the equations in (3), exhibits the set of states of the robot arm as a parametrized 3-manifold in $\mathbf{R}^6$. We leave it to you to check that $\mathbf{X}$ defines a smooth parametrized 3-manifold. ◆

Just like a parametrized surface, a parametrized k-manifold $M = \mathbf{X}(D)$ may or may not have a **boundary,** denoted, as usual, by ∂M. If M has a nonempty boundary, then ∂M is contained in the image under $\mathbf{X}$ of the portion of the boundary of the domain region D that is also part of D. Under suitable (and mild) hypotheses, ∂M, if nonempty, is, in turn, a union of finitely many $(k-1)$-manifolds (*without* boundaries).

EXAMPLE 4 Let $B \subset \mathbf{R}^3$ denote the closed unit ball $\{\mathbf{u} = (u_1, u_2, u_3) \mid u_1^2 + u_2^2 + u_3^2 \leq 1\}$, and define $\mathbf{X}\colon B \to \mathbf{R}^4$ by

$$\mathbf{X}(u_1, u_2, u_3) = (u_1, u_2, u_3, u_1^2 + u_2^2 + u_3^2).$$

Then $M = \mathbf{X}(B)$ is a portion of a "generalized paraboloid" having equation $w = x^2 + y^2 + z^2$; we have $M = \{(x, y, z, w) \in \mathbf{R}^4 \mid w = x^2 + y^2 + z^2,\ x^2 + y^2 + z^2 \leq 1\}$. In this case, $\partial M = \{(x, y, z, 1) \mid x^2 + y^2 + z^2 = 1\}$. Note that ∂M is a parametrized 2-manifold in $\mathbf{R}^4$, as we may see via the map

$$\mathbf{Y}\colon [0, \pi] \times [0, 2\pi) \to \mathbf{R}^4, \quad \mathbf{Y}(s, t) = (\sin s \cos t, \sin s \sin t, \cos s, 1).$$ ◆

Integrals over Parametrized k-manifolds

Now, we see how to define the integral of a k-form over a smooth parametrized k-manifold. Our definition generalizes those of Definitions 2.1 and 2.3.

■ **Definition 2.6** Let D be a bounded, connected region in $\mathbf{R}^k$ and $\mathbf{X}\colon D \to \mathbf{R}^n$ a smooth parametrized k-manifold. If ω is a k-form defined on an open set in $\mathbf{R}^n$ that contains $M = \mathbf{X}(D)$, then we define the **integral** of ω over M (denoted $\int_{\mathbf{X}} \omega$) by

$$\int_{\mathbf{X}} \omega = \int \cdots \int_D \omega_{\mathbf{X}(\mathbf{u})}(\mathbf{T}_{u_1}, \ldots, \mathbf{T}_{u_k})\, du_1 \cdots du_k.$$

(Here $\int \cdots \int$ refers to the k-dimensional integral over D.)

EXAMPLE 5 Let $\mathbf{X}\colon [0,1] \times [1,2] \times [-1,1] \to \mathbf{R}^5$ be the parametrized 3-manifold defined by

$$\mathbf{X}(u_1, u_2, u_3) = (u_1 + u_2, 3u_2, u_2 u_3^2, u_2 - u_3, 5u_3).$$

(See Example 2.) Let ω be the 3-form defined on $\mathbf{R}^5$ as

$$\omega = x_1 x_3\, dx_1 \wedge dx_3 \wedge dx_5 + (x_3 x_4 - 2x_2 x_5)\, dx_2 \wedge dx_4 \wedge dx_5.$$

We calculate $\int_{\mathbf{X}} \omega$.

Recall from Example 2 that the tangent vectors to the three coordinate curves are

$$\mathbf{T}_{u_1} = (1, 0, 0, 0, 0),$$
$$\mathbf{T}_{u_2} = (1, 3, u_3^2, 1, 0),$$

and

$$\mathbf{T}_{u_3} = (0, 0, 2u_2 u_3, -1, 5).$$

Then, from Definition 2.6,

$$\begin{aligned}
&\int_{\mathbf{X}} \omega \\
&= \int_{-1}^{1}\int_{1}^{2}\int_{0}^{1} \{(u_1+u_2)u_2u_3^2\, dx_1 \wedge dx_3 \wedge dx_5(\mathbf{T}_{u_1}, \mathbf{T}_{u_2}, \mathbf{T}_{u_3}) \\
&\qquad + (u_2u_3^2(u_2 - u_3) - 30u_2u_3)\, dx_2 \wedge dx_4 \wedge dx_5(\mathbf{T}_{u_1}, \mathbf{T}_{u_2}, \mathbf{T}_{u_3})\}\, du_1\, du_2\, du_3 \\
&= \int_{-1}^{1}\int_{1}^{2}\int_{0}^{1} \left\{ (u_1+u_2)u_2u_3^2 \begin{vmatrix} 1 & 1 & 0 \\ 0 & u_3^2 & 2u_2u_3 \\ 0 & 0 & 5 \end{vmatrix} \right. \\
&\qquad \left. + (u_2^2u_3^2 - u_2u_3^3 - 30u_2u_3) \begin{vmatrix} 0 & 3 & 0 \\ 0 & 1 & -1 \\ 0 & 0 & 5 \end{vmatrix} \right\} du_1\, du_2\, du_3 \\
&= \int_{-1}^{1}\int_{1}^{2}\int_{0}^{1} 5(u_1+u_2)u_2u_3^4\, du_1\, du_2\, du_3 = \tfrac{37}{6}.
\end{aligned}$$

◆

EXAMPLE 6 If ω is a 3-form on $\mathbf{R}^3$, then ω may be written as

$$\omega = F(x, y, z)\, dx \wedge dy \wedge dz.$$

(Why?) If D^* is a bounded region in $\mathbf{R}^3$ and $\mathbf{X}\colon D^* \to \mathbf{R}^3$ is a smooth parametrized 3-manifold, then Definition 2.6 tells us that

$$\begin{aligned}
\int_{\mathbf{X}} \omega &= \iiint_{D^*} \omega_{\mathbf{X}(u_1,u_2,u_3)}(\mathbf{T}_{u_1}, \mathbf{T}_{u_2}, \mathbf{T}_{u_3})\, du_1\, du_2\, du_3 \\
&= \iiint_{D^*} F(\mathbf{X}(\mathbf{u}))\, dx \wedge dy \wedge dz(\mathbf{T}_{u_1}, \mathbf{T}_{u_2}, \mathbf{T}_{u_3})\, du_1\, du_2\, du_3 \\
&= \iiint_{D^*} F(\mathbf{X}(\mathbf{u})) \begin{vmatrix} \partial x/\partial u_1 & \partial x/\partial u_2 & \partial x/\partial u_3 \\ \partial y/\partial u_1 & \partial y/\partial u_2 & \partial y/\partial u_3 \\ \partial z/\partial u_1 & \partial z/\partial u_2 & \partial z/\partial u_3 \end{vmatrix} du_1\, du_2\, du_3 \\
&= \iiint_{D^*} F(x(\mathbf{u}), y(\mathbf{u}), z(\mathbf{u})) \frac{\partial(x, y, z)}{\partial(u_1, u_2, u_3)}\, du_1\, du_2\, du_3 \\
&= \pm \iiint_{D} F(x, y, z)\, dx\, dy\, dz,
\end{aligned}$$

from the change of variables theorem for triple integrals (Theorem 5.5 of Chapter 5), where $D = \mathbf{X}(D^*)$. ◆

Orientation of a Parametrized k-manifold

We have seen that vector line integrals and vector surface integrals may be defined, respectively, over oriented curves and surfaces in a manner effectively independent of the parametrization used. We now see how it is possible to define the integral of a k-form over a parametrized k-manifold $\mathbf{X}\colon D \to \mathbf{R}^n$ so that it depends largely on the underlying manifold $M = \mathbf{X}(D)$, rather than on the particular map $\mathbf{X}$. To do this, we must consider how **reparametrization** of M affects the integral, and we must define what we mean by an **orientation** of M.

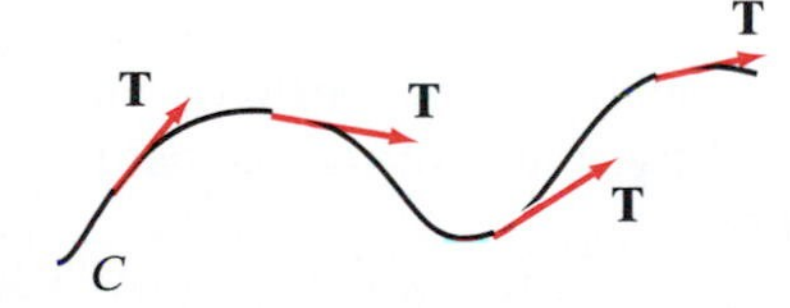

Figure 8.2 An orientation of the curve C shown is a choice of continuously varying unit tangent vector $\mathbf{T}$ along C.

First, we consider the notion of orientation. We have previously seen how parametrized curves and surfaces can be oriented by using some fairly natural geometric ideas. A smooth parametrized curve implicitly received an orientation from the parameter; typically, we orient a curve by indicating the direction in which the parameter variable increases. We may also think of an **orientation** of a curve as a choice of a unit tangent vector $\mathbf{T}$ at each point of the curve, made so that $\mathbf{T}$ varies continuously as we move along the curve. (See Figure 8.2.) An **orientation** of a smooth parametrized surface in $\mathbf{R}^3$, when it exists, is a choice of a continuously varying unit normal vector $\mathbf{N}$ at each point of the surface. (See Figure 8.3.)

To define notions of **orientation** and **orientability** for a parametrized k-manifold when $k > 2$, we will need to work more formally.

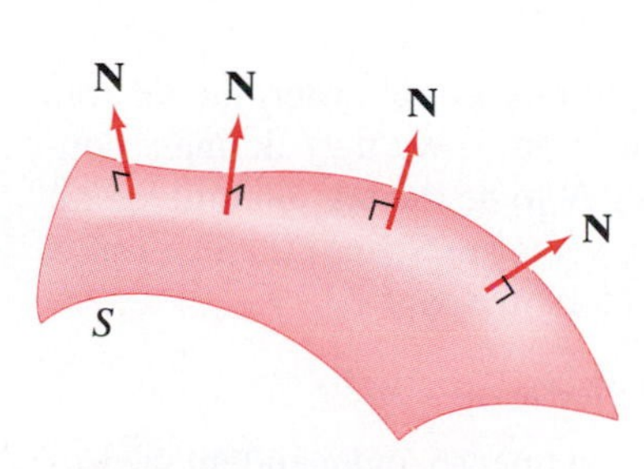

Figure 8.3 An orientation of the surface S is a choice of continuously varying unit normal vector $\mathbf{N}$ along S.

First, we need to introduce two related ideas from the linear algebra of $\mathbf{R}^n$. Thus, suppose $\mathbf{v}_1, \mathbf{v}_2, \ldots, \mathbf{v}_k$ are vectors in $\mathbf{R}^n$. By a **linear combination** of $\mathbf{v}_1, \ldots, \mathbf{v}_k$, we mean any vector $\mathbf{v} \in \mathbf{R}^n$ that can be written as

$$\mathbf{v} = c_1\mathbf{v}_1 + c_2\mathbf{v}_2 + \cdots + c_k\mathbf{v}_k$$

for suitable choices of the scalars $c_1, \ldots, c_k$. The set of all possible linear combinations of $\mathbf{v}_1, \ldots, \mathbf{v}_k$, called the **(linear) span** of $\mathbf{v}_1, \ldots, \mathbf{v}_k$, will be denoted $\operatorname{Span}\{\mathbf{v}_1, \ldots, \mathbf{v}_k\}$. That is,

$$\operatorname{Span}\{\mathbf{v}_1, \ldots, \mathbf{v}_k\} = \{c_1\mathbf{v}_1 + \cdots + c_k\mathbf{v}_k \mid c_1, \ldots, c_k \in \mathbf{R}\}.$$

Definition 2.7 Let $M = \mathbf{X}(D)$, where $\mathbf{X}: D \subseteq \mathbf{R}^k \to \mathbf{R}^n$, be a smooth parametrized k-manifold. An **orientation** of M is a choice of a smooth, nonzero k-form Ω defined on M. If such a k-form Ω exists, M is said to be **orientable** and **oriented** once a choice of such a k-form is made.

Although we cannot readily visualize an orientation Ω of a parametrized k-manifold when k is large, we can nonetheless see how the tangent vectors to the coordinate curves relate to it.

Definition 2.8 Let $M = \mathbf{X}(D)$ be a smooth parametrized k-manifold oriented by the k-form Ω. The tangent vectors $\mathbf{T}_{u_1}, \ldots, \mathbf{T}_{u_k}$ to the coordinate curves of M are said to be **compatible** with Ω if

$$\Omega_{\mathbf{X}(\mathbf{u})}(\mathbf{T}_{u_1}, \ldots, \mathbf{T}_{u_k}) > 0.$$

We also say that the parametrization $\mathbf{X}$ is **compatible** with the orientation Ω if the corresponding tangent vectors $\mathbf{T}_{u_1}, \ldots, \mathbf{T}_{u_k}$ are.

Note that if $\mathbf{T}_{u_1}, \ldots, \mathbf{T}_{u_k}$ are incompatible with the orientation Ω, then they are compatible with the *opposite* orientation $-\Omega$. Alternatively, we may change the parametrization $\mathbf{X}$ of M by reordering the variables $u_1, \ldots, u_k$ to, say, $u_2, u_1, u_3, \ldots, u_k$, so that $\mathbf{T}_{u_2}, \mathbf{T}_{u_1}, \mathbf{T}_{u_3}, \ldots, \mathbf{T}_{u_k}$ are compatible with Ω.

Definition 2.7 is consistent with the earlier definitions of orientations of curves and surfaces, as we now discuss. Suppose first that $\mathbf{x}: I \to \mathbf{R}^n$ is a smooth parametrized curve in $\mathbf{R}^n$ (where I is an interval in $\mathbf{R}$) and $\mathbf{T}$ is a continuously varying choice of unit tangent vector along $C = \mathbf{x}(I)$. Then we may define an orientation 1-form Ω on C by

$$\Omega_{\mathbf{x}(t)}(\mathbf{a}) = \mathbf{T} \cdot \mathbf{a}.$$

Conversely, given an orientation 1-form Ω, we may define a continuously varying unit tangent vector from it by taking $\mathbf{T}$ to be the unique unit vector parallel to $\mathbf{x}'(t)$ such that, for any nonzero vector $\mathbf{a}$ parallel to $\mathbf{x}'(t)$,

$$\mathbf{T} \cdot \mathbf{a} \text{ has the same sign as } \Omega_{\mathbf{x}(t)}(\mathbf{a}).$$

That $\mathbf{T}$ is uniquely determined follows because $\mathbf{T}$ must equal $\pm\mathbf{x}'(t)/\|\mathbf{x}'(t)\|$, so knowing $\mathbf{a}$ and the value of $\Omega_{\mathbf{x}(t)}(\mathbf{a})$ determines the choice of sign for $\mathbf{T}$.

Similarly, suppose $S = \mathbf{X}(D)$ is a smooth parametrized surface in $\mathbf{R}^3$ (i.e., a smooth parametrized 2-manifold). If we can orient S by a continuously varying unit normal $\mathbf{N}$, then we may define an orientation 2-form Ω on S by

$$\Omega_{\mathbf{X}(u_1,u_2)}(\mathbf{a}, \mathbf{b}) = \det \begin{bmatrix} \mathbf{N} & \mathbf{a} & \mathbf{b} \end{bmatrix},$$

where $\begin{bmatrix} \mathbf{N} & \mathbf{a} & \mathbf{b} \end{bmatrix}$ is the 3×3 matrix whose columns are, in order, the vectors $\mathbf{N}$, $\mathbf{a}$, $\mathbf{b}$. Conversely, given an orientation 2-form Ω on S, we may define a continuously varying unit normal $\mathbf{N}$ from it by taking $\mathbf{N}$ to be the *unique* unit vector perpendicular to $\mathbf{T}_{u_1}$ and $\mathbf{T}_{u_2}$ (and hence to every vector in $\operatorname{Span}\{\mathbf{T}_{u_1}, \mathbf{T}_{u_2}\}$) such that, for any pair $\mathbf{a}$, $\mathbf{b}$ of linearly independent vectors in $\operatorname{Span}\{\mathbf{T}_{u_1}, \mathbf{T}_{u_2}\}$,

$$\det \begin{bmatrix} \mathbf{N} & \mathbf{a} & \mathbf{b} \end{bmatrix} \text{ has the same sign as } \Omega_{\mathbf{X}(u_1,u_2)}(\mathbf{a}, \mathbf{b}).$$

To see that $\mathbf{N}$ is uniquely determined, note that, given linearly independent vectors $\mathbf{a}$, $\mathbf{b}$ in $\operatorname{Span}\{\mathbf{T}_{u_1}, \mathbf{T}_{u_2}\}$, the only possibilities for $\mathbf{N}$ are

$$\pm\frac{\mathbf{T}_{u_1} \times \mathbf{T}_{u_2}}{\|\mathbf{T}_{u_1} \times \mathbf{T}_{u_2}\|}.$$

Hence, we choose the sign for the normal vector $\mathbf{N}$ so that $\det\begin{bmatrix} \mathbf{N} & \mathbf{a} & \mathbf{b} \end{bmatrix}$ has the same sign as $\Omega_{\mathbf{X}(u_1,u_2)}(\mathbf{a}, \mathbf{b})$.

EXAMPLE 7 Consider the generalized paraboloid $M = \{(x, y, z, w) \in \mathbf{R}^4 \mid w = x^2 + y^2 + z^2\}$, which we may exhibit as a smooth parametrized 3-manifold via

$$\mathbf{X}\colon \mathbf{R}^3 \to \mathbf{R}^4, \quad \mathbf{X}(u_1, u_2, u_3) = (u_1, u_2, u_3, u_1^2 + u_2^2 + u_3^2).$$

We show how to orient M.

Note that the equation $x^2 + y^2 + z^2 - w = 0$ shows that M is the level set at height 0 of the function $F(x, y, z, w) = x^2 + y^2 + z^2 - w$. Hence, the gradient $\nabla F = (2x, 2y, 2z, -1)$ is a vector normal to M. If we employ the parametrization $\mathbf{X}$ and normalize the (parametrized) gradient, we see that

$$\mathbf{N}(u_1, u_2, u_3) = \frac{(2u_1, 2u_2, 2u_3, -1)}{\sqrt{4u_1^2 + 4u_2^2 + 4u_3^2 + 1}}$$

is a continuously varying unit normal. Moreover, the 2-form Ω defined on M as

$$\Omega_{\mathbf{X}(\mathbf{u})}(\mathbf{a}_1, \mathbf{a}_2, \mathbf{a}_3) = \det\begin{bmatrix} \mathbf{N} & \mathbf{a}_1 & \mathbf{a}_2 & \mathbf{a}_3 \end{bmatrix}$$

gives an orientation for M. Note that

$$\begin{aligned}
\Omega_{\mathbf{X}(\mathbf{u})}(\mathbf{T}_{u_1}, \mathbf{T}_{u_2}, \mathbf{T}_{u_3}) &= \det\begin{bmatrix} \mathbf{N} & \mathbf{T}_{u_1} & \mathbf{T}_{u_2} & \mathbf{T}_{u_3} \end{bmatrix} \\
&= \det\begin{bmatrix}
\dfrac{2u_1}{\sqrt{4u_1^2 + 4u_2^2 + 4u_3^2 + 1}} & 1 & 0 & 0 \\
\dfrac{2u_2}{\sqrt{4u_1^2 + 4u_2^2 + 4u_3^2 + 1}} & 0 & 1 & 0 \\
\dfrac{2u_3}{\sqrt{4u_1^2 + 4u_2^2 + 4u_3^2 + 1}} & 0 & 0 & 1 \\
\dfrac{-1}{\sqrt{4u_1^2 + 4u_2^2 + 4u_3^2 + 1}} & 2u_1 & 2u_2 & 2u_3
\end{bmatrix} \\
&= \sqrt{4u_1^2 + 4u_2^2 + 4u_3^2 + 1}.
\end{aligned}$$

Since this last expression is strictly positive, we see that $\mathbf{T}_{u_1}, \mathbf{T}_{u_2}, \mathbf{T}_{u_3}$ are compatible with Ω. ◆

EXAMPLE 8 We may generalize Example 7 as follows:

Suppose that $M \subseteq \mathbf{R}^n$ is the graph of a function $f\colon U \subseteq \mathbf{R}^{n-1} \to \mathbf{R}^n$, that is, suppose M is defined by the equation $x_n = f(x_1, \ldots, x_{n-1})$. Then M may be parametrized as an $(n-1)$-manifold via

$$\mathbf{X}\colon U \subseteq \mathbf{R}^{n-1} \to \mathbf{R}^n, \quad \mathbf{X}(u_1, \ldots, u_{n-1}) = (u_1, \ldots, u_{n-1}, f(u_1, \ldots, u_{n-1})).$$

Since M is also the level set at height 0 of the function

$$F(x_1, \ldots, x_n) = f(x_1, \ldots, x_{n-1}) - x_n,$$

a vector normal to M is provided by the gradient $\nabla F = (f_{x_1}, \ldots, f_{x_{n-1}}, -1)$. If we normalize ∇F and use the parametrization $\mathbf{X}$, we see that we have a continuously varying unit normal

$$\mathbf{N}(u_1, \ldots, u_{n-1}) = \frac{(f_{u_1}, \ldots, f_{u_{n-1}}, -1)}{\sqrt{f_{u_1}^2 + \cdots + f_{u_{n-1}}^2 + 1}},$$

from which we may define our orientation $(n-1)$-form Ω for M by

$$\Omega_{\mathbf{X}(\mathbf{u})}(\mathbf{a}_1, \ldots, \mathbf{a}_{n-1}) = \det\begin{bmatrix}\mathbf{N} & \mathbf{a}_1 & \cdots & \mathbf{a}_{n-1}\end{bmatrix}. \quad ◆$$

Now suppose that M is a smooth parametrized k-manifold in $\mathbf{R}^n$ with nonempty boundary ∂M. If M is oriented by the k-form Ω, then there is a way to derive from it an orientation for ∂M, which we describe in Definition 2.9. To set notation, let $\mathbf{X}\colon D \subseteq \mathbf{R}^k \to \mathbf{R}^n$ denote the parametrization of M and suppose $\mathbf{Y}\colon E \subseteq \mathbf{R}^{k-1} \to \mathbf{R}^n$ gives a parametrization of a connected piece of ∂M as a smooth $(k-1)$-manifold. Since ∂M is part of M, if $\mathbf{s} = (s_1, \ldots, s_{k-1}) \in E$, then there is some $\mathbf{u} = (u_1, \ldots, u_k) \in D$ such that $\mathbf{Y}(\mathbf{s}) = \mathbf{X}(\mathbf{u})$.

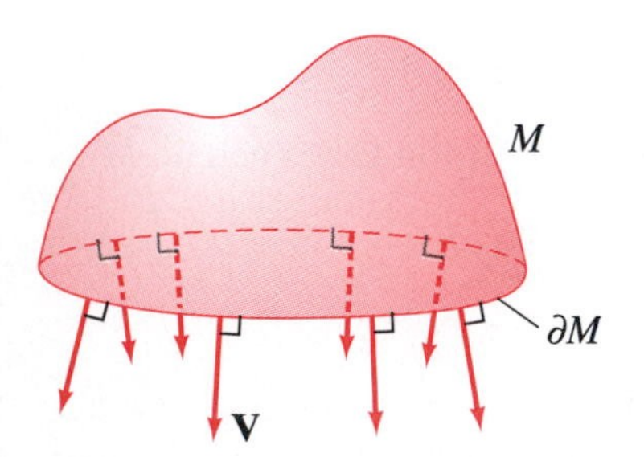

Figure 8.4 The outward-pointing unit vector **V** of Definition 2.9.

■ **Definition 2.9** Let M be a smooth parametrized k-manifold in $\mathbf{R}^n$ with boundary ∂M. Suppose M is oriented by the k-form Ω. Then the connected pieces of ∂M are said to be **oriented consistently** with M, or that ∂M has its **orientation induced** from that of M, if the orientation $(k-1)$-form $\Omega^{\partial M}$ is determined from Ω as follows. Let $\mathbf{V}$ be the unique, outward-pointing unit vector in $\mathbf{R}^n$, defined and varying continuously along ∂M, that is tangent to M and normal to ∂M. (See Figure 8.4.) Then $\Omega^{\partial M}$ is defined as

$$\Omega^{\partial M}_{\mathbf{Y}(\mathbf{s})}(\mathbf{a}_1, \ldots, \mathbf{a}_{k-1}) = \Omega_{\mathbf{X}(\mathbf{u})}(\mathbf{V}, \mathbf{a}_1, \ldots, \mathbf{a}_{k-1}),$$

where the map $\mathbf{X}\colon D \subseteq \mathbf{R}^k \to \mathbf{R}^n$ parametrizes M, the map $\mathbf{Y}\colon E \subseteq \mathbf{R}^{k-1} \to \mathbf{R}^n$ parametrizes a connected piece of ∂M, and $\mathbf{Y}(\mathbf{s}) = \mathbf{X}(\mathbf{u})$.

Note that, in particular, the vector $\mathbf{V}$ in Definition 2.9 must be such that

- $\mathbf{V} \in \operatorname{Span}\{\mathbf{T}_{u_1}, \ldots, \mathbf{T}_{u_k}\}$ (i.e., $\mathbf{V}$ is tangent to M);
- $\mathbf{V} \cdot \mathbf{T}_{s_i} = 0$ for $i = 1, \ldots, k-1$ (i.e., $\mathbf{V}$ is normal to ∂M);
- $\mathbf{V}$ points away from M.

These conditions are often not difficult to achieve in practice. Definition 2.9 will be very important when we consider a generalization of Stokes's theorem in the next section.

EXAMPLE 9 Consider the surface S in $\mathbf{R}^3$ consisting of the portion of the cylinder $x^2 + y^2 = 4$ with $2 \le z \le 5$. Note that the boundary of S consists of the two circles $\{(x, y, z) \mid x^2 + y^2 = 4,\ z = 2\}$ and $\{(x, y, z) \mid x^2 + y^2 = 4,\ z = 5\}$. We investigate how to orient ∂S consistently with an orientation of S.

The cylinder may be parametrized as a 2-manifold in $\mathbf{R}^3$ by

$$\mathbf{X}\colon [0, 2\pi) \times [2, 5] \to \mathbf{R}^3, \qquad \mathbf{X}(u_1, u_2) = (2\cos u_1, 2\sin u_1, u_2).$$

Then the tangent vectors to the coordinate curves are

$$\mathbf{T}_{u_1} = (-2\sin u_1, 2\cos u_1, 0)$$

and

$$\mathbf{T}_{u_2} = (0, 0, 1).$$

Since S is a portion of the level set at height 4 of the function $F(x, y, z) = x^2+y^2$, a unit normal $\mathbf{N}$ to S is given by

$$\frac{\nabla F}{\|\nabla F\|} = \frac{(2x, 2y, 0)}{\sqrt{4x^2+4y^2}} = \left(\frac{x}{2}, \frac{y}{2}, 0\right).$$

In terms of the parametrization $\mathbf{X}$, the normal $\mathbf{N}$ is also given by

$$\mathbf{N} = (\cos u_1, \sin u_1, 0).$$

Then we may define an orientation 2-form on S by

$$\Omega_{\mathbf{X}(u_1,u_2)}(\mathbf{a}_1, \mathbf{a}_2) = \det\begin{bmatrix}\mathbf{N} & \mathbf{a}_1 & \mathbf{a}_2\end{bmatrix}.$$

Hence,

$$\Omega_{\mathbf{X}(u_1,u_2)}(\mathbf{T}_{u_1}, \mathbf{T}_{u_2}) = \det\begin{bmatrix}\cos u_1 & -2\sin u_1 & 0\\ \sin u_1 & 2\cos u_1 & 0\\ 0 & 0 & 1\end{bmatrix} = 2 > 0.$$

Thus, $\mathbf{T}_{u_1}, \mathbf{T}_{u_2}$ are compatible with Ω.

We may parametrize ∂S by using two mappings:

$$\text{Bottom circle:}\quad \mathbf{Y}_1\colon [0, 2\pi) \to \mathbf{R}^3, \quad \mathbf{Y}_1(s) = (2\cos s, 2\sin s, 2)$$

and

$$\text{Top circle:}\quad \mathbf{Y}_2\colon [0, 2\pi) \to \mathbf{R}^3, \quad \mathbf{Y}_2(s) = (2\cos s, 2\sin s, 5)$$

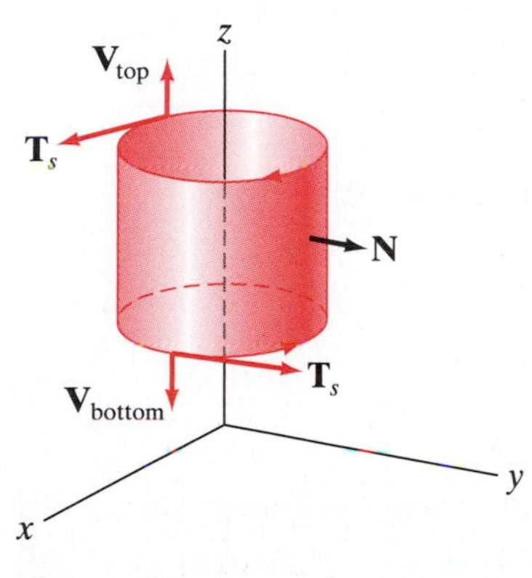

Figure 8.5 Orienting the boundary of the surface S of Example 9. Note the outward-pointing tangent vectors $\mathbf{V}_{\text{top}}$ and $\mathbf{V}_{\text{bottom}}$.

To use Definition 2.9 to orient ∂S, we must identify outward-pointing vectors tangent to S and normal to ∂S. From Figure 8.5, we see that along the top circle $\mathbf{V} = \mathbf{V}_{\text{top}} = (0, 0, 1)$ works, while along the bottom circle, $\mathbf{V} = \mathbf{V}_{\text{bottom}} = (0, 0, -1)$ suffices. Hence, Definition 2.9 tells us that, along the bottom circle,

$$\Omega^{\partial S}_{\mathbf{Y}_1(s)}(\mathbf{a}) = \Omega_{\mathbf{X}(s,2)}(\mathbf{V}_{\text{bottom}}, \mathbf{a}) = \det\begin{bmatrix}\mathbf{N} & \mathbf{V}_{\text{bottom}} & \mathbf{a}\end{bmatrix},$$

while along the top circle,

$$\Omega^{\partial S}_{\mathbf{Y}_1(s)}(\mathbf{a}) = \Omega_{\mathbf{X}(s,5)}(\mathbf{V}_{\text{top}}, \mathbf{a}) = \det\begin{bmatrix}\mathbf{N} & \mathbf{V}_{\text{top}} & \mathbf{a}\end{bmatrix}.$$

For both maps $\mathbf{Y}_1$ and $\mathbf{Y}_2$, we have that the coordinate tangent vector is $\mathbf{T}_s = (-2\sin s, 2\cos s, 0)$. Thus, along the bottom circle,

$$\Omega^{\partial S}_{\mathbf{Y}_1(s)}(\mathbf{T}_s) = \det\begin{bmatrix}\cos s & 0 & -2\sin s\\ \sin s & 0 & 2\cos s\\ 0 & -1 & 0\end{bmatrix} = 2,$$

so $\mathbf{T}_s$ is compatible with the orientation 1-form $\Omega^{\partial S}$. However, along the top circle,

$$\Omega^{\partial S}_{\mathbf{Y}_2(s)}(\mathbf{T}_s) = \det\begin{bmatrix}\cos s & 0 & -2\sin s\\ \sin s & 0 & 2\cos s\\ 0 & 1 & 0\end{bmatrix} = -2,$$

so $\mathbf{T}_s$ is incompatible with $\Omega^{\partial S}$. Therefore, we must orient the top circle *clockwise* around the z-axis and the bottom circle *counterclockwise*. ◆

The following example is the three-dimensional analogue of Example 9:

EXAMPLE 10 Consider the subset $M \subseteq \mathbf{R}^4$ given by $M = \{(x, y, z, w) \mid x^2 + y^2 + z^2 = 4,\ 2 \le w \le 5\}$. This set M is a portion of the cylinder over a sphere of radius 2. Note that the boundary of M consists of the two spheres $S_{\text{bottom}} = \{(x, y, z, 2) \mid x^2 + y^2 + z^2 = 4\}$ and $S_{\text{top}} = \{(x, y, z, 5) \mid x^2 + y^2 + z^2 = 4\}$.

We investigate M and ∂M as parametrized manifolds, orient M, and study the induced orientation on ∂M.

First, we note that M may be parametrized as a 3-manifold in $\mathbf{R}^4$ by

$$\mathbf{X}\colon [0, \pi] \times [0, 2\pi) \times [2, 5] \to \mathbf{R}^4,$$
$$\mathbf{X}(u_1, u_2, u_3) = (2 \sin u_1 \cos u_2, 2 \sin u_1 \sin u_2, 2 \cos u_1, u_3).$$

(This is the usual parametrization of a sphere using spherical coordinates $\varphi = u_1$, $\theta = u_2$, with an additional parameter u_3 for the "vertical" w-axis.) The tangent vectors to the coordinate curves are given by

$$\mathbf{T}_{u_1} = (2 \cos u_1 \cos u_2, 2 \cos u_1 \sin u_2, -2 \sin u_1, 0),$$
$$\mathbf{T}_{u_2} = (-2 \sin u_1 \sin u_2, 2 \sin u_1 \cos u_2, 0, 0),$$

and

$$\mathbf{T}_{u_3} = (0, 0, 0, 1).$$

Note that this parametrization fails to be smooth when u_1 is 0 or π, since then $\mathbf{T}_{u_2} = \mathbf{0}$ at those values for u_1. You can check that the parametrization is smooth at all other values of $\mathbf{X}(\mathbf{u})$ (i.e., for $\mathbf{u}$ in $(0, \pi) \times [0, 2\pi) \times [2, 5]$).

Because M is a portion of the level set at height 4 of the function $F(x, y, z, w) = x^2 + y^2 + z^2$, a unit normal $\mathbf{N}$ to M is given by

$$\frac{\nabla F}{\|\nabla F\|} = \frac{(2x, 2y, 2z, 0)}{\sqrt{4x^2 + 4y^2 + 4z^2}} = \left(\frac{x}{2}, \frac{y}{2}, \frac{z}{2}, 0\right).$$

In terms of the parametrization $\mathbf{X}$, the normal $\mathbf{N}$ is also given by

$$\mathbf{N} = (\sin u_1 \cos u_2, \sin u_1 \sin u_2, \cos u_1, 0).$$

We define an orientation 3-form Ω for M by

$$\Omega_{\mathbf{X}(\mathbf{u})}(\mathbf{a}_1, \mathbf{a}_2, \mathbf{a}_3) = \det \begin{bmatrix} \mathbf{N} & \mathbf{a}_1 & \mathbf{a}_2 & \mathbf{a}_3 \end{bmatrix}.$$

Then

$$\Omega_{\mathbf{X}(\mathbf{u})}(\mathbf{T}_{u_1}, \mathbf{T}_{u_2}, \mathbf{T}_{u_3}) = \det \begin{bmatrix} \sin u_1 \cos u_2 & 2 \cos u_1 \cos u_2 & -2 \sin u_1 \sin u_2 & 0 \\ \sin u_1 \sin u_2 & 2 \cos u_1 \sin u_2 & 2 \sin u_1 \cos u_2 & 0 \\ \cos u_1 & -2 \sin u_1 & 0 & 0 \\ 0 & 0 & 0 & 1 \end{bmatrix}$$
$$= 4 \sin u_1 > 0$$

for $0 < u_1 < \pi$ (which is where the parametrization $\mathbf{X}$ is smooth). Hence, $\mathbf{T}_{u_1}$, $\mathbf{T}_{u_2}$, $\mathbf{T}_{u_3}$ are compatible with Ω.

We parametrize the two pieces of ∂M with two mappings:

"Bottom" sphere S_{bottom}:

$$\mathbf{Y}_1\colon [0, \pi] \times [0, 2\pi) \to \mathbf{R}^4,$$
$$\mathbf{Y}_1(s_1, s_2) = (2 \sin s_1 \cos s_2, 2 \sin s_1 \sin s_2, 2 \cos s_1, 2),$$

and

"Top" sphere S_{top}:

$$\mathbf{Y}_2\colon [0, \pi] \times [0, 2\pi) \to \mathbf{R}^4,$$
$$\mathbf{Y}_2(s_1, s_2) = (2 \sin s_1 \cos s_2, 2 \sin s_1 \sin s_2, 2 \cos s_1, 5).$$

Note that both parametrizations $\mathbf{Y}_1$ and $\mathbf{Y}_2$ give the same tangent vectors to the corresponding coordinate curves, namely,

$$\mathbf{T}_{s_1} = (2\cos s_1 \cos s_2, 2\cos s_1 \sin s_2, -2\sin s_1, 0)$$

and

$$\mathbf{T}_{s_2} = (-2\sin s_1 \sin s_2, 2\sin s_1 \cos s_2, 0, 0),$$

and, by considering these tangent vectors, we see that the parametrizations are smooth whenever $s_1 \neq 0, \pi$.

To give ∂M the orientation induced from that of M, we identify outward-pointing unit vectors tangent to M and normal to ∂M. Thus, we need $\mathbf{V}$ such that

- $\mathbf{V} \in \operatorname{Span}\{\mathbf{T}_{u_1}, \mathbf{T}_{u_2}, \mathbf{T}_{u_3}\} \iff \mathbf{V}\cdot\mathbf{N} = 0$;
- $\mathbf{V}\cdot\mathbf{T}_{s_1} = \mathbf{V}\cdot\mathbf{T}_{s_2} = 0$;
- $\mathbf{V}$ points away from M.

It's not difficult to see that we must take $\mathbf{V} = \mathbf{V}_{\text{top}} = (0, 0, 0, 1)$ along S_{top} and $\mathbf{V} = \mathbf{V}_{\text{bottom}} = (0, 0, 0, -1)$ along S_{bottom}. Therefore, Definition 2.9 tells us that along S_{bottom},

$$\Omega^{\partial M}_{\mathbf{Y}_1(\mathbf{s})}(\mathbf{a}_1, \mathbf{a}_2) = \Omega_{\mathbf{X}(\mathbf{s},2)}(\mathbf{V}_{\text{bottom}}, \mathbf{a}_1, \mathbf{a}_2).$$

In particular,

$$\begin{aligned}\Omega^{\partial M}_{\mathbf{Y}_1(\mathbf{s})}(\mathbf{T}_{s_1}, \mathbf{T}_{s_2}) &= \det\begin{bmatrix}\mathbf{N} & \mathbf{V}_{\text{bottom}} & \mathbf{T}_{s_1} & \mathbf{T}_{s_2}\end{bmatrix} \\ &= \det\begin{bmatrix} \sin s_1 \cos s_2 & 0 & 2\cos s_1 \cos s_2 & -2\sin s_1 \sin s_2 \\ \sin s_1 \sin s_2 & 0 & 2\cos s_1 \sin s_2 & 2\sin s_1 \cos s_2 \\ \cos s_1 & 0 & -2\sin s_1 & 0 \\ 0 & -1 & 0 & 0 \end{bmatrix} \\ &= 4\sin s_1 > 0\end{aligned}$$

for $0 < s_1 < \pi$ (i.e., where the parametrization $\mathbf{Y}_1$ is smooth). Thus, $\mathbf{Y}_1$ is compatible with $\Omega^{\partial M}$. Along S_{top}, however, we have

$$\begin{aligned}\Omega^{\partial M}_{\mathbf{Y}_2(\mathbf{s})}(\mathbf{T}_{s_1}, \mathbf{T}_{s_2}) &= \det\begin{bmatrix}\mathbf{N} & \mathbf{V}_{\text{bottom}} & \mathbf{T}_{s_1} & \mathbf{T}_{s_2}\end{bmatrix} \\ &= \det\begin{bmatrix} \sin s_1 \cos s_2 & 0 & 2\cos s_1 \cos s_2 & -2\sin s_1 \sin s_2 \\ \sin s_1 \sin s_2 & 0 & 2\cos s_1 \sin s_2 & 2\sin s_1 \cos s_2 \\ \cos s_1 & 0 & -2\sin s_1 & 0 \\ 0 & 1 & 0 & 0 \end{bmatrix} \\ &= -4\sin s_1 < 0\end{aligned}$$

for $0 < s_1 < \pi$, so $\mathbf{Y}_2$ is incompatible with $\Omega^{\partial M}$. We must take care with this distinction when we consider the general version of Stokes's theorem. ◆

Next, we examine how the integral of a k-form ω can vary when taken over two different parametrizations $\mathbf{X}\colon D_1 \to \mathbf{R}^n$ and $\mathbf{Y}\colon D_2 \to \mathbf{R}^n$ for the same k-manifold $M = \mathbf{X}(D_1) = \mathbf{Y}(D_2)$.

■ **Definition 2.10** Let $\mathbf{X}\colon D_1 \subseteq \mathbf{R}^k \to \mathbf{R}^n$ and $\mathbf{Y}\colon D_2 \subseteq \mathbf{R}^k \to \mathbf{R}^n$ be parametrized k-manifolds. We say that $\mathbf{Y}$ is a **reparametrization** of $\mathbf{X}$ if there is a one-one and onto function $\mathbf{H}\colon D_2 \to D_1$ such that $\mathbf{Y}(\mathbf{s}) = \mathbf{X}(\mathbf{H}(\mathbf{s}))$; that is, such that $\mathbf{Y} = \mathbf{X} \circ \mathbf{H}$. If $\mathbf{X}$ and $\mathbf{Y}$ are smooth and $\mathbf{H}$ is of class C^1, then we say that $\mathbf{Y}$ is a **smooth** reparametrization of $\mathbf{X}$.

Since $\mathbf{H}$ is one-one, it can be shown that the Jacobian det $D\mathbf{H}$ cannot change sign from positive to negative (or vice versa). Thus, we say that both $\mathbf{H}$ and $\mathbf{Y}$ are **orientation-preserving** if the Jacobian det $D\mathbf{H}$ is nonnegative, **orientation-reversing** if det $D\mathbf{H}$ is nonpositive.

The following result is a generalization of Theorem 2.5 of Chapter 7 to the case of k-manifolds.

Theorem 2.11 Let $\mathbf{X}: D_1 \subseteq \mathbf{R}^k \to \mathbf{R}^n$ be a smooth parametrized k-manifold and ω a k-form defined on $\mathbf{X}(D_1)$. If $\mathbf{Y}: D_2 \subseteq \mathbf{R}^k \to \mathbf{R}^n$ is any smooth reparametrization of $\mathbf{X}$, then either

$$\int_{\mathbf{Y}} \omega = \int_{\mathbf{X}} \omega,$$

if $\mathbf{Y}$ is orientation-preserving, or

$$\int_{\mathbf{Y}} \omega = -\int_{\mathbf{X}} \omega,$$

if $\mathbf{Y}$ is orientation-reversing.

In view of Theorem 2.11, we can define what we mean by $\int_M \omega$, where M is a subset of $\mathbf{R}^n$ that can be parametrized as an oriented k-manifold and ω is a k-form defined on M. We simply let

$$\int_M \omega = \int_{\mathbf{X}} \omega,$$

where $\mathbf{X}: D \subseteq \mathbf{R}^k \to \mathbf{R}^n$ is *any* smooth parametrization of M that is compatible with the orientation chosen.

EXAMPLE 11 We evaluate $\int_C \omega$, where C is the (oriented) line segment in $\mathbf{R}^3$ from $(0, -1, -2)$ to $(1, 2, 3)$ and $\omega = z\,dx + x\,dy + y\,dz$.

Using Theorem 2.11, we may parametrize C in any way that preserves the orientation. Thus,

$$\mathbf{x}: [0, 1] \to \mathbf{R}^3, \quad \mathbf{x}(t) = (1-t)(0, -1, -2) + t(1, 2, 3) = (t, 3t-1, 5t-2)$$

is one way to make such a parametrization. Then $\mathbf{x}'(t) = (1, 3, 5)$, and, hence, from Definition 2.1, we have

$$\begin{aligned}\int_C \omega = \int_{\mathbf{x}} \omega &= \int_0^1 \omega_{\mathbf{x}(t)}(\mathbf{x}'(t))\,dt \\ &= \int_0^1 \{(5t-2)\cdot 1 + t\cdot 3 + (3t-1)\cdot 5\}\,dt \\ &= \int_0^1 (23t - 7)\,dt = \left(\frac{23}{2}t^2 - 7t\right)\Big|_0^1 = \frac{9}{2}.\end{aligned}$$

Note that if we parametrize C in the opposite direction, by using, for example, the map

$$\mathbf{y}: [0, 1] \to \mathbf{R}^3, \quad \mathbf{y}(t) = t(0, -1, -2) + (1-t)(1, 2, 3) = (1-t, 2-3t, 3-5t),$$

then we would have

$$\int_{\mathbf{y}} \omega = \int_0^1 \omega_{\mathbf{y}(t)}(\mathbf{y}'(t))\, dt$$
$$= \int_0^1 \{(3-5t)(-1) + (1-t)(-3) + (2-3t)(-5)\}\, dt$$
$$= \int_0^1 (23t - 16)\, dt = \left(\frac{23}{2}t^2 - 16t\right)\bigg|_0^1 = -\frac{9}{2}.$$

In light of Theorem 2.11, this result could have been anticipated from our preceding calculation of $\int_{\mathbf{x}} \omega$. ◆

Note on k-manifolds

The central geometric object of study in this section, namely, a parametrized k-manifold, is actually a rather special case of a more general notion of a k-manifold. In general, a ***k*-manifold** in $\mathbf{R}^n$ is a connected subset $M \subseteq \mathbf{R}^n$ such that, for every point $\mathbf{x} \in M$, there is an open set $U \subseteq \mathbf{R}^k$ and a continuous, one-one map $\mathbf{X}: U \to \mathbf{R}^n$ with $\mathbf{x} \in \mathbf{X}(U) \subset M$. (A k-manifold with nonempty boundary requires a somewhat modified definition.) That is, M is a (general) k-manifold if it is *locally* a parametrized k-manifold near each point. It is possible to extend notions of orientation and integration of k-forms to this more general setting, although it requires some finesse to do so. For the types of examples we are encountering, however, our more restrictive definitions suffice.

Exercises

1. Check that the parametrized 3-manifold in Example 3 is in fact a smooth parametrized 3-manifold.

2. A planar robot arm is constructed by using two rods as shown in Figure 8.6. Suppose that each of the two rods may *telescope,* that is, that their respective lengths l_1 and l_2 may vary between 1 and 3 units. Show that the set of states of this robot arm may be described by a smooth parametrized 4-manifold in $\mathbf{R}^4$.

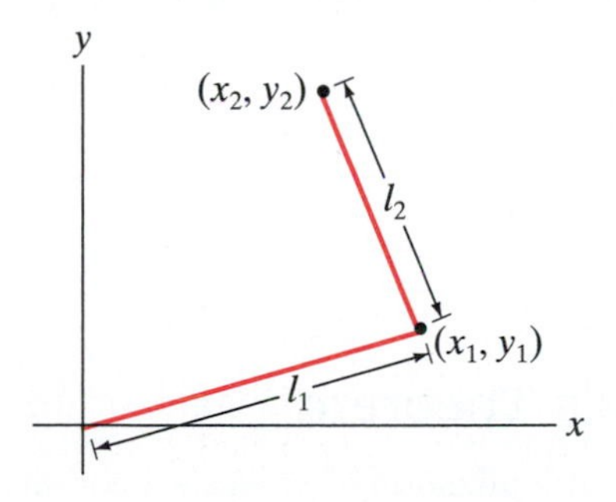

Figure 8.6 Figure for Exercise 2.

3. A planar robot arm is constructed by using a rod of length 3 anchored at the origin and two telescoping rods whose respective lengths l_2 and l_3 may vary between 1 and 2 units as shown in Figure 8.7. Show that the set of states of this robot arm may be described by a smooth parametrized 5-manifold in $\mathbf{R}^6$. (See Exercise 2.)

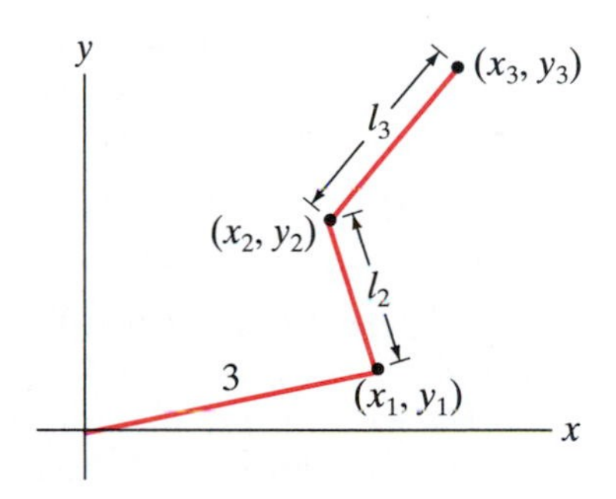

Figure 8.7 Figure for Exercise 3.

4. A robot arm is constructed in $\mathbf{R}^3$ by anchoring a rod of length 2 to the origin (using a ball joint so that the rod may swivel freely) and attaching to the free end of the rod another rod of length 1 (which may also swivel freely; see Figure 8.8). Show that the set of states of this robot arm may be described by a smooth parametrized 4-manifold in $\mathbf{R}^6$.

5. Suppose $\mathbf{v}_1, \ldots, \mathbf{v}_k$ are vectors in $\mathbf{R}^n$. If $\mathbf{x} \in \mathbf{R}^n$ is orthogonal to $\mathbf{v}_i$ for $i = 1, \ldots, k$, show that $\mathbf{x}$ is also orthogonal to any vector in $\operatorname{Span}\{\mathbf{v}_1, \ldots, \mathbf{v}_k\}$.

6. Let a, b, and c be positive constants and $\mathbf{x}: [0, \pi] \to \mathbf{R}^3$ the smooth path given by $\mathbf{x}(t) = (a\cos t, b\sin t, ct)$. If $\omega = b\,dx - a\,dy + xy\,dz$, calculate $\int_{\mathbf{x}} \omega$.

7. Evaluate $\int_C \omega$, where C is the unit circle $x^2 + y^2 = 1$, oriented counterclockwise, and $\omega = y\,dx - x\,dy$.

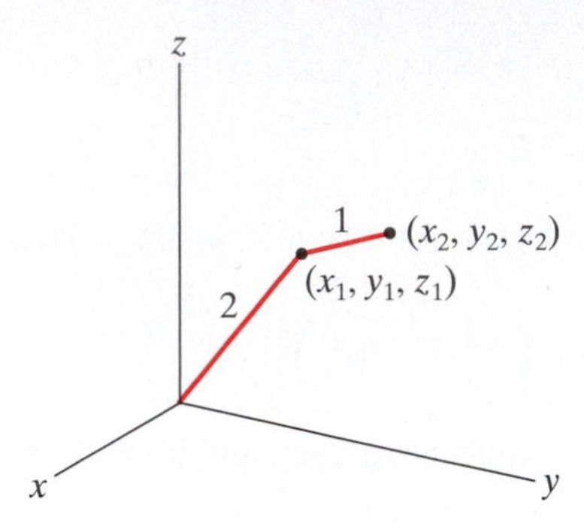

Figure 8.8 Figure for Exercise 4.

8. Compute $\int_C \omega$, where C is the line segment in $\mathbf{R}^n$ from $(0, 0, \ldots, 0)$ to $(3, 3, \ldots, 3)$ and $\omega = x_1\, dx_1 + x_2^2\, dx_2 + \cdots + x_n^n\, dx_n$.

9. Evaluate the integral $\int_{\mathbf{X}} \omega$, where $\mathbf{X}$ is the parametrized helicoid

$$\mathbf{X}(s, t) = (s \cos t, s \sin t, t), \quad 0 \le s \le 1, 0 \le t \le 4\pi$$

and

$$\omega = z\, dx \wedge dy + 3\, dz \wedge dx - x\, dy \wedge dz.$$

10. Consider the helicoid parametrized as

$$\mathbf{X}(u_1, u_2) = (u_1 \cos 3u_2, u_1 \sin 3u_2, 5u_2), \quad 0 \le u_1 \le 5, 0 \le u_2 \le 2\pi.$$

Let S denote the underlying surface of the helicoid and let Ω be the orientation 2-form defined in terms of $\mathbf{X}$ as

$$\Omega_{\mathbf{X}(u_1,u_2)}(\mathbf{a}, \mathbf{b}) = \det \begin{bmatrix} -5 \sin 3u_2 & a_1 & b_1 \\ 5 \cos 3u_2 & a_2 & b_2 \\ -3u_1 & a_3 & b_3 \end{bmatrix}.$$

(a) Explain why the parametrization $\mathbf{X}$ is incompatible with Ω.

(b) Modify the parametrization $\mathbf{X}$ to one having the same underlying surface S, but which is compatible with Ω.

(c) Alternatively, modify the orientation 2-form Ω to Ω' so that the original parametrization $\mathbf{X}$ is compatible with Ω'.

(d) Calculate $\int_S \omega$, where $\omega = z\, dx \wedge dy - (x^2 + y^2)\, dy \wedge dz$ and S is oriented using Ω.

11. Let M be the subset of $\mathbf{R}^3$ given by $\{(x, y, z) \mid x^2 + y^2 - 6 \le z \le 4 - x^2 - y^2\}$. Then M may be parametrized as a 3-manifold via

$$\mathbf{X}\colon D \to \mathbf{R}^3; \quad \mathbf{X}(u_1, u_2, u_3) = (u_1 \cos u_2, u_1 \sin u_2, u_3),$$

where

$$D = \{(u_1, u_2, u_3) \in \mathbf{R}^3 \mid 0 \le u_1 \le \sqrt{5},\ 0 \le u_2 < 2\pi,\ u_1^2 - 6 \le u_3 \le 4 - u_1^2\}.$$

(The parameters u_1, u_2, and u_3 correspond, respectively, to the cylindrical coordinates r, θ, and z. Hence, it is straightforward to obtain the aforementioned parametrization.)

(a) Orient M by using the 3-form Ω, where

$$\Omega_{\mathbf{X}(\mathbf{u})}(\mathbf{a}, \mathbf{b}, \mathbf{c}) = \det \begin{bmatrix} \mathbf{a} & \mathbf{b} & \mathbf{c} \end{bmatrix}.$$

Show that the parametrization, when smooth, is compatible with this orientation.

(b) Identify ∂M and parametrize it as a union of two 2-manifolds (i.e., as a piecewise smooth surface).

(c) Describe the outward-pointing unit vector $\mathbf{V}$, varying continuously along each smooth piece of ∂M, that is normal to ∂M. Give formulas for it in terms of the parametrizations used in part (b).

12. Calculate $\int_S \omega$, where S is the portion of the paraboloid $z = x^2 + y^2$ with $0 \le z \le 4$, oriented by upward-pointing normal vector $(-2x, -2y, 1)$, and $\omega = e^z dx \wedge dy + y\, dz \wedge dx + x\, dy \wedge dz$.

13. Consider the parametrized 3-manifold

$$\mathbf{X}\colon [0, 1] \times [0, 1] \times [0, 1] \to \mathbf{R}^4,$$
$$\mathbf{X}(u_1, u_2, u_3) = (u_1, u_2, u_3, (2u_1 - u_3)^2).$$

Find

$$\int_{\mathbf{X}} x_2\, dx_2 \wedge dx_3 \wedge dx_4 + 2x_1x_3\, dx_1 \wedge dx_2 \wedge dx_3.$$

8.3 The Generalized Stokes's Theorem

We conclude with a discussion of a generalization of Stokes's theorem that relates the integral of a k-form over a k-manifold to the integral of a $(k-1)$-form over the boundary of the manifold. Before we may state the result, however, we need to introduce the notion of the **exterior derivative** of a k-form.

The Exterior Derivative

The exterior derivative is an operator, denoted d, that takes differential k-forms to $(k+1)$-forms and is defined as follows:

■ Definition 3.1 The **exterior derivative** df of a 0-form f on $U \subseteq \mathbf{R}^n$ is the 1-form

$$df = \frac{\partial f}{\partial x_1}\,dx_1 + \frac{\partial f}{\partial x_2}\,dx_2 + \cdots + \frac{\partial f}{\partial x_n}\,dx_n.$$

For $k > 0$, the **exterior derivative** of a k-form

$$\omega = \sum F_{i_1 \dots i_k} dx_{i_1} \wedge \cdots \wedge dx_{i_k}$$

is the $(k+1)$-form

$$d\omega = \sum (dF_{i_1 \dots i_k}) \wedge dx_{i_1} \wedge \cdots \wedge dx_{i_k},$$

where $dF_{i_1 \dots i_k}$ is computed as the exterior derivative of a 0-form.

EXAMPLE 1 If

$$f(x_1, x_2, x_3, x_4, x_5, x_6) = x_1 x_2 x_3 + x_4 x_5 x_6,$$

then

$$df = x_2 x_3\,dx_1 + x_1 x_3\,dx_2 + x_1 x_2\,dx_3 + x_5 x_6\,dx_4 + x_4 x_6\,dx_5 + x_4 x_5\,dx_6. \quad \blacklozenge$$

EXAMPLE 2 If ω is the 1-form

$$\omega = x_1 x_2\,dx_1 + x_2 x_3\,dx_2 + (2x_1 - x_2)\,dx_3,$$

then

$$\begin{aligned} d\omega &= d(x_1 x_2) \wedge dx_1 + d(x_2 x_3) \wedge dx_2 + d(2x_1 - x_2) \wedge dx_3 \\ &= (x_2\,dx_1 + x_1\,dx_2) \wedge dx_1 + (x_3\,dx_2 + x_2\,dx_3) \wedge dx_2 + (2dx_1 - dx_2) \wedge dx_3. \end{aligned}$$

Using the distributivity property in Proposition 1.4 and the facts that $dx_i \wedge dx_i = 0$ and $dx_i \wedge dx_j = -dx_j \wedge dx_i$, we have

$$\begin{aligned} d\omega &= x_1\,dx_2 \wedge dx_1 + x_2\,dx_3 \wedge dx_2 + 2dx_1 \wedge dx_3 - dx_2 \wedge dx_3 \\ &= -x_1\,dx_1 \wedge dx_2 + 2\,dx_1 \wedge dx_3 - (x_2 + 1)\,dx_2 \wedge dx_3. \end{aligned} \quad \blacklozenge$$

Stokes's Theorem for k-forms

We now can state a generalization of Stokes's theorem to smooth parametrized k-manifolds in $\mathbf{R}^n$.

■ Theorem 3.2 (GENERALIZED STOKES'S THEOREM) Let $D \subseteq \mathbf{R}^k$ be a closed, bounded, connected region, and let $M = \mathbf{X}(D)$ be an oriented, parametrized k-manifold in $\mathbf{R}^n$. If $\partial M \neq \emptyset$, let ∂M be given the orientation induced from that of M. Let ω denote a $(k-1)$-form defined on an open set in $\mathbf{R}^n$ that contains M. Then

$$\int_M d\omega = \int_{\partial M} \omega.$$

If $\partial M = \emptyset$, then we take $\int_{\partial M} \omega$ to be 0 in the preceding equation.

We make no attempt to prove Theorem 3.2;[1] instead, we content ourselves for the moment by checking its correctness in a particular instance.

[1]For a full and rigorous discussion of differential forms and the generalized Stokes's theorem, see J. R. Munkres, *Analysis on Manifolds* (Addison-Wesley, 1991), Chapters 6 and 7.

EXAMPLE 3 We verify the generalized Stokes's theorem (Theorem 3.2) for the 2-form $\omega = zw\,dx \wedge dy$, where M is the 3-manifold $M = \{(x, y, z, w) \in \mathbf{R}^4 \mid w = x^2 + y^2 + z^2,\ x^2 + y^2 + z^2 \le 1\}$ oriented by the 3-form Ω corresponding to the unit normal

$$\mathbf{N} = \frac{(2x, 2y, 2z, -1)}{\sqrt{4x^2 + 4y^2 + 4z^2 + 1}}.$$

The manifold M is a portion of the 3-manifold given in Example 7 of §8.2 and may be parametrized as

$$\mathbf{X}\colon B \to \mathbf{R}^4, \quad \mathbf{X}(u_1, u_2, u_3) = (u_1, u_2, u_3, u_1^2 + u_2^2 + u_3^2),$$

where $B = \{(u_1, u_2, u_3) \mid u_1^2 + u_2^2 + u_3^2 \le 1\}$. Using this parametrization, we have

$$\begin{cases} \mathbf{T}_{u_1} = (1, 0, 0, 2u_1) \\ \mathbf{T}_{u_2} = (0, 1, 0, 2u_2) \\ \mathbf{T}_{u_3} = (0, 0, 1, 2u_3) \\ \mathbf{N} = \dfrac{(2u_1, 2u_2, 2u_3, -1)}{\sqrt{4u_1^2 + 4u_2^2 + 4u_3^2 + 1}} \end{cases},$$

so the orientation 3-form Ω is given by

$$\Omega_{\mathbf{X}(\mathbf{u})}(\mathbf{a}_1, \mathbf{a}_2, \mathbf{a}_3) = \det\begin{bmatrix} \mathbf{N} & \mathbf{a}_1 & \mathbf{a}_2 & \mathbf{a}_3 \end{bmatrix}.$$

Example 7 of §8.2 shows that the parametrization $\mathbf{X}$ is compatible with this orientation. Hence, we may use this parametrization without any adjustments when we calculate $\int_M d\omega$.

The boundary of M is $\partial M = \{(x, y, z, w) \mid x^2 + y^2 + z^2 = w = 1\}$ and may be parametrized as

$$\mathbf{Y}\colon [0, \pi] \times [0, 2\pi) \to \mathbf{R}^4, \quad \mathbf{Y}(s_1, s_2) = (\sin s_1 \cos s_2, \sin s_1 \sin s_2, \cos s_1, 1).$$

Then

$$\mathbf{T}_{s_1} = (\cos s_1 \cos s_2, \cos s_1 \sin s_2, -\sin s_1, 0)$$

and

$$\mathbf{T}_{s_2} = (-\sin s_1 \sin s_2, \sin s_1 \cos s_2, 0, 0).$$

An outward-pointing unit vector $\mathbf{V} = (v_1, v_2, v_3, v_4)$ tangent to M and normal to ∂M must satisfy

- $\mathbf{V} \cdot \mathbf{N} = 0$ along ∂M;
- $\mathbf{V} \cdot \mathbf{T}_{s_1} = \mathbf{V} \cdot \mathbf{T}_{s_2} = 0$.

Along ∂M, we have

$$\mathbf{N} = \frac{1}{\sqrt{5}}(2 \sin s_1 \cos s_2, 2 \sin s_1 \sin s_2, 2 \cos s_1, -1).$$

Thus, $\mathbf{V}$ must satisfy the system of equations

$$\begin{cases} (2 \sin s_1 \cos s_2)v_1 + (2 \sin s_1 \sin s_2)v_2 + (2 \cos s_1)v_3 - v_4 = 0 \\ (\cos s_1 \cos s_2)v_1 + (\cos s_1 \sin s_2)v_2 - (\sin s_1)v_3 = 0 \\ -(\sin s_1 \sin s_2)v_1 + (\sin s_1 \cos s_2)v_2 = 0 \end{cases}.$$

After some manipulation, one finds that the unit vector that satisfies these equations and also points away from M is

$$\mathbf{V} = \frac{1}{\sqrt{5}}(\sin s_1 \cos s_2, \sin s_1 \sin s_2, \cos s_1, 2).$$

Then the induced orientation 2-form $\Omega^{\partial M}$ for ∂M is given by

$$\Omega^{\partial M}_{\mathbf{Y(s)}}(\mathbf{a}_1, \mathbf{a}_2) = \Omega_{\mathbf{X(u)}}(\mathbf{V}, \mathbf{a}_1, \mathbf{a}_2),$$

where $\mathbf{X(u)} = \mathbf{Y(s)}$. In particular, we have

$$\begin{aligned}
&\Omega^{\partial M}_{\mathbf{Y(s)}}(\mathbf{T}_{s_1}, \mathbf{T}_{s_2}) \\
&\quad = \det \begin{bmatrix} \mathbf{N} & \mathbf{V} & \mathbf{T}_{s_1} & \mathbf{T}_{s_2} \end{bmatrix} \\
&\quad = \det \begin{bmatrix} \frac{2}{\sqrt{5}} \sin s_1 \cos s_2 & \frac{1}{\sqrt{5}} \sin s_1 \cos s_2 & \cos s_1 \cos s_2 & -\sin s_1 \sin s_2 \\ \frac{2}{\sqrt{5}} \sin s_1 \sin s_2 & \frac{1}{\sqrt{5}} \sin s_1 \sin s_2 & \cos s_1 \sin s_2 & \sin s_1 \cos s_2 \\ \frac{2}{\sqrt{5}} \cos s_1 & \frac{1}{\sqrt{5}} \cos s_1 & -\sin s_1 & 0 \\ -\frac{1}{\sqrt{5}} & \frac{2}{\sqrt{5}} & 0 & 0 \end{bmatrix} \\
&\quad = \sin s_1 > 0
\end{aligned}$$

for $0 < s_1 < \pi$. Hence, the parametrization $\mathbf{Y}$ of ∂M, when smooth, is compatible with the induced orientation, so we may use this parametrization to calculate $\int_{\partial M} \omega$.

Now we are ready to integrate. We first compute $\int_M d\omega$. Since $\omega = zw\, dx \wedge dy$, we have

$$\begin{aligned}
d\omega &= d(zw) \wedge dx \wedge dy = (z\, dw + w\, dz) \wedge dx \wedge dy \\
&= z\, dw \wedge dx \wedge dy + w\, dz \wedge dx \wedge dy.
\end{aligned}$$

Thus,

$$\begin{aligned}
\int_M d\omega &= \iiint_B d\omega_{\mathbf{X(u)}}(\mathbf{T}_{u_1}, \mathbf{T}_{u_2}, \mathbf{T}_{u_3})\, du_1\, du_2\, du_3 \\
&= \iiint_B \{u_3\, dw \wedge dx \wedge dy(\mathbf{T}_{u_1}, \mathbf{T}_{u_2}, \mathbf{T}_{u_3}) \\
&\qquad + (u_1^2 + u_2^2 + u_3^2)\, dz \wedge dx \wedge dy(\mathbf{T}_{u_1}, \mathbf{T}_{u_2}, \mathbf{T}_{u_3})\}\, du_1\, du_2\, du_3 \\
&= \iiint_B \left\{ u_3 \underbrace{\begin{vmatrix} 2u_1 & 2u_2 & 2u_3 \\ 1 & 0 & 0 \\ 0 & 1 & 0 \end{vmatrix}}_{=2u_3} \right. \\
&\qquad \left. + (u_1^2 + u_2^2 + u_3^2) \underbrace{\begin{vmatrix} 0 & 0 & 1 \\ 1 & 0 & 0 \\ 0 & 1 & 0 \end{vmatrix}}_{=1} \right\} du_1\, du_2\, du_3 \\
&= \iiint_B (u_1^2 + u_2^2 + 3u_3^2)\, du_1\, du_2\, du_3.
\end{aligned}$$

Since B is a solid unit ball, the easiest way to evaluate this iterated integral is to use spherical coordinates ρ, φ, and θ. Hence,

$$\begin{aligned}
\int_M d\omega &= \int_0^{2\pi}\int_0^{\pi}\int_0^1 (\rho^2 + 2\rho^2\cos^2\varphi)\,\rho^2\sin\varphi\,d\rho\,d\varphi\,d\theta \\
&= \int_0^{2\pi}\int_0^{\pi}\int_0^1 \rho^4\,(\sin\varphi + 2\cos^2\varphi\sin\varphi)\,d\rho\,d\varphi\,d\theta \\
&= \int_0^{2\pi}\int_0^{\pi} \frac{1}{5}\,(\sin\varphi + 2\cos^2\varphi\sin\varphi)\,d\varphi\,d\theta \\
&= \frac{1}{5}\int_0^{2\pi} \left(-\cos\varphi - \frac{2}{3}\cos^3\varphi\right)\Big|_{\varphi=0}^{\pi}\,d\theta \\
&= \frac{1}{5}\int_0^{2\pi} \frac{10}{3}\,d\theta = \frac{4\pi}{3}.
\end{aligned}$$

On the other hand,

$$\begin{aligned}
\int_{\partial M} \omega &= \iint_{[0,\pi]\times[0,2\pi)} \omega_{\mathbf{Y}(\mathbf{s})}(\mathbf{T}_{s_1}, \mathbf{T}_{s_2})\,ds_1\,ds_2 \\
&= \int_0^{2\pi}\int_0^{\pi} \cos s_1\,dx\wedge dy(\mathbf{T}_{s_1}, \mathbf{T}_{s_2})\,ds_1\,ds_2 \\
&= \int_0^{2\pi}\int_0^{\pi} \cos s_1 \begin{vmatrix} \cos s_1\cos s_2 & -\sin s_1\sin s_2 \\ \cos s_1\sin s_2 & \sin s_1\cos s_2 \end{vmatrix}\,ds_1\,ds_2 \\
&= \int_0^{2\pi}\int_0^{\pi} \cos s_1\,(\cos s_1\sin s_1)\,ds_1\,ds_2 \\
&= \int_0^{2\pi}\int_0^{\pi} \cos^2 s_1\sin s_1\,ds_1\,ds_2 \\
&= \int_0^{2\pi} \left(-\frac{1}{3}\cos^3 s_1\right)\Big|_{s_1=0}^{\pi}\,ds_2 = \int_0^{2\pi}\frac{2}{3}\,ds_2 = \frac{4\pi}{3}.
\end{aligned}$$

Therefore, the generalized Stokes's theorem is verified in this case. ◆

Besides being notationally elegant, the integral formula in Theorem 3.2 beautifully encompasses all three of the major results of vector analysis, as we now show.

First, let ω be a 1-form defined on an open set U in $\mathbf{R}^2$. Then,

$$\omega = M(x, y)\,dx + N(x, y)\,dy,$$

so that

$$\begin{aligned}
d\omega &= dM\wedge dx + dN\wedge dy \\
&= \left(\frac{\partial M}{\partial x}dx + \frac{\partial M}{\partial y}dy\right)\wedge dx + \left(\frac{\partial N}{\partial x}dx + \frac{\partial N}{\partial y}dy\right)\wedge dy \\
&= \left(\frac{\partial N}{\partial x} - \frac{\partial M}{\partial y}\right)dx\wedge dy.
\end{aligned}$$

The generalized Stokes's theorem (Theorem 3.2) says that if D is a 2-manifold contained in U and ∂D is given the induced orientation (see Figure 8.9), then

$$\int_D d\omega = \int_{\partial D} \omega,$$

or, in this instance, that

$$\iint_D \left(\frac{\partial N}{\partial x} - \frac{\partial M}{\partial y}\right) dx\, dy = \oint_{\partial D} M dx + N dy,$$

which is Green's theorem.

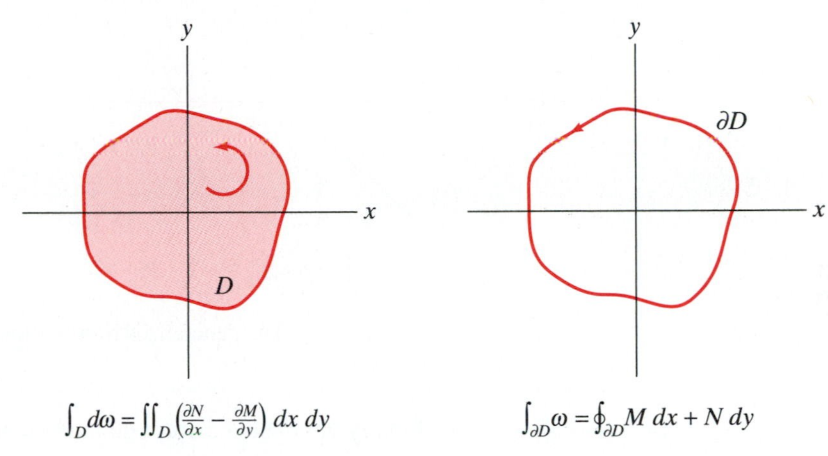

Figure 8.9 The generalized Stokes's theorem implies Green's theorem.

Next, suppose ω is a 1-form defined on an open set U in $\mathbf{R}^3$. Then

$$\omega = F_1(x, y, z)\, dx + F_2(x, y, z)\, dy + F_3(x, y, z)\, dz.$$

It follows that

$$d\omega = \left(\frac{\partial F_3}{\partial y} - \frac{\partial F_2}{\partial z}\right) dy \wedge dz + \left(\frac{\partial F_1}{\partial z} - \frac{\partial F_3}{\partial x}\right) dz \wedge dx + \left(\frac{\partial F_2}{\partial x} - \frac{\partial F_1}{\partial y}\right) dx \wedge dy.$$

Recall from Proposition 2.2 that if S is a parametrized 2-manifold (surface in $\mathbf{R}^3$), then

$$\int_{\partial S} \omega = \oint_{\partial S} \mathbf{F} \cdot d\mathbf{s},$$

where $\mathbf{F} = F_1\,\mathbf{i} + F_2\,\mathbf{j} + F_3\,\mathbf{k}$. From Proposition 2.4,

$$\int_S d\omega = \iint_S \mathbf{G} \cdot d\mathbf{S},$$

where

$$\mathbf{G} = \left(\frac{\partial F_3}{\partial y} - \frac{\partial F_2}{\partial z}\right)\mathbf{i} + \left(\frac{\partial F_1}{\partial z} - \frac{\partial F_3}{\partial x}\right)\mathbf{j} + \left(\frac{\partial F_2}{\partial x} - \frac{\partial F_1}{\partial y}\right)\mathbf{k} = \nabla \times \mathbf{F}.$$

Theorem 3.2 tells us, if S is oriented and ∂S is given the induced orientation, that

$$\int_{\partial S} \omega = \int_S d\omega,$$

or, equivalently, that

$$\oint_{\partial S} \mathbf{F} \cdot d\mathbf{S} = \iint_S \nabla \times \mathbf{F} \cdot d\mathbf{S},$$

which is the classical Stokes's theorem. (See Figure 8.10.)

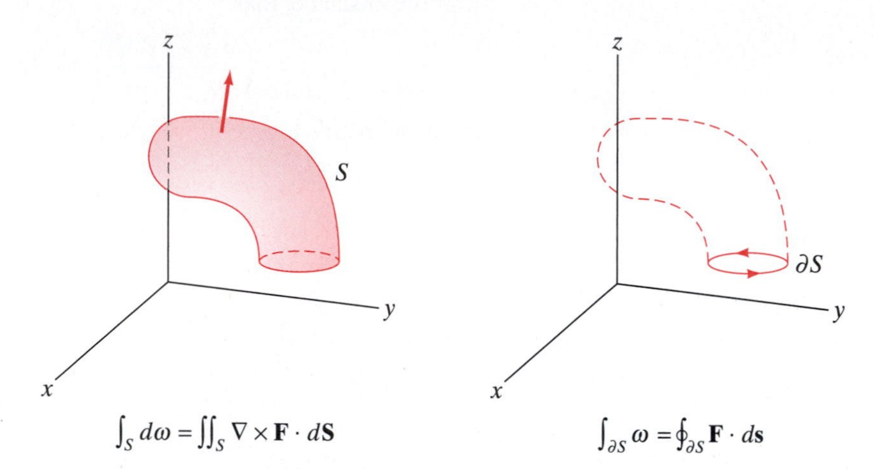

Figure 8.10 The generalized Stokes's theorem gives the classical Stokes's theorem.

Finally, let ω be a 2-form defined on an open set in $\mathbf{R}^3$. So

$$\omega = F_1(x, y, z)\, dy \wedge dz + F_2(x, y, z)\, dz \wedge dx + F_3(x, y, z)\, dx \wedge dy.$$

You can check that

$$d\omega = \left(\frac{\partial F_1}{\partial x} + \frac{\partial F_2}{\partial y} + \frac{\partial F_3}{\partial z} \right) dx \wedge dy \wedge dz.$$

If D is a region in $\mathbf{R}^3$, then D is automatically a parametrized 3-manifold, since the map $\mathbf{X}\colon D \to \mathbf{R}^3$, $\mathbf{X}(x, y, z) = (x, y, z)$ parametrizes D. (One can show that in this instance, D is always orientable as well.) If D is bounded and ∂D (which is a surface) is given the induced orientation (i.e., outward-pointing normal), then Proposition 2.4 states that

$$\int_{\partial D} \omega = \oiint_{\partial D} \mathbf{F} \cdot d\mathbf{S},$$

where $\mathbf{F} = F_1\,\mathbf{i} + F_2\,\mathbf{j} + F_3\,\mathbf{k}$. From Example 6 of §8.2,

$$\int_D d\omega = \int_D \left(\frac{\partial F_1}{\partial x} + \frac{\partial F_2}{\partial y} + \frac{\partial F_3}{\partial z} \right) dx \wedge dy \wedge dz = \iiint_D \nabla \cdot \mathbf{F}\, dV.$$

Theorem 3.2 indicates that $\int_{\partial D} \omega = \int_D d\omega$ or

$$\oiint_{\partial D} \mathbf{F} \cdot d\mathbf{S} = \iiint_D \nabla \cdot \mathbf{F}\, dV,$$

which is, of course, Gauss's theorem. (See Figure 8.11.)

In the foregoing remarks, we have implicitly set up a sort of "dictionary" between the language of differential forms and exterior derivatives and that of scalar and vector fields. To be explicit, see the table of correspondences shown in Figure 8.12.

The theorems of Green, Stokes, and Gauss all arise from Theorem 3.2 applied to 1-forms and 2-forms. The next question is, can the "dictionary" and

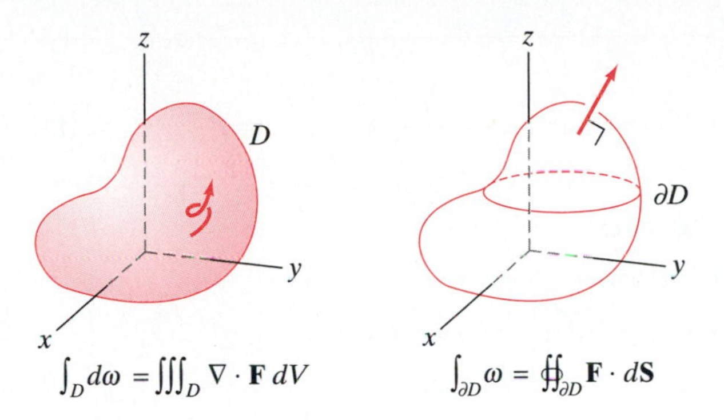

Figure 8.11 The generalized Stokes's theorem gives rise to Gauss's theorem.

k	Differential k-form	Field	Derivative
0	ω	Scalar field f	$d\omega \leftrightarrow \nabla f$
1	$\omega = F_1\, dx + F_2\, dy + F_3\, dz$	Vector field $\mathbf{F} = F_1\,\mathbf{i} + F_2\,\mathbf{j} + F_3\,\mathbf{k}$	$d\omega \leftrightarrow \nabla \times \mathbf{F}$
2	$\omega = F_1\, dy \wedge dz + F_2\, dz \wedge dx + F_3\, dx \wedge dy$	Vector field $\mathbf{F} = F_1\,\mathbf{i} + F_2\,\mathbf{j} + F_3\,\mathbf{k}$	$d\omega \leftrightarrow \nabla \cdot \mathbf{F}$

Figure 8.12 A differential forms–vector fields dictionary.

Theorem 3.2 provide a corresponding result for 0-forms? The generalized Stokes's theorem (Theorem 3.2) states, for a 0-form ω and an oriented parametrized curve C, that

$$\int_C d\omega = \int_{\partial C} \omega.$$

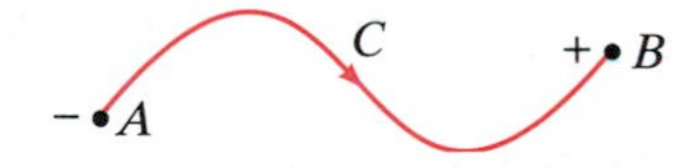

Figure 8.13 The orientation of the curve C induces an orientation of its boundary (i.e., the endpoints A and B).

Now, if C is closed, then ∂C is empty (and so $\int_{\partial C} \omega = 0$). But if C is not closed, then ∂C consists of just two points. In that case, what should $\int_{\partial C} \omega$ mean? In particular, to apply Theorem 3.2, we must orient ∂C in a manner that is consistent with the orientation of C, which can be done by assigning a "$-$" sign to the initial point A of C and a "$+$" sign to the terminal point B. (See Figure 8.13.) Then $\int_{\partial C} \omega$ is just $f(B) - f(A)$, where f is the function (scalar field) corresponding to ω in the table. Since $d\omega$ corresponds to ∇f, Theorem 3.2 tells us that

$$\int_C \nabla f \cdot d\mathbf{s} = f(B) - f(A), \tag{1}$$

the result of Theorem 3.3 in Chapter 6.

Finally, for the case $n = 1$, that is, the case of 0-forms (functions) on $\mathbf{R}$, the 0-form ω corresponds to a function f of a single variable, and ∇f is the ordinary derivative f'. Furthermore, a parametrized curve in $\mathbf{R}$ is simply a closed interval $[a, b]$. Then equation (1) reduces to

$$\int_a^b f'(x)\, dx = f(b) - f(a),$$

a version of the fundamental theorem of calculus. Thus, we can appreciate that the generalized Stokes's theorem is an elegant and powerful generalization of the fundamental theorem of calculus to arbitrary dimensions.

Exercises

In Exercises 1–7, determine $d\omega$, where ω is as indicated.

1. $\omega = e^{xyz}$
2. $\omega = x^3y - 2xz^2 + xy^2z$
3. $\omega = (x^2 + y^2)\,dx + xy\,dy$
4. $\omega = x_1\,dx_2 - x_2\,dx_1 + x_3x_4\,dx_4 - x_4x_5\,dx_5$
5. $\omega = xz\,dx \wedge dy - y^2z\,dx \wedge dz$
6. $\omega = x_1x_2x_3\,dx_2 \wedge dx_3 \wedge dx_4 + x_2x_3x_4\,dx_1 \wedge dx_2 \wedge dx_3$
7. $\omega = \sum_{i=1}^{n} x_i^2\,dx_1 \wedge \cdots \wedge \widehat{dx_i} \wedge \cdots \wedge dx_n$ (Note: $\widehat{dx_i}$ means that the term dx_i is omitted.)
8. Let $\mathbf{u}$ be a unit vector and f a differentiable function. Show that $df_{\mathbf{x}_0}(\mathbf{u}) = D_{\mathbf{u}}f(\mathbf{x}_0)$. (Recall that $D_{\mathbf{u}}f(\mathbf{x}_0)$ denotes the directional derivative of f at $\mathbf{x}_0$ in the direction of $\mathbf{u}$.)
9. If $\omega = F(x, z)\,dy + G(x, y)\,dz$ is a (differentiable) 1-form on $\mathbf{R}^3$, what can F and G be so that $d\omega = z\,dx \wedge dy + y\,dx \wedge dz$?
10. Verify the generalized Stokes's theorem (Theorem 3.2) for the 3-manifold M of Exercise 11 of §8.2, where $\omega = 2x\,dy \wedge dz - z\,dx \wedge dy$.
11. Verify the generalized Stokes's theorem (Theorem 3.2) for the 3-manifold
$$M = \{(x, y, z, w) \in \mathbf{R}^4 \mid x = 8 - 2y^2 - 2z^2 - 2w^2,\ x \geq 0\}$$
and the 2-form $\omega = xy\,dz \wedge dw$. (Hint: First compute $\int_{\partial M} \omega$. To calculate $\int_M d\omega$, study Example 3 of this section.)
12. (a) Let M be a parametrized 3-manifold in $\mathbf{R}^3$ (i.e., a solid). Show that
$$\text{Volume of } M = \frac{1}{3}\int_{\partial M} x\,dy \wedge dz - y\,dx \wedge dz + z\,dx \wedge dy.$$
(b) Let M be a parametrized n-manifold in $\mathbf{R}^n$. Explain why we should have
$$n\text{-dimensional volume of } M = \frac{1}{n}\int_{\partial M} x_1\,dx_2 \wedge \cdots \wedge dx_n - x_2\,dx_1 \wedge dx_3 \wedge \cdots \wedge dx_n + x_3\,dx_1 \wedge dx_2 \wedge dx_4 \wedge \cdots \wedge dx_n + \cdots + (-1)^{n-1}x_n\,dx_1 \wedge dx_2 \wedge \cdots \wedge dx_{n-1}.$$

8.4 True/False Exercises for Chapter 8

1. $(dx \wedge dy + dy \wedge dz)((1, 0, -2), (0, -1, 3)) = 0.$
2. $dx_1 \wedge dx_2 \wedge dx_3 \wedge dx_4 = dx_2 \wedge dx_4 \wedge dx_1 \wedge dx_3.$
3. There are 21 basic 5-forms in $\mathbf{R}^7$.
4. $dx_1 \wedge dx_2 = dx_2 \wedge dx_1.$
5. $(dx_1 \wedge dx_2) \wedge dx_3 = dx_3 \wedge (dx_1 \wedge dx_2).$
6. If ω is a 3-form on $\mathbf{R}^6$ and η is a 5-form on $\mathbf{R}^6$, then $\omega \wedge \eta = \eta \wedge \omega$.
7. If ω is a 2-form on $\mathbf{R}^8$ and η is a 3-form on $\mathbf{R}^8$, then $\omega \wedge \eta = \eta \wedge \omega$.
8. $dx \wedge dy \wedge dz(\mathbf{a}, \mathbf{b}, \mathbf{a}) = -dz \wedge dy \wedge dx(\mathbf{a}, \mathbf{a}, \mathbf{b}).$
9. $dx_i \wedge dx_j(\mathbf{a}, \mathbf{b}) = -dx_i \wedge dx_j(\mathbf{b}, \mathbf{a}).$
10. Let $D = [0, 2] \times [-1, 1]$ and let $\mathbf{X}\colon D \to \mathbf{R}^4$ be given by
$$\mathbf{X}(s, t) = (s - t,\ st^2,\ se^t,\ 4t).$$
Then $M = \mathbf{X}(D)$ is a smooth parametrized 2-manifold in $\mathbf{R}^4$.
11. Let $D = [-2, 2] \times [0, 5] \times [-3, 3]$ and let $\mathbf{X}\colon D \to \mathbf{R}^4$ be given by
$$\mathbf{X}(u_1, u_2, u_3) = (u_1u_3^2,\ u_2^2\cos u_3,\ u_1 - u_2,\ u_2^3u_3^4).$$
Then $M = \mathbf{X}(D)$ is a smooth parametrized 3-manifold in $\mathbf{R}^4$.
12. If $D = [0, 1] \times [0, 1]$, then the underlying manifolds of $\mathbf{X}\colon D \to \mathbf{R}^3$,
$$\mathbf{X}(s, t) = (s\cos 2\pi t,\ s\sin 2\pi t,\ s^2)$$
and $\mathbf{Y}\colon D \to \mathbf{R}^3$,
$$\mathbf{Y}(s, t) = (t\cos 2\pi s,\ t\sin 2\pi s,\ t^2)$$
are the same.
13. Let $\omega = dx \wedge dy$. Then $\int_{\mathbf{X}} \omega = \int_{\mathbf{Y}} \omega$, where $\mathbf{X}\colon D \to \mathbf{R}^3$,
$$\mathbf{X}(s, t) = (s\cos 2\pi t,\ s\sin 2\pi t,\ s^2),$$
and $\mathbf{Y}\colon D \to \mathbf{R}^3$,
$$\mathbf{Y}(s, t) = (t\cos 2\pi s,\ t\sin 2\pi s,\ t^2).$$
14. Let $B = \{\mathbf{u} \in \mathbf{R}^3 \mid u_1^2 + u_2^2 + u_3^2 \leq 1\}$. The generalized paraboloid $\mathbf{X}\colon B \to \mathbf{R}^4$ defined by
$$\mathbf{X}(u_1, u_2, u_3) = (u_1, u_2, u_3, u_1^2 + 2u_2^2 + 3u_3^2)$$
has as its boundary the ellipsoid $\mathbf{Y}\colon [0, \pi] \times [0, 2\pi) \to \mathbf{R}^4$,
$$\mathbf{Y}(s, t) = (\sin s\cos t,\ \tfrac{1}{\sqrt{2}}\sin s\sin t,\ \tfrac{1}{\sqrt{3}}\cos s,\ 1).$$

15. Let $M \subseteq \mathbf{R}^n$ be the graph of a function $f: U \subseteq \mathbf{R}^{n-1} \to \mathbf{R}^n$ parametrized by $\mathbf{X}: U \to \mathbf{R}^n$,

$$\mathbf{X}(u_1, \ldots, u_{n-1}) = (u_1, \ldots, u_{n-1}, f(u_1, \ldots, u_{n-1})).$$

If

$$\mathbf{N}(u_1, \ldots, u_{n-1}) = \frac{(f_{u_1}, \ldots, f_{u_{n-1}}, -1)}{\sqrt{f_{u_1}^2 + \cdots + f_{u_{n-1}}^2 + 1}}$$

is a unit normal, then the parametrization $\mathbf{X}$ is compatible with the $(n-1)$-form Ω defined by

$$\Omega_{\mathbf{X}(\mathbf{u})}(\mathbf{a}_1, \ldots, \mathbf{a}_{n-1}) = \det\left[\begin{array}{cccc} \mathbf{a}_1 & \cdots & \mathbf{a}_{n-1} & \mathbf{N} \end{array}\right].$$

16. If $\omega = x_1x_3\, dx_2 \wedge dx_4$, then $d\omega = x_2\, dx_2 \wedge dx_3 \wedge dx_4$.

17. If $\omega = x_1\, dx_3 - x_2\, dx_1 + x_1x_2x_3\, dx_3$, then

$$d\omega = (x_2x_3 + 1)\, dx_1 \wedge dx_3 + dx_1 \wedge dx_2 + x_1x_3\, dx_2 \wedge dx_3.$$

18. If $\omega = x_1x_2\, dx_1 \wedge dx_2 + x_2x_3\, dx_1 \wedge dx_3 + x_1x_3\, dx_2 \wedge dx_3$, then

$$d\omega = 2x_3\, dx_1 \wedge dx_2 \wedge dx_3.$$

19. If ω is an n-form on $\mathbf{R}^n$, then $d\omega = 0$.

20. If M is a parametrized k-manifold without boundary in $\mathbf{R}^n$ and ω is $(k-1)$-form defined on an open set containing M, then $\int_M d\omega = 0$.

8.5 Miscellaneous Exercises for Chapter 8

1. Let ω be a k-form, η an l-form. Show that

$$d(\omega \wedge \eta) = d\omega \wedge \eta + (-1)^k \omega \wedge d\eta.$$

This is accomplished by the following steps:

(a) Show that the result is true when $k = l = 0$, that is, when $\omega = f$ and $\eta = g$. (Here f and g are scalar-valued functions.)

(b) Establish the result when $k = 0$ and $l > 0$.

(c) Establish the result when $k > 0$ and $l = 0$.

(d) Establish the result when k and l are both positive.

2. Let M be the subset of $\mathbf{R}^5$ described as $\{(x_1, x_2, x_3, x_4, x_5) \mid x_5 = x_1x_2x_3x_4,\ 0 \le x_1, x_2, x_3, x_4 \le 1\}$.

(a) Give a parametrization for M (as a 4-manifold) and check that your parametrization is compatible with the orientation 4-form $\Omega = dx_1 \wedge dx_2 \wedge dx_3 \wedge dx_4$.

(b) Calculate $\int_M x_4\, dx_1 \wedge dx_2 \wedge dx_3 \wedge dx_5$.

3. (a) Let C be the curve in $\mathbf{R}^2$ given by $y = f(x)$, $a \le x \le b$. Assume that f is of class C^1. If C is oriented by the direction in which x increases, show that if $\omega = y\, dx$, then

$$\int_C \omega = \text{area under the graph of } f.$$

(b) Let S be the surface in $\mathbf{R}^3$ given by the equation $z = f(x, y)$, where $(x, y) \in [a, b] \times [c, d]$. Assume that f is of class C^1. If S is oriented by upward-pointing normal, show that if $\omega = z\, dx \wedge dy$, then

$$\int_S \omega = \text{volume under the graph of } f.$$

(c) Now we generalize parts (a) and (b) as follows: Suppose $f: D \to \mathbf{R}$ is a function of class C^1 defined on a connected region $D \subseteq \mathbf{R}^{n-1}$. Let M be the $(n-1)$-dimensional hypersurface in $\mathbf{R}^n$ defined by the equation $x_n = f(x_1, \ldots, x_{n-1})$, where $(x_1, \ldots, x_{n-1}) \in D$. If $\omega = x_n\, dx_1 \wedge \cdots \wedge dx_{n-1}$, show that

$$\int_M \omega = \pm(n\text{-dimensional volume under the graph of } f).$$

How can we guarantee a "+" sign in the equation?

4. Let M be the portion of the cylinder $x^2 + z^2 = 1$, $0 \le y \le 3$, oriented by unit normal $\mathbf{N} = (x, 0, z)$.

(a) Use $\mathbf{N}$ to give an orientation 2-form Ω for M. Find a parametrization for M compatible with Ω.

(b) Identify ∂M and parametrize it.

(c) Determine the orientation form $\Omega^{\partial M}$ for ∂M induced from Ω of part (a).

(d) Verify the generalized Stokes's theorem (Theorem 3.2) for M and $\omega = z\, dx + (x + y + z)dy - x\, dz$.

5. Use the generalized Stokes's theorem to calculate $\int_{S^4} \omega$, where S^4 denotes the unit 4-sphere $\{(x_1, x_2, x_3, x_4, x_5) \in \mathbf{R}^5 \mid x_1^2 + x_2^2 + x_3^2 + x_4^2 + x_5^2 = 1\}$ and $\omega = x_3\, dx_1 \wedge dx_2 \wedge dx_4 \wedge dx_5 + x_4\, dx_1 \wedge dx_2 \wedge dx_3 \wedge dx_5$.

6. (a) Let ω be a 0-form (i.e., a function) of class C^2. Show that $d(d\omega) = 0$.

(b) Now suppose that ω is a k-form of class C^2, meaning that when ω is written as

$$\sum_{1 \le i_1 < \cdots < i_k \le n} F_{i_1 \ldots i_k}\, dx_{i_1} \wedge \cdots \wedge dx_{i_k},$$

each $F_{i_1 \ldots i_k}$ is of class C^2. Use part (a) and the result of Exercise 1 to show that $d(d\omega) = 0$.

7. In this problem, show that the equation $d(d\omega) = 0$ implies two well-known results about scalar and vector fields.

(a) First, let ω be a 0-form (of class C^2). Then ω corresponds to a scalar field f. Use the chart on page 505 to interpret the equation $d(d\omega) = 0$.

(b) Next, suppose that ω is a 1-form (again of class C^2). Then ω corresponds to a vector field. Interpret the equation $d(d\omega) = 0$ in this case.

8. Let

$$\omega = \frac{x\,dy \wedge dz + y\,dz \wedge dx + z\,dx \wedge dy}{(x^2 + y^2 + z^2)^{3/2}}.$$

(a) Evaluate $\int_S \omega$, where S is the unit sphere $x^2+y^2+z^2 = 1$, oriented by outward normal.

(b) Calculate $d\omega$.

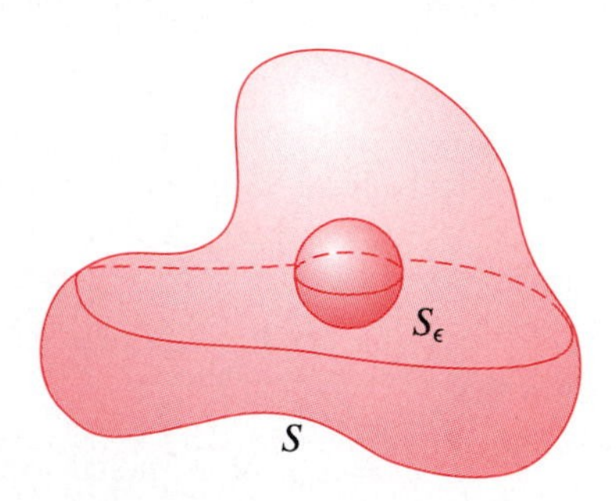

Figure 8.14 Figure for Exercise 8.

(c) Verify Theorem 3.2 over the region $M = \{(x, y, z) \mid a^2 \le x^2 + y^2 + z^2 \le 1\}$, where $a \ne 0$.

(d) Now let M be the solid unit ball $x^2 + y^2 + z^2 \le 1$. Does Theorem 3.2 hold for M and ω? Why or why not?

(e) Suppose that S is any closed, bounded surface that lies entirely outside the sphere $S_\epsilon = \{(x, y, z) \mid x^2 + y^2 + z^2 = \epsilon^2\}$. (See Figure 8.14.) Argue that if S is oriented by outward normal, then $\int_S \omega = 4\pi$.

9. Let M be an oriented $(k + l + 1)$-manifold in $\mathbf{R}^n$; let ω be a k-form and η an l-form defined on an open set of $\mathbf{R}^n$ that contains M. If $\partial M = \emptyset$, use Theorem 3.2 and Exercise 1 to show that

$$\int_M d\omega \wedge \eta = (-1)^{k+1} \int_M \omega \wedge d\eta.$$

10. Let M be an oriented k-manifold. Use Exercise 1 and the general version of Stokes's theorem to establish "integration by parts" for k-forms ω and 0-forms f:

$$\int_M f\,d\omega = \int_{\partial M} f\omega - \int_M df \wedge \omega.$$

Suggestions for Further Reading

General

Francis J. Flanigan and Jerry L. Kazdan, *Calculus Two: Linear and Nonlinear Functions*, 2nd ed., Springer, 1990. The essentials of vector calculus are presented from a linear-algebraic perspective.

John H. Hubbard and Barbara Burke Hubbard, *Vector Calculus, Linear Algebra, and Differential Forms: A Unified Approach*, 2nd ed., Prentice Hall, 2002. Treats the main topics of multivariable calculus, plus a significant amount of linear algebra. More sophisticated in approach than the current book, using differential forms to treat integration.

Jerrold E. Marsden and Anthony J. Tromba, *Vector Calculus*, 5th ed., W. H. Freeman, 2003. This text is probably the one most similar to the current book in both coverage and approach, using matrices and vectors to treat multivariable calculus in $\mathbf{R}^n$.

Jerrold E. Marsden, Anthony J. Tromba, and Alan Weinstein, *Basic Multivariable Calculus*, Springer/W. H. Freeman, 1993. Somewhat similar to Marsden and Tromba's *Vector Calculus*, but with less emphasis on a rigorous development of the subject. A good guide to the main ideas.

George B. Thomas and Ross L. Finney, *Calculus and Analytic Geometry*, 9th ed., Addison-Wesley, 1996. A complete treatment of the techniques of both single-variable and multivariable calculus. No linear algebra needed. A good reference for the main methods of vector calculus of functions of two and three variables.

Richard E. Williamson, Richard H. Crowell, and Hale F. Trotter, *Calculus of Vector Functions*, 3rd ed., Prentice Hall, 1972. A smooth and careful treatment of multivariable calculus and vector analysis, using linear algebra.

More Advanced Treatments

The following texts all offer relatively rigorous and theoretical developments of the main results of multivariable calculus. As such, they are especially useful for studying the foundations of the subject.

Tom M. Apostol, *Mathematical Analysis: A Modern Approach to Advanced Calculus*, 1st ed., Addison-Wesley, 1957. Look at the first edition, since the second edition contains much less regarding multivariable topics.

R. Creighton Buck, *Advanced Calculus*, 3rd ed., McGraw-Hill, 1978. Treats foundational issues in both single-variable and multivariable calculus. Uses the notation of differential forms for considering Green's, Stokes's, and Gauss's theorems.

Richard Courant and Fritz John, *Introduction to Calculus and Analysis*, Vol. Two, Wiley-Interscience, 1974. A famous and encyclopedic work on the analysis of functions of more than one variable (Volume One treats functions of a single variable), with fascinating examples.

Wilfred Kaplan, *Advanced Calculus*, 3rd ed., Addison-Wesley, 1984. A full treatment of advanced calculus of functions of two and three variables, plus material on calculus of functions of n variables (including some discussion of tensors). In addition, there are chapters on infinite series, differential equations, and functions of a complex variable.

O. D. Kellogg, *Foundations of Potential Theory*, originally published by Springer, 1929. Reprinted by Dover Publications, 1954. A classic work that ventures well beyond the subject of the current book. The writing style may seem somewhat old-fashioned, but Kellogg includes details of certain arguments that are difficult to find anywhere else.

James R. Munkres, *Analysis on Manifolds*, Addison-Wesley, 1991. A superbly well written and sophisticated treatment of calculus in $\mathbf{R}^n$. Requires a knowledge of linear algebra. Includes a full development of differential forms and exterior algebra to treat integration. For advanced mathematics students.

David V. Widder, *Advanced Calculus*, 2nd ed., originally published by Prentice-Hall, 1961. Reprinted by Dover Publications, 1989. Careful treatment of differentiation and integration of functions of one and several variables. Chapters on differential geometry, too.

Physics Oriented Texts

Mary L. Boas, *Mathematical Methods in the Physical Sciences*, 2nd ed., Wiley, 1983. A wide variety of topics that extend well beyond vector calculus. For the student interested in physics as well as mathematics.

Harry F. Davis and Arthur D. Snider, *Introduction to Vector Analysis*, 6th ed., William C. Brown, 1991. A detailed and relatively sophisticated treatment of multivariable topics. Includes appendices on classical mechanics and electromagnetism.

Edward M. Purcell, *Electricity and Magnetism*, 2nd ed., McGraw-Hill, 1985. This is a physics, not a mathematics, text. Provides excellent intuition regarding the meaning of line and surface integrals, differential operations on vector and scalar fields, and the significance of vector analysis.

Other

Alfred Gray, *Modern Differential Geometry of Curves and Surfaces with Mathematica®*, 2nd ed., CRC Press, 1998. Not a vector calculus text by any means, but rather a delightful library of geometric objects and how to understand them via *Mathematica*. Some of the differential geometric topics are somewhat remote from the subject of the current book, but the many illustrations are worth viewing. An outstanding aid for developing one's visualization skills.

H. M. Schey, *Div, Grad, Curl, and All That: An Informal Text on Vector Calculus*, 3rd ed., W. W. Norton, 1997. The classic "alternative" book on vector analysis, aimed at students of electricity and magnetism. Brief, but well done, intuitive account of vector analysis.

Answers to Selected Exercises

Chapter 1

Section 1.1

1. (a)

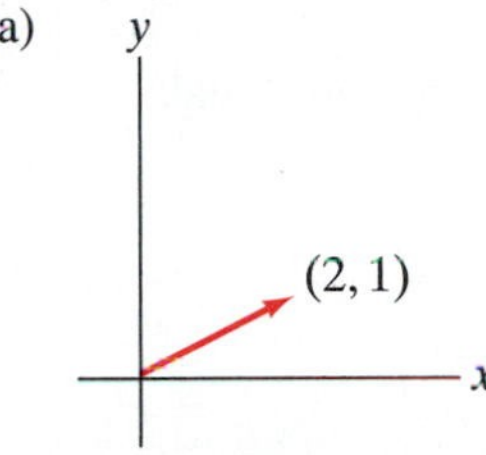

(b)

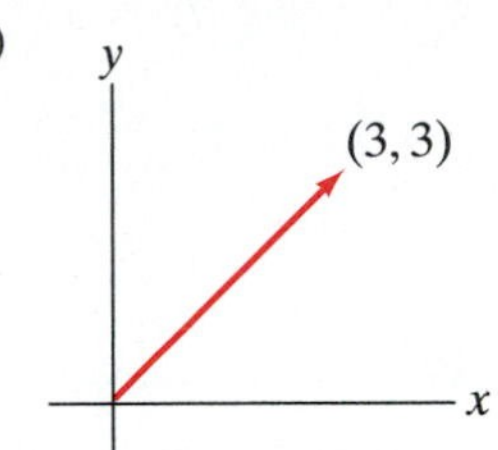

(c)

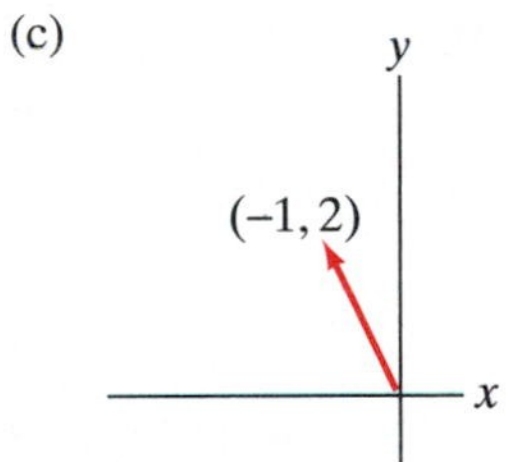

3. (a) $(2, 8)$
   (b) $(-16, -24)$
   (c) $(5, 15)$
   (d) $(-19, 25)$
   (e) $(8, -26)$

5.

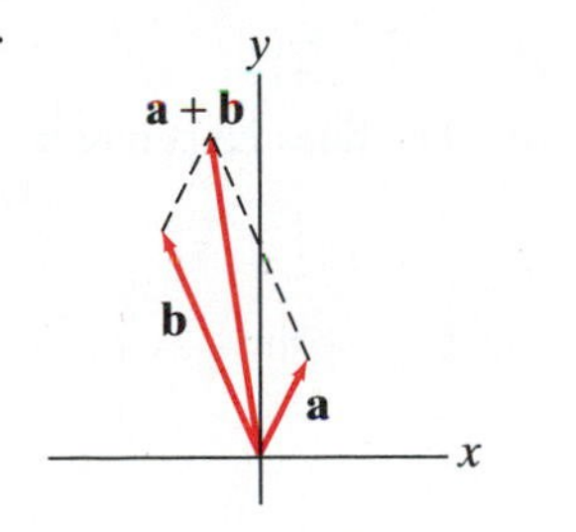

7. (a) $\overrightarrow{AB} = (-4, 3, -1)$, $\overrightarrow{BA} = (4, -3, 1)$
   (b) $\overrightarrow{AC} = (1, 1, 3)$, $\overrightarrow{BC} = (5, -2, 4)$, $\overrightarrow{AC} + \overrightarrow{CB} = (-4, 3, -1)$
   (c) In general, we have

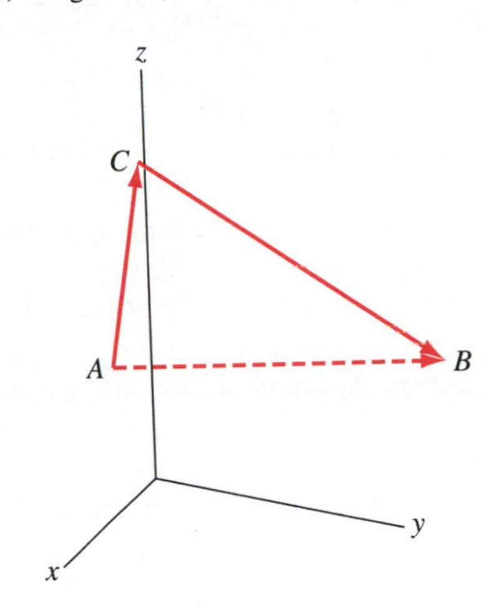

9. $x = 14, y = 16, z = 8$
11. Because $\mathbf{b} = 5\mathbf{a}$.
13. $(1, 2, 3, 4) + (5, -1, 2, 0) = (6, 1, 5, 4)$
    $2(7, 6, -3, 1) = (14, 12, -6, 2)$
    $\mathbf{a} + \mathbf{b} = (a_1 + b_1, \ldots, a_n + b_n)$; $k\mathbf{a} = (ka_1, \ldots, ka_n)$
15. $\overrightarrow{P_1P_2} = (1, -4, -1, 1)$
17. If your displacement vector is $\mathbf{a}$ and your friend's is $\mathbf{b}$, then $\mathbf{b} - \mathbf{a}$ is the displacement vector from you to your friend.
21. (b) The position vectors of points in the parallelogram determined by $(2, 2, 1)$ and $(0, 3, 2)$
23. (a) $\sqrt{5}$ units per minute
    (b) $(0, -4)$
    (c) 7 minutes
    (d) No
25. (a) $(5, 5, 4)$
    (b) $(-5, -5, -4)$

Section 1.2

1. $2\mathbf{i} + 4\mathbf{j}$
3. $3\mathbf{i} + \pi\mathbf{j} - 7\mathbf{k}$
5. $2\mathbf{i} + 4\mathbf{j}$
7. $(9, -2, \sqrt{2})$
9. $(\pi, -1)$
11. (a) $\mathbf{b} = 2\mathbf{a}_1 + \mathbf{a}_2$
    (b) $\mathbf{b} = -1\mathbf{a}_1 + 4\mathbf{a}_2$
    (c) Take $c_1 = (b_1 + b_2)/2$, $c_2 = (b_1 - b_2)/2$.
13. $x = t + 2, y = 3t - 1, z = 5 - 6t$
15. $x = t + 2, y = -7t - 1$
17. $x = t + 1, y = 4, z = 5 - 6t$

19. $x_1 = 1 - 2t, x_2 = 5t + 2, x_3 = 3t, x_4 = 7t + 4$

21. (a) $x = 2t - 1, y = 7 - t, z = 5t + 3$
(b) $x = 5 - 5t, y = 4t - 3, z = 5t + 4$
(c) One alternative for (a): $x = -2t - 1, y = t + 7, z = 3 - 5t$; one alternative for (b): $x = 5t, y = 1 - 4t, z = 9 - 5t$
(d) For (a): $(x + 1)/2 = (y - 7)/(-1) = (z - 3)/5$;
for (b): $(x - 5)/(-5) = (y + 3)/4 = (z + 4)/5$;
for (c): $(x + 1)/(-2) = (y - 7)/1 = (z - 3)/(-5)$ and $x/5 = (y - 1)/(-4) = (z - 9)/(-5)$.

23. Multiply the first symmetric form by $-\frac{1}{2}$. Then add $-\frac{1}{2}$ to each "side."

25. No. Setting $t = 0$ in the equations for l_1 gives the point $(2, -7, 1)$. To have $x = 2$ in l_2, we must take $t = \frac{1}{2}$. But $t = \frac{1}{2}$ gives the point $(2, -7, 2)$ on l_2, not $(2, -7, 1)$.

27. No, they do not. Note that $x \geq -1$, $y \geq 3$, and $z \leq 1$ only. The parametric equations define a *ray* with endpoint $(-1, 3, 1)$.

29. $(-14, 5, -18)$

31. With $x = 0$: $\left(0, \frac{13}{2}, \frac{7}{2}\right)$; with $y = 0$: $\left(-\frac{13}{3}, 0, \frac{17}{3}\right)$; with $z = 0$: $(7, 17, 0)$.

33. No

35. No

37. (a) Circle $x^2 + y^2 = 4$, traced once counterclockwise. If $0 \leq t \leq 2\pi$, circle is traced three times.
(b) Circle $x^2 + y^2 = 25$, traced once counterclockwise.
(c) Circle $x^2 + y^2 = 25$, traced once clockwise.
(d) Ellipse $x^2/25 + y^2/9 = 1$, traced once counterclockwise.

39. $x = (a + a\theta)\cos\theta, y = (a + a\theta)\sin\theta$

Section 1.3

1. $\mathbf{a}\cdot\mathbf{b} = 13$, $\|\mathbf{a}\| = \sqrt{26}$; $\|\mathbf{b}\| = \sqrt{13}$

3. $\mathbf{a}\cdot\mathbf{b} = -44$, $\|\mathbf{a}\| = \sqrt{50}$; $\|\mathbf{b}\| = 2\sqrt{14}$

5. $\mathbf{a}\cdot\mathbf{b} = 2$, $\|\mathbf{a}\| = \sqrt{26}$; $\|\mathbf{b}\| = \sqrt{3}$

7. $2\pi/3$

9. $\pi/2$

11. $\frac{5}{2}(\mathbf{i} + \mathbf{j})$

13. $(2\mathbf{i} - \mathbf{j} + \mathbf{k})/\sqrt{6}$

15. $\sqrt{3}(\mathbf{i} + \mathbf{j} - \mathbf{k})$

17. Yes, if $\mathbf{a}\cdot\mathbf{b} = 0$ or $\mathbf{a} = \mathbf{b}$.

21. (a) Work $= (\text{component of } \mathbf{F})\left\|\overrightarrow{PQ}\right\|$
$$= \|\mathbf{F}\|\cos\theta\left\|\overrightarrow{PQ}\right\| = \mathbf{F}\cdot\overrightarrow{PQ}$$
(b) 10,000 ft-lb

25. Hint: The diagonals are given by $\mathbf{d}_1 = \mathbf{a} + \mathbf{b}$ and $\mathbf{d}_2 = \mathbf{b} - \mathbf{a}$, where $\mathbf{a}$ and $\mathbf{b}$ determine the adjacent sides of the parallelogram.

Section 1.4

1. 2

3. -5

5. $(31, -5, 8)$

7. $5\mathbf{k}$

9. $-6\mathbf{i} + 14\mathbf{j} + 4\mathbf{k}$

11. $5\sqrt{30}$

13. $\mathbf{c}$ is parallel to the plane determined by $\mathbf{a}$ and $\mathbf{b}$ (or else $\mathbf{a}$ is parallel to $\mathbf{b}$).

15. $\sqrt{1002}/2$

17. $3\sqrt{34}/2$

19. 53

21. $$(\mathbf{a}\times\mathbf{b})\cdot\mathbf{c} = \begin{vmatrix} a_1 & a_2 & a_3 \\ b_1 & b_2 & b_3 \\ c_1 & c_2 & c_3 \end{vmatrix},$$
$$(\mathbf{b}\times\mathbf{c})\cdot\mathbf{a} = \begin{vmatrix} b_1 & b_2 & b_3 \\ c_1 & c_2 & c_3 \\ a_1 & a_2 & a_3 \end{vmatrix}.$$
Expand determinants to see they are equal.

23. (b) $\frac{1}{2}$

25. (a) $\mathbf{a}\times\mathbf{b}$
(b) $2\mathbf{a}\times\mathbf{b}/\|\mathbf{a}\times\mathbf{b}\|$
(c) $\left(\dfrac{\mathbf{a}\cdot\mathbf{b}}{\mathbf{a}\cdot\mathbf{a}}\right)\mathbf{a}$
(d) $\dfrac{\|\mathbf{b}\|}{\|\mathbf{a}\|}\mathbf{a}$
(e) $\mathbf{a}\times(\mathbf{b}\times\mathbf{c})$
(f) $(\mathbf{a}\times\mathbf{b})\times\mathbf{c}$

35. (a) $20\sqrt{3}$ ft-lb
(b) $30\sqrt{3}$ ft-lb

37. $75\sqrt{3}$ ft-lb

39. $400\pi/3$ in/min

41. Rotation about an axis that passes through the rigid body is very different from rotation about a parallel axis that does not pass through the body.

Section 1.5

1. $x - y + 2z = 8$

3. $5x - 4y + 3z = 25$

5. $5x - 4y + z = 12$

7. $3x - 2y - z = 3$

9. $x = 13t - \frac{23}{5}, y = -14t + \frac{24}{5}, z = -5t$

11. $-4/3$

13. $x = 2s + t - 1, y = 2 - 3s, z = s - 5t + 7$

15. $x = 2s + 5t - 1, y = 10t - 3s + 3, z = 4s + 7t - 2$

17. $x = 3s + 3t - 5, y = 10 - 3s - 6t, z = 2s - 2t + 9$

19. $19x - 16y + 7z = 59$

21. $31/\sqrt{34}$

23. $25/\sqrt{641}$

25. (a) 0
(b) The lines must intersect.

27. $\sqrt{14}/2$

31. $x + 3y - 5z = 2 \pm 3\sqrt{35}$

33. Hint: Consider Example 8 in §1.5 with A as P_1 and B as P_2.

Section 1.6

1. (a) $\mathbf{e}_1 + 2\mathbf{e}_2 + 3\mathbf{e}_3 + \cdots + n\mathbf{e}_n$
(b) $\mathbf{e}_1 - \mathbf{e}_3 + \mathbf{e}_4 - \mathbf{e}_6 + \cdots + \mathbf{e}_{n-2} - \mathbf{e}_n$

3. $(1, -2, 3, -4, \ldots, (-1)^{n+1}n)$

5. (a) $(3, -1, 11, -1, \ldots, 2n + (-1)^{n+1}2n - 1)$

(b) $(-1, 7, -1, \ldots, 2n + (-1)^n 2n - 1)$

(c) $(-3, -9, -15, \ldots, -6n + 3)$

(d) $\sqrt{1 + 9 + 25 + \cdots + (2n - 1)^2}$

(e) $2 - 12 + 30 + \cdots + (-1)^{n+1} 2n(2n - 1)$

9. Hint: Use the triangle inequality.

11. Hint: Square both sides of the equation and write as dot products.

13. Hyperplane in $\mathbf{R}^5$ passing through $(1, -2, 0, 4, -1)$ and perpendicular to the vector $\mathbf{n} = (2, 3, -7, 1, -5)$

15. (a) Total cost $=$ $(200, 250, 300, 375, 450, 500) \cdot (x_1, x_2, x_3, x_4, x_5, x_6)$

(b) $\{\mathbf{x} \in \mathbf{R}^6 \mid (200, \ldots, 500) \cdot (x_1, \ldots, x_6) \leq 100{,}000\}$. Budget hyperplane is $200x_1 + \cdots + 500x_6 = 100{,}000$.

17. $\begin{bmatrix} 5 & 8 & 8 \\ -2 & 5 & -2 \end{bmatrix}$

19. $\begin{bmatrix} -4 & 0 \\ 15 & -9 \\ 5 & 0 \end{bmatrix}$

21. -42

23. -240

25. (a) A lower triangular matrix is an $n \times n$ matrix whose entries above the main diagonal are zero.

(b) For an upper triangular matrix, use cofactor expansion about the first column or last row.

27. -289

31. $\begin{bmatrix} 1/2 & -1 & 1/2 \\ 0 & 1 & 0 \\ 0 & 0 & -1 \end{bmatrix}$

37. $\begin{bmatrix} \frac{3}{7} & -\frac{3}{14} & \frac{1}{7} \\ \frac{2}{7} & \frac{5}{14} & -\frac{4}{7} \\ -\frac{1}{7} & \frac{1}{14} & \frac{2}{7} \end{bmatrix}$

39. $(32, 46, 19, -35)$

Section 1.7

1. $(1, 1)$

3. $(3, 0)$

5. $(2\sqrt{2}, 3\pi/4)$

7. $(2\cos 2, 2\sin 2, 2)$

9. $\left(-1/2, \sqrt{3}/2, -2\right)$

11. $\left(0, 3\sqrt{3}/2, 3/2\right)$

13. $(0, 0, -2)$

15. $(2, 2\pi/3, 13)$

17. $(2\sqrt{2}, \pi/6, 7\pi/4)$

19. (a)

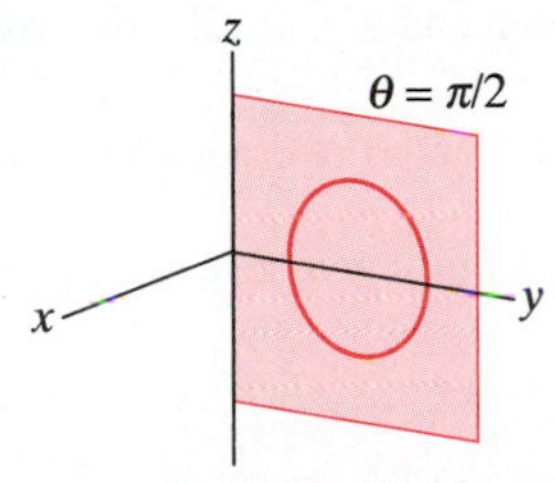

(b)

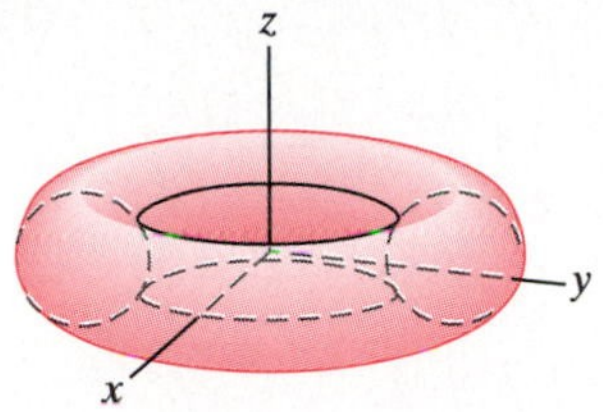

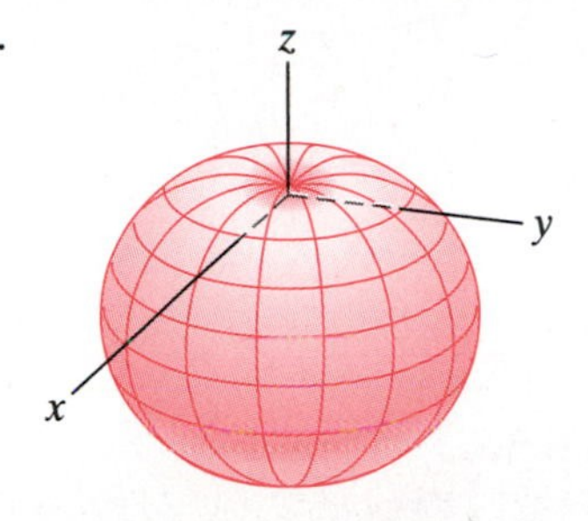

21.

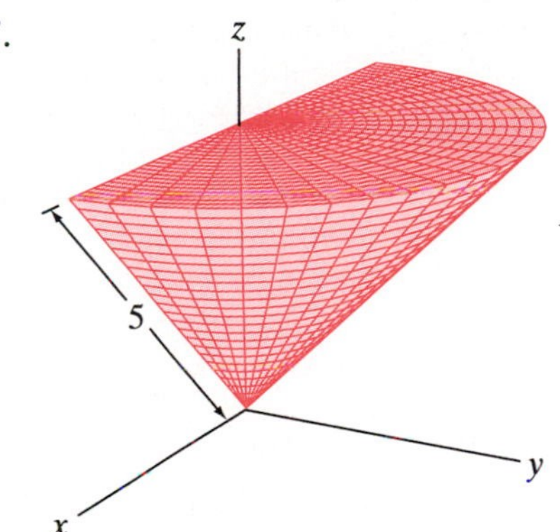

23. Cartesian: $y = 2$; cylindrical: $r \sin\theta = 2$; surface is a vertical plane.

25. Cartesian: $x = y = 0$; spherical: $\varphi = 0$ or π; object is the z-axis.

27.

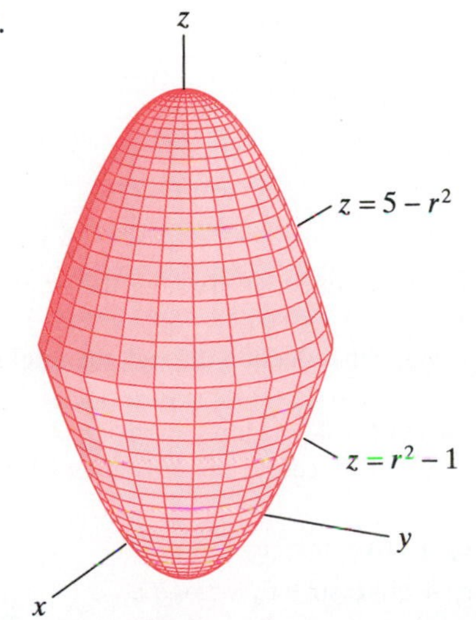

29.

31.

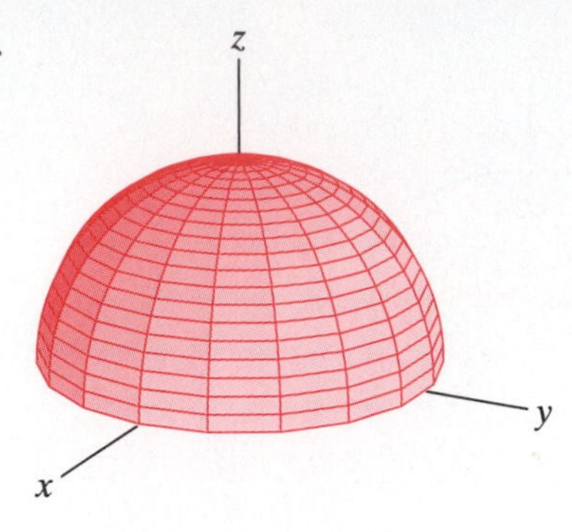

33.

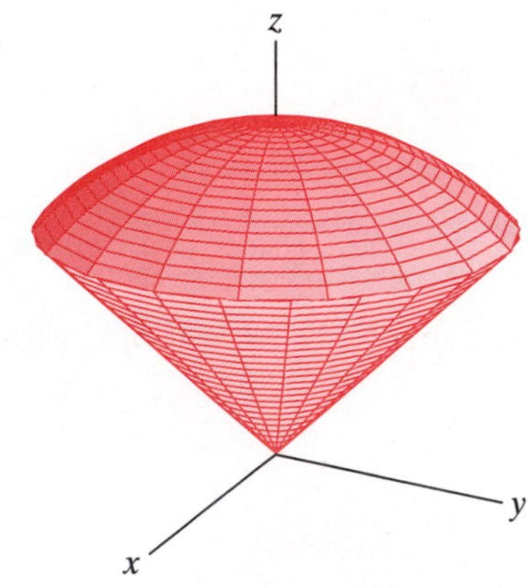

35.

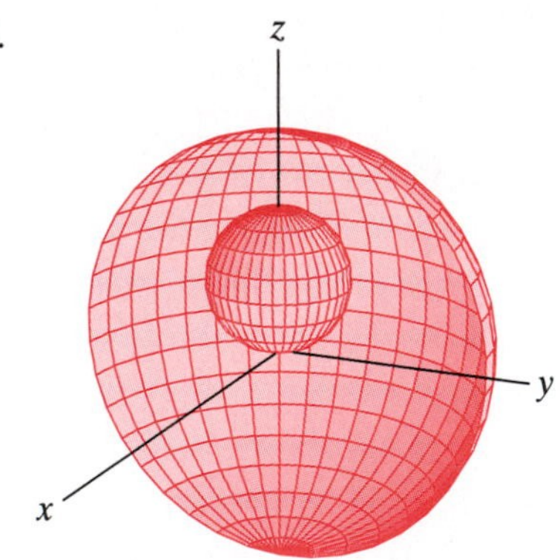

37. (a) $r = -f(\theta)$ is the reflection through the origin of $r = f(\theta)$.
(b) $\rho = -f(\varphi, \theta)$ is the reflection through the origin of $\rho = f(\varphi, \theta)$.
(c) $r = 3f(\theta)$ is a threefold magnification of $r = f(\theta)$.
(d) $\rho = 3f(\varphi, \theta)$ is a threefold magnification of $\rho = f(\varphi, \theta)$.

41. $\mathbf{i} = \sin\varphi\cos\theta\,\mathbf{e}_\rho + \cos\varphi\cos\theta\,\mathbf{e}_\varphi - \sin\theta\,\mathbf{e}_\theta$
$\mathbf{j} = \sin\varphi\sin\theta\,\mathbf{e}_\rho + \cos\varphi\sin\theta\,\mathbf{e}_\varphi + \cos\theta\,\mathbf{e}_\theta$
$\mathbf{k} = \cos\varphi\,\mathbf{e}_\rho - \sin\theta\,\mathbf{e}_\varphi$

True/False Exercises for Chapter 1

1. False. (The corresponding components must be equal.)
3. False. ($(-4, -3, -3)$ is the displacement vector from P_2 to P_1.)
5. False. (Velocity is a vector, but speed is a scalar.)
7. False. (The particle will be at $(2, -1) + 2(1, 3) = (4, 5)$.)
9. False. (From the parametric equations, we may read a vector parallel to the line to be $(-2, 4, 0)$. This vector is not parallel to $(-2, 4, 7)$.)
11. False. (The line has symmetric form $\dfrac{x-2}{-3} = y - 1 = \dfrac{z+3}{2}$.)
13. False. (The parametric equations describe a *semicircle* because of the restriction on t.)
15. False. ($\|k\mathbf{a}\| = |k|\|\mathbf{a}\|$.)
17. False. (Let $\mathbf{a} = \mathbf{b} = \mathbf{i}$, and $\mathbf{c} = \mathbf{j}$.)
19. True
21. True. (Check that each point satisfies the equation.)
23. False. (The product BA is not defined.)
25. False. ($\det(2A) = 2^n \det A$.)
27. False. (The surface with equation $\rho = 4\cos\varphi$ is a sphere.)
29. True

Miscellaneous Exercises for Chapter 1

3. $x = 63t + 1$, $y = 148t$, $z = 847t - 2$
5. (a) $x = 5t + 2$, $y = 3t - 2$
(b) $x = (b_2 - a_2)t + (a_1 + b_1)/2$, $y = (a_1 - b_1)t + (a_2 + b_2)/2$
7. (a) $2x_1 + 4x_2 - 2x_3 + x_4 - x_5 = 5$
(b) $(b_1 - a_1)x_1 + \cdots + (b_n - a_n)x_n = \frac{1}{2}(b_1^2 - a_1^2 + \cdots + b_n^2 - a_n^2)$
9. (a) No (b) No
11. (a) $\pi/3$
(b) $x = t$, $y = 1 - t$, $z = t$
13. The dot product measures the agreement of the answers.
15. Hint: $\mathbf{a} \times \mathbf{b}$ is normal to the plane determined by $\mathbf{a}$ and $\mathbf{b}$.
17. (a) $5\sqrt{21}$ (b) 10
19. (a) 1
21. Hint: Note that to determine $\mathbf{x}$, it suffices to determine $\|\mathbf{x}\|$ and the angle between $\mathbf{a}$ and $\mathbf{x}$. Consider the cases $c = 0$ and $c \neq 0$ separately.
23. (a) Hint: Let $\mathbf{a}$ and $\mathbf{b}$ denote two adjacent sides of the parallelogram.
(b) Hint: You should simply give a vector equation.
25. (b) $A^n = \begin{bmatrix} 1 & n \\ 0 & 1 \end{bmatrix}$

27. (a) $H_6 = \begin{bmatrix} 1 & \frac{1}{2} & \frac{1}{3} & \frac{1}{4} & \frac{1}{5} & \frac{1}{6} \\ \frac{1}{2} & \frac{1}{3} & \frac{1}{4} & \frac{1}{5} & \frac{1}{6} & \frac{1}{7} \\ \frac{1}{3} & \frac{1}{4} & \frac{1}{5} & \frac{1}{6} & \frac{1}{7} & \frac{1}{8} \\ \frac{1}{4} & \frac{1}{5} & \frac{1}{6} & \frac{1}{7} & \frac{1}{8} & \frac{1}{9} \\ \frac{1}{5} & \frac{1}{6} & \frac{1}{7} & \frac{1}{8} & \frac{1}{9} & \frac{1}{10} \\ \frac{1}{6} & \frac{1}{7} & \frac{1}{8} & \frac{1}{9} & \frac{1}{10} & \frac{1}{11} \end{bmatrix}$;

$\det H_2 = \dfrac{1}{12}$, $\det H_3 = \dfrac{1}{2160}$,

$\det H_4 = \dfrac{1}{6048000}$, $\det H_5 = \dfrac{1}{266716800000}$,

$\det H_6 = \dfrac{1}{186313420339200000}$

(b) $\det H_{10} \approx 2.16418 \times 10^{-53}$

(c) Most likely, AB and BA will *not* equal I_{10}.

29. $x = (a+b)\cos t - b\cos\left(\dfrac{a+b}{b}t\right)$,

$y = (a+b)\sin t - b\sin\left(\dfrac{a+b}{b}t\right)$

31. Hypotrochoid: $x = (a-b)\cos t + c\cos\left(\dfrac{a-b}{b}t\right)$,

$y = (a-b)\sin t - c\sin\left(\dfrac{a-b}{b}t\right)$;

Epitrochoid: $x = (a+b)\cos t - c\cos\left(\dfrac{a+b}{b}t\right)$,

$y = (a+b)\sin t - \sin\left(\dfrac{a+b}{b}t\right)$

33. (a)

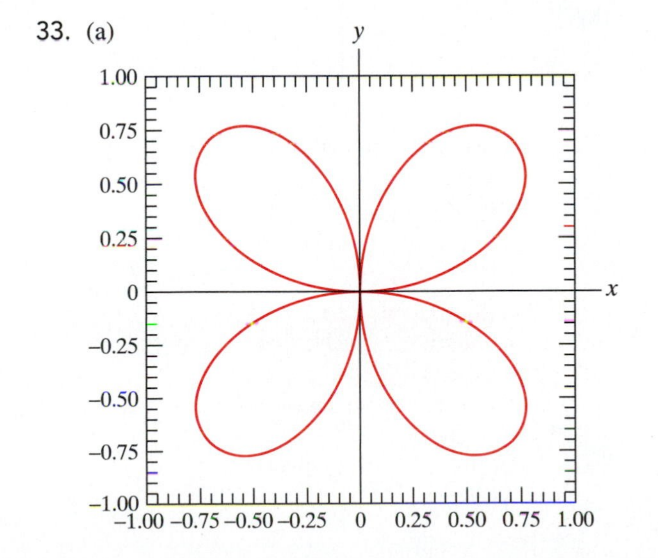

(b)

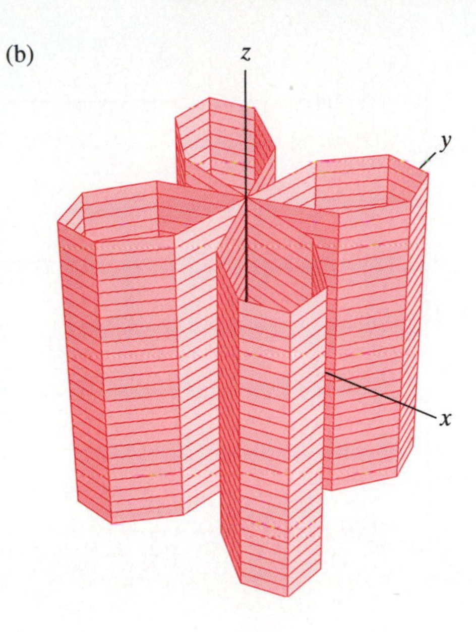

(c)

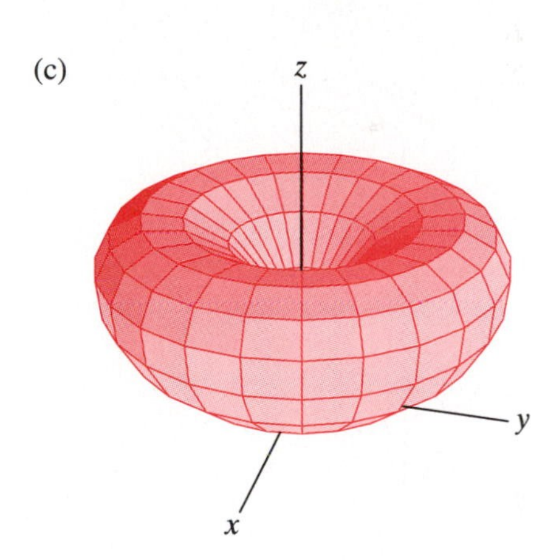

(d)

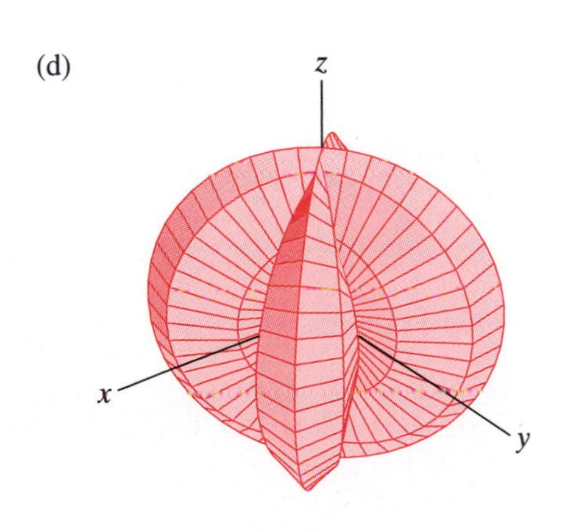

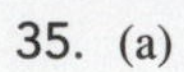

35. (a)

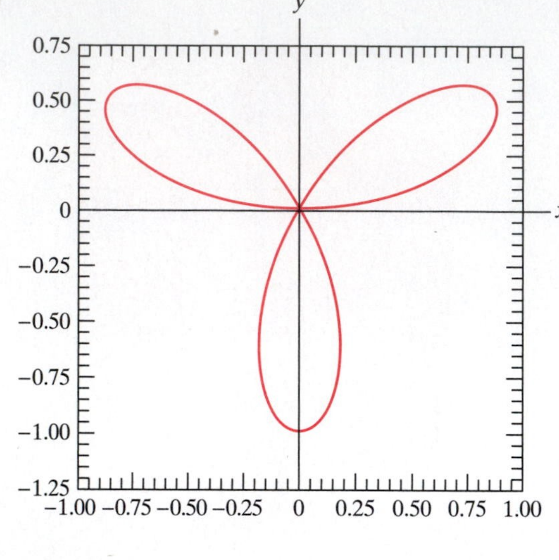
y
x
0.75
0.50
0.25
0
−0.25
−0.50
−0.75
−1.00
−1.25
−1.00 −0.75 −0.50 −0.25 0 0.25 0.50 0.75 1.00

(b)

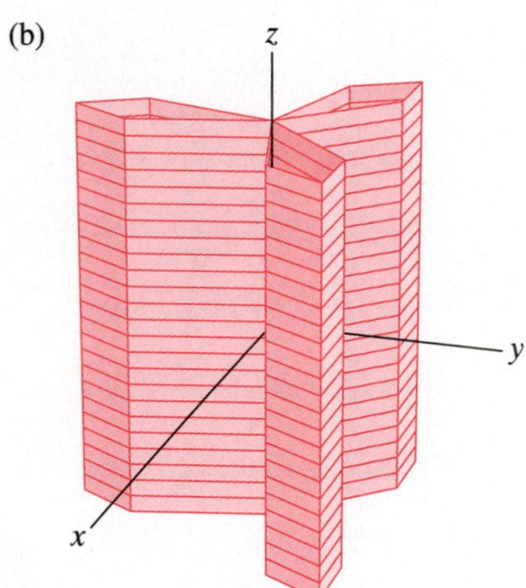
z
y
x

(c)

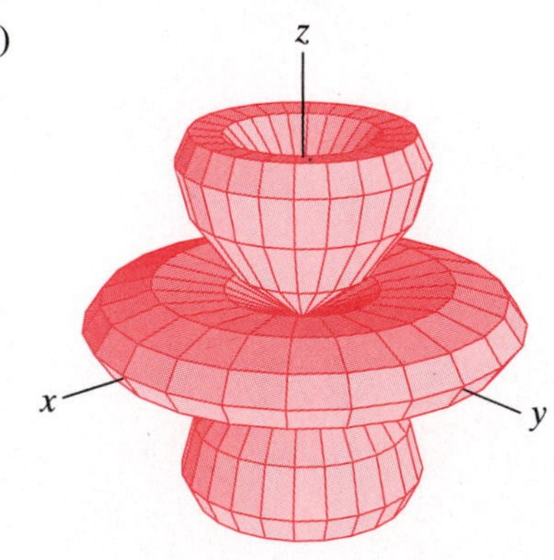
z
x
y

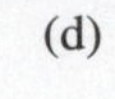

(d)

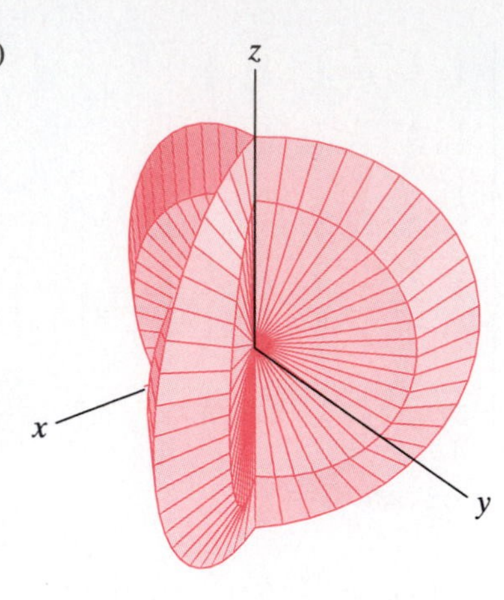
z
x
y

37. (a)

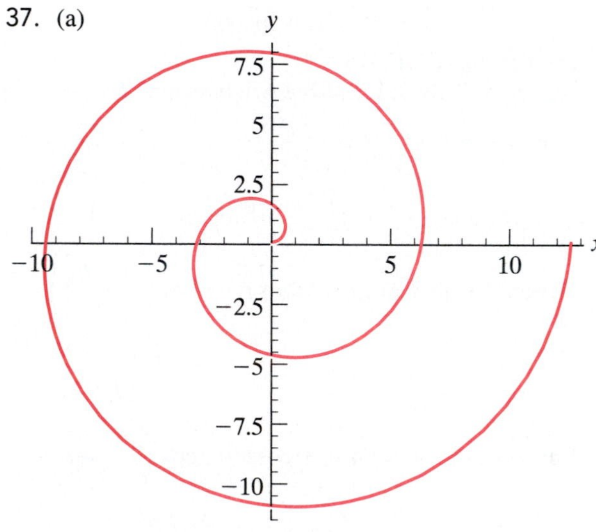
y
x
7.5
5
2.5
−2.5
−5
−7.5
−10
−10 −5 5 10

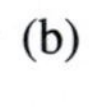

(b)

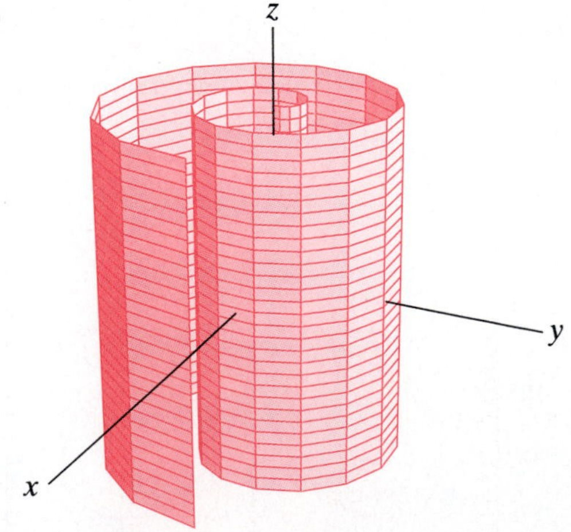
z
y
x

(c)

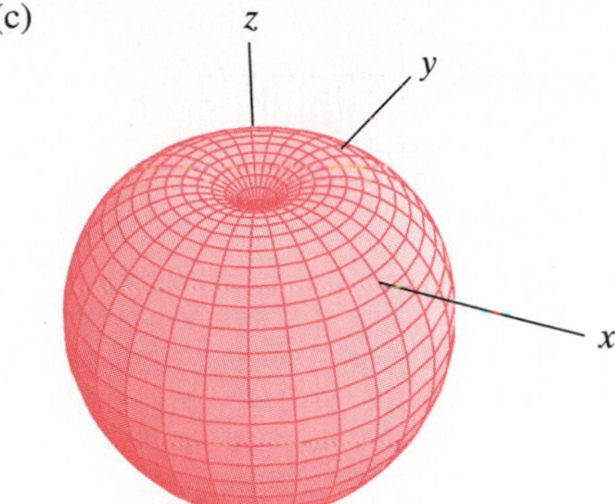

(d)

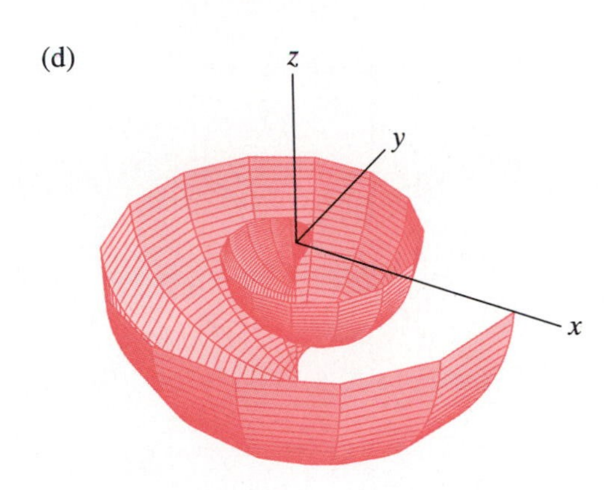

39. (a) $\{(r, \theta, z) \mid 0 \leq r \leq 3, 0 \leq \theta \leq 2\pi, 0 \leq z \leq 3\}$ if the cylinder is positioned so that the center of the bottom is at the origin and its axis is the z-axis.
    (b) Using the same positioning as in part (a), $\{(\rho, \varphi, \theta) \mid 0 \leq \rho \leq 3 \sec\varphi, 0 \leq \varphi \leq \pi/4, 0 \leq \theta \leq 2\pi\} \cup \{(\rho, \varphi, \theta) \mid 0 \leq \rho \leq 3 \csc\varphi, \pi/4 \leq \varphi \leq \pi/2, 0 \leq \theta \leq 2\pi\}$.

Chapter 2

Section 2.1

1. (a) Domain = $\mathbf{R}$; range = $\{x \mid x \geq 1\}$
   (b) No
   (c) No
3. Domain = $\{(x, y) \mid y \neq 0\}$; range = $\mathbf{R}$
5. Domain = $\mathbf{R}^3$; range = $[0, \infty)$
7. Domain = $\{(x, y) \mid y \neq 1\}$;
   range = $\{(x, y, z) \mid y \neq 0, y^2z = (xy - y - 1)^2 + (y + 1)^2\}$

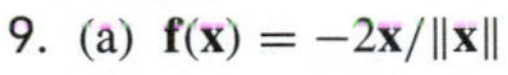

9. (a) $\mathbf{f}(\mathbf{x}) = -2\mathbf{x}/\|\mathbf{x}\|$
   (b) $f_1(x, y, z) = -2x/\sqrt{x^2 + y^2 + z^2}$,
   $f_2(x, y, z) = -2y/\sqrt{x^2 + y^2 + z^2}$,
   $f_3(x, y, z) = -2z/\sqrt{x^2 + y^2 + z^2}$

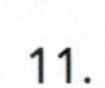

11.

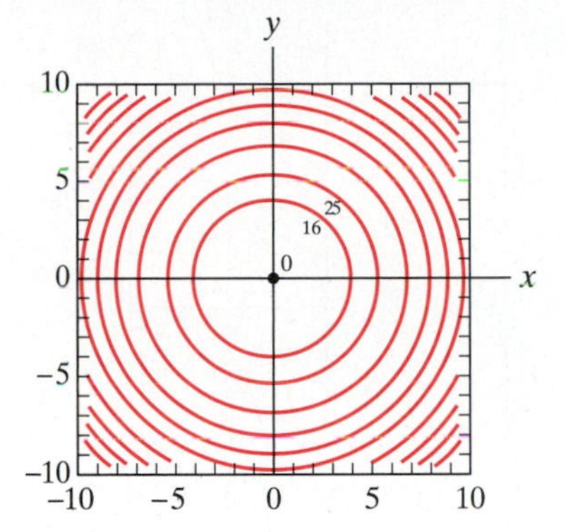

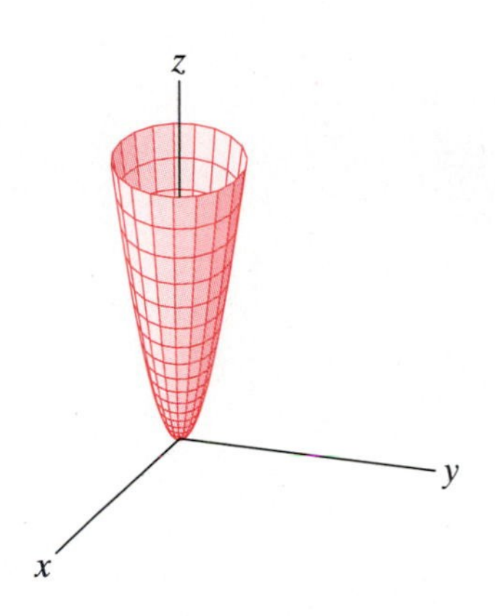

13.

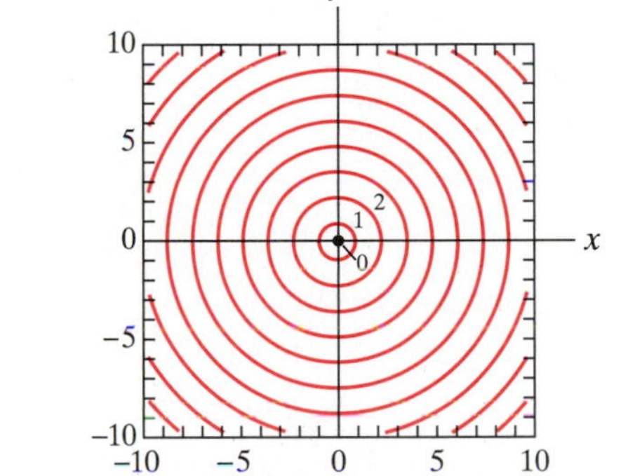

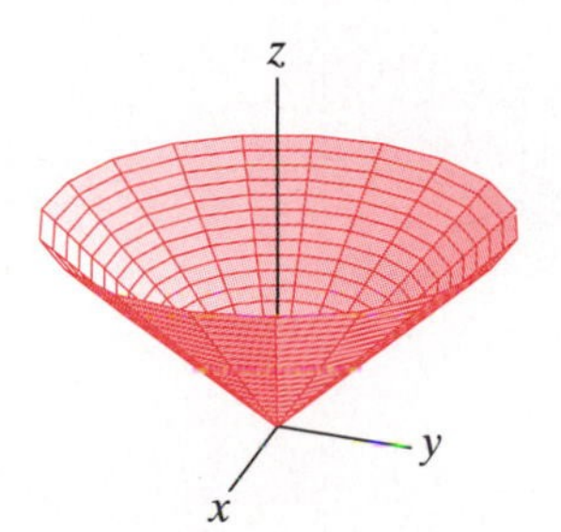

15.

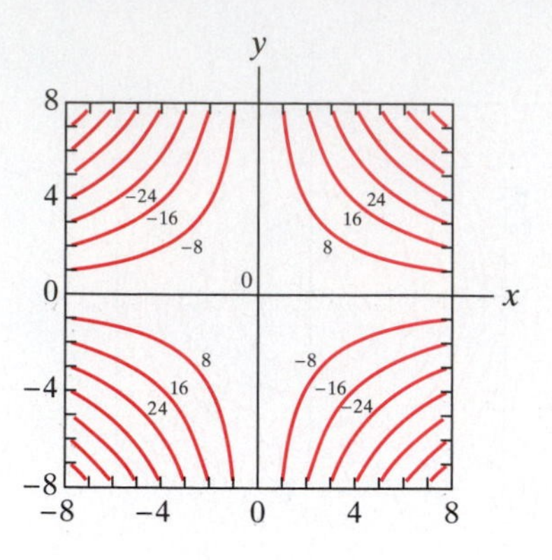

z
y
x

19.

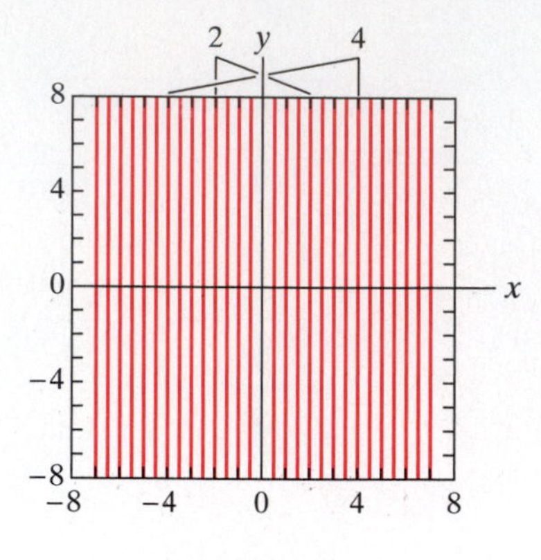

z
y
x

17.

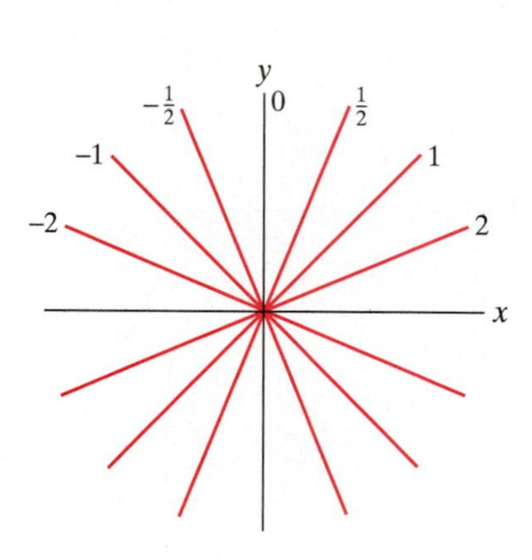

z
y
x

21.

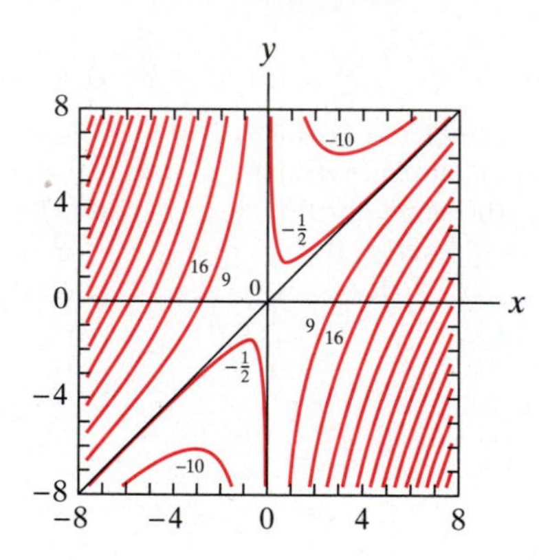

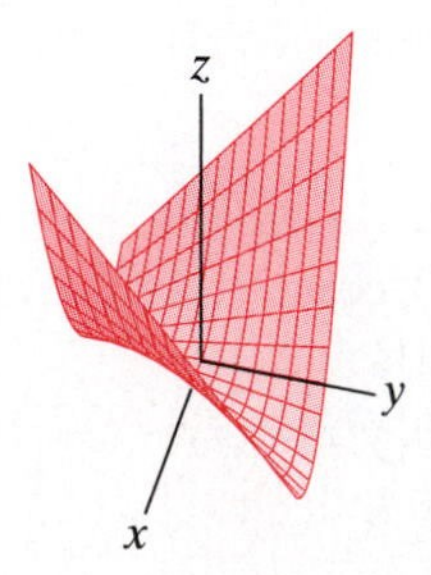

23.

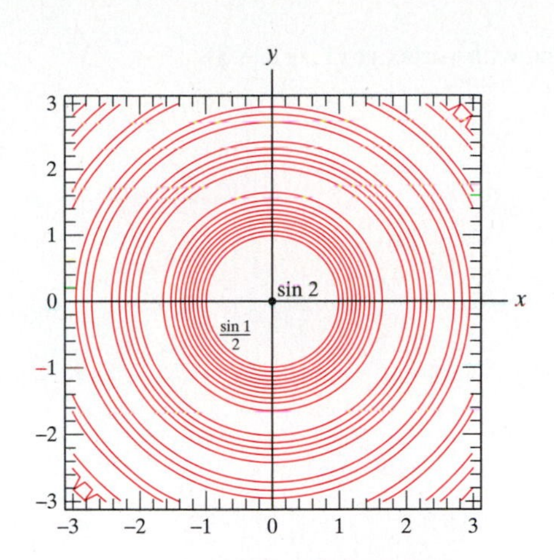

25. (a)

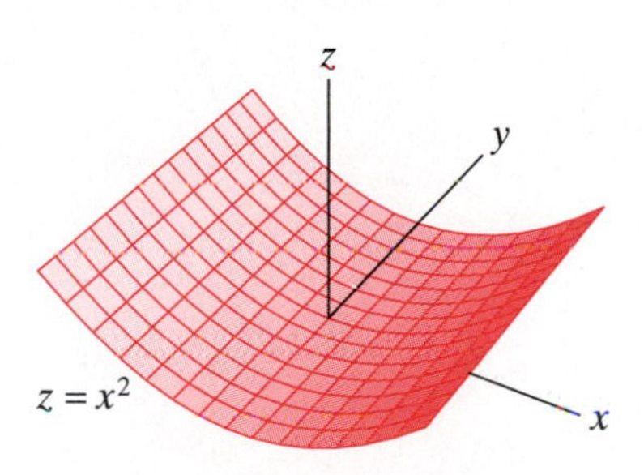

$z = y^2$

(c)

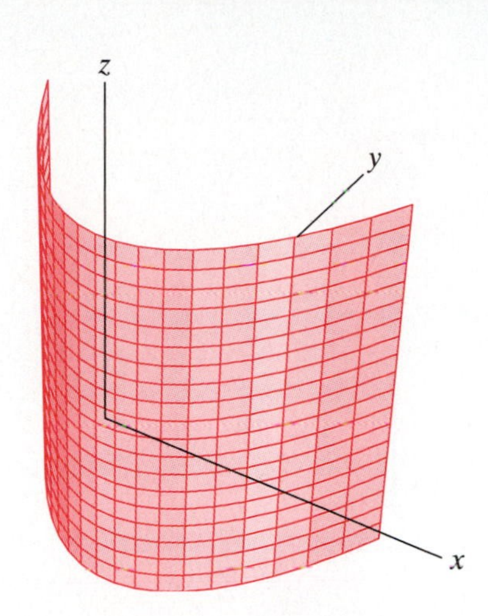

27. No

29.

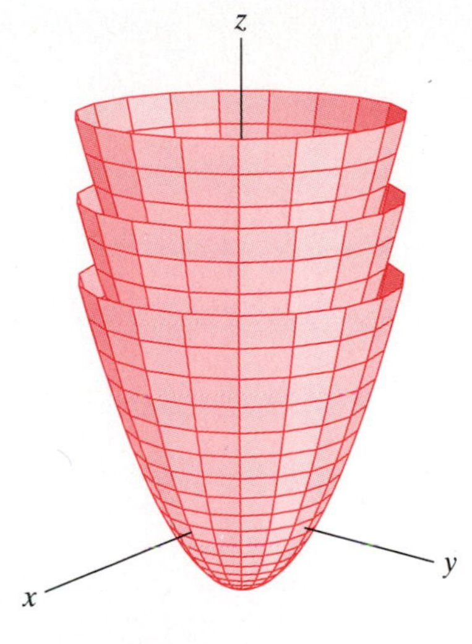

31.

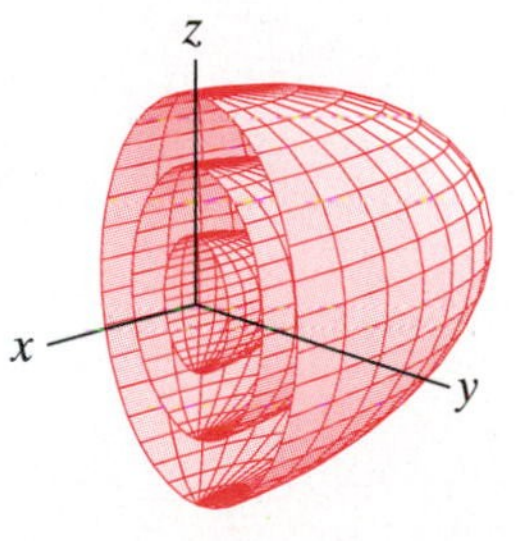

33. (a)

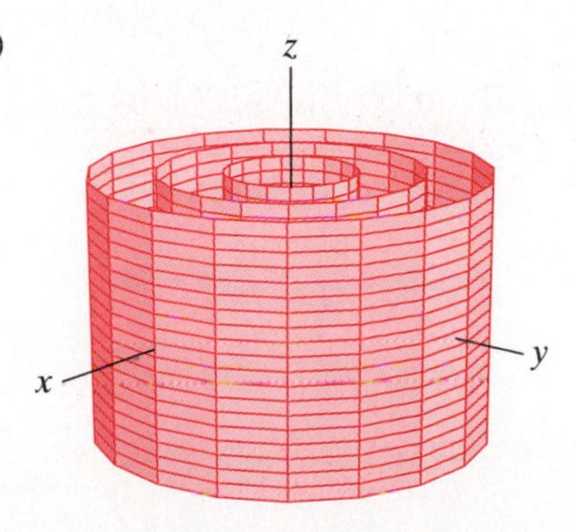

35.

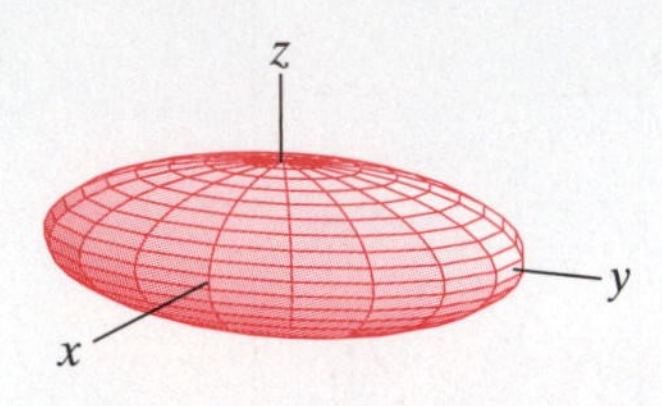

37.

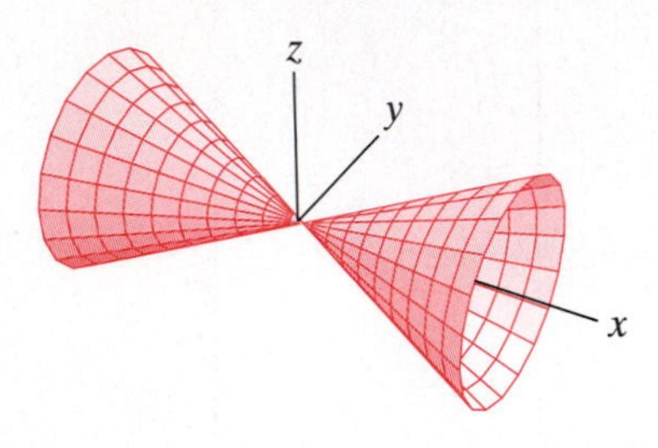

39.

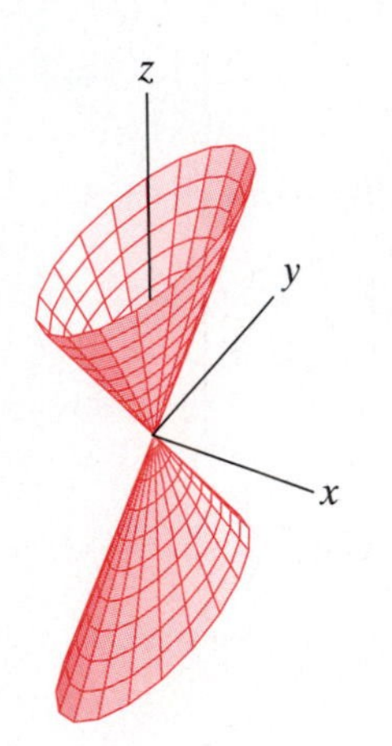

41.

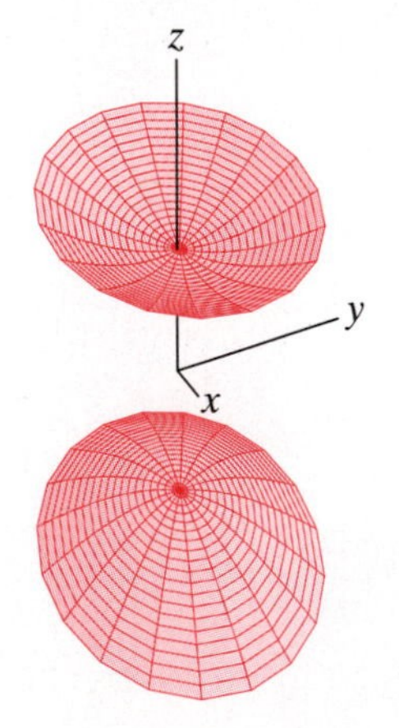

43. Cone with vertex at $(1, -1, -3)$

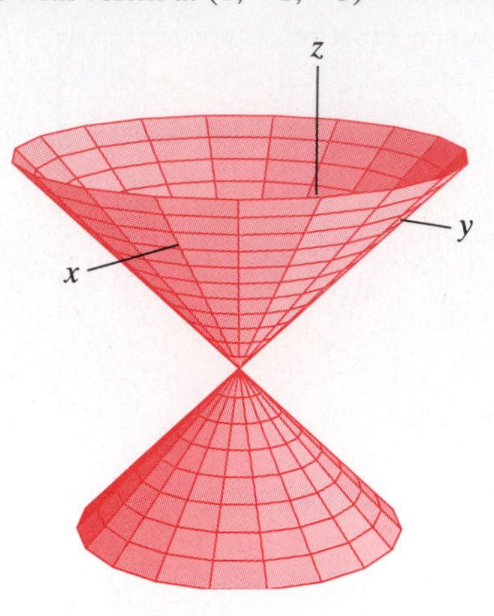

45. Ellipsoid centered at $(-1, 0, 0)$

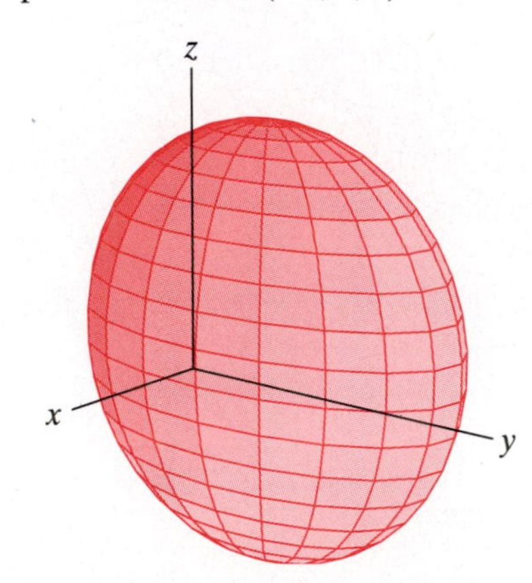

47. Paraboloid with vertex at $(3, 0, 1)$

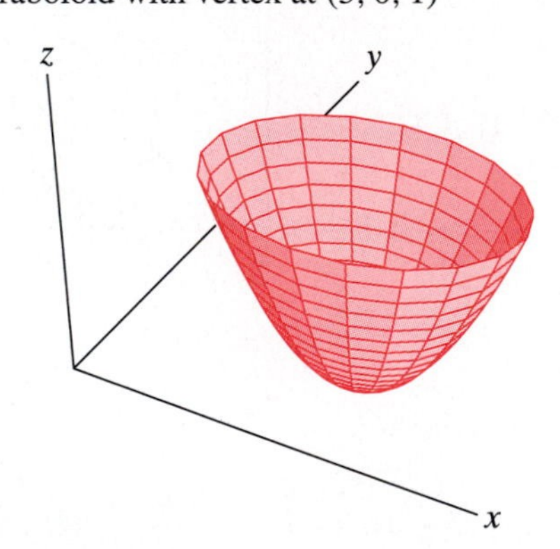

Section 2.2

1. Open
3. Neither
5. Neither
7. 2
9. Does not exist
11. Does not exist
13. 0
15. 0
17. 0
19. -1
21. Limit does not exist.
23. Limit does not exist.

25. Limit does not exist.

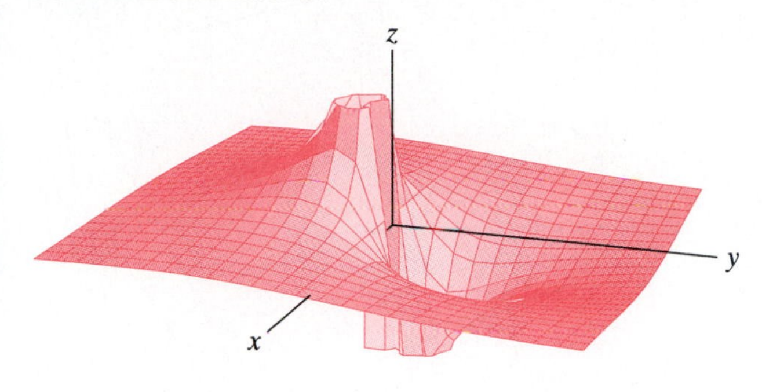

27. Limit is 0.

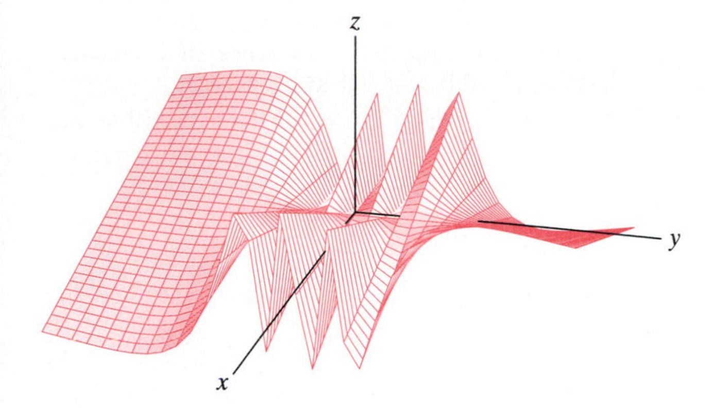

29. Limit does not exist.

31. 0

33. Limit does not exist.

35. Continuous

37. Continuous

39. Not continuous at (0, 0)

41. Continuous

43. Hint: Write $f(\mathbf{x})$ in terms of the components of $\mathbf{x}$.

Section 2.3

1. $f_x(x, y) = y^2 + 2xy$, $f_y(x, y) = 2xy + x^2$

3. $f_x(x, y) = y\cos xy - y\sin xy$,
$f_y(x, y) = x\cos xy - x\sin xy$

5. $f_x(x, y) = 4xy^2/(x^2 + y^2)^2$,
$f_y(x, y) = -4x^2y/(x^2 + y^2)^2$

7. $f_x(x, y) = -3x^2y\sin x^3y$,
$f_y(x, y) = -x^3\sin x^3y$

9. $F_x = x/\sqrt{x^2 + y^2 + z^2}$,
$F_y = y/\sqrt{x^2 + y^2 + z^2}$,
$F_z = z/\sqrt{x^2 + y^2 + z^2}$

11. $F_x = \dfrac{1 - 2x^2 - 3xy + y^2 - 3xz + z^2}{(1 + x^2 + y^2 + z^2)^{5/2}}$,
$F_y = \dfrac{1 + x^2 - 3xy - 2y^2 - 3yz + z^2}{(1 + x^2 + y^2 + z^2)^{5/2}}$,
$F_z = \dfrac{1 + x^2 + y^2 - 3xz - 3yz - 2z^2}{(1 + x^2 + y^2 + z^2)^{5/2}}$

13. $F_x = \dfrac{x^4 - 2xyz + 3x^2z^2 + 3x^2}{(x^2 + z^2 + 1)^2}$,
$F_y = \dfrac{z}{x^2 + z^2 + 1}$,
$F_z = \dfrac{x^2y - 2x^3z - yz^2 + y}{(x^2 + z^2 + 1)^2}$

15. $-\frac{1}{6}\mathbf{i}$

17. $-\mathbf{i} + (2\pi + 1)\mathbf{j} - 2\mathbf{k}$

19. $\mathbf{i} + \mathbf{j} - 2\mathbf{k}$

21. $\begin{bmatrix} 0 & -2 & 0 \\ 1/\sqrt{5} & 0 & -2/\sqrt{5} \end{bmatrix}$

23. $\begin{bmatrix} 3 & -7 & 1 & 0 \\ 5 & 0 & 2 & -8 \\ 0 & 1 & -17 & 3 \end{bmatrix}$

25. $\begin{bmatrix} -2 & 0 \\ 1 & -1 \\ 0 & 2 \end{bmatrix}$

29. (a) The function has continuous partial derivatives.
(b) $z = 3x + 8y + 3$

31. $z = e(x + y)$

33. $x_5 = -4x_1 + 12x_2 - 4x_3 - 6x_4 + 34$

35. (a) $h(0.1, -0.1) = 1$
(b) $f(0.1, -0.1) = 1$

37. (a) $h(1.01, 1.95, 2.2) = 21.76$
(b) $f(1.01, 1.95, 2.2) = 21.6657$

39. (a) $f_x(x, y) = (3x^4 + 8x^2y^2 + y^4)/(x^2 + y^2)^2$,
$f_y(x, y) = -(x^4 + 4x^3y + 2x^2y^2 + y^4)/(x^2 + y^2)^2$
(b) $f_x(0, 0) = 3$, $f_y(0, 0) = -1$

41. (a) $(8 + 4\sqrt{2})x - 8y = \sqrt{2}(\pi - 4)$
(b)

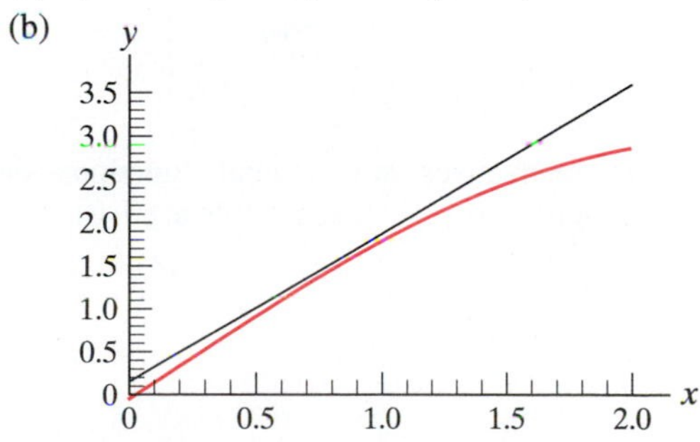

43. (a) $x + y = \ln 2 - 1$

(b)

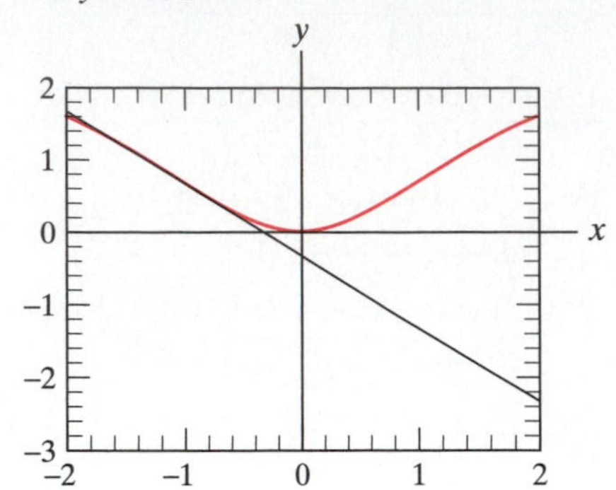

45. (a) $z = 11x - 15$

(b)

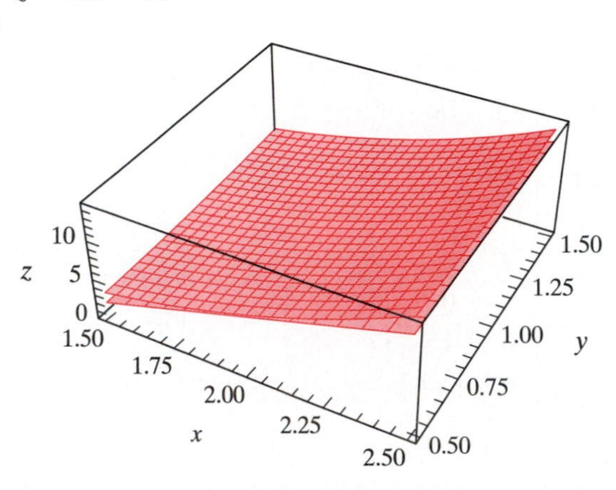

(c) Partial derivatives are polynomials—hence continuous—so f is differentiable at (2, 1).

47. (a) $z = 0$

(b)

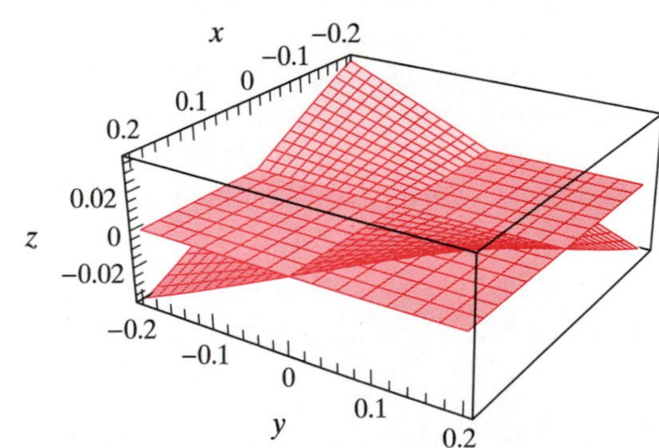

(c) Partial derivatives are rational functions—hence continuous—so f is differentiable at (0, 0).

49. (a) $z = -\frac{\pi}{288}[(9\sqrt{3}\pi - 96\sqrt{2})x - (16\sqrt{2}\pi + 72)y + (4\sqrt{2} - 3\sqrt{3})\pi^2 + (16\sqrt{2} + 9)\pi]$

(b)

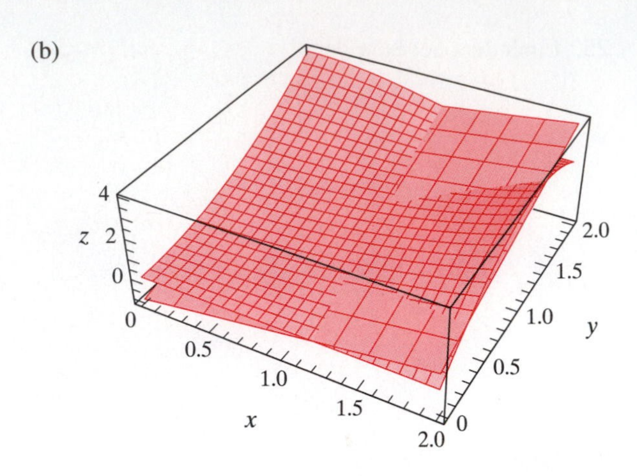

(c) Partial derivatives are products of polynomials and trigonometric functions and hence are continuous. Thus, f is differentiable at $(\pi/3, \pi/4)$.

51. $D\mathbf{f}(\mathbf{x}) = A$. Note that $f'(x) = a$ in the one-variable case.

Section 2.4

1. $D(f+g)$
$= \begin{bmatrix} y - \sin x + y\cos(xy) & x + 3y^2 + x\cos(xy) \end{bmatrix}$

3. $D(\mathbf{f}+\mathbf{g})$
$= \begin{bmatrix} \sin y + 3x^2\cos x - x^3\sin x & x\cos y & 1 \\ -6x + yz & e^z + xz & ye^z + xy \end{bmatrix}$

5. $D(fg) = \begin{bmatrix} 3x^2 + y^2 & 2xy \end{bmatrix}$,

$D(f/g) = \begin{bmatrix} y^2 - y^4/x^2 & 2xy + 4y^3/x \end{bmatrix}$

7. $D(fg)$
$= [12x^3y + 3x^2y^5 - 12xy^3 - 2y^7 \quad 3x^4 + 5x^3y^4 - 18x^2y^2 - 14xy^6]$,
$D(f/g)$
$= \begin{bmatrix} \dfrac{y(2y^6 - 6x^3 - 3x^2y^4)}{x^2(x^2 - 2y^2)^2} & \dfrac{3x^3 + 6xy^2 + 5x^2y^4 - 6y^6}{x(2y^2 - x^2)^2} \end{bmatrix}$

9. $f_{xx} = 6xy^7$, $f_{yy} = 42x^3y^5 + 6x$, $f_{xy} = f_{yx} = 21x^2y^6 + 6y - 7$

11. $f_{xx} = -ye^{-x} + 2yx^{-3}e^{y/x} + y^2x^{-4}e^{y/x}$, $f_{yy} = x^{-2}e^{y/x}$, $f_{xy} = f_{yx} = e^{-x} - x^{-2}e^{y/x} - yx^{-3}e^{y/x}$

13. $f_{xx} = \dfrac{4(8e^y\cos 2x + \cos 4x + 2\cos 2x - 3)}{(\cos 2x - 4e^y - 1)^3}$, $f_{yy} = \dfrac{8e^y(4e^y + \cos 2x - 1)}{(4e^y - \cos 2x + 1)^3}$, $f_{xy} = f_{yx} = \dfrac{4e^y\sin 2x}{(2e^y + \sin^2 x)^3}$

15. $f_{xx} = 2yz$, $f_{yy} = 2xz$, $f_{zz} = 2xy$, $f_{xy} = f_{yx} = 2xz + 2yz + z^2$, $f_{xz} = f_{zx} = 2xy + y^2 + 2yz$, $f_{yz} = f_{zy} = x^2 + 2xy + 2xz$

17. $f_{xx} = b^2e^{bx}\cos z + a^2e^{ax}\sin y$, $f_{yy} = -e^{ax}\sin y$, $f_{zz} = -e^{bx}\cos z$, $f_{xy} = f_{yx} = ae^{ax}\cos y$, $f_{xz} = f_{zx} = -be^{bx}\sin z$, $f_{yz} = f_{zy} = 0$

19. (a) p_x and p_y have degree 16; p_{xx}, p_{yy}, and p_{xy} all have degree 15.

(b) p_x and p_y have degree 3; p_{xx} has degree 2; p_{yy} has undefined degree; p_{xy} has degree 2.

(c) The degree of $\partial^k p/\partial x_{i_1} \cdots \partial x_{i_k}$ is $d-k$, where d is the highest degree of a term of p of the form $x_1^{d_1} x_2^{d_2} \cdots x_n^{d_n}$ such that, for $j = 1, 2, \ldots, n$, d_j is at least the number of times x_j occurs in the partial derivative. If p has no such term, then the degree of $\partial^k p/\partial x_{i_1} \cdots \partial x_{i_k}$ is undefined.

21. (a)

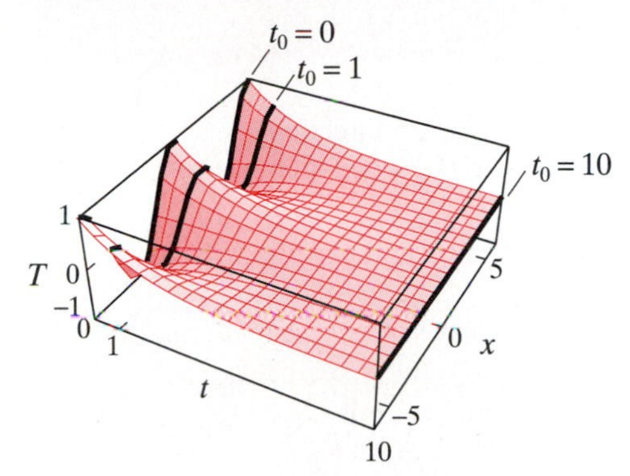

(b)

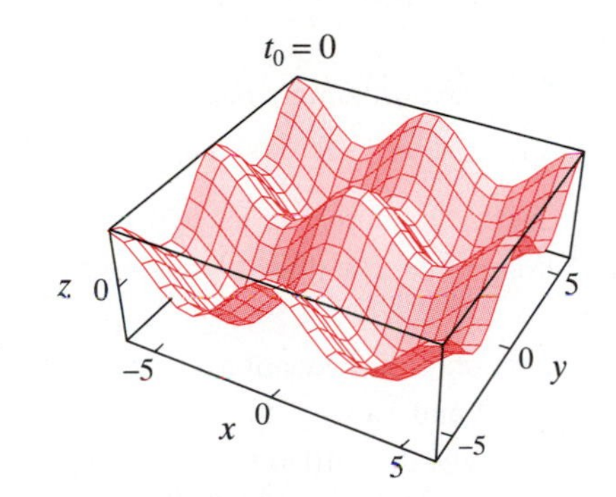

$t_0 = 1$

$t_0 = 10$

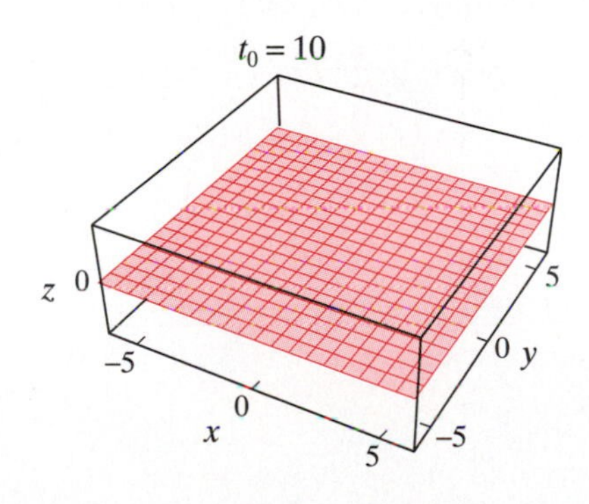

25. (a)

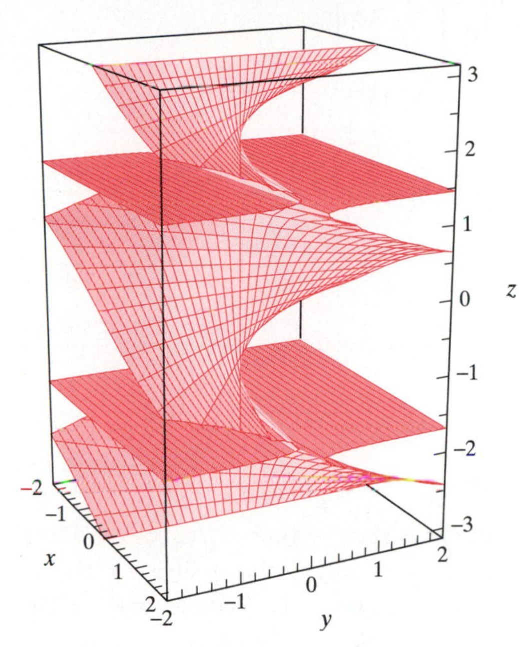

27. (a) $\mathbf{x}_0 = (-1, 1)$ leads to $(-1.2649111, 1.54919334)$

(b) $\mathbf{x}_0 = (1, -1)$ leads to $(1.26491106, -1.5491933)$, while $\mathbf{x}_0 = (-1, -1)$ leads to $(-1.2649111, -1.5491933)$

(c) It appears that an initial vector leads to an intersection point that lies in the same quadrant.

31. $(-0.9070154, -0.9070154)$

Section 2.5

1. $df/dt = (18t^2 + 14t)\sin 2t - 6\cos 2t \sin^2 2t + (12t^3 + 14t^2)\cos 2t + 72t + 84$

2. $\partial f/\partial s = (s^2 + t^2 + 2s^2 + 2st)\cos((s+t)(s^2+t^2))$,
$\partial f/\partial t = (s^2 + t^2 + 2st + 2t^2)\cos((s+t)(s^2+t^2))$

3. (a) $dP/dt = (36 - 27\pi)/\sqrt{2}$ atm/min
(c) Approximately $(.36 + (9 - .27)\pi)/\sqrt{2} \approx 19.6477$ atm

5. -6 in^3/min. Decreasing.

7. 0.766 units/month

11. Hint: $\partial w/\partial x = (dw/du)((y^3 - x^2 y)/(x^2 + y^2)^2)$ from the chain rule.

13. Hint: $\partial w/\partial x = -1/x^2(\partial w/\partial u + \partial w/\partial v)$ from the chain rule.

15. $D(\mathbf{f} \circ g) = \begin{bmatrix} 15(s-7t)^4 & -105(s-7t)^4 \\ 2e^{2s-14t} & -14e^{2s-14t} \end{bmatrix}$

17. $D(\mathbf{f} \circ \mathbf{g}) = \begin{bmatrix} 3s^2 - t^2 & -2st \\ 6s^5t^3 - s^{-2}t^{-2} & 3s^6t^2 - 2s^{-1}t^{-3} \end{bmatrix}$

19. $D(\mathbf{f} \circ \mathbf{g})$
$= \begin{bmatrix} t+u & s+u & s+t \\ 3s^2t^3 - tu^2e^{stu^2} & 3s^3t^2 - su^2e^{stu^2} & -2stue^{stu^2} \end{bmatrix}$

21. (a) $\begin{bmatrix} 7 & 10 \\ 31 & 44 \end{bmatrix}$

(b) $\begin{bmatrix} 1 & 13 \\ 2 & 31 \end{bmatrix}$

23. (a) $\frac{\partial^2}{\partial x^2} = \cos^2\theta \frac{\partial^2}{\partial r^2} + \frac{\sin^2\theta}{r}\frac{\partial}{\partial r} - \frac{2\sin\theta\cos\theta}{r}\frac{\partial^2}{\partial r\partial\theta} + \frac{2\sin\theta\cos\theta}{r}\frac{\partial}{\partial\theta} + \frac{\sin^2\theta}{r}\frac{\partial^2}{\partial\theta^2}$,

$\frac{\partial^2}{\partial y^2} = \sin^2\theta \frac{\partial^2}{\partial r^2} + \frac{\cos^2\theta}{r}\frac{\partial}{\partial r} + \frac{2\sin\theta\cos\theta}{r}\frac{\partial^2}{\partial r\partial\theta} - \frac{2\sin\theta\cos\theta}{r}\frac{\partial}{\partial\theta} + \frac{\cos^2\theta}{r^2}\frac{\partial^2}{\partial\theta^2}$

25. (a) $\frac{\partial}{\partial r} = \sin\varphi \frac{\partial}{\partial\rho} + \frac{\cos\varphi}{\rho}\frac{\partial}{\partial\varphi}$

27. $dy/dx = (2xy^7 - y\cos xy)/(x\cos xy - 7x^2y^6 + e^y)$

29. $\partial z/\partial x = 3x^2z^3/(yz^2\sin z + \sin y - x^3z^2)$, $\partial z/\partial y = (z^2\cos z + z\cos y)/(yz^2\sin z + \sin y - x^3z^2)$

31. (a) $\left(\frac{\partial w}{\partial x}\right)_{y,z} = 1, \left(\frac{\partial w}{\partial y}\right)_{x,z} = 7, \left(\frac{\partial w}{\partial z}\right)_{x,y} = -10; \left(\frac{\partial w}{\partial x}\right)_{y} = 1 - 20x, \left(\frac{\partial w}{\partial y}\right)_{x} = 7 - 20y$

(b) $\left(\frac{\partial w}{\partial x}\right)_{y} = \left(\frac{\partial w}{\partial x}\right)_{y,z}\left(\frac{\partial x}{\partial x}\right)_{y} + \left(\frac{\partial w}{\partial y}\right)_{x,z}\left(\frac{\partial y}{\partial x}\right)_{y} + \left(\frac{\partial w}{\partial z}\right)_{x,y}\left(\frac{\partial z}{\partial x}\right)_{y} = (1)(1) + (7)(0) + (-10)(2x)$

33. $\left(\frac{\partial s}{\partial z}\right)_{x,y,w} = xw - 2z$;

$\left(\frac{\partial s}{\partial z}\right)_{x,w} = xw - 2z + \frac{x^2y^3 - x^3}{xw - 3y^2z}$

Section 2.6

1. (a) The directional derivative of f at (x, y, z) in the negative z-direction.
(b) $\nabla f(x, y, z) \cdot (-\mathbf{k}) = -\partial f/\partial z$

3. $8\sqrt{5}/5$

5. $(2e - 9)\sqrt{5}/5$

7. $-4\sqrt{3}\,e^{-14}$

9. (a) $f_x(0, 0) = f_y(0, 0) = 0$
(b) $D_{\mathbf{v}}f(0, 0) = v|w|$ for all unit vectors $\mathbf{v} = v\,\mathbf{i} + w\,\mathbf{j}$.

(c)

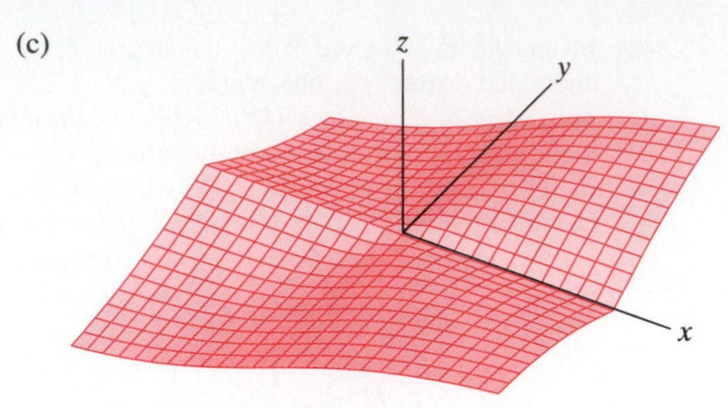

11. (a) $(-2\mathbf{i} + \mathbf{j})/\sqrt{5}$ direction
(b) $\pm(\mathbf{i} + 2\mathbf{j})/\sqrt{5}$ direction

13.

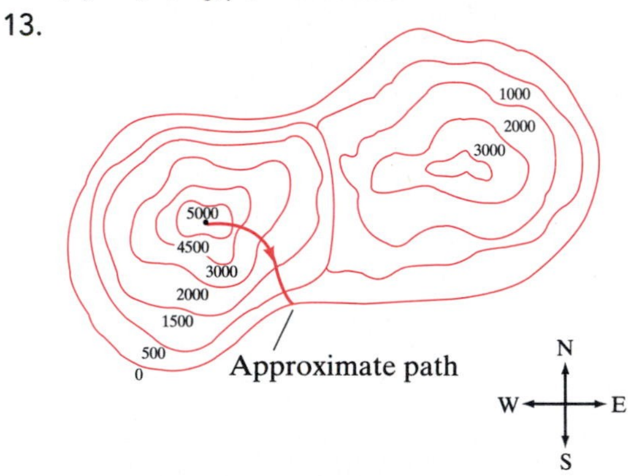

15. Travel along $y^3 = 27x$ (toward the origin)

17. $y - z = 1$

19. $x - 2y - 2z = 2$

21. The tangent plane has equation $x - \pi z = 2$.

23. $\left(\frac{5}{4}, -\frac{5}{4}, -\frac{9}{2}\right)$ and $\left(-\frac{5}{4}, \frac{5}{4}, \frac{9}{2}\right)$

27. (a) $3x + 12y + 2z + 10 = 0$
(b) There is no tangent plane at $(0, 0, 0)$.

29. Tangent line is $5x - 3\sqrt[3]{4}\,y = -1$

31. Parametric equations: $x = 5t + 5$, $y = 4t - 4$. Cartesian equation: $4x - 5y = 40$.

33. Parametric equations: $x = 14t + 2$, $y = t - 1$. Cartesian equation: $x - 14y = 16$.

35. $x = -2t - 1$, $y = t - 1$, $z = t - 1$

37. $x_5 - x_1 = \pi$

39. $x_1 + x_2 + \cdots + x_{n-1} - x_n = \sqrt{n}$

41. (a) Near all points of S such that $x \neq 0$. At such points $f(x, y) = \ln(1 - \sin xy - x^3y)/x$.
(b) All points $(0, y, z)$ (i.e., the yz-plane).
(c)

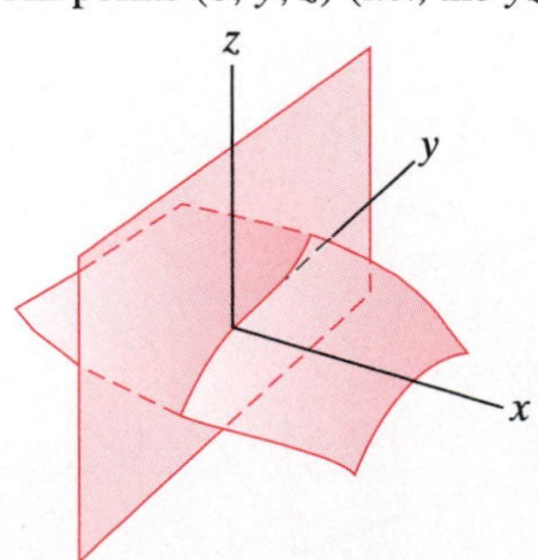

43. (b) Yes, $y = x^{3/2}$.

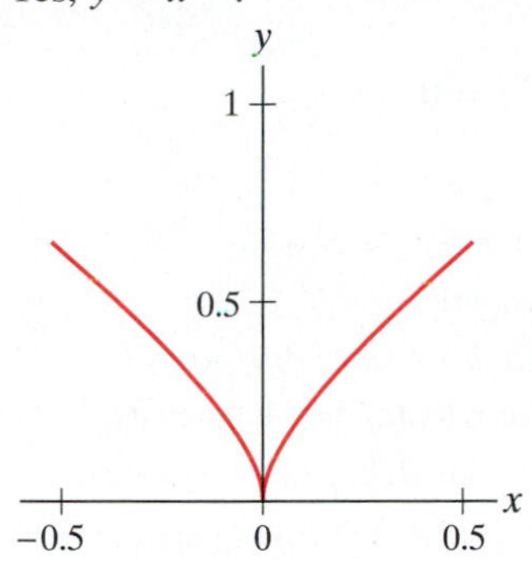

(c) The implicit function theorem suggests that we need not be able to solve for y in terms of x. Note that we can in this case, but that the function fails to be of class C^1 at $x = 0$.

45. (a) $G(-1, 1, 1) = F(-2, 1) = 0$
    (b) Hint: Use the chain rule to find $G_z(-1, 1, 1) = 30 \neq 0$.

47. (a) $\Delta(1, 0, -1, 1, 2) = -120 \neq 0$; apply the general implicit function theorem.
    (b) $\frac{\partial y_1}{\partial x_1}(1, 0) = -\frac{7}{5}$, $\frac{\partial y_2}{\partial x_1}(1, 0) = -\frac{1}{2}$, $\frac{\partial y_3}{\partial x_1}(1, 0) = -1$

49. (a) Anywhere except where $\varphi = 0$ or π.
    (b) We can solve anywhere except along the z-axis, which is where θ can have any value.

True/False Exercises for Chapter 2

1. False
3. False. (The range also requires $v \neq 0$.)
5. True
7. False. (The graph of $x^2 + y^2 + z^2 = 0$ is a single point.)
9. False
11. False. ($\lim_{(x,y)\to(0,0)} f(x, y) = 0 \neq 2$.)
13. False
15. False. ($\nabla f(x, y, z) = (0, \cos y, 0)$.)
17. True
19. False. (The partial derivatives must be continuous.)
21. False. ($f_{xy} \neq f_{yx}$.)
23. True. (Write the chain rule for this situation.)
25. False. (The correct equation is $x + y + 2z = 2$.)
27. True
29. False

Miscellaneous Exercises for Chapter 2

1. (a) $f_1(\mathbf{x}) = -x_2$, $f_2(\mathbf{x}) = x_1 - x_3$, $f_3(\mathbf{x}) = x_2$
   (b) Domain is $\mathbf{R}^3$; range consists of all vectors in $\mathbf{R}^3$ of the form $a\,\mathbf{i} + b\,\mathbf{j} - a\,\mathbf{k}$.

3. (a) Domain: $\{(x, y) \mid x \geq 0,\ y \geq 0\} \cup \{(x, y) \mid x < 0,\ y < 0\}$
   Range: $[0, \infty)$
   (b) Domain is closed.

5.

$f(x, y) =$	Graph	Level curves
$\frac{1}{x^2 + y^2 + 1}$	D	d
$\sin\sqrt{x^2 + y^2}$	B	e
$(3y^2 - 2x^2)e^{-x^2-2y^2}$	A	b
$y^3 - 3x^2y$	E	c
$x^2y^2e^{-x^2-2y^2}$	F	a
$ye^{-x^2-y^2}$	C	f

7. 0

11. (a) 15°F
    (b) 5°F

13. 1.3

15. (a) 9.57°F; 16.87°F; 7.75 mph
    (b) Effect of windspeed is greater in the Siple formula than in the table.
    (c) If $t < 91.4$°F or $s \geq 4$ mph, the Siple formula gives windchill values that are higher than the air temperature, which is unrealistic.

17. (a)

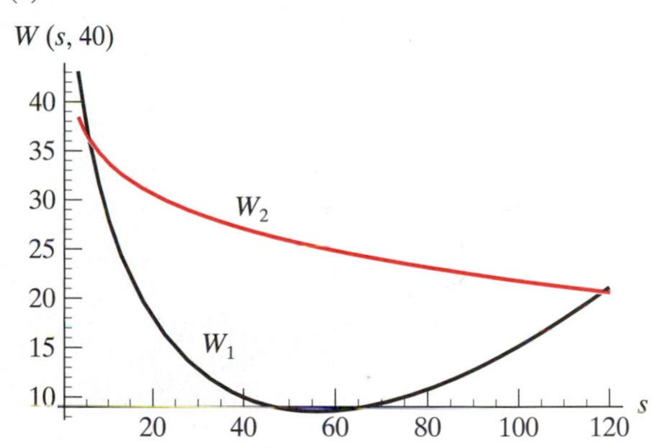

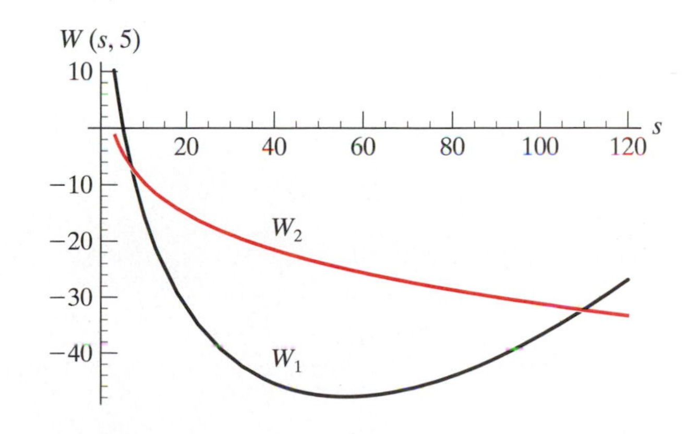

(b)

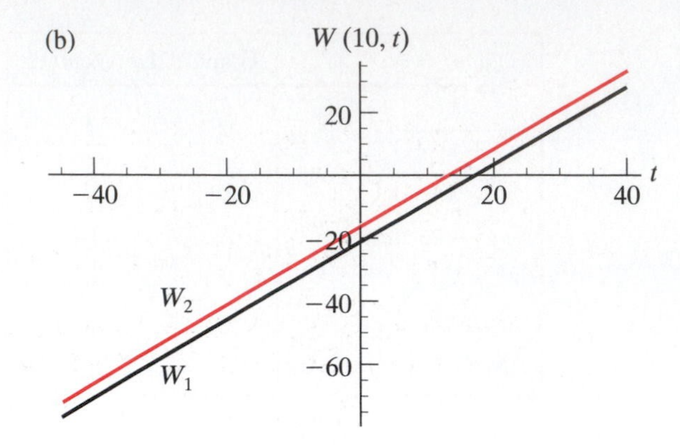

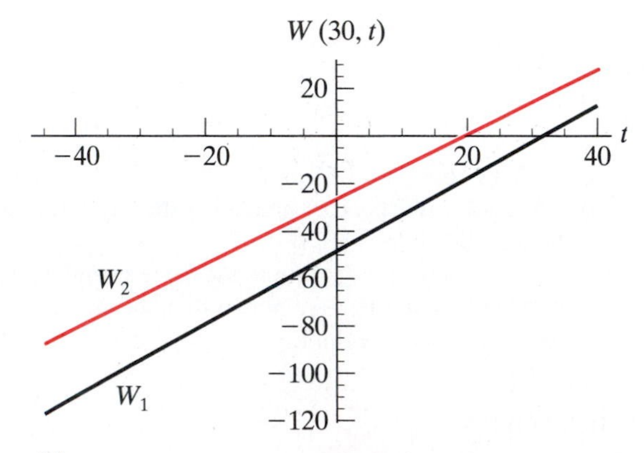

(c)

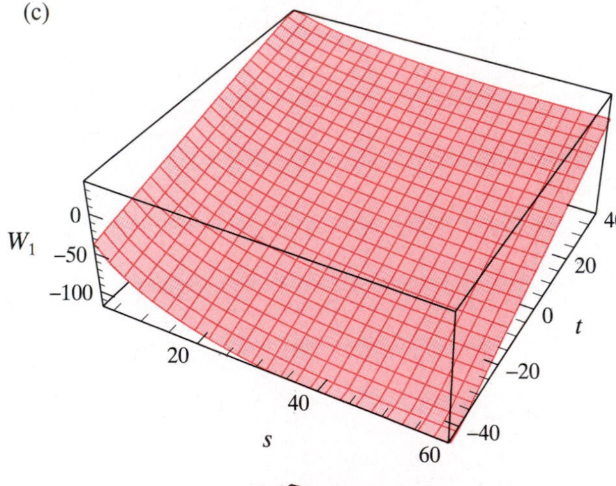

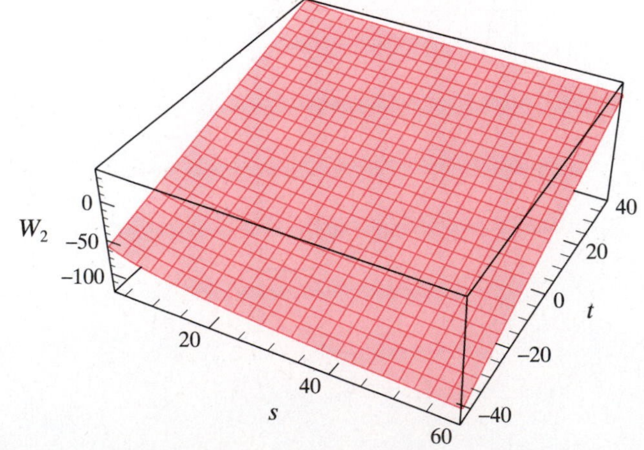

19. Hint: Compare the normal vector to the tangent plane with the vector $\overrightarrow{OP}$.

21. (a) $3x - 4y - 5z = 0$
    (b) $ax + by - cz = 0$

23. (a) $2x - 8y - z = 5$
    (b) $x = 1,\ y = t - 1,\ z = 5 - 8t$

25. 12, 201.4 units/month

27. (a) $\partial w/\partial \rho = 2\rho,\ \partial w/\partial \varphi = \partial w/\partial \theta = 0$

29. (a) $\partial z/\partial r = e^r(\cos\theta(\partial z/\partial x) + \sin\theta(\partial z/\partial y))$,
    $\partial z/\partial \theta = e^r(-\sin\theta(\partial z/\partial x) + \cos\theta(\partial z/\partial y))$;
    $\partial z/\partial x = e^{-r}(\cos\theta(\partial z/\partial r) - \sin\theta(\partial z/\partial \theta))$,
    $\partial z/\partial y = e^{-r}(\sin\theta(\partial z/\partial r) + \cos\theta(\partial z/\partial \theta))$
    (b) Hint: $\partial^2 z/\partial x^2 = e^{-2r}(\cos^2\theta(\partial^2 z/\partial r^2)$
    $+ (\sin^2\theta - \cos^2\theta)(\partial z/\partial r)$
    $+ 2\cos\theta\sin\theta(\partial z/\partial\theta)$
    $- 2\cos\theta\sin\theta(\partial^2 z/\partial r\partial\theta)$
    $+ \sin^2\theta(\partial^2 z/\partial\theta^2))$,
    $\partial^2 z/\partial y^2 = e^{-2r}(\sin^2\theta(\partial^2 z/\partial r^2)$
    $+ (\cos^2\theta - \sin^2\theta)(\partial z/\partial r)$
    $- 2\sin\theta\cos\theta(\partial z/\partial\theta)$
    $+ 2\sin\theta\cos\theta(\partial^2 z/\partial r\partial\theta)$
    $+ \cos^2\theta(\partial^2 z/\partial\theta^2))$

31. $u^u\left[u^{(u^u-1)} + u^{u^u}\ln u + u^{u^u}(\ln u)^2\right]$

35. (a) $z = f(x, y)$, where $f(x, y) = (x^3 - 3xy^2)/(x^2 + y^2)$ if $(x, y) \neq (0, 0)$; $f(x, y) = 0$ if $(x, y) = (0, 0)$
    (b) Yes
    (c) $\partial f/\partial x = (x^4 + 6x^2y^2 - 3y^4)/(x^2 + y^2)^2$,
    $\partial f/\partial y = -8x^3y/(x^2 + y^2)^2$ (if $(x, y) \neq (0, 0)$);
    $f_x(0, 0) = 1,\ f_y(0, 0) = 0$
    (d) $D_{\mathbf{u}}f(0, 0) = (\partial f/\partial r)|_{r=0} = \cos 3\theta$
    (e) $f_y(0, 0) = 0$ from (c), but along $y = x$, $f_y(x, x) = -2$, which does not have a limit of 0.
    (f)

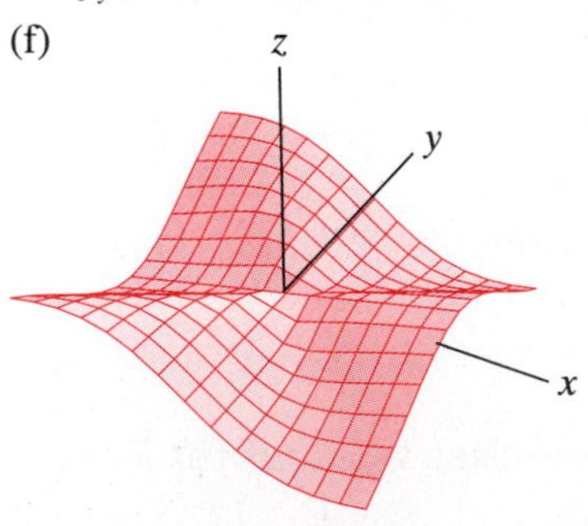

37. Homogeneous of degree 3

39. Homogeneous of degree 3

41. Homogeneous of degree 0

Chapter 3

Section 3.1

1.

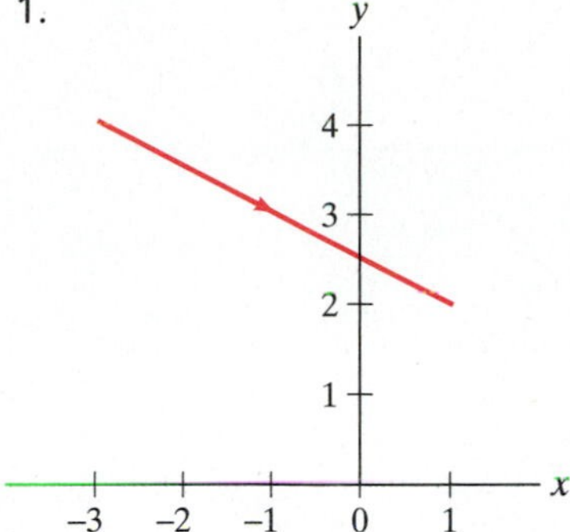

3.

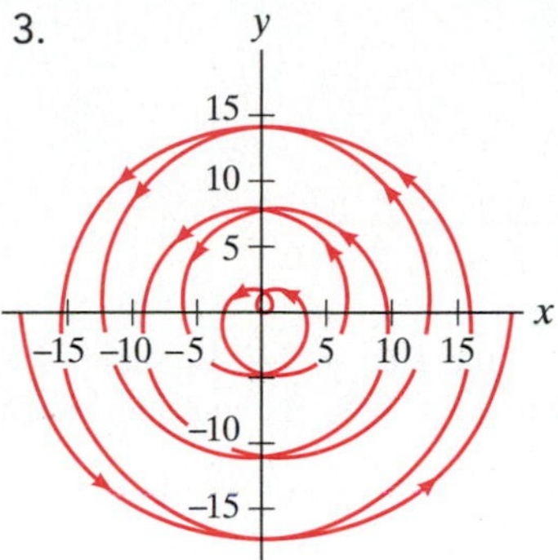

5.

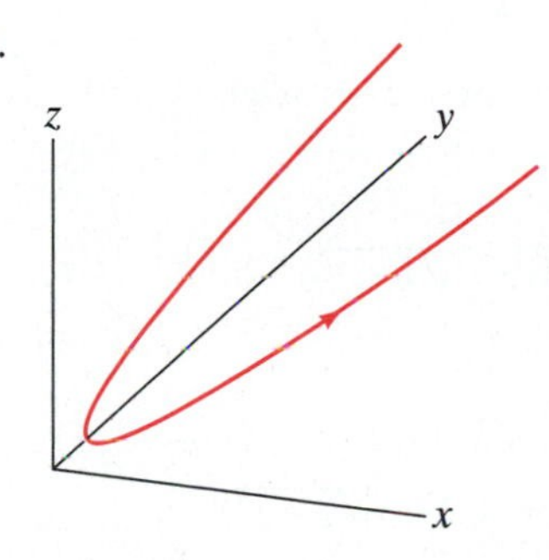

7. $\mathbf{v}(t) = 3\mathbf{i} + 2\mathbf{j}$, speed $= \sqrt{13}$, $\mathbf{a}(t) = \mathbf{0}$

9. $\mathbf{v}(t) = (\sin t + t\cos t, \cos t - t\sin t, 2t)$,
speed $= \sqrt{5t^2 + 1}$,
$\mathbf{a}(t) = (2\cos t - t\sin t, -2\sin t - t\cos t, 2)$

11. (a)

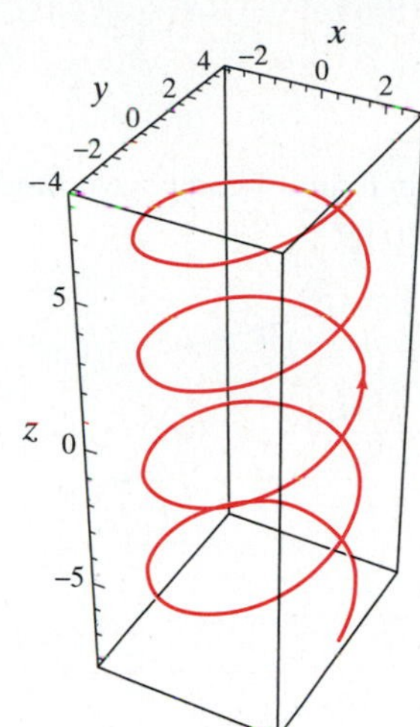

13. (a)

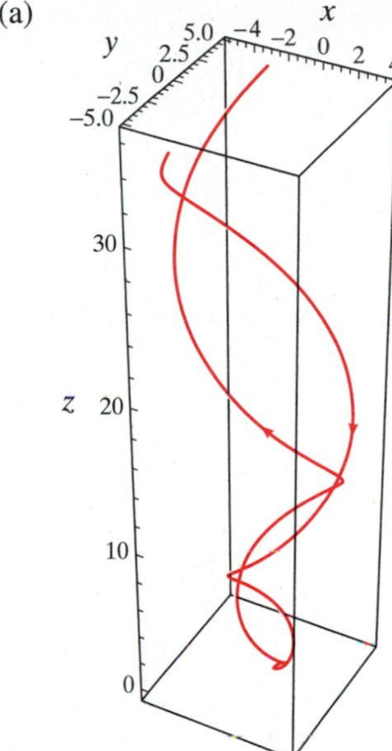

15. $\mathbf{l}(t) = t\,\mathbf{i} + (3t + 1)\mathbf{j}$

17. $\mathbf{l}(t) = (4t - 4, 12t - 16, 80t - 128)$

19. (a)

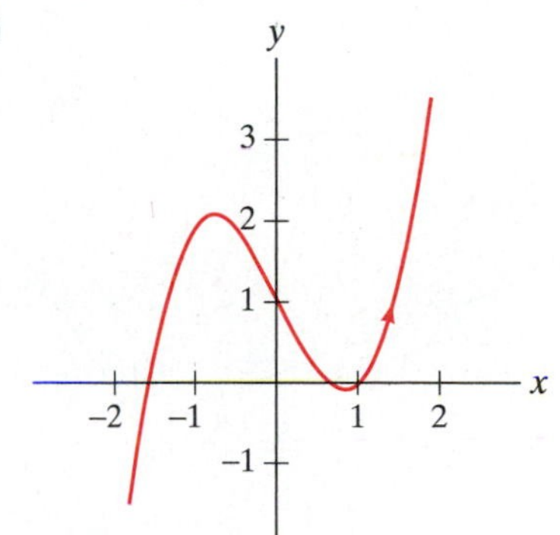

(b) $\mathbf{l}(t) = (t, 10t - 15)$

(c) $y = x^3 - 2x + 1$

(d) $y - 5 = 10(x - 2)$. (Let $x = t$ to verify agreement.)

21. 117.1875 ft
23. 50.17°
25. No
27. Write out both sides in terms of component functions.
29. Hint: Differentiate $\|\mathbf{x}(t)\|^2$.
31. (a) (i)

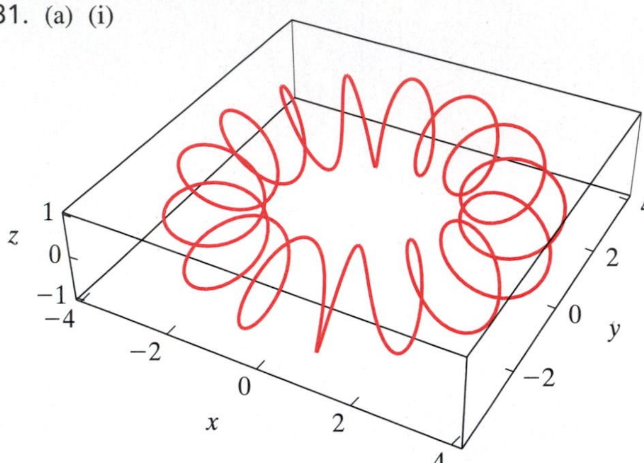

(ii)

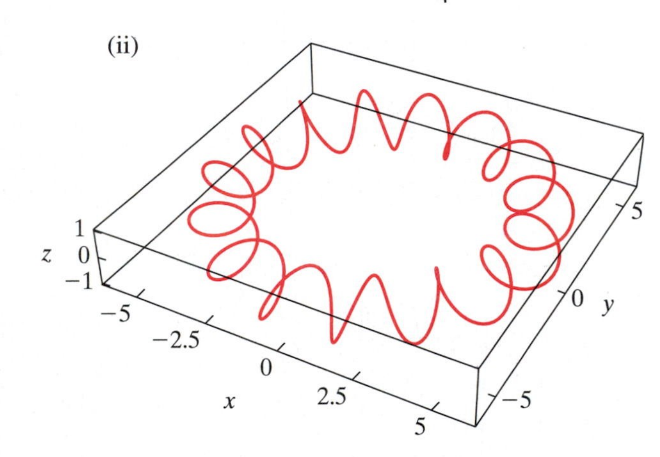

(iii)

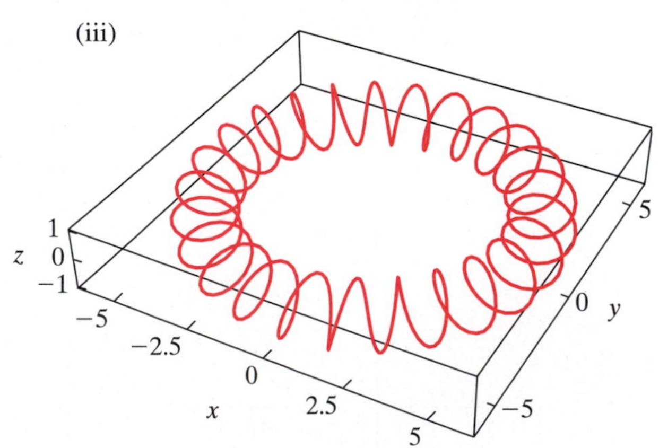

Section 3.2

1. $3\sqrt{13}$
3. $6\sqrt{3} - 2$
5. $\frac{15}{2} + \ln 4$
7. $6a$

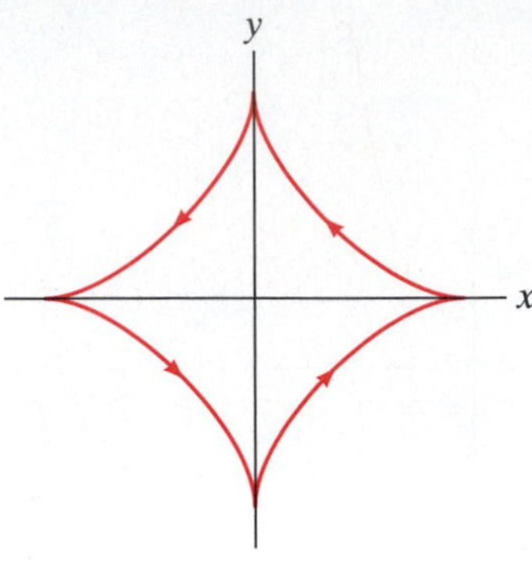

9. $(m^2 + 1)|x_0 - x_1|$
11. (a)

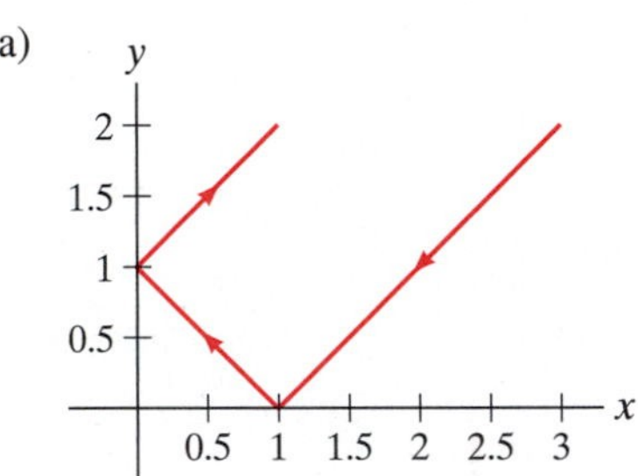

(b) The path consists of three C^1 pieces: one for $-2 \leq t \leq 0$, another for $0 \leq t \leq 1$, and another for $1 \leq t \leq 2$.

(c) $4\sqrt{2}$

13. $\mathbf{T} = \dfrac{1}{\sqrt{29}}(-5 \sin 3t\, \mathbf{i} + 2\mathbf{j} + 5 \cos 3t\, \mathbf{k})$,

$\mathbf{N} = -\cos 3t\, \mathbf{i} - \sin 3t\, \mathbf{k}$,

$\mathbf{B} = \dfrac{1}{\sqrt{29}}(-2 \sin 3t\, \mathbf{i} - 5\mathbf{j} + 2 \cos 3t\, \mathbf{k})$,

$\kappa = \frac{5}{29}$, $\tau = -\frac{2}{29}$

15. $\mathbf{T} = \sqrt{\frac{2}{3}}\left(1, \frac{1}{2}\sqrt{t+1}, -\frac{1}{2}\sqrt{1-t}\right)$,

$\mathbf{N} = 2\sqrt{2}\left(0, \frac{1}{4}\sqrt{1-t}, \frac{1}{4}\sqrt{t+1}\right)$,

$\mathbf{B} = \frac{1}{\sqrt{3}}\left(1, -\sqrt{t+1}, \sqrt{1-t}\right)$,

$\kappa = \sqrt{2}/(6\sqrt{1-t^2})$, $\tau = 1/(3\sqrt{1-t^2})$

17. (b) $\kappa = |\sin x|$
19. (a) $\dfrac{ab}{(a^2 \sin^2 t + b^2 \cos^2 t)^{3/2}}$

(b)

y
1.5
1.0
0.5
0
−0.5
−1.0
−1.5
−2
−1
0
1
2
x

21. (a) $\dfrac{3}{4a\sqrt{2+2\cos t}}$

(b)

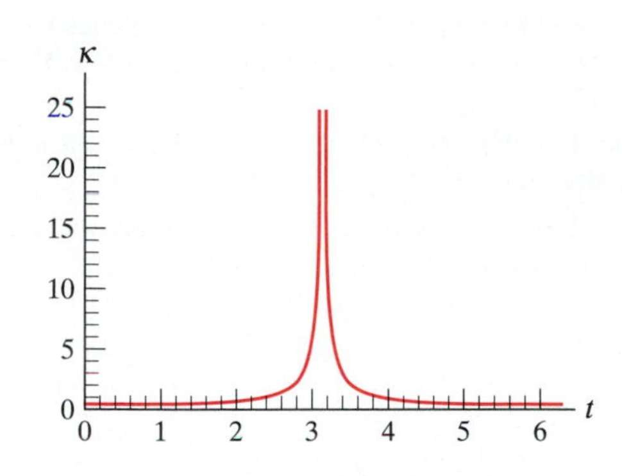

23. $a_{\text{tang}} = 4t/\sqrt{4t^2+1}$, $a_{\text{norm}} = 2/\sqrt{4t^2+1}$

25. $a_{\text{tang}} = \sqrt{5}e^t$, $a_{\text{norm}} = 2\sqrt{5}e^t$

27. $a_{\text{tang}} = 4t/\sqrt{2+4t^2}$, $a_{\text{norm}} = 2/\sqrt{1+2t^2}$

29. (b) $a_{\text{tang}} = 4t/\sqrt{4t^2+10}$, $a_{\text{norm}} = 2\sqrt{10}/\sqrt{4t^2+10}$

31. Hint: Write $\mathbf{v}$ and $\mathbf{a}$ in terms of $\mathbf{T}$ and $\mathbf{N}$ and use the Frenet–Serret formulas.

33. Hint: Use formula (17) and Exercise 31.

35. Hint: Recall that $\|\mathbf{x}-\mathbf{x}_0\|^2 = (\mathbf{x}-\mathbf{x}_0)\cdot(\mathbf{x}-\mathbf{x}_0)$.

37. Hint: Calculate $\mathbf{T}$ and show that $(\mathbf{x}(t)-(1,0,0))\cdot\mathbf{T} = 0$.

Section 3.3

1.

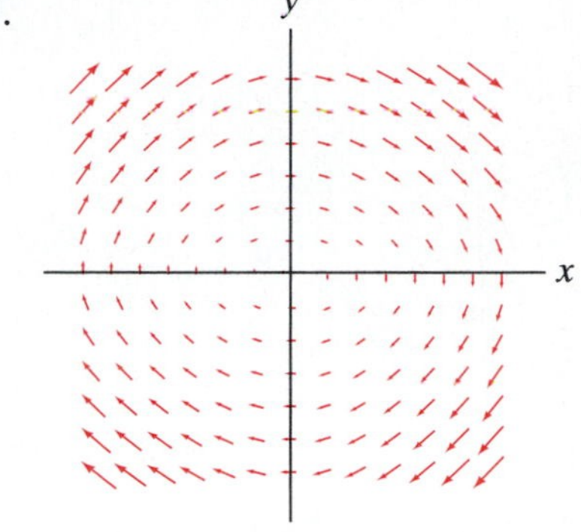

3.

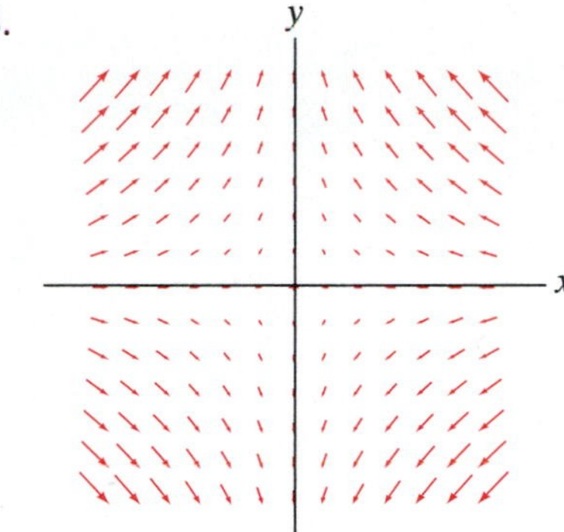

5.

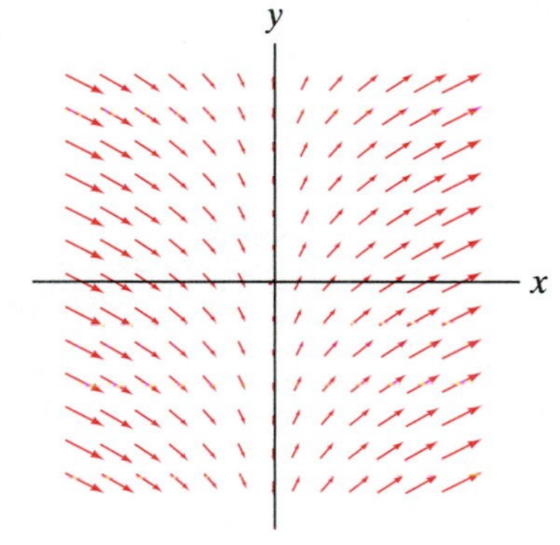

7.

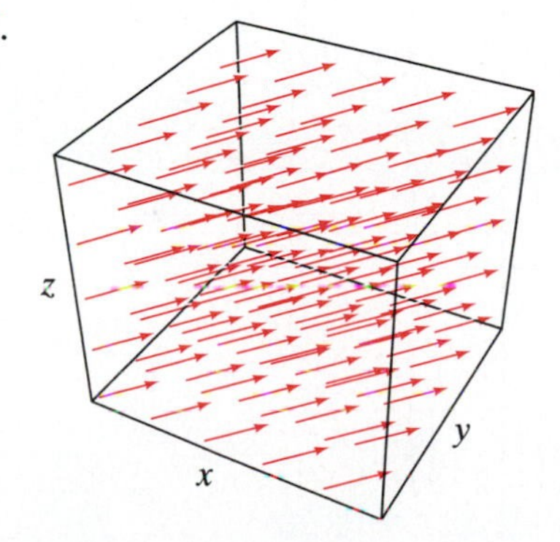

9.

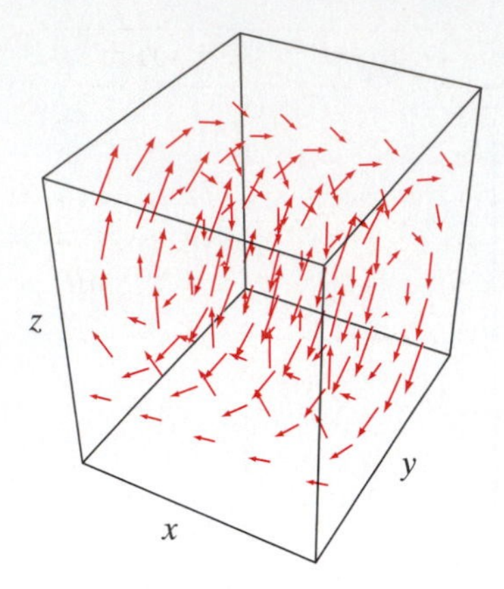

11.

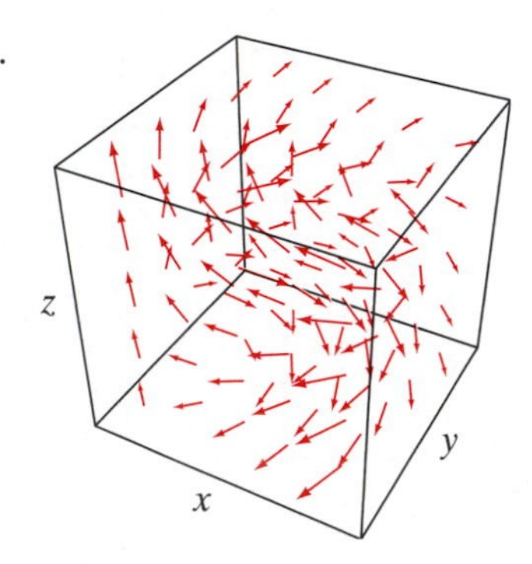

13.

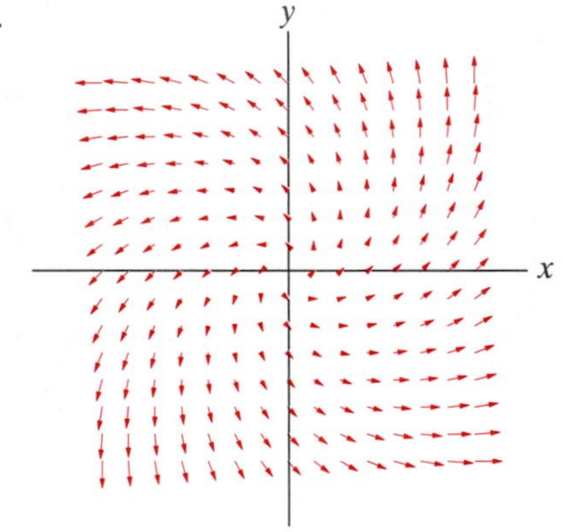

15.

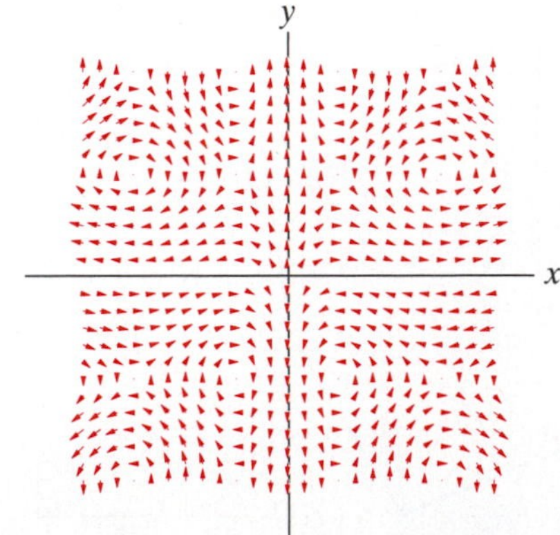

17. $\mathbf{F}(\mathbf{x}(t)) = (\cos t, -\sin t, 0) = \mathbf{x}'(t)$

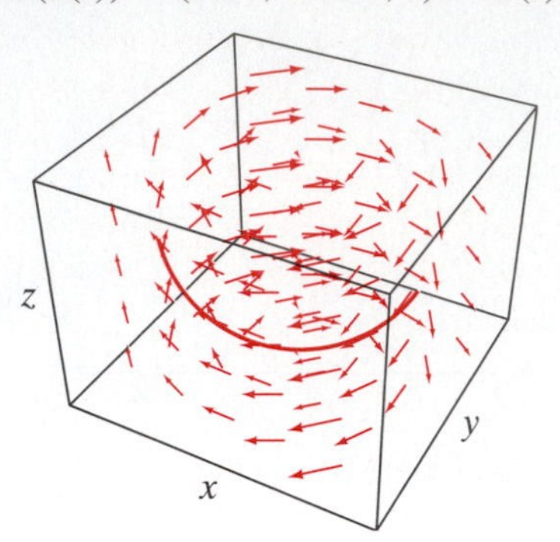

19. $\mathbf{F}(\mathbf{x}(t)) = (\cos t, -\sin t, 2e^{2t}) = \mathbf{x}'(t)$

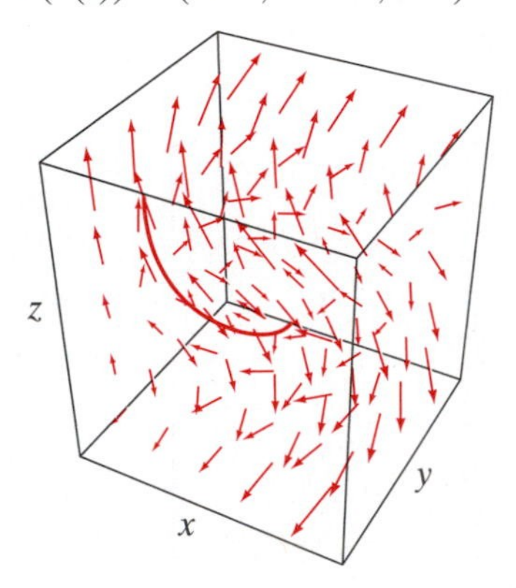

21. $\left(\dfrac{1}{2-t}, e^t\right)$

23. (a) $\mathbf{F} = \nabla f$, where $f(x, y, z) = 3x - 2y + z$
    (b) Equipotential surfaces are parallel planes with equation $3x - 2y + z = c$ (i.e., planes with normal $(3, -2, 1)$).

25. Hint: Use the chain rule and the facts that $\nabla f = \mathbf{F}$ and $\mathbf{x}$ is a flow line of $\mathbf{F}$.

31. Hint: Differentiate the defining differential equation for the flow with respect to $\mathbf{x} = (x_1, x_2, \ldots, x_n)$.

Section 3.4

1. $2x + 2y$
3. 3
5. $2x_1 + 4x_2 + 6x_3 + \cdots + 2nx_n$
7. $2xz\,\mathbf{i} - 2yz\,\mathbf{j} - e^y\,\mathbf{k}$
9. $\mathbf{0}$
11. $(x^2 - xye^{xyz})\mathbf{i} + (y^2 - 2xy)\mathbf{j} + (yze^{xyz} - 2yz)\mathbf{k}$
13. (a) div $\mathbf{F} > 0$ on all of $\mathbf{R}^2$
    (b) div $\mathbf{F} < 0$ on all of $\mathbf{R}^2$
    (c) div $\mathbf{F} > 0$ on $\{(x, y) \mid x > 0\}$, div $\mathbf{F} < 0$ on $\{(x, y) \mid x < 0\}$
    (d) div $\mathbf{F} > 0$ on $\{(x, y) \mid y > 0\}$, div $\mathbf{F} < 0$ on $\{(x, y) \mid y < 0\}$
21. Write out in terms of the component functions of $\mathbf{F}$ and $\mathbf{G}$.

23. Write out in terms of the component functions of $\mathbf{F}$.
25. Write out in terms of the component functions of $\mathbf{F}$ and $\mathbf{G}$.
27. First use the chain rule to replace occurrences of the Cartesian differential operators in ∇ by combinations of spherical differential operators. Then compute ∇f. Finally, replace $\mathbf{i}$, $\mathbf{j}$, and $\mathbf{k}$ by appropriate combinations of $\mathbf{e}_\rho$, $\mathbf{e}_\varphi$, and $\mathbf{e}_\theta$. (See also §1.7.)
29. Write out the components of $f\nabla f$.
33. $(1/\sqrt{3}, 4/\sqrt{3}, -1/\sqrt{3})$

True/False Exercises for Chapter 3

1. True
3. True
5. False. (There should be a negative sign in the second term on the right.)
7. True
9. False
11. True
13. True
15. False. (It's a scalar field.)
17. True
19. False. (It's a meaningless expression.)
21. True. (Check that $\mathbf{F}(\mathbf{x}(t)) = \mathbf{x}'(t)$.)
23. False. ($\nabla \times \mathbf{F} \neq \mathbf{0}$.)
25. False. (Consider $\mathbf{F} = y\,\mathbf{i} + x\,\mathbf{j}$.)
27. True
29. False. ($\nabla \cdot (\nabla \times \mathbf{F}) \neq 0$.)

Miscellaneous Exercises for Chapter 3

1. (a) D (b) F (c) A (d) B (e) C (f) E
3. Hint: Differentiate $(ds/dt)^2 = \|\mathbf{x}'(t)\|^2$.
7.
9. (a) Hint: Calculate $\mathbf{x}(0)$ and $\mathbf{x}(1)$.
   (b) Hint: Show that

$$\mathbf{x}(\tfrac{1}{2}) = \tfrac{w}{1+w}(x_2, y_2) + \tfrac{1}{1+w}\left(\tfrac{x_1+x_3}{2}, \tfrac{y_1+y_3}{2}\right).$$

11. (a) $\frac{1}{2(1+w)}\sqrt{(x_1 - 2x_2 + x_3)^2 + (y_1 - 2y_2 + y_3)^2}$
    (b) $\frac{w}{2(1+w)}\sqrt{(x_1 - 2x_2 + x_3)^2 + (y_1 - 2y_2 + y_3)^2}$
13. (a) Hint: Find where $\mathbf{x}'(t) = (0, 0)$.
    (b) Hint: The tangent line at $\mathbf{x}(t_0)$ is given by $\mathbf{l}(s) = \mathbf{x}(t_0) + s\mathbf{x}'(t_0)$. Find the y-intercept of this line.
15. Hint: Begin with $\mathbf{x}(\theta) = (f(\theta)\cos\theta, f(\theta)\sin\theta)$.
17. (a) $\mathbf{y}(t) = (a(\cos t + t\sin t), a(\sin t - t\cos t))$
    (b)

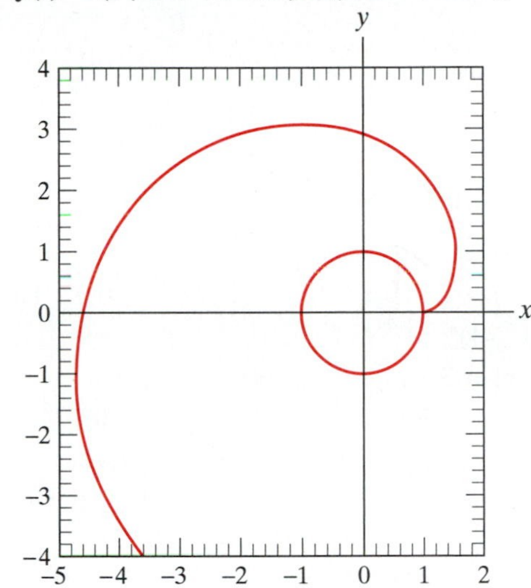

19. (a) Show that $\|\mathbf{x}(t) - \mathbf{y}(t)\| = s(t)$.
    (b) The vector difference $\mathbf{x}(t) - \mathbf{y}(t)$ is a tangent vector (to $\mathbf{x}$) $s(t)$ units long.
23. $\mathbf{e}(t) = (a(t - 3\sin t), 3a(1 - \cos t))$, which is a another type of cycloid.
25. Hint: Use the Frenet–Serret formulas.
27. (a) Show that $(x'(s))^2 + (y'(s))^2 = 1$ by means of the fundamental theorem of calculus.
    (b) $\kappa = |g'(s)|$
    (c) Set $g'(s) = \kappa(s)$. Find g by antidifferentiation and set $x(s) = \int_0^s \cos g(t)\,dt$, $y(s) = \int_0^s \sin g(t)\,dt$.
    (d) $x(s) = \int_0^s \cos(t^2/2)\,dt$, $y(s) = \int_0^s \sin(t^2/2)\,dt$
    (e)

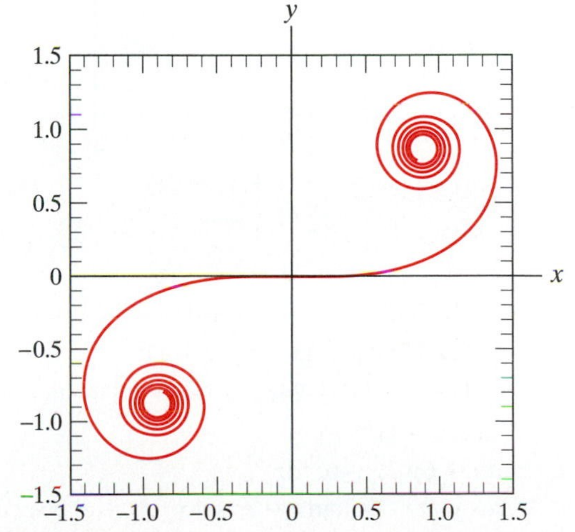

29. (a) Calculate **B** and remember that it is a unit vector.
(b) **B** is constant, so $\mathbf{B}' = \mathbf{0}$. Next use the Frenet–Serret formula.

31. (a) $\dfrac{a^2 + h^2/4\pi^2}{a}$ (b) 8.2771 ft

35. $\mathbf{N}' = -\kappa\mathbf{T} + \tau\mathbf{B}$. Argue that $\mathbf{N}' \neq \mathbf{0}$.

37. (a) $2\sqrt{2}\pi$ (b) $\mathbf{F} = \sqrt{2}(y\,\mathbf{i} - 2x\,\mathbf{j} + y\,\mathbf{k})$

41. Hint: Use Proposition 1.4.

43. No

Chapter 4

Section 4.1

1. $p_4(x) = 1 + 2x + 2x^2 + 4x^3/3 + 2x^4/3$

3. $p_4(x) = 1 - 2(x-1) + 3(x-1)^2 - 4(x-1)^3 + 5(x-1)^4$

5. $p_3(x) = 3 + (x-9)/6 - (x-9)^2/216 + (x-9)^3/3888$

7. $p_5(x) = 1 - (x-\pi/2)^2/2 + (x-\pi/2)^4/24$

9. $p_1(x, y) = \frac{1}{3} - 2(x-1)/9 + 2(y+1)/9$,
$p_2(x, y) = \frac{1}{3} - 2(x-1)/9 + 2(y+1)/9 + (x-1)^2/27 - 8(x-1)(y+1)/27 + (y+1)^2/27$

11. $p_1(x, y) = -1 - 2x$,
$p_2(x, y) = -1 - 2x - 2x^2 + 9(y-\pi)^2/2$

13. $p_1(x, y, z) = p_2(x, y, z) \equiv 0$

15. $\begin{bmatrix} 6 & 2 & 0 \\ 2 & 0 & -2 \\ 0 & -2 & 12 \end{bmatrix}$

17. $p_2(x, y) = 1 + \begin{bmatrix} 0 & 0 \end{bmatrix}\begin{bmatrix} x \\ y \end{bmatrix} + \frac{1}{2}\begin{bmatrix} x & y \end{bmatrix}\begin{bmatrix} -2 & 0 \\ 0 & -2 \end{bmatrix}\begin{bmatrix} x \\ y \end{bmatrix}$

19. (a) $Df(\mathbf{0}) = \begin{bmatrix} 1 & 2 & \cdots & n \end{bmatrix}$,
$$Hf(\mathbf{0}) = \begin{bmatrix} 1 & 2 & 3 & \cdots & n \\ 2 & 4 & 6 & \cdots & 2n \\ 3 & 6 & 9 & \cdots & 3n \\ \vdots & \vdots & \vdots & \ddots & \vdots \\ n & 2n & 3n & \cdots & n^2 \end{bmatrix}$$
(b) $p_1(x_1, \ldots, x_n) = 1 + x_1 + 2x_2 + \cdots + nx_n$
$p_2(x_1, \ldots, x_n) = 1 + \sum_{i=1}^n ix_i + \frac{1}{2}\sum_{i,j=1}^n ijx_ix_j$

21. (a) $2 - z + 3xy + x^2y - xz^2 + 2y^3$
(b) $-4 - 4(x-1) + 11(y+1) - z + \frac{1}{2}[4(x-1)^2 + 16(x-1)(y+1) - 12(y+1)^2 - 2z^2)] + \frac{1}{6}[18(x-1)^3 + 24(x-1)^2(y+1) - 6(x-1)z^2 + 12(y+1)^3]$

23. $2x\,dx + 6y\,dy - 6z^2dz$

25. $e^x \cos y\,dx + (e^y \sin z - e^x \sin y)\,dy + e^y \cos z\,dz$

27. (a) 388.08 (b) 0.2462 (c) 1.1

29. The (1, 1)-entry (upper left)

31. 0.0068 m

33. (a) $p_2(x, y) = 1 - \frac{1}{2}[x^2 + (y - \frac{\pi}{2})^2]$
(b) Accurate to at least 0.0360

Section 4.2

1. (a) (2, 3)
(b) $\Delta f = -h^2 - k^2 - k$
(c) There is a local maximum at (2, 3).

3. Local maximum at $\left(\frac{2}{9}, \frac{4}{9}\right)$

5. Local minimum at $\left(\frac{27}{2}, 5\right)$; saddle point at $\left(\frac{3}{2}, 1\right)$

7. Minimum at $\left(4, \frac{1}{2}\right)$

9. Saddle point at (0, 0); minimum at (0, 2)

11. Saddle point at $(0, \frac{1}{\sqrt{3}})$; local minimum at $(0, -\frac{1}{\sqrt{3}})$

13. Local maximum at $\left(-\frac{1}{2}, \frac{1}{3}\right)$

15. Saddle point at $(0, 6, -3)$

17. Local minimum at (0, 0, 0)

19. Saddle point at $\left(-1, \frac{1}{2}, \frac{1}{2}\right)$

21. (a) $(0, -2)$ and $(0, 3)$
(b) Local maximum at $(0, -2)$;
local minimum at (0, 3)

23. (a) Minimum if $a, b > 0$; maximum if $a, b < 0$; saddle point otherwise
(b) Minimum if $a, b, c > 0$; maximum if $a, b, c < 0$; saddle point otherwise
(c) Minimum if $a_1, \ldots, a_n > 0$; maximum if $a_1, \ldots, a_n < 0$; saddle point otherwise

25. Saddle points at (0, 0), $\left(\pm\sqrt{\frac{3}{2}}, 0\right)$; local maxima at $\left(\frac{1}{\sqrt{2}}, -\frac{1}{\sqrt{2}}\right), \left(-\frac{1}{\sqrt{2}}, \frac{1}{\sqrt{2}}\right)$

27. Saddle points at (0, 0, 0, 0), $(-\sqrt{2}, 2\sqrt{2}, 1, -\sqrt{2})$, $(\sqrt{2}, 2\sqrt{2}, -1, -\sqrt{2})$ $(-\sqrt{2}, -2\sqrt{2}, -1, \sqrt{2})$, $(\sqrt{2}, -2\sqrt{2}, 1, \sqrt{2})$

29. $\left(\frac{36}{13}, -\frac{48}{13}, -\frac{12}{13}\right)$

31. Maximum of 8 at (0, 0, 2);
minimum of $-\frac{191}{7}$ at $\left(-\frac{32}{7}, 3, \frac{8}{7}\right)$

33. Maximum of 1 at $(\pi/2, 0)$, $(\pi/2, 2\pi)$, $(3\pi/2, \pi)$;
minimum of -1 at $(3\pi/2, 0)$, $(3\pi/2, 2\pi)$, $(\pi/2, \pi)$

35. Maximum of e^6 at $(0, 1, -2)$; minimum of e at all (x, y, z) such that $x^2 + y^2 - 2y + z^2 + 4z = 0$

37. (b) Maximum

39. (b) Neither

41. (b) Maximum

43. (a) Local maximum at (0, 0, 0)
(b) $f(0, 0, 0) = e^2$ is a global maximum.

45. Global minimum of $2 + \ln 2$ at $\left(2, \frac{1}{2}\right)$. No global maximum.

47. (b) Critical points are (2, 1) and $(0, -1)$.

(c)

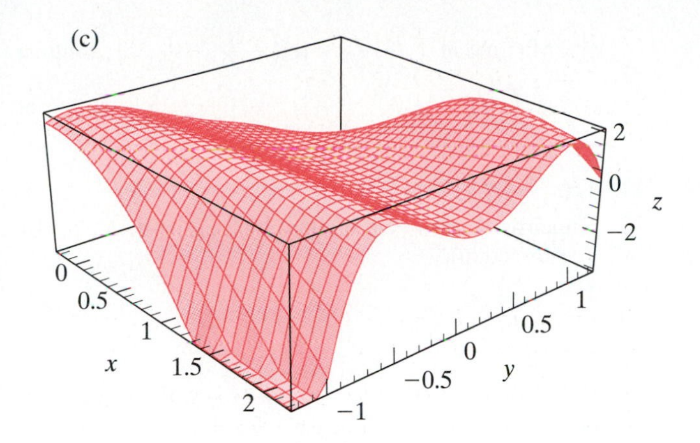

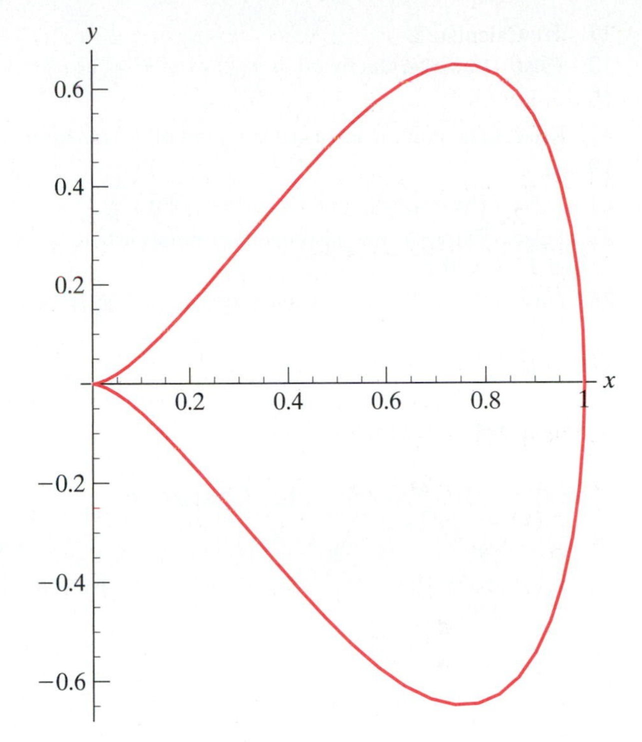

(c) $\nabla g = \mathbf{0}$ at $(0, 0)$

35. (a) Hint: Check that $\partial L/\partial l_i = c_i - g_i(\mathbf{x})$ for $i = 1, \ldots, k$ and

$$\frac{\partial L}{\partial x_j} = \frac{\partial f}{\partial x_j} - \sum_{i=1}^{k} l_i \frac{\partial g_i}{\partial x_j}$$

for $j = 1, \ldots, n$.

Section 4.3

1. (a) Minimize $f(x, y) = x^2 + y^2 + (2x - 3y - 4)^2$ to find that the closest point is $\left(\frac{4}{7}, -\frac{6}{7}, -\frac{2}{7}\right)$.
   (b) Minimize $f(x, y, z) = x^2 + y^2 + z^2$ subject to $2x - 3y - z = 4$.
3. $(\sqrt{2}, \sqrt{2})$ and $(-\sqrt{2}, -\sqrt{2})$
5. $\left(1, \frac{2}{3}, 2\right)$, $(3, 0, 0)$, $(0, 2, 0)$ and $(0, 0, 6)$
7. $\left(\frac{4}{3}, \frac{2}{3}, -\frac{4}{3}\right)$
9. (a) $\left(\pm\sqrt{\frac{7}{8}}, \frac{1}{4}\right)$, $\left(0, \pm\frac{1}{\sqrt{2}}\right)$
   (b) $\left(\pm\sqrt{\frac{7}{8}}, \frac{1}{4}\right)$ give maxima; $\left(0, \pm\frac{1}{\sqrt{2}}\right)$ give minima.
11. $\left(\sqrt{\frac{2}{3}}, \frac{1}{2}, \frac{12-\sqrt{6}}{8}\right)$
13. $(\pm 1, 0, 0)$, $(0, 1, 0)$, $(0, 0, \pm 1)$, $\left(\frac{2}{3}, -\frac{2}{3}, \frac{1}{3}\right)$, $\left(\frac{1}{8}\sqrt{\frac{11}{2}}, -\frac{3}{8}, -\frac{1}{8}\sqrt[3]{\frac{11}{2}}\right)$, $\left(-\frac{1}{8}\sqrt{\frac{11}{2}}, -\frac{3}{8}, \frac{1}{8}\sqrt[3]{\frac{11}{2}}\right)$
15. $\left(\frac{1}{\sqrt{2}}, \frac{1}{\sqrt{2}}, \frac{1-\sqrt{2}}{2}, \frac{1-\sqrt{2}}{2}\right)$, $\left(-\frac{1}{\sqrt{2}}, -\frac{1}{\sqrt{2}}, \frac{1+\sqrt{2}}{2}, \frac{1+\sqrt{2}}{2}\right)$
17. The numbers are 6, 6, 6.
19. Maximum value: 6. Minimum value: 0.
21. Height should be equal to diameter.
23. Locate at either $(-2, 2, 1)$ or $(-2, -2, 1)$.
25. Largest sphere has equation $x^2 + y^2 + z^2 = 2$.
27. $\left(\frac{9}{2}, 2, \frac{5}{2}\right)$
29. Highest point is $(-1, -1, 2)$; lowest point is $\left(\frac{1}{2}, \frac{1}{2}, \frac{1}{2}\right)$.
31. (a) $(\sqrt{6}, \sqrt{6})$ and $(-\sqrt{6}, -\sqrt{6})$
33. (a) Critical point at $(1, 0)$
    (b) There is a minimum of 0 at $(0, 0)$ and a maximum of 1 at $(1, 0)$.

Section 4.4

1. $35x - 49y + 98 = 0$
3. (a) $D(a, b) = \sum_{i=1}^{n}(y_i - (a/x_i + b))^2$
   (b) Minimize D with respect to a and b.
5. Hint: Let $D(a, b, c) = \sum_{i=1}^{n}(y_i - (ax_i^2 + bx_i + c))^2$.
7. (b) There is a single, stable equilibrium point at $\left(-\frac{1}{4}, -\frac{1}{4}\right)$.
9. Single equilibrium point at $\left(-1, \frac{3}{2}, \frac{3}{2}\right)$. There are no stable equilibria.
11. Produce 50,000 each of both the standard and executive models and 100,000 deluxe models.
13. (a) Invest \$120,000 for capital equipment and \$240,000 for labor.
    (b) Hint: Note that $L/K = 2$.

True/False Exercises for Chapter 4

1. True
3. True
5. True
7. False. (f is most sensitive to changes in y.)
9. False

11. True
13. False. (Consider the function $f(x, y) = x^2 + y^2$.)
15. True
17. False. (The point is not a critical point of the function.)
19. True
21. False. (The critical point is a saddle point.)
23. False. (Extrema may also occur at points where $g = c$ and $\nabla g = \mathbf{0}$.)
25. False. (You will have to solve a system of 7 equations in 7 unknowns.)
27. True
29. False. (The equilibrium points are the critical points of the potential function.)

Miscellaneous Exercises for Chapter 4

1. $r_0 = 2h_0$
3. Price the Mocha at \$2.70 per pound and the Kona at \$5 per pound.
5. Maximum value of 4 at $(1, -\sqrt{3}, 0)$. Minimum value of -4 at $(-1, \sqrt{3}, 0)$.
7. (e)

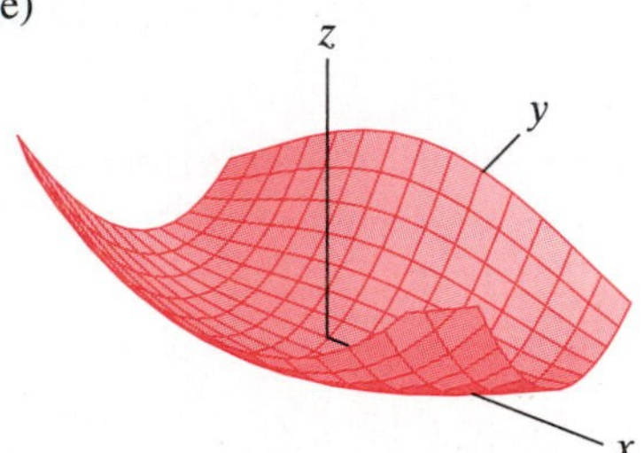

9. (a) $(0, 0, \pm 1)$, $\left(\frac{1}{\sqrt{2}}, \frac{1}{\sqrt{2}}, 0\right)$, $\left(\frac{1}{\sqrt{2}}, -\frac{1}{\sqrt{2}}, 0\right)$, $\left(-\frac{1}{\sqrt{2}}, \frac{1}{\sqrt{2}}, 0\right)$, $\left(-\frac{1}{\sqrt{2}}, -\frac{1}{\sqrt{2}}, 0\right)$

(b)

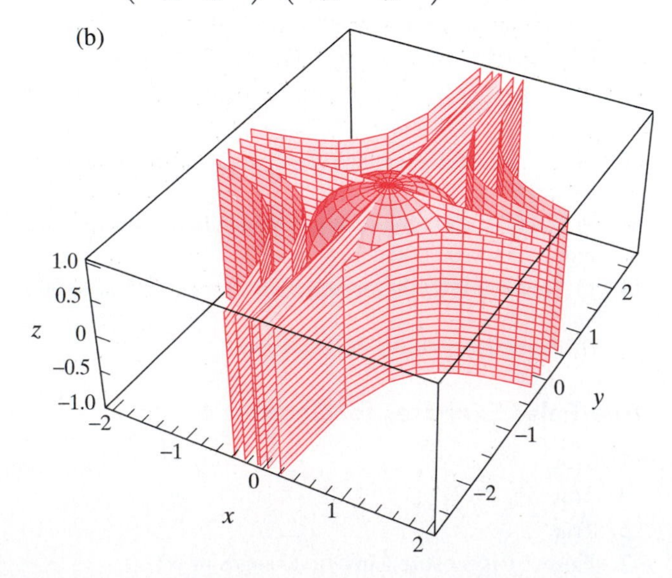

(c) Maxima at $\left(\frac{1}{\sqrt{2}}, \frac{1}{\sqrt{2}}, 0\right)$, $\left(-\frac{1}{\sqrt{2}}, -\frac{1}{\sqrt{2}}, 0\right)$; minima at $\left(\frac{1}{\sqrt{2}}, -\frac{1}{\sqrt{2}}, 0\right)$, $\left(-\frac{1}{\sqrt{2}}, \frac{1}{\sqrt{2}}, 0\right)$; saddle points at $(0, 0, \pm 1)$

11. $1/(a_1^2 + a_2^2 + \cdots + a_n^2)$
13. Dimensions are 4 (x-direction) by $2\sqrt{2}$ (y-direction) by 2 (z-direction).
15. 1
17. $a/3$ by $b/3$ by $c/3$
19. $3\sqrt{5}/8$
21. Hint: Minimize $D^2 = (x - x_0)^2 + (y - y_0)^2$, where (x, y) denotes a point on the line $ax + by = d$.
23. (a) Hint: Show that the maximum value occurs when $x^2 = y^2 = z^2 = a^2/3$.
    (b) Since $f(x, y, z) = x^2y^2z^2$ is maximized when $x^2 = y^2 = z^2 = a^2/3$, we must have $x^2y^2z^2 \le (a^2/3)^3 = ((x^2 + y^2 + z^2)/3)^3$.
    (c) Hint: Since $x_1, x_2, \ldots, x_n$ are assumed to be positive, we can write $x_i = y_i^2$ for $i = 1, \ldots, n$. Maximize $f(y_1, y_2, \ldots, y_n) = y_1^2 y_2^2 \cdots y_n^2$ subject to $y_1^2 + y_2^2 + \cdots + y_n^2 = a^2$.
25. (a) $\lambda_1, \lambda_2 = \dfrac{(a + c) \pm \sqrt{(a + c)^2 - 4(ac - b^2)}}{2}$
    (b) Rewrite as $\lambda_1, \lambda_2 = \dfrac{(a + c) \pm \sqrt{(a - c)^2 + 4b^2}}{2}$.

Chapter 5

Section 5.1

1. $\frac{40}{3}$
3. $6(e - 1)$
5. $e^3 - 2e^2 + e + 2\ln 2 - \frac{2}{3}$
7. (a) Volume $= \int_{-1}^{2} \int_{0}^{2} (x^2 + y^2 + 2)\, dy\, dx = 26$
   (b) Volume $= \int_{0}^{2} \int_{-1}^{2} (x^2 + y^2 + 2)\, dx\, dy$
9. $\int_0^1 \int_{-1}^{2} (2x^2 + y^4 \sin \pi x)\, dy\, dx = 2 + 66/5\pi$
11. The iterated integral gives the volume of the region bounded by the graph of $z = 16 - x^2 - y^2$, the xy-plane, and the planes $x = 1$, $x = 3$, $y = -2$, $y = 2$. The value of the integral is $\frac{248}{3}$.
13. The iterated integral gives the volume of the region bounded by the graph of $z = 4 - x^2$, the xy-plane, and the planes $x = -2$, $x = 2$, $y = 0$, $y = 5$. The value of the integral is $\frac{160}{3}$.
15. The iterated integral gives the volume of the region bounded by the graph of $z = 5 - |y|$, the xy-plane, and the planes $x = -1$, $x = 2$, $y = -5$, $y = 5$. The value of the integral is 75.

Section 5.2

1. (a) Check that $\int_{-2}^{2}\int_{0}^{4-x^2} x^3\,dy\,dx = 0$.
   (b) Hint: The region D is symmetric about the y-axis and x^3 is an odd function.

3. 4

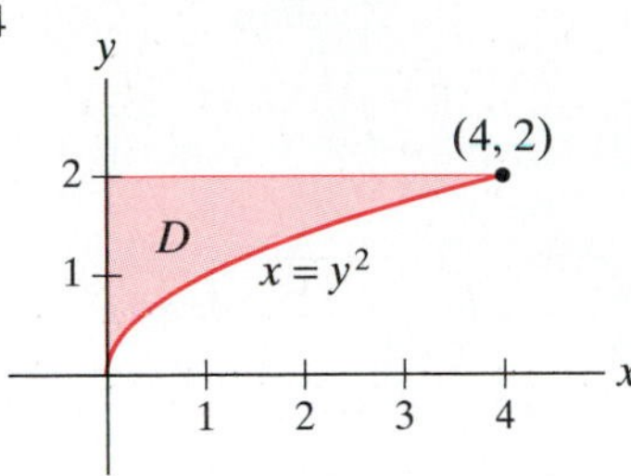

5. $\frac{152}{3}$

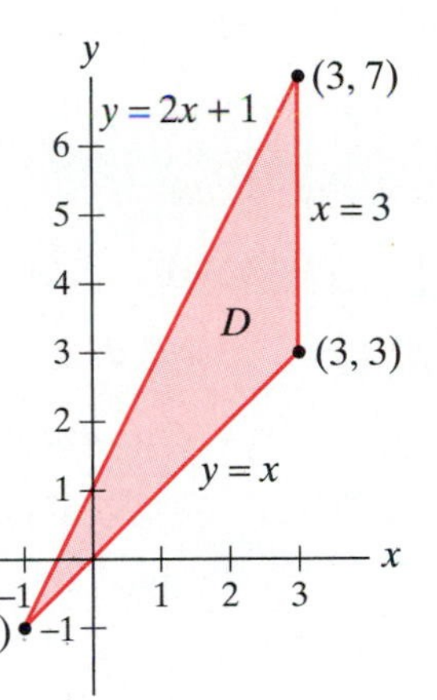

7. $3\pi/2$

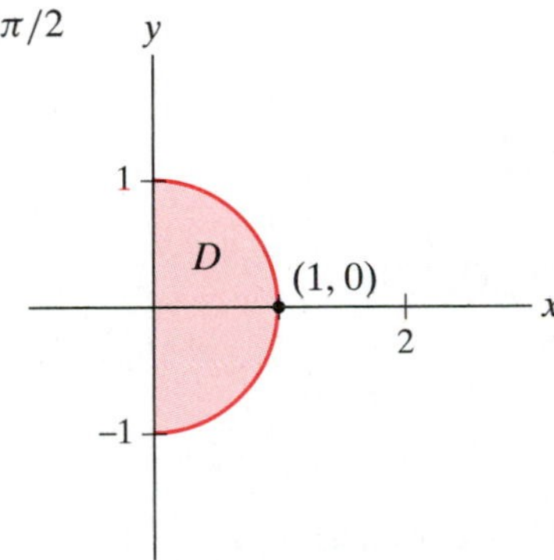

9. 0

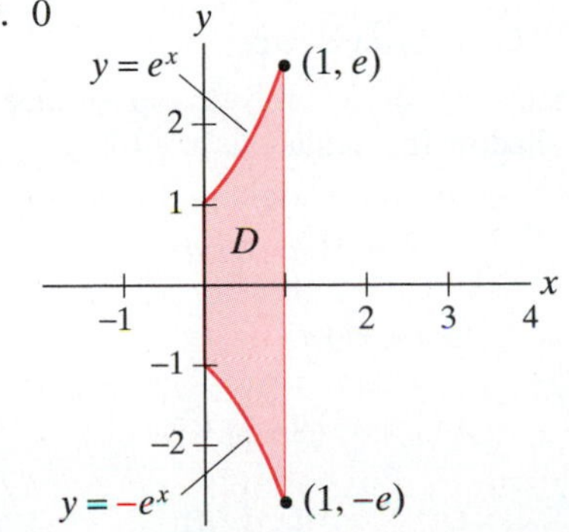

11. $\frac{4}{3}$

13. $\frac{99}{20}$

15. $-\frac{128}{5}$

17. Hint: Write $\iint_R cf\,dA$ and $\iint_R f\,dA$ as limits of appropriate Riemann sums.

19. Hint: Use the fact that $\left|\sum_{i=1}^{n} a_i\right| \le \sum_{i=1}^{n} |a_i|$.

21. $\frac{1}{12}$

23. Area is πab.

25. 11,664

27. $\frac{16}{3}$

29. (a) $\int_0^2 f(x, y)\,dy = 2$ regardless of whether x is rational or irrational.
    (b) Value is 2.
    (c) 2
    (d) Converges to 3.
    (e) The Riemann sum has no uniquely determined limit.

Section 5.3

1. (a) 4
   (b)

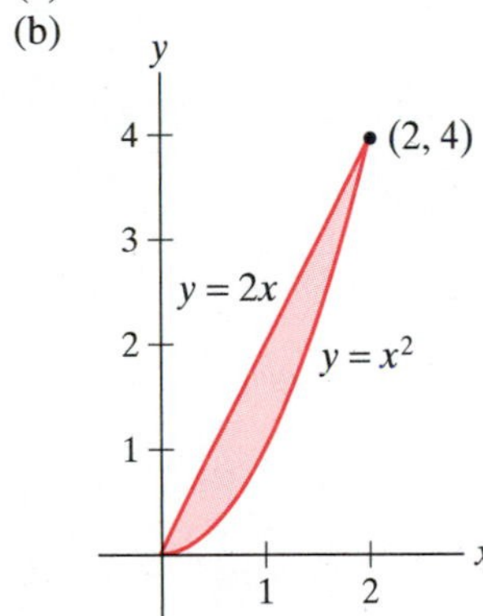

   (c) $\int_0^4 \int_{y/2}^{\sqrt{y}} (2x + 1)\,dx\,dy = 4$

3. $\int_0^4 \int_0^{2-y/2} y\,dx\,dy = \frac{16}{3}$

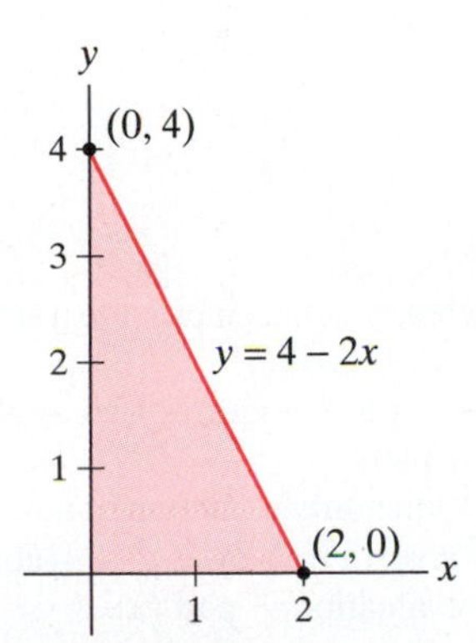

5. $\int_0^3 \int_0^{x^2} (x+y)\,dy\,dx = \frac{891}{20}$

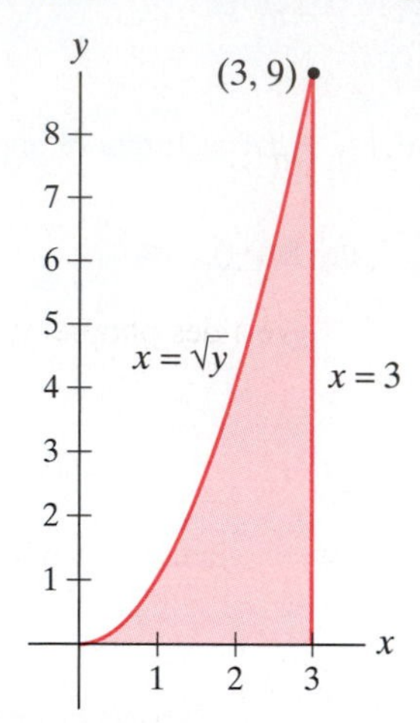

7. $\int_0^1 \int_{x/2}^x e^x\,dy\,dx + \int_1^2 \int_{x/2}^1 e^x\,dy\,dx = \frac{1}{2} + e^2/2 - e$

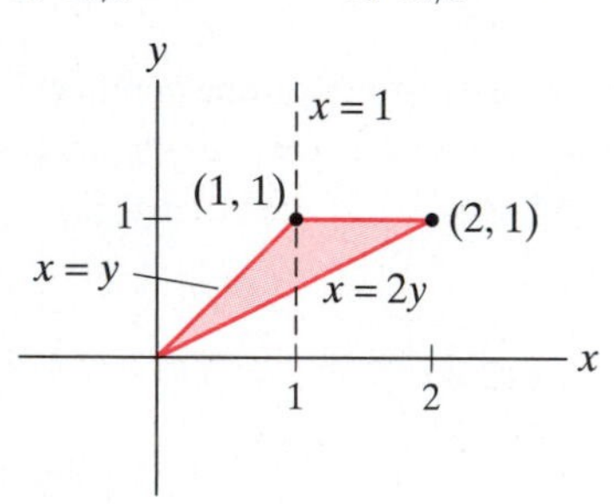

9. $\int_{-2}^2 \int_0^{\sqrt{4-x^2}} y\,dy\,dx = \frac{16}{3}$

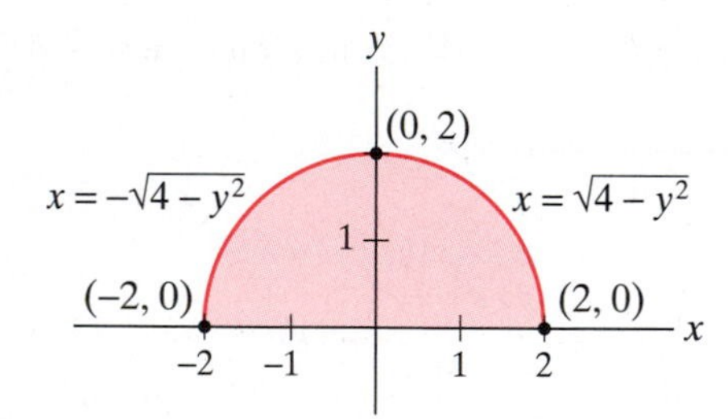

11. $\frac{625}{12}$
13. $\frac{896}{15}$
15. $(1 - \sin 1)/2$
17. $(e^{27} - 1)/6$
19. (a) If your computer algebra system can provide a simple answer, it should be $\frac{1}{4}(1 - \cos 2)$.
    (b) The integral with respect to y requires two applications of integration by parts.
    (c) It should take the computer only a fraction of a second to find that $\int_0^1 \int_0^{2y} y^2 \cos(xy)\,dx\,dy = \frac{1}{4}(1 - \cos 2)$, much faster than the evaluation in part (a).
21. (a) It is quite possible that the computer will be unable to make the evaluation.
    (b) With order reversed, the computer easily finds $\int_0^{\pi/2} \int_0^{\sin x} e^{\cos x}\,dy\,dx = e - 1$.

Section 5.4

1. 0
3. 1
5. $\frac{1539}{16}$
7. $-\frac{5}{24}$
9. Volume $= \int_{-a}^a \int_{-\sqrt{a^2-x^2}}^{\sqrt{a^2-x^2}} \int_{-\sqrt{a^2-x^2-y^2}}^{\sqrt{a^2-x^2-y^2}} dz\,dy\,dx$
11. $\frac{176}{15}$
13. 0
15. $\frac{1}{10}$
17. $81\sqrt{3}\pi/8$
19. $\sqrt{2}\pi/2$
21. $\int_0^1 \int_{-\sqrt{x}}^{\sqrt{x}} \int_0^{1-x} f(x,y,z)\,dz\,dy\,dx$
$= \int_0^1 \int_0^{1-x} \int_{-\sqrt{x}}^{\sqrt{x}} f(x,y,z)\,dy\,dz\,dx$
$= \int_0^1 \int_0^{1-z} \int_{-\sqrt{x}}^{\sqrt{x}} f(x,y,z)\,dy\,dx\,dz$
$= \int_{-1}^1 \int_0^{1-y^2} \int_{y^2}^{1-z} f(x,y,z)\,dx\,dz\,dy$
$= \int_0^1 \int_{-\sqrt{1-z}}^{\sqrt{1-z}} \int_{y^2}^{1-z} f(x,y,z)\,dx\,dy\,dz$

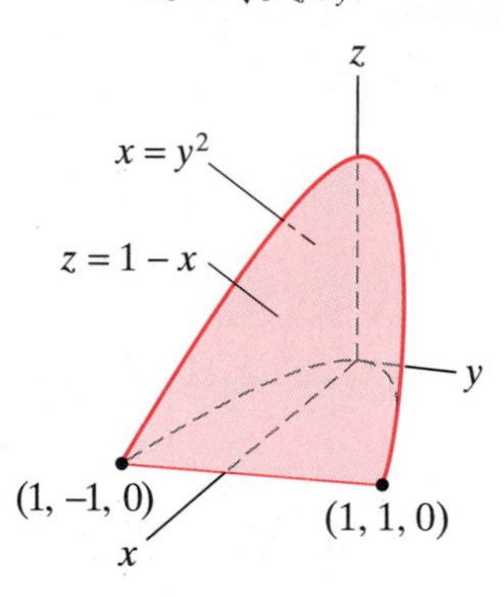

23. $\int_0^2 \int_y^2 \int_0^y f(x,y,z)\,dz\,dx\,dy$
$= \int_0^2 \int_0^x \int_z^x f(x,y,z)\,dy\,dz\,dx$
$= \int_0^2 \int_z^2 \int_z^x f(x,y,z)\,dy\,dx\,dz$
$= \int_0^2 \int_0^y \int_y^2 f(x,y,z)\,dx\,dz\,dy$
$= \int_0^2 \int_z^2 \int_y^2 f(x,y,z)\,dx\,dy\,dz$
25. (a) Bottom surface is $z = x^2 + 3y^2$, top surface is $z = 4 - y^2$; shadow in xy-plane is $x^2/4 + y^2 \le 1$, $y \ge 0$.
    (b) $\int_0^1 \int_{-2\sqrt{1-y^2}}^{2\sqrt{1-y^2}} \int_{x^2+3y^2}^{4-y^2} (x^3 + y^3)\,dz\,dx\,dy$
    (c) $\int_0^1 \int_{3y^2}^{4-y^2} \int_{-\sqrt{z-3y^2}}^{\sqrt{z-3y^2}} (x^3 + y^3)\,dx\,dz\,dy$
    (d) $\int_0^3 \int_0^{\sqrt{z/3}} \int_{-\sqrt{z-3y^2}}^{\sqrt{z-3y^2}} (x^3 + y^3)\,dx\,dy\,dz +$
    $\int_3^4 \int_0^{\sqrt{4-z}} \int_{-\sqrt{z-3y^2}}^{\sqrt{z-3y^2}} (x^3 + y^3)\,dx\,dy\,dz$

(e) $\int_{-2}^{2}\int_{x^2}^{3+x^2/4}\int_0^{\sqrt{(z-x^2)/3}}(x^3+y^3)\,dy\,dz\,dx +$
$\int_{-2}^{2}\int_{3+x^2/4}^{4}\int_0^{\sqrt{4-z}}(x^3+y^3)\,dy\,dz\,dx$

Section 5.5

1. (a) $\mathbf{T}(u,v)=\begin{bmatrix}3 & 0\\ 0 & -1\end{bmatrix}\begin{bmatrix}u\\ v\end{bmatrix}$
   (b) $D=[0,3]\times[-1,0]$
3. D is the parallelogram with vertices $(0,0)$, $(11,2)$, $(4,3)$, $(15,5)$.
5. $\mathbf{T}$ takes W^* onto the parallelepiped with vertices $(0,0,0)$, $(3,1,5)$, $(-1,-1,3)$, $(0,2,-1)$, $(2,0,8)$, $(3,3,4)$, $(-1,1,2)$, and $(2,2,7)$.
7. (a) $D=\{(x,y,z)\mid x^2+y^2+z^2\le 1\}$
   (b) $D=\{(x,y,z)\mid x^2+y^2+z^2\le 1,\ x,y,z\ge 0\}$
   (c) $D=\left\{(x,y,z)\mid \frac{1}{4}\le x^2+y^2+z^2\le 1,\ x,y,z\ge 0\right\}$
9. $\int_0^2\int_0^2 u^5ve^{v^2}\cdot\frac{1}{2}\,du\,dv=8(e^4-1)/3$
11. $7(e-1/e)$
13. 3π
15. $486\pi/5$
17. $3\ln(\sqrt{2}+1)$
19. $(16-3\pi)/12$
21. $2+\pi/4$
23. $(\pi\sin 1)/3$
25. 48π
27. $2\pi((1-a^2)e^{a^2}+(b^2-1)e^{b^2})$
29. $4\sqrt{2}\pi(5\sqrt{5}-8)/3$
31. $656\pi/5$

Section 5.6

1. (a) 80 cases
   (b) \$1.60
3. e^2-2e+1
5. (a) c, where c is the constant of proportionality.
   (b) $\{(x,y,z)\mid x^2+y^2+z^2=1\}$
7. $\dfrac{30-3\pi}{30-5\pi}$
9. 90 seconds
11. If the plate is located at $\{(x,y)\mid x^2+y^2\le a^2,\ y\ge 0\}$, then $(\bar{x},\bar{y})=(0,4a/3\pi)$.
13. $(\bar{x},\bar{y})=\left(\frac{27}{4},\frac{12}{7}\right)$
15. $(\bar{x},\bar{y})=\left((7\sqrt{3}+8\pi)/(3\sqrt{3}+4\pi),\ 15/(\sqrt{3}+4\pi)\right)$
17. $(\bar{x},\bar{y})=(21/20,0)$
19. (a) $(\bar{x},\bar{y},\bar{z})=\left(\frac{1}{2},0,\frac{9}{5}\right)$
    (b) $(\bar{x},\bar{y},\bar{z})=\left(\frac{43}{56},0,\frac{99}{49}\right)$
21. $(\bar{x},\bar{y},\bar{z})=\left(0,0,-\frac{17}{12}\right)$
23. $3a/8$.
25. (a) $I_x=I_y=I_z=\frac{1}{30}$
    (b) $r_x=r_y=r_z=1/\sqrt{5}$
27. (a) $I_z=6561\pi/4$; $r_z=(3\sqrt{3})/(2\sqrt{2})$
    (b) $I_z=8748\pi/35$; $r_z=(3\sqrt{3})/\sqrt{7}$
29. $1496/135$
31. $116\pi/3$
33. $V=-3GMm(b^2-a^2)/(2(b^3-a^3))$

True/False Exercises for Chapter 5

1. False. (Not all rectangles must have sides parallel to the coordinate axes.)
3. True
5. False
7. False
9. True
11. True
13. False. (The value of the integral is 3.)
15. True
17. True. (The inner integral is zero because of symmetry.)
19. True
21. False. (The integrals are opposites of one another.)
23. False. (A factor of r should appear in the integrand.)
25. False. (A factor of ρ is missing in the integrand.)
27. True
29. True

Miscellaneous Exercises for Chapter 5

1. 72π
3. (a) $\int_0^{\sqrt{3}}\int_{-\sqrt{9-3x^2}}^{\sqrt{9-3x^2}}\int_{2x^2+y^2}^{9-x^2}3\,dz\,dy\,dx$ and $\int_{-3}^{3}\int_0^{\sqrt{(9-y^2)/3}}\int_{2x^2+y^2}^{9-x^2}3\,dz\,dx\,dy$ are two possibilities.
   (b) $81\sqrt{3}\pi/4$
5. (a) $\int_{-1}^{1}\int_{-\sqrt{1-x^2}}^{\sqrt{1-x^2}}\int_0^{\sqrt{9-x^2-y^2}}dz\,dy\,dx$
   (b) $\int_0^{2\pi}\int_0^{\sin^{-1}1/3}\int_0^3\rho^2\sin\varphi\,d\rho\,d\varphi\,d\theta+\int_0^{2\pi}\int_{\sin^{-1}1/3}^{\pi/2}\int_0^{\csc\varphi}\rho^2\sin\varphi\,d\rho\,d\varphi\,d\theta$. Value is $2\pi\left(9-\dfrac{16\sqrt{2}}{3}\right)$.
7. $8\int_0^a\int_0^{\pi/4}\int_0^{\frac{a}{2}\sec\theta}r\,dr\,d\theta\,dz=a^3$
9. $(\sin 1+\sin 2)/6$
11. (a) $\int_{-a}^{a}\int_{-b\sqrt{1-x^2/a^2}}^{b\sqrt{1-x^2/a^2}}dy\,dx$
    (b) $E^*=\{(\bar{x},\bar{y})\mid \bar{x}^2+\bar{y}^2=1\}$
    (c) Area $=\int_0^{2\pi}\int_0^1 abr\,dr\,d\theta$
13. Area is π.
15. $-\frac{3}{5}\ln 4$
17. $\frac{3}{2}(\tan^{-1}4-\frac{\pi}{4})$

19. (a) $\int_{-a}^{a} \int_{-\sqrt{a^2-x^2}}^{\sqrt{a^2-x^2}} \int_{-\sqrt{a^2-x^2-y^2}}^{\sqrt{a^2-x^2-y^2}} \int_{-\sqrt{a^2-x^2-y^2-z^2}}^{\sqrt{a^2-x^2-y^2-z^2}} dw\, dz\, dy\, dx$
   (b) $\pi^2 a^4/2$
   (c) Five-dimensional ball has volume $8\pi^2 a^5/15$; six-dimensional ball has volume $\pi^3 a^6/6$. The pattern is not very clear from this information.
21. Within a disk of radius $\sqrt{5}a/3$ about the center, where a is the radius of the hemisphere.
23. (a) $4(\sqrt{1-\epsilon} - \sqrt{\epsilon})(\sqrt{1-\delta} - \sqrt{\delta})$
   (b) 4
25. Integral does not converge.
27. $-4\pi/9$
29. Converges when $p > 1$ and $q > 1$; value is $\dfrac{1}{(p-1)(q-1)}$.
31. Integral does not converge.
33. (a) Hint: Break up the integral into a sum of integrals from 0 to 1 and from 1 to ∞. Show convergence of the improper integral from 1 to ∞ by comparing it to $\int_1^\infty e^{-x}\, dx$.
   (b) Hint: Begin with $\iint_{\mathbf{R}^2} e^{-x^2-y^2}\, dA$ and use laws of exponents.
   (c) $\pi(1 - e^{-a^2})$
   (d) π
   (e) $\sqrt{\pi}$
35. (b) $\frac{3}{280}$
37. $C = ab/4$
39. (a) $C = \frac{4}{a^2+b^2}$
   (b) $\frac{1}{2}$
41. $1 - e^{-1/4} \approx 0.2212$

Chapter 6

Section 6.1

1. (a) 50 (b) 4
3. $(35\sqrt{35} - 17\sqrt{17})/27$
5. $(5^{3/2} + 87)/12$
7. 2
9. $4\pi + 16\pi^2/3$
11. 3π
13. $1 - e^{-4\pi}$
15. (a) $\int_{\mathbf{x}} \mathbf{F} \cdot d\mathbf{s} = \frac{1}{2}$, $\int_{\mathbf{y}} \mathbf{F} \cdot d\mathbf{s} = -\frac{1}{2}$
   (b) $\mathbf{y}(t) = \mathbf{x}(1 - 2t)$. Thus, the images of $\mathbf{x}$ and $\mathbf{y}$ are the same, although $\mathbf{y}$ traces the image in the opposite direction to that of $\mathbf{x}$. For these reasons, the results of part (a) could have been anticipated.
17. 0
19. $-\frac{137}{12}$
21. $-\frac{11}{3}$
23. Hint: Use formula (3).
25. (a) $2500\sqrt{13}$ ft-lb
   (b) 7500 ft-lb
33. (a) 7.65625 (b) 7.5

Section 6.2

1. $\oint_{\partial D} M\, dx + N\, dy = \iint_D (N_x - M_y)\, dA = 8\pi$
3. $\oint_{\partial D} M\, dx + N\, dy = \iint_D (N_x - M_y)\, dA = -4$
5. (a) 0
   (b) You will need to compute four separate line integrals—one for each edge of the square.
7. -2
9. Hint: Calculate $\frac{1}{2} \oint_C -y\, dx + x\, dy$, where C is the boundary of R, oriented counterclockwise.
11. -45
13. (a)

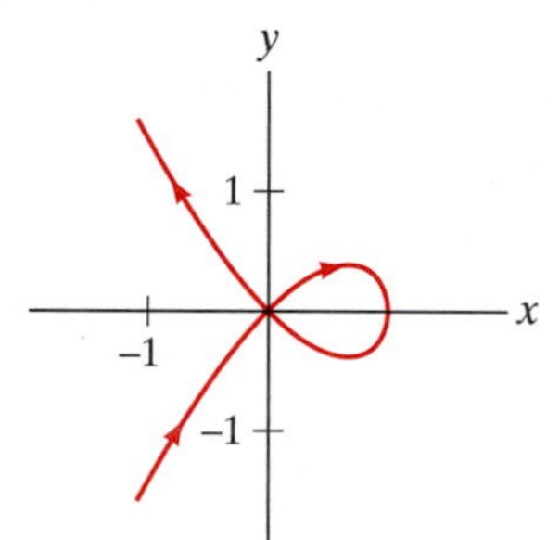

   (b) $\frac{8}{15}$
17. (a) Hint: Show $\nabla \cdot \mathbf{F} = 0$.
19. Directly apply Green's theorem.
21. The line integral is ± 3 times the area of the rectangle, where the sign depends on the orientation of the boundary.
23. Hint: $u\nabla v = (u\, \partial v/\partial x, u\, \partial v/\partial y)$. Then use Green's theorem.
25. Hint: Begin with $\displaystyle\oint_{\partial D} \frac{\partial f}{\partial n}\, ds$ and then use Green's theorem.

Section 6.3

1. (a) $\frac{5}{3}$
   (b) $\frac{11}{7}$
   (c) No. Line integrals are not path-independent.
3. Not conservative
5. Not conservative
7. Conservative; $f(x, y) = xe^{-y} + \cos xy$
9. Conservative; $f(x, y, z) = x^2 + xy + \sin yz$
11. Conservative; $f(x, y, z) = e^x \sin y + z^3 + 2z$
13. Not conservative
15. $N(x, y) = \frac{1}{2}e^{2x} + x^3 e^y + u(y)$, where u is any function of y of class C^2.

17. (a) Check that $\nabla \times \mathbf{F} = \mathbf{0}$. Since $\mathbf{F}$ is of class C^1 on all of $\mathbf{R}^3$, by Theorem 3.5, $\mathbf{F}$ is conservative. A scalar potential is $f(x, y, z) = \frac{1}{3}x^3 + \sin y \sin z$.
    (b) $\frac{7}{3} + \sin e \sin e^2 - (\sin 1)^2$
19. 0
21. (a) Conservative on $\{(x, y) \mid y > 0\}$ and on $\{(x, y) \mid y < 0\}$
    (b) $f(x, y) = (x^2y^2 + x^2 + 1)/(2y^2)$
    (c) 1
23. (a) $\mathbf{F}$ is conservative. (b) -2
25. Work $= GMm\left(\dfrac{1}{\|\mathbf{x}_1\|} - \dfrac{1}{\|\mathbf{x}_0\|}\right)$

True/False Exercises for Chapter 6

1. True
3. False. (It's negative.)
5. False. (The integral is 0.)
7. False. (There is equality only up to sign.)
9. True
11. True
13. True
15. False. (The line integral could be $\pm \int_C \|\mathbf{F}\|\, ds$, depending on whether $\mathbf{F}$ points in the same or the opposite direction as C.)
17. False. (Let $\mathbf{F} = y\mathbf{i} - x\mathbf{j}$ and consider Green's theorem.)
19. False. (Under appropriate conditions, the integral is $f(B) - f(A)$.)
21. True
23. False. (For the vector field to be conservative, the line integral must be zero for *all* closed curves, not just a particular one.)
25. False. (The vector field $(e^x \cos y \sin z, e^x \sin y \sin z, e^x \cos y \cos z)$ is not conservative.)
27. False. (The domain is not simply-connected.)
29. False. (f is only defined up to a constant.)

Miscellaneous Exercises for Chapter 6

1. Break up C into n segments, each of length Δs_k. By continuity, f will be nearly constant on each segment, so $[f]_{\text{avg}} \approx \sum_{k=1}^n f(\mathbf{c}_k)\Delta s_k / \sum_{k=1}^n \Delta s_k$. (Here $\mathbf{c}_k$ is any point in the kth segment.) The formula follows after taking limits as all $\Delta s_k \to 0$.
3. $2a/\pi$
5. 2
7. $\bar{x} = \bar{y} = \left(\frac{8-\sqrt{2}\pi-2\sqrt{2}}{4(\pi-2\sqrt{2})}\right) a$
9. (a) $\frac{36\pi-32}{3}$ (b) 2
11. $I_x = \frac{7769\sqrt{17}-1}{840}$, $\quad r_x = \sqrt{\frac{7769\sqrt{17}-1}{1190\sqrt{17}-70}}$
13. $I_z = 27\sqrt{11}$, $\quad r_z = \frac{9\sqrt{22}}{22}$
15. (a) $\int_a^b g(f(\theta)\cos\theta, f(\theta)\sin\theta)\sqrt{(f(\theta))^2 + (f'(\theta))^2}\, d\theta$
    (b) $\sqrt{10}\,[(e^{18\pi} - 1)/9 + 12(1 - e^{12\pi})/37]$
17. 6π
19. $K = 2\pi$
21. $\frac{6}{5} - \cos 1 - \sin 1$
23. Hint: Use the formula Area $= \frac{1}{2}\int_{\partial D} -y\, dx + x\, dy$.
25. Hint: $\bar{x} = \iint_D x\, dA/\text{area of } D$, $\bar{y} = \iint_D y\, dA/\text{area of } D$. Now apply Green's theorem to $\oint_{\partial D} x^2\, dy$ and $\oint_{\partial D} xy\, dy$.
27. $\bar{x} = \bar{y} = -\frac{1}{34}$
29. Hint: Use the result of Exercise 28 twice.
35. (a) Both the line integral and the double integral are zero.
    (b) No. The double integral is not defined properly over the disk because $\mathbf{F}$ is undefined at the origin.
    (c) $\iint_D [\frac{\partial}{\partial x}(\frac{x}{x^2+y^2}) - \frac{\partial}{\partial y}(\frac{-y}{x^2+y^2})]\, dA = 0 = \oint_C \mathbf{F}\cdot d\mathbf{s} + \oint_{C_a} \mathbf{F}\cdot d\mathbf{s}$
37. (a) 0
    (b) Apply the divergence theorem.
39. $\int_{\mathbf{x}} \mathbf{F}\cdot d\mathbf{s} = \int_{\mathbf{x}} -\nabla V \cdot d\mathbf{s} = -V(B) + V(A)$. Now use Exercise 38.

Chapter 7

Section 7.1

1. (a) $-\mathbf{i} - 4\mathbf{j} + 2\mathbf{k}$
   (b) $x + 4y - 2z = 5$
3. $y + 4z = 4$
5. (a)

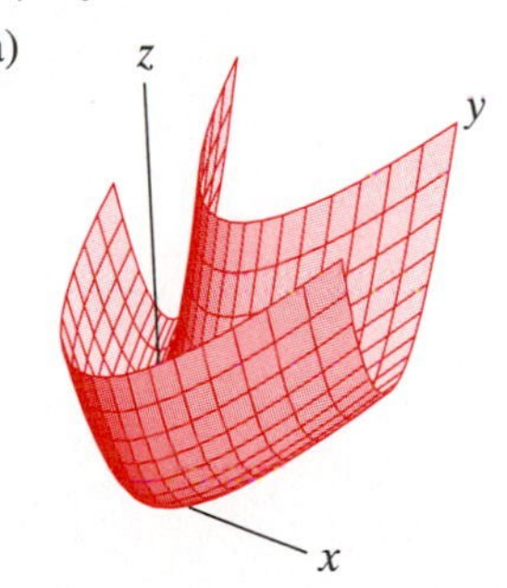

   (b) Yes
   (c) $4x - 2y - z = 3$
7. (a) Smooth except at $(0, 0, 0)$. Tangent plane at $(1, \sqrt{3}, 4)$ has equation $2x + 2\sqrt{3}y - z = 4$.
   (b) S is a paraboloid.
   (c) $z = x^2 + y^2$
   (d) Yes it does. At $(0, 0, 0)$, the tangent plane has equation $z = 0$.
9. Hint: Consider $x^2 + y^2$.

11. (a)–(c) All versions give the equation $-x + y + \sqrt{2}z = 1$.

13. (a) $y^2 z = t^2 s^2 = x^2$
(b) No
(c)

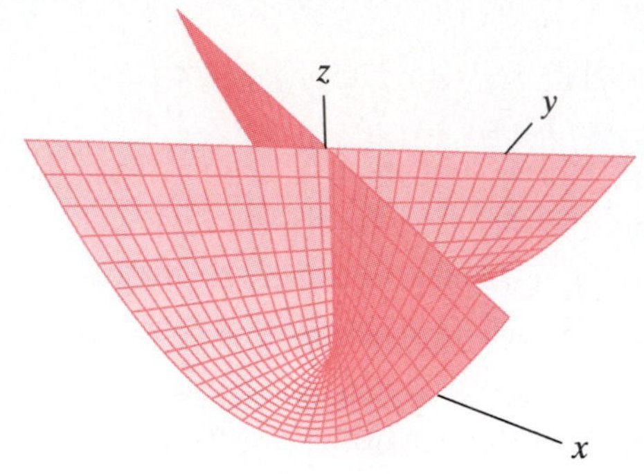

(d) Points on the positive z-axis
(e) $-4x + 8y + z = 4$
(f) $(0, 0, 1) = \mathbf{X}(\pm 1, 0)$; tangent planes have equations $x \pm y = 0$.

15. (a) At the point $(a, g(a, c), c)$, the tangent plane has equation $g_x(a, c)(x-a)-(y-g(a, c))+g_z(a, c)(z-c) = 0$.
(b) At the point $(h(b, c), b, c)$, the tangent plane has equation $-(x - h(b, c)) + h_y(b, c)(y - b) + h_z(b, c)(z - c) = 0$.

17. $\{\mathbf{x} \in \mathbf{R}^3 \mid \mathbf{x} = (1, 0, 1) + (s - 1)(1, 2, 0) + (t + 1)(0, 1, -2)\}$ or $x = s$, $y = 2s + t - 1$, $z = -2t - 1$. To verify consistency with Exercise 5(b), check that the points given by the parametric equations all lie in the plane determined by the equation in Exercise 5(b).

19. $\sqrt{6}\pi$

21. $4\pi a\sqrt{a^2 - b^2}$

23. $(65^{3/2} - 17^{3/2})\pi/24$

25. $\sqrt{1+a} \cdot$ area of D

27. $16a^2$

Section 7.2

1. $26\sqrt{3}/3$

3. 1

5. $\frac{640}{3}$

7. (a) $4\pi a^4$ (b) $4\pi a^4/3$

9. (a) Parametrize the cylinder as $x = 2\cos t$, $y = 2\sin t$, $z = s$ with $-2 \le s \le 2$, $0 \le t < 2\pi$. The integral evaluates to -64π.
(b) The integral is $-4 \cdot$ (surface area of S) $= -64\pi$.

11. 0

13. $297\pi/2$

15. 36π

17. 0

19. $2\pi a^3/3$

21. $-\pi a^2$

23. (a)

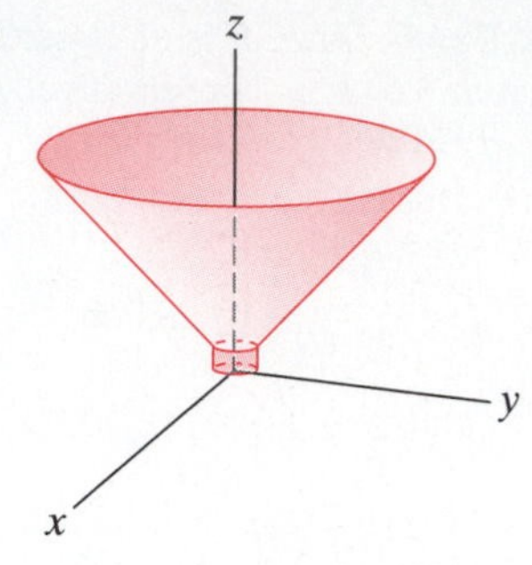

(b) $\mathbf{n}_1 = \left(x/(\sqrt{2}z), y/(\sqrt{2}z), -1/\sqrt{2}\right)$, $\mathbf{n}_2 = (x, y, 0)$
(c) $-1456\pi/3$

25. (a)

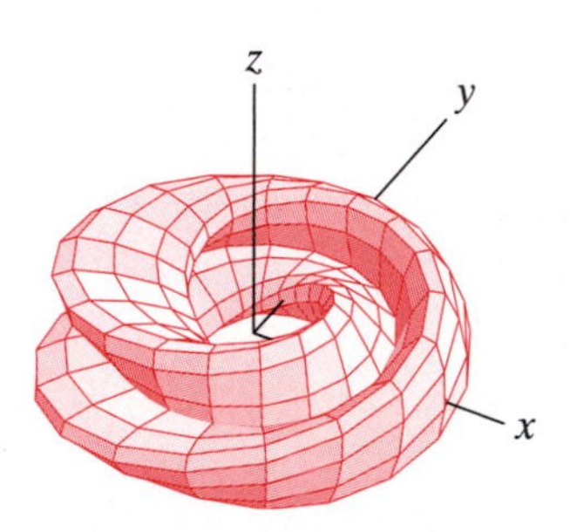

(b) $(a\cos s, a\sin s, 0)$
(c)
$$\mathbf{N}(s, 0) = (a\cos s\,(2\cos(s/2) + \sin(s/2)),\ a\sin s\,(2\cos(s/2) + \sin(s/2)),\ a(2\sin(s/2) - \cos(s/2))).$$

From this, we see that $\mathbf{N}(0, 0) = (2a, 0, -a)$, while $\mathbf{N}(2\pi, 0) = (-2a, 0, a)$. Therefore, the Klein bottle cannot be orientable, since the normal vector along the s-coordinate curve at $t = 0$ changes direction.

Section 7.3

1. 0

3. 0

5. 0

7. 0

9. $4\pi(b^2 - a^2)$

11. 45π

13. (a) Hint: Use the double angle formula.
(b) $-\frac{3\pi}{4}$

19. (a) πa
(b) πa
(c) The answers in parts (a) and (b) are the same; the three flat quarter-circles that are part of ∂D do not contribute anything to $\iint_S \nabla f \cdot \mathbf{n}\, dS$.

23. (a)
$$\begin{aligned}\oiint_S \mathbf{F} \cdot d\mathbf{S} &\approx F_z(r, \theta, z + \Delta z/2) r\, \Delta\theta\, \Delta r \\ &\quad - F_z(r, \theta, z - \Delta z/2) r\, \Delta\theta\, \Delta r \\ &\quad + F_r(r + \Delta r/2, \theta, z)(r + \Delta r/2)\Delta\theta\, \Delta z \\ &\quad - F_r(r - \Delta r/2, \theta, z)(r - \Delta r/2)\Delta\theta\, \Delta z \\ &\quad + F_\theta(r, \theta + \Delta\theta/2, z)\Delta r\, \Delta z \\ &\quad - F_\theta(r, \theta - \Delta\theta/2, z)\Delta r\, \Delta z\end{aligned}$$

Section 7.4

1. Hint: Use Gauss's theorem and the product rule for $\nabla \cdot (f \nabla g)$.
3. (a) Hint: Use Green's first formula with $f = g$.
   (b) Hint: Use part (a) and the fact that $\nabla f \cdot \nabla f = \|\nabla f\|^2$.
7. Hint: Use the argument in Exercise 6 and the product rule for $\nabla \cdot (k \nabla T)$.
9. (a) Hint: Use Gauss's theorem and Exercise 8.
   (b) Heat flows into D from the inner sphere and out through the outer sphere at the same rate.
11. Hint: Use Ampère's and Gauss's laws.
15. (a) In each case $k = \mu_0 \epsilon_0$.
19. Hint: Apply Gauss's theorem to $\mathbf{P} = \mathbf{E} \times \mathbf{B}$, then Faraday's and Ampère's laws.
21. Hint: Apply the arguments used in Exercise 20 to each component integral of $\mathbf{B}$.

True/False Exercises for Chapter 7

1. True
3. True. (Let $u = s^3$ and $v = \tan t$.)
5. False. (The limits of integration are not correct.)
7. False. (The value of the integral is 24.)
9. True
11. False. (The integral has value 32π.)
13. False. (The value is 3π.)
15. False. (The surface must be connected.)
17. True. (The result follows from Stokes's theorem.)
19. True. (Use Gauss's theorem.)
21. False. (Gauss's theorem implies that the integral is *at most* twice the surface area.)
23. True
25. False. (Should be the flux of the *curl* of $\mathbf{F}$.)
27. True. (Apply Green's first formula.)
29. False. (f is determined up to addition of a harmonic function.)

Miscellaneous Exercises for Chapter 7

1. (a) C (b) E (c) A
   (d) D (e) F (f) B
3. (a) $(\sqrt{z^2+1}\cos\theta, \sqrt{z^2+1}\sin\theta, z)$
   (b) $(a\sqrt{s^2+1}\cos t, b\sqrt{s^2+1}\sin t, cs)$
5. (a) $(a \sin\varphi \cos\theta, b \sin\varphi \cos\theta, c\cos\varphi)$
   (b) $\int_0^{2\pi}\int_0^{\pi} \sqrt{b^2c^2\sin^4\varphi\cos^2\theta + a^2c^2\sin^4\varphi\sin^2\theta + a^2b^2\cos^2\varphi\sin^2\varphi}\, d\varphi\, d\theta$
7. (a) $(s\cos t, f(s), s\sin t)$
9. (b) $\frac{98}{3}$
11. $\frac{11}{10}$
13. $(a/2, a/2, a/2)$
15. $(0, 0, a/3)$
17. (a) $15\sqrt{5}\pi/2$
    (b) $\sqrt{5/2}$
    (c) $I_z = 62\sqrt{5}\pi k/5$, $r_z = \sqrt{93/35}$
19. (a) $I_x = I_y = 2\pi ab\delta(3a^2+2b^2)/3$
    (b) $r_x = r_y = \sqrt{(3a^2+2b^2)/6}$
23. Hint: Use Stokes's theorem.
25. Use Exercise 24.
31. Hint: Show $\nabla \cdot (\mathbf{x}/\|\mathbf{x}\|^3) = 0$ where defined.
35. $\nabla \cdot \mathbf{F} \neq 0$, so there is no vector potential.
39. Hint: Calculate $\nabla \times (\mathbf{E} + \partial\mathbf{A}/\partial t)$.

Chapter 8

Section 8.1

1. -2
3. 6
5. -370
7. $-6a_1b_3 + 6a_3b_1 + a_2b_4 - a_4b_2$
9. $14\cos x - 7\sin z + 11y^2 + 33$
11. $2xy\, dx \wedge dy \wedge dz$
13. $(x_1e^{x_4x_5} - x_1x_2x_3\cos x_5)\, dx_1 \wedge dx_2 \wedge dx_3 \wedge dx_4 \wedge dx_5$

Section 8.2

1. Hint: Show linear independence of $\mathbf{T}_{\theta_1}, \mathbf{T}_{\theta_2}, \mathbf{T}_{\theta_3}$ by solving the vector equation $c_1\mathbf{T}_{\theta_1} + c_2\mathbf{T}_{\theta_2} + c_3\mathbf{T}_{\theta_3} = \mathbf{0}$ for c_1, c_2, c_3.
3. $\mathbf{X}\colon [0, 2\pi) \times [0, 2\pi) \times [0, 2\pi) \times [1, 2] \times [1, 2] \to \mathbf{R}^6$, $\mathbf{X}(\theta_1, \theta_2, \theta_3, l_2, l_3) = (x_1, y_1, x_2, y_2, x_3, y_3)$, where $x_1 = 3\cos\theta_1$, $y_1 = 3\sin\theta_1$, $x_2 = 3\cos\theta_1 + l_2\cos\theta_2$, $y_2 = 3\sin\theta_1 + l_2\sin\theta_2$, $x_3 = 3\cos\theta_1 + l_2\cos\theta_2 + l_3\cos\theta_3$, $y_3 = 3\sin\theta_1 + l_2\sin\theta_2 + l_3\sin\theta_3$
7. -2π
9. $4\pi^2$
11. (a) $\Omega_{\mathbf{X}(\mathbf{u})}(\mathbf{T}_{u_1}, \mathbf{T}_{u_2}, \mathbf{T}_{u_3}) = u_1 > 0$ for $0 < u_1 \leq \sqrt{5}$ which is where the parametrization is smooth.
    (b) Parametrize ∂M in two pieces as
    $\mathbf{Y}_1\colon [0, \sqrt{5}] \times [0, 2\pi) \to \mathbf{R}^3$,
    $\mathbf{Y}_1(s_1, s_2) = (s_1\cos s_2, s_1\sin s_2, s_1^2 - 6)$ and
    $\mathbf{Y}_2\colon [0, \sqrt{5}] \times [0, 2\pi) \to \mathbf{R}^3$,
    $\mathbf{Y}_2(s_1, s_2) = (s_1\cos s_2, s_1\sin s_2, 4 - s_1^2)$.

(c) $\mathbf{V}_1 = \dfrac{(2s_1 \cos s_2, 2s_1 \sin s_2, -1)}{\sqrt{4s_1^2+1}}$,

$\mathbf{V}_2 = \dfrac{(2s_1 \cos s_2, 2s_1 \sin s_2, 1)}{\sqrt{4s_1^2+1}}$

13. $\frac{3}{2}$

Section 8.3

1. $d\omega = e^{xyz}(yz\,dx + xz\,dy + xy\,dz)$
3. $d\omega = -y\,dx \wedge dy$
5. $d\omega = (x + 2yz)\,dx \wedge dy \wedge dz$
7. $d\omega = 2(x_1 - x_2 + x_3 - \cdots + (-1)^{n+1}x_n)\,dx_1 \wedge dx_2 \wedge \cdots \wedge dx_n$
9. $F(x, z) = xz + Cz + D_1$, $G(x, y) = xy + Cy + D_2$, where C, D_1, D_2 are arbitrary constants.
11. $\int_{\partial M} \omega = \int_M d\omega = 0$

True/False Exercises for Chapter 8

1. True
3. True
5. True
7. True
9. True
11. False. ($\mathbf{X}(1, 1, -1) = \mathbf{X}(1, 1, 1)$, so $\mathbf{X}$ is not one-one on D.)
13. False. (The agreement is only up to sign.)
15. False. (This is only true if n is even.)
17. True
19. True. ($d\omega$ would be an $(n+1)$-form, and there are no nonzero ones on $\mathbf{R}^n$.)

Miscellaneous Exercises for Chapter 8

5. 0
7. (a) $\nabla \times (\nabla f) = \mathbf{0}$
   (b) $\nabla \cdot (\nabla \times \mathbf{F}) = 0$
9. Hint: Consider $\int_M d(\omega \wedge \eta)$.

Index

D

K

L

M

N

Q

R

S

T